Energy and Power Generation Handbook

Established and Emerging Technologies

Editor
K. R. Rao

Library of Congress Cataloging-in-Publication Data

Energy and power generation handbook : established and emerging technologies / editor K.R. Rao.
   p. cm.
Includes bibliographical references.
ISBN 978-0-7918-5955-1
   1. Electric power production — Handbooks, manuals, etc. I. Rao, K. R., 1933–
TK1001.E56 2011
621.31'21—dc22
                2011009559

DEDICATION

This *ENERGY AND POWER GENERATION HANDBOOK* is dedicated to:

The late Dr. Baira Gowda, Pittsburgh, PA for introducing me to ASME, in the late 1980s;

Dr. Robert Toll Norman and Dr. Liane Ellison Norman, staunch supporters of the "Green Peace Movement" and Clean Energy at Pittsburgh, PA, where I was in the 1970s and 1980s, in whom I saw firsthand what these movements symbolize;

Mr. VRP Rao, Fellow-IE for encouraging in me interest in actually taking up of this project to cover energy generation sources "other than nuclear," especially *renewable energy generation*, and finally;

Victims and Site Staff of the Fukushima Daiichi Nuclear Plants at Japan devastated by the Tohoku-Taiheiyou-Oki Earthquake and Tsunami of March 11, 2011. This publication is especially dedicated to these and other victims of Japan for the "Fortitude of Japan as a Nation," that shows national strength in their hour of an utterly tragic accident.

ACKNOWLEDGEMENTS

The editor is indebted to several individuals who had directly or indirectly helped in coming up with this handbook.

My thanks are due to all of the 53 contributors whose dedicated efforts made this possible by their singular attention to detail, presentation of graphics, procuring the copyrights for the "artwork" and taking time to research the references to complement the write-up. Even while they succinctly conveyed the wealth of information and knowledge they acquired during their professional career, they followed the guidelines provided for adhering to the page length.

It was challenging to enlist 53 experts from around the world to address varied power and energy generation topics. The editor contacted professionals who knew the worldwide ensemble of energy and generation experts. Efforts were made not to miss any power generation sources, and this was largely facilitated by contacting over 100 practicing professionals and academia before settling down with 53 authors. The formidable task of acquiring the correct authors took over six months and was amply rewarded.

While it is difficult to chronicle everyone the editor contacted, worthy of particular mention are Dr. Gregory J. Walker, University of Tasmania, Hobart, Tasmania, Australia; Mr. V. R. P. Rao F-IE, Hyderabad, India; Dr. Hardayal Mehta of GE Hitachi Nuclear Energy, San Jose, CA; Mr. Richard Bunce of Siemens Energy, Inc., Orlando, FL; Mr. Roger Reedy of REEDY Engineering, Inc., Cambell, CA; Ms. Katherine Knurek Martin, NASA Glenn Research Center, Cleveland, OH; Mr. Clifford Wells of Structural Integrity Associates, San Jose, CA; Dr. Bob Swindemann, Oak Ridge, TN; Mr. Roger Bedard formerly of EPRI; and Dr. E. V. R. Sastry, Osmania University College of Engineering, Hyderabad, India.

This publication was ably supported by the staff of ASME Technical Publishing. My appreciation and thanks to them for their cooperation.

Finally, all of this enduring effort, spread over 18 months, would have never been possible had it not been for the constant help and untiring zeal provided by my wife, Dr. Indira Rao, that included all of the sundry-editorial chores associated with this project.

CONTRIBUTOR BIOGRAPHIES

AGRAWAL, RAVI K.

Dr. Ravi K. Agrawal is a Senior Process Manager at KBR. He is currently the process manager and work group leader for a 600-MW Kemper County IGCC Project (formerly Mississippi Gasification Project). He has over 25 years of experience in a wide variety of technologies including gasification, syngas production, gas-to-liquids, coal-to-liquids, biomass conversion, bioethanol, carbon capture, acid gas removal, combustion, sour water treatment, and specialty chemicals. He is the inventor of two patents and six patent applications. He is also the author of over 60 technical publications in refereed journals.

Previously, he was with ETEC, Fluor Daniel, Woodward Clyde Consultants, and Argonne National Laboratory. As a principal at ETEC, he developed and executed marketing plans to increase sales that resulted in ETEC being recognized by the Houston Business Journal as the sixth fastest growing small business in 2002. He has been responsible for sales, engineering, construction, and startup of over 41 combustion and air pollution control systems installed at several utilities and refineries. He is a registered Professional Engineer in the states of Texas and Pennsylvania. Dr. Agrawal holds a PhD and a MS degree in chemical engineering from Clarkson University, as well as a B.Tech from Osmania University (Hyderabad, India).

ANDREONE, CARL F.

Carl F. Andreone, PE, Fellow of the ASME is registered in Massachusetts. He was President of Heat Transfer Consultants, Inc. until 2001, but now practices as an individual contractor. He was a Staff Consultant at Stone & Webster Engineering Corporation, Boston, MA, and held several other positions from 1970 to his retirement in 1991. Before joining Stone & Webster, he was a heat exchanger specialist with Badger America and Crawford & Russell Inc.

Mr. Andreone gained broad experience as a maintenance engineer on refinery exchangers at Aramco. His career includes nearly a decade with Lummus Heat Exchanger Division (now Yuba Heat Transfer Corporation) as an application and product engineer on power and process heat exchangers. Mr. Andreone has been continuously active in the heat exchanger industry since 1951. His work in this field has involved specification, design, maintenance, and repair of more than 3000 power and process heat exchangers.

From 1981 to the present, he has assisted in trouble-shooting, failure analysis, repair, modification, and replacement of more than 400 feedwater heaters at various power stations.

From 1982 through 2007, he and Stanley Yokell presented annual seminars on Closed Feedwater Heaters and Inspection, Maintenance and Repair of Tubular Exchangers. He is the author of numerous papers on feedwater heaters and tubular heat transfer equipment. With Mr. Yokell, he has written, *Tubular Heat Exchanger Inspection, Maintenance and Repair*, published by McGraw-Hill Book Company 1997.

Mr. Andreone served on the ASME Power Division Heat Exchanger Committee. He has served on the ASME Boiler and Pressure Vessel Code Committee's Special Working Group on Heat Transfer Equipment, and the ASME Codes and Standards Committee for the ASME/ANSI Performance Test Code 12.1, Closed Feedwater Heaters. Mr. Andreone received the B.Ch.E. from Villanova University.

BAILEY, SHEILA GAYLE

Sheila G. Bailey has been a Senior Physicist working in photovoltaics at NASA Glenn Research Center for over 25 years. Her most recent projects include nanomaterials and nanostructures for space photovoltaics, quantum wire IIIV solar cells and quantum dot alpha-voltaics. She has authored or co-authored over 165 journal and conference publications, nine book chapters and two patents.

Dr. Bailey is on the Editorial Board of "Progress in Photovoltaics". She is a member of the American Physical Society and a speaker for the American Institute of Physics Visiting Scientist Program. She is a member of AIAA Aerospace Power Systems technical committee and the IEE Electron Devices Society Photovoltaic Devices Committee. She was the chair of the 4th World Conference on Photovoltaic Energy Conversion in 2006. She is executive vice president of the Lewis Engineers and Scientists Association.

Dr. Sheila Bailey was an adjunct professor at Baldwin Wallace College for 27 years and is currently an associate faculty member of the International Space University. She has a BS degree in physics from Duke University, a MS degree in physics from the University of NC at Chapel Hill, and a PhD in condensed matter physics from the University of Manchester in England. She spent a post-doctoral year at the Royal Military College (part of the University of New South Wales) in Canberra, Australia.

Dr. Bailey is the recipient of the faculty excellence award from Baldwin Wallace College and the Federal Women's Program award. She is an Ohio Academy of Science "Exemplar". She was awarded the NASA Exceptional Service Medal for her work in space photovoltaics in 1999. She has completed the Office of Personnel Management's Executive Potential Program. She was inducted into the Ohio Women's Hall of Fame in 2003 by Governor Taft.

BALDWIN, THOMAS L.

Thomas L. Baldwin, PE, PhD, IEEE Fellow, is a senior engineer at the Idaho National Laboratory. He conducts engineering studies and research in electrical power generation and transmission for the U.S. Department of Energy, U.S. Navy, and EPRI. His research interests are in distribution energy system design, industrial power systems, grounding issues, transformers, and the analysis of power quality problems.

Dr. Baldwin also holds the rank of professor at the FAMU-FSU College of Engineering at Florida State University, Tallahassee, FL, and has conducted research at the Center for Advanced Power Systems since 1999. He is a Registered Professional Engineer in the State of North Carolina.

Thomas L. Baldwin received the BSEE and MSE.E degrees from Clemson University, Clemson, SC, and the PhD degree in electrical engineering from Virginia Polytechnic Institute and State University, Blacksburg, VA, in 1987, 1989, and 1993, respectively.

Dr. Baldwin is a member of the IEEE Power and Energy Society and the Industrial Applications Society and serves on several committees and working groups including Power System Grounding and the IEEE Green Book.

BANNERJEE, RANGAN

Rangan Banerjee is a Professor of the Department of Energy Science and Engineering and currently the Dean of Research and Development at the Indian Institute of Technology Bombay. He was Associate Dean (R & D) of IIT Bombay from 2003 to 2006 and Head of the Department of Energy Science and Engineering (2006 to 2009).

Dr. Banerjee is a Convening Lead Analyst for Industrial End Use Efficiency and a member of the executive committee for the Global Energy Assessment (2008 to 2010) coordinated by the International Institute for Applied Systems Analysis. He is also an Adjunct faculty (Honorary) in the Department

of Engineering & Public Policy, Carnegie Mellon University. He was a member of the Planning Commission's Integrated Energy Policy (2004 to 2005) Committee and on the working group for renewable energy and energy efficiency and DSM for the Eleventh Five Year Plan.

Dr. Banerjee has coauthored a book on Planning for Demand Side Management in the Power sector, a book on Energy Cost in the Chemical Industry, and a book on Engineering Education in India. He has been involved in industrial projects with organizations like Essar, Indian Chemical Manufacturers Association, KSIDC, HR Johnson, Tata Consulting Engineers, BSES, Sterlite, International Institute of Energy Conservation and sponsored projects with the Department of Science & Technology, UN, MERC, PCRA, MNES, Hewlett Foundation.

Dr. Banerjee's areas of interest include energy management, modeling of energy systems, energy planning and policy, hydrogen energy, and fuel cells. He has conducted two international training programs on solar energy and several National programs on renewable energy and Energy Management.

BOEHM, ROBERT F.

Robert F. Boehm is a Distinguished Professor of Mechanical Engineering and Director of the Energy Research Center at the University of Nevada, Las Vegas (UNLV). His work has been primarily in the area of renewable and conventional energy conversion. He was on the faculty of the University of Utah Department of Mechanical Engineering prior to coming to UNLV. He holds a PhD in Mechanical Engineering from the University of California at Berkeley.

Dr. Boehm is a registered professional engineer, a Fellow of the American Society of Mechanical Engineers, and has received several awards, including the Harry Reid Silver State Research Award, the UNLV Distinguished Teaching Award, and the Rudolf W. Gunnerman Silver State Award for Excellence in Science and Technology from DRI. He has been an invited lecturer at many institutions here and abroad, and he has published over 400 papers in heat transfer, design of thermal systems, and energy conversion topics. He is the author or coauthor of the ten books. He serves as a technical editor for *Energy—the International Journal.*

BOYCE, MEHERWAN P.

Professor Meherwan P. Boyce, PhD, PE, C.Eng (UK), is the managing Partner of The Boyce Consultancy Group, LLC. He has 50 years of experience in the field of TurboMachinery in both industry and academia. Dr. Boyce is a Fellow of the American Society of Mechanical Engineers (USA), National Academy of Forensic Engineers (USA), the Institute of Mechanical Engineers (UK), and the Institution of Diesel and Gas Turbine Engineers (UK), and member of the Society of Automotive Engineers (SAE), and the National Society of Professional Engineers (NSPE), and several other professional and honorary societies such as Sigma Xi, Pi Tau Sigma, Phi Kappa Phi, and Tau Beta Phi. He is the recipient of the ASME award for Excellence in Aerodynamics and the Ralph Teetor Award of SAE for enhancement in Research and Teaching. He is also a Registered Professional Engineer in the State of Texas and a Chartered Engineer in the United Kingdom.

Industrial experience of Dr. Boyce covers 10 years with The Boyce Consultancy Group, LLC., 20 years as Chairman and CEO of Boyce Engineering International Inc., founder of Cogen Technologies Inc. His academic experience covers a 15-year period, which includes the position of Professor of Mechanical Engineering at Texas A&M University and Founder of the TurboMachinery Laboratories and The TurboMachinery Symposium, which is now in its Fortieth year.

Dr. Boyce is the author of several books such as the Gas Turbine Engineering Handbook (Third Edition, Elsevier), Handbook for Cogeneration & Combined Cycle Power Plants (Second Edition, ASME Press), and Centrifugal Compressors, A Basic Guide (PennWell Books). He is a major contributor to Perry's Chemical Engineering Handbook Seventh and Eight Editions (McGraw Hill) in the areas of Transport and Storage of Fluids, and Gas Turbines. Dr. Boyce has taught over 150 short courses around the world attended by over 3000 students representing over 300 Corporations. He is chair of ASME PTC 55 Aircraft Gas Turbine Committee on testing of aircraft gas turbines and a member of the ASME Ethics Review Board, Past Chairman of the following ASME Divisions Plant Engineering & Maintenance, the Conferences Committee and the Electric Utilities Committee.

Dr. Boyce has authored more than 150 technical papers and reports on Turbines, Compressors Pumps, and Fluid Mechanics.

Dr. Boyce received a BS (1962) and MS (1964) degrees in Mechanical Engineering from the South Dakota School of Mines and Technology and the State University of New York, respectively, and a PhD (Aerospace & Mechanical engineering) in 1969 from the University of Oklahoma.

CHORDIA, LALIT

Lalit Chordia is the Founder, President and CEO of Thar Technologies, Inc., Pittsburgh, PA, USA. Prior to starting Thar Technologies, Dr. Chordia founded two other companies: Superx Corporation and Visual Symphony. He also holds an Adjunct Research Scientist position at Carnegie Mellon University. He is a world-renowned expert in supercritical fluid technology and has pioneered its applications. Through his technological and commercial leadership, Dr. Chordia took Thar from one employee to a global leader in its field, with four technology groups in Pittsburgh and two international subsidiaries.

Dr. Chordia's company was the recipient of two National Institute of Technology's Advanced Technology Program (ATP) awards. He has been featured in several international publications, including Fortune. Dr Chordia has numerous patents and publications to his credit. He has won numerous awards, including being cited as the 2002 National Small Business Exporter of the Year by the Bush Administration and the 2009 IIT Madras Distinguished Alumnus Award. Dr. Chordia has a BS degree from IIT Madras and a PhD from Carnegie Mellon University.

EDEN, TIMOTHY J.

Dr. Timothy J. Eden joined the Applied Research Laboratory in 1990. He received a PhD from the Pennsylvania State University in 1996. His research interests include development and transition of the Cold Spray process, development and application of high performance aluminum alloys produced using Spray Metal Forming, design and fabrication of functionally tailored ceramic and composite structures, material characterization, process improvement, and material structure–performance relationships.

Dr. Eden is currently head of the Materials Processing Division in the Materials and Manufacturing Office at the Applied Research Laboratory. The Materials Processing Division includes the Advanced Coatings, Metals and Ceramic Processing, High Pressure Laboratory, and an Electronics Materials Initiative. He has participated in several large multidiscipline research programs and has helped transferred Cold Spray Technology to the U.S. Army, Navy, and to industry.

BRATTON, ROBERT

Robert Bratton is a principle investigator for graphite qualification for the NGNP at the Idaho National Laboratory. He has degrees in Nuclear Engineering and Applied Mechanics and has been employed at the INL for 18 years. He has worked on the NPR MHTGR, Light Water Tritium Target Program, and the National Spent Nuclear Fuel Program. He is a member of the ASME Project Team on Graphite Core Supports, which is developing future design codes for graphite core design.

EECEN, PETER

Peter Eecen is research manager of the group Rotor & Farm Aerodynamics at the Wind Energy department of the Energy research Centre of the Netherlands (ECN). His responsibilities are to establish the research strategy and priorities and manage a group of 20 researchers. In collaboration with the ECN Wind Industrial Support group (EWIS), he is responsible for the business development; an important aspect of ECN is to bring to the market the developed knowledge and technologies in renewable energy.

Peter holds a Master's degree in theoretical physics and did his PhD in the field of nuclear fusion at the FOM Institute for Plasma Physics, after which he worked for 3 years at TNO in the field of underwater acoustics.

Dr. Eecen joined the Wind Energy department of ECN in 2000 as manager of the group "Wind and Waves," concerning the wind and wave descriptions for turbine loading and wind resource assessments. After that, he led the experimental department for a year. He has been working in the field of Operation and Maintenance of large offshore wind farms. He started the project of Operation and Maintenance Cost Estimator (OMCE). Since 4 years, Peter is heading the group Rotor & Farm Aerodynamics. This group aims to optimize the aerodynamic performance of the wind turbine rotor and of the wind farm as a whole and reduce the uncertainties in modeling rotor aerodynamics, wake aerodynamics, boundary layers by development of CFD technology, development of aerodynamic design tools for wind turbines and wind farms.

During his career in wind energy, Dr. Eecen performed research in a variety of areas, which include modeling wind and waves, resource assessments, uncertainties in wind measurements, remote sensing, operation, and maintenance. He was responsible for measurements on full-scale wind farms and the ECN scale wind farm.

Peter is active in international organizations like IEA, MEASNET, TPWind, and European projects. He is coordinating the subprogram Aerodynamics of EERA-Wind, the European Energy Research Alliance. Ten leading European Research Institutes have founded EERA to accelerate the development of new energy technologies by conceiving and implementing Joint Research Programs in support of the Strategic Energy Technology (SET) plan by pooling and integrating activities and resources.

GAMBLE, SIMON

Simon Gamble is an accomplished leader in the Australian renewable energy industry, with over a decade of practical experience in the technical, commercial, strategic, and managerial aspects of the renewable energy development business. This leadership was recently recognized through Simon's selection as a Fulbright Scholar, through which Simon will spend 4 months with the National Renewable Lab in Colorado investigating emerging renewable energy technologies and their application in remote power systems.

Currently, as Manager Technology and Commercialization with Hydro Tasmania, Simon is responsible for the development and implementation of Hydro Tasmania's Renewable Energy and Bass Strait Islands Development Strategies; for the assessment of new and emerging renewable energy technologies; for Hydro Tasmania's Research and Development program; and for the preparation of remote area power system project feasibility assessments, project approvals, and business cases.

Simon has a Bachelor of Civil Engineering and a Masters of Engineering Science degrees from the University of Adelaide. He also has a Master's of Business Administration degree from the University of Tasmania. Simon sits on the advisory board for UTAS Centre for Renewable Energy and Power Systems and the Clean Energy Council Emerging Technology Directorate.

GONZÁLEZ AGUILAR, JOSÉ

Dr. Jose González is Senior Researcher in the R&D Unit of High Temperature Processes at the IMDEA Energia Institute. He received his PhD in Physics from the University of Cantabria (Spain) in 1999 and his Habilitation à Diriger des Recherches from the University Paul Sabatier, Toulouse (France) in 2007. Between 2000 and May 2009, he worked as R&D engineer — Project manager at the Center for Energy and Processes — MINES ParisTech. In September 2006, he became associate professor at MINES ParisTech (or Ecole nationale supérieure des mines de Paris, ENSMP).

The main research area of Dr. González is focused on the study and development of high temperature processes for energy and environmental issues, with special emphasis in concentrating solar systems and plasma technologies. His expertise includes process simulation from systems analysis (flow sheeting) to computational fluid dynamics. José González has participated in 15 national and international research projects, published 26 papers in peer review journals, two international patents, and a French patent, and he is author of more than 50 communications in national and international conferences.

HASEGAWA, KUNIO

Dr. Kunio Hasegawa graduated from Tohoku University with a Doctor of Engineering degree in 1973. He joined Hitachi Research Laboratory, Hitachi Ltd., over 30 years back. During his term at Hitachi, he was also a visiting professor of Yokohama National University and Kanazawa University for several years. Since 2006, Dr. Hasegawa serves as a principal staff in Japan Nuclear Energy Safety Organization (JNES).

Dr. Kunio Hasegawa is a member of Japan Society of Mechanical Engineers (JSME) and is a past member of the JSME Fitness-for-Service Committee for nuclear facilities. He is also a member of ASME and is involved in ASME Boiler and Pressure Vessel Code Section XI Working Group, Subgroup and Subcommittee activities.

He has been active for 3 years as a Technical Program Representative of Codes and Standards Technical Committee in ASME PVP Division. He has been involved with structural integrity for nuclear power components, particularly, leak-before-break, fracture and fatigue strengths for pipes with cracks and wall thinning, and flaw characterizations for fitness-for-service procedures. Dr. Hasegawa has published over 100 technical papers in journals and conference proceedings.

HEDDEN, OWEN F.

Owen F. Hedden retired from ABB Combustion Engineering in 1994 after over 25 years of ASME B&PV Committee activities with company support. His responsibilities included reactor vessel specifications, safety codes and standards, and interpretation of the B&PV Code and other industry standards. He continued working part-time for that organization into 2002. Subsequently, he has been a part-time consultant to the ITER project and several other organizations. Prior to joining ABB, he was with Foster Wheeler Corporation (1956 to 1967), Naval Nuclear program. Since 1968, Mr. Hedden has been active in the Section XI Code Committee, Secretary (1976 to 1978), Chair (1991 to 2000).

In addition to Section XI, Owen has been a member of the ASME C&S Board on Nuclear Codes and Standards, the Boiler and Pressure Vessel Committee, and B&PV Subcommittees on Power Boilers, Design, and Nondestructive Examination. He is active in ASME's PVP Division. Mr. Hedden was the first Chair of the NDE Engineering Division, 1982 to 1984. He has presented ASME Code short courses in the United States and overseas. He was educated at Antioch College and Massachusetts Institute of Technology. His publications are in the ASME Journal of Pressure Vessel Technology, WRC Bulletins and in the Proceedings of ASME PVP, ICONE, IIW, ASM, and SPIE. He is an ASME Fellow (1985), received the Dedicated Service Award (1991), and the ASME Bernard F. Langer Nuclear Codes and Standards Award in 1994.

HOFFELNER, WOLFGANG

Wolfgang Hoffelner is currently manager of the High Temperature Materials project at the Swiss Paul Scherrer Institute. He represents Switzerland in the Generation IV System Steering Committee and in the Project Management Board for VHTRs. He supports as PSI volunteer in the current ASME Sect III Div. 5 Code development. He is also Managing Director of RWH consult LlC, a Swiss-based consulting entity for materials and energy-related consultancy. In this function, he acts as task advisor and materials data analyst for ASME LlC. Wolfgang has been Senior Lecturer for High Temperature Materials at the Swiss Federal Institute of Technology since 1986, and he is currently responsible for the materials education within the Swiss Master of Nuclear Engineering Program.

Wolfgang received his PhD in Physics and has an MS in mathematics at the University of Vienna. He started his work as a research fellow at the same place. He improved his skills in structural materials and mechanics during his time at ABB (formerly BBC), where he was working in different positions ranging from a scientist in the Research Laboratory, Group Leader in the Laboratory, and Head of Section Mechanics and Materials for Gas Turbines and Combined Cycle Plants. In 1990, he joined the Swiss Company MGC-Plasma Inc. as a Board member, where he was responsible for technology of metallurgical and environmental applications (including low-level nuclear waste) of thermal plasma.

Dr. Hoffelner is member of ASME, ASM, and TMS, and he has published more than 120 papers in scientific and technical books and journals.

JACOBSON, PAUL T.

Paul T. Jacobson is the Ocean Energy Leader and a Senior Project Manager at the Electric Power Research Institute. Dr. Jacobson is also a faculty member in the Zanvyl Krieger School of Arts and Sciences, Johns Hopkins University, where he teaches a graduate-level course in ecological assessment. He holds a bachelor's degree in biology from Cornell University and MS and PhD degrees in oceanography and limnology from the University of Wisconsin-Madison. Dr. Jacobson has been engaged in assessment of electricity-generation systems and living resources for more than 30 years. Much of his work over this period has addressed the effects of electricity generation on aquatic ecosystems.

JENNER, MARK

Dr. Mark Jenner is a biomass systems economist with the consulting firm, Biomass Rules, LLC and the California Biomass Collaborative. Jenner creates and adds value to biomass through his expertise in biomass production and conversion technologies, as well as environmental and energy policies.

Since 2009, Mark Jenner has been studying the adoption economics of purpose grown energy crops with the California Biomass Collaborative, located at the University of California Davis. In 2003, he began his consulting firm Biomass Rules, LLC, which conducts feasibility studies on value-added biomass projects and biomass inventories. In 2006, Mark wrote the BioTown, USA Sourcebook for the State of Indiana. Since 2007, Mark Jenner has written the biomass energy outlook column for BioCycle Magazine.

Mark Jenner has a PhD in agricultural economics in production systems, two MS degrees in manure management, a BS in agronomy, and 30 years of professional biomass experience spanning three continents.

LYONS, KEVIN W.

Kevin W. Lyons is a Senior Research Engineer within the Manufacturing Engineering Laboratory (MEL), National Institute of Standards and Technology (NIST). His current assignment involves supporting the Sustainable Manufacturing Program in formalizing manufacturing resource descriptions, manufacturing readiness modeling, and simulation. His research interests are design and manufacturing processes for sustainable manufacturing, simulation and modeling, and nanomanufacturing. From 2004 through 2006, he served as Program Director for the Nanomanufacturing Program at the National Science Foundation (NSF). From 2000 to 2004, he served as Program Manager of the Nanomanufacturing Program at NIST. From 1996 to 2000, he served as Program Manager with the Defense Advanced Research Projects Agency (DARPA), where he managed advanced design and manufacturing projects. From 1977 to 1992, he worked in industry in various staff and supervisory positions in engineering marketing, product design and analysis, factory automation, and quality engineering.

MARTIN, HARRY F.

Harry F. Martin retired from Siemens Energy Corp as an Advisory Engineer at the Orlando Florida Facility. He has over 40 years of engineering experience in the power industry. Most of this related to turbo machinery. His engineering career started at Westinghouse Electric Corporation in Lester, PA. At Westinghouse, most of this experience was related to steam turbines. However, he also had assignments relating to gas turbines and heat exchangers. Harry held positions of various levels engineering responsibility and management. With Siemens, his efforts focused primarily on steam turbine design and operation.

Mr. Harry Martin has a Bachelor of Mechanical Engineering degree and Masters of Science Degree in Mechanical Engineering. His experiences include design, product and technology development, and operation of steam turbines. He has published 16 papers. These have included the subjects of turbine design, blading development and operation of steam turbines including transient analysis. He has ten patents. His technical specialization is in thermodynamics, fluid mechanics, and heat transfer. He is past Chairman of the Turbines, Generators and Auxiliaries Committee of the Power Division of the American Society of Mechanical Engineers.

MCDONALD, DENNIS K.

Denny McDonald is a Technical Fellow, Advanced Technology Development & Design, The Babcock & Wilcox Company.

Denny is currently responsible for the technical development and design of oxycombustion within B&W. He has led the conversion of B&W's 30 MWth Clean Environment Development Facility for oxy-coal testing and is deeply involved in oxycombustion performance and cost studies, process, and equipment design improvements, and emerging associated technologies. In addition to advancing the technology, he provides technical support for demonstration and commercial opportunities.

Mr. McDonald joined B&W in 1972 and has worked in various engineering capacities through his career. Up to 1985, he worked in various positions of increasing responsibility in the fields of mechanical design of boilers, field problem resolution including involvement in startup of a large utility PC plant, and development of design standards. From 1985 to 1995, he managed B&W's New Product Engineering department and had technical responsibility for B&W's scope of American Electric Power's CCT-I Tidd PFBC Demonstration Project. From 1995 until assuming his present position in late 2006, he served as Manager of Functional Technology responsible for development of B&W's core performance analysis and thermal hydraulic technologies including design standards and software, boiler performance testing, advanced computational modeling, and technical support of contract engineering and advanced coal-fired technologies including ultra-supercritical boilers. In recent years, he has contributed significantly to B&W's mercury removal program as well as oxycombustion development.

Denny McDonald holds BS and MS degrees in Engineering and is a licensed Professional Engineer in the State of Ohio. He has published over 40 technical papers, authored chapters in the 40th and 41st editions of B&W's "STEAM — its Generation and Use" and holds eight U.S. patents.

MEHTA, HARDAYAL S.

Dr. Mehta received his BS in Mechanical Engineering degree from Jodhpur University (India), MS and PhD from University of California, Berkeley. He was elected an ASME Fellow in 1999 and is a Registered Professional Engineer in the State of California.

Dr. Mehta has been with GE Nuclear Division (now, GE-Hitachi Nuclear Energy) since 1978 and currently holds the position of Chief Consulting Engineer. He has over 35 years of experience in the areas of stress analysis, linear-elastic and elastic-plastic fracture mechanics, residual stress evaluation, and ASME Code-related analyses pertaining to BWR components. He has also participated as principal investigator or project manager for several BWRVIP, BWROG, and EPRI sponsored programs at GE, including the Large Diameter Piping Crack Assessment, IHSI, Carbon Steel Environmental Fatigue Rules, RPV Upper Shelf Margin Assessment, and Shroud Integrity Assessment. He is the author/coauthor of over 40 ASME Journal/Volume papers. Prior to joining GE, he was with Impell Corporation, where he directed various piping and structural analyses.

For more than 25 years, Dr. Mehta has been an active member of the Section XI Subgroup on Evaluation Standards and associated working task groups. He also has been active for many years in ASME's PVP Division as a member of the Material & Fabrication Committee and as conference volume editor and session developer. His professional participation also included several committees of the PVRC, specially the Steering Committee on Cyclic Life and Environmental Effects in Nuclear Applications. He had a key role in the development of environmental fatigue initiation rules that are currently under consideration for adoption by various ASME Code Groups.

MILES, THOMAS R.

Thomas R. Miles is the President and Owner of T.R. Miles Technical Consultants, Portland, Oregon, which designs, develops, installs, and tests agricultural and industrial systems for fuel handling, air quality, and biomass energy. Energy projects include combustion and gasification of biomass fuels such as wood, straws, stalks, and manures. Mr. Miles conducts engineering design and feasibility studies and field tests for cofiring wood, straw, and coal. He has sponsored and hosted internet discussions on biomass energy since 1994 (www.trmiles.com).

MORTON, D. KEITH

Mr. D. Keith Morton is a Consulting Engineer at the Department of Energy's (DOE) Idaho National Laboratory (INL), operated by Battelle Energy Alliance. He has worked at the INL for 35 years. Mr. Morton has gained a wide variety of structural engineering experience in many areas, including performing nuclear piping and power piping stress analyses, completing plant walk downs, consulting with the Nuclear Regulatory Commission, developing life extension strategies for the Advanced Test Reactor, performing full-scale seismic and impact testing, and helping to develop the DOE standardized spent nuclear fuel canister. His most recent work activities include performing full-scale drop tests of DOE spent nuclear fuel canisters, developing a test methodology that allows for the quantification of true stress-strain curves that reflect strain rate effects and supporting the Next Generation Nuclear Plant (NGNP) Project.

Mr. Morton is a Member of the ASME Working Group on the Design of Division 3 Containments, is the Secretary for the ASME Subgroup on Containment Systems for Spent Fuel and High-Level Waste Transport Packagings, a Member of the ASME Working Group on High Temperature Gas-Cooled Reactors, a Member of the Subgroup on High Temperature Reactors, a Member of the Section III Subgroup on Strategy and Management, and is a Member of the ASME BPV III Standards Committee. He has coauthored over 25 conference papers, one journal article, coauthored an article on DOE spent nuclear fuel canisters for *Radwaste Solutions*, and coauthored Chapter 15 of the third edition of the *Companion Guide to the ASME Boiler & Pressure Vessel Code.*

Mr. Morton received a BS in Mechanical Engineering degree from California Polytechnic State University in 1975 and a Masters of Engineering in Mechanical Engineering from the University of Idaho in 1979. He is a Registered Professional Engineer in the state of Idaho.

NOTTINGHAM, LAWRENCE (LARRY) D.

Lawrence D. (Larry) Nottingham is a Senior Associate, Structural Integrity Associates (SI), Inc. at Charlotte, NC. From June 1995 to the present, he has been with Structural Integrity Associates. From 1993 to 1995, he was Founder, President, and Managing Director of AEA Sonomatic, Inc., Charlotte, NC. From 1986 through 1993, Larry was with Electric Power Research Institute at Nondestructive Evaluation Center, Charlotte, NC. From 1972 through 1986, Larry was with Westinghouse Electric Corp. as Manager at Steam Turbine Generator Division, Orlando, FL and Senior Development Engineer, at Large Rotating Apparatus Division, Pittsburgh, PA.

Mr. Nottingham graduated with a BS Mechanical Engineering degree in the University of Pittsburgh in 1971. His Professional Associations and Certifications include Nondestructive Evaluation (NDE) Level III Certification in Ultrasonic Testing, Penetrant Testing (PT), and Magnetic Particle Testing (MT).

Mr. Nottingham has been involved in design, design analysis, maintenance, and nondestructive evaluation of turbines, generators, and other power plant equipment and components since 1972. His experience covers all aspects of design and design analysis including finite element stress analysis, fracture mechanics, materials testing and characterization, failure modes and mechanisms, metallurgy, and nondestructive evaluation. He has extensive experience in the development and delivery of advanced nondestructive evaluation systems and procedures for numerous power plant applications, with emphasis on turbine and generator components, including boresonic and turbine disk rim inspection systems.

At Structural Integrity (SI), Mr. Nottingham continues to provide broad-based engineering expertise. Until 2007, he managed all engineering development efforts for both nuclear and fossil plant inspection services. In his role as Development Manager, he developed SI as a recognized technological leader in the power generation NDE services community. He also has remained active on a number of EPRI projects involving fossil power plant components, generator retaining rings, generator rotors, boiler tube, and most recently developing guideline document for inspection and life assessment of turbine and valve casings. In 2007, Mr. Nottingham assumed responsibility for SI's turbine and generator condition assessment efforts, with aggressive growth objectives.

Mr. Nottingham has published over 70 technical papers, reports, and articles. He has been an invited presenter at numerous conferences, workshops, and seminars on NDE and lecturer at a number of training courses. He provides training for advanced non-destructive evaluation technologies and has been invited lecturer at the United States Naval Academy. He currently holds 14 U.S. patents related to turbine and generator component designs and NDE systems. He is also a member of SI's Board of Directors.

O'DONNELL, WILLIAM J.

Bill O'Donnell has Engineering Degrees from Carnegie Mellon University and the University of Pittsburgh. He began his career at Westinghouse Research and Bettis, where he became an Advisory Engineer. In 1970, Bill founded O'Donnell and Associates, an engineering consulting firm specializing in design and analysis of structures and components. The firm has done extensive work in the evaluation of structural integrity, including corrosion fatigue, flaw sensitivity, crack propagation, creep rupture, and brittle fracture.

Dr. O'Donnell has published 96 papers in engineering mechanics, elastic-plastic fracture mechanics, strain limits, and damage evaluation methods. He is Chairman of the Subgroup on Fatigue Strength and a Member of the Subcommittee on Design of the ASME Code. He has patents on mechanical processes and devices used in plants worldwide. He is a recognized expert in Failure Causation Analyses.

Dr. O'Donnell has given invited lectures at many R&D laboratories, design firms, and universities. He is a registered Professional Engineer. He received the National Pi Tau Sigma Gold Medal Award "For Outstanding Achievement in Mechanical Engineering" and the ASME Award for "Best Conference Technical Paper" in 1973 and 1988. The Pittsburgh Section of ASME named Bill "Engineer of the Year" (1988). He was awarded the ASME PVP Medal (1994).

Dr. O'Donnell received the University of Pittsburgh Mechanical Engineering Department's Distinguished Alumni Award (1996) and Carnegie Mellon University's 2004 Distinguished Achievement Award for distinguished service and accomplishments in any field of human endeavor. He is a Fellow of the ASME and is listed in the Engineers Joint Council "Engineers of Distinction," Marquis "Who's Who in Science and Engineering," and "Who's Who in the World."

PIEKUTOWSKI, MARIAN

Marian Piekutowski is a recognized leader in the field of power system planning and analysis including transmission, generation, and economic analysis. He has been instrumental in development of wind integration strategies for Hydro Tasmania. With more than 30 years of working experience, he has developed extensive understanding of regulatory environment of electricity markets and of the requirements for efficient management and operation of an electricity utility.

During the last 2 years, Marian has been involved in demand, supply, and regulatory aspects of frequency control ancillary services. This work includes improvement of dynamic response of hydro turbines, improvement governing, new modes operation of hydro machines, and impact of low inertia on the frequency control.

Marian has been involved in integration of renewable energy sources to remote power systems with an aim to minimize diesel consumption and reduce emission of GHG. His recent work included energy storage (VRB, flywheels, diesel UPS), applications of power electronics to maximize penetration of wind generation, control strategies for operation of small islanded systems, and improvement of grid stability and reliability.

Marian has been the lead technical advisor for the connection of Woolnorth stages 1 and 2 wind farms in Tasmania as well as Cathedral Rocks in South Australia. Marian has also been instrumental from the developers perspective in development of fault ride through capability on wind turbines. He has led studies to determine maximum viable wind penetration in Tasmanian power systems with the main sources of limitations being low inertia and low fault level.

POLING, CHRISTOPHER W.

Mr. Poling is currently the Program Director for The Babcock & Wilcox Company's Post-Combustion Carbon Capture product development project. He joined B&W in 2002 and has worked in areas of increasing responsibility including Lead Proposal Engineer for large Flue Gas Desulfurization projects and as a Principal Engineer in B&W's Technology department. Prior to his experience at B&W, Mr. Poling worked for 7 years at Ceilcote Air Pollution Control in Strongsville, Ohio as a Product Manager for industrial wet scrubber systems. Mr. Poling earned his bachelor's degree in chemical engineering from the University of Toledo. Mr. Poling also earned his Executive Master's in Business Administration from Kent State University.

MUTHYA, RAMESH PRANESH RAO

Mr. Ramesh holds a Master's degree in Mechanical Engineering from the Indian Institute of Science from Bengaluru, India, and a Master of Science from Oldenburg University, Germany, in Application of Renewable Energy Technologies.

Mr. Ramesh started his career in wind energy at the National Aeronautical Laboratory, Bangalore, in 1979. He has many firsts to his credit starting from development of indigenous wind power battery chargers and transferred know-how to industry. He made performance measurements on wind turbines for the first time in India, which resolved many issues of underperformance. In order to extend utility of measured wind information to a larger area, he came up with the idea of localized wind maps superimposed on scaled survey maps. Over a hundred such reports were created and were the basis for wind farming. He holds a patent on passive speed control of windmills with Dr. S. K. Tewari. He introduced small wind chargers in Indian Antarctic station.

Mr. Ramesh took over the position of the first full time Executive Director, Centre for Wind Energy Technology in 2002. He was instrumental in maximizing the benefits of the DANIDA-funded project. He went on to create ground rules for effective, orderly, and sustainable growth of the field by continuous interactions with the industry, the Government, and other stake holders.

After a short stint in Canada as the scientific director/advisor to GP CO, Varrenes, Mr. Ramesh returned to India to help the Indian wind industry. As the founding Managing Director of Garrad Hassan India, Mr. Ramesh brought in some of the most sought-after consulting practices to India that spans across the resource quantification to asset management techniques. One of the least attended to areas in terms of design documentation capabilities in India was effectively addressed by Mr. Ramesh.

He has served as Chairman, Electro Technical committee #42 (ET-42), set up by Bureau of Indian Standards to interact with International Electrotechnical committee on all wind energy-related standardization efforts. He also chaired the committee that brought out draft grid code for wind turbine grid interconnection.

Presently, as the President, Wind Resource and Technology, at Enercon India, Mr. Ramesh is actively involved in bringing in the best practices on all aspects of wind energy deployment in India.

RAO, K. R.

K.R. Rao retired as a Senior Staff Engineer with Entergy Operations Inc. and was previously with Westinghouse Electric Corporation at Pittsburgh, PA, and Pullman Swindell Inc., Pittsburgh, PA. KR got his Bachelors in Engineering degree from Banaras University, India, with a Masters Diploma in Planning from School of Planning & Architecture, New Delhi, India. He completed Post Graduate Engineering courses in Seismic Engineering, Finite Element and Stress Analysis, and other engineering subjects at Carnegie Mellon University, Pittsburgh, PA. He earned his PhD, from University of Pittsburgh, PA. He is a Registered Professional Engineer in Pennsylvania and Texas. He is past Member of Operations Research Society of America (ORSA).

KR was Vice President, Southeastern Region of ASME International. He is a Fellow of ASME, active in National, Regional, Section, and Technical Divisions of ASME. He has been the Chair, Director, and Founder of ASME EXPO(s) at Mississippi Section. He was a member of General Awards Committee of ASME International. He was Chair of Codes & Standards Technical Committee, ASME PV&PD. He developed an ASME Tutorial for PVP Division covering select aspects of Code. KR is a member of the Special Working Group on Editing and Review (ASME B&PV Code Section XI) for September 2007 to June 2012 term.

Dr. Rao is a recipient of several Cash, Recognition, and Service Awards from Entergy Operations, Inc. and Westinghouse Electric Corporation. He is also the recipient of several awards, certificates, and plaques from ASME PV&P Division including Outstanding Service Award (2001) and Certificate for "Vision and Leadership" in Mississippi and Dick Duncan Award, Southeastern Region, ASME. Dr. Rao is the recipient of the prestigious ASME Society Level Dedicated Service Award. KR is a member of the Board of ASME District F Professional & Educational Trust Fund for 2008 to 2011.

Dr. Rao is a Fellow of American Society of Mechanical Engineers, Fellow of Institution of Engineers, India, and a Chartered Engineer, India. Dr. Rao was recognized as a "Life Time Member" for inclusion in the Cambridge "Who's Who" registry of executives and professionals. Dr. Rao was listed in the Marquis 25th Silver Anniversary Edition of "Who's Who in the World" as "one of the leading achievers from around the globe".

RAYEGAN, RAMBOD

Rambod Rayegan is a PhD candidate in the Department of Mechanical and Materials Engineering at Florida International University. Since January 2007, he has worked as a research assistant at FIU in the sustainable energy area. He is also the president of ASHRAE FIU Chapter since May 2009. He has been a member of prestigious Honor Societies like Tau Beta Pi, Phi Kappa Phi, Sigma Xi, and Golden Key. He has published a number of conference and journal papers in energy and sustainability area. Raised in Tehran, Iran, Rambod now lives with his wife in Miami. He has served as an instructor at Semnan University, Iran, for 5 years. He was selected as the best teacher of the Mechanical Engineering Department by students during the 2002 to 2003 academic year and the best senior project supervisor in the 2003 to 2004 academic year. He has served as a consultant in three companies in the field of air conditioning and hydraulic power plants.

REEDY, ROGER F.

Roger F. Reedy has a BS Civil Engineering degree from Illinois Institute of Technology (1953). His professional career includes the U.S. Navy Civil Engineering Corps, Chicago Bridge and Iron Company (1956 to 1976). Then, he established himself as a consultant and is an acknowledged expert in design of pressure vessels and nuclear components meeting the requirements of the

ASME B&PV Code. His experience includes design, analysis, fabrication, and erection of pressure vessels and piping components for nuclear reactors and containment vessels. He has expertise in components for fossil fuel power plants and pressure vessels and storage tanks for petroleum, chemical, and other energy industries. Mr. Reedy has been involved in licensing, engineering reviews, welding evaluations, quality programs, project coordination, and ASME Code training of personnel. He testified as an expert witness in litigations and before regulatory groups.

Mr. Reedy has written a summary of all changes made to the ASME B&PV Code in each Addenda published since 1950, which is maintained in a computer database, RA-search. Mr. Reedy served on ASME BP&V Code Committees for more than 40 years being Chair of several of them, including Section III for 15 years. Mr. Reedy was one of the founding members of the ASME PV&P Division. Mr. Reedy is registered.

Mr. Reedy is a Professional Engineer in seven states. He is a recipient of the ASME Bernard F. Langer Award and the ASME Centennial Medal and is a Life Fellow of ASME.

ROBINSON, CURT

Curt Robinson is the Executive Director of the 1770-member Geothermal Resources Council (GRC), headquartered in Davis, California. Since 1970, the GRC has built a solid reputation as the world's leading geothermal association. The GRC serves as a focal point for continuing professional development for its members through its fall annual meeting, transactions, bulletin, outreach, information transfer, and education services. GRC has members in 37 countries.

Prior to his work at GRC, he held executive assignments in higher education and government and has twice worked in energy development. He has also taught at six universities and colleges.

He earned his PhD and MA degrees in geography and a BA with honors, all at the University of California, Davis.

RICCARDELLA, PETER (PETE) C.

Pete Riccardella received his PhD from Carnegie Mellon University in 1973 and is an expert in the area of structural integrity of nuclear power plant components. He co-founded Structural Integrity Associates in 1983 and has contributed to the diagnosis and correction of several critical industry problems, including:

- Feedwater nozzle cracking in boiling water reactors
- Stress corrosion cracking in boiling water reactor piping and internals
- Irradiation embrittlement of nuclear reactor vessels
- Primary water stress corrosion cracking in pressurized water reactors
- Turbine-generator cracking and failures.

Dr. Riccardella has been principal investigator for a number of EPRI projects that led to advancements and cost savings for the industry. These include the FatiguePro fatigue monitoring system, the RRingLife software for turbine-generator retaining ring evaluation, Risk-Informed Inservice Inspection methodology for nuclear power plants, and several Probabilistic Fracture Mechanics applications to plant cracking issues. He has led major failure analysis efforts on electric utility equipment ranging from transmission towers to turbine-generator components and has testified as an expert witness in litigation related to such failures.

He has also been a prime mover on the ASME Nuclear Inservice Inspection Code in the development of evaluation procedures and acceptance standards for flaws detected during inspections. In 2002, he became an honorary member of the ASME Section XI Subcommittee on Inservice Inspection, after serving for over 20 years as a member of that committee. In 2003, Dr. Riccardella was elected a Fellow of ASME International.

ROCAFORT, LUIS A. BON

Luis A. Bon Rocafort graduated in 1999 from Purdue University with a Bachelor's of Science in Mechanical Engineering degree. During his undergraduate career, he received the National Action Council for Minorities in Engineering Scholarship, which provided for tuition and a stipend, as well as work experience as a summer intern at the sponsoring company's facilities. His work with BP Amoco, at their Whiting, IN, refinery involved plant facilities, cooling tower design and analysis, and pipe fluid flow analysis and modeling to optimize use of cooling water and eliminate bottlenecks.

In 2001, Luis A. Bon Rocafort graduated with a Master's of Science in Mechanical Engineering degree from Purdue University, having received the Graduate Engineering Minority Fellowship, to cover his graduate school as well as provide work experience with DaimlerChrysler. The knowledge gained in the advanced vehicle design group, as well as the concept and modeling group would prove useful in the modeling and analysis realm that he is currently working in. During his time with DaimlerChrysler, he characterized fluid flows inside an automatic transmission engine, wrote data capture modules for a real-time driving simulator, programmed autonomous vehicles for real-time driving simulator, and compared FEA stress analysis results to stress paint-treated parts to determine viability of two methods to real-world tests of manufactured parts.

Having graduated in 2001, Luis A. Bon Rocafort became a field service engineer for Schlumberger Oilfield Services, performing as a drilling service engineer. As a cell manager, providing services to ExxonMobil in the Bass Strait of the Southeast coast of Australia, he performs logging while drilling services and assists directional drilling efforts in order to fully develop a known field that has been producing oil for more than three decades. Using advanced tools and drilling techniques, undiscovered pockets of oils are identified, and drilling programs are developed.

Luis A. Bon Rocafort joined O'Donnell Consulting Engineers, Inc. in 2006. While working with OCEI, he performed static and transient finite element analysis using a variety of elements and methods available through the ANSYS program. Other analyses include, modal analyses, harmonic analyses, vibrations, fatigue life analyses, inelastic analyses, creep analyses, among others. These analyses were done to evaluate vessels or structures to ASME, AISC, and IEEE codes and standards.

ROMERO, MANUEL ALVAREZ

Manuel Romero received his PhD in Chemical Engineering in 1990 at the University of Valladolid for his research on the solarization of steam reforming of methane. At present, he is Deputy Director and Principal Researcher of the High Temperature Processes R&D Unit at IMDEA Energía. Dr. M. Romero has received the "Farrington Daniels Award" in 2009, the most prestigious award in the field of solar energy research, created in 1975 by the International Solar Energy Society, conferred for his intellectual leadership, international reputation, and R&D contributions to the development of high-temperature solar concentrating systems.

In June 1985, Dr. Romero joined CIEMAT, Spain's National Laboratory for Energy Research, working as Project Manager until 2002 with responsibilities on R&D for solar thermal power plants and solar hydrogen. In 2002, he became Director of the Plataforma Solar de Almería, largely recognized R&D facilities for testing and development of solar concentrating technologies, and Director of the Renewable Energy Division of CIEMAT since June 2004 until August 2008 with R&D activities on solar thermal power, photovoltaics, biomass, and wind energy.

During his career, Dr. Romero has participated in more than 45 collaborative R&D projects in energy research, 15 of them financed by the European Commission, with special emphasis on high-temperature solar towers. He is coauthor of the European Technology Roadmap on High Temperature Hydrogen Production Processes INNOHYP, contracted by the EC in 2005, and coauthor of the European Technology Roadmap for Solar Thermal Power Plants, ECOSTAR, contracted by the European Commission in 2004. He acted as member of the experts' committee of the Energy R&D Program of the VI and VII Framework Program of the EC until August 2008.

Dr. Romero is Associate Editor of the ASME Journal of Solar Energy Engineering since January 2007 and at the International Journal of Energy Research (IJER) published by Wiley & Sons since December 2009. He was Associate Editor of the International Journal of Solar Energy since January 2002 until January 2007.

SEIFERT, GARY D.

Gary D. Seifert, PE, EE, is senior program manager at the Idaho National Laboratory. He has responsibility for multiple technical tasks for the U.S. Department of Energy, Department of Homeland Security, the U.S. Air Force, U.S. Navy, and NASA, as well as various power systems upgrades at the Idaho National Laboratory. Renewable projects have included the Ascension Island Wind Project and Ascension Island Solar Power projects, which have displaced a significant amount of diesel generation resulting in major financial and emissions savings.

Gary has been involved in multiple projects improving control systems and adding automation. Other support tasks include Wind Powering America, wind anemometer loan program, wind radar integration, power system distribution upgrades, high reliability power systems, relay system updates, smart substation upgrades, fiber optic communication systems installations, National SCADA Testbed, and the design of process control systems.

Gary is also currently involved in studies for multiple Department of Defense government wind projects and is leading a technical wind radar interaction project for the U.S. DOE and supporting wind prospecting activities in Idaho and surrounding regions.

Gary holds patents in thermal photovoltaic and Electro Optical High Voltage (EOHV) sensor designs. He was awarded a Research and Development top 100 award in 1998 for his work on the EOHV sensor and was instrumental in the implementation of the DOD's first island wind farm at Ascension air station.

Gary has a Bachelor of Science Electrical Engineering degree from the University of Idaho in 1981. He is an Adjunct Instructor Department of Engineering Professional Development, University of Wisconsin since 1991.

SHEVENELL, LISA

Lisa Shevenell was awarded a B.A. in geology at New Mexico Institute of Mining and Technology in 1984 and a PhD in Hydrogeology at the University of Nevada, Reno in 1990. Shevenell conducted geothermal exploration in Central America in the mid-1980s as part of a USGS-Los Alamos National Laboratory (LANL) team. Additional basic and applied research was conducted while with LANL at numerous sites throughout the western United States. Work at Mt. St. Helens evolved into her PhD research on the geothermal systems that formed after the 1980 eruption. Following her PhD, she worked at Oak Ridge National Laboratory for 3 years. Shevenell has been a faculty member at the Nevada Bureau of Mines and Geology since 1993, where she has led numerous geothermal-related research projects and teams in Nevada. She is currently a member of the Nevada Geothermal Technical Advisory Panel to NV Energy, the Science Advisory Board to the National Geothermal Data Center initiative being led by the Arizona Geological Survey, Geothermal Energy Association Technical Advisory Committee, and member of the Blue Ribbon Panel on Renewable

Energy formed by Senator Harry Reid and former Board of Directors member to the Geothermal Resources Council, former member of the Renewable Energy Task Force reporting to the Governor and Nevada Legislature, and former Director of the Great Basin Center for Geothermal Energy.

SINGH, K. (KRIS) P.

Dr. K.P. (Kris) Singh is the President and Chief Executive Officer of Holtec International, an energy technology company that he established in 1986. Dr. Singh received his Ph.D. in Mechanical Engineering from the University of Pennsylvania in 1972, a Masters in Engineering Mechanics, also from Pennsylvania in 1969, and a BS in Mechanical Engineering from the Ranchi University in India in 1967.

Since the mid-1980s, Dr. Singh has endeavored to develop innovative design concepts and inventions that have been translated by the able technology team of Holtec International into equipment and systems that improve the safety and reliability of nuclear and fossil power plants. Dr. Singh holds numerous patents on storage and transport technologies for used nuclear fuel and on heat exchangers/pressure vessels used in nuclear and fossil power plants. Active for over 30 years in the academic aspects of the technologies underlying the power generation industry, Dr. Singh has published over 60 technical papers in the permanent literature in various disciplines of mechanical engineering and applied mechanics. He has edited, authored, or coauthored numerous monographs and books, including the widely used text "Mechanical Design of Heat Exchangers and Pressure Vessel Components," published in 1984. In 1987, he was elected a Fellow of the American Society of Mechanical Engineers. He is a Registered Professional Engineer in Pennsylvania and Michigan and has been a member of the American Nuclear Society since 1979 and a member of the American Society of Mechanical Engineers since 1974.

Over the decades, Dr. Singh has participated in technology development roles in a number of national organizations, including the Tubular Exchange Manufacturers Association, the Heat Exchange Institute, and the American Society of Mechanical Engineers. Dr. Singh has lectured extensively on nuclear technology issues in the United States and abroad, providing continuing education courses to practicing engineers, and served as an Adjunct Professor at the University of Pennsylvania (1986 to 1992).

Dr. Singh serves on several corporate boards including the Nuclear Energy Institute and the Board of Overseers, School of Engineering and Applied Science (University of Pennsylvania), Holtec International, and several other industrial companies.

SPAIN, STEPHEN D.

Stephen D. Spain, PE, PEng, is Vice President of HDR's Northwest Region, Hydropower Department and Director of Hydromechanical Engineering for all HDR and Project Manager, Project Engineer, and Lead Mechanical Engineer for numerous hydroelectric projects throughout North America. Previously, he was the Northwest Regional Manager for Devine Tarbell & Associates (DTA), Duke Engineering & Services (DE&S), Department Manager of Hydro Mechanical and Electrical Engineering for Northrop Devine & Tarbell (ND&T), and Hydroelectric Project Engineer at the E.C. Jordan Company in Portland, Maine. Stephen has served as the chair for the American Society for Mechanical Engineers' (ASME) Hydropower Committee from 2006 to 2009.

SRIRAM, RAM D.

Ram D. Sriram is currently leading the Design and Process group in the Manufacturing Systems Integration Division at the National Institute of Standards and Technology, where he conducts research on standards for sustainable manufacturing and interoperability of computer-aided design systems. He also holds a part-time appointment in the Information Technology Laboratory, where he conducts research on bioimaging and healthcare informatics. Prior to that, he was on the engineering faculty (1986 to 1994) at the Massachusetts Institute of Technology (MIT) and was instrumental in setting up the Intelligent Engineering Systems Laboratory. At MIT, Sriram initiated the MIT-DICE project, which was one of the pioneering projects in collaborative engineering. Sriram has coauthored or authored nearly 250 publications in computer-aided engineering and healthcare informatics, including several books. Sriram was a founding coeditor of the International Journal for AI in Engineering. In 1989, he was awarded a Presidential Young Investigator Award from the National Science Foundation, USA. Sriram is a Fellow of the American Society of Mechanical Engineers, a Senior Member of the Institute of Electrical and Electronics Engineers, a Member (life) of the Association for Computing Machinery, a member of the American Society of Civil Engineers, and a Fellow of the American Association of Advancement for Science. Sriram has a BS from IIT, Madras, India, and an MS and a PhD from Carnegie Mellon University, Pittsburgh, USA.

TANZOSH, JIM M.

James Tanzosh is employed at the Babcock & Wilcox Company as the Manager of Materials and Manufacturing Technology for the Power Generation Group in Barberton, Ohio. He has worked for B&W for 37 years in a number of technical areas involved with nuclear and fossil-fueled power generation including commercial and defense reactor programs, fast breeder reactor development, and a large range of utility and industrial boilers covering a wide range of fossil fuels, solar power, and biomass and refuse. He is presently responsible for research and development and all aspects of materials and welding technology for the Power Generation Group. He has been involved for the last 8 years with materials and manufacturing development of materials and designs of the advanced ultrasupercritical boiler. He has been a member of the ASME Boiler and Pressure Vessel Code and a member of a number of subgroups and committees in the area of materials, welding, and fired boilers and is presently Chairman of the Subgroup on Strength of Weldments.

TAO, YONG X.

Dr. Yong X. Tao is PACCAR Professor of Engineering and Chairperson of the Department of Mechanical and Energy Engineering at the University of North Texas (UNT). He is an ASME Fellow and Editor-in-Chief of Heat Transfer Research with more than 20 years of research and teaching experience. Prior to joining UNT, he was the Associate Dean of the College of Engineering and Computing at Florida International University in Miami and a Professor of Mechanical and Materials Engineering. An internationally known researcher in fundamentals of thermal sciences, refrigeration system performance, and renewable energy applications in buildings, he was also Director of the Building Energy, Environment, and Conservation Systems Lab (BEECS) and Multi-Phase Thermal Engineering Lab (MPTE) at FIU.

Dr. Tao has produced a total of more than 154 journal publications, book chapters, edited journals and proceedings, and peer-reviewed technical conference papers over the course of his career and holds two patents. He has received more than 12.2 million dollars of research funding as a single PI or Co-PI in multidisciplinary teamwork projects from the NSF, NASA, Air Force, DSL, DOE, ASHRAE, and various industries. He was the Associate Editor of the Journal of Science and Engineering Applications.

Dr. Tao is also an active member of the American Society of Heating Refrigeration and Air-Conditioning Engineers (ASHRAE) and member of Executive Committee of the Heat Transfer Division of ASME, and Editor of ASME Early-Career Technical Journal. He has served on many technical committees for ASME, ASHRAE, and AIAA. He was also the Program Chair for the 2009 Summer Heat Transfer Conference of ASME and, as the Founding Chair, established the first US-EU-China Thermophysics Conference on Renewable Energy held in Beijing in May 2009.

In 2005, he was the faculty leader of the award-winning FIU Solar Decathlon entry sponsored by the United States. In 2008, as Project Director of the Future House USA project, he led a consortium of academics, builders, industry sponsors, and lobbyists to represent the United States in a ten-country, international demonstration project of renewable energy and environmentally friendly construction that resulted in a 3200-sq ft zero-net-energy American House in Beijing, China. On July 16th, 2009, Dr. Tao hosted a visit from the U.S. Secretary of Commerce Gary Locke and Secretary of Energy Steven Chu in the American House and was praised by both Secretaries as playing "vital role in building better collaboration between the United States and China in the area of energy-efficient buildings."

Dr. Tao has a PhD in Mechanical Engineering degree from the University of Michigan, and a BS and MS in Mechanical Engineering from Tongji University in Shanghai, China.

THAREJA, DHARAM VIR

Dr. Dharam Vir THAREJA is the Director — Technical, SNC-Lavalin Engineering India Pvt. Ltd. since 2009 to date. Previously. he was a consultant with SNC-LAVALIN Engineering India Pvt. Ltd., Institute for Defence Studies and Analysis (IDSA), J&K Power Development Corporation and HP Power Development Corporation.

Dr. Thareja held various senior assignments from 1990 through 2008, chronologically the most recent to the last are Chairman Ganga Flood Control Commission (GFCC), Ministry of Water Resources, Govt. of India (GOI); Member (Design & Research Wing), CWC, Ministry of Water Resources; Commissioner, Indus Wing, Ministry of Water Resources; Chief Engineer, CWC, Ministry of Water Resources; and Chief Project Manager, WAPCOS, Ministry of Water Resources. Dr. Thareja worked previously from 1973 to 1999 in several capacities in CWC and WAPCOS. Dr. Thareja was responsible for several publications that include 21 notable ones.

He earned his PhD IIT, Delhi, India, MSc and BSc in Civil Engineering from the College of Engineering, University Of Delhi, India. Dr. Thareja's professional affiliations include Fellow, Institution of Engineers (India), Member, Indian Water Resources Society, Indian Geotechnical Society, Indian Society for Rock and Mechanics & Tunneling Technology. He attended several institutions including UN Fellowship; USBR, Denver, Colorado (USA); University of California, Berkeley (USA); University of Arizona, Tucson (USA); University of Swansea, Swansea (UK); and Hydro Power Engineering with M/s. Harza Engineering Co., USA under the World Bank program.

Dr Thareja's professional work country experience includes the Philippines, Burma, Vietnam, Afghanistan, Bhutan, and India.

TOUSEY, TERRY

Terry Tousey, an Independent Consultant at Alternative Fuels & Resources, LLC and President of Rose Energy Discovery, Inc., has a diverse background in the alternative energy, resource recovery, environmental and chemical industries. He has over 22 years of experience in the development, implementation, and management of hazardous and nonhazardous waste fuel projects and substitute raw material programs within the cement industry. Mr. Tousey has spent most of the last 5 years working on the commercialization of renewable energy technologies including gasification and anaerobic digestion of waste biomass materials for the production of heat and power.

Mr. Tousey was a key member of the management team at two startup resource recovery companies where, among other things, he directed the business development strategy for sourcing waste materials into the alternative fuels and raw materials programs. He has reviewed the quality and quantity of numerous waste streams for use as an alternative fuel or substitute raw material and has researched a number of technologies for processing these materials into a useable form. Mr. Tousey has extensive expertise in managing these programs from concept through startup including permitting, design, construction, operations, logistics, marketing, and regulatory compliance. His work on a wide range of highly innovative alternative energy projects, both captive and merchant, over the course of his career, has made him uniquely knowledgeable in the dynamics of resource recovery and the mechanics of the reverse distribution chain of waste from the generator to the processor.

Mr. Tousey is an alumnus of Purdue University, where he earned his BS Degree in General Science with a major concentration in Chemistry and minor concentrations in Biology and Mathematics. He is a member of the Water Environment Association, New England Water Environment Association, and Missouri Water Environment Association and a past member of the National Oil Recyclers Association and the National Chemical Recyclers Association. He has been an active participant in the Environmental Information Digest's Annual Industry Round Table where Mr. Tousey has given a number of presentations on the use of waste as fuels in the cement industry.

VITERNA, LARRY

Dr. Larry Viterna is a loaned executive to Case Western Reserve University from NASA. At Case, he serves as the Technical Director of the Great Lakes Energy Institute, leading the formation of technology development efforts in renewable energy and energy storage. Most recently at NASA, Dr. Viterna was Lead for Strategic Business Development at the Glenn Research Center, a major federal laboratory with a budget of over $600M and a workforce of 2500. Previously, he was assigned to the NASA Deputy Administrator in Washington, DC, where he coordinated the development of the implementation strategy for Agency-wide changes

following the Space Shuttle Columbia Accident. Dr. Viterna was on the team that created the world's first multimegawatt wind turbines starting in 1979. He is the recipient of NASA's Blue Marble Award for aerodynamic models, now named for him, that are part of international design tools for wind turbines and that helped enable passive aerodynamic power control in the wind energy industry. Dr. Viterna has also been recognized with NASA Glenn's highest Engineering Excellence award for his pioneering work in fuel efficient hybrid vehicles. He is certified for the Senior Executive Service and received his PhD in Engineering from the University of Michigan. He has also completed executive education in business administration at Stanford University, public policy at the Harvard Kennedy School of Government, and international management at the National University of Singapore.

WEAKLAND, DENNIS P.

Mr. Weakland has over 28 years of experience in materials behavior and structural integrity of major nuclear components at an operating nuclear power plant. He is recognized to have a broad understanding of materials issues in the Industry by the leadership positions held in several organizations. He specializes in ASME Code compliance, technical and program review, evaluation of Industry technologies for degradation mitigation, evaluation of material degradation concerns, and the oversight of fabrication activities for new or replacement components. His experience with Inservice Inspection and materials programs has provided him with a thorough understanding of nondestructive examination techniques and applications.

Mr. Dennis Weakland has served in several industry leadership roles, including the Chairman of the Pressurized Water Reactor Owners Groups Materials Sub-committee and the Chairman of the EPRI Materials Reliability Project (MRP). He currently is a member of the ASME Working Group — Operating Plant Criteria and Task Group Alloy 600. He is an alumnus of Carnegie-Mellon University where he earned a BS in Metallurgy and Material Science degree. He also has earned a MBA for the University Of Pittsburgh Katz School of Business.

Prior to joining the nuclear industry in 1982, Mr. Weakland spent 13 years in the heavy fabrication industry in the production of river barges, towboats, and railroad cars.

WEISSMAN, ALEXANDER

Alexander Weissman is a doctoral student in the Department of Mechanical Engineering and the Institute for Systems Research at the University of Maryland. He is currently working on research in design-stage estimation of energy consumption in manufacturing processes. His broader research goals include sustainable manufacturing and design for environment. Prior to this, he worked as a software engineer and developer for an automated analysis and process planning system for water-jet machining. He completed a Bachelor of Science (BS) degree in computer engineering in 2006 at the University of Maryland.

WEITZEL, PAUL S.

Paul Weitzel is employed by the Babcock and Wilcox Company as a Technical Consultant and Team Leader for New Product Development, Advanced Technology Design and Development, Technology Division, Power Generation Group at Barberton, Ohio. His involvement with B&W spans 42 years, beginning as a Service Engineer at Kansas City starting up boiler equipment and is currently responsible for the Advanced Ultra Supercritical steam generator product development. Early on in his career, there was a time out to serve in the U.S. Navy as an Engineering Duty Officer aboard the USS Midway as the Assistant Boilers Officer and at Hunters Point Naval Shipyard as a Ship Superintendent for repair and overhauls, primarily for the main propulsion plant — always on ships with B&W boilers. Primary assignments with B&W have been in engineering and service roles with a strong technical interest in thermodynamics, fluid dynamics, and heat transfer supporting performance and design of steam generators. He is the author of Chapter 3, Fluid Dynamics, Steam 41, The Babcock and Wilcox Company. He is a member of ASME.

WILLEMS, RYAN

Ryan is a renewable energy engineer working in the Technology and Commercialisation group of the Business Development division of Hydro Tasmania. Ryan joined Hydro Tasmania as an intern in 2005 prior to graduating from Murdoch University with a Bachelor of Renewable Energy Engineering degree in early 2006. Ryan has gained extensive knowledge in the field of Remote Area Power Supply (RAPS) systems in his time at Hydro Tasmania.

Ryan has been extensively involved in RAPS on both King and Flinders Islands and has a considerable level of understanding of the complexities and control of each power station. Ryan has also developed tools for the analysis of energy flows in RAPS systems utilizing a range of control philosophies and has applied his knowledge of renewable energy generation technologies and their integration in the development of this simulation tool. Ryan has also been involved in the King Island Dynamic Resistive Frequency Control (DRFC) project since its inception and has provided significant technical assistance in the design of control logic and troubleshooting during commissioning.

WILLIAMS, JAMES (JIM) L.

James L. Williams is owner and president of WTRG Economics. He has more than 30 years experience analyzing and forecasting energy markets primarily as a consulting energy economist. He publishes the Energy Economist Newsletter and is widely quoted on oil and natural gas issues in the national and international media. His clients and subscribers include major oil companies, international banks, large energy consumers, brokerage firms, energy traders, local, state, and U.S. government agencies.

Jim has an MSc degree in mathematics with additional postgraduate work in math and economics.

Mr. James Williams has taught forecasting, finance, and economics at the graduate and undergraduate level at two universities, testified on energy issues before Congress, and served as an expert witness in state and federal courts. His analysis of oil prices in Texas identified weaknesses in the method the state used to collect severance taxes on oil and contributed to a revision in the system that resulted in higher revenues to the state as well as royalty owners.

Williams' first work in the oil and gas industry was as senior economist with El Paso Company, where he analyzed and forecast petrochemical prices and markets. He modeled and forecast the financial performance of El Paso Petrochemicals division as well as new plants and acquisition targets.

He regularly analyses and forecasts exploration activity and its impact on the performance of oil and gas manufacturing and service companies. His experience ranges from the micro to macro level.

Williams' current research interests include global supply and demand for petroleum and natural gas, country risk, and the influence of financial markets and instruments on the price of energy.

WOLFE, DOUGLAS E.

Dr. Douglas E. Wolfe's research activities include the synthesis, processing, and characterization of ceramic and metallic coatings deposited by reactive and ion beam assisted, electron beam physical vapor deposition (EB-PVD), sputtering, plating, cathodic arc, cold spray, thermal spray, and hybrid processes. Dr. Wolfe is actively working on nanocomposite, nanolayered, multilayered, functionally graded, and multifunctional coatings and the enhancement of coating microstructure to tailor and improve the properties of vapor-deposited coatings such as thermal barrier coatings, transition metal nitrides, carbides, and borides, transition and rare-earth metals, for a variety of applications in the aerospace, defense, tooling, biomedical, nuclear, and optical industries, as well as corrosion-resistant applications. Other areas of interest include the development of advanced materials and new methodologies for microstructural enhancement, design structures/architectures, and coatings/thin films with improved properties. Dr. Wolfe received his PhD in Materials (2001), his MS degree in Materials Science and Engineering with an option in

Metallurgy (1996), and his BS degree in Ceramic Science and Engineering (1994) from The Pennsylvania State University. He has been a member of The Pennsylvania State University Faculty since May of 1998 and currently has a dual title appointment as Advanced Coating Department Head for the Applied Research Laboratory and Assistant Professor in the Department of Materials Science and Engineering. Dr. Wolfe has developed a short course entitled, "Determination, Causes and Effects of Residual Stresses on Coating Microstructure and Properties" and established a world class state-of-the-art Coatings Research Facility at the Pennsylvania State University. His expertise include the development and processing of vapor-deposited coatings as well as materials characterization using a variety of analytical techniques including: X-ray diffraction (XRD), scanning electron microscopy (SEM), optical microscopy (OM), energy dispersive spectroscopy (EDS), tribology, electron probe microanalysis (EPMA), X-ray photoelectron spectroscopy (XPS), secondary ion mass spectroscopy (SIMS), transmission electron microscopy (TEM), etc. Other research interests/topics include defining and developing structure-property-processing-performance relationships. Dr. Wolfe published more than 40 research manuscripts and technical memorandums.

Each year from 1981 through 2007, Mr. Yokell presented two or three 4-day short, intensive courses on Shell-and-Tube Heat Exchangers-Mechanical Aspects at various locations in the United States, Canada, South America, and Europe. During this period, he has also presented, in collaboration with Mr. Andreone, annual seminars on Closed Feedwater Heaters and Inspection, Maintenance and Repair of Tubular Exchangers. In addition, he has provided in-plant training to the maintenance forces of several oil refineries, chemical plants, and power stations.

Mr. Yokell is the author of numerous papers on tubular heat transfer equipment including tube-to-tubesheet joints, troubleshooting, and application of the ASME Code. He is the author of *A Working Guide to Shell-and-Tube Heat Exchangers*, McGraw-Hill Book Company, New York, 1990. With Mr. Andreone, he has written *Tubular Heat Exchanger Inspection, Maintenance and Repair*, McGraw-Hill Book Company, New York 1997. He holds two patents.

Mr. Yokell is a corresponding member of the ASME Code Section VIII's Special Working Group on Heat Transfer Equipment and is a member of the AIChE, the ASNT, the AWS, and the NSPE. Mr. Yokell received the BChE degree from New York University.

YOKELL, STANLEY

Stanley Yokell, PE, Fellow of the ASME is registered in Colorado, Illinois, Iowa, and New Jersey. He is President of MGT Inc., Boulder, Colorado, and a Consultant to HydroPro, Inc., San Jose, California, manufacturers of the HydroPro® system for heat exchanger tube hydraulic expanding, the BoilerPro® system for hydraulically expanding tubes into boiler drums and tubesheets, and the HydroProof® system for testing tube-to-tubesheet and tube-to-boiler drum joints.

From 1976 to 1979, Mr. Yokell was Vice President of Ecolaire Inc. and President and Director of its PEMCO subsidiary. From 1971 to 1976, he was President of Process Engineering and Machine Company, Inc., (PEMCO) of Elizabeth, New Jersey, a major manufacturer of heat exchangers and pressure vessels, where he held the position of Vice President and Chief Engineer from its founding in 1953. Previously, he held the positions of Process Engineer and Sales Manager at Industrial Process Engineers, Newark, New Jersey, and Shift Supervisor at Kolker Chemical Works, Newark, New Jersey.

Mr. Yokell works in analyzing and specifying requirements, construction and uses, troubleshooting, and life extension of tubular heat transfer equipment. He is well-known as a specialist on tube-to-tubesheet joining of tubular heat exchangers and maintenance and repair of tubular heat exchangers.

He renders technical assistance to attorneys and serves as an Expert Witness. Mr. Yokell's more than 48 years of work in the field has involved design and construction of more than 3000 tubular heat exchangers, design and manufacture of process equipment, consulting on maintenance and repair of a variety of process heat exchangers and pressure vessels, feedwater heaters and power plant auxiliary heat exchangers. From 1979 to the present, he has assisted in troubleshooting, failure analysis, repair, modification, and replacement of process and power heat exchangers.

ZAYAS, JOSE

Jose Zayas is the senior manager of the Renewable Energy Technologies group at Sandia National Laboratories. His responsibilities in this role include establishing strategy and priorities, defining technical and programmatic roles, business development, and performing management assurance for the renewable energy-related activities of the laboratory. He manages and develops programs to:

• Bring together key renewable energy technology capabilities to consistently implement a science-based reliability and systems approach

• Leverage Sandia's broader predictive simulation, testing/evaluation, materials science, and systems engineering capability with expertise in renewable energy technologies

• Expand and accelerate Sandia's role in the innovation, development, and penetration of renewable energy technologies

Mr. Zayas joined Sandia National Labs in 1996 and spent the first 10 years of his career supporting the national mission of the labs wind energy portfolio as a senior member of the technical staff. During his technical career, he had responsibilities for several programmatic research activities and new initiatives for the program. Jose's engineering research contributions, innovation, and outreach spanned a variety of areas, which include active aerodynamic flow control, sensors, dynamic modeling, data acquisition systems, and component testing.

After transition to the position of program manager in 2006, Jose has engaged and supported a variety of national initiatives to promote the expansion of clean energy technologies for the nation. Most recently, Jose has continued to lead the organization's clean energy activities and has coordinated and developed the laboratories cross-cutting activities in advanced water power systems. This program focuses on developing and supporting an emerging clean energy portfolio (wave, current, tide, and conventional hydro energy sources). Through developed partnerships with key national labs, industry, and academia, Sandia is supporting and leading a variety of activities to accelerate the advancement and viability of both wind energy and the comprehensive marine hydrokinetics industry. Additionally, Jose is currently leading a Federal interagency research program to address barriers affecting the continued deployment and acceptance of wind energy systems across the nation.

Jose holds a bachelors degree in Mechanical Engineering from the University of New Mexico and a Master's degree in Mechanical and Aeronautical Engineering from the University of California at Davis.

CONTENTS

CHAPTER 18   Carbon Capture for Coal-Fired Utility Power Generation: B&W's Perspective
D. K. McDonald and C. W. Poling **18-1**

CHAPTER 19   Petroleum Dependence, Biofuels — Economies of Scope and Scale; US and Global Perspective
James L. Williams and Mark Jenner **19-1**

CHAPTER 20   Coal Gasification
Ravi K. Agrawal **20-1**

VI. NUCLEAR ENERGY

CHAPTER 21   Construction of New Nuclear Power Plants: Lessons to Be Learned from the Past
Roger F. Reedy **21-1**

CHAPTER 22   Nuclear Power Industry Response to Materials Degradation — A Critical Review
Peter Riccardella and Dennis Weakland **22-1**

PREFACE

Energy and Power Generation Handbook: Established and Emerging Technologies, edited by K.R. Rao, and published by ASME Press, is a comprehensive reference work of 32 chapters authored by 53 expert contributors from around the world. This "Handbook" has 705 pages and contains about 1251 references and over 771 figures, tables, and pictures to complement the professional discussions covered by the authors. The authors are drawn from different specialties, each an expert in the respective field, with several decades of professional expertise and scores of technical publications.

This book is meant to cover the conventional technical discussions relating to energy sources as well as why(s) and wherefore(s) of power generation. The critical element of this book will thus be balanced and objective discussions of one energy source vis-à-vis another source, without making any recommendations or judgments which energy source is better than another.

A primary benefit of these discussions is that readers will learn that neither this nor that source is better, but together they complete the energy supply for this planet. This perhaps could be obvious even without going through the compendium of energy sources covered comprehensively in a book. However, a unique aspect of this publication is its foundation in the scholarly discussions and expert opinions expressed in this book, enabling the reader to make "value judgments" regarding which energy source(s) may be used in a given situation.

This book has the end user in view from the very beginning to the end. The audience targeted by this publication not only includes libraries, universities for use in their curriculum, utilities, consultants, and regulators, but is also meant to include ASME's global community. ASME's strategic plan includes Energy Technology as a priority. Instead of merely discussing the pros and cons of "energy sources," this publication also includes the application of energy and power generation.

Thus, the book could be of immense use to those looking beyond the conventional discussions contained in similar books that provide the "cost–benefit" rationale. In addition to which energy source is better than the other and to which geographic location, the discussions on economics of energy and power generation will portray the potentials as well.

Instead of picturing a static view, the contributors portray a futuristic perspective in their depictions, even considering the realities beyond the realm of socio-economic parameters to ramifications of the political climate. These discussions will captivate advocacy planners of global warming and energy conservation. University libraries, the "public-at-large," economists looking for technological answers, practicing engineers who are looking for greener pastures in pursuing their professions, young engineers who are scrutinizing job alternatives, and engineers caught in a limited vision of energy and power generation will find this publication informative.

Equally important is that all of the authors have cited from the public domain as well as textbook publications, handbooks, scholastic literature, and professional society publications, including ASME's Technical Publications, in addition to their own professional experience, items that deal with renewable energy and non-renewable energy sources. Thus, ASME members across most of the Technical Divisions will find this book worth having.

The discussions in this Handbook cover aspects of energy and power generation from all known sources of energy in use around the globe. This publication addresses energy sources such as solar, wind, hydro, tidal, and wave power, bio energy including biomass and bio-fuels, waste-material, geothermal, fossil, petroleum, gas, and nuclear. Experts were also invited to cover role of NASA in photovoltaic and wind energy in power generation, emerging technologies including efficiency in manufacturing and the role of NANO-technology.

The 32-chapter coverage in this Handbook is distributed into nine (IX) distinct sections with the majority addressing power and energy sources. Depending upon the usage, solar, wind, hydro, fossil, and nuclear are addressed in more than a single chapter. Renewable Energy Resources are covered in Sections I through IV, and Non-Renewable in Sections V and VI; Sections VII through IX cover energy generation-related topics.

Cost comparison with conventional energy sources such as fossil and fission has been made to ascertain the usage potential of renewable resources. This aspect has been dealt by authors while emphasizing the scope for increased usage of renewable energy. Authors therefore dwell on measures for promoting research and development to achieve the target of being cost-comparable.

Preceding all of Sections I through IX, biographical information pertaining to each of the authors is provided followed by Chapter Introductions. This information provides readers a fairly good idea of the credentials of the experts chosen to treat the chapter topic and a glimpse of the chapter coverage.

Section I, Chapters 1 through 6, deals with *Solar Energy* in 114 pages addressed by 10 experts from academia, NASA, and practicing professionals from the U.S., Europe, and India. Global interest in solar energy is apparent not only from the current usage but also from the untapped resources and its potential for greater usage.

The last chapter of Section I, Chapter 6, is authored by experts from NASA who elucidate NASA's efforts in both Solar and Wind energy sectors. This is appropriate since both of these energy sources constitute the most popular of the renewable energy resources.

In addition to the potential of *Wind Energy* already covered in Chapter 6, it is covered in detail in Chapters 7 through 10 of Section II. The increase in usage of wind energy in the past few years in the U.S. as well as in Asia and Europe surpasses any other energy resource. Thus, the potential, like solar energy, is enormous yet is vastly untapped. Global interest in wind as energy resource, although confined to countries uniquely located with wind potential, is limited by technological consequences. Authors from Sandia and Idaho National Laboratories, a research laboratory in the Netherlands, and a practicing professional from India discuss in 71 pages all of the ramifications of wind energy including the public perceptions and ways to technologically overcome environmental considerations including noise and visual aspects.

Section III deals with Hydro and Tidal Energy and has three chapters, Chapters 11, 12, and 13, devoted to *Hydro Power in the USA and Asia* in 40 pages. These three chapters are authored by three expert practicing professionals at the helm of their organizations and EPRI. Potential for this energy source is considerable in the U.S. and developing world, and lessons of experience with considerable "know-how" in hydro power are valuable for use in rest of the world. *Tidal and Wave Power* is unique and knowledge based, a privilege of the developed nations even though rest of the world have enormous potential for this energy source. This is addressed with abundant reference material by an expert from Electrical Power Research Institute (EPRI).

Section IV covers diverse modes of energy and power generation such as Bio Energy, Energy from Waste, and Geo Thermal Energy addressed in 56 pages in Chapters 14, 15, and 16 by practicing professionals and academia.

Bio Energy including Biomass and Biofuels is not exclusive to developed world. Even developing nations are aware of it although not dependent upon this source of energy. Bio-energy technology has been discussed by a practicing professional with expertise in this field in the U.S. and overseas. The author covers the potential of bio energy's future usage and developments, especially co-firing with coal.

Waste Energy has been addressed by a practicing professional with knowledge of municipal and industrial waste in both developed as well as underdeveloped or developing economies. Urbanization and concomitant suburban sprawl with demands for alternative sources of energy generation can release gasoline for automobiles. With the help of several schematics, the benefits and challenges of utilizing waste are covered including waste cycle, the regulatory perspective, business risks, and economic rationale.

This book that has as its target to investigate all "known" energy sources and *Geothermal Power* cannot be discounted now as well as in the immediate future. Even though confined in its application to a few isolated locations in the world such as Iceland, USA, Australia, Asia, and Europe, its contribution for solving global energy and power problems can be considerable, if this partially tapped resource of this planet can be harnessed to the fullest extent. Technological intricacies of this topic are addressed by two authors, an expert from the academia and in-charge of a professional organization in U.S.

In Section V, as part of *Non Renewable Fuels for Power and Energy Generation—Fossil Power Generation* comprising of *Coal, Oil, Gas, and Coal Gasification* is addressed by U.S. experts in Chapters 17 through 20 in 86 pages. The cutting edge of technology concerning the impact of CO_2 emissions, climate change, and coal gasification is addressed by U.S. industry experts in this Section. Both the U.S. and global economy are impacted by energy and power generation from petroleum and gas. This issue is also addressed in this Section by two U.S. economists. Chapters of this section will cover ongoing issues as well as the state-of-the-art technology.

While contributors cover the existing generation methods and technology, they also expound facets that deserve unique treatment. For example, the fossil power generation industry, responsible for 40 percent of carbon emissions, can be addressed with minimal socio-economic impact largely by technological advances. Whereas longer chimney heights and scrubbers were considered adequate technology for coal-fired units, technology has moved far ahead, and there are items worth attention of the readers. 'The Devil's in the Details of these technological advances'!

A discussion about *Fossil Power Generation* is incomplete without an understanding of "global warming," "climate change," and the Kyoto Protocol for dealing with carbon emissions. Authors of Chapters 14 and 15 associated with a premier fossil generation enterprise bring the wealth of their experience in covering the cutting edge of technology related to carbon emissions. If the abundant coal in the U.S. has to continue for coal-fired power plants as a blessing instead of a bane, it has to transform the technology for the use of coal. Authors aware of the efficiency of coal for power generation, to meet the global competition, have, with the help of impressive schematics and examples, implicitly demonstrated the U.S. dominance in this field.

A unique aspect of this handbook is the inclusion of a chapter by two U.S. economists who provide economic rationales for both petroleum and bio fuels. With the help of abundant schematics, authors drive home the point that a value judgment has to include beyond technical considerations economic parameters as well. Scope of coverage will include U.S. and developed economies such as Australasia, Europe, and North Americas and developing economies including countries of Asia, South America, Africa, and Middle East.

Previously, coal was converted to make gas that was piped to customers. Recently, investigation has been progressing for "BTU Conversion." Technological advancement has prompted Coal Gasification, methanation, and liquefaction. Author addressed these state-of-the-art-technologies in Chapter 20 including design issues and cost impacts.

The oil rig exploration on April 20, 2010 in the Deepwater Horizon 40 miles off the coast of Louisiana was the largest accident in the Gulf of Mexico, according to the U.S. Coast Guard. This has not been addressed in the discussions of Section V, since this will distract from the main theme of the subject matter.

In Section VI titled *Nuclear Energy,* seven U.S. authors and one each from Japan and Switzerland cover Chapters 21 through 24 in 67 pages. Throughout the world, the nuclear industry is experiencing a renaissance. The aspects addressed in this Section will be *self-assessment* of the current generation of Nuclear Reactors as much as covering salient points of the *next generation* of Nuclear Reactors. These and other issues of *Nuclear Power Generation* are taken up by these nine authors with a cumulative professional and nuclear-related experience of over 300 years.

Previous generations of Nuclear Reactors built in the U.S. were criticized for the costs, time taken, and security concerns. All of these factors were instrumental in stalling the pace of construction of nuclear reactors in this country. Self-assessment by owners, regulators, and consultants with the help of professional organizations such as ASME has largely addressed several or most of the items, so that if we were to build nuclear reactors, we are much wiser now than ever before. Several of the issues are technical, whereas some are pseudo-management issues. The authors in Chapters 21 through 24 of this Section VI succinctly chronicle the items for helping the future generation of reactors that will be built. Technological advances such as 3-D FEA methods, alloy metals used in the construction, and several other factors have made it possible by even a slight reduction in safety factors without reconciling the safety concerns; likewise, thinking process on the lines of predesigned and modular constructions has alleviated the time from the initiation through the construction stages up to the completion of a nuclear reactor; the regulatory perspective has also gone beyond the U.S. bounds to countries that use the ASME Stamp of Approval for their Nuclear installations.

The future of the nuclear industry holds immense promise based on strides made in the U.S., Europe, and Asia. ASME Codes and Standards are used globally in building Nuclear Reactors. A discussion about Nuclear Power Generation is never complete without an understanding about the country's energy regulatory structure and decision-making process. In the first chapter of this section, Chapter 21, *A Perspective of Lessons Learned,* has been addressed by an author with several decades of experience in the U.S. nuclear industry. Hopefully this could be useful in building new reactors.

In Chapter 22, two experts with nuclear background provide a critical review of the "Nuclear Power Industry Response to Materials Degradation" problems, especially as it relates to the new plants. Authors discuss the fleet-wide recognition of these issues.

Experts from Switzerland, Idaho National Laboratories, General Electric, and Japan Nuclear Safety with knowledge of the next generation of nuclear reactors have contributed Chapter 23 summarizing global efforts. Authors provided an assessment of the existing generation and potential for new projects. These recognized experts with several decades of professional and Code experience have addressed the ramifications of the past and current constructions while providing their perspectives for the next generation of nuclear reactors.

An ASME Code expert succinctly addresses in the last chapter of Section VI (Chapter 24) the future of nuclear reactors that seems to be at the crossroads. It is most appropriate that the author provides an open window to look at the current concerns, future challenges, and most importantly the unfinished business to revive nuclear power generation in the U.S.

Recent events such as at the *Fukushima Daiichi Nuclear Plants at Japan* devastated by the Tohoku-Taiheiyou-Oki Earthquake and Tsunami of March 11, 2011 have not been addressed by the authors, since these require a separate treatment and will distract from the main theme of discussions.

Section VII is titled *Steam Turbines and Generators* and has two chapters, Chapters 25 and 26, authored by two industry experts in 52 pages. Interdependency of all the energy sources needs to be addressed, especially as it relates to energy sources that are intermittent, and this has been done in Section VII.

In Section VII, Chapters 25 and 26 will be dedicated to *Turbines and Generators*, since they are a crucial and integral part

of power generation, especially as they relate to Wind, Solar, Fossil, and Nuclear Power Generation. Discussions pertain to types of Turbine Configurations, their design, performance, operation, and maintenance. Turbine components, disks, and rotors including non-destructive methods have been covered in the discussions. In Chapter 25, the author discusses generators and crucial components such as retaining rings and failures. Material properties are briefly addressed. In both Chapters 24 and 25, the authors dwell upon the advanced technology and next generation of turbines.

In Section VIII of the book, Selected Energy Generation Topics have been covered in Chapters 27 through 30 in 79 pages. Topics selected for this Section stem from the importance of the topics for Renewable as well as Non-Renewable Energy Generation. The topics include Combined Cycle Power Plants, A Case Study, Heat Exchangers, and Water Cooled Steam Surface Condensers.

A recognized authority in *Combined Cycle Power Plants* with a Handbook on the subject has authored Chapter 27 that covers gas and steam turbines. The author has addressed the availability, reliability, and continuity of energy and power by using the combined cycle power plants.

In Chapter 28, *Hydro Tasmania—King Island Case Study* has been authored by three professional engineers of Hydro Tasmania, Australia, who address the renewable energy integration project. The discussions cover benefits including the development project.

Heat Exchangers are crucial components of Power Generation discussed in Chapter 29 by two recognized authorities with several decades of professional experience. The discussions rally around design aspects, performance parameters, and structural integrity.

A well-recognized authority in nuclear industry with global experience has authored the role of *Water Cooled Steam Surface Condensers* in Chapter 30. The author has covered design aspects, the construction details, and the related topics with schematics and a technical discussion with the help of 55 equations.

Whereas the *preceding groups* can be considered as the "core" of the book, the future of energy sources cannot be overlooked. Indeed, ignorance cannot be considered bliss in overlooking the energy and power generation potentials of the world. Ultimately, this planet's very existence depends on augmenting the energy and power generation resources. This could also imply conservation of energy (also covered in several of the preceding chapters) and harnessing methods that could improve known techniques.

In the last section of this handbook (Section IX), Emerging Energy Technologies have been addressed in 36 pages, in two chapters, by six authors. Use of untapped energy sources and peripheral items such as *Conservation Techniques, Energy Applications, Efficiency, and suggestions for Energy Savings "inside the fence"* is worthy of consideration.

In pursuit of the above statements, Chapter 31, *Toward Energy Efficient Manufacturing Enterprises,* has been addressed by two authors from the U.S. government, an expert from industry and an author from academia. Energy efficiency is implied in conservation and saving of energy, and this has been dealt with by authors in this chapter.

The cutting edge of technology by the use of *Nano-Materials and Nano Coatings* has been dealt with in Chapter 32 by two authors from academia. These experts deal in this chapter the use of Nano Technology in Fuel Cells, Wind Energy, Turbines, Nano-structured Materials, Nano-coatings, and the Future of Nano-technology in power generation.

A publication such as this with over 53 contributors from around the world and nearly 700 pages with rich reference material documenting the essence of the contributors' expertise can be a valuable addition to university libraries, as well as for consultants, decision makers, and professionals engaged in the disciplines described in this book.

For the reader's benefit, brief biographical sketches as mentioned before are included for each contributing author. Another unique aspect of this book is an Index that facilitates a ready search of the topics covered in this publication.

K. R. Rao Ph.D., P.E.
(Editor)

INTRODUCTION

This handbook has been divided into nine ("IX") sections with each section dealing with a similar or identical energy and power generation topic.

Section I deals with Solar Energy, which includes Chapters 1 to 6.

Chapter 1, "Some Solar-Related Technologies and Their Applications" is addressed by Robert Boehm. In this chapter, the source of energy that has been available to humankind since we first roamed the earth is discussed. Some of the general concepts are not new, and several particular applications of these technologies are enhancements of previous concepts.

The discussion begins with the special effects that are possible with the use of concentration. For locations that have a high amount of beam radiation, this aspect allows some very positive properties to be employed. This yields a lower cost, more efficient way of generating electricity. Limitations to the use of concentration are also outlined.

Another aspect discussed is the current situation of solar thermal power generation. This approach has been in use for many years. Previously designed systems have been improved upon, which results in more efficient and more cost-effective means of power production. While trough technology has been more exploited than other approaches (and is still a leader in the field), several other systems are gaining interest, including tower technology. Thermal approaches are the most convenient to add storage into solar power generation.

Photovoltaic approaches are described. New developments in cells have both decreased costs and increased performance. Both high- and low-concentration systems, as well as flat plate arrangements in tracking or non-tracking designs, offer a variety of application modes, each with certain benefits and shortcomings.

The use of solar-generated hydrogen is discussed. This offers an approach to a totally sustainable mobile or stationary fuel source that can be generated from the sun.

The solar resource can be used for lighting, heating, cooling, and electrical generation in buildings. The concept of zero energy buildings is discussed. These are buildings that are extremely energy efficient and incorporate a means of power production that can result in net zero energy use from the utility over a year's period. Locations with a moderate-to-high solar resource can use this to make up for the energy used. Both solar domestic water heating (a concept that has been applied in the United States for well over a century) and building integrated photovoltaic (PV) are also discussed. South-facing windows that incorporate thin film PV could generate power and allow lighting to penetrate the building. Finally, some exciting direct solar lighting concepts (besides windows) are discussed. The author uses 28 references along with 24 schematics, figures, pictures, and tables to augment his professional and scholastic treatment of the subject.

Chapter 2 by Yong X. Tao and Rambod Rayegan deals with "Solar Energy Applications and Comparisons." The authors focus on energy system applications resulting from the direct solar radiation including:

- Utility-scale solar power systems that generate electricity and feed to the electricity grid. There are PV systems and solar thermal power systems; the latter can also produce heat for hot water or air, which is often referred to as the combined solar power and heat systems.
- Building-scale solar power systems, also known as distributed power systems, which generate electricity locally for the building, and may be connected to the grid, or may be stand-alone systems, which require batteries or other electricity storage units. They are primarily photovoltaic systems.
- Solar heating systems for buildings, which are either used as hot water systems or hot air heating systems.
- Solar high-temperature process heat systems for industrial applications, which involve concentrated solar collectors and high-temperature furnaces for producing high-temperature heat for chemical processing of materials.
- Other special solar heating systems for desalination plants and hydrogen production.

There are additional solar energy applications in either the appliance category or even much smaller scales such as solar cooking, solar lighting products, and instrument-level solar power sources (watches, backpacks, etc.) The discussion of those applications is beyond the scope of this chapter. Outer space applications of solar energy technology are also excluded. Investigations primarily undertaken in the United States of America are presented, although some examples from global applications are also discussed to address the potentials and needs for wider applications of solar energy in the United States. The authors use 57 references along with 46 schematics, figures, pictures, and tables to augment the professional and scholastic treatment of the subject.

Next is Chapter 3 dealing with "Solar Thermal Power Plants: From Endangered Species to Bulk Power Production in Sun Belt Regions," by Manuel Romero and José González-Aguilar.

Solar thermal power plants, due to their capacity for large-scale generation of electricity and the possible integration of thermal storage devices and hybridization with backup fossil fuels, are meant to supply a significant part of the demand in the countries of the solar belt such as in Spain, the United States of America, India,

China, Israel, Australia, Algeria, and Italy. This is the most promising technology to follow the pathway of wind energy in order to reach the goals for renewable energy implementation in 2020 and 2050.

Spain, with 2400 MW connected to the grid in 2013, is taking the lead on current commercial developments, together with the United States of America, where a target of 4500 MW for the same year has been fixed and other relevant programs like the "Solar Mission" in India recently approved for 22-GW solar, with a large fraction of thermal.

Solar Thermal Electricity or STE (also known as CSP or Concentrating Solar Power) is expected to impact enormously on the world's bulk power supply by the middle of the century. Only in Southern Europe, the technical potential of STE is estimated at 2000 TWh (annual electricity production), and in Northern Africa, it is immense.

The energy payback time of concentrating solar power systems will be less than 1 year, and most solar-field materials and structures can be recycled and used again for further plants. In terms of electric grid and quality of bulk power supply, it is the ability to provide dispatch on demand that makes STE stand out from other renewable energy technologies like PV or wind. Thermal energy storage systems store excess thermal heat collected by the solar field. Storage systems, alone or in combination with some fossil fuel backup, keep the plant running under full-load conditions. This capability of storing high-temperature thermal energy leads to economically competitive design options, since only the solar part has to be oversized. This STE plant feature is tremendously relevant, since penetration of solar energy into the bulk electricity market is possible only when substitution of intermediate-load power plants of about 4000 to 5000 hours/year is achieved.

The combination of energy on demand, grid stability, and high share of local content that lead to creation of local jobs provide a clear niche for STE within the renewable portfolio of technologies. Because of that, the European Commission is including STE within its Strategic Energy Technology Plan for 2020, and the U.S. DOE is launching new R&D projects on STE. A clear indicator of the globalization of such policies is that the International Energy Agency (IEA) is sensitive to STE within low-carbon future scenarios for the year 2050. At the IEA's Energy Technology Perspectives 2010, STE is considered to play a significant role among the necessary mix of energy technologies needed to halving global energy-related CO_2 emissions by 2050, and this scenario would require capacity additions of about 14 GW/year (55 new solar thermal power plants of 250 MW each).

In this chapter, the authors discuss, with the help of 21 figures, schematics, and tables along with 72 references, the Solar Thermal Power Plants — Schemes and Technologies, Parabolic-Troughs, Linear-Fresnel Reflectors, Central Receiver Systems (CRS), Dish/Stirling Systems, Technology Development Needs and Market Opportunities for STE. The authors use 72 references along with 27 schematics, figures, pictures, and tables to augment the professional and scholastic treatment of the subject.

Chapter 4 has been written by Rangan Banerjee and deals with "Solar Energy Applications in India." India has a population of 1.1 billion people (one-sixth of the world population) and accounts for less than 5% of the global primary energy consumption. India's power sector had an installed capacity of 159,650 MW as on 30th April, 2010. The annual generation was 724 billion units during 2008 to 2009 with an average electricity use of 704 kWh per person per year. Most states have peak and energy deficits. The average energy deficit is about 8.2% for energy and 12.6% for peak.

About 96,000 villages are un-electrified (16% of total villages in India) and a large proportion of the households do not have access to electricity.

India's development strategy is to provide access to energy to all households. Official projections indicate the need to add another 100,000 MW within the next decade. The scarcity of fossil fuels and the global warming and climate change problem has resulted in an increased emphasis on renewable energy sources. India has a dedicated ministry focusing on renewables (Ministry of New and Renewable Energy, MNRE). The installed capacity of grid-connected renewables is more than 15,000 MW. The main sources of renewable energy in the present supply mix are wind, small hydro- and biomass-based power and cogeneration. In 2010, India has launched the Jawaharlal Nehru Solar Mission (JNSM) as a part of its climate change mission with an aim to develop cost-effective solar power solutions.

Most of India enjoys excellent solar insolation. Almost the entire country has insolation greater than 1900 kWh/m^2/year with about 300 days of sunshine. Figure 4-2 shows a map with the insolation ranges for different parts of the country. The highest insolation (greater than 2300 kWh/m^2/year) is in the state of Rajasthan in the north of the country. The solar radiation (beam, diffuse, daily normal insolation) values are available at different locations from the handbook of solar radiation data for India and at 23 sites from an Indian Meteorological Department (IMD) MNRE report.

Rangan Banerjee discusses in this chapter, with the help of 24 schematics, pictures, graphics, figures, and tables, the chapter that deals with Status and Trends, Grid-Connected PV Systems, Village Electrification Using Solar PV, Solar Thermal Cooking Systems, Solar Thermal Hot Water Systems, Solar Thermal Systems for Industries, Solar Thermal Power Generation, Solar Lighting and Home Systems, Solar Mission, and Future of Solar Power in India. The author uses 35 references and 24 schematics, figures, pictures, and tables to augment his professional and scholastic treatment of the subject.

Chapter 5, "Solar Energy Applications: The Future (with Comparisons)" is covered by Luis A. Bon and W.J. O'Donnell. This chapter traces the roots of solar energy from 1838 through current technologies from an engineering perspective. Numerous diagrams and photographs are included, illustrating the technical concepts and challenges. Methods of concentrating solar power are described including parabolic troughs, Fresnel reflectors, solar towers, and sterling engine solar dishes. Methods of storing solar energy to provide continuous power are described, including batteries, fly-wheel energy storage, water energy storage, compressed air, and superconducting magnetic energy storage. Current energy use and production in the United States of America and worldwide are quantified. Solar energy's potential future is illustrated by the fact that it would require less than 1% of the land area of the world to produce all of the energy we need. Of course, solar energy's future lies in its integration into the residential and commercial infrastructure. This challenge is expected to limit the contribution of solar energy to <0.1% of the USA energy consumption over the next 25 years.

The final chapter of this section is Chapter 6 "Role of NASA in Photovoltaic and Wind Energy" by Sheila G. Bailey and Larry A. Viterna. Since the beginning of NASA over 50 years ago, there has been a strong link between the energy and environmental skills developed by NASA for the space environment and the needs of the terrestrial energy program. The technologies that served dual uses included solar, nuclear, biofuels and biomass, wind, geothermal, large-scale energy storage and distribution, efficiency and heat utilization, carbon mitigation and utilization, aviation and ground transportation systems, hydrogen utilization and infrastructure, and advanced energy technologies such as high-altitude wind, wave and hydro, space solar power (from space to earth), and nanostructured photovoltaics. NASA, in particular, with wind and solar energy, had extensive experience dating back to the 1970s and 1980s and continues today to have skills appropriate for solving our nation's energy and environmental issues that mimic, in fact, those needed for space flight. This chapter encompasses the historical role that NASA and, in particular, NASA Glenn Research Center, GRC, have played in developing solar and wind technologies. It takes you through the programs chronologically that have had synergistic value with the terrestrial communities. It ends with pointing out the possibilities for future NASA technologies that could impact our Nation's energy portfolio. The technologies that it has developed for aeronautical and space applications has given GRC a comprehensive perspective for applying NASA's skills and experience in energy on the problems of developing a sustainable energy future for our nation. Authors use 70 references along with 32 schematics, figures, and pictures to augment the professional and scholastic treatment of the subject.

Section II dealing with Wind Energy is covered in Chapters 7 through 10.

Chapter 7 is authored by Jose Zayas who addresses "Scope of Wind Energy Generation Technologies." The energy from the wind has been harnessed since early recorded history all across the world, and it has been a viable and dependable resource to support our ever changing needs (pump water, grind grains, and now produce cost-effective electricity).

When the price of oil skyrocketed in the 1970s, so did worldwide interest in wind turbine generators. The sudden increase in the price of oil stimulated a number of substantial government-funded programs of research, development, and demonstration, which led to many of the technology that drove the designs and the industry that can be seen today.

Since the early 1980s through today, wind farms and wind power plants have been built throughout the world, and now wind energy is the world's fastest-growing clean energy source that is powering our industry as well as homes with clean, renewable electricity.

Although the United States experienced a large influx of installations during the 1980s, it is not until recent years that wind energy in the United States has achieved large market installations and continued market acceptance. Through the 3rd Quarter of 2010, the United States has approximately 37,000 MW of installed capacity, which approximately represents 2% of our energy consumption.

Since the beginning, Sandia National Laboratories has had a key role in developing innovations in areas such as aerodynamics, materials, design tools, rotor concepts, manufacturing, and sensors, and today, through continued partnerships with industry, academia, and other national labs, Sandia continues to develop and deliver the next set of technology options that will continue to improve the reliability and efficiency of wind systems. It is difficult to predict what the next generation of technologies will bring to this industry, but we can be certain that as it continues to mature and leverage technologies from other sectors, the resulting turbines will be smarter, more efficient, and they will represent a significant percentage of our energy mix. The author uses 11 references along with 37 schematics, figures, and pictures to augment the professional and scholastic treatment of the subject.

Thomas Baldwin and Gary Seifert cover "Wind Energy in the U.S." in Chapter 8. Idaho National Laboratory (INL) is a science-based, applied engineering national laboratory supporting the U.S. Department of Energy (DOE). INL's mission includes ensuring the nation's energy security with safe, competitive, and sustainable energy. INL's Renewable Energy Program, consisting of wind, hydro, and geothermal energy systems, has conducted wind energy resource assessments, system integration, feasibility studies, turbine selection, and array designs since the mid-1990s. Engineering support has been provided for the U.S. Department of Defense, Wind Powering America, State of Idaho, commercial industries, and regional entities. INL is an international clearing house supporting private parties, industry, and government agencies as an independent subject matter expert on wind power, resource assessment, and renewable energy systems siting. INL has extensive wind analysis expertise, having collected and analyzed wind data for more than 10 years. INL has installed met towers at more than 60 sites and collected data with SODARs. Wind analysis modeling tools experience includes WAsP, WindSim, Windographer, NRG, SecondWind, as well as MS Excel-based models which have been developed internally over several years.

Thomas Baldwin, Gary Seifert, and the engineering team they work with bring a combined total of over 75 years of expertise to the power system integration, wind, solar, and renewable energy field. Gary Seifert has been involved for over 30 years in electrical and power systems projects for the Idaho National Laboratory. He is a Sr. Program Manager in INL's Power and Renewable Energy and Power Technologies Department. His background includes extensive power plants and total energy plants for the DOE and the DOD. These plants have included heat recovery and desalination plants to produce potable water and challenging power system integrations.

Thomas Baldwin has been involved in power system design and energy storage systems for 25 years at ABB, Florida State University, and the INL. His background includes earthing and grounding systems, power quality assessments, applications of uninterruptible power supplies and superconducting energy storage systems, and industrial power system protection. Authors use 34 references along with 26 schematics, figures, pictures, and tables to augment the professional and scholastic treatment of the subject.

The next, Chapter 9, is by Peter Eecen which deals with "Wind Energy Research in the Netherlands." The chapter describes the developments within The Netherlands with regard to the wind energy research since the first funding was organized by the National Wind Energy Research Program in the period 1976 to 1985. Wind energy research activities in the Netherlands have been and are predominantly performed at the wind energy department of the Energy Research Centre of the Netherlands ECN and the interfaculty wind energy department DUWIND at Delft University of Technology. Both institutes are involved in wind energy research since the start of the modern wind turbines. These institutes match their research programs with each other so that a consistent research program in The Netherlands is in place.

The research activities in wind energy have a strong focus on international cooperation, where the cooperation was organized through among others the International Energy Agency (IEA), European Wind Energy Association (EWEA), European Academy of Wind Energy (EAWE), the International Electrotechnical Commission (IEC), the European Energy Research Alliance (EERA), and European research projects.

In the Netherlands, the wind energy research is supported by an extensive experimental infrastructure. The Knowledge Centre WMC that has been founded by the DUT and ECN is a research institute for materials, components, and structures. WMC is performing blade tests for large wind turbines to 60 m in length. ECN made available a research wind farm where prototype wind turbines are tested, where a research farm of five full-scale turbines are used for research activities, and where a scale wind farm is located for research on farm control and wind farm aerodynamic research. At DUT, a large selection of experimental facilities is being used for wind energy applications. The most prominent facilities are the wind tunnels, of which the Open Jet Facility is the most recent addition.

The historic overview of the wind energy research activities in the Netherlands is written from the perspective of the research community and provides alternative insights as would be provided by existing historic overviews that focus on the implementation of wind energy and the development of support mechanisms. The description of research activities, the developed advanced design tools, developed knowledge, and intellectual property may provide an alternative source for further activities to reduce the cost of energy of wind power. The author uses 22 references along with ten schematics, figures, pictures, and tables to augment the professional and scholastic treatment of the subject.

The last chapter of Section II, Chapter 10, is by M. P. Ramesh covering "Role of Wind Energy Technology in India and Neighboring Countries." The interest in wind as a source of power in the Asian region had an early start in India. Owing to a variety of inhibiting factors, it was a low key activity for a long time. Ramesh, Muthya with his long and close association with the field has treated the subject of wind energy development as a source of power with a thorough understanding of development cycle in dismissive environments that are stronger in developing countries.

Indian engagement with wind power started as early as the late 1950s but was mostly for nonelectric applications. Origins of usage of wind for grid connection started only in the mid-1980s with few small turbines installed at known windy areas. Danida aided 20 MW wind farms paved way for a steady and sustained growth of the field. This is treated with the backdrop of the grid situation and programmatic approach of the support from the government. Approach to resource estimation is also treated rather cautiously in India.

China had been a little slow in taking to large-scale development in renewable energy except for perhaps the biogas plants, which score over the Indian design in many ways. Slowly but surely, the Indian floating collector design is being replaced by fixed dome Chinese design for community-based biogas plants. On wind power use over the last few years, there has been a drastic turn around in the approach to using re-technologies. It has taken just 3 years for China to graduate to the second position in the top ten lists in terms of wind power capacity addition. If one goes by the projections, the number one position may be achieved. In 2007, it was at about 4 GW, and by 2009, the installations touched an astounding 26 GW. With scores of companies undertaking development on a war footing, there is so much happening in China on wind power. A spate of wind turbine designs has been developed, and prototypes are being built and tested. Another paradoxical position is that though it is a country that produces about 40% of all solar photovoltaic devices, domestic utilization is quite small.

Sri Lanka has good wind resource. However, it cannot be expected to reach market sizes that India or China have. The potential and limitations have been briefly described.

In this paper, Ramesh Muthya has attempted to capture some of the salient aspects of technology development and deployment in India in the context of power supply systems management. Main RE technologies dealt with are wind energy sources. Small hydro-power adds considerable value for localized grids. Biomass sources work more or less like thermal stations though for limited periods in a year. The author uses 14 references along with 16 schematics, figures, pictures, and tables to augment the professional and scholastic treatment of the subject.

Section III deals with Hydro and Tidal Energy and has three chapters: Chapters 11, 12, and 13.

Chapter 11, "Hydro Power Generation: Global and U.S. Perspective" is by Stephen D. Spain. The development of dams on rivers, with associated benefits of water storage for flood control, irrigation, and "hydropower" has played a vital role in advancing civilization throughout history. Of these, hydropower ingeniously, and yet so simply, combines two of the most fundamental components of nature on planet Earth — water and gravity — to help sustain our survival and improve our lifestyle.

This chapter describes the role of hydropower from past to present and into the future. Hydropower has been demonstrated to be a safe, reliable, and renewable energy resource worldwide, essential to the overall power and energy mix, both traditionally from rivers. Recent and growing development of pumped energy storage from lower to upper water reservoirs and evolving in the future with tidal and wave energy from the oceans has also been covered by the author.

Stephen D. Spain discusses, with the help of 18 schematics, figures, pictures, and tables, the History of Hydropower, including in the United States, Hydropower Equipment, Hydropower for Energy Storage, Ocean and Kinetic Energy, and Hydropower Organizations. The discussions also include Hydropower Owners, Worldwide Hydropower, and The Future of Hydropower. The chapter made extensive references to 27 publications.

Chapter 12, "Hydro Power Generation in India — Status and Challenges" is authored by Dharam Vir Thareja. Power generation in India has come a long way from about 1000 MW at the time of independence (August, 1947) to about 160,000 MW as on 31st March 2010 (end of Financial Year). The share of hydropower in the past six decades has also been impressive as it increased from about 500 MW at the time of independence to about 37,000 MW as of March 2010. But the present level of hydropower is only about 25% of the ultimate installed capacity estimated at 150,000 MW.

The hydroproject implementation and ownership remained with the State Governments or with the Power Corporations owned by States or Central Government for about four and a half decades. Owing to slow pace of development and also in line with international stress for liberalization of economy, the Government of India reviewed the policy of power development. In 1992, the power sector was opened up allowing private capital participation in its development along and in parallel with continued development under public sector.

India now possesses the needed financial resources and professional capabilities to utilize the in-house and global capacities for implementation of hydropower projects. The 50,000 MW initiatives leading to identification of 162 projects in a span of 2 years (2004 to 2006) has been recognized as a laudable initiative. The achievement of commissioning of about 8000 MW in the 10th plan period (2002 to 2007), the target of over 15,000 MW in the running 11th plan period (2007 to 2012) and 20,000 MW proposed during the 12th plan period (2012 to 2017) speaks of the ambitious plan of hydropower development. To achieve the targets of 11th and 12th plans, the fund requirement would be of the order of US$ 30 billion of which about 30% would be the share of private sector. The participation of private sector in the investment projections for the 11th plan for infrastructure, which includes power sector, is estimated to be at 30%.

The hydropower that is yet to be tapped, projects with more than 75% of installation, are located in Himalayan region. This region is known for intense seismicity, wide range of geotechnical variability, and extensive hydrologic pose challenges for infrastructure development.

It has been planned to exploit the bulk of balance of hydropotential in the next about two decades. To meet the target, the hydrosector would require investment of US$100 billion requiring an average of US$4 billion per year for another 25 years and, thus, would remain attractive for financial institutions, project developers, contractors, and consultancy organizations.

The role of hydropower in the energy scenario; potential and status of development; small hydro- and pump-storage development; transmission setup and status; constitutional and regulatory provisions; resettlement and rehabilitation policies; techno-economic appraisal procedures; hydropower development in the neighboring countries; response and achievement of private sector; the issues, constraints, and challenges in development; and innovations for future projects are covered in this chapter. The author uses 16 schematics and 48 references to supplement the textual discussions.

The last chapter of this section, Chapter 13, is "Challenges and Opportunities in Tidal and Wave Power" by Paul T. Jacobson. Power generation from waves and tidal currents is a nascent industry with the potential to make globally significant contributions to renewable energy portfolios. Further development and deployment of the related, immature technologies present opportunities to benignly tap large quantities of renewable energy; however, such development and deployment also present numerous engineering, economic, ecological, and sociological challenges. A complex research, development, demonstration, and deployment environment must be skillfully navigated if wave and tidal power are to make significant contributions to national energy portfolios during the next several decades.

A striking feature of the wave and tidal power technologies in various stages of development is their number and diversity. Standardized classification of these technologies, as described here, will facilitate their development and deployment. The principal engineering challenge facing development of wave and tidal power devices is design of devices that can survive and operate reliably in the harsh marine environment. A significant advantage of tidal and wave energy conversion, compared to wind and photovoltaic generation, is the ability to forecast the short-term resource availability.

Environmental considerations play a large role in ongoing development of the wave and tidal energy industry. The number and novelty of device types, in combination with the ecological diversity among potential deployment sites, creates a complex array of ecological impact scenarios. Efficient means of addressing ecological concerns are in need of further development, so that the industry can advance in an environmentally sound manner. Adaptive management offers a means of moving the industry forward in the face of ecological uncertainty; however, the potential benefits of adaptive management will be realized only if it is implemented in its more scientifically rigorous form known as active adaptive management.

The ecological assessment challenge facing the wave and tidal energy industry is to acquire and apply the information needed to ensure that systems, sites, and deployment scales are protective of ecological resources. Many reports have identified the range of environmental issues associated with wave, tidal, and other renewable ocean energy projects. The remaining challenge is to prioritize the issues and address the most pressing ones in a cost-effective manner. For individual projects, this requires definition and evaluation of questions required for site-specific permitting and licensing. For the industry as a whole, this requires identification and acquisition of information that is transferable across projects. Several nontrivial activities are outlined that can be taken to address the large array of outstanding issues. The author uses 70 references along with eight schematics, figures, pictures, and tables to augment the professional and scholastic treatment of the subject.

Section IV covers diverse modes of energy and power generation such as Bio, Waste, and Geo Thermal energies.

In the leading chapter of Section IV, Chapter 14, "BioEnergy Including BioMass and Biofuels" is addressed by T. R. Miles. He describes biomass fuels and discusses technologies that are suitable for biomass conversion. T. R. Miles draws from more than 35 years experience in the design and development of biomass systems, from improved cooking stoves in developing countries, to industrial boilers, independent power plants, utility cofiring and the development of biofuel processes.

Oil shortages in the 1970s stimulated the development of new technologies to convert biomass to heat, electricity, and liquid fuels. Combined firing of biomass with coal reduces emissions, provides opportunities for high efficiency, and reduces fossil carbon use. Cofiring and markets for transportation fuels have stimulated global trading in biomass fuels. Pyrolysis of biomass to liquid fuels creates opportunities for coproducts, such as biochar, which can help sequester carbon and offset emissions from fossil fuels. This chapter provides an overview of biomass fuels and resources, the biomass power industry, conventional and new technologies, and future trends. Advances in combustion and gasification are described. Thermal and biological conversions to liquid fuels for heat, power, and transportation are described with implications for stationary use.

Topics in Chapter 14 include biomass fuels and feedstocks, heat and power generation, cofiring biomass with coal, and conversion technologies such as gasification, torrefaction, pyrolysis, carbonization, and biofuels. Biomass fuel properties that are important to energy conversion are compared with coal. The effect of moisture, energy density, volatile content, particle size, and ash are related to the selection, design, and operation of biomass boilers. Moisture, particle size, and density are shown to limit cofiring in existing boilers. New technologies can offset these properties enabling more biomass to be cofired with coal. The transformation of ash in biomass boilers is shown graphically in Figure 14-1. Biomass types, sources, and supplies are evaluated including woody biomass, wood pellets, urban residues, and agricultural residues. The infrastructure and logistics of biomass are described. The current state of harvesting systems for crop residues and herbaceous crops is shown to highlight the need for higher fuel density and lower costs. Life cycle analysis is shown to be a common method for evaluating biomass sustainability.

Technologies for heat and power generation include combustion, domestic heat, district energy, small scale, industrial, and utility boilers for power generation. Industrial and utility systems use spreader stokers, and bubbling and circulating fluidized bed boilers. Methods to improve combustion efficiency and emissions are explained. Technology needs are identified such as boilers for low-quality biomass fuels like poultry litter.

Pyrolysis and carbonization are discussed as methods to change the form of biomass to enable increased use. Biochar is considered as a coproduct of pyrolysis or gasification that can be used to sequester carbon and improve soil fertility.

Biofuel development is outlined with attention to the potential benefits of using existing infrastructure in the pulp and paper industry. The author concludes that future developments in biomass energy will depend on public policy decisions regarding the use of biomass resources and on the development of biomass energy as a means of offsetting carbon emissions from fossil fuels. The author uses 65 references along with 16 schematics, figures, pictures, and tables to augment the professional and scholastic treatment of the subject.

Chapter 15, "Utilizing Waste Materials as a Source of Alternative Energy: Benefits and Challenges" is addressed by T. Terry Tousey. There are numerous waste streams, both industrial and residential, that contain recoverable energy. However, many of them, in their "as-generated" form are not suitable to be used directly as a fuel. Either they have contaminants that reduce their energy value, or they are not in the proper physical form and need to be processed in order to recover their energy value. The question then becomes, can the energy value of these materials be recovered economically?

In this chapter, we will explore some of the benefits and challenges associated with using wastes and industrial by-products as a source of energy. We will look at how different waste materials are classified by the regulatory agencies and how this affects the economics of using them as a source of energy. We will review the economic and regulatory drivers for recovering the energy value of waste materials, and we will examine a few technologies being used to convert these materials into a usable form of energy, including anaerobic digestion and gasification. We will also address the energy efficiency issues of using the calorific value of the waste to produce electricity versus using it directly in the production of heat. Finally, we will identify some specific examples of industrial and post-consumer waste streams that are currently being used for energy production and look at their fuel characteristics. As part of this, we will explore the benefits and challenges associated with municipal waste to energy projects and we will look at a case study of the cement industry's experience with using hazardous waste fuels.

This chapter strives to give the reader a broad overview of all the aspects of implementing a waste to energy program. This includes not only dealing with the potential operational, regulatory, and community relations issues, but also with the issues associated with sourcing materials and the concept of reverse distribution. It is important to understand that for a waste to energy program to be successful, there must be an efficient mechanism in place to collect, process, and transport the material to the ultimate energy consumer.

Waste to energy will continue to play an increasing role in our future energy needs. These projects are unique in that their economics are almost always driven by disposal cost avoidance or regulatory compliance, and therefore, they are not completely dependent on the price of fossil fuels or government support to make them viable. For this reason, the economics are generally more favorable for these types of projects than a pure renewable energy project. Hopefully, this chapter will give the reader the tools to make an informed decision. The author uses 55 references along with ten schematics, figures, pictures, and tables to augment the professional and scholastic treatment of the subject.

Chapter 16 is authored by Lisa Shevenell and Curt Robinson who cover "Geothermal Energy and Power Development." The chapter covers the basic geology of the types of geothermal systems, where they are located, why they are there, and how geothermal systems in nature behave. Brief historical perspectives of geothermal development are presented as background. The authors discuss the three different types of power plants used to generate electricity and which type of power plant should be used to develop the particular type of geothermal systems most efficiently.

Current trends in geothermal development in the United States are discussed. An overview of geothermal exploration methods, historically successful, and new methods is presented. Environmental benefits and consequences and sustainability of geothermal power development are also outlined. Additionally, there is a comparison to other types of renewable energy sources and how geothermal is able to compete in energy markets.

Although the majority of the chapter focuses on power production from natural hydrothermal systems, the authors briefly introduce direct use applications, heat pumps, and enhanced geothermal systems. The authors use 78 references along with 25 schematics, figures, pictures, and tables to augment the professional and scholastic treatment of the subject.

In Section V, Fossil and Other Fuels are addressed in Chapters 17 to 20.

In Chapter 17, authors Paul S. Weitzel and James M. Tanzosh cover "Development of Advanced Ultra Supercritical Coal-Fired Steam Generators for Operation above 700°C." The chapter provides a view into the development work advancing the steam conditions to above 700°C for coal-fired steam generation. The term for this technology is Advanced Ultra Supercritical (A-USSC). The European Union, China, India, and the United States, with large reserves of coal providing a major portion of the fuel used in those nations' electric generating capacity, have committed and undertaken programs to solve the materials issues for this fuel. The program in the United States has been ongoing for over 10 years and is sponsored by the Department of Energy and the Ohio Coal Development Office, along with four steam generator vendors that shared costs. This program, "Boiler Materials for Ultrasupercritical Coal Power Plants," addressed what is called the precompetitive data development that was needed for ASME Code qualification of new materials and the supporting technology to fabricate and construct the higher-temperature boiler components.

The service life of materials has been a key factor in the historical development of this generator technology in order to provide satisfactory availability and reliability, and this advancement was built on following these previous lessons learned. The outline of the program and discussion of results, along with an example design, are presented. A summary of the materials selected and the weights of tubing and piping are included. Background on the technology and terminology regarding efficiency and fuel impact comparisons are provided.

The program work has provided a wealth of information that will help assure the future introduction of A-USC technology. It is important to carry forward the advancement of 700°C steam generation and meet the goal of lower cost of electric power generation and achieve the significant reduction of carbon dioxide emissions. The authors use 20 references along with 17 schematics, figures, pictures, and tables to augment the professional and scholastic treatment of the subject.

In Chapter 18, "Carbon Capture for Coal-Fired Utility Power Generation: B&W's Perspective" is written by D. K. McDonald and C. W. Poling. Changing climate and rising carbon dioxide (CO_2) concentration in the atmosphere have driven global concern about the role of CO_2 in the greenhouse effect and its contribution to global warming. Since CO_2 has become widely accepted as the primary anthropogenic contributor, most countries are seeking ways to reduce emissions in an attempt to limit its effect. This effort has shifted interest from fossil fuels, which have energized the economies of the world for over a century, to non-carbon emitting or renewable technologies.

For electricity generation, the non-carbon technologies include wind, solar, hydropower, and nuclear, while low-carbon technologies use various forms of biomass. Unfortunately, all current options are significantly more expensive than current commercial fossil-fueled technologies. Although use of wind and solar for power generation are increasing, they are incapable of supplying base load needs without energy storage capacity, which is currently impractical at the scale required. The only technologies capable of sustaining base load capacity and potential growth are coal, natural gas, and nuclear.

As the world grapples with CO_2 management, it is becoming increasingly clear that for the sake of the infrastructure in many developed countries and to provide a low carbon emissions option for developing countries relying heavily on coal, some form of carbon capture and storage (CCS) will be necessary. In a carbon-constrained world, the long-term viability of coal depends on the technical and economic success of emerging technologies for capture and storage. This may first be in the form of retrofit for the existing fleet, but considering the growth rate in emerging countries such as China and India, new plants will also be necessary as soon as practical.

Deployment of CCS is currently hindered by regulatory and social issues related primarily to long-term CO_2 storage. However, capture technologies are in the demonstration phase and are expected to be commercially available in the next decade. Carbon capture technologies are being developed by several companies for use with fossil fuels, especially coal. This chapter provides an overview from B&W's perspective of oxycombustion and post-combustion technologies for coal-fired steam and electricity generation, which are being developed for near-term deployment. The authors use eight references along with 20 schematics, figures, and pictures to augment the professional and scholastic treatment of the subject.

In Chapter 19, James L. Williams and Mark Jenner discuss "Petroleum Dependence, Biofuels — Economies of Scope and Scale; U.S. and Global Perspective." Our purpose is to examine petroleum dependence, identify risks of dependence, methods to mitigate dependence and risks, and provide a methodology for comparison of alternative sources of energy presenting a disciplined approach to the analysis of fuels and distinguishing between the fuel and its carrier or storage device.

The economic behavior of all fuels share many common characteristics. Principles concerning one fuel can be applied to others as the need arises. When comparing fuels, we focus on the cost per Btu and Btu/lb and Btu/ft^3 and emphasize the importance of efficiency in converting Btu's to mechanical or electrical forms of energy.

Fossil fuel economics are profoundly global in nature, and bioenergy economics are local. The two meet and compete directly at the consumer level. For example, biodiesel must be price competitive with petroleum diesel at the pump. The policy questions about whether it is worth subsidizing biofuels to reduce dependence on imported petroleum are central to the issue.

An international shortage of oil, which causes a price spike, can result in as much damage to an oil-dependent economy as an interruption in imports. Dependence can be lowered by either increasing domestic production or decreasing domestic consumption of petroleum.

Substitution of non-petroleum energy resources for petroleum is evident in the changes in the fuel mix in the post WWII era. Substitution is most difficult in the transportation sector because the high Btu content of petroleum relative to its weight and volume makes it an efficient fuel for the purpose.

The economics of bioenergy is much the same as the energy economics of fossil fuels. The frontier nature of the fledgling bioenergy industry adds enormous complexity. Biomass energy feedstocks can be wet, with different levels of moisture, dry, little to no water, or liquid without any water. Fuels can be produced specifically for energy, or they can be a by-product residual from the production of a higher-valued product. Waste-derived fuels bring with them many layers of environmental regulation, further adding to the complexity.

The economic viability of many biofuels depends upon feedstock transportation costs as well as coproducts in the production process. Ultimately, the use of biofuels is dependent on technological innovation, but the rate of innovation and implementation of new technologies is heavily dependent on public policy. Public policy can add incentives to reduce the costs and risks of supplying biomass fuels. It can also add incentives to enhance the demand and market prices for biomass-derived energy. The authors use 61 references along with 92 schematics, figures, pictures, and tables to augment the professional and scholastic treatment of the subject.

In the final chapter of Section V, Ravi K. Agrawal discusses "Coal Gasification" in Chapter 20. Fossil fuels supply almost all of the world's energy and feedstock demand. Among the fossil fuels, coal is the oldest and most abundant form of fossil fuel. Coal is widely available with reserves estimated to last over 140 years. Coal is the lowest-cost fuel, and projections of the Energy Information Administration (EIA) indicate that coal is likely to remain the cheapest fuel in the foreseeable future and will likely become cheaper as the price of other preferred energy sources rise. Based on the energy content in terms of price per BTU, electricity commands the highest premium, followed by oil and gas. This price differential is the key driver that determines the interest in coal and its conversion into other energy substitutes.

Since electricity trades at the highest premium, the cheapest source of energy, coal, has been extensively used to generate electricity. About 60% of global electric power is generated from coal, and about two-thirds of coal produced is used for power generation. Unfortunately, coal has the largest carbon footprint when compared to conventional fuels derived from oil and natural gas. Historically, most coal-based power plants are considered to be "dirty" as only a few had controls to lower the emissions of SO_2, NOx and particulates, and even to date, virtually none have controls on mercury and CO_2 emissions. Tightening of environmental regulations in countries like the United States, Europe, and Japan has prompted installation of emission control devices. Future regulations on greenhouse gas emissions will have a major impact on traditional coal-fired power plants. Another drawback of traditional coal-fired power plants is efficiency, which is low, about 25% to 30%. Increasing costs of fuel and emission controls have led to the development of coal gasification as an efficient method to generate power from coal, while minimizing the environmental impact. Coal gasification involves conversion of coal into gas. Gas generated via gasification is referred to as syngas and is rich in CO and H_2. Gasification of coal is typically conducted at high temperature and high pressure with the aid of gasification agents such as air/oxygen and steam. This chapter reviews gasification methods and technologies that are now being actively pursued for coal. The state-of-the-art technology review highlights strengths and weakness of gasification technologies that are currently being marketed.

In addition to generating power, syngas generated from coal can be used for a wide variety of products including chemicals, fuels, and metals. Typical chemicals that can be produced from syngas include hydrogen, methanol, ammonia, acetic acid, and oxygenates. Liquid fuels such as naphtha, diesel, ethanol, dimethyl ether, and methyl tertiary butyl ether can also be produced from syngas. This chapter discusses applications of gasification to produce these products and identifies specialized processing steps necessary to achieve the desired end product. The flexibility of coal gasification plants to coproduce multiple slates of products makes it more attractive than traditional power plants.

Technologies to control contaminants in syngas generated from coal are discussed with emphasis on end-use application. Removal of contaminates from syngas ensures that the emissions from coal gasification plants are significantly lower than those from a traditional coal-fired steam generation units. Future regulations on greenhouse gas emissions will have a major impact on traditional coal-fired power plants. This chapter also provides a review of the greenhouse gas removal technologies from syngas.

The gasification process is complex and requires more capital than traditional coal plants. This chapter discusses technologies available to treat the raw syngas for further processing. Raw syngas gas processing in Gasification Island is discussed as a combination of blocks that combines several unit operations and processes. Several options available to process the raw syngas within the block are discussed identifying efficiency improvements and cost associated with these processing steps.

Even for a known feedstock and a known end application, deciding on gasifier type and the design of the gasification island is a complex issue. The most cost-effective path is not always straightforward. This chapter provides a discussion on options available to resolve conflicting issues between capital cost, operating cost, and efficiency.

Most of the future energy and chemical demands are projected to come from countries such as the United States, Russia, China, India, Australia, S. Africa, and Eastern Europe, and all of these nations are rich in coal. Therefore, incentive for coal usage via gasification as an alternative to meet energy and chemical demand will be strong. This chapter provides an overview of the impacts of global energy consumption pattern and consumption growth areas on coal utilization.

Since 2000, several gasification plants have become operational; most of these plants have been installed in China. Chinese coal gasification activity is due to the encouragement of the Chinese government to develop coal-based fuels and chemicals as a strategic issue due to their reluctance to become more reliant on foreign oil. In recent years, interest in using syngas to reduce iron ore has picked up, especially in India. This chapter provides an outlook and information on recent global activities in coal gasification. The author uses 49 references along with 34 schematics, figures, pictures, and tables to augment the professional and scholastic treatment of the subject.

In Section VI titled "Nuclear Energy," four chapters are covered in Chapters 21 to 24.

Roger F. Reedy discusses in Chapter 21 "Construction of New Nuclear Power Plants: Lessons to be Learned from the Past." Without question, the costs associated with the construction of nuclear power plants in the 1970s and later escalated unreasonably. Although this increase was due to many factors, there are important factors that are still prevalent in the nuclear industry today. If these

factors are not understood and properly addressed now, the costs associated with new plants will escalate as it did in the past. In order to address the issues of concern, technical changes must be made to codes and standards, but even more importantly, changes must be made to the mindsets that caused the unjustified costs in the past.

The largest mindset modification required pertains to quality assurance. Quality assurance has been implemented on engineering projects since before the time of the pyramids. In simple terms, quality assurance is the method used to assure that dimensions are correct, that calculations are checked, that proper materials are used, that processes used in construction are appropriate, and that work is appropriately inspected. The Egyptians did a great job of controlling their work because the pyramids have stood for thousands of years. The design might be called an overkill, but who knows? They were designed to last forever!

Has the quality control used in the post-1960s nuclear plants enhanced quality any more than achieved by the use of commercial codes and standards? With regard to the technical aspects of the hardware, the answer is "probably not!" However, the cost difference between nuclear and commercial quality assurance is extremely significant. The largest cost has been the cost of administration of the QA program. Most of the administrative controls have resulted in wasted time and money. The nuclear quality assurance programs have concentrated on documentation, which in most cases is subjective evidence, not objective evidence. Very little time or effort was devoted to hardware evaluation. In the future, a more practical approach is required. Chapter 21 addresses this issue.

It is a fact that more than one-half of the nuclear plants operating in this country were designed and constructed prior to the requirements for the nuclear quality assurance programs mandated in the early 1970s. Are the units designed and fabricated after 1971 any safer than the earlier plants using commercial quality control? The record indicates that the answer is no.

There were a lot of other costly issues that arose after 1971 that also increased cost without enhancing quality. These were:

1. Inappropriate interpretation of codes and standards by unqualified personnel
2. Lack of understanding of the necessity of basic engineering judgments
3. Welding rejected by personnel unfamiliar with welding practices
4. Inappropriate record keeping
5. Lack of understanding of tolerances used in design and fabrication

The author addresses these and other issues that must be understood to avoid the mistakes and excessive costs of the last generation of nuclear power plants. The author uses 47 references to augment the professional and scholastic treatment of the subject.

In Chapter 22, Peter Riccardella and Dennis Weakland cover "Nuclear Power Industry Response to Materials Degradation — A Critical Review." In this article, the authors summarize several materials and structural integrity issues in operating nuclear power plants in which they have personally been involved, through their employment in technical and management positions in the industry for over 40 years. The issues include degradation and cracking of pressure vessels and piping in the plants that have collectively cost the industry hundreds of millions of dollars. The article looks retrospectively at the root causes of these problems, itemizes the lessons learned, and recommends an approach going forward that will anticipate and hopefully allows the industry to proactively address degradation issues in the future, for both the operating fleet as well as new plants that are currently in the licensing stages. The authors use 18 references along with five schematics and figures to augment the professional and scholastic treatment of the subject.

Chapter 23 is titled "New Generation Reactors." It is authored by Wolfgang Hoffelner, Robert Bratton, Hardayal Mehta, Kunio Hasegawa, and D. Keith Morton, with Mr. Morton coordinating the chapter write-up. The history of mankind repeatedly provides examples where a need is recognized and creative thinking is able to determine appropriate solutions. This human trait continues in the field of energy, especially in the nuclear energy sector, where advanced reactor designs are being refined and updated to achieve increased efficiencies, increased safety, greater security through better proliferation control of nuclear material, and increased use of new metallic and nonmetallic materials for construction.

With minimal greenhouse gas emissions, nuclear energy can safely provide the world with not only electrical energy production but also process-heat energy production. Examples of the benefits that can be derived from process-heat generation include the generation of hydrogen, the production of steam for extraction of oil-in-oil sand deposits, and the production of process heat for other industries so that natural gas or oil does not have to be used. Nuclear energy can advance and better the lives of mankind, while helping to preserve our natural resources. This chapter provides the reader greater understanding on how advanced nuclear reactors are not a "pie-in-the-sky" idea but are actually operational on a test scale or are near term.

In fact, efforts are currently underway in many nations to design and build full-scale advanced reactors. Many countries including Japan, Canada, China, Korea, the United States, Germany, Russia, France, India, and more have significant ongoing efforts associated with advanced nuclear reactors. This chapter briefly describes the six Generation IV concepts and then provides additional details, focusing on the two near-term viable Generation IV concepts. The current status of the applicable international projects is then summarized. These new technologies have also created remarkable demands on materials compared to light water reactors. Higher temperatures, higher neutron doses, environments very different from water, and design lives of 60 years present a real engineering challenge. These new demands have led to many exciting research activities and to new Codes and Standards developments, which are summarized in the final sections of this chapter. The reader is encouraged to enjoy this chapter, for the future is just around the corner. The authors use 98 references along with 42 schematics, figures, pictures, and tables to augment the professional and scholastic treatment of the subject.

The last chapter of this section is Chapter 24, authored by Owen Hedden and deals with "Preserving Nuclear Power's Place in a Balanced Power Generation Policy." In considering the future of nuclear power for electricity production in the United States, it is necessary to consider the present public perception of nuclear power. It is also necessary to consider public perceptions of the various competing sources of electricity production. These include coal, natural gas, and the several "green" or "renewable" sources, including hydro, wind, and solar.

The most recent nuclear power plant was completed in 1994. No new nuclear power plants have gone on line since then, but nearly 20% of our electric power is still provided by nuclear power.

In 2010, the media view had changed to accept resumption of new reactor construction. Now, however, in 2011, the Fukushima disaster will reverse that. While the number of U.S. nuclear power plants at risk of tsunamis is very small, there will be demands for additional safeguards for emergency diesel generators and their fuel supplies and the need for more flexible switchgear. Consideration of the location and operation of spent fuel pools will also be required. New plant construction will be suspended.

However, the demand for additional supply of cheap reliable electric power capacity will continue. So will the demand for reduction in power plant emission of carbon dioxide. Natural gas is little help in reducing carbon dioxide emissions. Wind and solar facilities do not have the capacity or reliability. That leaves hydro and nuclear. As noted in this chapter, all the good locations for hydro have been taken.

Nuclear power advocates will have to carefully refute claims of damage and radiation risk at Fukushima. At least to date, Fukushima is no Chernobyl. Release of radioactive material at Fukushima is about one-tenth of that at Chernobyl. The cores at three of the six reactors, and the spent fuel rods at one, have been damaged, but no cores have exploded. No immediate deaths due to radiation have been reported at Fukushima versus 30 at Chernobyl. This must be equated to deaths due to other power-generation sources. Consider coal mining and transport, refinery and pipeline explosions, and long-term effects of air pollution caused by coal-burning plants.

There is a greater risk to the public from the other major energy producers than from nuclear. With electric energy needs increasing, nuclear power production must be part of that increase. The author uses 30 references along with one table to augment the professional and scholastic treatment of the subject.

Section VII is titled "Steam Turbines and Generators" and has two chapters: Chapters 25 and 26.

Chapter 25 "Steam Turbine and Generator Inspection and Condition Assessment" is authored by Lawrence D. Nottingham. Modern turbines and generators are large, complex machines that utilize a vast array of materials, are often exposed to high stress levels and a number of potentially hostile environments and which suffer operative damage mechanisms as a direct result. While failures of the massive rotating components are relatively rare events, the consequences of such failures are severe and extremely costly to the owner/operator. Consequently, rigorous programs have evolved for assessing current conditions, for detecting and quantifying existing flaws, and for predicting the remaining lives of the primary components, with an overall objective of providing effective run/repair/replace decision making. In this chapter, the author, who has spent most of his 39-year professional career dealing with turbine and generator condition assessment issues, presents an overview of the state-of-the-art for inspection and remaining life assessment of critical rotating T/G components.

The chapter additionally provides background information on machine design, materials, stress sources, applicable nondestructive evaluation methods, and analytical processes available for predicting remaining life, as related to and appropriate for an overall understanding of the assessment process. The author uses four references along with 23 schematics, figures, and pictures to augment the professional and scholastic treatment of the subject.

Chapter 26 Steam Turbines for Power Generation is written by Harry F. Martin. Steam turbines have historically been the prime source of power for electric power generation. This chapter will focus on steam turbines currently being applied to power generation. The steam conditions will include those currently applied to fossil-fired power plants, combined cycle and nuclear power units. In addition, future cycle pressures and temperature requirements are discussed along with the material requirements for these applications.

The chapter includes steam turbine fundamentals, equipment configurations, various design types and technology applications, performance levels, performance testing, and operation and maintenance. Blading fundamentals are included along with current technology improvements for increased performance. New sealing concepts including brush seals, retractable seals, and abradable seals are discussed. The effects of low pressure turbine exhaust size and diffuser design on performance is included.

This chapter should appeal to many types of readers. The presentation makes use of many references to permit the reader to expand his knowledge on a specific subject while limiting the size of the chapter.

Some of the topics included are seldom found in an overview type presentation. These include moisture and solid particle erosion, turbine heat transfer analysis, stalled and unstalled blade flutter, and limit load.

Concepts of automated turbine control are discussed including rotor stress modeling concepts applied in control systems. The pros and cons of various operating modes such as the use of partial arc and full arc of admission high-pressure turbine designs and sliding pressure are reviewed. The author covers the topic with 22 equations, 42 references along with 35 schematics, figures, and pictures to augment the professional and scholastic treatment of the subject.

Section VIII: Selected Energy Generation Topics and has four chapters: Chapters 27 to 30.

Chapter 27 is titled "Combined Cycle Power Plant" and is authored by Meherwan P. Boyce. Combined cycle power plants are most efficient power plants available today for the production of electric Power with an efficiency ranging between 53% and 57%. A short summary of other power plants such as steam, diesel, and gas turbine plants are given. Combined cycle power plants come in all sizes. In this chapter, we are emphasizing the larger plants ranging in size from 60 MW to 1500 MW. These combined cycle plants have as their core the gas turbine, which acts as the Topping Cycle and the steam turbine, which acts as the Bottoming Cycle. In between the gas turbine and the steam turbine is a Waste Heat Recovery Steam Generator (HRSG), which takes the heat from the exhaust of the gas turbine and generates high-pressure steam for the steam turbine.

The Brayton (Gas Turbine) and Rankine (Steam Turbine) cycles are examined in detail as they are the most common cycles used in a combined cycle power plant. Various modifications of the basic Brayton Cycle taking into account various cooling effects, such as refrigerated cooling and evaporative and fogging effects are examined. Also examined are intercooling, regenerative, and reheat effects.

The Rankine Cycle is also examined in detail, taking into accounts the effects of regeneration, reheat, backpressure and condensation on the performance, power, and efficiency of the steam turbine.

The Brayton Rankine Cycle, which is the basic cycle for a combined cycle, is examined, and the effect of the above-described effects, as they are applied to the combination cycle, is discussed. Various techniques to improve Power and Efficiency, on such a cycle are presented.

The combined cycle power plant comprises of the Gas Turbine, Heat Recovery and Steam Generator (HRSG), the Steam Turbine, and Condensers. Each of these major components is examined in detail, and furthermore, each of the major components are closely detailed to describe their operations.

The gas turbine section examines the various types of gas turbines such as the Aeroderivative Gas Turbine, as well as the large Frame-Type Gas Turbines. In this section, new trends on gas turbine compressors, combustors, and the hot section expander turbine are discussed, and the effect each of these components have on the efficiency and environment are carefully examined to ensure that they are operated as designed.

The HRSG system is a critically important subsystem in a combined cycle power plant. Under this section, the efficiency of energy transfer in multiple pressure steam sections is closely examined as are supplementary-fired HRSGs. Also examined are the Once Through Steam Generators (OTSG) and compared to the traditional drum-type steam generators. Also discussed in this section are the effects of deaerator, economizers, evaporators, superheaters, attemperators, and desuperheaters.

The steam turbine section deals with the reheat extraction and condensing steam turbines with various sections, which match the multiple pressure in the heat recovery steam generators. Various steam turbine characteristics are described as to their effect on power and efficiency.

The chapter closes with a short analysis of some of the major cost centers in a Combined Cycle Power Plant and the Reliability and Availability of such power plant systems. The author covers the topic with 31 equations, two references along with one table, 37 schematics, figures, and pictures to augment the professional and scholastic treatment of the subject.

Simon Gamble, Marian Piekutowski, and Ryan Willems authored Chapter 28 and discuss "Hydro Tasmania — King Island Case Study." Hydro Tasmania has developed a remote island power system in the Bass Strait, Australia, that achieves a high level of renewable energy penetration through the integration of wind and solar generation with new and innovative storage and enabling technologies. The ongoing development of the power system is focused on reducing or replacing the use of diesel fuel while maintaining power quality and system security in a low inertia system. The projects completed to date include:

- Wind farm developments completed in 1997 and expanded to 2.25 MW in 2003;
- Installation of a 200-kW, 800-kWh Vanadium Redox Battery (2003);
- Installation of a two-axis tracking 100-kW solar photovoltaic array (2008), and
- Development of a 1.5-MW dynamic frequency control resistor bank that operates during excessive wind generation (2010).

The results achieved to date include 85% instantaneous renewable energy penetration and an annual contribution of over 35%, forecast to increase to 45% postcommissioning of the resistor. Hydro Tasmania has designed a further innovative program of renewable energy and enabling technology projects. The proposed King Island Renewable Energy Integration Project, which recently received funding support from the Australian Federal Government, is currently under assessment to be rolled out by Hydro Tasmania (including elements with our partners CBD Energy) by 2012. These include:

- Installation of short-term energy storage (flywheels) to improve system security during periods of high wind;
- Reinstatement or replacement of the Vanadium Redox Battery (VRB) that is currently out of service;
- Wind expansion — includes increasing the existing farm capacity by up to 4 MW;

- Graphite energy storage — installation of graphite block thermal storage units for storing and recovering spilt wind energy;
- Biodiesel project — conversion of fuel systems and generation units to operate on B100 (100% biodiesel); and
- Smart Grid Development — Demand Side Management — establishing the ability to control demand side response through the use of smart metering throughout the Island community.

This program of activities aims to achieve a greater than 65% long-term contribution from renewable energy sources (excluding biodiesel contribution), with 100% instantaneous renewable energy penetration. The use of biodiesel will see a 95% reduction in greenhouse gas emissions. The projects will address the following issues of relevance to small- and large-scale power systems aiming to achieve high levels of renewable energy penetration:

- Management of low inertia and low fault level operation;
- Effectiveness of short-term storage in managing system security;
- Testing alternative system frequency control strategies; and
- Impact of demand side management on stabilizing wind energy variability.

The authors cover the topic with 24 equations, 12 references along with two tables, 22 schematics, figures, and pictures to augment the professional and scholastic treatment of the subject.

Chapter 29, titled "Heat Exchangers in Power Generation" has been authored by Stanley Yokell and Carl F. Andreone describes shell-and-tube and plate and frame types of power plant heat exchangers and tubular closed feedwater heaters and the language that applies to them. The chapter briefly discusses Header Type Feedwater Heaters and their application and use. It defines the design point used to establish exchanger surface and suggests suitable exchanger configurations for various design-point conditions and the criteria used to measure performance. It does not cover power plant main and auxiliary steam surface condensers because of the differences in how they are designed and operated The chapter briefly discusses the effects on the exchanger of normal and abnormal deviations from design point during operation. The authors have several schematics and references to supplement their discussion.

The last chapter of this section, Chapter 30, "Water Cooled Steam Surface Condensers" has been authored by K. P. (Kris) Singh. Chapter 30 is devoted to providing a comprehensive exposition to the technologies underlying the design and performance of steam surface condensers used to condense the exhaust steam from the low-pressure turbine in a power plant. The surface condenser is an indispensable component in any power station and is also one whose performance directly affects both the thermodynamic efficiency and the service life of the plant. In terms of sheer size, the surface condenser is the largest of any equipment in the power cycle, which alone makes it an important subject matter in the field of power plant technology.

The specific class of surface condensers considered in this chapter is of the so-called water-cooled type, wherein the condensing of the exhaust steam occurs outside the tube bundle by extraction of its latent heat by the cooling water circulated through the inside of the tubes.

Empirical information on the design and operation of classical surface condensers is provided in the standards published by the Heat Exchange institute, which is an industry consensus document that contains valuable practical guidance. Accordingly, the material in this chapter is not a substitute for a design manual;

rather, it is intended to illuminate the thermal-hydraulic concepts underlying the state-of-the-art condenser design technology to possibly serve as the technical beacon for developing designs for new challenging situations such as condensing of noncondensible-laden geothermal steam. Inadequate removal of noncondensibles, flow-induced vibration, entrainment of (corrosive) oxygen in the condensate, excessive fouling, erosion of the tubes from flashing action in the condensate are among the many ailments that can afflict a surface condenser that require an in-depth understanding of the thermal and hydraulic phenomena that are present in its operation. The mission of this chapter is to provide the necessary cerebral understanding to the designer to enable him (her) to navigate through the many challenges that lie on the path to a successful design.

Special emphasis is placed on explaining the efficacy of a type of tube support , called nonsegmental baffles that has not been used to the extent in the industry as its thermodynamic merit would warrant. The mathematical means to quantify the advantage of the nonsegmental baffle configuration over the conventional plate-type supports is provided.

The author covers the topic with 55 equations, 18 references along with seven tables, 11 schematics, figures, and pictures to augment the professional and scholastic treatment of the subject.

Section IX is titled "Emerging Energy Technologies" and has two chapters: Chapters 31 and 32.

Chapter 31: "Toward Energy Efficient Manufacturing Enterprises" has been authored by Kevin W. Lyons, Ram D. Sriram, Lalit Chordia, and Alexander Weissman. Industrial enterprises have significant negative impacts on the global environment. Collectively, from energy consumption to greenhouse gases to solid waste, they are the single largest contributor to a growing number of planet-threatening environmental problems. According to the Department of Energy's Energy Information Administration, the industrial sector consumes 30% of the total energy, and the transportation sector consumes 29% of the energy. Considering that a large portion of the transportation energy costs are involved in moving manufactured goods, the energy consumption of the industrial sector could reach nearly 45% of the total energy costs. Hence, it is very important to improve the energy efficiency of our manufacturing enterprises.

A product's energy life cycle includes all aspects of energy production. Depending on the type of material and the product, energy consumption in certain stages may have a significant impact on the product energy costs. For example, 1 kg of aluminum requires about 12 kg of raw materials and consumes 290 MJ of energy. Several different strategies can be used to improve the energy efficiency of manufacturing enterprises, including reducing energy consumption at the process level, reducing energy consumption at the facilities level, and improving the efficiency of the energy generation and conversion process. While the primary focus of this chapter is on process level energy efficiency, the authors also briefly discuss energy reduction methods and efficient energy generation process through case studies.

In this chapter, the authors introduce the concept of unit manufacturing processes, which are formal descriptions of manufacturing resources at the individual operations level (e.g., casting, machining, forming, surface treatment, joining, and assembly) required to produce finished goods. They classify these processes into mass-change processes, phase-change processes, structure-change processes, deformation processes, consolidation processes, and integrated processes. Several mechanisms used to determine energy consumption for these processes are described. This is followed by a method that describes how to improve the efficiency of this energy consumption through improved product design for injection molding. HARBEC Plastics, Inc. is an innovator in implementing sustainable manufacturing practices. This company, which makes high-quality injection-molded parts, has made a considerable commitment to being green. Various techniques employed by HARBEC Plastics to improve energy efficiency are briefly described. A specific technique — using supercritical fluids — that can be effectively used to improve energy efficiency and to improve processes that generate energy from nontraditional sources is also outlined. Finally, the authors point out how best practices, regulations, and standards can play an important role in increasing energy efficiency.

The authors cover the topic with 58 references along with three tables, 12 schematics, figures, and pictures to augment the professional and scholastic treatment of the subject.

Chapter 32: "The Role of Nano-Technology for Energy and Power Generation: Nano-Coatings and Materials" has been authored by Douglas E. Wolfe and Timothy J. Eden. Chapter 32 discusses a variety of nano-coatings and materials used in the energy and power generation fields. Nano-coatings, nanocomposite coatings, nanolayered coatings, functional graded coatings, and multifunctional coatings deposited by a wide range of methods and techniques and their roles in assisting to generate energy and power for the fuel cell, solar cell, wind turbine, coal, combustion, and nuclear industries are discussed. Chapter 32 provides a brief description of the past and present state-of-the-art nanotechnology within different industrial areas, such as turbine, nuclear, fuel cell, solar cell, and coal industries, that is used to improve efficiency and performance. Challenges facing these industries as pertaining to nanotechnology and how nanotechnology will aid in the improved performance within these industries are also discussed.

The role of coating constitution and microstructure including grain size, morphology, density, and design architecture is presented with regard to the science and relationship with processing-structure-performance relationships for select applications. The impact of nanostructured materials and coatings and their future in energy are also discussed with regard to nanostructured configuration, strategy, and conceptual design architectures. Coating materials and performance are also discussed related to materials in extreme environments such as power generation, erosion environment, and hot corrosion. The role of nanotechnology, nanocoatings, and materials for power and energy varies depending on the working environment but, in general, can be classified into the following categories of improved wear resistance, corrosion resistance, erosion resistance, thermal protection, and increased surface area for energy storage. Since not all environments are the same, slight modifications may be required to optimize nano-material and nano-coatings for a particular application. These include composite coatings, functional gradient coatings, superhard coatings, superlattice coatings, metastable multifunctional, solid solution, nano-crystalline, multilayer coatings, and mixed combinations. With all the material choices that we have ranging from binary, ternary, and quaternary nitride, boride, carbide, oxide, and mixed combinations, choosing the optimum coating material and design architecture can be challenging. The approach to materials solutions starts with understanding the system performance, operating environment, maintenance issues, material compatibility, cost, and life cycle. Chapter 32 concludes with a summary of the future role of nanotechnology and nano-coatings and materials in the fields of power generation and energy.

The authors cover the topic with nine equations, 73 references along with four tables, 20 schematics, figures, and pictures to augment the professional and scholastic treatment of the subject.

SOME SOLAR RELATED TECHNOLOGIES AND THEIR APPLICATIONS

Robert Boehm

1.1 SOLAR POWER CONVERSION, UTILITY SCALE

1.1.1 Introduction

The use of solar energy for generating power is a concept that has been around for a long time. Some early examples of this, to be dealt with in the sections that follow, include dishes and troughs that were developed in a flurry of creative activity between the late 1800s and the early 1900s. There was somewhat of a hiatus from much development in first half of the 20th century until work again resumed after the Second World War. This was driven by interest in exploring space as well as other thrusts. An excellent summary of the early history of solar applications has been given by Butti and Perlin [1].

Momentum to examine renewable energy generally picked up considerably with the oil embargo of the 1970s. At that time, President Jimmy Carter outlined the challenge the US particularly, and the world more generally, faced in terms of dealing with energy issues. He warned that the situation was "the moral equivalent of war" in a 1977 speech. In this same year, he was responsible for forming the US Department of Energy, which combined some existing agencies, including the US Energy Research and Development Administration. He also initiated the Solar Energy Research Institute, in Golden, Colorado, that later became the National Renewable Energy Laboratory (NREL).

Following that time, interest again waned as a result of falling energy prices. President Ronald Reagan removed the solar panels that were placed on the White House under President Carter. Oil again became the primary focus in energy. Various starts and stops to solar development followed, including the start of the large development of the Luz systems in the Mohave Desert area of California, which was later stopped when existing tax credits went away.

Although many countries of the world, particularly Israel and Japan, had active programs in solar water heating for many years, solar power generation had taken a low profile existence. About the turn of the 21st century this began to change. Not only was there a heightened awareness of problems with continuing the reliance on Middle East oil, as had been the case all along, but this became more strikingly apparent as various types of strife broke out there, including the wars in Iran and the terroristic activities of many Middle East groups. Also, more and more people came to appreciate the concerns about peak oil. Simultaneously, concern grew about climate change and the impact of burning of fossil fuels on that. Worldwide, there is a growing awareness of the greenhouse gas increases in the atmosphere. At the present time, there is a rebirth of solar power generation energy activity. Some of Europe has reacted with lucrative feed-in tariffs that have encouraged development, particularly in Spain and Germany. In the US, much of the encouragement is being given by individual states. This aspect is quite different than the wave of development under Carter, which was primarily driven from the Federal level, and thus was easily changed by a new president with a difference in philosophy. Now we find mandates, tax incentives, and rebate programs interwoven in a great deal of society, including at the state and local levels.

Generally, new power sources require some types of financial incentives when bringing them to market. There is usually a need for some type of government support. Nuclear power has received massive subsidies, but these are masked somewhat by the fact that both power applications and weapons development were at play at the same time. Solar development has been plagued by the fact that the costs of related equipment are high because the market has been small, and the market has been small because the costs are high. In the photovoltaic (PV) arena, one where a great deal of progress has been made with generally more financial support from the government than other types of solar power generation, historically a 20% reduction in cost has been realized for each doubling of production.

The development story is complicated by the overlaying economic situations as well as political reasons. Solar installations tend to be high capital investment items because, in a sense, the fuel is being paid for upon construction. In good economic times, this high cost, while still a concern, can be covered by loans. During economic hard times, with the usual tightening of credit during those periods, the solar market can become severely constrained. Such has been the situation in the latter half of the first decade of the 21st century.

Two major ways of producing power are either by thermal means, where high temperatures from solar heating can be used in some type of heat engine, or by photovoltaic approaches. Because these are two quite different approaches, each will be touched upon separately in what follows.

1.1.2 The Use of Concentration

Concentration opens a number of opportunities for solar power systems. In thermal systems, concentration allows much higher temperatures to be generated than would be possible with simple flat plate systems. This, in turn, results in higher temperatures within the associated thermal engine. As is well known from the Second Law of Thermodynamics, higher temperatures result in higher engine efficiencies. Another benefit of concentration that goes hand-in-hand with this one is that the profile of the heated element is relatively smaller than flat plate systems (comparing energy harvested per unit area of collector). This allows special attention to be given to minimizing the heat loss from the receiver elements.

In PV systems, concentration permits tradeoffs between solar cells, which are generally more expensive, against tracking mechanisms and support structure including lenses or mirrors, which are generally less expensive in comparison to the PV elements. Concentration allows for each PV cell to produce proportionally more power per unit, and concentration may also increase the cell's efficiency. One possible concern is that some types of cells (silicon particularly) can have a significant negative temperature effect on efficiency compared to others. High ambient temperatures can result in noticeable decreases in production efficiency.

Direct beam radiation is required for concentrating systems to be effective. Diffuse components of the radiation are not of value in these approaches unless the concentration is small. In addition, large amounts of flux are preferred to make the capital investment in these systems more profitable. Several locations in the world meet this requirement. For example, solar radiation resource in the US is shown in Figure 1.1. This resource is enormous, as shown in Figure 1.2 which is a representation that a square area of land approximately 100 miles (161 km) on a side could furnish with current technology all of the US's electricity needs. Of course the transmission of this generated power to the whole country poses a significant problem. Storage is another with current technology.

However, several issues come into play when the choices of locations for concentrating solar plants are selected. For these plants, only the beam radiation is to be considered. Also, rugged landscape is to be shunned. Generally a relatively flat area is preferred. When these factors are considered, the areas where solar plants can be located are far less numerous. See Figure 1.3. Here, all land areas with greater than a 3% grade have been eliminated. Another factor that comes into play that is depicted here is that the plant has to be near a major power grid corridor. Construction of long lengths of major power lines can be extremely expensive.

Concentration can be accomplished in a variety of ways in two general categories: single axis (line focus) and dual axis (point focus) tracking. There are many kinds of single axis approaches because of a fixed axis (which can be fixed in a variety of ways) in one dimension and a moveable one in the other. In all of the concentrating systems, the beam can be developed either by refracting or reflecting the incident energy. Generally applications of refracting systems are less frequently used in practice, and these are usually for PV systems.

Another issue in concentration is the way this takes place in terms of the imaging aspects of it [2]. This involves the concept of imaging optics vs. non-imaging optics. In general, the latter has a great deal to offer in the design of concentrating systems over the former. For one thing, the amount of light concentrated in non-imaging optical systems can be much greater than the comparable situation in imaging systems. Secondly, non-imaging approaches allow much better control of radiation, giving, say, more uniform irradiance or controlling the distribution better.

Lastly, but certainly not the least, cleaning is required for the concentrating elements of all CSP systems. Some geographic areas are more prone to lens or mirror soiling than others, but cleaning is a necessity. Usually, water is used for this purpose, so some water must be available for these kinds of systems even if it is not used for cooling purposes. Some treatments are available that facilitate rain running off surfaces quickly if they are oriented in a direction that gravity can be of assistance. It is desirable to have surfaces that stow facing downward during periods of darkness and inclement weather to minimize the cleaning requirements, but this may complicate the design requirements.

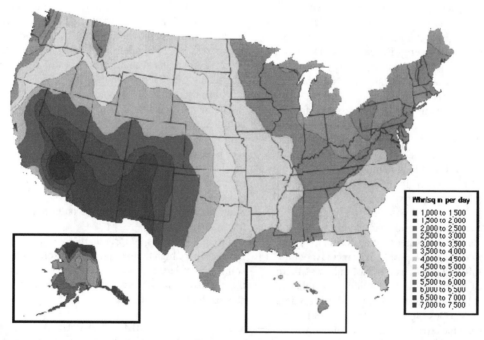

FIG. 1.1 THE MAP OF THE US INDICATING THE AVERAGE SOLAR RADIATION PER YEAR IN UNITS OF W-HR/SQ M AREA/DAY IS SHOWN (Source: US Department of Energy)

FIG. 1.2 IN THIS REPRESENTATION FROM THE NATIONAL RENEWABLE ENERGY LABORATORY, THE FACT IS DEPICTED THAT THE TOTAL ELECTRICITY REQUIREMENTS OF THE US COULD BE FURNISHED FROM SOLAR ENERGY FALLING AND CONVERTED ON A SQUARE 100 MILE X 100 MILE (161 KM X 161 KM) AREA OF THE STATE OF NEVADA

1.1.3 Thermal Power Generation Approaches

1.1.3.1 Rankine Cycles Many solar thermal power generation schemes (particularly the trough and tower systems to be discussed below) involve Rankine cycles to generate power from the heat collected by the solar field. There is both good and bad news about these kinds of plants.

Comparisons of the resulting operational levels of thermal power plants when using concentration are given in Table 1.1 [3]. Here the performance characteristics of trough and tower plants

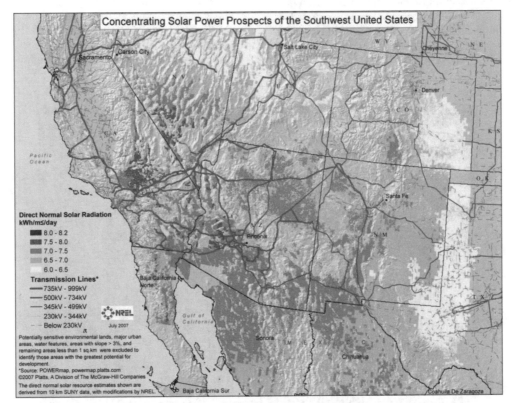

FIG. 1.3 THIS PLOT OF THE SOUTHWESTERN US AND NORTHWESTERN MEXICO SHOWS THE DIRECT NORMAL SOLAR FLUX (UNITS OF KW-HR/SQ M AREA/DAY) ONLY ON LAND THAT HAS LESS THAN A 3% GRADE. MAJOR TRANSMISSION LINES ARE ALSO SHOWN (Source: NREL)

TABLE 1.1 CYCLE PERFORMANCE COMPARISON [3]

Type of CSP	Maximum steam temperature expected, °C	Potential cycle efficiency ranges, %*
Flat panels	110	7–10
Fresnel collector	310	20–26
Parabolic trough	395	28–32
Solar tower	550	35–42

*Depends upon many factors of plant design.

are compared. Flat plate performance, which is not really a viable approach in utility scale solar thermal power plants, is shown also for completeness.

On the plus side, the power block of these plants is in many respects common to conventional gas-fired or coal-fired units, although there may be design differences due to the transient nature of the solar source. This commonality greatly simplifies their design and operation. These plants can utilize conventional steam plant components, from the turbine to the condenser. Any major power plant component manufacturer can supply the power block for these plants.

The other side of the coin is that both the high and low temperatures of these types of plants may be limited. For example, solar plants typically operate at high temperatures that are lower than those of conventional plants. See Table 1.1 for typical values for solar plants. This poses implications for the condenser end. While the condensing temperature affects the operation of all Rankine plants, when the high temperature is moderate, the condenser temperature has a relatively higher impact on plant efficiency. In general, wet cooling offers lower condensing temperatures than does dry cooling. Since most solar thermal power plants will be located in arid regions of the world, scarce water for cooling could be a significant limitation.

1.1.3.2 Trough Technology This approach uses a fluid circulating through a field of collectors where energy is reflected from a very large dimension parabolic trough (often aluminized or mirrored in some way on the inner side) to a small tubular arrangement along the axis. The small tube is often a metallic absorber where the fluid flows and a glass tube (usually evacuated) is concentrically located around it to increase the resistance to heat loss. A significant array of piping is used to connect all of the troughs together to bring the heated fluid to the main power plant and to return the cooled fluid back to the field.

The concentrator elements in the plant can offer a large wind profile. It is the case that plants of this type, like most tracking kinds of plants, have some arrangement for stowing their solar collection elements during high wind to give as low of a wind profile as possible.

Between 1984 and 1990 the Israeli company Luz International built several plants using this technology in the Mojave Desert region of California. An interesting analysis of this development is given in the book by Berger [4]. In all, a total of 364 MWe of these plants was constructed. Each of these series of plants was over 15% more cost efficient than its predecessor [5].

Solargenix (now Acciona Solar Power), recently made further improvements to this technology and built plants that include a 1 MWe organic Rankine cycle plant in Arizona and a 64 MWe (75 MWe maximum capacity) plant in Southern Nevada. The latter is shown in Figure 1.4. This plant includes 760 parabolic troughs with more than 180,000 mirrors, 18,240 4-m long tubes, and heats the field fluid to 390°C (735°F).

It is usually the case that a synthetic oil circulates through the field returning hot to a heat exchanger where the heat is transferred to steam. The steam then traverses a fairly conventional Rankine cycle engine.

Several other companies have begun to develop potentially large solar thermal plants that also use a line focus approach. One of the approaches that may be used to replace the parabolic trough includes a segmented set of nearly flat mirror facets that can in-

FIG. 1.4 NEVADA SOLAR ONE, A 64 MWE (75 MWE MAXIMUM) PLANT BUILT BY ACCIONA SOLAR POWER BEGAN ELECTRICAL PRODUCTION NEAR LAS VEGAS DURING 2007 (Courtesy Acciona Solar Power)

dividually follow the sun. This is denoted as Fresnel collector in Table 1.1. One such manufacturer is Areva (Palo Alto, California) that is using a technology that was developed in Australia. See Figure 1.5. The benefits from this approach are claimed in the field design. The facets are lower to the ground and offer a smaller profile to the wind than do parabolic troughs. Also, the receiver tubes are stationary, so no flexible connections are required between trough segments as is the case with the parabolic troughs, and thus the new design is claimed to be able to handle high-pressure steam in the field. No heat exchanger would then be required to remove heat from the field fluid for the Rankine cycle fluid for these types of designs.

1.1.3.3 Power Tower (Central Receiver) Systems In the power tower system, a large number of ground-mounted mirrors are made to follow the sun on a nearly instantaneous fashion, reflecting the sunlight to the top of a centrally located tower where a receiver section is located. The receiver design could take a variety of forms, but one involves a series of parallel tubes that cover the outside of the structure or are located in the back of a cavity-type receiver. Depending upon the type of fluid circulating through the receiver (both steam and molten salt have been used in various experiments that have been performed previously), there may or may not be a need for a heat exchanger to transfer the collected heat into steam for the conventional power block. Heat rejection systems for these types of units are either wet or dry cooling towers. Of course, the latter use no water for cooling (ideal in desert areas), but operate at a slight deficiency in engine efficiency compared to wet cooled systems.

These kinds of systems can be large. The original unit was Solar One, located just outside of Barstow, California, which is also in the Mohave Desert region. See Figure 1.6. The tall tower was illuminated by a surround field of 1818 heliostats. Each of the heliostats had an area of 40 m^2. The unit produced 10 MWe when the receiver fluid was steam. A later modification to the plant enlarged the field somewhat and used molten salt mixture of 60% sodium nitrate and 40% potassium nitrate as the receiver fluid, with a heat exchanger to transfer the heat to steam for the power block. Some degree of thermal storage was incorporated into both configurations. When steam was the receiver fluid, an organic heat transfer fluid was used as the storage medium. When molten salt was used as the receiver fluid, it served as the storage medium.

Abengoa Solar has two tower projects in Spain. See Figure 1.7. One, PS10, is a tower that generates 11 MW of electricity. To do this, it uses 624 heliostats that are each 120 m^2 in area. Unlike Solar One in California, this plant does not use a surround field. Rather it has a north-side field, which Abengoa Solar feels is more appropriate for their application. Next to PS10 is PS20 that is designed to generate 20 MWe, and it started producing power in 2009. The field consists of 1,255 mirrored heliostats, each with the same area as those on PS10. PS20's tower is 531 feet tall.

More recently some smaller configurations of these types of systems have been proposed. When smaller towers are considered, the fields are smaller and several units might be located in adjacent spaces, perhaps more than one feeding a given power block.

1.1.3.4 Dish Stirling Systems These units are usually found on a dish arrangement where both the mirror facets and the engine/generator follow the sun throughout the day. Unlike the trough and the power tower arrangement which can be constructed for large

FIG. 1.5 A CONCEPT OF A FRESNEL (LINE FOCUS) SOLAR THERMAL PLANT HAS BEEN PUT FORTH BY AREVA. IN THIS APPROACH, LONG AND TRANSVERSELY NEARLY FLAT FACETS EACH ROTATE ON THE GROUND TO FOCUS THE DIRECT BEAM RADIATION ONTO A FIXED RECEIVER OVERHEAD (Courtesy of AREVA)

FIG. 1.6 SHOWN IS A SOLAR CENTRAL RECEIVER PLANT THAT OPERATED DURING THE 1980S AND 1990S NEAR BARSTOW, CALIFORNIA. COURTESY OF SANDIA NATIONAL LABORATORIES

power ratings from a single engine, this approach requires many smaller units (typically around 25 kW each) to be constructed.

The basis of this approach is the Stirling cycle, which is known to be one of the more efficient thermodynamic engines because it has nearly constant temperature heat transfer processes. In these devices the working fluid (a gas) is contained in the engine housing and is usually either hydrogen or helium. A low molecular weight fluid is preferred for this application because of its good heat transfer characteristics and low-pressure drop behavior. Ambient air is used for heat rejection purposes. The dish maintains a high solar flux on the

FIG. 1.7 THIS PHOTO SHOWS PS10 IN OPERATION IN THE BACKGROUND AND PS20 UNDER CONSTRUCTION IN THE FOREGROUND. PS20 STARTED POWER PRODUCTION IN 2009 (From ABENGOA SOLAR)

heat input area of the engine, and air is circulated over a heat rejection area, usually through one or more compact heat exchangers.

Two different designs of dish-Stirling systems are shown in Figure 1.8. The one in the right foreground is a design developed by McDonald-Douglas in the 1980s that later sold the terrestrial application rights to Stirling Energy Systems (SES). For many years this type of system held the highest demonstrated efficiency conversion rates of sunlight to electricity for any type of system of around 30%. The SES unit used fixed mirror facets and an engine that was developed by the Scandinavian firm Kockums. SAIC International developed the unit shown on the left side, and it used stretched membrane facets that could be focused by exerting a slight vacuum on the chamber behind the membrane in each facet. It used an engine developed by Stirling Thermal Motors (STM). Both units produced an output of nominally 25 kW. When installing these kinds of systems, a significant pedestal has to be constructed that penetrates the ground a distance appropriate to counteract wind loads, but these types of units are designed to move to stow position in very high winds.

More recently SES in conjunction with Sandia National Laboratories has redesigned several aspects of its system that they call the Suncatcher™. They have moved toward a more production-oriented design, and several of these units are shown in Figure 1.9. Even more recently a 1.5 MWe solar field of this design has been installed outside Phoenix, Arizona. The exclusive development sister company of SES is Tessera.

As is the case with both the trough and power tower technologies, several companies are developing dish systems of this design. Although much interest in this approach has been expressed, the author is not aware of any systems of this type in actual commercial production at the present time other than the one noted above.

1.1.3.5 Thermal Storage It is often said that solar systems cannot be used to furnish a constant rate of power generation because the sun shines only a fraction of the hours in a day. While this is obviously true about the sun, it does not have to be true about the power plant. The last big frontier in solar power generation is the use of effective means of storage. This would allow solar energy collected during sunny periods to be used in periods when the solar flux is low or zero in magnitude.

When the solar thermal system operates on a Rankine cycle, the solution is clearly a thermal storage element in the plant design. A larger field than is needed to drive the plant during the day would be used, storing some of the energy to be used at a later time when it is needed. Traditionally storage has not been used to any great extent in solar thermal power plants because there were no incentives for this to happen. However, with the possibilities of solar being called upon to furnish 24-hour-a-day power, the issue becomes quite important.

Two approaches have been used in the past with solar thermal plants. As noted previously in this presentation, earlier designs have used either molten nitrate salts or high temperature heat transfer fluids. When the latter is used, the storage tanks may have a filler inserted, such as rock or something that is of lower cost than the fluid being used. Molten salt has also been used for storage applications. In salt flowing systems, the latter cannot be allowed to solidify (cool below the freezing/melting temperature, but still greatly above ambient temperature) unless extensive heat tracing is incorporated in the design. Some newer designs have other fluids adding or removing heat from non-flowing salt systems. These kinds of systems may be able to take advantage of phase change phenomena (solid/liquid) in the salt, offering additional storage capacity per unit volume of salt.

Designers often favor the two-tank approach in flowing systems. One tank is for hot fluid and the other is for cold fluid. A typical evening will begin with the hot tank fully charged and the cold tank relatively depleted. As the hot fluid is used during the night to

FIG. 1.8 TWO DIFFERENT DESIGNS OF DISH-STIRLING SOLAR DEVICES ARE SHOWN ON THE UNIVERSITY OF NEVADA, LAS VEGAS (UNLV) CAMPUS IN THE EARLY 2000S. THE UNIT ON THE RIGHT IS A DESIGN OWNED BY SES AND ON THE LEFT IS A DESIGN FROM SAIC INTERNATIONAL (Courtesy of UNLV)

FIG. 1.9   UNITS OF A RECENT DESIGN OF THE SES DISH-STIRLING SYSTEM (THE SUNCATCHER™) ARE SHOWN HERE AT SANDIA LABORATORIES UNDERGOING EVALUATION (From SANDIA Laboratories)

generate steam, it is returned in a cooled condition to the cold tank. Efforts have examined the use of stratified tanks with hot fluid on the top and cold fluid on the bottom, with a moving boundary between them (thermocline approach).

For dish-Stirling systems, as well as all PV systems discussed below, storage presents a more difficult problem. Perhaps the future will bring some viable possibilities either including or beyond storage batteries, but only time will tell.

1.1.4   Photovoltaic Approaches to Utility Scale Power Generation

1.1.4.1 Overview Comments The photovoltaic effect has been known since Becquerel identified it in 1839. Many incremental developments have occurred at various time intervals after that. One was the discovery of a method of monocrystalline silicon production by Czochralski in 1918. However, it was not until 1941 that the first monocrystalline cell was constructed. In the 1950s photovoltaic cells received a great deal more attention where applications were apparent to the growing space exploration interests. In 1955 Hoffman Electronics-Semiconductor Division introduced a commercial product that demonstrated 2% efficiency with 14 mW peak power for $25/cell. Both the types of cells and their associated efficiencies have grown with time. This is illustrated in Figure 1.10. Now there are many companies producing a wide range of PV products for both utility applications as well as residential installations.

Silicon has been the mainstay of the industry for over a half century, has proven to be quite a durable product, and has shown steady increases in efficiency and decreases in cost. In addition to the monocrystalline form that has all of these good characteristics, silicon cells are also produced in an amorphous form that, although it has lower efficiency, can be applied in thin film form with cor-

respondingly lower cost than the monocrystalline variety. A good review of solar cell development through the date of publication has been given [6]. One type of cell given special attention to here is CdTe. While cadmium is toxic, and the amount of tellurium on the earth's surface is not well known and may be limited, this cell is still one to be kept in mind because of its relatively low cost. A presentation at the 2010 Renewable Energy World Conference estimated that these types of cells (less than $1/peak W at the time of this writing and decreasing) would result in PV having grid parity for 97% of the US customers by 2015 [7]. In this same presentation, it was noted that a thin film plant of 12.6 MW_{DC} was completed in Boulder City, Nevada, in 2008 for $3.2/$W_{DC}$, the lowest price of all plants investigated in that presentation.

Another type of cells that will be given some brief attention here are those of the multi-junction (often triple junction) types that have currently been used either on spacecraft or high concentration PV systems (see more discussion on the latter below). One of the factors that limits PV cells to the typically low efficiency that most demonstrate is that a given semiconductor material is sensitive to only a limited range of the solar spectrum. By using thin layers of carefully selected, but differing, materials, this sensitivity range is greatly expanded. The result is higher efficiency. From Figure 1.10, it can be seen that triple junction cells show the highest of all efficiencies indicated there. Another positive aspect of many types of multi-junction cells is that their efficiency is less sensitive to increases in operational temperature than are those of the silicon type.

In more recent times, attention has been brought to bear on concentrating systems applied to photovoltaic systems [8]. In the period when this paper was written, CPV (concentrating photovoltaic) systems had not lived up to their promise of large-scale applications, partially because the cost of alternative fuels was relatively low, and flat plate PV approaches were being developed quite rapidly. The latter had the ability to be applied to roof structures, thus

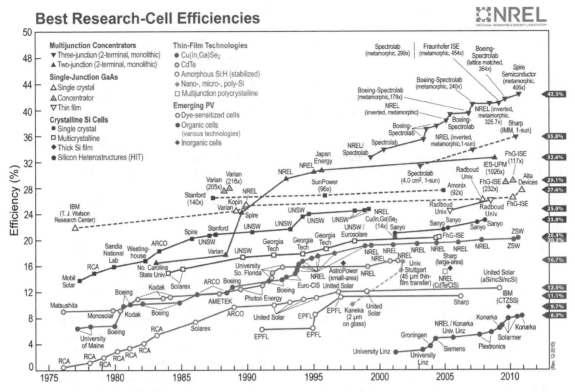

FIG. 1.10 THE EFFICIENCIES AND TYPES OF PHOTOVOLTAIC CELLS HAVE BEEN GROWING YEAR BY YEAR (Courtesy L. Kazmerski, NREL.)

opening a large market for them. Concern about tracker reliability was also a factor that held back CPV development. Since the market was small, the companies involved in this technology could not realize the economies of mass production and all of the related benefits that this implies.

This situation is obviously changing. For one thing, cell efficiencies have increased substantially over the years as noted above. Also, we are facing rapidly escalating conventional fuel prices, driven by higher costs of oil. Swanson [8] indicated that 10% of the output efficiency goes to pay for everything but the cells, so if efficiency increases, it encourages the application of CPV approaches.

One additional benefit of CPV systems compared to fixed flat plate systems is their ability to capture more of the sun's energy because of their tracking nature, compared to fixed flat plate PV systems. When more efficient cells can be applied to the CPV systems, this can greatly boost the energy generated for this approach. In all, Swanson [8] discusses the following benefits for CPV compared to fixed flat plate systems.

- Lower cost
- Superior efficiency
- High annual capacity factor
- Less materials' availability issues
- Less toxic material use (compared to some thin film systems)
- Ease of recycling
- Ease of rapid manufacturing capacity scale-up
- High local manufacturing content

For the purposes of this presentation we will put CPV into two categories with an imprecise division point: low concentration and high concentration. This is a result, in general, of the temperature of operation and the corresponding quality of cell required to appropriately utilize the concentrated beam. Low concentration will use cells that can be one-sun units, and passive cooling could always be used. On the other hand, high-concentration units will require improved fill factors and other design features in addition to some well designed cooling system, either passive or active.

For example, very high concentration systems for PV applications have been reported rendering up to 10^4 suns [9]. These are obviously not production systems, but rather special experimental systems. At the other end of the concentration range, JX Crystals is producing a 3× system [10]. Hence a wide range of concentrations is potentially available for PV systems.

The cells used in these kinds of systems can range from conventional one-sun types up to multi-junction (mj), very high efficiency, types. In between lie high-efficiency single-junction cells. It had been argued that high concentration systems were more appropriate using high efficiency single junction than mj cells [11], but that is changing in favor of the mj cells as the latter become more developed for this market.

Unlike concentrating thermal systems, where virtually all concentrators are of the reflecting type, CPV systems can have refracting or reflecting configurations. Usually the distinction is if irradiation is desired on single cells in tandem or a densely-packed array. Typically Fresnel lenses are used for the former and mirror-like reflectors are used for the latter.

Concentration not only boosts the cell incident flux, it also generates more heat within the cell compared to one-sun situations. When concentrations are sought at very high levels, enhanced heat removal from the cell is almost always called for. This is because most types of cells suffer a decrease in efficiency when the cell temperature increases. See, for example, Ref. [12].

1.1.4.2 High Concentration Photovoltaic (HCPV) Systems

High concentration systems might be defined as using concentrations above 10, but this is not a precise definition. Many of the systems now being marketed have design concentrations in the range of 100× to 500×, but 1000× does not seem too far from reality.

One of the companies that has been quite active in this field is Amonix, which is located in the Los Angeles, California, area. For quite some time, Amonix has been marketing a 25-kW system and a 35-kW system with a 250× concentration. More recently they have moved to 500× systems with multi-junction cells that have nearly doubled their units' output. Multi-junction cells are now being included in designs from several manufacturers [13]. A building block for Amonix is a Megamodule™ (typically 5 or 7 of these are included in a given concentrating unit depending upon the power output sought) that consists of many dispersed cells each with a Fresnel concentrating lens mounted on a configuration that assures structural integrity. Cooling of the cells is performed passively using a simply configured but quite effective heat sink on the rear side of the system. As is the case with any high concentration system, a two-axis tracker is used to follow the sun across the sky. In a similarity to operational aspects of dish Stirling systems noted above, the system must be mounted on a sunken support structure, and their system goes into stow position (horizontal) when local winds reach high values. A photo of several Amonix high concentration PV (HCPV) systems is shown in Figure 1.11.

Another approach to HCPV systems is the use of a dish. Some investigation of this type of system has taken place in the past (see Figure 1.12). In the figure, a design developed by SAIC is shown. The nominal output of this device was also 25 kW, using fixed facet mirrors. The unit at the focus of the beam looks very much like the Stirling engines shown in Figure 1.8, but in this case, it is entirely a cooling unit for the PV module that is located on the side closest to the facets. In fact, it is made up of four radiators that are similar to those found in trucks, and air is forced through them

from a large fan located on the side opposite from the PV array. The latter had high flux cells that were very similar to the ones used on the Amonix units, but in this application they are mounted in a dense array at the focal point of the beam from the dish. They were mounted to a copper plate that had cooling water flowing over the back. A small pump forced the water through passages in the copper plate and then through the radiators to give up the heat to the ambient air. Performance of this system is reported in a thesis by Newmarker [14].

Two distinct differences are present between the two HCPV systems shown here. First, in the dish system the flux comes from a large number of facets to a particular point on the PV array. This simplifies to the concentration assembly, but it also can cause hot spots to exist. A similar kind of behavior is sometimes demonstrated on dish Stirling units. A second issue that differs between the Amonix unit and the SAIC unit is the cooling approaches used. In the first case the cooling is accomplished in a passive manner, while the latter is performed actively using the system denoted in Figure 1.13 [15]. There are plusses and minuses for each of the approaches, but presumably the active system could offer better control in varying conditions.

1.1.4.3 Low Concentration PV Systems The other end of the range of concentration involves the use of low concentration designs. As was noted earlier, this is not a precisely defined category, but it might include concentration ratios of less than 10. For these kinds of systems, the precision of the tracking can be reduced significantly. Also, the heat that needs to be removed from the cells is greatly reduced and may all be done by passive interaction with the ambient conditions. Finally, many of these systems use one-sun cells. In one such design by JX Crystals, a 3× concentration ratio is used. As was stated by the developer, the bulk of the PV cells are replaced by reflecting aluminum that is on the order of 20 times less expensive than the cells [10]. This design uses aluminum

FIG. 1.11 AN ARRAY OF 25-KW AMONIX DESIGNS IN SPAIN AT THE GUASCOR FOTON INSTALLATION LICENSE PARTNER OF AMONIX. NEWER UNITS HAVE MUCH HIGHER POWER OUTPUT BECAUSE OF THE USE OF MULTIJUNCTION CELLS (From AMONIX)

FIG. 1.12 A DISH-PV CONCENTRATING SYSTEM IS SHOWN OPERATING ON SUN. THE CONCENTRATION IS APPROXIMATELY 250X (From UNLV)

"valleys" with conventional cells mounted at the base of the valley. One axis tracking is possible with a carousel type of arrangement in this application. See Figure 1.14. This unit was installed flat on a roof without penetrations. So the assembly can be installed without compromising any roofing construction.

1.1.4.4 Tradeoffs between One-Sun, Low Concentration, and High Concentration Systems

- One-sun systems, particularly fixed, flat plate assemblies, are simple to install and maintain.
- One-sun systems take up more ground area for the same power out than do concentrating types of systems, in general. This is particularly true if cells like the lower cost but lower efficiency CdTe type are used.
- One-axis tracking on flat plate systems can be adapted that increase the power output but do not increase significantly the initial cost and maintenance expenses.
- Tracking capabilities for high concentration systems must be more precise than are needed in low concentration systems.
- Low concentration systems can use one-sun cells and passive cooling designs.
- Some low concentration systems can be mounted on a flat surface without penetrating the latter.
- Each cell in a high concentration system produces significantly more power than does each in a low concentration or one-sun system.
- Cell cooling is a key concern in high concentration systems.
- High efficiency cells that are of high cost will be more cost effective in high concentration compared to low concentration systems.
- To minimize the costs of the tracking system per unit of power produced, high concentration systems must be on the order of 25 to 50 kW each to minimize the cost of the tracking system

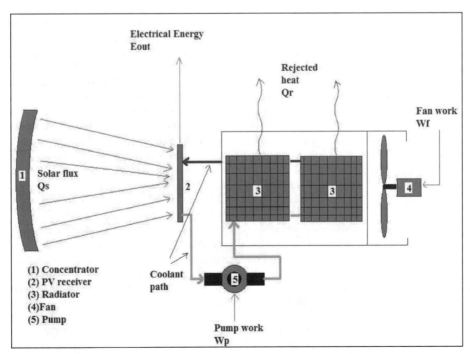

FIG. 1.13 THE SCHEMATIC DESIGN OF THE ASSEMBLY AT THE SNOUT OF THE DISH ARRAY SHOWN IN FIGURE 12 IS DEPICTED HERE [15]

FIG. 1.14   THE JX CRYSTALS ONE-AXIS TRACKING 3X CONCENTRATION PV SYSTEM IS SHOWN (Courtesy UNLV)

in the overall cost of power. These types of systems must have some careful design to minimize wind effects on the stability of the installation and dissipate heat generated in the cell.

1.2   HYDROGEN PRODUCTION AND USE

1.2.1   Background

Hydrogen is the most commonly-occurring element in the universe. It can be used for a variety of applications including manufacturing, electricity generation, and motive power. In addition, it can be produced from solar and wind energy systems, and it makes an excellent storage medium for these renewable resources. NASA (National Aeronautical and Space Administration) and the oil refining business have used hydrogen for years in achieving their goals with few hydrogen accidents. In spite of all of these good qualities, why has it not been used more as a core fuel in the world? For one thing, it is not a fuel in the sense as, say, natural gas and gasoline are fuels. Those two examples can be found in reservoirs at various places around the globe. Hydrogen, on the other hand, is an energy carrier. It has to be produced, much like electricity has to be generated.

Another characteristic of hydrogen is that it is the smallest atom in existence. The result of this is that it typically takes up a large volume to store, and it is prone to leaking unless the containing system is extremely well sealed. It can cause stress corrosion and hydrogen embittlement in some common materials that might be used in contact with it. Finally, it has a wide flammability range compared to most common fuels.

None of these drawbacks is insurmountable. What is clear, though, is that the infrastructure for generating, distributing, and utilizing hydrogen on a country-wide scale is not in place. However, it remains an intriguing option for future developments.

Hydrogen has been used to some extent over a large number of years. It is not our purpose here to cover all aspects of its history. Instead we will give some insights on how this can be generated from solar energy, and how it might be utilized in the modern world.

1.2.2   Generation of Hydrogen

Most of the world's supply of hydrogen is generated from a process called steam-methane reforming (SMR). In this technique, methane (normally as natural gas) and steam are brought together at elevated temperatures (in the range of 700°C to 1100°C). A major hydrogen distribution system is located on the Gulf coast area of the US, and this is a significant part of a US hydrogen industry that produced 10.6 million metric tons of hydrogen in 2006, most of it generated using SMR [16]. This is enough hydrogen to furnish 20 to 30 million cars with fuel or for energy uses in 5 to 8 million homes. SMR is currently the least expensive way to produce hydrogen. Its drawbacks, however, from a sustainability point of view include that it uses what is essentially a non-renewable energy supply (natural gas) and it generates greenhouse gases (primarily CO_2).

A cleaner approach, both environmentally as well as in hydrogen purity, can be electrolysis where water is broken into its molecular components using electricity. This is the second largest source of hydrogen, but the amount generated from this is much smaller than from SMR. This approach can be accomplished with greenhouse gases liberation, but this depends how the electrical input is generated. If the latter is generated from dirty fossil-fired plants, that fact becomes part of the hydrogen's carbon footprint. On the other hand, electrolysis produces very pure hydrogen. So when very high purity of hydrogen is needed, electrolysis is the desired source.

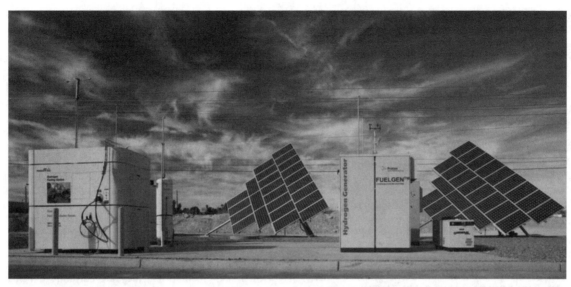

FIG. 1.15 THE UNLV/LAS VEGAS VALLEY WATER DISTRICT SOLAR POWERED HYDROGEN GENERATION AND FILLING STATION IS SHOWN. TO THE RIGHT IN THE FIGURE IS THE ELECTROLYZER SYSTEM, TO THE LEFT ARE THE COMPRESSOR, STORAGE AND DISPENSING STATION, AND IN THE REAR TWO OF THE FOUR PV PANELS ARE SHOWN THAT POWER THE SYSTEM

Because hydrogen can be made from electricity, if the electricity is made from renewably generated power, then the hydrogen becomes renewable also. This approach is a good addition to a time-variable renewable power generation source like solar or wind. Like "making hay while the sun shines," it is possible to save solar energy by the use of electrolytic hydrogen that is generated from it. Certainly one of the drawbacks to this approach, currently, is the high cost of generating hydrogen. When the cost of the photovoltaic, or other means of power generation from the sun, is factored in with the cost of the electrolysis unit, this does equate to higher cost hydrogen than that available from SMR approaches. However, PV systems are decreasing markedly in cost, and there are new approaches called photoelectrochemical that could bring the costs down significantly. An example of this has been reported [17].

Several solar powered hydrogen generation stations have been developed over the last several years in the US. One of these types of hydrogen generation and filling stations has been built by UNLV, Proton Energy, Air Products, and the Southern Nevada Water Authority at the Las Vegas Valley Water District headquarters in Las Vegas. This system uses four one-axis tracking flat plate PV units for power. This, or grid power, can then furnish the electricity to an electrolyzer. Water is converted there to hydrogen and oxygen, where the hydrogen is piped to a compressor and the oxygen is vented. Storage of the hydrogen takes place at 6000 psi for refueling of hydrogen vehicles. A photo of the filling station with two of its four PV panels showing is given in Figure 1.15.

One of the big issues in developing a hydrogen infrastructure for vehicle powering is how to transport the hydrogen from where it is generated to where it is used. Various options like using tanker trucks or pipelines have obvious drawbacks. One possibility is to have dispersed generation filling stations that make hydrogen on site. This could be used in locations where the solar resource is very good, such as the Southwest portion of the US. In locations that might have substantial wind availability, that resource could be used. Consideration of local renewable resources to generate hydrogen on the sites where it is dispensed could furnish hydrogen for a significant amount of the US land area.

1.2.3   Utilization of Hydrogen

Hydrogen can be used in the various ways any fuels are used: combustion processes, engine fuels, electricity generation in fuel cells, and others. Here a focus on transportation applications will be discussed. Not all transportation applications will be covered, however. It is well known that hydrogen can be used to power rockets as has been shown over the years. It can also be used to fuel aircraft, although this application has not been demonstrated to any large extent to date. In what follows, we will discuss road vehicle applications. Generally the motive power for hydrogen road vehicles will be internal compression ignition or spark ignition combustion engines and fuel cells.

First consider using some type of internal combustion engine (ICE) for motive power. In this application, the ICE is a machine that has been around significantly over a century so a great deal of infrastructure has been developed for both the manufacture as well as the maintenance of these devices. If they can be used for hydrogen applications, then one piece of the required hydrogen utilization infrastructure would be pretty much in place. What is required to convert a conventional (say gasoline powered) engine into a hydrogen power engine? It is actually quite simple. For one thing the spark timing must be changed quite significantly because of the ease of ignition of hydrogen as well as the resulting flame speed being so high. Also the fuel injection/air-fuel-ratio needs to be controlled quite carefully. For all of these reasons, the engine computer control must be appropriately calibrated. One of the bigger concerns using hydrogen in ICE's is to minimize the NOx formed. This can be done with high air fuel ratio. This however, causes minimal power with normally aspirated engines.

Of course, almost all modern engines use some type of fuel injection, and some of them (in addition to Diesel engines) use direct cylinder injection. Adaptation of direct injectors to hydrogen engines is important for a number of reasons. One of them is that the flame speed in hydrogen is extremely fast, much faster than typical hy-

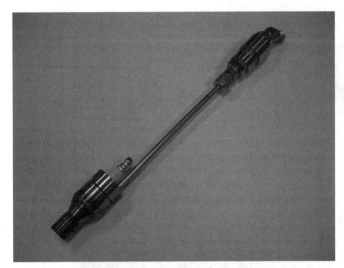

FIG. 1.16 AN INTEGRAL IN CYLINDER FUEL INJECTOR AND SPARK PLUG ASSEMBLY IS SHOWN. THIS ASSEMBLY REPLACES THE CONVENTIONAL SPARK PLUG OF A CONVENTIONAL SPARK IGNITION ENGINE [18]

surfaces work well. Hydrogen is a drying fuel, and sliding metal surfaces are to be shunned in hydrogen systems. Work has been ongoing related to special in-cylinder fuel injection systems at several locations. One such design that combines an in-cylinder fuel injector integral with a spark plug, and which can be used to replace the conventional spark plug is shown in Figure 1.16 [17]. Something like this greatly simplifies the conversion of a hydrocarbon spark ignition engine to a hydrogen in-cylinder fuel injected engine.

Another way that hydrogen can be used in vehicles is in conjunction with fuel cells. Good insights into the development of fuel cells have been recently given [19]. These devices are essentially the opposite process of electrolyzers. In the latter, water and electricity are used to make hydrogen and oxygen. In fuel cells, hydrogen and oxygen (the latter usually from atmospheric air) combine to make water and electricity. Hence fuel cell vehicles are really electric vehicles that use hydrogen fuel to generate the electricity. These devices have a number of very desirable characteristics, including quiet operation and higher efficiency than ICE's. Down sides include that they are currently expensive and their membranes can be poisoned with impurities in the air.

The most typical type of fuel cell to use for vehicular propulsion is of the proton exchange membrane type. These cells operate at relatively low temperatures (about 175°F or 80°C), have high power density, can vary their output rather quickly (but not as fast as automobile engines) to meet shifts in power demand, and are suited for applications–such as in automobiles–where quick startup is required. The proton exchange membrane is a semi-permeable membrane that allows hydrogen ions to pass through it. The membrane is coated on both sides with highly dispersed metal alloy particles (mostly platinum) that are active catalysts. This type of fuel cell is sensitive to fuel impurities. Cell outputs generally range from 50 to 250 kW, although they are also available in smaller sizes.

drocarbon fuels. Also, even more critically than HC fueled engines, the combustion process may need to be controlled to minimize NOx formation. Direct in-cylinder fuel injection can offer a significant improvement over carburetion or port injection. However, hydrogen fuel injectors, whether they be for port or direct cylinder injection of hydrogen need to be designed more carefully because of the drying nature of hydrogen fuel compared to hydrocarbon fuels. This is because hydrocarbon fuels have lubricity, so sliding metal

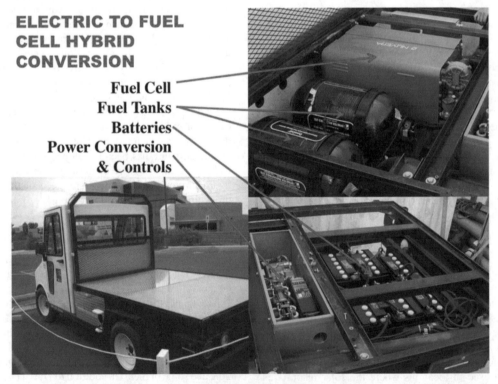

ELECTRIC TO FUEL CELL HYBRID CONVERSION

Fuel Cell
Fuel Tanks
Batteries
Power Conversion & Controls

FIG. 1.17 AN ELECTRIC UTILITY VEHICLE THAT HAS BEEN CONVERTED TO A FUEL-CELL-ELECTRIC HYBRID CONFIGURATION IS SHOWN. THE SPACE ORIGINALLY USED FOR HALF OF THE ORIGINAL BATTERY PACK HAS BEEN USED FOR A FUEL CELL UNIT AND HYDROGEN STORAGE TANKS (Courtesy of UNLV)

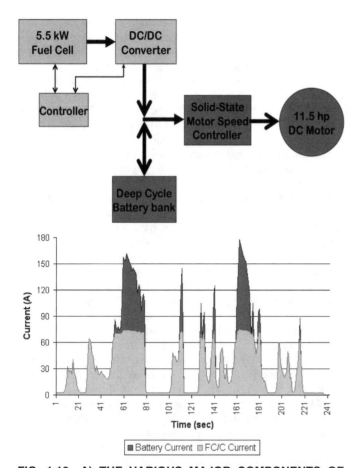

FIG. 1.18 A) THE VARIOUS MAJOR COMPONENTS OF THE POWER SYSTEM DEPICTED FOR THE VEHICLE CONVERSION SHOWN IN FIGURE 1.17. B) SOME EXAMPLES OF THE POWER DRAW DURING A DRIVING CYCLE ARE SHOWN DENOTING WHAT AMOUNT COMES FROM THE FUEL CELL AND WHAT COMES FROM THE ENERGY STORED IN THE BATTERIES (Courtesy of UNLV)

Because fuel cells generate electricity, the vehicles that they propel are electric. Most of the major automobile manufacturers have been working on the development of fuel cell vehicles. The first one of these vehicles to appear on the market is the Honda Clarity. This is a model that succeeds the Honda FCX vehicles that were evaluated on an experimental basis for several years.

Depending upon the function of the vehicle, it could be a straightforward task to convert an all-electric vehicle to a fuel cell hybrid design that uses hydrogen fuel. One such example is shown in Figure 1.17. This Taylor–Dunn electric utility vehicle was converted to be a hybrid hydrogen fuel cell electric vehicle. In doing this, approximately 50% of the original batteries were removed, and the space they had taken was used to install a 5.5-kW fuel cell and two high-pressure hydrogen storage tanks. Necessary power control aspects were added to allow the system to operate off of a base power from the fuel cell and to tap the storage battery reserve power if needed for high load periods. When the fuel cell is actually producing more power than the vehicle needs for its motion, the extra energy generated can be used to recharge the batteries. A block diagram of the system used for this conversion is shown in Figure 1.18a. In the companion curve (Figure 1.18b), the green denotes the amount of power directly from the fuel cell, while the red indicates the power that is from energy stored in the battery.

Whatever is used as the prime mover, one significant issue that needs to be addressed is how the hydrogen is stored on board. As noted earlier, because hydrogen has the lowest molecular weight of all molecules, considerable space is needed to store it. It is not unusual for gaseous hydrogen storage tanks to be designed to hold 5000 to 6000 psi, and in spite of this, the tanks have to be large for reasonable vehicle range. Other means of storage have also been applied. BMW, for example, uses liquefied hydrogen storage, which is more efficient space-wise than gaseous hydrogen storage. However, the liquid has to be stored at very low temperatures. Some boil-off of hydrogen due to heat transfer through the insulated tank walls will occur. This might cause substantial loss of hydrogen if the vehicle is not used for a long period of time. Another way to store hydrogen is with metal hydride beds. These are like a metal sponge to hydrogen, soaking up gas when the bed is cooled,

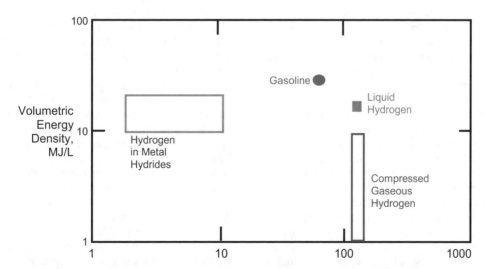

FIG. 1.19   AN APPROXIMATE COMPARISON OF THE THREE WAYS OF STORING HYDROGEN IN TERMS OF THE ENERGY STORED PER UNIT OF WEIGHT AND PER UNIT OF VOLUME. THESE ARE CONTRASTED TO GASOLINE STORAGE. NOTE THAT LIQUEFIED HYDROGEN, ALTHOUGH REQUIRING VERY LOW TEMPERATURES FOR STORAGE, ACHIEVES CHARACTERISTICS ON THIS MAP THAT ARE NOT TOO DIFFERENT THAN GASOLINE

and releasing the hydrogen when the bed is heated. The downside is that the beds are heavy, and this approach has typically been used only in large vehicles like buses and trucks. The upside is that hydrogen is stored in these beds at relatively low pressure (say 100 to 200 psi). An approximate map of the storage characteristics of hydrogen is shown compared to gasoline in Figure 1.19.

A key element in designing with hydrogen in any application is to allow venting to the atmosphere in case of some type of malfunction. Since the gas is so light, it will rapidly go straight up in the atmosphere in the area of the hydrogen leak as long as it has a clear passage upward.

1.3 BUILDING APPLICATIONS OF SOLAR ENERGY

1.3.1 General Philosophy

Knowledge of interaction of solar energy with buildings is very important for the planning for a low energy future. Sometimes these interactions need to be minimized as much as possible, and at other times they need to be accepted to the full extent; but at all times, the building needs to be designed for smart solar interactions. These may include the use of the photons in their basic form (lighting and PV), and at other times the photon conversion to thermal energy is desired (space and water heating).

Of course the environmental conditions will have an impact on how all this is done. Hot, dry climates will have to be handled differently than temperate climates. It could be that different parts of the building may be designed to achieve different kinds of outcomes. Also, the orientation of the building compared to the sun path in the different seasons of the year should be carefully incorporated in the design.

By far, the most important aspect to general building design is cost-effective energy conservation. Forms that this may take vary, but they almost always involve carefully chosen insulation and high performance windows. In addition, high electrical use devices in the house (particularly air conditioners in warm or hot climates) need to be chosen for high efficiency.

A critical tool in good design is the use of building energy simulation (BES). Like many types of simulation software packages, BES codes tend to range between codes that are extremely comprehensive and often hard to master and easier codes that may not be as detailed in their ability to analyze. BES codes that allow "what if" kinds of analyses to examine the changes to performance with change in design are particularly good to use. One such example that is quite easy to apply is Energy10 [20]. There may be some aspects of the design that are not easily accounted for in this code, but the major elements can be. It also allows changes to the building design within a given sized and oriented structure. Codes can be used with detailed measurements to estimate the cost-effectiveness of conservation components on actual buildings [21].

However, it is also critical that component costs of various options be considered in the analysis, and these are sometimes more difficult to gather without direct access to the local construction industry data. One simple way to handle this (if cost information is available) is to consider a base case design, assess its performance and cost (both capital cost as well as operating costs), and then consider "what if" options. Looking at simple payback may be sufficient to see what the long-term cost implications of the design may be.

After the basic structure has been optimized for performance and cost, then various solar applications can be considered for their impact on the overall building. One of the significant problems in all of this is incorporating good estimates of cost increases of conventional energy. If the latter remain stable (or nearly so) compared to current costs, a different set of conclusions may be drawn than if the conventional energy greatly increases in cost over time. One way to handle this is to consider two cost estimation scenarios. One can be considering no increase in cost. The other can be for cost escalation at the average historical cost rate of increase. Each state's public utility commission, or even the local utility, should be able to furnish the latter information.

Human comfort and possible indoor environmental pollution needs to be considered in energy efficient design. Many of these factors have been discussed [22].

In the next section a concept that is attracting a great deal of interest, particularly in the US and Europe, is discussed: Zero Energy Buildings. Generally these buildings have a great deal of energy saving design incorporated in their construction, and they also include some means of alternative energy for generation to make up for the energy used in the house. Following this, we discuss the two most-common types of solar add-ons for buildings. These are solar domestic water heating, and photovoltaics. In both of those applications, a south-facing, unshaded orientation is preferred. Some inclinations that are nearly southerly might offer less solar energy harvest than direct south, but the deficit may be small. Inclinations to be used vary somewhat because of ease of application or desire to achieve some particular goal. For good all year performance, an optimal angle of orientation is to face south with a tilt angle equal to the local latitude. If more winter input is desired, it can be achieved by using a higher angle, but this will be at the expense of summer input. Oftentimes solar units on homes will be mounted at whatever the roof inclination is. Of course, for flat roof installation, any orientation can be achieved with a mounting system.

Cleaning is another issue. Rain, if it occurs periodically like it does in many parts of the nation, will generally do a good job in cleaning solar collection units. Many solar collection systems have not been cleaned during their entire lifetime. If the atmosphere is particularly dirty, and rain is particularly sparse, it may be that a periodic cleaning may improve system performance.

1.3.2 Net Zero Energy Buildings

This is a concept that has, on one hand, been around for a long time, but, on the other hand, is only now starting to catch on in the US. In buildings of this sort, very good energy-conserving design is coupled with some means of generating energy from a renewable energy source. If the customer matches the net energy use from the grid by an excess generation over the year, this is often called a net site zero energy building. In some quarters this is referred to as a zero site energy building. Another distinction sometimes made is to denote a building as being net zero source energy. This is related to the site definition, but also accounts for losses in generating the energy and bringing it to the consumer. A similar definition can work for net zero energy costs. In some service areas, utilities pay local generators for the renewable energy they send to the grid. If the payment to the utility matches the payment from the utility, this is denoted as net zero energy costs. A similar definition can be based upon emissions.

On net zero energy buildings in a good solar climate, both a solar domestic water heater and a photovoltaic array will be applied. Each of these topics is addressed separately in sections that follow. Of course the electricity demand for a building will typically be greater than the energy demand for heating of water. As a result, the size of the two collection systems is relatively much larger for the PV and smaller for the solar domestic water heating (SDWH)

FIG. 1.20 TWO EXAMPLES OF ZERO ENERGY HOMES ARE SHOWN WHERE THE SIZE OF THE PV ARRAY IS MUCH LARGER IN PHYSICAL SIZE COMPARED TO THE SDWH SYSTEM (Credits: above, NREL #14164; below, UNLV)

system. See Figure 1.20 where some examples of the size of the PV system can be compared to that of the SDWH system. It should be noted that the first of these buildings uses a SDW heating system that has freeze protection, while the second uses an integrated collector storage (ICS) system. Both of these are described in the next section.

1.3.3 Solar Domestic Water Heating

A great deal of the world has solar applied domestic water heating more than we have in the US. Europe, Asia, and the Middle East are among those areas where this is the case.

In considering solar domestic water heating, several aspects should be considered. Of course, a south or near-south exposure for the solar collection unit is desired. Careful attention to the climatic conditions is also required. Climates that have freezing conditions at some points in the year will require freeze protection that may not be required in more balmy regions. Some types of systems may require a quite-heavy installation on or under the roof, and the basic building design may not allow this without significant strengthening of the structure. Water quality can have an important effect on some types of systems. Also important for cost-effective installations is the cost of the competing conventional energy used for water heating. A rule of thumb that is almost always true is that conventional DWH (domestic water heating) systems that use electricity for heat input are usually good candidates for SDWH replacement.

Typical elements of a SDWH system include some type of solar collection element, some means of storing the water being heated,

and a way of connecting these two together. Finally, it is, of course, necessary to merge this between the domestic water source and the hot water system of the building. It is quite common to install the SDWH system between the water input to the building and the conventional water heating system. This allows the latter to serve as "back-up." If significant periods of inclement weather occur minimizing the SDWH output, the conventional system can furnish the necessary heating service.

Two types of SDWH systems will be described here, but it should be remembered that many more options than those described are available on the market. One of the two types of systems considered is the integrated collector-storage (ICS) configuration. The other is a freeze-protected type that either allows the collector to drain when heating is not taking place or uses an anti-freeze loop. An ICS system and a collector for a freeze-protected system are shown in Figure 1.21.

First consider the ICS system. It includes a combination collection/storage unit typically located on the roof of the building. The collection/storage unit is often constructed of several very large

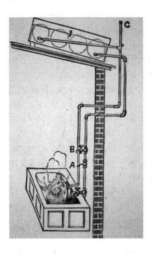

FIG. 1.21 TWO OF THE MANY TYPES OF SDWH SYSTEMS ARE REPRESENTED HERE: A) THE INTEGRATED COLLECTOR-STORAGE SYSTEM AS IT IS CALLED TODAY IS SHOWN IN THIS SKETCH THAT ACCOMPANIED A 1893 US PATENT, AND B) A CUTAWAY VIEW OF A COLLECTOR TYPE THAT MIGHT BE USED WITH A DRAIN-DOWN OR AN ANTIFREEZE PROTECTED SYSTEM

diameter copper tubes, perhaps 4 to 6 in. in diameter, that are connected with small tubing in a serpentine pattern. The hot water supply source for the building is connected directly to the inlet of this unit, and the outlet of the unit is connected to the conventional DWH system of the building. When hot water is used, the DWH system supplies the water, the ICS system supplies the DWH, and the water supply source feeds the ICS unit. Through a day, water use will allow the supply water to flow slowly (because the ICS unit has a very large volume relative to the demand use) through the ICS unit, ultimately heating it to whatever the climatic conditions allow. This results in a system that has very little impact on floor space in the building. It operates in a totally passive manner. That is: no pump or special piping is required other than to connect the unit to the water source and the conventional heater. No second fluid, like anti-freeze solution, is required. Two downsides include the fact that the unit is heavy, and the location where it is located, if on the building, may need structural reinforcement. Another concern is that there is no positive freeze protection other than the thermal inertia of the large unit filled with water. If severe freezing conditions occur in the local area, freeze damage would be an issue. Finally another drawback is that the storage of the hot water occurs out in the weather and the night. This influences the overall efficiency of the system because of heat losses from the water at those times.

Many approaches have been put forth to make a system less susceptible to freezing. One of these uses an anti-freeze loop and is thus preferred over the ICS system where severe freezing conditions can occur. The system may use two tanks (if so, the first is called the preheat tank, and the second is called the back up tank, the latter typically being the conventional water heater), or one tank which serves both purposes. If only one tank is used, a heat exchanger is incorporated, and care should be used to try to achieve stratification in that tank. This is not always easy to accomplish. A distinct loop with a pump, thermostatic control system, expansion tank, and back flow preventer is filled with an anti-freeze solution, typically a mixture of water and propylene glycol. When the anti-freeze solution in the collector is warmer than the tank, the pump is turned on. The anti-freeze solution moves from the collector to the heat exchanger with the result of heating the top of the storage tank. The cooled anti-freeze solution circulates back to the collector. The heat exchanger that is used can be one in the tank (perhaps a bayonet unit), or a separate loop that brings water from the tank to an external heat exchanger where it exchanges heat with the anti-freeze solution. Obviously, this type of approach is more complicated than the ICS approach, but it does offer positive freeze protection. This can be an extremely important characteristic. Every few years, the anti-freeze solution needs to be drained, and the system recharged with a new solution.

1.3.4 Photovoltaic Applications on Buildings

Most photovoltaic systems on buildings are hooked in parallel with the local electrical grid. This will mean the system must interface to the local utility, and usually the latter will have fairly strict rules about how this takes place. This, in a sense, allows the local grid to serve as "storage" for the local PV system, furnishing power when generation is not sufficient for the building requirements. Of course, if it is not desirable or if it is impossible to hook to a grid because there is not one locally, then typically, storage batteries will be needed. This off-grid type of system will not be explored here, but there are certainly many of them throughout the US. The placement of PV systems on buildings is normally referred to as "distributed generation," and this is in distinction from

"utility scale generation" addressed earlier in this review. Both distributed generation issues as well as electrical storage issues have been reviewed [23].

Usually, PV panels are placed on the roof of the building and oriented to the south or nearly south (depending upon the roof orientation). Since PV modules generate DC power, this needs to be converted to AC by use of an inverter. Various meters and disconnect switches are usually required by the local utility for final hook up to the grid. Some locations have what is called "net metering." This means that the power meter for grid power can turn forwards (when the building is taking power from the grid) or backwards (when the PV system on the building is making more power than is being used locally). These systems allow the owner to sell PV generated power at the same cost as power is purchased. Other locations have a difference between the price that power is purchased from the utility compared to the price at which it purchases power. In these cases, the latter is less. Some locations may not have a selling price to the utility.

In recent years, more interest in building-integrated photovoltaics (BIPV) has developed. BIPV means that the PV array becomes much more a part of the building design, rather than simply panels that are mounted quite obviously on the roof. This can mean, for example, that the PV units could be of the same size and shape as the roofing tiles. An example of this is shown in the lower photograph in Figure 1.20 of the zero energy house. In fact this set of pictures shows the distinction between panels (upper picture) and BIPV (lower picture). BIPV could also mean that photovoltaic ele-

FIG. 1.22 BUILDING INTEGRATED PHOTOVOLTAICS (BIPV) ARE INCORPORATED IN A NUMBER OF THE FACADES OF THIS BUILDING [24]

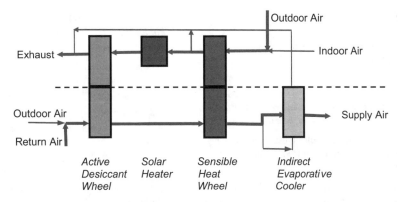

FIG. 1.23   A POSSIBLE SOLAR DRIVEN DESICCANT DEHUMIDIFICATION/COOLING SYSTEM

ments might be used for building siding, skylights, or windows. In some of these applications, the mounted elements may have to be at least partially light transmitting. The upside of these arrangements is that power can be generated by almost any element of the building. However, the downside includes that the sun may not have access to these types of elements for all times of the day; and when it does, the angles might be quite obtuse. Of course, when the element is light transmitting, this means that the power harvest per unit of active area may be quite low compared to conventional arrays located in optimal harvesting orientations. An example of an application is shown in Figure 1.22.

Potential shading is of particular concern in PV applications on buildings. Depending upon the design of the system, it could be the case that partial shading will completely shut down the production in the panel. While there are design features that can minimize this potential problem, shading is undesirable.

The current situation is that some lower cost PV panels (usually based upon cost per peak W generated) have lower efficiency than higher cost panels. Depending upon the limitations of the mounting location (say a smaller roof), higher-performance panels may be desirable.

1.3.5   Other Solar Applications in Buildings

Several other solar applications in buildings are in various stages of development. Only a few will be noted here.

Solar lighting is a concept where actual sunlight is brought into the building to provide a natural source. Of course, windows are an example of this; and when the window has a good view from it, this may be irreplaceable. However, windows can be the source of large heat gains or losses. In situations where this is not desirable, some enhancements to the windows may be required. Included is a variety of shading devices (usually preferred on the outside of the window), some type of insulation (like drapes or louver blinds), window tinting, and many other products on the market, as well as design aspects like overhangs and light shelves. If only good quality natural lighting is desired, this might be accomplished with products like light conducting ducts through the roof [25] or solar hybrid lighting [26]. These devices are designed to minimize the thermal impact of bringing sunlight into buildings. A review of the operational aspects of many of these kinds of systems has been given [27].

Solar building space cooling is another technology that is moving more toward the market. One of these applications is that of a solar absorption cooling unit that is driven by heat from the sun.

Oftentimes these approaches require higher temperatures than are typical from solar water heating systems. Currently the prices of these systems are quite high compared to conventional air conditioning units.

Another approach to cooling is to use a system with desiccants. These materials will soak up water from moisture-laden air when the latter is circulated through the material. Both cooling and removal of water can occur without the use of a vapor compression or absorption cooling system. Desiccant cooling work has been reviewed by Daou et al. [28]. A possible cooling system using this kind of an approach is shown in Figure 1.23.

1.4   CLOSING COMMENTS

A summary has been given of some of the ways that solar energy may be used in power generation, transportation, and building applications. While solar energy may be more useful in some areas of the country than others, some applications can be adapted in virtually all areas.

The technology has, by and large, been expensive because a sizeable market has not developed in the past, which, in turn, kept the prices up. Now, between a great deal of encouragement in terms of incentives on the Federal level as well as state and local levels, and some technical breakthroughs, the markets are developing and prices are distinctly decreasing. Better means of storing energy are also coming along, so this periodic resource will be able to deliver useful and economical energy 24 hours a day in the not-too-distant future.

Developments like this can assist the US in becoming much more energy self sufficient in a sustainable way. These are goals that all should desire.

1.5   ACKNOWLEDGMENTS

I particularly want to credit the excellent assistance by the various students over the years in the UNLV Center for Energy Research, without whom the large scope of work we have carried out, some of it touched upon here, would not have been possible. I also appreciate the collaboration of several companies who funded and/or assisted us in a large number of the projects. A great deal of support was received from the US Department of Energy and the National Renewable Energy Laboratory, and that made the whole operation possible.

1.6 REFERENCES

1. Butti, K., and Perlin, J, 1980. *Golden Thread: 2500 Years of Solar Architecture & Technology*, Cheshire Books, Palo Alto, California, p. 29.

2. Welford, W., 1983. "Connections and Transitions between Imaging and Nonimaging Optics," Proceedings of the SPIE International Conference on Nonimaging Concentrators, Volume 441, August, San Diego.

3. Zachary, J., 2010. "CSP-Downstream of the Solar Field: Design and Optimization of the Power Plant," presentation at Renewable Energy World Conference, Austin, Texas.

4. Berger, John J., 1997. *Charging Ahead: The Business of Renewable Energy and What It Means for America,* University of California Press, Berkeley, California, Chapter 4.

5. Ricker, C, 2008. "Is Solar Thermal Power the Answer?" *Photonics Spectra,* July, pp. 58-59.

6. Perlin, J., 1999, *From Space to Earth—The Story of Solar Electricity,* Aatec Publications, Ann Arbor, Michigan.

7. Jaffe, S., 2010. "Race to the Quarter; Photovoltaic Pricing Trends," paper presented at the Renewable Energy World Conference, Austin, TX.

8. Swanson, R, 2000. "The Promise of Concentrators," *Progress in Photovoltaics: Research and Applications.* 8, pp. 93-111.

9. Gordon, J. M., E. A. Katz, D. Feuermann, and M. Huleihil, 2004. "Toward Ultrahigh-Flux Photovoltaic Concentration," *Applied Physics Letters,* 84, May 3.

10. Fraas, L.M, Avery J, Huang H, Minkin L, Corio R, and Fraas J, 2007. "Start-Up of First 100 kW System in Shanghai with 3-Sun PV Mirror Modules," *Proceedings of the International Conference on Solar Concentrators (ICSC 4),* El Escorial, Spain, March.

11. Slade, A., Stone K, Gordon R, and Garboushian V, 2005. "High Efficiency Solar Cells for Concentrator Systems: Silicon or Multi-Junction," SPIE Optics and Photonics Conference, August, San Diego, California.

12. King, D.L., Kratochvil JA, and Boyson WE, 1997. "Temperature Coefficients for PV Modules and Arrays: Measurement Methods, Difficulties, and Results," *Proceedings of the 26th IEEE Photovoltaic Specialists Conference,* Anaheim, California, September 29-October 3.

13. Zubi, G., Bernal-Agustin J, and Fracastoro G, 2009. "High Concentration Photovoltaic Systems Applying III-V Cells," *Renewable and Sustainable Energy Reviews,* 13, pp. 2645-2652.

14. Newmarker, M., 2007. *Development and Testing of an Advanced Photovoltaic Receiver,* MS Thesis, Mechanical Engineering, UNLV, p. 30, December.

15. Mahderekal, I., Halford C.K., and Boehm R.F., 2006. "Simulation and Optimization of a Concentrated Photovoltaic System," *Journal of Solar Energy Engineering,* 128, p. 139.

16. EIA, 2008. "The Impact of Increased Use of Hydrogen on Petroleum Consumption and Carbon Dioxide Emissions, Appendix C: Existing Hydrogen Production Capacity," EIA Report #: SR-OIAF-CNEAF/2008-04.

17. Wang, H., Deutsch T., and Turner J., 2008. "Direct Water Splitting under Visible Light with a Nanostructured Photoanode and GaInP2 Photocathode," *ECS Transactions,* 6 (17).

18. Wilson, Max, 2010. "Development of Hydrogen Direct Injection For Conversion of Gasoline Engines," MS Thesis, Mechanical Engineering, UNLV, June.

19. Andujar, J, and Segura F. [2009]. "Fuel Cells: History and Updating. A Walk along Two Centuries," *Renewable and Sustainable Energy Reviews,* 13, pp. 2309-2322.

20. ENERGY-10, 2010. Website: http://www.sbicouncil.org/displaycommon.cfm?an=1&subarticlenbr=112

21. Zhu, L., Hurt R, Correa D, and Boehm R, 2009. "Comprehensive Energy and Economic Analyses on a Zero Energy House Versus a Conventional House," *Energy,* 34, 9, pp. 1043-1053.

22. Omer, A., 2008. "Renewable Building Energy Systems and Passive Human Comfort Solutions," *Renewable and Sustainable Energy Reviews,* 12, pp. 1562-1587.

23. Toledo, O.M., Filio FO, and Diniz A, 2010, "Distributed Photovoltaic Generation and Energy Storage Systems: A Review," *Renewable and Sustainable Energy Reviews,* 14, pp. 506-511.

24. Sinopuren, 2009. "Experiencing the Dreams" brochure, Sinopuren Energy Group Ltd., Shenzhen, China.

25. Solatube, 2010. Website: http://www.solatube.com/

26. Sunlight Direct, 2010. Website: http://www.sunlight-direct.com/index.php

27. Kandilli, C., and Ulgen K, 2009. "Review and Modelling the Systems of Transmission Concentrated Solar Energy via Optical Fibres," *Renewable and Sustainable Energy Reviews,* 13, pp. 67-84.

28. Daou, K., Wang RZ, Xia Z.Z, 2006. "Desiccant Cooling Air Conditioning: A Review," *Renewable and Sustainable Energy Reviews,* 10, pp. 55-77.

SOLAR ENERGY APPLICATIONS AND COMPARISONS

Yong X. Tao and Rambod Rayegan

2.1 INTRODUCTION

Strictly speaking, all the practical energy sources for applications to human activities on the Planet Earth are from the sun. Even fossil fuels resulted from millions and millions years of layered deposition of once living plants (and animals eating plants), which obtained their energy intakes and conversion directly and indirectly from solar rays and reached their maturity before they were buried. From ancient times to today, humans have used various ways to harness direct solar radiation, from using it as a heating source to today's electricity generation systems. Solar energy is also responsible for such renewable energy sources as wind and wave power, hydroelectricity and biomass. Today, the total energy consumption in the United States still predominantly comes from fossil fuels, although recent interests in investing in wind and solar electricity have been accelerating.

In this chapter, only energy system applications resulting from the direct solar radiation will be discussed and are limited to commercial applications. The applications of solar energy systems can be categorized as follows:

- Utility-scale solar power systems that generate electricity and feed to the electricity grid. There are photovoltaic (PV) systems and solar thermal power systems; the latter can also produce heat for hot water or air, which are often referred as the combined solar power and heat systems.
- Building-scale solar power systems, also known as distributed power systems, which generate electricity locally for the building, may be connected to the grid or may be stand-alone systems that require batteries or other electricity storage units. They are primarily PV systems.
- Solar heating systems for buildings, which are either used as hot water systems, or hot air heating systems.
- Solar high-temperature process heat systems for industrial applications, which involve concentrated solar collectors and high-temperature furnaces for producing high-temperature heat for chemical processing of materials.
- Other special solar heating systems for desalination plants and hydrogen production.

There are additional solar energy applications in either the appliance category, or even much smaller scales such as solar cooking, solar lighting products, and instrument-level solar power sources (watches, backpacks, etc.) The discussion of those applications is beyond the scope of this Chapter. Outer space applications of solar energy technology are also excluded. Investigations primarily undertaken in the United States are presented, although some examples from global applications are also discussed to address the potentials and needs for wider applications of solar energy in the United States.

2.2 LARGE-SCALE SOLAR ENERGY PLANTS FOR POWER GENERATION

One of the fastest-growing applications is in utility scale solar power generation systems. Under increasing pressure to reduce domestic dependence on foreign sources of energy, it has been recognized [1] that solar energy represents a huge domestic energy resource for the United States, particularly in the Southwest where the deserts have some of the best solar resource levels in the world. For example, an area approximately 12% the size of Nevada (15% of federal lands in Nevada) has the potential to supply all of the electric needs of the United States. In addition, solar power often complements other renewable power sources such as hydroelectric and wind power. Solar resources are typically higher during poor hydroelectric periods, and solar output peaks during the summer, whereas wind power typically peaks in the winter. Solar can complement fossil power sources as well. Eskom, a coal-dominated power utility in South Africa with one of the lowest power costs in the world, has identified large-scale solar power technologies as a good intermediate load power source for its grid. Although some renewable power technologies provide an intermittent energy supply, large-scale thermal electric solar technologies can provide dispatchable power through the integration of thermal energy storage. Thermal energy storage allows solar thermal energy collected during the day to be used to generate solar electricity to meet the utility's peak loads, whether during the summer afternoons or the winter evenings. Although solar energy is abundant and free, it is a diffuse energy source, so the cost to harness (or harvest) it with solar collectors can be significant. As a result, electricity generated from solar energy is currently more expensive than power from conventional fossil-power plants. However, the Western Governors' Association has determined that even at moderate levels of deployment, large-scale solar power can potentially compete directly with conventional fossil generation [1].

There are two major types of systems. One is solar thermal power generation systems, which can be further categorized by solar collector geometric characteristics as trough and tower systems. The other type is large-scale PV (LSPV) systems. In this section, we first describe the main technical features of the systems, followed by discussion to compare those two systems in terms of their performance, cost structure and barriers to wide applications.

2.2.1 Solar Thermal Power Generation Systems

Solar thermal power generation technology generally refers to a power generation system that involves collecting solar radiation through concentrated collectors to an absorber surface, which will heat a carrier fluid to a high temperature. Through a piping and boiling system, the hot fluid will be able to generate steam to power a turbine by means of a standard Rankine cycle. A generator connected to the turbine will then generate electricity. In general, such a system can be seen as similar to a coal-burning power plant except that the equipment component for steam generation with coal combustion is replaced by a solar heating system. Currently, two types of major solar thermal power systems have been in commercial-scale operation. One is parabolic trough (or trough in short) technology, and the other is the tower configuration (Fig. 2.1). Parabolic dish with Stirling engine is another technology that has not been developed enough yet to be available for large-scale operation.

2.2.1.1 Parabolic Trough Solar Power Technology Although many solar technologies have been demonstrated, parabolic trough solar thermal electric power plant technology represents one of the major renewable energy success stories of the last two decades. The main reason is that it has been recognized as one of the lowest-cost solar-electric power options available today and has significant potential for further cost reduction. Nine parabolic trough plants, totaling over 350 megawatts (MW) of electric generation, have been in daily operation in the California Mojave Desert for up to 18 years [2]. These plants provide enough solar electricity to meet the residential needs of a city with 350,000 people. They have demonstrated excellent availabilities (near 100% availability during solar hours) and have reliably delivered power to help California meet its peak electric loads, especially during the California energy crisis of 2000 to 2001. Several new parabolic trough plants have been built or are currently under development. Growing interest in green power and CO_2-reducing power technologies have helped to increase interest in this technology around the world. New parabolic trough plants are currently

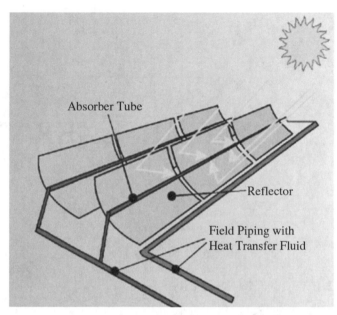

FIG. 2.2 PARABOLIC TROUGH SCHEMATIC [1]

under construction or in the early stages of operation in support of solar portfolio standards in Nevada and Arizona and a solar tariff premium in Spain.

Parabolic trough power plants use concentrated sunlight, in place of fossil fuels, to provide the thermal energy required to drive a conventional power plant. As shown in Figs. 2.1a and 2.2, these plants use a large field of parabolic trough collectors, made of non-optical mirrors for low cost, which track the sun during the day and concentrate the solar radiation onto a receiver tube located at the focus of the parabolic shaped mirrors. A heat transfer fluid (HTF) passes through the receiver and is heated to temperatures required to generate steam and drive a conventional Rankine cycle steam power plant. The largest collection of parabolic systems in the world is the Solar Energy Generating Systems (SEGSs) I through IX plants in the Mojave Desert in southern California [2]. The SEGS plants were built in the 1985 to 1991 time frame. Figure 2.3 shows one of five 30-MW SEGS plants in the Kramer Junction site, California. The largest of the SEGS plants, SEGS IX, located at Harper Lake, is rated 80 MW. All of the SEGS plants are "hybrids," using fossil fuel to supplement the solar output during periods of low solar radiation. Each plant is allowed to generate

A
B

FIG. 2.1 TWO TYPES OF SOLAR THERMAL SYSTEMS FOR POWER GENERATION: (A) PARABOLIC TROUGH AND (B) POWER TOWER [1]

FIG. 2.3 SOLAR TROUGH FARM IN KRAMER JUNCTION, CALIFORNIA [1]

25% of its energy annually using fossil fuel. With the use of the fossil hybrid capability, the SEGS plants, in Southern California Edison on peak hours, have exceeded 100% capacity factor for more than a decade, with greater than 85% from solar operation.

2.2.1.2 Power Tower Systems consist of a field of thousands of sun-tracking mirrors which direct insolation to a receiver atop a tall tower (Figure 2.1b). A molten salt heat-transfer fluid is heated in the receiver and is piped to a ground based steam generator. The steam drives a steam turbine-generator to produce electricity. Because trough and power tower systems collect heat to drive central turbine generators, they are best suited for large-scale plants: 50 MW or larger [2]. Trough and tower plants, with their large central turbine generators and balance of plant equipment, can take advantage of economies of scale for cost reduction, as cost per kilowatt goes down with increased size. Additionally, these plants can make use of thermal storage or hybrid fossil systems to achieve greater operating flexibility and dispatchability. This provides the ability to produce electricity when needed by the utility system, rather than only when sufficient solar insolation is available to produce electricity, for example, during short cloudy periods or after sunset. This capability has significantly more value to the utility and potentially allows the owner of the solar power plant to receive additional credit, or payment, for the electric generating capacity of the plant.

In the summer of 2009, the first and only commercial power tower plant in North America, Sierra SunTower plant, started to operate and interconnected to the grid. It is a 5 MW, commercial facility located in Lancaster, California. There are several more

power tower systems around the world that have been built or were under construction during 2009-2010 with general confidence that uncertainty in the cost, performance, and technical risk of this technology is decreasing. A 2004 predictive analysis [3] shows that, assuming the technology improvements are limited to current demonstrated or tested systems and a deployment of 2.6 GWe of installed capacity by the year 2020, tower costs could drop to approximately 5.5¢/kWh (see Fig. 2.4), or better than trough systems. However, the data to confirm this prediction has yet to come. Owing to the limited studies for power tower systems, our discussion therefore mainly focuses on parabolic trough systems of the following.

There are several technical challenges for the solar thermal power systems that have been recently investigated. The overall goal is to reduce the plant cost:

- air or water cooling for Rankine cycles
- optical durability and high-temperature selective coating
- optimal piping design
- performance evaluation technique

Air Versus Water Condenser Cooling. Kelly [4] conducted a comparison study between the air cooled Rankine system and water cooled one. He selected an 80-MWe trough system located in Barstow, California, for both cases but with different modes of heat rejection. The first case has a dry-air cooled condenser, whereas the second case uses a wet cooling tower. The system is modeled using GateCycle software, developed by GE Energy [5].

Figure 2.5 is a GateCycle [5] flow diagram and shows the system schematic of an 80-MWe Rankine cycle in which the heat generated from the condensers is removed by the dry air flow. The main consideration for this application is to utilize dry air in the areas where supply water is limited, such as in the desert. The Rankine cycle design closely followed that developed by Fichtner for the 55-MWe AndaSol project in Spain. The cycle is a conventional, single reheat design with five closed and one open extraction feedwater heaters. The live steam pressure and temperature are 1,450 lbf/in^2 and 703°F, respectively, and the reheat steam temperature is 703°F. The independent parameters studied include dry-bulb ambient temperatures, preferred initial temperature difference, which is defined as the temperature difference between the dry-bulb temperature and steam condensation temperature, and

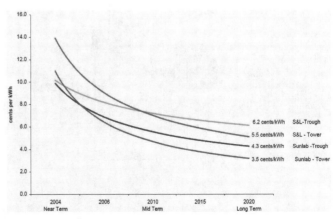

FIG. 2.4 ENERGY COST ANALYSIS FOR TROUGH AND TOWER TECHNOLOGY CONDUCTED (S&L: SARGENT & LUNDY LLC; SUNLAB: DOE NATIONAL LABS) [3]

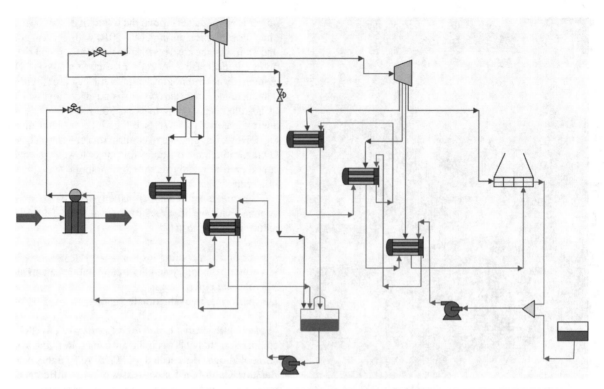

FIG. 2.5 FLOW DIAGRAM FOR AN 80 MWE RANKINE CYCLE WITH DRY HEAT REJECTION [4]

energy value in terms of dollar per mega watt hours electrical. The output parameters to be investigated are the power output, overall efficiency, and capital cost and net cost benefit.

The second model uses the water cooling tower as a means for discharging the condenser heat. Figure 2.6 shows that cycle. In this case, in addition to the parameters discussed above for the dry air heat rejection case, the sensitivity of relativity humidity is also studied.

It was found that for the dry heat rejection case with the range of initial temperature difference studied (24°F to 49°F), the gross

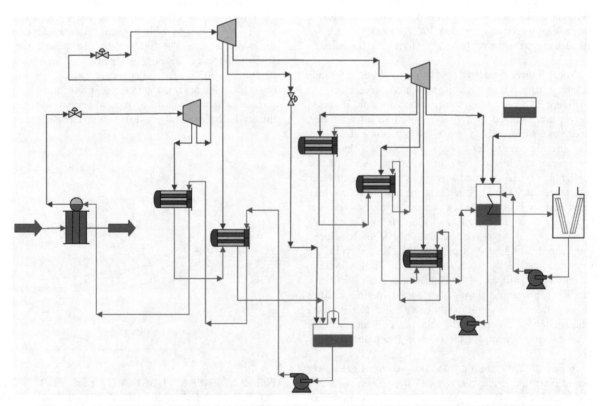

FIG. 2.6 FLOW DIAGRAM FOR AN 80 MWE RANKINE CYCLE WITH WET HEAT REJECTION [4]

energy output always increases as the initial temperature difference decreases, no matter what the ambient temperature is. In the meantime, the fan energy consumed in the air cooler also increases monotonically with the ambient temperature. As a result, the highest net output (gross minus fan energy) occurs with an initial temperature difference of 29°F because the incremental increase in the fan power demand at 24°F compared with 29°F is higher than the incremental increase in gross output. For a typical bin distribution of the dry-bulb temperature in Barstow, California, the ambient temperature that yields both the longest available annual hours and highest net energy is 83°F, although lower ambient temperature and lower initial temperature difference give the highest power output, from the thermodynamic point of view. The effect of ambient temperature on the capital cost, on the other hand, shows a different trend. Because the air-cooled condenser heat transfer area increases inversely with the initial temperature difference, the highest cost for condenser cost corresponds to the lowest initial temperature difference. For a reference initial temperature difference point set at the highest value studied, i.e., the lowest energy output point, D, an allowable incremental capital cost in dollars, is defined as follows:

$$D[\$] = \frac{(\textit{Net incremental output,MWHe})(\textit{Energy value,} \frac{\$}{\textit{MWHe}})}{\textit{Fixed Charge rate}}$$

(2.1)

This leads to the conclusion that the difference between the above-defined allowable cost and actual incremental capital cost yields an optimal cost benefit at an initial temperature difference in the range of 35°F to 40°F. It is also found that this value is relatively insensitive to the selling price of electricity.

For wet heat rejection, the modeling results show that the gross power output of the turbine is nearly invariant with the ambient temperature (within less than 0.56%). However, the consumption of water in the cooling tower to make up the evaporation loss is a strong function of ambient temperature, increasing more than twice when the ambient temperature increases from 40°F to 120°F. The effect of relative humidity on the power plant output is not significant for desert areas studied and may not be neglected for high-humidity regions.

Overall, the 80-MW system with wet heat rejection will yield a net energy generation that is 4% to 6% higher than that with the dry air cooling tower. The resulting annual solar-to-electricity

efficiency will be 13.4% for the wet heat rejection case compared with 12.8% to 12.9% for dry heat rejection. However, the wet heat rejection requires raw water usage that is 13 times higher than that of the dry heat rejection scenario.

Optical Durability. One of the challenges in expanding the applications of concentrating solar power (CSP) technologies for trough configuration is to reduce the cost of reflector materials while maintaining their high specular reflectance for long lifetimes, even under severe outdoor environments. A target of cost reduction of up to 50% with a lifetime of 10 to 30 years to the solar concentrator has been determined by the US DOE's Solar Program [6]. These goals may be achieved with lightweight front-surface reflectors that include anti-soiling coatings. Kennedy et al [6] conducted a study to identify new, cost-effective advanced reflector materials that are durable with weathering. The CSP official program goals were set as follows:

- 90% into a 4-mrad half-cone angle
- greater than 10 years under outdoor service conditions
- large-volume manufacturing cost of less than $1/ft^2 ($10.8/m^2)

The following more aggressive goals have been pursued in the research community:

- 95% reflectivity
- 15- to 30-year lifetime
- cost goal of $2.50/ft^2 ($27/m^2) for structural mirrors (e.g., self-supporting mirrors)
- reflectors themselves (not self-supporting) at $1.44/ft^2 ($15.46/m^2)

In their study, several mirror materials were tested using various UV-VIS-NIR spectrophotometers with wavelengths from 250 to 2500 nm, infrared (IR) spectrophotometer (2.5 to 50 μm) and specular reflectometer (7-, 15-, and 25-mrad cone angle at 660 nm). Accelerated exposure testing was done by employing Weather Ometers, solar simulators, UV lamps, heating chamber, and gas oven at high-temperature exposures. The study focuses on three materials that are close to the target criteria:

(1) thick glass (see Fig. 2.7a): Flabeg 4- to 5-mm silvered, slumped glass mirrors with proprietary multi-layer paint system commercially deployed at nine California SEGS plants,

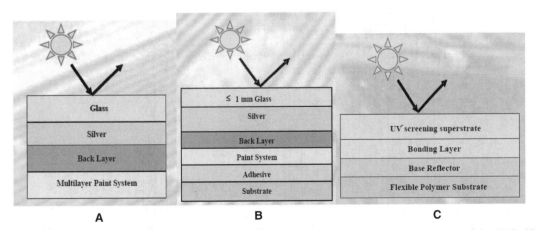

FIG. 2.7 CONFIGURATION OF THREE SOLAR REFLECTORS: (A) THICK GLASS, (B) THIN GLASS, AND (C) SILVERED POLYMERS [6]

or Pilkington (UK) (4-mm) and "Spanish" (Cristaleria Espanola S.A — Saint-Gobain Spanish branch) glass mirrors (3 mm) with copper-free and lead-free paint for possible use at Solar Tres

(2) thin (1 mm) lightweight glass (Fig. 2.7b) made by wet-silvered, copper-free processes, painted with commercially lead-free paints

(3) silvered polymer with laminated UV screening film to provide outdoor durability commercially available from ReflecTech

It was found that all the above three commercially available types of materials (glass, ReflecTech, and Alanod mirrors) may meet the 10-year lifetime goals based on accelerated exposure testing. However, a real outdoor lifetime may not be reliably predicted solely based on accelerated exposure testing. Currently, the commercially available solar reflectors have been in outdoor real-time exposure testing for less than 6 years, and their actual durability needs to be further determined.

Kennedy and Price [7] further modeled a high-temperature selective coating for improving the material properties and achieving high efficiency of parabolic trough collectors while reducing the cost of solar electricity. This may be achieved by increasing the operating temperature above the current operating limits of 400°C. Current coatings such as $Mo-Al_2O_3$ cermet solar coating (cermets are highly absorbing metal-dielectric composites consisting of fine metal particles in a dielectric or ceramic matrix), or multi-layer Al_2O_3-based cermet, do not have the stability and performance necessary to move to higher operating temperatures. Therefore, it was proposed to develop high-temperature solar-selective coatings using multiple cermet layers by physical vapor deposition. A model using Essential Macleod™ [8] was developed to design the

proposed multi-layer coating and successfully predicted, optically, a coating with $\alpha = 0.959$ and $\varepsilon = 0.061\mu$ at 400°C composed of materials stable at high temperature. Further modeling to the entire heat-collection element (HCE: air/glass/anti-reflective (AR) coating, vacuum, AR coating/solar-selective coating/stainless steel) structure will advance the design in this aspect.

Optimal Piping Design. A main part of solar power plant is the solar steam system [9]. The main elements include the solar collectors, control system, HTF piping system, HTF pump system, and solar heat exchangers. The piping system consists of header piping, valves, and fittings, and its cost can constitute up to 10% of the total solar system cost. The piping system design also affects performance. For example, the pumping power required to circulate the HTF through the system is a significant contributor to the plant parasitic power requirement. Furthermore, the piping heat loss reduces the useful heat delivered by the solar field to the power plant. Therefore, it is very important to obtain a reasonably accurate estimate through an optimal design and analysis of the piping system. Two useful design tools are an internal solar field piping model, SolPipe, developed by Flabeg Solar International (Cologne, Germany), and a model by NREL. Figure 2.8 shows an example of such design [9].

The key design variables that will lead to the final design through optimization are piping diameter, number of loops, field configuration, cold fluid and hot fluid temperatures, and collector field power ratings. Header diameters will be determined through the optimization model based on the cost criteria. Table 2.1 compares the feature of these two models.

In general, for the case of a 30-MW plant, similar to the SEGS VI solar plant, both NREL and SolPipe models satisfactorily compare the cost results of the headers as a whole (NREL: $16.6/m^2$ versus SolPipe: $16.5/m^2$). The results, however, differ in header

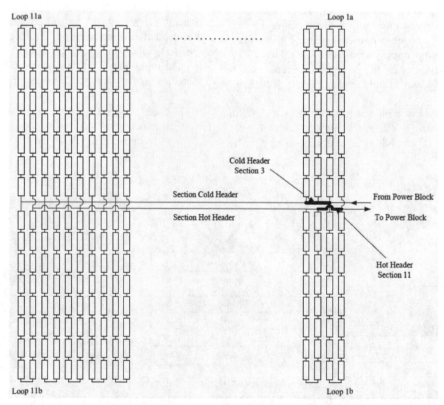

FIG. 2.8 SECTIONAL PIPING (H) CONFIGURATION FOR A 30-MW SOLAR TROUGH SYSTEM [9]

TABLE 2.1 COMPARISON BETWEEN SOLPIPE AND NREL MODELS [9]

Feature	SolPipe	NREL
Overall layout	H (80MW) or (30MW) with loops; also straight no-loop layout	Presently H and I configurations
Pipe sizing	Sized based on design velocity set by user	Optimizes each piping section (defined as header piping between loop connections). See below.
Pipe wall thickness	Uses Sked 40 piping	Wall thickness is no thicker than needed for required pressure
Piping/fitting capital costs	Table lookup from vendor data	Table lookup from vendor data
Piping/fitting labor costs	Table lookup from vendor data	Base on Bechtel experience
Insulation costs	Table lookup from vendor data	Table lookup from Solar Two data based on both ID and thickness of insulation
Pumping power cost	Not used	Calculated for typical year
Heat cost	Not used	Calculated for typical year
Optimization for pipe size	Not done	Per section pipe D assumed→model calculates wall thickness, then capital costs. equiv. heat loss cost equiv. pumping cost. Then increases D to next std size and recalculates cost for comparison. Within this process, thickness of insulation is optimized.
Expansion loops	Based on SEGS design; between every 2 loops	Same
Loops	Design based on SEGS	Same
Valves/fittings	Specified	Same
Calculation method	Simple arithmetic within cells	Uses macros
Input data	Entered in Input worksheet	Entered in Input worksheet

piping sizes, which result in differences in cost estimate and installation labor hours as well as fitting and insulation costs. Those differences are primarily due to understandable differences in the model assumptions and configurations. For example, for the HTF pressure drop for solar team systems, the NREL code is more inclusive because it includes all elements of the system (i.e., header piping, solar field loops, and solar heat exchangers), whereas Sol-Pipe has a less sophisticated pressure drop calculation.

Performance Evaluation Technique. It is important to measure the field performance of a SEGS after years of service in order to determine the extent of system degradation compared with the original design. Price et al [10] presented a technique that uses an IR camera to evaluate the in situ thermal performance of parabolic trough receivers at operating solar power plants. Through an analytical model, it was shown that the glass temperature measured with the IR camera correlates well with modeled thermal losses from the receiver. The work presented the results of a field survey that used this technique to quickly sample a large number of receivers to develop a better understanding of how both original and replacement receivers are performing after up to 17 years of operational service.

The key component to the system performance is the HCE, as shown in Fig. 2.9. It consists of a stainless steel tube with a cermet solar-selective absorber surface, which is surrounded by an AR evacuated glass tube. The HCE incorporates conventional glass-to-metal seals and metal bellows to achieve the necessary vacuum-tight enclosure and also to ensure for thermal expansion difference between the steel tubing and the glass envelope. The vacuum enclosure serves primarily to significantly reduce heat losses. The multi-layer cermet coating is sputtered onto the steel tube to result in excellent selective optical properties with high solar absorptance of direct beam solar radiation and a low thermal emittance at operating temperature to reduce thermal re-radiation. The outer glass cylinder has an AR coating on both surfaces to reduce Fresnel reflective losses from the glass surfaces, thus maximizing the solar transmittance. Getters, which are metallic compounds designed to absorb gas molecules, are installed in the vacuum space to absorb hydrogen and other gases that permeate into the vacuum annulus over time. The receivers include an evaporable barium getter that is used to monitor the vacuum in the receiver. One issue is hydrogen buildup in the vacuum owing to the substantial thermal decomposition of the HTF. A special hydrogen removal (HR) membrane made from a palladium alloy was attempted to remove excess hydrogen from the vacuum annulus [11, 12]. The reliability problems proved that this design was not feasible [13].

The IR camera methodology presented here provides a useful technique for rapid evaluation of parabolic trough receivers in an

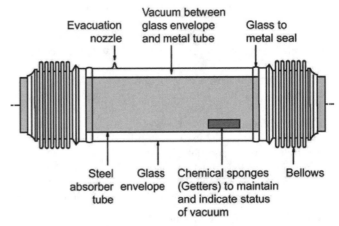

FIG. 2.9 HEAT COLLECTOR ELEMENT (HCE) [10]

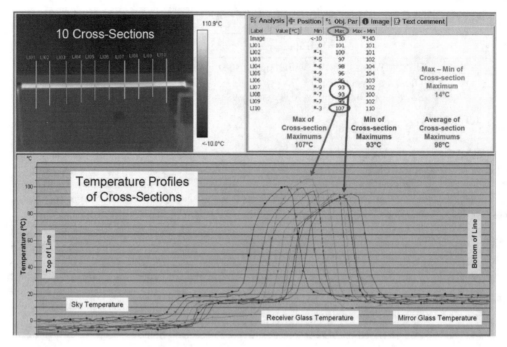

FIG. 2.10 ANALYSIS OF IR TEMPERATURE DATA ACROSS THE HCE [10]

operating solar field. A large number of receivers can be evaluated in a short period with the IR camera as shown in Fig. 2.10 where both the IR image and temperature profile are analyzed. Wind velocity has a significant effect on glass temperature, therefore it is important to do the measurements when there is very little wind. The field-testing helped identify that hydrogen buildup appears to be an issue in some receivers, but not others. The initial Luz cermet receivers with HR appear to be performing adequately without a noticeable buildup of hydrogen. However, a number of replacement receivers without the HR show signs of elevated glass temperature that is indicative of hydrogen buildup. The recent new designs that increase the hydrogen capacity of the getters addressed this problem. Additional receiver testing is being conducted in controlled laboratory experiments on a new receiver thermal loss test stand at NREL. The testing will further explore the relationship between thermal losses and receiver glass envelope temperature as a function of receiver type and gas composition in the annulus. A new non-destructive instrument has been developed [14] to identify gas composition and pressures in the receiver vacuum annulus. Preliminary test results on a number of receivers removed from the solar field show that hydrogen is present in receivers that exhibit elevated glass temperatures. The findings could be used to develop the correlation between the glass temperature and hydrogen level.

2.2.1.3 Parabolic Dish and Stirling Engine Technology Parabolic dish systems use a dish-shaped arrangement of mirror facets to focus energy onto a receiver at the focal point of the collector. A working fluid such as hydrogen is heated in the receiver and drives a turbine or Stirling engine. Most current dish applications use Stirling engine technology because of its high efficiency. The Sandia National Laboratories (SNL) of the US Department of Energy has been pursuing the aggressive deployment of 25-kW dish-Stirling systems for bulk power [15]. The immediate objectives of the development are to

- improve reliability and reduce cost of dish/engine components and systems;

- test, evaluate, and improve performance of dish/engine components and systems; and
- develop tools for industry to characterize their systems and components.

In 2009, the SNL development team, along with many industrial partners, continued to operate, maintain, and improve the Stirling Energy Systems six-dish model power plant, cataloging more than 100 development areas. They developed a real-time mirror characterization prototype system for 100% inspection of mirrors on assembly line, engine simulator for development of modern engine control hardware and software. SNL and Infinia together further conducted optical and systems design of a 3-kW free piston Stirling engine dish system, which is hermetically sealed and requires no maintenance. Figure 2.11 illustrates one of the comprehensively developed, tested and analyzed Stirling Energy Systems. The system consists of the following main components:

- optical mirror and structure assembly
- sun ray tracking mechanism and controller
- hydrogen storage unit
- power conversion unit, which is further comprised of Stirling engine, heat receiver, radiator with cooling fan, frame, and control unit.

The development team has conducted a number of tasks such as:

- mirror loading analysis
- velocity profiling and wind load analysis using computational fluid dynamics models
- dish alignment optimization modeling and validation
- test facility development
- design tool development and validation
- rigorous field testing

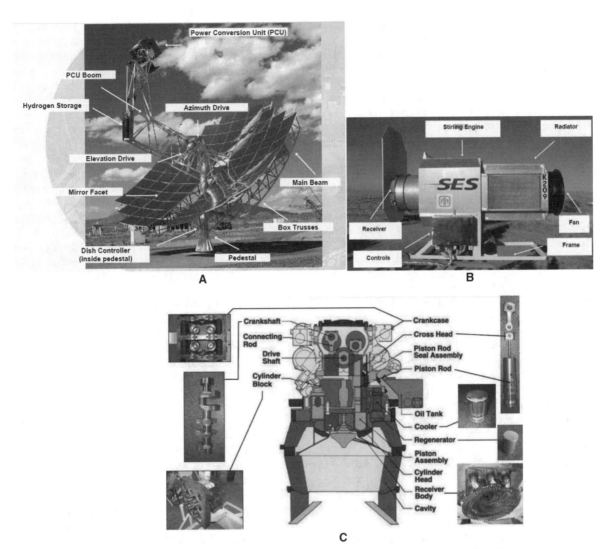

FIG. 2.11 ILLUSTRATION OF A DISH/STIRLING ENGINE SYSTEM: (A) WHOLE SYSTEM; (B) POWER CONVERSION UNIT (PCU), AND (C) STIRLING ENGINE [16]

It was reported that, in 2007 [16], the test unit, shown in Fig. 2.11 (A) had been in operation for 100,000 on-sun hours, which is equivalent to 28 years of daily solar operation. The one shown in Fig. 2.11 (B) was running for 161,000 on-sun or test cell hours, which is equivalent to 45 years of daily solar operation. The initial success further advanced the continued federal funding for accelerating the development. As of today, no commercially proved units or systems are available for wide deployment yet.

2.2.2 Large-Scale Photovoltaic Systems

There has been a surge of installing large-scale flat panel PV array systems for utility power productions in the United States and globally during last 5 years. The capacity of a single plant has reached as high as 60 MW in Spain. Fig. 2.12 shows the Wyandot Solar Energy Facility, a 12.6 megawatt (DC) solar PV facility located in Upper Sandusky, Ohio. Starting power production in May 2010, the system has 159,200 ground-mounted, thin-film solar panels on a 77-acre plot of land, the largest solar project in Ohio. Projects of similar size have been completed or are under construction in California, Florida, Nevada, Illinois, and other states.

Large-scale ground based PV systems face a fundamental question of land-use impacts. Denholm and Margolis [18] investigated

this issue. When deployed horizontally, the PV land area needed to meet 100% of an average US citizen's electricity demand is about 100 m^2. This requirement roughly doubles to about 200 m^2 per person when using 1-axis tracking arrays. By comparison, golf courses and airports each currently occupy about 35 m^2 per person

FIG. 2.12 WYANDOT SOLAR FACILITY IN UPPER SANDUSKY, OHIO, USA (COURTESY OF JUWI SOLAR, INC [17])

in the United States, whereas land used to grow corn for ethanol production exceeds 200 m² per person, although this land is concentrated in a fairly small number of states. They also pointed out another factor of disrupting local ecosystems. Deploying tightly packed PV arrays would create the most disruption but would require the least amount of land area. In contrast, the use of pole-mounted 2-axis arrays would require significantly more land but could be substantially less disruptive. Although the land-use requirements for wide-scale deployment of PV are modest when considering both the large area of rooftop availability and when compared with other uses of land in the United States, additional studies are needed to better understand the potential opportunities and impacts of PV on land use. For example, a more comprehensive estimate of rooftop availability is needed on a national basis for both residential and commercial buildings. Actual shading impacts of ground based PV arrays are also important. Actual ecosystem impacts of LSPV deployment need to be investigated to arrive at solutions of growing shade-tolerant native and beneficial species under LSPV arrays. This also includes evaluation of best practices to minimize the use of herbicides and other chemicals and the use of installation and maintenance techniques to provide minimum impacts to the environment.

One of the alternative solutions to provide LSPV power generation systems is using concentrated Photovoltaic (CPV) systems, which have a higher energy conversion efficiency than flat panels — currently 40% [19] with potential up to 50 %. If deployed on a large scale, they will require less land to produce the same amount of energy. Although no proven commercial scale applications have been installed, the commercialization of this technology has been accelerated.

Concentrated Photovoltaic systems use either parabolic dish mirror systems or a large array of flat Fresnel lenses to focus energy on PV cells. Unlike the dish and Stirling system, which is a solar thermal power system, in the dish and CPV systems, the solar PV cells generate direct current electricity, which is converted to alternating current (AC) using a solid state inverter. Dish and CPV systems are modular in nature, with single units producing power in the range of 10 to 35 kW. Thus, dish and CPV systems could be used for either distributed or remote generation applications,

which will be discussed in a later section, or in large arrays of several hundred or thousand units to produce power on a utility scale. Dish and CPV systems have the potential advantage of mass production of individual units, similar to the mass production of automobiles or wind turbines, yet they can be integrated to a utility scale solar power plant.

The key technology challenges for increasing the efficiency of CPV systems are optical elements and cell development. Spectrolab, Inc., has developed since 2007 metamorphic multi-junction solar cells up to 40.7% efficiency and lattice-matched 3-junction terrestrial cells with 40.1% efficiency. These efforts were partially supported by the High-Performance Photovoltaics program through DOE NREL [19]. The efficiency also benefits from high-band-gap, disordered GaInP top cells and wide-band-gap tunnel junctions under the terrestrial solar spectrum at high concentration.

Under a contract by US Department of Energy, SolFocus, Inc., has developed a 1.6-MW system using the technology illustrated in Fig. 2.13. It possesses the following characteristics:

- optical efficiency: 74%
- power unit efficiency: 27.0%
- module efficiency: 25.3%
- acceptance angle: greater than 1 deg.
- cell temperature: 50.8°C to 53.9°C
- module degradation: 1.2% per Kelvin

The system uses primary and secondary mirrors (Fig. 2.13a) and optical rods to focus the sun's rays to highly efficient multi-junction PV cells. Two examples of high-efficiency PV cell architectures are shown in Fig. 2.14, resulting in a high-efficiency inverted metamorphic 3-junction cell structure [21]. The combined mirrors, rod, and cells form a power unit, 20 of which are integrated into one panel. Twenty-eight such panels are then mounted and combined on a dual-axis tracking support as shown in Fig. 2.13 (b). Eventually, such a system will could be scaled up to a utility scale capacity.

The target of this development is to provide an operational rate of greater than 3 MW with module manufacturing cost of less than $5/W. The developed automated assembly line was validated by a third party to have exceeded the cost target and was expanded to

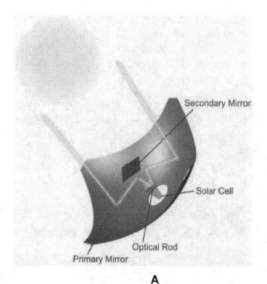

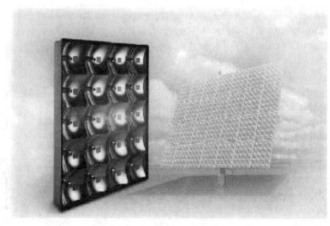

A   B

FIG. 2.13 SCHEMATIC OF CONCENTRATED PHOTOVOLTAIC (CPV) ARRAYS: (A) POWER UNIT, AND (B) 20-UNIT PANEL AND 28-PANEL TRACKER ON A SINGLE POLE [20]

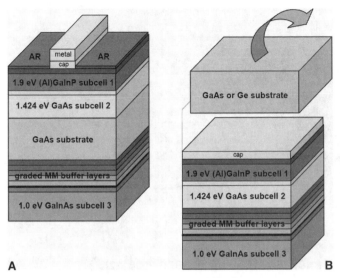

FIG. 2.14 SCHEMATIC OF 3-JUNCTION GaInP/Ga(In)As/ GaInAs CELLS USING TRANSPARENT METAMORPHIC (MM) GRADED BUFFER LAYER, AND AN INVERTED MM GaInAs SUBCELL 3: (A) WITH GROWTH ON BOTH SIDES OF A GaAs SUBSTRATE AND (B) WITH GROWTH ON ONE SIDE OF A GaAs OR Ge SUBSTRATE, FOLLOWED BY SUB-STRATE REMOVAL [21]

other manufacturing partners. The projected manufacturing capacity is to have a 100 MW capacity in 2010 [20].

2.2.2.1 Energy Storage Issues Unlike in solar thermal power systems where the harvested solar thermal energy can be stored in molten salt or other media and produce steam to run turbine and generate electricity during nights or cloudy days, neither the dish-Stirling nor CPV systems use storage or hybrid fossil capabilities to provide a firm resource, although CPV systems could, in principle, make use of battery energy storage. However, present battery storage technology is comparatively inefficient and cost-ineffective. A grid-tie back feed capability has to be in place in order to make economic sense in the short run. More discussion is given in Section 2.2.3 on this topic.

2.2.3 Energy Storage With Renewable Electricity Generation

In general, renewable energy sources such as PV and wind energy have variable and uncertain (also referred to as intermittent) power output. They are categorized as undispatchable sources compared with the conventional electricity generation used in the United States. The variability of those sources has led to concerns regarding the reliability of an electric grid if it derives a large fraction of its energy from renewable energy (and variable) generation into the electric grid. Denholm et al [22] investigated the impact of those uncertainties. In their report, three classes of energy storage were defined in Table 2.2.

The first two categories in Table 2.2 correspond to a range of ramping and ancillary services, but do not typically require continuous discharge for extended periods of time for conventional electricity. In the case of renewable energy-driven applications, this could require discharge time up to an hour to allow fast-start thermal generators to come online in response to forecast errors. Bridging power typically refers to the ability of a storage device to "bridge" the gap from one energy source to another. The third category (energy management) corresponds to energy flexibility, or the ability to shift bulk energy over periods of several hours or more. Figure 2.15 illustrates a range of available energy storage technologies used for both conventional and renewable energies, excluding the thermal energy storage applications. It can be seen that pumped storage hydro has the highest capacity and longest discharge time, and Nickel-cadmium has the least capacity and a relatively short discharge time. Double-layer capacitors have the shortest discharge time suitable for power quality management. In general, bridging the gap in types of energy storage classes are the ones suitable for renewable energy systems. However, when the time comes for renewable energy systems to be of a significant fraction of the overall electricity portfolio, the conventional power plants can serve as a backup system to the variable generation (VG) system in addition to the energy storage systems of each renewable energy plant.

The technical and economic limits to how much of a system's energy can be provided by renewable energy are based on at least two factors: coincidence of VG supply and demand and the ability to reduce output from conventional generators. There is no simple answer, primarily because the availability and cost of grid flexibility options are not well understood and vary by region. Although it is clear that high penetration of VG increases the need for all flexibility options including storage, historically, storage has been difficult to sell to the market. Both high costs and its services that it provides face challenges in quantifying the value of those services. The value of energy storage is best captured when selling to the entire grid, instead of any single source. Therefore, evaluating the role of storage with renewable energy requires continued analysis, improved data, and new evaluation techniques.

TABLE 2.2   THREE CLASSES OF ENERGY STORAGE

Common name	Example applications	Discharge time required	Technologies (Fig. 2.15)
Power quality	Transient stability, frequency regulation	Seconds to minutes	Flywheels, capacitors, and superconducting magnetic energy storage (SMES)
Bridging power	Contingency reserves, ramping	Minutes to ~1 hr	Lead-acid, nickel-cadmium, nickel-metal hydride, and (more recently) lithium-ion
Energy management	Load leveling, firm capacity, T&D deferral	Hours	High-energy batteries (sodium-sulfur, sodium-nickel chloride — ZEBRA, vanadium redox, zinc-bromine, polysulfide-bromine), pumped hydro storage (PHS), compressed air energy storage (CAES), and thermal energy storage (TES)

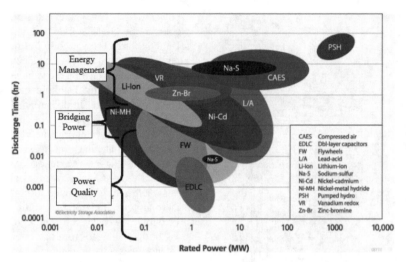

FIG. 2.15 INSTALLED ENERGY STORAGE CAPACITIES AS OF NOVEMBER 2008 [22]

2.2.4 Comparisons

Among all the large-scale solar power systems, parabolic trough systems and power tower systems have been proved to be best suited for large-scale plants of 50 MW or larger. Trough and tower plants, with their large central turbine generators and balance of plant equipment, can take advantage of economies of scale for cost reduction, as cost per kilowatt goes down with increased size. Additionally, these plants can make use of thermal storage or hybrid fossil systems to achieve greater operating flexibility and dispatchability, accounting for situations in which sufficient solar insolation is unavailable. Large-scale flat panel PV array systems for utility power productions are also widely used because they are easy to scale up or down. When tied to the electric grids, they provide flexible, variable power generation options. However, their costs are much higher compared with parabolic trough systems. Concentrating PV systems and parabolic dish systems with Stirling engine technology have relatively high efficiency. Dish and CPV systems are modular in nature, and a single unit could produce power in the range of 10 to 35 kW. They could be used for either distributed or remote generation applications, or in large arrays of several hundred or thousand units for a utility scale application. They also have the potential advantage of mass production of individual units.

The largest group of solar systems in the world is the SEGSs I through IX parabolic trough plants in the Mojave Desert in southern California, built between 1985 and 1991 and have a total capacity of 354 MW. These plants have generally performed well over their 15 to 20 years of operation. There are no operating commercial power tower or dish-Stirling power plants, although some commercial purchase agreements have been in place to pursue those options [1].

2.2.5 Reducing the Cost of Parabolic Trough Solar Power

Parabolic trough technology has continued to advance in recent years as a result of research and development efforts by the operators of the existing trough plants, the parabolic trough industry, and government-sponsored laboratories around the world. Key advances during the last 10 years include

- reduction in operation and maintenance costs,
- development of improved trough receivers,

- development of improved parabolic trough concentrators,
- reduction of solar field pumping parasites, and
- development of a thermal energy storage technology for parabolic trough plants.

Although parabolic trough technology is the least-cost solar power option available today, it is still more expensive than power from conventional fossil-fueled power plants. Recent increases in the price of natural gas have helped reduce the gap between parabolic trough solar electricity and fossil energy in the United States.

2.2.6 Cost Reduction Potential

At current fossil energy prices ($5 to $7/MMBtu), large-scale central solar generation must achieve costs in the range of $0.08 to $0.10/kWh (nominal) to directly compete with fossil power alternatives. The Western Governors' Association has shown that a significant reduction in the cost of energy is possible for parabolic trough solar power. Major cost reductions are possible through the following efforts:

- plant scale-up: increasing the size of plants to 200 MW or larger
- development of advanced technologies: improved thermal storage, concentrator, and receiver designs
- learning curve: cost reductions through plant deployments.

2.2.7 Combined Solar Power and Heat System

For all the solar thermal power systems, there always are subsystems that can be integrated to recover heat at various stages. They not only produce additional hot water or air for heating but also improve the overall plant efficiencies.

2.3 DISTRIBUTED PHOTOVOLTAIC SYSTEMS FOR BUILDINGS

Buildings have a significant impact on energy consumption. Annually, buildings account for 40% of the energy consumption of the world. Energy efficient design and quality construction can drive the cost of powering a home down by more than 50%. But to become zero energy buildings, i.e., where the net annual elec-

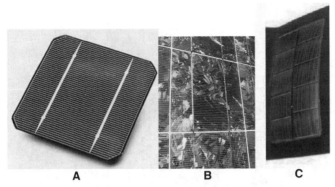

FIG. 2.16 THREE TYPES OF SOLAR CELLS: (A) MONO-CRYSTALLINE SILICON, (B) POLY-CRYSTALLINE SILICON, AND (C) THIN FILMS MADE FROM A VARIETY OF MATE-RIALS (Source: Wikipedia, Solar Cell)

tricity consumption from utility is zero, homes must incorporate some type of on-site energy generation. Photovoltaic systems, as the only available solar electricity production technologies for buildings, are reasonably developed to the extent that numerous companies warranty their building integrated solar products for 20 to 25 years. Most of recent building solar energy system manufacturers combine reliability, functionality, and aesthetics to remain in the solar market competition. Photovoltaic shingles and slates are examples of these modern products. Some PV companies provide ongoing Web-based monitoring of PV electric output, notifying consumers if problems arise, to make the system highly reliable.

There is an enormous potential for basing solar energy on buildings. Today, there is enough residential and commercial rooftop space to site more than 500 GW of PV capacity, equivalent to placing 4-kW PV systems on more than 125 million homes [23]. Current US electric capacity is about 1000 GW. A PV power system mainly consists of a few essential elements: PV panels with solar cells, inverters, and disconnects.

2.3.1 System Components

2.3.1.1 Cells and Panels The majority of PV panels available in the market right now fall into three categories depending on the materials and manufacturing methods of the cells, the element device that converts the energy of sunlight directly into electricity by the PV effect: cells made from a mono-crystalline silicon wafer, poly-crystalline PV cells laminated to backing material in a module, and thin films (Fig. 2.16).

2.3.1.2 Inverters Photovoltaics produce direct current (DC) electricity in the same way as batteries. Inverters, which can be called the most complex parts of a residential PV system,

take DC electricity and convert it to AC for powering typical household appliances as shown in Fig. 2.17. Residential PV systems can be connected to the electric utility distribution grid, so power can be sold to the utility when on-site power generation is more than demand. These grid-connected PV systems are the most common and simplest PV systems installed in houses. Unless a home is being built on a site more than half a mile away from any electric utility, or there is a very specific regular need for battery backup, the Department of Energy [24] recommends grid-tied systems. Essentially, grid-tied systems use the grid as a battery. Facilities that offer emergency services during disasters are good candidates for battery backup systems. Candidates would include clinics, fire stations, police departments, and dispatch facilities.

During electricity conversion from DC to AC heat generation will be occurred. This heat should be dissipated in some way. Inverters can be installed inside or outside the house, but indoor installation is more common, and the location of the inverter should be close to the main electrical panel.

2.3.1.3 Disconnects All homes need one main disconnect. National Electrical Code [25] requires that each piece of PV equipment have disconnect switches in addition to the main feeder disconnect, allowing service providers to disconnect the equipment from all sources of power. Some inverters have incorporated a disconnect switch into a box attached to the inverter. Disconnects may be circuit breakers or switches. The number of disconnects in grid-tied systems changes with the system's complexity. In a simple system, there would be disconnects on both sides of the inverter. In more complex systems, there may be disconnect switches for each string of arrays, sub-array disconnects, main PV disconnects for each inverter, sub-panel and AC disconnects. If there are multiple inverters, they must be combined in a dedicated sub-panel that feeds one AC disconnect and a main feed AC disconnect.

2.3.1.4 Aesthetic Efforts Manufacturers have produced PVs that can be installed in some roofing systems to make them closely invisible. Some PVs look like a part of the roof rain-shedding system. Some PV systems are mounted very near the roof, parallel to it, making them resemble skylights. Thin film PVs can be adhered directly to the flat part of raised seam metal roofs without changing the roof profile.

2.3.1.5 Performance Measures There are numerous efforts to measure the performance of PV systems in buildings. Measurement procedures are often inconsistent with each other. The Performance Metric Project [26] conducted by NREL is an attempt to resolve differences among the various approaches in building performance monitoring. The project has determined which performance metrics have the greatest importance in building auditing. Other NREL-conducted research [27] expands on previous

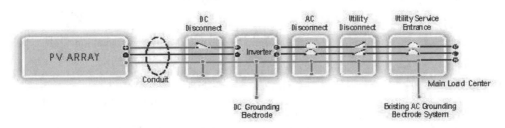

FIG. 2.17 SCHEMATIC DIAGRAM OF A TYPICAL RESIDENTIAL PV SYSTEM [24]

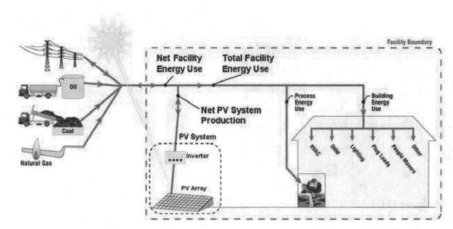

FIG. 2.18   ENERGY FLOW DIAGRAMS FOR PV SYSTEMS IN BUILDINGS [27]

works in performance evaluation of grid-tied PV systems by determining proper long-term energy performance metrics and providing a standard procedure for measuring and characterizing them. The procedure may be used to evaluate the power supplied to a building's electrical system from a grid-tied PV system and the implications for the building's energy use. Figure 2.18 depicts the energy flow diagram and shows the highest level metrics determined in the procedure. Table 2.3 shows all possible performance metrics in a grid-tied PV system. Based on project goals and the questions that should be answered by performance measuring, performance metrics will be selected. A simpler level of analysis called tier 1 yields monthly and annual results for the system, mostly based on utility meter readings and building and PV system drawings. Tier 2 provides small time interval (i.e., 15 minutes) results, in addition to monthly and annual results based on a data acquisition system with sub-metering. A flow chart representing the grid-tied PV system performance measuring procedure is shown in Fig 2.19.

2.3.1.6 Ongoing Developments for Thin Films Developed in 1954, thin-film solar cells have increasingly become a major market force in the PV industry, although still having a relatively small market share compared with crystalline silicon modules. A thin-film PV cell is a solar cell made by depositing one or more thin layers of PV material on a substrate. The thickness range of such a layer is wide and varies from a few nanometers to tens of micrometers. Three main types of thin film solar cells are amorphous silicon (a-Si), cadmium telluride (CdTe), and copper indium gallium diselenide (CIGS). Other types include dye-sensitized and organic cells.

a-Si solar cells are fabricated on substrates such as low-cost soda lime glass, stainless steel, and polyimide. The plasma-enhanced chemical vapor deposition is used for the deposition process. The typical superstrate structure is p-i-n (intrinsic semiconductor region between a p-type semiconductor and an n-type semiconductor region) on glass or substrate structure of n-i-p (intrinsic semiconductor region between an n-type semiconductor and a p-type semiconduc-

TABLE 2.3   HIGH-LEVEL AND RELATED METRICS [27]

High-level metrics	Related metrics
Net PV system production	Total PV system production
	PV system standby use
	Equivalent daily hours of peak rated PV production
	Equivalent annual hours of peak rated PV production
	Total incident solar radiation
	PV system AC electricity generation effectiveness
	PV system performance ratio
	Maximum time-series net PV production
	Average daily time-series PV production profiles
Total facility energy use	Total facility electricity use
	Peak demand of total facility electricity use without PV system
	Facility electricity costs without PV system
Net facility energy use	Facility's electrical load offset by PV production
	Facility's total load met by PV production
	Total electricity delivered to utility
	Peak demand of net facility electricity use
	Reduction of peak demand resulting from the PV system
	Facility electricity costs
	Energy cost savings accruing from PV system

I. Project Definition

A. Identify project goals.
B. List specific questions to be answered.
C. Determine boundaries of the site to be analyzed.
D. Select tier 1 or tier 2 analysis.
E. Specify desired accuracy of results.
F. Develop estimated budget for performance analysis.
G. Identify period of analysis.
H. Gather basic site and PV system data.
I. Obtain pre-existing performance data.

II. Measurement System Design

A. Select performance metrics to be measured.
B. Identify data required for each metric.
C. Specify physical location of each measurement.
D. Specify frequency of each measurement.
E. Specify measurement equipment.
F. Determine feasibility of measurements.
G. Estimate cost of DAS equipment and operation.
H. Calculate uncertainty of measurements.
I. Resolve cost, uncertainty, and practicality with expectations (steps I.E and I.F).

III. Data Collection and Analysis

A. Validate data for quality control.
B. Assemble data for the period of analysis.
C. Calculate monthly metrics.
D. Calculate annual metrics.

IV. Reporting Results

FIG. 2.19 GRID-TIED PV SYSTEM PERFORMANCE MEASURING PROCEDURE [27]

tor region) on foil with a transparent conducting oxide as a top contact and Al along with ZnO as the bottom contact [28]. To enhance the short circuit current, the ZnO/Al serves as a back reflector. In this way, the short-circuit current may increase by approximately 15%. With recent developments in tandem devices, the performance of ZnO/Ag or ZnO/Ag/Al has been quite reliable for stable back contacts. There is a great effort worldwide to introduce tandem devices using a-Si as the top cell and microcrystalline absorbers as the bottom cells. Some of the major challenges to produce a-Si solar cells are [28] to increase the a-Si solar cell efficiency (a 12% to 13% range has been demonstrated for small-area laboratory solar cells now), reduce the light-induced changes in the devices, develop higher deposition rates for the microcrystalline bottom cell without compromising on the opto-electronic properties of the a-Si tandem devices, which could potentially have negative effect on the solar cell performance, and ultimately reduce the manufacturing cost.

Thin-film CdTe solar cells with a perfect match with the solar spectrum are one of the most promising thin-film PV devices. Theoretical efficiencies for these devices are about 26%. Several deposition processes have been developed for the growth of the absorber layer. Most of deposition processes results in 10% or higher efficiency for thin film CdTe solar cell. Five of these

processes have demonstrated prototype power modules, namely, close-space sublimation, electrodeposition, spray, screen printing, and vapor transport deposition [28]. Recent developments are under way to produce CdTe solar modules with increased cell efficiency (laboratory efficiencies of 16.5% for thin-film CdTe solar cell has been demonstrated by NREL scientists [29]), standardization of equipment for deposition of the absorber layer, back-contact stability, reduction of absorber layer thickness to less than 1 μm, and controlled film and junction uniformity over a large area [28].

Thin-film copper indium selenide (CIS) is a direct band-gap semiconductor and has a band-gap of ~0.95 eV. When gallium (Ga) is added to CIS, the band-gap increases to ~1.2 eV, depending on the amount of Ga added to the CIGS film. This material has demonstrated the highest total-area conversion efficiency for any thin-film solar cells. Some of the major developments for CIS modules are undertaken to increase the CIGS solar cell efficiency to scale up the laboratory range of 19.3% to 19.9%, prevent moisture ingress for flexible CIGS modules, decrease absorber layers to less than 1 μm, and investigate CIGS absorber film stoichiometry [28].

Furthermore, developing approaches for using transparent wide band-gap in CIGS and alloyed CdTe to increase the cell efficiency up to 25% is also one of the ongoing research topics [30]. For CIGS's case, the relative Ga/In and S/Se compositions play the key role in changing the thin film band-gap. CIGS thin films can be prepared by thermal co-evaporation under different uniform and sequential processes to elucidate the film formation and composition control. For CdTe solar cells, improving cell performance by controlling chemistry and materials processing during film deposition, post-deposition treatments, and contact formation has the main focus.

2.3.2 Roof-Mounted BIPV Systems

2.3.2.1 Cement Tile Systems This type of roof mounted system is the most common type of Building Integrated Photovoltaics (BIPV) system. In this system, some cement tiles are replaced by PV panels that are sized and mounted with the same overall dimensions (Fig. 2.20). Usually, PV panels are lighter than the cement tiles they replace; therefore, structural assessments to add any supporting structure should be conducted before installing PVs. Photovoltaic panels follow the contour of the roof in exactly the

FIG. 2.20 CEMENT TILE SYSTEMS [24]

FIG. 2.21 THIN-FILM PV LAMINATE FOR STANDING SEAM METAL ROOFS [24]

same way as the cement tiles. Since PV tiles are replacing roofing materials, they should be compliant with local and national roofing regulations. Electrical connections are made between each tile. In some cases, a single module will replace a set of three or four tiles, reducing the number of connections.

2.3.2.2 Thin-Film PV Laminate for Standing Seam Metal Roofs These types of BIVPs are common and easy to install with no additional structural support required. Solar laminate PV is installed on metal raised seam roofs between the seams as shown in Fig. 2.21. It is directly attached with glue during installation of the roof without replacing the metal on the roof. The roof ridge, as an easily accessible place for maintenance, is used for electrical wiring and connections. Since they have no glass components, they have higher durability than other types of BIPVs. Thin-film PV laminates are approximately half as efficient as the mono- or multi-crystalline modules, but also currently cost about half as much.

2.3.2.3 Shingle Systems Photovoltaics can replace shingles in two ways. In the first method, shingled roofs take advantage of thin-film PV. This lightweight plastic replacement for shingles comes in relatively long strips to replace courses of asphalt shingles and reduce the number of electrical connections. In the second method roof, shingles powered by solar energy — PV shingles — serve to replace ordinary shingles. Electrical lead wires extend from the underside of each shingle and pass through the roof deck, allowing interior roof space connections. The sun's heat helps bond the shingles together, forming a weather-resistant seal.

2.3.3 Non-Roof-Mounted Systems

South-facing walls have the potential to generate electricity using BIPVs. They can be covered in PV vertically or can be slanted to act as PV and window shading at the same time. A filtered window or filtered skylight can be made by setting thin-film PV between two sheets of tempered glass. This type of PVs may cost the owner three times more than regular PVs, but can make a powerful architectural impact.

Mono- or poly-crystalline PV can also be set between sheets of glass to create a dappled effect, blocking the majority of sunlight to make electricity, but allowing shaded light through. This reduces solar gain to the interior of the building while producing electricity.

Ground-mounted PV systems may be applicable for buildings with large land holdings mostly located in rural areas. Commonly, solar ground-mounted systems involve steel or aluminum frames attached to a concrete foundation. The lower edge of ground-mounted arrays should be high enough to clear vegetation and accumulated dirt or snow.

Either crystalline or thin-film PV technologies can be utilized in these systems. In cold climates of northerly latitudes, sun-tracking systems are used to maximize the solar energy harvest of ground mounted arrays. It is only worth installing trackers in regions with mostly direct sunlight. In diffuse light, tracking has no noticeable effect on solar gain by PV panels. Ground-mounted PV systems are more suitable with respect to rooftop systems for high-wind regions. Some areas have restrictions on installing ground-mounted PV systems for aesthetic or agricultural reasons.

2.3.4 Integration Issue

Because solar energy is not available all the time, the continuous usage of electricity by buildings requires power to be drawn from either an energy storage facility charged by the building solar energy system during the production period or from the conventional electricity grid. Energy storage such as battery banks is expensive and inefficient unless they are used in remote areas where the utility grid is not available. The best solution is for PV systems to have an interconnection with the grid, a scheme of net energy metering that allows a PV system to export excess electricity to the utility and import it back when needed. Most electric utilities in the United States have adopted regulations and guidelines to help design PV systems that operate in parallel with the utility systems. In residential applications, because each of the distributed systems usually has a very small capacity such as less than 5 KW, the impact of back feed due to net metering to the grid is negligible. For commercial-scale applications, the impact could be significant.

Coddington et al (2009) [31] investigated four investor-owned utilities in the United States that operate secondary network distribution systems and that have allowed PV systems to be interconnected to those networks. Six PV systems in four cities were examined with their capacities ranging from 17 to 676 KW. Several approaches were adopted by the utilities. There are two types of electric distribution systems: radial and secondary network distribution system. In the radial system, which comprises more than 90% of the distribution system in North America, single feeders serve single transformers, each of which serves multiple electricity consumers. In other words, the radial system contains no closed loop. In contrast, the secondary network distribution system (or more simply, network) uses multiple feeders that serve multiple transformers, which in turn serve multiple electricity consumers. It means the consumer is simultaneously served from more than one primary feeder, which increases reliability of the system. Therefore, customers served from a network system will not experience an outage if any of the network feeders are interrupted. Alternatively, customers served from radial systems will experience outages when the radial feeder is interrupted. For this reason, networks are preferred electrical distribution systems for areas with high-load concentrations such as high-rise commercial buildings, to provide excellent service reliability.

Networks incorporate an important design feature at each network transformer called "network protectors" (NPs), which ensure

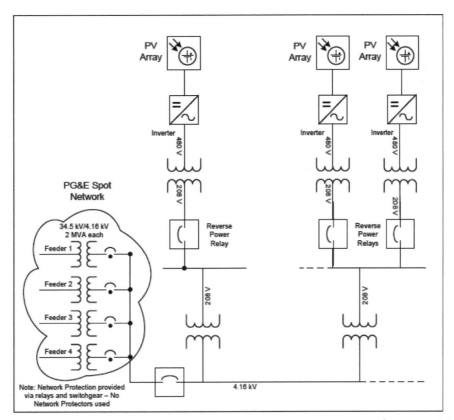

FIG. 2.22   SIMPLIFIED ELECTRICAL DIAGRAM FOR THE MOSCONE CENTER PV SYSTEM [31]

reliability and continuous operation if one or more feeders are lost due to a device failure or if the utility must conduct maintenance activities. The normal direction of the current flow on network feeders is from the utility toward the loads. An NP is installed on the low-voltage side of each network transformer and detects current flow direction. If the direction of the current flow reverses during a short-circuit condition on the high-voltage side of one of the network transformers, NP detects this reversal and interrupts the current from flowing back toward the utility system by opening its contacts. This design prevents undesired current flow and allows the network to maintain uninterrupted service to utility customers.

Figure 2.22 shows an example of such system for the Moscone Center PV System investigated by Coddington et al [31]. That example system generates approximately only 3.18% of the energy for the entire building uses in the course of a year. Because the minimum daytime load is typically much greater than the PV system generation capability, there is no chance of any power flowing out toward the utility system. This is the case for almost all the cases studied in [31]. However, perceivably a commercial scale PV system may produce a flow-back current when its electricity generation is higher than customer demand. In these cases, back-flow current actuates the corresponding NP and may cause the device to open unnecessarily. In addition, the NP is designed to re-close for a pre-defined forward flow condition, and it is possible to have a setting where the NP might try to re-close out of synch on an islanded distributed generation. Islanding is a condition in which a portion of the utility system that contains both load and distributed resources remains energized while isolated from the remainder of the utility system. Distributed resource islanding is an islanding condition in which the distributed resource(s) supplying

the loads within the island are not within the direct control of the power system operator [32].

The accepted interconnection of the PV system on the secondary network happens when it is ensured that the electricity produced by the PV system is not fed back toward the utility system. The general recommendations are to minimize, reduce, or eliminate the possibility of back feed from the PV system through the NPs [31]. As more and more commercially installed distributed PV systems increases, it will be imperative to have a comprehensive solution for seamless integration of distributed solar power systems with the utility grid. Recent research and development in smart grid technology have been emphasized to address this issue.

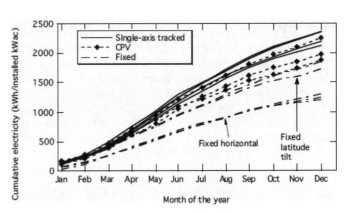

FIG. 2.23   KILOWATT-HOURS GENERATED PER INSTALLED KWAC FOR SINGLE-AXIS TRACKED FLAT-PLATE, CONCENTRATOR, AND FIXED ROOFTOP SYSTEMS SITED IN ARIZONA [33]

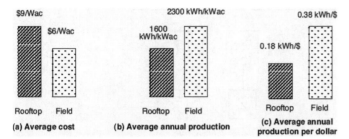

FIG. 2.24 COMPARISON OF FLAT-PLATE ROOFTOP AND FIELD SYSTEM COSTS AND AVERAGE ANNUAL PRODUCTION FOR RECENT INSTALLATIONS IN ARIZONA [33]

2.3.5 Large- and Small-Scale PV Systems Comparison

The cost per watt of installing and operating distributed PV systems is higher than the costs of similar utility systems because of the economics of scale. In the case of a new, large building where the builder might choose to integrate and install PV systems at the time of construction installation, costs may be low, but, in general, there are added costs associated with rooftop systems; especially for retrofits, there can be permit, architectural, structural, and electrical issues.

On the other hand, distributed systems avoid transmission losses through delivering power where it is needed, and residential and commercial systems can be financed along with the rest of a building. Results of an NREL-conducted research [33] show that the installation and operation of PV systems for buildings may cost more than twice the installation and operation cost in utility approaches with the same capacity. Figures 2.23 and 2.24 illustrate a comparison of rooftop and field PV systems. Concentrating PV has often been presented as a lower-cost approach to utility scale power and could be a major player in this market as discussed in earlier sections.

The kilowatt rating in Fig. 2.23 was determined from the DC electricity generated at 1000 W/m² for flat-plate or 850 W/m² for concentrator systems. The AC rating was assumed to be 85% of the DC STC rating. There are several reasons for lower performance of fixed rooftop systems with respect to CPV systems. In most rooftop systems, PV panels are installed in a horizontal configuration that is not an optimal position for year-round sunlight. The efficiency of a solar cell decreases by increasing its temperature. If the temperature of the panel increases because of poor ventilation from close contact with the roof, performance of the panel is negatively affected. In practical cases, panels may be shaded by a nearby building or tree.

In Fig. 2.24 relative costs, kWh electricity output per installed kWac and kilowatt-hour electricity output per dollar spent have been compared for rooftop and field PV systems. In this analysis, tracked PV systems represent utility scale solar systems. Results show that the field PV systems generated nearly twice as many kWh per dollar spent, compared with the rooftop PV systems. Therefore, given a fixed budget, the field approach results in higher electricity generation in comparison to retail approach. This also implies that public support for utility market incentives would yield a higher rate of return for the investment.

2.4 SOLAR THERMAL SYSTEMS FOR BUILDINGS

Solar thermal systems are among the most cost-effective renewable energy options that have a great potential for use in building energy systems. In contrast with PV systems, solar thermal systems for buildings do not generate electricity. These systems employ the heating potential of the sun to heat water or air to be used for domestic consumption, pool heating, or space conditioning. Solar water heating systems for domestic consumption or pool heating are well-established technologies, whereas space conditioning with solar heated water or air is not yet mature enough for building applications.

2.4.1 Solar Water Heating Systems

Solar water heating accounts for 11.7% of the energy used in the US residential sector [34]. It is a well-developed, highly effective, and environmentally friendly technology. Key functions of solar water heating systems are collecting solar heat and protecting the system against freezing. Different anti-freezing mechanisms affect the total heating system and its components. First, solar collectors with water heating application will be discussed. Systems will be analyzed in the next section.

2.4.1.1 Collectors Solar collectors for buildings have been available in the market for over a century and have been greatly developed and enhanced in these years. Improved materials and modern testing and commissioning significantly raise efficiency, durability, and reliability of solar collectors and the whole thermal system as well.

There are three types of solar collectors for building water heating: flat-plate collector, evacuated tube collector, and batch heater.

Flat-Plate Collector (Glazed or Unglazed). This collector is the most common solar collector for use in solar water heating. A glazed flat-plate collector is typically 2 in. or 3 in. thick and consists basically of an insulated metal box with a glass or plastic cover (the glazing) and a dark-colored absorber. It resembles a skylight when it is installed on the roof. The glazing reduces heat loss and traps heat inside the collector. Absorber coating plays a significant role in energy efficiency of the collector. Selective coatings like black chrome, black nickel, and aluminum oxide with nickel enable the conversion of a high proportion of solar radiation into heat. Flat-plate collectors heat the circulating fluid to a temperature noticeably less than that of the boiling temperature of water and are best suited to applications where the demand temperature is 86°F to 158°F (30°C to 70°C).

Evacuated Tube Collector. An evacuated-tube collector contains several rows of glass tubes interconnected with a manifold, along the top of the collector. Each tube is evacuated to eliminate heat loss through convection. Inside the glass tube, a flat or curved selective coated aluminum or copper fin is attached to a metal pipe.

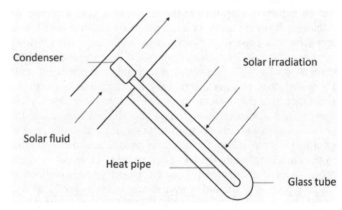

FIG. 2.25 SCHEMATIC FLUID FLOW IN THE HEAT PIPE EVACUATED TUBE COLLECTOR

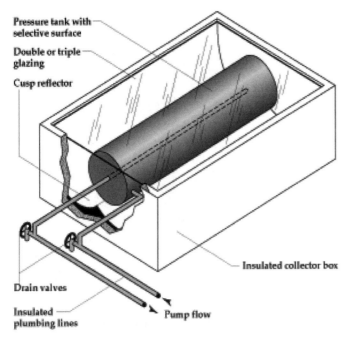

Pressure tank with selective surface

Double or triple glazing

Cusp reflector

Insulated collector box

Drain valves

Insulated plumbing lines

Pump flow

FIG. 2.26 BATCH HEATER (*Source*: US Department of Energy)

The heat-carrying fluid circulates through this metal pipe. Evacuated tube collectors can provide higher temperatures in comparison to flat-plate collectors. There are three types of evacuated tube solar collectors for buildings: direct flow, heat pipe, and thermosiphon evacuated tube solar collectors. (Thermosiphoning is the upward motion of heated fluid by natural convection.) A direct-flow evacuated tube collector has two pipes that run down and back, inside the tube. One pipe is for inlet fluid, and the other is for outlet fluid. In this type, collector tubes are not easily replaced, and all of the fluid could be pumped out if a tube breaks. A heat pipe collector employs a fluid in the metal pipe that has a low boiling point and is resistant to freezing. Because of the vacuum, the liquid boils at a lower temperature than it would at atmospheric pressure. The vapor rises to the top of the pipe. Heat-carrying fluid flows through the manifold and absorbs the heat. The fluid in the heat pipe condenses and flows back down to the tube. Fig 2.25 demonstrates the schematic fluid flow in the heat pipe evacuated tube collector. Thermosiphon evacuated tube solar collectors are preferred where temperatures are not likely to drop into the freezing zone. They use an integrated tank and work passively with heated liquid from the evacuated tubes rising into the tank. With the tank included in the system, water flow is controlled by the household water pressure. Water pumps, expansion tanks, and electronics are reduced in this system, which makes the system very economical. Thermosiphon evacuated tube collectors will be described in more detail in the Systems section.

Batch Heater: Batch heaters are also known as integrated collector storages (ICSs). The solar collector, combined with the storage tank in the batch heater, is shown in Fig. 2.26. This is the cheapest and oldest solar water heating system, which one can trace back to the late 1800s. In the batch heater, water pressure in the plumbing system pushes the water through the pipes and the tank. This system should not be used where there is a chance of freezing. A 40-gallon batch heater can place more than 500 lb on the roof, which should be considered in the structural design of the building's roof.

2.4.1.2 Systems There are four fundamental terms that help categorizing solar water heating systems: Active and passive systems, and direct and indirect systems. Active systems employ pumps to move fluid through pipes in contracts to passive systems that use warm fluid buoyancy and gravity for fluid circulation. In direct heating systems, potable water is the HTF; but in indirect systems, an independent piping system and heat exchangers isolate HTF from potable water.

Whether the system is active or passive, direct or indirect, solar heating system components may be combined in different ways to achieve efficient water heating. All non-obsolete combinations are as follows:

Batch Heater (or ICS) Passive Direct System. Since the tank and collector are combined in ICS, potable water is the HTF in this system. Plumbing system pressure helps to fill the tank with cold water when the hot water is consumed. The system shows the best performance if water is consumed during late afternoon and evening. When the heat transfer direction changes from hot water to the ambient, a check valve should be considered to avoid reverse thermosiphoning. Reverse thermosiphoning may be stopped by the pipe run's arrangements.

Thermosiphon Passive Systems. Thermosiphon systems are also known as convection heat storage units. No pumps are used to enforce circulation. The unit relies on natural convection and gravity to transfer the fluid between the collector and the tank. The tank is located in a higher level than the collector to permit thermosiphoning. Both direct and indirect mechanisms can be applied on thermosiphon passive systems (Fig 2.27). Some direct units use collectors with built-in tanks. In these prepackaged units, the tank is mostly mounted on the roof. In other units, tanks are located in the attic spaces above the collectors. Direct systems are not recommended in climates with severe freezing. Indirect systems are equipped with heat exchangers within their tanks and may use non-freezing fluids for freeze protection. However, water pipes and tanks containing water must be in conditioned spaces in cold climates. Stagnant conditions, which can lead to overheating fluid, should be considered in warmer regions.

Drainback Active Indirect System. Drainback systems use water as the HTF most of the time; however, a mix of water and propylene glycol has been also recommended. The collector and pipes are discharged when temperature falls near the freezing point. To avoid overheating, these systems also drain back when the water temperature is near a pre-specified high temperature. High temperatures occur during warm and low-demand days. No electric valve is used in this system. Therefore, there is no concern of decreasing the system capacity during a power outage. A schematic of a drainback active indirect system is illustrated in Fig 2.28(a).

Controller-Based Active Direct Systems. Controller-based mechanisms can be applied only to active systems that use a pump to circulate water between the storage tank and collectors. These systems operate only when a pump is on. A differential pump controller directs system operation. The controller detects the temperature of the solar collector as well as that of the storage tank. The controller turns the pump on whenever the collector is hotter than the stored water (by 4°F to 8°F). These differential temperatures are pre-set by the installer in accordance with the specific characteristics of the installation. Another approach is to use a PV module to power a DC pump. When the sun shines, the pump runs. In this system, check valves are needed to stop reverse thermosiphoning and freeze prevention valves are used to protect the system against freezing like many other solar heating systems. However, direct systems are never recommended for climates with severe freezing

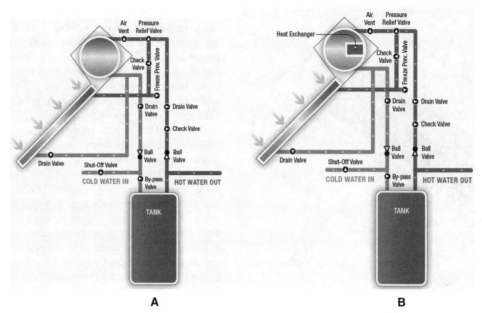

FIG. 2.27 SCHEMATIC DIAGRAM OF A THERMOSIPHON PASSIVE SYSTEM: (A) DIRECT SYSTEM AND (B) INDIRECT SYSTEM [24]

conditions. A schematic of a controller based active direct system is illustrated in Fig 2.28(b).

Pressurized Glycol Active Indirect System. A non-toxic glycol or a mix of the glycol with water is used as the HTF in pressurized glycol systems. They are suitable for regions with temperatures below the freezing point of water during some periods of the year. Glycol is circulated in the system using a pump that may be powered by a PV module. Because these systems are pressurized, fill and drain valves must be incorporated to add or change the collector fluid. As depicted in Fig 2.29, an expansion tank is needed to absorb excess glycol pressure caused by thermal expansion as glycol is heated. Since glycol is used as the heat transfer medium, a heat exchanger is needed to transfer solar heat absorbed by glycol to the water. Stagnant conditions should be controlled by adding a

shunt load to the system, which can be an outdoor radiator or a buried un-insulated pipe. Glycol must be changed every 10 to 15 years.

2.4.1.3 Freezing Issue In northern regions where freezing days occur, different anti-freezing mechanisms are employed, which has a major effect on the system's components and connections. There are five methods for preventing freezing in the solar water heating system: draindown, drainback, water flow, freeze prevention valves, and pressurized glycol. In the draindown system, when the collector inlet temperature falls to a pre-specified temperature, a draindown valve isolates the collector inlet and outlet from the tank. At the same time, it opens a valve that allows water in the collector to drain away. This method is no longer in use and is not certified by the Solar Rating and Certification Corporation. In the

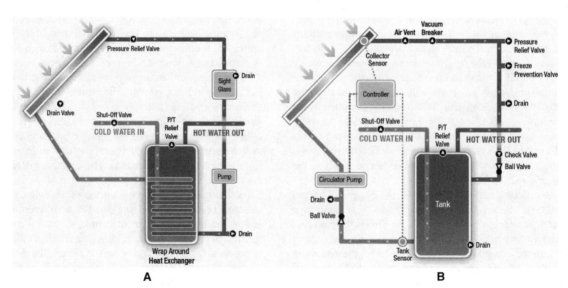

FIG. 2.28 SCHEMATIC OF (A) A DRAINBACK ACTIVE INDIRECT SYSTEM AND (B) CONTROLLER-BASED ACTIVE DIRECT SYSTEM [24]

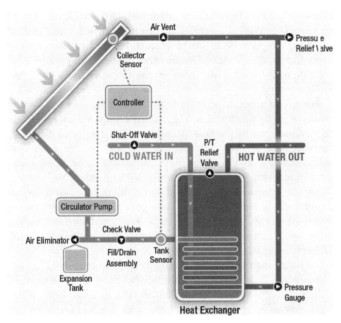

FIG. 2.29 PRESSURIZED GLYCOL ACTIVE INDIRECT SYSTEM [24]

drainback system, fluid that drains from the solar collector is collected in a drainback tank, which is located in a conditioned space. In the water flow method, water is circulated through the collector by a pump with a freeze switch. This method needs electricity to work. Therefore, it is unreliable when there is a power outage. Freeze prevention valves allow water to drain out of collectors when the weather is cold. Pressurized glycol systems are more complex than other systems. They use a mix of a non-toxic glycol (usually propylene or ethylene glycol) and water as working fluid.

2.4.1.4 Further Development Use of polymer material for constructing solar water heating systems has recently drawn the attention of some experts. A noticeable manufacturing cost and weight reduction may be maintained by using polymers instead of glass, copper, and steel in the collectors and the piping. There are some challenges to overcome in order to ensure the successful use of polymers in solar water heater systems: the thermal conductance of polymers is much lower than that of metals, which results in lower efficiency. The polymer industry is developing new high-conductance polymer materials manufactured with chemical additives. The durability of some polymers under intense sun exposure is uncertain. Additional UV-resistant coatings may help the durability issue of polymers. Non-accelerated experiments in the future may provide a better picture of the durability of polymer solar heating systems. The use of plastic heat exchangers, thin-film plastic absorbers and glazings, plastic collectors, and plastic storage vessels are high on the research agenda of some laboratories. If results from the field experiments of these systems are promising, the work will be continued toward final product development and manufacturing.

2.4.2 Solar Heating and Cooling for Air-Conditioning Systems

2.4.2.1 Space Heating Solar space heating systems are usually employed as auxiliary space heaters, especially in residential buildings. Technical limitations, weather condition requirements,

and high capital costs are main reasons for keeping them as auxiliary heaters. One of the most advantageous solutions is using a solar system that serves both water and partial space heating purposes. There are passive and active methods to employ solar heat for space heating. Active solar space heating systems take advantage of the same solar collectors as solar water heating systems. The only difference is their high demand for collector and storage capacity, which makes them complex and expensive.

2.4.2.2 Passive Systems In solar passive space heating systems, the sun's heat is transferred to the building through design features such as large sun-facing windows and thermal mass in walls and floors, to absorb heat during the day and release it during the night. Passive systems may have direct gain, indirect gain or isolated gain designs. Direct gain refers to the heat gained by a building element like tile or concrete through direct solar radiation. Indirect gain refers to the heat gained by the building through an intermediate medium such as the walls or roof. Isolated gain refers to the heat gained by an area that is not a primary living area. A sunroom attached to a house that absorbs some heat from the warmer air flowing to the primary living area can be a good example of isolated gain in passive solar system design.

2.4.2.3 Active Systems There are three different approaches in active solar space heating:

Direct Air-Space Heating. Direct air-space heating is the simplest solar space heating system. An air-based solar collector that employs air as the HTF is used in this system. Air is drawn by a fan into the air-based collector, and the heated air directly goes to the building. A simple controller activates the fan when air temperature in the collector is higher than that in the house. Direct air-space heating systems should be controlled properly to avoid overheating. In some systems, air enters a storage medium such as rock in order to stabilize the supply temperature. In other cases, this system is used as a pre-heating system for ventilation air in commercial buildings. A special type of air-based collector called a perforated plate collector is commonly used for air pre-heating purposes (Fig. 2.30).

Solar-Powered Radiators. Radiators can be fed by liquid-based solar collector systems. Collectors in this system are exactly the same as water heating collectors that comprise flat-plate and evacuated tube solar collectors. Baseboards and other radiators need high volumes of water at temperatures around 150°F (65°C). Solar systems cannot provide that, unless they use over-large collector arrays and storage systems. For this reason, a boiler must be added to the system. In this combined system, the solar part works as a pre-heating system for the boiler to reduce its oil or gas.

Solar-Powered Radiant Floors. Radiant floor systems require temperatures around 100°F (38°C), which are easily achievable by using solar collectors. Therefore, solar-powered radiant floors can operate as stand-alone heating systems. Radiant floor heating systems are compatible with solar hot water systems. These systems are expensive in comparison to solar-powered radiators. Evacuated tube collectors are suitable for cold climates, and flat-plate collectors can operate in mild climates. Modern radiant floor systems use a lightweight durable cross-linked polyethylene network called PEX. Some manufacturers present packages integrating various components.

2.4.2.4 Solar Absorption Technology Solar absorption is a space cooling system that demands high temperatures. Therefore, evacuated tube and concentrating type collectors are suitable for

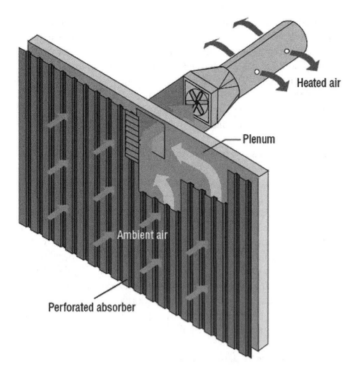

FIG. 2.30 A PERFORATED SOLAR COLLECTOR (*Source:* US Department of Energy)

this purpose. A schematic diagram of a solar absorption cooling system is illustrated in Fig. 2.31. Solar absorption systems, which are used for space cooling, are relatively low-capacity chillers. These low-capacity chillers are often single-effect water lithium bromide, driven by hot water. Solar absorption systems can usually provide 30% to 60% cooling load of the building and are usually equipped with an auxiliary heat source and a storage unit. Since solar absorption is an expensive technology, it is best to use a solar system that serves more than just the cooling needs of a house to maximize the return on the investment and not leave the system idle when cooling is not required. Water heating and/or space heating can be provided with the same equipment used for the solar cooling system. Different arrangements of solar collectors, absorption chillers, storage unit, and the load result in different system performance [35]. The auxiliary source can be

connected, either in series, or in parallel, to the solar absorption system. The series configuration allows use of solar heat even if the fluid temperature in the storage tank is not high enough to drive the cooling system. On the other hand, in the parallel configuration, the fluid in the storage tank may be heated up by the auxiliary sources.

For optimum operation of the system, using a double-hot-storage tank system as illustrated at Fig. 2.32 is highly recommended. In this system, one of the tanks (tank b in Fig. 2.32) provides heat-carrying fluid at about 10°C to 15°C lower than the other one. When the temperature of the fluid reaches the value of the lower temperature storage, the pump is activated with valves Va_1 and Va_2 open. If it exceeds a pre-specified value, the pump stops for a few minutes. If the collector temperature goes over the hotter tank set point during this period, the pump is reactivated with valves Vb_1 and Vb_2 open. If not, the operation starts again with the colder tank recovering the load of the system.

To have long continuous working periods, using a cold storage is necessary. In addition, using a cold storage reduces the required chiller capacity to meet the peak load. The load can be in series with the chiller and the cold storage or in parallel. The series arrangement is not recommended as the cold storage, and the chiller cannot maintain the load independently. The parallel arrangement is a good alternative that provides three different options for the system to supplement the load: by the cold storage, directly by the chiller, and indirectly by the chiller through cooling the storage only. The double-cold-storage tank configuration as depicted in Figure 2.33 is recommended to avoid stratification in the cold storage. Only a couple of degrees temperature gradient in the cold storage can already adversely affect the COP of the system. Chilled water is taken from tank b and returned to tank a. The tank b level is changed by the flow rate of the two pumps. At close flow rates of the pumps, the fluid level in tank b remains stable.

2.4.2.5 PV/Thermal Hybrid Systems A significant amount of research and development work on the PV/thermal hybrid technology has been carried out since 1970. Chow [37] recently published a review outlining theoretical models of the technology and their validation. The integrated design of both PV and thermal fluid channels such as those shown in Fig. 2.34 allows the harvesting of heat from a PV panel that is otherwise lost to the ambient; thus maximizing utilization of the total surface area and increasing the overall energy conversion efficiency. Despite significant development efforts, however, its commercial applications have been limited [37].

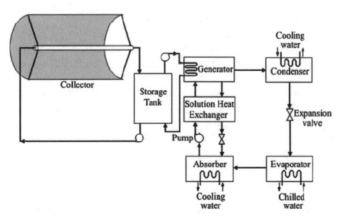

FIG. 2.31 SCHEMATIC DIAGRAM OF A SOLAR ABSORPTION COOLING SYSTEM [36]

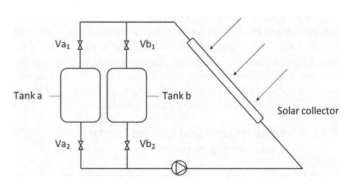

FIG. 2.32 SCHEMATIC OF A DOUBLE-HOT-STORAGE TANK SYSTEM FOR SOLAR ABSORPTION COOLING

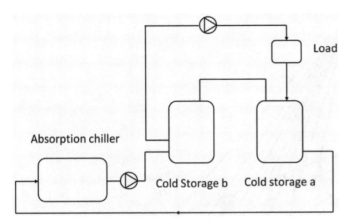

FIG. 2.33 CONFIGURATION FOR DOUBLE-COLD-STORAGE TANK SYSTEM

2.5 SOLAR PROCESS HEAT FOR MANUFACTURING APPLICATIONS

The industrial sector of US economy consumes a significant amount of energy, particularly in manufacturing processes. The primary metals industry is one of the most energy-intensive in the manufacturing sector [38]. Throughout evolution of manufacturing technology, heating through combustion is the dominant, traditionally economical mode of delivering the required energy. It naturally contributes to tremendous climate-changing gas emissions. During the past decades, investigations have shown that solar energy might provide an alternative solution to the high-temperature heat requirement. However, the main reason for the investigation still lies on economic and process requirements. For example, the replacement of electrolysis or electrothermal processes with direct reduction processes for aluminum production using high-temperature solar process heat may well be economical, especially when the costs of CO emission are included in the analysis. In particular, aluminum production by carbothermal reduction is a very high-temperature, energy-intensive process. The temperature required for the process is in the range of 2300 to 2500 K and therefore is too high for practical process heat addition from combustion sources alone. Only electric-arc furnaces or highly-concentrated solar reactors are capable of supplying process heat at these high temperatures. The unique opportunities for such industrial implementation of solar process heat may make possible a direct thermal route from the ore to metal. One of the in-depth investigations of such scenarios was presented by Murray [38]. Aluminum Company of America (Alcoa) demonstrated the following direct reduction process:

$$3SiO_2 + 9C \rightarrow 3SiC + 6CO \quad 1500\text{--}1600°C \quad (2.2)$$
$$2Al_2O_3 + 3C \rightarrow Al_4O_4C + 2CO \quad 1600\text{--}1900°C \quad (2.3)$$
$$Al_4O_4C + 3SiC \rightarrow 4Al + 3Si + 4CO \quad 1950°C \text{ to } 2200°C \quad (2.4)$$

This process makes the solar-thermal processes a good candidate to achieve the desired temperature. Figure 2.35 illustrates the solar process heat system that could be used for such an application [39].

Although the investigation focuses on the complexity of the manufacturing processes, not on the solar heat system itself, it points out the great potential of solar thermal systems to be an alternative energy for process heat applications.

The solar process heat for direct aluminum production may be an extreme case in terms of temperature range. Many industrial processes require heating at a lower temperature range such as between 80°C and 250°C. Therefore, all the following solar collector types could be the candidates for supplying the process heat [40]:

- advanced flat-plate collectors
- evacuated tube collectors
- compound parabolic concentrator (CPC) collectors
- parabolic trough collectors
- linear concentrating Fresnel collectors
- concentrating collectors with stationary reflector

2.5.1 Advanced Flat-Plate Collectors

Advanced flat-plate collectors for process heat are based on those widely used in building hot water solar systems. In order to achieve a temperature higher than 80°C, it is necessary to reduce the collector heat losses mainly on the front side of the collector, but without sacrificing too much of the optical performance at the same time. The improvements include hermetically sealed collectors with inert gas fillings, double covered flat-plate collectors, vacuum flat-plate collectors, and combinations of the above mentioned. For example, if the standard glass in the collector is replaced by AR glass, it has been experimentally confirmed that the efficiency of a 2-AR collector is more than 33% better than a standard collection, as illustrated in Figure 2.36.

2.5.2 Evacuated Tubes

Evacuated tubes have been used for years in the main markets globally [41]. Because of the increasing quantity of vacuum, currently vacuum tube collectors have been a standard component of solar thermal systems working at higher temperatures. The main advantage is that they can also be used with less collector area compared with a standard flat-plate collector. In the direct-flow vacuum collectors, the stagnation temperature can reach 300°C; however, owing to the limit of working fluid, such

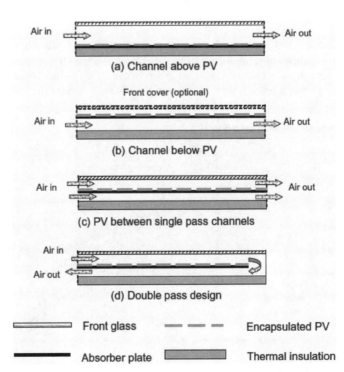

FIG. 2.34 LONGITUDINAL CROSS-SECTIONS OF COMMON PVT/A COLLECTOR DESIGNS [37]

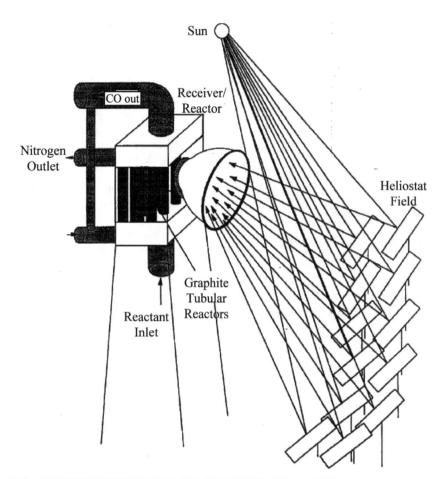

FIG. 2.35   ENGINEER'S VIEW OF A SOLAR POWER FOR ALUMINUM PROCESSING [39]

as glycol, in a region where winter conditions could be below freezing, the maximum temperature has to be kept at less than 170°C [42].

2.5.3   Compound Parabolic Concentrator Collectors

Compound parabolic concentrator collectors use a CPC to concentrate solar radiation on an absorber. Because they are not focusing (non-imaging), they are a natural candidate to bridge the gap between the lower temperature solar application field of flat-plate collectors (T < 80°C) to the much higher-temperature applications field of focusing concentrators (T > 200°C).

As shown in Fig. 2.37, an ideal CPC concentrator can concentrate isotropic radiation incident on an aperture "a" within a solid angle Θ, with the normal to this aperture without losses. The minimal possible aperture "b" on to which the radiation can be delivered is shown in Figure 2.37. A full CPC can achieve the maximum concentration C_{max} within the angle Θ, where C_{max} = 1/sin(Θ). In practice, cost reduction often yields a design of CPC with truncated configuration to eliminate the upper part of the

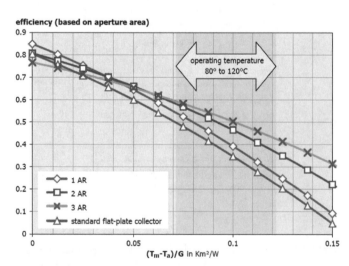

FIG. 2.36   EFFICIENCY CURVES OF A SINGLE-, DOUBLE-, OR TRIPLE-GLAZED AR COLLECTOR [40]

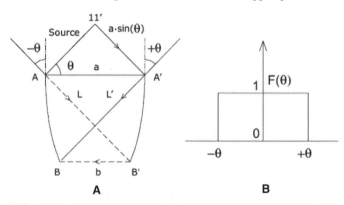

FIG. 2.37   APERTURE OF A CPC CONCENTRATOR: (A) FLAT ABSORBER PARALLEL TO THE FLAT ENTRANCE APERTURE A AND (B) ACCEPTANCE ANGLE AS A FUNCTION OF HALF ACCEPTANCE ANGLE Θ [40]

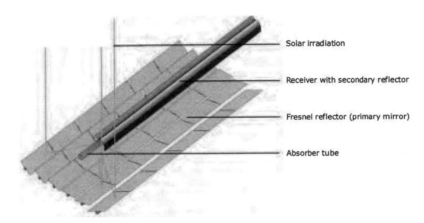

FIG. 2.38   SCHEMATIC OF A LINER CONCENTRATING FRESNEL COLLECTOR [40]

mirrors. Developments have been undertaken since 1970 to achieve a high temperature between 150°C and 200°C. The most recent one utilized evacuated tubes with CPC reflectors.

2.5.4   Parabolic Trough Concentrators

Parabolic trough concentrators as discussed in Section 2.2 can be scaled to meet the requirements of process heat in industrial applications. For example, to achieve a required temperature of 100°C to 130°C, a parabolic trough system can generate steam either in a direct (loop) or indirect (through a heat exchanger) mode. The steam can then be distributed through the process plant.

2.5.5   Linear Concentrating Fresnel Collectors

Linear concentrating Fresnel collectors are based on the simple geometric configuration shown in Fig. 2.38. It consists of an array of uniaxially tracked mirror strips to reflect sunlight onto a stationary thermal receiver. In addition to their relatively simple construction, linear concentrating Fresnel collectors have low-wind loads. The stationary receiver can easily be integrated with a high-ground application such as covered parking lots.

Similar to parabolic trough systems, linear Fresnel collectors can be scaled down from large-scale power generation applications to a thermal capacity of 50 kW to several megawatts for industrial process heat applications. Commercially, there are several demonstration projects in Germany and Italy for industrial process heat

and absorption chiller applications, respectively. The operation temperature range is 100°C to 400°C [41].

2.5.6   Concentrating Collectors With Stationary Reflector (CCStaR)

Concentrating collectors with stationary reflector (CCStaR) is also referred as a fixed-mirror solar collector [43, 44]. The fixed reflector can be linear mirrors like Fresnel geometry (Fig. 2.39) or a smooth parabolic segment. The key requirement to reach the highest possible efficiency for CCStaR is for the receiver to rotate and form an angle to the aperture.

Theoretically, it is possible to reach concentrations of 40 to 50 suns. The capability of these collection systems to concentrate solar energy is described in terms of their mean flux concentration ratio $\tilde{C}$ over a targeted area A at the focal plane, normalized with respect to the incident normal beam insolation I,

$$\tilde{C} = \frac{Q_{\text{solar}}}{I\,A}$$

where Q_{solar} is the solar power intercepted by a collector surface. $\tilde{C}$ is often expressed in units of "suns" when normalized to $I = 1$ kW/m^2. The solar flux concentration ratio typically obtained is at the level of 100, 1000, and 10,000 suns for trough, tower, and dish systems, respectively.

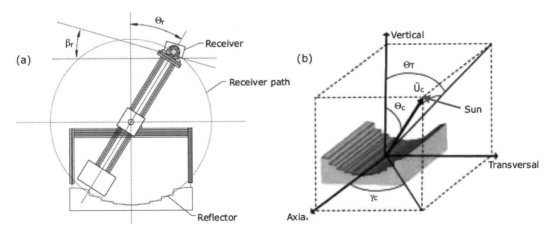

FIG. 2.39   BASIC CONFIGURATION OF A CCSTAR [44]: (A) ROTATING RECEIVER AND (B) DEFINITION OF THE SOLID ANGLE, Θ

The averaged optical efficiency in this range of concentration ratios falls to about 60%. Therefore, the current development rests on a more realistic target of 15 suns with the working temperature range between 80°C and 140°C. This technology will serve as a good candidate for the food industry for pasteurizing, boiling and sterilizing and textile industry for bleaching and dyeing. Other process applications could be found in chemical industry and absorption refrigeration systems. Again, only a handful of governmental and industrial demonstration projects have been established for CCStaR systems, and wide commercialization of this application has yet to be realized.

2.6 OTHER SOLAR ENERGY APPLICATIONS

2.6.1 Photovoltaic-Driven Desalination Systems

The electricity produced from PV systems can be used for desalination applications by running pumps in either reverse osmosis (RO) or electrodialysis (ED) desalination units. In RO applications, AC is required for the pumps; therefore, a DC/AC inverter has to be installed. In contrast, ED uses DC for the electrodes at the cell stack, and hence, it can use the energy supplied from the PV panels directly with some minor power conditioning. For sustained operation, the system requires energy storage capacity.

Reverse osmosis is a physical process that occurs under the osmotic pressure difference between salt water and pure water. In this process, a pressure greater than the osmotic pressure is applied on salt water (feedwater), and clean water passes through the synthetic membrane pores separated from the salt. A concentrated salt solution is retained for disposal (Fig. 2.40). The RO process is effective for removing total dissolved solid (TDS) concentrations of up to 50,000 parts per million (ppm), which can be applied for both brackish-water (1500 to 10,000 ppm) and seawater (33,000 to 45,000 ppm). It is currently the most widely used process for seawater desalination [45]. A pump running on electricity is required to provide the pressure difference.

Photovoltaic-powered RO is widely viewed as one of the most promising forms of renewable-energy-powered desalination, especially used in remote areas. Small-scale PV-RO has received much attention in recent years, and numerous demonstration systems have been built. Two types of PV-RO systems are available in the market: brackish-water and seawater PV-RO systems. Different membranes are used for brackish water, and much higher recovery ratios are possible, which makes energy recovery less critical [46].

The ED process uses electromotive force applied to electrodes adjacent to both sides of a membrane to separate dissolved minerals in water. The separation of minerals occurs in an individual membrane called a cell pair, which consists of an anion transfer membrane, cation transfer membrane, and two spacers. The complete assembly of the cell pairs and electrode is called a membrane stack (Fig. 2.41). Electrodialysis reversal (EDR) is a similar process, except that the cations and anions reverse to routinely alternate the current flow. In design applications, the polarity is reversed four times per hour, which creates a cleaning mechanism and decreases the scaling and fouling potential of the membrane. Electrodialysis and EDR are best used in treating brackish water with TDS of up to 5000 ppm and are not economical for higher concentrations. A DC voltage difference is required to provide the potential to the separation process occurring across the membrane.

Electrodialysis using conventional power has been used in commercial applications since 1954, more than 10 years before RO. Production of potable water seems mostly the focus of this type of systems. Because of its modular structure, ED can be available in a variety of dimensions with capacities from small (<2 m³/day) to large (145,000 m³/day). In the United States, ED systems have 31% of the total installed capacity. In Europe and the Middle East, ED has 15% and 23% of the total installed capacity, respectively. The EDR process was developed in the early 1970s. Today, the process is used in 1100 installations world-

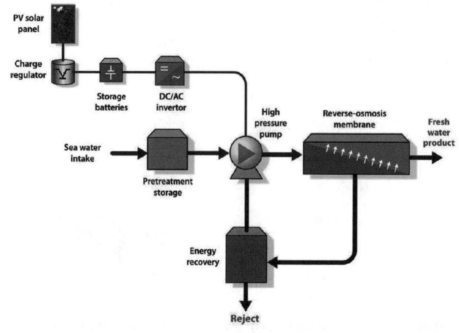

FIG. 2.40 SCHEMATIC OF A PV-RO SYSTEM [45]

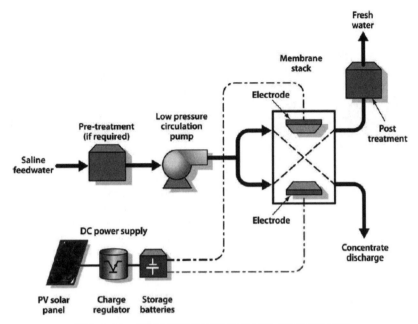

FIG. 2.41   SCHEMATIC OF A PV-ED SYSTEM [45]

wide. The installed PV-ED units are only of small capacity and are mainly used in remote areas. In general, desalination systems that incorporate renewable energy resources are still far from achieving their full potential.

2.6.2   Other PV-Driven Applications

During the last decade, the continuous decline in manufacturing costs for PV has rapidly increased the pace of development in the variety of applications. Examples include solar powered cars [47] and boats [48]. However, the commercial applications of such products still have a long way to go until achieving market-ready status. Of course, solar-powered space exploration vehicles, satellites, space stations, and instruments have been in applications for a long time, which are not part of the discussion in this chapter.

2.6.3   Solar Chemical Reactors for Hydrogen Production

There have been significant investigations in the literature during the last decade to investigate the method of hydrogen production using solar energy. Three basic pathways have been utilized for producing hydrogen with solar energy: electrochemical, photochemical, and thermochemical [49-51]. The thermochemical process is based on the use of concentrated solar radiation as the energy source of high-temperature process heat for driving an endothermic chemical transformation. Large-scale concentration of solar energy is mainly based on three optical configurations using parabolic reflectors, namely, trough, tower, and dish systems [51]. As discussed above, the capability of these collection systems to concentrate solar energy is described in terms of their mean flux concentration in units of

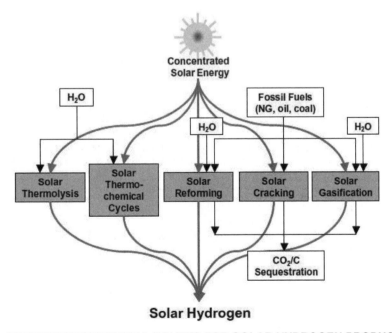

FIG. 2.42   FIVE THERMOCHEMICAL ROUTES FOR SOLAR HYDROGEN PRODUCTION [50]

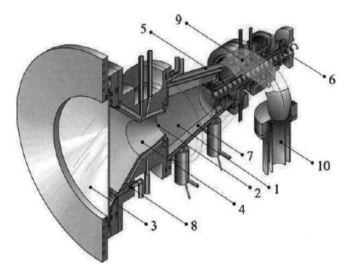

FIG. 2.43 SCHEMATIC OF THE "ROTATING-CAVITY" SOLAR REACTOR CONCEPT FOR THE THERMAL DISSOCIATION OF ZnO TO Zn AND O_2 AT 2300 K [50]

"suns." Higher concentration ratios correspond to lower heat losses from smaller areas and, consequently, higher attainable temperatures at the receiver. Solar thermochemical applications, although not as far developed as solar thermal electricity generation, employ the same solar concentrating technologies [51].

In Fig. 2.42, five thermochemical routes for solar hydrogen production are depicted. The chemical source of H_2 is water for the solar thermolysis and the solar thermochemical cycles, fossil fuels for the solar cracking, and a combination of fossil fuels and H_2O for the solar reforming and solar gasification. All of those routes involve endothermic reactions that can make use of concentrated solar radiation as the energy source of high-temperature process heat [52].

No matter which route is taken, the reactors designed for the processes basically consist of an irradiation receiver and a reaction chamber with a feeder of fuel (H_2O or fossil fuels). Figure 2.43 shows one type of solar reactor, the rotating-cavity solar reactor, which is used for thermal dissociation of ZnO to Zn and O_2 based on the following reaction:

$$\text{First step (solar): } ZnO \rightarrow Zn + \frac{1}{2} O_2 \qquad (2.5)$$

$$\text{Second step (non-solar): } Zn + H_2O \rightarrow ZnO + H_2 \qquad (2.6)$$

As shown in Fig. 2.43, the reactor consists of a rotating conical cavity-receiver (1), which contains an aperture (2) for access of concentrated solar radiation through a quartz window (3). The solar flux concentration is further enhanced by incorporating a CPC (4) in front of the aperture. Both the window mount and the CPC are water cooled and integrated into a concentric (non-rotating) conical shell (5). A screw powder feeder continuously feeds ZnO particles from the rear of the reactor (6). The centripetal acceleration therefore forces the ZnO powder to the wall, where it forms a thick layer of ZnO (7) that also insulates the inner cavity walls and reduces the thermal load. A purge gas flow enters the cavity-receiver tangentially at the front (8) and keeps the window cool and clear of particles or condensable gases. The gaseous products Zn and O_2 continuously exit via an outlet port (9) to a quench device (10) [53]. A 10-kW system has been tested for this design and reached 4000 suns.

There are other types of reactor designs such as the two-cavity type for ZnO carbothermal reduction where the inner cavity acts as the solar absorber and the outer one as the reaction chamber, and the vortex type that can be used for the combined ZnO reduction and CH_4 reforming [54]. Those solar hydrogen production systems have been assessed for the economics [55-57]. The assessments indicate that the solar thermochemical production of hydrogen can be competitive with the electrolysis of water using solar-generated electricity and, under certain conditions, might become competitive with conventional fossil-fuel based processes at current fuel prices, even before the application of credit for CO_2 mitigation and pollution avoidance. The weaknesses identified from these economic evaluations are primarily related to the uncertainties in the viable efficiencies and investment costs of the various components owing to their early stage of development and their economy of scale. It was recognized that further development and large-scale demonstration are needed.

2.7 SUMMARY

In this chapter, a number of commercial applications of solar energy are presented. By far, PV-driven applications are dominant in the present markets in terms of annual productions of cells, panels, and systems and the number of installations in buildings and other applications. The impacts of distributed PV and thermal systems are tremendous simply because of the sheer volume of applications. While solar hot water systems are relatively less expensive, the challenge to significantly bring down the cost of PV systems to compete against the conventional fossil fuel-based electricity requires both technological breakthroughs and non-technical measures.

From the most active perspective of emerging applications, large-scale solar thermal power plants have been gaining momentum in both governmental and private investments and developments. This is mainly because of the potential economical benefits of such systems among the renewable energy portfolios at the utility scale. Recent acceleration in project implementation and increase in funding opportunities from various governments indicate a global expansion in this field.

The development in solar process heat technology has been relatively slow. The motivations for further development will most likely rely on new methods or requirements in material processing or manufacturing processes, which could be achieved only by using solar energy. Also, regulations that account for the economic values of environmental impacts may speed up the adoption of solar energy systems for process heat requirements as an alternative.

As one of the special solar process heat options, systems that produce hydrogen using solar energy have been attracting much attention. So far, they have been mainly in the research and development stages. No commercial-scale demonstration project has been reported. The application potential for this technology is, however, very high.

2.8 REFERENCES

1. Stoddard, L., Abiecunas, J., and O'Connell, R. (2006). Economic, Energy, and Environmental Benefits of Concentrating Solar Power in California, Black & Veatch Overland Park, Kansas, *NREL Report* SR-550-39291.

2. DOE NREL (2006). *Parabolic Trough Solar Thermal Electric Power Plants*. Publication FS-550-40211; DOE/GO-102006-2339.

3. Sargent and Lundy LLC. (2003). Assessment of Parabolic Trough and Power Tower Solar Technology Cost and Performance Forecasts, *NREL Report* SR-550-34440.

4. Kelly, B. (2006). Nexant Parabolic Trough Solar Power Plant Systems Analysis; Task 2: Comparison of Wet and Dry Rankine Cycle Heat Rejection, *NREL Report* SR-550-40163, pp. 1-22.

5. GateCycle Program, Version 5.20, GE Enter Software, Inc. and the Electric Power Research Institute.

6. Kennedy, C., Terwilliger, K., and Warrick, A. (2007). Optical Durability of Candidate Solar Reflector Materials, the 2007 Parabolic Trough Technology Workshop, 8-9 March 2007, Golden, Colorado, *NREL Report* PO-550-41428.

7. Kennedy, C. E., and Price, H. (2006). Progress in Development of High-Temperature Solar-Selective Coating. Proceedings of the 2005 International Solar Energy Conference (ISEC2005), 6-12 August 005, Orlando, Florida, Paper No. ISEC2005-76039, pp. 749-755. American Society of Mechanical Engineers (ASME), New York, *NREL Report* CP-520-36997.

8. Macleod, H. A. (2005). *Essential Macleod: Optical Coating Design Program*, 8.10.37, Tucson, AZ: Thin Film Center.

9. Kelly, B., and Kearney, D. (2006). Parabolic Trough Solar System Piping Model: Final Report, 13 May 2002-31 December 2004. *NREL Report* SR-550-40165.

10. Price, H., Forristall, R., Wendelin, T., Lewandowski, A., Moss, T. and Gummo, C. (2006). Field Survey of Parabolic Trough Receiver Thermal Performance, the ASME International Solar Energy Conference (ISEC2006), 8-13 July 2006, Denver, Colorado, *NREL Report* CP-550-39459, pp. 1-11.

11. Kearney, D., Price, H. (2005). Advances in Parabolic Trough Solar Power Technology, Advances in Solar Energy. Vol 16, Kreith, F., Goswami, D.Y. (Eds.), ASES, Boulder, Colorado, 2005.

12. Labaton, I., Y. Harats, Hydrogen Pump, United States Patent 4,886.048, Dec. 12, 1989.

13. Price, H., M. Hale, R. Mahoney, C. Gummo, R. Fimbres, R. Cipriani, Development in High Temperature Parabolic Trough Receiver Technology, Proceeding of ISEC: Solar 2004, Portland, Oregon, July 11-14, 2004.

14. Latent Structures. (2006). Spectral Evaluation Experiments of HCEs, NREL Subcontract ACO-5-55519-01, Boulder, CO, Feb-Mar 2006.

15. DOE Solar Energy Technologies Program FY 2008 Annual Report (2009). pp. 157-158, *NREL Report* MP-840-43987.

16. Stirling Energy Systems, Inc. (2007). Solar Dish Stirling Systems Report for NREL CSP Technology Workshop, March 7, 2007 (www.nrel.gov/csp/troughnet/pdfs/2007/liden_ses_dish_stirling.pdf).

17. Juwi solar Inc. Press Release: Boulder, Colorado, June 14, 2010. http://www.juwisolar.com/wyandot-solar.

18. Denholm, P., Margolis, R. M. (2008). Impacts of Array Configuration on Land-Use Requirements for Large-Scale Photovoltaic Deployment in the United States, SOLAR 2008 — American Solar Energy Society (ASES), 3-8 May 2008, San Diego, California, pp. 1-7. *NREL Report* CP-670-42971.

19. King, R. R. (2010). Ultra-High-Efficiency Multijunction Cell and Receiver Module, Phase 1B: High Performance PV Exploring and Accelerating Ultimate Pathways, Spectrolab, Inc., Subcontract Final Report, *NREL Report* SR-520-47602.

20. Horne, S., McDonald, M., Hartsoch, N., Desy, K. (2009). Reflective Optics CPV Panels Enabling Large Scale, Reliable Generation of Solar Energy Cost Competitive with Fossil Fuels, *NREL Report* SR-520-47310.

21. King, R. R. (2010). Ultra-High-Efficiency Multijunction Cell and Receiver Module, Phase 1B: High Performance PV Exploring and Accelerating Ultimate Pathways, *NREL Report* SR-520-47602.

22. Denholm, P., Ela, E., Kirby, B., and Milligan, M. (2010). Role of Energy Storage with Renewable Electricity Generation. *NREL Report* TP-6A2-47187, pp. 1-66.

23. Solar America Initiative–In Focus: The Building Industry. (2007). 2 pp., *NREL Report* DOE/GO-102007-2389.

24. Baechler, M., Gilbride, T., Ruiz, K., Steward, H., Love, P. (2007). High-Performance Home Technologies: Solar Thermal & Photovoltaic Systems. Volume 6 Building America Best Practices Series. 159 pp., *NREL Report* No. TP-550-41085.

25. Wiles, J. (2005). *Photovoltaic Power Systems and the 2005 National Electrical Code: Suggested Practices*.

26. Deru, M., Torcellini, P. (2005). Performance Metrics Research Project — final report. 22 pp, *NREL Report* No. TP-550-38700.

27. Pless, S., Deru, M., Torcellini, P., Hayter, S. (2005). Procedure for Measuring and Reporting the Performance of Photovoltaic Systems in Buildings. 61 pp., *NREL Report* No. TP-550-38603.

28. Ullal, H. S. (2008). Overview and Challenges of Thin Film Solar Electric Technologies. 7 pp., *NREL Report* No. CP-520-43355.

29. Wu, X. (2004). High-efficiency polycrystalline CdTe thin-film solar cells, *Solar Energy*, Vol.77, n 6, p 803-814,

30. Shafarman, W., McCandless, B. (2008). Development of a Wide Bandgap Cell for Thin Film Tandem Solar Cells: Final Technical Report, 6 November 2003 — 5 January 2007. 40 pp., *NREL Report* No. SR-520-42388.

31. Coddington, M., Kroposki, B., Basso, T., Lynn, K., Sammon, D., Vaziri, M., Yohn, T. Photovoltaic Systems Interconnected onto Network Distribution Systems–Success Stories. *NREL Report* TP-550-45061.

32. IEEE Std 929-2000. (2000). *IEEE Recommended Practice for Utility Interface of Photovoltaic Systems*.

33. Kurtz, S., Lewandowski, A., Hayden, H. (2004). Recent Progress and future potential for concentrating photovoltaic power systems: preprint. 9 pp. *NREL Report* No. CP-520-36330.

34. 2008 *Buildings Energy Data Book, March 2009, Prepared for the Buildings Technologies Program, Energy Efficiency and Renewable Energy*, US Department of Energy, by D&R International, Ltd.

35. Lazzarin, R. M. (2007). Solar cooling plants: How to arrange solar collectors, absorption chillers and the load. *International Journal of Low Carbon Technologies*, **2**, pp. 376-90.

36. Mazloumi, M., Naghashzadegan, M., and Javaherdeh, K. (2008). Simulation of solar lithium bromide–water absorption cooling system with parabolic trough collector. *Energy Conversion and Management*, **49**, pp. 2820-2832.

37. Chow, T. T. (2010). A review on photovoltaic/thermal hybrid solar technology, *Applied Energy*, **87**, pp. 365-379.

38. Murray, J. (1999). Aluminum Production Using High-Temperature Solar Process Heat, *Solar Energy*, **66**, pp. 133-142.

39. Murray, J. (2006). Investigation of opportunities for high-temperature solar energy in aluminum industry. *NREL Report* SR-550-39819.

40. IEA. (2008). Process Heat Collectors, State of the Art Within Task 33/IV, edited by Weiss, W., and Rommel, M., Report by Solar Heating and Cooling Executive Committee of the International Energy Agency (IEA), published by AEE INTEC, Gleisdorf, Feldgasse 19, Austria, 2008. pp. 1-55.

41. Weiss, W., and Bergmann, I. (2008). *Solar Heat Worldwide 2006*, Gleisdorf.

42. Rommel, M. (2005). Medium Temperature Collectors for Solar Process Heat up to 250°C. *Proceedings of European Solar Thermal Energy Conference (ESTEC)*.

43. Russell, J.L, Jr. (1773). Investigation of a Central Station Solar Power Plant, *General Atomic Company Report GA-A12759*, August 31, 1973.

44. Ravinder Kumar Bansal. (1974). *Theoretical Analysis of Fixed Mirror Solar Concentrator*. A thesis presented in partial fulfillment of the requirements for a Master of Science in Engineering. Arizona State University.

45. Al-Karaghouli, A., Renne, D., and Kazmerski, L. L. (2009). Technical and economic assessment of photovoltaic-driven desalination systems, *Renewable Energy*, **35**, pp. 323-328.

46. Thomson, M., Gwillim, J., Rowbottom, A., Draisey, I., Batteryless, M. M. (2000). Photovoltaic reverse osmosis desalination system, S/P2/00305/REP, ETSU, DT1, UK.

47. Hampson, C. E., Holmes, C., Long, L. P., Piacesi, R. F.D., and Raynor, W. C. (1991). The pride of Maryland: A solar powered car for GM Sunrayce USA, *Solar Cells*, **31**, pp. 459-495.

48. Steve Barrett. (2006). Fuel cell boat at Dutch solar yacht race, *Fuel Cells Bulletin*, Issue 7, p. 9.

49. Rodat, S., Abanades, S., Sans, J.-L., and Flamant, G. (2009). Hydrogen production from solar thermal dissociation of natural gas: Development of a 10 kW solar chemical reactor prototype, *Solar Energy*, **83**, pp. 1599-1610.

50. Steinfeld, A. (2005). Solar thermochemical production of hydrogen — a review, *Solar Energy*, **78**, pp. 603-615.

51. Tyner, C.E., Kolb, G.J., Geyer, M., Romero, M. (2001). Concentrating solar power in 2001 to — an IEA/SolarPACES summary of present status and future prospects. SolarPACES 2001 (www.solarpaces.org).

52. Steinfeld, A. and Meier, A. (2004). Solar fuels and materials, *Encyclopedia of Energy. Volume 5.* C. Cleveland ed., Elsevier Inc., pp. 623-637.

53. Steinfeld, A., Spiewak, I. (1998). Economic evaluation of the solar thermal co-production of zinc and synthesis gas. *Energy Conversion Management*, **39**, pp. 1513-1518.

54. Spath, P., Amos, W.A. (2003). Using a concentrating solar reactor to produce hydrogen and carbon black via thermal decomposition of natutal gas: feasibility and economics. *J Solar Energy Eng.* **125**, pp. 159-164.

55. Spiewak, I., Tyner, C.E., Langnickel, U. (1992). Solar reforming applications study summary. In: *Proceedings of the 6th International Symposium on Solar Thermal Concentrating Technologies, Mojacar, Spain, Sept. 28-Oct. 2*, pp. 955-968.

56. Diver, R.B., Fletcher, E.A. (1985). Hydrogen and sulfur from H2S-III. The economics of a quench process. *Energy*, **10**, pp. 831-842.

57. Steinfeld, A. (2002). Solar hydrogen production via a 2-step water-splitting thermochemical cycle based on Zn/ZnO redox reactions. *International Journal of Hydrogen Energy*, **27**, 611-619.

SOLAR THERMAL POWER PLANTS: FROM ENDANGERED SPECIES TO BULK POWER PRODUCTION IN SUN BELT REGIONS

Manuel Romero and José González-Aguilar

3.1   INTRODUCTION

Solar thermal power plants, due to their capacity for large-scale generation of electricity and the possible integration of thermal storage devices and hybridization with backup fossil fuels, are meant to supply a significant part of the demand in the countries of the solar belt [1]. Nowadays, the high-temperature thermal conversion of concentrated solar energy is rapidly increasing with many commercial projects taken up in Spain, USA, and other countries such as India, China, Israel, Australia, Algeria, and Italy. This is the most promising technology to follow the pathway of wind energy in order to reach the goals for renewable energy implementation in 2020 and 2050.

Spain with 2400 MW connected to the grid in 2013 is taking the lead on current commercial developments, together with USA where a target of 4500 MW for the same year has been fixed and other relevant programs like the "Solar Mission" in India recently approved and going for 22 GW-solar, with a large fraction of thermal [2].

Solar Thermal Electricity or STE (also known as CSP or Concentrating Solar Power) is expected to impact enormously on the world's bulk power supply by the middle of the century. Only in Southern Europe, the technical potential of STE is estimated at 2,000 TWh (annual electricity production) and in Northern Africa it is immense [3]. Worldwide, the exploitation of less than 1% of the total solar thermal power plant potential would be enough to meet the recommendations of the United Nations' Intergovernmental Panel on Climate Change for long-term climate stabilization [4]. One MW of installed concentrating solar thermal power avoids 688 tons of CO_2 compared to a Combined Cycle conventional plant and 1360 tons of CO_2 compared to a conventional coal/steam plant. A 1-m^2 mirror in the primary solar field produces 400 kWh of electricity per year, avoids 12 tons of CO_2 and contributes to a 2.5-ton savings of fossil fuels during its 25-year operation lifetime. The energy payback time of concentrating solar power systems is less than 1 year, and most solar-field materials and structures can be recycled and used again for further plants.

But in terms of electric grid and quality of bulk power supply, it is the ability to provide dispatch on demand that makes STE stand out from other renewable energy technologies like PV or wind. Thermal energy storage systems store excess thermal heat collected by the solar field. Storage systems, alone or in combination with some fossil fuel backup, keep the plant running under full-load conditions. This capability of storing high-temperature thermal energy leads to economically competitive design options, since only the solar part has to be oversized. This STE plant feature is tremendously relevant, since penetration of solar energy into the bulk electricity market is possible only when substitution of intermediate-load power plants of about 4000–5000 hours/year is achieved.

The combination of energy on demand, grid stability, and high share of local content that lead to creation of local jobs provide a clear niche for STE within the renewable portfolio of technologies. Because of that, the European Commission is including STE within its Strategic Energy Technology Plan for 2020, and the US DOE is launching new R&D projects on STE. A clear indicator of the globalization of such policies is that the International Energy Agency (IEA) is sensitive to STE within low-carbon future scenarios for the year 2050. At the IEA's Energy Technology Perspectives 2010 [5], STE is considered to play a significant role among the necessary mix of energy technologies needed to halving global energy-related CO_2 emissions by 2050, and this scenario would require capacity additions of about 14 GW/year (55 new solar thermal power plants of 250 MW each).

STE systems consist of a large reflective surface collecting the incoming solar radiation and concentrating it onto a solar receiver with a small aperture area. The solar receiver is a high-absorptance radiative/convective heat exchanger that emulates as closely as possible the performance of a radiative black body. An ideal solar receiver would thus have negligible convection and conduction losses. In the case of a solar thermal power plant, the solar energy is transferred to a thermal fluid at an outlet temperature high enough to feed a heat engine or a turbine that produces electricity. The solar thermal element can be a parabolic trough field, a linear Fresnel reflector field, a central receiver system or a field of parabolic dishes, normally designed for a normal incident radiation of 800–900 W/m². Annual normal incident radiation varies from 1600 to 2800 kWh/m², allowing from 2000 to 3500 annual full-load operating hours with the solar element, depending on the available radiation at the particular site [6].

The first generation of commercial STE projects is mainly based on technological developments and concepts that matured after more than two decades of research. Nevertheless, the current solar thermal power plants are still based on conservative schemes and technological devices which do not exploit the enormous potential of concentrated solar energy. Commercial projects use technologies of parabolic troughs with low concentration in two dimensions and linear focus, or systems of central tower and heliostat fields, operating with thermal fluids at relatively modest temperatures, below 400°C [7]. The most immediate consequence of these conservative designs is the use of systems with efficiencies below 20% nominal in the conversion of direct solar radiation to electricity; the tight limitation in the use of efficient energy storage systems; the high water consumption and land extension due to the inefficiency of the integration with the power block; the lack of rational schemes for their integration in distributed generation architectures and the limitation to reach the temperatures needed for the thermochemical routes used to produce solar fuels like hydrogen.

In the first commercial projects involving parabolic trough technology, some improvements are being introduced, such as the use of large molten salt heat storage systems able to provide high degrees of dispatch for the operation of the plant, such as the plants Andasol 1 and 2 in Guadix, Spain, with 7.5 hours of nominal storage, or the use of direct steam generation loops to replace thermal oil at the solar field. Central towers are opening the field to new thermal fluids like molten salts (Gemasolar tower plant in Seville, Spain) and air, and new solar receivers like volumetric absorbers.

In parallel, a new generation of concentrating solar thermal power systems is starting deployment. This new generation is characterized by its modularity and higher conversion efficiencies. The design strategy is based on the use of highly compact heliostat fields, using mirrors and towers of small size, and looking for integration into high-temperature thermodynamic cycles. There are currently some initiatives with prototypes at an experimental stage and announcing large commercial projects like the one proposed by BrightSource with a prototype of 6 MWth at the Neguev desert in Israel, the 100 kWe prototype promoted by the AORA company and researchers of the Weizmann Institute in Israel, and the 5 MWe prototype built by the company eSolar in California.

3.2 SOLAR THERMAL POWER PLANTS: SCHEMES AND TECHNOLOGIES

3.2.1 Solar Concentration: Fundamentals for STE Plants

Solar energy has a high exergy value since it originates from processes occurring at the sun's surface at a black-body equivalent temperature of approximately 5777 K. Because of this high exergetic value, more than 93% of the energy may be theoretically converted to mechanical work by using thermodynamic cycles [8], or to Gibbs free energy of chemicals by solarized chemical reactions [9]. According to thermodynamics and Planck's equation, the conversion of solar heat to mechanical work or Gibbs free energy is limited by the Carnot efficiency, and, therefore, to achieve maximum conversion rates, the energy should be transferred to a thermal fluid or reactants at temperatures close to that of the sun.

Even though solar radiation is a source of high temperature and exergy at origin, with a high radiosity of 63 MW/m², sun-to-earth geometrical constraints lead to a dramatic dilution of flux and to irradiance available for terrestrial use; only slightly higher than 1 kW/m² with a consequent supply of low temperatures to the thermal fluid. It is, therefore, an essential requisite for solar thermal power plants and high-temperature solar chemistry applications to make use of optical concentration devices that enable the thermal conversion to be carried out at high solar fluxes and with relatively low heat losses. A simplified model of a STE plant is depicted in Fig. 3.1.

Solar reflective concentrators follow the basic principles of Snell's Law of reflection [10]. In a specular surface, like the mirrors used in solar thermal power plants, the reflection angle equals the angle of incidence. The most practical and simplest primary geometrical concentrator typically used in STE systems is the parabola. Even though there are other concentrating devices like lenses or Compound Parabolic Concentrators [11], the reflective parabolic concentrators and their analogues are the systems with the greatest potential for scaling up at a reasonable cost. Parabolas are imaging concentrators able to focus all incident paraxial rays onto a focal point located on the optical axis (see Fig. 3.2).

The paraboloid is a surface generated by rotating a parabola around its axis. The parabolic dish is a truncated portion of a pa-

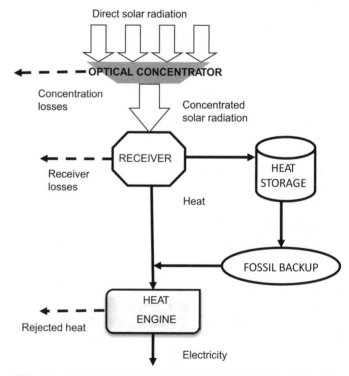

FIG. 3.1 FLOW DIAGRAM FOR A TYPICAL SOLAR THERMAL POWER PLANT

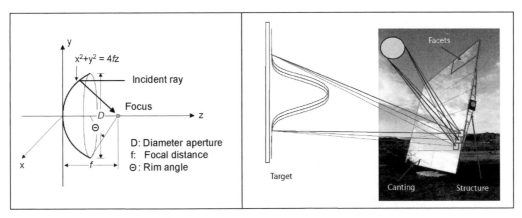

FIG. 3.2 CONFIGURATION OF IDEAL PARABOLIC CONCENTRATOR (LEFT) AND FLUX PROFILE REFLECTED BY A TYPICAL CONCENTRATOR INCLUDING SUNSHAPE (RIGHT)

raboloid. For optimum sizing of the parabolic dish and absorber geometries, the geometrical ratio between the focal distance, f, the aperture diameter of the concentrator, d, and the rim angle, Θ, must be taken into account. The ratio can be deduced from the equation describing the geometry of a truncated paraboloid, $x^2 + y^2 = 4fz$, where x and y are the coordinates on the aperture plane and z is the distance from the plane to the vertex. For small rim angles, the paraboloid equals to a sphere and in many cases spherical facets are used, therefore, in most solar concentrators, the following correlation is valid:

$$f/d = \frac{1}{4\tan(\Theta/2)} \quad (3.1)$$

For example, a paraboloid with a rim angle of 45 deg has f/d of 0.6. The ratio f/d increases as the rim angle decreases. A parabolic concentrator with a very small rim angle has very little curvature and the focal point far from the reflecting surface. Because of that, STE systems making use of cavity receivers with small apertures should use small rim angles. Conversely, those STE systems using external or tubular receivers will make use of large rim angles and short focal lengths.

The thermodynamic limit or maximum concentration ratio for an ideal solar concentrator would be set by the size of the sun. By applying the geometrical conservation of energy in a solar concentrator, the following expressions are obtained for 3D and 2D systems (for a refraction index $n = 1$, and assuming a flat aperture area):

$$C_{max,3D} = \frac{1}{\sin^2\theta_S} \le 46,200 \quad (3.2)$$

$$C_{max,2D} = \frac{1}{\sin\theta_S} \le 215 \quad (3.3)$$

Then the semi-angle subtended by the sun is $\theta_S = 4.653 \times 10^{-3}$ rad (16'), and the maximum concentration values 46,200 for 3D and 215 for 2D. For real concentrators, the maximum ratios of concentration are much lower because of microscopic and macroscopic, tracking and mechanical, sunshape, and other errors.

In a real mirror with intrinsic and constructional errors, the reflected ray distribution can be described with "cone optics" and,

as depicted in Fig. 3.2, the flux profile obtained onto the target can be described by a Gaussian shape. The reflected ray direction has an associated error that can be described with a Normal Distribution Function. The errors of a typical reflecting solar concentrator may be either microscopic (specularity) or macroscopic (waviness of the mirror and error of curvature). All the errors together end up modifying the direction of the normal compared to the reference reflecting element. However, it is necessary to discriminate between microscopic and macroscopic errors. Microscopic errors are intrinsic to the material and depend on the fabrication process, and can be measured at the lab with mirror samples. Macroscopic errors are characteristic of the concentrator and the erection process, therefore, they should be measured and quantified with the final system in operation [12]. It is relatively easy to quantify all that random errors onto the target by using the standard deviation of the reflected rays ($\sigma^2 = \Sigma \sigma_i^2$) or by fitting the solar concentrated flux to a normal distribution function with a standard deviation σ, commonly known as beam quality. Typical beam qualities of solar reflective concentrators used in STE plants are of a few milliradians and their values determine the size of the absorber or cavity. Beam quality convolves with the sunshape leading to an increment of the acceptance angle of the absorber, receiver aperture, or cavity from $2.\theta_S$ (ideal value corresponding to the size and errors of the sun) to $2.\beta$ (real value of the acceptance angle including all sources of errors).

The concentrator with more extended use is the parabolic trough it is a 2D parabolic reflector with a linear focus and single axis tracking. The concentration ratio C for this system is determined by the ratio between the aperture of the parabolic surface and the total area of the tubular receiver (Eq. 3.4).

$$C = \frac{A_C}{A_R} = \frac{L_A \cdot L}{2\pi R \cdot L} = \frac{L_A}{2\pi R} \quad (3.4)$$

Where A_C is the aperture area of the concentrator, A_R is the area of the receiver or absorber, L_A is the width of the aperture of the concentrator, L is the length of the concentrator and R is the radius of the receiver tube (Fig. 3.3). According to this, the following relations are obtained:

$$\frac{R}{r} = \sin\beta \quad (3.5)$$

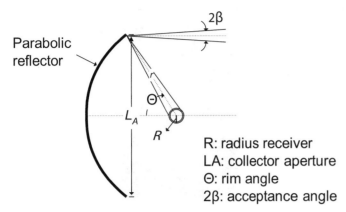

Parabolic reflector

R: radius receiver
LA: collector aperture
Θ: rim angle
2β: acceptance angle

FIG. 3.3 RIM ANGLE AND ACCEPTANCE ANGLE IN A PARABOLIC TROUGH CONCENTRATOR

$$\frac{L_A}{2r} = \sin\Theta \qquad (3.6)$$

Where r is the distance from the edge of the parabola to the center of the receiver, 2β is the acceptance angle and Θ is the rim angle of the concentrator.

Then the Concentration Ratio C can be expressed also as:

$$C = \frac{\sin\Theta}{\pi\sin\beta} = \frac{\sin\Theta}{\pi}C_{2D\,max} \qquad (3.7)$$

The concentration ratio for a parabolic trough can reach a maximum value of 67.7 (for a refractive index n = 1), considering the best case when the rim angle $\Theta = \pi/2$ rad and the acceptance angle is $\beta = \theta s = 4.7 \times 10^{-3}$ rad. In practice and because of imperfections of the reflective surface, errors on tracking and absorber tube positioning for large collector assemblies of 5.8 m of aperture and 100 m length, the typical values of C are between 20 and 30.

If we repeat the same analysis of C for a 3D parabolic reflector, the expression leading to the maximum concentration is:

$$C = \frac{\sin^2\Theta \cdot \cos^2(\Theta + \beta)}{\sin^2\beta} \qquad (3.8)$$

Figure 3.4 represents C versus the rim angle Θ obtained from Eq. 8. As it can be observed the maximum concentration ratio is reached for a rim angle of 45 deg. and, according to equation 1, for a focal ratio of $f/D = 0.6$. This optimum concentration is independent of the acceptance angle of the system. The acceptance half-angle β is the sum of the half-angle subtended by the Sun, θ_s, and the optical error associated to the concentrator system, ψ, as $\beta = \theta_s + \psi$. Typical values of C are 1000 to 3000 for parabolic dishes between 5 and 10 m diameter.

3.2.2 Solar Thermal Power Plant Technologies

Four concentrating solar power technologies are today represented at pilot and commercial-scale [13]: parabolic trough collectors (PTC), linear Fresnel reflector systems (LFR), power towers or central receiver systems (CRS), and dish/engine systems (DE). All the existing pilot plants mimic parabolic geometries with large mirror areas and work under real operating conditions (Fig. 3.5).

PTC and LFR are 2-D concentrating systems in which the incoming solar radiation is concentrated onto a focal line by one-axis tracking mirrors. They are able to concentrate the solar radiation flux 30 to 80 times, heating the thermal fluid up to 450ºC, with power conversion unit sizes of 30 to 80MW, and therefore, they are well suited for centralized power generation at dispatchable markets with a Rankine steam turbine/generator cycle. CRS optics is more complex, since the solar receiver is mounted on top of a tower and sunlight is concentrated by means of a large paraboloid that is discretized into a field of heliostats. This 3-D concentrator is, therefore, off-axis and heliostats require two-axis tracking. Concentration factors are between 200 and 1000 and unit sizes are between 10 and 200 MW, and they are, therefore, well suited for dispatchable markets and integration into advanced thermodynamic cycles. A wide variety of thermal fluids, like saturated steam, superheated steam, molten salts, atmospheric air or pressurized air, can be used, and temperatures vary between 300ºC and 1000ºC. Finally, DE systems are small modular units with autonomous generation of electricity by Stirling engines or Brayton mini-turbines located at the focal point. Dishes are parabolic 3D concentrators with high concentration ratios (1000–3000) and unit sizes of 5–25 kW. Their current market niche is in both distributed on-grid and remote/off-grid power applications.

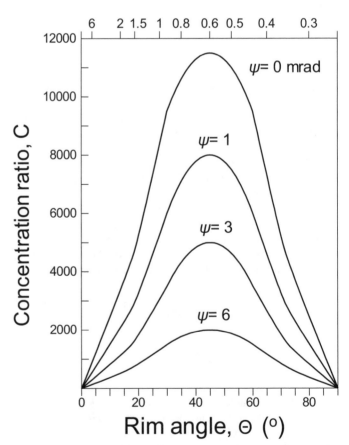

FIG. 3.4 CONCENTRATION RATIO VERSUS RIM ANGLE AT DIFFERENT BEAM QUALITIES OF THE CONCENTRATOR ($\psi = 0$ CORRESPONDS TO AN IDEAL CONCENTRATOR OF $\beta = \theta$ S)

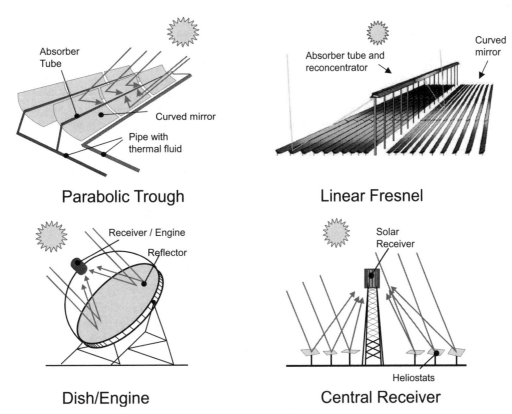

FIG. 3.5 SCHEMATIC DIAGRAMS OF THE FOUR STE SYSTEMS CURRENTLY SCALED UP TO PILOT AND DEMONSTRATION SIZES

Typical solar-to-electric conversion efficiencies and annual capacity factors are listed in Table 3.1 below [6]. The values for parabolic troughs, by far the most mature technology, have been demonstrated commercially. Those for linear Fresnel, dish and tower systems are, in general, projections based on component and early commercial projects and the assumption of mature development of current technology. With current investment costs, all STE technologies are generally thought to require a public fi-

nancial support strategy for market deployment. At present direct capital costs of STE and power generation costs are estimated to be 2–3 times those of fossil-fueled power plants, however, industry roadmaps advance 60% cost reduction before 2025 [7]. In fact governments at some countries like Spain are already accelerating the process of drastic tariff reduction with the goal of STE, PV, and wind energy becoming tariff-equivalent in less than one decade.

TABLE 3.1   CHARACTERISTICS OF SOLAR THERMAL ELECTRICITY SYSTEM (ADAPTED FROM REF. [15]). FRESNEL SYSTEMS ARE NOT INCLUDED SINCE PERFORMANCE DATA AVAILABLE ARE NOT CONCLUSIVE FOR A COMPARATIVE ASSESSMENT

	Parabolic troughs	**Central receiver**	**Dish–Stirling**
Power Unit	30–80 MW*	10–200 MW*	5–25 kW
Temperature operation	390°C	565°C	750°C
Annual capacity factor	23–50%*	20–77%*	25%
Peak efficiency	20%	23%	29.4%
Net annual efficiency	11–16%*	7–20%*	12–25%
Commercial status	Mature	Early projects	Prototypes-demonstration
Technology risk	Low	Medium	High
Thermal storage	Limited	Yes	Batteries
Hybrid schemes	Yes	Yes	Yes
Cost W installed			
$/W	3.49–2.34*	3.83–2.16*	11.00–1.14*
$/Wpeak**	3.49–1.13*	2.09–0.78*	11.00–0.96*

* Data interval for the period 2010–2025.

** Without thermal storage.

Every square meter of STE field can produce up to 1200 kWh thermal energy per year or up to 500 kWh of electricity per year. That means a cumulative savings of up to 12 tons of carbon dioxide and 2.5 tons of fossil fuel per square meter of CSP system over its 25-year lifetime [14]. After two decades of frozen or failed projects, approval in the past few years for specific financial incentives in Europe, the US, India, Australia and elsewhere, is now paving the way for launching of the first commercial ventures. Spain with 2400 MW connected to the grid in 2013 is taking the lead on current commercial developments, together with USA where a target of 4500 MW for the same year has been fixed and other relevant programs such as the "Solar Mission" in India has been recently approved for 22 GW-solar, with a large fraction being thermal [2].

3.3   PARABOLIC-TROUGHS

3.3.1   The parabolic-trough collector (PTC)

Parabolic-trough collectors are linear-focus concentrating solar devices suitable for working in the 150°C–400°C temperature range [16]; current research with new thermal fluids intends to increase the operating temperature up to 500°C [17]. The concentrated radiation heats the fluid that circulates through the receiver tube, thus transforming the solar radiation into thermal energy in the form of the sensible heat of the fluid. Figure 3.6 shows a typical PTC and its components.

Collector rotation around its axis requires a drive unit. One drive unit is usually sufficient for several parabolic-trough modules connected in series and driven together as a single collector. The type of drive unit assembly depends on the size and dimensions of the collector. Drive units composed of an electric motor and a gearbox combination are used for small collectors (aperture area < 100 m²), while powerful hydraulic drive units are required to rotate large

collectors. A drive unit placed on the central pylon is commanded by a local control unit in order to track the sun. At present, all commercial PTC designs use a single-axis sun-tracking system [18].

Thermal oils are commonly used as the working fluid in these collectors for temperatures above 200°C, because at these operating temperatures, normal water would produce high pressures inside the receiver tubes and piping. This high pressure would require stronger joints and piping, and thus raise the price of the collectors and the entire solar field. However, the use of demineralized water for high temperatures/pressures is currently under investigation in Spain at the Plataforma Solar de Almería (PSA) and the feasibility of direct steam generation at 100 bar/400°C in the receiver tubes of parabolic trough collectors has already been proven in an experimental stage [19]. For temperatures below 200°C, either a mixture of water/ethylene glycol or pressurized liquid water can be used as the working fluids because the pressure required in the liquid phase is moderate.

The oil most widely used in parabolic-trough collectors for temperatures up to 395°C is VP-1, which is a eutectic mixture of 73.5% diphenyl oxide/26.5% diphenyl. The main problem with this oil is its high-solidification temperature (12°C), which requires an auxiliary heating system when oil lines run the risk of cooling below this temperature. Since the boiling temperature at 1013mbar is 257°C, the oil circuit must be pressurized with nitrogen, argon, or some other inert gas when oil is heated above this temperature. Blanketing of the entire oil circuit with an oxygen-free gas is a must when working at high temperatures because high-pressure mists can form an explosive mixture with air. Though there are other suitable thermal oils for slightly higher working temperatures with lower solidification temperatures, they are too expensive for large solar plants.

The typical PTC receiver tube is composed of an inner steel pipe surrounded by a glass tube to reduce convective heat losses from the hot steel pipe [6]. The steel pipe which has a selective high-absorption (>90%), low-emission (<30% in the infrared)

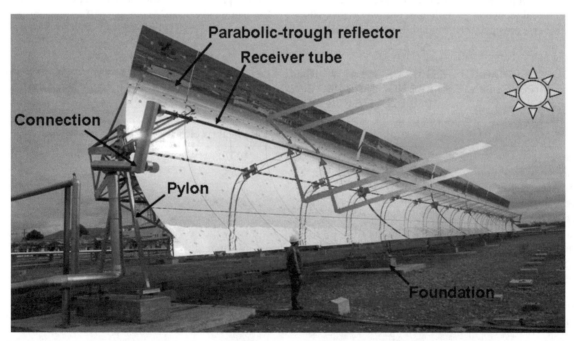

FIG. 3.6   TYPICAL PARABOLIC-TROUGH COLLECTOR (PTC TEST FACILITY AT THE PLATAFORMA SOLAR DE ALMERÍA, SPAIN) (*Source*: Eduardo Zarza, Plataforma Solar de Almería)

coating that reduces radiation thermal losses. Receiver tubes with glass vacuum tubes and glass pipes with an anti-reflective coating achieve higher PTC thermal efficiency and better annual performance, especially at higher operating temperatures. Receiver tubes with no vacuum are usually for working temperatures below 250°C because thermal losses are not so critical at these temperatures. Due to manufacturing constraints, the maximum length of a single receiver pipe is less than 6 m, so that the complete receiver tube of a PTC is composed of a number of single receiver pipes welded in series up to the total length of the PTC. The total length of a PTC is usually within 25–150 m.

Figure 3.7 shows a typical PTC vacuum receiver pipe. The outer glass tube is attached to the steel pipe by means of flexible metal differential expansion joints which compensate for the different thermal expansion of glass and steel when the receiver tube is working at nominal temperature. At present there are only two manufacturers of PTC vacuum absorber tubes: the German company Schott and the Israeli company SOLEL. The flexible expansion joint used by these two manufacturers is shown in Fig. 3.7. The glass-to-metal-welding used to connect the glass tube and the expansion joint is a weak point in the receiver tube and has to be protected from the concentrated solar radiation to avoid high thermal and mechanical stress that could damage the welding. An aluminum shield is usually placed over the joint to protect the welds.

As seen in Fig. 3.7, several chemical getters are placed in the gap between the steel receiver pipe and the glass cover to absorb gas molecules from the fluid that get through the steel pipe wall to the annulus.

Parabolic-trough collector reflectors have a high specular reflectance (>88%) to reflect as much solar radiation as possible. Solar reflectors commonly used in PTC are made of back-silvered glass mirrors, since their durability and solar spectral reflectance are better than the polished aluminum and metallized acrylic mirrors also available in the market. Solar spectral reflectance is typically 0.93 for silvered glass mirrors and 0.87 for polished aluminum. Low-iron glass is used for the silvered glass reflectors and the glass to improve solar transmission.

Two PTC designs specially conceived for large solar thermal power plants are the LS-3 (owned by the Israeli company SOLEL Solar Systems) and EuroTrough (owned by the EuroTrough Consortium), both of which have a total length of 100 m and a width of 5.76 m, with back-silvered thick-glass mirrors and vacuum absorber pipes. American Solargenix design has an aluminum structure. However, other collector designs are recently becoming commercially available in the short-to-medium term like the ones developed by the companies Solargenix, Albiasa, or Sener [17]. The main constraint when developing the mechanical design of a PTC is the maximum torsion at the collector ends, because high torsion would lead to a smaller intercept factor and lower optical efficiency.

In a typical parabolic-trough collector field, several collectors connected in series make a row, and a number of rows are connected in parallel to achieve the required nominal thermal power output at design point. The number of collectors connected in series in every row depends on the temperature increase to be achieved between the row inlet and outlet. In every row of collectors, the receiver tubes in adjacent parabolic-trough collectors have to be connected by flexible joints to allow independent rotation of both collectors as they track the sun during the day. These flexible connections are also necessary to allow the linear thermal expansion of the receiver tubes when their temperature increases from ambient to nominal temperature during system start-up. Two main types of flexible connections are available: flexible hoses and ball joints.

Flexible connections are also needed to connect the row to the main field pipe header inlet and outlet.

3.3.2 Electricity Generation with Parabolic-Trough Collectors

The suitable parabolic-trough collector temperature range and their good solar-to-thermal efficiency up to 400°C make it possible to integrate a parabolic-trough solar field in a Rankine water/steam

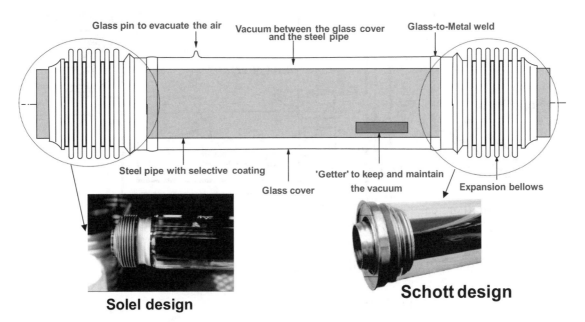

FIG. 3.7 EVACUATED RECEIVER TUBE DESIGN WITH GLASS-TO-METAL WELDS FOR PARABOLIC TROUGH COLLECTORS, AND DETAILS OF COMMERCIAL TUBES SUPPLIED BY THE COMPANIES SOLEL AND SCHOTT

power cycle to produce electricity. The simplified scheme of a typical solar thermal power plant using parabolic-troughs integrated in a Rankine cycle is shown in Fig. 3.8. The technology commercially available at present for parabolic-trough power plants is the HTF (Heat Transfer Fluid) technology, which uses oil as the heat carrier between the solar field and the power block. The solar field collects the solar energy available in the form of direct solar radiation and converts it into thermal energy as the temperature of the oil circulating through the receiver tubes of the collector's increases. Once heated in the solar field, the oil goes to the steam generator, which is an oil-water heat exchanger where the oil transfers its thermal energy to the water that is used to generate the superheated steam required by the turbine. The steam generator is, therefore, the interface between the solar system (solar field + oil circuit) and the power conversion system (PCS) itself. Normally, the steam generator used in these solar power plants consists of three stages: preheater, evaporator, and superheater.

Though parabolic-trough power plants usually have an auxiliary gas-fired heater to produce electricity when direct solar radiation is not available, the amount of electricity produced with natural gas is always limited to a reasonable level. This limit changes from one country to another: 25% in California (USA), 15% in Spain, and no limit in Algeria. Typical solar-to-electric efficiencies of a large solar thermal power plant (>30 MWe) with parabolic-trough collectors is between 15% and 22%, with an average value of about 17%. The yearly average efficiency of the solar field is about 50%.

The maturity of PTC systems is confirmed by the SEGS plants. The plants SEGS (Solar Electricity Generating Systems) II to IX, which use thermal oil as the working fluid (HTF technology), were designed and implemented by the LUZ International Limited company from 1985 to 1990. All the SEGS plants are located in the Mojave Desert, Northwest of Los Angeles (California, USA). With their daily operation and over 2.2 million m^2 of parabolic-trough collectors, SEGS plants are this technology's best example of commercial maturity and reliability. Their plant availability is over 98% and their solar-to-electric annual efficiency is in the range of 14–18%, with a peak efficiency of 22% [16].

The electricity produced by the SEGS plants is sold to the local utility, Southern California Edison, under individual 30-year contracts for every plant. To optimize the profitability of these plants, it is essential to produce the maximum possible energy during peak-demand hours, when the electricity price is the highest. The gas boilers can be operated for this, either to supplement the solar field or alone. Nevertheless, the total yearly electricity production using natural gas is limited by the Federal Commission for Energy Regulation in the United States to 25% of the overall yearly production.

Peak-demand hours are when electricity consumption is more and, therefore, the tariff is the highest. Off-peak and super off-peak hours are when electricity consumption is low, and the electricity price is, therefore, also lower. At present, 16% of the SEGS plants' annual net production is generated during summer peak-demand hours, and the revenues from this are in the order of 55% of the annual total. These figures show how important electricity generated during peak-demand hours is for the profitability of these plants. Thanks to the continuous improvements in the SEGS plants, the total SEGS I cost of $0.22/kWh$_e$ for electricity produced was reduced to $0.16/kWh$_e$ in the SEGS II and down to $0.09/kWh$_e$ in SEGS IX [20].

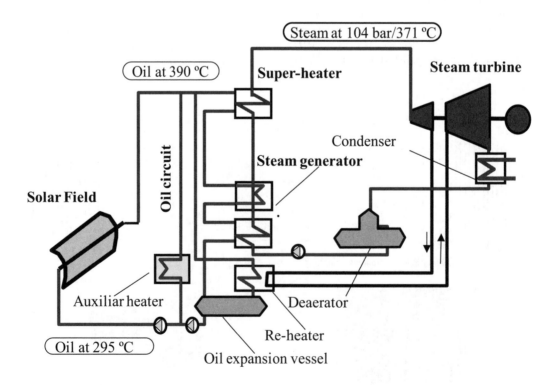

SEGS Plant

FIG. 3.8 TYPICAL FLOW DIAGRAM OF A SEGS PLANT WITH OIL AS HEAT TRANSFER FLUID IN THE SOLAR FIELD (*Source*: Eduardo Zarza, Plataforma Solar de Almería)

TABLE 3.2   MAIN CHARACTERISTICS OF NEVADA SOLAR ONE PLANT

Solar field		Power block	
Solar collector assemblies	760	Turbine generator gross output	75 MWe
Aperture area (m)	5	Net output to utility	72 MWe
Aperture area (m^2)	470	Solar steam inlet pressure	86.1 bar
Length (m)	100	Solar steam reheat pressure	19.5 bar
Concentration ratio	71	Solar steam inlet temperature	371°C
Optical efficiency	0.77		
Number of mirror segments	182,400		
Number of receiver tubes	18240		
Field aperture (m^2)	357,200		
Site area (km^2)	1.62		
Field inlet temperature (°C)	300		
Field outlet temperature (°C)	390		

With the revival of commercial STE projects since 2006 in US and Spain, a new generation of SEGS-type plants has come to the arena. This is the case of the Nevada Solar One project of 75 MWe in US, the Ibersol project in Puertollano, Spain or the Shams One 100 MW plant in Abu Dhabi [21]. Nevada Solar One started grid-connected operation in June 2007 and it is considered a milestone in the opening of the second market deployment of PTC technology in the world after SEGS experience. Since then, more than 40 PTC plants (about 50 MWe each) are constructed and started operation between the period 2007 and 2013 and more than 2 GW on track in the US [2]. Main characteristics of NSO Plant are shown in Table 3.2.

In spite of their environmental benefits, there are still some obstacles to the commercial use of this technology. The main barriers at present are the high investment cost (2500–4000 $/kW, depending on plant size and thermal storage capacity) and the minimum size of the power block required for high thermodynamic efficiency. However, these barriers are shared by all the solar thermal power technologies currently available [22].

One of the strategies to mitigate risk perception and increment the capacity factor of the power block is to integrate a parabolic-trough solar field in the bottoming cycle of a combined-cycle gas-fired power plant. This configuration is called the Integrated Solar Combined Cycle System (ISCCS). Though the contribution of the solar system to the overall plant power output is small (10%–15% approximately) in the ISCCS configuration, it seems to be a good approach to market penetration in some developing countries, which is why the Government of Algeria has promoted an ISCCS plant and the World Bank, through its Global Environment Facility (GEF), is promoting ISCCS plants in Morocco and Egypt [23]. Figure 3.9 shows the schematic diagram of a typical ISCCS plant.

3.3.3   Thermal Energy Storage and New Thermal Fluids

In many countries, the market penetration of STE systems is based on feed-in-tariffs or green certificates linked to significant restrictions or regulations regarding the use of hybrid concepts like ISCCS schemes. Because of that the use of thermal energy storage systems with an oversized solar field is pursued to optimize economics and dispatchability of PTC plants.

In this case, the solar field has to be oversized so that it can simultaneously feed the power conversion system and charge the storage system during sunlight hours. Thermal energy from the storage system is then used to keep the steam turbine running and producing electricity after sunset or during cloudy periods. Yearly hours of operation can be significantly increased and plant amortization is thus enhanced when a storage system is implemented. However, the required total investment cost is also higher.

For temperatures of up to 300°C, thermal mineral oil can be stored at ambient pressure, and is the most economical and practical solution. Synthetic and silicone oils, available for up to 410°C, have to be pressurized and are expensive. Then, molten salts can be used between 220 and 560°C at ambient pressure, but require parasitic energy to keep them liquid.

Two pioneer projects introducing thermal storage are the plants Andasol-I and Andasol-II. This PTC plants installed in Guadix, Spain, have a nominal power of 50 MWe each and an oversized solar field (510,120 m^2 mirrors surface area) with an integrated 1010 MWth molten-salt thermal storage system to extend the plant's full-load operation 7.7 hours beyond daylight hours leading to a capacity factor of 41%. Basically, the system is composed of [24]:

- 2 tanks (cold and hot) containing 28,500 tons of nitrate molten salts at 2 different temperature levels (292 and 386°C). Molten salt composition: 60% $NaNO_3$ + 40% KNO_3.
- 6 heat exchangers (series connected) located between the tanks, which allow storage charging and discharging from and to the Heat Transfer Fluid (HTF) of the plant, in this case, Dowtherm A.
- 3 cantilever pumps in each tank to transfer the molten salt from one tank to the other through the 6 heat exchangers (counterflow). There is a spare pump unit in the cold tank.
- 6 electrical heating devices in the bottom of each tank to maintain salt temperature above 292°C during long maintenance periods (salts crystallize at 238°C and solidify at 221°C).
- Electrical heat tracing in piping and heat exchangers parts to keep temperatures at adequate levels.
- A natural convection system through tanks foundations in order to decrease concrete temperature.
- A drainage tank to ease draining of piping and heat exchangers.

Figure 3.10 shows Andasol plant flow diagram and Fig. 3.11 an aerial view of the two tanks, pumps and salt circuit. As it can be

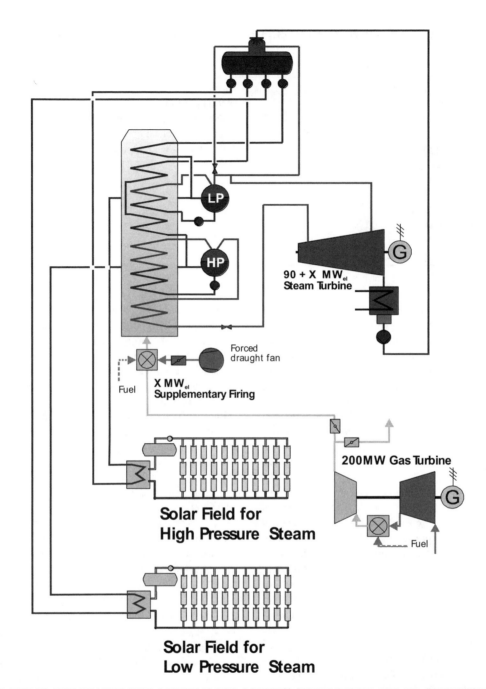

FIG. 3.9 HYBRID STE PLANT MAKING USE OF THE ISCCS SCHEME (INTEGRATED SOLAR COMBINED CYCLE SYSTEM). THE SOLAR FIELD IS INTEGRATED IN THE BOTTOMING CYCLE OF A COMBINED-CYCLE GAS-FIRED POWER PLANT (*Source*: SolarPACES)

observed, the introduction of a third circuit with molten salts adds more complexity with a number of new heat exchangers and heat tracing to avoid the salts thawing. It is still early to know the final result in terms of energy management and operational robustness. Early operational data demonstrate the capacity of the plant to supply electricity several hours after sunset.

The company ACS-Cobra is constructing three more 50-MWe plants similar to the Andasol projects, Extresol 1 (in collaboration with the company Sener) and Extresol 2 in Torre de Miguel Sesmero (Badajoz, Spain), and Manchasol 1 in Alcázar de San Juan (Toledo, Spain). This company is also planning 2 more plants, Extresol 3 in Badajoz, and Manchasol 2 in Ciudad Real [17]. Other

countries such as Italy are planning further steps with the use of molten salts both for thermal storage and solar field, like the 5-MW demonstration project Archimede [25].

Even though molten salts are nowadays the preferred option for demonstration and first commercial projects with storage in the US and Spain, there are other options under development and assessment like the use of concrete or other solid bed materials [26], and the use of Phase change media (PCM) [27]. PCM provide a number of desirable features, e.g., high volumetric storage capacities and heat availability at constant temperatures. Energy storage systems using the latent heat released on melting eutectic salts or metals have often been proposed, but never carried out on a large

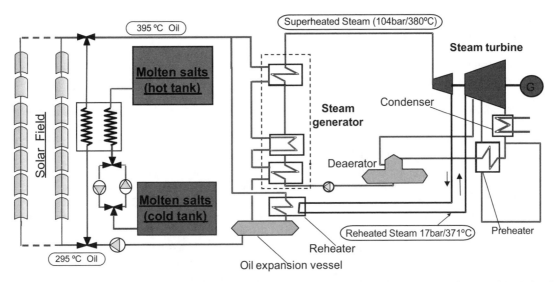

FIG. 3.10 ANDASOL PLANT SCHEME INTRODUCING A MOLTEN-SALT CIRCUIT WITH TWO STORAGE TANKS TO INCRE-MENT CAPACITY FACTOR (*Source*: Eduardo Zarza, Plataforma Solar de Almería)

scale due to difficult and expensive internal heat exchange and cycling problems. Heat exchange between the heat transfer fluid (HTF) and the storage medium is seriously affected when the storage medium solidifies. Encapsulation of PCMs has been proposed to improve this. Combining the advantages of direct-contact heat exchange and latent heat, hybrid salt-ceramic phase change storage media have recently been proposed. The salt is retained within the submicron pores of a solid ceramic matrix such as magnesium oxide by surface tension and capillary force. Heat storage is then

accomplished in two modes: by the latent heat of the salt and also the sensible heat of the salt and the ceramic matrix.

PCM are extremely suitable for direct steam generation PTC plants. If the superheated steam required to feed the steam turbine in the power block were produced directly in the receiver tubes of the parabolic-trough collectors, the oil would be no longer necessary, and temperature limitation and environmental risks associated with the oil would be avoided. Simplification and cost reduction of overall plant configuration is then evident since only one fluid is

FIG. 3.11 AERIAL VIEW OF POWER ISLAND AND DETAIL OF MOLTEN-SALT STORAGE TANKS AND HEAT EXCHANGERS OF ANDASOL PLANT. THIS 50 MW PLANT STORES THERMAL ENERGY EXCESS IN 28,500 TONNES OF NITRATE MOLTEN SALTS ABLE TO PROVIDE UP TO 7.7 EQUIVALENT HOURS OF OPERATION AT NOMINAL CAPACITY (ANDASOL-I PLANT, GUADIX, SPAIN) (*Source*: ACS/Cobra Energía (Spain))

used. This effect combined with the increment of efficiency after removing intermediate heat exchanger might lead to a reduction of 15% of the cost of the electricity produced. The disadvantages of this concept originate from the thermo-hydraulic problems associated with the two-phase flow existing in the evaporating section of the solar field. Nevertheless, experiments performed at in a 2-MWth loop at the Plataforma Solar de Almería (PSA) in Spain have proven the technical feasibility of direct steam generation with horizontal parabolic-trough collectors at 100 bar/400°C within the framework of two European projects, Direct Solar Steam (DISS) and Integration of DSG Technology for Electricity Production (INDITEP) [28]. Two pre-commercial projects, Puertollano GDV and Real-DISS, are under development in Spain to demonstrate the technical feasibility of direct steam generation combined with a power block [17].

3.4 LINEAR-FRESNEL REFLECTORS

3.4.1 Historical Evolution of Linear-Fresnel Reflector Systems

Conceptually, Linear-Fresnel Reflectors (LFR) are optical analogues of parabolic troughs. They are 2-D concentrating reflectors with linear focus, where the parabolic reflective surface is obtained by an array of linear mirror strips which independently move and collectively focus on absorber lines suspended from elevated towers [13, 29]. They are fix focus reflectors where the absorber is static. Reflective segments are close to the ground and can be assembled in a compact way up to 1 ha/MW. The objective is to reproduce the performance of parabolic troughs though with lower costs. However, optical quality and thermal efficiency is lower because of a higher influence of the incidence angle and the cosine factor, and therefore, the temperature obtained at the working fluid is also lower (150–350°C). Because of that, LFR are mainly oriented to produce saturated steam via direct in-tube steam generation, and application into ISCCS or in regenerative Rankine cycles, though current R&D is aiming at higher temperatures above 400°C.

In contrast, LFRs typically use

- lower-cost non-vacuum thermal absorbers where the stagnant air cavity provides significant thermal insulation
- light reflector support structures close to the ground
- low-cost flat float glass reflector
- low-cost manual cleaning, because the reflectors are at human height.

The LFRs also have much better ground utilization, typically using 60–70% of the ground area compared to about 33% for a trough system, and lower O&M costs due to more accessible reflectors.

After some pioneering experiences [29], the first serious development began on the CLFR Australian design at the University of Sydney in 1993, in which a single field of reflectors used multiple linear receivers and reflectors which change their focal point from one receiver to another during the day in order to minimize shading in the dense reflector field. These are called the Compact Linear Fresnel Reflector systems, or CLFR systems and cover about 71% of the ground compared with 33% for trough systems [30]. In 1994, a company called Solel Europe entered into a commercial

in-confidence agreement with Sydney University regarding CLFR technology. This company was later reconstituted as Solarmundo and it built a 2400 m² LFR prototype collector field at Liege in Belgium 2000, but test results were not reported. Later the company moved to Germany and was renamed the Solar Power Group. SPG signed an exclusivity cooperation agreement with DSD Industrieanlagen GmbH a company of the MAN/Ferrostaal group. A 800kW Linear Fresnel pilot operating at 450°C has already been tested in Plataforma Solar de Almería with components developed by Solar Power Group, and the technology is expected to be commercially available by 2012 [7, 31].

Back in Australia, in early 2002, a new company, Solar Heat and Power Pty Ltd (SHP), was formed after acquiring the patents covering the Australian CLFR work. SHP immediately made extensive changes to the engineering design of the reflectors to lower cost and has become the first to commercialize LFR technology. SHP initiated in 2003 for Macquarie Generation, Australia's largest generator, a demonstration project of 103 MWth (approximately 39 MWe) plant with the aim of supplying preheat to the coal fired Liddell power station. Phase 1 of the project, completed in 2004, resulted in a 1350 m² segment not connected to the coal fired plant, and was used to trial initial performance, and it first produced steam at 290°C in July 2004. The expansion to 9 MWth was completed by 2008. Activities of the company moved to the US and were continued by Ausra, Palo Alto. Ausra established a factory of components, tubular absorbers and mirrors in Las Vegas and built the Kimberlina 5-MW demonstration plant in Bakersfield at the end of 2008. In 2010, Aura has been purchased by AREVA that is committed to the commercial deployment of this technology.

The third technology player after SPG/MAN and AREVA is the German company NOVATEC Biosol [7]. The technology of NOVATEC is based upon its collector Nova-1 aimed to produce saturated steam at 270°C. They have developed a serial production factory for pre-fabricated components and a 1.4 MW small commercial plant in Puerto Errado, Murcia, Spain. This plant is grid connected since March 2009 [32]. NOVATEC is promoting 50 MW plants mixing PTC and LFR fields, where LFR provides pre-heating and evaporation and PTC field takes over superheating. The company claims that this hybridization results in 22% less land use and higher profitability.

3.4.2 Future Technology Development and Performance Trends

Even though some solid commercial programmes are underway on LFR, still it is early to have consolidated performance data with respect to electricity production. The final optimization would integrate components development to increment temperature of operation and possible hybridization with other STE systems like parabolic troughs.

There are many possible types of receiver, including evacuated tube and PV modules, but the most cost-effective system seems to be an inverted cavity receiver. In the case of SHP technology, the absorber is a simple parallel array of steam pipes at the top of a linear cavity, with no additional redirection of the incoming light from the heliostats to minimise optical losses and the use of hot reflectors. In the case of SPG technology, the absorber is a single tube surmounted by a hot non-imaging reflector made of glass, which must be carefully manufactured to avoid thermal stress un-

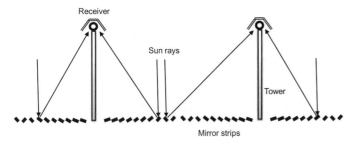

FIG. 3.12 SCHEME OF COMPACT LFR SYSTEM WITH A MULTITOWER ARRAY AND DYNAMIC AIMING STRATEGY OF MIRROR STRIPS

der heating, and exhibit some optical loss. Both systems can produce saturated steam or pressurized water. At present, AREVA and NOVATEC are looking for new absorbers able to work at temperatures above 450°C. By 2015, according to developers, Linear Fresnel can be expected to be operating with superheated steam at 500°C yielding an efficiency improvement of up to 18.1% relative to current saturated steam operation at 270°C [7].

For reflectors, automation is a key issue that has been demonstrated by NOVATEC. Additional effort should be given to the optimized demonstration of multi-tower arrays to maximize ground coverage ratios (Fig. 3.12).

However, it is the lack of reliable information regarding annual performance and daily evolution of steam production that should be targeted as a first priority. Still, some concerns remain regarding the ability to control steam production, because of the pronounced effect of cosine factor in this kind of plants. This

dynamic performance would also affect the potential integration with other STE systems such as PTC or CRS. Until now most comparative assessments vis-à-vis parabolic troughs are not economically conclusive, revealing the need to use much larger fields to compensate lower efficiencies [33]. In order to achieve break even costs for electricity with current LFR technology, the cost target for the Fresnel solar power plants need to be about 55% of the specific costs of parabolic trough systems [33].

3.5 CENTRAL RECEIVER SYSTEMS (CRS)

In power towers or central receiver systems, incident sunrays are tracked by large mirrored collectors (heliostats), which concentrate the energy flux onto radiative/convective heat exchangers called solar receivers, where energy is transferred to a thermal fluid, mounted on top of a tower (Fig. 3.13). Plant sizes of 10 to 200 MW are chosen because of economy of scale, even though advanced integration schemes are claiming the economics of smaller units as well [34]. The high solar flux incident on the receiver (averaging between 300 and 1000 k W/m^2) enables operation at relatively high temperatures of up to 1000°C and integration of thermal energy into more efficient cycles in a step-by-step approach. CRS can be easily integrated in fossil plants for hybrid operation in a wide variety of options and has the potential for generating electricity with high annual capacity factors through the use of thermal storage. With storage, CRS plants are able to operate over 4500 hours per year at nominal power [35].

The heliostat field and solar receiver systems distinguish solar thermal tower power plants from other STE plants, and are,

FIG. 3.13 AERIAL VIEW OF THE FIRST COMMERCIAL SOLAR TOWERS IN THE WORLD. PS20 WITH 20 MW AT THE FRONT AND THE PS10 WITH 11 MW AT UPPER LEFT. HELIOSTAT FIELD LAYOUT IS FORMED BY HUNDREDS OF TRACKING MIRRORS FOCUSING CONCENTRATED LIGHT ONTO THE RECEIVER APERTURE. PS20 AND PS10 HAVE BEEN DEVELOPED AND THE TECHNOLOGY IS OWNED BY ABENGOA SOLAR S.A. AND ARE LOCATED IN SANLUCAR LA MAYOR, SEVILLE, SPAIN (*Source*: Abengoa Solar, S.A.)

therefore, given more attention below. In particular, the heliostat field is the single factor with the most impact on plant investment. Collector field and power block together represent about 72% of the typical solar-only plant (without fossil backup) investment, of which heliostats represent 60% of the solar share. Even though the solar receiver impacts the capital investment much less (about 14%), it can be considered the most critical subsystem in terms of performance, since it centralizes the entire energy flux exchange. The largest heliostat investment is the drive mechanism and reflecting surface, which alone are almost 70% of the total.

3.5.1 Technology of Heliostats and Solar Receivers

The collector field consists of a large number of mirrors, called heliostats, with two-axis tracking and a local control system to continuously focus direct solar radiation onto the receiver aperture area. During cloud passages and transients the control system must defocus the field and react to prevent damage to the receiver and tower structure. Heliostats fields are characterized by their off-axis optics. Since the solar receiver is located in a fixed position, the entire collector field must track the sun in such a way that each and every heliostat individually places its surface normal to the bisection of the angle subtended by sun and the solar receiver. Each and every heliostat has its own elevation and azimuthal angle updated about every 4 sec.

Heliostat field optical efficiency includes the cosine effect, shadowing, blocking, mirror reflectivity, atmospheric attenuation and receiver spillage [36]. Because of the large area of land required, complex optimization algorithms are used to optimize the annual energy produced by unit of land, and heliostats must be packed as close as possible so the receiver can be small and concentration high. Since the reflective surface of the heliostat is not normal to the incident rays, its effective area is reduced by the cosine of the angle of incidence; the annual average cosine varies from about 0.9 at two tower heights north of the tower to about 0.7 at two tower heights south of the tower. Of course, annual average cosine is highly dependent on site latitude. Consequently, in places close to the Equator a surround field would be the best option to make best use of the land and reduce the tower height. North fields improve performance as latitude increases (South fields in the Southern hemisphere), in which case, all the heliostats are arranged on the North side of the tower.

The combination of all the above-mentioned factors influencing the performance of the heliostat field should be optimized to determine an efficient layout. There are many optimization approaches to establish the radial and azimuthal spacing of heliostats and rows [37]. One of the most classic, effective and widespread procedures is the "radial staggered" pattern, originally proposed by the University of Houston in the 1970s [38]. Integral optimization of the heliostat field is decided by a tradeoff between cost and performance parameters. Heliostats, land and cabling network must be correlated with costs. Cost and performance also often have reverse trends, so that when heliostats are packed closer together, blocking and shadowing penalties increase, but related costs for land and wiring decrease. A classical code in use since the 1980s for optimization of central receiver subsystems is DELSOL3 [39].

Mature low-cost heliostats consist of a reflecting surface, a support structure, a two-axis tracking mechanism, pedestal, foundation and control system (Fig. 3.14). The development of heliostats shows a clear trend from the early first generation prototypes, with

FIG. 3.14 LATERAL VIEW OF A TYPICAL GLASS-METAL HELIOSTAT WITH T-SHAPE SUPPORTING STRUCTURE FORMED BY PEDESTAL, TORQUE TUBE AND TRUSSES. REFLECTIVE SURFACE IS MADE OF FACETS WITH A METALLIC FRAME AND GLASS MIRROR. LOCAL CONTROL BOX IS LOCATED AT THE BOTTOM OF THE PEDESTAL (*Source*: Abengoa Solar, S.A.)

a heavy, rigid structure, second-surface mirrors and reflecting surfaces of around 40 m^2 [40] , to designs with large 100–120 m^2 reflecting surfaces, lighter structures and lower-cost materials [41]. Since the first-generation units, heliostats have demonstrated beam qualities below 2.5 mrad (standard deviation of reflected rays including all heliostat errors but not including intrinsic to the solar disk) that are good enough for practical applications in solar towers, so the main focus of development is directed at cost reduction.

In Spain, some developments worthy of mention are the 105-m^2 GM-100 [42] and more recently, the 90-m^2 and 120-m^2 Sanlúcar heliostats finally adopted for the first commercial tower power plant PS10 promoted by the company Abengoa Solar [43]. Recently, the company SENER has developed a similar 115-m^2 heliostat for its Gemasolar plant. Estimated production costs of large area glass/metal heliostats for sustainable market scenarios are around \$130–200/m^2. Large-area glass/metal units make use of glass mirrors supported by metallic frame facets.

Recently, some developers are introducing substantial changes in the conception of heliostat design. A number of projects based upon

Cylindrical

Cavity

Billboard

Volumetric

FIG. 3.15 DIFFERENT CONFIGURATIONS OF SOLAR RECEIVERS. FROM LEFT TO RIGHT AND TOP TO BOTTOM: A) EXTERNAL TUBULAR CYLINDRICAL, B) CAVITY TUBULAR, C) BILLBOARD TUBULAR AND D) VOLUMETRIC

the paradigm of maximum modularity and mass production of components are claiming small-size heliostats as a competitive low-cost option. Companies like Brightsource, eSolar, Aora, or Cloncurry are introducing heliostat units of only a few square meters. Brightsource with an ambitious program of large projects is making use of single-facet 7.3-m^2 heliostats [44] and the company eSolar with a multi-tower plant configuration presents a highly innovative field with ganged heliostats of extremely small size (1.14 m^2 each) that implies the large number of 12,180 units for a single 2.5 MW tower [45]. If such small heliostats may reach installed costs below 200 \$/m^2, it can only be understood under aggressive mass production plans and pre-assembly during manufacturing process by reducing on-site mounting works. Annual performance and availability of those highly-populated fields are still under testing.

In a solar power tower plant, the receiver is the heat exchanger where the concentrated sunlight is intercepted and transformed into thermal energy useful in thermodynamic cycles. Radiant flux and temperature are substantially higher than in parabolic troughs, and therefore, high technology is involved in the design, and high-performance materials should be chosen. The solar receiver should mimic a black body by minimizing radiation losses. To do so, cavities, black-painted tube panels, or porous absorbers able to trap incident photons are used. In most designs, the solar receiver is a single unit that centralizes all the energy collected by the large mirror field, and therefore, high availabilities and durability are a must. Just as cost reduction is the priority for further development in the collector field, in solar receivers, the priorities are thermal efficiency and durability. Typical receiver absorber operating temperatures are between 500°C and 1200°C and incident flux covers a wide range between 300 and over 1000 kW/m^2 [6, 46].

There are different solar receiver classifications criteria depending on the construction solution, the use of intermediate absorber materials, the kind of thermal fluid used, or heat transfer mechanisms. According to the geometrical configuration, there are basically two design options, external and cavity-type receivers. In a cavity receiver, the radiation reflected from the heliostats passes through an aperture into a box-like structure before impinging on the heat transfer surface. Cavities are constrained angularly and subsequently used in North-field (or South-field) layouts. External receivers can be designed with a flat plate tubular panel or are cylindrically shaped. Cylindrical external receivers are the typical solution adopted for surround heliostat fields. Figure 3.15 shows examples of cylindrical external, billboard external and cavity receivers.

Receivers can be directly or indirectly irradiated depending on the absorber materials used to transfer the energy to the working fluid [47]. Directly irradiated receivers make use of fluids or particle streams able to efficiently absorb the concentrated flux. Particle receiver designs make use of falling curtains or fluidized beds. In many applications, and to avoid leaks to the atmosphere, direct receivers should have a transparent window. Windowed receivers are excellent solutions for chemical applications as well, but they are strongly limited by the size of a single window, and therefore, clusters of receivers are necessary.

The key design element of indirectly heated receivers is the radiative/convective heat exchange surface or mechanism. Basically, two heat transfer options are used, tubular panels and volumetric surfaces. In tubular panels, the cooling thermal fluid flows inside the tube and removes the heat collected by the external black panel surface by convection. It is, therefore, operating as a recuperative heat exchanger. Depending on the heat transfer fluid properties and incident solar flux, the tube might undergo thermo-mechanical stress. Since heat transfer is through the tube surface, it is difficult to operate at an incident flux above 600 kW/m^2 (peak). Table 3.3 shows how only with high thermal conductivity liquids like sodium it is possible to reach operating fluxes above 1 MW/m^2. Air-cooled receivers have difficulties working with tubular receivers because of the lower heat transfer coefficients. To improve the contact surface, a different approach based on wire, foam or appropriately shaped materials within a volume are used. In volumetric receivers, highly porous structures operating as convective heat exchangers absorb the concentrated solar radiation. The solar radiation is not absorbed on an outer surface, but inside the structure "volume." The heat transfer medium (mostly air) is forced through the porous structure and is heated by convective heat transfer [58]. Volumetric absorbers are usually made of thin heat-resistant wires (in knitted or layered grids) or either metal or ceramic (reticulated foams, etc.) open-cell matrix structures. Good volumetric absorbers are very porous, allowing the radiation to penetrate deeply into the structure. Thin substructures (wires, walls or struts) ensure good convective heat transfer. A good volumetric absorber produces the so-called volumetric effect, which means that the irradiated side of the absorber is at a lower temperature than the medium leaving the absorber. Under specific operating conditions, volumetric absorbers tend to have an unstable mass flow distribution [49]. Receiver arrangements with mass flow adaptation elements (e.g., perforated plates) located behind the absorber can reduce this tendency, as well as appropriate selection of the operating conditions and the absorber material [53]. A number of initiatives have formulated air-cooled volumetric schemes for STE plants, both for atmospheric pressure ([51], [52]) and for pressurized systems [46], though still commercial plants with these technologies are missing. Though air-cooled open volumetric receivers are a promising way of producing superheated steam, the modest thermal efficiency at the receiver (74% nominal and 61.4% annual average) must still be improved [53]. At present, all the benefits from using higher outlet temperatures are sacrificed by radiation losses at the receiver, leading to low annual electricity production, so it is clear that volumetric receiver improvements must reduce losses.

Selection of a particular receiver technology is a complex task, since operating temperature, heat storage system and thermodynamic cycle influence the design. In general, tubular technologies allow either high temperatures (up to 1000ºC) or high pressures (up to 120bar), but not both [54]. Directly irradiated or volumetric receivers allow even higher temperatures but limit pressures to below 15 bar.

TABLE 3.3 OPERATING TEMPERATURE AND FLUX RANGES OF SOLAR TOWER RECEIVERS

Fluid	Water/ steam	Liquid sodium	Molten salt (nitrates)	Volumetric air
Flux (MW/m^2)				
- Average	0.1–0.3	0.4–0.5	0.4–0.5	0.5–0.6
- Peak	0.4–0.6	1.4–2.5	0.7–0.8	0.8–1.0
Fluid outlet temperature (ºC)	490–525	540	540–565	(700–1000)

3.5.2 Experience in central receiver systems

Although there have been a large number of STE tower projects, only a few have culminated in the construction of an entire experimental system. Table 3.4 lists systems that have been tested all over the world along with new early commercial plants. In general terms, as observed, they are characterized as being small demonstration systems between 0.5 and 10 MW, and most of them were operated in the 1980s ([6], [36], [55]). The thermal fluids used in the receiver are liquid sodium, saturated or superheated steam, nitrate-based molten salts and air. All of them can easily be represented by flow charts, where the main variables are determined by working fluids, with the interface between power block and the solar share.

The set of experiences referred to has served to demonstrate the technical feasibility of the CRS power plants, whose technology is sufficiently mature. The most extensive experience has been collected by several European projects located in Spain at the premises of the Plataforma Solar de Almería [55] and the 10 MW Solar One [56] and Solar Two plants [57] in the US. At present, water/steam and molten salts are the heat transfer fluids being selected for the first generation of commercial plants.

3.5.3 Water/Steam Plants from PS 10 Project to Superheated Steam

Production of superheated steam in the solar receiver has been demonstrated in several plants, such as Solar One, Eurelios, and CESA-1, but operating experience showed critical problems related to the control of zones with dissimilar heat transfer coefficients like boilers and superheaters [55]. Better results regarding absorber panel lifetime and controllability have been reported for saturated steam receivers. The good performance of saturated steam receivers was qualified at the 2-MW Weizmann receiver that produced steam at 15bar for 500 hours in 1989 [58]. Even though technical risks are reduced by saturated steam receivers, the outlet temperatures are significantly lower than those of superheated steam, making applications where heat storage is replaced by fossil fuel backup necessary.

PS10, the first commercial CRS plant in the world, adopted the conservative scheme of producing saturated steam to limit risk perception and avoid technology uncertainties. The 11 MW plant, located near Seville in South Spain, was designed to achieve an annual electricity production of 23 GWh at an investment cost of less than 3500€/kW. The project made use of available, well-proven technologies like the glass-metal heliostats developed by the Spanish INABENSA company and the saturated steam cavity receiver developed by the TECNICAL company to produce steam at 40 bar and 250ºC [43]. The plant is a solar-only system with saturated steam heat storage able to supply 50 minutes of plant operation at 50% load [6]. The system makes use of 624 heliostats of 121 m^2 each, distributed in a North-field configuration, a 100-m high tower, a 15 MWh heat storage system and a cavity receiver with four 4.8 × 12 m tubular panels. The basic flow diagram selected for PS10 is shown in Fig. 3.16. Though the system makes use of a saturated steam turbine working at extremely low temperature, the nominal efficiency of the power block (30.7%) is relatively good. This efficiency is the result of optimized management of waste heat in the thermodynamic cycle. As summarized in Table 3.5, the combination of optical, receiver and power block efficiencies lead to a total nominal efficiency at design point of 21.7%. Total an-

TABLE 3.4   POWER TOWERS IN THE WORLD (ONLY PS10, PS20 AND GEMASOLAR ARE COMMERCIAL GRID-CONNECTED PROJECTS)

Project	Country	Power (MWe)	Heat transfer fluid	Storage media	Beginning operation
SSPS	Spain	0.5	Liquid Sodium	Sodium	1981
EURELIOS	Italy	1	Steam	Nitrate salt/water	1981
SUNSHINE	Japan	1	Steam	Nitrate salt/water	1981
Solar One	U.S.A.	10	Steam	Oil/rock	1982
CESA-1	Spain	1	Steam	Nitrate salt	1982
MSEE/Cat B	U.S.A.	1	Nitrate salt	Nitrate salt	1983
THEMIS	France	2.5	Hitec salt	Hitech salt	1984
SPP-5	Russia	5	Steam	Water/steam	1986
TSA	Spain	1	Air	Ceramic	1993
Solar Two	U.S.A.	10	Nitrate salt	Nitrate salt	1996
Consolar	Israel	0.5*	Pressurized air	Fossil hybrid	2001
Solgate	Spain	0.3	Pressurized air	Fossil hybrid	2002
PS10	Spain	11	Water/steam	Saturated steam	2006
PS20	Spain	20	Water/steam	Saturated steam	2009
Gemasolar**	Spain	17	Nitrate salt	Nitrate salt	2011

* Thermal
** Project under construction.

nual efficiency decreases to 15.4%, including operational losses and outages. PS10 is a milestone in the CRS deployment process, since it is the first solar power tower plant developed for commercial exploitation. Commercial operation started in June 21st 2007. Since then the plant is performing as designed. The construction of PS20, a 20 MWe plant with the same technology as PS10, followed. PS20 started operation in May 2009. With 1,255 heliostats (120 m^2 each) spread over 90 ha and with a tower of 165 m high, the plant is expected to produce 48.9 GWh/year.

Saturated steam plants are considered a temporary step to the more efficient superheated steam systems. Considering the problems found in the 1980s with superheated steam receivers, the current trend is to develop dual receivers with independent absorbers, one of them for the preheating and evaporation and another one for the superheating step. The experience accumulated with heuristic algorithms in central control systems applied to aiming point strategies at heliostat fields allows achieving a flexible operation with multi-aperture receivers. The company Abengoa Solar, developer of PS10 and PS20, is at present designing a superheated steam receiver for a new generation of water/steam plants [59].

China also has selected superheated steam for the first tower construction (Dahan) of 1 MW located in Yanqing, 65 km North of Beijing. It is a small project promoted by the Chinese Ministry of Science and Technology and the Chinese Academy of Science. The water/steam receiver is a joint development with Korea.

But the most advanced strategy is the program announced by the BrightSource Industries (Israel) Ltd (BSII), is that it has already built a demonstration plant of 6 MWth located at the Neguev dessert in June 2008 [44]. The final objective of BSII is to promote plants producing superheated steam at 160 bar and temperature of

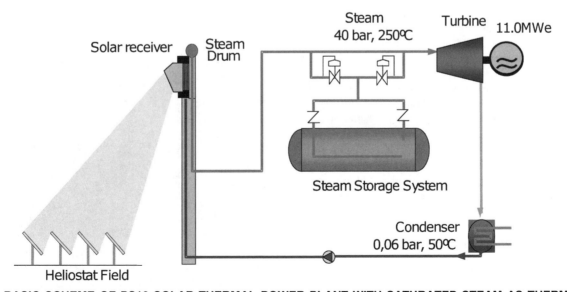

FIG. 3.16   BASIC SCHEME OF PS10 SOLAR THERMAL POWER PLANT WITH SATURATED STEAM AS THERMAL FLUID
(*Source*: Abengoa Solar, S.A.) (All rights reserved on Figure 3.16 to Abengoa Solar, S.A. © 2010)

TABLE 3.5   ANNUAL ENERGY BALANCE FOR THE PS10 PLANT AT NOMINAL CONDITIONS

Nominal rate operation		
Optical efficiency	77.0%	67.5 MW → 51.9 MW
Receiver and heat handling efficiency	92.0%	51.9 MW → 47.7 MW
Thermal power to storage		11.9 MW
Thermal power to turbine		35.8 MW
Thermal power/electric power efficiency	30.7%	35.8 MW → 11.0 MW
Total efficiency at nominal rate	**21.7%**	
Energetical balance in annual basis		
Mean annual optical efficiency	64.0%	148.63 GWh → 95.12 GWh
Mean annual receiver & heat handling efficiency	90.2%	95.12 GWh → 85.80 GWh
Operational efficiency (start-up/stop)	92.0%	85.80 GWh → 78.94 GWh
Operational efficiency (breakages, O&M)	95.0%	78.94 GWh → 75.00 GWh
Mean annual thermal energy /electricity efficiency	30.6%	75.00 GWh → 23.0 GWh
Total annual efficiency	**15.4%**	

565°C (named DPT550). With those characteristics, they expect up to 40% conversion efficiency at the power block for unit sizes between 100 and 200 MW. The receiver is cylindrical, dual and with a drum. The first commercial Project under development in California is Ivanpah Solar in California of 400 MW and expected to commission in 2012, followed by a 1.8 GW unit more in California and 400 MW unit in Nevada [2]. Planned conversion efficiency from solar to electricity is 20%.

The combination of recent initiatives on small heliostats, compact modular multi-tower fields and production of superheated steam may be clearly visualized in the development program of the company eSolar. This company proposes a high degree of modularity with power units of 46 MW covering 64 ha, consisting of 16 towers and their corresponding heliostat fields sharing a single central power block. With replication, modularity sizes up to 500 MW and upward may be obtained [60]. Within the development program of STE in the US, eSolar has four plants totalizing 334 MW [2]. Figure 3.17 depicts the first two modules of 2.5 MW each installed in 2009 by eSolar in Lancaster, California, and already in operation. It can be observed that each receiver has two independent cavities and the heliostat layout consists of identical arrays to maximize replication and modularity that clearly penalizes optical efficiency and cosine factor. Each tower is associated with 12,180 flat heliostats of 1.14 m² each and a single 46 MW commercial plant would contain 194,880 heliostats [45]. The receivers are dual-cavity, natural-circulation boilers. Inside the cavity, the feedwater is preheated with economizer panels before entering the steam drum. A downcomer supplies water to evaporator panels where it is boiled. The saturated water/vapor mixture returns to the drum where the steam is separated, enters superheater panels, and reaches 440°C at 6.0 MPa. Each receiver absorbs a full-load power of 8.8 MWth. Overall plant efficiency expected by eSolar would be 23% solar to electricity [60].

3.5.4   Molten Salt Systems: Solar Two and Gemasolar

For high annual capacity factors, solar-only power plants must have an integrated cost-effective thermal storage system. One such thermal storage system employs molten nitrate salt as the receiver heat transfer fluid and thermal storage media. To be usable, the operating range of the molten nitrate salt, a mixture of 60% sodium nitrate and 40% potassium nitrate, must match the operating temperatures of modern Rankine cycle turbines. In a molten-salt power tower plant, cold salt at 290°C is pumped from a tank at ground level to the receiver mounted atop a tower where it is heated by concentrated sunlight to 565°C. The salt flows back to ground level into another tank. To generate electricity, hot salt is pumped from the hot tank through a steam generator to make superheated steam. The superheated steam powers a Rankine-cycle turbine. The collector field can be sized to collect more power than is demanded by the steam generator system, and the excess salt is accumulated in the hot storage tank. With this type of storage system, solar power tower plants can be built with annual capacity factors up to 70%. As molten salt has a high-energy storage capacity per volume (up to 500–700 kWh/m³), they are excellent candidates for solar thermal power plants with large capacity factors. Even though nitrate salt has a lower specific heat capacity per volume than carbonates, they still store 250 kWh/m³. The average heat conductivity of ni-

FIG. 3.17 FIRST EXPERIMENTAL EXAMPLE IN THE WORLD OF A MULTITOWER SOLAR ARRAY CORRESPONDING TO ESOLAR'S SIERRA SOLAR GENERATING STATION WITH TWO FULL-SIZED HELIOSTAT FIELDS IN LANCASTER, CALIFORNIA (*Source*: eSolar)

trates is 0.52 W/mK and their heat capacity is about 1.6 kJ/kgK. Nitrates are a cheap solution for large storage systems.

Several molten salt development and demonstration experiments have been conducted over the past two decades in the US and Europe to test entire systems and develop components. The largest demonstration of a molten-salt power tower was the Solar Two project, a 10-MW power tower located near Barstow, CA. The purpose of the Solar Two project was to validate the technical characteristics of the molten-salt receiver, thermal storage, and steam generator technologies, improve the accuracy of economic projections for commercial projects by increasing the capital, operating, and maintenance cost database, and distribute information to utilities and the solar industry to foster a wider interest in the first commercial plants. The Solar Two plant was built at the same site as the Solar One pilot plant and reused much of the hardware including the heliostat collector field, tower structure, 10 MW turbine and balance of plant. A new, 110 MWh$_t$ two-tank molten-salt thermal storage system was installed, as well as a new 42 MWth receiver, a 35 MWth steam generator system (535°C, 100 bar), and master control system [61].

The plant began operating in June 1996. The project successfully demonstrated the potential of nitrate-salt technology. Some of the key results were: receiver efficiency was measured at 88%, the thermal storage system had a measured round-trip efficiency of over 97%, and the gross Rankine-turbine cycle efficiency was 34%, all of which matched performance projections. On one occasion, the plant operated around-the-clock for 154 hours straight [62]. On April 8, 1999, testing and evaluation of this demonstration project was completed and subsequently was shut down.

To reduce the risks associated with scaling up hardware, the first commercial molten-salt power tower should be approximately three times the size of Solar Two. One attempt to prove scaled-up molten-salt technology is the Gemasolar project being promoted by Torresol Energy, a joint venture between the Spanish SENER and MASDAR initiative from Abbu Dhabi ([63], [64]). Table 3.6 summarizes the main technical specifications of Gemasolar project.

With only 17 MWe, the plant is predicted to produce 112 GWh$_e$/year. A large heliostat field of 304,750 m^2 (115 m^2 each heliostat) is oversized to supply 15-hour equivalent heat storage capacity.

TABLE 3.6 TECHNICAL SPECIFICATIONS AND DESIGN PERFORMANCE OF THE GEMASOLAR PROJECT

Technical specifications	
Heliostat field reflectant surface	304,750 m^2
Number heliostats	2650
Land area of solar field	142 ha
Receiver thermal power	120 MWth
Tower height	145 m
Heat storage capacity	15 hours
Power at turbine	17 MWe
Power NG burner	16 MWth
Operation	
Annual electricity production	112 MWhe
Production from fossil (annual)	15%
Capacity factor	74%

The plant is designed to operate around-the-clock in summertime, leading to an annual capacity factor of 74%. Fossil backup corresponding to 15% of annual production will be added. The levelized energy costs are estimated to be approximately $0.16/kWh. Gemasolar might represent a breakthrough for STE technology. Construction works started in November 2008 and commercial operation is scheduled for May 2011.

3.6 DISH/STIRLING SYSTEMS

3.6.1 Dish–Stirling Technology: Subsystems and Components

Dish–Stirling systems track the sun and focus solar energy onto a cavity receiver, where it is absorbed and transferred to a heat engine/generator. An electrical generator, directly connected to the crankshaft of the engine, converts the mechanical energy into electricity (AC). To constantly keep the reflected radiation at the focal point during the day, a sun-tracking system continuously rotates the solar concentrator on two axes following the daily path of the sun. With current technologies, a 5 kW$_e$ Dish–Stirling system would require 5.5 m diameter concentrator, and for 25 kW$_e$, the diameter would have to increase up to 10 m. Stirling engines are preferred for these systems because of their high efficiencies (40% thermal-to-mechanical), high-power density (40–70 kW/liter), and potential for long-term, low-maintenance operation. Dish–Stirling systems are modular, i.e., each system is a self-contained power generator, allowing their assembly in plants ranging in size from a few kilowatts to tens of megawatts [65].

The concentrator is a key element of any Dish–Stirling system. The curved reflective surface can be manufactured by attached segments, by individual facets or by stretched membranes shaped by a continuous plenum. In all cases, the curved surface should be coated or covered by aluminum or silver reflectors. Second-surface glass mirrors, front surface thin-glass mirrors or polymer films have been used in various different prototypes.

First-generation parabolic dishes developed in the 1980s were shaped with multiple, spherical mirrors supported by a trussed structure [66]. Though extremely efficient, this structure concept was costly and heavy. The introduction of automotive industry concepts and manufacturing processes has led to optimized commercial versions like the 25-kW SunCatcher system developed by the company Stirling Energy Systems (Fig. 3.18) that has earned the world's highest recorded peak efficiency rating, 31.25%, for converting solar energy-to-grid quality electricity. Another good example of the latest developments in Dish–Stirling systems is the 10-kW Eurodish prototype. The Eurodish concentrator consists of 12 single segments made of fiber glass resin. When mounted, the segments form an 8.5-m diameter parabolic shell. The shell rim is stiffened by a ring truss to which bearings and the Stirling support structure are later attached. Thin 0.8-mm-thick glass mirrors, are glued onto the front of the segments for durable and high reflectivity around 94% [67].

As in central receivers and parabolic trough absorbers, the receiver absorbs the light and transfers the energy as heat to the engine's working gas, usually helium or hydrogen. Thermal fluid working temperatures are between 650°C and 750°C. This temperature strongly influences the efficiency of the engine. Because of the high operating temperatures, radiation losses strongly penalize

FIG. 3.18   ROW OF DISH-STIRLING SYSTEMS (STIRLING ENERGY SYSTEMS SUNCATCHER) OF 25 KW EACH AT THE MARICOPA SOLAR PLANT IN ARIZONA (*Source*: Stirling Energy Systems)

the efficiency of the receiver; therefore, a cavity design is the optimum solution for this kind of system.

Two different heat transfer methods are commonly used in parabolic dish receivers [68]. In directly illuminated receivers, the same fluid used inside the engine is externally heated in the receiver through a pipe bundle. In indirect receivers, an intermediate fluid and a heat pipe is used to decouple solar flux and working temperature from the engine fluid. The phase change guarantees good temperature control, providing uniform heating of the Stirling engine [69].

Stirling engines solarized for parabolic dishes are externally heated gas-phase engines in which the working gas is alternately heated and cooled in constant-temperature, constant-volume processes (Fig. 3.19). This possibility of integrating additional external heat what makes it an ideal candidate for solar applications. Since the Stirling cycle is very similar to the Carnot cycle, the theoretical efficiency is high. High reversibility is achieved since work is supplied to and extracted from the engine at isothermal conditions. The clever use of a regenerator that collects the heat during constant-volume cooling and heating substantially enhances the final system efficiency. For most engine designs, power is extracted by kinematics of the rotating crankshaft connected to the pistons by a connecting rod. An example of a kinematic Stirling engine is shown in Fig. 3.20. Though theoretically, Stirling engines may

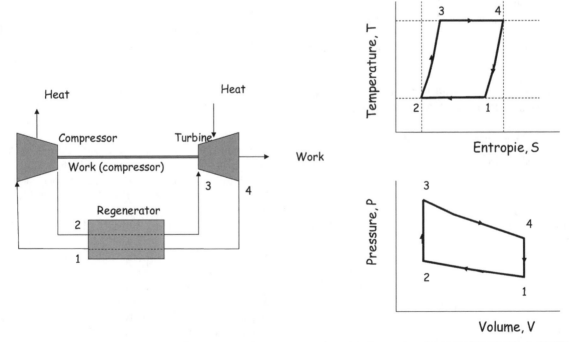

FIG. 3.19   IDEAL STIRLING CYCLE PRESENTS ISOTHERMAL COMPRESSION AND EXPANSION PLUS A REGENERATOR THAT LEADS TO A HIGH THERMODYNAMIC EFFICIENCY

have a high life-cycle projection, the actual fact is that today their availability is still not satisfactory, since an important percentage of operating failures and outages are caused by pistons and moving mechanical components. Availability is, therefore, one of the key issues, since it must operate for more than 40,000 hours in 20-year lifetime, or ten times more than an automobile engine. One option to improve availability is the use of free-piston designs. Free-piston engines make use of gas or a mechanical spring so that mechanical connections are not required to move reciprocating pistons. Apparently, they are better than kinematic engines in terms of availability and reliability. The most relevant program in developing dishes with free-piston technology is promoted by the company Infinia.

3.6.2   Dish–Stirling Developments

Like the other CSP technologies, practical Dish–Stirling development started in the early 1980s. Most development has been concentrated in the US and Germany, though developed for commercial markets they have been tested in a small number of units [70].

The first generation of dishes was a facet-type concentrator with second-surface mirrors that already established concentration records ($C = 3000$), and had excellent performances, though their estimated costs for mass production were above \$300/m^2. Their robust structures were extremely heavy, weighing at 100 kg/m^2 [55]. The 25 kW Vanguard-1 prototype built by Advanco was operated at Rancho Mirage, California, in the Mojave desert in production mode for 18 months (February 1984 to June 1985). This system was 10.7 m in diameter with a reflecting surface of 86.7 m^2 and a 25 kW PCU made by United Stirling AB (USAB) model 4-95 Mark II. This engine had four cylinders with a 95 cm^3 cylinder displacement. The working gas was hydrogen at a maximum pressure of 20 MPa and temperature of 720°C. Engine power was controlled by varying the working-gas pressure. McDonnell Douglas later developed another somewhat improved dish system making use of the same technology and the same engine. The dish was 10.5 m and 25 kW. The 88 m^2 parabolic dish consisted of 82 spherically curved glass facets. Reported performances and efficiencies were

similar to those of Advanco/Vanguard [66]. The project was frozen for several years until in 1996, Stirling Energy Systems (SES) acquired the intellectual and technology rights to the concentrator and the US manufacturing rights to what is now called the Kockums, 4-95 Stirling engine-based PCU [65]. The re-designed 25-kWe system named SunCatcher is being qualified in a commercial basis at the Maricopa Solar Plant in Arizona (Fig. 3.20). The plant totalizes 1.5 MW with 60 dishes and is in operation since January 2010. Commercialization is expected in California with two large plants of about 800 MW each.

In Europe the most important development is the EuroDish system. The EuroDish project is a joint venture undertaken by the European Community, German/Spanish Industry (SBP, MERO, Klein+Stekl, Inabensa), and research institutions DLR and CIEMAT. The engine used in the EuroDish is SOLO Kleinmotoren 161. Two new 10-kW EuroDish units were installed at the Plataforma Solar de Almeria, Spain, early in 2001 for testing and demonstration. In a follow-up project called EnviroDish, additional units were deployed in France, India, Italy and Spain to accumulate operating experience at different sites. The peak system efficiency was first measured at 20%. The estimated annual production of a EuroDish system operating in Albuquerque, New Mexico, is 20,252 kWh of electricity with 90% availability and an annual efficiency of 15.7% [65]. SBP and the associated EuroDish industry have performed cost estimates for a yearly production rate of 500 units per year (5 MW/year) and 5000 units per year, which corresponds to 50 MW/year. The actual cost of the 10-kW unit without transportation and installation cost and excluding foundations is approximately US\$10,000/kW. The cost projections at production rates of 500 and 5000 units per year are US\$2,500/kW and US\$1,500/kW, respectively based upon replication of automotive mass-production processes.

3.7   TECHNOLOGY DEVELOPMENT NEEDS AND MARKET OPPORTUNITIES FOR SOLAR THERMAL ELECTRICITY (STE)

From the 1970s to the 1990s the development of solar thermal electricity technologies remained restricted to a few countries and only a few, though important, research institutions and industries were involved. The situation has dramatically changed since 2006 with the approval of specific feed-in-tariffs or power purchase agreements in Spain and the US. Both countries with more than 6 GW of projects under development and more than 1 GW in operation at the end of 2010 are undoubtedly leading the commercialization of STE [2]. Other countries such as India, China, Australia and Italy adopted the STE process. Subsequently a number and variety of engineering and construction companies, consultants, technologists and developers committed to STE are rapidly growing.

A clear indicator of the globalization of STE commercial deployment for the future energy scenario has been elaborated by the International Energy Agency (IEA). This considers STE to play a significant role among the necessary mix of energy technologies for halving global energy-related CO_2 emissions by 2050 [5]. This scenario would require capacity addition of about 14 GW/year (55 new solar thermal power plants of 250 MW each). However, this new opportunity is introducing an important stress to the developers

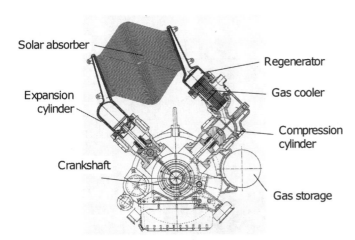

FIG. 3.20   KINEMATIC STIRLING ENGINE V-161 OF 10 KWE MANUFACTURED BY SOLO KLEINMOTOREN WITH PISTONS SITUATED IN V AND CONNECTED TO A TUBULAR ARRAY HEAT EXCHANGER (*Source*: Solo Kleinmotoren)

of STE. In a period of less than 5 years, in different parts of the world, these developers of STE are forced to move from strategies oriented to early commercialization markets based upon special tariffs, to strategies oriented to a massive production of components and the development of large amounts of projects with less profitable tariffs. This situation is speeding up the implementation of second generation technologies even though in some cases still some innovations are under assessment in early commercialization plants or demonstration projects. The projected evolution of Levelized Electricity Costs of different STE technologies is depicted in Fig. 3.21. LEC value may be reduced an additional 30% when moving to future sites with very high Direct Normal Irradiance.

The reduction in electricity production costs should be a consequence, not only of mass production but also of scaling-up and R&D. A technology roadmap promoted by the European Industry Association ESTELA [7] states that by 2015, when most of the improvements currently under development are expected to be implemented in new plants, energy production boosts greater than 10% and cost decreases up to 20% are expected to be achieved. Furthermore, economies of scale resulting from plant size increase will also contribute to reduce plants' CAPEX per MW installed up to 30%. STE deployment in locations with very high solar radiation further contributes to the achievement of cost competitiveness of this technology by reducing costs of electricity up to 25%. All these factors can lead to electricity generation cost savings up to 30% by 2015 and up to 50% by 2025, reaching competitive levels with conventional sources (e.g., coal/gas with stabilized Electricity Costs <10€c/kWh). Similar projections are published in another recent roadmap issued by the IEA [71]. Other roadmaps coordinated by R&D centers expect larger influence of innovations (up to 25%) in cost reduction [72]. Some of the key general topics on medium to long-term R&D proposed by the STE community are [22]:

Build confidence in the technology through:

- pilot applications based on proven technologies;
- high reliability of unattended operation;
- increased system efficiency through higher design temperatures;
- hybrid (solar/fossil fuel) plants with small solar share.

Reduce costs through:

- improved designs, materials, components, subsystems and processes;
- exploitation of economies of scale.

Increase solar share through:

- suitable process design;
- integration of storage.

In all cases, R&D is multi-disciplinary, involving optics, materials science, heat transfer, control, instrumentation and measurement techniques, energy engineering and thermal storage.

3.7.1 Trough and Linear Fresnel Power Plants

To further reduce costs and increase reliability in next generation PTC and LFR technology, the following are expected:

- Lighter and lower-cost structural designs including front surface mirrors with high solar-weighted reflectivity of about 95%.
- Development of high-absorptance coatings for tube receivers (96% and higher) able to work efficiently at over 500°C.
- Development of medium temperature thermal energy storage systems (Phase Change Materials, molten salts, concrete) suitable for solar-only systems;
- Continued improvement in overall system O&M, including mirror cleaning, integral automation and largely unattended control;
- System cost reductions and efficiency improvements from substituting water for synthetic oil as the heat-transfer fluid (Direct Steam Generation technology)

3.7.2 Power Tower Plants

Power tower R&D in the United States, Europe, and Israel is concentrated in the two most relevant subsystems with regard to costs: heliostat field and solar receiver. The following improvements are expected:

Improvements in the heliostat field as a result of better optical properties, lower cost structures, and better control. Improvements in materials should be analogous to those for trough collectors.

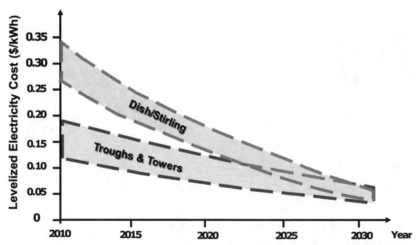

FIG. 3.21 EVOLUTION OF LEVELIZED ELECTRICITY COST FOR STE TECHNOLOGIES BASED UPON TECHNOLOGY ROADMAPS AND INDUSTRY

In general terms, optical performance and durability of existing heliostats is acceptable (95% availability and beam quality below 2.5 mrad), therefore, R&D resources should focus basically on cost reduction.

Development of water/superheated steam and advanced air-cooled volumetric receivers using both wire-mesh absorbers and ceramic monoliths is the subject of various projects. Dual-aperture receivers for water/steam and volumetric receivers for air still need further development for scale-up, materials durability and thermal efficiency.

Heat storage is another key issue for CRS development. The new developments in air-cooled receivers have led to the development of advanced thermocline storage systems making use of packed-bed ceramic materials. This system has shown excellent performance for small units of a few MWh but pressure losses and design restrictions appear when size is increased.

Finally, more distributed control architectures, system integration and hybridization in high-efficiency electricity production schemes should be developed as already mentioned for trough systems.

3.7.3   Dish/Engine Power Plants

Several dish/engine prototypes have operated successfully during the last 20 years in the US and Europe, but there is no large-scale deployment yet. This situation may change soon with some existing commercialization initiatives. The use of dishes for stand-alone or grid-support installations will reach near-term markets as costs drop to less than 12¢/kWh. This lower cost can be achieved through:

- improvements in mirrors and support structures, improvements in hybrid heat-pipe and volumetric receivers coupled to Stirling and Brayton engines, and development of control systems for fully automatic operation; and
- improvements in system integration by reduction of parasitic loads, optimization of startup procedures, better control strategies, and hybrid Stirling-Brayton operation.

3.8   NOMENCLATURE AND UNIT CONVERSIONS

bar	Unit of pressure equal to 100 kilopascals
CLFR	Compact Linear Fresnel Reflector
CRS	Central Receiver System
CSP	Concentrating Solar Power
DE	Dish Engine
GEF	Global Environmental Facility
GW	Gigawatt (10^9 Watts)
GWh	Gigawatt-hour (10^9 Wh)
Ha/MW	Hectares per MW
HTF	Heat Transfer Fluid
IEA	International Energy Agency
ISCCS	Integrated Solar Combined Cycle System
kJ/kgK	kiloJoule per kilogram and Kelvin
kW	kiloWatt (10^3 Watts)
kWh/m^3	kiloWatt-hour per cubic meter
LFR	Linear Fresnel Reflector
m	meter

m^2	square meter
mbar	millibar
mrad	milliradian
MW	Megawatts (10^6 Watts)
MWe	Megawatts electric (10^6 Watts)
MWth	Megawatts thermal (10^6 Watts)
NSO	Nevada Solar One
O&M	Operation and Maintenance
PCM	Phase Change Material
PTC	Parabolic Trough Collector
PV	Photovoltaics
R&D	Research and Development
SEGS	Solar Electricity Generating Systems
SHP	Solar Heat and Power
SPG	Solar Power Group
STE	Solar Thermal Electricity
TWh	TeraWatt hour (10^{12} Wh)
W/mK	Watts per meter and Kelvin

3.9   REFERENCES

1. Becker M., Meinecke W., Geyer M., Trieb F., Blanco M., Romero M., Ferriere A. (2002), "Solar Thermal Power Plants". In: "The future for Renewable Energy 2: Prospects and Directions". Eurec Agency. Pub. James & James Science Publishers Ltd., London, UK. pp. 115-137.

2. Herring G. "Concentrating solar thermal power gains steam in Spain, as momentum builds for major projects in the US, North Africa, the Middle East, Asia and Australia". Photon International, December 2009, 46-52.

3. Nitsch, J. and Krewitt, W. "Encyclopedia of Energy", Elesevier Inc., Volume 5, 2004.

4. Philibert C. (2004). International Energy Technology Collaboration and Climate Change Mitigation, Case Study 1: Concentrating Solar Power Technologies, OECD/IEA Information Paper, Paris. COM/ENV/EPOC/IEA/SLT(2004)8.

5. IEA (2010) Energy Technology Perspectives 2010 – Scenarios and strategies to 2050. ISBN 978-92-64-08597-8.

6. Romero M., Zarza E. (2007) "Concentrating Solar Thermal Power". In: Handboook of Energy Efficiency and Renewable Energy. F. Kreith and Y. Goswami (Eds.) Chapter 21. pp. 1-98. CRC Press Taylor & Francis Group, Boca Raton, Florida.

7. Kearney A.T. (2010) "Solar thermal electricity 2025". STE industry roadmap for the European Solar Thermal Electricity Association (ESTELA). Available at: www.atkearney.com.

8. Winter C.J., Sizmann R.L., Vant-Hull L.L. (Eds) (1991), "Solar Power Plants", Springer-Verlag, Berlin, ISBN 3-540-18897-5.

9. Kodama T. (2003), "High-temperature solar chemistry for converting solar heat to chemical fuels", *Progress in Energy and Combustion Science* **29**, 567–597.

10. Rabl, A. Active Solar Collectors and Their Applications. New York, Oxford University Press, 1985: pp. 59-66. ISBN: 0-19-503546-1.

11. Welford W.T., Winston R. (1989) "High Collection Non-Imaging Optics". New York; Academic Press.

12. Biggs F., Vittitoe C. (1979). "The Helios model for the optical behavior of reflecting solar concentrators". Report Sandia National Labs. SAND76-0347. March 1979.

13. Mills D. (2004) "Advances in solar thermal electricity technology". *Solar Energy* 76 (2004) 19–31.

14. Geyer M. (2002), Panel 1 Briefing Material on Status of Major Project Opportunities. The Current Situation, Issues, Barriers and Planned Solutions. International Executive Conference on Expanding the Market for Concentrating Solar Power (CSP) – Moving Opportunities into Projects; 19 – 20 June 2002; Berlin, Germany.

15. DeMeo E.A., Galdo J.F. (1997), "Renewable Energy Technology Characterizations", TR-109496 Topical Report, December 1997, U.S. DOE-Washington and EPRI, Palo Alto, California.

16. Price H., Luepfert E., Kearney D., Zarza E., Cohen G., Gee R., Mahoney R. (2002), "Advances in Parabolic Trough Solar Power Technology", Int. J. Solar Energy Eng., Vol. 124, pp. 109-125.

17. Fernandez-Garcia A., Zarza E., Valenzuela L., Perez M. "Parabolic-trough solar collectors and their applications". Renewable and Sustainable Energy Reviews 14 (2010) 1695–1721.

18. Rabl, A. Active Solar Collectors and Their Applications. New York, Oxford University Press, 1985: pp. 59-66. ISBN: 0-19-503546-1.

19. Eck, M. Zarza, E. Eickhoff, M. Rheinländer, J. Valenzuela, L. Applied research concerning the direct steam generation in parabolic troughs. Solar Energy, no. 74, 2003: pp. 341-351.

20. Kearney, D.W and Cohen, G.E. (1997) "Current experiences with the SEGS parabolic trough plants". In: Proceedings of the 8th International Symposium on Solar Thermal Concentrating Technologies. Vol. 1. Cologne, Germany, 1996. Becker, M.; Böhmer, M. eds. Heidelberg, Alemania, C.F. Müller, 1997: pp. 217-224.

21. Goebel O. "Shams One 100 MW CSP Plant in Abu Dhabi"; Proceedings SolarPACES 2009 (CD). Ref. manuscript: 15523; Berlín, Germany; 15-18 September 2009. Ed. DLR, Stuttgart, Germany. ISBN 978-3-00-028755-8.

22. Romero M. (2004)."Solar Thermal Power Plants". In: "Report on research and development of energy technologies". Edited by IUPAP working group on energy; October 6, 2004. pp 96-108.

23. Horn M, Führing H, Rheinländer J. "Economic analysis of integrated solar combined cycle power plants. A sample case: The economic feasibility of an ISCCS power plant in Egypt". Energy 2004;29:935–1011.

24. Relloso S., Delgado E. "Experience with molten salt thermal storage in a commercial parabolic trough plant. Andasol-1 commissioning and operation. Proceedings SolarPACES 2009 (CD). Ref. manuscript: 11396; Berlin, Germany; 15-18 September 2009. Ed. DLR, Stuttgart, Germany. ISBN 978-3-00-028755-8.

25. Falchetta M., Mazzei D., Crescenzi T., Merlo L. "Design of the Archimede 5 MW molten salt parabolic trough solar plant". Proceedings SolarPACES 2009 (CD). Ref. manuscript: 11608; Berlín, Germany; 15-18 September 2009. Ed. DLR, Stuttgart, Germany. ISBN 978-3-00-028755-8.

26. Laing, D., Steinmann, W.-D., Tamme, R., and Richter, C., 2006, "Solid Media Thermal Storage for Parabolic Trough Power Plants", Solar Energy, 80(10), pp. 1283-1289.

27. Laing D., Bahl C., Bauer T., Lehmann D., Steinmann W.D. "Thermal energy storage for direct steam generation". Proceedings SolarPACES 2009 (CD). Ref. manuscript: 12055; Berlin, Germany; 15-18 September 2009. Ed. DLR, Stuttgart, Germany. ISBN 978-3-00-028755-8.

28. Zarza, E., Valenzuela, L., Leon, J., Hennecke, K., Eck, M., Weyers, H. D., and Eickhoff, M., 2004, "Direct steam generation in parabolic troughs: Final results and conclusions of the DISS project", Energy, 29(5-6), pp. 635-644.

29. Kalogirou S.A. "Solar thermal collectors and applications". Progress in Energy and Combustion Science. Volume 30, Issue 3, 2004, 231-295.

30. Mills D.R and Morrison G.L.(1999). Compact linear Fresnel reflector solar thermal powerplants. Solar Energy. V68, pp 263 – 283.

31. Bernhard R., Hein S., de Lalaing J., Eck M., Eickhoff M., Pfänder M., Morin G., Häberle A. (2008). Linear Fresnel Collector Demonstration on the PSA, Part II – Commissioning and first Performance Tests, Proceedings of the 14th Solar Paces Symposium, Las Vegas, USA. http://www.nrel.gov/docs/gen/fy08/42709CD.zip.

32. Hautmann G., Selig M., Mertins M., "First European Linear Fresnel power plant in operation – Operational experience&outlook". Proceedings SolarPACES 2009 (CD). Ref. manuscript: 16541; Berlin, Germany; 15-18 September 2009. Ed. DLR, Stuttgart, Germany. ISBN 978-3-00-028755-8.

33. Dersch J., Morin G., Eck M., Häberle A.: Comparison of Linear Fresnel and Parabolic Trough Collector Systems – System Analysis to determine break even Costs of Linear Fresnel Collectors. Proceedings SolarPACES 2009 (CD). Ref. manuscript: 15162; Berlin, Germany; 15-18 September 2009. Ed. DLR, Stuttgart, Germany. ISBN 978-3-00-028755-8.

34. Romero M., Marcos M.J., Téllez F.M., Blanco M., Fernández V., Baonza F., Berger S. (2000), "Distributed power from solar tower systems: A MIUS approach", Solar Energy, 67 (4-6) 249-264.

35. Kolb G.J. (1998). Economic evaluation of solar-only and hybrid power towers using molten-salt technology, Solar Energy. 62, 51-61.

36. Falcone P.K. (1986), "A handbook for Solar Central Receiver Design", SAND86-8009, Sandia National Laboratories, Livermore, (USA). http://prod.sandia.gov/techlib/access-control.cgi/1986/868009.pdf.

37. Sánchez M., Romero M. (2006) "Methodology for generation of heliostat field layout in central receiver systems based on yearly normalized energy surfaces", Solar Energy, 80, pp 861-874.

38. Lipps F.W., Vant-Hull L.L. (1978), "A Cellwise method for the optimization of large central receiver systems" Solar Energy 20, 505-516.

39. Kistler B.L. (1986), "A User's Manual for DELSOL3: A Computer Code for Calculating the Optical Performance and Optimal System Design for Solar Thermal Central Receiver Plants". Sandia Report, SAND-86-8018. Sandia National Laboratories, USA. http://prod.sandia.gov/techlib/access-control.cgi/1986/868018.pdf.

40. Mavis, C.L. (1989), "A description and assessment of heliostat technology", SAND87-8025, Sandia Nat. Labs., January 1989.

41. Romero M., Conejero E. and M. Sánchez (1991) "Recent experiences on reflectant module components for innovative heliostats", Solar Energy Materials 24, 320-332.

42. Monterreal, R., Romero, M., García, G. and Barrera, G. (1997)," Development and testing of a 100 m² glass-metal heliostat with a new local control system". Solar Engineering 1997, pp. 251-259, Eds. D.E. Claridge and J.E. Pacheco, ASME, New York, 1997. ISBN: 0-7918-1556-0.

43. Osuna R., Fernández V., Romero M., Sanchez M. (2004). "PS10: A 11-MW solar tower power plant with saturated steam receiver", Proceedings 12th SolarPACES International Symposium, 6-8 October 2004, Oaxaca, Mexico. S3-102. CD-Rom. Eds. C. Ramos and J. Huacuz. ISBN: 968-6114-18-1.

44. Silberstein E., Magen Y., Kroyzer G., Hayut R., Huss H. (2009) "Brightsource solar tower pilot in Israel's Negev operation at 130 bar @ 530°C Superheated Steam". Proceedings SolarPACES 2009 (CD). Berlín, Germany; 15-18 September 2009. Ed. DLR, Stuttgart, Germany. ISBN 978-3-00-028755-8.

45. Schell S. (2009). "Design and evaluation of eSolar's heliostat fields". Proceedings SolarPACES 2009 (CD). Berlín, Germany; 15-18 September 2009. Ed. DLR, Stuttgart, Germany. ISBN 978-3-00-028755-8.

46. Romero M., Buck R., Pacheco J.E. (2002), "An Update on Solar Central Receiver Systems, Projects, and Technologies.", Int. J. Solar Energy Eng., Vol. 124, pp. 98-108.

47. Becker M. and Vant-Hull L.L. (1991). "Thermal receivers". In "Solar Power Plants", Winter C.J., Sizmann R.L., Vant-Hull L.L. (Eds), Springer-Verlag, Berlin, pp. 163-197. ISBN 3-540-18897-5.

48. Agrafiotis C.C., Mavroidis I., Konstandopoulos A.G., Hoffschmidt B., Stobbe P., Romero M., Fernández-Quero V. (2007) "Evaluation of porous silicon carbide monolithic honeycombs as volumetric receivers/collectors of concentrated solar radiation". Solar Energy Materials & Solar Cells 91 (2007) 474–488.

49. Palero S., Romero M., Castillo J.L. (2008) "Comparison of Experimental and Numerical Air Temperature Distributions Behind a Cylindrical Volumetric Solar Absorber Module". Journal of Solar Energy Engineering 130, 011011-1-8.

50. Marcos M.J., Romero M., Palero S. (2004) "Analysis of air return alternatives for CRS-type open volumetric receiver". Energy, 29, 677–686.

51. Hoffschmidt B., Fernandez V., R. Pitz-Paal, M. Romero, P. Stobbe, F. Téllez (2002). "The Development Strategy of the HitRec Volumetric Receiver Technology - Up-Scaling from 200kWth via 3MWth up to 10MWel –". 11th SolarPACES International Symposium on Concentrated Solar Power and Chemical Energy Technologies. September 4-6, 2002. Zurich, Switzerland. pp. 117-126. ISBN: 3-9521409-3-7.

52. Romero M., Marcos M.J., Osuna R. and Fernández V. (2000), "Design and Implementation Plan of a 10 MW Solar Tower Power Plant based on Volumetric-Air Technology in Seville (Spain)". SOLAR ENGINEERING 2000-Proceedings of the ASME International Solar Energy Conference, Madison, Wisconsin, June 16-21, 2000. Ed.: J.D. Pacheco and M.D. Thornbloom, ASME, New York, 2000. ISBN: 0791818799.

53. Tellez F., Romero M., Heller P., Valverde A., Reche J.F., Ulmer S., Dibowski G. (2004). "Thermal Performance of "SolAir 3000 kWth" Ceramic Volumetric Solar Receiver", Proceedings 12th SolarPACES International Symposium, 6-8 October 2004, Oaxaca, Mexico. S9-206. CD-Rom. Eds. C. Ramos and J. Huacuz. ISBN: 968-6114-18-1.

54. Kribus A. (1999). "Future directions in solar thermal electricity generation". In: Solar thermal electricity generation. Colección documentos CIEMAT. CIEMAT, Madrid, Spain. pp. 251-285. ISBN: 84-7834-353-9.

55. Grasse W., Hertlein H.P., Winter C.J. (1991), "Thermal Solar Power Plants Experience" . In "Solar Power Plants", Winter C.J., Sizmann R.L., Vant-Hull L.L. (Eds), Springer-Verlag, Berlin, pp 215-282. ISBN 3-540-18897-5.

56. Radosevich L.G., Skinrood A.C.(1989), "The power production operation of Solar One, the 10 MWe solar thermal central receiver pilot plant", J. Solar Energy Engineering, 111, 144-151.

57. Pacheco J.E., Gilbert R. (1999), "Overview of recent results of the Solar Two test and evaluations program". In Renewable and Advanced Energy Systems for the 21st Century RAES'99 April 11–15, 1999 — Maui, Hawaii, pp. RAES99-7731, Eds.R. Hogan, Y. Kim, S. Kleis, D. O'Neal and T. Tanaka; ASME, New York, 1999. ISBN: 0-7918-1963-9.

58. Epstein M, Liebermann D., Rosh M, Shor A. J (1991), "Solar testing of 2 MW (th) water/steam receiver at the Weizmann Institute solar tower", Solar Energy Materials, Vol. 24, pp. 265-278.

59. Fernandez-Quero V., Osuna R., Romero M., Sanchez M., Ruiz V., Silva M., (2005) EURECA: Advanced Receiver For Direct Superheated Steam Generation In Solar Towers, As An Option For Increasing Efficiency In Large Low Cost Direct Steam Generation Plants. Proceedings of the 2005 Solar World Congress ISES-2005. 6-12 August, Orlando, Florida. Ed. By D.Y. Goswami, S. Vijayaraghaveng, R. Campbell-Howe. Pub. American Solar Energy Society, Boulder, Colorado, USA .ISBN 0-89553-177-1.

60. Tyner CE., Pacheco J.E. (2009). "eSolar's power plant architecture". Proceedings SolarPACES 2009 (CD). Berlín, Germany; 15-18 September 2009. Ed. DLR, Stuttgart, Germany. ISBN 978-3-00-028755-8.

61. Kelly, B., Singh, M. (1995) "Summary of the Final Design for the 10 MWe Solar Two Central Receiver Project," Solar Engineering: 1995, ASME, Vol. 1, p. 575.

62. Pacheco, J. E., H. E. Reilly, G. J. Kolb, C. E. Tyner (2000), "Summary of the Solar Two Test and Evaluation Program". Proceeding of the Renewable Energy for the New Millennium, Sydney, Australia, March 8-10, 2000. pp.1-11.

63. Ortega J.I., Burgaleta J.I., Tellez F. (2006) "Central Receiver System (CRS) solar power plant using molten salt as heat transfer fluid". Proceedings of the 13th International Symposium on Concentrated Solar Power and Chemical Energy Technologies. June 20, 2006, Seville, Spain. M. Romero, D. Martínez, V. Ruiz, M. Silva, M. Brown (Eds.) (2006). Pub. CIEMAT. ISBN: 84-7834-519-1.

64. Burgaleta J.I., Arias S., Salbidegoitia I.B. (2009) "Operative advantages of a central tower solar plant with thermal storage system". Proceedings SolarPACES 2009 (CD). Ref. manuscript: 11720; Berlin, Germany; 15-18 September 2009. Ed. DLR, Stuttgart, Germany. ISBN 978-3-00-028755-8.

65. Mancini T., Heller P., Butler B., Osborn B., Schiel W., Goldberg V., Buck R., Diver R., Andraka C., Moreno J. (2003), "Dish–Stirling Systems: An Overview of Development and Status", Int. J. Solar Energy Eng., Vol. 125, pp. 135-151.

66. Lopez C., Stone K., (1992), "Design and Performance of the Southern California Edison Stirling Dish," Solar Engineering, Proc. of ASME Int. Solar Energy Conf., Maui, HI, ISBN 0-7918-762-2, pp. 945–952.

67. Keck T., Heller P., Weinrebe G., (2003). "Envirodish and Eurodish — system and status". Proc. ISES Solar World Congress, Göteborg, Sweden, June 2003. ISBN: 91-631-4740-8.

68. Diver, R.B., (1987), J. Solar Energy Engineering, 109 (3), 199-204.

69. Moreno J. B., Modesto-Beato M., Rawlinson K. S., Andraka C. E., Showalter S. K., Moss T. A., Mehos M., and Baturkin V., (2001), "Recent Progress in Heat-Pipe Solar Receivers," SAND2001-1079, 36th Intersociety Energy Conversion Engineering Conf., Savannah, GA, pp. 565–572.

70. Stine W., Diver R.B (1994), "A Compendium of Solar Dish/Stirling Technology", report SAND93-7026; Sandia National Laboratories, Albuquerque, New Mexico.

71. IEA (2010) Technology Roadmap – Concentrating Solar Power. Available free at http://www.iea.org.

72. Pitz-Paal R., Dersch J., Milow B., Ferriere A., Romero M., Téllez F., Zarza E., Steinfeld A., Langnickel U., Shpilrain E., Popel O., Epstein M., Karni J. (2005). ECOSTAR Roadmap Document for the European Commission; SES-CT-2003-502578. Edited by: Robert Pitz-Paal, Jürgen Dersch, Barbara Milow. Deutsches Zentrum für Luft- und Raumfahrt e.V., Cologne, Germany. February 2005. http://www.vgb.org/data/vgborg_/Forschung/roadmap252.pdf.

SOLAR ENERGY APPLICATIONS IN INDIA

Rangan Banerjee

4.1   INTRODUCTION

India has a population of 1.1 billion people (1/6th of the world population) and accounts for less than 5% of the global primary energy consumption. India's power sector had an installed capacity of 159,650 MW [1] as on 30th April, 2010. The share of installed capacity from different sources is shown in Fig. 4.1.

The annual generation was 724 Billion units during 2008–2009 with an average electricity use of 704 kWh/person/year. Most states have peak and energy deficits. The average energy deficit is about 8.2% for energy and 12.6% for peak [1]. About 96,000 villages are unelectrified (16% of total villages in India) and a large proportion of the households do not have access to electricity.

India's development strategy is to provide access to energy to all households. Official projections indicate the need to add another 100,000 MW within the next decade. The scarcity of fossil fuels and the global warming and climate change problem has resulted in an increased emphasis on renewable energy sources. India has a dedicated ministry focussing on renewables (Ministry of New and Renewable Energy, MNRE). The installed capacity of grid connected renewables is more than 15,000 MW. The main sources of renewable energy in the present supply mix are wind, small hydro- and biomass-based power and cogeneration. In 2010, India has launched the Jawaharlal Nehru Solar Mission (JNSM) as a part of its climate change mission with an aim to develop cost-effective solar power solutions.

Most of India enjoys excellent solar insolation. Almost the entire country has insolation greater than 1900 kWh/m^2/year with about 300 days of sunshine. Figure 4.2 shows a map with the insolation ranges for different parts of the country. The highest insolation (greater than 2300 kWh/m^2/year) is in the state of Rajasthan in the north of the country. The solar radiation (beam, diffuse, daily normal insolation) values are available at different locations from the handbook of solar radiation data for India [2] and at 23 sites from an Indian Meteorological Department (IMD) MNRE report [3].

4.2   STATUS AND TRENDS

Figure 4.3 shows the different end uses for energy. For solar energy applications the two distinct routes are solar photovoltaics (PV) and solar thermal. An overview of solar PV [4, 5] and solar thermal [5–7] reveals steady growth and developments in applications.

The Indian PV industry started in 1976 with the research and development efforts of Central Electronics (CEL). In the 1980s, public sector undertakings like CEL, Bharat Heavy Electricals and Udaya Semiconductors manufactured solar cells in India. Since 1992, the manufacture of solar cells was liberalized. At present, the solar cell and module manufacture is predominantly driven by large private sector companies like Tata BP Solar, Moser Baer [8]. Figure 4.4 shows the trend in growth of production capacity of cells and modules. A large part of the PV manufacturing capacity (70% or more) caters to the international market. At present, in the established PV plants, the capacity utilization of installed production capacity is above 90%. During 1999–2009, India's PV module production has increased from about 11 MW/year to about 240 MW/year (a compound annual growth rate of 36 % per year).

Figure 4.5 shows the cumulative deployment of PV modules produced in India. It can be seen that exports accounted for 525 MW (66%) of the total module production. PV applications include solar lanterns, solar home lighting and street lighting, solar power plants (isolated-off grid) and grid-connected systems, and PV for telecom applications. Almost all the commercial PV capacity is based on crystalline silicon. Recently, Moser Baer has set up an 80-MW capacity line for manufacture of amorphous silicon modules.

Solar thermal systems are indigenously available for water heating, drying and cooking. Figure 4.6 shows the trend in the growth of the cumulative installed capacity of flat plate collectors in India. The actual installation of solar hot water systems was about 3.5 million m^2 in March 2010 [10]. The installed solar hot water collectors in India are about 3.5 m^2/1000 people in India as compared to the world average of 30 m^2/1000 people. Cities like Bangalore and Pune have a large number of solar hot water installations in domestic buildings.

4.3   GRID CONNECTED PV SYSTEMS

The progress of grid connected PV systems in India was slow till 2008. The capacity of the largest grid connected PV system in 2008 was 239 kW. The average capacity factor of grid connected PV systems was 14% in 2002–2003. Recent efforts to provide a preferential tariff for solar PV and the launch of the JNSM has resulted in an increased emphasis on grid connected PV systems. In December 2009, two MW scale grid connected PV plants were installed at Jamuria, Asansol, West Bengal (Fig. 4.7) [9] and Amritsar in Punjab resulting in an increase in the present installed grid connected PV power to 6 MW. In 2009–2010, a total of 8 MW of grid connected PV plants were installed taking the total grid connected PV capacity to 10 MW.

The largest grid connected system in India is a 3 MW system at Yalesandra in Kolar district in Karnataka (Fig. 4.8) (commissioned

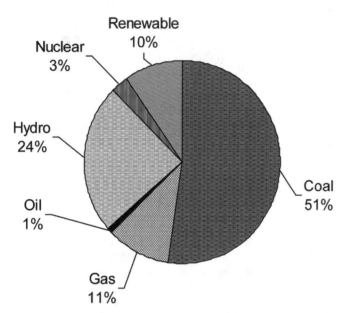

FIG. 4.1 SHARE OF POWER GENERATION INSTALLED CAPACITY 2010 [1]

in June 2010) [11] at a cost of Rs. 590 million (14 million US $). The system has twelve 250 kW inverters supplying power to the grid at 11 kV level. The cost of electricity generation is estimated to be Rs 16.9 /kWh and the plant occupies an area of 15 acres. A simulation of the output of 1 MW peak rated power plant located in different regions of India reveals the effect of varying solar insolation and ambient temperature [12]. The annual generation from a 1 MW peak plant varies from 1752 MWh in Rajasthan (cost of generated electricity Rs13 or US 26 ¢/kWh) to 1462 MWh in Kolkata (Rs 16.5 or US 33 ¢/kWh) [12].

4.4 VILLAGE ELECTRIFICATION USING SOLAR PV

It is difficult to obtain accurate data on the number of village electrification systems using solar PV that are operational in the country. Table 4.1 provides an estimate of different systems in select states. Many systems are 4- to 6-hour grids only meeting the residential loads during the evening. Systems can be single phase AC or three-phase AC. Table 4.2 provides an estimate of the capacity factor of a few sample village electrification systems. An analysis of sample systems [7] reveals that the average cost of supply can be reduced by 25% (average value Rs. 32/kWh) or more through optimal sizing and distribution planning.

There are several examples of successful solar PV distributed generation systems in the islands. For example, the nine solar minigrids in the Sunderbans installed by the West Bengal Renewable Energy Development Agency have a total peak rating of 345 kW and meets the needs of 1750 consumers. An example of a 26 kW peak system in the Sundarbans (battery capacity 2 V, 800 Ah) had a capital cost of Rs 10.6 million Rupees and supplied 120 users [13]. The users paid a fixed monthly connection charge of Rs1000. The model used in the Sunderbans was to have a fixed monthly charge irrespective of use but couple it with loads limiters for each connection. Village co-operatives would be responsible for plant

operation and bill collection. Some of these systems are trying out prepaid metering [13].

Figure 4.9 shows a schematic of a solar PV power system along with typical voltage values. One of the main issues in an isolated system is matching the supply with the demand. Figure 4.10 shows the load profile measured on a typical PV village electrification project. Several attempts have been made to improve the capacity factor and the economic viability by linking with a constant/base load. In many of the island grids solar photovoltaic systems have been hybridised with the existing diesel micro-grids. BHEL has installed a 50 kW (peak) PV system along with a diesel engine generator and battery storage (two 75-kVA diesel generators) at Bangaram island in Lakshwadeep in 2006 [15]. The system operates as a 24 hour power supply for the island.

4.5 SOLAR THERMAL COOKING SYSTEMS

In India, the first solar cookers were developed by the National Physical Laboratory in 1954 with a circular parabolic reflector focussing on to a cooking pot. Subsequently, box type solar cookers were developed in 1961. It is estimated that about 0.5 million solar cookers have been installed in the country [17, 18]. Most of these cookers are used for supplementary cooking.

In the early 1990s, solar dish cooker for community application (Scheffler cooker) was developed and manufactured in India. The aperture area of the dish is about 7 m². The automatically tracked parabolic reflector is located outside the kitchen and reflects the

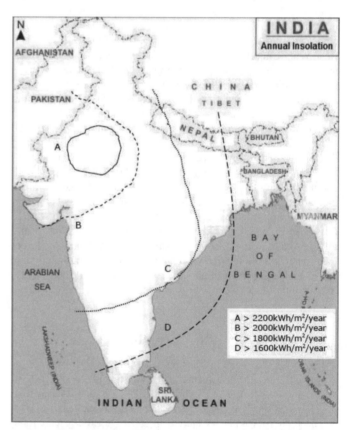

FIG. 4.2 AVERAGE ANNUAL DAILY NORMAL INSOLATION FOR INDIA Data Source [3]

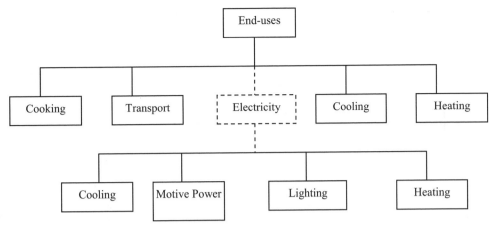

FIG. 4.3 ENERGY END-USES [4]

incident solar energy through an opening in its mouth wall. A secondary reflector is used to focus the solar radiation on to the cooking vessel. Figure 4.11 shows a schematic of a Scheffler cooker.

Table 4.3 lists some of the large community installations of solar cooking in India. The cooking system installed at Tirumala Tirupati Devasthanam at Tirumala in Andhra Pradesh (see Fig. 4.12) is considered to be one of the world's largest solar cooking system, consisting of 106 automatically track parabolic concentrator's generates steam. This system was installed in 2002 with a capital cost of Rs. 11 million. The system generates 4 tonnes of steam per day 180°C (10 kg/cm^2 pressure). Government provided 50% subsidy for this plant. It is estimated that the savings are around 118,000 L of diesel per year.

In 2006, 300 community solar dish cookers (2.3 m diameter) were installed in tribal school in Maharashtra. The project cost was Rs. Eight million and estimated LPG savings was 110 kg annually resulting in a payback period of 6 years [19].

Solar concentrator systems generating steam seem suitable for community kitchens, hotels. The technology is available though the payback period for LPG replacement is about 6 years.

4.6 SOLAR THERMAL HOT WATER SYSTEMS

Solar thermal hot water heaters are a manufactured, installed and serviced by indigenous manufacturers. There are 56 Bureau of Indian Standards (BIS) approved manufacturers of flat plate collectors and 23 MNRE approved manufacturers (assemblers) of evacuated tube-based systems in the country. The actual installation is about 3.5 million m^2 in 2010 [9]. The cost of a typical 100 litres per day household system is about Rs 20,000 (US 500$). Most of the manufacturers are small and medium size companies with no single manufacturer having more than 15 % share [22]. Figure 4.13 shows a 25,000 lpd solar water heating system installed at a student hostel at the Indian Institute of Technology Bombay at Mumbai 23]. The residential sector accounts for 80% of the total solar water heater demand. Other potential users are hotels and hospitals.

A framework was proposed for estimation of potential of solar water heating systems for a target area and a country [24, 25]. The framework has been applied to Pune and extrapolated to get the potential of solar water heating in India. The estimated potential

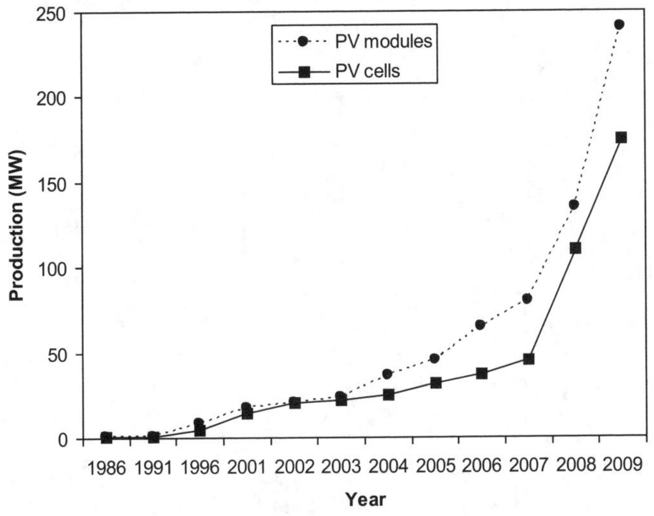

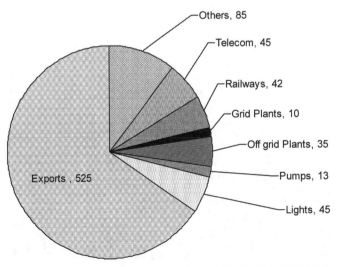

FIG. 4.4 GROWTH OF PRODUCTION CAPACITY OF CELLS AND MODULES Data Source [9, 4]

FIG. 4.5 CUMULATIVE PRODUCTION (IN MW) OF SOLAR PV IN INDIA Source: Mnre Presentation, Solar Energy Conclave, January 2010

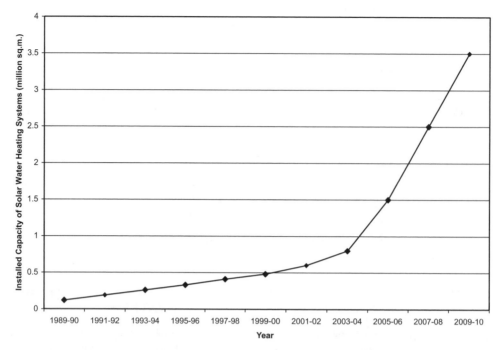

FIG. 4.6   GROWTH OF SOLAR WATER HEATING SYSTEMS IN INDIA [4]

of solar water heating in India is 60 million m² leading to annual electricity savings of 12.2 billion kWh, The potential estimated is used for the synthesis of S-curve for solar water heating systems (Fig 4.14). The payback periods for hotels and hospitals range between 2 and 4 years while the payback is higher for single households due to the discrete sizes of solar water heaters available, morning peak hot water usage patterns.

Figure 4.15 shows an estimation of the impact of electric water heaters on the load curve for Pune in Maharashtra [26]. It can be seen that electric water heating contributes significantly to the morning peak of the utility. Solar water heaters can effectively reduce morning peak. The MNRE provides an interest subsidy for end-users and a capital subsidy for government institutions. Some state governments (Delhi) also offer a capital subsidy. Some municipalities like Bangalore, Thane, and Kalyan are making solar water heaters compulsory for new housing units, hotels, and hospitals.

4.7   SOLAR THERMAL SYSTEMS FOR INDUSTRIES

Many industries have low-temperature process heat requirements (hot water, steam) which is ideal for the application of solar thermal systems. In India, there have been several installations of flat plate and evacuated installations of flat plate and evacuated tubular collectors for preheating boiler feedwater and low-temperature process heating.

FIG. 4.7   1 MW SPV POWER PLANT INSTALLED BY WEST BENGAL GREEN DEVELOPMENT CORPORATION LIMITED AT JAMURIA VILLAGE IN ASANSOL, WEST BENGAL [9]

FIG. 4.8   3 MW SYSTEM AT YALESANDRA IN KOLAR DISTRICT IN KARNATAKA [11]

TABLE 4.1   SAMPLE ISOLATED SOLAR PV POWER PLANTS IN INDIA

Name of the plant	PV capacity (kWp)	Inverter capacity (kVA)	Battery capacity (Ah)	Battery voltage (V)	Connected load (kW)	Number of households	Plant capacity factor (%)
Dound-II, Chattisgargh	1	1.5	200	48	0.3	25	5.8
Latdadar, Chattisgargh	2	3	400	48	0.7	30	7.1
Chatal, Chattisgargh	3	5	400	48	0.7	24	4.4
Gudagarh, Chattisgargh	4	5	800	48	1.2	60	6.3
Sura, Udaipur, Rajasthan	17.25	15	1200	120	5.0	50	7.3
Nurda Village, Jharkhand	28	20	1200	120	9.5	350	8.5
Anandgarh, Bikaner, Rajasthan.	34.5	2*15	2*1200	120	10	50	7.5

Source: http://mapunity.org/projects/gvep and Rajasthan Renewable Energy Corporation (RREC).

A solar pond of 6000 m^2 was constructed at Bhuj in Gujarat to provide process heat to a dairy in 1990 to 1991. Table 4.4 provides a few examples of industrial solar flat plate water heating systems in the country.

A system of imported line focussing parabolic concentrators for producing process steam at 150°C and 100 kg/hour was installed at a silk factory in Mysore in 1987 [27]. Scheffler dishes have been used to generate saturated steam at 5 bar (20 dishes of 12.64 m^2) in a demonstration project supported by the Ministry of New and Renewable Energy at the Global Hospital and Research Centre, Mount Abu. The overall efficiency reported was about 28%.

A Fresnel paraboloid dish (ARUN160) [28] with two-axis tracking has been developed by Clique and has been installed at Mahananda dairy at Latur in Maharashtra (Fig. 4.16). This is a 160-m^2 dish that is used to replace steam fed from a 1-tonne/hour boiler. Figure 4.17 shows a schematic of its integration into the dairy utility system. The boiler can be used as a backup for days where there is inadequate energy from the solar system, resulting in a reliable heat supply system without any need for storage. Solar dryers have been used in India for different crops, food products and timber. It is estimated that more than 10,000 m^2 of dryers are operational in the country [21]. Solar cooling systems have also been demonstrated based on vapour absorption cycles with lithium bromide-water. A 25-TR (tonnes of refrigeration) solar air conditioning plant with a collector area of 280 m^2 has been installed in Ahmedabad [21]. The capital cost is Rs. 5.8 million with the solar component accounting for Rs. 4 million.

TABLE 4.2   ISOLATED SPV POWER PLANTS IN INDIA [14]

S. no.	State	PV Capacity (kWp)	No. of power plants
1	Maharashtra	5	2
2	Chattisgargh	1–6	108
3	Rajasthan	5–34	83
4	Jharkhand	28	1
5	Orissa	2	11
6	Haryana	10	–
7	Mizoram	25	1
8	UP (NTPC)	11.9 kW	1
9	West Bengal	25 kW	15

Installed capacity (As on 31/01/2009): 2.8 MWp
Total number of SPV power plants: 230 nos.

A municipal hospital of the Thane Municipal Corporation in Kalwa, near Mumbai has installed a 160 tonnes of Refrigeration plant based on solar thermal energy has been commissioned in 2010. This has 184 Scheffler dishes to generate 700 kg/hour of steam that runs vapor absorption refrigerator to generate chilled water at 7°C. The project is developed by Sharada Inventions Nasik at a cost of Rs. 40 million. The system has agro residue-based briquette fired boilers as a back up to generate steam on cloudy days or days with low solar insolation. The estimated annual electricity savings is about 1 million kWh with money savings of Rs. 4.5 million. The payback period is about 9 years. The Government has provided a capital subsidy of Rs. 12.4 million reducing the payback period to about 6 years. The system also has a solar dehumidifier developed by IIT Bombay.

4.8   SOLAR THERMAL POWER GENERATION

India has limited experience in solar thermal power generation. There was a demonstration 50 kW line focussing parabolic collector demonstration plant (imported) at the solar energy centre at Gwalpahari (near Delhi) by the Ministry of New and Renewable Energy. Due to maintenance problems the unit was derated and then combined with a biomass gasifier system.

An integrated solar combined cycle plant with a 35-MW solar component (based on line focussing parabolic collectors) was planned at Mathania in Rajasthan in early 1990s. However, this plant was not installed due to problems in obtaining multi-lateral funding. Demonstrations units based on imported paraboloid dishes have been installed near Hyderabad (20 kW rating) and at Vellore (10 kW rating). At present, there are no grid connected solar thermal power plants in operation in the country. The launch of the JNSM has resulted in an increased interest in solar thermal power generation.

There are several indigenous manufacturers who are developing solar concentrator systems that generate steam either through parabolic trough concentrators (KIE Solatherm), Fresnel paraboloid dish (Clique), Scheffler (Gadhia solar) or Compact Linear Fresnel Reflectors (KG Design Services, Coimbatore). An energy sector manufacturing company Thermax also has planned demonstration projects for solar power generation based on CLFR and parabolic concentrator technology.

MNRE has funded a national solar thermal power plant cum testing facility. This is planned as a consortium led by the Indian

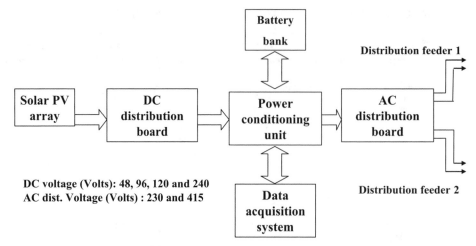

FIG. 4.9 SCHEMATIC OF SPV POWER PLANT [16]

Institute of Technology Bombay, Mumbai with the Solar Energy Centre and several industry partners (Tata Power, TCE, L&T, KIE Solatherm, Clique). The facility is to be located at the Solar Energy Centre at Gwalpahari (near Delhi) and will supply 1 MW of electricity to the Haryana grid (during sunshine hours). The plant is expected to be operational in 2011 and will enable testing of solar components and systems. It is planned to develop a simulator for a solar-based power plant and validate with the prototype plant. The unique feature of this facility is its ability to integrate and test different medium and low-temperature solar concentration technologies. The objective of this facility is to promote the development of cost-effective solar thermal power plants in the future in India.

There is an ongoing research project at IIT Bombay, supported by MNRE, to develop 3 kW Stirling engine for solar applications. At present, there are no indigenous Stirling engine suppliers. In Pune, there is a group working on integration of solar dishes with a steam engine at the 10-kW level.

4.9 SOLAR LIGHTING AND HOME SYSTEMS

Solar lighting systems in India were developed in the 1970s. As per the MNRE, there are about 580,000 solar home lighting systems, 790,000 solar lanterns (Fig. 4.18) and 88,000 solar street lights installed in the country. It is estimated that there are 72 million households in India which use kerosene for lighting in rural India. This represents the target potential for solar lighting systems in the country. Table 4.5 provides specifications of some of the solar PV lanterns and home systems available commercially in India. Innovative financing schemes that enable rural households and low-income users to afford solar lighting and home systems have been implemented by companies like SELCO and a few non-government organisations. At present, a 90% capital subsidy is available for solar home systems in remote areas. In India, it is estimated that 3600 million liters of kerosene are used for domestic lighting [29]. Replacing kerosene-based lighting with solar lighting systems will result in significant savings in annual kerosene subsidy and carbon dioxide emissions.

4.10 SOLAR MISSION AND FUTURE OF SOLAR IN INDIA

The Jawaharlal Nehru Solar Mission (JNSM) was announced by the Indian government in 2009 and formally launched in January

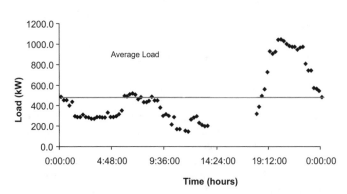

FIG. 4.10 LOAD PROFILE TAKEN OVER A DAY (5 KW SYSTEM) [14]

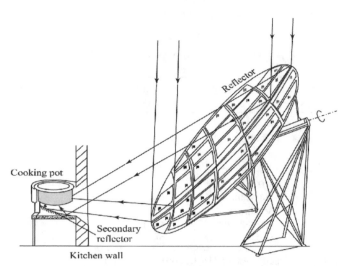

FIG. 4.11 SCHEMATIC OF A SCHEFFLER COOKER [21]

TABLE 4.3 LARGE SOLAR COOKING INSTALLATIONS IN INDIA [4,20]

S. no.	Location	Cooking capacity people/day	Collector area
1.	Sri Sai Baba Sansthan, Shirdi, Maharashtara	40,000	73 reflective dishes of 16 m^2 each (1168 m^2)
2.	Brahmakumaris training centre, Hubli	600	75 m^2
3.	Brahmakumaris Ashram, Gurgaon, Haryana	2000	265 m^2
4.	Brahmakumaris Ashram, Talleti, near Mount Abu	10,000	800 m^2
5.	Rishi Valley School, Chitoor District, Andhra Pradesh	500	94 m^2
6.	Tirupalli in Andhra Pradesh	15000	106 collectors of 9.2 m^2 each (975 m^2)
7.	Auroville, Pondicherry, Tamil Nadu	1000	15 m diameter solar bowl concentrator

FIG. 4.13 SOLAR WATER HEATING SYSTEM AT IIT BOMBAY, MUMBAI [23]

2010. Table 4.6 shows the targets set by the JNSM. It is expected that 20 GW of grid connected solar power will be established in India till 2022. 20 million solar lighting systems is the target set for 2022. Incentives have been announced for rooftop solar PV systems including a generation-based incentive. For off grid systems, a capital subsidy and a 5% interest soft loan have been offered. The capital subsidy is about Rs 70 to 90/W peak and about Rs 2100 to 6000/m^2 for different solar thermal concentrator systems. The subsidy is estimated to be 30% of the market cost. An amount of Rs 43370 million has been approved for implementation of the mission till 2013 (1 billion US$) for 1000 MW of grid connected solar (at the 33 kV level), 100 MW of solar plants at the low tension/11 kV level and 200 MW of off-grid solar plants.

The National Thermal Power Corporation (NTPC) one of the largest public sector generation companies in the world has a fully owned subsidiary the NTPC Vidyut Vyapar Nigam Ltd. (NVVN). NVVN has been designated as the nodal agency to purchase solar power fed to 33 kV and above and is expected to bundle solar power with coal-based power. The power sold through NVVN can be used to meet the renewable purchase obligations of different Indian utilities. The Central Electricity Regulatory Commission has notified a renewable based preferential tariff that is based on a capital cost of Rs 170 million/ MW for solar PV and a capacity utilization factor of 19% and a capital cost of Rs 130 million/MW for solar thermal and a capacity utilization factor of 23%. This results in an effective preferential tariff of Rs 18.4/kWh for solar photovoltaics and a preferential tariff of Rs 15/ kWh for solar thermal.

Solar water heaters for low-grade heating are already viable for residential and commercial applications. These are now ready for

FIG. 4.12 SOLAR COOKING SYSTEM AT TIRUMALA TIRUPATI DEVASTHANAM, ANDHRA PRADESH [20]

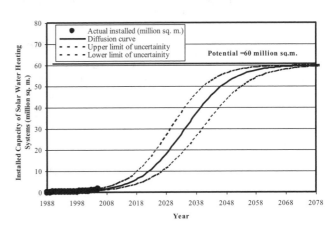

FIG. 4.14 DIFFUSION CURVES FOR SOLAR WATER HEATING SYSTEMS IN INDIA [4]

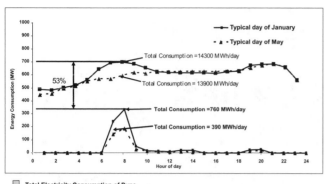

FIG. 4.15 LOAD CURVE REPRESENTING ENERGY REQUIREMENT FOR WATER HEATING [26]

FIG. 4.16 ARUN160 AT MAHANANDA DAIRY, LATUR, MAHARASHTRA [28]

widespread diffusion and market penetration. Application engineering and optimal sizing is needed to have a larger number of installations in hotels and hospitals. Changes in the building codes and a regulatory push will enable their faster adoption. A recent study supported by the MNRE has proposed that utilities provide a rebate of Rs 75/m^2 for solar collectors installed to offset electric water heating [35] since it results in electrical energy savings and peak power savings.

Solar thermal systems in industry have a significant potential to replace/reduce oil-fired steam generation. At present a few demonstration systems based on dish (ARUN160) and Scheffler technologies have been installed. It is expected that these technologies along with CLFR would effectively diffuse into the industrial steam market. Technology development and cost reduction efforts are needed in order to penetrate the industrial steam market.

For solar thermal power generation a large number of agreements have been signed with many of the international suppliers in the states of Gujarat and Rajasthan. By 2012, it is expected that India will have several grid connected solar thermal plants. Indigenous technology development and operating experience is likely

to promote the next generation of solar thermal power plants in the country. MNRE and Department of Science and Technology are supporting solar research and training efforts that will enable development of cost-effective solar thermal systems.

In solar photovoltaics, the incentives provided in the JNSM are likely to see the growth of the domestic market — both grid connected — large and rooftop and off grid applications. India is likely to significantly increase its production capacity of cells and modules. Based on the applications and announcements made by solar PV companies, it is expected that the cell and module capacity would increase to more than 1000 MW per year by 2012. India has a research based on capabilities in materials, amorphous silicon, thin films, and new solar cell concepts. This is being supported by the MNRE and the Solar Energy Centre. A National Centre for Photovoltaics Research and Education has been planned at IIT Bombay to facilitate and enhance PV research. The entry of companies like Moser Baer who have a track record of innovation and cost reduction is likely to drive down costs.

The Indian government has announced its intent to support solar energy in country through the JNSM and has provided financial support for solar implementation. There exists a manufacturing base in the country that can take up this mission. This needs to be supported by a critical mass of researchers and technology developers. It is important to assess the actual performance of implementation and cost reduction of solar technologies supported by the JNSM. The mainstreaming of solar technologies will depend on the innovations and technology solutions provided that will reduce costs and improve reliability. An enabling policy environment has been provided by the JNSM. India has a rapidly growing energy sector. If technology developers, researchers, and industry take up the challenge and develop cost-effective solar solutions, India can have an increasing solar share in the future energy mix. Under such a scenario, India has the potential to be a solar manufacturing hub for the world.

TABLE 4.4 INDUSTRIAL APPLICATIONS OF SOLAR FLAT PLATE SYSTEMS [22]

		Comments
Panchmahal dairy, Godha	20,000 lpd 80°C preheating boiler feed water	236 flat plate collectors storage 20,000 litres savings 110 litres/day furnace oil
Synthokem Labs Pvt. Ltd., Hyderabad	10,000 lpd boiler feed water preheat	16% reduction in energy cost, 2 years payback
Kangaroo Industries, Ludhiana	Electroplating 100 lpd	
GFTCL Kakinada	120,000 lpd Boiler feedwater preheating	1309 collectors, Payback period 5 years

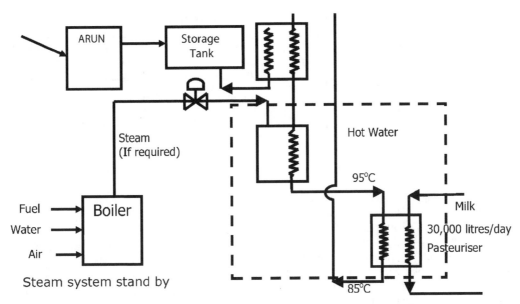

FIG. 4.17   SOLAR SYSTEM FOR A DAIRY IN MAHARASHTRA

4.11   CONCLUSION

The conditions for solar energy in India are similar to the conditions prevalent in other countries of South Asia — Nepal, Bangladesh, Pakistan and Sri Lanka. All these countries have electricity shortages and a large proportion of the population without access to convenient energy sources. In Bangladesh, about 70% of the households are not connected to the grid. There have been a growth in the deployment of solar home systems (solar PV systems 40 W to 75 W) and solar flat plate water heating systems in Bangladesh and Nepal. In Bangladesh, an NGO Gramee Shakti has been instrumental in innovative financing for Solar Home systems resulting in a total deployment of 220,000 Solar Home Systems. In Sri Lanka, the Japanese Government has provided assistance of 860 million Yen (INR 350 million) to set up a 400 kW grid connected PV plant at Hambantota.

In India, there is a well-developed indigenous industry for solar flat plate heater and solar PV cells and modules. The capability for concentrated solar thermal and solar thermal-based power generation is being developed through a few R&D projects. The attractive tariffs being provided under the JNSM is expected to encourage leading international suppliers to set up solar thermal power plants

FIG. 4.18   COMMERCIALLY AVAILABLE SOLAR LANTERN [33]

TABLE 4.5   SPECIFICATIONS OF SOLAR PV LANTERNS AND SOLAR HOME SYSTEMS [30–32]

S. no.	PV Module	Light	Battery	Cost
Solar lanterns				
1.	10 Wp	7 W CFL	12 V, 7 Ah	Rs 3300
2.	5 Wp	5 W CFL	6 V, 5 Ah	Rs. 1600
3.	3 Wp	2.5 W LED	6 V, 4.5 Ah	Rs. 1600
Small home systems				
1.	18 Wp	1 CFL (9 W/ 11 W)	12 V, 20 Ah	Rs. 8000
2.	37 Wp	2 CFL (9 W/ 11 W) or 1 CFL and DC Fan	12 V, 40 Ah	Rs. 13000
3.	50 Wp	9 W and 11 W CFL, portable 4-B/W TV	80 Ah	Rs. 18000
4.	74 Wp	2 CFL, 1 DC Fan, 1 B/W TV	12 V, 75/80 Ah	Rs. 28000

TABLE 4.6   JNSM TARGETS [34]

S. no.	Application segment	Target for Phase I (2010–2013)	Target for Phase II (2013–2017)	Target for Phase III (2017–2022)
1.	Solar collectors	7 million m^2	15 million m^2	20 million m^2
2.	Off grid solar applications	200 MW	1000 MW	2000 MW
3.	Utility grid power, including roof top	1000–2000 MW	4000–10000 MW	20000 MW

in India. Several collaborative projects in solar energy are being proposed. In India, there is an Indo-UK collaborative research project with multi-institutions in both countries to study the stability and field performance of PV cells, modules and systems in both countries. A joint Indo-EU research programme plans to support collaborative research projects in solar energy between EU, and Indian research organisations and industry. The US has planned to set up a Joint Centre for Solar Energy research between leading US and Indian research agencies and companies and is currently evaluating different proposals submitted by interested organisations in both countries.

The potential growth and untapped demand for energy and the availability of good solar insolation makes the Indian sub-continent an emerging solar market. The challenge however is to devise cost-effective solution that fit the customers' ability to pay.

4.12   ACKNOWLEDGMENTS

The author is grateful to Mr. Balkrishna Surve for assistance in preparing this chapter.

4.13   REFERENCES

1. *www.powermin.nic.in/indian_electricity_scenario/introduction. htm* Last accessed on June 22, 2010.

2. Mani, A., 1981. 'Handbook of solar radiation data for India', Allied Publishers.

3. Solar radiant energy over India, Indian Meteorological Department (IMD) - MNRE available at *http://www.mnre.gov.in/sec/solarradiant-energyoverIndia.pdf* Last accessed on July 6, 2010.

4. Pillai, I. R. and Banerjee, R., 2009. 'Renewable energy in India: Status and potential', Energy 34, 970-980.

5. Bhargava, B., 2008. Photovoltaic Technology Development in India: an overview, 25 years of renewable energy in India, Ministry of New and Renewable Energy, New Delhi, 45-73.

6. Kumar, A., Development and Promotion of low temperature solar thermal energy technologies in India, 74-95.

7. Manoj, M.V. and Banerjee, R., 2010. 'Analysis of Isolated Power Systems for Village Electrification', accepted for publication in Energy for Sustainable Development.

8. India Semiconductor Association (ISA) report on the 'Solar PV Industry 2010: Contemporary Scenario and Emerging Trends', supported by office of the Principal Scientific Adviser (PSA) to the Government of India.

9. Annual Report 2009, Ministry of New and Renewabale *Energy Source* (http://www.mnre.gov.in)

10. Ministry of New and Renewable Energy cumulative achievements as on 31st March, 2010; http://www.mnre.gov.in/; Last accessed on July 5, 2010.

11. The Kerala State Electricity Regulatory Commission (KSERC) (http://www.kseboa.org/kseb/)

12. Doolla, S. and Banerjee, R., "Diffusion of Grid-Connected PV in India: An Analysis of Variations in Capacity Factor", Proceedings of IEEE PhotoVoltaic Specialist Conference, 20-25 June 2010, Hawaii, USA.

13. Moharii, R.M. and Kulkarni, P.S. 2009 'A case study of solar photovoltaic power system at Sagardeep Island, India', Renewable and Sustainable Energy Reviews, Volume 13, Issue 3, 673-681.

14. Sivapriya, M. B., 2008. M.Tech Dissertation 'Optimal Placement and Sizing of Distributed Generators in Microgrid', Department of Energy Science and Engineering, Indian Institute of Technology Bombay.

15. *http://www.bhel.com/images/pdf/ENJune2006.pdf;* last accessed on July 7, 2010

16. Manoj, M. V., 2009. M.Tech Dissertation 'Design of Isolated Power Systems for Village Electrification', Department of Energy Science and Engineering, Indian Institute of Technology Bombay.

17. Pohekar, S.D., Kumar, D. and Ramachandran, M., 2005. 'Dissemination of cooking energy alternatives in India- a review.' Renewable and Sustainable Energy Reviews, Volume 9, Issue 4, 379-393.

18. Purohit, P., Kumar, A., Rana, S., and Kandpal, T. C., 2002. 'Using renewable energy technologies for domestic cooking in India: a methodology for potential estimation', Renewable Energy, Volume 26, Issue 2, 235-246.

19. Chandak, A., Dubey D. and Kulkarni R., 'Development of 2.3 m dia. Solar community dish cooker', presented at Asia Regional Workshop on Solar Cooking and Food Processing" Kathmandu, Nepal, 16 - 17 April 2007.

20. Gadhia Solar, http://www.gadhia-solar.com/ Last accessed on October 15, 2010.

21. Sukhatme, S. P. and Nayak, J. K., 2008. Principles of Thermal Collection and Storage, Solar Energy, 3rd Edition, Tata McGraw Hill Publishing company, New Delhi.

22. Report on 'Solar water heaters in India: Market assessment studies and surveys for different sectors and demand segments' submitted to MNRE by Greentech Knowledge Solutions (P) Ltd., available at *mnre.gov.in/pdf/greentech-SWH-MarketAssessment-report.pdf*, (2010). Last accessed on July 6, 2010.

23. Personal communication, Executive Engineer, Electrical, IIT Bombay, Powai. Mumbai 2010.

24. Pillai, I. R. and Banerjee, R., 2006. 'Potential of solar water heating systems for India', Proceedings of Renewable Energy 2006, Chiba, Japan, 557-560.

25. Pillai, I. R. and Banerjee R., 2007. 'Methodology for estimation of potential for solar water heating in a target area', Solar Energy 2007; 81 (2): 162-172.

26. Pillai, I. R., 2008. Ph.D. Thesis on 'Diffusion of Renewable Energy Technologies', Department of Energy Science and Engineering, Indian Institute of Technology Bombay, 2008.

27. Thomas, A., 1996. 'Design methodology for a small solar steam generation system using the flash boiler concept', Energy Conversion and Management, Volume 37, Issue 1, 1-15.

28. Presentation on 'Capacity Building for Solar Thermal Energy in India presentation' at Two day Indo -German Dialogue on Accelerated Dissemination of Solar Energy Technologies in India – 5 March, 2010, Kochi.

29. Deshmukh, R., Gambhir, A., Sant, G., 'Need to Re-align India's National Solar Mission', Prayas (Energy Group), Pune, 2010.

30. *http://www.geda.org.in/solar/so_slr_hmlight_spec.htm* Last accessed on July 6, 2010.

31. Chaurey, A. and Kandpal, T. C., 2009. 'Solar lanterns for domestic lighting in India: Viability of central charging station model', Energy Policy, 37, 4910–4918.

32. Mukerjee, A.K., 2000. 'Comparative study of solar lanterns', Energy Conversion & Management 41 621-624.

33. Tata BP Solar India Limited, http://www.tatabpsolar.com Last accessed on October, 18, 2010.

34. Towards Building Solar India – JNSM Document available at *http://mnre. gov.in/pdf/mission-document-JNNSM.pdf* Last accessed on July 6, 2010.

35. Report on 'Scheme and Framework for Promotion of Solar Water Heating Systems by Utilities and Regulators' submitted to MNRE by ABPS Infrastructure Advisory Private Limited, January 2010 available at *http://www.mnre.gov.in/pdf/abps-SWHS-u&r-report.pdf* Last accessed on July 6, 2010.

SOLAR ENERGY APPLICATIONS: THE FUTURE (WITH COMPARISONS)

Luis A. Bon Rocafort and W.J. O'Donnell

5.1 HISTORY

5.1.1 Awakening

Solar energy can trace its roots to the early 19th century, when in 1838 French physicist Edmund Becquerel [1],[2] published his findings about the nature of materials being able to turn light into energy. He discovered the photovoltaic effect while experimenting with an electrolytic cell made up of two metal electrodes. Becquerel found that certain materials would produce small amounts of electric current when exposed to light. At the time this was an interesting discovery that was not appreciated.

Twenty years passed before Auguste Mouchout [1], a French mathematics teacher, designed and patented the first machine that generated electricity using the sun. Mouchout began his work with solar energy in 1860. He produced steam by heating water using a glass-enclosed, water-filled iron cauldron. Mouchout then added a reflector to concentrate additional radiation onto the cauldron, thus increasing the steam output. He succeeded in using his apparatus to operate a small steam engine.

At the 1878 Paris Exhibition, he demonstrated a solar generator that powered a steam engine, similar to the one shown in Figure 5.1. This engine included a mirror and a boiler that drove an ice-maker that produced a block of ice. Later in 1869, Mouchot wrote one of the first books devoted to solar energy: "Le Chaleur Solaire et les Applications Industrielles." Mouchout's work help lay the foundation for our current understanding of the conversion of solar radiation into mechanical power driven by steam.

The next promising discovery concerning solar technology came from an Englishman who while developing a method for continually testing an underwater telegram cable used selenium and noted that the conductivity of the selenium rods decreased significantly when exposed to strong light. Willoughby Smith [2], an electrical engineer, discovered the photoconductivity of selenium, which led to the invention of photoelectric cells.

Shortly after, William Adams [5], [6] wrote the first book about Solar Energy called: "A Substitute for Fuel in Tropical Countries." With the use of mirrors, Adams and his team were able to power a 2.5 horsepowered steam engine, bigger than Mouchout's 0.5 horsepowered steam engine. His design, known as the Power Tower concept is shown below in Figure 5.2 in a more current setting.

Charles Fritts [2] created the first working solar cell in 1883 turning the sun's rays into electricity. Fritts coated the semiconductor material selenium with a thin layer of gold. The resulting cells had a conversion efficiency of about 1% due to the properties of selenium, which in combination with the material's cost precluded the use of such cells for energy supply.

The first solar energy system for heating household water on rooftop was developed by Charles Tellier [2] in the late 1880's. This concept is shown in Figure 5.3. He used a non-concentrating solar motor for refrigeration much like a solar heat pump. The solar water heater that is employed today largely in warm climates originated in the late 19th century. Further advancements in solar refrigeration at that time were halted, Tellier's efforts concentrated on refrigeration while transporting across the oceans.

At the turn of the 20th century, in 1904, Henry Willsie recognized one of the fundamental limitations of solar power generation as being the inability to generate power without sunlight. He developed a concept to store generated power and use it at night. His method consisted keeping the water warm at night by storing it in an insulated basin. Tubes were then inserted into the heated water, and sulfur dioxide flowed through the tubes, transforming it into a high-pressure vapor, which operated an engine. Two small power plants were built using this method.

The next big advancement for solar energy came at the hands of Calvin Fuller, Gerald Pearson and Daryl Chaplin [2] of Bell Laboratories who accidentally discovered the use of silicon as a semi-conductor, which led to the construction of a solar panel with an efficiency rate of 6% in 1954. The first practical means of collecting energy from the sun and turning it into a current of electricity was at hand. The invention of the solar battery resulted in a major improvement in the ability to harness the sun's power into electricity.

In 1958, Vanguard I [8] was launched; it was the first satellite that used solar energy to generate electricity. Photovoltaic silicon solar cells provided the electrical power to the 6.4-inch, 3.5-pound satellite, demonstrating the potential for solar energy to generate reliable power. An illustration showing the solar cells used in the satellite is shown in Figure 5.4.

Throughout history discoveries of little consequence have a way of becoming more as time passes on. Solar energy has been on a quest for a long time, to demonstrate to us as a society the potential there is in harnessing the sun's rays for power, and ultimately for our survival. Solar energy has to become part of our solution in the grand scheme of our energy supply make up.

FIG. 5.1 SOLAR STEAM ENGINE [4]

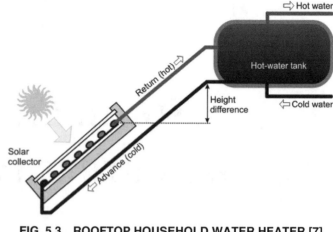

FIG. 5.3 ROOFTOP HOUSEHOLD WATER HEATER [7]

5.1.2 Revival

Every time a shortage of fuels is encountered, be it raw materials, processing capacity shortage, or transportation interruptions, a push for renewable energy sources and technologies emerge until the short term problem is solved. This cycle is not new, and will continue until permanent solutions are found. It seems that every time the cycle emerges, the duration of the push for new technology is prolonged. In the future, this cycle will be long enough to achieve a satisfactory solution that does not involve finite fuel resources.

The cost of producing energy using solar radiation has come down significantly over the last century, but the biggest hurdle, the availability of sunlight will always limit solar energy technology until an adequate advancement in energy storing technologies is found. Improving efficiencies in production, transmission and delivery systems, along with improvements on appliances and electrical equipment will help reduce our power needs. Using an improved infrastructure will help increase the efficiency and reli-

ability of the power grid. Some reexamination of the traditional concepts concerning power generation and delivery must be revised in order to take advantage of the non-traditional methods of power generation (solar, wind, wave, etc.). Local on-site generation and storage must be a part of any plan that will succeed in the future. Better use of energy by appliances and other equipment must be implemented to reduce energy losses due to inefficiencies. Site specific power generation plans that take advantage of the local strengths available must be considered and a system designed that can help lower the load on the national grid system.

Solar electric systems are now used to power many homes, businesses, holiday cottages, even villages in Africa. Solar cells can be used to power anything from household appliances to cars and satellites. Solar technology is becoming increasingly cost-effective as more distributors enter the market and new technologies continue to offer more choices and new products. Technologies that can be used to advance solar energy into the future are discussed in the next section.

FIG. 5.2 POWER TOWER (Courtesy of DOE/NREL, Credit — Sandia National Labs)

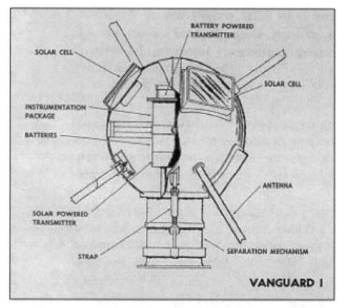

FIG. 5.4 VANGUARD I SATELLITE ILLUSTRATION

5.2 CURRENT TECHNOLOGIES

Solar technology can be divided into two categories; passive, and active. Passive technologies tend to rely on scientific concepts and phenomena to convert solar irradiation to power, while active technologies use mechanical systems to augment the power production. This enhancement can come at a cost to the efficiency of the system. Passive technologies use the sun's rays and scientific concepts of thermal heat and mass transfer to either cool or heat mediums like air or water. While the goal of active technologies is to produce measurable power, which can then be transmitted, used on demand, or stored in a battery for use when power can not be generated.

5.2.1 Passive

Passive solar energy technology uses sunlight to generate energy without the aid of a mechanical system. The main goal behind these types of technology is to convert sunlight or solar radiation into usable heat (water, air, thermal mass), causing thermally induced ventilation, or stored for future use. Passive solar technologies include direct and indirect solar gain for space heating, solar water heating systems based on the natural convection, use of thermal mass and phase-change materials for slowing indoor air temperature swings, solar cookers, the solar chimney for enhancing natural ventilation, and earth sheltering. Passive solar technologies also include the solar furnace and solar forge, but these typically require some external energy to power auxiliary systems that help align their concentrating mirrors or receivers, which has shown over time to be impractical and not cost effective for wide-spread use. Energy used for space and water heating, however, have demonstrated to be a good use of passive use solar energy.

5.2.1.1 Thermosiphon This method of passive heat exchange is based on natural convection. Natural convection causes the circulation of the fluid within a loop when the fluid is heated, which causes it to expand and become less dense. The denser fluid, the cool water, moves to the bottom of the loop, while the less dense fluid, the hot water, will rise to the top of the loop. Convection moves heated liquid in the system as it is replaced by cooler liquid returning by gravity. This type of system could be used in moderate temperature regions to pre heat incoming cold water for an instant water heater. This would reduce the electrical demand on the instant water heater and increase the overall efficiency of the system. A similar system is shown in Figure 5.3.

5.2.1.2 Thermal Collectors Solar collectors can be non-concentrating and concentrating. In non-concentrating collectors, the collector area is the same as the absorber area. In these types the whole solar panel absorbs the solar energy. Flat plate and evacuated tube solar collectors are two types of non-concentrating solar collectors that are used to collect heat for space heating or domestic hot water. Flat plate collectors consist of a dark flat-plate absorber, a transparent cover that allows solar energy to pass through but reduces heat losses, a heat-transport fluid (air, anti-freeze, or water) flowing through tubes to remove heat from the absorber, and an insulating backing. Fluid is circulated through the tubing to transfer heat from the absorber to an insulated water tank. This may be achieved directly or through a heat exchanger.

Evacuated tube collectors consist of evacuated glass tubes which heat up a fluid in order to heat water, or to provide space heating. The tubes are evacuated, that is under vacuum, such that convection and conduction heat losses are reduced. This allows them to heat up to temperatures that surpass those of flat plate collectors. Tube collectors have an advantage over flat-plate collectors with regards to the shape of the tube; since it is cylindrical the collector surface will always be perpendicular to the sun.

5.2.1.3 Thermal Mass A thermal mass is a solid or liquid body that has the capacity store heat during the day and then releases the heat slowly when the heat source is removed. The use of thermal mass has become popular in the world of building design as an alternative to passively heating an interior space. Using a thermal mass will prevent extreme temperature fluctuations during the day by serving as a thermal inertia, in other words, softens the temperature fluctuation experienced, by absorbing heat during the day, and slowly releasing this heat at cooler temperatures of the night, effectively heating the space. A thermal mass will absorb heat from the surroundings, as long as the surrounding temperature is hotter than the mass, once the temperature around the mass gets cooler, the thermal mass releases the stored heat to the surroundings. This phenomenon is illustrated below in Figure 5.5.

The use of thermal mass in building design, construction, and rehabilitation has increased due to renewed interest in green construction. Thermal masses can reduce the load on heating and cooling systems, while increasing individual comfort.

Leadership in Energy and Environmental Design (LEED [10]) is a rating system for buildings that use criteria such as energy savings, water management, among others to score how sustainable a building is. The LEED rating systems was created by the U.S. Green Building Council (USGBC), and are internationally accepted benchmarks for design, construction, and operation of high performance green buildings.

5.2.1.4 Solar Cooker A solar cooker is a cooking apparatus that uses sunlight to cook food. Solar cookers by nature require sunlight to perform and are usually reserved for outdoor use, but with careful planning and design could be adapted to be used indoors, concept shown in Figure 5.6.

Use of optics and appropriate materials can ensure every home in the future can prepare meals using this form of passive solar

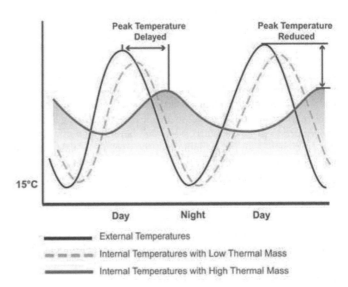

FIG. 5.5 THERMAL MASS TEMPERATURE FLUCTUATION [9]

FIG. 5.6 INDOOR SOLAR COOKER DESIGN (Courtesy DOE/NREL, Credit — Tsuo, Simon)

technology whenever possible, reducing the demand on the electrical grid as well as on natural gas.

In order for a solar cooker to perform satisfactorily, they must use some form of concentrated sunlight, by way of mirrors or reflective metals, to direct the energy used to the cooking area. They must also convert sunlight to usable heat. A solar cooker must also be thermally insulated, that is once the heat is trapped inside the cooker, and it should be insulated so that the heat does not escape.

Solar cookers can reach temperatures of 300°F. Although this is significantly less than what can be achieved with your stovetop, or conventional oven, it will still cook food. The catch is that it takes longer to cook the food. Improvements in concentrating materials and amplification of the suns rays could reduce the amount of time required to cook the food using a solar cooker. Another limitation is that the cooker must be used around the times when the sun is highest in sky to take full advantage of the solar power.

The cooker can be used to warm food and drinks and can also be used to pasteurize water or milk. Unlike cooking on a stove or over a fire, which may require more than an hour of constant supervision, food in a solar cooker is generally not stirred or turned over, both because it is unnecessary and because opening the solar cooker allows the trapped heat to escape thereby slowing the cooking process. Air temperature, wind, and latitude also affect cooking efficiency. Careful planning must be used when using a solar cooker to ensure the food is prepared properly as to take advantage of the cooking method.

5.2.1.5 Solar Chimney Using passive solar energy, improvements can be made to the natural ventilation of buildings through the use of convection heat transfer concepts. A solar chimney uses solar energy to accomplish this. The solar chimney is not a new concept, it is not even recent, and it has been implemented in the past by Persians and Romans.

A solar chimney works by absorbing solar energy during the day and heating the chimney and the air in it. This heated air wants to

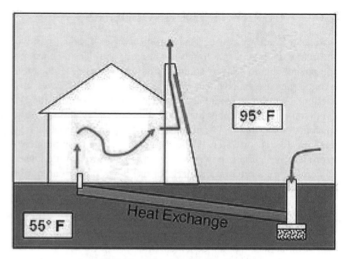

FIG. 5.7 SOLAR CHIMNEY [11]

move to higher elevations creating an updraft through the chimney (similar to a thermosiphon). The movement of the air in the chimney can be used to suck in air at the base of the chimney, which would cause air drafts along the building to which the chimney is attached. This can be used to circulate cooler air inside the building, as shown in Figure 5.7. Although this system is not ideal in every location or climate, its concept can be applied on most new construction to improve natural ventilation using simple heat convection concepts.

A solar chimney can also be used in colder climates to circulate hot air inside the building by reversing the flows into the structure, or even devising a closed loop system that will help reduce the load on your heating appliance. In hotter climates combining the solar chimney technology with water can increase the cooling effect by using evaporation cooling.

5.2.1.6 Solar Furnace A solar furnace is a structure used to harness the rays of the sun in order to produce high temperatures, usually for industrial applications. This is achieved using a parabolic reflector, concentrating direct sun light, also known as direct insolation, onto a focal point. The solar furnace consists of an array of plane mirrors which in turn reflects sun light onto a large curved mirror. After the rays bounce off both set of mirrors they are then focused onto a small area that can reach more than 6,000°F, useful for some industrial processes. The solar furnace at Odeillo in the Pyrenees-Orientales in France is shown in Figure 5.8.

5.2.2 Active

Active solar technologies are used to convert solar energy into light, heat, ventilation, cooling, or to store heat for future use. This type of solar energy generation uses electrical or mechanical equipment, i.e., pumps and fans, to increase system efficiencies. Solar hot water systems that use pumps or fans to circulate a working thermal fluid through solar collectors are one type of active solar technology. Another type of active solar technology used to convert sunlight to electric power is by use of photovoltaic, or with concentrating solar power. The latter focuses the sunlight to boil water which is then used to provide power. Another concept that uses concentrating solar power is the Stirling engine dishes which use a Stirling cycle engine to power a generator.

FIG. 5.8 SOLAR FURNACE IN FRANCE [12]

5.2.2.1 Photovoltaic Photovoltaic technology is an evolving one. Currently, there are many competing technologies, such as thin film, monocrystalline silicon, polycrystalline silicon, amorphous cells among others. The efficiency for these different technologies can vary from 5% – 18% depending on the technology. A PV Power Plant using this technology is shown in Figure 5.9.

The earliest significant application of solar cells was as a back-up power source to the Vanguard I satellite in 1958, which allowed it to continue transmitting for over a year after its chemical battery was exhausted. The successful operation of solar cells on this mission was duplicated in many other Soviet and American satellites, and by the late 1960s, PV had become the established source of power for them.

After the successful application of solar panels on the Vanguard I satellite (see Figure 5.4), in the 1970s, photovoltaic solar panels became more than back up power systems on spacecraft. Photovoltaic went on to play an essential part in the success of early commercial satellites such as Telstar, and they remain vital to the telecommunications infrastructure today. Branching out from the aerospace industry, PV was adopted by electronics manufacturers, builders and others.

The use of PV in construction has led to some advances in system integration, increase in efficiency of domestic appliances and a shift in the design and architecture paradigm. Consideration of how the different systems interact with each other and how to reuse waste heat, waste water and refuse to generate power or reduce power consumption is generating interest in a recent field known as Green Construction. Some concepts in green construction use building-integrated photovoltaic to cover the roofs or sides of an increasing number of buildings to generate some of the power required for these buildings (see Figure 5.10 and Figure 5.11). If sized correctly an energy neutral building could be achieved. An oversized system could sell the extra power generated to the power company or trade it for power at night when no power is generated by the PV in the absence of energy storage methods.

The high cost of solar cells limited terrestrial uses throughout the 1960s. This changed in the early 1970s when prices reached levels that made PV generation competitive in remote areas without grid access. Early uses included powering telecommunication stations, offshore oil rigs, navigational buoys and railroad crossings. The 1973 oil crisis had the effect of increasing the production of PV during the 1970s and 1980s. Economies of scale which resulted from increasing production along with improvements in system performance brought the price of PV down tenfold in less than 15 years.

FIG. 5.9 PV PLANT (courtesy of DOE/NREL, Credit — Steve Wilcox)

FIG. 5.10 BUILDING INTEGRATED PV (Courtesy of DOE/NREL, Credit-BP Solarex)

FIG. 5.11 PV RESIDENTIAL SOLAR ARRAY (Courtesy of DOE/NREL)

5.2.2.2 Concentrating Solar Power Concentrating photovoltaic systems use sunlight concentrated onto photovoltaic surfaces for the purpose of electrical power production. Solar concentrators of all varieties may be used, and are often mounted on a solar tracker to make sure the cell faces the sun as it moves across the sky. Solar tracking increases flat panel photovoltaic output. The concentrated sunlight coming from a concentrating mirror has many uses, like the ancient legend that claims Archimedes used polished shields to concentrate sunlight on the invading Roman fleet and repel them from Syracuse. Something of a non-lethal crowd control, before its time.

5.2.2.2.1 Parabolic Trough Auguste Mouchot used a parabolic trough to produce steam for the first solar steam engine in 1866. Concentrating Solar Power systems use lenses or mirrors and tracking systems to focus a large area of sunlight into a small beam or focal point as shown in Figure 5.12 and Figure 5.13. The concentrated heat is then used as a heat source for a conventional power plant. A wide range of concentrating technologies exists; the most developed are the parabolic trough, the concentrating linear Fresnel reflector, the Stirling dish and the solar power tower.

Various techniques are used to track the Sun and focus sunlight. In all of these systems, a working fluid is heated by the concentrated sunlight, and is then used for power generation or energy

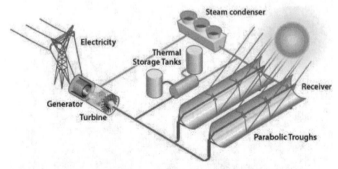

FIG. 5.12 A LINEAR CONCENTRATOR POWER PLANT USING PARABOLIC TROUGH COLLECTORS (Courtesy of DOE/NREL)

FIG. 5.13 CONCENTRATING SOLAR TROUGH (Courtesy of DOE/NREL)

storage. A parabolic trough is made up of a parabolic reflector, a receiver, and a working fluid. Sunlight hits the parabolic reflector surface which concentrates light onto the receiver positioned along the reflector's focal line. The receiver is usually a tube located at the parabola's focal point, and it is filled with a working fluid. The reflector usually tracks the sun during the daylight hours thus increasing its solar input. Parabolic trough systems provide the best land-use factor of any solar technology.

5.2.2.2.2 Fresnel Reflectors Concentrating Linear Fresnel Reflectors are Concentrating Solar Power plants which use many thin mirror strips instead of parabolic mirrors to concentrate sunlight onto two tubes with working fluid. In this configuration, flat mirrors can be used and a cheaper alternative to the parabolic concept can be implemented on less space. A sample of this technology is shown in Figure 5.14.

5.2.2.2.3 Stirling Engine A Stirling solar dish consists of a stand-alone parabolic reflector that receives sunlight onto a receiver positioned at the reflector's focal point. The reflector may track the Sun along one or two axes. The concentrated sunlight is used to heat a working fluid in a Stirling Engine. It drives a Stirling Cycle, although it could be used to boil water and produce steam to drive a steam generator. Power output of this system depends on the working fluid and the process of power generation. The Stirling solar dish combines a parabolic concentrating dish with a Stirling heat engine which normally drives an electric generator as shown in Figure 5.15. The advantages of Stirling solar over PV cells are higher efficiency of converting sunlight into electricity and longer lifetime.

5.2.2.2.4 Solar Tower The Solar Tower concept uses solar collectors to warm the air near the surface which is then channeled up the central tower. Turbines are placed at the bottom of the tower to make electricity from the updraft. The solar tower is an active version of a solar chimney, shown in Figure 5.7, an old technique for providing cooling to a home by creating a natural updraft from heated air inside a chimney. The downside is the large amount of space needed for this concept to function properly.

One of the most effective configurations has an 800 to 1,000 meter tower with a canopy of 2.5 km radius on the ground similar to one shown below in Figure 5.16.

The land use factor for this technology is high, thus making this alternative more attractive as the prices for traditional energy generation and delivery climbs. This technology can also be appropri-

FIG. 5.14   LINEAR FRESNEL REFLECTORS (Courtesy of DOE/NREL)

ate for rural communities where the delivery of power is restrictive in either cost or feasibility.

5.2.2.2.5 Solar Bowl A solar bowl is a spherical dish mirror that is fixed in place like the one shown below in Figure 5.17. The receiver follows the line focus created by the dish (as opposed to a point focus with tracking parabolic mirrors). The receiver is on trackers moving along the bowl maximizing the amount of direct insolation it can absorb. It is similar to a solar engine, except that the tracking mechanism is on the generator, thus making it smaller than the one used on the dish. This setup is also less prone to weather issues, and does not require a structure to support the dish.

5.3   STORING ENERGY

Solar energy is not available at night, making energy storage an important issue in order to provide continuous power. Solar power is an intermittent energy source, meaning that all available output must be used as it is generated, either by transmitting it for immediate consumption, or storing it for later use when no output is generated. Using Solar Energy to generate power can be done in many ways, as shown in Section 5.2, the problem with Solar Energy Generation, is that when no sun is available, little or no power can be produced. It is possible, however, to generate more power than it is needed when sunlight is available, and that power can either be stored for use at night, or cloudy days, or it can be sold back to the utility by way of net metering as shown in Figure 5.18.

Designing a system that optimizes storage involves many assumptions, particularly on energy use, number of days without sun, and design life of the system. Consider a traditional battery system to be used in conjunction with PV panels to provide the same level of power to a couple of secluded homes like the one shown in Figure 5.19. For the secluded home in a location with a low number of sunny days, the energy storing system could be very expensive, compared to a secluded home whose location has a high number of sunny days. The land use factor dedicated to the solar power plant will be higher in a location with a low number of

FIG. 5.15   STIRLING SOLAR ENGINE (Courtesy of DOE/NREL, Credit — Infinia Corporation)

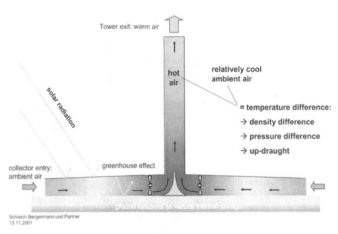

FIG. 5.16   SOLAR TOWER CONCEPT [13]

FIG. 5.17   SOLAR BOWL [14]

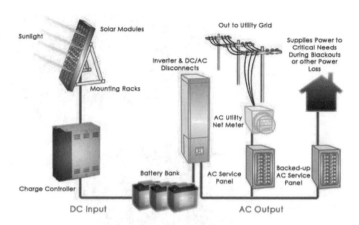

FIG. 5.19   SOLAR PV AND BATTERY SYSTEM [15]

sunny days. A higher land use factor is necessary to accommodate more solar PV panels to generate extra power to store in the extra battery storage to maintain the power throughout the longer period without sunlight.

There are many energy storing methods in use today. Some of these methods, like potential energy have been around for more than a thousand years, others like batteries are more recent and both are still being refined to improve system efficiencies. Methods for storing energy include; Batteries, Flywheel, Water (Hydro Reservoir), Compressed Air, and Superconducting Magnetic Energy Storage.

5.3.1   Batteries

Batteries use a controlled chemical reaction that takes place inside a series of cells, which have a positive and negative electrodes divided by a conductive electrolyte separator. When the battery is connected to a load, positively charged ions will travel from the negative electrode to the positive electrode, providing an electric current that can be used to provide electrical power. Reverse this process, and you can recharge a battery. The chemicals used vary and impact the characteristics of the battery. For instance, a car battery is usually made with lead acid, while a laptop battery uses

nickel cadmium, and electrical cars use some variation of the nickel metal hydride form.

The chemical makeup of the battery and its size have been studied extensively by scientists to determine what combination works best for specific uses. The lithium ion battery is currently the newest battery technology around that packs a lot of energy into a lightweight form. Although care must be taken so that this battery does not overheat and explode, a phenomenon many early laptop computer users faced when using lithium ion batteries.

5.3.2   Flywheel Energy Storage

Accelerating a rotor (flywheel) to high speeds and maintaining that speed to create rotational kinetic energy is a method of storing generated power. This would be the equivalent of a mechanical battery. The energy is released when the flywheel turns a generator which creates electricity using the rotational energy stored in the flywheel. To store energy in the flywheel, the generator is reversed thus creating a motor that spins the flywheel. This method of storing power is simple in concept, but has its difficulties in reducing all the friction losses in the system to enable the appropriate use of this technology. To reduce the friction losses, a combination of a vacuum chamber and magnetic bearings are used. This energy

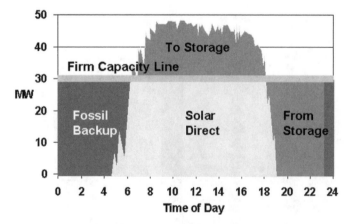

FIG. 5.18   SAMPLE ENERGY STORAGE USAGE [16]

FIG. 5.20   FLYWHEEL ENERGY STORAGE SYSTEM [17]

FIG. 5.21 LA MUELA PUMPED STORAGE FACILITY [19]

storage has the advantages of safety, little maintenance and providing regular output of electricity (see Figure 5.20).

5.3.3 Water Energy Storage

Using water to store solar energy might seem cheap, but it really depends on how you want to store the energy and then reclaim the saved energy later. Using water as a thermal mass is cheap, but limited in the uses for the reclaimed energy. Instead consider using the generated power to pump the water to a higher reservoir, now we have a flexible solution, although costly.

Using potential energy to generate power is not new; we have had hydroelectric dams for a long time, and are proven very effective at generating power. Using excess solar energy to pump water to a higher reservoir, which can later be used to turn generators on its way down to the lower reservoir, is a unique system of energy storage. The land use factor for this type of system could be very large, and the efficiencies are somewhat lower than other energy storing technologies, but for some locations and large-scale deployments it may make sense (see Figure 5.21).

5.3.4 Compressed Air

Another concept that uses potential energy is using compressed air to turn turbines and produce electricity. The advantage of using compressed air is that natural gas is used sparingly since the compressed air drives the turbine. The air can be compressed using extra power generated from an oversized system and stored underground, under a cap rock. The compressed air is used later by releasing to the surface where it can be used to drive a power generating system (see Figure 5.22).

5.3.5 Superconducting Magnetic Energy Storage

This type of energy storing system stores energy in a magnetic field. This type of system is composed of three parts: superconducting coil, power conditioning system and cryogenically cooled refrigerator. One of the advantages of this system is that once the coil is charged the current does not decay, meaning the magnetic energy can be stored for prolonged periods of time. The stored energy can easily be used by discharging the coil. Superconducting Magnetic Energy Storage is the most efficient method of storing and using stored energy, although it is more suited for short duration energy storage due to the energy requirements of the process, and the high cost of its materials, especially the superconducting wire. As such, it is mainly used to improve power quality.

5.4 HOW CAN SOLAR ENERGY HELP

According to the Energy Information Administration (EIA), a governmental agency within the Department of Energy (DOE), the residential energy use breakdown shown in Figure 5.23, illustrates

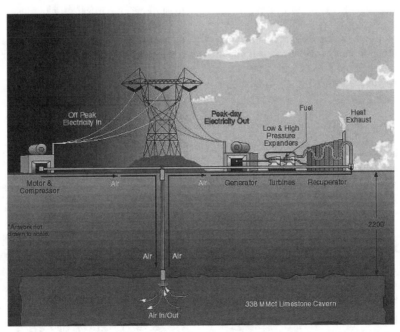

FIG. 5.22 COMPRESSED AIR STORAGE DIAGRAM (Courtesy of DOE/NREL)

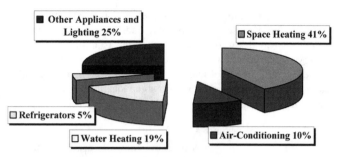

FIG. 5.23 HOUSEHOLD ENERGY USED, DATA (Courtesy of EIA)

that the majority of the household energy use, relates to comfort level of the occupant. More than half of the energy used in the home is used to handle heating air and water and Air Conditioning. Household energy demand can be reduced substantially by improving natural lighting during the day, making appliances more energy efficient, along with improvements in insulation, and using energy efficient windows, and doors.

Building energy use breakdown is shown in Figure 5.24 is similar to residential energy use breakdown with the exception of water heating, since not as much heated water is needed in commercial buildings. Lighting use in buildings is much higher also, to accommodate all of the buildings' occupants and interior offices.

Also, there is a greater use of computers and electronics in a building than in a residential home. Buildings and residences can also benefit from the methods described above. Buildings operate mainly during the day, whereas residential energy use occurs mainly in the evening hours as shown in Figure 5.25. A solar PV system could be readily used by the building as power is generated, although the space available to mount a PV system does not readily match the ratio of a residential home, where the ratio of energy needed to possible solar energy generation through PV is closer to unity. Energy used in buildings is greater than in residential homes, although there are more residential homes than buildings. Hence the discrepancy in energy use by sectors shown in Figure 5.26. Energy use in buildings is close to constant from the start of business until the end of the day, and then there are slow drop-offs for a smaller workforce during the early to late evening shift. Residential energy use is low and constant during the day hours where space and water heating, or air conditioning maintains the home at some preset level. Then, when the home is occupied, energy use increases, as shown at the end of the day, when hot showers, dinner and home entertainment is used at length.

5.4.1 Energy Usage and Production

If we consider energy consumption in the USA we can look at the different sectors that consume energy to determine where the

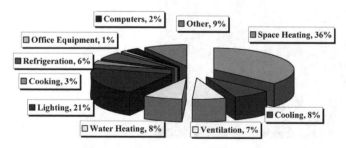

FIG. 5.24 BUILDING ENERGY USE, DATA (Courtesy of EIA)

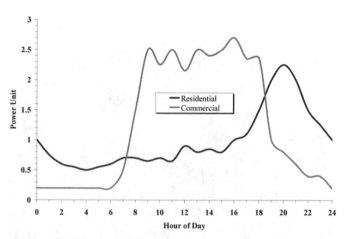

FIG. 5.25 SAMPLE LOAD PROFILE — RESIDENTIAL VS. COMMERCIAL

highest consumption occurs. A historical trend is shown below in Figure 5.26. Residential and commercial energy usage is less than half of the total energy usage in the USA; also evident in this figure is the growth of energy use in these sectors, more than 300% in 60 years. Residential energy usage is more than commercial energy usage, and both are less than industrial and transportation energy usage. The percentage of each sector compared to the total has been constant across all sectors over time; in essence they have grown at the same rate over the last 60 years. Solar energy technologies, passive and active, can be used as an added energy reservoir to spur growth and opportunity across all sectors, as well as make up for losses due to non-production related problems associated with traditional fuel sources, interruptions in transportation, refining, processing and overall delivery schedule issues of some of these fuels, namely, oil and gas.

Over the last 60 years the bulk of our energy has been produced by limited resources, namely fossil fuels. Fossil fuels currently account for 84% of the energy consumed by the USA. Nuclear power currently provides 8.5% of our energy, and has only contributed to our energy portfolio for the last 5 decades. Renewable energy, which has contributed as much as 9.3% in 1949, currently contributes 7.4% as shown in Figure 5.27. The bulk of renewable energy came from hydroelectric power and biomass.

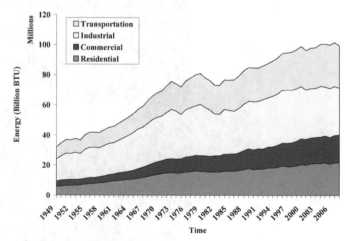

FIG. 5.26 ENERGY USAGE BY SECTOR, DATA (Courtesy of EIA)

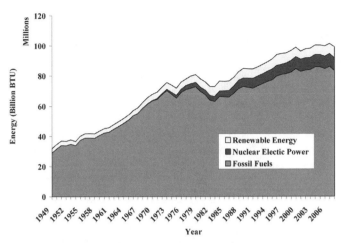

FIG. 5.27 HISTORICAL ENERGY CONSUMED MAKEUP (Data courtesy of EIA)

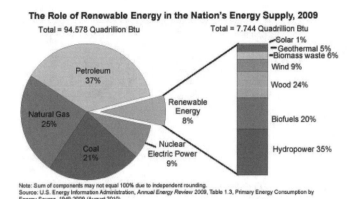

FIG. 5.29 2009 TOTAL ENERGY BREAKDOWN FOR THE USA (Courtesy of EIA)

5.4.2 Solar Energy's Role

The renewable energy portfolio for the USA is shown in Figure 5.28. The majority of renewable energy is attributed to hydroelectric power and biomass. An increase in the last 20 years from other energy sources shows promise, specifically for geothermal, wind, and solar energy. Passive solar energy can help reduce fossil energy consumption across many categories in residential and commercial energy use, as well as industrial processes. Local on-site active solar energy power generation can help reduce industry's energy demand by having local power plants that run on active solar technologies, like a Stirling engine, or one of the many solar PV configurations. Efforts like these can provide power cheaply when it is needed the most as shown in Figure 5.25. Using solar forges and furnaces can also reduce industry's reliance on fossil fuel for high temperature processes.

For 2009, the total energy used in the USA is estimated at 94.6 Quadrillion BTU, with 7.7 Quadrillion BTU provided by renewable energy as seen in Figure 5.29. These figures indicate that for 2009 8.2% of the total energy used in the USA came from renewable sources. Solar Energy provided 1% of the 8.2%, which is a 0.08% of the total energy needed in the USA. Figure 5.30 and Figure 5.31 show the room for improvement in solar energy infrastructure as there is ample solar irradiation in the USA and the world.

5.5 WHAT THE FUTURE HOLDS

Energy Information Administration publishes annually information on energy use and trends it expects in the near future. Figure 5.32 illustrates their expectation that renewable energy will double in the next 25 years, although they expect much of the growth from wind and biomass generated power.

Solar power is expected to increase more in proportion to wind power. A paradigm shift needs to occur that will include solar power as a greater contributor if we want to supplant the 84% energy produced by fossil fuels in the long term future. The USA with an estimated population of over 300 million, according to the U.S. Census Bureau, uses 21% of the world's energy. China follows the USA in energy consumption with a 17% portion of the world's energy. Next is Russia with 5.8% followed by India with 5.2%. Other countries are shown in Figure 5.33, this data obtained from World Bank data. While the energy use of the USA, China and India are increasing, most of the world's other economies have lived with a more constant usage level for the better part of 50 years. China and India, the world's two largest populations, currently use only 1,484 and 529 kg_{oil}/capita respectively (see Figure 5.34). These

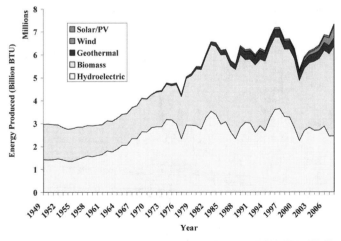

FIG. 5.28 HISTORICAL RENEWABLE ENERGY MAKEUP (Data courtesy of EIA)

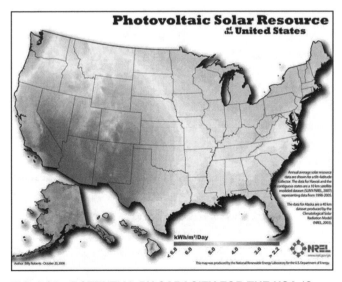

FIG. 5.30 POTENTIAL PV CAPACITY FOR THE USA (Courtesy of NREL)

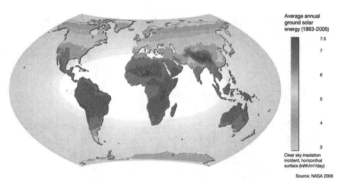

FIG. 5.31 POTENTIAL PV CAPACITY FOR THE WORLD (Courtesy of NREL)

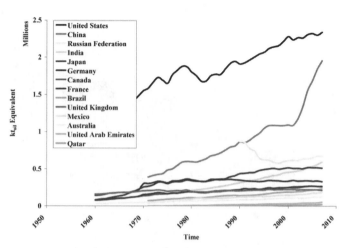

FIG. 5.33 WORLD ENERGY USE EQUIVALENT, SELECTED COUNTRIES, WORLD BANK

two nations are positioned to be economic superpowers, and in their ascension will become major energy consumers.

There is a great opportunity for solar energy; passive and active, to become part of the growth that these emerging nations need to satisfy their energy demand without adding further strain to current worldwide energy supplies.

5.5.1 Solar's Potential

Solar irradiation on Earth's surface varies from 200 W/m^2 in the late morning hours, or early evening hours to 1,000 W/m^2 at peak, and the available hours of usable sunlight range from less than 2 hours to 7 hours depending on the region. The amount of solar radiation converted to usable power varies from 5% to 30% dependent on the type of generating technology used. Using current PV technology, a 230-W panel that occupies 1.67 m^2 receiving 1000 W/m^2 solar irradiation would be 13.8% efficient at peak production. This same panel at 800 W/m^2 will only produce 165.6 W, making it 12.4% efficient.

Assuming we need a global power level consumption of 20 terawatts (TW; 1 TW = 1e12 W), a theoretical land use can be found to determine the feasibility of solar providing 20 TW of power. Assuming we use the 230 W panel discussed above, which occupies 1.67 m^2 of area, and that this panel receives 5 hours on sun per day average, we can expect this panel to deliver 1.15 kWh/day. To satisfy world energy usage of 20 TW, on average this would be equivalent to 480 TWh/day, we would need about 417.4 billion

panels, which would equate to 697e9 m^2. The world's land surface area is just north of 148e12 m^2; this would be less than 0.5% of the world's total land area. Figure 5.35 illustrates world land use required versus power demand, for different PV configurations. This figure does not take into account facilities, wiring, and installations. Also not considered in these land use calculations are cost and efficiency losses due to storing, transmitting, converting and using the power generated. No credit is taken for future improvements to PV technology and power delivery methods. Figure 5.36 shows the area required for PV installations depending on overall system efficiency. Improvements in certain areas of the system will offset losses in others and it is expected that efficiencies will improve over time, thus requiring less land use to provide the same power, or the ability to increase the power generation to promote growth.

Considering USA's possible energy usage of 120e15 BTU, we would need 96.4e9 kWh/day average of energy production to meet our demand. Using the same solar panel from above that produces 1.15 kWh/day, we would need 83.8 billion panels that would occupy about 140e9 m^2. The land mass of the USA is listed at 9.8 million km^2 by the CIA. The USA would have to use at least

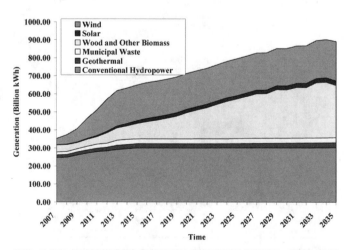

FIG. 5.32 EXPECTED ENERGY MAKEUP IN USA, ENERGY INFORMATION ADMINISTRATION (Data courtesy of EIA)

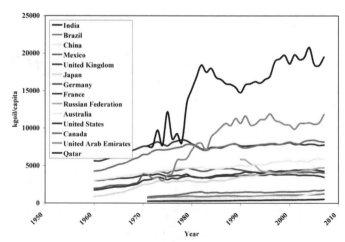

FIG. 5.34 WORLD ENERGY USE PER CAPITA EQUIVALENT, SELECTED COUNTRIES, WORLD BANK

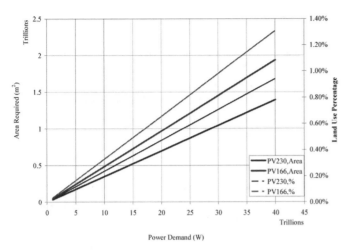

FIG. 5.35 AREA USE BY PV REQUIRED TO MEET EXPECTED POWER DEMAND

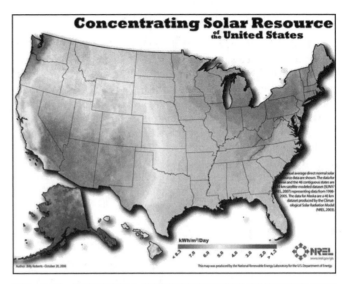

FIG. 5.37 POTENTIAL CONCENTRATING CAPACITY FOR THE USA (Courtesy of NREL)

1.5% of its land mass to become energy independent using solar PV technology at its current level.

Another technology that could be used to generate the power needed, would be concentrating solar power, this gives us the ability to get more energy per unit area used thus lowering the system size. Figure 5.37 shows the concentrating solar energy potential similar to Figure 5.30, which shows PV energy potential for the USA. These figures show that in the USA we can expect a maximum above 8.3 kWh/m^2/day to a minimum of less than 1.3 kWh/m^2/Day for concentrating solar heat flux compared to a maximum above 6.8 kWh/m^2/day to a minimum of less than 2.2 kWh/m^2/day for PV solar heat flux.

Assuming we can build our power plant at a region with the maximum solar irradiation of 8.3 kWh/m^2/day and a 30% efficiency, which many solar engines can achieve at today's standards, we have the potential of 2.5 kWh/day, which would require a total area of 38.6e9 m^2. This means that we can build many plants that are composed of both PV and concentrating solar engines to take advantage of each technologies' strengths. For instance, PV can produce energy under cloud cover, which for concentrating systems is not possible, and while there is no cloud cover under direct

sun concentrating solar can produce twice as much as a PV array if configured properly.

There are, however, some clarifications that must be made regarding these calculations. We used a yearly energy demand number and averaged it over the whole year, as opposed to the actual demand per month, which varies according to region and season. We also assumed we could store energy not used during generation indefinitely, and without a loss. Both of these assumptions lead to a bigger system, or development of more efficient technologies at converting solar energy to usable power.

5.6 CONCLUSION

The future of solar energy lies in the successful integration of solar technologies into our current residential and commercial infrastructure. This will allow the extension of our finite resources, essentially giving us time to solve the long term problem of supplanting these sources with renewable energy sources without sacrificing current lifestyle and comfort levels. We have to supplement the current 84% energy from finite resources.

Active solar energy for general consumption for residences, businesses and industrial complexes has proven effective in recent years and is gaining widespread adoption as backup energy.

Passive solar energy, like thermal collectors, or solar water heaters, can pre-heat water that is being circulated to traditional water heaters. This could reduce the 19% of residential energy currently used for water heaters. Implementing other passive solar energy techniques can promote growth and prosperity.

Active solar technologies generate power that can be inserted to our power grid immediately. Any location can have a small power plant that provides excess power produced to the national grid to help meet demand at peak times, which generally occur during the day, when solar power production is viable.

The expansion of active solar power generation is inhibited by the low 0.08% of USA energy that it provides now, and the < 0.1% that is expected over the next 25 years. This can be dramatically changed by ongoing improvements in the technology of solar power generation and storage.

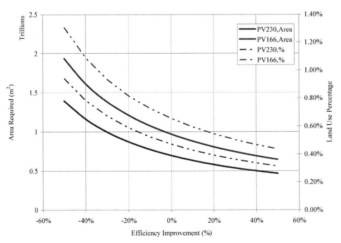

FIG. 5.36 AREA REQUIRED FOR 20 TW DEMAND, AS A FUNCTION OF EFFICIENCY IMPROVEMENT

5.7 ABOUT THE AUTHORS

Luis A. Bon Rocafort and W.J. O'Donnell are with O'Donnell Consulting Engineers, Inc, located in Bethel Park, PA. Their work is in related energy fields (renewable, nuclear, gas, etc.), high temperature, and high pressure component design. They have worked on solar energy projects with Stirling engines, as well as frame construction for PV panels. OCEI has solved structural problems in frames, poles, towers, etc. for flow induced vibrations, harmonic, random vibration, wind, wave and other loading phenomena that these renewable technologies may experience. Their extensive work with high temperature materials in the creep regime solves advanced solar energy system problems.

5.8 REFERENCES

1. Sam Elliot, Alex Nies (Lead Author); Dawn Wright (Topic Editor) . "Solar Energy." In: Encyclopedia of Earth. Eds. Cutler J. Cleveland (Washington, D.C.: Environmental Information Coalition, National Council for Science and the Environment). [First published in the Encyclopedia of Earth May 5, 2010; last revised date May 5, 2010; retrieved October 11, 2010 <http://www.eoearth.org/article/Solar_Energy>.

2. "This Month in Physics History. April 25, 1954: Bell Labs Demonstrates the First Practical Silicon Solar Cell." American Physical Society. 2009. http://www.aps.org/publications/apsnews/200904/physicshistory.cfm.

3. Cutler Cleveland (Lead Author);Peter Saundry (Topic Editor). "Mouchout, Auguste." In: Encyclopedia of Earth. Eds. Cutler J. Cleveland (Washington, D.C.: Environmental Information Coalition, National Council for Science and the Environment). [First published in the Encyclopedia of Earth August 18, 2006; Last revised Date August 18, 2006; Retrieved October 11, 2010 <http://www.eoearth.org/article/Mouchout,_Auguste>.

4. Wolf, Martin, I. N. (Year). Solar Energy Utilization By Physical Methods. Science, Volume 184, page 383.

5. Bradford, Travis. Solar revolution: The economic transformation of the global energy industry. 2006.

6. Sukhatme, S. P. Solar Energy: Principles of Thermal Collection and Storage. 2nd Edition. Tata McGraw-Hill. Delhi, India. 1996.

7. Quaschning, Volker. Solar thermal water heating. Renewable Energy World, 02/2004.

8. Naval Research Laboratory. "Vanguard Satellite Marks 45 Years in Space." NRL Press Release 21-03r. 3/10/2003.

9. New4Old. 2010. http://www.new4old.eu/Default.aspx.

10. National Resources Defense Council. http://www.nrdc.org/buildinggreen/about.asp.

11. "A Theory of Power" ISBN 0595330304 Vail, Jeff (2005-06-28). Passive Solar & Independence. Retrieved on 2007-03-10.

12. PROMES. 2010. http://www.promes.cnrs.fr/.

13. EnviroMission Limited. 2010. http://www.enviromission.com.au/.

14. Auroville Universal Township. 2010. http://www.auroville.org/research/ren_energy/solar_bowl.htm.

15. Astralux Solar. 2010. http://www.AESsolarenergy.com.

16. Power from the Sun. 2010. http://www.powerfromthesun.net/chapter1/Chapter1.htm.

17. Beacon Power. 2010. http://www.beaconpower.com/products/about-flywheels.asp.

18. Iberdrola. 2010. http://www.iberdrolarenovables.es.

19. U.S. Energy Information Administration. 2008. http://www.eia.gov.

20. U.S. Census Bureau. 2010. http://www.census.gov.

21. World Bank. 2010. http://www.worldbank.org.

ROLE OF NASA IN PHOTOVOLTAIC AND WIND ENERGY

Sheila G. Bailey and Larry A. Viterna

6.1   INTRODUCTION

Since the beginning of NASA over 50 years ago there has been a strong link between the energy and environmental skills developed by NASA for the space environment and the needs of the terrestrial energy program. The technologies that served dual uses included solar, nuclear, biofuels and biomass, wind, geothermal, large-scale energy storage and distribution, efficiency and heat utilization, carbon mitigation and utilization, aviation and ground transportation systems, hydrogen utilization and infrastructure and advance energy technologies such as high-altitude wind, wave and hydro, space solar power (from space to earth) and nanotech photovoltaics. NASA, in particular with wind and solar energy, had extensive experience dating back to the 1970's and 1980's and continues today to have skills appropriate for solving our nation's energy and environmental issues that mimic in fact those needed for space flight. NASA has long been interested in maturing new laboratory level technologies into industry products and has a well-founded system of interactions with Universities for research leading to Proof-of-Concept and then to prototype demonstrations and mission applications with industry. The capability to assess and then test the feasibility for future commercial development is a well-honed NASA skill. This has certainly promoted new businesses and job growth as a result. That is particularly evident in the case of photovoltaic energy but also relevant for wind energy as well.

6.2   PHOTOVOLTAIC ENERGY

6.2.1   The Early Years

In 1954, Chapin reported a solar conversion efficiency of 6% for a silicon single-crystal cell marking the beginning of modern day photovoltaics [1]. At approximately the same time the first thin film solar cells of CdS/Cu were being developed by the US Air Force Laboratory in Dayton, Ohio [2]. These cells had an efficiency of ~1.5%. In 1955 the first III-V cells (GaAs, InP) were made and by 1956 GaAs had a reported efficiency of 6% [3]. By 1958, small area Silicon solar cells had reached an efficiency of 14% under terrestrial sunlight. The big push to develop solar power, however, came from its obvious space application. On March 17, 1958 the world's first solar powered satellite was launched, Vanguard 1 (see Figure 6.1) [4].

It carried two separate radio transmitters to transmit scientific and engineering data concerning, among other things, performance and lifetime of the 48 p/n silicon solar cells on its exterior. The battery powered transmitter operated for 20 days; the solar cell powered transmitter operated until 1964, at which time it is believed that the transmitter circuitry failed. The solar cells were fabricated by Hoffman Electronics for the U.S. Army Signal Research and Development Laboratory at Fort Monmouth. Until the use of those silicon cells on Vanguard 1 solar cells were really just a novelty – powering toys mostly. As photovoltaic's pioneer Martin Wolf observed, the success of solar cells on Vanguard "was the salvation of the solar cell industry." Setting a record at the time for satellite longevity, Vanguard 1 proved the merit of space solar cell power.

Silicon cells supported the emerging commercial satellite industry and provided power for multiple NASA and military missions. The differences between the functionality of silicon cells in space as opposed to on the ground gradually became apparent. The Van Allen trapped radiation belts were discovered in 1958 by the Explorer 1 satellite that transmitted data for 22 days. However, there were many failures both in the Russian Sputnik series of satellites and in the U.S. Explorer and Vanguard series of satellites. As a direct result of the Sputnik crisis, the Congress and the President of the United States created the National Aeronautic and Space Administration (NASA) on October 1, 1958. NASA absorbed into itself the earlier National Advisory Committee for Aeronautics intact: its 8,000 employees, an annual budget of $100 million, three major research laboratories-Langley Aeronautical Laboratory, Ames Aeronautical Laboratory, and Lewis Flight Propulsion Laboratory-and two smaller test facilities. It quickly incorporated other organizations into the new agency, notably the space science group of the Naval Research Laboratory in Maryland, the Jet Propulsion Laboratory, JPL, managed by the California Institute of Technology for the Army, and the Army Ballistic Missile Agency in Huntsville, Alabama, where Wernher von Braun's team of engineers were engaged in the development of large rockets. The latter became the George C. Marshall Space Flight Center in 1960. In 1959 NASA established Goddard Space Flight Center (GSFC), in 1961 the Johnson Space Center (JSC) and in 1962 the Launch Operations Center that became, along with Cape Canaveral Auxiliary Air Force Station, the Kennedy Space Center (KSC) in 1963. There are now ten NASA centers.

The Vanguard served as the progenitor of practical photovoltaic applications from Space to Earth, demonstrating the importance of

FIG. 6.1 VANGUARD 1 SATELLITE MODEL AT THE PARADE OF PROGRESS SHOW AT THE PUBLIC HALL IN CLEVELAND, OH

Space in developing the simplest, most benign source of energy yet conceived. In 1961 many of the staff from the silicon cell program at Fort Monmouth transferred to NASA Lewis Research Center, LeRC, (now John H. Glenn Research Center at Lewis Field, GRC) in Cleveland, Ohio. From that time to the present, the Photovoltaic Branch at NASA Glenn has served as the research and development base for NASA's solar power needs. Impressed by the lightweight and reliability of photovoltaics, almost all communication and military satellites and scientific space probes have been solar powered. NASA's space flight centers, especially JPL and GSFC, were also involved in testing, understanding the space environment and technology development specifically for the solar powered missions that they supported.

At the beginning of the 1960s, a large number of organizations under the Department of Defense and NASA supported research and development on energy conversion devices. Upon the recommendation of the Solar Working Group of the Interservice Group for Flight Vehicle Power a conference was organized by the Institute for Defense Analysis to include research from Universities, industry and government on photovoltaic devices. That conference on April 14, 1961 was the 1st Photovoltaic Specialist Conference (PVSC). Since the 4th conference in 1964 the PVSC has been supported by the IEEE Electron Devices Society. The 37th IEEE PVSC will be held in Seattle in June of 2011. The conference is notable for the record of photovoltaic research not just in the U.S. but worldwide. A Space Photovoltaic Research and Technology (SPRAT) conference was started in 1974 and has always been held at NASA GRC sponsored by the Photovoltaic Branch. The 22nd SPRAT will be held in the fall of 2011.

The early NASA missions (Pioneer, Beacon, Vanguard, and Explorer) were Earth orbiting and included many failures. Tiros 1, see Figure 6.2, gave us our first TV image from space.

Echo 1 was the first passive communication satellite that relayed voice and TV signals in August, 1960. Many of the NASA early satellites focused on learning more about the Earth and its space environment. However, in 1959 Pioneer 4 passed within 60,000 km of the moon and in 1962 Mariner 2 successful flew by Venus. NASA's project Mercury, beginning with Alan Shepard in 1961, had two suborbital and four successful human orbital flights ending with Gordon Cooper in 1963. The Gemini program followed in 1965 with the first manned orbital maneuvers and the first extra – vehicular activity (EVA). The manned lunar program (Apollo) began in 1964 resulting in the first man on the moon in 1969. The program ended with Apollo 17 in 1972 after landing 12 astronauts on the moon. Skylab, our first orbital space station, followed in

FIG. 6.2 FIRST TELEVISION PICTURE FROM TIROS 1 SATELLITE, APRIL 1, 1960

FIG. 6.3 SKYLAB

1973, and was the first NASA human mission to be powered by photovoltaic arrays, see Figure 6.3.

During these early years, unmanned missions continued to use solar cells for primary power, and there were improvements in the space silicon solar cell developed by NASA: Use of n-on-p silicon semiconductor type (rather than p-on-n) for superior radiation resistance (1963); Use of shallow junction silicon cells for increased blue response and current output (1971); Use of back-surface fields (BSF) and low/high junction theory for increased silicon cell voltage output (1972). Research was focused on understanding and mitigating the factors that limited cell efficiency (e.g., minority carrier lifetime, surface recombination velocity, series resistance, reflection of incident light, and non-ideal diode behavior).

While both the terrestrial and the space markets were focused in improving cell efficiency, a new set of requirements arose in addressing the needs of the space community. Early satellites needed only a few watts to several hundred watts. The power source must be available, reliable and ideally have a high specific power, watts per kg, (W/kg) since early launch costs were ~ $10K/kg or more. The cost of the power system for these satellites was not of paramount importance since it was a small fraction of the satellite and launch cost. The size of the array was important for many early satellites due to the body-mounted array design, therefore limiting total power. The launch of Telstar in 1962 created new markets for space photovoltaics (i.e., terrestrial communications) [5]. Telstar's beginning of life (BOL) power was 14 W but high radiation caused by a nuclear weapon test reduced the power output. There was a great deal of both theoretical and experimental research in the '60s. The early CdS/Cu solar cells were found to degrade over time. CdTe cells were developed reaching efficiencies of ~ 7.5%. However, the higher efficiency and stability of the silicon solar cells assured their preeminence in satellite power for the next 3 decades.

Research on thin film cells, because of their higher specific power and projected lower costs, was also funded at lower levels by the space community. As the first photovoltaic devices were being created there were corresponding theoretical predictions emerging citing ~20% as the potential efficiency of Si and 26% of an optimum bandgap material (1.5eV) under terrestrial illumination [6]. In addition the concept of a tandem cell was proposed to enhance the overall efficiency. An optimized three-cell stack was soon to follow with a theoretical optimum efficiency of 37% [7]. Aside from the cell response to a radiation environment, the goals of both the terrestrial and space community were the same. In the sixties, researchers at NASA Lewis, Goddard, Langley and JPL contributed to the growing base of knowledge about photovoltaics in general and particularly their use in space.

Although the price of silicon solar cells dropped by about 300 percent between 1956 and 1971 the cells still cost $100/watt compared to the price of electricity of $.05/watt [8]. A new industry was born in that time span to provide cells for the space markets and also a growing Department of Defense (DOD) surveillance market.

6.2.2 The '70s and '80s: A Strong Link between Space and Terrestrial Photovoltaics

The '70's involved interaction between NASA and the Energy Research and Development Administration (ERDA), that was the predecessor of the Department of Energy (DOE) founded in 1977. A Terrestrial Photovoltaic Measurements Workshop under the joint sponsorship of ERDA and NASA was held at the NASA Lewis Research Center in March 1975 [9]. Nearly 100 people at-

tended from all segments of the solar cell community. The workshop was divided into three separate sessions:

(1) Solar Intensity and Spectrum Conditions for Terrestrial Photovoltaics
(2) Terrestrial Sunlight Simulation
(3) Methodology for Measurements and Calibration of Solar Cells

A set of procedures for testing solar cells for terrestrial applications resulted from that workshop. The issues that the workshop addressed were equally important in the space program. The intensity and appropriate spectral distribution for solar simulation and standard test procedures were concerns for both communities.

ERDA had established, as a goal of the National Photovoltaic Program, that low-cost solar cell arrays be developed with a lifetime of 20 years. As part of this program, JPL purchased a quantity of solar cell modules from various manufacturers. Some these modules were installed in the ERDA Photovoltaic Systems Test Facility, STF, located outdoors at NASA Lewis (Glenn) Research Center in Cleveland, OH, see Figure 6.4 [10].

The first of these modules were installed in early 1976. The effects of the environment on modules installed in the Systems Test Facility (STF) were of particular interest for several reasons: first, because JPL needed data to begin evaluating the capability of different module designs towards attaining the ERDA lifetime goal, and second, because the modules in the STF were utilized in arrays whose voltage output is relatively high (approx 200 V dc). With respect to the latter point, little or no data existed regarding environmental effects on solar cell modules used in arrays delivering these relatively high output voltages. The objective of the work was to acquire data under standard, easily reproducible, laboratory conditions for modules both before and after exposure to the environment. The data so obtained were intended to serve as the beginning of a database that was used to evaluate the effects of environmental exposure and use on representative terrestrial solar cell modules used in the STF.

In the latter years of the '70s the National Photovoltaic Program run by DOE had several projects, one of which was the Tests and Applications Project that was managed for DOE by NASA Lewis

FIG. 6.4 DR. LOU ROSENBLUM DURING A TELEVISION INTERVIEW AT THE STF IN 1977

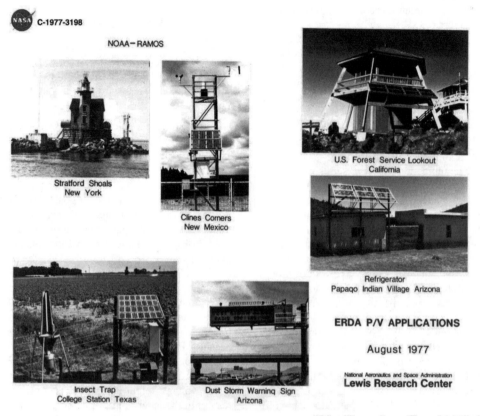

FIG. 6.5   ERDA/NASA PV APPLICATIONS

Research Center (LeRC). The testing part of that project was discussed previously. The Applications effort was intended to introduce PV power systems to a wide range of potential users with near-term applications in order to accelerate their entry into the commercials market, see Figure 6.5 [11].

Applications which indicated near-term cost-effectiveness, technical feasibility, solar cell promotional value, and/or substantial use multiplication were considered for joint cost shared experiments. By the end of 1978 seven different applications comprising a total of 16 systems were brought to operational status. In addition at that time four additional experiments were being developed: an air pollution monitor in New Jersey, two seismic sensors in Hawaii and a village power system in Arizona. In addition to the DOE sponsored projects, LeRC installed a PV-powered water pump and grain grinder in the remote African village of Tangaye, Upper Volta, see Figure 6.6.

This project was sponsored by the U.S. Agency for International Development (AID). Photovoltaic systems applications approaching near-term cost-effectiveness in the late '70s were almost exclusively associated with remote applications where the user's loads were powered by batteries, thermoelectric generators, or small engine-generators without any utility connections or back-up power supplies. During implementation of these experiments LeRC personnel worked closely with the user in selection of loads and load profiles as well as designing array structures that would integrate well with the environment of the site. LeRC personnel also provided the installation and checkout plus at least one site visit per year. The systems ranged in size from 25 watts to 3.5 kW. Operating data was collected. The installed systems included PV powered refrigerator at remote sites, U.S. Forest Service forest lookout tow-

ers, highway dust storm warning signs, remote automatic weather stations for NOAA, U.S. Dept of Agriculture insect survey traps, a water chiller for a remote drinking fountain, and a PV power system for a Papago Indian village.

The Upper Volta project mentioned above was the beginning of several projects sponsored by AID. The intent was to stimulate interest and awareness on the part of planners and decision makers from Third World countries and within the international development assistance community. The AID/NASA/DOE projects included a large village power system in Gabon (four sites) and the Utirik Island in the Marshall Islands, PV powered vaccine refrigerator systems in 18 countries, a farmhouse system and two water pumping-irrigation systems in Tunisia, five medical clinic systems in Guyana, Ecuador, Kenya and Zimbabwe and a PV powered remote earth station in Indonesia [12]. LeRC personnel installed two of the projects and then managed the contracts to U.S. companies such as Solar Power Corp., Solavolt, Solarex and Hughes Aircraft for the remaining projects. The last project was completed in 1984.

In 1982 a space station task force was established by NASA to provide focus and direction for Space Station planning activities. It also provided Congress and the current Administration with sufficient information to allow them to make an informed decision on whether the U.S. should proceed with a Space Station as the next major national initiative. Space station planning actually began when NASA was created in 1958 and was given a higher priority in the early years over a lunar landing program. However President's Kennedy's announcement in 1961 to land a man on the moon postponed the possibility of a Space Station to the post-Apollo era. Skylab, America's first space station, was launched in 1973. A

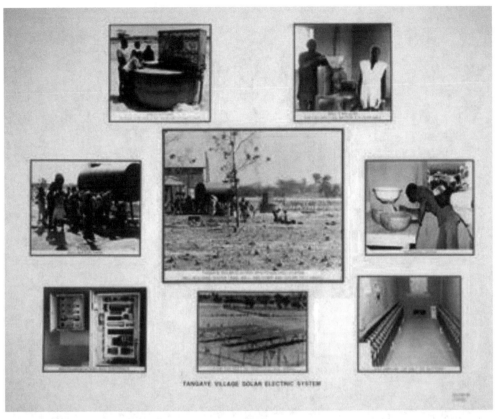

FIG. 6.6   TANGAYE UPPER VOLTA VILLAGE AFRICA SOLAR ELECTRIC SYSTEM

decision was made in 1972 to construct a reusable transportation system and a great deal of NASA's resources were focused on building and testing a Space Shuttle that first flew in 1981. In 1984 Pres. Reagan directed NASA to develop a permanently manned space station (SS) within a decade. NASA embarked on a definition phase to completely define cost, capabilities, and components of the initial Space Station [13]. The original design for the SS was to provide 75 kW of continuous electrical power (60 kW for the customer and 15 kW for SS needs) that could grow to 150 kW. The requirements of a permanent system, the need for evolutionary growth, human presence, operational flexibility, and multiplicity of use drove the power technology long operational lifetime, maintainability, growable/modular subsystems, safety and crew accommodations, ground independence including on-board monitoring and control, and user friendliness. Because of the extremely large arrays, the high power level technology may include high voltage power distribution whether or not the solar array power is generated at high voltage. That, in turn, would require space plasma compatibility. At that point in time a number of alternative energy sources were also considered. These included solar planar or concentrator arrays of either silicon or gallium arsenide solar cells. The arrays could be either erectable or deployable or some combination of both. Solar dynamic systems were also considered for SS power. The idea that the power system should by designed to incorporate technology that can later evolve to an operational readiness state and replace the technology used on the initial SS was incorporated in these first designs of SS.

The work that was accomplished in the evolution of the silicon space solar cell design can be seen in Figure 6.7.

Accomplishments included: Development of wrap around contacts for high efficiency silicon cells to enable automated array as-

sembly and reduced costs (1977); Development of large-area solar cells to reduce array cell and assembly costs (1983); A variety of array blanket materials and laminating technologies, welding methods, and large area encapsulants were developed and evaluated for thermal cycling durability for increased orbital lifetime (1972-1990); Developed materials and methods to protect arrays against the effects of atomic oxygen and damaging plasma interaction effects in low Earth orbit (1985-1995).

The major goals of NASA in the mid '80s were essentially the same as today. High efficiency, low mass, increased radiation tolerance and low cost continue as the primary focus of the photovoltaic program [14]. However, it was already recognized that III-V solar cells offered higher efficiency and greater radiation tolerance than silicon solar cells. NASA focused on development of a cascade III-V cell structure and also on ultralightweight CLEFT technology. The CLEFT process permitted the growth of thin single-crystal films on reusable substrates, thus reducing both cell cost and weight [15]. NASA also began to explore InP cells that had substantially greater radiation resistance than GaAs.

6.2.3   The '90s to 2010

It should be noted that after the large increase in the DOE Solar Program budget during Pres. Carter's years that reached almost $700M, the PV budget, including concentrated solar power, fluctuated above and below $100M from 1986 to 2006. NASA's budget during this time span was about 10% of the DOE budget. DOE focused on reducing PV module costs and increasing production while NASA focused primarily on increasing solar cell efficiency. Both programs were remarkably successful although it is important to acknowledge that DOD also invested more dollars than NASA

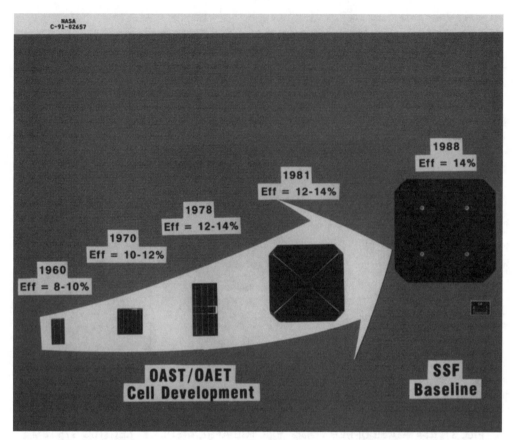

FIG. 6.7 EVOLUTION OF THE SILICON SPACE STATION SOLAR CELL TECHNOLOGY

in space photovoltaics primarily on technology development while NASA invested in research.

In the '90s the gap between theoretical efficiencies and experimental efficiencies for silicon, gallium arsenide and indium phosphide became almost non-existent. Thin film cells of amorphous silicon, $CuInSe_2$ and CdTe renewed the enthusiasm for the promise of lower costs for the terrestrial community and the potential for increasing the thin film efficiency and making them on flexible substrates excited the space community. Satellites grew in both size and power requirements and structures were designed to deploy large solar arrays during these two decades. However, the mass and fuel penalty for attitude control still dictated a move to more efficient cells. The first GaAs solar cells reached flight status in 1983 on the Living Plum Shield (LIPS)-II satellite [16]. In 1997 the first dual-junction GaInP/GaAs cells were launched into space, a technology that quickly supplanted single-junction GaAs, and then was itself supplanted by triple-junction GaInP/GaAs/Ge cells. NASA JPL completed in 1991 the Advanced Photovoltaic Solar Array (APSA) program that demonstrated a lightweight deployable flexible array wing using 55micron silicon cells or 100micron GaAs/Ge solar cells. As might be expected the space commercial community was and still is risk adverse in moving laboratory cells into space. DOD had sufficient funds to develop cell production and array assembly capabilities concurrently while NASA's cell/array development has generally followed a very different path, governed by limited funds and lack of a specific mission need [17]. Program needs have followed from the demonstration of a space-ready component or device. While this does leverage funding both in the terrestrial and military communities it does not always address NASA specific issues.

The New Millennium Program was created in 1994 to accelerate the insertion of advanced space-related technologies into future science missions using deep-space and Earth-orbiting technology validation spacecraft. This included the use of new solar cells and array designs. Both GRC and JPL were involved in the technology development for solar arrays and solar cells for this program.

FIG. 6.8 DEEP SPACE 1 SPACECRAFT REPRESENTED PICTORIALLY

Almost all of the photovoltaic power systems used in space have been flat-panel arrays. The Deep Space 1 mission, the first of the New Millennium missions launched in October of 1998, tested the use of a Fresnel lens concentrator with dual-junction solar cells on an asteroid/comet fly-by mission, see Figure 6.8.

The Solar Concentrator Array with Refractive Linear Element Technology (SCARLET) concentrator array [18] on this mission performed flawlessly and produced 2.5 kW of power at Air Mass Zero (AM0), with a power density of 200 W/m2 and a specific power of 45 W/kg [19]. During a highly successful primary mission, Deep Space 1 tested 12 advanced, high-risk technologies in space. In an extremely successful extended mission, it encountered Comet Borrelly and returned the best images and other science data ever from a comet. During its fully successful hyperextended mission, it conducted further technology tests. The spacecraft was retired on December 18, 2001.

The largest solar power system ever flown in space is that of the International Space Station (ISS), see Figure 6.9.

The SS Project Office at LeRC (Work package 4) had a major role in the early SS project with the goal of selecting, designing, building, and verifying the performance of flight hardware including: Performed and managed early trade studies of the power system design and technology selection (1983-1991); Managed contract with Rocketdyne and subcontractors to design, build and test SS hardware (1990-1995); Tested key flight components such as the beta gimbal roll rings, the radiators, the power electronics boxes, etc. (1994-1999); Developed, built and tested the hollow cathode assembly for the plasma contactor (based on research performed in the Ion Thruster Branch) to protect SS from plasma damage (1993-1999). A key technology development effort for the space station arrays was to reduce the cost involved in the interconnection and lay-up of solar arrays. This was done by developing a larger area solar cell and developing the "wrap-through" contact to allow all of the cell electrical interconnection to be done from the back side of the solar cell.

The ISS array uses silicon cells, one of the first uses of large area (8cm by 8cm) cells, with wrap-through contacts allowing all the electrical connections to be made from the backside. The front surface of this cell is shown in Figure 6.4. ISS cells were assembled

FIG. 6.10   ISS SOLAR ARRAY WING

into panels of 200 cells, with the panels assembled into accordion-folded 107 ft by 38 ft (34 m by 12 m) wings (Figure 6.5), producing ~32 kW of power per wing. The final array was delivered to the Space Station in March, 2009. ISS electrical power system hardware was sized to provide a total of 75 kW of continuous orbital power, during both the sunlit and eclipse portion of the orbit, with Nickel-hydrogen batteries providing the storage capacity for the eclipse operations. The system generates about 110 kW of average power, that after battery charging, life support, and distribution, supplies ~46 kW of continuous power for research experiments. The Russians also supply an additional 20 kW of solar power to the ISS. A total of 32,800 cells are bonded to the flexible panels of each wing, see Figure 6.10.

The total array assembly has a total of 262,400 silicon solar cells that could produce ~250 kW of photovoltaic array power (8 wings at 32 kW each). ISS is expected to have an operational lifetime extending to 2020. The solar array system for ISS was managed by NASA GRC and built by Lockheed Martin. Today GRC still provides monitoring and problem solving for the power system of ISS.

6.2.4   Planetary Missions

Dr. Geoffrey Landis kindly contributed this section on planetary missions. Solar arrays were used to power the Surveyor spacecraft, the first US missions to land on the moon, and also powered the

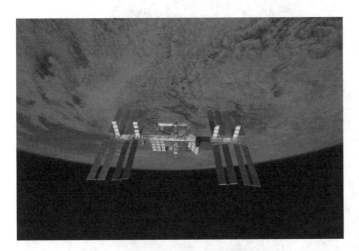

FIG. 6.9   BACKDROPPED BY THE BLACKNESS OF SPACE AND EARTH'S HORIZON, THE INTERNATIONAL SPACE STATION IS SEEN FROM SPACE SHUTTLE DISCOVERY AS THE TWO SPACECRAFT BEGIN THEIR RELATIVE SEPARATION ON MARCH 25, 2009

FIG. 6.11   THE PATHFINDER LANDER (left) AND SOJOURNER ROVER (right) ON MARS

FIG. 6.12   THE SOLAR ARRAYS ON THE MARS EXPLORATION ROVER "SPIRIT" VIEWED IN TRUE COLOR ON THE 1229TH DAY OF OPERATION ON MARS (IMAGE COURTESY NASA/JPL AND CORNELL UNIVERSITY)

Apollo Lunar Surface Experiments Package, ALSEP, scientific package left on the moon by the Apollo-11 mission, although later ALSEP packages used radioisotope power.

The first lander on Mars, the 1976 Viking mission, was powered by a radioisotope source, but in 1997 NASA launched the Mars Pathfinder Lander and Sojourner rover (see Figure 6.11), the first use of solar power on the surface of another planet [20]. The cells that powered Pathfinder were single-junction GaAs on Germanium

cells [21]. Dust accumulation was monitored on that mission by the Materials Adherence Experiment (MAE) instrument. and was thought to be the limiting problem with using solar arrays on Mars. The MAE measurements indicated a steady dust accumulation at a rate of ~0.29% per day [22].

In 2003 the Mars Exploration Rovers (MER) mission (see Figure 6.12) were launched.

At over six years of operations on Mars, it is an understatement to comment that the performance of the MER solar arrays is quite unlike that of any other space array. The MER rovers, Spirit and Opportunity, have now been on the surface of Mars for 6.5 years. They demonstrated the first use of triple-junction GaInP/GaAs/Ge solar cells on Mars [23]. A significant finding of the MER mission was that in the Martian spring, wind events remove dust from the solar arrays, and hence dust accumulation was not an insuperable limit to solar array lifetime on Mars [24]. The Opportunity rover has now passed its 13th mile, and continues to drive. After 4.8 miles of traverse, the Spirit rover has become stuck at the feature named "Troy." As a result of this immobility, it was unable to change its tilt to optimize solar energy during the winter. It is now the middle of the southern-hemisphere Martian winter, and on March 22, 2010, the Spirit Rover stopped communicating with Earth due to a low-power fault. Although it is possible that as Mars approaches the southern-hemisphere summer, the amount of incident solar power available may increase to a level sufficient that the solar arrays might gradually recharge her batteries and allow the spacecraft computer to reboot. At the moment, it is unclear whether Spirit's mission will resume.

The success of a PV power system on Mars has been remarkably demonstrated. As opposed to the more common mode of spacecraft operation the MER power generation is so minimal and variable that the mission power profile must be adjusted to the expected power generated. The impact of unpredictable factors such as atmospheric conditions and dust accumulation and removal on the solar panels limits the accurate prediction of array power.

FIG. 6.13   PHOENIX LANDER, SEEN DURING SPACECRAFT INTEGRATION

FIG. 6.14   PHOENIX ULTRAFLEX ARRAYS, AS DEPLOYED ON MARS

In 2007 the Mars Phoenix mission was launched, arriving in May of 2008. The Phoenix mission was the first chosen in NASA's Scout program, sent to the polar region of Mars. It used a lander that was intended for the 2001 Mars Surveyor project that was cancelled. The lander was powered by two 1.8m diameter solar arrays and lasted a little more than five months in the Martian northern plains. A picture of the spacecraft during assemble with extended arrays can be seen in Figure 6.13 and deployed on Mars in Figure 6.14. The Phoenix lander uses the UltraFlex array that had been chosen as the array for NASA's Orion program. The mission was not designed to survive the polar winter, during which it was expected to be covered with several meters of dry ice, and as expected, it fell silent due to low-power fault in early Martian autumn, on November 2, 2008, and was not subsequently recovered. Following the winter, a recent flyover by Mars Reconnaissance Orbiter suggested there was severe ice damage to the lander's solar panels over the course of the Martian winter.

Launched in 1999, the Stardust mission was a probe to gather dust from the comet Wild-2 and return the material to Earth. Because the mission would operate at low temperatures, 10-Ohm-cm silicon cells outperformed the state of the art (at the time) GaAs solar cells, and hence Si cells (16.3% efficient at AM0, 28°C) were used for the mission [25]. Stardust broke the record for the farthest spacecraft from the sun powered by solar energy, at a farthest distance of 2.72 Astronomical Units (AU) from the sun. This record was subsequently broken by the Dawn mission to the asteroids Vesta and Ceres [26], launched in launched in July 2007, with a power of 10.5 kW at AM0 1-sun conditions, and a power at 3 AU estimated at 1.4 kW. The distance record will be set again by the Juno mission to Jupiter .[27]

NASA has also been interested in missions operating at distances closer to the sun. Starting in 1962, NASA sent a number of probes inward toward Venus, and one probe, Mariner 10, to fly past Venus on the way to a series of three encounters past Mercury. This was the first NASA mission that came closer than half an astronomical unit to the sun, encountering Mercury on March 29, 1974 [28]. To avoid overheating the solar panels at solar intensity approaching ten times the intensity at Earth orbit, the solar panels were rotated up to 76° away from normal to the incident sunlight. By this technique, the solar cells were kept at a temperature under about 115°C (239°F) at the closest approach of the spacecraft to the Sun.

Two other spacecraft also flew at the high-intensity region, Helios A, launched on 10 December 1974, which reached 0.31 AU; and Helios B, launched on 15 January 1976, which reached 0.29 AU. Both of these spacecraft used silicon cells that were slightly modified for high-intensity use in conjunction with surface mirrors to reduce the intensity on the array.

Following Mariner and Helios, there were no further missions inward of Venus until the ambitious MESSENGER mission to orbit Mercury, launched August 3, 2004. The spacecraft, built by Johns Hopkins Applied Physics Laboratory for the NASA mission, is designed to operate under solar power at a distance as close as 0.3 astronomical units from the sun, where the solar intensity is 11 times that seen in Earth orbit. To minimize temperature, the same technique as used on the Helios mission was adapted to the MESSENGER array, and much of the surface of the solar cell array was populated with Optical Solar Reflector (OSR) mirrors, see Figure 6.15,

FIG. 6.15   SOLAR PANEL FOR THE MESSENGER MISSION TO MERCURY, SEEN IN THE ASSEMBLY PHASE. NOTE THE OPTICAL REFLECTORS COVERING MUCH OF THE SURFACE OF THE ARRAY

with a cell to OSR ratio of 1:2, to limit the solar absorbance and avoid overheating the panels. The arrays are also operated at a tilt angle to the sun, to reduce the projected area illuminated limiting the array temperature below the maximum of 150°C (actual maximum temperatures seen on the mission are closer to 100°C). The solar cells are 0.14-mm thick, 3 cm by 4 cm Advanced Triple Junction (ATJ) cells with a minimum efficiency of 28% at 1-sun AM0. This mission is operating successfully at distances close to the sun, and will reach Mercury early in 2011 [29].

The Solar Dynamics Observatory mission was launched Feb. 11, 2010. Although this gets no closer to the sun than 1 AU, spectacular images of the dynamic nature of our sun are being recorded. The total available power is 1450 W from 6.6 m2 of solar arrays operating at an efficiency of 16%. The cells are Ultra Triple Junction (UTJ) cells with a nominal efficiency rating of 28.3% AM0 at 28°C.

6.2.5 Programmatic Accomplishments

Both GSFC and JPL have been involved in array development and dealing with issues concerning the solar cells related to their specific NASA missions. Due to the efforts of Dr. Edward Gaddy, GSFC became experts in electrostatically clean arrays, necessary for their Earth monitoring missions and also developed expertise in photovoltaic power systems to be used near the sun. JPL work, supported primarily by Paul Stella, developed cells and arrays unique to the needs of missions to Mars and other planets.

From the '90s until 2001 GRC had a diverse research program to develop solar cell technology that spanned advanced high efficiency III-V solar cells, advanced thin film technology, space environmental effects, measurement and calibration, and advanced photovoltaic concepts. The three programs with the most overlap with the terrestrial community were the advanced thin film technology program, the high-efficiency III-V program, and the advanced photovoltaic concepts program. The other programs were focused solely on space applications.

The Glenn program in thin film solar cells worked on approaches to reduce the mass of solar arrays. The best thin film cells required processing temperatures in excess of 600°C, which prohibit the use of current polyimide substrates. A low-cost flexible substrate would also benefit the terrestrial community by replacing the expensive and fragile heavy glass structures. Thin-film research led by Dr. Aloysius Hepp focused on developing the capability to produce polymer or inorganic or hybrid solar cells on lightweight polymer substrates by finding a lower temperature deposition process for CIGS cells, for example, or by finding a high temperature survivable polymer substrate [30]. The thin film group worked on both approaches. A new single-source precursor for CIGS cells was patented [31]. Additionally the ability to interconnect the thin films cells and produce an integrated array was part of the Glenn Small Business Innovative Research (SBIR) program. This is now a commercial product of Ascent Solar.

The high efficiency III-V program led by Henry Curtis and later David Wilt included efforts to enhance the efficiency of space cells by texturing the surface of GaAs [32] and InP [33] (see Figure 6.16).

Later efforts involved the Epitaxial Lift-Off (ELO) of these cells [34] and a patent for the ELO of InP [35]. Today a commercial company, MicroLink Devices, uses that technology to market both GaAs and InP based devices. The III-V effort also focused on the possibility of fabricating the multi-junction (MJ) III-V cells on lower cost substrates, both silicon [36] and polycrystalline germanium [37]. Both of these technologies resulted in the formation of

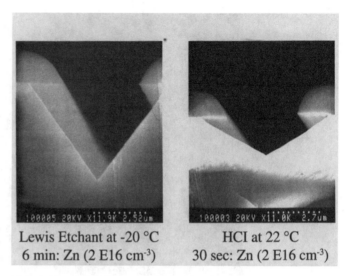

Lewis Etchant at -20 °C	HCl at 22 °C
6 min: Zn (2 E16 cm⁻³)	30 sec: Zn (2 E16 cm⁻³)

FIG. 6.16 PROFILE ETCHING OF InP

commercial companies, Amberwave and Wakonda Technologies. The lure of being able to produce a high efficiency III-V cell on a flexible substrate that potentially could be used in a low-cost roll to roll manufacturing process drove the efforts of the program.

The Advanced Photovoltaic Concepts effort in the GRC program led by Dr. Sheila Bailey included investigating possible methods for direct conversion of photons to electrical energy using other than the conventional II-VI, III-V or silicon and germanium semiconductor p/n junction solar cells. Approaches included organic semiconductors, nanostructured materials, quantum dots, quantum wells, and optical antennae/rectifier devices, among other potential concepts. This effort also involved investigating integrated power and propulsion solar sails for low-cost microsatellites for near Earth space and deep space applications and integrated thin film power also applicable to dirigibles, balloons, and reusable space flight vehicles. One of the achievements from this effort was the development of the first SiC solar cell [38]. These cells were developed for high temperature, high light intensity, and high radiation missions such as experienced by solar probes. The project was funded by the JPL Solar Probe Program. The cells were developed and tested by GRC (Dr. Sheila Bailey) and Rochester Institute of Technology, RIT (Dr. Ryne Raffaelle). The performance was especially noteworthy at high temperature (300 °C) and under concentration (150 suns) [39].

In 2001-2002 a technology review committee from NASA, DOE and AFRL was formed to assess solar cell and array technologies required for future NASA science missions. After consultation with mission planning offices, solar cell and array manufacturers, universities and research laboratories, an assessment of the state of the art of solar cells and arrays was made and compared with the projected needs. A technology development program was proposed in high efficiency cells, electrostatically clean arrays, high temperature solar arrays, high power arrays for solar electric propulsion, low-intensity/low-temperature array conditions for deep space mission, high radiation missions, and Mars arrays that operate in dusty environments [40]. The results of that study were remarkably prophetic in predicting the directions that high efficiency multi-junction cells would take and the recommendations are still valid today.

The Advanced Photovoltaic Concepts effort was split into three more focused programs in 2002: An Extended Temperature project

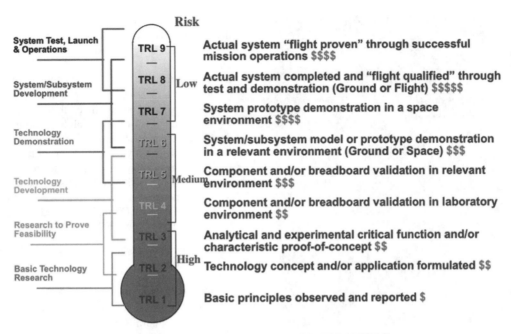

FIG. 6.17 TECHNOLOGY READINESS LEVELS

led by Dr. Geoffrey Landis; An Advanced Blanket and Array Technology project led by Michael Piszczor; and a Quantum Dot Technology project led by Dr. Sheila Bailey. At about the same time frame there was competition based internal funding (Director's Discretionary Fund, DDF) for the projects: Quantum Dot Alpha Voltaics, Size-graded Quantum Dot Thin Film Solar Cells and Quantum Wire III-V Photovoltaics. The entire program was entitled Energetics and ran through 2004. The Energetics program was reviewed by an outside panel from the National Research Council (NRC) in 2002 and then again by a technical peer review panel in 2003. The Photovoltaic program was commended for the quality of their projects, the technical achievements and the technical direction. It was noted that the choice of direction in any truly advanced research enterprise is difficult. In part this is due to the fluid nature of long-range mission plans that in many cases are beyond the horizon of budget timelines, a particularly severe problem for NASA. Thus, any specific research goals can only address potential future missions in general terms and must retain the flexibility to adjust to changes in mission plans. Even to a greater part, the research projects in themselves generate enablers for advanced missions. They can drive the technology, cost, schedule and risks and in some cases the entire feasibility of advanced, highly demanding missions. Several areas of research were noted as impressive by the peer review panel: Transition of lattice-mismatched semiconductor growth to industry, MJ Thin-Film PVs at 20%, Quantum Dot diagnostics and fabrication, and extended-temperature solar cell development that was directed at a specific NASA mission.

NASA has a set of Technology Readiness Levels that are illustrated in Figure 6.17.

The Energetics Program had two major thrusts relative to photovoltaics: Advanced Space Technology at TRL levels 2-5 and Technology Maturation at TRL levels 3-6. While focused on mission uses for NASA the program had a great deal of synergy with terrestrial photovoltaics [41].

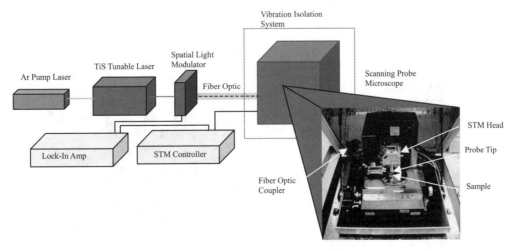

FIG. 6.18 SCHEMATIC OF STORM SETUP

FIG. 6.19   A 6 µM × 6 µM STORM IMAGE OF ALTERNATING DOPED AND UNDOPED EPILAYERS OF INP

The quantum dot project and three DDF projects all had value for the terrestrial community. Some of the accomplishments of this program are listed below:

1. A new diagnostic technique was developed which has the capability to interrogate the optical bandgap as a function of position within a semiconductor microstructure by Scanning Tunneling Optical Resonance Microscopy or STORM. This technique involves the use a broadly tunable continuous wave Coherent 899-01 Ti:Sapphire laser. Output from the laser is intensity modulated by a high-speed ferro-electric liquid crystal modulator and fed into the single mode fiber. The output from the optical fiber is used to illuminate the tip-sample junction in a Digital Instruments D3100 scanning probe microscope. The change in the STM tunneling current due to the photo- induced carriers is measured. The change in this photo-enhanced portion of the tunneling current as a function of the laser frequency can be used to determine optoelectronic bandgaps, see Figures 6.18 and 6.19 [42], [43].

The tunneling current due to photoexcited carriers is clearly discernable. STORM can be used to investigate quantum dot structures, enabling analysis of both the optical and electrical characteristics.

2. A theoretical analysis of photovoltaic materials and quantum dots (1., 2., & 3.) yielded a potential quantum dot cell structure for an amorphous silicon thin film cell with an intermediate band of $CuInSe_2$ or $CuInS_2$ quantum dots [44].

3. An intensive effort to synthesize quantum dots while controlling size and yield, providing suitable cladding, and dispersion onto appropriate substrates resulted in the following accomplishments: (See Figures 6.20 and 6.21).

4. A polymeric quantum dot solar cell was fabricated using 1% w/w $CuInS_2$ in P3OT on ITO coated PET flexible lightweight substrate, see Figure 6.22 [45].

Three patent disclosures were filed as a result of this work [46, 47, 48]

In 2004 Dr. Bailey's proposal "Nanomaterials and Nanostructures for Space Photovoltaics" was selected for funding from the Intramural Call for Proposals Human and Robotic Technology (H&RT). The program was for 4 years and 6 million dollars. It funded senior researchers at GRC, JPL, RIT, Penn State, Univ. of Toledo, and Univ. of Houston. Research would be conducted on developing new nano-materials and nano-structures in three different systems which offer the potential for radical advances in space power generating capability by enabling higher efficiency and/or higher specific power, better radiation tolerance, favorable temperature coefficients, and lower cost for future space solar arrays.

Three different cell structures were explored with the following generic goals:

- Identify suitable nanomaterials (matching the bandgaps, electron affinities, and compatibility of various PV materials to the desired properties).
- Selecting a compatible structure commensurate our developed ability to manufacture the nanomaterials (i.e., colloidal synthesis, laser ablation, or Stranski-Krastanow growth).
- Incorporation of nanomaterials into a complete device structures.
- Characterization of cell structure for intended H & RT space utilization.

CuInS₂ and CuInSe₂
Single-source precursor (SSP)
$(PPh_3)_2Cu(m\text{-}SEt)_2In(SEt)_2$
$(PBu_3)_2Cu(m\text{-}SEt)_2In(SEt)_2$
$(PPh_3)_2Cu(m\text{-}SePh)_2In(SePh)_2$

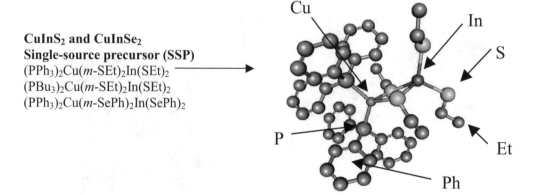

FIG. 6.20   SSP IS HEATED IN A MIXTURE OF DOP (DIOCTYLPHTHALATE) AND HEXANETHIOL

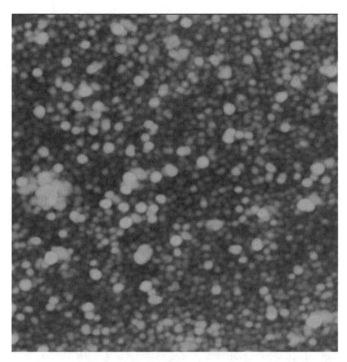

FIG. 6.21   A 1 μM × 1 μM TAPPING MODE AFM IMAGE OF COLLOIDAL CuInS₂ (MERCAPTOACETIC ACID) QUANTUM DOTS ON GLASS

The three solar cell material areas targeted were:

- Epitaxially grown III-V nanostructures (quantum dots and quantum wires) inserted into State of the Art, SOA, multi-junction space solar cells,
- Thin film amorphous silicon solar cells with silicon or chalcopyrite quantum dots, and
- Thin film, flexible polymeric solar cells incorporating quantum dots and/or carbon nanotubes.

The actual program ran for 11 months until Nov. of 2005 with a remarkable number of publications, 37, by the principal scientists involved in the project.

In 2005, a new NASA administrator changed the NASA direction to focus efforts on developing a new rocket to address the challenges in Bush's 2004. "Vision for Space Exploration." To accomplish this Dr. Griffin completely cut many of NASA's research and technology programs. The photovoltaic research budget

for NASA went to essentially zero in 2006, resulting in significant losses to the Photovoltaic Branch at GRC. Small research efforts have continued based on Space Act Agreements (SAA) with Universities, businesses and other Government Agencies. A SAA with the Univ. of Toledo has continued the quantum dot cell funding utilizing a new cell design. A SAA with AlphaMicron is to create a window that has a self-regulating controllable tint that is powered using solar cells developed by NASA GRC. Some NASA funding existed under the Constellation program that has been cancelled in President Obama's 2011 budget

In recent years two new terrestrial installations have been installed at GRC. A 2 KW system was installed connected to the grid and a solar concentrator system developed by Greenfield Solar utilyzing vertical junction silicon solar cells patented by Bernard Sater who is currently a Distinguished Research Associate at NASA GRC.

There has been a low-funded thermophotovoltaic (TPV) program at GRC for many years. Dr. Donald Chubb has provided expertise in selected emitters and testing for multiple programs and David Wilt and Eric Clark have designed and built III-V cells for these systems. Dave Wolford has provided expert testing. Thermophotovoltaics converts heat energy from the sun or a nuclear, combustion or radioisotope heat source to electric energy. Thermal energy from any of the thermal sources is supplied to an emitter. Infrared radiation from the emitter is directed to photovoltaic (PV) cells where the radiation is converted to electrical energy. In order to make the process efficient, the energy of the photons reaching the PV array must be greater than the bandgap energy of the PV cells. Shaping of the radiation or *spectral control* is accomplished by using a selective emitter that has large emittance for photon energies above the bandgap energy of the PV cells and small emittance for photon energies less than the bandgap energy or by using a gray body emitter (constant emittance) and a band pass filter. The bandpass filter should have large transmittance for photon energies above the PV cells bandgap energy and large reflectance for photon energies below the PV cells bandgap energy. A back surface reflector on the PV array can also be used to reflect the low-energy photons back to the emitter. It is also possible to combine all three of these methods [49]. There are terrestrial uses for TPV systems both in waste heat energy conversion and also as a primary low-maintenance system for remote locations.

6.2.6   The Future for Photovoltaic Energy

The President's budget for FY 2011 proposed to invest an additional $6 billion in NASA over the next 5 years. The intent is to

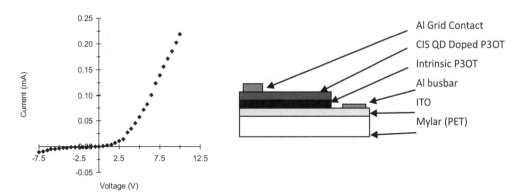

FIG. 6.22   CURRENT VERSUS VOLTAGE CHARACTERISTICS OF CUINS₂ QUANTUM DOT DOPED P3OT ON ITO JUNCTION

reinvigorate research and technology development. Areas of interest to the PV community are a focus on transformative technology development and flagship technology demonstrations to pursue new approaches to space exploration, robotic precursor missions to multiple destinations in the solar system, cross-cutting technology development aimed at improving space capabilities and accelerating the next wave of Climate change research and observations spacecraft. These were discussed in the May 24, 2010 Space Exploration Workshop [50]. It should be noted that the following programs are still under discussion and formulation and are therefore subject to change.

In the Enabling Technology Development and Demonstration, ETDD, there will be a combination of short duration projects to test concepts or provide in-space technology demonstrations, near-term development and demonstration of prototype systems, long-range development of technologies critical to future human exploration and an infusion path for promising, game-changing technologies. Some of the proposed areas relevant to PV are high-efficiency space power systems and advanced in-space solar propulsion. The ETDD program also encompasses human exploration telerobotics that will, of course, require power although there are no details as yet concerning that.

The ETDD program is expected to feed the Flagship Technology Demonstration Program. This program demonstrates the technologies needed to reduce the cost and expand the capability of future space exploration activities. They are large-scale demonstrations in space of technologies that could be transformation and improve the capability and reduce the cost of future exploration missions. The intent is to provide $400M to $1B each, including launch vehicle, for a project lifetime of no longer than five years. The proposed demonstration most interesting to the PV community is the Advance In-Space Propulsion Demonstration. The intent of this is to deliver revolutionary benefits by combining advanced space propulsion with efficient, lightweight, array technology. The top-level requirements is to have a 10 to 70 kg/kW specific mass, including the power system but excluding propellant and propellant tankage. There are numerous missions that could be enabled by Solar Electric Propulsion, SEP.

The Exploration Precursor Robotic Missions, xPRM, also offer opportunities for PV power in some cases. The intent is for "a steady stream of robotic precursor missions". The effort would consist of two Programs, the xPRP and xScout. xPRP would provide a set of linked flight mission, instrument developments, and research and development for the purpose of acquiring applied precursor knowledge of human spaceflight. These are estimated to be in the $500M to $800M range. The xScout missions are focused, less-expensive, higher-risk missions, with a cost cap of $200M including launch. All of these missions are designed to quantify the engineering boundary conditions associated with space environments beyond LEO, to identify hazards and ensure safety, to identify resources and to facilitate sustainability, lower launch mass and "living off the land", and to provide knowledge to inform the selection of Human Exploration destinations. The missions would have an average launch rate of one mission every 18 months, with a goal of one every year. A number of missions have been proposed but it will take time for the details to emerge of the power systems requirements.

There is also an effort being made at NASA Headquarters to look at Breakthrough Capabilities that will, within the next 40 years, provide technology breakthroughs both within and outside of the space industry that will have a large impact on human exploration programs. One of these is entitled "Ubiquitous Access to Abundant Power" [51]. A report will be published by the end of 2010 with possible power technology candidates. The scope is space-focused beyond Low Earth Orbit, LEO, with a timeline through 2050. The technology concepts must remain within the laws of physics, be referenced and sourced from academically-respected sources and be realistic and achievable based on current experimental results or demonstrated technologies. There are thirteen listed breakthrough capabilities including many that would require power. However, "Ubiquitous Access to Abundant Power" is focused on energy generation, storage, and distribution. The workshop was an exercise in discussing an array of technologies and deciding if the concepts are "risky" enough to be included in the study or if they are too risky. Concepts were characterized and the challenges and hurdles to technology development will be addressed. There is a time assessment such as more near term or closer to the 40 year end of the timeframe. Some of the supporting technology concepts relevant to PV included Multifunctional Photovoltaic Materials, Microradioisotope Power Sources (alpha or beta voltaic), Power Beaming/Wireless Power Transfer, etc.

The Multifunctional PV Materials refers to providing energy generation in addition to one or more other primary function. These materials often consist of nanoscale PV, such as quantum dots, embedded in the primary material. Multifunctional PV can be developed for structures, corrosion-resistant paint, light-filtering windows, or radiation protection. An assumption made here was that additional research would be necessary to qualify nanoscale PV and develop composite materials. Quantum dot solar cells are particularly attractive for space applications due to inherent radiation resistance, which may significantly increase solar cell lifetime [52]. Several research organizations have developed techniques for dispersing quantum dots in a solvent, potentially leading to a multifunctional paint that provides energy generation and aesthetics [53]. With additional nanotechnologies, this paint could provide corrosion resistance, self-healing properties, or dust mitigation coating. Structural materials that provide energy may result from a combination of carbon nanotubes and photovoltaic technologies. Carbon nanotubes increase the conductivity and improve the performance of many photovoltaics, and they may provide structural stability for rigid, thin-film cells, reducing the mass of photovoltaic arrays [54]. Over the next 40 years, new photovoltaic materials will likely be developed that could significantly improve these applications, in addition to providing new, multifunctional capabilities. Terrestrial challenges overlap with space exploration challenges; however, the space environment is more extreme, and space systems require a higher level of ruggedness. Exploration-specific research will be necessary to develop and validate terrestrial technologies for lightweight, radiation-hard, long-duration solar arrays. If multifunctional solar cells can be incorporated into a space exploration architecture with minimal additional cost, risk, and weight, then power generation can be provided by distributed and ubiquitous systems. With these technologies, power generation will be limited by surface area and proximity to the Sun; however, with improvements in low-power electronics and energy storage, many architecture elements could become power self-sufficient, potentially including suits, rovers, portable electronics, sensors, and other distributed systems.

The Microradioisotope Power Sources use energetic particles from radioactive decay to create electron-hole pairs in semiconductors. Here there are assumed to be engineering challenges with efficiently converting energy released by radioactive decay with lifetime-limiting damage to the conversion unit. These are several examples of the technology concepts: alphavoltaics [55] betavoltaics. thermoelectric conversion. The conversion process can be direct or indirect. There are several challenges to microradioisotope power sources that will require engineering and design solutions, or in some cases new materials and physical models. These systems are relatively complex, and several of the potential technologies have multi-stage energy conversion processes. Improvements in efficiency, lifetime, and mass are necessary at each of these stages. New nanomaterials may be required for efficient conversion of radiated particles without material damage. With microradioisotope power sources, NASA could replace batteries, fuel cells, and power generation technologies in all portable electronics, medical technologies, sensors, spacesuit power, and some satellites and rovers with radioisotope power sources that last for years or decades.

In the President's new approach to Space Exploration there is a focus on the development of technology both to benefit space exploration and terrestrial applications. There is a belief that, rather than delaying human space exploration, the resulting technology benefits will actually speed the process for accomplishing human missions. The new office of NASA Chief Technologist has been formed. The strategy has focused on the foundation of early stage innovation and basic research efforts. Providing a steady stream of technology demonstrations that will enable a flexible path of capabilities that will begin with a set of crewed flights early in the next decade. After the early missions the focus will shift to a 2025 human mission to a near-Earth asteroid and 2035 human mission to Mars orbit and return and a later Mars surface mission. The technology development program seeks to change the game by expanding the alternatives available for human exploration through timely, strategic and significant technology investment.

6.2.7   Beamed Power from Space to Earth

The concept of generating electrical power in space and beaming it to Earth was first proposed by Dr. Peter Glaser in 1968 [56]. Glaser proposed putting a large solar array in geosynchronohnus orbit (GEO) and beaming the power to the ground using microwaves. Since that time numerous studies have investigated the possibilities of beamed power from space to earth. These studies usually evolved around assessing the current technologies, systems concepts, and terrestrial markets that could provide a customer base. A major study led by DOE with support from NASA was conducted from 1976-1980 [57], [58]. This study proposed providing the primary electrical power to the U.S. with as many as 60 space power satellites (SPS) in GEO with a base power of 5 to 10 GW of continuous energy per satellite. Both the U.S. National Research Council (NRC) and the Congressional Office of Tecnology Assessment (OTA), in reviewing the proposed SSP concept in 1980-1981, concluded that although the solar power satellites were technically feasible they were economically unachievable [59]. The report suggested revisiting the concept in ten years and continuing related research to further develop the needed technologies. In 1995, a "Fresh Look" study was performed by NASA [60]. This study looked at new marketplace conditions, different orbital possibilities, and different modes of power transmission encompassing about 30 SSP system concepts. While finding the prospect for power from space more viable technically than in 1980 it was still exceptionally challenging. In 1999 and 2000 NASA conducted the SSP Exploratory Research and Technology (SERT) program [61]. The program defined new systems concepts, better defined technical challenges, and initiated strategic research projects involving businesses, other agencies and laboratories, universities and in-house resources. In 2002 the National Science Foundation (NSF), the Electric Power Research Institute (EPRI), and NASA jointly issued a broad area

announcement that yielded a number of high-leverage, high-risk research studies targeting some of the key challenges facing future SSP systems. The major challenges are the system mass and launch costs, device efficiencies (solar array, convertor to microwave or laser, receiver), power management and distribution, thermal management, assembly, integration, maintenance and repair, and low-cost manufacturing of space-qualified systems.

In 2006 Dr. Geoffrey Landis reexamined the economics of SSP [62]. He pointed out the synergy with terrestrial solar systems that would integrate solar and microwave receivers on the ground and use SPS to beam to receivers when ground solar was unavailable. In 2009 Dr. Landis examined the SSP concept on a physics basis [63]. Some of his conclusions were that, although a space location for solar panels gets more sun than a ground location, it is not that much more than the best ground locations; electromagnetic beam diffraction means that SPS is inherently large and switching to laser transmission reduces size but loses on efficiency; solar arrays in space produce 3.5 times more power than non-tracking arrays on the ground or 2.2 times more than a tracking array (accounting for transmission losses, this reduces to 1.63 times more power/solar-cell area; if the cost of solar arrays is less than 61% of the total SPS cost, the array is better on the ground; GEO is only reasonable orbit choice; can space cell cost equal cost of cheap terrestrial cells? John Mankins also discusses new directions for space solar power in a 2009 publication [64]. He discusses some of the key issues associated with cost-competitive space solar power in terrestrial markets and ends with an example of the kind of novel architectural approach for space solar power that is needed.

Space Solar Power makes the most sense in reaching isolated communities where the cost of terrestrial based power is very high. It might also make sense for military applications if the ground system was designed to be portable. A major stumbling block for space-based system is the high cost to launch a kilogram. However the idea of beaming power in space from SPS to other satellites, for example, might provide one of the first tests for SSP and also provide a potential useful commodity in space.

6.2.8   The People

It is perhaps important here to note that although most of NASA's photovoltaic work has been discussed here without names there were indeed people behind these accomplishments and leaders who greatly contributed to and indeed made these efforts possible. In the early years of the Photovoltaic Branch at LeRC, contributors to the photovoltaic community included Americo Forestieri, Cosmo Barona, Henry Curtis, Dr. Irv Weinberg, Victor Weizer, Bernard Sater and others. NASA GRC played a pivotal role in the history of the PVSC. The 4th conference was actually held in Cleveland and five chairs (Americo Forestieri, Dr. Henry Brandhorst, Cosmo Baraona, Dr. Dennis Flood and Dr. Sheila Bailey) of the conference have been from the Photovoltaic Branch. Chairs of both the 1st, 2nd and 4th World Conferences were from GRC.

In the '70s and early '80s Dr. Henry Brandhorst led the PV group followed by Dr. Dennis Flood from 1985 to 2000. Dr. Flood had the enviable capabilities of in-depth technical knowledge, effective communication skills and the ability to motivate and guide subordinates. His leadership created a productive, capable and knowledgeable group of PV researchers and technology developers. Lead researchers in the PV branch at GRC over the years included Dr. Irving Weinberg who began the InP effort in the branch, Henry Curtis who participated in many space flight projects, Victor Weizer who was an expert in silicon solar cells, Russel Hart, Dr. Dave Brinker, Phillip Jenkins, Dave Scheiman and

Dr. Dave Snyder who were experts in measurement and calibration techniques, and Dave Wilt who was an expert III-V metal-organic vapor phase epitaxy (MOVPE) crystal grower . Dr. Geoffrey Landis is a renown expert in planetary photovoltaics and serves on the Science team of the Mars Exploration Rovers (MER). Dr. Sheila Bailey has led the nano-technology efforts. Michael Pisczcor is an expert in array technology and is now chief of the Photovoltaic and Power Technologies Branch. Dr. Donald Chubb and Dave Wolford are TPV experts. Dr. Edward Gaddy, who recently retired from GSFC and now is at John Hopkins Applied Physics Laboratory, and Dr. Bruce Anspaugh who has retired from JPL and Paul Stella currently with JPL, were icons in the Space PV world.

6.3   WIND ENERGY

In the decade from 1975 through the 1985 the United States government worked with industry to advance the technology and help enable large commercial wind turbines. The impetus for this effort was two events on the world stage (1) an oil embargo by the Organization of Petroleum Exporting Countries (OPEC) in 1973 followed by (2) the fall of the Shah of Iran in 1979. In response to the first crises, in 1975, the NASA Lewis Research Center in Ohio (now named the Glenn Research Center) took on the role of leading the technology development of large horizontal-axis wind turbines, the dominant type of wind turbine in use today. The following pages document the development of major wind turbine systems and technologies, their impact on the industry and future development areas that could benefit the industry.

As a major national laboratory, NASA's Lewis Research Center (now named the Glenn Research Center) in Ohio had the facilities and engineering capabilities necessary to advance wind turbine technology. At that time, as it is still true today, NASA was able to successfully apply its capabilities in areas such as propeller and turbine design, structural analysis, advanced materials, aerodynamics, instrumentation and power system control to the design of wind energy systems. During the 1970's and '80s, with funding from the National Science Foundation, the Department of Energy (DOE), and the Department of Interior (DOI) a total of 13 experimental wind turbines were put into operation including four major wind turbine designs. This program pioneered many of the multi-

megawatt turbine technologies in use today, including: steel tube towers, variable-speed generators, composite blade materials, high lift-to-drag airfoils. In addition, key engineering analysis, design and measurement capabilities were developed such as models for aerodynamic performance, structural dynamics, fatigue life, wind measurement and acoustics. The outstanding success of wind energy can be seen today in the growth of the installed power capacity. In 2008 and 2009 the U.S. increased its installed capability for wind energy by 45% and achieved the largest installed capacity of any country in the world. Nearly all of this increased capacity was from horizontal-axis wind turbines that generate electrical energy. Large horizontal-axis wind turbines are in fact the dominant type in use throughout the world and owe much of their characteristics to the NASA technologies and the DOE investment.

The large wind turbines developed under the NASA-led technology development effort set several world records for diameter and power output. Figure 6.23 shows the size and power output of the designs developed.

All turbines shown were built and operated in single or multiple units with the exception of the Mod-5A that was a competitive design to the Mod-5B, the winner in a down selection process. Each of the remaining turbines will be discussed below.

NASA designed, constructed and operated its first experimental wind turbine, the Mod-0, in Sandusky Ohio (see Figure 6.24).

The Mod-0 had a rotor diameter of 38 meters and a power rating of 100 kilowatt. The size of the Mod-0 was nearly ideal because it was large enough to be representative of electric utility wind turbines but not too large to be overly costly to make changes to the hardware. As further explanation, a critical parameter for aerodynamic performance is the Reynolds number. The size scale of the Mod-0 resulted in the airfoils operating above the critical value of 1 million for modern airfoils. Alternatively, the relatively small size of the turbine allowed NASA to select off the shelf commercial products for key components such as the gearbox and generator. This allowed for many technologies to be tested and developed on the Mod-0 at much lower cost than on a larger scale turbine.

Given that it was the first wind turbine design for NASA, the Mod-0 operated as mostly expected with regard to most design goals. Power /speed control were excellent and synchronization with the utility grid's line frequency was found to be easier than

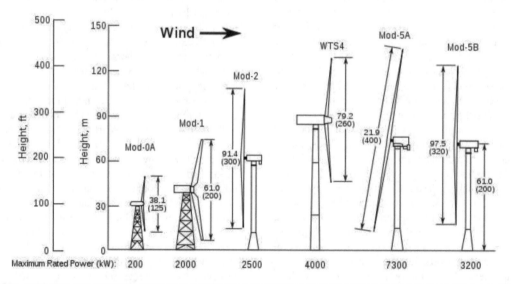

FIG. 6.23   SIZE AND RATED POWER OUTPUT OF WIND TURBINES DEVELOPED UNDER THE NASA-LED PROGRAM

FIG. 6.24 NASA MOD-0 WIND TURBINE, INSTALLED IN SANDUSKY, OHIO IN 1975 BECAME THE TEST BED FOR DEVELOPING NEW TECHNOLOGIES FOR LARGE HORIZONTAL AXIS WIND TURBINES

anticipated. One important exception to the Mod-0's excellent performance was the blade stress and life. Within months of beginning operation, cracks were found in the aluminum blades near the hub of the rotor. Even though this was a significant challenge, it fostered the development of reliable wind turbine blades and improved system design. It took two years and the best of NASA's capabilities in materials, structures and aerodynamics testing and analysis to resolve the problems. The causes were several. First, analytical methods to model the coupled aeroelastic structural behavior needed to be improved. The rotor used a rigid hub causing blade loads to be resolved primarily by internal stress rather than dissipated into acceleration of blade mass. The cyclic bending load out of the rotor plane was higher than expected. The problem was further defined through NASA wind tunnel testing that showed that the truss tower and access stairs in the center blocked much of the wind passage. The Mod-0 was designed with the rotor downwind of the tower. As the blades passed by the tower each revolution, the aerodynamic forces changed significantly resulting in damaging fatigue stresses. Removal of the stairs resolved much of this factor. Finally, the riveted aluminum blades were found to not be a good choice for a fatigue driven machine such as a wind turbine. A

significant alternative blade development program began and over the next year successful blades were developed that were made from wood or fiberglass.

With the Mod-0 design proven reliable, four other turbines of a more powerful version of the Mod-0 were built and operated throughout the US. The generator was doubled in size to 200kw, but most all other characteristics remained unchanged. These wind turbines were designated the Mod-0As were built in partnership with Westinghouse. They were installed and operated at Culebra Island, Puerto Rico; Clayton, New Mexico; Block Island, Rhode Island and Oahu, Hawaii. These turbines were instrumental in defining the grid interaction and electrical characteristics for small utilities (soft grids) and large utilities (hard grids).

The next step was increasing the size of wind turbines. Under the technical management of NASA in Cleveland, the General Electric Company (Space Division/Valley Forge) scaled up the Mod-0 wind turbine in Sandusky Ohio to achieve 2 megawatts of power output from a 61 meter diameter rotor [65]. The Mod-1 wind turbine shown in Figure 6.25 was the world's first multi-megawatt wind turbine under partnership with commercial industry.

It was installed and operated in Boone North Carolina and helped develop the transportation and erection methods for large wind turbines. The Mod-1 location next to residential housing however uncovered two unexpected environmental impacts – low-frequency acoustic emissions and electromagnetic television inter-

FIG. 6.25 THE NASA/GENERAL ELECTRIC MOD-1 WIND TURBINE WAS THE WORLD'S FIRST MULTI-MEGAWATT COMMERCIAL PROTOTYPE WIND TURBINE (Derived from the Mod-0 in Sandusky, Ohio)

ference. The same truss tower and downwind rotor configuration as the Mod-0 gave the Mod-1 a similar "tower shadow" effect on the blade aerodynamics. This manifested itself as a sound wave that was propagated and focused towards local residential housing in the mountainous site. NASA developed a new analytical model [66] to predict the low-frequency sound waves due to the rotation of the blades through the wake of the tower. Studies using this analysis tool showed that decreasing the rotational speed could reduce the sound levels by 10dB. This change was implemented and noise complaints from residents came to an end. The electromagnetic interference proved more difficult to solve at this mountainous site. Residents were provided cable television as a local solution. Challenging as the world's first multi-megawatt wind turbine was, it was instrumental in developing location standards for wind turbines near populated areas.

In an effort to further build the wind industry, NASA and DOE awarded the design and construction of the next generation of lower cost wind turbines to Boeing Engineering and Construction using new technology proven out at the NASA Mod-0 wind turbine in Sandusky, designated the Mod-2 (Figure 6.26), the 300 foot diameter rotor produced 2.5 megawatts.

One of the new technologies was the steel monopole tube tower and the upwind rotor configuration. The steel tube was unproven at these large sizes and a primary concern was the possibility of a mechanical vibration occurring at a resonant natural frequency that could destroy the wind turbine. The tube tower, referred to by NASA as the "soft tower" had a natural frequency that would be excited by the rotor as it increased in speed during startup.

FIG. 6.27   THE NASA / DOI / HAMILTON STANDARD WTS-4 (4 MEGAWATT) WIND TURBINE IN WYOMING TYPIFIED THE "COMPLIANT" DESIGN APPROACH AND HELD THE WORLD RECORD FOR POWER OUTPUT FOR OVER 20 YEARS

Development of the steel tube tower required NASA to advance the structural analysis models and create electronic control systems to quickly increase the rotational speeds before resonance vibrations damage the wind turbines. This "soft" steel tube tower is used throughout the industry today. Five Mod-2 wind turbines were put into operation. Another technology introduced on the Mod-2 was partial span (tip) pitch control as seen in Figure 6.26. This technology reduced the weight and cost of the aerodynamic power control compared to full span pitch control used on the Mod-1. The cluster of three Mod-2 wind turbines at Goodnoe Hills in the state of Washington produced a total of 7.5 megawatt of power in 1981. Two other Mod-2 turbines were built and operated in California and Wyoming.

Success in moving from the early rigid towers to lighter and "softer" towers provided a basis for applying this "compliant" design approach throughout the wind turbine system. The 4 megawatt WTS-4 wind turbine shown in Figure 6.27 is an example of this compliant design approach.

Designed by Hamilton Standard, under the technical management of NASA and with funding this time from the Department of Interior, the WTS-4 was placed into operation in Wyoming in 1982. It featured a "soft" steel tube tower, fiberglass blades, torsional springs and dashpots in the drivetrain, and a flexible teetered hub (to be discussed later). To this day, the WTS-4 is the most

FIG. 6.26   THE NASA / DOE/ BOEING MOD-2 (2.5 MEGAWATT) WIND TURBINE WAS A SECOND GENERATION DESIGN THAT PIONEERED THE LOWER COST "SOFT" STEEL TUBE TOWERS USED IN THE INDUSTRY TODAY

powerful wind turbine to have operated in the US and it held the world record for power output for over 20 years. A second commercial prototype with a smaller generator (3 megawatts) designated the WTS-3 was constructed and operated in Sweden.

The last generation of wind turbines developed under the NASA-led program was the Mod-5. Two competing designs were developed as a parallel effort, the 7.3 megawatt Mod-5A design in partnership with the General Electric Company, and the 3.2 megawatt Mod-5B design in partnership with Boeing Engineering and Construction. Government funding limitations allowed only one to be funded - the Mod-5B. With a rotor diameter of nearly 100 meters, the Mod-5B (Figure 6.28) was the world's largest operating wind turbine into the 1990's.

It demonstrated an availability of 95 percent, an unparalleled level for a new first-unit wind turbine. The Mod-5B had the first large-scale variable speed drive train using a double fed induction generator, the industry standard today. It also featured a sectioned, two-blade rotor that enabled easy transport of the very large blades.

The significant government and industry effort led by NASA begs the question of what happened to these large commercial prototypes in terms of the development of the utility scale wind turbine market in the 1990's. The story is an important case study of public policy and business decisions that set the course of the emerging wind industry in the US. In 1982 the price of oil decreased by a factor of 3. By 1986 the US government began to phase out large

FIG. 6.28   THE NASA/DOE/BOEING MOD-5 IN HAWAII PIO-NEERED VARIABLE SPEED GENERATORS AND WAS THE WORLD'S LARGEST OPERATING WIND TURBINE INTO THE 1990S

wind turbine development and support. The original orders of 100 commercial units of the Mod-5B for Hawaii were cancelled and the industry partners left the US wind turbine market. European countries, in particular Denmark, on the other hand developed a long-term public policy and technology program to address their nation's oil dependency. The effect of the change in US policy in 1986 can be seen in Figure 6.29 [67] where the Europeans came from behind to take the lead in installed wind turbine capacity.

Denmark specifically went on to capture the dominant position in a market that exceeded $15 billion a year by 2004. Fortunately, US wind industry manufacturers such as General Electric are now back in the market and have recovered substantial position. In 2010, the US took the lead again in the world for total installed wind energy capacity.

The success of both the US and European manufacturers owe much to the NASA program and the technologies already discussed including the "soft" steel tube towers, variable speed generators, structural dynamic and acoustic emission modeling. Other technologies not discussed include the NACA (predecessor to NASA) 6 series airfoils, used widely throughout the industry and pioneered on the NASA Mod-0. Also, more accurate modeling of stall aerodynamics [68] that helped enable the early industry's dominant use of passive fixed pitch power control. These are just a few of the examples of impact on this important industry.

Some technologies developed under the NASA program have yet to be fully duplicated by industry. One example is the ability to produce multi-megawatt power using a flexible two blade rotor. Two blades can capture essentially the same energy in the wind as three blade turbines but offer several cost and reliability advantages. First, there is a direct cost reduction of not having to buy one blade, a very expensive component. Second, the reduction in weight at the rotor permeates throughout the entire system and usually yields about a 25% reduction in system weight. Weight is a surrogate for cost because it affects material, manufacturing and transportation costs. Third it is easier to assemble – some two blade turbines have been assembled by lifting the rotor by the tower. Fourth, two blade rotors require a smaller gearbox ratio because the typically operate at higher tip speed ratios. This also reduces the torque in the drive train for the same amount of power. Fifth, two blades easily allow for flexibility in the rotor hub that can effectively uncouple the rotor plane from the motion of the tower. The simplest is the teetered hub used on the NASA Mod-2, WTS-4 and Mod-5B turbines (Figure 6.30).

The uncoupling of the cyclic loads into the drivetrain also greatly increases reliability of key components in the drivetrain. Today's wind turbine industry uses a rigid hub and has experienced a broad and critical problem of gearbox failures in the drivetrain. None of the multi megawatt wind turbines NASA developed had gearbox problems, and some ran for over a decade. The WTS-4 (4MW) and WTS-3 (3 MW) still hold records for operating time and total energy produced – and they were prototypes. The ability of NASA to successfully design and construct these flexible rotor systems was the result of its expertise in helicopter and rotorcraft technology, something very few other countries in the world possess.

Another technology that could be adopted by today's industry is automated blade manufacturing. Today all large wind turbine blades in the world are manufactured using hand layup of the composite materials, a very labor intensive process. Thus most blades are imported from countries where labor is relatively inexpensive. Figure 6.31 shows an automated blade manufacturing system developed on the NASA-led program that could be implemented in to enhance the US wind industry manufacturing capabilities.

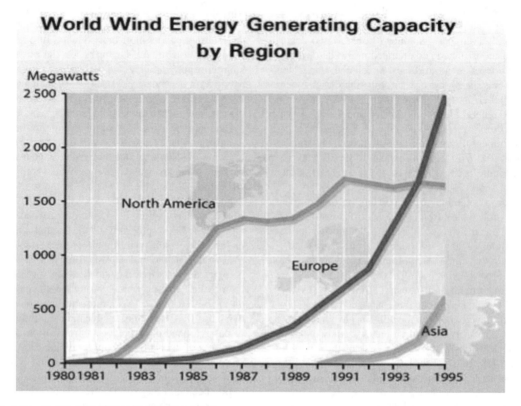

FIG. 6.29 A US PUBLIC POLICY CHANGE IN 1986 POSTPONED US LEADERSHIP IN WIND ENERGY

Several automated designs were developed and blades were fabricated and tested in lengths up to 50 meters.

Details of the all the efforts and results by NASA and its partners are too broad to do justice within this chapter. An extensive list of reports on the technology development conducted during the Large Horizontal Axis Wind turbine program can be found in the references [69]. It should assist those active in the field of wind energy in locating the technical information they need on wind power planning, wind loads, turbine design and analysis, fabrication and installation, laboratory and field testing, and operations and maintenance. That bibliography contains approximately 620 citations of publications by over 520 authors and co-authors. Sources are:

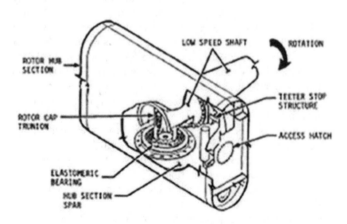

FIG. 6.30 FLEXIBLE ROTOR HUBS SUCH AS THE MOD-5B TEETERED HUBS COULD REDUCE COSTS AND INCREASE THE RELIABILITY OF TODAY'S DRIVETRAINS

FIG. 6.31 NASA'S BLADE TECHNOLOGY PROGRAM DEVELOPED 50 METER BLADES USING AUTOMATED MANUFACTURING TECHNIQUES THAT COULD BE ADOPTED TODAY

FIG. 6.32 NASA'S LIGHT WEIGHT COMPLIANT DESIGN APPROACH AND TECHNOLOGIES MAY ENABLE LOW COST AND RELIABLE OFFSHORE WIND TURBINES IN THE FUTURE

(1) NASA reports authored by government, grantee, and contractor personnel, (2) papers presented by attendees at NASA-sponsored workshops and conferences, (3) papers presented by NASA personnel at outside workshops and conferences, and (4) outside publications related to research performed at NASA/DOE wind turbine sites.

6.3.1   The Future of Wind Energy

Today and into the future, NASA will continue to develop new technologies for the aeronautics and space industry that will find application in the future wind industry as well. As a vision for this future, perhaps the emerging offshore wind industry is a case in point [70]. Currently no offshore wind turbines exist in the US, but there are a number of wind farms in the shallow waters off of Europe. A primary challenge to this new industry is getting the weight and cost of deployment reduced. Deeper waters will require floating wind turbines where weight and reliability will be even more important. Some of the design characteristics of the compliant wind turbines pioneered by NASA such as flexible towers, articulated blades and drive trains as well as downwind rotors and free yaw (Figure 6.32) may offer significant advantage to this new offshore industry (U.S. Offshore Wind Collaborative) for many years into the future.

6.4   CONCLUSIONS

In the wake of Apollo, GRC, then named Lewis Research Center, helped the predecessor to DOE by making major foundational

contributions in wind energy, solar power, batteries, and fuel cells. Many of the people who spearheaded these achievements still work at GRC, augmented with new talent in a broad range of aerospace technologies. GRC also currently serves as the lead center for power and propulsion technology within NASA. The technologies that it has developed for aeronautical and space applications has given GRC a comprehensive perspective for applying NASA's skills and experience in energy on a problem with national and global implications.

Perhaps what is needed is an "Apollo-type" effort to solve our nation's energy crises. One could envision a national effort directed at renewable energy and the environment that would combine not only the DOE's program in renewable energy but also would encompass contributions from multiple other agencies (NASA, DOD, EPA, Commerce, Transportation, etc.) NASA GRC could provide a lead role in such an endeavor based on its history of projects, ranging from concept studies and fundamental research efforts, to proof-of-concept experiments and commercial prototype demonstrations. NASA's ability to work cooperatively with other agencies in developing technologies, it's long-standing interest in alternative/renewable energy research, demonstrating technologies, and infusing these products into commercial applications and it's capability to lead and coordinate projects makes it an ideal choice to lead the way to developing a sustainable energy future for our nation.

6.5   ACKNOWLEDGMENTS

Dr. Bailey would like to acknowledge the significant contributions of Dr. Geoffrey Landis of the Photovoltaic and Power Technologies Branch, George Schmidt, Deputy Director of Research and Technology Directorate, and Dr. Ryne Raffaelle, Director of the National Center for Photovoltaics, DOE.

6.6   ACRONYMS

AFM	Atomic Force Microscopy
AFRL	Air Force Research Laboratory
ALSEP	Apollo Lunar Surface Experiments Package
AMO	Air Mass Zero
APSA	Advance Photovoltaic Solar Array
AU	Astronomical Unit
CIGS	Copper Indium Gallium Diselenide
CLEFT	A Peeled Film Technique
D3100	Model Number
DDF	Director's Discretionary Fund
DOD	Department of Defense
DOE	Department of Energy
DOI	Department of the Interior
ELO	Epitaxial Liff-Off
EPA	Environmental Protection Agency
ERDA	Energy Research and Development Administration
ETDD	Enabling Technology Development and Demonstration
GaAs	Gallium Arsenide
GRC	Glenn Research Center
GSFC	Goddard Space Flight Center
H&RT	Human and Robotic Technology
IEEE	Institute of Electrical and Electronics Engineers

InP	Indium Phosphide
ISS	International Space Station
ITO	Indium Tin Oxide
JPL	Jet Propulsion Laboratory
KSC	Kennedy Space Center
LED	Light Emitting Diode
LeRC	Lewis Research Center
MAE	Materials Adherence Experiment
MER	Mars Exploration Rovers
MOD	Module
MSFC	Marshall Space Flight Center
NACA	National Advisory Committee for Aeronautics
NASA	National Aeronautics and Space Administration
NRC	National Research Council
OPEC	Organization of Petroleum Exporting Countries
P30T	Poly 3-Octyl-Thiophene
PET	Polyethylene Terephthalate
PV	Photovoltaics
PVSC	Photovoltaic Specialists Conference
RIT	Rochester Institute of Technology
SAA	Space Act Agreement
SEP	Solar Electric Propulsion
SPRAT	Space Photovoltaic Research and Technology
SSP	Space Solar Power
STF	Systems Test Facility
STM	Scanning Tunneling Microscopy
STORM	Scanning Tunneling Optical Resonance Microscopy
TPV	Thermophotovoltaic
UTJ	Ultra Triple Junction
WTS	Wind Turbine System
xPRM	Exploration Precursor Robotic Missions
xPRP	Exploration Precursor Robotic Program
xSCOUT	Exploration Scout Missions

6.7   REFERENCES

1. Chapin D.M., Fuller C.S., Pearson, G.L. (1954), "A new silicon photocell for converting solar radiation into electrical power". J. Appl Phys. 25, 676.

2. Reynolds D.C., Leies G., Antes L., Marburger R.E. ,"*Photovoltaic effect in cadmium sulfide"*, Phys. Rev. 96, 1954, p533.

3. Jenny D.A., Loverski J.J., Rappaport P., *Photovoltaic effect in GaAs p-m-junctions and solar energy conversion,* Phys. Rev. 101, 1956, p 1208.

4. Easton R.L., Votaw M.J., *Vanguard I IGY Satellite,* Rev. of Scientific Instruments, 30, 2, 1958, pp. 70-75.

5. *The Telstar Experiment,* Bell Syst. Tech. J., 1963, p. 42.

6. Loeferski J., *Theoretical considerations governing the choice of the optimum semi-conductor for the photovoltaic solar energy conversion,* J. Appl. Phys. 27, 1956, p. 777.

7. Jackson E.D., *Areas for improvement of the semiconductor solar energy converter.* Trans. Of the Conf. On the Use of Solar energy, Tucson, Arizona, 5, 1955, p. 122.

8. Perlin J., *From Space to Earth,* aacet publications, 1999, p.50.

9. Terrestrial Photovoltaic Measurments, *Workshop Proceedings,* March 19-21, Cleveland, OH, 1975.

10. I Weinberg, H. Curtis and A. Forestieri, *Effects ofF OutdooR Exposure on Solar Cell Modules in the ERDA/NASA LewisS Research Center Systems TestT Facility,* NASA TM X-73657, *1977.*

11. A. Ratajczak, W. Bifano, J. Martz, and P. O'Donnell, *NASA Lewis Research Center Photovoltaic Application Experiments,* AIAA/ASERC Conference on Solar Energy: Technology Status, Phoenix, Ariz./ Nov. 27-29, 1978.

12. W. Bifano, R. DeLobard, A. Ratajczak and L. Scudder, *Status of DOE and AID Stand-Alone Photovoltaic System Field Tests,* Proc. Of the 17th Photovoltaic Specialist Conference, pp1159-1167, 1984.

13. A. Forestieri and C. Baraona, *Space Station Power System,* Proc. Of the 17th Photovoltaic Specialist Conference, pp7-11, 1984.

14. J. Mullin, J. Loria, and H. Brandhorst, *The NASA Photovoltaic Terchnology Program,* Proc. Of the 17th Photovoltaic Specialist Conference, pp12-16, 1984.

15. J. Fan, R. McClelland, and B. King, *GaAs CLEFT Solar Cells for Space Applications,* Proc. of the 17th Photovoltaic Specialist Conference, 1984, pp32-35.

16. T. Trumble and F. Betz, *Evaluation of a Gallium Arsenide Solar Panel on the LIPS-II Satellite,* Proc. Of the 17th Photovoltaic Specialist Conference, 1984, pp12-16.

17. P. Stella and R. Kurland, "Thin Film GaAs for Space – Moving Out of the Laboratory", Proc. Of the 23rd IEEE Photovoltaic Specialist Conference, 1993, pp 21-26.

18. P. A. Jones, *et al.,* "The SCARLET Light Concentrating Solar Array," *25th IEEE Photovoltaic Specialists Conference,* Washington DC, 1996.

19. J. Stubstad, *et al.,* "SCARLET and Deep Space 1: Successfully Validating Advanced Solar Array Technology," *AIAA paper AIAA-99-4487,* Space Technology Conference & Exposition, Albuquerque NM September 28-30 1999.

20. G. Landis, P. Jenkins, and D. Scheiman, "Photovoltaic Cell Operation on Mars," *19th European Photovoltaic Science and Engineering Conf.,* Paris, France, June 7-11 2004, 3674-3677.

21. R. Ewell and D.R. Burger, "Solar Array Model Corrections from Mars Pathfinder Lander Data," 26th IEEE Photovoltaic Specialists Conference, Sept.-Oct. 1997. 1019-1022.

22. G. Landis and P. Jenkins, "Measurement of the settling rate of atmospheric dust on Mars by the MAE instrument on Mars Pathfinder, *Journal of Geophysical Research, 105, No. El,* 2000, pp. 1855-1857.

23. P. Stella *et al.,* "Managing PV Power on Mars – MER Rovers", 35th IEEE Photovoltaic Specialists Conf., 2009.

24. G. Landis, "Exploring Mars with Solar-Powered Rovers," *31st. IEEE Photovoltaic Specialists Conference,* Orlando FL, Jan 3-7 2005, 858-861.

25. Gasner, *et al.,* "The Stardust Solar Array," World Conf. Photovoltaic Energy Conversion, Osaka Japan, May 12-18, 2003.

26. P. Stella and S. DiStefano, "Early Mission Power Assessment of the Dawn Solar Array," *34th IEEE Photovoltaic Specialists Conf.,* Philadelphia PA, June 7-9, 2007.

27. M. Piszczor, *et al.,* "Advanced Solar Cell and Array Technology for NASA Deep Space Missions," 33rd IEEE Photovoltaic Specialists Conf., San Diego CA, May 12-16 2008.

28. J. Dunne and E. Burgess, *The Voyage of Mariner 10,* NASA Special Publication SP-424, 1978.

29. G. Dakermanji, J. Jenkins, and C. J. Ercol, "The MESSENGER Spacecraft Solar Array Design and Early Mission Performance," *4th World Conf. Photovoltaic Energy Conversion,* Waikoloa, Hawaii, May 7-12, 2006, 1919-1922.

30. A. F. Hepp, J. S. McNatt, S. G. Bailey, and J. E. Dickman, R. P. Raffaelle, B. J. Landi, A. Anctil, and R. DiLeoM. H.-C. Jin, Chun-Young Lee and T.J. Friske, S.-S. Sun, C. Zhang, S. Choi, A. Ledbetter, K. Seo, C. E. BonnerK. K. Banger, S. L. Castro, David Rauh"Ultra-Lightweight Hybrid Thin-Film Solar Cells: A Survey of Enabling Technologies for Space Power Applications", Proceedings of the 5th IECEC, 2007, pp 179-200.

31. A.N. MacInnes, M.B. Power, A.R. Barron, A.F. Hepp, and P.P. Jenkins, "Chemical Vapor Deposition from Single Organometallic Precursors" , *United States Patent* 5,300,320, April 5, 1994. United States Patent.

32. S.G.Bailey, G.A.Landis, and D.M.Wilt, "Effect of Crystal Orientation on Anisotropic Etching and MOCVD Growth of Grooves on GaAs," *J. Electrochemical Soc.*, 136, No.11, 1989, pp 3444-3449.

33. "Textured InP for Device Applications", S. Bailey, N. Fatimi, and G. Landis, *Chemical Surface Preparation, Passivation and Cleaning for Semiconductor Growth and Processing,* Vol. 259, 341-346 (1992).

34. S. Bailey, D. Wilt, F. DeAngelo, and E. Clark, "Preferentially Etched Epitaxial Liftoff of InP", , *Proceedings of the 23rd IEEE PVSC*, 1993, pp. 783-785.

35. S. Bailey, D. Wilt and F. DeAngelo, U.S. Patent No. 5,641,381, "Preferentially Etched Epitaxial Liftoff of InP Material", issued on June 24, 1997.

36. J. Carlin, M. Hudait, S. Ringet, D. Wilt, E. Clark, C. Leitz, M. Currie, T. Langdo, E. Fitzgerald, "High Efficiency GaAs on Si Solar Cells", *Proceedings of the 28th IEEE PVSC*, pp. 1006-10011(2000).

37. S. Bailey, D.Wilt,, R. Raffaelle, S. Hubbard, "Thin Film Poly III-V Space Solar Cells", C. Bailey *Proceedings of the 33rd IEEE PVSC*, 2008.

38. S. Bailey, R. Raffaelle, P. Nuedeck, S. Hubbard, "Optical and Electrical Characterization of SiC Devices", *Proceedings of the 28th IEEE PVSC*, 2000, pp. 1257-1260.

39. R. Raffaelle, S. Kurinec, D. Scheiman, P. Jenkins, S. Bailey, " Silicon Carbide Cells for Space Applications", *Proceedings of the the 17th European PVSEC, Volume 1, 2001, pp 281- 283.*

40. S. Bailey, H. Curtis, M. Piszczor, R. Surampudi, T. Hamilton, D. Rapp, P. Stella, N. Mardesich, J. Mondt, R. Bunker, B. Nesmith, E. Gaddy, D. Marvin, and L. Kazmerski, "Photovoltaic Cell and Array Technology Development for Future Unique NASA Missions", *Proceedings of the 29th IEEE PVSC 2002,*, pp, 799-803.

41. S. Bailey, R. Raffaelle, and Keith Emery, "Space and Terrestrial Photovoltaics – Synergy and Diversity", *Progress in Photovoltaics*, Vol 10 Number 6, 2002, pp. 399-406.

42. S.G. Bailey, R.P. Raffaelle, J. E. Lau, P. Jenkins, S.L. Castro, P. Tin, D.M. Wilt, A.M. Pal, and S. Fahey. "Scanning Tunneling Optical Resonance Spectroscopy", *Proceedings of the 1st International Energy Conversion Engineering Conference,* 2003.

43. S.G. Bailey, R.P. Raffaelle, J. E. Lau, P. Jenkins, S.L. Castro, D. Scheiman, P. Tin, D.M. Wilt, "Scanning Tunneling Optical Resonance Spectroscopy", *Photonics Tech Briefs,* 2a-6a, October 2003.

44. S.L. Castro, S.G. Bailey, R.P. Raffaelle, K.K. Banger, A.F. Hepp, "Nanocrystalline Chalcopyrite Materials (CuInS2 and CuInSe2) via Low-Temperature Pyrolysis of Molecular Single-Source Precursors", *Chem. Mater.,* American Chemical Society, August 2003.

45. S. Bailey, S. Castro, R. Raffaelle, "Nano-structured Materials for Solar Cells", *Proceedings of the 3rd World Conference on Photovoltaic Energy Conversion,* (2003).

46. R. Raffaelle, T. Gennett, S. Bailey, D. Wilt, P. Tin, J. Lau, P. Jenkins, D. Scheiman, "Scanning Tunneling Optical Resonance Spectroscopy (STORM)", 3/28/02, LEW-17344-1.

47. R. Rafaelle, T. Gennett, G. Landis, S. Bailey, J. Lau, "Carbon Nanotube Thermal Power Converter", 8/9/02 . LEW-17413-1.

48. A. Hepp, S. Castro, K. Banger, S. Bailey, "Preparation of Ternary Nanostructured Crystallites, Via Ternary Single source Precursors", 10/31/02, LEW-17446-1.

49. Chubb, D., "Fundamentals of Thermophotovotlaic Energy Conversion", Elsevier Science, 2007, pp. 1-530.

50. NASA Space Exploration workshop, May 24 2010; see http://www.nasa.gov/exploration/new_space_enterprise/home/workshop_home.html.

51. "Ubiquitous Access to Abundant Power," *Technology Frontiers* workshop, hosted by NASA Directorate Interrogation Office, (Washington, DC: June 16, 2010.

52. C. Cress, C. Bailey, S. Hubbard, D. Wilt, S. Bailey, R. Raffaelle, "Radiation Effects on Strain Compensated Quantum Dot Solar Cells", *Proceedings of the 33rd IEEE PVSC,* 2008.

53. Stephen Markley, "MIT Prints Solar Cells on Paper; Could It Work as Car Paint?," *Kicking Tires,* May 6, 2010, accessed September 24, 2010, http://blogs.cars.com/kickingtires/2010/05/mit-prints-solar-cells-on-paper-could-it-work-as-car-paint.html; U.S. Department of Energy, Solar Energy Technologies Program, "Paint-On Solar Cell Captures Infrared Radiation," news release, January 19, 2005, accessed September 24, 2010, http://www1.eere.energy.gov/solar/news_detail.html?news_id=8785.

54. S. Bailey, B. Landi, R. Raffaelle, "Carbon Nanotubes Synthesized and Assessed for Space Photovoltaics", , *GRC 2005 Research and Technology,* 2006, pp78-80.

55. Sheila Bailey, Dave Wilt, Stephanie Castro & Ryne Raffaelle, "Alpha-Voltaic Power Source Designs Investigated", *GRC 2005 Research and Technology 2006,* pp 76-77.

56. P. E. Glaser, "Power from the Sun: Its Future," *Science Vol. 162, 1968, pp.* 957-961.

57. "The Final Proceedings of the Solar Power Satellite Program Review", NASA TM-94183, July, 1980, pp. 1-330.

58. NASA, *Satellite Power System Concept Development and Evaluation Program System Definition Technical Assessment Report*, U.S. Department of Energy, Office of Energy Research, Dec. 1980, Report DOE/ER/10035-03.

59. U. S. Office of Technology Assessment, *Solar Power Satellites,* 1981.

60. J. Mankins, "A Fresh Look At Space Solar Power: New Architecture, Concepts, and Technologies," paper IAF-97-R.2.03, 48th International Astronautical Conference, October 6-10 1997, Turin, Italy.

61. J. Mankins, J. Howell, "Overview of the space solar power exploratory research and technology program", AIAA 2000-3060, 35th Intersociety Energy Conversion Engineering Conference, July, 2000, pp. 24-28.

62. G. Landis, Re-evaluating Satellite SolarPower Systems for Earth", IEEE 4th World Conference on Photovoltaic Energy Conversion, Waikoloa, HI, May 7-12, 2006.

63. Landis, G., "Solar Power from Space: separating speculation from reality", XXIth Space Photovoltaic Research and Technology Conference, Cleveland, OH, October 6-8, 2009.

64. Mankins, J., "New Directions for Space Solar Power", Acta Astronautica 65, 2009, pp. 146-156.

65. Spera, D.A., Viterna, L.A., Richards, T.R. and Neustadter, H. E. (1979) *Preliminary analysis of performance and loads data from the 2-megawatt Mod-1 wind turbine generator.* Cleveland OH: NASA Lewis Research Center., NASA TM-81408 http://ntrs.nasa.gov/archive/nasa/casi.ntrs.nasa.gov/19800008234_1980008234.pdf.

66. Viterna, L. A. , 1981, The NASA-LeRC Wind Turbine Sound Prediction Code, NASA TM-81737, Cleveland, Ohio, NASA Lewis Research Center, http://gltrs.grc.nasa.gov/reports/1981/TM-81737.pdf.

67. UNEP *United Nations Environment Programme.* http://maps.grida.no/go/graphic/wind_power_generation_per_region.

68. Viterna. Larry A., 1981, *Fixed Pitch Rotor Performance of Large Horizontal Axis Wind Turbines*, NASA CR- 195462, http://ntrs.nasa.gov/search.jsp?R=194076&id=9&as=false&or=false&qs=Ne%3D35%26Ns%3DHarvestDate%257c1%26N%3D245%2B4294888887%2B33.

69. Spera, David A., (1995) *Bibliography of NASA-Related Publications on Wind Turbine Technology 1973-1995*, NASA CR-195462, http://ntrs.nasa.gov/archive/nasa/casi.ntrs.nasa.gov/19950020314_1995120314.pdf.

70. "U.S. Offshore Wind Energy: A Path Forward", U.S. Offshore Wind Collaborative, 2009-10-16. http://www.usowc.org/pdfs/PathForwardfinal.pdf.

SCOPE OF WIND ENERGY GENERATION TECHNOLOGIES

Jose Zayas

The author wishes to acknowledge many figures in this chapter that were provided by Sandia National Laboratories, and which are used here with permission.

7.1 INTRODUCTION: WIND ENERGY TREND AND CURRENT STATUS

The energy from the wind has been harnessed since early recorded history all across the world. There are proofs that wind energy propelled boats along the Nile River around 5000 B.C. The use of wind to provide mechanical power came somewhat later in time — by 200 B.C. simple windmills started pumping water in China, and vertical-axis windmills with woven reed sails were grinding grain in the Middle East. The Europeans got the idea of using wind power from the Persians who introduced it to the Roman Empire by 250 A.D. By the 11th century, a strong focus on technical improvements enabled wind power to be leveraged by the people in the Middle East extensively for food production. Returning merchants and crusaders carried this idea back to Europe where the Dutch refined the windmill and adapted it for draining lakes and marshes in 1300s.

In the late 19th century settlers in America began using windmills to pump water for farms and ranches, and later, to generate electricity for homes and industry applications. Although the industrial revolution influenced the propagation of wind energy, larger wind turbines generating electricity continued to appear. The first one was built in Scotland in 1887 by prof. James Blyth from Glasgow. Blyth's 33-ft. tall, cloth-sailed wind turbine was installed in the garden of his holiday home and was used to charge accumulators that powered the lights, thus making it the first house in the world to have its wind power supplied electricity. At the same time across the Atlantic, in Cleveland, Ohio, a larger and heavily engineered machine was constructed in 1888 by Charles F. Brush. His wind turbine had a rotor 17 m in diameter and was mounted on an 18-m tower. Although relatively large, the machine was only rated at 12 kW. The connected dynamo had the ability to charge a bank of batteries or to operate up to 100 incandescent light bulbs, three arc lamps, and various motors in Brush's laboratory. The machine was decommissioned soon after the turn of the century. In the 1940s, the largest wind turbine of the time began operating on a Vermont hilltop known as Grandpa's Knob. This turbine, rated at 1.25 MW fed electric power to the local utility network for several months during World War II.

In Denmark, wind power has played an important role since the first quarter of the 20th century, partly because of Poul la Cour who constructed wind turbines. In 1956, a 24-m diameter wind turbine had been installed at Gedser, where it ran until 1967. This was a three-bladed, horizontal-axis, upwind, stall-regulated turbine similar to those used through the 1980s and into the 1990s for commercial wind energy development, see Fig. 7.1. The popularity of using the wind energy has always fluctuated with the price of fossil fuels. When fuel prices fell in the late 1940s, interest in wind turbines decreased, but when the price of oil skyrocketed in the 1970s, so did worldwide interest in wind turbine generators.

The sudden increase in the price of oil stimulated a number of substantial government-funded programs of research, development, and demonstration. In the USA, this led to the construction of a series of prototype turbines starting with the 38-m diameter 100-kW Mod-0 in 1975 (Fig. 7.2) and culminating in the 97.5-m diameter 2.5-MW Mod-5B in 1987.

Similar programs were pursued in the UK, Germany, and Sweden. There was considerable uncertainty as to which architecture

FIG. 7.1 EARLY WIND FARM IN TEHACHAPI, CA
(*Source:* webcoist.com)

FIG. 7.3   SNL 34-m VAWT

FIG. 7.2   NASA MOD-0 WIND TURBINE

might prove most cost-effective, and several innovative concepts were investigated at full scale. In Canada, a 4-MW vertical-axis Darrieus wind turbine was constructed and this concept was also investigated by one of the Department of Energy's (DOE) National labs, Sandia National Laboratories (Sandia). The 34-m diameter Sandia Vertical Axis Test Bed was rated at 500 kW and was tested at the USDA-ARS site in Bushland, TX, see Fig. 7.3.

In the UK, an alternative vertical-axis design using straight blades to give an 'H' type rotor was proposed by Dr Peter Musgrove and a 500-kW prototype was constructed. In 1981, an innovative horizontal-axis 3 MW wind turbine was built and tested in the USA. This machine used a hydraulic transmission and, as an alternative to a yaw drive, the entire structure was orientated into the wind. The best choice for the number of blades remained unclear for some time and the industry and research entities experimented with large turbines constructed with one, two, or three blades, eventually converging with three.

Since the early 1980s through today, wind farms and wind power plants have been built throughout the country, and now wind energy appears to be the world's fastest-growing energy source that will power our industry as well as homes with clean, renewable electricity, see Fig. 7.4.

7.2   SANDIA'S HISTORY IN WIND ENERGY

Sandia National Laboratories' (Sandia) roots lie in World War II's Manhattan Project and its history reflects the changing national security needs of postwar America. Sandia's original emphasis on ordnance engineering — turning the nuclear physics packages created by Los Alamos and Lawrence Livermore National Laboratories into deployable weapons — expanded into new areas as national security requirements changed. In addition to ensuring the safety and reliability of the stockpile, Sandia applied the expertise it acquired in weapons work to a variety of related areas,

FIG. 7.4   MODERN WIND FARM IN NEW MEXICO (GE 1.5 MW)

such as energy research, supercomputing, treaty verification, and nonproliferation.

That expertise both in terms of capabilities and facilities was applied to wind energy during the mid-1970s, when the price of oil rose to unprecedented levels, and the nation began a commitment in identifying alternative, clean, and affordable energy generation. For the last 35 years, the laboratory has been committed to this mission and has contributed key technology advancements targeted at reducing the cost of delivered wind energy, while improving the reliability and efficiency of wind system. Historical contributions are captured below and are represented in time by the SNL Vertical Axis Wind Turbine (VAWT) Program, Rotor Innovations, Material and Manufacturing Program, to today's diverse wind research portfolio structured to meet the industry's needs and develop the next generation of components that will continue to improve the efficiency, reliability, and cost effectiveness of wind turbines.

7.2.1 Sandia's VAWT Program

7.2.1.1 History: Transition to The Modern Vertical Axis Wind Turbine
French inventor Georges Jean Marie Darrieus filed the first patent for a modern type of vertical axis wind turbine (VAWT) in France in 1925, then in the U.S. in 1931. His idea received little attention at that time, so little in fact that two Canadian researchers re-invented his concept in the late 1960s for the National Aeronautics Establishment of Canada without knowing Darrieus' patent. They later learned of the French inventor, and today's VAWT is known as Darrieus-type wind turbine.

In 1973, the Atomic Energy Commission, a predecessor to the current DOE, asked Sandia National Laboratories, a national laboratory devoted to engineering research and development, to investigate and develop alternative energy sources. Using their extensive experience in aerodynamics and structural dynamics from years of work with delivery systems for weapons, Sandia's engineers began to look into the feasibility of developing an efficient wind turbine that the industry could manufacture. During this time, the Canadians shared their re-invention with Sandia, and interest in the VAWT concept began in earnest.

7.2.1.2 R&D Beginning: From Desktop to Rooftop
The first Darrieus-type VAWT in America was actually only 12 in. tall and was constructed on top of an engineer's desk. To demonstrate that the VAWT concept worked, Sandia's engineers used a fan to create wind for the miniature turbine and a blackboard to perform their calculations — using these simple means, they converted non-believers.

Darrieus' concept appeals to engineers because it works on the principle of aerodynamic lift. Lift is what keeps an airplane in the sky — the wind actually pulls the blades along. In contrast, the traditional Holland-type windmill operates on the principle of drag, meaning the wind has to push a manmade barrier, such as a blade. Modern vertical- and horizontal-axis wind machines both use lift, which makes them more efficient compared with traditional windmills.

Sandia's original modal VAWT combined Darrieus' design with another concept for a wind turbine, called Savonius after its inventor, a Swede. Because the Darrieus VAWT could not start itself, some researchers thought it might be at a disadvantage. The turbine Savonius design used some lift, but its theoretical advantage was in using cups or vanes to trap the wind — employing the principle of drag — and so it was able to catch the wind and start itself in motion. However, the Savonius element was soon abandoned because

of the blade size required for it to work: instead, engineers opted to start up the turbine manually and to use the Darrieus design.

To test the aerodynamics of the turbine, a larger working model was built on the rooftop of the main administration building at Sandia. This model measured 5 m across the outer edges of the two bowed blades, each constructed out of a shank of steel covered by foam and fiberglass, then molded into the characteristic teardrop airfoil shape commonly used in the aircraft industry. Putting the test turbine in motion was no easy feat — researchers patiently waited for the wind to begin to blow, strapped themselves onto the roof of the building and spun the blades by hand. Whenever a thunderstorm, with its accompanying high winds, would blow into Albuquerque — night or day — the engineers rushed to the laboratory, climbed to the roof, and began turning the blades.

Starting the blades was not the only problem, however. To sustain their rotation, the blades had to be turning at least two to four times faster than the wind so that lift could work properly. At this early stage in the turbine's development, the blades required certain wind conditions, which did not occur on a daily basis. In the spring of 1974, however, the winds cooperated, the VAWT blades rotated smoothly on their own, and the demonstration phase began.

Another factor engineers had to consider was that under certain conditions, wind turbines can literally spin apart; they go into what is known as a runaway condition. Researchers knew that if their VAWT had a load to power — a generator for example — the load would act as a brake against runaway, but at that time, there was no load in the test system. For this reason, they built a disc brake consisted of a commercially available automobile disk caliper clamped onto a machined disk.

7.2.1.3 Tech Transfer: Moving to Industry
Some two years after constructing the rooftop model, Sandia built a second, larger wind turbine- this one on the ground. With a blade span of 17 m, the turbine's main purpose was to show that it could compete in cost with the more traditional horizontal-axis machines. An economic study from 1976 supported the research: vertical and horizontal axis wind turbines, or HAWTs, should indeed be comparable in performance and price if some improvements were made to the VAWT's design.

The 1976 study suggested these improvements. First, two blades would be better — the earliest design had three. Next, slimming down the shape of the turbine would improve its design, and the turbine's efficiency could be improved with better airfoil shapes. Finally, the study also found that a blade span of at least 17 m was best. During its first year of operation, 1976, this experimental machine was the largest VAWT in existence, and its performance compared favorably with that of a horizontal-axis machine.

The first VAWT blades were expensive because they were made of aluminum, fiberglass, and a man-made honeycomb-like material, all of which had to be carefully fitted together. Alcoa Industries was interested in reducing manufacturing costs of VAWT blades and in the mid-1970s developed an extrusion process in which partially molten bars of aluminum are forced into a die cut into the shape of airfoil. The aluminum is under such pressure that it melts and flows through the die, where it cools and resolidifies. The result is a uniformly manufactured airfoil in the required shape. The process dramatically reduces the cost to manufacture VAWT blades, and it continues to be used today.

Alcoa won a DOE contract a few years later, in 1979, to construct four low cost VAWTs, each to have a 17-m blade span and to deliver 100 kW of electricity. Construction lasted from January 1980 until March 1981; however, because of DOE budget constraints, only three of the units were installed. Each of the sites was chosen for a specific application: Bushland, Texas, to demonstrate an agricultural application, Rocky Flats, Colorado, to confirm structural and performance tests, and Martha's Vineyard, Massachusetts, to demonstrate the VAWT's applicability to the utility grid.

Their successful operation — more than 10,000 hours for the Bushland machine — convinced two companies to commercialize this design. VAWTPOWER and FloWind each manufactured VAWTs for use in California, where weather conditions favor using the wind's power for electricity. The result was more than 500 VAWTs were operating in California and producing electricity by the mid-1980s (Fig. 7.5).

7.2.1.4 Using the Information for a New, Larger Machine
Because the 17-m VAWTs showed such success, the DOE Wind Program directed Sandia to develop an expanded research machine. System studies indicated 34 m was a good size for the blade diameter to test the new airfoils, and the size made economic sense. In cooperation with the Department of Agriculture, the culmination of planning began in 1984.

Called simply the 34-m Test Bed, this VAWT is a research tool for testing and developing advanced concepts. It can produce 500 kW of electricity, more than half of the local community's normal

FIG. 7.5 FLOWIND COMMERCIAL 19-m VAWT COMMERCIALIZED IN COOPERATION WITH SANDIA

power needs, but its purpose is research, not power production. For this reason, instruments are strategically mounted on the VAWT to measure its parameters, especially stress on the blades. Weather conditions that affect the VAWT's performance are also recorded, including the wind direction and speed, ambient temperature, and barometric pressure.

A special feature of the Test Bed is that it can run over a continuously variable range of rotor speeds, from 25 to 40 rpm, whereas most wind turbines are designed to turn at a constant speed. The large, bowed aluminum blades are made of sections of specially designed airfoils that are bolted together; three different sizes and designs increase efficiency and regulate power through stall.

The work at Sandia and its Test Bed includes validating computer models, testing airfoil designs, and developing various control strategies. The work is part of improving the first-generation design, which has been commercialized in California, as well as developing next-generation VAWTs. Transferring technology from its national laboratories to the commercial sector is a major goal of the DOE, and Sandia's development of the VAWT and its subsequent adoption by industry is a good example of such a program.

7.2.1.5 The Future of VAWT Research
Within the DOE's Office of Energy Efficiency and Renewable Energy (EERE) is the Wind and Waterpower program, which oversees the current federal wind energy program, including wind research and development supported by the national laboratories. The DOE supported Sandia's efforts to develop VAWT technology, which serves as the basis for private industry to develop new generations of VAWTs with greater efficiency and longer life expectancy than any machine produced in the past. To this end, the Department supports its laboratories' forming cooperative research agreements with commercial firms to improve wind turbine designs.

The DOE's program for the vertical axis wind turbine came a long way since Sandia built its 30-cm-tall desktop version, and many of the elements which we see today on utility scale horizontal axis wind turbines (HAWT) were developed during this time.

In the mid-early 1990s, it was apparent that the industry had chosen a new path, and that it would convert primarily to the three-bladed HAWT. There are many reasons why that path was taken; in particular, the pursuit of higher wind resources at higher elevations, but it is difficult to quantify where VAWTs would be today if that decision would have been different.

7.3 SNL'S TRANSITION TO HAWT'S IN THE MID 1990s

Although VAWT technology had proven its feasibility to compete as a viable wind energy architecture, there was a fundamental shift in the early to mid-1990s that ended the investment of utility-scale VAWTs. Additionally, the industry in the U.S. had diminished given an expiration of the production tax credit, see Fig. 7.6. During this period, designers were seeking larger machines that could sweep a larger area and take advantage of the more benign higher velocity wind found at higher altitudes, see Fig. 7.7.

During this time, Sandia began to focus its research activities in HAWT technology and take the capabilities and core-competencies of the laboratory and apply them synergistically to HAWTs. Although the industry in the U.S. had dwindled, Sandia transitioned and began applying their 20-year wind energy experience to wind rotors. Since that time and continuing today, Sandia

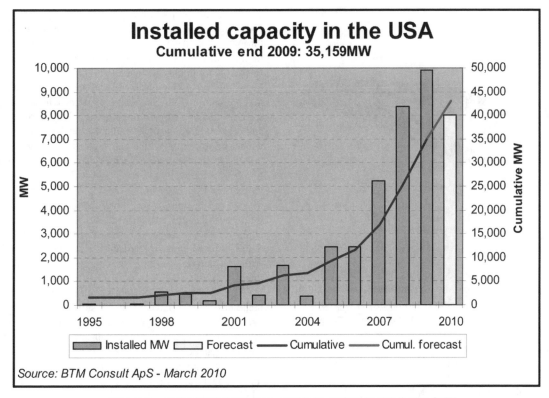

FIG. 7.6   BTM CONSULT U.S. ANNUAL INSTALLED CAPACITY

has been engaged in developing next-generation blades that are designed to be innovative, low-cost, reliable, and maximize energy capture. Programs in aerodynamics, structural dynamics, materials and manufacturing, and testing and evaluation provided the foundation for the research program.

7.3.1   Rotor Innovation

Wind turbine blades are designed to maximize energy capture and survive structurally the stochastic wind input for a 20-year design life. Although these structures appear quite simple from the exterior, there is immense innovation that has been applied over

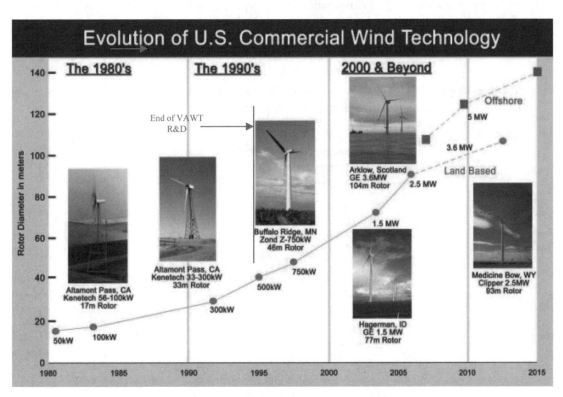

FIG. 7.7   WIND TURBINE EVOLUTION

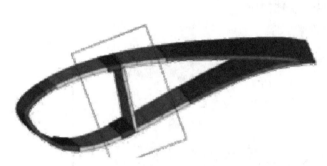

FIG. 7.8   ANSYS FEA WIND BLADE CROSS SECTION

the history that have enabled blades to be efficient, reliable, and cost-effective. In order to maximize the efficiency of the rotor, designers focus on balancing structural requirements and aerodynamic efficiency to maximize the operational coefficient of performance, C_p.

Equation 7.1:

$$\text{Wind Power} = \frac{1}{2}\rho A C_p V^3,$$

where ρ = air density, A = rotor swept area, C_p = coefficient of performance, and V = wind velocity. All utility-scale rotors today are comprised of three lift-based blades, which theoretically have a collective maximum efficiency of 59%, known as the Betz limit [1,2]. Advancements in computational fluid dynamic modeling coupled with airfoil evaluation and testing have enabled operational rotors today to have C_p in the high 40s to low 50s. That is quite a remarkable engineering accomplishment, given the random nature of the wind input and the fact that there is limited control authority in the system, variable speed and pitch.

Structurally, wind turbine blades are driven and designed to survive high fatigue cycles [4], and survive the environment conditions throughout the design life. Given these design constraints, composite materials lends themselves well for this application, and today, fiberglass dominates the market given its low cost and ease of manufacturing. Most wind turbine blades are designed and manufactured in three sections, a high and low pressure skin, and one or two shear webs as the main support member (Fig. 7.8). In order to save weight and prevent large unsuspended panel buckling, the panels are sandwich type structures with a core material, balsa wood or foam.

A large challenge for structural designers is the non-linear relationship as the weight of the blade scales to the third power of the length, see Fig. 7.9.

As wind turbine blades have gotten larger (30 to 60 m today) innovative designs and utilization of advanced materials have enabled rotors to scale and remain competitive. A large portion of Sandia's research is targeted at evaluating the utilization of lighter-stronger materials, such as carbon fiber to optimize the structural integrity of the blade.

Over the past decade, Sandia's blade program has developed three-blade designs that have evaluated strategic methods for optimizing structural design, aerodynamics, and weight. All designs have taken into account economics, manufacturing, and performance to validate the next generation of blades for the industry (Fig. 7.10).

As an example, in 2002, Sandia developed a blade design utilizing "flatback" airfoils for the inboard section of the blade to achieve a lighter, stronger blade. Flatback airfoils are generated by opening up the trailing edge of an airfoil uniformly along the camber line, thus preserving the camber of the original airfoil. This process is in distinct contrast to the generation of truncated airfoils, where the trailing edge the airfoil is simply cut off, changing the camber and subsequently degrading the aerodynamic performance. Compared to a thick conventional, sharp trailing-edge airfoil, a flatback airfoil with the same thickness exhibits increased lift and reduced sensitivity to soiling [7].

Today, several manufacturers incorporate carbon fiber in their blade designs and are evaluating the utilization of inboard structurally efficient flatback type airfoils.

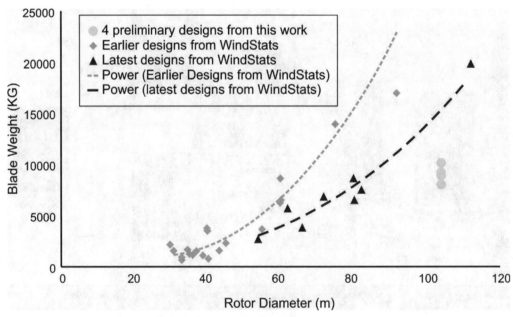

FIG. 7.9   WINDSTATS BLADE WEIGHT-VS-ROTOR DIAMETER

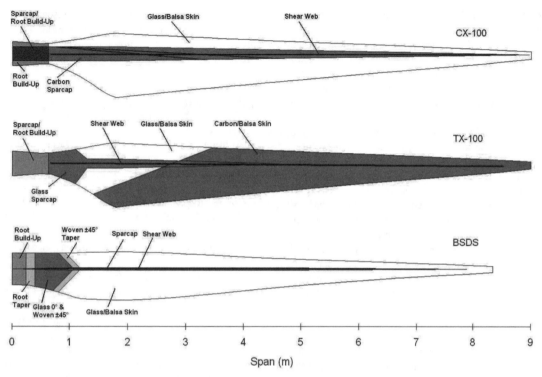

FIG. 7.10 SNL'S CARBON FIBER INNOVATIVE BLADE DESIGNS. TOP TO BOTTOM: CX-100 — CARBON SPAR BLADE, TX-100 — OFF-AXIS CARBON SKINS FOR AEROELASTIC TAILORING, AND BSDS — OPTIMIZED STRUCTURAL/AERO BLADE DESIGN

7.3.2   Manufacturing Research

Typical utility-scaled wind turbine blades being manufactured today can range between 30 and 60+ m in length, but given that the majority of the installations are land-based, the range is between 30 and 45+ m (Fig. 7.11). Wind turbine blades pose manufacturing and supply chain challenges given their large size, large amount of raw materials, and significant labor content associated to the various accepted manufacturing processes. Additionally, in order to meet the demand and support large and emerging global markets, some utility scaled turbine manufacturers have their own blade manufacturing, while others have chosen to purchase them from

FIG. 7.11   PICTURE OF UTILITY SCALE BLADE MANUFAC-TURING (Courtesy of TPI)

component suppliers to displace risk and large capital investment in manufacturing infrastructure.

As an example, focusing on a record year, 2009, where approximately 10,000 MW were installed across the U.S. and assuming an average machines being 1.5 MW in size (~40-m blades), ~20,000 blades were manufactured just to meet the U.S. installations. A typical 40-m blade weighs approximately 12,500 lbs and is composed of fiberglass, some OEMs have carbon fiber on spar cap, core material (balsa wood or foam), and a resin system (epoxy, polyester, or vinylester) and is primarily manufactured through an infusion process. Out of the total weight of a blade, the dry fiberglass can represent 70% of the total weight, the resin 25%, and the rest is the coring material. The raw material supply and delivered quality is crucial to manufacturing a high-quality product that can not only meet the certified requirements, but can survive the industry average design life of 20 years.

In manufacturing, Sandia, through the support from DOE, has embarked on a manufacturing program to address the challenges and opportunities of manufacturing high-quality cost-effective wind blades. The program is multi-disciplinary in nature, where quality, reliability, and cost-effectiveness are the primary metrics for success.

As blade length increases, the associated increase in blade weight places additional loads on both the rotor and the supporting structure. This increase in blade length has also resulted in scaling issues for structural aspects like bond lines, root attachments, and thick laminate infusion. In addition to gravitational loads, wind turbines also experience tens of millions of fatigue cycles during their operational lifetime due to turbulence in the wind, making fatigue resistant materials necessary for design. Wind turbines also often operate in difficult and harsh environments, which necessitate the use of coatings for protection. Finally, since wind must compete with other generation resources, there is a cost constraint on the blades of around $5 to $7/lb. These three factors create a uniquely challenging design problem for wind engineers.

To address and ensure quality, the program targets improvement opportunities in robust and lean manufacturing techniques to minimize human errors, given the labor intensiveness in manufacturing, and nondestructive inspection techniques (NDT) to indentify and address issues in the finish product prior to shipment and delivery. Nondestructive techniques typically used for wind blades, ultrasonic and thermography, provide mixed results and vary in applicability given the complex geometry and internal architecture. Through experience and design, knowledgeable manufacturers inspect critical regions and developed guidelines for acceptable flaws. SNLs manufacturing program evaluates all available applicable NDT techniques to develop a portfolio of options that will minimize false-positive inspection results, which can lead to field problems where cost of repair grows exponentially.

Given expert projections or the results of industry studies, such as the DOE 20% by 2030 scenario where the analysis documents the viability and improvements needed to achieve 20% wind energy by the year 2030, it is clear that a robust, reliable, and high-quality wind blade supply chain is needed for the industry.

7.3.3 Materials Research

Sandia National Laboratories has performed research in the area of wind turbine materials for over 20 years. A primary effort of that work has been a partnership with Montana State University to produce the DOE/SNL/MSU Composite Material Fatigue Database. The database features the results of over 10,000 mechanical tests of wind turbine blade materials and is the largest publically available data set of its kind in the world [3]. The focus of much of this research has been in the area of high-cycle composite fatigue. This research has been broad in focus, with investigations of resins, fibers, resin–fiber interfaces, fabrics, adhesives, and design/manufac-

turing implementations. Through this research, the turbine OEMs have been able to discover material solutions to challenging design problems, and material suppliers have been able to evaluate their products and fast-track them into the industry.

7.4 MOVING FORWARD: STATE OF THE INDUSTRY

Although the U.S. experienced a large influx of installations during the 1980s, it is not until recent years that wind energy in the U.S. has achieved large market installations and continued market acceptance. Over the past few years, the Federal government has continued to provide a production tax credit (1.8 ¢/kW/h), and it is the combination of large amount available land with adequate resource, renewable portfolio standards, tax credit, renewed market pull for clean energy, and technology viability that has spurred this growth.

As can be seen in Fig. 7.6, the U.S. industry has experienced an exponential growth over the last five years, and many states have chosen to have a significant percentage of wind in their system (Fig. 7.12). Today, the U.S. has the largest installed capacity, but there is significant competition from emerging countries, such as China, which installed over 13 GW in 2009.

Through the third quarter of 2010, the U.S. has approximately 37,000 MW of installed capacity, which approximately represents 2% of our energy consumption [5]. Additionally, the growth of the industry has enabled wind energy to emerge as the leader of the new-generation clean energy portfolio (Fig. 7.13).

The growth and acceptance of the industry has disrupted the installation trends of the more traditional forms of generation, such as natural gas. Over the past three years, wind energy has represented on average 35% of all new-generation installations (Fig. 7.14).

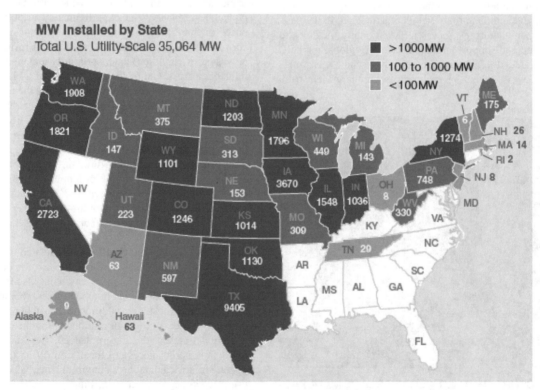

FIG. 7.12   WIND ENERGY INSTALLED BY STATE — END OF 2009

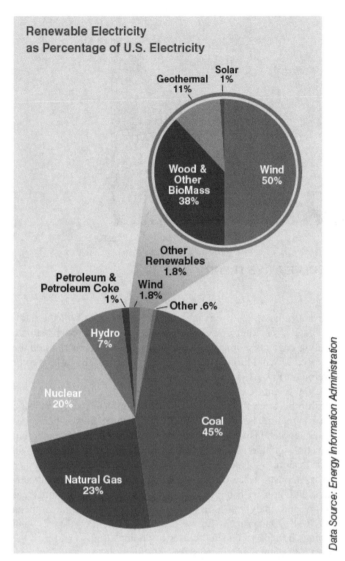

FIG. 7.13 EIA U.S. INSTALLED CAPACITY BY GENERATION SOURCE

In 2010, overall energy demand is down and combined with low natural gas prices, it is expected to be a low year for wind.

7.4.1 DOEs 20% by 2030 Scenario

In May of 2008, the Department of Energy (DOE) sponsored a National Study to analyze the feasibility of 20% of the U.S. energy by 2030 delivered from wind energy [6]. The 20% by 2030 scenario report outlines the required advancements in technology, as well as improvements in the siting process, the manufacturing scale-up, and the integration needs in order to achieve the goal.

Since being published, the report has provided the wind industry with a target that has become a unifying objective that is both recognized and acknowledged as an achievable goal. The analysis results captured a scenario that predicts the need for 305 GW to be online by 2030 (Fig. 7.15). Although mostly land-based technology, the capacity target does include 54 GW of offshore wind as well. In order to achieve the 305 GW, the analysis projects a scale-up to a steady state ~16 GW annual installation by 2016. As an example of feasibility, in the record year in 2009 ~9.9 GW were installed; showing the ability of the industry to scale to meet the demand.

Unfortunately, 2010 is not continuing the growth of the last few years, and as a matter of fact, it is expected that 2010 will be a difficult year for the industry (Fig. 7.16). This change is not solely due to the deep economic crisis being experienced, it is also a function of lower energy demand, lower-cost energy sources, and available transmission.

The 20% by 2030 report also has serves as key document to drive DOE's activities and investment in research and development. Through a balance program portfolio, Sandia, in coordination with other laboratories, academia, and industry, supports the industry and develops the next generation of technology targeted at improving the efficiency and reliability and promote a larger market acceptance.

7.4.2 Aeroacoustics

As wind turbines continue to be deployed across the nation, the likelihood of wind farms being sited near inhabited areas increases. An important constraint on wind turbine placement arises due to the consideration of wind turbine noise. As a key design metric, the noise generated by a turbine can determine its required setback distance from residences or buildings and depends on local community noise regulations. Noise is typically measured on a logarithmic or decibel scale [9, 10]. As an example, a six-decibel increase in the noise of a turbine would double the required turbine setback distance; likewise, a six-decibel decrease in noise may allow the turbine to be half as far away. Wind developers seek to place turbines in locations with the optimal wind resource, but as installations encroach populated areas, the noise constraint can prevent the optimal placement and adversely impact the economics of a wind farm.

Noise involves several distinct elements, including the source, the propagation through the atmosphere, and the perception, all of which are relevant to wind turbine acoustics and design. It is important to recognize that not all noise is the same, and that not all noise is perceived in the same way. Tones, or noise at discrete frequencies, tend to be perceived as more bothersome to humans than broadband noise, which is spread over a continuous range of frequencies. Low-frequency noise propagates through the atmosphere more efficiently than high-frequency noise; hence, it can travel over large distances.

There are two primary sources of noise generated by wind turbines: mechanical noise, and aero-acoustic noise. Mechanical

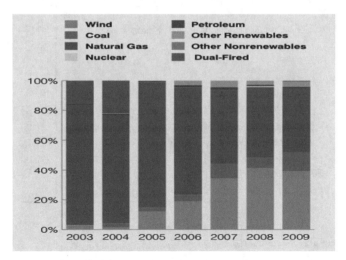

FIG. 7.14 EIA ANNUAL INSTALLED CAPACITY BY GENERATION SOURCE

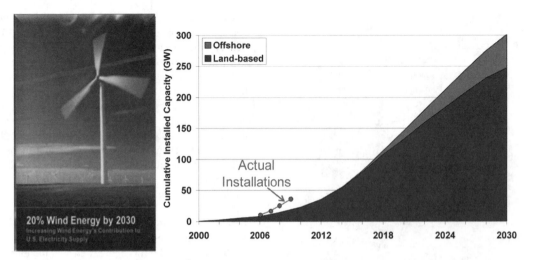

FIG. 7.15   DOE 20% BY 2030 CUMULATIVE TREND

noise involves machinery-generated noise from the gearbox, bearings and generator. This noise can directly radiate from the machinery components and cause vibration in the surrounding structures, such as the nacelle and tower (called "structure-borne" noise). Mechanical noise often occurs at well-defined tones associated with the rotational frequencies of the machinery components, such as gears and individual gear teeth. Unlike aero-acoustic noise, mechanical noise sources are often easier to isolate since the source and location is well known and can lend themselves to effective mitigation through the use of insulating material in the nacelle and vibration isolation to prevent structure-borne noise.

Aero-acoustic noise is the noise created due to the motion of the rotating turbine blades relative to the surrounding air. Aero-acoustic noise is the result of several complex fluid dynamical phenomena that occur on a wind turbine blade and is usually broadband in nature, meaning that the noise signal is spread over a continuous range of frequencies. A particularly important aero-acoustic noise source is trailing edge noise, which results from the flow of air past the aft or trailing edge of a blade. For an observer on the ground near a turbine, this noise is modulated by the passage of the rotating blades, resulting in a characteristic "swoosh, swoosh" sound. Trailing edge noise imposes a rather strict design constraint on the tip speed of wind turbine rotors, limiting how fast the turbine rotor can rotate (Fig 7.17).

A key scientific challenge involves the fact that the precise relationship between the shape of a blade design and its aeroacoustic noise signature is not well understood, which makes blade designers apprehensive to large changes that could result in a higher acoustic signature. This constraint tends to limit innovation in blade design.

Key acoustic research being conducted at national labs, universities, and industry is targeted at developing the underpinning technology and analytical tools to better understand the phenomena (Fig. 7.18). Once successful, we can expect that not only will wind turbines be able to be sited closer to populated areas but also the overall efficiency of wind systems will increase.

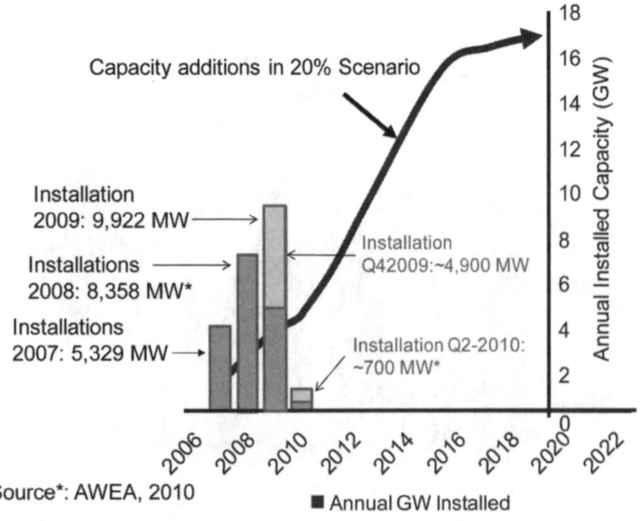

FIG. 7.16   DOE 20% BY 2030 ANNUAL CAPACITY ADDITION AND U.S. ANNUAL WIND INSTALLATIONS

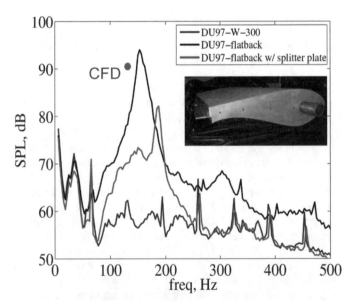

FIG. 7.17   MEASURED SOUND PRESSURE LEVELS (DU97 AIRFOIL — SHARP TE, FLATBACK AND FLATBACK WITH SPLITTER PLATE)

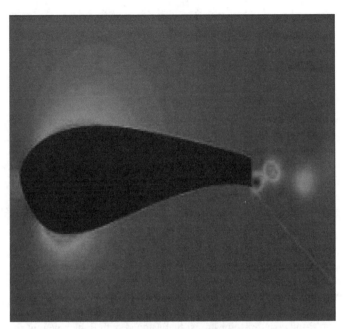

FIG. 7.18 FLATBACK CFD ANALYSIS: NOTE THE ASYM-METRIC VORTEX SHED OF THE TRAILING EDGE WHICH RESULTS IN HIGH ACOUSTIC EMISSION

7.4.3 Aerodynamics

The discipline of aerodynamics plays a critical role in wind turbine design for two main reasons: first, aerodynamic blade lift forces are responsible for creating the torque on the rotor necessary to drive the generator; and second, aerodynamic forces create the primary loads that drive the structural design of the turbine. The aerodynamic forces are the result of complex fluid-dynamical processes occurring in the wind, over the blades themselves, and in the wake of the turbine. The fluid dynamical system surrounding a turbine is multi-scale in nature, creating a significant modeling challenge for aerodynamicists (Fig. 7.19). The scales range from rotor-scale fluctuations in the atmospheric boundary layer down to micron-scale turbulent fluctuations in the boundary layer surrounding the rotor blades.

The turbine inflow, or oncoming wind seen by the rotor, is not uniform; in fact, it varies both in time and space in a stochastic

fashion that is typical of turbulent flows. This unsteady, stochastic operating environment is especially important when considering both fatigue and extreme loads on the turbine rotor, tower, and drive-train. The aerodynamic response to the wind inflow is determined by the blade aerodynamic characteristics.

Modern HAWT blades are comprised of a continuous sequence of two-dimensional airfoil sections defined along the span of the blade. Blades are typically designed with non-uniform twist, taper, and chord distributions along the span in order to maximize energy capture. Power-generating torque is generated by lift forces, while drag acts to decrease torque and power. Thus, airfoil sections with high lift-to-drag ratios are critical for good turbine performance. However, other considerations also govern the selection and design of airfoils for blades, such as structural constraints on the thickness of the airfoils and performance of the airfoils under soiled conditions.

Under design conditions and steady, uniform inflow, air flows smoothly over the wind turbine blades, leading to predictable performance and loads and optimal energy capture. However, various phenomena, such as sudden wind gusts or insect and dirt build-up on the blade leading edge, can lead to dramatic changes in the flow over the blades that have important consequences for both loads and energy capture. Parts of the blade may enter stall, or an unsteady form of airfoil stall known as dynamic stall. In either case, the flow detaches, or "separates," from part of the blade, leading to large changes in lift and drag forces. Separated flow regions may become three-dimensional, such that near the blade surface air moves radially along the blade instead of flowing from the leading to trailing edge. These complex phenomena are difficult to measure experimentally at full scale, as well as difficult to model (Fig. 7.20). However, modeling approaches based on Computational Fluid Dynamics (CFD) offer the ability to model these off-design aerodynamic phenomena important to wind turbine design.

Wind turbine rotors slow down the wind as they extract kinetic energy from it, resulting in a region of low momentum air downwind of the turbine known as the wake. The flow in the wake is also quite complex and three-dimensional. Flow perturbations caused by fluid motion in the wake are felt by the blades, affecting rotor performance and loads. Thus, any engineering model for a wind turbine rotor must consider both the blade forces as well as the behavior of the wake. Wakes tend to persist for large distances downwind from turbines (Fig. 7.21). They interact with and modify the inflow seen by turbines placed downwind of turbines in a wind farm array. This can have important implications for energy capture and reliability of large wind farms.

7.4.4 Sensors and Condition Health Monitoring

Although most machines today share the same architecture as the older machines, advancements in sensors, controls, and power

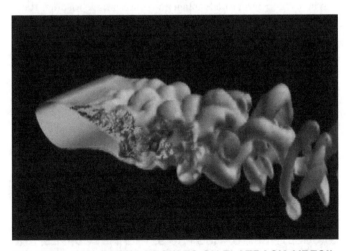

FIG. 7.19 CFD-BASED RESULTS ON FLATBACK AIRFOIL

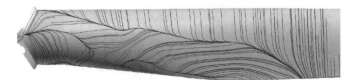

FIG. 7.20 CFD SOLUTION SHOWING NEAR-SURFACE FLOW STREAMLINES OVER THE INBOARD REGION OF A UTILITY-SCALE WIND TURBINE BLADE, SHOWING A REGION OF THREE-DIMENSIONAL SEPARATED FLOW

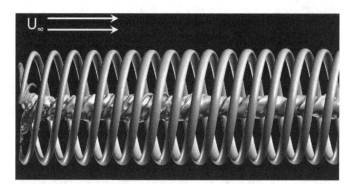

FIG. 7.21 CFD SOLUTION VISUALIZING THE HELICAL STRUCTURE OF THE WAKE OF A WIND TURBINE OPERATING IN A UNIFORM WIND, DEMONSTRATING THE PERSISTENCE OF THE WAKE WELL DOWN-WIND FROM THE ROTOR, WHICH IS LOCATED AT THE LEFT END OF THE FIGURE

electronics have provided opportunities for designers to develop algorithms and operational strategies that continually attempt to maximize energy capture, load management, and reliability.

A typical turbine today relies on hundreds of sensors for their effective operation and survivability. The role of those sensors vary from control observers (i.e., wind speed, high-speed shaft RPM, pitch position, etc.), fault detections (generator over temp, cable twist, etc.), to conditional health monitoring (gearbox lubricant quality, vibration levels, etc.). The effective operation of these sensor systems or networks is crucial for the safe operation of the machine and must operate reliably throughout the design life, which is typically 20 years. This strategy is increasingly important for offshore deployed systems, as machines are more complicated and have limitations in access when compared to land-based systems.

In the future sensor systems may play an even larger role on wind turbines. Currently, Sandia as well as other European laboratories are all engaged in the development and application of sensor and operational measurement methods. Some of the key objectives include: determination of inflow loads and damage state (Sandia National Laboratories), advanced condition monitoring of gearboxes (National Renewable Energy Laboratory), and monitoring of localized aerodynamic flow conditions (Risø DTU National Laboratory) [7a]. These technologies are all targeted at building a smarter wind turbine that can itself, identify the loads being applied by the wind, the damage created by these loads, and deploy control strategies to mitigate the loads while maintaining optimal power productions.

In order for newer, higher fidelity sensors to be adopted, there are several challenges/observations that must be addressed: sensor arrays and interrogator must have minimal cost, simple installation, and an operational life on the order of years and tens of years. Over these long durations of application, the sensor must also maintain calibration and sensitivity, otherwise Type 1 and Type 3 errors (false positive and false negative) will reduce the reliability and usefulness of the technology. Sandia's sensor program is focused on identifying sensor technologies that can potentially fulfill these design requirements. Currently, Fiber Bragg strain sensors interrogated over fiber optic lines, ruggedized accelerometers, hot-film aerodynamic sensors, and aerodynamic surface pressure taps are all simultaneously being investigated (Fig. 7.22). Each sensor technology is evaluated to determine the relative cost which is dictated by the number of sensors required to accurately monitor the

rotor blade, the cost of the interrogator used to measure the sensor signal, and the optimal/reliable method for integrating and protecting the sensor to maximize survivability.

Examples of adoption of new sensor technologies can be seen in several commercial machines today [8].

As an example, several wind manufacturers rely on fiber optic networks on the blades to enhance operation and control strategy. These sensors offer the flexibility in that many sensors can be placed in a single fiber line and can be incorporated in the manufacturing process.

As we foresee future designs, it is important to acknowledge that innovation will continue to play a key role in making wind systems more reliable and cost-effective. Sensor technologies are just one of those key elements that will continue to contribute to turbine optimization. It is conceivable that sensors will not only contribute to single turbine improvements in the future, and can be utilized for wind plant operations, as machines could have the ability to adapt to address real-time conditions.

7.4.5 Advanced Control Strategies

In order to continue to reduce wind turbine costs, future large multi-megawatt turbines must be designed with lighter-weight rotors, potentially implementing active controls strategies to mitigate fatigue loads while maximizing energy capture and adding active damping to maintain stability for these dynamically active structures operating in a complex inflow environment. Development, evaluation, and testing of advanced controls to mitigate fatigue loads caused by complex turbulent inflow are crucial for future designs.

The wind turbine is a highly non-linear dynamic machine that operates over a large turbulent wind regime. Current conventional designs are limited to linearized models about nominal wind speed operating points that require gain scheduling to transition between each nominal wind speed operating point [7]. Today, commercial machines rely on either classical single-input–single-output (SISO) controllers or state-space multiple-input–multiple-output (MIMO) controllers based on linearized models. While adequate for controlling the "stiff" machines of the past, these methods are not ideal for stabilizing future large multi-megawatt turbines that will experience greater dynamic coupling due to greater flexibility and lower rotor speeds. To meet these future challenges, Sandia focuses on advanced control methods and paradigms that are designed to meet multiple control objectives with a single unified control loop, where multiple control actuators and multiple sensors can be used to greatest advantage to reduce fatigue loading, stabilize the complex structure, and maximize power.

Moving forward, the possibility of designing full non-linear dynamical system in a non-linear/adaptive control design may allow for the potential to capture more energy in below rated-power conditions, efficient transition between below and rated-power conditions, and for above rated-power conditions to mitigate and reduce fatigue loads on turbine components and blades. This results in longer operational life for the wind turbine components (gearboxes, blades, etc.).

FIG. 7.22 SANDIA'S SENSOR BLADE—INCLUDES STRAIN GAUGES, FIBER OPTICS FBG'S, THERMOCOUPLES, AND ACCELEROMETERS

7.4.6 Advanced Architecture

Recent technology innovation in rotor technology — including individual blade pitch control, passive bend-twist and sweep-twist coupling (aero-elastic tailoring), and fast-acting active aerodynamic load control — offer the potential for further enhancing turbine energy capture and decreasing turbine cost of energy (COE). There is a significant amount of research domestically and globally that showcase the value of these innovation, and ongoing research in both controls and sensing will provide the operational architecture to make them a reality.

Advanced control architectures that fully take advantage of these innovations can provide the technology pathway to continue to refine these large machines and ensure that safety, efficiency, economics, and reliability metrics are fully realized.

With today's computational capabilities and the researchers' ability to model the integrated system, multidisciplinary approaches are a key to improving the technology and ensuring that the maximum efficiencies are attained (Fig. 7.23).

7.4.7 Revitalizing U.S. Clean Manufacturing

In order to continue to support the growth of the wind industry, local and cost-competitive manufacturing is an important aspect to provide sustainability and the green economy. The wind industry provides a series of jobs that will require a trained a robust workforce. Like any other energy industry, the opportunities span from engineering, service and operation, and manufacturing. A big challenge for the U.S. in the manufacturing area is the labor cost when compared to other parts of the world. While the industry in the U.S. was ramping up, a large percentage of the components were being imported. Today given the persistence of the U.S. market, many companies have established U.S. manufacturing, but in order to continue to capitalize on this opportunity, a cost-effective manufacturing strategy must be developed to displace the higher labor costs. Unlike the semiconductor industry, which has predominantly left the U.S., wind turbine components have the opportunity to stay given their large size, the countries manufacturing capabilities, and transportation and logistics cost (Fig. 7.24).

Sandia's Advanced Manufacturing Initiative (AMI) is targeted at evaluating the manufacturing process, optimizing the process flow, identifying opportunities for automation, and improving quality and plant output. The initial program is targeting wind

FIG. 7.24 2 MW WIND TURBINE BLADE BEING TRANSPORTED

blade manufacturing due to its high labor content, the size of the components, and need for improve quality and reliability.

7.4.8 Testing and Evaluation

As wind turbines have grown in size and capacity over the last three decades, the importance of reliability and technology innovation have been quite apparent. Even though, these "Gentle Giants" look much like their predecessors of the 1980s — three-bladed, upwind configuration — technology improvements have been vital for the success of this vibrant industry. Every component and subcomponent of the turbine including airfoils, materials, structures, and sensor and control systems have to be tested and evaluated prior to being deployed and accepted.

All engineered components and systems must go through a testing phase in order to validate the engineering assumptions made in the design, analysis and manufacturing processes. The key question for testing is why, when, and is some cases how? As the wind industry has gone through its maturity phase, the trial-and-error days of going straight to the field and patching flaws real-time are hopefully long gone. This has been driven and enabled by both the fidelity of today's engineering tools and computing capacity, as well as the requirement to certify components and systems to standards and the shear cost of building and testing structures of this magnitude (Fig. 7.25).

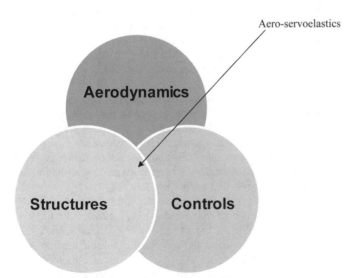

FIG. 7.23   SNL INTEGRATED AEROSERVOELASTIC STRATEGY

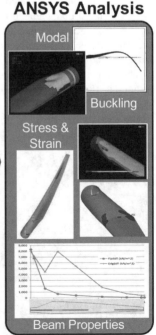

FIG. 7.25   ANSYS FEA BLADE MODEL AND CALCULATED ANALYSIS

Wind energy components pose many challenges when it comes to testing and evaluation. Not only does wind have a series of unique components, the size of the components and the measurement requirements can make testing quite costly and challenging. Take for example a typical utility-scale wind turbine blade, which is 30 to 60 m in length, weighs 10 to 20 tons, and is quite complex in shape. As previously explained, modern blades are predominately made of a combination of fiberglass and carbon fibers, resin, and balsa or foam. Each one these materials has to be certified and tested by their respective manufacturer and must be brought together to design a blade which itself has to be certified. Airfoils are now optimized for aerodynamic and acoustic performance and are tested rigorously in wind tunnels to ensure this performance under a variety of simulated field conditions. Moreover, the shape must lend itself to not only localized aerodynamic performance, but must also be coupled to the structural and manufacturing design, where the internal structure is designed to take advantage of the materials available and how they will be organized or stacked to develop a structurally efficient blade. Also, as part of the design phase, the manufacturability must be continually evaluated to ensure that the blade can be manufactured to the specifications without the introduction of defects.

Although, there will be several sub-scale testing phases throughout the engineering process, the entire completed blade must also be tested for structural strength and aerodynamic performance to validate the computer models and receive certification. These tests can be both complex and costly due to the size of the structure and the complex testing and loading requirements, which must replicate such things as, in the case of fatigue loading, the number of and magnitude of loading cycles that a blade will see over its 20-year life (Fig. 7.26). It is key at this point that the tested article is close to the final design, as it can cause significant cost increases and project delays to redesign, rebuild, and retest.

Coincidently, these complex testing processes must also be completed as applicable for other components of a wind turbine, including gearboxes, generators, controllers, etc. In order to completely be sure that the system will operate as predicted, all components should be evaluated together to validate that the systems will work reliably. Sandia's technology program relies on this approach and evaluates the technology at the sub-scale test site in Bushland, TX (Fig. 7.27).

The tools used to develop and evaluate these designs are only as good as the data used to validate and improve the fidelity of the

FIG. 7.26 UTILITY SCALE STATIC TESTING AT NREL'S NWTC

FIG. 7.27 SANDIA'S SUB-SCALE TEST SITE IN BUSHLAND, TX

code. Additionally, the tools are only able to model the article to a certain degree of fidelity within a certain operational envelope, and many practical elements can diverge from the model in the as-manufactured, as-installed final product. The lessons learned and data gathered from lab and field testing, at both full-scale and sub-scale enables engineers to continually improve their designs, the result of which can be clearly seen in the viability and resilience of the wind industry.

7.4.9 Design Tools

Wind turbines are designed and optimized to capture as much energy as possible in a given wind resource. Research into innovative wind turbine design improvements is being performed on all components of the wind turbine in order to improve the overall reliability and efficiency of the machines in the field. Example research topic areas include towers, generators, drive trains, and blades. With the component size of today, the ability to build and test these components is cost prohibited and would delay the product cycle significantly.

Additionally, changes to individual component technology affect system behavior and result in both costs and benefits for any given innovative idea. It is vital to understand not only the benefits resulting from an innovation but also to understand the nature and magnitude of resulting costs that may be present elsewhere in the system. Common system costs may include the following: increased forces and moments elsewhere in the system, increased complexity or decreased energy capture. Use of system dynamic models enables researchers to assess overall cost and benefit of new ideas even to the point that effects on final cost of electricity (COE) may be determined.

The system dynamics model of a wind turbine includes physics representing three major areas as seen in Fig. 7.23. It includes elements to describe aerodynamic and structural aspects as well as wind turbine controls interactions. The combination of these elements is known as an *aeroelastic*, or *aeroservoelastic*, problem due to the coupled interactions of the aerodynamics, structural deflections, and controls that are involved.

The system model is used to assess the wind turbine design with respect to typical design requirements (Fig. 7.28). Wind turbines are designed for a 20-year life and are subjected to stochastic turbulent wind input, extreme loads, and operation through component faults. Potential failures are assessed in terms of ultimate loads, fatigues loads, and functional requirements, such as blade-tower clearance. Finally, the system model can be used to simulate the efficiency of energy capture during normal operations.

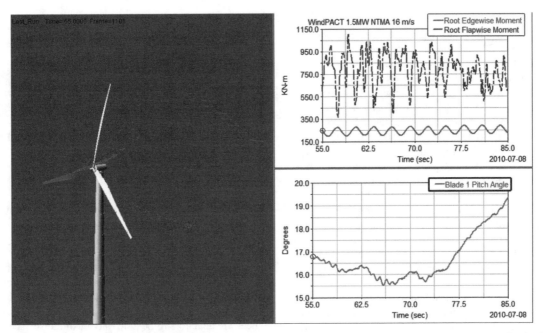

FIG. 7.28 GRAPHICAL REPRESENTATION OF A SYSTEM DYNAMICS SIMULATION AND RESPONSES

7.5 FUTURE TRENDS

Although the wind energy industry has experienced a large growth over the last decade, both nationally and globally, technology improvements continue to be required in order for the industry to be competitive and continue to be a key part of the energy mix of today and the future. Technology innovations that balance efficiency improvements and reliability are especially attractive, and national labs, industry, and academia are engaged to develop and implement these technologies.

Additionally, although the U.S. has an immense resource in particular in the Great Plains region (Fig. 7.29), technology that is viable in lower resource sites and offshore is needed.

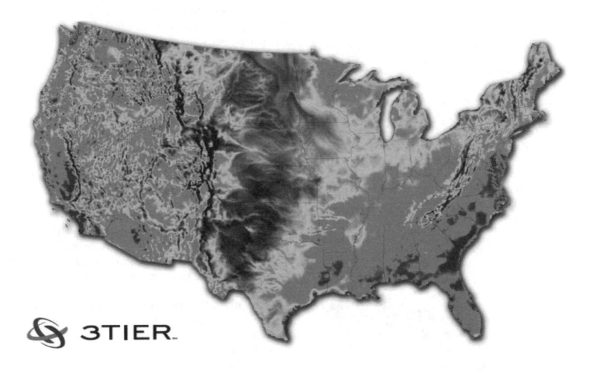

FIG. 7.29 U.S. 80-m WIND MAP

7.5.1 SMART Rotor

Since the global acceptance of the utility-tied, three-bladed up-wind configuration over 30 years ago, engineers and scientist across the globe have developed innovative techniques and improvements to increase the efficiency, availability, and reliability of wind turbines. Efficiency improvement options are always attractive given the strong coupling to cost-of-energy (COE) and because of the ease of calculating the return on the investment. Revolutionary examples of innovation over the last 30 years include the use of laminar airfoils, the transition to variable speed and pitch from stall-regulated designs, and many more. These innovations have enabled the wind industry to become globally cost-competitive and to install products that are designed for a 20-year lifespan.

As we look to the future, the large "low-hanging fruit" of efficiency improvement areas for land-based deployment are no longer there, and designers, engineers, and scientist at national labs, universities, and manufacturers are evaluating, designing, and implementing concepts that are focused on refining and improving the technology.

Ongoing research — taking place both domestically and internationally — focused on next-generation concepts has identified the viability and feasibility of both passive and active aerodynamic surfaces on wind turbine blades.

There are two ways to implement passive aeroelastic coupling on a wind turbine blade: off-axis materials and geometric sweep (Fig. 7.30).

A key advantage of the passive methods is the simplicity versus authority that the design enables. By passively modifying the blade, the rotor is able to change the incident angle resulting in an overall load reduction over the entire power curve spectrum. The result is the ability of designing a larger rotor that captures more energy and manages the system loads. The disadvantage is that the wind is not constant and is quite random in nature. Since the system is passive, designers must identify the primary design point and try to maximize performance throughout the operational envelope. Initial field evaluation studies of both methodologies have shown energy capture improvements ranging from 5% to 10% [11].

Sandia's active aerodynamics program focuses on designing and implementing low energy-consuming, fast-acting, and simple aero surfaces that can modify the localized flow in order to affect the high-frequency content in the wind (Fig. 7.31). This capability will provide designers with a new set of actuators that can be managed to fine-tune the performance of machines and will be able to adapt to local atmospheric phenomena that are difficult to resolve with current actuation.

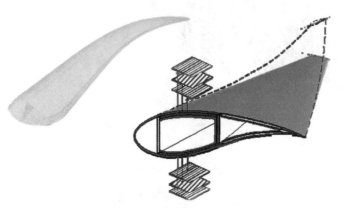

FIG. 7.30 LEFT: SWEPT K&C STAR DESIGN AND RIGHT: SNL OFF AXIS DESIGN

FIG. 7.31 MICROTAB ACTIVE AERO CONCEPT AND CFD CALCULATION

Initial results from Sandia's SMART rotor program have shown the ability of these methodologies to significantly reduce system loads and enable designers to increase the rotor size for a given architecture, yielding a net annual energy increase (Fig. 7.32).

In order to fully realize the benefits of active aerodynamics, the localized conditions must be understood. Cost-effective sensor technologies that provide the necessary information (load, pressure, etc.) and have the appropriate resolution must be collocated near the actuators to be able to control the surface effectively and efficiently. Although the initial results are quite promising, it is important to keep cost, reliability, and maintainability in mind in order to ensure the implementation and acceptance by the industry.

Innovation, such as the SMART program, will always be a part of technology development. Although cost-competitive, the wind industry must continue to identify improvement areas to increase viability and ensure that wind can compete in a diverse energy sector of the future.

7.5.2 Offshore Wind

Although developers will continue to pursue economically feasible land-based sites to install projects, concerns or limitations in our transmission system, NIMBY, and limited land on most coastal states are all reasons why a strong interest has been spurred in exploiting our coastlines for offshore wind installations. There are many attractive reasons why offshore wind installations should be pursued on the Gulf coast, the Atlantic or Pacific coast, and/or the Great Lakes. The U.S. has an excellent offshore wind resource, and approximately 78% of our total energy consumption is consumed by the 28 states with a coastline (Fig. 7.33). Currently, there are no installed offshore projects in the U.S., but there are 13 projects being proposed, totaling to 2.4 GW.

Several European countries have already leveraged their coastlines for offshore projects; there are currently 39 installed offshore projects totaling over 2 GW (Fig. 7.34). Although the modern wind industry has many decades of experience, the offshore wind industry is quite young, and several of the initial projects have experienced several technical premature reliability challenges since their installation.

Fundamentally, offshore machines are quite similar to the land-based systems that we have become accustomed to, but both are driven by economic and environmental differences. Offshore machines must be equipped with systems and an operational architecture that provide accessibility and enable them to survive the challenges of the offshore environment (Fig. 7.35).

From a resource perspective, it is well understood that wind over the water is often more consistent and less turbulent than on land. This typically translates to higher capacity factors and more-predictable electrical output. Coincidently, the design of the machine from the foundation to the rotor must take these differences

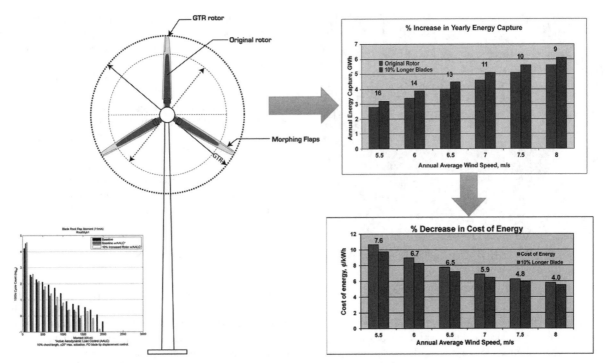

FIG. 7.32   ACTIVE AERODYNAMIC STRATEGY — IMPROVEMENTS IN ENERGY CAPTURE AND COST OF ENERGY

into account, as well as the impact of the hydrodynamic loading induced by the ocean in order to design a machine that is efficient, reliable, and cost-effective. All current offshore installations in Europe today have been installed in fairly shallow waters (<30 m), which have provided the opportunity to leverage well-known foundation designs (primarily monopole) from other industries. Unfortunately, if the U.S. wants to capitalize on its vast offshore resource, research and development must be performed for deeper water depths where jacketed or floating structures will be needed (Fig. 7.36).

There are many advantages as to why offshore projects should be pursued in the U.S. Outside of a key advantage of proximity to large load centers, offshore machines can be significantly larger than the typical land-based machines being installed today (1 to 2.

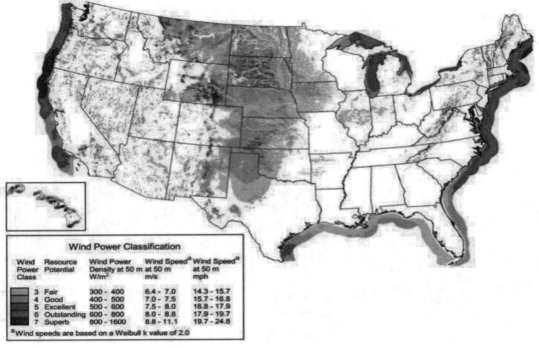

FIG. 7.33   LAND-BASED AND OFFSHORE WIND RESOURCE MAP

FIG. 7.34 SIEMENS OFFSHORE MACHINES — COPENHA-GEN HARBOR (*Source:* Wikimedia Commons)

5MW) since the limitations in both infrastructure and transportation can be mitigated by having coastal manufacturing and barging the components to the installation sites. Typical offshore machines today range from 2 to 5 MW, but larger turbines are being designed and tested. There are several challenges associated with offshore wind as well. In comparison to the land-based machines, the cost

balance of an offshore project is not the same, as the turbine represents a much smaller percentage of the total cost (~25%). Cost associated with the support structure, the electrical infrastructure, and operations and maintenance (O&M) are significantly higher for offshore projects; hence which is why current offshore projects come in at a higher cost (Fig. 7.37).

To outweigh these challenges, future designs must be smarter and able to operate and report upcoming failures and service requirements prior to a catastrophic system failure. As an example, new machines could incorporate a sensor network that increases the fidelity in operation and condition health monitoring. Given that the turbine cost does not dominate the total cost of the installation as the land-based system does, innovation in this area is critical and a cost-effective way to enable reliable offshore turbines.

7.6   CONCLUSION

The last 30 years of investment in the wind industry have transformed this old technology into the fastest-growing clean energy source in the world. Through record setting years, wind energy deployment both worldwide and in the U.S. continues to show the effectiveness of policies and incentives, coupled with a clean, affordable, and reliable energy supply.

There is still an immense set of technology options that can potentially improve wind systems, such as condition health monitoring systems, distributed sensor networks, advance materials

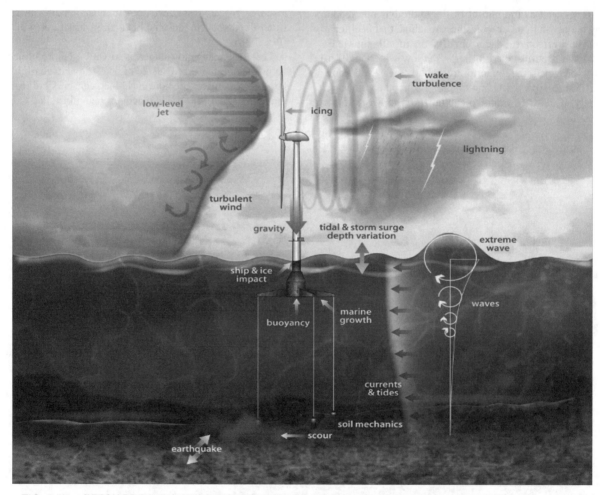

FIG. 7.35   OFFSHORE WIND TURBINE WITH COMPLEX DESIGN CONDITION SHOWN (Courtesy of NREL)

FIG. 7.36 OFFSHORE WIND FOUNDATION/PLATFORM DESIGNS (Courtesy of NREL)

options, such as nanoparticles to strengthen local areas, etc., but evaluating and balancing these technologies on an economic basis is the key. Sandia National Laboratories in conjunction with other labs, academia, and the industry will continue to explore these options in order to continue to innovate.

Like the leadership and direction that DOE's 20% by 2030 report outlined, there are several new ongoing studies targeted at evaluating and calculating the feasibility of large penetrations of renewable in the future energy mix, with all studies showing that wind energy will represent a significant percentage of the clean energy portfolio. Additionally, with current administration goals, which aggressively suggest up to 10% renewable energy by 2012 and 25% by 2025, it is important to understand that not only clean technology viability and feasibility is needed, but also a robust supply chain and manufacturing sector is imperative to meet these goals.

As we foresee, a future for the wind industry where there will be offshore and land-based machines available and installed globally, it is important to acknowledge that technology innovation will need to play a key role in making wind systems more reliable and cost-effective. As the leader of the clean energy portfolio, the wind industry must continue to find ways to improve the technology and pave the road for other upcoming technologies. It is hard to predict what the future energy picture will look like, but there is a high probability that if this industry continues to innovate, grow, and lead, it will have a key role in our energy future.

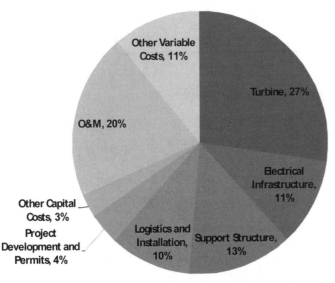

FIG. 7.37 OFFSHORE LIFE-CYCLE COST BREAKDOWN

7.7 ACRONYMS

AMI: Advanced Manufacturing Initiative
CFD: Computational Fluid Dynamics

COE:	Cost of Energy
DOE:	Department of Energy
FBG:	Fiber Bragg Grating
FEA:	Finite Element Analysis
GW:	gigaWatt
HAWT:	Horizontal Axis Wind Turbine
MIMO:	Multiple-Input Multiple-Output
MW:	megawatt
NDT:	Non-Destructive inspection Technology
NREL:	National Renewable Energy Laboratory
O&M:	Operations and Maintenance
SISO:	Single-Input Single-Output
SMART:	Structural Mechanical Adaptive Rotor Technology
SNL or Sandia:	Sandia National Laboratories
USDA:	United States Department of Agriculture
VAWT:	Vertical Axis Wind Turbine

7.8   REFERENCES

1. http://en.wikipedia.org/wiki/Betz%27_law

2. Gijs A.M. van Kuik, *The Lanchester–Betz–JoukowskyLimit*, Wind Energy Journal. 2007; 10:289–291

3. John F. Mandell and Daniel D. Samborsky, *DOE/MSU Composite Material Fatigue Database:Test Methods, Materials, and Analysis*, SAND97-3002 UC-1210, Sandia National Laboratories

4. H. J. Sutherland and John F. Mandell, *Application of the U.S. High Cycle Fatigue DataBase to Wind Turbine Blade Lifetime Predictions*, Energy Week 1996, Book VIII: Wind Energy, ASME, January-February, 1996, pp. 85-92.

5. http://www.awea.org

6. http://www1.eere.energy.gov/windandhydro/wind_2030.html

7a. http://www.risoe.dtu.dk/?sc_lang=en

7. Rush D. Robinett, III and David G. Wilson, *Maximizing the Performance of Wind Turbines with Nonlinear Aeroservoelastic Power Flow Control*, AWEA 2009

7. Dale E. Berg and Jose R. Zayas, *Aerodynamic and Aeroacoustic Properties of Flatback Airfoils*, AIAA Symposium 2008

8. Mark A. Rumsey and Joshua A. Paquette, *Structural Health Monitoring of Wind Turbine Blades*, SPIE Conference 2008

9. M. Barone, D. Berg, W. Devenport, and R. Burdisso, *Aerodynamic and Aeroacoustic Tests of a Flatback Version of the DU97-W-300 Airfoil*. SAND2009-4185, Sandia National Laboratories, August 2009.

10. S. Wagner, R. BareiB, G. Guidati, *Wind Turbine Noise*. Springer-Verlag, Berlin, 1996.

11. D. Berry, T. Ashwill, *Design of 9-Meter Carbon-Fiberglass Prototype Blades: CX-100 and TX-100*, SAND07-0201, Sandia National Laboratories, 2007

WIND ENERGY IN THE U.S.

Thomas Baldwin and Gary Seifert

8.1 INTRODUCTION

Humans have been harnessing the energy in the wind for several thousand years. Early uses included sailboats and windmills used to pump water and grind grain into flour — hence the term wind *mill*. In the 1800s, settlers in the western United States used wooden windmills to pump water, and many are still standing. In the late 1800s, the windmill was connected to an electric generator to produce electricity — hence, wind *turbine*.

In the 1970s and 1980s, wind turbines were clustered into wind farms and connected to the electric grid in California, which marked the first commercial, utility-scale use of wind energy. The size of those wind turbines were 100 kW and smaller. In the following decades, wind turbine technology progressed quickly, and by 2010, grid-connected wind turbines were typically in the 2-MW range and turbines as large as 6-MW have been deployed [17].

Adequate wind speeds are essential to the success of wind energy facilities. The potential energy in wind is determined as a function of the cube of the speed. Given the cubic relationship between power and wind speed, when wind speed doubles there is eight times more power available.

In 2009, wind supplied about 1.8% of the electricity in the United States [1]. However, experience shows that wind can provide up to one fifth of a power system's electricity, and that amount can increase with optimized system designs. For example, wind power currently provides more than 20% of the electricity distributed in northern Spain and in Denmark. A goal is to increase wind generationin the United States to 20% by 2030 [2].

Utility transmission lines carry wind-generated electricity from vast and sparsely populated areas where the wind is most abundant, like the Great Plains, to large cities where demand for electricity is high. At the moment, there is insufficient transmission infrastructure connecting the windiest parts of the country with large cities. Enhancing the transmission line power capacity for utility-scale wind plants is a key issue that must be resolved in the coming years [35].

Most regions of the United States are served by "power pools" of utilities that join together to generate electricity and transmit it to where it is needed. The name "power pool" is descriptive of the electric power coming from many different sources (a coal-fired power plant, a hydro plant, and others) flows into a "pool" from which it is distributed to thousands of end users. A power pool can easily absorb the electric power from a wind plant and add it to the generation mixture up to a penetration level of around 20%. Wind penetration greater than 20% requires advancements in transmission capacity, forecasting accuracy, and energy storage as addressed in the Wind Energy Research section. Wind plants could be installed in many locations, providing income, jobs, and electricity for homes and businesses.

Wind energy is a particularly appealing way to generate electricity because it is essentially pollution-free once the energy used to manufacture the turbine is offset. More than half of all the electricity that is used in the United States is generated from burning coal, and in the process, large amounts of toxic metals, air pollutants, and greenhouse gases are emitted into the atmosphere. Developing just 10% of the wind potential in the ten windiest U.S. states would provide more than enough energy to displace emissions from many of the nation's fossil-fuel power plants and offset their growth for many years.

Wind farms and plants can also revitalize the economy of rural communities, providing steady income through lease or royalty payments to farmers and other landowners. Although leasing arrangements can vary widely, in 2010, a reasonable estimate for income to a landowner from a single utility-scale turbine was approximately $4,000 a year, depending on the wind resource, the size of the turbine, and other factors. For a typical farm, income from wind comes with little interruption in farming activities with about one acre removed from agricultural production per turbine. Farmers can grow crops or raise livestock next to the towers. While wind farms may extend over a large geographical area, their actual "footprint" covers only a very small portion of the land, making wind development a cooperative way for farmers to earn additional income.

8.2 WIND TURBINE TECHNOLOGIES

Wind energy systems transform the kinetic energy of wind into mechanical and then into electrical energy, which is distributed for industrial, commercial, and residential use.

8.2.1 Types of Wind Turbines

There are two basic designs of wind electric turbines: horizontal-axis (propeller-style) machines and vertical-axis, or "egg-beater" style as shown in Fig. 8.1. The horizontal-axis wind turbines are commonly used in the United States, constituting nearly all of the "utility-scale" (100 kW, capacity and larger) turbines. The use of vertical turbines continues to lag at the utility scale because of the high cost to place the rotors into the higher wind speeds at elevations of 80 to 100 m above ground. However, use for residential units is gaining popularity with regular new entries in the market.

Wind turbines vary substantially in size. Table 8.1 depicts a variety of historical turbine sizes and the rated amount of electric power they are capable of generating. In 2010, offshore wind

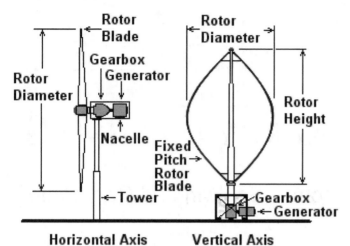

FIG. 8.1 TYPICAL WIND TURBINE CONFIGURATIONS

turbines operated at the 5-MW range with rotor diameters reaching 120 m. Technology developments foresee growth to the 10-MW range projected by 2015. Onshore turbines are operating up to the 3-MW range with 100 m rotor diameters.

8.3 WIND RESOURCES IN THE U.S.

There are three parts that make up a successful wind energy project. They include a wind resource, an energy sale contract, and a transmission/grid interconnection. Successful wind projects have tended to cluster along transmission grid corridors that have adequate wind. At the beginning of a project, a survey of wind resources must be conducted to ensure that a desired location has adequate energy supply. Wind resource maps, developed by the U.S. Department of Energy (DOE), provide general wind characteristics for areas of interest and are used to screen locations for suitable investigation. A review of these maps provides an initial decision point whether further exploration is warranted. Wind resource at the local level can vary significantly. Local wind measurements and evaluations should be conducted for the specific site of interest, a process known as micro-siting and validation. An evaluation characterizes the wind resource and predicts the turbine performance and economic feasibility.

The Wind Powering America organization also provides high-resolution wind maps and estimates of wind resource potential. The maps show the predicted mean annual wind speeds with a spatial resolution of 2.5 km for a height of 80-m above the terrain, the typical wind turbine hub height. Locations with annual average wind speeds greater than 6.5 m/s at the 80-m height are considered to have suitable wind resource for development of most areas.

The U.S. DOE's National Renewable Energy Laboratory (NREL) has identified practical locations for wind energy explora-

TABLE 8.2 DEFINITIONS OF WIND POWER CLASSES

Wind class	Description
1	$V < 5.9$ m/s
2	$5.9 \ V < 6.9$ m/s
3	$6.9 \ V < 7.5$ m/s
4	$7.5 \ V < 8.1$ m/s
5	$8.1 \ V < 8.6$ m/s
6	$8.6 \ V < 9.4$ m/s
7	$V \ 9.4$ m/s

tion. In the wind energy resource maps, wind velocity characteristics are categorized and presented in terms of wind classes. Table 8.2 lists the definitions of the wind classes, which range from class 1 (the lowest) to class 7 (the highest). Each class represents a range of mean wind power density or approximate mean wind speed at specified heights above the ground. Areas designated as class 4 or greater are suitable for most wind energy applications. Class 3 areas are transitional, where the power purchase costs become important. Class 2 areas are marginal and class 1 sites are generally not suitable for development.

NREL conducted a preliminary review and validated the AWS Truepower's (formerly AWS Truewind) 80-m map estimates for 19 selected states (6 Western states, 6 Midwestern states, and 7 Eastern states) using tower measurements at heights of about 50 m and greater from more than 300 locations.

8.3.1 Wind Resource Potential

In a joint effort between NREL and AWS Truepower of Albany, New York, a national dataset was produced of estimated gross capacity factors. These resource maps identify potential energy resources, both on and offshore. The data is interpreted to a spatial resolution of 200 m for heights at 80 m and 100 m. Wind power is greatly impacted by the turbine height, as illustrated by Table 8.3.

Mean wind speed is based on the Rayleigh speed distribution of equivalent mean wind power density. The power available in the wind depends on air density as well as wind speed. In Table 8.3, wind speed is reported for standard sea-level conditions. To maintain the same power density at higher elevations, wind speed must increase 3% per 1000 m (5%/5000 ft) of elevation gain.

Each wind power class is designed to span a range of power densities. For example, Wind Power Class 3 represents a wind power density range between 150 W/m^2 and 200 W/m^2. The offset cells in the first column of Table 8.3 illustrate this concept.

8.3.2 Potential Energy Supply for the U.S.

Useful wind energy resources for generating electricity can be found in nearly every state. In the 20% Wind Energy by 2030 report issued by the U.S. DOE in 2008, a scenario has been pre-

TABLE 8.1 CHANGES IN WIND TURBINE ROTOR DIAMETERS, POWER RATINGS AND ESTIMATED ENERGY OUTPUT OVER TIME

Typical dimensions	1981	1985	1990	1996	1999	2000	2010
Rotor diameter (m)	10	17	27	40	50	71	120
Power rating (kW)	25	100	225	550	750	1,650	5000
Annual energy (MWh)	45	220	550	1,480	2,200	5,600	17,500

TABLE 8.3 CLASSES OF WIND POWER DENSITY AT 10-M AND 50-M; VERTICAL EXTRAPOLATION OF WIND SPEED BASED ON THE 1/7 POWER LAW

Wind power class	10 m (33 ft)		50 meter (164 feet)	
	Power density [W/m^2]	Mean speed [m/s (mph)]	Power density [W/m^2]	Mean speed [m/s (mph)]
1	0	0	0	0
2	100	4.4 (9.8)	200	5.6 (12.5)
3	015	5.1 (11.5)	300	6.4 (14.3)
4	200	5.6 (12.5)	400	7.0 (15.7)
5	250	6.0 (13.4)	500	7.5 (16.8)
6	300	6.4 (14.3)	600	8.0 (17.9)
	400	7.0 (15.7)	800	8.8 (19.7)
7	1000	9.4 (21.1)	2000	11.9 (26.6)

sented by which 20% of the nation's electricity could be produced by wind energy by the year 2030. [2] By comparison, wind energy accounted for less than 2% of the nation's electricity in 2009 [1].

U.S. wind resource potential is significantly greater than 20% of U.S. demand in 2009. North Dakota alone is theoretically capable of producing enough wind-generated energy to exceed 25% of the nation's 2009 electricity energy demand (3,950 TW/h [3]). In 2010, however, electric transmission constraints would prevent much of that electricity from being distributed to load centers throughout the county; transmission will be discussed later in this chapter. The theoretical potentials of the windiest states are shown in Table 8.4.

These wind potential estimates resulted from a collaborative project between the NREL and AWS Truepower. This is the first comprehensive update of the wind energy potential by state since 1993. NREL has worked with AWS Truepower for almost a decade updating wind resource maps for 36 states and producing validated maps for 50-m height above ground level. U.S. DOE's Wind Powering America project supported the mapping efforts, and the results are publicly available. Table 8.4 indicates that the wind resources available in the United States vastly exceed current electric energy demand. As discussed later, careful selection of sites is essential to ensure adequate power generation.

Wind power facilities tend to develop in areas with adequate local transmission infrastructure and capacity factors of 35% or greater. Figure 8.2 indicates that much more wind power capacity could be installed if sites with capacity factors as low as 25% were developed.

The two lines shown in Fig. 8.2 plot the installed power capacity versus capacity factor at two different hub heights. As Table 8.3 indicates, taller hub heights would result in greater power capacity. Turbine manufacturers tend to design turbines for sites with capacity factors of 30% or greater. Research and design of lower-wind speed turbines would support the development of 25% capacity factor sites.

8.3.2.1 Wind Maps NREL provides both a national wind resource assessment of the United States and high-resolution wind data as a service to the public. The national wind resource assessment was created for the U.S. DOE in 1986 by the Pacific Northwest Laboratory and is documented in the Wind Energy Resource Atlas of the United States, October 1986, and subsequently updated with modern 80 m maps available on NREL and DOE websites. According to NREL:

The wind resource assessment was based on surface wind data, coastal marine area data, and upper-air data, where applicable. In data-sparse areas, three qualitative indicators of wind speed or power were used when applicable: topographic/meteorological indicators (e.g., gorges, mountain summits, sheltered valleys); wind deformed vegetation; and eolian landforms (e.g., playas, sand dunes). The data was evaluated at a regional level to produce 12 regional wind resource assessments; the regional assessments were then incorporated into the national wind resource assessment.

TABLE 8.4 THE TOP 20 STATES FOR WIND ENERGY POTENTIAL, MEASURED IN BILLIONS OF KWH, FACTORING IN ENVIRONMENTAL AND LAND USE EXCLUSIONS FOR WIND CLASSES OF 3 AND GREATER

Ranking	State	TWh/Yr	Ranking	State	TWh/Yr
1.	North Dakota	1,210	11.	Colorado	481
2.	Texas	1,190	12.	New Mexico	435
3.	Kansas	1,070	13.	Idaho	73
4.	South Dakota	1,030	14.	Michigan	65
5.	Montana	1,020	15.	New York	62
6.	Nebraska	868	16.	Illinois	61
7.	Wyoming	747	17.	California	59
8.	Oklahoma	725	18.	Wisconsin	58
9.	Minnesota	657	19.	Maine	56
10.	Iowa	551	20.	Missouri	52

Source: An Assessment of the Available Windy Land Area and Wind Energy Potential in the Contiguous United States, Pacific Northwest Laboratory, August 1991. PNL-7789.

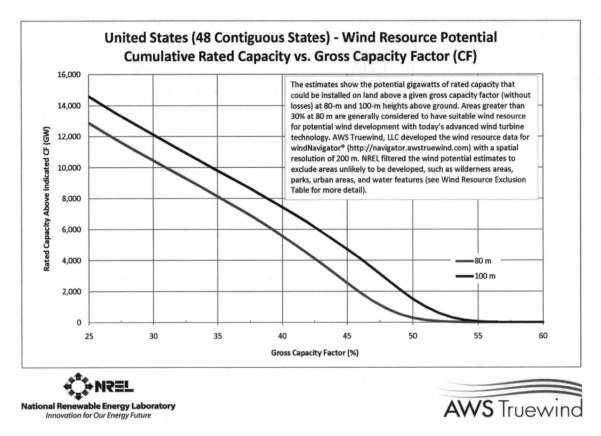

FIG. 8.2   U.S. CUMULATIVE RATED CAPACITY VS. GROSS CAPACITY FACTOR [5]

The conterminous United States was divided into grid cells ¼ degree of latitude by ⅓ degree of longitude. Each grid cell was assigned a wind power class ranging from 1 to 6, with 6 being the windiest. The wind power density limits for each wind power class are shown in Table 8.1. Each grid cell contains sites of varying power class. The assigned wind power class is representative of the range of wind power densities likely to occur at exposed sites within the grid cell. Hilltops, ridge crests, mountain summits, large clearings, and other locations free of local obstruction to the wind will be well exposed to the wind. In contrast, locations in narrow valleys and canyons, downwind of hills or obstructions, or in forested or urban areas are likely to have poor wind exposure.

Wind classes are defined in Table 8.2. The degree of certainty with which the wind power class can be specified depends on three factors: the abundance and quality of wind data; the complexity of the terrain; and the geographical variability of the resource. A certainty rating was assigned to each grid cell based on these three factors, and is included in the Wind Energy Resource Atlas of the United States [36] (Fig. 8.3).

8.3.2.2 Local Terrain Effects When using public wind maps, it is important to consider the effect of local terrain, which may not be well represented on such data sources. A site's topography affects its power density (W/m²), which will impact both the number of turbines that can be placed on a site and turbine spacing (see the Wind Farm Development section of this chapter). In areas with relatively flat terrain, such as the Great Plains, power density tends to be greater than in mountainous areas. In areas with complex, mountainous terrain, wind turbine placement is

typically limited to areas that are well exposed, such as ridgetops and tables. In heavily wooded areas, turbines are best placed in clear spaces with limited wind obstructions. In any region, measuring a potential site's *actual* wind resource (micro-siting) is an essential step when developing a wind farm and designing turbine arrays.

8.3.2.3 High-Resolution Wind Data High-resolution datasets are geographic shape-files generated from the original raster data. According to NREL, the original raster data varied in resolution from 200 m to 1000 m cell sizes and provide an estimate of annual average wind resource for specific states or regions. The data is separated into two distinct groups: NREL produced and AWS Truepower produced/NREL validated [36].

The NREL-produced map data only applies to areas of low surface roughness (i.e., grassy plains), and excludes areas with slopes greater than 20%. For areas of high surface roughness (i.e., forests), the values shown may need to be reduced by one or more power classes. The AWS Truepower-produced resource estimates factor surface roughness into their calculations and do not exclude areas with slopes greater than 20%. This data was produced in cooperation with U.S. DOE's Wind Powering America program and have been validated by NREL and other wind energy meteorological consultants. To download state wind resource maps, visit the Wind Powering America website and the Idaho National Laboratory's (INL) Virtual Renewable Energy Prospector (VREP) website. It is important to emphasize that all of these publicly available data is intended to be used as initial screening tools only. Site-specific measurements, as described in the Wind Farm Development section, are required to make sound decisions.

United States - Annual Average Wind Speed at 80 m

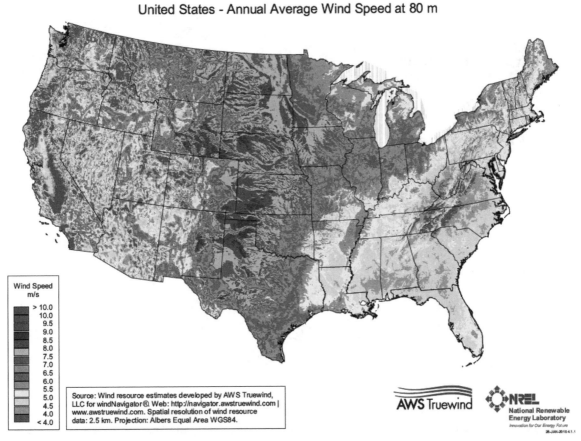

Wind Speed
m/s

> 10.0
10.0
9.5
9.0
8.5
8.0
7.5
7.0
6.5
6.0
5.5
5.0
4.5
4.0
< 4.0

Source: Wind resource estimates developed by AWS Truewind, LLC for windNavigator®. Web: http://navigator.awstruewind.com | www.awstruewind.com. Spatial resolution of wind resource data: 2.5 km. Projection: Albers Equal Area WGS84.

AWS Truewind

NREL
National Renewable Energy Laboratory
Innovation for Our Energy Future

FIG. 8.3   MAP OF U.S. WIND RESOURCE AT 80-M HEIGHT ABOVE GROUND LEVEL [6]

8.4   WIND PLANT ECONOMICS

From 1990 to 2010, the cost of electricity from utility-scale wind systems dropped by more than 80%. In the early 1980s, when the first utility-scale turbines were installed, wind-generated electricity cost as much as 30 cents per kilowatt-hour. Now, state-of-the-art wind power plants can generate electricity between 3 and 8 cents/kWh with tax incentives, a price that is competitive with and often lower than new coal- or gas-fired power plants.

The DOE is working with the wind industry to develop a next generation of wind turbine technology. The products from this program are expected to generate electricity at prices that will be lower still and allow expansion to lower wind speed requirements around the nation.

The most important factors in determining the cost of wind-generated electricity from a wind farm are (1) size of the wind farm; (2) wind speed at the site; (3) cost of installing the turbines; and (4) grid interconnection and transmission costs. Each of these factors can have a major impact. Generally speaking:

- The larger the wind farm, all other factors being equal, the lower the cost of energy;
- The higher the wind speed, the lower the cost of energy;
- The less expensive construction costs are, the lower the cost of energy;
- The closer to transmission, the lower the cost of energy.

On isolated ridgelines, for example, wind farms are likely to be smaller and cost more to install than in the flat terrain of northern Plains states. While larger wind farm power may cost less than 5 to 6 cents/kWh in the northern Plains, it may cost 6 to 7 cents/kWh in small sites.

In the case of offshore wind farms, the distance that power must be transmitted to shore and water depth are potentially significant cost elements, yet offshore farms can be significantly closer to population and load centers, changing transmission impacts.

8.4.1   Energy Subsidies and Incentives

In 2008, the Nuclear Energy Institute (NEI) asked Management Information Services Inc. (MISI) to prepare an independent assessment [7] of the amounts and types of federal incentives provided from 1950 to 2006 and the energy sources targeted with each type of incentive. As summarized in Fig. 8.4, the largest beneficiaries of federal energy incentives have been oil and gas, receiving more than half of all incentives provided since 1950. Carbon-emitting sources (oil, natural gas, and coal) combined account for approximately 73% of all incentives over this time period.

8.4.1.1 Wind Power Plants Cost Comparison to Other Renewable Energy Sources Wind is the lowest-cost renewable energy resourceto develop in the beginning of the 21st century and consistently ranks low in utility integration resource planning.

8.4.1.2 Production Tax Credit for Wind Energy A 1.5-cent per kilowatt-hour [1] production tax credit (PTC) for wind energy was included in the Energy Policy Act of 1992. Passage of the credit reflected recognition of the important role that wind energy

(Billions of 2006 Dollars[1])

TYPE OF INCENTIVE	ENERGY SOURCE							SUMMARY FOR INCENTIVE TYPE	
	Oil	Nat. Gas	Coal	Hydro	Nuclear	Renewable[2]	Geothermal	Total	Share
Tax Policy	173	88	31	12		20	2	326	45%
Regulation	116	3	7	5	11			142	20%
R&D	7	6	32	1	67	19	3	135	19%
Market Activity	5	2	2	59		2	2	72	10%
Gov't Services	31	1	14	1	1	2		50	7%
Disbursements	3		8	2	-14	2		1	~0%
Total	335	100	94	80	65	45	7	726	
Share	46%	14%	13%	11%	9%	6%	1%		100%

[1] All estimates quoted are in constant 2006 dollars, unless otherwise noted, and refer to actual expenditures in the relevant fiscal year, rounded to the nearest billion. Totals and percentages may differ slightly due to independent rounding.

[2] Renewables are primarily wind and solar energy sources.

FIG. 8.4   SUMMARY OF FEDERAL ENERGY INCENTIVES FROM 1950 THROUGH 2006

can and should play in the nation's energy mix. It also was intended to partially correct an existing bias of the federal energy tax code, which has historically favored conventional fossil energy technologies, such as oil and coal, as shown in Fig. 8.4.

Generally, the credit is a business credit that applies to electricity generated from wind plants for sale at "wholesale" (i.e., to a utility or other electricity supplier which then sells the electricity to customers at "retail"). It applies to electricity produced during the first 10 years of a wind plant's operation. The company that owns the wind plant subtracts the value of the credit from the business taxes that it would otherwise pay.

As part of the American Recovery and Reinvestment Act 2009 (ARRA), legislation (H.R. 1) expanded the eligibility criteria by removing the wind turbine size capacity limitation. This allowed all Renewable Electricity Production Tax Credit (PTC) eligible facilities to use the 30% Business Energy Investment Tax Credit (ITC) on any size of wind turbine project. The 30% credit is applied to the total project cost. Confirmation of the tax credits and their application to this specific project were confirmed by three methods. The following documents show the details of these different incentives:

Department of the Treasury Internal Revenue Service FORM 8835 [8] (Department of the Treasury, Internal Revenue Service, 2009),

Renewable Energy Production Tax, FORM 3468 [9] (Department of the Treasury, Internal Revenue Service, 2009),

Investment Credit Tax, FORM 3800 [10] (Department of the Treasury, Internal Revenue Service, 2009)

Under the ARRA program, the federal government will give a cash rebate for 30% of the cost of building a renewable-energy facility, awarded 60 days after an application is approved, which is contingent upon the first energy date. Investors are also given valuable accelerated depreciation deductions, which help offset taxes. Critics have used the government subsidies as a way to draw negative attention to renewable energy, but in a report by the Environmental Law Institute (ELI) for the years 2002 to 2008 the government provided more subsidies to fossil fuels than to renewable energy sources. Approximately $72 billion dollars in fossil fuel subsidies represented a direct cost to tax paying Americans, whereas only $29 billion were given to the renewable fuels. The numbers explain the need for government subsidies for all types of fuels. Fossil fuels have been well established as the prime energy generation source, and the incentives are written in as permanent provisions. Renewable fuels incentives are mostly time-limited and are not as useful or effective as they could be if they were permanent. On a final note from the study done by ELI, more than half of the subsidies for renewables are for the processing of corn into ethanol (Environmental Law Institute, 2009) [11].

8.4.1.3 Green Pricing Programs There are several reasons for the cost premium (typically 2 to 3 cents per kilowatt-hour) that most green marketers charge for wind-generated electricity. Green pricing varies from state to state and is regulated by state public utility commissions.

Assume wind energy costs 2 cents more per kilowatt-hour (2 cents/kWh) than the rest of the electricity a utility is generating or buying — a conservative estimate. If the utility were to decide to use wind energy to generate 10% of its electricity (more than nearly all utilities in the United States), then the added cost would be 0.2 cents/kWh. An average U.S. home uses about 800 kWh per month, so a ratepayer would pay an extra $1.60 per month or about a nickel a day.

At times when the prices of natural gas, oil, and other fossil fuels are increasing, such as 2005, wind energy is more economical. A study of wind integration into the New York State electric power system, looking at a 10% addition of wind generation (3,300 MW of wind in a 34,000-MW system), projected a reduction in payments by electricity customers of $305 million in one year. [12] Natural gas prices are lower in 2010 than they were in 2005, but utility companies are still interested in adding wind.

8.4.1.4 Energy Payback Time for a Wind Turbine The "energy payback time" is a term used to measure the net energy value between production and the construction of a wind turbine or other power source. It is expressed in the number of years required for a plant to operate in order to generate the total energy required for its manufacture, construction, operation, and retirement. Several studies over the years have examined this quantity and have concluded that wind energy has one of the shortest energy payback times of any energy technology. Wind turbines typically take between 3 and 8 months, depending on the local average wind speeds to reach the payback point. From this point forward, the wind turbine will generate emission-free electricity. By comparison, conventional fossil-fuel power plants require several years to pay back the infrastructure's embedded energy and continue to generate emissions for their entire life.

8.4.1.5 Cost of Non-Dispatchable Energy Some critics claim that since wind power cannot be dispatched to address corresponding load demand, a utility should have available equal proportions of power producing capability from dispatchable fossil fuel generation as connected wind power. For example, 100 MW of fossil-fuel generation is needed for every 100 MW of wind turbines to allow for the times when the wind is calm.

Utilities use complicated statistical models to determine the value in added energy and power capacities that each new generating plant adds to the system. According to those models, the energy capacity value of a new wind plant can be equal to the product of its capacity and capacity factor. Thus, adding a 100-MW wind plant with an average capacity factor of 35% to the system is approximately the same as adding 40 MW of conventional fueled generating capacity with an 87.5% availability factor.

The exact answer depends on, among other factors, the correlation between the time that the wind blows and the time that the utility sees peak demand and diversity of multiple wind farms and rate structures and agreement determined by the PUC/PSC. Thus, wind farms whose output is highest in the spring months or early morning hours will generally have a lower capacity value than wind farms whose output is high on hot summer evenings or cold winter days.

8.4.2   Turbine Performance

8.4.2.1 Capacity Factor The Capacity Factor is a commonly used metric employed in measuring the productivity of a wind turbine or any other power production facility. It compares the plant's actual production over a given period of time with the amount of power the plant would have produced if it had run at full capacity for the same amount of time.

$$\text{Capacity factor} = \frac{A}{B} \qquad (8.1)$$

where:

A = Actual amount of power produced over time.

B = Power that would have been produced if the turbine operated at maximum output 100% of the time.

A conventional utility power plant uses fuel, so it will normally run much of the time unless it is idled by equipment problems or for maintenance. A capacity factor of 60% to 95% is typical for conventional fossil fuel and nuclear power plants and 30% to 50% for hydropower plants [13].

A wind turbine's prime mover is the wind, which blows steadily at times and not at all at other times. Modern utility-scale wind turbines typically operate 65% to 90% of the time, often at less than full capacity. Therefore, annual capacity factors ranging between 25% and 40% are common, although they may operate at much higher capacity factors during windy times.

It is important to note that while capacity factor is almost entirely a matter of reliability and market demand for a fueled power plant, it is not for a wind plant. For a wind plant, it is a matter of economical turbine design and wind resource. With a large rotor and a small generator, a wind turbine could run at full capacity whenever the wind blew and have a 60% to 80% capacity factor, but it would produce very little electricity compared to investment. The optimal electricity per dollar of investment is gained by using a larger generator with high wind resource and optimizing the capacity factor and revenue. Wind turbines are fundamentally different from fueled power plants in this respect and their inability to be dispatched by grid operators.

8.4.2.2 Availability Factor The Availability Factor is a measurement of the reliability of a wind turbine or other power plant. It refers to the percentage of time that a plant is ready to generate (i.e., not out of service for maintenance or repairs). Modern wind turbines have an availability of more than 95%, higher than most other types of power plant.

8.5   TECHNICAL ISSUES

8.5.1   Grid Operations with Wind Power

Modern wind turbine systems can provide very good voltage regulations, but the output from the power plant cannot be controlled by grid operators, rather it is dependent on the environment.

Wind power fluctuations are stochastic in nature, but not completely random. Since the wind does not blow all the time a wind farm's contribution to a utility's generating capacity is the subject of debate.

Utilities must maintain enough power plant capacity to meet expected customer electricity demand at all times, plus an additional reserve margin. They schedule many resources, including wind, to meet those loads. All other things being equal, utilities generally prefer plants that can generate as needed (that is, conventional plants) to plants that cannot (such as wind plants).

However, despite the fact that the wind is variable and sometimes does not blow at all, wind plants do increase the overall statistical probability that a utility system will be able to meet demand requirements while reducing the cost of fossil fuel input. A rough rule of thumb is that the capacity value of adding a wind plant to a utility system is about the same as the wind plant's capacity factor multiplied by its capacity. Thus, a 100-MW wind plant with a capacity factor of 35% would be similar in capacity value to a 35-MW conventional generator. For example, in 2001 the Colorado Public Utility Commission found the capacity value of a proposed 162-MW wind plant in eastern Colorado (with a 30% capacity factor) to be approximately 48 MW [14].

The exact amount of capacity value that a given wind project provides depends on a number of factors, including average wind speeds at the site and the match between wind patterns and utility load (demand) requirements. It also depends on how dispersed geographically wind plants on a utility system are, and how well-connected the utility is with neighboring systems that may also have wind generators. The broader the wind plants are scattered geographically, the greater the chance that some of them will be producing power at any given time, increasing wind power's contribution to the electric grid.

8.6   ENVIRONMENTAL ISSUES

8.6.1   Wildlife Considerations

Birds occasionally collide with wind turbines, as they do with other tall structures such as buildings. Avian deaths have become a concern at Altamont Pass in California, which is an area of extensive wind development and also high year-round raptor use. Detailed studies and monitoring following construction at other wind development areas indicate that this is a site-specific issue that will not be a problem at most potential wind sites. However, it is an important issue which should be considered during planning and layout with site specific studies. Overall, wind turbines' impact on birds is low compared with other human-related sources of avian mortality. See "Avian Collisions with Wind Turbines," for more information. Figure 8.5 presents results of that study.

No matter how extensively wind is developed in the future, bird deaths from wind energy are unlikely to ever reach as high as 1% of those from other human-related sources such as hunters, house cats, buildings, and autos (house cats, for example, are believed to kill 1 billion birds annually in the United States alone). Wind turbine causes are proportionally insignificant. Areas that are commonly used by threatened or endangered bird species should be regarded as less suitable for wind development. The wind industry is working with environmental groups, federal regulators, and other interested parties to develop methods of measuring and mitigating wind energy's effect on birds and accommodating special issues.

Wind energy can also negatively impact birds and other wildlife by fragmenting habitat, both through installation and operation of wind turbines themselves and through the roads and power lines that may be needed. This has been raised as an issue in areas with unbroken stretches of prairie grasslands or of forests. More research is needed to better understand these impacts.

Bat collisions at wind plants generally tend to be low in number and to involve common species which are quite numerous. Human disturbance of hibernating bats in caves is a far greater threat to species of concern. Still, a surprisingly high number of bat kills at a new wind plant in West Virginia in the fall of 2003 has raised concerns, and research at that plant and another in Pennsylvania in 2004 suggests that the problem may be a regional one. The wind industry has joined with the U.S. Fish and Wildlife Service, the U.S. DOE's NREL, and Bat Conservation International to form the Bats and Wind Energy Cooperative (BWEC), which funded the 2004 research program and is continuing to explore ways to avoid or reduce bat kills.

8.6.1.1 Additional Information Sources

- Comparative Impacts of Wind and Other Energy Sources on Wildlife
- Avian Collisions with Wind Turbines: A Summary of Existing Studies and Comparisons to Other Sources of Avian Collision Mortality in the United States
- Wind Turbine Interactions with Birds and Bats: A Summary of Research Results and Remaining Questions, National Wind Coordinating Committee.

8.7   RADAR IMPACTS

In the first decade of the 21st Century, deployment of wind power systems around the world grew significantly. This growth has led to many interaction issues between wind farms, military, aviation and weather radar systems. The UK RAF issued three studies in 2005, which illustrated new interactions and the need for additional review. The U.S. government is also concerned about the same issues and initiated a military readiness review in 2006, and issued interim policies that had the unfortunate effect of delaying over 4 billion dollars of wind projects in 2005 and over 600 billion since. Diligent efforts reduced this impact significantly and established new and reasonable policies.

Mitigation and coexistence is the goal for wind energy developers, yet defense entities are focused on eliminating wind radar interaction, creating impasses in many locations. Many mitigation methods have been successfully implemented, including wind farm/turbine site adjustments, radar (software or hardware) upgrades, and improved assessments. Early communication between affected parties in a wind farm development is a key to identifying possible problem areas and solutions to those problems.

The use of wind farms in the United States to generate electricity has increased dramatically from 2000 to 2010 [17]. Such installations can have over 100 turbines with blade-tip heights over 120 m (400 ft) above ground level (AGL). Blade-tip heights of over 180 m (600 ft) AGL are expected within a few years. Building a wind farm involves many considerations, including consultation with various aviation interests, both civil and military, and also with weather radar installations. These parties may raise objections to a proposed wind farm for a variety of reasons.

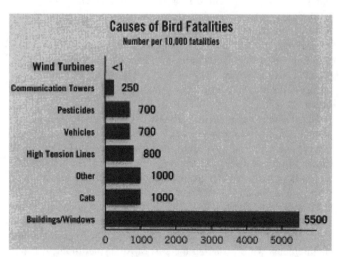

FIG. 8.5   CAUSES OF BIRD FATALITIES [15]

Data from the national network of Weather Surveillance Radar-1988, Doppler (WSR-88D) systems are a key component in the decision making process of issuing weather forecasts and severe weather warnings, and supporting air traffic safety. Experience has shown that when wind farms are located "close" to weather radar systems, the turbine towers, rotating blades, and the wake turbulence induced by the blades negatively impact data quality and can degrade the performance of radar algorithms particularly due to Doppler return from the turbine blades. Another example, the wind turbines might appear as echoes on the display of radar used in air traffic control (ATC) and reduce radar sensitivity near and over wind farms. These echoes may distract the air traffic controller from the aircraft echoes which are his main interest and can reduce the effectiveness of the radar by masking genuine aircraft returns.

Recognizing the radar concerns relating to the Wind Radar Issue is the first step. Supporting a resolve to such concerns as quickly as possible is the next step. Protecting our borders, to provide air traffic support, forecasting the weather, and preceding with rapid development of domestic sustainable energy resources are all important. They are all vital to national security. The wind industry is in a confusing and uncertain position because it is unclear how many projects will be affected by Wind Radar. The broad-brush approach being taken by some stops all development over a multi-state region pending outcome of a study and often downplays industry and military operational experience. A number of U.S. government installations have both wind turbines and functional radar co-existing, and the British military has a track record of successfully addressing this challenge. Policies on wind turbine impacts to radar should explore the solutions used elsewhere in the world and look at ways of mitigating the problem and developing new technologies, rather than limiting wind development in large areas. Wind turbines are installed at U.S. Air Force bases and near airports in the United States and elsewhere, and the experience at those sites demonstrate this is a solvable challenge. Several radar manufacturers are confident that the majority of radar systems can perform well in presence of wind turbines, albeit with some optimization, modifications, and upgrades. Impacts, costs of mitigation technologies and DOD and FAA guidelines should be able to help streamline the development and integration processes and integrate correction technologies. Systems updated to ASR-9 have had good experience mitigating wind systems and others are having good luck with their ASR-11 radar in presence of turbines.

The DHS and DOD have contested many proposed wind turbines, which stalled development of several thousands of MW of wind energy nationwide (billions of dollars of investment). The large number of denials is a serious impediment to the nation's growth of sustainable energy.

Technology can help, as Radar is basically designed to filter out stationary objects and display moving ones, and moving wind turbine blades create radar interference. It is possible to modify a radar installation to eliminate this problem, according to a consulting firm that has studied it for the British government [16]. That study concludes that " . . . radars can be modified to ensure that air safety is maintained in the presence of wind turbine farms. Individual circumstances will dictate the degree and cost of modification required, some installations may require no change at all while others may require significant modification."

If a wind project is proposed near an airport or military airfield, this issue will likely require further technical investigation. The interference is generally limited to objects (airplanes) that are physically obscured over and nearby turbines. Radar

manufacturers are aggressively upgrading the radar system to allow discrimination between airplanes and turbines and foster co-existence.

8.8   LOCAL IMPACTS

Local opposition to proposed wind farms sometimes arises because some people perceive that the development will spoil the view that they are accustomed to. It is true that a large wind farm can be a significant change, but while some people express concern about the effect wind turbines have on the beauty of our landscape, others see them as elegant and beautiful, or symbols of a better, less polluted future.

The visual effect of wind farms is a subjective issue. Most of the other criticisms made about wind energy today are exaggerated or untrue and simply reflect attempts by groups to discredit the technology, worry local communities, and turn them against proposed projects. In the electronic age, myths and misinformation about wind power spread at lightning speed and due diligence are needed to dispassionately inform decision makers of the facts.

8.8.1   Visual Impacts

Visual impacts can be minimized through careful design of a wind power plant. Using turbines of the same size and type and spacing them uniformly generally results in a wind plant that satisfies most aesthetic concerns. Computer simulation is helpful in evaluating and demonstrating visual impacts before construction begins. Public opinion polls show that the vast majority of people favor wind energy, and support for wind plants often increases after they are actually installed and operating. The bibliography references [30, 31], and [32] provide more information on public attitudes toward wind.

Shadow flicker is occasionally raised as an issue by close neighbors of wind farm projects. A wind turbine's moving blades can cast a moving shadow on a nearby residence, depending on the time of the year (which determines how low the sun is in the sky) and time of day. It is possible to calculate very precisely whether a flickering shadow will in fact fall on a given location near a wind farm, and how many hours in a year it will do so. Therefore, it is easy to determine the time and duration of the flicker. Normally, it should not be a problem in the United States, because at U.S. latitudes (except in Alaska) the sun's angle is not very low in the sky. Further, the appropriate setbacks for noise will be sufficient to prevent shadow flicker problems from becoming longer-term impacts.

8.8.2   Acoustic Noise

Impacts and levels of audible noise emanating from a wind turbine depend on its surroundings and proximity to civilization.

Noise was an issue with some early wind turbine designs, but it has been largely eliminated as a problem through improved engineering and through appropriate use of setbacks from nearby residences. Aerodynamic noise has been reduced by changing the thickness of the blades' trailing edges, adjusted tip speed and by making machines "upwind" rather than "downwind" so that the wind flows through the rotor blades first, then the tower (on turbines with the rotor "downwind" of the tower, a thumping noise occurs each time a blade passes behind the tower). A small amount of noise is generated by the mechanical components of the turbine. To put this into perspective, a wind turbine 300 m away is no noisier than the reading room of a library.

8.9 ADDRESSING NEEDS FOR WIND TO REACH ITS FULL POTENTIAL IN THE U.S.

8.9.1 Consistent Policy Support

Since 1999, the federal production tax credit has been extended four times, but three times (2000, 2002, and 2004) Congress allowed the credit to expire before acting, and then only approved short durations. Lapses in the PTC have a significant effect on wind turbine installations as evidenced in Fig. 8.6. These expiration-and-extension cycles inflict a high cost on the industry, cause large lay-offs, and hold up investments. Long-term, consistent policy support would help unleash the industry's pent-up potential.

8.9.2 Electric Transmission

8.9.2.1 Nondiscriminatory Access to Transmission Lines To provide a stable electric power grid, system operators must maintain the balance between power consumption by customer loads and power production by the generation facilities. For each hour of the day, the operators dispatch a sufficient amount of generation to meet the expected load. Transmission grid operators typically charge generation providers substantial penalty fees if they fail to deliver electricity when it is scheduled to be transmitted. One purpose of these fees is to deter power providers from using transmission scheduling as a "gaming" technique to raise spot prices or gain advantage against competitors.

Because the wind is variable, wind plant owners naturally cannot guarantee delivery of electric power for transmission at a scheduled time. Wind energy needs a modified policy that recognizes the different nature of wind plants and allows these plants to compete on a fair basis.

8.9.2.2 New Transmission Lines The transmission grid of the wind-rich Great Plains, which covers the central one third of the United States, was designed to meet the power needs of a less densely populated region. In order to capture and transmit the wind energy, extensive grid develop is required. At present, this system consists mostly of smaller-power rural transmission service. A series of extra-high-voltage transmission lines is needed

to transportpower from wind plants to population centers. Such a redevelopment will be expensive, but it will also benefit consumers and national security by making the electrical transmission system more reliable, reducing dependence on fossil energy, and reducing shortages, sensitivity and price volatility due to natural gas and other fossil fuels. National energy policy addressing transmission grids will be a key component for the development of the wind industry's future over the next two decades.

Wind is an intermittent energy source and therefore adds complexity for grid operators. At current levels of use, wind integration has not yet become a significant issue on most utility systems. Current best practice is to permit wind production to operate up to the point where wind generates about 10% of the electric power that the system is delivering in any given hour of the day. At this operating level, there is enough flexibility in the form of spinning reserve built into the system for generation reserve backup, varying loads, etc. Studies show, that there is effectively little difference between a system with 10% wind production and a system with 0% wind generation. Observations of actual wind generation at 10% indicate that the power variations introduced by wind are smaller than routine variations in customer load demand.

At the point where wind is generating 10% to 20% of the power it can become a larger concern for grid operators. Significant improvement can be achieved with wind forecasting and spatial diversity. Both are the subjects of current research and involve the application of statistical tools. Concepts of an energy grid overlay to the power grid are possible solutions to creating a virtual super-wind plant where the power production across vast distances is integrated together to form very predictable wind power generation.

Once wind is generating more than about 20% of the electricity, the system operator can incur significant additional expense because of the need to procure additional equipment that is solely related to the system's increased variability and/or managing load differently.

These figures assume that the utility system has an "average" resource portfolio that is complementary to wind's variability (e.g., hydroelectric dams and natural gas peaking plants) and an "average" amount of load that can vary quickly (e.g., electric arc furnace steel mills). Actual utility systems can vary quite widely in their ability to handle as-available output resources like wind farms, especially when they are part of a larger grid

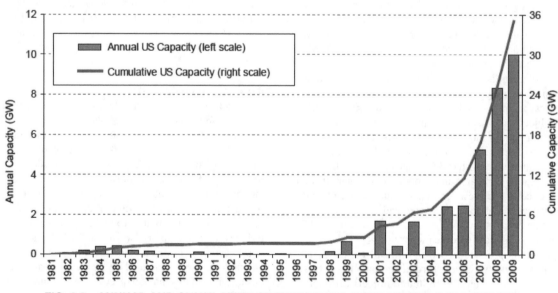

FIG. 8.6   ANNUAL AND CUMULATIVE GROWTH IN U.S. WIND POWER CAPACITY [17]

system. However, as wholesale electricity markets grow, fewer, larger utility systems are emerging. Therefore, over time, more and more utility systems will be able to integrate larger ratios of renewable power.

For detailed information on this topic, see "Grid Impacts of Wind Power: A Summary of Recent Studies in the United States," Milligan et al, NREL [33].

Industry and utilities note that one of the biggest constraints to the expanded growth of wind power in the United States will be the ability of the transmission grid to deliver large amounts of wind energy to customers. AWEA, DOE, and WPA work together to transform electric industry practices, including the area of transmission, through technology, planning advocacy and outreach.

Transmission is an important factor in the development of wind energy because some of the best wind resources in the country are located in areas remote from the large electric markets. By expanding and upgrading transmission systems, the nation could have better access to wind energy, which could be more easily moved from distant areas to population centers where electricity demand is greatest. Moreover, by facilitating the expansion and geographical dispersion of wind power across a wide area, an upgraded transmission grid improves both the reliability and capacity of wind. When the wind is not blowing at one location, it is usually blowing somewhere else. Thus, dispersed wind power compensates for short-term fluctuations and increases capacity value.

Electric utilities originally developed their electric transmission grids over time to connect local load with local generation, and then gradually those local utility grids were interconnected in furtherance of reliability benefits, into a "national electric grid." That national grid was never intended to function, and never has functioned, to move truly large amounts of power from one region of the country to another. America's electric grid could be designed to function much like the interstate highway system functions to efficiently transport goods across the country. This "electric superhighway" would allow low-priced clean energy to reach consumers across the country. Such a national grid would require state, regional, and national rethinking in terms of transmission line construction processes, yet provide vastly improved transfer capability.

8.10   WIND FARM DEVELOPMENT

To be viable, a candidate wind plant site needs to have the following components:
- Favorable economics
  - o Ability to generate income
    - Sufficient wind resource
    - Buyer/PPA
  - o Reasonable construction costs
    - Roads
    - Cost effective turbine
    - Interconnection fees
- Transmission/path to market
  - o Proximity to transmission lines
  - o Cost effective interfacing substation
- To be legally permissible
  - o Siting/permitting
    - Environmental impacts
    - Radar impacts
    - Local concerns
- Access to financing

Typically, mechanical engineers are not involved in all aspects of these steps, but need to be aware of them in order to properly assess a candidate site's viability. A given site may have an excellent wind resource, but other factors (i.e., if it is near an airport or lacks adequate transmission) may prevent its project development. The following sections address the project development steps typically handled by engineers.

8.11   WIND RESOURCE ASSESSMENT

A site's wind resource is a key factor in determining the ability to generate income. There are many steps involved in resource assessment, but the end goal is estimating how much annual energy (MWh) can be generated at a site. Wind resource metrics include average wind speed at a certain height, frequency distribution of wind speeds at the site, and wind shear. Each metric will be discussed individually.

8.11.1   Power Available in the Wind

Average wind speed is one of the most frequently discussed metrics of a farm or plant site. Because it is a single number, it is easy to discuss and is the metric used to define wind classes (refer to Table 8.1). However, relying on a site's average wind speed can be very misleading because of the cubic relationship between instantaneous wind speed, instantaneous power available, and the total annual energy production. It is important to understand the nature of the power available in the wind.

The following equation describes the amount of power, P, available in the wind as a function of air density, cross-sectional or "swept" area, A, of the rotor, and the wind speed, v.

$$P = \frac{1}{2}\rho A v^3 \tag{8.2}$$

Note that this equation does not describe the power delivered to the grid. To calculate grid power, one must account for generator efficiency and line losses. The total available energy is obtained from the time integral of the power.

$$E = \int P(t)\mathrm{d}t \tag{8.3}$$

Consider for example the impact of the wind characteristics on the annual energy available at two sites having the same annual average wind speed, but very different distribution of wind speeds. It is assumed that each project has four discrete wind speeds evenly distributed over the year (a quarter of the time is equal to 2190 hours). The site data is provided in Table 8.5. For both sites, the simplified wind speed distributions yield the same annual average wind speed of 10 (m/s). However, Site A's average cubic wind speed, which is proportional to power, is 2020 (m/s)³, nearly double that of Site B. If one assumes the turbines placed at each site have equal rotor diameters (i.e., equal swept area) and each site experiences the same air density over the course of a year, then site A will generate approximately double the energy that Site B generates. This over-simplified example demonstrates the importance of a site's wind speed frequency distribution in addition to its average velocity and the importance of having actual long-term wind speed data to assist in decisions and optimization.

8.11.2   Wind Speed Frequency Distribution

The best way to determine a site's wind speed frequency distribution is to measure the wind speed over the course of a full year at

TABLE 8.5   AVERAGE WIND SPEED

Site A			Site B		
Time (hours/yr)	Wind speed (m/s)	Wind speed cubed (m/s³)	Time (hrs/yr)	Wind speed (m/s)	Wind speed cubed (m/s)³
2190	2	8	2190	8	512
2190	8	512	2190	9	729
2190	12	1728	2190	11	1331
2190	18	5832	2190	12	1728
Averages	10	2020	Averages	10	1075

a minimum. In professional practice, many investors are requiring wind sampling over the course of 2 years to gain more confidence in a site's wind resource. Anemometers placed on meteorological towers have been the most common instrument used to collect wind speeds up to the early part of this decade. Sonar and laser type instruments capable of measuring wind speed and direction at various heights without use of tower (SODARs and LIDARs) have become more widely used since 2005.

Regardless of which instrument is used to measure wind speed and direction, data collection is a fairly uniform process. Most data loggers used in 2010 sample wind speed and direction every 2 seconds and report wind speed and direction in 10-minute averages. Wind resource engineers use this data to develop wind speed frequency distribution curves similar to the one shown in Fig. 8.7.

The wind speed frequency distribution can be used to estimate the annual energy (kWh) production as plotted in the figure. Although 12 mph wind is the most frequent, occurring at 6% of the time, the 22 mph winds generate the greatest amount of energy annually. This reflects the cubic relationship between wind speed and power.

8.11.3   Wind Shear

As with other flows, wind blowing across land has velocity profiles. In general, wind nearest the ground has the least velocity and wind at higher levels above the ground has the greatest velocity

(i.e., free stream). The variation of wind speeds at different heights is known as wind shear.

Wind shear is an important factor in determining the optimal tower height for wind turbine performance. To optimize tower height, two competing factors are considered. First, in general, wind speeds increase with height above the ground, which can lead to greater energy production and ultimately greater income. Conversely, taller towers increase a project's capital and operations and maintenance (O&M) costs. Increased capital costs include the cost of the towers themselves, transportation costs to the site, taller cranes required to erect the turbines, stronger/larger foundations, and roads suitable to carry the loads. O&M costs increase because of the increased nacelle height, which will affect everything from routine maintenance (due to the increased time and energy required for technicians to gain access) to replacing large components, especially blades and gearboxes.

In order to optimize tower height, a clear understanding of the site's wind shear is crucial. Experience has shown that wind shear varies daily, seasonally and with changes in wind direction (especially for sites with mountainous topography). The best way to understand wind shear is to directly measure wind speed and direction at multiple heights, preferably including the hub height and rotor diameter. Figure 8.8 shows the hub height, top and bottom of the rotor of a Suzlon S-88 wind turbine with an 80-m tower and 88-m rotor diameter.

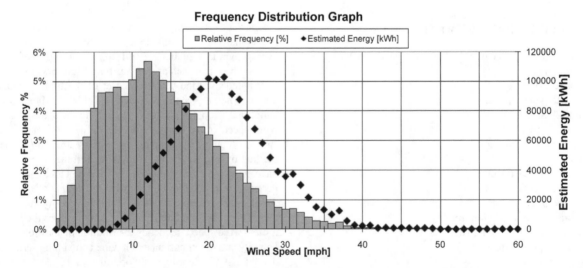

Frequency Distribution Graph

FIG. 8.7   FREQUENCY DISTRIBUTION GRAPH OF WIND SPEEDS COLLECTED OVER A 5-MONTH PERIOD IN 2009 FROM A SITE IN IDAHO. THE X-AXIS INDICATES WIND SPEED. THE BARS INDICATE PERCENT OF TIME THAT A GIVEN WIND SPEED OCCURRED DURING THE PERIOD (LEFT VERTICAL AXIS). THE BLACK DIAMONDS INDICATE ENERGY (RIGHT VERTICAL AXIS) [18]

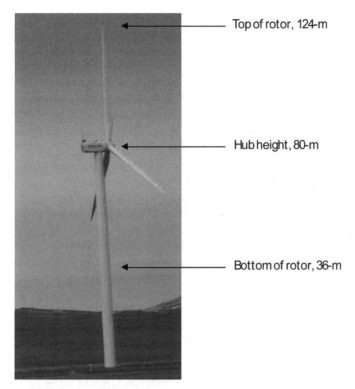

Top of rotor, 124-m

Hub height, 80-m

Bottom of rotor, 36-m

FIG. 8.8 WIND TURBINE NEAR MOUNTAIN HOME, ID (Photo Courtesy of Todd Haynes, 2008)

In practice, few meteorological towers are as tall as wind turbine towers (i.e., they cannot measure wind speed at hub height). As a result, wind speeds are measured at various heights (typically 10-m intervals) along the met-towers. The measured wind speeds are used to create a velocity profile which can be extrapolated to the turbine's hub height and maximum rotor height. Meteorologists have developed relationships to estimate wind speed at one height given a known wind speed at another height. In North America, the power law is most often used.

$$\frac{v}{v_0} = \left(\frac{h}{h_0} \right)^a \qquad (8.4)$$

In the power law, v_0 is a known velocity at height v_0 and v is an unknown velocity at height h. The exponent is the wind shear exponent. Under very precise meteorological conditions, the wind shear exponent has been shown to be $\alpha = 1/7$. In practice, these conditions are not often present. Inexperienced wind resource engineers and developers will assume that $\alpha = 1/7$ in the absence of measured data. This can lead to inaccurate estimates of annual energy generation, and thus inaccurate estimates of income potential. Wind resource engineers should exercise caution when estimating wind shear, and reminded that for a given site, wind shear will vary under different atmospheric conditions. The best way to know a site's wind shear is to measure it. A discussion of various instruments available follows in the Micro-siting section.

8.11.4 Prospecting

Given the costs associated with installing met-towers or other instrumentation, wind farms developers typically engage in "prospecting" sites before actually measuring data. Prospecting involves collecting cursory information in a low cost manner to determine if a site warrants a more detailed resource assessment.

During the 1990s and 2000s, the amount and quality of public data available for prospecting has increased considerably. Low-tech prospecting may be as simple as observing tree growth. If the wind blows consistently from one direction, a tree will "flag" as it grows. That is, the trunk will bend away from the prevailing wind, and the branches will tend to grow on the downwind side as shown in the photograph of Fig. 8.9.

There are many sources of public information that can be useful for prospecting. The U.S. DOE's Wind Powering America (DOE–WPA) program has produced publicly available wind maps for the nation (refer to Fig. 8.3) and many individual states.

In addition to wind maps, DOE-WPA initiated a state anemometer loan program in the early 2000s. In this program, DOE-WPA provided dozens of met-towers to participating states. The met-towers are loaned on an annual basis to interested landowners — both public and private. If the site shows promise, it is possible for the tower to remain a second year and continue collecting data. If not, the tower is removed and installed at another location in the state. DOE-WPA does charge the landowners to participate, but requires that the data collected be made publicly available for the common good. More than 20 states participate in the anemometer loan program.

Much of this data has been collected on 20-m and 30-m towers, which are considered short by 2010 wind resource assessment standards. These data are not meant to be relied upon for wind project development. Instead, they are meant to be indicators of sites which show sufficient promise that a developer or other interested party might invest in installing taller met-towers or other instrumentation. State of the art wind resource instrumentation is discussed in the Micro-siting/Data Collection section. Links to the publicly available met-data can be found by clicking on the respective states at the website as shown in Fig. 8.10.

For example, by clicking on the Idaho link on the website leads to websites hosted by the INL and Boise State University, both of which contain further links to wind data collected at individual sites in Idaho. Other states with anemometer loan programs (shown in Fig. 8.10) have similar public data sources. The Idaho National Lab map contains data going back to the program's

FIG. 8.9 GERALD FLEISCHMAN, WIND ENERGY ENGINEER AT THE IDAHO OFFICE OF ENERGY RESOURCES, STANDS NEAR A FLAGGING TREE (Photo Courtesy of Kurt Myers, 2002)

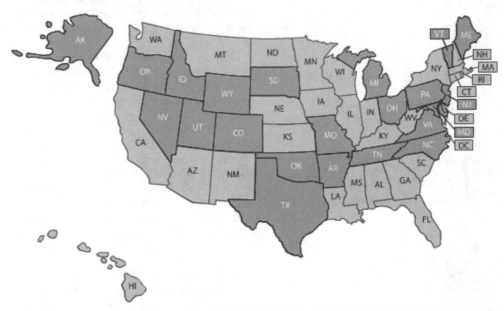

State Anemometer Loan Programs
The U.S. map below shows which states have anemometer loan programs. Click on a state to read more about a program and eligibility.

FIG. 8.10 MAP OF STATES WHICH PARTICIPATE IN THE DOE-WPA ANEMOMETER LOAN PROGRAM [19]

inception in 2002; see Fig. 8.11. The Boise State University College of Engineering became involved in 2008 and contains data collected and analyzed since that time. By clicking on one of the dots on the map at the website, the user will find raw data and summary data for that site, similar to the data shown in Fig. 8.12.

8.11.5 Micro-siting

Once a candidate site has been identified through the prospecting process, the next step is to collect "investment grade" data. This data is typically collected with a tall tower (60-m or greater) or a sensor which measures multiple heights without a tower (SODAR/LIDAR).

8.11.5.1 Met-Towers In 2010, the most commonly installed met-towers are 60-m tall. Taller towers are available, but it can be difficult to obtain permits for towers taller than 200 ft (60 m) because of FAA regulations regarding lights and the can be cost prohibitive. Regardless of height, met-towers typically contain multiple instruments. This can include anemometers at 10-m height intervals, a direction vane and temperature sensor.

Most counties in the U.S. require that met-towers taller than 33 ft (10-m) obtain conditional use or special use permits. Additionally, in 2010, the wind industry was made aware of concerns from crop dusters over met-towers being installed on or near farmland. These issues typically do not prevent met-towers from being installed. They are addressed here to alert the reader to be cognizant of the need to inform appropriate officials, especially since most met-towers do not have lights at the top.

Many companies manufacturer met-towers in North America. Typical packages include the tower, appropriate instrumentation and a choice of data loggers. A typical met-tower is shown in Fig. 8.13.

Depending on the geographic footprint of the project being considered, more than one met-tower may be necessary. This is especially true in complex (i.e., mountainous) terrain. When placing multiple met-towers, it is advisable to locate some towers at points

expected to have good wind (i.e., ridge tops or other high points) and others at places which are expected to have less wind. This will enable the wind resource engineer to create a more realistic model of the wind speeds across the site than would be the case if measurements were only taken on ridge tops.

8.11.5.2 SODARs and LIDARS In addition to MET-towers, other instruments have gained popularity in the wind resource assessment field over the last decade. Sonic Detection and Ranging (SODAR) instruments use sound to measure wind speed and direction, taking advantage of the doppler shift. Light Detection and Ranging (LIDAR) instruments use light to measure wind speed and direction. Both SODARs and LIDARs are able to measure wind speed and direction at multiple height intervals without using a tower. This allows an engineer to collect measurements higher above the ground than can be done with most towers, which can lead to more accurate estimates of wind shear at a given site. In practice, SODARs and LIDARs are often used in conjunction with met-towers to create very high resolution wind maps of a potential site. Figure 8.14 shows a typical SODAR setup.

SODARs and LIDARs have both advantages and disadvantages as compared to met-towers. Advantages include the ability to measure at more heights as already discussed. Another advantage is the fact that they can be moved to different locations much easier than towers. One disadvantage to these technologies is that they tend to cost more than towers. However, the costs are coming down and nearing convergence with tall towers in 2010. Another disadvantage to SODARs is the noise emitted. They are generally not suitable for placement near residences or places of business. This is often not an issue with wind resource assessment, because wind farms also tend to be located away from homes and businesses.

8.11.6 Data Analysis

Data is typically collected at 2-second sample rates and averaged over 10-minute intervals. This data is used to create frequency distribution curves (similar to Fig. 8.7), calculate a sites average

Idaho National Laboratory and Boise State's Wind Application Center have partnered together to provide data collected and analyzed across the state. For a complete data set, check both INL's Idaho Wind Data site and Boise State's Idaho Wind Data site.

Please click Boise States Logo to visit their Wind Data:

FIG. 8.11 MAP OF INDIVIDUAL ANEMOMETER LOAN PROGRAM SITES IN IDAHO. IDAHO NATIONAL LABORATORY AND BOISE STATE UNIVERSITY ARE PARTNERS WHICH ANALYZE AND POST THESE DATA [20]

wind speed, and also to create wind roses. A wind rose is helpful in determining the orientation of a wind turbine array. It indicates the direction of the prevailing wind(s), the percentage of time that wind comes from any particular direction and the percentage of energy associated with a wind direction. The site represented in Fig. 8.15 has two prevailing wind directions: the northwest and the southeast. It is interesting to note that although the wind comes from the northwest much more frequently than the southeast (the

green slice), there is only slightly more energy from the northwest than from the southeast. This suggests that southeasterly winds are typically higher velocity than those from the northwest.

8.11.6.1 Data Analysis Tools There are many software packages available to analyze wind data and create frequency distribution curves, wind roses and the like. Some are offered by the met-tower manufacturers. Others are offered as services along with data

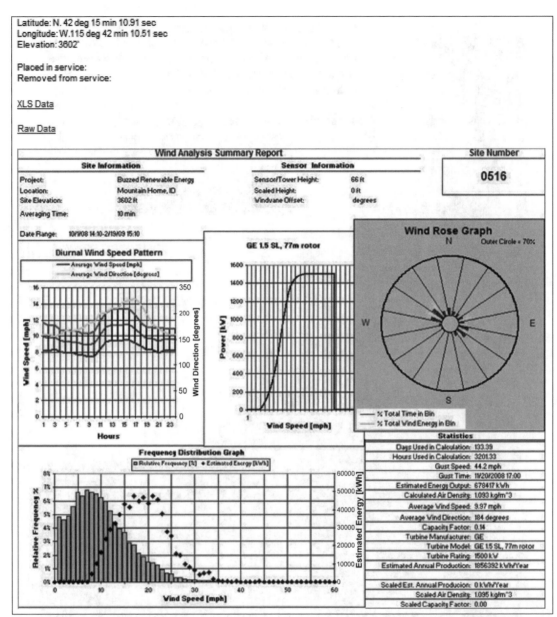

FIG. 8.12 EXAMPLE OF THE SUMMARY SHEET AND LINKS TO ANALYZED (XLS) AND RAW DATA AVAILABLE FROM THE ANEMOMETER LOAN PROGRAM [21]

loggers. Some are similar to Computational Fluid Dynamics (CFD) software codes in that they use measured wind speeds at one or more locations on a site to create a high resolution wind map of the candidate site. Some create visualizations of how various wind turbines will look from a distance when placed at a given site. The sophistication of these packages in increasing rapidly in 2010 and wind developers and engineers are encouraged to become familiar with several options to determine which will best suit a project's needs.

The U.S. DOE's INL has created free data analysis software tools which are based in MS Excel. This software allows a user to input wind data which has been measured at a site (regardless of the instrument). It contains power curves for a variety of commercial wind turbines. The user is able to compare the performance of a variety of wind turbines for a given site. Figure 8.12 was created using the INL software, which is available for download at bibliographic reference [34].

8.12 WIND FARM DESIGN

After a candidate site is deemed viable in the wind resource assessment process, the design process may be initiated. There are many steps involved in the design of a wind power facility. The steps below are typically handled by a combination of mechanical, electrical and civil engineers.

8.12.1 Turbine Selection

At any given site, different turbines will perform differently. It is typically the job of a mechanical engineer to select an optimal turbine given a site's wind speed and turbulence characteristics. The selection process consist of (i) the selection of the number and type of turbine, including the manufacturer and power capacity, (ii) the tower height, and (iii) the blade length. The cost optimization process considers both income from energy generation and the costs for construction, operation, and maintenance.

FIG. 8.13   A 50-M MET-TOWER IN SOUTHERN IDAHO (Photo Courtesy of Todd Haynes, 2009)

8.12.2   Turbine Placement and Layout

The layout of a turbine array is a critical step in wind farm design. Proper spacing between turbines is necessary to prevent one turbine from being in another turbines wake, which can decrease energy output, increase turbulence, and increase O&M costs. Although each site should be carefully modeled to precisely space turbines, some rules of thumb or best practices exist. Side-to-side spacing of turbines, perpendicular to the prevailing wind direction, should be separated by 3 to 4 rotor diameter lengths. For front to back spacing, parallel to prevailing wind direction, turbines should be spaced 8 to 12 rotor diameters apart. In sites with a singular or predominant prevailing wind (such as indicated in the rose graph of Fig. 8.15), the low-end of the range can be used. In sites with frequent winds from multiple directions and having a more balanced wind rose pattern, the greater spacing distances should be used. Turbine layouts can be optimized using computer models.

One advantage to wind energy is that it allows a site to continue with its traditional land use. For wind turbines placed on active farmland, it may be necessary to consider the land's current use when creating a turbine array. For example, a landowner may prefer that some income from energy be forfeited in order to continue farming the majority of the land. In other words, rather than placing all turbines in the optimal spots for energy production, they may be moved to areas not farmed, such as irrigation pivot corners.

8.12.3   Access Road Design

Access roads must have adequate width andbases to support trucks transporting tower sections and blades as well as the cranes

necessary to erect the turbines. When roads are built,curves must be of sufficient radius to accommodate the long blades of modern turbines, as shown in Fig. 8.16. Crane pads must be sufficient to support cranes as shown in Fig. 8.17. Civil engineers typically design a wind farm's roads and crane pads.

8.12.4   Electrical Collection System

Typical wind farm design in the United States includes many electrical power systems aspects. Wind farm engineers need to be cognizant of electrical grid and transmission issues when looking for suitable locations and planning their developments. Available transmission capacities, voltage levels, proximity to loads, and many other factors need to be considered. The voltage level of the proposed interconnect with the utility's transmission will have a direct impact on the cost of the interconnection equipment and the economics of the wind project. Most wind projects above 20 MW in size typically attempt to tie in to transmission voltage levels between 69kV and 230 kV, while only very large projects above 500 MW attempt interconnects at voltages above 230 kV due to substation costs.

Once an interconnect application is filed with a utility, both the utility and the wind developer's electrical engineers start more in-depth analyses of the power system from both perspectives. The utility will model its system and the wind farm addition with software programs addressing items such as load flow, system voltages and voltage support equipment, potential harmonics, system dynamics and stability, and any other power quality areas of concern. The wind developer's engineers will analyze the wind system with similar modeling tools and calculations and will refine the wind farm electrical design based on this interaction with the utility and other design factors. Layout and spacing of the wind turbines drives much of the design for theelectric power collector system, along with other aspects such as soil thermal conductivity, electrical resistivity, and underground soil and rock characteristics. Therefore, the electrical engineers must work closely with the meteorologists doing the wind turbine layouts to put together an appropriate design.

Transformers and cables are sized to keep losses and voltage dropsbelow acceptable levels (typically less than 5%) for the

FIG. 8.14   BOISE STATE UNIVERSITY ENGINEERING STUDENT KEN FUKUMOTO STANDS NEXT TO A TRITON™ SODAR NEAR MOUNTAIN HOME, ID (Photo Courtesy of Todd Haynes, 2009)

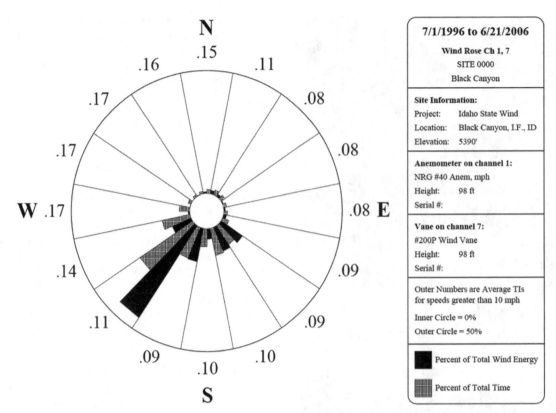

FIG. 8.15 EXAMPLE OF A WIND ROSE GRAPH (TOTAL 10-MINUTE INTERVALS: 524592. INTERVALS USED IN CALCULATIONS: 217906. PERCENT DATA USED: 41.5) [20]

project and to keep cables from being damaged due to overheating. The latter concern is related to the soil thermal conductivity determined in testing and number of cables in parallel, and designs should consider appropriate cable sizing and deratings based on applicable electrical codes and good design practices. Wind farm collector systems are typically run underground to reduce potential bird and environmental impacts, and once the electrical system is far enough away from the turbines or back to the substation, it usually transitions back to overhead power lines to keep costs down. Typical collector system voltages are from the 25 kV and 35 kV voltage classes with 34.5 kV being a common nominal voltage. Some smaller projects may have systems at 25 kV or lower distribution voltages. Typical wind turbine output voltages are 575 V to 690 V, so step-up transformers are located either in the turbine nacelle or on the ground next to the turbine to increase the voltage to the collection voltage level and reduce transmission losses. Grounding of the wind turbine and electrical system is also a significant component of the electrical design, and is related to the

electrical resistivity tests and wind turbine grounding requirements that must be part of every wind farm design.

Wind turbines are designed with circuit breakers and/or contactors and fault protection devices that monitor voltages, frequency, current and other parameters in order to protect the turbines. Additional fault protective systems to protect the wind farm transformers, cables and other equipment should be designed in the electrical system and substation at the point of interconnect. These systems usually include high and low side transformer breakers and switches, electronic protective relays, feeder breakers and relays, switch-

FIG. 8.17 CRANES USED TO ERECT TOWER AND ROTOR SECTIONS AT A WIND FARM NEAR BLISS, ID (Photo Courtesy of Todd Haynes, 2008)

FIG. 8.16 TRUCK TRANSPORTING A 44-M BLADE TO A WIND FARM NEAR BLISS, ID (Photo Courtesy of Todd Haynes, 2008)

gear, grounding transformers, current and potential transformers, and related equipments. Other items that may be incorporated in the substation include metering equipment, control building and backup batteries, voltage support equipment, such as switched capacitor banks or reactors, and other equipment determined necessary during the project design.

8.12.5 Permitting

Engineers typically support the permitting process steps by providing:

- topographical maps describing the proposed turbine array,
- visual simulations of the turbines from a distance (such as how they might be visible from a nearby city or town),
- estimates of sound levels at the property boundary, and
- technical details about the proposed turbines.

During the permitting process, wind farm developers must consider potential environmental impacts, radar proximity, and local ordinances regarding sound, set back, and flicker.

8.13 WIND ENERGY RESEARCH

From 1998 through 2009, installed wind power capacity grew very rapidly. At the end of 2009, the cumulative total capacity of wind power in the United States was greater than 35,000 MW [17]. Given wind power's intermittent nature, there is evidence that this rapid growth is adding complexity to grid operators and electric utility companies. Grid operators must constantly balance load demand and generation supply on the electric grids. Since typical users do not schedule their demand ahead of time, grid operators must forecast demand ahead of time and adjust generation to meet the forecast. Forecasts and scheduling are done hourly in many markets and sub-hourly in other markets. The grid is kept in balance by adjusting generation power levels. Output from most traditional energy sources thermal generators and hydropower can be controlled, or "dispatched," by grid operators to meet the load forecasts. Intermittent energy sources, such as wind, cannot be controlled by grid operators. In fact, the power produced by intermittent sources is accounted as negative loading on the grid. As currently configured, this fact increases the complexity of keeping electric grids in balance.

8.13.1 Grid Integration

During some periods, the output from wind plants is increasing or decreasing at the same time as load, which helps keep the grid in balance. At times when wind is increasing and load is decreasing, or vice-versa, the wind output makes balancing the grid much more difficult.

The Bonneville Power Administration's (BPA) balancing control area in the Pacific Northwest has the largest penetration of wind generation in the nation at approximately 30% of peak load (3000 MW of 10,500 MW) [22]. The U.S. DOE has a stated goal of 20% wind energy nationally by the year 2030 [1], and BPA was well ahead of this goal in 2009. Its experiences are considered a leading indicator of wind power's effects (both positive

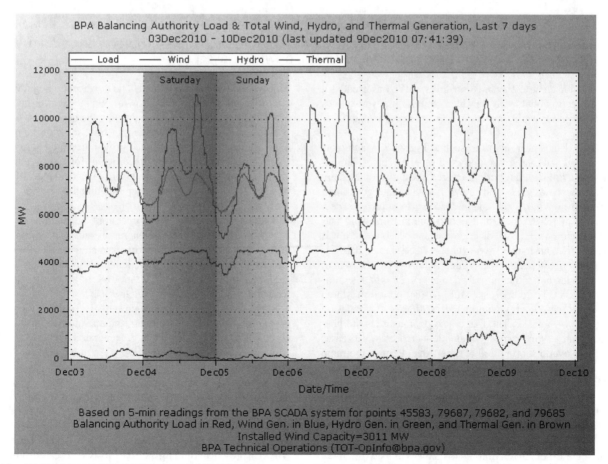

FIG. 8.18   PLOT OF 7-DAY PERIOD IN DECEMBER 2010, SHOWING LOAD (RED LINE) AND GENERATION CONTRIBUTIONS FROM SEVERAL ENERGY SOURCES ON THE BONNEVILLE POWER ADMINSTRATION GRID [23]

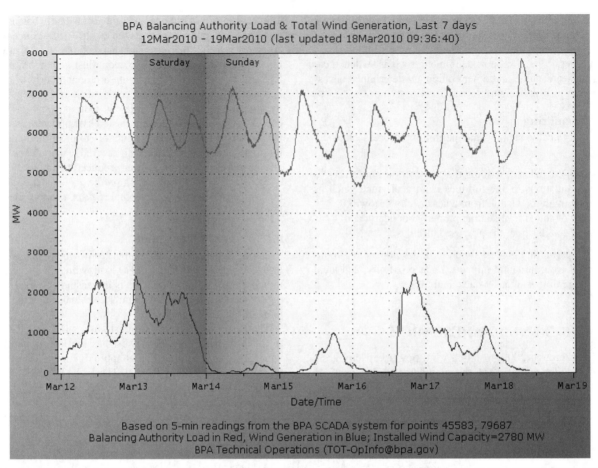

FIG. 8.19 PLOT OF 7-DAY PERIOD IN MARCH 2010 SHOWING LOAD AND WIND GENERATION ON BPA'S GRID, WHICH DEMONSTRATES SEVERAL WIND RAMP, BOTH UPWARD AND DOWNWARD [22]

and negative) on electric transmission grids. BPA has created a Wind Power section on its website which contains many resources regarding wind power's grid impacts (http://www.bpa.gov/corporate/WindPower/).

Some of the data is near real-time, such as Fig. 8.18, which plots BPA's total load and total generation from several sources, including wind, over a rolling 7-day period. The red line indicates the daily load cycles; the blue, green and brown lines indicate generation contributions from wind, hydro and thermal energy sources respectively. During this 7-day period, the contribution from wind was near-zero, which is not unusual during the winter season. There was around 1000MW contribution from wind on December 9, 2010.

While there is often little wind generation during the winter months, the spring and fall months tend to have much more significant contribution. Figure 8.19, showing a 7-day period in March 2010, indicates wind energy contributed 2000 MW or greater on March 12, 13, 14, and 17. The figure also demonstrates the phenomena known as "wind ramps," dramatic increases or decreases in wind power over a short period of time. An extreme wind ramp occurred on March 17, when wind power generation increased from 0 MW around noon to greater than 2000 MW around 4 p.m. This up-ramp was followed by a down-ramp, which was nearly as extreme. From approximately 8 p.m. March 17 to 3 a.m. March 18, wind power generation decreased from around 2500 MW to around 1000 MW. While these ramps are interesting to study alone, it is important to view them in relationship to changes in load (red line). The March 17 up-ramp provided a more difficult situation than the down-ramp. This is because most of the up-ramp was coinci-

dent with a down-ramp in load. The down-ramp beginning around 8 p.m. was coincident with a down-ramp in load; this had the effect of assisting grid operators in balancing the grid.

Wind ramps such as these demonstrate some of the most pressing research needs for continued growth in wind energy. Ongoing research in grid integration, specifically forecasting, storage and development of hybrid energy systems, is intended to allow greater percentages of intermittent generation on electric grids.

8.13.1.1 Forecasting Grid operators balance the grid by forecasting load and adjusting generation output accordingly. Since utility companies cannot directly dispatch wind, the ability to forecast wind output would be beneficial by allowing them to properly adjust output from other generation — to account for changes in load and wind generation simultaneously.

State-of-the-art monitoring and statistical computation technologies allow wind farm developers and operators to fairly accurately predict how much energy can be generated annually at a facility. However, knowing when that energy will be available is a much more difficult problem. Utility companies are continually working to keep the grid in balance, and rely on short- and medium-term load forecasts, ranging from hour-ahead to 24-hours-ahead. The ability to accurately forecast wind plant outputs in those time frames has been identified by BPA as a high priority [24]. Similarly, the U.S. DOE Energy Efficiency and Renewable Energy Office (EERE) provided funding in 2010 to two research projects with the goal of improving short-term wind forecasting. The research teams include collaborators from private companies, universities, the DOE, and National Oceanic and Atmospheric Administration (NOAA) [25].

8.13.1.2 Storage To allow high percentages of intermittent energy generation on transmission grids, economical grid-scale energy storage is arguably more important than forecasting. Such storage would allow energy from wind, solar, and other intermittent sources to be generated when available and stored until it was needed to meet load. Although many forms of energy storage are readily available (batteries, reservoirs, capacitors, etc.), few are economical, large-scale and ubiquitous. In 2010, pumped-hydro storage is both large-scale and relatively cost effective, but it is not ubiquitous; that is, the number of candidate sites is limited by topography, availability of water and environmental considerations. Batteries (including chemical and flow) are ubiquitous, but are not economical on large scale. Capacitors and ultra-capacitors are ubiquitous and relatively economical, but cannot store large quantities of energy over time.

The U.S. DOE recognizes the need to develop larger scale energy storage for improving grid integration. Through the Advanced Research Projects Agency-Energy (DOE ARPA-E) 12 grid-scale, rampable energy storage research projects were initiated in 2010 [26]. The project teams include private companies, universities and national laboratories. Figure 8.20 indicates some of the ARPA-E targets, with respect to cost, energy and power, for the research.

8.13.1.3 Hybrid Energy Systems While wind, solar, and other renewable energy sources tend to be intermittent, other energy sources, such as nuclear, tend to operate most efficiently at steady state. Electric load tends to be cyclic in nature, varying daily and seasonally. Figure 8.21 shows the electric demand (load) in California on November 12, 2007. The load is lightest at night and in the morning, and peaks in mid/late afternoon. The shape of such curves will vary regionally and seasonally, but all have peaks and troughs.

Neither intermittent sources nor steady-state base-loading energy sources naturally lend themselves to match such cyclic load

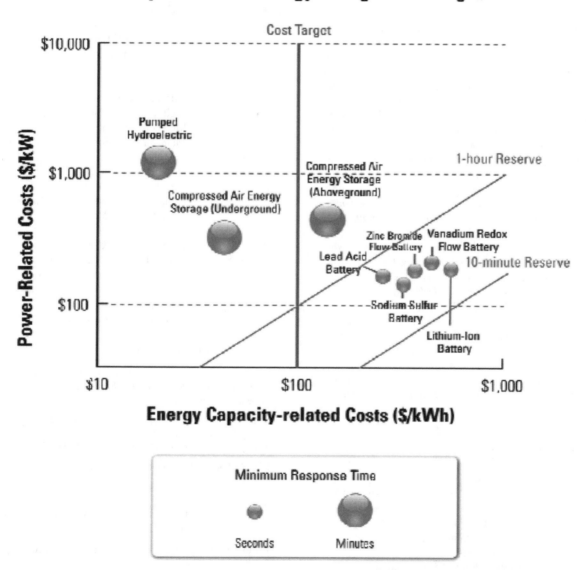

Capital Costs of Energy Storage Technologies

FIG. 8.20 POWER COSTS AND ENERGY COSTS ARE DISPLAYED FOR A RANGE OF ENERGY STORAGE TECHNOLOGIES [27]

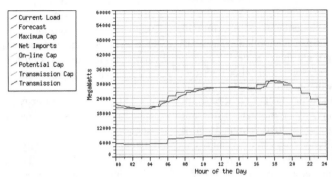

FIG. 8.21 EXAMPLE OF A 24-HOUR LOAD CYCLE ON THE CALIFORNIA ELECTRICAL SYSTEM [29]

patterns. Hybrid Energy Systems (HES) offer a possible solution by combining steady-state and intermittent energy sources to meet electrical demand when needed and create liquid fuels at times of low-demand. The INL is actively researching HES. The following excerpt introduces the HES concept.

Broadly described, HES are energy product production plants that take two or more energy resource inputs (typically includes both carbon and non-carbon based sources) and produce two or more energy products (e.g., electricity, liquid transportation fuels, industrial chemicals) in an integrated plant. Nuclear energy integration into HES offers intriguing potential, particularly if smaller (<300 MWe) reactors are available. Although the concept of using nuclear energy in a variety of non-electrical process applications is certainly not new, renewed interest in more tightly coupled energy product plants (such as HES) that meet the objectives outline above have gained additional interest recently, an interest likely sparked by sharpening energy security concerns. Studies have shown that non-nuclear integrated (hybrid) energy systems can have appealing attributes in terms of overall process efficiency, enhanced electric grid stability, renewable energy integration, and economic performance, and lifecycle greenhouse gas emissions. These attributes seem to be sufficiently compelling that several significant commercial investments in fossil-renewable HES are being made in the United States while the U.S. Defense Advanced Research Projects Agency (DARPA) has openly solicited information regarding nuclear energy integration schemes. In testimony before the U.S. Senate, a senior researcher at Rand Corporation summed up the potential value of hybrid systems well, stating "... the combined use of fossil and solar or nuclear technologies may make for cost-effective and environmentally superior approaches" [28].

8.14 ACRONYMS

AGL	Above ground level
ARRA	American Recovery and Reinvestment Act of 2009
ASR-11	Raytheon Terminal Radar
ASR-9	Grumman Terminal Radar
ATC	Air Traffic Control
BPA	Bonneville Power Administration
BWEC	Bats and Wind Energy Cooperative
CFD	Computational Fluid Dynamics
DARPA	Defense Advanced Research Projects Agency
DHS	Department of Homeland Security
DOD	Department of Defense
DOE	Department of Energy
DOE-WPA	Department of Energy Wind Powering America
FAA	Federal Aviation Administration
HES	Hybrid Energy Systems
INL	Idaho National Laboratory
ITC	Investment Tax Credit
kVa	Kilovolt amps
kW	Kilowatts
kWh	Kilowatt-hours
LIDAR	Light Detection and Ranging
m/s	Meters per second
MISI	Management Information Services Inc.
MS Excel	Microsoft Excel
MW	Megawatts
NEI	Nuclear Energy Institute
NREL	National Renewable Energy Laboratory
O&M	Operations and Maintenance
PPA	Power Purchase Agreement
PTC	Production Tax Credit
SODAR	Sonic Detection and Ranging
USDOE	United States Department of Energy
VREP	Virtual Renewable Energy Prospector
W/m^2	Watts per meter squared
WSR-88D	Weather Surveillance Radar-1988, Doppler

8.15 REFERENCES

1. (U.S. Energy Information Administration, http://www.eia.doe.gov/cneaf/electricity/epm/table1_1.html,http://www.eia.doe.gov/cneaf/electricity/epm/table1_1_a.html, accessed 12/13/10)

2. 20% Wind Energy by 2030, US Department of Energy Office of Energy Efficiency and Renewable Energy (DOE-EERE) Wind and Hydropower Technologies Program (WHTP), July 2008

3. (U.S. Energy Information Administration, http://www.eia.doe.gov/cneaf/electricity/epa/epa_sum.html, accessed 12/3/10)

4. Source: An Assessment of the Available Windy Land Area and Wind Energy Potential in the Contiguous United States, Pacific Northwest Laboratory, August 1991. PNL-7789

5. http://www.windpoweringamerica.gov/pdfs/wind_maps/us_wind_potential_chart.pdf

6. http://www.windpoweringamerica.gov/pdfs/wind_maps/us_windmap_80meters.pdf

7. Analysis of Federal Expenditures for Energy Development, Management Information Services, Inc., September 2008.

8. Department of the Treasury, Internal Revenue Service. (2009). Form 8835

9. Department of the Treasury, Internal Revenue Service. (2009). Investment Credit, Form 3468

10. Department of the Treasury, Internal Revenue Service. (2009). General Business Credit, Form 3800

11. Environmental Law Institute. (2009). Estimating U.S. Government Subsidies to Energy Sources: 2002-2008.

12. "The Effects of Integrating Wind Power on Transmission System Planning, Reliability, and Operations: Report on Phase 2: System Performance Evaluation," Executive Summary, p. 2.13.

13. http://www.jcmiras.net/surge/p130.htm

14. http://www.nrel.gov/docs/fy01osti/30551.pdf

15. Source: Erickson, et. al, 2002. Summary of Anthropogenic Causes of Bird Mortality

16. http://www.bwea.com/aviation/ams_report.html

17. Wiser, Ryan and Bolinger, Mark, "2009 WindTechnologies Market Report", August 2010, pp 3. http://www.windpoweringamerica.gov/filter_detail.asp?itemid=2788

18. This frequency distribution graph was taken from publicly available wind data posted on the Idaho National Laboratory website.http://www.inl.gov/wind/idaho/

19. http://www.windpoweringamerica.gov/anemometer_loans.asp

20. http://www.inl.gov/wind/idaho/

21. http://coen.boisestate.edu/WindEnergy/WindData/index.asp

22. DOE Bonneville Power Administration, Wind Generation Capacity http://www.transmission.bpa.gov/business/operations/Wind/WIND_InstalledCapacity_current.xls

23. http://transmission.bpa.gov/Business/Operations/Wind/baltwg.aspx

24. Haynes, Todd; Dawson, Paul; and Nuss, Kevin, Forecasting for Wind Energy, February 2010, prepared for Bonneville Power Administration.

25. http://www1.eere.energy.gov/windandhydro/news_detail.html?news_id=16316

26. http://arpa-e.energy.gov/ProgramsProjects/GRIDS.aspx

27. U.S. Department of Energy Advanced Research Projects Agency-Energy (DOE ARPA-E), Grid-Scale Rampable Intermittent Dispatchable Storage Funding Opportunity Announcement (DE-FOA-0000290), page 12, March 2010.

28. Aumeier, Steven, Cherry, Robert, Boardman, Richard and Smith, Joseph, Nuclear Hybrid Energy Systems: Imperatives, Prospects, and Challenges, pp. 2, October 2010

29. http://currentenergy.lbl.gov/ca/index.php

30. S. Krohn and S. Damborg, "On public attitude towards wind power," Renewable Energy, vol. 16, no. 1/4, pp. 954-960, 1999, Pergamon Press, Inc.

31. M. Wolsink, "Wind power implementation: The nature of public attitudes: Equity and fairness instead of backyard motives," Renewable and Substanable Energy Reviews, vol. 11, no. 6, pp. 1188-1207, Aug. 2007, Elsevier Ltd.

32. I. Bishop "Determination of thresholds of visual impact: the case of wind turbines" Environment and Planning B: Planning and Design, vol. 29, no. 5, pp. 707–718, 2002, Pion Ltd.

33. http://www.nrel.gov/docs/fy03osti/34318.pdf

34. http://www.inl.gov/wind

WIND ENERGY RESEARCH IN THE NETHERLANDS

Peter Eecen

9.1   INTRODUCTION

Interest in the application of modern wind energy grew in the Netherlands in the 1970s when the limit of fossil fuels became clear. Wind energy has been an important source of energy in The Netherlands for centuries and the country was known for the many wooden wind mills especially in the coastal regions. At the same time, research activities in the modern wind energy were started, which have led to the relatively large wind energy research community in the Netherlands today.

Wind energy research activities in the Netherlands are predominantly performed at the Energy research Centre of the Netherlands (ECN) and the Delft University of Technology (DUT). Both institutes are involved in wind energy research since the start of the modern wind turbines, the 1970s. These institutes match their research programs with each other as close as possible. ECN Wind Energy has a research staff of 55 scientists and DUT has a research staff of 15 permanent researchers and more than 35 PhD students. Another institute dedicated to wind energy research is the foundation Knowledge Centre WMC that has been founded by the DUT and ECN in 2003 with an additional research staff of 25 scientists. Although the major part of the wind energy research is concentrated in these institutes, many other universities and scientific institutes contribute to the research with dedicated and specialized research.

The current wind energy research and associated industrial activities are taking place in an international context, mostly the European context, therefore the research activities not only take account of the long-term energy research program of the Dutch government, such as the long-term energy research program EOS [1], but also of the R&D priorities defined in the international context, such as the Strategic Research Agenda (SRA) of the wind energy sector [2]. The three wind energy research organizations are well represented in international bodies such as European Wind Energy Association (EWEA), European Academy of Wind Energy (EAWE), International Energy Agency (IEA), International Electrotechnical Commission (IEC), International Network for Harmonised and Recognised Measurements in Wind Energy (MEASNET), European Wind Energy Technology Platform (TPWind) and the European Energy Research Alliance (EERA).

9.2   WIND ENERGY IN THE NETHERLANDS

In the last 20 years, the amount of wind energy has grown at a rate of approximately 30% per year. In Europe, at the end of 2008 a total amount of 64.5 GW wind energy has been installed onshore and 1.5 GW offshore. In the Netherlands, the installed wind energy capacity amounts 1921 MW onshore and 228 MW offshore. During 2009, the Netherlands decommissioned 106 turbines (total capacity 34.8 MW) and installed new turbines totaling 101.4 MW to have a total of 2216 MW installed wind power which generated 4% of the national electricity demand. Wind energy has an important role in the extension and replacement of electricity production capacity.

The official target for the Netherlands for the implementation of renewable energy, set in a European context, is 14% in 2020. However, the Netherlands has the ambition to reach the share of renewable energy of 20% in 2020. In order to reach that ambition, the share of renewable electricity generation in the total electricity production should reach 35%. Based on this ambition, the Netherlands has set the target for installed wind energy capacity in 2020 at 4000 MW onshore and 6000 MW offshore.

Currently, the cost of electricity production using onshore wind turbines reaches the cost of fossil Electricity production (6 - 8 ct/kWh). Offshore, the cost of wind energy electricity production is strongly depending on the complexity of the wind farm, such as distance to the coast, water depth, soil, etc. The cost offshore is in the order of 12 to 18 ct/kWh. However, it is expected that the relatively young offshore wind technology will experience a strong learning curve, leading to significant cost reductions. This is one of the reasons that the Dutch government balances its funding between supporting the implementation and funding further innovation by supporting wind energy research with a focus on the further development of offshore wind energy.

Geographically, the Netherlands has a central position with regards to the North Sea. In the next decades, offshore wind energy will be a significant part of the total wind energy capacity in the countries surrounding the North Sea. The conditions are relatively favorable: the average wind speed is high, the water depths are moderate (20 to 50 m), and there are many harbors to access the farms. In order to realize the ambitious targets, substantial R&D

effort is required. The wind energy research program in the Netherlands has therefore been focused on offshore applications since the beginning of the century. Not only because most capacity must be installed offshore, but in addition, the offshore application has the highest challenge in knowledge, technology, reliability, installation and maintenance that is also applicable to onshore applications. The knowledge institutes in the Netherlands active in the field of wind energy are collaborating intensively. The long-term collaboration between the Energy research Centre of the Netherlands ECN and Delft University of Technology DUWIND has led to among others the common foundation Wind turbine Materials and Constructions WMC where blade tests are performed.

In the 1990s, the Netherlands had several wind turbine manufacturers, among which Lagerwey, Nedwind en Windmaster. Although these manufacturers did not develop to large global players, the Netherlands still has significant knowledge in the field of development and manufacturing of wind turbines. Currently, in the Netherlands wind turbines are being developed by for instance XEMC Darwind, 2B-energy, VWEC, Lagerwey Wind, STX-Harakosan and EWT. In addition, the Netherlands has a strongly developed industrial sector in the field of offshore technology, which is heavily involved in the installation of offshore wind energy. The activities range from offshore engineering to the development of installation and maintenance vessels and foundations. Large Dutch players are among others Ballast Nedam, Mamoet van Oord, Heerema, IHC Merwede, Fugro, and MSC Gusto. Because of the strong Dutch knowledge position in offshore wind energy by the collaborative applied research activities, the Dutch industry is involved in the Construction of offshore wind farms in The Netherlands, Denmark, Germany and the United Kingdom.

9.3   HISTORIC VIEW TO 1990

The first proposals for producing electricity from wind in The Netherlands date from the 1920s. After some experiments in the 1930s, there were more experiments until the 1970s without too much success. Especially in the 1960s, the fossil fuels were cheap and the general idea was that nuclear energy would provide cheap energy for a long time. The Energy research Centre of the Netherlands in that time was called Reactor Centre of the Netherlands (RCN). In 1975, a national committee reported that wind energy is the preferred option for large scale renewable energy, and remarkably, these large numbers would not be installed only onshore, and specifically offshore wind energy in the North Sea should be investigated. A committee guided by RCN had the assignment to define the first Dutch national research program on wind energy.

The first funding of Dutch wind energy research was organized by the National Wind Energy Research Program (NOW) phase in the period 1976 to 1985 which was followed by the 1986 to 1990 NOW phase to stimulate technological, economical and environmental research related to wind energy. The intention of these programs was to give The Netherlands a leading role in the development of wind energy given the rich wind history. And indeed, these programs facilitated the development of wind energy as a realistic energy option for the next decades. The 1986 to 1990 Integral Wind Energy Program (IWP) was intended to initiate the large-scale application of wind energy in the Netherlands and to provide an incentive for the further development of cost-effective wind turbines. In 1990, these programs led to a large involvement of Dutch industry in wind energy research and contributed to the accumulation of wind technology know-how in the Netherlands.

Examples are the know-how developed in the field of noise emission reduction, material expertise and fatigue properties that led to significant improvements of wind turbines. However, while The Netherlands significantly invested in wind energy research, the development of large scale wind energy was not realized. The national realization of wind turbines always was far below the targets and also the national industry was largely overtaken by international competition.

In the first phase of the NOW research program, the comparison was made between horizontal and vertical axis wind turbines. Technical options were investigated and no choice could be made then. Both options were considered for the exploratory phase and experimental turbines of both types were realized. Although there were no solid arguments to not further develop the vertical axis wind turbine and given that the horizontal axis turbine was the dominant design in the international market, a decision was made to stop the further development of vertical axis wind turbines. In the mean time Polymarin built a floating 15 m Darrieus turbine in the Gaasperplas in Amsterdam in the early 1980s.

From 1977 to 1978, a larger turbine with horizontal axis was designed, the HAT25. The name indicated the rotor diameter of 25 m of this turbine. ECN, the National Aerospace Laboratory NLR and Eindhoven university of Technology were responsible for parts of the research program. The 25-m HAT (horizontal axis turbine) experimental wind turbine was erected at the site of ECN and the turbine had the typical half concrete, half steel tower. It has been operated by ECN since August 1981 and extensive measurement campaigns have been performed at this turbine. The experiments have been used for validation of codes, testing of control strategies, analyses of load cases, etc. Alternative rotors have been installed, such as the FLEXHAT rotor and many more. Some prominent measurements on the 25-m HAT turbine are the axial force measurements to verify the computer program PHATAS, which calculates static and dynamic loads in the large wind turbine components and dynamic loads in the large turbine components. These were used to determine excitation and response frequency spectrums for different modes of operation and to determine the fatigue characteristics of the machines by means of the rain flow counting method.

In 1985 a prototype of the advanced 1 MW NEWECS-45 was built and in 1986 a start was made with the experimental wind farm Sexbierum, near the city of Sexbierum in the Netherlands. This wind farm consisted of 18 turbines with rotor diameter of 18 m rated at 300 kW totaling 5.4 MW. Seven towers were installed for meteorological measurements. Especially the wake measurements in this wind farm were unique for that time. Also the effect of the wind farm on the environment, and especially birds, was part of the research, where it was concluded that wind farms do not pose an excessive threat to birds.

In the Netherlands, the research activities on wind energy always did have a large international focus and was based on international cooperation. Already in the IEA R&D WECS [3] program that was initiated in 1977, the Netherlands participated with the objective to perform cooperative research, development and demonstration, and exchange of information in the field of wind energy utilization. From the early 1970s, the Netherlands, just as many other countries were exploring the use of wind energy. After the oil crisis, an urge to collaborate between R&D institutes in wind energy led to the start of the International Energy Agency Implementing Agreement for Co-operation in the Research and Development of Wind Turbine Systems [4]. A manual for structural safety analysis of wind turbines was prepared and technical feasibility studies were carried out on multi-unit offshore wind farms. Already in 1981, it was concluded that it was technical feasible to install a large, multi-unit offshore wind farm with 50 to 200 turbines on individual support structures, with rotor diameters between 50 and 110 m in water depths of 10 to 45 m. The study group on recommended practices completed a study of power performance testing and studied recommended practices for wind turbine costing, evaluation of fatigue, acoustics, safety and reliability, quality of power, etc. As early as 1980, a new task was initiated by ECN on wake effects of wind turbines. The objective to better estimate the performance of arrays of wind turbines led to the theoretical and experimental modeling of turbine clusters, among others experiments in the Dutch TNO wind tunnels [5, 6]. The measured data still are used for analyses and validations today. In addition, generated turbulence levels and associated dynamic effects were part of the studies. The Dutch TNO wake interaction model MILLY was tested against wind tunnel experiments. In 1983 the Netherlands joined the IEA agreement for cooperation in the development of Large-Scale Wind Energy Systems [3] with ECN as the contracting party. One of the activities was to prepare the specifications for the 3 MW 80 m GROHAT turbine with an industrial consortium.

9.4   HISTORIC VIEW 1990 TO 2000

The wind energy research activity in The Netherlands in the period from 1990 to 2000 period was concentrated at ECN Wind Energy and the Delft University of Technology. Additional activities were carried out by the Dutch organization for Applied Scientific Research TNO and research and development by the industry. ECN focused on aerodynamics, durability and fatigue, electrical conversion and regulation, criteria for certification and standards, and offshore applications. The primary focus of the ECN program is the improvement of the institute's research potential in order to be able to react adequately to possible assignments from industry and therefore the program is oriented towards the development of measuring methods, simulation models, test procedures, etc. The development of measuring methods is facilitated by the availability of the HAT25 turbine at the Petten site. In the same period, the Delft University of Technology plays an important role in research and researcher training, with predominant focus on fundamental aspects, such as aerodynamics and loads. The 1991 to 1995 TWIN program administered by NOVEM was followed by the 1996 to 2000 program. The main focus in these programs are further research in rotor aerodynamics, especially dynamic and 3D effects and development of engineering rules; design tools for rotor development, o.a. buckling and optimization through cost functions; inventory of extreme wind conditions; reduction of emission noise, o.a. through design and field testing of rotor blades with serrated edges and empirical research in which serrated edges and tip shapes are tested in wind tunnels; standards and certification; and design and construction of light weight turbine concepts such as the FlexHat research [7]. The FlexHat program aimed at diminishing internal peak loads, by employing the dynamics of the rotor instead of withstanding them. Flexibility is introduced in the four main degrees of freedom of the rotor, resulting in a variable speed conversion system with soft characteristics, rotor control by means of passive activation of the tip in combination with variable speed, an elastomeric teeter with elastomeric teeter limiters (bumpers), and a flex beam which is moderately flexible in flap wise direction and stiff in torsional direction. The components were tested on the HAT25 research turbine, located at ECN Petten.

In 1997, NOVEM carried out a feasibility study of a demonstration project of a near shore wind farm. The wind farm is meant to gain experience and knowledge of offshore installation, construction and operations. The idea to have a demonstration offshore wind farm was steadily further developed and the wind farm became operational in 2007. The realization of the farm included an extensive monitoring and evaluation program. The farm is highlighted further in this Chapter.

Around the same time, engineering design codes came available, like the wind field generator for aeroelastic codes SWIFT, an aerodynamic correction method for 3D effects TIDIS, an optimization tool for blade design BLADOPT, and a tool for calculations on noise emissions from blades SILANT. In 1999 the DUT developed the method NewGust to quickly generate extreme turbulence gusts for load calculations. These codes have been used by industry and most of these codes are still in use today.

The objective of the project 3D effects in stall was to improve methods that are used in order to predict the effect of rotation on sectional aerodynamic coefficients, particularly the lift coefficients. The project was a cooperation between ECN, DUT and the National Aerospace Laboratory NLR. In the first stage, following the ideas of Herman Snel, boundary layer equations in a rotating frame of reference were developed by NLR and subsequently put in integral form. In parallel, the aerodynamic analysis program XFOIL was purchased and improved by ECN and DUT resulting in a more accurate prediction of lift coefficients near stall. In the second stage, the integral form of boundary layers in the rotating frame of reference was implemented in the improved XFOIL by NLR. The resulting program called RFOIL was validated using the experimental aerodynamic field data from ECN's HAT25 experimental wind turbine. In addition, power curve predictions were validated by experimental data. In the third stage, an engineering method was developed by ECN, allowing the effect of rotation on the lift coefficient to be obtained without having to perform a calculation with RFOIL. The main result of the project was the code RFOIL, which is more accurate than its predecessor in rotating cases like wind turbines. The RFOIL code is to date still freely distributed by ECN and is in use with most wind turbine manufacturers and wind energy researchers.

ECN coordinated the JOULE II project Dynamic Inflow [8], research on the validation of 3D effects in stall. These effects describe the wake induced unsteadiness and non-uniformity of the flow in the rotor plane. Aim of the project was the definition and implementation of engineering models within computer codes for dynamic load calculations. These models have been qualified by means of comparison with both existing sophisticated models (dynamic vortex wake calculations) and experiments on turbines with a large range of rotor diameters. Important dynamic inflow effects were found at fast pitching transients and at yawed flow conditions. The effects were predicted well with the newly developed engineering methods. The projects provided many insights in the behavior of the induced velocities in the rotor plane under several conditions. Measurements and calculations showed important effects of dynamic inflow on the mechanical loads at fast pitching transients and at yawed conditions. At fast pitching transients large overshoots in the loads were apparent. At yawed conditions, the influence of the skewed wake on the induced velocities effected the phase and amplitude of the azimuthally binned averaged flat wise moments. For partial span pitch conditions and for wind gusts, the dynamic inflow effects were much less. Direct evidence of dynamic inflow was found in the wake flow measurements in the wind tunnel at pitching transients and yawed conditions. An interesting result was that under yawed conditions, the skewed wake does not only effect the axial induction (which was expected from helicopter aerodynamics), but also the in plane velocities. The dynamic inflow effects were predicted well with the engineering methods, developed in the project.

Delft University of Technology coordinated the Joule project "Structural and Economic Optimization of Bottom Mounted Offshore Wind Energy Converters" (Opti-OWECS) with the aim to reduce cost of electricity by extending the technology and demonstrating practical solutions for offshore wind turbines. An innovative integral design methodology was developed, the OWECS design approach, considering all components of an offshore wind farm. The design solution is then considering criteria like levelized production costs, adaptation to local site conditions, dynamics of the system, installation effort as well as availability of the turbines. This also required a novel offshore wind farm cost model which was developed at DUT.

During this period, the Dutch wind energy research was strongly involved in international cooperation, like numerous European projects and participation in IEA Annexes. The Dutch researchers have been heavily involved in the development of design guidelines and IEC International Standards.

For instance, TASK XVIII — Enhanced Field Rotor Aerodynamics Database has been operated by ECN [9]. In 1998, the ExCo approved Task XVIII to extend the database developed in Task XIV and to disseminate the results. The objective of the ANNEX-XIV was the coordination of full-scale aerodynamic test programs on wind turbines, in order to acquire the maximum of experimental data at minimum costs. In these full-scale aerodynamic test programs local aerodynamic quantities (forces, inflow velocities, inflow angles) are measured at several radial positions along the blade. The supply of local aerodynamic data, is a major step forward in understanding the very complicated aerodynamic behavior of a wind turbine. In conventional test programs only blade (or rotor) quantities are measured. Usually these quantities are integrated over the rotor blade and they are not only influenced by aerodynamic effects, but also by mass effects. Then the local aerodynamic properties of the blade can only be derived indirectly, introducing an uncertainty. Since aerodynamic field experiments are typically very time consuming, expensive and complicated because of the large volumes of data and the extensive data reduction it is very advantageous to cooperate. And because specific turbine configurations that are investigated experimentally may exhibit a very different aerodynamic response characteristics, the combination of measurement data on very different facilities will provide much more insight about the general validity of aerodynamic phenomena.

In 1999 a new Dutch wind energy research strategy was formulated by the government, and with input from Dutch wind turbine and blade manufacturers, engineering firms and end users such as utilities, insurance companies and certifying institutes the following priority subjects were set:

(1) New developments: offshore, innovative materials and recycling.
(2) Testing and measuring: condition monitoring systems, wind turbine test facilities.
(3) Databases: failure statistics of wind turbines and components.
(4) Design tools: reliability, wind turbine control and aerodynamics.

The research programs at ECN and DUT were shaped after these priorities and the research projects were defined accordingly.

In 1999 several projects were started under the Ecology, Economy and Technology funding scheme: One of the projects researches the possibilities of producing large blades from ecologically friendly materials. The other is the DOWEC project.

The Dutch Offshore Wind Energy Converter project (DOWEC) included NEG Micon, LM Glassfiber, Van Oord ACZ, DUT and ECN and was started as a concept study for offshore wind turbines [10]. The goal of the DOWEC Concept Study was to make an inventory of all wind turbine concepts in order to select the most optimal concept for a 5 to 6 MW offshore wind turbine. In the first phase the DOWEC concept study aimed at the choice of the optimal wind turbine concept. These turbines should be able to withstand the severe wind and wave conditions at the Dutch North Sea. In the design process, the wind turbine was not treated as an isolated system, but the designs of different wind turbine concepts were evaluated as an integral part of a complete large-scale offshore wind farm. All significant properties like the structural loads, the power performance, the system reliability, the costs of the electric infrastructure, maintenance costs and installation costs were determined for the optimized designs. The concept study resulted in five feasible concepts for far offshore wind turbines of 5 to 6 MW. These concepts were quantitative ranked based on the cost of generated energy. Furthermore, qualitative criteria like development risk and market potential were taken into consideration when finalizing the choice of concept. From these concepts, one was selected by industry for further development. The industrial tasks resulted in the engineering, purchase, and construction of a 2.75 MW wind turbine. The turbine has been erected and tested at the ECN Wind turbine Test site Wieringermeer. The final result of the DOWEC project should have been the realization of a 6 MW offshore wind turbine which is optimized for exploitation in large wind farms in the North Sea addressing all design and operational aspects in a cost-effective way. The realization of this turbine never happened, but the design of the 6 MW turbine was delivered in 2003. The design is publicly available and the information has been used extensively. For instance the 5 MW reference turbine defined by NREL is largely derived from it and many aspects of the DOWEC 6 MW design can be found in modern offshore wind turbines installed today (Fig. 9.1).

At the end of the decade, ECN developed and patented a novel technique to visualize the stall behavior of rotor blades. The power performance of stall regulated turbines in some cases varied considerably and to quantify the cause, a measurement technique was required, which could be applied on large commercial wind turbines, was fast enough to monitor the dynamic changes of the stall pattern, and was not influenced by the centrifugal force and most importantly did not disturb the flow. The technique developed and patented at ECN was called Stall Flag Method [11] and was composed by small reflecting sheets covering with a hinged non-reflecting sheet. When glued to the blade the hinged non-reflecting blade will open upon stall and the reflecting part becomes visible. During night time, the reflecting parts can be recorded by camera if the turbine is lit by a light source. With this technique problems can be searched for, optimum locations for stall strips and vortex generators can be found and insight can be obtained into the problem of multiple stall. The statistical stall behavior is characterized from the sequences of thousands of analyzed images. This leads to the visualization of the stall pattern, the blade azimuth angles and the rotor speed. It also measures the yaw error and the wind speed from the optical signals of other sensors, which are recorded simultaneously. For example, the delay in the stall angle by vortex generators can be measured with an accuracy of one degree from

FIG. 9.1 THE NM90 2.75 MW WIND TURBINE (ALSO KNOWN AS DOWEC TURBINE) AT THE ECN TEST SITE IN THE WIERINGERMEER AS SEEN FROM THE MEASURE-MENT MAST (109 M) (*Source*: [ECN, taken by R. Nijdam])

the stall flag signals. Many experiments have confirmed the independence of stall flags from the centrifugal force and that stall flags respond quickly to changes in the flow.

9.5 HISTORIC VIEW 2000 TO 2010

In 2000, almost 40 MW of wind capacity was installed in the Netherlands. This is a continuation of the slow but steady installation rate of 40 to 50 MW in the last 4 years. At the time, the national target was 2750 MW installed wind power in 2020 and therefore it was necessary to go offshore. In 2000, preparations for the necessary step to offshore were taken at the technical and administrative level. The research in wind energy became focused on offshore application and a demonstration offshore wind farm was tendered. The consortium which was called Noordzeewind and consisted of Shell Renewables, energy company Nuon International, bank ING Bank, engineering firm Jacobs Compimo, and NUON owned project developer WEOM won the tender for the Near Shore Wind farm. According to the requirements in the environmental effect report for the Near Shore Wind farm, a monitoring and evaluation program had to be carried out. The outlines of this program, which cover environmental, economical and technical aspects were drafted early 2000. According to these outlines, this program started with an evaluation of the undisturbed situation. The entire program aimed at learning as much as possible from the near shore demonstration wind farm in order to better realize the future larger offshore wind farms that are necessary to reach the national targets.

As a follow-up to the JOULE project "Database on Wind Characteristics," Sweden, Norway, United States, The Netherlands, Japan, and Denmark formulated IEA Task XVII. There was an urgency to initiate an international effort to collect, describe and store high-quality wind data measured by several wind energy

related projects. The unique database on wind characteristics consists of quality controlled and documented wind field time series supplemented with tools that enable easy access and a simple analysis through an internet connection. The statistics database contains a list of all time series, the derived statistics together with other parameters. The raw time series (74.6 GB) consist of files of 10-minute periods in a common format. The accessibility of the database has been developed during the operational years of the database and is quite easy to use and it is comprehensive. From numerous areas over the world, controlled and high quality data have been added to the database. Important data consists of wake measurements, three component offshore measurements, extreme complex terrain measurements and long-term data sets covering time spans of at least 20 years. Reports [12, 13] are available on description of data that have been added by ECN to the international wind database [14].

An interesting spin off of the research at DUT is the development of the Ampelmann [15]: a motion compensated platform to safely access offshore wind turbines. The initial feasibility study, preliminary design, and proof of concept of an offshore access system named Ampelmann was supported by SenterNovem. The Ampelmann is a ship-based self stabilizing platform that actively compensates all vessel motions to make offshore access safe, easy and fast. The base of the Ampelmann is mounted on the deck of the vessel or barge and takes care of the motion compensation, keeping the top side of the Ampelmann stationary. The gangway, which is mounted on the top side, can now be easily deployed onto the offshore structure enabling safe offshore crew transfer. To achieve this, the vessel is equipped with a set of motion sensors and a Stewart platform. The motions of the ship are continuously registered in six degrees of freedom by the motion sensors on the deck of the vessel and instantaneously fed into a control system in order to keep the top plate of the Stewart platform motionless compared to the fixed world. On this top plate, a transfer deck is installed. By extending a gangway between the transfer deck and the offshore structure, the structure can be accessed in an easy and safe way, even in high waves. The Demonstrator project was sponsored by We@Sea, Delft University of Technology, Shell, Smit, Heerema fabrication, SMST, and Ecofys with the common goal to further develop and test its design, compensation performance, operational procedures and safety systems in the harsh offshore conditions of the North Sea. Currently, various Ampelmann systems are in use for daily delivery of personnel to fixed platforms in the North Sea.

In 2005, ECN completed the wind resource atlas of the Dutch part of the North Sea for several heights. ECN developed the atlas by combining data from two sources: the numerical weather prediction model Hirlam and the meteorological stations at the North Sea, thus combining a source with dense spatial sampling and a source with accurate (measured) values. At the ECN website, maps with the mean wind speed at 60, 90, 120, and 150 m above sea level are published. ECN can for every location at the North Sea and for each height produce a time series of 10 years based on the Hirlam data which is validated by using the measurement stations. The wind resource in such a location is expressed in terms of the wind speeds and wind directions, the turbulence intensities, and the stability classes (Fig. 9.2).

9.5.1 UpWind

UpWind is a large European project funded under the EU's Sixth Framework Programme (FP6). The project looks towards the wind power of tomorrow, more precisely towards the design of very large wind turbines (8 to 10 MW), both onshore and offshore.

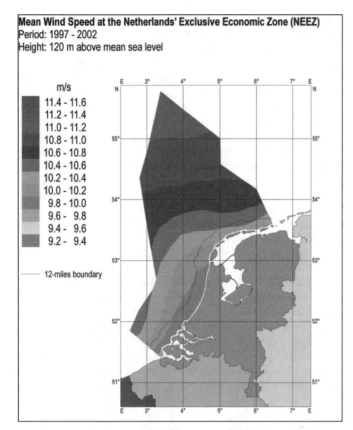

Mean Wind Speed at the Netherlands' Exclusive Economic Zone (NEEZ)
Period: 1997 - 2002
Height: 120 m above mean sea level

m/s
11.4 - 11.6
11.2 - 11.4
11.0 - 11.2
10.8 - 11.0
10.6 - 10.8
10.4 - 10.6
10.2 - 10.4
10.0 - 10.2
9.8 - 10.0
9.6 - 9.8
9.4 - 9.6
9.2 - 9.4

········· 12-miles boundary

FIG. 9.2   MEAN WIND SPEED AT THE DUTCH PART OF THE NORTH SEA AT 120 M IN HEIGHT (*Source*: [ECN, Arno Brand])

Furthermore, the research also focuses on the requirements to the wind energy technology of 20 MW wind turbines. The challenges inherent to the creation of wind farms of several hundreds MW request the highest possible standards in design, complete understanding of external design conditions, the design of materials with extreme strength to mass ratios and advanced control and measuring systems geared towards the highest degree of reliability, and critically, reduced overall turbine mass. The aim of the project is to develop the accurate, verified tools and component concepts the industry needs to design and manufacture new, cost effective type of turbines. UpWind focuses on design tools for the complete range of turbine components addressing the aerodynamic, aero-elastic, structural and material design of rotors. The UpWind consortium, composed of 40 partners, brings together the most advanced European specialists of the wind industry. In most work packages, researchers from the Netherlands are involved and the following work packages have coordinators from the Netherlands: Metrology, Upscaling and Rotor structure and materials.

9.5.2   Smart Rotors

For many years DUT is carrying out research work on smart rotors in the context of the Dutch INNWIND and the EU UPWIND projects. The objective of smart structures for rotor blades is to alleviate significant blade loads by applying spanwise-distributed load control devices without incurring lower reliability or higher maintenance. DUT concentrates on the investigation of concepts, feasibility, and integrated design; aerodynamics; structural integration; control, identification, and experimental investigation of

developed models. DUT proved the concept in their wind tunnel. Researchers observed load reductions between 70% and 90% and concluded a significant amount of fatigue blade load alleviation is possible. This was confirmed by research at ECN that is concentrating on the application of synthetic jets for this purpose. DUT has carried out experiments in the Open Jet Facility with a rotating 2 m diameter wind tunnel model.

9.5.3   WE@SEA

The 4-year WE@SEA program on offshore wind energy is funded from the Dutch national natural gas fund. This program is run by a consortium consists of companies in offshore technology, wind energy technology, offshore wind farm development, logistics, investors, energy consultants, environment, and other stakeholders and was operated from 2004 to 2008. The program of 26 million concentrated on medium-term research in the research lines: offshore wind power generation; spatial planning and environmental aspects; energy transport and distribution; energy market and finance; installation, operation and maintenance, and dismantling; education, training, and knowledge dissemination; and the PhD@ Sea project at DUT. The PhD@Sea research subjects were divided over the research lines of the overall program. Its topics are: large blades; wind turbine concepts; morphology of the North Sea bed; grid stability of large-scale integration of wind energy in electrical power systems; park-grid interaction; offshore access through the Ampelmann; reliability, availability, maintainability, and serviceability analyses and scenarios.

9.5.4   Controlling Wind and Heat & Flux

Over the years 1999–2003, ECN invented and patented two wind farm control techniques, the one called Heat & Flux, the other Controlling Wind. The concept Heat & Flux aims at maximizing the power output of a wind farm by adjusting the axial induction of the windward turbines below their individual optimum for power production, which means making them more transparent for the wind than usual by realizing an axial induction factor below the Lanchester–Betz optimum of 1/3. This will reduce the velocity deficit in the wake and increase the output of the downwind turbines. Other benefits are decreased average loading of the upwind turbines and decreased fatigue loading of the turbines in the wake. ECN quantified the effects of "Heat and Flux" farm control. For this purpose models have been developed for calculating the power and energy production with "Heat and Flux" operation. The models have been developed in close interaction with wind tunnel experiments on model turbines and farms and have been tested in full-scale experiments on a row of five 2.5-MW turbines. Qualitatively, the wind tunnel tests and the field experiments revealed positive effects of simple "Heat and Flux" control settings on several occasions. Quantitatively, the accuracy and reliability of the wind tunnel measurements are questionable because of the large scatter and the conditions that deviate from full scale in various aspects. The full-scale experiments point at comparable optima for "Heat and Flux" control settings as model predictions of the model developed by ECN. This method has been implemented in ECN's Control Tool for designing wind turbine control algorithms.

9.5.5   Wind Tunnel Experiment Mexico and Mexnext

In the past, the accuracy of wind turbine design models has been assessed in several validation projects [16]. They all showed that the modeling of a wind turbine response which can be either the power or the loads is subject to large uncertainties. These uncertainties mainly find their origin in the aerodynamic modeling where several phenomena such as 3-D geometric and rotational effects, instationary effects, yaw effects, stall, and tower effects, amongst others, contribute to unknown responses, particularly at off-design conditions. The availability of high quality measurements is considered to be the most important pre-requisite to gain insight into these uncertainties and to validate and improve aerodynamic wind turbine models. For this reason, the European Union project "MEXICO: Measurements and EXperiments In COntrolled conditions has been carried out [17]. The project was coordinated by ECN and was carried out with ten institutes from six countries. They co-operated in doing experiments on an instrumented, 3-bladed, 4.5-m diameter wind turbine placed in the 9.5 m² open section of the Large Low-speed Facility (LLF) of German Dutch Wind Tunnel (DNW) in the Netherlands (see Fig. 9.3). The LLF is described later in this Chapter. The measurements were performed in December 2006 and resulted in a database of combined blade pressure distributions, loads, and flow field measurements that can be used for aerodynamic model validation and improvement.

A similar previous experiment was performed by the National Renewable Energy Laboratory (NREL) in the National Aeronautics and Space Administration (NASA) Ames wind tunnel [18] on a 10-m diameter turbine. An obvious difference between the two types of experiments lies in the larger size of turbine diameter the latter experiment. But in addition to the NASA-Ames experiment, the Mexico experiment also included flow field measurements. Especially the flow field of inflow and wake are interesting, where the flow field is important for the understanding of discrepancies between calculated and measured blade loads. This is mainly because the load calculations are done in two steps. First, the flow field around the blade, the induction, is calculated, and second from that the loads are derived. Where in conventional experimental programs, only blade loads are measured, it is not possible to distinguish between these two sources of discrepancies. The addition of flow field measurements should open up this possibility.

At the end of the MEXICO project, due to budget reductions, the database was still in a rather rudimentary form and only limited analyses had been carried out. Especially since it is a huge amount of data which requires a lot of effort to analyze. Therefore it is beneficial to organize the analysis of the MEXICO data in a joint project under IEA Wind in which various countries share this task. Given the platform for discussion and interpretation of the results the outcome of the data analysis shall be better than the summed result from individual projects.

ECN is the Operating Agent of the IEA Wind Task 29 called MEXNEX(T) which has the objective to improve aerodynamic modes used for wind turbine design. In this task the access of the Mexico data is provided and a shared thorough analysis of the data takes place. This includes an assessment of the measurement uncertainties and a validation of different categories of aerodynamic models. The insights are compared with the insights that were gained within IEA Wind Task 20 on the NASA-Ames experiment and other wind tunnel experiments.

Special attention is paid to yawed flow, instationary aerodynamics, 3-D effects, tip effects, non-uniformity of the flow between the blades, near wake aerodynamics, turbulent wake, standstill, tunnel effects, etc. These effects are be analyzed by means of different categories of models like CFD, free wake methods, engineering methods, etc. As such, the Task will provide insight on the accuracy of different types of models and the following activities are performed.

(1) Processing and presentation of the measured data and the associated uncertainties. High quality measurement data are provided to the partners to facilitate and compare calculations.

(2) Analysis of tunnel effects. Since the 4.5-m diameter wind turbine model was placed in the open jet section of the LLF facility (9.5 m x 9.5 m) the ratio of turbine diameter over tunnel size may make the wind tunnel situation not fully representative of the free stream situation. The tunnel effects have been studied with advanced CFD models. Supporting information on tunnel effects will also be obtained from eight additional pressure measurements, which were measured with taps in the collector entrance. These pressures measure the speedup in the outer flow (outside the wake) needed for the mass conservation of the tunnel flow.

(3) Comparison of calculational results from different types of codes of the various partners with the MEXICO measurement data. This is proving to be a thorough validation of different codes and it provides insights into the phenomena that need further investigation.

(4) Deeper investigation into the observed phenomena. A deeper investigation of different phenomena using isolated sub-models, simple analytical tools, or by physical rules. These phenomena that are investigated include 3-D effects,

FIG. 9.3 THE MEXICO WIND TUNNEL EXPERIMENT IN THE DNW WIND TUNNEL (*Source*: [ECN, photo Toon Westra]).

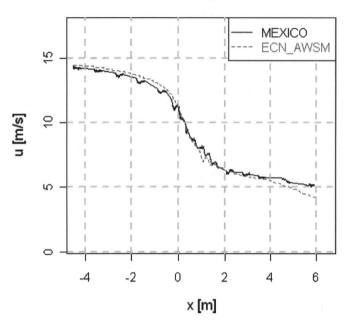

u [m/s], r=1.8 m, U∞=15 m/s

FIG. 9.4 MEXNEXT RESULT: AXIAL TRAVERSE FROM PIV MEASUREMENTS COMPARED TO AWSM PREDICTIONS (*Source*: [ECN])

instationary effects, yawed flow, non-uniformity of the flow between the blades (i.e., tip corrections), and the wake flow at different conditions.

(5) Comparison with results from other (mainly NASA-Ames) measurements. It is investigated whether these findings are consistent with results from other aerodynamic experiments, particularly the data provided within IEA Wind Task 20 by NREL from the NASA-Ames experiment.

A large variety of results have been obtained, the data are analyzed on quality and tunnel effects. PIV measurements in the wake have been analyzed, and a comparison has been made between the observations from the NASA-Ames and MEXICO experiments. The effects from airfoil imperfections have been estimated which is important for the comparison with models. A typical example of the comparison between calculated and measured results are shown in Fig. 9.4 where the results are compared to predictions by the ECN code AWSM, a numerical code based on the Generalized Prandtl's Lifting Vortex Line Method.

9.5.6 DU Airfoil Data

In the last 15 years DUT has developed many airfoils for wind turbine application [19]. These airfoils are called the DU airfoils and are applied world-wide in wind turbine blades ranging from 6 to 60 m. The development of the DU airfoils has been supported by the European Commission, the Dutch Ministry of Economic affairs and the DUT in the framework of the educational and research program of the section Wind Energy of the Faculty of Civil Engineering and Geosciences, which is at present the Wind Energy section of the Faculty of Aerospace Engineering. The two-dimensional aerodynamic characteristics of the DU airfoils have been extensively verified in the Delft University low-speed low-turbulence wind tunnel for various Reynolds numbers. In addition to the clean configuration for Reynolds numbers ranging from

1 million to 3 millions, the characteristics of several airfoils with vortex generators, zigzag tape, trip wires, trailing edge wedges or Gurney flaps of 1% and 2% chord heights have been established. In some cases a stethoscope was used to determine the transition location. The DU airfoils are included in the ATG software package of ECN.

9.6 RESEARCH PROGRAMS

The research programs described here concern the programs of ECN Wind Energy, Knowledge Centre WMC and the Wind Energy group at DUT, DUWind.

9.6.1 Energy Research Centre of the Netherlands ECN and Knowledge Centre WMC

The Energy Research Centre of the Netherlands is dedicated to the research in efficient use of energy and infrastructure, deployment of renewable energy sources, clean conversion of fossil fuels and development of energy analyses and policies. While targets for the near and medium term have been set in Europe, substantial further acceleration of new technologies is required to meet the long-term goals of reducing the dependence on fossil fuels and the CO_2 emissions. ECN aims to carry out groundbreaking research that will have a major influence on energy transition and brings technologies to every stage of development. The strength of ECN lies in its portfolio, which enables the development of new generations of technologies. Many technologies developed by ECN have reached maturity in recent years, increasingly resulting in third-party economic activity. ECN focuses on the needs of government and industry. In addition, it conducts contract research for companies and governmental institutions and to a large extent for the European Union. ECN collaborates intensively with other knowledge institutes and universities.

ECN Wind Energy has the mission to develop high-quality knowledge and technology for large-scale cost effective application of wind energy and transfer these to the market. ECN Wind Energy has the ambition to contribute substantially to lowering the production costs of offshore wind energy to a level at which it is competitive with fossil-fuel generation in 2020.

The long-term research program of ECN Wind Energy is combined with the long-term research program of Knowledge Centre WMC because of the close co-operation between the two organizations and the fact that the programs are complementary and dependent on each other. ECN Wind Energy and WMC organize their research in four priority areas:

(1) Rotor and Farm Aerodynamics
(2) Integrated Wind Turbine Design
(3) Operation and Maintenance
(4) Materials and Structures

The research is supported by extensive experimental facilities and an Experiments & Measurements group that is MEASNET and ISO17025 accredited.

In the field of **Aerodynamic** research, ECN Wind Energy aims at optimizing the aerodynamic performance of the wind turbine rotor and of the wind farm as a whole. By developing new knowledge, methods and design software, the results of the long-term research can directly be transferred to the industry. An additional result of the research is that Intellectual Property is generated. The tools and Intellectual Property are used for research purposes of turbine and blade manufacturers and project developers or are directly applied.

A unique feature of its research is the combination of theoretical and numerical research with the various experimental test facilities: wind tunnels, scale wind farm and full-scale test wind farm in the Wieringermeer. The aerodynamic research is divided in rotor aerodynamics and wind farm aerodynamics.

Research in the field of rotor aerodynamics aims at the development of advanced aerodynamic design tools. ECN has a tradition in developing BEM, but also developed the aerodynamic free wake model based on lifting line panel method AWSM. A new code is being developed where also the boundary layer is being modeled. This method where the external pressure field of the rotating panel method that includes the wake is combined with a three dimensional unsteady boundary layer model will lead to an accurate method that is significantly faster than full Navier–Stokes field solvers. It is called ROTORFLOW. In a Dutch consortium ECN has developed the RFOIL 3D code, a modified version of the XFOIL software. The progressively advanced aerodynamic design tools are used to design and model future large wind turbine rotors with increased accuracies.

Research in the field of wind farm aerodynamics aims at the understanding of the flow field in and around wind farms, which is especially relevant for the construction and operation of large offshore wind farms. Insight in wake effects in wind farms is essential. Computational fluid dynamics (CFD) are applied for the further understanding of flow phenomena in wind farms. The research comprises the development of validated wake models and CFD wind farm models using commercial solvers as well as the development of dedicated wind farm aerodynamics CFD solvers. Interesting model challenges are combined, such as the modeling of the atmospheric boundary layer, the flow inside the wind farm, the modeling of the turbine including its control, the wind farm control as well as the optimization of this system.

The FarmFlow program is a validated tool developed by ECN to calculate the wake effects of large offshore wind farms. It is unique that it calculates both wake losses and added turbulence levels with an accuracy currently unmatched. The FarmFlow tool allows the user to accurately optimize the lay-out of the wind farm with respect to power output and, in conjunction with aero-elastic codes, the possibility to perform design calculations on wind turbines that are placed in wind farms (Fig. 9.5).

In the field of **Integrated Wind Turbine Design** research, ECN Wind Energy and WMC aim at the improvement of the design process in industry by developing new, improved models, concepts and tools for the integrated system design. ECN focuses on aero-elasticity, control and concept studies. WMC is dedicated to the development of the integrated design tool and the blade design module. The research in the field of integrated design is among others.

Aero-elasticity: rotor aerodynamic models are coupled to structural dynamic models and the resulting aero-elastic models are verified and validated against measurements and used for integrated design. Improvements are needed for simulations concerning oblique inflow, individual pitch, large deformations and complex drive train and support structure geometries. Coupling the improved aerodynamic models to structural dynamic models (multi-body or FEM) improves the results obtained from simulations. To enable a straightforward coupling the ECN develops the Aero-module. This module couples the aerodynamic codes to aeroelastic codes, such as TURBU and PHATAS which have been developed by ECN.

Research in wind turbine control aims at the improvement of the present control algorithms, and the development of new ones, mainly targeting reduction of the cost of energy. To this end, the following objectives are pursued:

- optimal power and rotor speed control,
- load reduction in the mechanical components of the wind turbine,
- high availability and reliability of the wind turbine,
- easy and robust commissioning of the controller,
- coping with all grid requirements.

ECN developed the user-friendly ECN Control Design Tool which provides the wind turbine control engineer with an excellent set of tools in an open-source MATLAB environment, covering the complete process of turbine modeling, control design, stability analysis and compilation of the final controller into executable code for linking to advanced simulation software for load calculations or to dedicated hardware. Furthermore, the following four major control components of the integral approach are investigated

- Optimized feedback control (OFC), aiming at removing limitations for up-scaling of wind turbines, such as high turbine loads and stability problems,
- Extreme event control (EEC), dealing with prevention of unnecessary standstill and/or increased loads due to extreme events,
- Fault-tolerant control (FTC), having the task of detection of minor (sensor) faults followed by recovery actions,
- Optimal shutdown control (OSC), for preventing the accumulation of damage in cases of a turbine shut-down caused by a severe failure.

ECN develops design tools for large wind turbines and these tools are constantly improved in order to address aeroelastic stability issues and to minimize cost of energy. The advanced tools are developed to reduce uncertainties in design load calculations, through an improved modeling of the physics involved. ECN upholds the quality and validity of its computer programs through participation in European benchmark studies and through feedback from field experiences. Aeroelastic and hydromechanical behavior of a wind turbine are modeled using sophisticated tools like, e.g., AeroModule (including BEM and AWSM for rotor aerodynamics) and PHATAS and TURBU for dynamic load modeling of the turbine and support structure in time domain and frequency domain respectively. Wind and wave fields for use in the time domain code are generated with the SWIFT and ROWS code respectively. For quick assessments of aeroelastic turbine behavior the code BLADMODE is used and development and testing of custom made commercial control algorithms for wind turbines is done with the Control Design Tool. The BLADOPT: program optimizes rotor blade geometry for lowest cost of energy. WMC has developed the FOCUS code which includes several of these design tools.

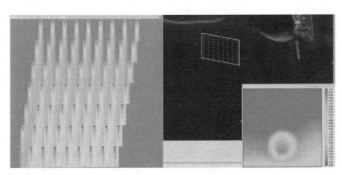

FIG. 9.5 ECN FARMFLOW WIND FARM CALCULATION MODEL (*Source*: [ECN])

Research in the field of Electrical Systems and Components aims at the design and optimization of electrical systems in a wind turbine that is connected to the wind farm grid; statically and dynamically. ECN developed, in collaboration with TUD Electrical Power Processing, two types of models: EeFarm and DynFarm. These programs are used with the developed FarmFlow code that calculates the wake losses in wind farms to optimize offshore wind farm designs.

For more than 15 years ECN Wind Energy is developing methods and tools to optimize the **Operations and Maintenance** (O&M) of offshore wind farms and in fact, ECN is one of the few institutes carrying out long-term R&D on this topic. The scope of work comprises the development of models and software (Decision Support Tools) to analyze the O&M aspects, to determine costs and downtime, identify cost drivers, and to optimize the O&M strategy and the development of diagnostics and measurement techniques for those situations where condition based maintenance is preferred instead of preventive or corrective maintenance. ECN's leading role in the field of O&M for offshore wind farms has been recognized for many years. Major project developers make use of ECN's knowledge and models to develop O&M strategies for their offshore wind farms. Turbine designers apply the knowledge in the design process of offshore wind turbines.

WMC aims with the research on **Materials and Structures** at increasing its knowledge of the structural behavior of the blade and its materials under the complex loading and the environmental conditions and to make this knowledge applicable for wind turbine and blade manufacturers. With this knowledge current designs and the design of the next generation of blades are further optimized.

9.6.2 Delft University of Technology DUT

Research on wind energy at the Delft University of Technology began 30 years ago, starting with the tip vane project, an aerodynamic research project at the faculty of Aerospace Engineering. Nowadays DUWIND is the wind energy research organization of the Delft University of Technology. Since an increasing number of research questions require a multi-disciplinary approach, DUWIND was established in August 1999 as an interfaculty research organization, specifically for wind energy. World-wide, DUWIND now is one of the largest academic research groups in wind energy. Its research program covers almost all aspects of modern wind turbine technology, and is undertaken across 5 faculties in 13 groups. The five faculties are 1. Faculty of Aerospace Engineering, 2 Faculty of Civil Engineering and Geosciences 3.Faculty of Electrical Engineering, Mathematics and Computer Science, 4 Faculty of Mechanical and Materials Engineering and 5. Faculty of Technology, Policy and Management. Each of the research groups at these faculties has its own specific expertise therefore the DUWIND research group covers a wide area of wind energy research.

The research program of DUWIND has been reformulated and is published in October 2008. The program encompasses approximately 60 PhD positions, which implies a doubling in size of DUWIND compared to 2007. From the start the core of the research program is the technology development of wind turbines and wind farms, including the underlying basic research. The program focuses on long-term research efforts, mainly performed by PhD researchers. Long term is to be understood as leading to useful results in a 5- to 10-year time frame after starting the PhD research. The program is made up and executed by the research groups at the five TU-Delft faculties, and is "bottom-up": a new topic is included only when it can be embedded in an existing expertise group, and when funding is found. The PhD students will have activities in five sub-programs:

- Unsteady aerodynamics
- Smart structure rotors
- Design methods
- Offshore components and design
- Dutch wind energy in Europe.

The focus of DUWIND program is on the development of turbine and wind farm technology, ranging from basic research through technology development to design support for the industry. Being a university, DUWIND provides courses for students and for professionals in the wind energy industry. As is indicated before, DUWIND closely co-operates with the wind energy group of ECN. Many projects are performed jointly, and mutual use is made of the research facilities. DUWIND and ECN together form the Dutch node of the European Academy of Wind Energy EAWE. DUWIND is part of the international wind energy research community. It is one of the founding members of the European Academy for Wind Energy and has started the Academy conference series 'The Science of making Torque from Wind'. DUWIND contributes to the Wind Energy Technology Platform of the European Commission, the European Energy Research Alliance (EERA) and to the wind energy programs of the International Energy Agency. It has participated in all Framework Programs of the European Commission.

9.7 EXPERIMENTAL RESEARCH INFRASTRUCTURE IN THE NETHERLANDS

9.7.1 Blade Testing and Material Research at WMC

Knowledge Centre WMC is a research institute for materials, components and structures. The major activities are fundamental and applied research on fiber reinforced plastics and wind turbine structures. WMC is active in research projects for the European and Dutch governments and industry and performs blade and material tests on contract basis for the international industry. WMC develops design tools, the FOCUS software, that are being used worldwide by many of the largest wind turbine manufacturers. Results of the research work are published and presented internationally.

The Knowledge Centre WMC has a history from 2000, when it was started as "WMC-Group" in the Delft University of Technology doing quite a lot of full-scale testing work on rotor blades. However, the test facility at Delft University of Technology became too small for the next generation of large rotor blades. Therefore, in order to keep track with the market's needs another, larger location needed to be found to continue the activities. A larger test facility was built in Wieringerwerf, in the northern part of the Netherlands where a new industrial site and yacht-basin "Waterpark Wieringermeer" was developed.

For the new laboratory a location was chosen close to open water, on the borders of the IJsselmeer. Now the sometimes very large structures that are brought for testing can also be transported by water. Starting 2003, as the centre moved from Delft to Wieringerwerf, the former "WMC-Group" of the DUT continued its work as a new foundation with a new name: Knowledge Centre WMC, established by the Delft University of Technology and the Energy research Centre of the Netherlands ECN. With its links to both organizations the new Knowledge Centre WMC can continue to combine fundamental and applied research on wind turbine and fiber reinforced plastics structures.

With the background of continuous testing and development of tests, WMC is actively involved in international standardization committees. For the experimental research WMC has a laboratory for material research and a large test area with a dedicated strong floor enabling testing of structures of over 60 m in length, such as rotor blades for large wind turbines. Testing machines up to 300 tons are available for material and component testing, both static and fatigue. The facility is one of the largest of its kind and has its own mechanical, electro technical and hydraulic workshops for development and maintenance of equipment and test rigs.

The research topics for WMC in national and European research projects are amongst others:

- Material behavior under complex and fatigue loading
- New materials
- Condition monitoring
- Reliability of design codes
- Reliability of full-scale testing
- Connection methods
- Design recommendations and standards on full-scale testing of wind turbine rotor blades

On the other hand, WMC is performing more fundamental research activities in the context of many international projects. WMC initiated the EU project OPTIMAT BLADES to investigate the behavior of Fiber Reinforced Plastics and WMC is work package leader for materials research in the UPWIND project. The projects executed for industrial research purposes include:

- Testing of materials
- Numerical analyses
- Design evaluation
- Verification tests on mechanisms
- Strength and fatigue behavior of components and connections
- Development of design software: the integral wind turbine design software FOCUS
- Full-scale verification tests on (sub)components such as:
  - Wind turbine rotor blades (or parts thereof)
  - Wind turbine pitch bearings and hubs
  - Connections

For larger wind turbines, the potential power yields scales with the square of the rotor diameter, but the blade mass scales to the third power of rotor diameter (square-cube law). With the gravity load induced by the dead weight of the blades, this increase of blade mass can even prevent successful and economical deployment of very large wind turbines. In order to meet this challenge and allow for the next generation of larger wind turbines, higher demands are placed on materials and structures. This requires more thorough knowledge of materials and safety factors, as well as further investigation into new materials with a higher strength to mass ratio. Furthermore, a change in the whole concept of structural safety of the blade is required. WMC is performing research activities in this field in order to improve both the empirical and fundamental understanding of materials, extend the material database, study effective blade details, establish of tolerant design concepts and probabilistic strength analysis and establish material testing procedures and design recommendations.

9.7.2 ECN Wind Turbine Test Site Wieringermeer

At the Petten site, ECN has operated a wind turbine test site for many years. The size of the turbines allowed at the site is limited and therefore it is not in use anymore since the year 2000.

FIG. 9.6 ECN WIND TURBINE TEST SITE PETTEN
(*Source*: [ECN, Picture by Aris Homan])

From 2003, ECN has expanded its wind turbine facilities with a wind turbine test site in the municipality of Wieringermeer. The site is located just south of the village Kreileroord, at about 30 km distance from ECN's main offices in Petten (Fig. 9.6). This unique facility is a combination of:

Four locations for prototype wind turbines up to 6 MW, recently expanded with a fifth location for prototype wind turbines up to 10 MW.
Five Nordex N80/2500 research wind turbines; these wind turbines are equipped for experimental research.
Three meteorological masts, 108 m, 108 m, and 100 m high equipped with atmospheric measurement equipment.
A scaled wind farm consisting of ten turbines rated 10 kW and 14 meteo masts up to 19 m height.
A measurement pavilion with offices, a computer centre and advanced glass fibre based data acquisition system.

The research wind farm — consisting of five Nordex N80 wind turbines — enables ECN to perform wind farm specific research and development programs. The site also comprises five prototype locations. These locations enable manufacturers to test, optimize and certify prototypes together with ECN. Supporting facilities are three meteo towers, a 36 MVA grid connection, data collection equipment and a test site control center. A fourth meteo mast of 100 m height is expected early 2011. The test site also accommodates a unique facility, which allows ECN to perform accurate wind field measurements in a scale wind farm. This Scaled Wind Farm — consisting of ten Aircon P10 turbines plus fifteen measurement towers — allows for the development and testing of wind farm specific control strategies. The site has a favorable wind climate: the average wind speed at 100 m height is 8.3 m/s. With this wind climate, not only do the five Nordex N80 turbines produce about 30,000,000 kWh per year from which the income of the green electricity gives a solid base for the financial exploitation of the site, the wind conditions allow for fast fulfillment of the experimental capture matrices of the test turbines.

Many research programs have been carried out at the experimental facility of ECN, also many European programs. The fact that the research turbines are owned by ECN and a proper research agreement has been reached with the manufacturer, allows ECN to optimally use these research turbines for its R&D programs. What was obvious from many defined programs is that it is very difficult to execute these programs in commercial wind farms since wind farm operators are reluctant to facilitate research programs that might conflict with maximum energy production.

ECN investigates aspects like operation and maintenance strategies; wake effects, noise effects, effect on birds and condition monitoring. Other important activities are the development of improved wind farm control strategies and new advanced measurement techniques.

Currently, in 2010, plans are made to extend the research wind farm. Some fourteen turbines in the range from 5 MW to 8 MW are to be set up over the coming years, at least seven will be dedicated offshore turbines. ECN intends to purchase seven turbines for further research activities and a further seven proto-type test locations will be made available. A fifth proto-type location has already been created and a XEMC-Darwind turbine is being built and will be tested in 2011.

9.7.3 ECN Scale Wind Farm

A recent addition to the ECN experimental research infrastructure is the ECN scale wind farm. It has been designed to further advance the knowledge in wind fields in and around wind farms, including the understanding of wakes and turbine-turbine interaction. The development of large-scale offshore wind power implies the construction of even larger wind farms. At the moment, the large uncertainties connected to the wind field in the wind farm leads to financial risks when investing in these large wind farms. Furthermore, the cost of wind energy should be reduced even further. Therefore, improved models are required describing the flow within and around wind farms so that optimized wind farm control strategies can be developed. The high-quality data of the ECN scale wind farm are used for the development and validation of wind farm aerodynamic models and wind farm control strategies (Fig. 9.7).

The understanding of unsteady wind fields within and around wind farms could be greatly increased when more detailed models or measurements of the wind field in a wind farm would be available. The same applies to understanding the response of many turbines within a wind farm on the mutual wakes. One reason for this is that adequate measurements are lacking and the models are not (yet) accurate enough. In full-scale wind farms meteorological masts are very expensive and the number of masts is thus limited. On the other hand, the value of modeling wind farms in wind tun-

nels is limited due to scaling effects. Therefore, ECN has overcome this problem by building the scaled wind farm facility which consists of relatively small wind turbines together with many measurement masts that measure the wind conditions in the wind farm and above the wind farm. The scale of this wind farm is not too small to alleviate the dominant scaling effects and the scale is not too large to permit the building of sufficient meteorological masts.

The ECN scaled farm consists of ten permanent magnet, direct drive, pitch-controlled wind turbines. The turbines have 10 kW rated power, a rotor diameter of 7.6 m and a hub height of 7.5 m. It is essential that the researchers have full access to the hardware and software of the wind turbines. The scaled farm has been designed in a way that allows ECN performing experiments without any risks for the environment as well as the turbines themselves. As a result, ECN is able to adapt the controllers as well as the turbines for the dedicated experiments. A dedicated wind farm controller has been installed. Inside and around the wind farm a network of fourteen measurement masts has been installed, which measure the wind velocity field from 3.6 m to 19 m height. This covers the rotor area and up to one rotor diameter above the rotor. The large number of meteorological masts within the wind farm permits to measure at the same time single, double, triple and quadruple wakes while simultaneously measuring the external conditions with three nearby 108 m meteorological masts. The unusually densely spaced wind measurements gives the unique possibility to capture the complete wind field, which gives valuable additional information compared to the usual measurement of the wind speed at a single location. Furthermore, most of the wind measurements will be performed using 3D sonic anemometers thus capturing the three wind velocity vectors of the wind field.

The scaled farm is located in ECN's large wind turbine test field in Wieringermeer in between the prototype turbines. The scaled wind farm and its surroundings are characterized by flat terrain, consisting of mainly agricultural area, with single farmhouses and rows of trees. The lake IJsselmeer is located at a distance of 1 km East of the scaled wind farm. Great care has been taken to ensure undisturbed inflow of the wind in the scaled wind farm.

This worldwide unique research facility shall give further insights in the field of wind farm aerodynamics, wake interaction and wind farm control. The high quality data are used for the development and validation of wind farm aerodynamic models and wind farm control strategies. This will allow operating a wind farm at maximum efficiency while guaranteeing at the same time a maximum in reliability and a minimum in mechanical and electrical loads. A unique feature installed in the wind farm controller is that the controller can adjust the yaw of the turbines to exact positions, as well as the pitch angle. Since it has been measured that the yaw angles of the ECN research turbines

FIG. 9.7   THE SCALE WIND FARM IN BETWEEN TWO PROTOTYPE TURBINES (*Source*: [ECN])

FIG. 9.8   THE SCALE WIND FARM (*Source*: [ECN])

are seldom equal when the wind is along the row, this feature will make comparison with models more easy and the experimental results more accurate. The side by side comparison of the two rows allows the demonstration of small differences due to changes in turbine or wind farm control (Fig. 9.8).

9.7.4 Experimental Facilities at Delft

DUWIND has access to all experimental facilities of TU-Delft. The most important ones that have an important role in the wind energy research are:

- The wave tank of the faculty of Civil Engineering and Geosciences, where the scale model of the motion-compensating offshore access system, the Ampelmann, has been tested.
- The Low Speed Low Turbulence Tunnel of the faculty of Aerospace Engineering, where all of the Delft University airfoils have been measured. The test section is 1.25*1.80 m, with a max wind speed of 120 m/s. The LST has an very low turbulence level: <0.1%.
- Structures and Materials Laboratory of the faculty of Aerospace Engineering, where innovative composite materials, such as thermoplastic composites and smart materials, are produced and tested.
- The Smart Structures laboratory at the Delft Centre for Systems and Control, faculty of Mechanical, Maritime and Materials Engineering is set up to run real time control on structures and to test the performance of active laminates produced at the materials laboratory.

Especially for wind energy research, the wind tunnels are important and are elaborated in the next section.

9.7.5 Wind Tunnels at TU Delft

The Delft University of Technology has a large selection of wind tunnels available with a wide range in wind speed from low speed to large speed. The most recent addition, the Open Jet Facility, is very relevant for wind energy application. The tunnels are applied for fundamental research as well as for applied research and many students use the tunnels for their practical. The wind tunnels available at DUT are indicated in Table 9.1.

The Delft Open Jet Facility

In 2009, at the Delft University of Technology, a new octagonal 500 kW 30 m/s open jet facility began operation that can test model rotors up to 1.8 m for concepts such as flexible smart dynamic rotors and controls. For this kind of experiments the large advantage of the OJF over the other wind tunnels at DUT is that it has an open jet and an outlet diameter of almost three meters and can handle very large models that may obstruct the airflow quite considerably. The new wind tunnel offers more possibilities than ever before for teaching, such as laboratory courses involving model rotors, and research; parameter studies can be executed effectively and efficiently. The initial plans for the OJF were made in the 1980s in response to growing interest in wind energy. The development was delayed because of several reasons, one of them the large budget that was needed for the construction. When the wind group at DUT became part of the Faculty of Aerospace Engineering in 2003 it was possible to seriously proceed with the construction of the facility. The requirement was that the OJF should be versatile and also be used for other research than research into wind turbines. Construction of the OJF started in November 2006 and the opening of the tunnel was in 2009.

The dimensions of the OJF are very impressive. A large fan powered by a 500 kW electric motor enables it to achieve a maximum speed of around 120 km per hour. Air is rotated 180 degrees through a long diffuser and two rows of corner vanes. It then passes through a short diffuser before entering the "settling chamber." Here, five fine-mesh screens reduce the turbulence and velocity deviations in the airflow. Via a contraction the air is then blown into the test section as an even jet stream and cooled at the end by an enormous cooling radiator and guided back to the fan. Some of the many unique technical features in this project are the high

TABLE 9.1   LIST OF WIND TUNNELS AT DELFT UNIVERSITY OF TECHNOLOGY

			Low-speed windtunnels		
Name	Type	#	Dimensions test section ($W \times H$)	Cross-section	V_{max} (m/s)
BXF	oj	4	5 × 10 cm	rectangular	25
M-tunnel	oj/cc	1	40 × 40 cm	rectangular	35
W-tunnel	oj	1	40 × 40 cm	rectangular	35
BLT	cc	1	125 × 25 cm (×540 cm)	rectangular	38
LTT	cc	1	180 × 125 cm	octagonal	120
OJF	cc	1	285 × 285 cm	octagonal	35
			High-speed windtunnels		
Name	Type	#	Dimensions test section ($W \times H$)	Cross-section	M-Range
TST-27	bd	1	280 × 270 mm	rectangular	0.5–4.2
ST-15	bd	1	150 × 150 mm	rectangular	0.7–3.0
ST-3	co	1	30 × 30 mm	rectangular	1.5–3.5
HTFD	bd	1	350 mm	circular	6.0–11.0

FIG. 9.9   THE ROTOR OF THE WIND RACE CAR OF ECN IN THE OJF (*Source*: ECN, Picture by P. Eecen)

power motor, the modular adjustable frequency drive, the thermal sensors, and the large cooling fan (Fig. 9.9).

The OJF is mainly used by PhD students, graduates and members of the permanent academic staff. It fulfils an important role in research into the aerodynamic effects that wind can have on buildings and ships as well as in the field of sports. The vast majority of the models are made in the faculty's own workshop. To accommodate these research activities, the OJF is constructed to be extremely versatile. The wind energy group at DUT uses the OJF extensively, for instance for the verification and validation of the various calculation models. In the context of the large scale European research project Upwind experiments in the OJF contribute to the definition of new concepts for wind turbine blades, including smart rotors.

9.7.6   German–Dutch Wind Tunnels at DNW

DNW, the German–Dutch Wind Tunnels, is a non-profit foundation and was established by the German Aerospace Center DLR and the Dutch National Aerospace Laboratory NLR. Its headquarters are in the Noordoostpolder in the Netherlands and it has wind tunnels situated in a number of locations in the Netherlands and Germany. DNW operates its own large, low-speed facility and the aeronautical wind tunnels of DLR and NLR. DNW provides solutions for the experimental simulation requirements of aerodynamic research and development projects. These projects can originate in the research community (universities, research establishments or research consortia) or in the course of industrial development of new products. Most of the industrial development projects originate in the aeronautical industry, but the automotive, civil engineering, shipbuilding and sports industries have also benefited from DNW's capabilities.

The wind tunnel of DNW used for the experiments in the European project Mexico is the large low-speed facility (LLF). The DNW LLF wind-tunnel facility allows testing over a wide range of conditions and flight regimes. Due to the modular design of the wind tunnel a number of test sections can be used. These include open jet, 9·5 m × 9·5 m ($0 < V < 62$ m/s), 8 m × 6 m ($0 < V < 116$ m/s) and 6 m × 6 m ($0 < V < 152$ m/s) closed test sections. The open jet wind-tunnel configuration allows large model heights above the ground at zero to low speeds. The closed jet test sections enable testing to

be conducted over a higher range of airspeeds. The Mexico experiment, described earlier has been performed in this wind tunnel.

9.7.7   The Offshore Wind Farm Egmond aan Zee (OWEZ)

In the beginning of the century, the government formulated a target for wind energy that required the development of offshore wind farms due to the restricted possibilities onshore. Besides the focus of the wind energy research on offshore developments, an offshore demonstration wind farm should be built relatively close to shore in order to demonstrate the feasibility, but more importantly acquire knowledge and subsequently decrease the cost of offshore wind energy. The government decided on the final location for the demonstration project 100-MW Near Shore Wind farm. The location is situated in Dutch territorial waters of the North Sea between 10 and 18 km from the coast near the village of Egmond aan Zee. After a tender procedure, the Egmond Building Combination (EBC), a joint venture of Ballast Nedam and Vestas, built the 108 MW wind farm on the order of Shell Renewables and Nuon. The Offshore Wind farm Egmond aan Zee (OWEZ) comprises 36 Vestas V90 wind turbines of 3 MW each and associated support systems. Each wind turbine is connected by a transition piece to a steel monopole foundation, piled to a penetration depth of about 30 m. The power generated is transmitted through three 34 kV cables to shore, which land north of IJmuiden harbour. A substation, located near Wijk aan Zee, transforms the voltage from 34 kV to 150 kV and transmits the power into the national grid. Investment costs are around 200 million and financed on balance by Nuon and Shell. The wind farm initially has been operated by EBC under a 5-year warranty, operations, and maintenance contract. The wind farm produced approximately 350 GWh per year.

In the context of realizing the wind farm, an extensive Monitoring and Evaluation Program (MEP) was carried out. In that context, a meteorological mast was installed and data from the turbines were collected. The meteorological mast has a height of 106 m above sea level and measures the wind conditions at three levels. Data collected in the MEP includes contractual issues, project organisation; permits; technical description of the design, support structure, wind turbines, and electrical design; assembly and installation; planning versus execution; budget; health, safety, security, and environment management; risk management; financing, insurance, and power purchase agreements; quality assurance management; requirements and qualifications; monitoring and evaluation program; and lessons learned [20]. This information has been made available [21]. Commercially sensitive data that has been collected during the execution of the MEP is on the subjects: corrosion and lightning; dynamics of turbines; aero-elastic stability; scour protection; electricity production, disruptions, failure data, availability, maintenance, and reliability; power quality, grid stability, and power forecasts; and wind turbine P–V curve and wake effects. ECN and DUT have executed several projects with selections of this data under an NDA agreement with NoordzeeWind.

9.7.8   The Offshore Wind Farm 'Prinses Amalia Windpark'

The second offshore wind farm in Dutch National waters of the North Sea is Prinses Amaliawindpark. This wind farm is situated some 23 km offshore from IJmuiden, in block Q7 of the Dutch continental shelf. This is just outside the 12-mi zone south-west of OWEZ. The wind farm consists of 60 Vestas V80 wind turbines of 2 MW each with a total capacity of 120 MW. The water depth at the site is between 19 m and 24 m. The project is owned and developed by a group of companies of ENECO Holding NV, Econcern BV, and Energy Investment Holdings. It was built by Vestas Wind Systems A/S, Van Oord Dredging, and Marine Contractors BV under separate construction contracts. Initially, the windfarm is operated by Vestas Offshore, an affiliate of Vestas, under a 5-year warranty, operations, and maintenance contract. During its operation, more and more research is carried out at the 'Prinses Amaliawindpark', also in the context of the FLOW program.

9.8 SUMMARY

This Chapter is an attempt to provide insight in the wind energy research of the Netherlands. The result is not complete and must be regarded as a personal selection of the activities that have been performed in all the years of wind energy research in the Netherlands. A more extensive overview can be found in Verbong [22], written in Dutch. The research in wind energy is concentrated at the wind energy department of the Energy research Centre of the Netherlands ECN and the interfaculty wind energy department DUWIND at Delft University of Technology. Both institutes have been involved in wind energy research from the start in the 1970s and closely match their research programs. ECN Wind Energy has a research staff of 55 scientists and DUT has a research staff of 15 permanent researchers and more than 35 PhD students. Together ECN and DUT belong to the top 5 of the international wind energy research groups.

In the Netherlands, the wind energy research is supported by an extensive experimental infrastructure. The Knowledge Centre WMC that has been founded by the DUT and ECN has a research staff of 25 scientists and is a research institute for materials, components and structures. WMC is performing blade tests for large wind turbines to 60 m in length. ECN has a research wind farm where proto-type wind turbines are tested, where a research farm of 5 full-scale turbines are used for research activities and where a scale wind farm is located for research on farm control and wind farm aerodynamic research. At DUT a large selection of experimental facilities are being used for wind energy applications. The most prominent facilities are the wind tunnels, of which the Open Jet Facility is the most recent addition.

9.9 ACRONYMS AND INTERNET

2B-energy	www.2-benergy.com
AgencyNL	www.agentschapnl.nl, formerly known as SenterNovem www.senternovem.nl
DUT	Delft University of Technology — www.tudelft.nl
EAWE	European Academy of Wind Energy — www.eawe.eu
ECN	Energy research Centre of the Netherlands — www.ecn.nl
EERA	European Energy Research Alliance — www.eera-set.eu
EWEA	European Wind Energy Association — www.ewea.org
EWT	Emergya Wind Technologies — www.ewtinternational.com
IEA	International Energy Agency — www.iea.org
IEC	International Electrotechnical Commission — www.iec.ch
Lagerwey Wind	www.lagerweywind.nl
LSEO	Landelijke Stuurgroep Energie Onderzoek
MEASNET	International Network for Harmonised and Recognised Measurements in Wind Energy — www.measnet.org
MEP	Monitoring and Evaluation Program executed at the Offshore Wind farm Egmond aan Zee
NLR	Nationaal Lucht- en Ruimtevaartlaboratorium — www.nlr.nl
OWEZ	Offshore Wind farm Egmond aan Zee — www.noordzeewind.nl
RCN	Reactor Centre of the Netherlands
STX-Harakosan	www.stxwind.com
TNO	Dutch organization for Applied Scientific Research
TPWind	European Wind Energy Technology Platform — www.windplatform.eu
VWEC	VWEC Wind Energy Consult
WMC	The Knowledge Centre Wind turbine Materials and Constructions — www.wmc.eu
XEMC Darwind	www.xemc-darwind.com

9.10 REFERENCES

1. Energie Onderzoek Subsidie (EOS) carried out by Agentschap NL (www.agentschapnl.nl, www.senternovem.nl/eos).

2. Strategic Research Agenda, EWEA 2005.

3. IEA Large-Scale WECS Annual Reports, 1978 to 1990.

4. IEA R&D WECS Annual Reports, 1978 to 2010.

5. P.E.J. Vermeulen, Report of the IEA Technical Meeting on Wakes and Clusters,, February 2-3 1981, MT-TNO 81-04676.

6. P.E.J. Vermeulen, P.J.H. Builtjes, J.B.A. Vijge, Mathematical modeling of wake interaction in wind turbine arrays, MT-TNO 81-01473.

7. G. A. M. van Kuik, J. W. M. Dekker, The FLEXHAT program, technology development and testing of flexible rotor systems with fast passive pitch control, Journal of Wind Engineering and Industrial Aerodynamics, Volume 39, 1992, Pages 435-448.

8. J.G. Schepers, H. Snel, Final results of the EU JOULE projects 'DYNAMIC INFLOW', presented at the ASME '96, ECN-RX-95-062.

9. J.G. Schepers et al., Final report of IEA ANNEX XIV. Field Rotor Aerodynamics, ECN-C-97-027.

10. H.B. Hendriks and M. Zaaijer, DOWEC Dutch Offshore Wind Energy Converter 1997 – 2003. Executive summary of the public research activities.

11. G.P. Corten and H.F. Veldkamp, Insects can halve wind-turbine power, NATURE, VOL 412, 42-43, 5 July 2001.

12. P.J. Eecen, S.A.M. Barhorst, 'Data added to International Wind Database, IEA Annex XVII', ECN-C-03-128.

13. P.J. Eecen, S.A.M. Barhorst, 'Extreme wind conditions, Measurements at 50m Meteorological Mast at ECN, Petten', ECN-C-04-019.

14. G.C. Larsen and K.S. and Hansen, Structure and Philosophy, RISO-R-1299.

15. www.ampelmann.nl.

16. J.G. Schepers et al. Verification of European Wind Turbine Design Codes, VEWTDC. European Wind Energy Conference, EWEC, Copenhagen, July 2001.

17. H. Snel, J.G. Schepers and B. Montgomerie. The Mexico project the database and first results of data processing and interpretation In 'The Science of Making Torque from the Wind', August 2007.

18. S. Schreck. IEA Wind Annex XX: HAWT Aerodynamics and Models from Wind Tunnel Measurements. NREL/TP-500-43508, NREL, December 2008.

19. W.A. Timmer and R.P.J.O.M. van Rooij, Summary of the Delft University wind turbine dedicated airfoils, AIAA-2003-0352.

20. Offshore Windfarm Egmond aan Zee, General report OWEZ_R_141_20080215.

21. www.senternovem.nl/offshorewindenergy/technology_nsw/ www.monitoring_mep-nsw/index.aspwww.noordzeewind.nl.

22. G. Verbong, Kwestie van lange adem, de geschiedenis van duurzame energie in Nederland (history of renewable energy in the Netherlands).

ROLE OF WIND ENERGY TECHNOLOGY IN INDIA AND NEIGHBORING COUNTRIES

M. P. Ramesh

ABSTRACT

Renewable energy technologies except for large hydroelectric power have, until recently, played a marginal role in most of the countries in terms of energy supply mix and it is no different in developing nations such as India. Global concerns about excessive consumption and mitigation efforts have put Renewable Energy Technologies on centre stage not merely because of their novelty but out of necessity. In this paper we attempt to put in perspective the opportunities and challenges that these emerging technologies face in a wide range of situations afforded by different countries in the region. Approach taken by different countries is analyzed with a peep through the recent historic information and its influence on present and future developments. An attempt is made to estimate future role of the renewable energy technologies in the energy market.

10.1 INTRODUCTION

Renewable energy technology applications in India have been of recent origin if we discount the traditional use of solar drying, use of biomass, and agricultural waste for heating purposes. First wind mills for pumping were designed at National Aeronautical Laboratory (NAL), Bangalore during 1959-1961 (Fig. 10.1) and deployed in different parts of the country. It was perhaps a little ahead of the times. During this time, it was generally felt that we have electrical power to be able to provide free electricity to agricultural pump sets in rural areas. With this back drop, there was no way large-scale deployment of wind mills could be financed. However, the designs developed were quite good, and until 2000-2001, some of these windmills were still seen around (Fig. 10.2).

Once NAL stopped working on these mills, there was a long holiday for any effort in the country on any scale. The oil shock of 1972-1973 rekindled the interest to some extent, and it was strengthened during 1976 again due to global shortage of oil. This time around, there were some sustained efforts to introduce renewable energy technologies for low-grade applications, such as solar hot water systems, solar cookers, improved stoves (chulas) for biomass, water pumping windmills, and other such devices which could be used in a decentralized manner.

China had been a little slow in taking to large-scale development in renewable energy except for perhaps the biogas plants which score over the Indian design in many ways. Slowly, but surely, the Indian floating collector design is being replaced by fixed dome Chinese design for community-based biogas plants. On wind power use over the last few years, there has been a drastic turn around in the approach to using re-technologies. It has taken just 3 years for China to graduate to the second position in the top ten lists in terms of Wind power capacity addition. If one goes by the projections, the number one position may be achieved. In 2007, it was at about 4 GW, and by 2009, the installations touched an astounding 26 GW. With scores of companies undertaking development on a war footing, there is so much happening in China on wind power. A spate of wind turbine designs has been developed, and prototypes are being built and tested. Another paradoxical position is that though it is a country that produces about 40% of all solar photovoltaic devices, domestic utilization is quite small.

In this paper, an attempt is made to capture some of the salient aspects of technology development and deployment in India in the context of power supply systems management. Main RE technologies dealt with are wind and solar energy sources. Small hydro power adds considerable value for localized grids. Bio-mass sources work more or less like thermal stations though for limited periods in a year.

10.2 INDIA

India, with its agro-based economy has always been a low per capita energy consumer. With a large population. the total energy consumed could reach fairly high levels in decades to come. On the ground, the fact remains that there will be a steady growth in the per capita consumption, but it may not be at the expected rates due to a variety of reasons.

10.2.1 Electricity Generation, Distribution and Management

India is among the fast-developing nations with significant growth rates over the past 5 years. It has also been a country with one of the lowest per capita energy consumption. With an economy

FIG. 10.1 ONE OF THE 70 WINDMILLS (MODEL WP 2) INSTALLED IN RURAL INDIA DURING EARLY 1960S

on the upswing, this is expected to change in the coming decades. In a situation where there is a deficit, the targets sets are also somewhat conditioned by knowledge of available resources. Table 10.1 demonstrates that there is a trend of about 6% to 7% growth though the year 2008 to 2009 showed a slowing down. This is attributed to the general economic crisis around the world including India. It is seen that the growth rate has reached earlier levels during 2009 to 2010.

India with an installed capacity of about 153 GW (2010) of conventional power generation has a peak capacity shortage of about 13% and about 10% in terms of delivered energy (Table 10.2). It should be noted that it is the restricted demand that is projected and not the full demand. There are a number of circumstances under which the demand is to be deferred and this will be done in consultation with the consumer in advance.

With a country that is on a fast track development, projections have it that the demand may touch 300 GW by 2022. Table 10.3 gives the existing and projected installed capacities till 2022. These projections assume that there will be space for private and private–public funded power generation projects in the years to come.

It is expected that most of this would come from Coal based generating stations for base load. Although nuclear power is sometimes considered as an alternate source for baseload, the progress is rather slow, and a look at the achieved plant load factors leaves some doubts about its becoming as significant as is expected. Major hydroelectric projects seem to be no longer feasible looking at the number of impediments that need to be overcome. Apart from being expensive if one considers all costs involved, they are also fraught with population relocation and associated human problems.

Growth rate of thermal and other installations is given in Fig. 10.3. It can be seen that thermal installations are maintaining steady growth and in fact over the last 2 years there is an upward trend. Hydroelectric and nuclear powers have tabled off to some extent and wind has shown a stunted growth rate.

Management of electricity is shared between the State Government and the Central Government. Policy and major infrastructure related issues are handled at the central level through Ministry of Power (MOP). MOP has set up public sector undertakings, such as National Thermal Power Corporation (NTPC) to build, own, and operate huge generating stations in strategically located places and has arranged for evacuation of power through the Power Grid Corporation of India Limited (PGCIL). With central and state government-run entities managing the system, economics associated was relatively easy so long as the states concerned were able to come to an understanding. For example, if there is surplus power available in a given state at a given time and a neighbor state had a need, the two distribution companies (both owned by Government) would come to an agreement and bills would be settled in due course after the supplies are completed.

With a view to enhancing public–private partnership reforms were brought into the electricity market by allowing Independent Power Producers (IPPs) to build and operate generating stations. The state would purchase the electricity at a pre-agreed price under a Power Purchase Agreement (PPA). Based on this scheme, several coal- and oil-based hydroelectric stations were set up. They were used for peaking requirements and served as reserve capacities. IPPs were largely restricted to respective state boundaries. Due to the nature of financing such projects, IPPs were able to obtain a minimum guaranteed grid access for supply of power. Even if at a given time, the distribution company was not able to use the energy and has asked the IPP to back down, they would be eligible to charge for deemed generation. Provisions were also put in place wherein electricity could be purchased from IPPs at negotiated prices in case of urgent need.

In order to resolve any issues that may arise in the overall management, a quasi-judicial body known as the Central Electricity Regulatory Commission (CERC) has been set up at the central Government level and each state has its own State Electricity Regulatory

FIG. 10.2 NAL'S WINDMILL (WP-2) FOUND TO BE OPERATIONAL IN 2001 AT PONDICHERRY

TABLE 10.1 ELECTRICITY CONSUMPTION AND SOURCES (BU) [1]

Year	2004–2005	2005–2006	2006–2007	2007–2008	2008–2009	2009–2010
Target	586.40	621.50	623.00	710.00	774.10	781.55
Achievement	587.40	617.50	662.50	704.50	723.60	771.55
% of target	100.20	99.40	99.90	99.20	93.40	97.73
% growth	5.20	5.10	7.30	6.30	2.70	6.60
Thermal	486.10	497.20	527.50	559.00	589.90	640.90
Nuclear	16.80	17.20	18.60	16.80	14.70	18.60
Hydro	84.50	103.10	116.40	128.70	113.00	106.70
Wind[#]		6.00	9.60	11.40	13.00	18.00
Biomass and other[#]		3.20	4.80	7.20	7.00	8.00

[#]MNRE.

Commission(SERC). One of the primary functions of SERC is to take into account the energy mix available in the state, the demand and paying capacities of different sections of the society, and to fix a just price for the electricity. It is possible to bring to the regulators any grievances for redress. SERC in consultation with CERC also mandates the Renewable Energy Purchase Obligations (RPOs) for their state and provide a preferential tariff. The SERC would also define the conditions of supply and evacuation to the bodies who are managing the generating stations and distribution companies. The system in place gives ample room for negotiation. The flip side of negotiation is that it inhibits rapid development.

System operators are familiar with generating stations whose output could be controlled and scheduled in advance. Fairly good estimates of loading patterns are also available. Therefore, mostly, the demand side management in terms of available generating power in short term is employed. The state entities are required to give their day ahead, week ahead, and month ahead requirements to load dispatch centers. In order to prevent over drawl from any specific state, a system known as Availability Based Tariff (ABT) has been put in place. Under this system, for any over draw of power when grid frequency is low, the specific state will have to pay higher tariff. Similarly surplus power supplied when frequencies are higher attracts lower tariff. This mechanism is termed Unscheduled Interchange charges (UI). The UI charges are settled everyday and monthly bills are reconciled among participating agencies, i.e., the Central pool, concerned state distribution companies and other stake holders. State utilities or distribution companies are required to pay the IPPs and also honor the inter-state and Centre-State commitments. With decades of experience behind them, they normally are able to balance the grid in an optimal manner.

Distribution companies will have no issues with renewable energy technology-based systems getting into the system if the magnitudes are small. However, when the proportion of the wind or other renewable energy gets larger, problems with grid management become serious. A case in point is the Tamil Nadu grid. The state has a firm power in terms of thermal and Hydro power apart from the Nuclear power coming from Central sector in the range of 9000 MW. The state presently has over 4500 MW wind power installed in various parts of the state. Incidentally Tamil Nadu Electricity Board has considerable experience with wind power in their grid. Based on anticipated wind power generation levels during high wind season, the distribution company could set limits to draw from central pool. This will impact savings so that in case of sudden wind power drops to a low grid frequency, the board has to import energy at a much higher tariff. With a must run status for wind power generators, the board would have to pay for energy supplied by wind generators at negotiated fixed price and also pay the central pool at higher tariff. When the state exports surplus power while frequency approaches rated frequency of 50 Hz, they would get lower tariff from other consumers on the grid. The distribution companies would also have to provide nearly free electricity to agricultural pump sets and get grants from the State Government to make good the expenses it would incur while managing the system. This would make the distribution companies wary of higher penetration levels of wind or any other RE technology powered generation. Presently, the size of solar power in grid interactive mode is very small and therefore most of the complexity envisaged and experienced is from wind power on the grid. The Small hydro accounts for a sizeable potential particularly in the northern and north eastern states of the country.MOP posts 161351.80 MW as the installed capacity in India out of which 16,429.42 MW comes from RE technologies. Grid codes, tariff fixation, and planning of grid extension address the issues connected with RE power. Technological situation with the renewable energy application in India are dealt with in the following paragraphs with particular emphasis on grid connected wind power.

TABLE 10.2 ENERGY REQUIRED AND FULFILLED [2]

Year	Energy				Peak			
	Requirement	Availability	Surplus/deficits (−)		Peak demand	Peak met	Surplus/deficits (−)	
	(MU)	(MU)	(MU)	(%)	(MW)	(MW)	(MW)	(%)
2007–2008	7,39,345	6,66,007	73,338	−9.9	1,08,866	90,793	18,073	−16.6
2008–2009	7,77,039	6,91,038	86,001	−11.1	1,09,809	96,785	13,024	−11.9
2009–2010	8,30,594	7,46,644	83,950	−10.1	1,19,166	1,04,009	15,157	−12.7

TABLE 10.3   PROJECTION OF ELECTRICITY CONSUMPTION IN INDIA [3]

Year ending	2007	2012	2017	2022
Peak load at substation busbar [GW]	100	153	218	298
Anticipated consumption [GWh]	690	969	1392	1915
Anticipated CUF (capacity utilization factor)	0.7876	0.723	0.729	0.734

10.2.2   Policy Support

The trends indicate that the demand and supply gap would not narrow down over time. Therefore, there are huge opportunities for additional capacities and energy savings. Government of India set up an independent Ministry of New and Renewable Energy (MNRE) quite early to nurture and promote use of renewable energy on a large scale. This ministry at the federal level plays a catalytic role and has set up laboratories and knowledge centers that cater for needs of the industry. As the nodal ministry it deals with other relevant ministries and other governmental bodies which have common areas of interest, such as finance ministry or power ministry with a focus on development of all RE technologies. The Indian energy supply systems be it fuel or grid power, has been traditionally dominated by Governmental bodies or public sector undertakings. This includes distribution systems. Therefore, it is useful to look at the way electricity market is presently managed to have the right perspective.

Initially, when the exposure was small, India went along the subsidy route for most of the RE devices. It was administered through State nodal agencies which are institutions set up by local governments. Some specific models which had been demonstrated to work satisfactorily were enlisted by MNRE. When a wind mill of a specific design was purchased under intimation to the respective state nodal agency, subsidy was released to the buyer upon successful commissioning. The amount of subsidy was decided upon the basic cost of the equipment and its affordability by an average user. For example, a water pumping windmill will get INR 25,000 (US$500). A wind solar hybrid system will attract nearly 80% cost of the equipment because of the very high cost of the solar panels. Administration of these subsidies was fraught with considerable delays and not so mature technologies make it less attractive. One of the more successful program was the solar water heater subsidy.

Indian Renewable Energy Development Agency (IREDA), a financing agency under MNRE, provides an interest subsidy through nationalized banks.

Large grid connected systems do not have any subsidy attached. The projects had to be financed based purely on their merits after taking into consideration the depreciation benefit. In the initial years, it was possible to write off 100% of the cost of equipment in the first year against the gross profit. In the recent times, it has been reduced to 80%. There will be a 5-year electricity tax holiday in addition to this. In most of the states, the private wind turbine deployments were for captive consumption. The distribution companies permitted using the grid with a banking arrangement. There is a charge attached to it as a percentage. Apart from these enabling policy measures, some local governments introduced additional benefits to attract private investments. One of the most attractive features was the sales tax write off in the states of Madhya Pradesh and Maharashtra. With this, marginal sites became attractive enough to merit projects coming up. It did give a good boost to wind farming in these two states but the scheme was not renewed.

Once the tariff fixation came under the purview of the CERC/SERCs purchase price from RE technologies, it became a fully transparent process. It was possible to seek re-consideration. With a capacity addition of around 1500/2000 MW annually, for about 3 years, there was a very slight slowdown in the capacity addition. Though this was to some extent attributed to general recession, the outlook remained optimistic for wind power.

10.2.3   New RE-Based Incentives

The field which was doing very well with income tax-based rebates and captive generation attracted mostly profit-making concerns from manufacturing and to some extent from venture capitalists.

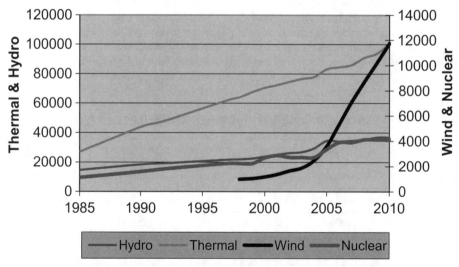

FIG. 10.3   GROWTH OF INSTALLED POWER IN INDIA (MW) [4]

There was an interest from quite a few financial institutions, infrastructure developers to enter the electric power industry in a major way. With global concerns about excessive dependence on fossil fuels and consequent threat to energy security, government started to consider green energy as a means of mitigating the risks. Presently, there are only few options to get sizeable capacities dependent on RE technologies. One is wind power which can give aggregated capacities of 100, 200 MW of installed power against 1000 or 2000 MW thermal units. Mini and micro-hydro units and biomass-based generation units are the other options. Under the regulatory authorities' directions, each state was required to have a certain percentage of electricity generated using greener technologies.

10.2.3.1 Generation-Based Incentive

Investor bases were not growing at the same rate as was envisaged and this engendered a need to find ways of roping in more investors into the field. In 2008, MNRE moved a proposal to encourage investors with no possibility of claiming accelerated depreciation to enter the market. A pilot scheme of 50 MW on a first come first served basis for GBI at a 0.5 INR was proposed by MNRE. There was considerable enthusiasm to get things underway. However, the scheme did not take off due to a number of administrative issues. However, a proposal was pushed through the planning commission to finance ministry to allow up to 4000 MW to come under the GBI scheme. Projects were commissioned between 17.12.2009 and 31.03.2012. Driving considerations were encouraging large Independent Power Producers (IPPs) and attracting Foreign Direct Investments. GBI would encourage higher efficiency of operation thereby increasing productivity.

Some of the preconditions and limitations to avail the GBI are that one cannot avail accelerated depreciation for the project. It is also not possible to sell power to third party. The INR 0.50 is payable over and above what the SERC would fix as the feed in tariff. This can somewhat be negated if the SERC takes into account the GBI component while fixing the tariff. The incentive period is limited to 10 years during which first 4 years maximum pay out shall not exceed INR 1.55 million/MW in any given year with an overall cap of INR 6.2 million. There will be no upper or lower limit to the number of MW that gets registered under the scheme. Till October 6th, 2010, a total of 261 MW of projects were registered under the GBI scheme [5]. Taken as a percentage of the MW commissioned during more or less the same period, this is about 20% of the projects. This percentage is likely to grow over the next 2 years, rapidly considering that quite a few international companies are seriously perusing projects.

- The allocation is for 4000 MW installed capacity to be used up before 2012.
- The 0.5 INR/kWh would be provided for first 4 to 10 years depending on the discretion of SERC concerned.
- Either this benefit could be availed or the 80% depreciation could be availed, but not both.
- Scheme cannot be used for third party sale of electricity or self consumption.

SERCs may take into account GBI element while fixing feed in tariff. This may result in some erosion of additional income anticipated under this scheme.

This is being administered by IREDA in conjunction with state distribution companies. Investors would naturally go to high wind areas so that their ROIs are better. With a view to encourage wind farming in not so windy areas, CERC and SERCs are relating the

TABLE 10.4   MERC FEED IN TARIFF ORDER [6]

Annual Mean Wind Power Density (W/m^2)	CUF	Tariff fixed (INR/kWh)
200–250	20%	4.29
250–300	23%	3.73
300–400	27%	3.18
> 400	30%	2.86

wind power densities inversely with the tariff. That is, wind farmable areas with lower wind power densities attract higher tariff. Following the CERC recommendations, Maharashtra Electricity Regulatory Commission (MERC) has announced a tariff structure given in the Table 10.4.

They have also indicated a range of anticipated plant load factors. Though there is considerable debate regarding this aspect, the wind industry welcomes it as a welcome change in the policy. Other wind active states are following up on similar schemes. However, there will be considerable public consultations, hearings, and opposing views from various stake holders in front of the State Electricity Regulatory commissions before passing an order on the feed in tariff takes typically 6 months to 1 year. Law provides for appeal against orders, there will always be some additional delays in implementation of such orders.

10.2.3.2 Clean Development Mechanism

Clean Development Mechanism implementation (CDM) in India with respect to wind power projects was considered as a way of making wind power projects an attractive option in terms of return on investment, particularly in areas with moderate potential. Till date, about 5543 MW [7] of wind power have been registered under 357 projects. Most of the wind power feed-in tariff orders have a provision for the transmission company/electricity board to get a share of the CDM revenue. This makes it much less attractive to avail CDM benefits, considering the amount of paper work associated with such claims. For smaller projects where depreciation was the main driver, the preparation of Project Design Document (PDD), validation, and maintenance would sometimes cost more than the costs incurred. In addition, part of CDM revenue would become payable to the Transco/electricity board. These issues make the idea of going in for CDM benefits for IPPs set up with the sole purpose of making projects viable by sale of electricity attractive. This is because such IPPs would have a team of professionals who would be very carefully monitoring everything that happens in the wind farm and necessary documentation and follow up will be part of portfolio management.

10.2.3.3 Renewable Energy Certificates

India, with its geological spread, has an uneven distribution of renewable energy availability. While South, Southwestern, Western states have good potential, eastern states of the country, small states, such as Delhi, have to almost entirely depend on conventional resources for their energy requirements. A financial instrument similar to CDM mechanism has been devised to increase share of RE in the national perspective. The idea is that there is an obligation on the part of distribution companies, captive generators, and open access consumers who constitute obligated entities. They would have to meet a part of their consumption through use of renewable energy. The quantum of such an obligation is decided by the SERC and obligated entities should top up their obligation by purchase of renewable energy certificates.

In states where renewable energy projects are concentrated, the RPO can be easily fulfilled. Once this obligation is met with, the state may have little incentive to encourage signing up for more projects. Renewable Energy Certificate (REC) is another route through which the surplus RE could be traded. This is facilitated by first registering RE projects with a centrally administered registry which would enable selling of RECs. The generation information would be compiled and sent to the registry. Each MWH will constitute a unit. The Load Dispatch Centre (LDC) keeps track of the energy generated. A mechanism has been formulated to get this information to the registry for registered projects. This would be verified by National Load Dispatch Centre (NLDC) and REC would be issued. These certificates can be traded on the Power Exchanges. They cannot be sold or purchased outside. There would be a floor price and a ceiling revised from time to time by CERC. This scheme fosters higher utilization of renewable energy in the country as a whole. The scheme is in an advanced state of becoming operational. The mechanism has been debated for nearly a year since being formally announced. Ultimately, it has to be implemented through SERC through specific tariff orders for actual implementation. This may take some time as each state will have to consider the REC mechanism in the state's own perspective. There are of course issues of grid congestion and management, which will continue to be debated by the distribution companies and the load dispatch centers where component of renewable energy is high. Being engineering problems, it they are likely to get sorted out by suitable measures, such as good quality forecasting and grid management techniques.

10.2.4 Growth of Wind Turbine Field

Among all new renewable energy sources, wind energy continues to remain at forefront and is likely to remain so in the foreseeable future. With the first machines making their appearance way back in 1985 to 1986 with 55 kW units a steady growth was seen over the past two decades. In the recent times, notwithstanding its nature of being an unsteady source of power, wind and other renewable energy technologies are listed alongside conventional sources like thermal power. Major part of the RE technologies come from Wind (11807 MW). The development showed a steady growth over last 5 years, installations did not show any significant impact of global economic slowdown (Fig. 10.4).

Eight states in India — Rajasthan, Gujarat, Maharashtra, Karnataka, Tamilnadu, and Andhra Pradesh have perused wind power with varying degrees of success (Fig. 10.5). It is seen that Tamilnadu has 4566 MW of installed wind power and has been consistently leading the pack. It is followed by Maharashtra with 2004 MW installed. Gujarat, from where the grid connected wind power utilisation started takes the third place with 1667 MW. Karnataka is similar to Maharashtra in terms of site conditions. The wind farmable sites are mostly in mountaneous regions. Karnataka has 1396 MW installed. Rajasthan has picked up installations in the recent times and presently has 846 MW of installed power. Kerala has just started recently with a total of 27 MW. The figures shown are as of 31.03.2010 [8]. Though mountains of Kerala show high potential, the access, infrastructural development-related difficulties have resulted in a slow development relative to other states. For example, Ramakkalmedu, a complex terrain site with one of the country's highest wind speeds has no grid access for evacuation on Kerala side. Though evacuation through Tamil Nadu can be done, the interstate power transmission rules were not as yet available for private sector. It was an administrative bottleneck which had no easy solution.

Indian forays in the wind turbine field are two decades old and till today few of the first machines installed are working reasonably well. There is an issue with the nature of winds in India. It is a seasonal phenomenon, and averages are not very high as may be desired. On the other hand, there are cyclonic conditions that can push the wind turbines to limits of their design and that needs to be taken into account at the design level as the exact place where the turbines get deployed cannot be controlled. With a peak deficit and energy deficit of over 10% at present, the supply and demand gap is unlikely to be closed presently or in the future, unless all available resources are deployed to the fullest extent and that includes wind power.

At around this time, the European certification procedures were getting consolidated. The system of certification was in an evolutionary mode. Notwithstanding this, India adopted a policy that the wind turbines that get installed in India shall have a valid type certificate. This could be the cause of moderate growth rates that have also been sustainable. With wind power equipment getting installed more often than not far away from load centers, there was always an issue with grid extension. Enhancing evacuation facilities is a time-consuming process. With moderate growth rates of installed capacities, the mismatch between machine deployment and grid connection has fewer problems. There are very few instances

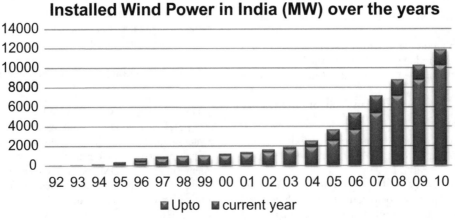

Installed Wind Power in India (MW) over the years

■ Upto ■ current year

FIG. 10.4 GROWTH OF INSTALLED WIND POWER IN INDIA [4]

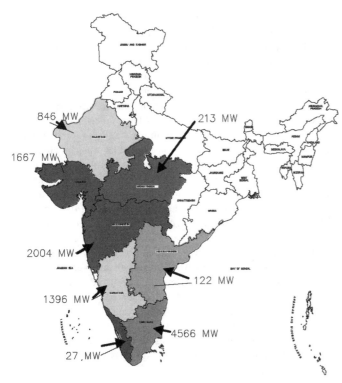

FIG. 10.5 INDIAN STATES WHERE WIND IS BEING ACTIVELY EXPLOITED (INSTALLED CAPACITY AS ON 31.03.2010)

locations for collecting long-term data. India, notwithstanding this major effort, does not possess long-term reference data sought by most of the consultants for reducing uncertainties in energy calculations. However, it has to be recognized that if the data from this program is rejected, there is no other data in public domain. For over a decade, this was the only source of wind information based on which wind farms were set up. In the recent times, manufacturers and developers carry out measurements prior to setting up projects. In fact, most of the major wind farm developers have their own private data bases created through measurements. This information is not available in public domain due to commercial considerations.

The primary purpose of establishing these masts has been largely justified by installation of wind farms in the area in large numbers. For example, initially, there were just two to three masts in Aralvoimozi pass in southern tip of India. Presently, there are over 2000 MW of wind power equipment deployed in and around this single location. However, there is a price to pay. Present-day wind farming practice looks for long-term data from near the sites proposed, and this is impossible to get.

Monsoon-based wind system makes the western and south western part of the country windier than the eastern and north eastern parts of the country. Southwest monsoon winds start from June and the movement would continue till September, and Northeast monsoon movement would start by October and continue till January. Southwesterly movement would be far stronger compared to the receding north easterlies. Therefore, there would occur low wind

when grid availability is an issue for the wind farms already set up or the ones which are being set up.

Over a period of the next 10 to 12 years, the turbines grew in size and largely turbines of Danish origin made their way into the Indian soil. Tax breaks, soft loans from World Bank facilitated rapid growth of installed capacity. After the initial rush of early 1990s and soft loan options were exhausted, rate of growth started falling for next 2 to 3 years. By this time, there was some operational experience gained, and it was feasible to get an idea of what to expect. On a parallel information on the resource availability, grid integration issues were better understood.

10.2.5 The Wind Resource

Initial years of wind energy development largely depended on the data Indian Meteorological Department (IMD) collected through a countrywide network of observatories. This data was not very helpful in getting a good estimate of the resource availability. It became essential to start an independent measurement program which was independent of the normal meteorological observations. Under this project, initially during 1986–1988, a network of wind monitoring stations was established initially in the states of Tamil nadu, Karnataka, Orissa, and Gujarat. The locations were so chosen that wind turbines could be established at a future date. Twenty-meter tall masts were equipped with anemometers and direction vanes installed at 10 and 20 m agl and solid state data loggers were used to store data automatically. Figure 10.6 shows one such installation in a complex terrain.

Over the years, more than five hundred such stations were established and the data collected for periods of 1 to 5 years. It is perhaps one of the longest sustained efforts by MNRE to gather resource information on a national level. Presently, the winds are measured at 50 m agl, and there are four 120-m masts installed at strategic

FIG. 10.6 WIND MONITORING STATION IN THE EARLY YEARS

months during October/November and again in March/April. June, July, and August months normally exhibit high winds with latter part of May and early part of September showing some winds.

While the location-specific measurements have been instrumental in bringing about wind farming in a significant manner, there has been a felt need to get an assessment of wind power potential at a global level. Under a joint program between Riso National Laboratory-Danish Technical University and CWET collaboration, a meso-scale modeling exercise for the entire country was undertaken. To the extent feasible, the map has been validated. The results have not changed the earlier perceptions about windy and non-windy areas except some portions. Figure 10.7 shows results of the modeling and mapping effort for winds at 80 m agl. The country has, by and large, class II wind zones mostly in the western and southern parts. East coast has some patches where higher wind power densities can be expected. East coast line is also fraught with cyclonic conditions with regularity. Some portions of the hills of Uttaranchal show some promise. Hills of Kashmir show high wind activity according to this meso-scale modeling. However, detailed measurements are required for validation of the estimates made. There are quite a few locations across the country in complex terrain that has not been captured in totality by the model because of the coarse grid size. However, it is possible to create maps with higher resolution for specific places.

Effect of monsoon can be seen in Fig. 10.8. The wind activity strengthens during these 3 months and areas with mean sea elevations above 600 to 700 m experience sufficiently high winds. Consequently, most of the outputs from wind turbines installed in India produce nearly 70% of their annual outputs during these 3 months.

Many of the wind farmable areas are to be found in Southern Tamilnadu, parts of Karnataka, hills of Maharashtra, vast plains of Gujarat and Rajasthan. Andhra Pradesh and Madhya Pradesh have some interesting pockets. Looking at the high capital cost and return on investment, there was a consensus that areas in wind class II and above are wind farmable. With this stipulation, the given area must have an annual WPD of over 200 W/m^2 at an extrapolated height of 50 m agl to merit installation of grid connected wind turbines. There is a debate about how do we arrive at the long-term WPD with limited measurements on site. An estimated 48,564 MW potential has been indicated for India. This estimate is based on an interpolated wind power density map of India. Source for this estimation was the recorded wind speeds from over 500 on-site measurements carried out in different parts of the country. It was assumed that 1% of the area coming under 200 W/m^2 and above would become available for wind farming and a generous spacing is provided between the wind turbines. With class II winds, it is a reasonable assumption. Wind industry estimates that the potential is higher and can reach 100,000 MW. One basis could be that the

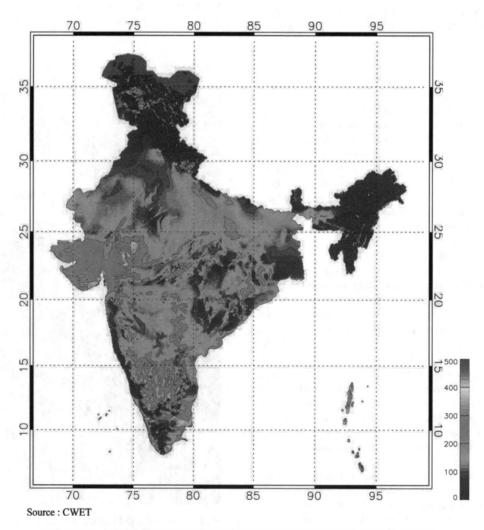

Source : CWET

FIG. 10.7   MESO-SCALE WIND POWER DENSITY MAP OF INDIA [9]

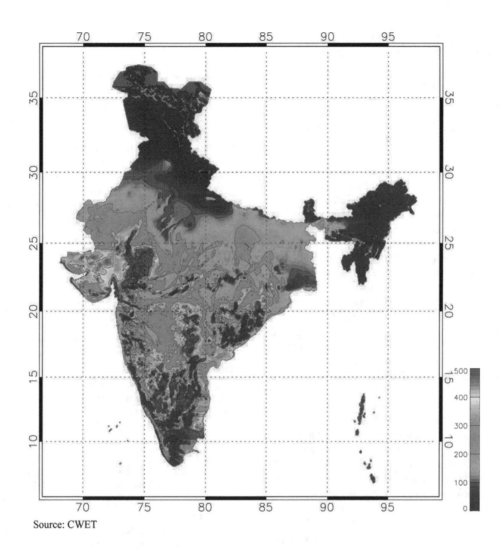

Source: CWET

FIG. 10.8   WIND POWER DENSITY MAP DURING MONSOONS [9]

1% assumption is strictly not correct and the procedure adopted to arrive at wind farmable areas was not sufficiently rigorous. However, the electrical infrastructure itself can be an inhibitor for very rapid development even in known windy areas.

10.2.6   Technology Deployment and Certification

Europe, notably Denmark and Holland, had just introduced commercial versions of grid connected wind turbines of 40 to 55 kW size during early 1980s. The first grid connected wind turbine in India was established in 1985 in Gujarat in a factory campus. The 40 kW rated wind turbine was connected to the Gujarat Electricity Board's grid and worked for a few years. Due to O & M-related difficulties, the turbine could not be kept working for long. Subsequently, a number of 55 kW turbines were introduced to India under demonstration projects. Installed at known windy locations, they gave the electricity boards hands-on experience and were seen as something actually feasible. At the same time, the effort to collect wind data at a height of 20 m agl was initiated because that was about the height of wind turbines. By 1989, Denmark came up with a novel idea of bringing 200/225 kW wind turbines adding up to 20 MW under a bilateral arrangement with the Government of India. Accordingly, by 1991, three wind farms of 10, 6, and 4 MW were established in the states of Gujarat and Tamil Nadu. The systems came with central monitoring systems. For the first time it was seen that the turbine operation and control could be done on a computer, data collected for off line analysis. After 20 years, many of these turbines are still functional.

With much interest generated in wind turbine technology, quite a few industrialists and venture capitalists started getting into the field to manufacture, install, and operate wind turbines in collaboration with turbine manufacturers, notably from Denmark. There were joint ventures between Indian companies and companies from the Netherlands, Japan, and United States. Two important events took place after the demonstration wind farms were set up. First factor was the Income Tax Rule by which the cost of the wind turbines could be written off by the company installing them as depreciation in the first year. This was later reduced to 80% in the first year. Therefore, any profit-making company could use this provision to set up wind farms. The second important factor was that the World Bank made available soft loans through the Government of India. These two key factors promoted wind power deployment in a major way. By 1993 to 1994, around twenty-seven companies registered with the Government as suppliers of wind turbines. Quite quickly, the World Bank Loan was exhausted. At the same time, the wind turbines that were installed without sufficient analysis were doing badly. Quite a few models started having spare part-related issues, lack of back office support from technology providers became a serious bottleneck. In Europe, quite a few companies were taken over by other companies, and traceability of necessary documentation became big challenges. Around this time, type certification systems were being evolved in Europe. Though there was some interest in Indian market for American companies, their own internal markets were attractive enough for them to pursue Indian clientele. With many complaints about problems with turbines, fear of second-hand machines finding their way to India eventually reached Government of India. In 1997, there was a hurricane in Gujarat and about 149 wind turbines suffered major damages. It was then decided by the Government that type certification should be available for a given model to permit it to be connected to a grid. It also became a pre-condition for availing accelerated depreciation. Lending agencies also made it a policy that the turbines with type certification alone will get financing. The type certification was somewhat country specific and as a temporary safety measure, GOI started insisting on permitting only those wind turbines which possessed a valid type certificate. MNRE issued guidelines for selection of wind turbines for deployment and set forth the criteria in detail. The original guidelines issued in 1995 has seen quite a few amendments and modifications based on the need of the industry. The lack of India-specific certification system was creating some confusion as to which system to follow to ensure some safety for large investments that were being made in the field.

The Ministry set up Centre for Wind Energy Technology (CWET) as an autonomous institution in 1999 under an Indo Danish collaboration. One of the primary objectives of the center was to create an India-specific certification system. The Riso National Laboratory, Denmark was appointed as the technical consultant for the project. With their assistance, Type Approval Provisional Scheme was prepared by the year 2000 (TAPS-2000) [10]. This was based on the International Electro-technical Committee (IEC) standards. It was formulated such that certification carried out elsewhere could also be taken up for recertification under which the machine suitability for the so-called Indian working conditions. There are three categories of type certificates that could be obtained by manufacturers under this scheme.

1. Simplest system accepts the existing type certificate and checks the following:

    a. Effect of lower grid frequency,
    b. Frequent interruptions in power — 350 grid outages per year as against 20 grid outages as per IEC,
    c. An extra load case for extreme wind which was not part of the IEC,
    d. No changes in the design which has been verified by an accredited certification body,
    e. A safety and function test is carried out.

2. In case the manufacturer wishes to introduce any changes in the design that would affect the loading on wind turbine components, it is possible under TAPS-2000 to take up a certification that will investigate effects of such changes on the integrity of the system as per accepted methods of evaluation. This would also require a type measurement as per IEC.

3. Provisions are also made for taking up a full certification for a new design. Under this, the manufacturer would submit the entire documentation for design verification, provide a typical turbine for type measurements, and have a manufacturing quality system in place. Though ISO 9001-2000 was not insisted upon, a verification of manufacturing system against a checklist was essential. Slowly, the industry found

it easier to get their manufacturing facilities certified under 9001-2000.

With a view to have good quality of supplied equipment and ensure safe operation for 20 years or more, a system has been put in place which is independent of Indian type certification. It is possible to market any wind turbine in India with a valid type certificate. The validity of a type certificate is verified on a quarterly basis in order to ensure that there are no delays in admitting a new model. CWET verifies conformity and validity of type certificates all models by a careful study of the type certificates and related documents and creates a list of models which have all the satisfactory documents. This list is widely published and available on the web pages of CWET. Under the original rules, there was a provision to allow marketing of models which were under type certification at CWET. In order to have more models and therefore more competition, the rules were modified to extend this facility to models registered for certification under any recognized and accredited certification body.

10.2.7 Design of Wind Turbines in India

With liberalization the need to design and develop new turbines was not felt in India. It was easy to get manufacturing rights under technology transfer arrangements from European countries under joint venture or agency basis or as subsidiary manufacturing units. Original estimates of wind farming potential were around 20,000 to 25,000 MW. At this level, there was no serious interest in investing time and effort to develop Indian designs. Indian industrialists were content to get technologies from elsewhere and concentrate on manufacture, deployment, and maintenance. There were a few isolated attempts but were abandoned after a few trials. Some of the notable ones are 200 kW turbines by Himalaya Machinery Co, Gujarat, and Bharat Heavy Electrical Limited (BHEL, a public sector undertaking). After a few prototypes, Himalaya Machinery Company abandoned the project. BHEL surprisingly went into collaboration with Nordex A/S for a 250-kW design.

A more recent trend is toward getting wind turbines designed by consultants and concentrate on manufacture of these designs. Under this arrangement, the manufacturer is not bound by the usual technology transfer related restrictions. This affords a freedom of making any decisions about the wind turbine without having to seek concurrence from the joint sector partners. There are quite a few models of this type which are undergoing prototype testing and certification. The via media here is that this will not come in the way of commercial deployment. Presently, turbines of up to a rated capacity of 2.5 MW are under testing in India.

With much hands-on experience with the wind turbine behavior in India, several locally relevant modifications have been implemented in many already deployed wind turbine models. One of the most notable changes was inclusion of blade extenders in many models to give higher swept area and, therefore, higher energy capture. This was quite a popular method with wind turbines in the capacity range of 200 to 300 kW. Though this may have had an impact on the maintenance of type certificates, such modifications occurred during a time when the rules for validation of type certification were getting formulated. In the recent times, blade extenders, if used, would be evaluated as part of design evaluation. CWET as a certification body was not able to take up any design-related activity in house. However, a number of attempts were made to nurture knowledge creation centers in laboratories. This has lead to some indigenous understanding of the design process. However, with influx of fully developed designs and customer comfort with such an approach inhibits very serious attempts at having to develop fully

TABLE 10.5 POWER SUPPLY MIX IN CHINA [11]

Source	Capacity (million kW)	% of total capacity
Thermal	652.00	74.60
Nuclear	9.08	1.00
Hydro	197.00	22.50
Wind	16.13	1.80
Total	874.21	

indigenous designs as at the end of proto-type development there could be no takers. Besides this, with the continuous interaction with Indian wind turbine manufacturers the designers overseas have a clearer design check lists. There is also a marked difference in the approach in the recent times. Presently many of the Indian wind turbine manufacturers have research facilities and design centers established in Europe. Since they are owned by Indian companies, the designs made by them would become indigenous to India.

10.3 CHINA

China has today a total installed capacity of 874.21 Million kW installed power (Table 10.5). It has been a slow starter in the field of wind energy. A variety of driving factors were instrumental in the sudden burst of interest in a field that is undeniably linked with pollution free power generation. The apparently impulsive change of attitude was followed up with massive planning and implementation phases that catapulted China as the next destination for wind power. China is poised to take on United States of America in terms of capacity additions. It is a clear demonstration of what political will can achieve in any given situation. After the initial rush of the way things were put in place, it is possible to perceive a well thought out strategy to make things work on ground at a pace that was difficult to comprehend initially.

Notwithstanding the fact that China has the second largest installed capacity of grid infrastructure in the world, there are shortages of power and has a reputation of having largest carbon foot print in the world. Looking at a need-based and realistic scenario, the sector development will be largely based on coal and other fossil fuel-based systems. Twenty-four Million kW nuclear power stations are under construction. Apart from the new thermal stations, there are major initiatives to de-commission old and inefficient thermal stations in order to improve overall system efficiency.

There is a good case for RE technologies in China from every point of view. Legendary infrastructure for manufacture, availability of very large tracts of land that can be employed for any other productive use, Government's commitment to make use of every possible resource to the fullest extent to meet its own internally set goals of reducing dependence of fossil fuels are just a few factors which make China destination next for wind power community.

Enactment of Renewable Energy Law (REL) followed up by policies and programs has made development of wind power one of the fastest growing field in China speeding past Germany in terms of installed capacity. Like all things Chinese, magnitude of operations is enormous. The speed at which things are happening makes information presented always appear to be a little dated. GWEC's wind power installation count for wind lists 25,802 MW by the end of 2009.

10.3.1 The Resource

There are a number of estimates of wind power potential made for China. Initial years of wind farming depended almost entirely on the meteorological information available from weather service and used the 10-m data. Based on this data the northern and eastern parts of the country appeared to possess relatively higher potential. Chinese Meteorological Administration put the figures at 4350 GW of reserve and of that 297 GW was expected to be exploitable. Subsequently, under a United Nations Environment Program, numerical simulations put the number at 1400 GW as the exploitable potential. The National Climate center carried out its own simula-

FIG. 10.9 NEAR TERM POTENTIAL AREAS WHERE VERY LARGE-SCALE INSTALLATIONS ARE PLANNED

tions and has put the potential at 2548 GW at 10 m agl. Thus, it is seen that there are different estimates available on the potential [12].

There are at least seven locations (Fig. 10.9) which can support 10 GW of wind power in the northern and north eastern parts of the country in the near term.

a. East and western inner Mongolia,
b. Kumul in Xinjiang,
c. Juquan in Gansu,
d. Bashang in Hebei,
e. western part of Jilin and
f. shallow seas off Jiangsu.

Keeping in view the extraordinary growth plan, a major initiative of setting up exclusive measuring stations was drawn up with masts of 50 m, 70 m, 100 m and 120 m height to match with the proposed hub heights at four hundred locations spread across known windy areas in 2006. Results of these measurements used in conjunction with actual generation from various wind farms already running would give a better estimate of the potential in foreseeable future.

One of the difficult situations is the distances between load centers and high wind potential areas. With load centers mostly along Eastern coast, potential areas to the North and North West, the grid extension can become quite a challenge. However, the problem has been addressed at a planning level.

10.3.2   The Growth of the Sector

There is a certain incredulity associated with the growth of wind power in China. Chinese wind power development started together with India with lower end grid connected wind turbines getting installed under trial basis. Initial forays by wind power failed to impress the leadership for quite a few years. Around 2004, there was a big turnaround in the government stand toward wind technology. By 2006, the changed scenario toward wind started to show results. Installed capacity of wind power equipment started doubling every year (Fig. 10.10). Government's enthusiasm toward wind power could be understood when one sees that there are 70 odd manufacturers of wind turbines in China.

Being a fast-developing country, the need for power is one of the highest in the world. It is also under internal and external pressure to use less polluting resources. Cities have reportedly high pollution levels. Gradual advancement of desertification in the windswept

arid zones in Mongolia is a reality. China boasts of one of the largest forestation programs in the world. One of the suggested ideas to conservation of humidity and prevention of desertification was to have wind breakers. Farmers would plant suitable trees along boundaries of farms to prevent crops from getting damaged due to winds simultaneously serving as barriers to dehumidification. It has been suggested in the past that a series of wind turbines may be installed to serve as wind breakers. They would, in addition, provide electricity. At a policy level, China mooted the RELin 2004 and it was implemented by 2005. It became essential for the utilities to have a certain percentage of RE technologies on the grid and was mandated to buy all energy from non-hydro renewable energy technologies at a preferential tariff. There were penalties to be paid for non-compliance. As is normal for developing countries during the initial years, it was mandatory to have 70% of the components to be sourced from within the country. This was later amended to encourage investments. It is mandatory for the power generators to source 3% of their electricity from non-hydro renewable sources and by 2020 the percentage share is to reach 8%.

By 2009, the REL was amended to include grid access, clear mandate for the transmission companies to arrange to absorb all the energy produced by RE technologies. An RE fund was set up to finance extra costs that would be required to integrate unsteady inputs from wind and other RE technologies. The amendment also addressed the issue of the price difference between coal generated electricity and renewable energy would have to be shared by all stake holders. An additional cess of 0.04 €/kWh is charged to fund additional integration costs. Wind energy also commands a higher tariff ranging between 0.56 and 0.65 €/kWh to be paid over 20 years of the service life of the turbine [13]. This constancy of policy for considerable time period would send right signals to the investors. If one looks at the consistent development of wind power in Tamil Nadu in India was partly because there was a certainty on ROI though it was somewhat lower compared to other competing states.

While such positive law and its implementation have fostered a galloping installation rate, there appears to be a gap between expectation and the realities. For many wind power companies who set up shop in China keeping in view booming market a need to recalibrate their expectations seem to be imminent. Frequently bidding processes for project implementation have gotten into a difficult sit-

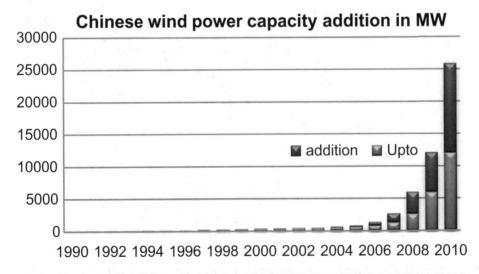

FIG. 10.10   GROWTH OF INSTALLED CAPACITIES OF WIND POWER IN CHINA

uation due to the fact that they were based on the price bids for sale of energy. Obviously, lowest cost per kWh supplied would get to do the projects. With so much competition, project financing gets into very difficult situations. Under the circumstances, established wind turbine manufacturers from world markets could not match the local prices and were disadvantaged to some extent.

10.3.3   Technology Development

China has always been perceived as a master where mass production and heavy engineering are concerned. Their planning and execution capabilities are almost unparalleled when it comes to mega projects. When they set their mind to any such ventures, the development path is never cluttered by any impediments. When it came to wind energy development, the same principles were applied. There is so much encouragement for developing indigenous designs that there are now over 70 wind turbine designs being designed, tested, and sold. Some of the leading turbine models have started making their way into foreign markets. Most notable ones are the Goldwind, Sinowel, Dongfang have a combined manufacturing capacity of about 8.2 GW annually. With targets set at 13.5 GW/year requirements, there is bound to be a shortfall of hardware if the installations have to go at the expected rates.

The similarity with Indian situation ended with experimental phase. Though China started off in the initial years with imported wind turbines, looking at the enormity of possibilities, development of indigenous design became a priority area for development. Most of the European and many American wind turbine companies including a lone Indian turbine manufacturer started establishing facilities for indigenous manufacture. With the newly found access for component manufacture at lower costs, quite a few Chinese component manufacturers became exporters of WT components and subassemblies. There was a need for having 70% of wind turbine components to be sourced from within the country that has also been done away with.

With a need for indigenous designs and the freedom it affords the manufacturers, there was a push to have designs carried out both in house and through design consultancy became quite an attractive route both for the consultancies and large number of state owned companies. It is notable that Europe, by this time had a growing body of wind turbine design consulting houses and China was requiring such quick design know how. Therefore, there was a matching of requirements and availability. It is a normal practice for a wind turbine company which has its proprietary rights on its designs and would always have a say in everything to do with the particular model either under a joint venture or a licensed production. A design consultant makes no such pre-conditions. This, coupled with the need for type certified turbines not being a primary concern, gave an impetus to the approach taken by most of the newly formed design teams in publicly held companies. This is in complete contrast to Indian situation where much importance is placed on a type certification for the models deployed in commercial space.

The National Bureau of Quality and Technology set up the Chinese Wind Turbine Standardization Technical Committee (CWSTC) and a national technical committee TC 50 was set up with constituents drawn from industry, academia, and relevant institutions. The committee interacts with IEC TC 88 on matters relating to wind turbine testing and certification standards. A number of equivalent standards which correspond to IEC standards have been published. With much indigenization, the turbines produced with in China get a type certificate. However, there is a lack of clarity as to the accreditation status of the body issuing the type certificate. With so much planned, there are considerable number of designers and certification bodies working on Chinese projects. In fact, the pace at which work gets ordered and executed, most of the independent design companies engaged do not have time for any new assignments perhaps for next few years. Keeping in view export market potential type, certification as per international standards is also pursued by a large number of manufacturers.

10.4   SRI LANKA

Sri Lanka has been pursuing wind energy for over 15 years. Table 10.6 gives a synopsis of installed capacity as of 2009.

As can be seen, there is considerable hydro power availability in the country which can be profitably used by installing sizeable wind turbine capacity thereby reducing dependence on the thermal power generation. NREL wind resource assessment shows high potential along the northwestern part of the country and central Sri Lanka where hills are to be found. Southeastern coast line also has few areas which can support wind farm activity. It is estimated that about 24000 MW wind power installations could be planned with about 10% of land made available from the identified windy areas.

With the grid capacity available presently, there is a smaller possibility of major development. A number of prospective IPPs are approaching the opportunity with a view to develop large wind farms for generation and sale to the Ceylon Electricity Board. Prospective Independent Power Producers entering wind market are known to be setting up measurement systems to gather authentic information on the resource availability.

10.5   SUMMARY AND CONCLUSION

It is expected that next decade on shore wind power would be the most exploited resource world over and Asian region would see sizeable installations. New ways of handling the variable nature of the wind power availability will cease to be an issue with the grid managers. Infrastructure development will continue to be a source of concern from evacuation point of view. In a country like India grid extension can take considerable effort and time because of right of way issues and net work congestion at certain times could adversely affect the stability of grid. The engineering problems associated with such issues have solutions.

Field being highly capital intensive will have to have sops either in kind or financing mechanisms to make them viable. It cannot be assumed that increased cost of energy would result in making wind more attractive as the input costs would also go up with increased energy costs. Electricity distribution companies are now willing to discuss issues. This is a big improvement in the approach. Forecasting wind power has caught imagination of the grid managers and in India there is a commercial and policy push to get short-

TABLE 10.6   INSTALLED POWER IN MW AS OF 2009 [14]

	CEB
HYDRO	1207 MW
THERMAL	548 MW
WIND	3 MW
	PPP
SMALL HYDRO	184 MW
THERMAL	742 MW
TOTAL	2684 MW

term forecasting introduced formally and this could pave the way for higher penetration levels. Therefore, the outlook appears to be quite optimistic.

10.6   ACRONYMS

ABT	Availability Based Tariff
BU	Billion Units (Billion kWh)
CEA	Central Electricity Authority
CEB	Ceylon Electricity Board
CERC	Central Electricity Regulatory Commission
CUF	Capacity Utilization Factor
CWET	Centre for Wind Energy Technology
EPS	Electrical Power Survey
GBI	Generation Based Incentive
GOI	Government of India
GW	Giga Watts
Hz	Hertz
IEC	International Electrotechnical Committee
INR/kWh-	Indian Rupees/ kilowatt-hour
IPPs	Independent Power Producers
IREDA	Indian Renewable Energy Development Agency
IWTMA	Indian Wind Turbine Manufacturer's Association
LDC	Load Dispatch Centre
magl	meters above ground level
MNRE	Ministry of New and Renewable Energy
MOP	Ministry of Power
MU	Million Units
MW	Mega Watts
NAL	National Aerospace Laboratories
NTPC	National Thermal Power Corporation
O&M	Operation and Maintenance
PGCIL	Power Grid Corporation India Limited
PPA	Power Purchase Agreement
PPP	Private Public Partnership
Ps/kWh	Paise/kilowatt hour (100 paise = INR 1)
RE	Renewable Energy
REC	Renewable Energy Certificate
REL	Renewable Energy Law
ROI	Return On Investment
RPO	Renewable energy Purchase Obligation
SERC	State Electricity Regulatory Commission
TAPS-2000	Type Approval Provisional Scheme – 2000 (year of adoption)
TWh	Terra Watt hour
UI	Unscheduled Interchange
W/m^2	Watts/ square meter
WPD	Wind Power Density

10.7   REFERENCES

1. Ministry of Power Annual Reports

2. http://www.powermin.nic.in/indian_electricity_scenario/pdf/Annual_Report_2009-10_English.pdf

3. Ministry of Power http://www.powermin.nic.in/generation/pdf/17th%20EPS.pdf

4. Ramesh. M.P. (2007) 'Wind Resource Assessment in India: A program with a difference' International Journal of Environmental Studies,64:695-708 figures updated till March 31st , 2010

5. IREDA http://119.82.68.103/IREDawindmill/form/frmuserlogin.aspx

6. *MERC_Final RE Tariff Order(SuoMotu)_14July_2010*

7. http://cdmindia.nic.in/

8. State wise installed capacity www.iwtma.co.in

9. Srivalsan E. et al, CWET, Indian Wind Atlas (2010)

10. TAPS 2000 amended 2003 http://www.cwet.tn.nic.in/Docu/TAPS-2000 amended.pdf

11. http://english.gov.cn/2010-01/07/content_1505403.htm

12. Li Junfeng, Shi Pengfei, Gao Hu, 2010 China Wind Power Outlook.

13. http://www.gwec.net/index.php?id=125

14. http://www.ceb.lk/generation/Digest%20Report%202009/Brouchure.pdf

HYDRO POWER GENERATION: GLOBAL AND US PERSPECTIVE

Stephen D. Spain

11.1   INTRODUCTION TO HYDROPOWER

The development of dams on rivers, with associated benefits of water storage for flood control, irrigation, and "hydropower" has played a vital role in advancing civilization throughout history. Of these, hydropower ingeniously and yet so simply combines two of the most fundamental components of nature on planet Earth — water and gravity — to help sustain our survival and improve our lifestyle.

This chapter will describe the role of hydropower from past to present and potentially into the future. Hydropower will be demonstrated to be a safe, reliable, and renewable energy resource worldwide, essential to our overall power and energy mix, both traditionally from rivers, through recent and growing development of pumped energy storage from lower to upper water reservoirs and evolving in the future with tidal and wave energy from the oceans.

11.2   HISTORY OF HYDROPOWER

For nearly 5000 years humans have in the cradles of civilization Babylonia, Egypt, India, Persia, and the Far East constructed dams on rivers for drinking water, irrigation and flood control. The remains and upgrades to these ancient structures exist throughout the planet. Some of the outstanding waterworks of antiquity eventually declined into disuse because the knowledge of their designers and builders was not preserved by the generations who inherited them. Although history does not record exactly when irrigation systems and dams were first constructed, archeology in ancient China, India, Iran, and Egypt does reveal that attempts at harnessing rivers began thousands of years ago and provided lifelines on which civilizations depended. Menes, the first Pharaoh of Egypt, ordered irrigation works to draw from the River Nile. In China, dams were constructed on the Min River for flood control and diversion of water to nearby farm lands. The sacred books of India cite the very early operation of dams, channels, and wells; evidence that this land may have been a birthplace of this science and art. Persians of ancient times also recognized the importance of irrigation to the sustenance of civilization, excavating underground water tunnel systems and constructing many dams [1].

As for "hydropower," dams storing water also created an abrupt difference from upstream to downstream water elevations to concentrate the force of gravity and then canals diverted flow to turn water wheels. The Greek engineer Philo of Byzantium first documented in "Parasceuastica" circa 200 BC that hydropower from Tympanum and Noriatype waterwheels was used to lift ore from mines, to grind grain and to saw lumber. There is also evidence that Egyptians were using the combination of water and gravity to turn water wheels for grinding wheat into flour in early centuries BC. And water wheels mills have been documented in China since 200 AD. Figure 11.1 [2] illustrates an example of an early waterpower development.

After another thousand years, the so-called Dark Ages, fast forward to the industrial revolution, when waterpower development and technology increased dramatically, with transition from wood to cast iron waterwheels. That invention was attributed to the British engineer John Smeaton, who in the 19th century wrote "An Experimental Enquiry Concerning the Natural Powers of Water and Wind to Turn Mills and Other Machines Depending on Circular Motion." Ironically, the industrial revolution was also when coal became readily available and steam power was developed, presenting hydropower with competition, which was less reliable seasonally, because of low river flows during the summer and icing during winter.

Water power technology advanced again, when the French engineer, Benoit Fourneyron, developed a higher efficiency water turbine, directing water into the center of the wheel and then radially outward. French countryman Jean Poncelet also contributed by designing an outward to inward flowing turbine using these same "reaction turbine" principles. In the United States, S. B. Howd obtained a U.S. patent for a similar design in 1838. Within 10 more years James B. Francis had perfected these designs to create a modern hydro turbine that now bears his name, achieving a remarkable milestone of 90% efficiency.

Meanwhile, Cragside, Rothbury, England was the site of the world's first "hydroelectric" power project, developed in 1870. The first "industrial" use of hydroelectric power in the United States occurred in 1880, when 16 brush-arc lights were powered using a water turbine at the Wolverine Chair Factory in Grand Rapids, Michigan. A hydroelectric dynamo also powered city street lamps in Niagara Falls, New York. And in 1882, the Vulcan Street plant in Appleton, Wisconsin became the first hydroelectric station to use the "Edison system."

The United States saw further hydroelectric developments throughout the late 19th and into the 20th centuries. The first alternating current (AC) hydroelectric plant was installed at Willamette

FIG. 11.1 WATERPOWER DEVELOPED IN EUROPE [2]

fixed blade propeller, and the invention of the hydracone draft tube by W. M. White. By the 1920s, hydroelectric plants were being developed for peak demand power and pumped storage was used to store energy at Candlewood Lake near New Milford, Connecticut.

Throughout the 1930s and 1940s, hydropower continued to be developed internationally as a viable means of power. And with the development of electricity and ready generation by hydropower and other technologies, the growth of world energy consumption consequently has been dramatic during the past 100 years.

Hydroelectric capacity became to be rated in two ways: by a facility's installed power "capacity" (typically in watts, kilowatts, and for larger projects in MW's or one million watts) and rated in energy production over time (typically in kw-hours or MW-hours). However, a hydro-electric plant rarely operates at its full power rating all the time, over a full seasonal year. Thus, the relationship between installed power capacity and actual annual energy generated is known as the plant capacity factor. This ratio between potential and actual energy varies by site, river and hydrology, but typically ranges from less than 40% to more than 60%.

11.3 HYDROPOWER HISTORY OF THE UNITED STATES

In the colonies of North America, and specifically in the United States since its inception in 1776, hydropower was initially used to provide energy for growing industries. Population centers and industries have always tended to be located along rivers, for drinking, irrigation and process water, and later for power. Before electricity, water driven wheels initially powered shafts, connected directly or by belts to machinery, to produce flour, textiles, and other manufactured goods. With electricity, power could be generated by water wheels and transmitted wherever needed remotely, beginning with street lighting and trolleys. By 1890, there were approximately 200 electricity generating plants in the United States and Canada using water for some or all their power. This generation industry accelerated between 1895 and 1915, with the advent of new technologies, and after World War I hydroelectric equipment and plant designs grew further and began to be standardized. By the early 1900s, hydroelectric power accounted for more than 40% of the United States' supply of electricity [3].

In the United States, as the population saw street lights, electric trolleys and learned the benefits of electricity for the home and industry, demand continue to increase. The Federal Water Power Act was created in 1901, encouraging developers but requiring permission from the government to build hydroelectric plants on rivers large enough to carry boat traffic. In 1902, the Bureau of Reclamation was created to manage water resources and have the authority to build hydropower plants at dams in the western U.S. Demand was so great that by 1920, the Federal Power Act was created by the Federal Power Commission (later becoming the Federal Energy Regulatory Commission) to issue licenses for hydropower development on all public lands.

Throughout the 1920s, urban Americans received more access to power, while rural dwellers were left behind. Rural customers battled with public utility commissions for access to electricity throughout the 1920s and 1930s. When President Roosevelt was elected, only 10% of rural Americans had electricity. In 1933, he created the Tennessee Valley Authority, primarily as a means to invigorate the economy and control flood waters along the Tennessee and Mississippi Rivers, but also as a means to provide electricity to rural Americans. Eight years after TVA's inception,

Falls in Oregon City, Oregon in 1889. That plant produced hydropower electricity transmitted at a revolutionary 4000 volts, through a single phase power line for 13 miles to Portland, where it was then transformed down to 50 volts for distribution and local use.

Later in Europe, a 175-km-long, 25,000 volt demonstration line between Frankfort on Main, Germany was used to connect with the first three-phase hydroelectric system. As hydropower gained credibility as a viable energy source, other countries worldwide began to plan and develop hydroelectricity. The first publicly-owned hydro-electric plant in the Southern Hemisphere was completed in Tasmania in 1895 to supply street lighting to the city of Launceston.

Other technology developments or that era included: high head reaction turbines in 1901 at Trenton Falls, New York; a low head plant with direct connected vertical shaft turbines and generators in Michigan; a fully submerged hydroelectric plant in Maryland in 1906; and a commercially installed vertical thrust bearing in Pennsylvania. Numerous inventors contributed turbines and generator and related component developments, including: S. J. Zowski's Francis reaction turbine for low head applications, Forrest Nagler's

household electricity in the region grew from 6000 homes to early 500,000. By 1936, the issue had gotten so contentious that the Rural Electrification Act (REA) was created to mitigate demand. The REA enabled the government to make low-cost loans to non-profit cooperatives such as farmers to assist with electrifying their communities [4].

Prompted by the Great Depression in the 1930s and World War II in the 1940s, the United States federal government began massive public work programs to develop many large hydropower projects. These projects put people to work, while both creating and meeting a growing demand for electricity. Public work projects and organizations included the Tennessee Valley Authority and Bonneville Power Administration, two federal agencies with missions to develop hydropower projects in the mid-Atlantic and Northwest respectively. Notable new hydropower projects included Grand Coulee dam on the Columbia River and Hoover dam on the Colorado. By the 1940s, hydropower provided about 75% of all the electricity consumed in the West.

TABLE 11.1 CONCISE HISTORY OF AMERICAN HYDROPOWER [5]

1879	Thomas Edison demonstrates incandescent lamp, Menlo Park, New Jersey.
1880	Brush arc light dynamo driven by water turbine in Grand Rapids Michigan.
1889	American Electrical Directory lists 200 electric companies that use waterpower.
1891	Ames, Colorado; Westinghouse alternator driven by Pelton waterwheel.
1891	60 cycle AC system introduced in United States.
1893	Mill Creek, California; First American three-phase hydroelectric plant.
1895	Niagara Falls, New York; 5000 horsepower, 60-cycle, three-phase generators.
1897	Minneapolis, Minnesota; hydroelectric plant at St. Anthony's Falls on the Mississippi.
1901	Federal Water Power Act
1902	Reclamation Act of 1902 establishes Bureau of Reclamation for water and hydropower.
1911	R. D. Johnson invents differential surge tank and Johnson hydrostatic penstock valve.
1914	S. J. Zowski develops reaction (later refined by "Francis") turbine runner for low head.
1931	Construction begins on Hoover Dam, Colorado River, in Arizona, Nevada.
1937	Bonneville Project Act creates Bonneville Power Administration.
1941	First power generated at Grand Coulee Powerplant, Washington, later 6800 MW
1969	National Environmental Policy Act — ensures environmental considerations for power.
1978	Public Utility Regulatory Policies Act — encourages small-scale hydropower production.
1986	Electric Consumers Protection Act — gives equal consideration to non-power values.
1994	National Hydropower Association — established to promote value of hydropower.

Later, the FPC was granted authority to authorize all hydroelectric projects built by utilities that were engaged in interstate commerce. Boulder (now Hoover) Dam started operating in 1936 on the Colorado River, with the capacity to produce at the time a significant 130,000 kW of electricity. In 1937, the U.S. Army Corps of Engineers (USACE) completed Bonneville Dam on the Columbia River, and the Bonneville Power Administration (BPA) was established. By 1940, the United States had over 1500 hydroelectric facilities, which produced almost one-third of power in America.

After the rush of the 1930s and 1940s, fewer hydropower plants have been developed in the United States. Several legislative policies have made the construction of dams less feasible, including the Wild and Scenic Rivers Act, National Environmental Policy Act, and Fish and Wildlife Coordination Act.

The major historical developments in hydroelectricity in the United States are chronicled in the following Table 11.1 [5].

11.4 HYDROPOWER EQUIPMENT

Besides the essential civil works of dams, waterways, and powerhouse structures, the mechanical and electrical hydropower equipment basically consists of turbines, generators, controls, transformers, and transmission systems. A typical layout of a hydropower project is shown in Fig. 11.2 [6] below, with its major equipment.

Hydro turbines are effective rotary engines that extract potential and dynamic energy from a fluid flow and convert it into usable power. Historically, there were four basic types of waterwheels, described briefly below.

- An "overshot" waterwheel, which operates by having water supplied from above, flowing down one side of buckets to create a weight imbalance, causing the wheel to turn, as illustrated in Fig. 11.3 [7]. Ideally, the bottom of the wheel should be above the lower water level to eliminate churning resistance.
- The "breastshot" waterwheel was similar, but positioned at the same level as the water source.
- An "undershot" waterwheel was positioned above the water source, with flat panels for the water to paddle underneath.
- The "pitchback" is similar to the overshot, positioned below the water source, but turns inward rather than out.

Efficiencies for these early water wheels varied greatly. The pitchback wheel was considered to have the highest rating at over 80% the overshot at 70%; breastshot 50%; and then the undershot at only 20% [7].

Modern turbines were developed during the past few hundred years, based largely on fundamentals of the waterwheels just described.

11.4.1 Francis Turbine

The first modern reactive water turbines were developed from the 1700s through the 1800s by Jan Andrej Segner, Jean-Victor Poncelet, and Benoit Fourneyron. Then in 1849, James B. Francis, while working as chief engineer of the Locks and Canals company in the water-powered factory city of Lowell, MA, improved on these designs to create a turbine with a 90% efficiency. Francis applied scientific principles, mathematics, graphical methods, and testing methods to produce these improvement and those techniques greatly improved the state of the art of turbine design and engineering for the future. The modern Francis turbine is shown below in Fig. 11.4 [8].

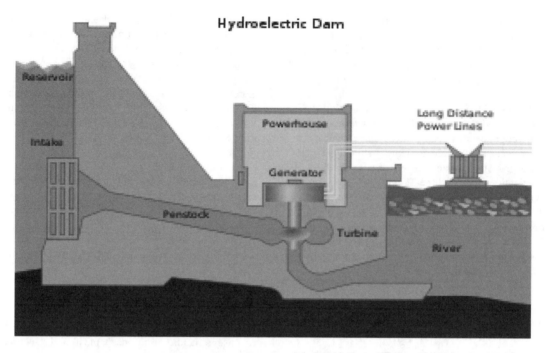

FIG. 11.2   TYPICAL LAYOUT OF A HYDROPOWER PROJECT [6]

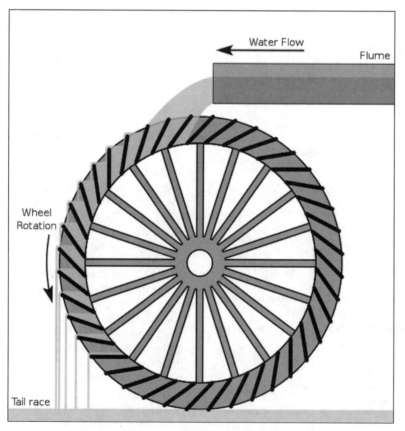

FIG. 11.3   OVERSHOT WATERWHEEL (Courtesy of Whitemill) [7]

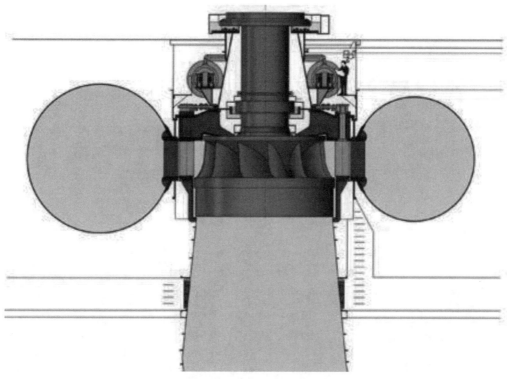

FIG. 11.4   FRANCIS TURBINE [8]

11.4.2   Kaplan Turbine

In 1912, Viktor Kaplan living in Brno, Moravia, developed and obtained a patent for an adjustable blade propeller turbine. Later Mr. Kaplan installed a demonstration of his unique design at Pod brady, Czechoslovakia. In 1922, the Voith Hydro company of Germany adopted and introduced a standard 1000 horsepower size version of the new "Kaplan" turbine. Soon after, a larger 8 MW Kaplan unit went on line at Lilla Edet, Sweden. This marked the commercial success and widespread acceptance of the design, which is still very popular, because of its relatively high efficiency over a wide range of head and flow (Fig. 11.5) [9].

11.4.3   Pelton Turbine

The Pelton wheel is another efficient water turbine, specialized for higher heads. It was invented by Lester Allan Pelton in the 1870s. The Pelton wheel extracts energy from the impulse or momentum of moving water, as opposed to its weight like traditional overshot water wheel. Although many variations of impulse turbines existed prior to Pelton's design, they were less efficient than Pelton's design; the water leaving these wheels typically still had high speed, and carried away much of the energy. Pelton' paddle geometry was designed so that when the rim runs at half the speed of the water jet, the water leaves the wheel with very little speed, extracting almost all of its energy, and allowing for a very efficient turbine (Fig. 11.6) [10].

11.4.4   Pump Turbine

A hybrid of these types is a pump turbine, used in modern bulk energy storage projects. The concept is to store energy in the form of flow pumped from a lower to an upper reservoir during low demand and then generating from upper to the lower during peak loads. A turbine crossed between a pump and Francis turbine is

used for this purpose. The result is a compromise in performance and net cycle losses, but provides renewable, load balancing, and economical power because of price differentials from off to peak.

11.4.5   Hydroelectric Generators

With the discovery of electricity and the invention of the generator, hydro power evolved from water wheels driving only shaft-connected mechanical equipment to the ability to transmit that power across long distances for use elsewhere. In 1821, Michael Faraday's experiments with electromagnetism led to the Maxwell in 1868 and subsequently applied to modern machinery including hydroelectric units [13]. Present day hydroelectric governors can be digital processor controlled. The algorithms are described below. Invention of the first rotating field "dynamo." If an electric current field drives a dynamo, it is a motor. If the field is driven by external mechanical means, the dynamo becomes a generator. The combination of this new dynamo and traditional water wheels resulted in hydro-electricity!

Dynamo's evolved into two basic types of modern hydroelectric generators: synchronous and induction.

- "Synchronous generators" produce power with rotating electromagnetic fields that are surrounded by stationary coils which then generate three phase-alternating currents. The "rotors" draw field excitation from an external power source and the "stators" can self-regulate their voltage, produce "reactive power, and generate electricity independently from the electrical transmission grid.

- "Induction generators" are driven slightly faster than synchronous speed by the prime mover, or hydro turbine. These machines rely upon the grid for excitation, cannot produce generate voltage until the generator is connected to the grid, and are not self regulating. They are simpler and less

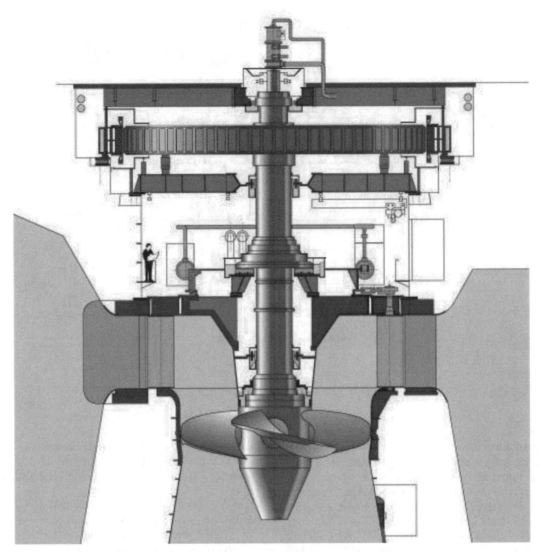

FIG. 11.5 VERTICAL KAPLAN TURBINE (Courtesy Voith-Siemens) [9]

expensive than synchronous generators, but now are typically used only in relatively small units [11].

11.4.6 Hydroelectric Controls

Turbine generators require controls and ancillary equipment. These include turbine flow governors, generator field exciters, electrical power and support system controls, and electrical fault and safety protection. The following are brief descriptions of each.

11.4.6.1 Governors A hydroelectric governor controls the speed and power of the turbine by regulating water flow. The theory is to sense and control output by sensing and controlling input. The first centrifugal flyball speed controlling governor was invented in 1788 by James Watt [12]. Although invented originally for the purpose of controlling steam turbines, the flyball mechanical governor was soon adapted to control hydro turbines. The flyball design consists of a rotating governor speed sensing shaft, which is either directly driven, belt connected or motor driven by the hydro turbine. The governor shaft is arranged with flyballs that rotate and extend outwards with increasing centrifugal speed. Through levers, valves, hydraulic oil power systems, and operating gates,

blades and jets, depending on the type of turbine, the governor then increases or decreases water flow to the unit. This speed sensing and control loop forms the basis of governor control.

There are other variations of the speed sensing governor with more complicated feedback loops. Those include speed droop, to damping response and prevent overcompensating and overshooting, especially in system of multiple units operating together at a plant to follow grid frequency and load. The servomotor's speed is directed by its position and cannot operate outside of the position's parameters. Speed droop, by definition, is the governor characteristic that requires a decrease in speed to produce an increase in gate opening. A compensating dashpot provides an additional method of balancing. It moves proportionally in the opposite direction to droop. The dashpot also adds temporary droop to the governor system and provides compensation for the effects of inertia of the unit and the water column.

A flyball and mechanical governor arrangement is shown in Fig. 11.7.

Dynamic control theory was later refined mathematically by James

- Proportional — This simplest compensation is based on the premise that the controller output is directly proportional to an

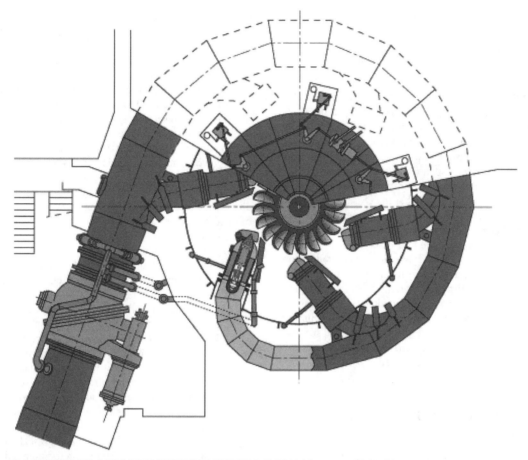

FIG. 11.6 PLAN VIEW OF A PELTON TURBINE INSTALLATION (Courtesy Voith Siemens Hydro Power Generation [10]

output. The correction signal is proportional to the difference between the process variable and set point.
- Integral — accumulative correction based compensation.
- Differential — rate of correction based compensation.
- "PID" — combined proportional, integral, and derivative correction-based compensation.

A schematic of turbine generator control loop is shown in Fig. 11.8.

Hydroelectric governor types have evolved as follows:

- *Flyball* — flyball weights mounted as rotating pendulums sensitive to speed changes.
- *Gate shaft* — flyballs, with oil hydraulic leveraging, and mechanical gate flow control.
- *Mechanical cabinet* — central hydraulic relay, with motor driven flyballs, oil pilot and servo hydraulic leveraging, and mechanical gate flow control.

Speed-Sensing Governor

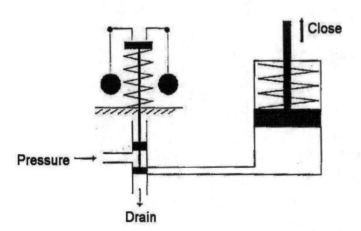

FIG. 11.7 FLYBALL AND SPEED SENSING GOVERNOR SCHEMATIC (*Source:* Woodward Governor)

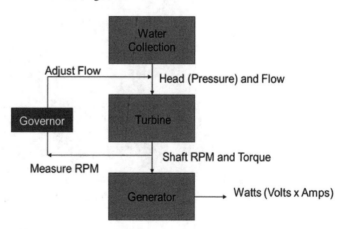

FIG. 11.8 HYDROELECTRIC GOVERNOR CONTROL LOOP (Woodward Governor)

- **Electrical cabinet** — electrical analog, hydraulically leveraged, mechanical.
- **Electronic** — solid state analog and mechanical.
- **Digital** — modern digital processor and mechanical.

11.4.7 Exciters

Exciters provide the direct current (DC) to energize the magnetic fields within a generator. The DC field on the hydraulically driven rotor then induces alternating current (AC) power output in the generator stator. Original exciters were the rotating type, either driven directly from the turbine shaft or driven by a motor, while modern exciters are typically solid-state or "static." Rotating exciters are either brush or brushless, to transfer the DC excitation current to the generator field. Static excitation for the generator fields is supplied by field-flashing voltage from storage batteries or from solid-state rectifiers powered by station service.

Hydroelectric turbines, generators, governors, and exciters are typically connected through transformers to a large electrical transmission grid. These electrical power systems operate at a nearly constant frequency, at 60 Hz or 50 Hz depending on the location. Usually, a single unit or plant cannot affect the larger transmission system frequency. Once synchronized, the generating unit speed is essentially locked to system frequency and follows load demand. This is how most units and plants are operated.

11.5 HYDROPOWER FOR ENERGY STORAGE

The modern development of wind power and solar power provide excellent sources of renewable energy. These renewable resources consume no fossil fuels, emit zero pollution, and are driven by free and practically limitless wind and sunshine. The main challenge of the renewable wind and solar energy is that their sources of power and therefore outputs are variable. Wind comes and goes with the weather. And sunshine comes and goes with the rotation of the earth and clouds. This makes these energy sources variable and somewhat unreliable for the purposes of power planning and meeting demand. Because both residences and industry loads are also variable, a variable supply will require power reserves to match and balance changes in generation and demand.

Therefore, other sources of reliable, flexible, and dispatchable generation are required to provide large electrical system capacity and balancing reserves on a daily, hourly, and real-time basis. In addition to system reserves, there is also a need for energy storage to balance potentially excess generation, for example, shifting wind power at night, when demand is low, to peak demand hours during the day. Conventional hydropower projects do this by shutting down units and storing energy in the form of water, and that is already the most common, although perhaps not well known, form of energy storage in the world. As variable renewable energy and the associated ratio of wind and solar generation to system load grows, so too will additional energy storage be required to balance supply and demand, and assure grid reliability.

A typically modern variable wind energy supply and grid demand is shown in Fig. 11.9.

Hydroelectric generation, and specifically hydroelectric pumped storage, is uniquely positioned to provide large-scale integration of variable renewable energy resources. Hydropower is itself a sustainable resource, which can also balance other renewable energy resources, by providing relatively large capacity energy storage and reserves. Hydropower is already the preferred technology providing system reserves throughout the world. While there are many potential solutions such as batteries to absorb excess energy and maintain a balanced energy system, pumped storage in particular is a proven, successful technology on a large scale in terms of power capacity and time duration. Hydroelectric pumped storage can provide other valuable ancillary benefits in addition to energy storage, including hydrologic seasonal storage, immediate electrical load balancing, frequency control, and incremental and decremental power reserves. It has historically been used to provide reserve capability to balance system load and allow large, thermal generating sources to operate at optimum conditions. And as variable genera-

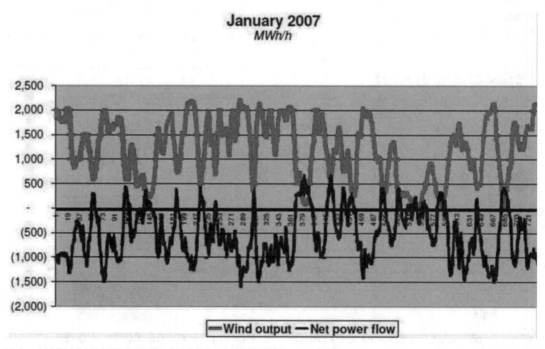

January 2007
MWh/h

— Wind output — Net power flow

FIG. 11.9   VARIABLE WIND ENERGY SUPPLY AND GRID DEMAND (*Source:* Bonneville Power Administration)

tion resources increase, pumped storage is being chosen worldwide as the preferred means to integrate all of these resources.

Europe has led the world in renewable energy and in addressing the associated challenge of balancing reserves for over a decade. The United States is now on track to significantly develop renewable wind and solar energy comparable to the proportional levels with the electrical grid of the European Union, now the benchmark for successfully integrating variable renewable energy. For example, Denmark's experience shows that introducing greater variable supply into the generation mix can very likely lead to a greater demand for system reserves. Norway and Sweden, with their predominately hydropower-supplied grids and strong interconnections with other countries, are generally able to accommodate power inrushes during periods of high-wind generation and then return that energy during low-wind periods. Those experiences, now given the rapid deployment of thousands of new wind turbines and solar arrays in the United States, indicate that large-scale hydroelectric pump storage must be planned and implemented as soon as possible or existing power and storage reserves will become overwhelmed, unbalanced, and renewable energy resources underutilized.

11.6 OCEAN AND KINETIC ENERGY

Over 70% of the Earth's surface is covered by oceans. Renewable energy resources associated with the world's oceans now include wave, tidal, undersea geothermal, and offshore wind. Kinetic energy machines developed for tidal currents are also being applied to convert energy from free-flowing river reaches without dams [14].

The available power and energy at a tidal energy site is dependent on the differential range of tide levels, local tidal current velocities, volume of the tidal basin, cross sectional geometry of the channel, and channel bottom roughness. The physical processes driving tidal exchanges are well understood and highly predictable, because of the primary gravitational pull and rotation of the moon around the earth. As a result, the potential of tidal energy can be extraordinarily predicable and reliable over the life of a project. For a free-flowing river, capturing hydro kinetic energy is dependent on the river hydrology over time, flow volume, velocity profile, and channel gradient, cross section and roughness.

Harnessing energy from waves, similar to the weather, is less predictable, more complex and potentially more powerful. Potential wave energy available is directly related to the heights and time intervals between wave peaks. The distribution of ocean surface energy is often described by a wave energy spectrum. The energy in the spectrum, in turn, can be characterized using the significant wave height and peak period. The wave height, peak period, and shape of the energy spectrum will combine to establish the design criteria, type of machine, and energy output of a wave energy conversion system.

11.6.1 Hydrokinetic Turbine Types

Horizontal axis hydrokinetic turbines utilize a fixed or variable pitch rotor with its rotational axis parallel to the flow. Due to variations in flow direction at some sites, horizontal axis turbines may require yaw control to keep the rotor aligned with the flow direction. Horizontal axis turbines can be rigidly mounted to the channel bottom or suspended from a floating structure. Depending on the flow speed and diameter of the rotor, horizontal axis turbines can either directly drive a generator or connect to a generator via a reduction gear.

Open rotor, horizontal axis machines are very similar to modern wind-turbines, consisting of a rotor that is driving a horizontally aligned drive-shaft. These machines can further be divided into rotors that use variable pitch blades and rotors that use fixed pitch blades. In addition, many machines of these types make provisions for either active or passive yaw of the turbine blades to orient with the flow direction. A gearbox is often incorporated into the nacelle of the turbine to increase the shaft speed and reduce the size of the generator required (Fig. 11.10).

Ducted rotor horizontal axis turbines utilize a steel or fiberglass duct to align the direction of the flow with the rotor. This flow alignment allows the turbine to operate efficiently over a range of flow directions without yawing the rotor. The rotor directly drives a generator that is housed within the support structure for the rotor and duct. The diameter of the generator allows for a large number of generator poles, enabling the relatively slow speed of the turbine to be converted into electricity. A large diameter bearing is required between the armature and the stator to maintain the alignment of the rotor (Fig. 11.11).

Vertical-axis hydrokinetic turbines consist of a rotor that turns around a vertical axis of rotation. Vertical axis machines can accommodate fluid flows from all directions without having to reverse the direction of turbine rotation. Vertical axis turbines are

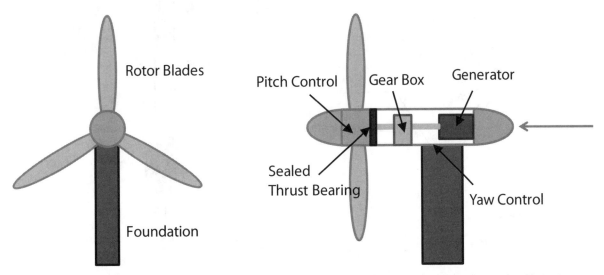

FIG. 11.10 OPEN ROTOR HORIZONTAL AXIS TIDAL TURBINE [ASME Hydromechanical Guide]

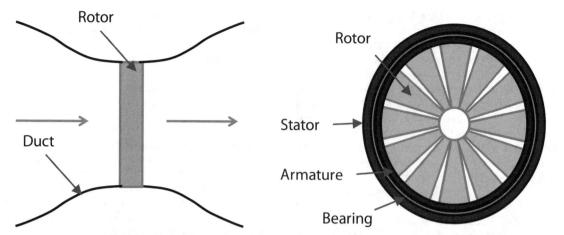

FIG. 11.11 DUCTED ROTOR HORIZONTAL AXIS TIDAL TURBINE (*Source:* ASME Hydromechanical Guide)

typically suspended from a structure on the water surface with the generator and associated electrical equipment above water. A series of vertical axis turbines can be installed across a channel forming a "tidal fence."

11.6.2   Wave Energy Converters

Wave energy conversion systems are best deployed offshore in deep water where the waves are generally linear and long crested. Those devices are not subjected to the non-linear forces applied to systems operating in shallow water, where wave breaking may occur. The wave energy available in deep water is also greater than that in shallow water due to avoiding the effects of bottom friction, as their waves begin to lose energy.

A point absorber is a wave energy device that is smaller than the length of the waves from which it extracts energy. Point absorbers extract energy from the waves by oscillating in one or more degrees of freedom, creating destructive interference as they move in the waves. Symmetric point absorbers generally extract energy from the waves by moving vertically (heaving) in the water column. The efficiency of a symmetric point absorber is independent of the wave direction and generally performs well over a wide range of wave frequencies (Fig. 11.12).

Typically, asymmetrical point absorbers extract energy from the waves by pitching back and forth about an enclosed horizontal axis point, with a wheel or pendulum. The device pitches in the water, while an internal mechanism swings, converting the wave power

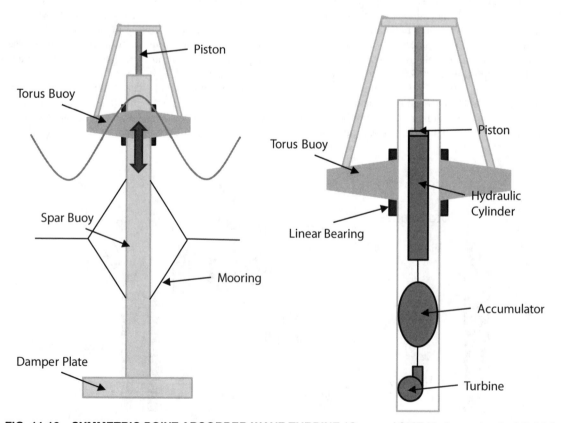

FIG. 11.12 SYMMETRIC POINT ABSORBER WAVE TURBINE (*Source:* ASME Hydromechanical Guide)

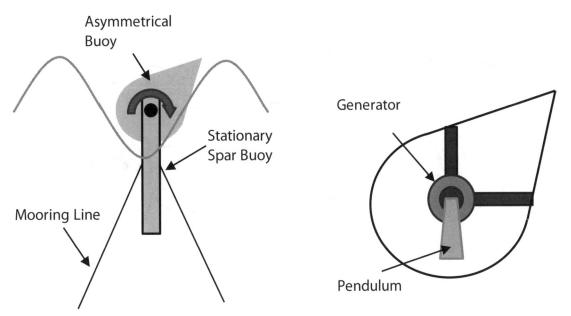

FIG. 11.13   ASYMMETRIC POINT ABSORBER WAVE TURBINE (*Source:* ASME Hydromechanical Guide)

into mechanical power. Electrical energy is then produced through the use of hydraulic cylinders and turbines or through a linear-to-rotary mechanism coupled to a rotary electric generator. The efficiency of an asymmetric point absorber is sensitive to the direction of the waves, therefore provisions must be made in the mooring system to orient the device toward the predominant wave direction. The shape of asymmetric point absorbers is also usually optimized for a single-wave frequency. As such, the efficiency is reduced for wave lengths away from the design point (Fig. 11.13).

An attenuator is a floating multiple-segment device that is arranged and moored in-line with the principal wave direction. Wave crests run along the length of the attenuator, causing flexing between segments. The flexing extracts kinetic energy through hydraulic pumps or similar mechanical–electrical converters. This type of wave energy conversion system must always be aligned with the predominant wave direction (Fig. 11.14).

Shallow water wave energy conversion systems are located near shore, reducing the need for expensive subsea transmission

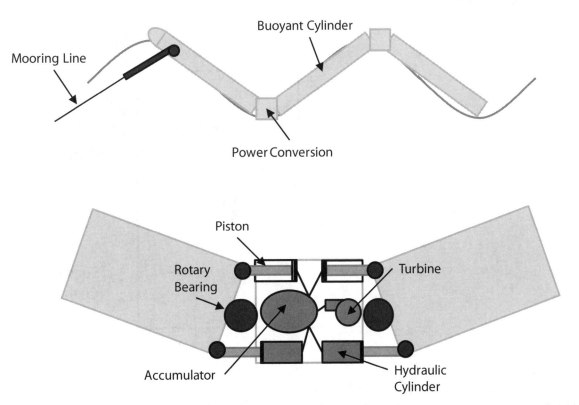

FIG. 11.14   ATTENUATOR WAVE ENERGY CONVERSION SYSTEM (*Source:* ASME Hydromechanical Guide)

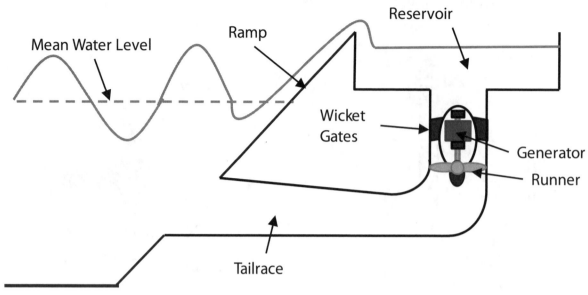

FIG. 11.15 OVERTOPPING WAVE ENERGY SYSTEM (*Source:* ASME Hydromechanical Guide)

cables and deep water mooring systems. As waves approach the shore, they begin to "feel" the effect of the bottom and generally the height of the waves increase and the wavelengths decrease. The effect of bottom friction also removes energy from the waves as they approach the shore, and when the steepness of the wave face reaches a critical value, the wave will break, dissipating a large portion of its energy. Shallow water wave energy systems harness the energy of these shallow water waves either just outside the breakpoint or, if the bottom is steep, on the shoreline.

An overtopping wave energy device typically consists of an enclosed basin into which waves spill. The water inside the enclosed basin is elevated over the sea-level, creating a low hydraulic head, which is converted into electricity using a low-head vertical hydraulic turbine. Overtopping systems can be installed on shore or deployed on floating platforms. Overtopping systems can use either traditional fixed blade propeller or Kaplan turbines to covert the stored wave energy in the reservoir into electricity (Fig. 11.15).

Oscillating water column wave energy conversion systems consist of a sealed chamber open near the bottom to allow waves to enter and exit under an air chamber. The wave action inside the structure compresses the air-column into a nozzle. A bi-directional air-turbine can then be used to convert the pressure differential into

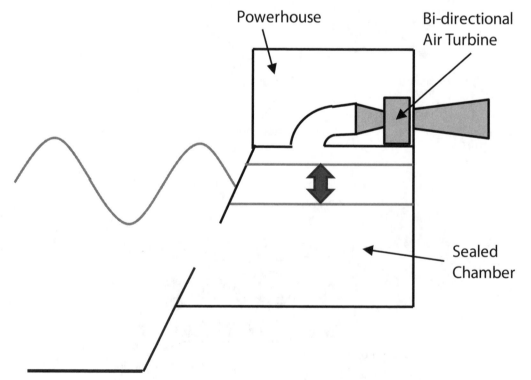

FIG. 11.16 OSCILLATING WAVE TO AIR COLUMN ENERGY SYSTEM (*Source*: ASME Hydromechanical Guide)

electricity. The most common type of turbine used in oscillating water column systems is the wells turbine, a slow speed reaction device designed to spin in the same direction regardless of the flow direction. Oscillating water column systems can be built on existing civil structures such as breakwaters or installed on fixed or floating platforms (Fig. 11.16).

There are many other promising types of wave, tidal and kinetic energy converters being developed. Some of those designs are yet unproven while the ocean energy industry matures technologically and commercially, similar to what the now more established river hydroelectric industry was only a hundred years ago.

11.7 HYDROPOWER ORGANIZATIONS AND OWNERS

For reference to the established and still evolving industry of hydropower, the following hydroelectric resource organizations and U.S. owners are active and listed below.

11.7.1 Organizations

- The American Society of Mechanical Engineers (ASME) is, as its name implies, a professional organization of mechanical engineers, founded in 1880. The ASME develops mechanical codes, standards, guidelines and books for industry and reference. Its vision is to be the premier organization to promote the art, science, and practice of mechanical and multidisciplinary sciences throughout the world [16]. Related to hydropower, the ASME includes a Hydro Power Technical Committee (HPTC) and a Power Test Committee (PTC).
- The American Society of Civil Engineers (ASCE) is a professional organization founded in 1852 to represent civil engineers internationally. It is the oldest national engineering body in the United States [17].
- International Committee on Large Dams (ICOLD) is an international non-governmental organization dedicated to sharing professional information on the design, construction, and maintenance of large dams. It was founded in 1928 with headquarters in France.
- The Institute of Electric and Electronics Engineers (IEEE) is an international non-profit professional body founded to promote the advancement of technology related to electricity. It was incorporated in New York from the merger of the American Institute of Electrical Engineers and the Institute of Radio Engineers [18].
- International Electro-technical Committee (IEC) is an international non-governmental standards organization that prepares and publishes international standards for all electrical, electronic, and related technologies. It was founded in 1906, and presently has over 130 countries subscribing to its practices [19].

11.7.2 Owners

- In the United States, the Department of Interior, Bureau of Reclamation is an agency which oversees water resource management as it relates primarily to water diversion, delivery, storage, and associated hydroelectric power generation throughout the western United States. The Bureau was founded under the Reclamation Act of 1902, with its original purpose to study potential water development projects in each western state with federal lands. Reclamation's hydroelectric Technical

Service Center is located in Denver, Colorado. Examples of hydropower projects include Grand Coulee Dam in Washington and Hoover Dam in Colorado [20].

- Also in the United States, the Department of Defense, Corps of Engineers is a agency that has developed and now manages many federal hydroelectric projects throughout the U.S. The Corps' Hydroelectric Design Center is located in Portland, Oregon. Hydropower projects include Chief Joseph, Bonneville, the Dalles, and many other dams on the Columbia River, as part of the Federal Columbia River Power System (FCRPS), in collaboration with the Bonneville Power Administration and Reclamation [21].
- State and Public Utility Districts and Municipals PUDs are jurisdictions throughout the U.S. created to manage public generation and demand for electricity, natural gas, sewage treatment, water, and communications.
- Utilities also include large and small publically and privately held companies. There are hundreds of utilities for thousands of hydroelectric projects owned, operated, and maintained throughout the country.
- In the United States, the larger non-federal hydroelectric projects are licensed by the Federal Energy Regulatory Commission (FERC). FERC hold licenses for about 1700 hydroelectric projects, totaling approximately 55,000 MW of installed capacity, or about half of all U.S. hydroelectric capability [22].

11.8 HYDROPOWER WORLDWIDE

In 2010, world energy usage has grown to a total of approximately 500 quadrillion (10^{15}) BTU per year. Of that, renewable energy sources, including hydropower, wind, and solar energy contribute about 50 quadrillion BTU, or 10%, compared to oil, coal, gas, nuclear, and consumable sources of energy. In terms of "electrical" energy, world annual generation and usage is presently approximately 19 trillion (10^{12}) kilowatt-hours. Of that world total electrical energy production, renewable energy sources contribute approximately 3.5 trillion watt-hours, or 18%. Hydroelectric energy comprises most of that renewable energy at about 3 trillion watt-hours, from about 822,000 MW of installed capacity, providing about 16% of the world's electricity [23].

To put this number in perspective, the hydropower plants on the mighty Columbia River and its tributaries in the Pacific Northwest of United States, including Grand Coulee Dam, have a total installed capacity of about 27,000 MW, which is about 28% of the United States hydroelectric capacity and 3% of hydroelectric capacity worldwide [24].

Brazil, Canada, Norway, Switzerland, and Venezuela are the only countries in the world that rely on hydroelectric power for the majority of their internal electric energy production. Paraguay produces all of its electricity from hydroelectric dams, but then exports 90% of its production to Brazil and to Argentina. Norway produces 99% of all its electricity from hydropower [25].

For now, the world's largest hydroelectric project is Three Gorges Dam in China, which has a design capacity production of approximately 22,000 MW. Next is Brazil and Paraguay's Itaipu Dam, built in 1984 with a capacity of 14,000 MW. Guri Dam in India is third at approximately 10,000 MW. The largest hydroelectric project in the United States is Grand Coulee, begun in 1934, now with an installed capacity of approximately 7000 MW.

These large-scale hydroelectric power stations are the largest power producing plant of any kind on the planet. In fact, many of

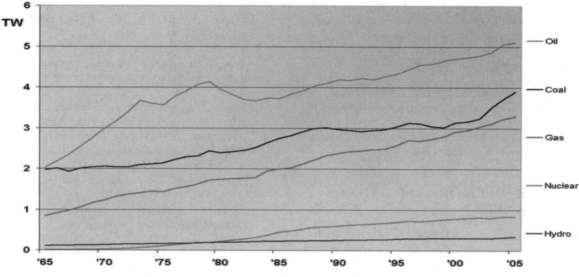

FIG. 11.17   GRAPHIC OF HYDROPOWER WORLDWIDE OVER HISTORICAL TIME [25]

very large hydroelectric facilities are capable of generating over several times the installed capacities of the world's largest nuclear or fossil powered power plants.

At the other extreme in size is small hydro, which refers to a plant developed on a scale that serves a small community or industrial plant. The definition of a small hydro project varies but a generating capacity of up to 10 (MW) is generally accepted as the upper limit of what can be termed small hydro. Small-scale hydroelectricity production grew during the industrial revolution and has resurged during the past decade. Small-scale hydro worldwide has a capacity that is not so small after all, in total estimated to be about 85,000 MW.

11.9   THE FUTURE OF HYDROPOWER

11.9.1   Renewable Portfolio Standards

A Renewable Portfolio Standard (RPS) is a governmental policy which requires electricity providers to obtain a minimum percentage of their power from renewable sources, such as wind, solar, biomass, and geothermal by a certain date. An RPS typically places an obligation on electricity supply companies to produce a specified percentage of their electricity from renewable energy sources. Certified renewable energy generators earn certificates for every unit of electricity they produce and can sell these along with their electricity to supply companies. Supply companies then typically pass the certificates to a state-level regulatory body to demonstrate their compliance with their regulatory obligations. As the RPS is market-mandated, it relies almost exclusively on the private market for its implementation. Supporters of RPS mechanisms state that market implementation will result in competition, efficiency, and innovation that will drive down costs and make them more capable of competing with fossil fuel energy sources. The Renewable Electricity Standard (RES) is the proposed national policy in the United States and seeks to have 25% of electricity produced by renewable resources by 2025. In the United Kingdom, it is known as Renewables Obligation, and supports the large-scale generation of renewable energy [26].

11.9.2   Energy Trends

Over the past 100 years, world electrical energy demand and supply have risen dramatically. Hydroelectric supply grew initially in the late 1800s and first half of the 1900s, but has not kept pace with overall energy demand, at least in the U.S., as can be seen from Fig. 11.17 [27] below.

So what is the future for hydropower, for the next ten, hundred, and thousand years? Will other renewables such as solar, wind, and ocean wave energy increase dramatically? Or will nuclear fission and fusion come back and along with other yet undiscovered exotic energy sources? And meanwhile, will hydropower on the simple dependable renewable workhorse on rivers for millennia continue to have value for future "generations"? We think so, but only time will tell!

11.10   REFERENCES

1. Dams and Public Safety (Part I) by Robert B. Jansen, U.S. Department of the Interior, Bureau of Reclamation (1980). http://ussdams.com/ussdeducation/Media/damsfrombegin.doc.

2. Wikipedia, water wheel, illustration released to public domain. http://upload.wikimedia.org/wikipedia/en/0/0d/Agricola1.jpg.

3. U.S. Bureau of Reclamation. http://www.usbr.gov/power/edu/history.html and the U.S.

4. "Tennessee Valley Authority Act (1933)." Major Acts of Congress. Ed. Brian K. Landsberg. Macmillan-Thomson Gale, 2004. eNotes.com. 2006. 15 Nov, 2010. http://www.enotes.com/major-acts-congress/tennessee-valley-authority-act.

5. U.S. Bureau of Reclamation. http://www.usbr.gov/power/edu/history.html and the U.S.

6. http://en.wikipedia.org/wiki/File:Hydroelectric_dam.svg.

7. http://www.whitemill.org/z0028.htm.

8. http://upload.wikimedia.org/wikipedia/commons/5/5f/M_vs_francis_schnitt_1_zoom.jpg.

9. http://upload.wikimedia.org/wikipedia/commons/1/16/S_vs_kaplan_schnitt_1_zoom.jpg.

10. http://upload.wikimedia.org/wikipedia/commons/6/65/S_vs_pelton_schnitt_1_zoom.png.

11. http://www.powergeneratorinfo.com/synchronous-generator/synchronous-generator.php.

12. http://en.wikipedia.org/wiki/Flyball_governor.

13. http://en.wikipedia.org/wiki/Control_theory.

14. David Elwood, excerpted from ASME HPTC, Hydromechanical Handbook, updated draft manuscript 2010.

16. http://en.wikipedia.org/wiki/American_Society_of_Mechanical_Engineers.

17. http://en.wikipedia.org/wiki/American_Society_of_Civil_Engineers.

18. http://en.wikipedia.org/wiki/IEEE.

19. http://en.wikipedia.org/wiki/International_Electrotechnical_Commission.

20. http://www.usbr.gov/power/edu/history.html.

21. http://www.nwp.usace.army.mil/hdc/home.asp.

22. http://www.ferc.gov/industries/hydropower.asp.

23. International Energy Outlook, by the U.S. Energy Information Administration (2010). http://www.eia.doe.gov/oiaf/ieo/pdf/0484(2010).pdf.

24. Federal Columbia River Power System, http://www.bpa.gov/power/pgf/hydrpnw.shtml.

25. http://www.economist.com/node/12970769?story_id=12970769.

26. http://apps1.eere.energy.gov/states/maps/renewable_portfolio_states.cfm.
http://www.decc.gov.uk/en/content/cms/what_we_do/uk_supply/energy_mix/renewable/policy/renew_obs/renew_obs.aspx.

27. http://en.wikipedia.org/wiki/World_energy_resources_and_consumption.

HYDRO POWER GENERATION IN INDIA – STATUS AND CHALLENGES

Dharam Vir Thareja

12.1   INTRODUCTION

Power generation in India has come a long way from about 1000 MW at the time of independence (August, 1947) to about 160,000 MW as on 31st March 2010 (end of Financial Year). The share of hydro power in this growth, in these over six decades, has also been impressive as it increased from about 500 MW at the time of independence to about 37,000 MW as of March 2010. But the present level of hydro power exploitation is only about 25% of the ultimate installed capacity estimated at 150,000 MW.

The demand of power is increasing rapidly (at the rate of over 8% per annum) so has been the realization that hydro power energy, being a renewable source, be exploited to its full available potential. In this endeavor, of the balance over 113,000 MW of power, i.e., yet to be commissioned, projects with about 14,000 MW are under construction and about 100,000 MW are under various stages of implementation.

The hydro project implementation and ownership remained with the State Governments or with the Power Corporations owned by States or Central Government for about four and a half decades. Due to slow pace of development and also in line with international stress for liberalization of economy, the Government of India reviewed the policy of power development. In 1992, the power sector was opened up allowing private capital participation in its development along and in parallel with continued development under public sector (under the Five-Year Plan system) [1]. With inherent constraints associated with hydro power development, construction activity in the private sectors did not show expected results in the initial period of over one decade. But it is picking up now and at present projects aggregating to about 25,000 MW are allocated to private developers on build, own, operate, and transfer basis (BOOT). Of these, projects with installed capacity of 20,000 MW are under various stages leading to their implementation including survey investigation, DPR preparation and clearances, about 4000 MW are under construction and over 1400 MW are in operation.

India now possesses the needed financial resources and professional capabilities to utilize the in-house and global capacities for implementation of hydro power projects in the challenging environment. 50,000 MW initiatives leading to identification of 162 projects and completion of their Pre-feasibility Reports (PFRs) in a span of 2 years (2004 to 2006) has been recognized as a laudable initiative [2,3]. The achievement of commissioning of about 8000 MW in the 10th Plan period (2002 to 2007), the target of over 15,000 MW in the running 11th Plan period (2007 to 2012) and 20,000 MW proposed during the 12th Plan period (2012 to 2017) speaks of the ambitious plan of hydro power development [4]. To achieve the targets of 11th and 12th plan, the fund requirement would be of the order of US $ 30 billion of which about 30% would be the share of private sector. The participation of private sector in the investment projections for 11th plan for infrastructure, which includes power sector also, is estimates at 30% [5].

The hydro power that is yet to be tapped, projects with more than 75% of installation are located in Himalayan region. This region is known for intense seismicity, wide range of geo-technical variability even in short stretches, extensive hydrologic variations over and within the year both for water and sediment flows, which pose challenges for infra-structure development, during project construction and their operation and maintenance stages. With about 70,000 MW of installation that is lined up for implementation in the Himalayan region, India will continue to remain a source of professional grooming with innovations in project planning, investigation, analysis, hydrological and sedimentation studies, structural design, construction methodologies and judicious construction equipment decisions. The expert consultancy, construction, and project management organizations will get opportunities to plan projects, which can be implemented in time frame of five years. The constraints of infra-structure will require to be integrated in the project formulation. The experience of projects that have been completed and those under construction will have to be pooled to decide the way forward for implementation of future projects without time and cost over-runs.

It has been planned to exploit the bulk of balance hydro potential in the next about two decades [6]. To meet the target, the hydro-sector would require investment of US$ 100 billion requiring an average of US$ 4 billion per year for another 25 years and thus would remain attractive for financial institutions, project developers, contractors, and consultancy organizations.

Role of hydro power in the energy scenario; potential and status of development; small hydro- and pump-storage development; transmission set-up and status; constitutional and regulatory provisions; resettlement and rehabilitation policies; techno-economic appraisal procedures; hydro power development in the neighboring countries; response and achievement of private sector; the issues, constraints and challenges in development; innovations for future projects; are covered in this Chapter.

12.2 ENERGY SCENARIO AND ROLE OF HYDRO

12.2.1 Energy Scenario

12.2.1.1 Installed Capacity and Share of Hydro India has a total installed capacity of about 1,59,400 Megawatt (MW) of power from all conventional energy sources, as on 31st March 2010. Fuel-wise breakup is covered in Table 12.1.

In terms of percentages, the distribution is 64.3% thermal, 23.1% hydro, 9.7% renewable, 2.9% nuclear-based, and is depicted in Fig. 12.1.

Hydro energy is a renewable source but for accounting purposes only small hydro projects (projects up to 25 MW capacity) are considered in the renewable category and their regulation is controlled by Ministry of New and Renewable Energy (MNRE), Government of India whereas hydro and thermal energy development is governed by Ministry of Power and nuclear energy by Atomic Energy Commission and Nuclear Power Corporation.

The share of hydro generation in the total generating capacity of the country remained at over 40% in the first three decades of power development. Thereafter, it has shown a downward trend to its present level of about 23%. This can be attributed to the fact that from mid-1970s, the Government started relatively favoring thermal power development vis-à-vis hydro with the objective of increasing the pace of development as in its view hydro projects were taking unduly long time in completion and thus affecting the pace of development. This caused relative set back to the pace of hydro development and in the process adversely affecting optimum desired thermal-hydro mix of 60:40 for smooth operation of system. This policy was not a prudent one. Hydro — being a perennial source of power gifted by nature with many other advantages over other resources as detailed in the subsequent paragraphs, deserved to be given a rather high priority. Regarding the long time and delays in completion of hydro projects one had to take cognizance of the fact that unlike thermal, hydro development-particularly of medium and large projects—would, in certain cases (depending upon geography, location, type of development envisaged), involve some inherent problems, including resettlement and rehabilitation, delay in land acquisition, interstate disputes, law and order problem, delay in investment decision, environmental clearances, contractual disputes, geological surprises, etc. The only way to enhance the share of hydro is to accept that these problems are inherent part of hydro development and the way forward is in the prompt resolution by mobilizing requisite resources/inputs.

12.2.1.2 The Country's Rising Energy Needs and Necessity of Hydro Power Exploitation India's per capita electricity consumption is among the lowest in the world at about 1/5th of the

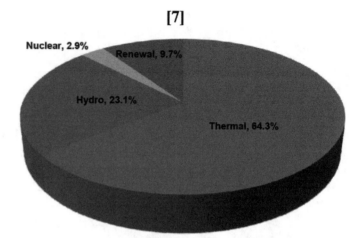

FIG. 12.1 INSTALLED CAPACITY VS. SHARE OF HYDRO-SOURCE: [7]

global average. With impending economic growth, a sharp rise in consumption is inevitable. Per capita power consumption will rise from about 670 kWh at present and is expected to go to 2000 kWh in next two decades. The National Electricity Policy announced in 2005 [8] aims at access of electricity by all households and per capita availability of electricity to be increased to 1000 units by 2012.

India's grid connected power generation capacity will need to go up by 3 to 4 times from the present about 160 GW to 400 to 600 GW if India is to meet the growing demand and to attain status of economic superpower. Since hydro power is one of the cleanest and most reliable sources of energy India has targeted to exploit bulk of its potential in the next two decades.

12.2.2 Role of Hydro

12.2.2.1 Distinct Economic and Social Advantages Hydro power is a renewable and non-polluting source of energy. Development of hydro has a long-term economic advantage as the annual operating costs are a tiny fraction of the initial capital cost. Hydroelectric projects have long useful life extending over 50 years and help in conserving scarce fossil fuels. Hydro's autonomy from the fuel price is a single distinct advantage. The flexibility of storage hydro (using reservoirs) also makes it a compelling partner to ensure security in mixed power systems. Multi-purpose hydro reservoirs can service the increasing need for water management and thus bring security of water supply as well as power. These projects results in far greater socio-economic benefits by bringing in the dimensions of irrigation, providing potable drinking water, flood mitigation, drought protection, water resource management and larger community-based economic development of remote, and backward areas, in addition to power generation.

12.2.2.2 Synergy with other Renewable Energies Storage hydro and, in regions where the quantity of water is limited, pumped storage can solve a plethora of system challenges. It can follow load fluctuations, so that fossil-fuel plants can continue to operate at their best efficiency. Wind power produces a variable and intermittent supply and hydro can provide the firming capacity to ensure both security and quality in the system. Similar synergies with solar, bio-generation, and marine power are being viewed when In-

TABLE 12.1 FUELWISE BREAKUP OF INSTALLED CAPACITY. SOURCE: [7]

Fuel	MW
Thermal	102453.98
Hydro	36863.40
Renewable	15521.11
Nuclear	4560.00
Total	159398.49

dia is aiming toward generating significant additional power from renewable sources in the next two decades. National Solar Mission has set a target of generating 20,000 MW of solar power by 2020 [9]. The wind energy potential is estimated at 50,000 MW [10].

12.2.2.3 Favorable Impact on and Resilient to Climate Change
In terms of climate change, hydro power tends to have a very low greenhouse gas footprint. As water carries carbon in the natural cycle, scientists have investigated the extent to which a new reservoir might accelerate carbon emissions. Many reservoirs around the world have been monitored and it has been established that hydro power is one of the cleanest methods of power generation. By employing hydro power generation in place of fossil-fuel technologies, a significant offset of greenhouse gas emissions can be achieved. Storage-based hydro power projects can absorb the fluctuations in the inflows that are expected to increase on account of climate change and ensure availability of designed power output. In India, tendency of converting storage based hydro development to run-of-river schemes is being discouraged.

12.2.2.4 Proven Efficient, Flexible, and Reliable Power Source
In terms of efficiency, hydro power shows the best conversion rate (90%) due to the direct transformation of hydraulic forces to electricity. It has the most favorable energy payback ratio considering the amount of energy required to build, maintain, and fuel a power-plant compared with the energy it produces during its normal life span.

In respect of flexibility, the storage of potential electricity in reservoirs, hydro power has the capacity to provide base and peak-load. It is the ideal back-up source for intermittent electricity sources such as wind, solar and optimizes efficiency of less flexible fossil or nuclear generating options. Hydro power has the capacity to follow demand fluctuations almost instantly and offers a quick response to failings in power grids.

Regarding reliability hydro power is a proven and well-advanced technology based on more than a century of experience. It is a backbone of an integrated renewable grid, a clean source of renewable energy with the capacity to make a significant contribution to the world's ever-growing need for electricity.

12.2.3 Policies Conducive to Emphasis on Hydro Power

12.2.3.1 Hydro Policy
The new Hydro Policy announced by Government of India in 2008 [11] has liberal provisions that would provide the private sector players a level playing field with transparent Selection criteria for awarding projects and induce large private investments in development of hydro power projects. The objectives of the hydro power policy are:

(i) Inducing private investment in Hydro Power Development
(ii) Harnessing the balance hydro-electric potential on priority
(iii) Improving resettlement and rehabilitation to generate support for hydro development
(iv) Facilitating financial viabilities
(v) Enable developers to recover his additional costs including upfront investments through merchant sale of power up to a maximum of 40% of the saleable energy. There is a deterrent for delay as the merchant power will be reduced by 5% for every 6 months of delay in commissioning of the project.

12.2.3.2 Electricity Act 2003
With the liberalization of the economy, the Government of India has been encouraging and inviting private sector for investment in the power sector. Accordingly, a conducive policy environment has been created by modifying the Electricity Act repealing the Indian Electricity Act, 1910, the Electricity (Supply) Act, 1948 and the Electricity Regulatory Commission Act, 1998. The new Electricity Act, 2003 [12] deals with the laws relating to generation, transmission, distribution, trading, and use of electricity. The Act has specific provision for the promotion of renewable energy including hydro power and cogeneration. It has been made mandatory that every state would specify a percentage of electricity to be purchased from renewable by a distribution licensee. This act has permitted direct commercial relationships between generating companies and consumers/ traders. The Act has provided a generating company the right to open access through state/central transmission facilities.

12.2.3.3 National Electricity Policy
The Policy announced in 2005 [8] underlines that hydro energy potential needs to be exploited and private sector would be encouraged through suitable promotional measures. The policy stipulates that progressively the share of electricity from hydro would need to be increased.

12.2.3.4 National Water Policy
The government has brought out a national water policy in the year 2002 [13]. It has been stipulated in the policy that in the planning and operation of system, water allocation priority should broadly be in the order of drinking water, irrigation, hydro power, ecology, agro industries and non-agriculture industries, navigation and other uses. The priority for hydro-development is very high in the list of water uses for different purposes.

12.2.3.5 Mega Power Project Policy
Keeping in view the requirements of power projects located in certain special category States of J&K, Sikkim and the seven states of North East the minimum qualifying capacity of hydro power plants to avail mega project benefits, has been reduced from 500 MW to 350 MW in the revised Mega Power Projects policy so that they can avail the advantages of reduced duties and taxes [14].

12.2.4 National Level Resolve

Hydroelectricity being a clean and renewable source of energy with highest energy pay back ratio, maximum emphasis would have to be laid on the full development of the feasible hydro potential in the country. Slow pace of development for some years should not be viewed with discouragement instead more focused approach with national level resolve and matching resources needs to be channelized in harnessing hydro potential speedily, as it is also one of the easiest and best way to facilitate economic development of States, particularly, North-Eastern States, Sikkim, Uttarakhand, Himachal Pradesh and J&K, where a large proportion of our hydro power potential is located. Hydro projects call for comparatively larger capital investment and therefore, debt financing of longer tenure would need to be made available. Central Government has to remain committed to policies that ensure financing of viable hydro projects. State Governments need to review procedures for land acquisition, and other approvals/clearances for speedy implementation of hydroelectric projects. Working for future projects will demand innovations in planning, design and construction and for sustaining the pace of development, the Central Government will have to be more vigorous in extending its support to the State Governments, power corporations and private developers by offering expert services of Central Public Sector Corporations, Central Government Departments and also facilitating timely availability of global expertise as and when required.

12.3 BASIN WISE POTENTIAL AND DEVELOPMENT SCENARIO

12.3.1 Hydro Potential

India with a total geographical area of 329 million hectares is a land of diversity blessed with a large number of rivers as well as mountains. As these rivers flow from their source in mountains and hills, they provide plenty of scope for large-scale hydro power development. In the peninsular rivers there is a very wide disparity between the discharge in the monsoon period and the non-monsoon period, with the variation being as much as 25 times. Therefore, in case of hydro projects on these rivers, large storage capacity is often required to balance out the flows in order to increase the firm benefits from hydroelectric power. The Himalayan Rivers being snow fed have a more equitable distribution of water flow and therefore, large numbers of run-of-the-river type or small storage capacity hydroelectric projects yielding substantial firm power are feasible. In addition, a few reasonably large to very large capacity storage based sites are available to provide the needed stability and security to the hydro power system. Considering the topography and heavy to moderate rainfall in most of the regions/areas, there are extremely large numbers of possible sites for development of pumped storage schemes for providing much needed peaking capacity. Only attractive scheme with high heads, proximity to load centers and economy in cost of civil structures has been targeted for development. Accordingly, Himalayan Region and in particular Arunachal Pradesh & Sikkim, has been kept out of the focus of identifying pumped storage schemes.

12.3.1.1 Systematic Assessment The first systematic and comprehensive study to assess the hydro-electric resources in the coun-

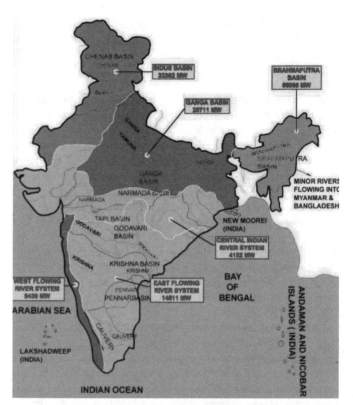

FIG. 12.2 BASIN-WISE MAP OF INDIA WITH THEIR HYDRO POTENTIAL. SOURCE: [15, 17]

try was undertaken during the period 1953 to 1959 by the Power Wing of the erstwhile Central Water and Power Commission [15]. This study placed the economical and utilizable hydro power potential of the country at 42100 MW from 250 schemes, at 60% load factor (corresponding to an annual energy generation of 221 billion units).

The Reassessment studies (1978 to 1987) of Hydro Electric Potential were undertaken by CEA for providing up-to-date data of hydro electric potential of the country and facilitate development of hydro power capacity [15, 16].

The Reassessment study completed in 1987 revealed that the Hydro Electric Power Potential of the country was of the order of about 84,000 MW at 60% load factor (with probable installed capacity of about 1,50,000 MW corresponding to an annual energy generation of 442 billion units) from a total of 845 Nos. schemes. With seasonal energy, the total energy potential is assessed to be around 600 billion units per year. In addition, 56 Nos. sites for development of Pumped Storage Schemes with likely aggregate installed capacity of about 94,000 MW were also identified in various regions of the country as detailed under a separate section of this chapter.

The basin-wise hydroelectric power potential of the country as per Reassessment Studies is covered in Table 12.2 and is also depicted in the map of India in Fig. 12.2.

12.3.2 Development of Hydro Power

12.3.2.1 Pre-independence Development Hydro-electric Development in India started in 1897 with the commissioning of the first station of 200 KW capacity at Sidrapong near Darjeeling in West Bengal. Tapping of hydro power generation was generally an open field, subject to obtaining permission from local authori-

TABLE 12.2 BASIN-WISE PROBABLE INSTALLED CAPACITIES. SOURCE: [15]

River Basin	No. of Schemes	Probable installed capacity (MW)
Himalayan Rivers		
Indus	190	33,832
Brahmaputra	226	66,065
Ganga	142	20,711
Sub-total	558	120,158
Peninsular Rivers		
Central Indian River System	53	4152
West Flowing Rivers of Southern India	94	9430
East Flowing Rivers of Southern India	140	14,511
Sub-total	287	28,093
Total	845	148,701
Pumped Storage Schemes	56	94,000

ties, especially in case of private developers. A few private utility companies, princely States and industrial houses constructed hydro power stations of varying capacities, from the micro/mini to a few tens of megawatts according to their requirements. It took quite some time before some of the provincial governments undertook the development. Shiva Samudaram Power House on river Cauvery with initial capacity of 7.92 MW in 1902 increasing to 47 MW in 1938 came up with the initiative of Mysore State. Governments in the State of Himachal, Punjab, Madras, and Kerala also took lead in hydro power construction. Major contribution was made by renowned industrial house of Tata's in bringing up three major hydro-stations in the Western Ghats in Maharashtra, namely; 40 MW Khopoli (5 × 8 MW) in 1915, 48 MW Bhivpuri (4 × 12 MW) during 1922 to 1925, and 90 MW Bihra (5 × 18 MW) in 1927.

Total installed hydro capacity at the time of independence (August 1947) grew up to 508 MW in a span of about 50 years since the first development. There were many reasons for such a small growth in power development such as absence of concerted efforts and funding at the centre level, adverse effects of two world wars, lack of interest from financial institutions and private enterprise, and apprehension of lack of demand/slow demand growth.

12.3.2.2 Post-independence Development After 1947, the Government of India accorded a very high priority to development of power sector and decided to undertake projects under public funding in a pattern of centrally planned economy — the 5-Year Plan Model. Plan-wise growth of hydro power up to the part of on-going 11th Plan and its share as a percentage of total power installation is presented in Table 12.3. Staring from 1951, the launching of 1st Plan, 5-year hydro power development up till 4th plan, remained mainly with States. From 5th Plan onward, the Central Government entered in a big way by way of setting-up power corporations.

TABLE 12.3   PLAN-WISE GROWTH OF INSTALLED CAPACITY OF HYDRO POWER. SOURCE: [4, 18]

Plan-wise growth of power and share of hydro power			
Plan period	**Installed capacity at the end of plan (MW)**		
	Hydro	**Total**	**Hydro as % of total**
1st Plan (1951–1956)	1061	2886	36.78
2nd Plan (1956–1961)	1917	4653	41.19
3rd Plan (1961–1966)	4124	9027	45.68
Three Annual Plans (1966–1969)	5907	12957	45.58
4th Plan (1969–1974)	6966	16664	41.80
5th Plan (1974–1979)	10833	26680	40.60
Annual Plan (1979–1980)	11384	28448	40.01
6th Plan (1980–1985)	14460	42585	33.96
7th Plan (1985–1990)	18307	63636	28.77
Two Annual Plans (1990–1992)	19194	69065	27.79
8th Plan (1992–1997)	21658	85795	25.46
9th Plan (1997–2002)	26269	105046	25.40
10th Plan (2002–2007)	34654	132329	26.19
11th Plan (2007–2012)*	36863	159398	23.13

*11th plan is ongoing; status is given up to 31st March 2010.

TABLE 12.4   SECTOR-WISE INSTALLED CAPACITY AT THE END OF THE COMPLETED 10TH PLAN (2002–2007). SOURCE: [7]

Sector	Installed Capacity
Central	7562
State	25789
Private	1306
Total	34654

The National Hydro-electric Power Corporation (NHPC) was set up in 1975. North-Eastern Electric Power Corporation (NEEPCO) was set up in 1976 to implement the regional power projects in the North-East. Subsequently two more power generation corporations were set up jointly by Central and State Governments in 1988 viz. Tehri Hydro Development Corporation (THDC) and Nathpa Jhakri Power Corporation (NJPC).

It would be seen from the above that hydro share was more than 45% at the end of 3rd Plan which reduced to about 26% at the end of 10th Plan. The sector-wise breakups of installed capacity at the end of 10th Plan is given in Table 12.4. The contribution of private sector up-till the year 2007 has been only about 4%.

12.3.2.3 Plan-wise Status of Achievements Vs. Targets of Hydro Capacity The details of hydro capacity actually added to the grid vs. targeted for addition from 4th Plan onward is presented in Fig. 12.3.

A review of this status reveals that in none of the plan periods targets could be achieved. The shortfall has been relatively less up-till 7th Plan whereas in 8th Plan the shortfall had reached 75%. In the 9th Plan and the last completed Plan, i.e., 10th Plan only about 50% of the planned hydro capacity could be achieved.

The Working Group of power for the 10th Plan had recommended hydro capacity addition of 14,393 MW comprising of 8742 MW in central sector, 4481 MW in state sector and 1170 MW in Private Sector. However, hydro capacity of 7886 MW comprising of 4495 MW in Central Sector, 2691 MW in State Sector and 700 MW in Private Sector could actually be commissioned during the 10th Plan. The contribution of private sector in this capacity addition has been about 9%. The main reasons for slippages in 10th Plan are delay in supplies / erection by suppliers/contractor, delay in award of works, delay in clearances/investment decisions, law and order problems and such other reasons like delay in environmental clearances,

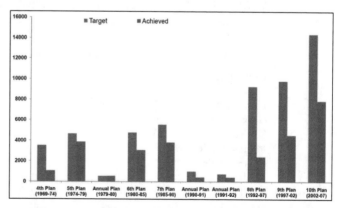

FIG. 12.3   PLANWISE HYDRO CAPACITY (MW), TARGETS VS. ACHIEVEMENTS. SOURCE: [4, 6, 18]

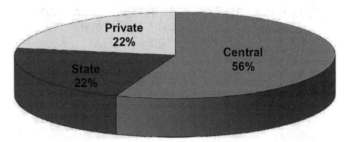

FIG. 12.4 PLANNED SHARE OF PRIVATE SECTOR IN 11TH PLAN. SOURCE: [4]

geological surprises, natural calamities, R&R issues, delay in signing of MOU, court cases, etc.

In the continuing 11th Plan, hydro power capacity addition of 15627 MW was targeted. The share committed to Private Sector as depicted in Fig. 12.4 shows their increasing role. In the first 3 years of this running plan, projects with 2210 MW of installation have been commissioned and up-till the completion of this plan an additional about 6000 MW is likely to be loaded. The expected achievement would thus be 8237 MW against a target of 15,627 MW, which again would be short by 50%.

12.3.3   Acceleration of Development

To give necessary fillip for development of the balance hydro electric schemes and with a view to prioritize the large number of identified schemes to harness vast untapped hydro resources in the order of their attractiveness for implementation, ranking studies were carried out by CEA in October, 2001. The Ranking Study gives inter-se prioritization of the projects which could be considered for further implementation including their survey & investigation so that hydro power development is effected in an appropriate sequence.

The ten major aspects which play vital role in the implementation of the hydro projects were adopted in the criteria considered for ranking study. These were R&R aspects, international aspects, interstate aspects, potential of the scheme, type of scheme, height of dam, length of tunnel / channel, accessibility to site, status of the project and status of upstream or downstream hydroelectric development. For each of the criteria, certain marks with weightage

FIG. 12.5 MAP OF INDIA SHOWING STATES. SOURCE: [20]

were allotted on its applicability to each individual project and 400 schemes with probable installed capacity of about 1,07,000 MW were prioritized under categories A, B & C. The categorization A implies ease whereas C indicates difficulties in implementation. River Basin-wise summary of categorization of the schemes is given in Table 12.5.

12.3.3.1 50,000 MW Hydroelectric Initiative A further step forward toward accelerated development of hydro power has been when Honorable Prime Minister of India launched a two year program for preparation of Preliminary Feasibility Reports (PFRs) in respect of 162 hydroelectric schemes, located in 16 States, (Fig. 12.5 shows Map of India with States) with installed capacity of

TABLE 12.5   BASIN-WISE CATEGORIZATION FOR ACCELERATED DEVELOPMENT. SOURCE: [19]

River System	Category A		Category B		Category C		Total	
	Nos.	MW	Nos.	MW	Nos.	MW	Nos.	MW
Himalayan Rivers								
Indus	11	4088	51	8811	17	6080	79	18979
Brahmaputra	52	7800	97	42574	19	12954	168	63328
Ganga	20	2023	54	9616	1	600	75	12239
Sub Total	83	13911	202	61001	37	19634	322	94546
Peninsular Rivers								
Central Indian	3	283	9	1425	1	186	13	1894
West Flowing	1	35	10	958	14	1508	25	2501
East Flowing	11	1412	26	6469	2	88	39	7969
Sub Total	15	1730	45	8852	17	1782	77	12364
Total	98	15641	247	69853	54	21416	399	106910

TABLE 12.6 STATE-WISE NUMBER OF PRELIMINARY FEASIBILITY REPORTS AND THEIR INSTALLATION. SOURCE: [3]

Sl. no.	State	Number of Schemes	Installed Capacity (MW)
1.	Andhra Pradesh	1	81
2.	Arunachal Pradesh	42	27293
3.	Chhattisgarh	5	848
4.	Himachal Pradesh	15	3328
5.	Jammu & Kashmir	13	2675
6.	Karnataka	5	1900
7.	Kerala	2	126
8.	Madhya Pradesh	3	205
9.	Maharashtra	9	411
10.	Manipur	3	362
11.	Meghalaya	11	931
12.	Mizoram	3	1500
13.	Nagaland	3	330
14.	Orissa	4	1189
15.	Sikkim	10	1469
16.	Uttarakhand	33	5282
Total		**162**	**47930**

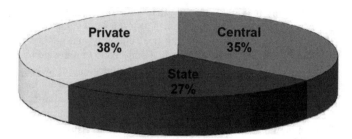

FIG. 12.7 SHARE OF PRIVATE SECTOR IN THE CAPACITY ADDITION PLANNED FOR 12TH PLAN (2012–2017). SOURCE: [18, 21]

about 50,000 MW. The scheme was formulated by CEA and sanctioned/funded by the Ministry of Power on 31st March, 2003. The state-wise details of Preliminary Feasibility Report prepared under the above program are given in Table 12.6.

Based on the projected tariff of the various schemes, 77 projects with installed capacity of 33951 MW were selected for preparation of Detailed Project Report and development.

12.3.4 Strategy for Hydro Development during 12th Plan period (2012 to 2017)

To achieve the ambitious program of hydro capacity addition of over 20,000 MW during 12th Plan (2012 to 217), a shelf of 109 candidate hydro projects aggregating to 30920 MW, having higher level of confidence for realizing benefits during 12th Plan, based on their status of preparedness, had been finalized [18].

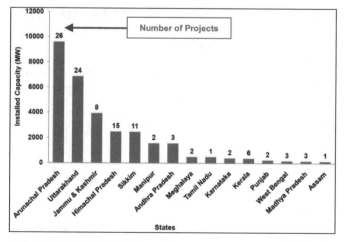

FIG. 12.6 STATE-WISE NUMBER OF PROJECTS AND INSTALLATION IDENTIFIED FOR 12TH PLAN (2012–2017). SOURCE: [18, 21]

The state-wise distribution of schemes with their total installation is depicted in the bar diagram in Fig. 12.6. Since long period is required for development of DPRs, obtaining various clearances like Environment & Forest clearances, CEA clearance, investment decision and achieving financial closure, regular monitoring is being done to ensure start of their construction in the 11th Plan itself.

As of now, excluding the projects that may slip from 11th Plan, a shelf of 87 projects with installation aggregating to over 20,000 MW that have the probability of their being amongst the list of projects for development in 12th Plan have been short listed. Their sector-wise distribution covered in Fig. 12.7 shows a further increasing role of private sector.

12.4 SMALL HYDRO DEVELOPMENT

12.4.1 The Aim and Focus of Small Hydro Program

In India, about 70% of the population lives in villages. These villages are scattered all over the country, including those in remote and hilly areas. To achieve 100% electrification target by 2012 is a stupendous task. Keeping in view the large investment required for transmission and distribution system, as against grid connectivity, the decentralized energy access to these villages and semi-urban centers is being aimed at through renewable energy including small hydro sources. India is blessed with an abundance of sunlight, water and biomass and vigorous efforts of last two decades in tapping these sources are now bearing fruits. India has the world's largest program to harness all viable renewable energy sources to produce electricity, for local supply as well as for feeding into the larger grid. India is the only country, which has an exclusive Ministry (MNRE) dealing with New and Renewable Energy Sources. As per MNRE, hydro power projects up to 25 MW station capacity have been categorized as Small Hydro Power (SHP) projects, which includes Micro (up to 100 KW) and Mini (101 to 2000 KW) projects. These are connected to the intra-State system and come under the direct jurisdiction of the respective State Electricity Regulatory Commission.

The MNRE has taken a decision that out of the total grid interactive power generation capacity that is being installed, 2% should come from small hydro. Today, the SHP program is essentially private investment driven. Private sector entrepreneurs are finding attractive business opportunities in small hydro and State Governments have also realized that the private participation may be necessary for tapping the full potential of rivers and canals for power generation. The focus of the SHP program is to lower the cost of equipment, increase its reliability, and set up projects in areas, which give the maximum advantage in terms of capacity utilization. A series of steps have been taken to promote development of SHP in a planned manner and improve reliability and quality of the projects. Concurrently, efforts are being made to renovate and modernize old SHP projects and complete languishing projects.

The procedure for allotment of sites has also been streamlined and made transparent. While some states continued with allotment of already identified sites, some other even allowed identification of potential sites by the entrepreneurs themselves. Number of financial institutions and banks are financing the projects.

12.4.2 Renewable Energy Policy of Central/State Government

The Government of India have announced various policies, i.e., the National Electricity Policy in 2005, Tariff Policy in 2006 [22] and Hydro Policy in 2008 and amended the Electricity Act in 2003 to create a conductive atmosphere for investments in the power sector including development of SHPs.

Section 86(1) (e) of Electricity Act, 2003 empowers Electricity Regulatory Commission to promote co-generation and generation from renewable sources of Energy by providing suitable measures of connectivity with the grid and sale of electricity to any person, and also specify percentage of renewable energy to be procured as "Renewable Purchase Obligation" for distribution licensees in the States.

Section 61(h) of Electricity Act, 2003 further stipulates that the Appropriate Regulatory Commission shall subject to provisions of the Act, specify the terms and conditions for determining of tariff, and in doing so, shall be guided by the promotion of co-generation and generation of electricity from renewable sources.

In compliance to the mandate as indicated above in the provisions of Electricity Act, 2003 the Central and State Electricity Regulatory Commissions issue from time to time specific regulations/ orders on Tariff and other important issues relating to small hydro power projects. The State Governments have also announced specific policies for SHP projects through private participation. The facilities available in the State include wheeling of power produced, banking, buy-back of power, facility for third party sale, etc. Government of India and the State Governments are providing fiscal and financial incentives like concessional import duty, custom and excise duty relief, sales tax relief, capital subsidy, and tax holiday.

For promoting small hydel projects the hydro policy lists out the following measures/incentives.

(i) Incentives to Private Sector during the execution of the project in the form of capital subsidy.
(ii) Special incentives for execution of small hydro projects in the North Eastern Region by the Government departments/ SEB/State agencies.
(iii) Financial support for renovation and modernization and up-rating of old small hydro power stations.

12.4.3 Projects Under MNRE Subsidy Schemes

The Ministry of New and Renewable Energy is giving financial subsidy, both in public and private sector to set-up SHP projects. In order to improve quality and reliability of projects, it has been made mandatory to get the project tested for its performance by an independent agency and achieving 80% of the envisaged energy generation before the subsidy is released. In order to ensure project quality/performance, the ministry has been insisting to adhere to IEC/International standards for equipment and civil works. The subsidy available from the Ministry is linked to use of equipment manufactured as per IEC or other prescribed international standards. MNRE is providing financial support to state governments/ agencies for overall estimation of potential in a state, identification of new potential SHP sites and preparation of state perspective plan.

It has been made mandatory that all newly commissioned SHP projects should be tested for performance and quality to avail subsidy. Accordingly, necessary onsite testing facilities have been created at the Alternate Hydro Energy Centre (AHEC), IIT Roorkee with 3 sub-centers at Jadavpur University, Kolkata, National Institute of Technology, Thiruchirapally and Maulana Azad National Institute of Technology, Bhopal. The entire testing is being coordinated by AHEC [23, 24].

12.4.4 Potential and Status of Development

The Estimated Potential for power generation in the country from small hydro power development is over 15,000 MW. The actual potential of small hydro power (SHP) is likely to be much more than that assessed as different states are offering sites for self-identification by prospective bidders for installation of the small hydro power stations. Out of 15,000 MW small hydro power in natural streams, canal falls and dam toes mostly in Himalayan and other hilly regions were fast flowing and perennial streams can easily be exploited to tape this renewable source of energy, 5415 potential sites with an aggregate capacity of 14,305 MW have been identified. As of now, about 700 hydro power schemes with an installed capacity of 2550 MW have been completed; schemes with an installed capacity of about 500 MW are under implementation [24]. Thus, there is great scope for exploitation of the balance capacity.

12.4.5 Economic Viability of SHP and Comparison with Other Sources

SHP projects are low cost development schemes, which cannot justify investment on transmission system. The primary responsibility to evacuate power from SHP's rests with the distribution licensees. As such, the cost of installation per MW as well as cost of per unit generation is high for SHP when compared with conventional sources. A comparison between conventional and non-conventional sources to have an idea of the order of cost in India is given in Table 12.7.

TABLE 12.7 SOURCE-WISE COST OF INSTALLATION PER MW AND ENERGY GENERATION PER UNIT. SOURCE: [24]

Source	Cost per MW installation (Rs. in Crores)	Per unit energy generation (Rs.)
Renewable		
Small hydro*	5–7	2–4
Solar PV	17	12–17
Solar thermal	13	10–13
Biomass	4.5	2–4
Wind	5–6	2–4
Conventional		
Thermal	4.5	2.5 to 3
Gas based	3.5	
Hydro	5–7.5	

* Highest efficiency and longest life amongst renewable sources.

12.4.6 Issues Constraints, Appraisal, and Clearances

The Constraints perceived in development of the small hydro segment are technical, procedural and cost-related in nature. The technical barriers include factors such as accessibility to the sites and risks involved in transporting heavy equipments to the sites. The procedural issues primarily relate to the number of clearances required before taking the project. Typically, a developer is required to get a project allotment from the state nodal agency, obtain clearance from MOEF where forest land is involved (in projects costing more than Rs.100 crores), clearance from the Irrigation/Water Resources Department, clearance from the state government on land availability, etc.

There are hardly any databanks available on water flow from small streams and rivulets and there is uncertainty risk like hydrology, power tariff, natural calamities and geological surprises.

12.5 PUMPED STORAGE DEVELOPMENTS

12.5.1 Role of Pumped Storage Schemes

In India the base load requirements would generally be met by thermal and nuclear generating plants, whereas peaking loads would be met from conventional hydro projects along with pumped storage schemes. The demand for peak power is growing faster compared to the average demand. The large untapped conventional hydro potential located in Himalayan River basins would need to be developed to meet the peaking power requirements of Northern and North-eastern regions. The eastern region, though, deficit in hydro resources (rich in thermal) would benefit from the large hydroelectric potential in the north eastern region to meet the peaking requirement. However, the development of pumped storage schemes in Western and Southern region could be visualized in a significant magnitude to meet the peaking requirements due to lack of availability of sufficient conventional hydro potential resources in these regions. Due to inter-state disputes involved on harnessing of water and thereby in harnessing some of the conventional hydro potential in Southern region and also due to decrease in the availability of water in river systems on account of upstream consumptive uses, pumped storage schemes are becoming inevitable for meeting the peak loads.

The hydrological uncertainties with hydro power projects are much more, compared to PSS and in a dry year when the discharges is abnormally low, the hydro power projects may fail to even peak load. In such conditions, PSS is very helpful as it re-circulates the stored water only and loss of water is due to seepage and evaporation losses only. The discharge of peninsular Indian rivers has significant seasonal fluctuation, during monsoon, the discharge is very high, so the hydro power projects are able to take based load also. However, in non-monsoon period discharge is very low and therefore one has to depend upon the thermal plants. Under such condition, the peak power becoming available from PSS would be the best way forward.

Further, if hydro power project is available to supply power at the cost lower than PSS, it is definitely preferable to get power from hydro power project than PSS. In fact, if the hydro power is so much abundant, there is no need to run thermal power plants. But there are two problems being experienced with hydro power plants. The first problem is with the run-of-the-river scheme and the second with multi-purpose projects. In these plants, the generation of power depends upon the discharge availability, which is not consistent with the demand. If discharge is surplus, it can be spilled through spillway. But when there is shortfall in discharge,

the frequency goes down and load shedding becomes necessary. In case of storage type plants such fluctuating demand can be met by storing surplus water and releasing water during the shortfall. In case of multi-purpose schemes, the water release schedule is more often dictated by other demands like irrigation, and power generation is not a priority item. Therefore, in general, power generation is not according to the demand and the requirement of frequency regulation is being met in a very limited way by hydro power plants. The option in such cases is the implementation of pumped storage schemes. When compared with conventional hydro power projects, in the case of pumped storage schemes, the environmental impacts are very much limited. Also, because of re-circulation of water, the problem of silt is absent.

In order to provide peaking power the use of natural gas and DG sets for power generation has been increasing over the years in some of the regions of our country. In view of this, an utmost need for such a power generation system, which can provide peak power in a shorter gestation period utilizing renewable energy resources at comparable economics and which can be installed relatively nearer to the load center so that the transmission and distribution losses can be considerably reduced, is being felt. Pumped storage schemes are being considered as the best option to fulfill these requirements.

12.5.2 Status of Development and Potential

In India, the need for developing pumped storage schemes was realized as early as 1960 and the first pumped storage plant could be commissioned at Nagarjuna Sagar in Andhra Pradesh in the year 1980 with an installed capacity of 700 MW. Since then pumped storage plants with their aggregate capacity of 4804 MW are in operation. The operational pumped storage plants are: Kadamparai (400 MW) in Tamilnadu; Bhira (150 MW), Ghatghar (250 MW), Paithan (12 MW) and Ujjaini (12 MW) in Maharashtra; Srisailam Left Bank (900 MW) in Andhra Pradesh; Purulia (900 MW) in West Bengal; Kadana (240 MW) and Sardar Sarovar (1200 MW) in Gujarat; Panchet Hills (40 MW) in Bihar [25].

In view of the increasing role of pumped storage schemes for providing much needed peaking capacity and also bringing improvement in the power system, the systematic identification and preparation of inventory of attractive pumped storage schemes was attempted in the country for the first time along with reassessment study (1978 to 1987) carried out by CEA to assess the hydroelectric potential of the country. Fifty-six potential sites have been identified for pumped storage schemes with total installation of about 94,000 MW with individual installed capacity varying from 600 MW to 2800 MW. Region wise/ State wise distribution of potential sites for installation of pumped storage schemes is given in Table 12.8.

12.5.3 Economic Viability

Although these schemes do not provide additional energy and infact consume some energy from the existing power system, increasing emphasis of their use in Indian power system is inevitable in view of high peak load demand and the pumped storage projects possess the unique capability to re-deliver the energy as and when system requires. Their reservoir is usually of a smaller size compared to those of conventional hydro power station with similar capacity. This benefit is increased by adopting pumped storage with higher operating head since the size of reservoir can be made smaller. Also, their reservoir can be located on a small stream which would not pass any significant impact on the surrounding environment. With the decline of available sites for conventional hydroelectric generation, in times to come, the development of pumped storage schemes would attract additional attention. The

TABLE 12.8 POTENTIAL SITES IDENTIFIED FOR INSTALLATION OF PUMPED STORAGE SCHEMES. SOURCE: [26]

Region/State	Number of Schemes identified	Portable total installed capacity (MW)
Northern Region		
Jammu and Kashmir	1	1650
Himachal Pradesh	2	3600
Uttar Pradesh	2	4035
Rajasthan	2	3780
Sub total	**7**	**13065**
Western Region		
Madhya Pradesh	7	11150
Maharashtra	18	27070
Sub total	**25**	**38220**
Southern Region		
Andhra Pradesh	1	1650
Karnataka	4	7900
Kerala	2	4400
Tami Nadu	1	2700
Sub total	**8**	**16650**
Eastern Region		
Bihar	1	2800
Orissa	1	2500
West Bengal	4	3785
Sub total	**6**	**9085**
North-Eastern Region		
Manipur	2	4350
Assam	1	2100
Mizoram	7	10450
Sub total	**10**	**16900**
All India Total	**56**	**93920**

pumped storage schemes improve the economics of fuel as has been demonstrated by the recently implemented Purulia Pumped Storage Scheme in West Bengal. It is desirable to make pumped storage scheme as a part of thermal power plant just like the DG backup of thermal power plant and absorb the cost of pumped storage scheme in the cost of thermal power plant and accordingly the tariff of thermal plant requires to be structured instead of working out separate tariff for pumped storage scheme. Also, in cities and towns of India, consumers are investing heavily in captive generators and invertors, to avoid inconvenience due to abrupt breakdown. The savings on account of this cost and to the environment can be accounted toward the economic viability of PSS.

Today, because of continuous shortage of power, off peak power for pumping operation is costly making the pumped storage look unviable. But this situation may not continue in times to come. A sizable part of pumped storage potential would need to be exploited at some point of time in the future.

12.6 TRANSMISSION, SET-UP, AND STATUS

12.6.1 Implementation Set-up and Requirements

In order to optimally utilize the dispersed sources for power generation it was decided right at the beginning of the 1960s that the country would be divided into five (5) regions (Northern, Southern, Western, Eastern and North Eastern) and the transmission planning process would aim at achieving regional self sufficiency. The planning was so far based on a Region as a unit and accordingly the power systems have been developed and operated on regional basis. Today, strong integrated grids exist in all the five regions of the country and the energy resources developed are widely utilized within the regional grids. Presently, the Eastern and North-Eastern Regions are operating in parallel.

To construct, operate and maintain the inter-State and interregional transmission systems the National Power Transmission Corporation (NPTC) was set up in 1989. The corporation was renamed as POWER GRID in 1992. It has a network of about 75,000 circuit km of transmission line and wheels almost half of the total power generation of India [27]. With the proposed inter-regional links being developed, it is envisaged that it would be possible for power to flow anywhere in the country with the concept of National Grid becoming a reality during 12th Plan Period. The intra-state transmission system is built by state transmission and distribution utilities.

Evacuation of power from hydro projects has been planned basin-wise. Comprehensive transmission system for each basin has been evolved by CEA for development in a phased manner and components of the transmission system associated with specific generation and transmission system from the pooling stations to the de-pooling stations have been identified. As the transmission system for evacuation of power from the hydro projects would benefit the region where these projects are located, as well as other regions which will avail the power, the probable buyers who would have a long-term commitment for the transmission charges for each of these transmission systems, have also been identified.

For power evacuation from North-Eastern Region, where bulk of untapped hydro potential is available, planning of appropriate transmission system through the chicken neck area is of utmost importance. Transmission corridor through the chicken neck has been planned considering the evacuation not only from North-Eastern Region, but also hydro power from Sikkim and Bhutan to Northern and Western Regions in the country. Transmission system comprising of 800 kV HVDC bi-pole lines and 400 kV double circuit AC lines in the hybrid systems have been planned for evacuation of power through the chicken neck area.

The Electricity Laws have been passed with a view to make transmission as a separate activity for inviting greater participation in investment from public and private sectors. The participation by private sector in the area of transmission is proposed to be limited to construction and maintenance of transmission lines, whereas operation remains under the supervision and control of Central Transmission Utility (CTU)/State Transmission Utility (STU). On selection of the private company, the CTU/STU would recommend to the CERC/SERC for issue of transmission license to the private company.

All the project developers would be required to obtain Long Term Open Access (LTOA) for evacuation of power and for transmission of power to the identified beneficiaries. If exact buyers/States could not be identified, to start with, the project developer should specify the target Regions where the power would be absorbed, and accordingly, apply for LTOA for firming up of the transmission system. While planning the dedicated evacuation system from hydro power projects constraint of Right of Way (ROW) specially in hilly areas needs to be kept in view.

12.6.2 Status of Development and the Challenges

In the ongoing 11th Plan, the main focus of transmission system development is the formation of the National Power Grid together with strengthening of regional and state grids. The aggregate inter regional transmission capacity which was 14,000 MW at end of 10th Plan is proposed to be increased to about 38,000 MW by the end of 11th Plan and about 75,000 MW by the end of 12th Plan.

The 11th Plan has a generation capacity addition program of around 78,700 MW. A perspective transmission plan has been drawn in order to evacuate this power and transmit it to the load centers. Accordingly it is envisaged that about 5400 ckm (circuit kilometer) of 765 kV lines, 49,200 ckm of 400 kV lines, 35,300 ckm of 220 kV lines and 5,200 ckm of 800 kV / 500 kV HVDC lines would be added during the 11th Plan. Uptill the end of 10th Plan 765 kV, 400 kV and 220 kV lines were 2184 ckm, 75,722 ckm and 114,629 ckm, respectively.

For the 12th Plan involving 100,000 MW of capacity addition which includes over 20,000 MW of hydro, the expected transmission requirement would be 25,000 to 30,000 ckm of 765 kV lines; 50,000 ckm of 400 kV lines; 40,000 ckm of 220 kV lines and 4,000 to 6,000 ckm of HVDC 800 kV lines. Total fund requirement would be of the order of US$ 50 billion [28].

The challenges and the areas of concern in the development of transmission system are: obtaining of forest clearances, conserving of right of way, land acquisition, minimizing the impact of natural resources, and cost-effectiveness in evacuation.

12.7 CONSTITUTIONAL PROVISIONS OF WATER AND POWER RESOURCES

The Constitution of India was adopted on January 26, 1951 [29]. In it, the subject of "Power" was kept as a concurrent subject with both Centre and State empowered to legislate on it. The subject of "Water Resources" was kept primarily as a State subject, with Centre having some power to intervene in interstate aspects/issues. To carry out major economic development in a centrally planned manner — the 5-Year-Plan model was adopted under the overall supervision/watch of Central Planning Commission.

Electricity Act, 2003 provides an elaborate institutional frame work and financing norms of the performance of the electricity industry in the country. The Act envisaged Creation of State Electricity Boards (SEBs) for planning and implementing the power development programs in their respective States. The Act also provided for creation of central generation companies for setting up and operating generating facilities in the Central Sector. The Central Electricity Authority constituted under the Act is responsible for power planning at the national level.

12.8 REGULATORY AGENCIES

Government of India has promulgated Electricity Regulatory Commission Act, 1998, now a part of Electricity Act, 2003, for setting up of Independent Regulatory bodies both at the Central level and at the State level namely The Central Electricity Regulatory Commission (CERC) and the State Electricity Regulatory Commission (SERCs). The main function of the CERC are to regulate the tariff of generating companies owned or controlled by the Central Government, to regulate the tariff of generating companies, other than those owned or controlled by the Central Government, if such generating companies enter into or otherwise have a composite scheme for generation and sale of electricity in more than one State, to regulate the inter-state transmission of energy including tariff of the transmission utilities, to regulate inter-state bulk sale of power and to aid and advise the Central Government in formulation of tariff policy.

The main functions of the SERC would be to determine the tariff for electricity, wholesale bulk, grid, or retail to determine the tariff payable for use by the transmission facilities, to regulate power purchase and procurement process of transmission utilities and distribution utilities, to promote competition, efficiency and economy in the activities of the electricity industries, etc. Subsequently, as and when each State Government notifies, other regulatory functions would also be assigned to SERCs.

12.9 RESETTLEMENT AND REHABILITATION POLICIES

A National Policy on Resettlement and Rehabilitation (R&R) for Project Affected Families was formulated in 2003, and it came into force with effect from February, 2004. The provisions of the National R&R policy (INRRP) were revised in 2007. The policy provides for the basic minimum requirements, and all projects leading to involuntary displacement of people must address the rehabilitation and resettlement issues comprehensively. The State Governments, Public Sector Undertakings or agencies, and other requiring bodies shall be at liberty to put in place greater benefit levels than those prescribed in NRRP 2007 [30].

The Policy addresses the need to provide succor to the asset less rural poor, support the rehabilitation efforts of the resource poor sections, namely small and marginal farmers, SCs/STs and women who have been displaced. Besides, it seeks to provide a broad canvas for an effective dialogue between the Project Affected Families (PAF) and the Administration for Resettlement and Rehabilitation to enable timely completion of project with a sense of definiteness as regards costs and adequate attention to the needs of the displaced persons. The objectives of the Policy are to minimize displacement, to plan the R&R of PAFs including special needs of Tribal and vulnerable sections, to provide better standard of living to PAFs and to facilitate harmonious relationship between the Requiring Body and PAFs through mutual cooperation. Proper implementation of National Policy on Rehabilitation and Resettlement would be essential in this regard so as to ensure that the concerns of project-affected families are addressed adequately. Because of smaller requirement for land of SHP's, the no. of PAF is minimum. However, provisions of R&R policy need to be implemented wherever applicable.

Hydro power development is not being seen in isolation but as a "Comprehensive Project Development Initiative." The model for development includes benefit for the basin reserve area, incorporates the entire planning and development, accommodates the interests of the populace by making them stake holders. This Tehri Township is the best example and veritable role model worth replicating, which has been developed along with the construction of 1000 MW hydroelectric project in Uttarakhand State of India.

12.10 APPRAISAL AND TECHNO-ECONOMIC CLEARANCES

12.10.1 Appraisal of DPRs

12.10.1.1 Appraising Agencies All the multi-purpose water resource development project proposals with hydro power development component are first appraised by CWC, then by the Advisory Committee on Irrigation, Flood Control and Multi-purpose projects of MoWR. After clearance of the Advisory Committee, CEA examines and clears the hydro power component of the scheme. For single purpose hydro power projects, CEA act as a single agency for techno-economic clearance.

Section 8 of Electricity Act, 2003, provides that "any generating Company intending to set up a hydro-generating station shall prepare and submit to Authority for its concurrence, a scheme estimated to involve a capital expenditure exceeding such sum as may be fixed by Central Government, from time to time by notification." The Authority, on receipt of such schemes, examines them from techno-economic aspects before it accords its concurrence.

At present, the Central Government has fixed the following cost limits for submission of the schemes for concurrence of CEA.

i) In relation to a scheme for generating station prepared by a Generating Company and selected through a process of competitive bidding by the competent Government, rupees one thousand crores;

ii) In relation to a scheme for the Generating station prepared by a Generating company whose tariff for sale of electricity is determined by the Central Electricity Regulatory Commission or any State Electricity Regulatory Commission, Rs.2500 crores.

iii) In relation to a scheme for renovation and modernization of existing power generating stations, rupees five hundred crores and

iv) In relation to all other schemes, rupees two hundred and fifty crores.

v) All hydroelectric schemes utilizing water of inter-state rivers shall be submitted to the Authority for its concurrence.

For private sector HEPs, the clearance from the State Government would be required in the area of water availability, Rehabilitation and Resettlement (R&R) of displaced persons, land availability and E&F clearance. The Report duly accepted by State Government is to be submitted to CEA.

Techno-economic examination of project reports of hydro electric/multi-purpose project is an interactive and complex process and involves various disciplines like hydrology, civil design, electrical and mechanical formations with a view to finalize the features of the project based on the optimal plan development of water resources, and also considering techno-economic feasibility and requirements of system.

CEA coordinates examination of DPRs with CWC, CSMRS, and GSI on aspects pertaining to their respective specialized discipline.

12.10.2 Documents to be Submitted for Appraisal

The procedure given hereunder is the one in practice and not emerging from role assigned to CEA under the Act. The objective of such a procedure is to expedite the appraisal process so that developers get the clearance quickly from CEA. The under listed development stage is a part of DPR preparation stage but by complying with this prior to undertaking the rigor of DPR would help overall saving in time of getting concurrence from CEA.

12.10.2.1 Development Stage The Developer will prepare a Preliminary Project Report (PPR) of Hydro Power Project based on survey and investigation and Hydrology and submitted to CEA. CEA will forward PPR for examination and approval to CWC for hydrology and Inter-State/International aspects and to GSI for Geological aspect (in principle). PPR will contain basic concept of the project, basic data and planning, general design philosophy, project features, capacity basis (power Potential studies), construction planning, geological investigation details of inter-state/international implications; details of survey and investigations, hydrological details, etc. along with check list as per prescribed format. The chapters in PPR will be prepared based on CWC Guide Lines.

After approval of Hydrology, CEA will give approval (in principle) for installed capacity, number of units, energy generation, etc. after receipt of clarification, if any.

The Developer/State Government shall also submit Environment Impact Assessment Report together with R&R Plan, Environment and Disasters Management Plans along with broad cost estimates and application for forest clearance (if required to the Ministry of Environment and Forest. If the schedule tribe population is affected the R&R Plan will also be submitted to the Ministry of Social Justice and Empowerment/Tribal affairs. The concerned Ministries will accord relevant clearances after due appraisal and if required, obtain Defense clearance, from Ministry of Defense also.

12.10.2.2 Detailed Project Report (DPR) Preparation Stage Hydro electric projects are capital intensive involving high technology and relatively long gestation period. Due to their complex nature, a large amount of preparatory work is done by the project proponents. Preparatory work covers detailed field investigations, planning, assessment of benefits, design and engineering studies, detailed cost estimates based on analysis, cost of inputs and equipment, identification and tie up of inputs, project need, justification and economics and environment studies, safety aspects, etc.

The Developer after completing the preparatory work will prepare Detailed Project Report (DPR) of Hydroelectric Project in accordance with the "Guidelines for preparation of Detailed Project Report of Irrigation & Multipurpose Projects" published by Central Government [31]. After obtaining the required clearances as stated under Para on "Development Stage", shall submit to the Central Electricity Authority complete DPR for accord of Techno Economic Clearance as required under section 8 of Electricity Act 2003.

The civil cost estimates forming part of DPR shall be prepared based on "Guide lines for preparation of Estimates for River Valley Projects" published by Central Water Commission [32] and the Electrical & Mechanical cost estimates shall be prepared as per prescribed CEA Performa [33].

12.10.2.3 Finalization of Project Cost On receipt of the DPRs of hydro projects, copies of the same would be forwarded to CWC and GSI in addition to examination in the various divisions of CEA.

The Developer shall interact, provide clarifications/additional information and obtain approval for Hydraulic structures, Construction Machinery/Equipment, civil cost estimates, etc. from CWC, Construction Materials from CSMRS, Geological aspects from GSI and Electro-mechanical designs, cost estimates of electrical and mechanical works, etc. from CEA.

The total cost estimates of the project shall be finalized by CEA after approval of electrical and mechanical cost estimates and civil cost estimates and including cost of environmental and R&R/Tribal Welfare Works as cleared by Ministry of Environment and Forest and Ministry of Social Justice and Empowerment. The financial costs shall be submitted by the Developer based on the financial package and these shall be examined and finalized in CEA.

Two check lists, Check List-A and Check List-B have also been included to be filled in and submitted along with DPR. The gist of these checklists is as under. The complete details of these checklists are given in the document "Guidelines for Formulation of Project Reports for Hydro Power Projects" [33, 34].

Check List-A First Stage Check List for Hydro Electric Projects

a) Registration of the Company as per the Company Act and authorization of the State Government/Central Government as the case may be in accordance with Section 10 Electricity Act, 2003
b) Land and water availability certificate from State Government
c) Clearance from Ministry of Environment and Forest, Govt. of India
d) Justification of the Scheme from power supply–demand consideration
e) Completed cost, Present Day Cost (for SEBs) and both for Generating Companies in Public Sector as per CEA formats
f) Financial and commercial aspects as per CEA format
g) Defense clearance, if applicable
h) Salient features as per CEA format
i) Cost estimates as per CEA format

Legal Aspects (As applicable)

- Compliance under Indian Electricity Act
- Power Purchase Agreement (PPA)
- Tariff Notification
- Equity participation
- Change of name of implementing agency
- General/Specific conditions.

Check List-B Second Stage Check List for Hydro Electric Projects

a) General Data of Project (Location, State, District, etc.).
b) Registration of the Company as per the Company Act and authorization of the State Government/Central Government as the case may be in accordance with Section 10 of Electricity Act, 2003.
c) Land and water availability certificate from State Government.
d) Clearance from Ministry of Environment and Forest, Govt. of India.
e) Justification of the Scheme from power supply- demand consideration.
f) Completed cost, Present Day Cost (for SEBs) and both for Generating Companies in Public Sector as per CEA formats.
g) Financial and commercial aspects as per CEA format.
h) Rehabilitation and resettlement plan from State Revenue Department.
i) Defense Clearance, if applicable.
j) Section 8 and 9 of Electricity Act, 2003 (for private sector projects).
k) Competent Government's recommendation of DPR and cost (in case of private generating companies).
l) Detailed information regarding the following aspects of concerned Hydro Electric Project must be incorporated in the Detailed Project Report/Feasibility Report:
   i) Planning
   ii) Inter-state/International aspects
   iii) Survey (Topographical, construction material, etc.)
   iv) Investigations (Geological, Seismic, Foundation, Hydrological and Meteorological, etc.)
   v) Hydrology
   vi) Land acquisition and resettlement of ousters
   vii) Design and model studies carried out
   viii) Floor control and drainage
   ix) Power planning and related proposals regarding transmission system, energy charges, etc.
   x) Construction program and man power and plant planning.
   xi) Foreign exchange and financial resources
   xii) Estimates
   xiii) B.C. ratio and revenue
   xiv) Ecological aspects and Soil Conservation (if needed).

Details of Financial Package

a) Financial package should be in CEA prescribed Performa 1001–1004 filled up as under:
   1001 Financial package Summary.
      - Broad details of hard cost, interest during construction and financing charges.
   1001 Financial Package Abstract.
      - Financial structure i.e., Amount of foreign/domestic debt and equity (Foreign/ domestic/ promoters/ other partners), exchange rates, etc.
   1002 Financial Package Details.
      - Details of each debt package (Amount, source, interest rate, repayment period, moratorium period, financing charges, etc.)
   1003 - Phasing of expenditure and drawl of fund statement.
b) Commitment letters from foreign/ domestic lenders along with their terms and conditions.
c) Equity partner agreement
d) Package wise Interest during Construction (in respective currency) along with detailed calculations.
e) Package wise Financing Charges (in respective currency) including guarantee fee and commitment charges along with calculations.
f) In case of projects of SEBs / Central Sector the calculation of IDC and financing charges are required both at current and completed cost.

Break up of Cost Estimates

Break up of Cost Estimates of Electrical and Mechanical and Civil works of Hydro Electric Project as per CEA format / CWC Guidelines.

The Check List-A is for checking of legal and other major clearances at the time of receipt of DPR. If the requirement as per Check List-A are not met, the proposal would not be accepted for processing in CEA. The Check List-B is for checking of the completeness of the DPR. If the requirements as per Check List-B are not there, the proposal would be returned.

If the proposal is found to be meeting the requirements as per Check List-A and Check List-B, it would be further processed and clarifications/additional information, if necessary, may be sought.

12.10.3 Aspects Appraised by Different Agencies

12.10.3.1 Aspects Appraised by CWC Central Water Commission and wherever necessary in consultation with MOWR, examines and accords approval to the issues like hydrology of the project, design and cost estimates of civil components, inter-state/

international aspects, dam safety aspects, etc. A brief about the objective of appraising these aspects is given as under:

Hydrology: CWC appraise the hydrological inputs and hydrological studies. It includes assessment of quantities of available water at the project site and time variation, estimation of design flood, silt studies for estimation of life of the project. They play a vital role in the planning of hydro electric projects and the design of various hydraulic structures. An over estimate of water availability may lead to larger investment and project may become costlier resulting in a higher installation. On the other hand, lower estimates of water availability may result in a wastage of some hydro potential and non-utilization of selected site optimally.

Construction Machinery: CWC carry out appraisal of the construction planning machinery and construction methodology, number and type of equipment required, use rate of equipment, unit rates of work, etc. in order to have a realistic estimates of the cost of the project and the time required for construction.

Foundation Engineering, Dam Safety and Civil Design Aspects: CWC have specialized formations to examine foundation engineering aspects and civil design aspects of civil structures including dam safety aspects for various structures related with hydro electric schemes.

Inter-State/International Aspects: CWC also examines inter-State/international aspects related with a hydro electric project in consultation with Ministry of Water Resources, if considered necessary and provides necessary suggestions to CEA.

Cost Estimates of Civil Works: Before cost estimates of civil works are cleared by CWC, the project is evaluated from the angle of various aspects like hydrology, civil design, dams design, gates design, dam safety, foundation engineering, power plant engineering, barrage and canal design, construction machinery aspects, etc. by CWC. It is essential because the cost depends upon all the above mentioned clearances. A change anywhere in these aspects may affect the cost aspect.

Once all the aspects are finalized, the cost estimates included in DPR are verified. To verify the estimated cost of civil works, hourly use rates of equipment and analysis of rate of main works, like excavation, concreting, RCC works, stripping, filling, grouting, etc. are determined for each activity. Based on the construction designs, finalized, the quantities of the items required are worked out. Based on these, the estimated cost of civil works proposed in DPR is reviewed and finalized.

12.10.3.2 Aspects Appraised by CEA

Legal Aspects: CEA examines and ensures whether the project authorities have complied with all the legal provisions.

Justification of the Project: CEA carries out studies and forecasts the 'Power Supply Position' with and without the proposed project and examine the need/justification of the project from system demand point of view. Necessary inputs/information regarding future system demand in both peak demand and energy requirement for these studies is provided by Electric Power Survey Report published by CEA.

Hydro Power Planning Aspects: CEA examines the general layout of the hydro scheme as proposed by the project authorities and suggest modification, if any. CEA also examines the power potential studies carried out by the project authorities for the years for which hydrological data is available, proposed installed capacity and number and unit size, etc. CEA also examines the overall basin development and how the project proposal fits into it.

Designs Aspects of E & M Works: From the point of view of suitability of electro-mechanical design aspects, CEA examines

designs of turbine or pump/turbine (for PSSs), generator or motor/generator (for PSSs), main step-up transformer along with cooling arrangement/system, switch-yard equipment (conventional or gas insulated switchgear), single line scheme, control and protection equipment, auxiliary services and transportation, etc. included in the DPR.

Cost of E & M Works: For E & M Works, estimated cost is verified by CEA based on cost data of similar equipment available in CEA.

Evacuation of Power: CEA examines the adequacy of power evacuation system proposed by the project authorities to evacuate the power generated by the project and suggests necessary modifications, if any.

Construction Program: CEA examines activity-wise, item-wise and year-wise targets/schedule for construction for each of the major components of the project, which are based on detailed Bar/PERT Chart. The completion cost of the scheme is worked out based on detailed construction Program.

Financial and Commercial Aspects: Financial and commercial aspects of a hydro-electric projects are examined by CEA, which includes examination of financial package, calculation of interest during construction based on different financial packages, leveled tariff calculation with the objective of assessing as to whether the project is attractive enough to proceed further.

12.10.3.3 Aspects Appraised by GSI
The geological aspects are appraised by Geological Survey of India (GSI).

The objective of appraisal is to assess the adequacy of mapping drilling, drifting, geological investigation and impact of geological features on designs. The BIS and guidelines of CWC form the basis. The objective is that the rock type, weathering limit, the overburden depth projected and marked on the drawings is based on the input of investigations carried out at site. The drive of the checking is that there is no room for geological surprises during construction as this can be one of the main factor leading to time and cost over runs.

12.10.3.4 Aspects Appraised by CSMRS
Central Soil and Material Research Station (CSMRS) carries out appraisal of construction material aspects in regards to suitability of different materials and their source for use in the project.

12.10.3.5 Clearance from MOEF
Development of hydro-electric projects has also adverse impacts on the surrounding environment. Hydro-electric projects may involve submergence causing environmental and ecological aspects, rehabilitation and resettlement and forest land. This necessitates scrutiny and clearance from Ministry of Environment and Forest. In case the projects involve diversion of forest land, clearance is also required from forest angle from MOEF under Forest Conservation Act, 1980 [35].

12.10.3.6 Clearance from Defense
If a hydro-electric project involves Defense aspects, then the clearance is also required from Defense angle. The request is made to Ministry of Power to take up the case with the Ministry of Defense.

12.10.4 Techno-economic Clearance
After CEA satisfies itself about the technical and economic viability of the project and if necessary inputs/clearances for the scheme are tied up, it accords concurrence to the proposal as proposed or subject to some conditions as an interim step prior to its

occurrence. At times when, MoEF clearance is pending, CEA accords conditional clearance so that the developer could take advance action for initiating certain activities including financial tie-up. The intimation regarding accord of concurrence for hydro-electric projects are given to the Project Authorities, Ministry of Power, Planning Commission and other Government Departments for further action.

Techno Economic Concurrence of CEA: After all the specified statutory sanctions and clearance from various Government Departments/Organizations are obtained, CEA's concurrence is conveyed as per Section 8(2) of Electricity Act, 2003.

Final Financial Package: CEA also examines the final financial package (after the financial tie-up with lending financial institution has been done) submitted by the project authorities, and accords its approval.

Issues Related with Techno-economic Appraisal of Hydro Schemes. Often, it is found that the DPR submitted by the Project Authorities lack details required for proper examination and finalization of the project features. DPRs lack proper surveys and investigations studies, hydrological data/studies, design details, proper power potential studies, proper evaluation of quantities of civil works, detailed cost estimates, etc.

During the course of examination when deficiencies involving data/investigation, etc. are found, back references are made to the project authorities for obtaining complete information and it normally takes some time for them to attend to such observations. In case the DPRs of hydroelectric projects are prepared as per the guidelines of CWC and various quarries/clarifications raised by CEA/CWC/GSI are replied promptly by the Project Authorities, a scheme could be accorded concurrence by CEA within a short period after the receipt of DPR.

12.11 HYDRO-DEVELOPMENT IN THE NEIGHBOURING COUNTRIES

In the next over two decades when India would reach its targeted power installation of about 600 GW, with domestic hydro capacity of 150 GW at that time, the share of hydro would be at about 25% level only. To achieve the desirable thermal-hydro mix, one of the better option is to collaborate, at this stage itself, with the neighboring [20] hydro rich countries viz. Bhutan, Nepal, Myanmar, for joint exploitation of available hydro potential. At present, co-operation at national level, in the field of hydro power development, with these three countries, exist. In the case of Bhutan a good degree of success in joint co-operation for construction of hydro projects has been demonstrated which needs to be replicated with Nepal and Myanmar. The hydro potential of these three countries is estimated at 150 GW, i.e., equal to India's potential [10].

Bhutan with its major river systems comprising of Torsa, Wangchu (Raidak), Sankosh and Manas, has total hydro potential capacity of 30,000 MW. Through project specific agreements Chukha, (336 MW), Kurichu (60 MW), Tala (1200 MW) projects have been commissioned and Punatsangchu-I (1200 MW) and Mangdechu (600 MW) projects are under construction. Both countries signed an umbrella agreement in 2006 wherein both the governments have agreed to facilitate, encourage and promote development and construction of hydro projects and associated transmission system as well trade in electricity. Projects are to be developed both through public and private sector participation and would be governed by separate project implementation and power purchase agreements [36].

Nepal, with its major river systems comprising of Kosi, Gandaki, Karnali, Mahakali and Southern Rivers, has total hydro potential estimated at 83,000 MW. India and Nepal have jointly implemented projects that includes Pokhra (1 MW), Trisuli (21 MW), Western Gandak (15 MW) and Devi Ghat (14 MW) during the period 1968 to 1983. In case of Myanmar, with its major river systems comprising of Ayeyawady, Thanlwin, Sittaung, Chindwin, the total hydro potential is estimated at 39,720 MW. The existing Umbrella agreement on power exchange between India and Nepal and similar level of understanding between Myanmar and India are the instruments to be utilized for exploitation of the available hydro power resources.

The present stage of development in these three countries is not even at 5% level. Also their in-house power demand is far less than their hydro power potential. India being a major market of power, tremendous scope of inter-country cooperation for further exploitation of hydro power for the mutual benefit of the region is being seen as a way forward.

12.12 RESPONSE AND ACHIEVEMENT OF PRIVATE SECTOR

Unlike thermal, hydro development-particularly of medium and large projects-would, in certain cases (depending upon geography, location, type of development envisaged), involve some inherent problems such as (i) difficulty of investigation at detailed project preparation stage; (ii) environmental aspects/clearance at project approval/clearance stage; (iii) inter-state aspects/issues; (iv) rehabilitation of populaces/land acquisition; and (v) natural and geological risks and uncertainties particularly in the Himalayan regions. These problems need to be recognized as part of hydro development and to be promptly resolved by commandeering all requisite resources/inputs. Only a handful of private companies that could recognize the long-term advantages of hydro, remained in the field in first one decade of opening up of power sector to private companies. In this period, encompassing 8th (1992 to 1997) and 9th (1997 to 2002) 5-year plans, the private companies contribution in adding to hydro capacity has been less than 550 MW in a total hydro capacity addition of about 7000 MW.

The reforms which were initiated in the 1990s when the power sector was opened up to private sector got strengthened with the enactment of Electricity Act 2003, followed by open access regulations, national electricity policy and national tariff and integrated energy policy [37].

Private companies which had not shown much interest in undertaking hydro power projects because of high capital cost, long gestation periods, geological risks and high rate of royalty/free power demanded by the state governments in the first over one decade is no more the trend. The hydro sector which historically had mainly central and state utilities is now witnessing interest from over 50 new players looking at investment across the value chain.

In the second decade (encompassing 10th and the running 11th Five Year Plan up to 31st March 2010), a total of 900 MW is the share of private companies in a total hydro capacity addition of over 10,000 MW; whereas capacity under construction with them is 4000 MW. In the 12th Plan about 40 projects with installed capacity aggregating to about 8000 MW are expected to be commissioned by private companies, against a total of over 20,000 MW programmed hydro capacity addition. In addition projects with over 15,000 MW are under various stages of DPR preparation. The increasing interest shown by private sector with each successive

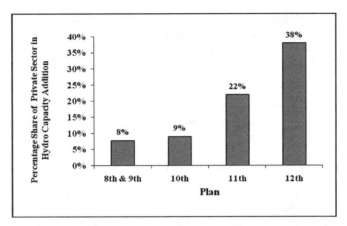

FIG. 12.8 TREND OF INCREASING PARTICIPATION BY PRIVATE COMPANIES IN SUCCESSIVE PLANS. SOURCE: [4, 21]

plan since the power sector reforms in 1992, is amply clear from the bar diagram in Fig. 12.8.

12.13 ISSUES, CONSTRAINTS, AND CHALLENGES IN DEVELOPMENT

12.13.1 Issues in Hydro Development

The issues which need to be favorably addressed by State and Central Government agencies to facilitate uninterrupted and accelerated development of hydro power potential include:

Unlike in the past when inadequate financial resources used to be the main cause, at present, some of the power generation schemes are not able to take off and some other on-going are facing hurdles on account of interstate disputes on water and power sharing, resettlement and rehabilitation issues, flood mitigation benefits to the lower riparians, etc. The power generation should be de-linked from river water at least for those schemes which are run of the river in nature as they do not hurt the riparian rights of the downstream states or are the cause of the said problems. This could help avoid water related politics plaguing hydro power sector. In the long run, to ensure that neither water nor power development is used as leverage for intra and inter-state political ends, centre need to seriously work toward enacting a legislation that would bring water into the Concurrent List.

Planning of the hydro electric project requires long-term hydro-meteorological data which are collected at specific locations by various State and Central Government agencies in the country. This hydrological data is at times not available in the reasonable vicinity of the project for adequate period, without inconsistencies, as required for optimum planning of hydro schemes especially for projects in the Himalayan region. This results in constraints in power potential estimation and confidence in project economics. There is need to pool the available wisdom and to do the best that is possible in the available circumstances.

The developers face difficulties in obtaining data related to hydrology, toposheets, geology, remote sensing, previous experiences of working in the project area and the projects in the vicinity. The procedure for obtaining related data that can help in better assessment of the cost and benefits needs to be streamlined.

For development of hydro electric projects located in the remote areas, infrastructural facilities like construction of bridges, strengthening of existing roads, efficient and reliable tele-communication links, better road transport/air services, etc. are required for their early implementation. Development of infrastructural facilities in hydro rich States needs to be planned and implemented jointly by Central and State agencies well in advance so as to facilitate hydro development in these States.

The royalty demanded by some to the States on use of water by hydro projects is unduly high. This requires co-operation of the central and state governments to arrive at a realistic royalty structure. Also, there is a need to take a fresh look on the varying practice of free power to the Host State. Security costs, common infrastructure facilities, R&R costs should be borne by the states in lieu of royalty received by them This will not only reduce the cost of generation but will make State Governments partner in ensuring that hydro-development proceeds uninterrupted [38].

Hydro electric projects involve submergence and often require reserved as well as unreserved forest land for their implementation. The impact on ecology, monuments, seismicity, resettlement and rehabilitation, catchment area treatment, flora and fauna are assessed in the Environment Appraisal of the project. Forest clearances also require compensatory afforestation on the non-forest lands. Identifying such land in view of availability of non-forest land in some of the hydro rich states like Arunachal Pradesh, Himachal Pradesh, is difficult. The resolution of these aspects is often time consuming and results in delays in the clearance of the projects. For the projects identified for implementation, basin-wise environmental clearance needs to be initiated by joint efforts of State and Central agencies rather than leaving it for developers to undertake project-wise such works.

Recent norms of Net Present Value and its upfront payment for assessing the cost of forest diversion need to be reviewed. Particularly, in case of storage schemes, huge financial burden, as at times it works out to be of the order of 20% of the project cost, makes them unviable. Such a step would encourage providing pondages more than the bare minimum needed for peaking in case of run-off-river type of hydro development.

The Ministry of Environment and Forests insist on treatment of very highly degraded areas of free draining catchment, i.e., up to the next project upstream, at the project cost, whereas degradation of catchment takes place due to various economic activities such as rising population pressure, developmental activities, over grazing and pressure of agricultural activities, wood requirement, etc. With a view to decrease not only the project cost but also discipline the contributing player, a review is required to be made to equitably apportion the cost amongst various beneficiary sectors like agriculture, irrigation and flood control, etc., along with power component.

Unfounded concerns of environmentalists often, hold up execution of well thought out and investigated projects thereby escalating the project cost manifold. Without making a distinction whether a project is good or bad, some NGOs are opposing all the hydro projects. Multi-lateral financial institutions fear backlash from these NGOs and are averse to support hydro projects. The Central and State Government agencies need to work jointly to ensure such dampening trends are not allowed to propagate.

In the recent times, State Governments are encouraging development of run-off-river (ROR) type hydro projects only and in some cases, possible storage projects are being converted into 2 or 3 ROR schemes. These are driven by short-term narrow gains even though the reason quoted is submergence of agricultural land besides R&R issues.

Active construction of sanctioned H.E. Schemes is held up in a number of cases simply on account of serious problems and difficulties in acquiring required land at various stages of construction viz., during pre-diversion and post diversion periods, for start of active construction, for compensatory afforestation and for resettlement and rehabilitation. In the case of Tehri hydro-electric project even religious sentiments by projecting that the sacred Ganga no more remains sacred once it meets and emerges from power house tail race, were excited. There is a need to handle these issues with sensitiveness so that the requirement of the country for exploiting the available hydro resources optimally, is not derailed.

It is well established that storage schemes provide necessary regulation to increase power generation during lean flow period and also provide flood moderation, irrigation, and drinking water supply benefits besides increasing benefits from downstream projects and improving power system stability. Therefore, any move to convert storage schemes into ROR schemes needs to be strongly discouraged as such sites being the gift of nature, once lost can never be retrieved.

There has also been slow progress on the tariff formulation to allow a premium on sale rate for hydro projects during peak period. Hydro power is most suited for peak generation and therefore early implementation of 'peak time tariff' would help mobilize investments for this sector. Hydro power being environmentally clean and a renewable source of power need to be given special incentives and hydro power basin developmental area linked concessional tax, which can be a big resource mobilization for R&R.

Concurrence of CEA is required under Electricity Act for setting up of a hydro power generation station estimated to involve a capital expenditure exceeding such sum as may be notified by the Central Government from time to time. The limit is presently Rs.500 crores. Since the cost of SHP's is lower than this stipulation, SHP's are generally not required to obtain concurrence of CEA. There is a need to ensure that appraisal agencies prove as facilitator.

The Ministry of Environment and Forest (MoEF) must be concerned to ensure that hydro power development is encouraged and not increase, beyond requirement, coal based power generation. It needs to champion the development of hydro including putting forth the suggestions for extension of benefits of renewal energy to it.

RTI Act at times becomes a deterrent with fears of decision being reviewed and questioned. Such an environment, particularly in public sector units, is leading to delaying development of hydro power projects. Contractual arrangements need to be revisited to facilitate mid-term review due to contractual difficulties and uncertainties.

The scheme of private sector participation in development of transmission projects will need to be expanded to cover larger number of projects. The procedure will need to be re-visited to ensure that developments of these projects are achieved as per the targeted schedules.

12.13.2 Constraints in Hydro Development

The constraints which are inherent to hydro power development projects, have to be understood by the developers and ways and means would need to be devised to ensure that constraints are converted to opportunities. The constraints include:

Features of the hydro-electric projects, being site specific, depend on the geology, topography and hydrology at the site. In spite of detailed investigations using state-of-the-art techniques that are carried out at the time of preparation of Detailed Project Report, there is still possibility of hydro projects facing some uncertainties in the sub-surface geology during implementation of the project. The Geological surprises during actual construction cannot be

completely ruled out. The success of mitigation measures not only depend upon the preparedness but also on the accessibility of the site. All these factors, inter-alia, result in cost escalations and time over runs. The construction time and the cost of a hydro project being greatly influenced by the geology of the project area and its accessibility, the developers and the financial institutions of the projects in Himalayan region should be mentally prepared to accept and accommodate such eventualities, then only innovative ways would emerge that would lead to successful completion of projects [39, 40].

Hydro-electric projects are generally located in remote hilly areas and construction of long transmission lines in difficult terrain for evacuation of power from these projects to the load centers takes considerable time. With State authorities taking time in exercising the option of buying whole power from the private developers, identification of power buyers and power delivery locations takes time, thus, delaying the finalization of transmission system associated with such hydro projects. The transmission system needs to be firmed up well in time so that completion of the transmission lines matches the commissioning schedule of such projects. Now, that a bulk of projects have been allocated, the project developers could at least indicate the region (s) to which they intend to supply power rather than the exact buyer(s) of the power to facilitate firming up of the transmission work at this stage.

Manpower constraint has been perceived as a major challenge for increasing capacity addition. Whereas, manpower at the higher level is being supplied by the market, the technician level manpower is a major constraint which has to be appropriately dealt by utility companies, academics and professional Institutions.

Community involvement is essential to know the requirements of the locals and accordingly decide the resource endowments. Skills development commensurate to technology deployment is desirable so that operation and maintenance can be done by locals.

Govt. of India had initiated 'adopt an ITI' scheme to build up skilled manpower around the project area. Under the scheme, the project developer is expected to adopt an Industrial Training Institute (ITI) or open a technical training centre in the neighborhood of the project well before commencement of construction activities, upgrade the facilities for training of students in different trades needed for the project. This would provide an opportunity to local residents to improve their skills to get employment opportunities in the project. Likewise, health and education facilities with remote areas can be improved utilizing the facilities created during project construction in a systematic plan of regional development.

Industry is also being motivated to take initiatives in Mission Mode to develop skills in the country like build up a pool of high pressure welders by arranging training and certification of their capability. These measures will not only help building up skilled manpower force in the country but also provide employment opportunity to youth in the project area benefiting the PAPs. It is necessary to ensure their implementation through appropriate monitoring mechanism. Involvement of power generating and transmission utilities will be helpful in ensuring the success of these initiatives.

There is a need to establish a Hydro Power Institute with expertise in Planning design, engineering, project management for promotion of new technologies in the hydro power projects, pooling and dissemination of experience of planning, design, construction, and operation maintenance including failures experienced during project implementation.

An independent agency for construction management, quality control is considered essential to ensure timely completion, quality and completeness of hydro power projects.

Better quality control at manufacturers' works is desirable as this would reduce the erection and operational problems. Manufacturers need to review the quality assurance program of the E&M and HM equipments.

A Central warehouse of latest construction machinery needs to be created for use by various developers to suit the site requirements.

There is a need to redesign Tunnel Boring Machines suitable for Himalayan geology as the experience of using TBM in Himalayan Geology has not been encouraging. But there are projects, which envisage about 5 km of tunnel from one face and the use of TBM may be unavoidable. For successful completion of these projects in a reasonable time, selection of the most suited type of the TBM based on the geo-technical investigation inputs is essential. Also for achieving the desired progress, fore-probing as the excavation progresses, will need to be made integral to the construction methodology. Also, the seismic prediction methodology needs improvement for assessing geological conditions in underground works to minimize geological surprise and the time and cost over-run.

To facilitate competition and to ensure that hydro development does not suffer for want of experienced contractors, qualifying requirements need to be rationalized to facilitate financially sound parties having experience in other civil infrastructure works to enter into hydro project construction work.

12.13.3 Challenges in Hydro Development

Hydro development projects involves working with the natural factors like topography, geology, hydrology and by virtue of their being spread over a large area, they inherently involve wide variety of field conditions, and managing these for successful tapping of this precious resource is a challenge. Some of the challenges with the way forward are:

Hydro sector has been assigned an important role in the power scenario and it is expected that the current imbalance of its reduced share vs. thermal, would gradually be set right. With use of advanced analysis and design tools along with modern construction equipment based methodologies and effective project management techniques, that are available in abundance, efforts would be made to establish hydro power plants in a shorter time frame so that the targets set for the Five Year Plan period are achieved [41, 42].

Geological surprises could be minimized by adopting suitable means during planning and investigation stage and their impact on time and cost over runs could be reduced during construction stage by timely decision and timely adoption of more suited/alternative construction methodologies and equipments. During planning stage, the suitable project specific program of investigations has to be planned and there should be regular interaction amongst planning, design and engineering formations with engineering, geology and geo-physical experts. Geological investigations should be taken up after certain preparatory works like accurate topographic surveys, procurement of good quality drilling machines, specialized manpower with drifting, etc., are complete. During construction stage, site engineer must ensure that geo-technical experts closely observes the formations after every round of excavation, Project specific data of projects in operation is available with the respective project authorities for reference and use. Underground civil works are properly secured to avoid unnecessary rock falls and hill slopes in proximity of major surface civil structures are properly stabilized. Also adequate supplies of alternative materials, spares and equipments are readily available to cater to varying site requirements [43].

The Challenge is to continuously improve hydro power technology in terms of environmental performance, materials, efficiency,

operating range, and costs. From the smallest to the largest, all developments have a footprint, especially evident in the cumulative effect of many small schemes. Smaller-scale hydro plays an important role in remote areas, in community developments, the large number of run-off river and pumped storage are required to meet the peaking demand, the storage based hydro cater to the power system stability and meeting multiple demand of water, mega schemes will continue to be the most environmentally benign in supporting grid systems and powering industrial and urban centers.

The least-cost option for producers desiring additional capacity is almost always to modernize existing plants, when this is an option. Equipment with improved performance can be retrofitted, often to accommodate market demands for more flexible, peaking modes of operation. Most of the hydro equipment in operation today will need to be modernized by 2030. The layout plans of civil, electro and hydro mechanical components need to account for this need of the future.

Most hydro power projects in the country suffer execution problems and delays due to the inability to assign and distribute risks amongst the stake-holders vis-à-vis the contractors, developers, banks and financial institutions and the governments (State/Centre). The risks assessment itself is a complex and continuous process and needs innovative treatment. Risk management can be achieved by logical risk allocation and sharing the cost of risk and overcoming the challenges posed by land acquisition problems; law and order issues; security of people involved with projects and the projects itself and issues of local public interest besides the inherent risks of disasters and disaster management. The key element in the direction of rendering the process of Risk Mitigation more effective is hydro power development accorded the Renewable Energy status so that the portfolio benefits starts become available [44].

A model contract document for hydro projects has recently been developed by a committee under the chairmanship of Chairperson, CEA. The document includes Risk Register indicating the various risks and proper allocation of the risk, between the employer and contractor. The model contract document is under discussion with the stake holders and after taking their inputs into consideration, the same may be ready for adoption. This is an attempt to minimize disputes in contracts.

The private sector role is increasing in the development of hydro power with more than a dozen groups engaged more actively in the development of hydro power projects besides keen interest shown by many. Developers may be having love for this sector, but the financial sector seems to be insensitive to the issues of hydro power development and the risks of investments. A major relook is required to bring in greater degree of predictability with elements in building greater certainly to factor like assessments of geology, time of completion and cost. The success of private sector depends on its will to learn to live and work through uncertainties and the regulatory institutions. It is expected that government policy makers would be more sensitive to the development issues of hydro power [45, 46].

12.14 INNOVATIONS FOR FUTURE PROJECTS

Many technically feasible hydro power projects are financially challenged because of their remote location or they are at such high altitude that their access road is available during non-snow period of 6 to 8 months only. High up-front costs for such projects are a deterrent for investment. There are yet another set of projects that are storage reservoir based with weak foundation conditions, i.e., overburden in the river bed extends for a large depth. Many a times when flood discharge for such projects is very high and abutments are steep, locating a concrete gravity type structure to support a spillway, in the river bed, is a challenge. The provision of spillway in the bank involve not only extensive slope stabilization measures during construction but at times become safety concerns during operation and maintenance phase of the project. Another category of projects are the ones that involve long lengths of tunnels as the water carriers and for some reaches, locating adits, to control length to be excavated from each face to be in the manageable limits, becomes a topographical constraint. Conventional construction methodologies need to be replaced with innovative solutions for their development. Stand alone, some of these projects may look economically unattractive but they need to be implemented to service the social good of the remote localities and at the times to meet the requirements of stability of the power system. The reservoir based storage projects do value addition in firming up the energy generation of the cascade of downstream projects. There is a need to take-up these projects by apportioning part of their cost to social development objectives and for the other set of projects, their cost can be off loaded to the downstream projects that are getting benefited. In this way the economic viability of projects gets justified and the hydro power development could proceed unhindered.

Now, that investor's confidence in hydro-sector is encouraging, the main challenge for the power utilities would be to ensure risk reduction and timely return. In the process, of advantages of low operating cost in comparison to high capital investment with lengthy lead time for planning, permitting and construction, would be physically experienced by developers. This is the only way that the investor will remain glued to this sector and in the process more appropriate financing model will develop that would ensure long-term sustainability of development with optimum role for public and private sectors. Green markets and trading in emissions reductions will undoubtedly provide incentives in some areas so also the interconnection between countries and the formation of power pools will build investor confidence.

India's massive program to develop Himalayan water resources for hydroelectricity has produced a large amount of data on rates of sedimentation of reservoirs and the sediment related damage to turbines and under water parts. Invariably the volume of incoming sediment has been much higher than hydroelectric engineers had predicted prior to project construction.

The natural processes in the Himalayan region are so predominant that there is no requirement to seek human intervention as the cause of siltation. In other words, artificial reservoirs will silt up rapidly in this very dynamic region regardless of human influences — negative or positive. Unless effective sediment management strategies suited to the type of development (ROR or storage) are integral to the project formulation, the project may suffer losses during operation and maintenance stage and returns may get affected.

The low strength rocks would be encountered in some of the projects in Himalayas. The aggregates produced from such rocks may not meet the required standards stipulated in the specifications. As against long distance transportation of acceptable quality aggregates which would make the project economics unviable, when locally available inferior quality aggregates are used to generate both coarse and fine fractions for concrete, innovative engineering and well researched construction material studies would have to be undertaken to justify the compliance to the safety norms.

12.15 CONCLUSIONS

An analysis of the energy scenario of India indicates that hydro power development has a very important role in meeting the energy needs of the country. To meet the ever increasing power demand, India would need to increase its installation from 160 GW at present to 4 times. The present stage of hydro exploitation at 25% level of its potential estimated at 150 GW, would also need to be accelerated to achieve the desired hydro thermal mix. An investment of over US$ 100 billion would be required in a period of about two decades to meet the planned target of hydro power projects. A significant role of private participation in exploiting the balance of over 100 GW of hydro is being envisaged. To meet the ambitious targets set for the coming Five Year Plans, pooling of resources from the Engineering Consultancy firms, Contractors, Equipment Manufacturers and Suppliers Groups is very essential. Fund requirements would be of the order of US$ 30 billion for hydro sector up to 12th Plan and US$ 50 billion for 12th Plan (2012 to 2017) for transmission sector.

Since balance of the project construction is pre-dominantly located in the Himalayan region, which is known for wide range of rock mass strengths with dry to very high quantity of water ingress, and when it comes to underground works, predicting their locations is a challenge, India would remain a hub for innovations in hydro power development. Many of these projects, with large installed capacities, would require mega-size companies to play the needed role in their development. With its ambitious plan of exploiting the vast potential of small hydro sector and the incentives declared by the Government to boost the construction activity, smaller groups could find attractive avenues for investment and engineering.

The Government of India and the State Governments have carried out amendments in the policies to attract developers to take initiative for construction of project on BOOT basis. There are well laid out procedures for appraisal and accord of techno-economic concurrence to the Detailed Project Reports, as a pre-requisite to meet the requirement of undertaking project construction.

The issues, which need further favorable consideration by the Government agencies have been discussed, the constraints in development that have been listed are required to be viewed as a challenge to be overcome by being partners in the development. The determination of developers to face challenges inherent to hydro power development of Himalayan rivers would lead to abundant advantages in terms of attractive revenue from the sale of energy, in the operational phase of the project.

The hydro power development in the neighboring hydro rich countries would also need to proceed at a faster pace to meet the growing energy need of the region. The security of consumption can be viewed with comfort as a large populace of India would serve as an assured market for all times to come.

The private participation, which has shown an increasing trend of their contribution, since expects timely return of their investments, the Governments cannot escape but to become partners and play a more aggressive role in resolving the hindrances at local and Government levels.

12.16 ACRONYMS

AHEC	:	Alternate Hydro Energy Centre
BOOT	:	Build Own Operate and Transfer
CEA	:	Central Electricity Authority
CERC	:	Central Electricity Regulatory Commission

CRORES : The number Ten Million is referred as one crore
CSMRS : Central Soil and Material Research Station
CTU : Central Transmission Utility
CWC : Central Water Commission
DPR : Detailed Project Report
GSI : Geological Survey of India
HE : Hydro Electric
HEP : Hydro Electric Project
ITI : Industrial Training Institute
J&K : Jammu and Kashmir (one of the state of India)
MNRE : Ministry of New and Renewable Energy
MoEF : Ministry of Environment and Forest
MoP : Ministry of Power
MoWR : Ministry of Water Resources
NEEPCO : North Eastern Electric Power Corporation
NGO : Non-Governmental Organization
NHPC : National Hydroelectric Power Corporation
NRRP : National Resettlement and Rehabilitation Policy
NTPC : National Thermal Power Corporation
PAF : Project Affected Families
PFC : Power Finance Corporation
PFR : Pre-Feasibility Report
PSS : Pumped Storage Scheme
R&R : Resettlement and Rehabilitation
REC : Rural Electrification Corporation
ROR : Run-of-the-River
Rs. : Indian Rupees (1US$ = about Rs.50)
RTI : Right to Information
SEB : State Electricity Board
SERC : State Electricity Regulatory Commission
SHP : Small Hydro Project
STU : State Transmission Utilities
TEC : Techno-Economic Clearances

12.17 REFERENCES AND GOVERNMENT OF INDIA WEBSITES (IN PUBLIC DOMAIN)

1. "Hydroelectric Power Stations in operation in India", Publication No. 288, Central Board of Irrigation and Power, New Delhi-110021, India.

2. "Future power scenario — policy and issues vis-a-vis hydro development" Proceedings of India Hydro — 2005, seminar organized by Indian National Hydropower Association (INHA) and International Hydropower Association (IHA), UK, New Delhi, India, 19th–21st February, 2005.

3. 50,000 MW initiative - http://www.cea.nic.in/hydro/Status of 50,000 MW Hydro Electric Initiative/Status of 50000 MW HydroElectric Initiative.pdf.

4. Hydro Development Plan for 12th five year plan (2012–2017) Central Electricity Authority, New Delhi, India - http://www.cea.nic.in/hydro/Hydro Development Plan for 12th Five Year Plan.pdf.

5. "Building Infrastructure : Challenges and opportunities, conference at Vigyan Bhawan, New Delhi, March 2010 - Projections in the Eleventh Five Year Plan- Investment in Infrastructure", Secretariat for the Committee on Infrastructure, Planning Commission, Government of India, New Delhi-110001, India.

6. "Hydro by the year 2025 — road ahead" Proceedings of India Hydro — 2005, seminar organized by Indian National Hydropower Association (INHA) and International Hydropower Association (IHA), UK, New Delhi, India, 19th -21st February, 2005.

7. "Statistical Data", published in Water and Energy, Vol. 67, No.2, April 2010, Central Board of Irrigation and Power, New Delhi — 110021, India.

8. National Electricity Policy, 2005 - http://www.powermin.nic.in/indian_electricity_scenario/national_electricity_policy.htm.

9. "Tata BP Solar expands solar manufacturing capacity by 62% to serve growing solar market in India", Enertia, Vol. 3, Issue-IV, April 2010, Falcon Media, Thane (Maharashtra), India.

10. "Vision Energy: Enhancing declining Hydro Power potential is essential", Enertia, Vol. 3, Issue-IV, April 2010, Falcon Media, Thane (Maharashtra), India.

11. New hydro policy 2008 - http://www.powermin.nic.in/whats_new/pdf/new_hydro_policy.pdf.

12. Electricity Act, 2003 - http://www.powermin.nic.in/acts_notification/electricity_act2003/pdf/The Electricity Act_2003.pdf.

13. National Water Policy, 2002 - http://wrmin.nic.in/writereaddata/linkimages/nwp20025617515534.pdf.

14. Mega Power Projects: Revised Policy Guidelines -http://www.powermin.nic.in/whats_new/pdf/revised_mail.pdf.

15. Systematic Assessment -http://www.cea.nic.in/hydro/Special_reports/Ranking_Study/preliminary_ranking_study_of_hyd.htm.

16. "Hydro Electric Power Potential of India", Central Electricity Authority, Ministry of Energy, Department of Power, Government of India, New Delhi, December 1988.

17. River Basins Map — http://cwc.gov.in/main/images/india_map.jpg.

18. Hydro capacity addition proposed for 12th Plan — http://www.cea.nic.in/planning/Conclave%2018-19%20Aug%2009/1_Over%20View%20of%20PS_Sh%20Bakshi.pdf.

19. Ranking Study Report for the Development of Hydro-Electric Potential-http://pib.nic.in/archieve/lreleng/lyr2001/rdec2001/05122001/r051220014.html.

20. Indian State Map — http://www.imd.gov.in/images/india_map_orig.gif.

21. "Development of Indian Power Sector — Present Status", Water and Energy, Vol. 67, No.2, April 2010 published by Central Board of Irrigation and Power, New Delhi — 110021, India, 2009.

22. Tariff Policy 2006 : http://www.powermin.nic.in/whats_new/pdf/Tariff_Policy.pdf.

23. "Manual on Development of Small Hydroelectric Projects", Central Board of Irrigation and Power, New Delhi — 110021, India.

24. "Development of Small Hydro Power in India", Water and Energy, Vol. 67, No.3, May 2010, Central Board of Irrigation and Power, New Delhi — 110021, India.

25. "Pumped Hydro Storage - India's perspective" Proceedings of India Hydro — 2005 (Supplementary Volume), seminar organized by Indian National Hydropower Association (INHA) and International Hydropower Association (IHA), UK, New Delhi, India, 19th -21st February, 2005.

26. "Development of Pumped Storage Schemes in India" Proceedings of India Hydro — 2005 (Supplementary Volume), seminar organized by Indian National Hydropower Association (INHA) and International Hydropower Association (IHA), UK, New Delhi, India, 19th -21st February, 2005.

27. "Interview — CMD Power Grid", published in Water and Energy, Vol. 67, No.2, April 2010 Central Board of Irrigation and Power, New Delhi — 110021, India.

28. "Development of Transmission System in India — An Overview", Water and Energy, Vol. 67, No.3, May 2010, Central Board of Irrigation and Power, New Delhi — 110021, India.

29. Constitution of India - http://india.gov.in/govt/constitutions_of_india.php.

30. National R&R Policy - http://www.indiaenvironmentportal.org.in/files/NRRP2007.pdf.

31. "Guidelines for preparation for of Detailed Project of Irrigation and Multipurpose projects, August 2010", Central Water Commission, New Delhi, India - http://www.cwc.gov.in/main/downloads/Revised Guidelines 2010.doc.

32. "Guidelines for Preparation of project estimates for river valley projects, 1997", Central Water Commission, New Delhi, India - http://www.cwc.gov.in/main/downloads/1997_Guidelines_preparation_project_estimates_river_valley_projects.pdf.

33. "Guidelines for Formulation of Detailed Project Reports for hydro-electric schemes, their acceptance and examination for concurrence, January 2007" Central Electricity Authority, New Delhi, India — http://www.cea.nic.in/hydro/Special_reports/GUIDELINES%20for%20formulationof%20DPR%20for%20HE%20Schemes.pdf.

34. "Best Practices in Hydro Power Generation" — http://www.cea.nic.in/hydro/Special_reports/best_practises.pdf.

35. Forest Conservation Act, 1980 — http://www.ielrc.org/content/e8003.pdf.

36. "Tala-A Shining Example of Indo-Bhutan cooperation", International Workshop on Experiences Gained in Design and Construction of Tala Hydroelectric Project Bhutan on 14th–15th June 2007, CSMRS, New Delhi, India organized by Indian Society for Rock Mechanics and Tunneling Technology, New Delhi, India.

37. "Building Infrastructure : Challenges and opportunities, conference at Vigyan Bhawan, New Delhi, March 2010 - Private Participation in Infrastructure", published by Secretariat for Infrastructure, Planning Commission, Government of India, New Delhi — 110001, January 2010.

38. "India's Power & Energy Economy — Version 2010–2011", Falcon Strategic Advisors (India) — FSAI, Dombivli, (Maharashtra), India.

39. "Integrating site specific geotechnical condition in design of Hydel Projects", Indian Rock Conference, ISRMTT, November, 2002, New Delhi, India.

40. "Geo-technical considerations in the planning and design of spillway for Sankosh multipurpose project", Symposium on modern practices in Geo-techniques, ISEG, November, 1996, Lucknow, India.

41. "Complexities of the future water resources projects and challenges for their planning and design", The Indian Journal of Public Administration, New Delhi, India, July-September, 2003.

42. "Contract Management of underground construction in Himalayas — a Challenge", Indian Rock Conference, ISRMTT, November, 2002, New Delhi, India.

43. "Speedy completion of Tunneling projects in Himalayas through innovative approaches", — seminar on "Productivity and Speed in Tunneling", 2003, Dehradun, India.

44. "3rd Hydro Vision India 2010 — Risk Mitigation in Hydropower Projects", Enertia, Vol. 3, Issue-IV, April 2010, Falcon Media, Thane (Maharashtra), India.

45. "Challenges of Power Capacity additions in Twelfth Plan", Enertia, Vol. 3, Issue-III, March 2010, Falcon Media, Thane (Maharashtra), India.

46. India's Electricity Sector — Widening Scope for Private Participation - www.powermin.nic.in/whats_new/pdf/brochure-revised.doc.

CHALLENGES AND OPPORTUNITIES IN TIDAL AND WAVE POWER

Paul T. Jacobson

ABSTRACT

Power generation from waves and tidal currents is a nascent industry with the potential to make globally significant contributions to renewable energy portfolios. Further development and deployment of the related, immature technologies present opportunities to benignly tap large quantities of renewable energy; however, such development and deployment also present numerous engineering, economic, ecological, and sociological challenges. A complex research, development, demonstration, and deployment environment must be skillfully navigated if wave and tidal power are to make significant contributions to national energy portfolios during the next several decades.

13.1 INTRODUCTION

Societal demand for energy continues to grow. Over the period 1980 to 2007, total global primary energy consumption grew at an average annual rate of 2.0% [1]. While improvements in the efficiency of energy use have the potential to slow this rate of growth, an ever-increasing world human population and industrial and economic development ensure that demand for energy will continue to grow into the foreseeable future. Thus, worldwide energy consumption is projected to grow at an average annual rate of 1.4% over the period 2007–2035 [2].

Societal concern regarding climate change resulting from carbon emissions associated with fossil fuel consumption is driving worldwide efforts to reduce net emissions of carbon dioxide and other greenhouse gases. A recent study by the Electric Power Research Institute (EPRI) [3] demonstrated that 58% reduction from 2005 emission levels can be achieved by 2030 for the US electricity sector (including electro-technologies and electric transportation) given aggressive deployment of a suite of energy technologies, including conservation and improved efficiency of energy use. A key point highlighted by this study is that deployment of a diverse array of technologies allows the target reductions in carbon emissions to be achieved in a more cost-effective manner.

Other considerations favoring renewable power generation include the environmental effects associated with extraction and consumption of fossil fuels, a desire to reduce reliance on foreign suppliers, and concerns regarding long-term supply of fossil fuels.

Waterpower constitutes a minor but significant contribution to meeting future energy demand with renewable sources. Currently, virtually all hydropower is generated by conventional means involving impoundment of water bodies and passage of water through turbines placed in water control structures. Conventional hydropower constituted 6% of all primary energy production and 15.8% of net electricity generation worldwide in 2007 [4]. There is substantial potential for expansion of conventional hydropower generation through construction of new hydropower dams, expanded generation at existing hydropower facilities, and installation of turbines at existing dams that currently lack generating capacity. The International Hydropower Association estimates that conventional hydropower can be expanded by 70% worldwide, mostly in developing countries [5]. The US-based National Hydropower Association has established the goal of doubling hydropower's contribution to the nation's energy portfolio, with capacity increases coming from development of new pumped storage projects, conversion of non-powered dams, new capacity and modernization at existing hydroelectric facilities, and hydrokinetic projects, including in-stream, wave, and tidal power projects. The EPRI [6] estimates that conventional hydropower capacity (exclusive of pumped storage) could be increased 18% by 2025, with just over half of that increase coming from new capacity at existing, non-powered dams.

New dam construction is constrained or precluded in many parts of the world, including the US, by siting constraints and concerns about the environmental effects of dam construction and operation. Furthermore, the presumed low carbon status of conventional hydropower is threatened by reports that reservoirs used for hydropower generation can be net sources of greenhouse gas emissions, most notably methane [7-9].

Emerging technologies for hydropower generation such as wave power and tidal power do not require impoundment of water bodies. These technologies have the potential to add significantly to hydropower generating capacity while avoiding many of the siting and environmental concerns associated with conventional hydropower and other forms of power generation.

13.2 THE RESOURCE

Tidal power and wave power are potential sources of significant quantities of domestic, renewable, low-carbon energy. Coastal areas are in closest proximity to these renewable resources. Tidal power is, by its nature, limited to coastal areas, whereas wave power tends to increase with distance from shore. According to the Census Bureau, 48.9% of the US population lived within 50 miles of a coastline in 2000. These figures include coastlines with very limited wave and tidal energy (such as in the Great Lakes and the Gulf of Mexico). Nonetheless, potential for local, ocean-based power generation has relevance for a very significant portion of the US population. Ocean-based power generation has the potential to supply renewable energy to areas of high demand, without occupying limited terrestrial open space or requiring long-distance transmission.

Quantification of wave and tidal energy resources is of value both for assessing potential for resource development and for siting of generating facilities.

13.2.1 Wave Energy

Power density of waves for a given sea state is:

$$P = kH_s^2 T_z \tag{13.1}$$

where P is the wave power flux (kW m^{-1}), H_s is the significant wave height (m), T_z is the mean wave period (s), and k is a constant usually in the range of 0.4 to 0.6, depending on the relative amounts of energy in the short-period, wind-driven component and the longer-period swell component of the given sea state.

Wave power develops over the unobstructed length of open water acted upon by the wind (called the fetch); thus, wave climate in the US tends to be best along the Pacific coast, which is exposed to prevailing westerly winds. Wave power is diminished by interaction of wave energy with the sea bottom. This effect becomes insignificant with water depths greater than half a wavelength; thus, wave power generally increases with distance offshore out to a depth of approximately 200 m, which corresponds to the approximate depth of the continental shelf edge.

The magnitude of the global wave energy resource is estimated to be 2 TW (18,000 TWh/yr) [10]. The Carbon Trust [11] estimates that approximately 13%-20% of this natural worldwide wave energy resource constitutes a practical resource (accounting for site availability and technical factors). Figure 13.1 shows an estimate of the global distribution of annual mean wave power based upon a prediction of wave climatology over the period 1997–2006. Annual mean wave power is greatest:

- In the North Atlantic south of Iceland and west of Scotland (90 kW/m) [12]
- In the North Pacific south of the Aleutian Islands and off the west coast of Canada, Washington, and Oregon (75 kW/m) [13]
- Between 40°S and 60°S in the Southern Hemisphere (140 kW/m, southwest of Australia) [12]

It should be pointed out that the high latitude areas exhibiting the highest annual mean wave power also exhibit the greatest seasonal variability. The maximum monthly average wave power is approximately 200 kW m^{-1} in both the Northern and Southern Hemispheres [12].

An ongoing US Department of Energy (USDOE)–sponsored study being conducted by EPRI has estimated that the total available energy flux across the continental shelf break for the United States is approximately 2640 TWh/yr (Table 13.1) [14]. Estimates of annual mean wave power density are being mapped along the coastlines of the United States extending out to 50 nautical miles from the shoreline (see Figure 13.2).

13.2.2 Tidal Energy

Hydraulic power density (P, W m^{-2}) is a function of the density of water (ρ, kg m^{-3}) and its speed (V, m s^{-1}), scaling with the third power of the water speed [15]:

$$P = \frac{1}{2}\rho |V|^3 \tag{13.2}$$

Given the density of sea water (1024 kg m^{-3}), power density can be high in tidal currents. Water speed of just 1.0 m s^{-1} yields a power density of 0.5 kW m^{-2}, while a tidal stream of 3.0 m s^{-1} yields a power density of 13.8 kW m^{-2} (Figure 13.3).

The total power extracted ($P_{\text{extracted}}$) from a tidal stream scales with cross-sectional area of the intercepted flow (A) and the efficiency with which the energy is extracted (η_e) [16]:

$$P_{\text{extracted}} = \frac{1}{2}\eta_e \rho A V^3 \tag{13.3}$$

The theoretical maximum limit on extraction of energy from the total kinetic flux in unconstrained flow (known as the Betz limit) is 59%; however, several authors have questioned the relevance of

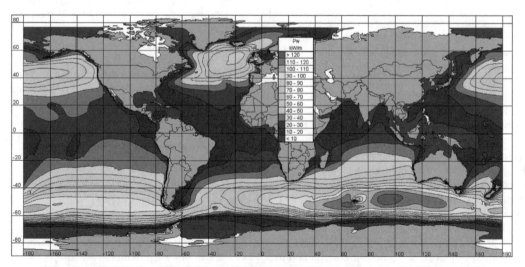

FIG. 13.1 ANNUAL AVERAGE WAVE POWER DENSITY (Source: CORNETT [13])

TABLE 13.1 MEAN ANNUAL WAVE ENERGY FLUX ESTIMATED FROM WAVEWATCH III MULTI-PARTITION HINDCAST SEA STATE PARAMETERS AND RECONSTRUCTED SPECTRA AND AGGREGATED BY REGION [14]

Region	Total Available Flux (TWh/yr)
West Coast	590
East Coast	200
Alaska	1,360
Bearing Sea	210
SE Atlantic	40
Gulf of Mexico	80
Puerto Rico	30
Hawaii	130
Total	2,640

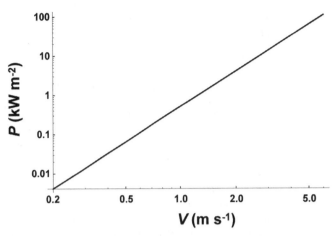

FIG. 13.3 HYDRAULIC POWER OF SEAWATER (P; kW m^{-2}) AS A FUNCTION OF CURRENT SPEED (V; m s^{-1}).

Betz's assumptions to a tidal scenario [17-19]. Resource assessment based on the energy flux in undisturbed flow is considered to be only preliminary by some (e.g., 18), and others (e.g., 20) consider such an approach to be flawed. True resource assessments should be based on energy extraction modeling that incorporates knowledge of the undisturbed kinetic energy flux and site-specific information on channel characteristics [18, 20-23].

As a consequence of these challenges, energy resources available from tidal streams are not as well characterized as is the wave energy resource. Global estimates of the tidal resource range from >100 TWh/yr [24] to >3500 TWh/yr [25]. Additionally, several published documents report a practical, global resource of approximately 800 TWh/yr [11, 26-28], although it is unclear if the latter value comprises multiple independent estimates.

EPRI examined the tidal energy resource at many, but not all, potential US tidal energy sites. The aggregate estimate at these sites (which excludes sites with annual average power densities less than 1 kW/m²) is 115 TWh/yr [29, 30]. This suite of sites includes:

- Alaska (many sites), 109 TWh/yr
- Puget Sound, Washington, 3.5 TWh/yr
- Golden Gate, San Francisco, California, 2 TWh/yr
- Western Passage, Maine, 0.5 TWh/yr

Since this estimate does not encompass all potential sites in the US, and it excludes sites with annual average power densities less than 1 kW/m² that may ultimately prove to be exploitable, the estimate of 115 TWh/yr likely is a lower bound on the total available US tidal energy resource. A USDOE-sponsored study presently is working toward a more comprehensive, rigorous estimate of the total available US tidal energy resource.

As of August 2009, worldwide installed capacity of tidal instream energy conversion devices (TISECs) totaled 2 MW, none of which was in the United States.

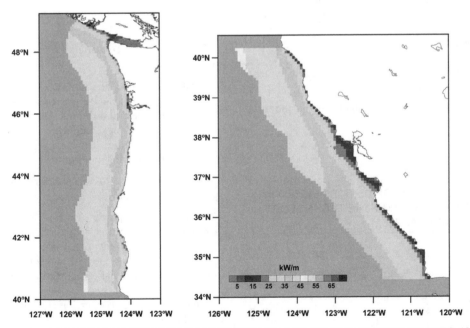

FIG. 13.2 ANNUAL AVERAGE WAVE POWER FOR THE WEST COAST OF THE UNITED STATES (Source: JACOBSON ET AL. [14])

13.3 ENGINEERING CHALLENGES AND OPPORTUNITIES

Electricity generation from ocean waves and tidal currents presents a rich array of engineering challenges and opportunities. While the earliest known patent for a wave energy conversion device was filed in 1799 [31], the technology to cost-effectively deploy wave energy conversions and TISECs still is in an early stage of development. A large number of device types compete for limited capital to support research, development, deployment, and demonstration activities (RDD&D); however, the corollary is that there is a notable absence of technology lock-in, and the door is open to new entrants and innovation. From among the thousands of concepts and patents that exist, hundreds have progressed to the stage of rigorous laboratory tow- or wave-tank tests of physical models, and only a few dozen have progressed to the stage of short-term tests in natural waters. Long-term testing (>1 year in duration) of prototype devices in natural waters is rarer still [29]. The USDOE Office of Energy Efficiency and Renewable Energy (EERE) maintains a database of marine and hydrokinetic technologies [32] that listed 129 technologies as of November 2008. The database classification of these technologies according to development stage is listed in Table 13.2. It should be pointed out that the USDOE-EERE database is only a partial listing, with poorest representation of the earlier technology stages. Furthermore, the database was last updated in late 2008, and since that time, there has been growth in the number of technologies in each stage and stage advancement for some technologies.

A striking feature of the technologies in various stages of development is their diversity [33]. This diversity has motivated development of various classification schemes. The most useful classification scheme may be that of Myers et al. [34], which classifies wave and tidal technologies separately at three levels corresponding to three major subsystems of the devices:

1. The hydrodynamic subsystem, which classifies the means by which the device captures energy from the environment and converts into motion that is suitable for extraction
2. The power take-off subsystem, which classifies the means by which mechanical work is converted to a form of energy useful for conveyance to shore (e.g., electricity, hydrostatic pressure)
3. The reaction subsystem, which classifies the means by which the device is held on station.

This system of classification is useful for engineering, economic, and environmental assessment.

TABLE 13.2 CLASSIFICATION OF DEVICES LISTED IN USDOE'S MARINE AND HYDROKINETIC TECHNOLOGY DATABASE [32]

Technology Stage	Number
Commercially available	1
Full-scale prototype	28
Scale model testing — sea trials	42
Scale model testing — tank testing	23
Detailed design	30
Concept design	5

Within level 1, tidal technologies are further classified according to

- the primary method of energy conversion (i.e., lift, drag, Venturi)
- principle motion of the hydrodynamic subsystem (i.e., linear oscillation, rotation, uni-directional [over a half tidal cycle])
- orientation of rotational axis (i.e., vertical; horizontal, parallel to flow; horizontal, perpendicular to flow)
- number of hydrodynamic subsystems per device
- number of lift or drag elements (i.e., blades) per hydrodynamic subsystem
- incorporation of free stream velocity augmentation (e.g., ducted versus unducted rotor)
- capture area of hydrodynamic subsystem(s) (i.e., aggregate area of the tidal stream intercepted or swept by the hydrodynamic subsystem[s])

Within level 2, the power take-off system is classified according to

- rated electrical power of device (i.e., "nameplate" or "nominal" output)
- rated flow speed (i.e., flow speed at the center of the capture area corresponding to the rated electrical power)
- generator type (i.e., singly fed asynchronous, singly fed synchronous, doubly fed asynchronous, doubly fed synchronous, linear)
- gearbox/differential specification (e.g., gear ratio, lever ratio)
- electrical/energy conversion and output (e.g., DC-AC, 33 kV; hydraulic-AC, 11 kV)

Within level 3, the reaction and control subsystems are classified according to

- the basic form of the foundation/anchor (i.e., monopole, multipile, single-point anchor, multipoint anchor, gravity base, pontoon)
- working water depth range
- presence/absence of an alignment mechanism for hydrodynamic subsystem and whether it is active or passive
- presence/absence of power regulation and whether it is active or passive

For wave energy converters (WECs), level 1 classifies the device according to

- the combination of exciting forces on the primary conversion component and the type or source of opposing force (i.e., buoyancy, inertia, or lift in combination with buoyancy, inertia, lift, seabed, or shoreline)
- axis of motion (i.e., heave, surge, or pitch)
- number of moving members

Level 2 for WECs provides a description of the power take-off system with respect to

- rated electrical power
- power matrix (power output as a function of significant wave height and wave energy period)
- generator type (see level 2 above for TISECs)
- gearbox/differential specification (see level 2 above for TISECs)
- electrical/energy conversion and output (see level 2 above for TISECs)
- energy storage/smoothing type (i.e., mechanical, electrical, other, none) and capacity (i.e., seconds at rated power)

Level 3 (reaction and control subsystems) characterizes

- the type of anchoring system (see Level 3 above for TISECs)
- the working depth range
- the alignment mechanism for the hydrodynamic subsystem (presence/absence and active/passive)
- presence or absence of power regulation and whether it is active or passive

The above classification/characterization scheme presented by Myers et al. [34] reflects the diversity of technologies and the early stage of development of the industry. This scheme or a similar one will be useful for system description and specification and for standards development. The scheme will also be useful for evaluation of environmental effects, because the scheme is structured according to the major components of the devices and their operation.

The principal engineering challenge facing development of tidal and wave power devices is design of devices that can survive and operate reliably in the harsh marine environment. Deployment sites are attractive precisely because they are high-energy sites. For economic reasons, generating capacity of the devices is less than peak extractable power at the site, but devices must be designed and built to withstand extreme events during which the environment is substantially more energetic and turbulent. This is true primarily for WECs, because wave action is more strongly affected by local weather conditions than are tidal currents.

A significant challenge that cannot be overlooked is that installation, operation, monitoring, maintenance, and repair must occur — at least in part — at sea. Working from a vessel and in water affects logistics, safety, and cost. These challenges are exacerbated by the high-energy characteristics of favorable project sites. The cost and logistical considerations favoring infrequent maintenance and repair are opposed by the harsh marine environment in which the devices must operate. Transport of the large, heavy devices by ship from manufacturing to deployment sites, however, is likely easier than it would be over land.

A significant advantage of tidal and WEC relative to other renewable energy technologies is the ability to forecast the short-term resource availability. Significant wave power at high-quality wave sites arises from large waves that develop and propagate over large expanses of open ocean. Given the time required for such waves to propagate over the ocean, wave power is forecastable 24–48 hours, or longer, in advance. Tides are even more predictable because they are governed by the motion of the Sun and the Moon relative to the Earth. Tide tables are computable centuries in advance, and while local and regional weather can alter tide predictions, significant effects on tidal currents are manifested at the time scale of hours and days rather than minutes or seconds. This degree of forecastability facilitates integration of wave and tidal power into the electric grid for which generation and load must be balanced in real time. By contrast, wind and photovoltaic generation are subject to unpredictable, short-time scale variation in energy reaching the device due to gusty winds and passing clouds.

Substantial enhancements to the efficiency of wave energy extraction can be achieved by tuning devices to near-term wave conditions. This can be accomplished through adjustments of buoyancy and power take-off characteristics. The complement to wave-by-wave tuning to improve efficiency of wave energy extraction is near-term prediction of quiescent periods during periods when the most energetic waves are potentially damaging to WECs. Successful prediction and adaptation to wave characteristics present opportunity for substantial gains in the capacity factor of WECs [35, 36];

thus, design and control of adaptive devices are a topic rich in engineering and economic opportunity.

As discussed, reliable resource assessment requires energy extraction modeling involving fairly detailed, site-specific factors. This encompasses device specification, device placement, and array design. Far-field effects of TISEC arrays on coastal hydraulics also have implications for environmental resources. Much work remains to be done to enhance our understanding of the reciprocal influences and effects of TISEC arrays and the coastal areas in which they are placed.

Design and siting of TISECs have been primarily focused on economical extraction of energy; however, siting of TISEC devices may be constrained by concerns about potential adverse effects on fish and other aquatic life — especially as projects use larger and more numerous devices. Some early results suggest that adverse effects on selected biota are minimal or non-existent [37-39]. However, many ecological questions remain unanswered (see ECOLOGICAL CONSIDERATIONS below), and incorporation of biological considerations in device design (such as rotational rate, tip speed ratio, leading edge thickness, and noise and vibration characteristics) may relax possible ecological constraints on device size, number, and placement.

13.4 SOCIOLOGICAL AND ECONOMIC FACTORS

While potential development of wave and tidal energy garners great enthusiasm from many because it is a renewable, domestic, emission-free means of generation, it also raises significant concerns on the part of others, and ambivalence in some, because it involves industrialization of sites that typically have not been industrialized. Such areas may be heavily transited by ships, may be used for recreational and commercial fishing, and perhaps are even degraded by other anthropic coastal activities; however, they generally do not have permanent, fixed industrial facilities. Public ownership of these apparently "unoccupied" sites allows for multiple, pre-existing, potentially conflicting uses [40, 41]. Thus, wave and tidal projects constitute novel human footprints on the marine landscape with the potential to displace pre-existing activities, and they are scrutinized as such.

Visibility and noise — two factors affecting public acceptance of onshore and offshore wind generation — are not a significant factor with regard to public acceptance of wave and tidal projects. The above-water profile of wave and tidal devices is much reduced compared with that of wind turbines, and noise effects are minimal or non-existent for the human passersby.

Wave and tidal energy development presents an opportunity to apply existing maritime industry support services to a new industry. Ocean engineering capabilities and experience are transferrable from ship building, offshore oil and gas development, and the emerging offshore wind industry. Local job creation related to manufacture, installation, operation, and maintenance of wave and tidal energy projects is often cited as an important benefit.

Capital costs for wave and tidal energy conversion devices are related to the engineering challenges noted above. Perhaps the most authoritative analysis of device costs for current devices is reported by Callaghan and Boud [42]. Their estimates of capital costs and cost of energy are presented in Table 13.3. Their analysis assumes projects are limited to 10 MW, which also limits economies of scale. These costs are higher than costs for other technologies, including more established renewables. However, costs can

TABLE 13.3   CURRENT CAPITAL COST AND COST OF ENERGY FOR WAVE AND TIDAL ENERGY CONVERSION DEVICES AND ARRAYS [42]

Device Type	Initial Device Capital Cost ($/kW)[1]	
	Prototype	Production Model
Wave	6600–14,000	2600–6600
Tidal	7400–12,000	2200–4600

Device Type	Initial Array Cost of Energy (¢/kWh)[1]	
	Central Range	Uncertainty Band
Wave	34–39	18–68
Tidal	18–23	14–28

[1]All values are converted from UK to US currency at a rate of $1.54/£.

be expected to decline with cumulative capacity installation (as has occurred with other renewable energy technologies) [43] and eventually equal or better the costs of other renewable energy technologies [44]. Organizations promoting development of the wave and tidal power industry and others consider government support and subsidies essential to acquiring the cumulative installation that will reduce unit costs and allow the industry to become cost-competitive [28, 45].

13.5   ECOLOGICAL CONSIDERATIONS

Wave and tidal power have received substantial scrutiny regarding their potential for causing adverse ecological effects. The novelty of the devices and their interaction with the environment have raised many questions that remain unanswered and have generated many other questions that are difficult to articulate and address in the context of specific project proposals. As a consequence, environmental review contributes substantial uncertainty and cost to the permitting and licensing process. Many of the engineering issues — as well as uncertainty regarding many of the ecological concerns — that must be resolved to create a viable wave and tidal power industry can be addressed only through deployment of prototype devices and demonstration projects in open-water settings. Environmental permitting, however, requires answers to the outstanding questions. This scenario creates a circular problem that precludes its own solution (see Figure 13.4a). Resolution of this dilemma requires careful definition of the questions to be answered regarding environmental effects, and relaxation of one or more of the requirements depicted in Figure 13.4a. Adaptive management (discussed below) is the primary means of resolving this dilemma.

Three major factors control the potential for adverse ecological effects (the three S's): system, siting, and scale.

13.5.1   System

System refers to the wave or tidal-generating equipment and all of its supporting structure including electricity transmission equipment. This is the technology side of the project-environment interaction. System characteristics are still poorly defined at the level of the ocean renewable industry. "Technology lock-on" does not yet exist in this industry; there are many modes of operation and

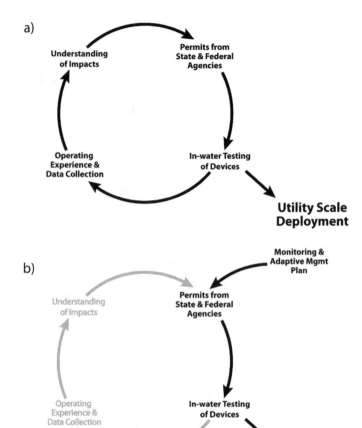

FIG. 13.4 (A) RELATIONSHIP OF SELECTED ACTIVITIES REQUIRED TO DEPLOY WAVE AND TIDAL DEVICES AT THE UTILITY SCALE, (B) THE ALTERED RELATIONSHIP OF ACTIVITIES WITH THE ADDITION OF ADAPTIVE MANAGEMENT

countless designs. The diversity of wave and tidal energy devices was noted above under engineering challenges and opportunities. It is unknown which types of devices will provide the best balance of reliability, cost-effectiveness, and environmental performance.

13.5.2   Siting

Siting refers to the environmental setting of the project and its components. Siting encompasses geographic scales ranging from the region to the estuary or coastal water body and the location within a specific water body. While most biological entities and physical/chemical environmental factors are broadly distributed, they come together in combination and proportion in unique, site-specific ways. As a consequence, organisms respond to variability in their environment on both large and small geographic scales [46]. For example, presence or absence of a particular species of marine mammal or fish may depend on the geographic region; abundance may vary among water bodies, and distribution within a water body will be variable but non-uniform over the long term. The environmental resources that are present to interact with a given technological system are thus highly site-specific. The energy resource to be exploited, however, may be extremely localized within a water body (see Figure 13.5), thereby limiting siting options.

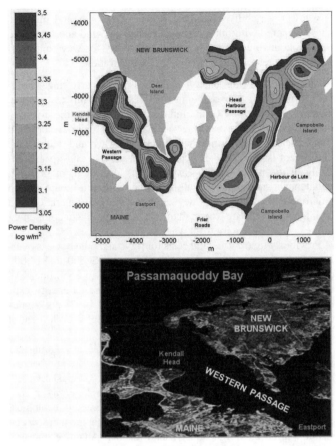

FIG. 13.5 GEOGRAPHY AND POWER TIDAL POWER DEN-SITY IN THE VICINITY OF WESTERN PASSAGE, EAST-PORT, MAINE, USA (Source: BEDARD [70])

13.5.3 Scale

At the level of the individual project, scale comprises factors such as the number and unit size of the devices, the spatial footprint of the deployment site, and the total generating capacity of the project or magnitude of energy it dissipates. Deployment scale also is relevant for water bodies and geographic regions, because the potential for cumulative effects across multiple sites increases with the aggregate scale of the industry. The relationship between ecological effects and deployment scale is likely to be non-linear — at least for some types of interaction; thus, extrapolation of effects across scales may be unreliable. For example, the behavioral response of fish migrating out of an estuary may differ, depending on the number and size of tidal turbines in a given tranche. Behavioral response — and the ability to respond — may be further affected by the number and spacing of tranches in an estuary. Behavioral response, in turn, may affect survival rate. Limiting the scale of demonstration and early-stage commercial projects reduces the potential for impacts, but it also compromises the ability to detect, quantify, and characterize specific impacts that may or may not occur at larger scales of deployment.

13.5.4 Adaptive Management

Adaptive management is a structured process of learning by doing [47-49]. It involves treating the management action or policy decision (in this case, deployment of the wave or tidal energy project) as a quasi-experiment. Monitoring is conducted prior to deployment to provide baseline information and following deploy-

ment during project operation to detect environmental effects. Adaptive management plans for wave and tidal projects have included requirements that the project will be modified or removed if adverse effects are detected. This process has been identified as the principal means of permitting deployment of wave and tidal energy devices in the face of substantial uncertainty regarding their environmental effects (Figure 13.4b) [50-53]. While incorporation of adaptive management into licensing and permitting of wave and tidal power projects allows the projects to go forward, project owners are at risk of having to substantially modify or remove the project if monitoring reveals unacceptable environmental effects. Given the limited scale and duration of wave and tidal projects under the Federal Energy Regulatory Commission's Hydrokinetic Pilot Project criteria, this appears to be an acceptable trade-off for at least some projects.

13.5.5 The Assessment Challenge

If we consider the number of wave and tidal device configurations that could be deployed (n_{system}), the number of sites energetically suitable for development (n_{site}), and a number of project size classes (n_{scale}), the industry-wide number of scenarios relevant to ecological effects assessment ($n_{assessment}$) can be seen to be an n-cubed problem:

$$n_{assessment} = n_{system} \times n_{site} \times n_{scale} \qquad (13.4)$$

Individual device and project developers find themselves alone in a metaphorical ocean of ecological assessment space. Given the novelty of the impact assessment scenarios they encounter, and an apparent lack of useful existing information, device and project developers face multifaceted, expensive environmental monitoring and assessment programs. To date, project developers have relied on government funding to help meet these costs.

Some have asserted that subsequent projects will benefit from the effort applied to initial projects and encounter lesser environmental assessment requirements. This will be true only if the questions implicit in the monitoring and assessment efforts are credibly answered by the initial stage projects, and only to the degree the results are transferrable to other projects. Adaptive management plans are not guaranteed to yield defensible, transferrable results. In its initial formulation, adaptive management was designed primarily to promote learning about managed natural resource systems [48, 49]. In its active form, adaptive management designs current management decisions to optimize learning about how the system responds to management; in its passive form, data available at each step is used to update current management, with the result that management and environmental effects are confounded and learning is compromised [54]. Among the countless initiatives that invoke adaptive management, few if any embody its defining characteristics, especially the attributes of active adaptive management [55]. Wave and tidal energy project applications of adaptive management are at risk of conforming to this pattern. Projects that are small in scale and limited in duration, are designed from the outset to minimize the potential for detectable adverse impacts, whereas the work of Walters and Holling [54] indicates that dramatic actions are more informative. Furthermore, since the adaptive management plans call for modification of the project if any adverse effects are detected through monitoring, the project-as-experiment is terminated prematurely. These attributes of adaptive management plans for wave and tidal energy projects are characteristic of evolutionary and passive adaptive management — weak forms of adaptive management that compromise learning.

The ecological assessment challenge facing the wave and tidal energy industry is to acquire and apply the information needed to ensure that systems, sites, and deployment scales are protective of ecological resources. Many reports have identified the range of environmental issues associated with wave, tidal, and other renewable ocean energy projects [52, 56-69]. The remaining challenge is to prioritize the issues and address the most pressing ones in a cost-effective manner. For individual projects, this requires definition and evaluation of questions required for site-specific permitting and licensing. For the industry as a whole, this requires identification and acquisition of information that is transferrable across projects. Several non-trivial activities can be taken to address the large array of outstanding issues:

1. Explicitly define conceptual models of the interaction of wave and tidal devices and arrays with the environment. These conceptual models can be used to identify commonalities among systems and sites and opportunities for information transfer. The conceptual models can also serve as peer-reviewed points of departure for design of project-specific assessment and monitoring programs.
2. Use risk-based approaches to identify the most significant environmental issues and prioritize them for detailed assessment. Although many quantitative values of risk are likely to be project-specific, generic analyses will still be useful for triage. The conceptual models will help identify opportunities for cross-cutting quantitative risk analyses.
3. Based on the conceptual models and risk analyses, identify overarching questions and information requirements that need not or cannot be effectively addressed in the context of individual projects.
4. Based on the conceptual models and the risk analyses, identify questions of broad relevance to regulators and resource managers. Parse the broadly relevant permitting and licensing questions into narrow questions suitable for rigorous assessment, and parse them in ways that maximize transferability across projects.
5. Develop generic protocols for active adaptive management of wave and tidal energy projects to serve as a template for project-specific adaptive management plans.

The foregoing steps provide a means of efficiently allocating monitoring and assessment activities among government agencies and laboratories, university researchers, project developers and industry collaboratives. The foregoing steps also provide opportunities for thorough review by the relevant academic, regulatory, and stakeholder communities while displacing duplicative but less thorough efforts.

13.6 SUMMARY

Wave power and tidal power have the potential to make significant contributions to emission-free renewable energy portfolios. Several factors enhance the value of the available wave and tidal energy resource, including

- proximity to the coastal areas where nearly half of the US population lives, which reduces the need for long-distance transmission
- high power density, which occurs at or below the waterline, thereby minimizing the spatial footprint and visibility of projects

- lower variability over time scales of seconds and minutes, compared with other renewables such as wind and solar, and 48-hour to centuries-scale forecastability for wave and tidal power, respectively, facilitate grid integration

The wave and tidal energy industry is in an early stage of development, and a large amount of RDD&D remains to be done. The nascent character of the industry is reflected in the diversity of device types in various stages of pre-commercial development. The principal engineering challenge facing wave and tidal device developers is design of devices that can survive and operate reliably in the harsh marine environment. Operation, monitoring, maintenance, and repair are all made more difficult by the marine environment in which the devices are deployed. Operation and maintenance costs are expected to be a very significant component of the cost of electricity from wave and tidal devices.

Previously reported capital costs for wave and tidal devices overlap substantially, ranging from approximately \$6600 to \$14,000 per kW for prototype devices and from \$2200 to \$6600 per kW for production devices. The corresponding cost of energy estimates range from 18 to 23 ¢/kWh (central range, 14–28 ¢/kWh uncertainty band) for tidal device arrays and 34–39 ¢/kWh (central range, 18–68 ¢/kWh uncertainty band).

Environmental considerations play a large role in ongoing development of the wave and tidal energy industry. The number and novelty of device types, in combination with the ecological diversity among potential deployment sites, creates a complex array of ecological impact scenarios. Efficient means of addressing ecological concerns are in need of further development so that the industry can advance in an environmentally sound manner. Adaptive management offers a means of moving the industry forward in the face of ecological uncertainty; however, the potential benefits of adaptive management will be realized only if it is implemented in its more scientifically rigorous form known as active adaptive management.

13.7 ACRONYMS

kV	kilovolt(s)
kW m^{-1}	kilowatt(s) per meter
N	lower case n number (count)
RDD&D	research, development, deployment, and demonstration
TISECs	tidal in-stream energy converters
TW	terawatt(s)
TWh/yr	terawatt-hour(s) per year
USDOE-EERE	United States Department of Energy - Office of Energy Efficiency and Renewable Energy
WEC	wave energy conversion/wave energy converter

13.8 REFERENCES

1. EIA (Energy Information Administration). 2010. U.S. Energy Information Administration: Independent Statistics and Analysis. U.S. Department of Energy. Available online at: www.eia.doe.gov.

2. EIA (Energy Information Administration). 2010. *International Energy Outlook 2010*. U.S. Energy Information Administration, Washington, DC. DOE/EIA-0484(2010). July 2010.

3. EPRI (Electric Power Research Institute). 2009. *Prism/MERGE Analyses: 2009 Update*. Electric Power Research Institute, Palo Alto, CA. 1019563. August 2009.

4. EIA (Energy Information Administration). 2010. *Annual Energy Review 2009*. U.S. Department of Energy. DOE/EIA-0384 (2009). August 19, 2010.

5. IHA (International Hydropower Association). 2003. *The Role of Hydropower in Sustainable Development*. February 2003.

6. EPRI (Electric Power Research Institute). in press. *Assessment of Waterpower Potential and Development Needs: 2010 Update*. Palo Alto, CA. 1020638.

7. UNESCO/IHA (United Nations Educational, Scientific and Cultural Organization/International Hydropower Association). 2010. *GHG Measurement Guidelines for Freshwater Reservoirs*.

8. UNESCO/IHA (United Nations Educational, Scientific and Cultural Organization/International Hydropower Association). 2008. *Scoping Paper: Assessment of the GHG Status of Freshwater Reservoirs*. IHP/GHG-WG/3. July 2008.

9. EPRI (Electric Power Research Institute). 2010. *The Role of Hydropower Reservoirs in Greenhouse Gas Emissions*. 1017971. May 2010.

10. Barstow, S., G. Mørk, D. Mollison, and J. Cruz. 2008. The Wave Energy Resource. In: *Ocean Wave Energy*. Springer Berlin Heidelberg. pp 93-132.

11. The Carbon Trust. 2003. *Building Options for UK Renewable Energy*. CT/2008/08. October 2003.

12. Barstow, S., G. Mørk, L. Lønseth, and J. P. Mathisen. 2009. World-Waves wave energy resource assessments from the deep ocean to the coast. *8th European Wave and Tidal Energy Conference*. Uppsala, Sweden.

13. Cornett, A. M. 2008. A Global Wave Energy Resource Assessment. In: 18th (2008) International Offshore and Polar Engineering Conference, July 6-11, 2008, Vancouver, British Columbia, Canada. The International Society of Offshore and Polar Engineers (ISOPE).

14. Jacobson, P. T., G. Hagerman, and G. Scott. 2010. Assessment and mapping of available U.S. wave energy resource. *HydroVision International 2010*. Charlotte, NC. July 29, 2010.

15. Hardisty, J. 2009. *The Analysis of Tidal Stream Power*. Wiley, Chichester, West Sussex, UK ; Hoboken, NJ.

16. Polagye, B. and M. Previsic. 2010. *Deployment Effects of Marine Renewable Energy Technologies: Tidal Energy Scenarios*. Prepared by RE Vision Consulting, LLC for DOE. RE Vision DE-002. June 2010.

17. Gorban, A. N., A. M. Gorlov, and V. M. Silantyev. 2001. Limits of the turbine efficiency for free fluid flow. *Journal of Energy Resour. Technol.* **123**: 311-317.

18. Bryden, I. G., S. J. Couch, A. Owen, and G. T. Melville. 2007. Tidal Current Resource Assessment. *Proc. IMechE Part A: Journal of Power and Energy* **221**: 125-135.

19. Garrett, C. and P. Cummins. 2004. Generating power from tidal currents. *Journal of Waterway, Port, Coastal, and Ocean Engineering* **130**(3): 113-118.

20. Polagye, B., and P. C. Malte. 2010. Far-field dynamics of tidal energy extraction in channel networks. *Renewable Energy* **in press**.

21. Bryden, I., and S. J. Couch. 2006. ME1 — marine energy extraction: tidal resource analysis. *Renewable Energy* **31**(2):133-139.

22. Garrett, C., and P. Cummins. 2008. Limits to tidal current power. *Renewable Energy* **33**(11), pp. 2485-2490.

23. Bryden, I. G., and S. J. Couch. 2007. How much energy can be extracted from moving water with a free surface: a question of importance in the field of tidal current energy? *Journal of Renewable Energy* **32**(11), pp. 1961-1966.

24. Fraenkel, P. L. 2007. Marine current turbines: pioneering the development of marine kinetic energy converters. *Proc. IMechE Part A: Journal of Power and Energy* **Volume 221**, pp. 159-169.

25. Gross, R., M. Leach, and A. Bauen. 2003. Progress in renewable energy. *Environment International* **29**, pp. 105-122.

26. Bhuyan, G. S. 2010. World-wide Status for Harnessing Ocean Renewable Resources. *IEEE PES General Meeting. Minneapolis, MN, July 25-29, 2010*.

27. Hammons, T. J. 1993. Tidal Power. *Proceedings of the IEEE* **89**(3), pp. 419-433.

28. DTI (Department of Trade and Industry). 2004. *The World Offshore Renewable Energy Report 2004-2008*. Department of Trade and Industry, United Kingdom. URN 04/393 CD.

29. EPRI (Electric Power Research Institute). 2008. *Prioritized Research, Development, Deployment and Demonstration (RDD&D) Needs: Marine and Other Hydrokinetic Renewable Energy*. Electric Power Research Institute, Palo Alto, CA.

30. Bedard, R., M. Previsic, and B. Polagye. 2009. Marine energy: How much development potential is there? *Hydro Review* **April 2009**, pp. 32-36.

31. Clément, A., et al. 2002. Wave energy in Europe: Current status and perspectives. *Renewable and Sustainable Energy Reviews* **6**(5), pp. 405-431.

32. USDOE-EERE (U.S. Department of Energy-Office of Energy Efficiency and Renewable Energy). 2010. Marine and Hydrokinetic Technology Database. Available online at: http://www1.eere.energy.gov/windandhydro/hydrokinetic/default.aspx.

33. Bedard, R., P. T. Jacobson, M. Previsic, W. Musial, and R. Varley. 2010. An overview of ocean renewable energy technologies. *Oceanography* **23**(2), pp. 22-31.

34. Myers, L. E., A. S. Bahaj, F. Gardner, C. Bittencourt, and J. Flinn. 2010. *Equitable Testing and Evaluation of Marine Energy Extraction Devices in Terms of Performance, Cost and Environmental Impact: Deliverable D5.2 Device classification template*. Available online: http://www.equimar.org/.

35. Belmont, M. R. 2009. A lower bound estimate of the gains stemming from quiescent period predictive control using conventional sea state statistics. *Journal of Renewable and Sustainable Energy* **1**(6), pp. 063104.

36. Belmont, M. R. 2010. Increases in the average power output of wave energy converters using quiescent period predictive control. *Renewable Energy* **35**, pp. 2812-2820.

37. NAI (Normandeau Associates, Inc.). 2009. *An Estimation of Survival and Injury of Fish Passes Through the Hydro Green Energy Hydrokinetic System, and a Characterization of Fish Entrainment Potential at the Mississippi Lock and Dam No. 2 Hydroelectric Project (P-4306), Hastings Minnesota*. Prepared for Hydro Green Energy, LLC by Normandeau Associates, Inc.

38. Amaral, S. V., G. Hecker, N. Perkins, D. Dixon, and P. T. Jacobson. 2010. Development of theoretical models for estimating hydrokinetic turbine strike probability and survival. *Symposium on Hydrokinetic Electricity Generation and Fish: Asking the Right Questions, Getting Useful Answers. Annual Meeting of the American Fisheries Society*. Pittsburgh, PA. September 15, 2010.

39. Amaral, S. V., B. J. McMahon, G. Allen, D. Dixon, and P. T. Jacobson. 2010. Laboratory evaluation of fish survival and behavior associated with passage through hydrokinetic turbines. *Symposium on Hydrokinetic Electricity Generation and Fish: Asking the Right Questions, Getting Useful Answers. Annual Meeting of the American Fisheries Society*. Pittsburgh, PA. September 15, 2010.

40. PEV (Pacific Energy Ventures). 2010. *Siting Methodologies for Hydrokinetics: Stakeholder Perspectives.*

41. Whittaker, D. and B. Shelby. 2010. *Hydrokinetic Energy Projects and Recreation: A Guide to Assessing Impacts. Public Review Draft (June 2010).* Prepared for National Park Service Hydropower Assistance Program, Department of Energy, and Hydropower Reform Coalition. June 2010.

42. Callaghan, J. and R. Boud. 2006. *Future Marine Energy, Results of the Marine Energy Challenge: Cost competiveness and growth of wave and tidal stream energy.* The Carbon Trust.

43. IEA (International Energy Agency). 2000. *Experience Curves for Energy Technology Policy.* ISBN 92-64-17650-0.

44. EPRI (Electric Power Research Institute). 2009. *Offshore Ocean Wave Energy: A Summer 2009 Technology and Market Assessment Update.* Electric Power Research Institute, Palo Alto, CA. EPRI-WP-013-OR. July 21, 2009.

45. Elefant, C. and S. O'Neill. 2010. *2009 Annual Report on the Marine Renewables Industry in the United States: A comprehensive review of legislation, regulations, executive policies and financial incentives.* Ocean Renewable Energy Coalition. February 17, 2010.

46. Levin, S. A. 1992. The problem of pattern and scale in ecology. *Ecology* **73**(6), pp. 1943-1967.

47. Williams, B. K., R. C. Szaro, and C. D. Shapiro. 2009. *Adaptive Management: The U.S. Department of the Interior Technical Guide.* Adaptive Management Working Group, Washington, DC. ISBN: 978-1-4133-2478-7.

48. Walters, C. J. 1986. *Adaptive Management of Renewable Resources.* Macmillan, New York.

49. Holling, C. S. 1978. *Adaptive Environmental Assessment and Management.* Wiley, New York.

50. Bald, J., et al. 2010. Protocol to Develop an Environmental Impact Study of Wave Energy Converters. *Revista de Investigación Marina* **17**(5), pp. 62-138.

51. Oram, C. and C. Marriott. 2010. Using adaptive management to resolve uncertainties for wave and tidal energy projects. *Oceanography* **23**(2), pp. 92-97.

52. USDOE (U.S. Department of Energy). 2009. *Report to Congress on the Potential Environmental Effects of Marine and Hydrokinetic Energy Technologies.* Energy Efficiency and Renewable Energy, Wind and Hydropower Technologies Program, Washington, DC. December 2009.

53. FERC (Federal Energy Regulatory Commission). 2008. Licensing Hydrokinetic Pilot Projects. Available online at: http://www.ferc.gov/industries/hydropower/indus-act/hydrokinetics.asp.

54. Walters, C. J. and C. S. Holling. 1990. Large-scale management experiments and learning by doing. *Ecology* **71**(6), pp. 2060-2068.

55. Gregory, R., D. Ohlson, and J. Arvai. 2006. Deconstructing adaptive management: criteria for applications to environmental management. *Ecological Applications* **16**(6), pp. 2411-2425.

56. Boehlert, G. W., G. R. McMurray, and C. E. Tortorici. 2008. *Ecological effects of wave energy development in the Pacific Northwest* National Oceanic and Atmospheric Administration. NOAA Tech. Memo. NMFS-F/SPO-92

57. Cada, G. F., et al. 2007. Potential impacts of hydrokinetic and wave energy conversion technologies on aquatic environments. *Fisheries* **32**, pp. 174-181.

58. EPRI (Electric Power Research Institute). 2006. *Instream Tidal Power in North America: Environmental and Permitting Issues.* EPRI-TP-007-NA.

59. Gill, A. B. 2005. Offshore renewable energy: Ecological implications of generating electricity in the coastal zone. *Journal of Applied Ecology* **42**(4), pp. 605-615.

60. Inger, R., et al. 2009. Marine renewable energy: potential benefits to biodiversity? An urgent call for research. *Journal of Applied Ecology* **46**, pp. 1145-1153.

61. Kramer, S., M. Previsic, P. Nelson, and S. Woo. 2010. *Deployment Effects of Marine Renewable Energy Technologies: Framework for Identifying Key Environmental Concerns in Marine Renewable Energy Projects.* Prepared by RE Vision Consulting, LLC for DOE. RE Vision DE-003.

62. Michel, J., et al. 2007. *Worldwide Synthesis and Analysis of Existing Information Regarding Environmental Effects of Alternative Energy Uses on the Outer Continental Shelf.* MMS OCS Report 2007-038.

63. MMS (Minerals Management Service). 2007. *Programmatic Final Environmental Impact Statement for Alternative Energy Development and Alternate Uses of Facilities on the Outer Continental Shelf. Volume I: Executive Summary through Chapter 4.* U.S. Department of Interior, Minerals Management Service. OCS EIS/EA MMS 2007-046. October 2007.

64. Nelson, P. A., et al. 2008. *Developing Wave Energy in Coastal California: Potential Socio-Economic and Environmental Effects.* California Coastal Commission, PIER Energy-Related Environmental Research PRogram and California Ocean Protection Council. CEC-500-2008-083.

65. Simas, T., A. Moura, R. Batty, D. Thompson, and J. Norris. 2009. *Equitable Testing and Evaluation of Marine Energy Extraction Devices in Terms of Performance, Cost and Environmental Impact: Uncertainties regarding environmental impacts. A draft.* EquiMar. Deliverable D6.3.1. Available online at: www.equimar.org. June 2009.

66. Boehlert, G. W., and A. B. Gill. 2010. Environmental and ecological effects of ocean renewable energy development. *Oceanography* **23**(2), pp. 68-81.

67. ABP MER (ABP Marine Environmental Research). 2009. *Wet Renewable Energy and Marine Nature Conservation: Developing Strategies for Management.* R.1451.

68. Faber Maunsell and METOC PLC. 2007. Scottish Marine Renewables: Strategic Environmental Assessment. Scottish Executive. Available online at: http://www.seaenergyscotland.net/SEA_Public_Environmental_Report.htm.

69. EMEC (European Marine Energy Center). 2008. *Environmental Impact Assessment (EIA). Guidance for Developers at the European Marine Energy Centre.* Orkney, Scotland. GUIDE003-01-03 20081106.

70. Bedard, R. 2007. *Overview of Tides and Tidal In-stream Energy Conversion Technology.* Electric Power Research Institute.

BIOENERGY INCLUDING BIOMASS AND BIOFUELS

T. R. Miles

14.1 INTRODUCTION

Oil shortages in the 1970s stimulated the development of new technologies to convert biomass to heat, electricity, and liquid fuels. Combined firing of biomass with coal reduces emissions, provides opportunities for high efficiency, and reduces fossil carbon use. Cofiring and markets for transportation fuels have stimulated global trading in biomass fuels. Pyrolysis of biomass to liquid fuels creates opportunities for co-products, such as biochar, which can help sequester carbon and offset emissions from fossil fuels. This chapter provides an overview of biomass fuels and resources, the biomass power industry, conventional and new technologies, and future trends. Advances in combustion and gasification are described. Thermal and biological conversions to liquid fuels for heat, power, and transportation are described with implications for stationary use.

14.2 BIOMASS FUELS AND FEEDSTOCKS

14.2.1 Biomass Properties

Combustion properties of biomass are apparent when compared with coal. The distinguishing features of biomass compared with coal are higher moisture content (MC), lower energy density, higher volatile matter content, variable particle size, and higher volatile ash components [1]. Tables 14.1 and 14.2 show the composition of biomass fuels compared with coal. The coal and switchgrass were cofired at a utility power plant. The coal is from the Powder River Basin (PRB), Wyoming, which is a major source of coal for North American utilities. The switchgrass is an herbaceous crop from Iowa [2]. The wood is an urban wood waste mix from a biomass power plant [3].

14.2.1.1 Moisture
Living biomass contains about 50% to 70% water. Except in very dry or very cold environments, wood fuels will stay moist (50% MC) when stored in log form or when chipped and stored in bulk piles. Crop residues are either harvested green (>70% MC), in damp field conditions (>35% MC), or in dry conditions (<15% MC). Dry grasses and crop residues are often 12% MC. The as-fired higher heating value (HHV) of woody biomass and wet field crop residues is often 50% to 65% of the dry higher heating value when adjusted for moisture.

When biomass fuels are burned, the moisture must first evaporate. Free water evaporates until the moisture remaining is chemically bound in the cells. This is called the fiber saturation point and is often at 20% MC wet basis. The remaining moisture will migrate through the cells to the surface and evaporate with volatile combustion gases. Since the chemical energy in the fuel cannot be converted to gas faster than the water evaporates, the moisture tends to regulate the rate of combustion. This feature is useful when burning wood with coal on a grate or when burning high-alkali fuels like agricultural crops and residues. Biomass combustion cools the coal flame which moderates the rate of combustion. The low-temperature biomass flame prevents high-temperature ash agglomeration.

Moisture in biomass is important to boiler design and efficiency. For example the efficiency, based on the higher heating value of the fuel (HHV), of a small wood boiler that generates 20,000 pph steam at 400 psig, with a stack temperature of 580°F, can be 69% with fuel at 20% MC. Efficiency drops to 57% at 50% MC. If an economizer is used to reduce the stack temperature to 325°F than efficiency is 78% at 20M% and 67% at 50% MC.

Europe has developed grades of biomass based on moisture content for use in heat and power generation [4, 5]. The U.S. has not yet developed similar standards.

14.2.1.2 Density and Energy Density
The dry bulk density of biomass fuels is often 7 to 10 lb/ft^3 (112 to 160 kg/m^3) compared with coal at 40 to 60 lb/ft3 (640 to 961 kg/m^3).This results in a low-energy density. Table 14.1 shows that the as-fired heating value (HHV) of PRB coal with high ash and moisture is similar to dry biomass at 10% MC. Wet wood chips at 50% MC are 4,200 Btu/lb and 24 lb/ft3 and have about one-third the energy density of PRB coal. When grass or wood are pulverized for suspension firing the energy density is also 20% to 30% of PRB coal. Because of this low-energy density, biomass systems that replace coal require larger facilities for fuel receiving, storage, and handling. It is often practical to design a separate system for the preparation, handling, and delivery of biomass fuel.

Biomass is densified into pellets to offset the difference in energy density between wood and coal for cofiring. Biomass pellets at 10% MC and 40 lb/ft^3 have an energy density of 300,000 Btu/ft^3 which is similar to the PRB coal. Torrefaction combined with densification is another method of increasing the energy density of biomass for cofiring with coal (see Section 14.6).

14.2.1.3 Volatile Content
Biomass has a higher volatile content than coal. Table 14.1 shows that when ash and moisture are removed biomass contains approximately 84% volatile carbon and

16% fixed carbon which means that more than 80% of the biomass solids convert to gas before burning. The rest remains as charcoal which burns in direct contact with air. When biomass is burned on a grate up to two-thirds of the air required for combustion is introduced under the grate in order to promote gasification and combustion. The gases, together with suspended fuel particles, are burned, with overfire air in a fireball above the grate. The coal represented in Table 14.1 contains about 50% fixed carbon and 50% volatile carbon. In a coal boiler about 80% of the combustion air is introduced through the grate where most of the combustion occurs. Coal combustion on a grate usually results in low flame height but volatiles in coal can produce streams of unburned hydrocarbons that are difficult to completely burn. By burning wood with coal the long volatile wood flame can help burnout products of incomplete combustion from coal.

High volatile matter causes challenges for burning biomass cleanly on fixed or travelling grates and in bubbling fluidized beds because it is difficult to mix the evolving combustible gases with sufficient air, and at a high enough temperature, to ignite and completely burn the carbon monoxide and hydrocarbons. Bubbling fluidized bed boilers have the advantage of delivering a uniform flow of combustion air up through the bed but if the fuel is not distributed evenly, then incomplete combustion occurs. A circulating fluidized bed boiler provides even better control of fuel and air resulting in higher combustion efficiency and lower emissions of unburned carbon.

14.2.1.4 Particle Size Biomass fuels can derive from either woody or non-woody sources. When wood is chipped or milled, it breaks into a wide range of particle sizes. Wood chips are usually thick particles (>3 mm, > 0.125 in.). Most wood particles take longer to burn than pulverized coal. Non-woody biomass such as straw can be thin (>0.25 mm or >0.01 in.) when it is milled. Thin particles volatilize rapidly so they burn easily in suspension with pulverized coal.

Biomass boilers are designed to provide long residence times to allow thick biomass particles sufficient time to burn. Boilers with fixed or travelling grates, and fluidized beds allow thick particles time to burn completely. When burned in suspension biomass is often dried and sized to ensure complete particle burnout. A 1-mm-thick wood particle takes approximately 3 seconds to burn:

one second to dry; one second to devolatilize; and one second for char burnout [6]. Pulverized coal (PC) boilers often have less than 2.5 seconds residence time available for complete particle burnout because coal is crushed and milled in a roller mill to sizes of 70 to 90 μm. Biomass particles are sized to ensure complete combustion when cofired with coal. Since wood is not as brittle as most coals, it takes more energy, and many more operations, to mill biomass to this fine particle size than coal. One commercial plant in Denmark has been pulverizing biomass separately for cofiring. Biomass is milled to less than ¼ in (6 mm) and compressed to high-density pellets. Pellets are then pulverized in a roller mill that was modified for biomass [7].

Processing biomass to a uniform dense format for thermal or chemical conversion is the object of current research funded by the U.S. Department of Energy [8]. Their goal is to reduce moisture from 50% MC to 10% MC and to increase bulk density from as low as 5 lb/ft^3 (80 kg/m^3) to 15 lb/ft^3 (240 kg/m^3) at an affordable cost. The final product will be similar to densified biomass pellets that have been pulverized. This research may result in a form of biomass that is more efficient to convert to heat and power.

14.2.1.5 Volatile Ash Biomass contains less ash than coal but it contains volatile components that can cause slagging, deposits, and corrosion in boilers. Clean wood that is used to make residential wood pellets contains less than 0.5% ash. Bark from wood processing residues contains 3% to 5% ash. Urban wood residues (UWW), which is composed of mixtures of clean wood and contaminants, contains 5% to 15% ash. Herbaceous energy crops and residues often contain 5% to 8% ash. The range 5% to 15% may be similar to coal but the composition of the biomass ash causes it to behave differently in combustion. The switchgrass in Tables 14.1 to 14.2 is a native prairie grass that has been considered for production as an energy crop. When heated to combustion temperatures, some of the inorganic components, such as potassium or sodium, phosphorous, sulfur, and chlorine, melt or vaporize [9, 10]. Potassium and sodium react with chlorine and sulfur to form salts that melt at low temperatures, or condense in cold parts of boilers to promote corrosion. The concentrations of these compounds are important when designing a biomass boiler or gasifier to prevent slagging and deposition on heat transfer surfaces. The transformations of mineral matter in biomass during combustion and cooling in a boiler are shown in Fig. 14.1.

Table 14.2 shows that wood and switchgrass have much higher concentrations of these alkali metals than coal. Alkali metals in coal are more tightly bound and do not usually melt or vaporize except at high temperatures [11]. Rice straw and animal manures are biomass fuels that can cause severe slagging, fouling, and corrosion. They contain sufficient alkali and silica to cause a eutectic, or low melting temperature, of about 1450°F (750°C) which is below combustion temperatures in a typical boiler. The partially molten ash can block fuel on grates, agglomerate fluidized bed media, erode boiler tubes, and deposit on boiler tubes which reduces heat transfer, blocks gas flow, and promotes corrosion.

An awareness of the behavior of these compounds in combustion has led to more efficient boiler designs which offset the effect of the volatile inorganic compounds. The choice of a combustor can reduce the effect of volatile alkali on boiler performance [12]. Corrosive elements like chlorine can be separated and volatile elements can be sequestered in the ash by controlling temperatures in the fuel bed [13]. While additives like calcium can be used to offset the effect of low melting fuels, it is an additional cost and operating expense for a boiler owner.

TABLE 14.1 COMPOSITION AND FUEL VALUES OF BIOMASS FUELS COMPARED WITH COAL

As received		Coal	Switchgrass	Wood
Moisture	%	33.52	6.34	16.67
Ash	%	5.51	6.34	4.62
Volatile	%	28.98	73.84	66.02
Fixed carbon	%	32.00	14.48	12.69
Btu/lb (HHV)	Btu/lb	7,774	7,458	6,967
Dry basis				
Ash	%	8.24	5.70	5.54
Volatile	%	43.60	78.84	79.23
Fixed carbon	%	48.16	15.46	15.23
Btu/lb (HHV)	Btu/lb	11,696	7,965	8,361
MAF (HHV)	Btu/lb	12,746	8,446	8,851

Source: Powder River Basin Coal and Iowa Switchgrass — Amos, 2002 [2], Urban Wood Waste — Miles et al [3]

TABLE 14.2 ULTIMATE ANALYSIS OF BIOMASS FUELS COMPARED WITH COAL

Ultimate (dry basis)		Coal	Switchgrass	Wood
C	%	67.98	48.41	48.77
H	%	4.48	4.80	5.76
N	%	1.16	0.22	0.27
S	%	0.55	0.12	0.07
Ash	%	8.24	5.70	5.54
O_2	%	17.58	40.16	39.59
Cl	%	0.02	0.14	0.06
Na_2O	%	0.067	0.003	0.177
K_2O	%	0.004	0.809	0.265
Alkali	lb/MMBtu	0.06	1.02	0.53
Ash	lb/MMBtu	7.05	7.16	6.63
SO_2	lb/MMBtu	1.18	0.38	0.21

Source: Powder River Basin Coal and Iowa Switchgrass — Amos, 2002 [2], Urban Wood Waste — Miles et al [3]

14.2.2 Biomass Types, Sources, and Supplies

Biomass is an abundant fuel in portions of North America. Recent studies in the U.S. have identified more than 1 billion tons of biomass available for conversion to heat, power, or liquid fuels [14]. These studies have identified 368 million dry tons per year of forest resources and 998 million tons per year of agricultural residues for a total resource potential of 1,366 million dry tons per year. This compares with about 1 billion tons of coal that are mined each year and primarily used for power. While coal is supplied to power plants primarily from Wyoming, biomass is distributed throughout the country.

In 2009, biomass, including biofuels, waste (landfill gas, municipal solid waste, biosolids), wood, and wood derived fuels, accounted for 3.884 quadrillion Btu, or 50% of the 7.745 quadrillion Btu of renewable energy consumed in the U.S. [15]. Biomass provided 3.3% of the energy for heat, power, and transportation. Biofuel production increased to 1.546 quadrillion Btu in 2009 while wood and wood derived fuels decreased to 1.891 quadrillion Btu.

14.2.2.1 Woody Biomass Consumption for fuel in the U.S. represents about 118 million dry tons of wood per year. According to the Energy Information Agency (EIA) about 9% (173 trillion Btu) of the 1.891 quadrillion Btu of wood energy consumed in 2009 was used for power generation and the rest was used for domestic heat [15].

Until the 1970s, biomass fuels used for heat and power were primarily wood residues from sawmills. Wood residues continue to be important sources of fuels. Following the first oil shortage, other sources of wood fuels were developed, such as urban wood waste (UWW). UWW became an important source of fuel when the expansion of independent power plants in California and in the Northeastern states overtook wood industry residues. UWW includes landscape prunings, pallets, packaging and construction, and demolition debris. Wood recovered and diverted from

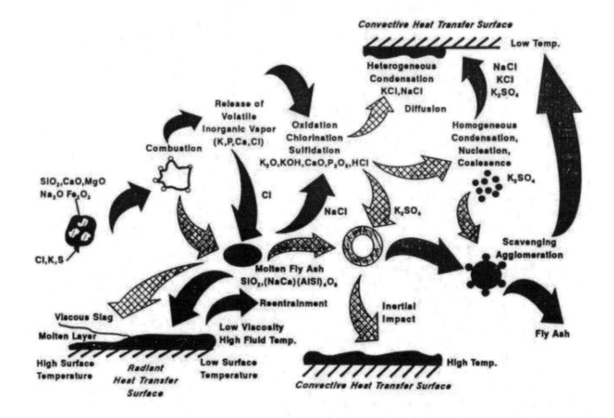

Transformation of Mineral Matter in Biomass

FIG. 14.1 TRANSFORMATIONS MINERAL MATTER IN BIOMASS (*Source*: Miles et al, Alkali Deposits Found in Biomass Power Plants: A Preliminary Investigation of Their Extent and Nature, 1996 [3])

landfills is milled to size at intermediate sites and delivered to power boilers. Regulators ensure that contaminants, such as heavy metals, do not enter the fuel or the waste from these power plants. Industrial consumers and independent biomass power plants pay up to $4/MMbtu ($60/dry ton) equal to about $60/ton of PRB coal.

Forest residues that should be removed to prevent forest fires are expected to be a second large source of woody biomass. This wood accumulates through re-growth or disease, such as infestation by the mountain pine beetle in the Rocky Mountains. While contracts to remove this wood often provide for residues to be mulched and redistributed on site some public agencies provide transportation incentives for contractors to haul forest residues for use as soil amendment, fuel, or fiber. Wood residues are usually sized by chippers and grinders in the field to a maximum length of 3 or 4 in. (10 cm). These fuels are usually available at a delivered cost of $2 to $6/MMBtu ($5 per cubic yard or $60/ton).

Orchard pruning and other woody agricultural residues are typically processed by the same contractors who process wood residues and are similar in delivered form and cost. Annual prunings are usually mulched. Orchard replacement, or tear-outs, generates large volumes of wood. Orchard prunings have become an important fuel for biomass power plants in California.

14.2.2.2 Wood Pellets Wood is the primary residential biomass heating fuel. EIA reports that 430 trillion btu of wood, equal to about 27 million tons of dry wood, was consumed for heating in 2009 [15]. Approximately 1.2 million tons of clean wood residues are densified to wood pellets for residential fuel [16]. Additional wood pellet capacity is under construction to reach a growing export market.

Wood pellets are characterized by their low moisture (<10% MC), and high density (40 lb/ft^3, 640 kg/m^3). Residential consumers demand low (<0.5%) ash content. Bagged fuels at the pellet plant command prices of about $10/MMBtu ($150/ton). Pellets are sold at retail outlets for $12 to 20/MMBtu ($200 to $300/ton). Clean wood residues for residential wood pellets are becoming scarce in many parts of the U.S. The Pellet Fuels Institute has instituted a "standard" grade of pellets for commercial use which allows a higher ash content. Pellets containing agricultural residues are made by a few companies but they find it difficult to sell them to existing power plants [17].

Global wood pellet production has reached 13 million tons. Consumption has grown to more than 8 million tons per year in Europe where they are used extensively for cofiring with coal [18]. In countries, like Sweden, domestic wood pellet heaters and boilers can compete with fossil fuels. Bulk pellets are sold for commercial and industrial use. Some U.S. and Canadian mills supply wood pellets to Europe for cofiring but low-cost oil and natural gas have limited the widespread use of commercial or industrial wood pellets in the U.S.

14.2.2.3 Urban Residues Municipal waste contributed 18 billion kilowatt hours (kWh) or one-third of the 54.3 billion kWh of net electrical generation from biomass energy in the U.S. in 2009 [15]. Landfill gas contributed 7.3 billion kWh or 40% of the total from waste. Wood and wood derived fuels such as black liquor contributed 36.3 billion kWh or two-thirds of net electrical generation from biomass. Increased recycling and composting and the difficulty of obtaining sites and permits have limited the growth of municipal waste-to-energy plants [19].

Biosolids from municipal wastewater treatment are often suggested for heat and power generation because disposal is a problem in the highly populated areas of the Eastern U.S. Heavy metals, mercury, and other contaminants pose practical, operational, and legal constraints to the use of biosolids in industry. Thermal conversion of biosolids is usually limited to municipalities. Many conversion and pollution control issues have not yet been resolved [20].

14.2.2.4 Agricultural Residues The current use of residues from annual crops, like wheat straw and corn stover, is limited to small quantities for bedding and mulch. They are disperse and not harvested in large concentrations. Few areas in the world use straws or residues for energy or fiber in quantities of 200,000 tons per year (tpy) or more. Denmark consumes more than 1 million tpy wheat straw for heat and power. Switchgrass, miscanthus, and other herbaceous energy crops are under development as potential energy crops but so far they are not in commercial production. Danish technology was incorporated in a cofiring demonstration of switchgrass in Iowa in 2006. In that project, it required several years to establish the crop, develop cultivation and harvesting techniques, and to organize businesses to supply 200,000 tpy to generate 37 MWe from biomass [2, 21]. Power production and cellulose to liquid fuel projects that intend to use large quantities of corn stover and wheat straw residues are still in development and negotiating contracts with potential suppliers.

Animal manures are often suggested as potential biomass fuels. Feed and bedding are transported to concentrated animal feed operations (CAFO). Wastes from CAFOs are often concentrated in areas that have low energy demand. Their use must be carried out in coordination with growers. They are still used as fertilizers and soil amendments. They usually cost about $10/ton or about $0.75/MMBtu. Litter and bedding from poultry and hog operations contain partially digested feed which carry high concentrations of nutrients which make them costly to convert to heat and power due to slagging, deposits, corrosion, and pollution control [22].

14.2.3 Infrastructure and Logistics of Biomass Supply

The infrastructure for delivering large quantities of woody biomass has been well developed by the pulp, paper, and wood product industries. Systems for harvesting, handling, processing, storing, and feeding woody biomass to new uses are similar to equipment already in industrial use [23]. Even a small, 5-MWe power plant, consuming 70,000 tpy, depends on several small contractors. New equipment like mobile grinders and screens help individual contractors supply 30,000 tons or more per year. Large consumers sometimes develop their own fuel supply companies. A 50-MWe facility may have more than 100 suppliers.

A few companies specialize in supplying both fiber and fuel. One company in the South supplies more than 14 million tons per year of chips and fuel to pulp mills, engineered wood products plants, pellet mills, and biomass power plants. Companies that handle residues in related industries, such as composting or soil amendments, use their fleets of trucks to supply plants with biomass.

Equipment and infrastructure to supply crop residues and herbaceous crops as fuels is less developed. Most equipment for harvesting, handling, baling, and hauling agricultural residues is designed for limited annual use of about 400 hours and less than 10,000 tons per year. Manufacturers are starting to develop advanced systems to supply large-scale biomass conversion facilities. Since no large biomass facilities have been built, this development is still funded by equipment used in feed and forage crops. Some companies that are experienced in the large volume export of hay and straw are positioning themselves to supply large power stations, and cellulose to liquid fuel plants.

Most of the existing harvesting and handling equipment for field crops and residues is designed for materials that contain less than 15% MC. Equipment must still be developed for harvesting, baling, storing, and reclaiming large volumes of wet (35% MC) herbaceous feedstocks, like sweet sorghum or wet corn stover.

14.2.4 Biomass Sustainability and the Environment

Development of biomass energy has been paralleled by an increase in the forest resource and a reduction in timber harvesting and processing. A large portion of wood fiber is now imported. The global trade in wood, wood products, and wood fiber, including solid and liquid biomass fuels, has raised public concern about the environmental impacts of increased biomass use for energy. Contracts from public agencies frequently require analyses of environmental impacts and sustainability. U.S. Department of Energy, U.S. Forest Service and other agencies are increasingly investigating impacts of biomass use. Life cycle analyses are employed to evaluate the potential for carbon capture and sequestration or storage (CCS or CSS) and the net cost of energy projects [24, 25].

14.3 HEAT AND POWER GENERATION

14.3.1 Industry

Biomass has been used for heat and power by the wood industry for many years. Systems are designed according to industry standard practices [26, 27]. Fuels have changed from the simple use of wood industry residues to fuels from many sources. Paper companies are now finding new and efficient ways to convert residues and waste lignin, in the form of black liquor, to heat, power and synthesis gas via gasification. Pulp mills are investigating the potential of becoming biorefineries to produce pulp heat, power, and liquid fuels as co-products.

Independent power producers and utilities have usually built wood burning facilities where there are financial incentives and long-term power purchase agreements. Early small-scale (<5 MWe, <200 tpd) biomass powerplants were built by large steam consumers with abundant wood wastes. Today these small systems are primarily in demonstration projects where capital costs are paid with public funds. A small number of companies supply this market with boilers and specialized wood chip handling systems.

Large biomass plants consume1000 tpd and generate 20-30 MWe. A few biomass plants generate 50-80 MWe. These plants are small compared with coal fired power plants but they are important to local economies since they process from 200,000 to 400,000 tpy biomass. Utilities, developers, and investors wanting to engage in biomass energy have many resources today. During the last 20 years, many engineering firms have designed and built biomass energy systems. Equipment is supplied by traditional vendors that supply the wood products industry and coal utilities. Private companies have taken over the business of promoting biomass energy projects in public conferences and publications. All states have incentives for biomass power through waste reduction and recycling. Local, state, and federal agencies regulate biomass installations. Air, water, solid waste, and building permits are usually the first contact with public agencies.

14.3.2 Combustion

14.3.2.1 Small-ScaleSystems Small-scalebiomasscombustion equipment includes direct fired heaters for dryers, small furnaces

and boilers, small chip, and pellet furnaces. Small scale is often defined by air quality regulations. Heating systems were considered small up to 25 MMBtuh (7.4 MW, 26.4 GJ). Proposed new regulations for boilers have reduced that level to10 MMBtuh (3 MW, 10.55 GJ) [28]. Above 10 MMBtuh new boilers will be required to employ more efficient pollution control devices such as ceramic filters or electrostatic precipitators to control fine particulates (e.g., less than 2.5 microns). Some states have special air quality regulations starting at 2.5 MMBtuh (2.64 GJ).

14.3.2.2 Domestic Heat Domestic heating appliances, such as cord wood and chip fired furnaces, include indoor and outdoor boilers that are regulated at the state and federal level. Manufacturers of these appliances must submit their equipment for certification by approved laboratories. Acceptable emissions levels are then established by either federal (e.g., EPA) regulations or state rules. Similar air quality regulations exist in Canada and Europe. Emissions from domestic boilers has improved in recent years. Future improvements may be guided by improved control over ash constituents during combustion. New devices are designed to promote incorporation of volatile ash components in the grate ash rather than being emitted from the stack. It is estimated that emissions in Europe could be improved 75% by replacing old equipment [29].

Many European wood burning appliances are now sold in North America. These include hot water and steam boilers. These systems are usually designed for clean fuels with little provision for ash removal. Wood pellet boilers are now installed in hospitals, schools, and small commercial applications. Since emissions are minimal, there is often no gas cleaning equipment but some industrial gas filtration devices, like ceramic filters, are now adapted to domestic appliances.

There are about ten times as many cordwood heating appliances in use in the U. S. as there are pellet appliances. Cordwood furnaces usually consist of a firebox with different arrangements for under fire and over fire air to control combustion and emissions. The largest of these are outdoor wood boilers (OWB) and air heaters for shops and greenhouses. Use of these appliances depends on local supply, demand, and regulations. Manufacturers are often members of the Hearth Products Association [30] or the Pellet Fuels Institute [16].

14.3.2.3 District Energy North American homes are disperse and there are few incentives for developing infrastructure for district heating networks so the U.S. has not used biomass for district heat to the extent that is found in Europe. Interest has increased as building, business, and academic campus arrangements seek environmental and economic benefits with public assistance.

14.3.2.4 Small Biomass Boilers The wood boiler in a sawmill traditionally powered a steam engine. When industry changed to oil, gas, and electricity these boilers were abandoned except to generate low-pressure steam or hot water for lumber dry kilns. Small sawmills are now supplied by companies that make small dry kilns, small industrial boilers, and small outdoor wood boilers. Small wood chip and pellet boilers have become increasingly popular for public schools and hospitals.

Small boilers use simple grates. Chipped or ground wood fuel is pushed onto the grate by an auger or ram (Fig. 14.2).

These boilers usually have fixed grates although some moving grates are now available. The grate usually has slots or pinholes to provide under grate or primary combustion air. In the case of staged combustors or gasifiers, the combustible gases are burned

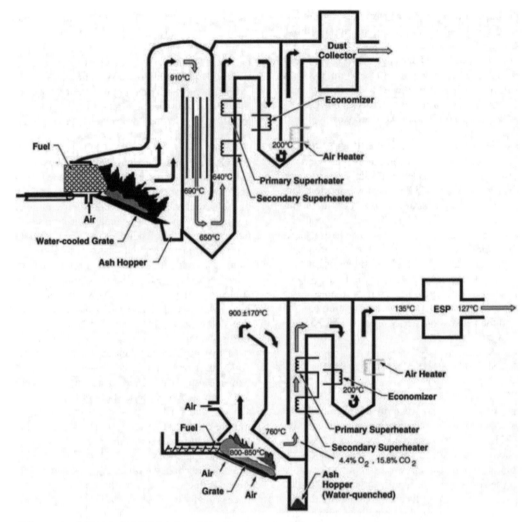

FIG. 14.2 SMALL RECIPROCATING GRATE BOILERS IN DENMARK (*Source*: Miles et al [3])

in a firetube boiler. Fuel requirements are 300 to 1100 lb/hour (130 to 500 kg/hour) or up to about a truckload per day. Ash is removed manually or by an auger that intermittently collects ash at the end of the grate and conveys it to an isolated pot. Vendors prefer not to develop travelling grates for continuous ash removal because of added cost in a competitive market. Increased use of urban wood and agricultural residues may cause suppliers to develop grates that will burn higher ash fuels.

Companies that specialize in small wood heating systems usually provide the fuel handling system (including a receiving bin and metering equipment), and the firebox, boiler, gas cleaning equipment, exhaust stack, and controls. They design and fabricate the fuel feeding and firebox. They usually outsource the firetube boiler, gas cleaning, fans, and stack. They integrate the system and controls and commission the installation. When a small boiler is installed in a separate building, the capital cost is usually comprised of one-third for the boiler and fuel equipment, one-third for the heat distribution or integration to an existing heating system, and one-third for the building and civil construction. It is more economic to install a small boiler in an existing building.

Particulate removal in small boilers is most often by cyclone or multi-cyclone before exhausting gases to a stack. Local regulations sometimes require baghouse filters or electrostatic precipitators for gas cleaning. Ceramic fiber filters can also be used. Proposed federal regulations may require more stringent gas and particulate emissions which may add 20% to 30% to the cost of a small boiler system.

Future developments in small boilers are likely to include improved combustion and emissions and designs for low-quality fuels. Small boilers for straws, corn stalks, and agricultural residues are used in Europe but have not been used in the U.S. Improvements in gas cleaning include electrostatic precipitators adapted from large biomass boilers. Emissions of nitrogen oxides (NOx), volatile organic carbons (VOCs), and other hazardous air pollutants (HAPS) for small boilers are usually controlled through best management practices (BMP). BMPs are issued through local environmental agencies.

14.3.2.5 Small-Scale Power Generation There is interest in small-scale (e.g., 25 to 250 kWe) power generation in locations like Alaska or Hawaii where off-grid power is expensive. Small steam turbines are inefficient and require special operators so they are often integrated with larger industrial processes. There have been many demonstrations for generating power at the small scale but few proven systems. Steam engines, Stirling engines, Organic Rankin Cycle (ORC) systems, and small-scale gasifiers have been demonstrated but are not in commercial use due to high cost. Biomass gasification with solid oxide fuel cells (SOFC) have been investigated [31]. Partial combustion of biomass to make a char-coal (biochar) co-product with the waste heat used to generate heat or power has been proposed but no commercial systems are in operation.

14.3.2.6 Small Industrial Boilers Direct fired wood burners are used to heat wood dry kilns, and wood dryers for pellets and shavings. These are either pile burners, staged combustors, cyclonic suspension burners, or register-style suspension burners. Suspension burners require clean dry fuel that has been milled to <0.250 in. (<6 mm). They are used where products of combustion do not contaminate the dried product. These burners are best used where the fuel is generated on site because of the very low density of fine dry fuels. Pollution control for these burners often depends on whether the heat is used in a boiler or dryer.

Small industrial boilers are similar to small institutional boilers with the addition of travelling grates for continuous ash removal. They typically provide process steam to food and fiber industries at rates of up to 20 MMBtuh or 600 HP (5MW, 21.1 GJ). Fuel requirements are 1 to 2 dry tons per hour. Boilers are pile burning combustion cells, underfed stokers, spreader stokers with travelling grates, and gasifiers. Wood boiler suppliers usually also supply boilers for coal, oil, and gas. Boiler designs often include improved combustion air control. While gasifiers and fluidized bed combustors have generally not been successful at the small scale, some new suppliers are developing efficient systems that comply with the more stringent emissions regulations. Installations often include the fuel preparation and handling equipment, furnace, boiler, multi-cyclone, baghouse, or electrostatic precipitator, induced draft fans, and stack. Some vendors also supply pollution control equipment. Small wet scrubbers may also be used if the fuel has the potential to emit acid gases from chlorine or sulfur. New regulations will require improved control of NOx and CO emissions.

Power is usually not generated in small-scale systems. Back pressure turbines are sometimes used to reduce pressure for large steady process heat consumers, such as lumber dry kilns. However, small turbines (500 KWe to 1,000 kWe) are expensive and require too much labor unless it is shared with other processes. It is also often difficult to optimize turbine operation with varying steam demand.

Small biomass power plants generate 25 MWe or less consuming up to 40 wet (20 dry) tons per hour and 1000 wet tons per day. Fuel sources can typically supply 200,000 to 300,000 tpy wet wood to power plants. No plants of this size in North America burn straw or crop residues. There are straw burning plants in Europe, notably in England and Denmark.

The U.S. has more than 11,353 MWe of installed biomass power capacity [15]. Many of these plants were installed in the 1980s when special power contracts were provided to utilities and independent power producers [32]. Most of those contracts have expired. In California, for example, there are 31 plants running, and 11 plants that are idle. The idle plants are in various states of disrepair. Many need major investments to restart. The total operating capacity is about 610 MW, and the idle capacity is about 122 MW [33]. New biomass plants are being built and more are expected as utilities are required to increase renewable energy.

14.3.2.7 Spreader Stoker Boilers Spreader stoker boilers with moving grates have been the preferred combustion technology for woody biomass in North America (Fig. 14.3). These include travelling grates and vibrating grates that are water or air cooled. Travelling and reciprocating grates must be designed so that ash loads are not excessive. Otherwise, fuels with ash that melt at low temperatures can cause clinkering. Large facilities that burn biomass with high ash, such as the 55-MWe boiler in Minnesota that burns poultry litter, use water cooled vibrating grates [34]. Combustion is staged in a tall furnace so that fuel gasifies on the grate and volatile gases are burned above the grate. Alkali compounds condense and deposit on the water walls where they are removed with soot blowers. Superheater inlet temperatures are limited and tube spacing in convection systems is adjusted to ease cleaning and limit corrosion. Similar boilers are used

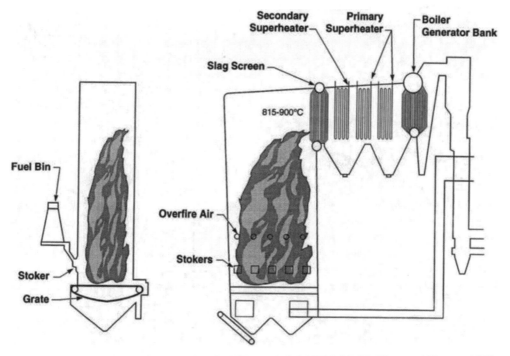

FIG. 14.3   A SPREADER STOKER TRAVELLING GRATE BOILER (*Source*: Miles et al [3])

for burning straw in Spain, Denmark, and England. These challenging fuels have stimulated major advances in moving grate boilers resulting in efficient air mixing for complete combustion, reduction of hazardous air pollutants, and improved particulate collection [35].

14.3.2.8 Fluidized Bed Boilers Fluidized bed boilers were introduced to wood, refuse, and bio-solids combustion in the U.S. in the 1970s by domestic and European companies (Fig. 14.4) Fluidized bed boilers afford uniform mixing of fuel and air for good control of combustion temperatures.

Low-peak combustion temperatures can be maintained which help prevent formation of nitrogen oxides (NOx) and melting of ash. Fluidized bed boilers are less sensitive to variations in particle size and moisture than travelling grates. Two plants in Europe burn poultry litter in bubbling fluidized bed boilers [36]. Alkali from these fuels can build up on the sand media in fluidized beds by forming a coating of alkali silicates which eventually causes

bed agglomeration [12]. A portion of the spent media is usually removed and refractory materials, such as calcium and magnesium oxides, are added to the bed to prevent agglomeration. When fuels such as urban wood waste that contain sand and calcium are burned, the media is often screened and re-circulated so they can operate for several years without completely replacing the bed media. Fluidized bed boilers have generally not been used for straws and agricultural residues.

Major components for large biomass plants, such as boilers and turbine systems, are supplied by several international companies. A typical project will attract suppliers from Europe and North America.

14.4 COFIRING BIOMASS WITH COAL

Combined firing with coal has long been seen as a good method to convert biomass to power. It has the highest energy yield and

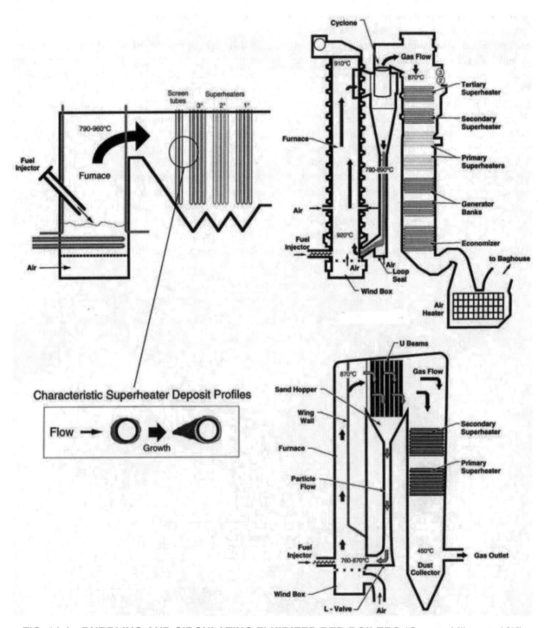

FIG. 14.4 BUBBLING AND CIRCULATING FLUIDIZED BED BOILERS (*Source*: Miles et al [3])

the greatest potential for carbon substitution and sequestration. It represents a near-term, low-risk, low-cost, and sustainable energy strategy. Cofiring can take advantage of the higher efficiency and heat rate of coal compared with wood. When cofiring up to 7% biomass the heat rate of a coal facility can be 10,000 to 12,000 Btu/kWh compared with 14,000 Btu/kWh for a wood plant [37]. Cofiring can take advantage of the large installed coal infrastructure that consumes more than 1 billion tons of coal per year. Cofiring combined with carbon sequestration at coal plants would make a significant impact on carbon emissions. Biomass-coal cofiring reduces CO_2 and SO_2 emissions and it may reduce NOx emissions. It is the most effective measure to reduce CO_2 emissions because it substitutes coal, which has the most intensive CO_2 emissions per kWh electricity production, with biomass, which has zero net CO_2 emissions [38].

Cofiring has been the subject of extensive research and development in Europe and North America [39–42]. It has been stimulated by the requirement for utilities in Europe to comply with obligations under the United Nations Framework Convention on Climate Change (UNFCCC). The International Energy Agency (IEA) Task 32 on Combustion and Cofiring reports more than 30 boilers that cofire with biomass [43]. Increased cofiring has caused millions of tons of biomass to be imported to Europe. Most of the biomass fuel has been pelletized for ease of use in the coal handling equipment.

Cofiring in the U.S. has been limited. In 2008, there were 66 plants with a biomass/coal cofiring capacity of 3,772 MWe. Total plant capacity was 6,147 MWe [44]. There have been many feasibility studies. New cofiring capacity is expected in areas like the Southeast where biomass supplies are abundant, where there is good potential for afforestation of range and forest lands, and where renewable portfolio standards (RPS) are becoming mandatory [45]. The cost of fuel and capital equipment have been the principal constraints for cofiring in the U.S. While legislation encourages the production of electricity with renewable fuels biomass fuels have been more expensive than alternatives.

A test cofiring coal with switchgrass in a PC boiler was conducted in 2006 [46]. A combination of public incentives was required to close the gap between the cost of coal and the cost of biomass. The delivered price of coal to the burner tip equaled $35/ton but the cost of biomass was $70/ton. While the cost of coal has increased, biomass is still expensive compared with coal.

It is often possible to fire only 5% biomass with coal in older stoker boilers due to the low density of biomass. Corn cobs, paper cubes, or wood from local sources have been added in small quantities on travelling grates to prevent agglomeration. The higher volatile flame cools the coal fire and reduces clinkering. Alkali in the biomass fuel reacts with the sulfur in the coal to reduce sulfur emissions which can also help prevent corrosion. In these applications cofiring is best suited to boilers that are operated at 60% of their rated capacity. Cofiring of biomass with coal can also be limited by the volume of a combustor since biomass requires more combustion volume and more water wall surface area than coal. Many older stokers are being retired and replaced with more efficient circulating fluidized bed boilers which can accept higher proportions of biomass with coal.

Circulating fluidized bed (CFB) boilers burn coal at the industrial and utility scales. Suppliers report that many new or retrofit facilities are being planned with systems that will allow biomass to be cofired when it becomes required or economic. In a few cases, CFBs have been modified to completely replace coal with biomass.

Biomass can be fed into the recycle loop of a CFB which is uniquely suited to staged combustion for temperature and emissions control. Recent studies show that when biomass is cofired up to 25% with coal in a fluidized bed the cofiring can have a beneficial effect on agglomeration and fouling. Alkali in biomass was found to react with the ash rather than the bed media and was removed rather than reacting with the chlorine and sulfur [18].

Biomass is often cofired in pulverized coal (PC) boilers. PC boilers are tangentially fired (corner-fired) or wall fired. Biomass is milled and injected into these boilers either by adding burners or by substituting biomass for coal in individual burners. Burners that are usually selected are low in the furnace so that the char particles have an opportunity to burn out before entering the convection sections of the boiler. A single burner or injector will consume 5 to 6 tph of biomass or approximately 90 MMBtuh.

Wall-fired boilers have an array of individual burners arranged on one wall or on opposing walls. Advanced combustion modeling has improved the design and efficiency of these burners to reduce emissions and increase efficiency.

Current research identifies density, moisture, and particle size as the principal limitations for biomass cofiring [47]. Methods to increase density, reduce moisture, and reduce particle size will significantly increase the potential for cofiring. Wood pellets have provided a suitable format for handling large volumes for automated, low-maintenance biomass firing. In Europe, wood pellets are added to the coal prior to milling. With modifications to the mill, up to 12% biomass can be added to coal. In the process of densification the biomass is sized, dried, resized, densified, and cooled. Wet biomass requires about 15% of the biomass raw material to be burned to provide heat for drying. Waste heat sources from the coal plant have been considered for drying wood fuels. Densification consumes about 150 kWh per ton of fuel. It can add $100 per ton to the cost of the biomass or about $6/MMBtu.

Wood has been the principal biomass fuel that has been cofired with coal. Straw cofiring has been limited to a few plants in Europe. One pellet plant in Denmark prepares 240,00 tpy straw pellets for cofiring with coal [7]. Two dedicated roller mills at one plant further reduce the pellets to a 90-μm size for firing. That plant can fire up to 100% straw.

Milling and direct firing biomass has been employed with sawdust, and with straw, in two locations in the U.S. The specific energy required for processing 1,000 lb straw bales in Iowa was 35 kWh per ton of biomass fired [48]. Another method of cofiring biomass with coal is to burn the high-alkali straw in a low-pressure boiler and superheat the steam in a wood or coal boiler that is less prone to fouling. This was demonstrated at a plant in Ensted, Denmark [38]. No new plants have been built using this technique.

Torrefaction has also been considered for cofiring. Torrefaction increases the energy density of biomass through drying and partial pyrolysis. Torrefaction reduces the overall energy required for densifying and burning biomass [49]. The first torrefaction plant will produce 60,000 metric tons per year for a utility in the Netherlands [50]. Delivered costs of torrefied fuel have not yet been validated by industrial use. Neither densification nor torrefaction are used for cofiring in the U.S.

Gasification is another method suitable for cofiring. Gasification is expected to be employed as a method of cofiring waste and high-alkali fuels where the contaminants can be removed with a hot gas filter before the producer gas is burned in the boiler. Removal of sulfur and other contaminates may be necessary to protect equipment for selective catalytic reduction (SCR) of nitrogen oxide emissions. Gasification of biomass for cofiring has been demonstrated at Lahti, Finland. No further plants have been built [51].

14.5 GASIFICATION

Gasification occurs early in combustion when fuels are heated and volatile solids convert to gases before combustion. Gasification emerged for industrial use in the late 1970s. At that time it, was considered as a means of retrofitting biomass fuels to existing wood boilers. Gasification technologies include fixed bed updraft, and fluidized bed or circulating fluidized bed reactors. Early systems were fixed bed updraft or downdraft gasifiers. They were based on the gasifiers that had been used to make methanol and ammonia from coal and biomass before these chemicals were made from coal or natural gas. Fluidized bed gasifiers were also developed. Commercial gasifiers and several demonstration facilities were built in Europe and North America. One, at Varnamo, Sweden, was an integrated combined cycle power plant [52]. Subsidies for these demonstrations ended between 1998 and 2000.

Gasification has not been applied to the extent that was once expected because it has had few advantages over direct combustion. Only a few industrial scale biomass gasifiers are in use in North America. Some gasifiers capable of converting wet fuels have been successful in direct heating applications. Gasification is now seen as a means of cofiring biomass with coal or gas, and a means of repowering district heating plants for efficient combined heat and power [53, 54]. Gasification could solve the principal problems associated with cofiring biomass: moisture, density, and particle size.

Companies installing gasification systems have focused on heat applications and reforming gas for use in engines and turbines. Turbine applications largely stopped in the 1990s and have not been actively pursued by industry in favor of direct combustion in boilers. While gasification has long been considered for small-scale power generation, there are few commercial situations where gasifiers can compete with industrial steam boilers and turbines. Installed costs for small-scale gasifiers with generators ($2,500 to $10,000/kWe) can be two to four times as much as comparable steam systems ($2,500 to $3,500/kWe). Even subsidized plants have not demonstrated commercial viability or long periods of availability. Nonetheless, several companies in Europe and North America continue to develop gasifiers for small-scale and industrial use.

The most suitable applications for gasification appear to be production heat and for synthesis gas, principally carbon monoxide and hydrogen, for conversion to ethanol, methanol, and other chemicals through thermal catalytic or biochemical processes [55].

14.6 TORREFACTION

Torrefaction has been promoted as a useful method for solving the challenges of biomass fuel. Biomass is heated to between 250°C and 300°C (482°F to 572°F) in the absence of air until the moisture and light volatiles are removed. Heating reduces the mechanical strength of the biomass so that less energy is required to densify or pulverize it for cofiring. No energy is gained in the process. Utilities consider torrified wood to be a good form of biomass for cofiring if it can be made economically.

Torrefaction processes are in development. Some industrial carbon producers that have attempted to make torrefied wood attest that it is difficult to control the torrefaction process. Above 285°C (545°F) the process can become exothermic and it is difficult to control the reactor temperature.

The first industrial torrefaction plant is being installed by RWE and Topell in the Netherlands [50]. At 60,000 metric tons per year, it is intended to demonstrate the economic viability of converting wood to carbonized wood for industrial use in coal fired boilers. The dried, torrified, and densified wood is expected to be sold at pelletized wood prices. Perceived benefits are increased energy density and reduced transportation costs. The best location for torrified biomass processing may be at a coal-fired facility where the product can be fired directly. The torrefaction plant would take advantage of the shared utilities and services in order to put biomass into a more convenient form while reducing the cost of processing.

14.7 PYROLYSIS AND CARBONIZATION

Heating wood to convert volatile and fixed carbon to vapors, oil, and gas has long been a goal for public and private development. Pyrolysis oil can be used to generate power in combustion turbines or engines. While there has been extensive development to produce pyrolysis oils, there are apparently no successful commercial pyrolysis plants for heat and power generation. Commercial developers of pyrolysis technologies point to several limits to commercialization including bio-oil handling, storage, utilization, and upgrading technologies [56].

Pyrolysis or gasification can also produce carbonized biomass as a co-product. Pyrolysis of three tons of dry biomass can produce approximately 16 MMBtu (16.9 GJ) of energy, that can be directly fired or cofired with coal, and one ton of charcoal. The charcoal, or biochar, can be applied as a soil amendment to improve plant growth, restore soil fertility, capture nutrients, and sequester carbon [57]. When applied to soils with low organic matter, the biochar can stimulate the growth of additional biomass, thereby increasing the sequestration of carbon in the soil. While markets are still in development, the production of biochar as a co-product can improve the feasibility of a gasification or pyrolysis process. The permanent sequestration of the carbon is attractive to utilities that have considered combined energy and biochar for disposal of prunings from power line right of ways.

As with cofiring or liquid fuels, policy decisions will be required to make co-production of energy and biochar feasible. Carbon currently has a low value in the U.S. market. Technology improvements are also needed. Current technology development is limited by a clear understanding of process conditions. Technologies must be specific to the feedstocks and location. Commercial producers need better knowledge of the science of pyrolysis, better technologies to refine the bio-oil, and experience with developing projects efficiently [58].

14.8 BIOFUELS

Biofuels include ethanol [59] and biodiesel, bio-oils, Fischer–Tropsch fuels [60], and products of liquefaction [61]. The biofuels industry that has been developed on ethanol and biodiesel from grains accounts for an increasing share of biomass energy produced in the U.S. [15]. Alcohols and derivatives produced through thermal and biological processes are expected to play a critical role for sustainable transportation fuels in the future [62]. These include biofuels from cellulosic ethanol, and advanced biofuels from ligno-cellulosic feedstocks, and algae. Several technologies and pathways are being explored including production of intermediate products for processing in existing refineries. Many challenges remain in the synthesis of these products [61, 63]. Biofuels may become the highest value products of biomass renewable energy [64]. While transportation markets may be able to afford biofuels, power generation is not likely to be a major consumer unless they can be delivered at competitive prices.

The pulp and paper industry can add biofuels to their current production of fiber and combined heat and power. Although the paper industry has been in decline in the U.S., it has the potential for being a suitable platform for biorefineries. Experts point out that the existing industry has several advantages [65].

- It is among the top10 manufacturers in 42 states.
- It generates nearly 70% of its own energy needs.
- It generates over 80% of all biomass power generated in the U.S. through combined heat and power.
- It maintains high operating rates with low cost typically 24/7.
- It is skilled at running and maintaining large integrated facilities.
- It handles more biomass than any other industry.
- It consistently improves its safety record and environmental footprint.
- It utilizes the most intensely recycled material in our society.

Biorefinery processes have been demonstrated at the pilot and industrial scale. The industry has made fuels through biological and thermal processes from biomass and black liquor. Processing options include adding a biorefinery, preprocessing fiber to extract hemicelluloses (termed Value Prior to Pulping), new value from spent pulping liquors, and biodiesel from tall oil. The typical plant capacity of 1,000 tpd would be similar to a 30-MWe boiler. Incremental investments for biorefining would cost from $90 to $250 million.

Studies estimate that the pulp and paper industry has the potential and infrastructure to support biorefineries for liquid fuels. The fitting of 90 biorefineries to existing pulp and paper mills would produce 17 billion gpy of oil or 44% of the current USDOE Targets. Large capital support will be required to start the industry at current oil prices of $3/gallon. The industry would need prices of $100/barrel for most options.

14.9   FUTURE DEVELOPMENTS

Future developments in biomass energy depend on policy decisions regarding the use of biomass resources and the development of biomass as an alternative fuel source. European countries have used income from oil and gas to subsidize the development of biomass technologies for combined heat and power. U.S. constituencies have passed legislation encouraging biomass use but private utilities will only build biomass power generation when they can obtain power purchase agreements that will allow them to finance new capacity. The lack of energy and carbon policies would suggest that biomass may remain a low priority and even coal CSS technologies will not be implemented until we have experienced significant climate change.

14.10   CONCLUSION

Abundant biomass resources are available for conversion to energy. as they become feasible. Continuing advances in conversion technologies make biomass technically feasible for power generation, conversion to transportation fuels, and for carbon sequestration. Cofiring, combustion, torrefaction, pyrolysis, gasification, and biochemical conversions are all suitable options. Biomass conversion can be combined with carbon strategies such as biochar and afforestation to increase carbon sequestration. Continued growth and development of biomass, development of improved resource management, infrastructure, and conversion technologies will depend on energy and carbon markets and public policy decisions.

14.11   ACRONYMS

BMP	Best Management Practices
Btuh	British thermal units per hour
CAFO	Concentrated Animal Feed Operation
CEC	California Energy Commission
CFB	Circulating Fluidized Bed
CO	Carbon Monoxide
CCS	Carbon Capture and Storage
CSS	Carbon Capture and Sequestration or storage
DOE	U.S. Department of Energy
EIA	Energy Information Agency, U.S. Department of Energy
EPA	Environmental Protection Agency
GPY	Gallons Per Year
HAPS	Hazardous Air Pollutants
HHV	Higher Heating Value, Btu/lb
IEA	International Energy Agency
kWh	Kilowatt Hours
LHV	Lower Heating Value, Btu.lb
MMbtuh	Million Btu per hour
Mt	Metric Mon
MW	Megawatt, Thermal
MWe	Megawatt, Electric
NOX	Nitrogen Oxides
NREL	National Renewable Energy Laboratory
ORC	Organic Rankin Cycle
OWB	Outdoor Wood Boiler
PC	Pulverized Coal
PPH	Pounds Per Hour
RPS	Renewable Portfolio Standard
SCR	Selective Catalytic Reduction
SOFC	Solid Oxide Fuel Cell
Tph	Short Tons Per Hour
Tpy	Short Tons Per Year
UNFCCC	United Nations Framework Convention on Climate Change
UWW	Urban Wood Waste
VOC	Volatile Organic Sompounds

14.12   REFERENCES

1. Jenkins BM, et al. Combustion Properties of Biomass. *Fuel Processing Technology* 1998;54:29.

2. Amos W. Summary of Chariton Valley Switchgrass Co-Fire Testing at the Ottumwa Generating Station in Chillicothe, Iowa. Golden, CO: National Renewable Energy Laboratory; 2002, p. 40.

3. Miles TR, et al. Alkali Deposits Found in Biomass Power Plants: A Preliminary Investigation of Their Extent and Nature. Golden, CO: National Renewable Energy Laboratory: 1996; p. 120.

4. EBI Association and BS Institute. Quality Standards for Solid Biomass CEN/TC 355. In: *Biomass Stanards*. British Standards Institute, 2010.

5. Alakangas E, Valtanen J, Levlin JE. CEN technical specification for solid biofuels--Fuel specification and classes. *Biomass and Bioenergy* 2006;30:908–914.

6. Bharadwaj A, Baxter LL, Robinson AL. Effects of Intraparticle Heat and Mass Transfer on Biomass Devolatilization: Experimental Results and Model Predictions. *Energy & Fuels* 2004;18:1021–1031.

7. Gjernes E. Fuel Flexibility at Amager Unit 1 Using Pulverized Fuels. Cologne, Germany: Power-Gen Europe Conference; 2006.

8. Hess JR, et al. Uniform-Format Solid Feedstock Supply System: A Commodity-Scale Design to Produce an Infrastructure-Compatible Bulk Solidfrom Lignocellulosic Biomass Executive Summary. INL/EXT-09-15423; TRN: US201004%%449; 2009.

9. Miles TR. Alkali Deposits in Biomass Power Plant Boilers. In: *Strategic Benefits of Biomass and Waste Fuels*. Washingon, D.C., USA: Electric Power Research Institute, 1993.

10. Knudsen JN, Jensen PA, Dam-Johansen K. Transformation and Release to the Gas Phase of Cl, K, and S during Combustion of Annual Biomass. *Energy & Fuels* 2004;18:1385–1399.

11. Baxter LL, et al. The Behavior of Inorganic Material in Biomass-fired Power Boilers: Field and Laboratory Experiences. *Fuel Processing Technology* 1998;54:47–78.

12. Zevenhoven-Onderwater M, et al. Bed Agglomeration Characteristics of Wood-Derived Fuels in FBC. *Energy & Fuels* 2006;20:818–824.

13. Obernberger I, Brunner T, Bärnthaler G. Chemical properties of solid biofuels—significance and impact. *Biomass and Bioenergy* 2006;30:973–982.

14. Perlack RD, et al. Biomass as Feedstock for a Bioenergy and Bioproducts Industry: The Technical Feasibility of a Billion-Ton Annual Supply. Oak Ridge, TN: Oak Ridge National Lab; 2005; p. 72.

15. Renewable Energy Consumption and Electricity: Preliminary Statistics 2009. Washington, DC: U.S. Energy Information Administration; 2010; p. 14.

16. Pellet Fuels Institute: Arlington, VA.

17. Obernberger I, Thek G. *The Pellet Handbook*. London: Earthscan Ltd, 2010; p. 549.

18. Pirraglia A., et al. Wood Pellets: An Expanding Market Opportunity. *Biomass Magazine* 2010; http://www.biomassmagazine.com/article.jsp?article_id=3853&q=&page=all.

19. Vehlow J, et al. (2007) Management of Solid Residues in Waste-to-Energy and Biomass Systems. *Bioenergy NoE;* www.bioenergynoe.org.

20. Elled AL, et al. The fate of trace elements in fluidised bed combustion of sewage sludge and wood. *Fuel* 2006;86:843–852.

21. Knudsen JN, et al. Corrosion and Deposit Investigations During Large Scale Co-combustion of Switchgrass at a Coal-fired Power Plant. Snowbird, UT: Impacts of Fuel Quality on Power Production Conference; 2006.

22. Miles TR, Bock BR. Energy and Nutrient Recovery from Combustion of Swine Solids and Turkey Litter. Research Triangle Park, NC: Animal Waste Management Symposium; 2005.

23. Webb, E., et al., A preliminary assessment of the state of harvest and collection technology for forest residues. Oak Ridge National Laboratory Report ORNL/TM-2007/195, 2008.

24. Plevin RJ, et al. Greenhouse Gas Emissions from Biofuels' Indirect Land Use Change Are Uncertain but May Be Much Greater than Previously Estimated. *Environmental Science & Technology* 2010.

25. Roberts KG, et al. Life Cycle Assessment of Biochar Systems: Estimating the Energetic, Economic, and Climate Change Potential. *Environmental Science & Technology* 2009;44:827–833.

26. Stutz SC, Kitto JB. *Steam Its Generation and Use*. 40 ed. Barberton, OH: Babcock & Wilcox Company, 1992.

27. Drbal LF, et al. *Power Plant Engineering*., Boston, MA: Kluwer Academic Publishers, 1996.

28. Kapur KB, International R. Regulatory Impact Analysis: National Emission Standards for Hazardous Air Pollutants for Industrial, Commercial, and Institutional Boilers and Process Heaters. U.S. Environmental Protection Agency, Office of Air Quality Planning and Standards (OAQPS), 2010.

29. Brunner TG, Barnthaler G, Obernberger I. Evaluation of Parameters Determining PM Emissions and their Chemical Composition in Modern Residential Biomass Heating Appliances. In: *International Conference World Bioenergy*. Stockholm, Sweden: Swedish Bioenergy Association; 2008.

30. Hearth Products Association, Arlington, VA.

31. Colpan CO, et al. Effect of gasification agent on the performance of solid oxide fuel cell and biomass gasification systems. *International Journal of Hydrogen Energy* 2010;35:5001–5009.

32. Turnbull JH. Use of biomass in electric power generation: the California experience. *Biomass and Bioenergy* 1993;4:75–84.

33. California Biomass Energy Alliance: Sacramento, CA.

34. Jackson P, Kelso P, Morrow R. Renewable Energy Power Station Utilizing Turkey Litter: The 55 MW Fibrominn Biomass Plant 1st Year of Commercial Operation. Las Vegas, NV: Renewable Energy World Conference & Expo, 2009.

35. Yin C, Rosendahl LA, Kær SK. Grate-firing of biomass for heat and power production. *Progress in Energy and Combustion Science* 2008;34:725–754.

36. Bolhàr-Nordenkampf, M. Operating Experiences From Combustion of BViomass at elevated Steam temperatures With The Focus On Challenging Biomass Fuels. In: *European Biomass Conference*. Lyon, France: International Energy Agency; 2010.

37. Wiltsee G. Lessons learned from existing biomass power plants. National Renewable Energy Laboratory Report NREL SR-570-26946, *24 Feb 2000*.

38. Al-Mansour F, Zuwala J. An evaluation of biomass co-firing in Europe. *Biomass and Bioenergy* 2010;34:620–629.

39. Sami M, Annamalai K, Wooldridge M. Co-firing of Coal and Biomass Fuel Blends. *Progress in Energy and Combustion Science* 2001;27:711–214.

40. Tillman DA. Biomass cofiring: the technology, the experience, the combustion consequences. *Biomass and Bioenergy* 2000;19:365–384.

41. Hus PJ, Tillman DA. Cofiring multiple opportunity fuels with coal at Bailly Generating Station. *Biomass and Bioenergy*, 2000;19:385–394.

42. Battista J, et al. Biomass cofiring at Seward Station. *Biomass and Bioenergy* 2000;19:419–427.

43. van Loo S, Koppejan J. *Handbook of Biomass Combustion and Co-Firing*. London: Earthscan, 2008.

44. Renewable Energy Trends in Consumption and Electricity 2008. Washington, DC: U.S. Energy Information Administration; 2010.

45. Kadyszewski J, et al. New Carbon Sequestration & Biomass Co-Firing. In: *22nd Annual International Pittsburgh Coal Conference (PCC)*. Pittsburgh, PA: University of Pittsburgh School of Engineering, 2005.

46. Knudsen JN, et al. Corrosion and Deposit Investigations During Large Scale Co-combustion of Switchgrass at a Coal-fired Power Plant. In: *Impacts of Fuel Quality on Power Production.* Snowbird, UT: Electric Power Research Institute; 2006.

47. Lu H, et al. Effects of particle shape and size on devolatilization of biomass particle. *Fuel* 2010;89:1156–1168.

48. Miles TR. Collection, Storage and Preprocessing of Switchgrass for Power Generation—the Ottumwa Co-firing Experience 1998–2006. Louisville, KY: Biomass and Bioenergy Technical Sessions Conference, American Society of Agricultural and Biological Engineers (ASABE); 2006.

49. Uslu A, Faaij APC, Bergman PCA. Pre-treatment technologies, and their effect on international bioenergy supply chain logistics. Techno-economic evaluation of torrefaction, fast pyrolysis and pelletisation. *Energy* 2008;33:1206–1223.

50. Maaskant E. Torrefaction. In: *High Cofiring Percentages in New Coal Fired Power Plants.* J Koppejan, ed. Hamburg: IEA Bioenergy Task 32, 2009.

51. Kymijärvi Biomass CFB Gasifier, Finland 2010. http://www.power-technology.com/projects/kymijarvi/.

52. Stahle K. *Varnamo Demonstration Plant: The Demonstration Programme 1996–2000.* Trelleborg, Sweden: Berlings Skogs, 2001.

53. Knoef HAM. *Handbook Biomass Gasification.* Enschede, Netherlands: BTG Biomass Technology Group, 2005.

54. Larson ED, Haiming JE, Celik, FE. Gasification-based fuels and Electricity Production from Biomass, without and with Carbon Capture and Storage. Princeton Environmental Institute; https://www.princeton.edu/pei/energy/publications/texts/LarsonJinCelik-Biofuels-October-2005.pdf.

55. Biagini E, Masoni L, Tognotti L. Comparative study of thermochemical processes for hydrogen production from biomass fuels. *Bioresource Technology* 2010;101:6381–6388.

56. Radlein D. The Past and Future of Fast Pyrolysis For the Production of Bio-Oil. Ames, IA: Iowa State UniversityTCS 2010 Symposium on Thermal and Catalytic Sciences for Biofuels and Biobased Products: 2010.

57. Lehmann J, Joseph S. *Biochar for Environmental Management: Science and Technology.* Sterling, England: Earthscan, 2009.

58. Garcia-Perez M, Lewis T, Kruger C. Feasible Pyrolytic Methods for Producing Biochar and Advanced Biofuels in the State of Washington. Pullman, WA: Washington State University Report, 2010.

59. Phillips S, et al. Thermochemical Ethanol via Indirect Gasification and Mixed Alcohol Synthesis of a Lignocellulosic Biomass. National enewable Energy Laboratory Report NREL/TP-510-41168, 2007.

60. Tijmensen M. *The Production of Fischer Tropsch Liquids and Power Through Biomass Gasification*, in *Science and Policy.* Unpublished dissertation, University of Utrecht: Utrecht, Netherlands, November 2000.

61. Zhang L, Xu C, Champagne P. Overview of recent advances in thermo-chemical conversion of biomass. *Energy Conversion and Management* 2010;51:969–982.

62. Foust TD. The Future of Biomass for Transportation. Ames, IA: Iowa State University TCS 2010 Symposium on Thermal and Catalytic Sciences for Biofuels and Biobased Products. 2010.

63. Hayes DJ. An examination of biorefining processes, catalysts and challenges. *Catalysis Today* 2009;145:138–151.

64. Lynd LR, et al. The role of biomass in America's energy future: framing the analysis. *Biofuels, Bioproducts and Biorefining* 2009;3: 113–123.

65. Thorp BA, Akhtar M. Is the Biorefinery for Real? *Paper360*, 2010; 5:8–12.

UTILIZING WASTE MATERIALS AS A SOURCE OF ALTERNATIVE ENERGY: BENEFITS AND CHALLENGES

T. Terry Tousey

15.1   INTRODUCTION

The generation of waste, whether industrial or residential, is a fact of life in our society today. Nearly everything we do creates some type of waste. It is estimated that the United States alone generates 7.6 billion tons per year of non-hazardous industrial waste [1], 48 million tons per year of hazardous waste [2], and 250 million tons per year of municipal solid waste (MSW) [3].

Many of the waste streams and industrial by-products we generate each year contain recoverable energy. Capturing and utilizing this energy can create a positive impact both economically and environmentally. It not only extends a material's life cycle, but it also reduces the volume of waste sent to landfills, conserves non-renewable resources, and helps reduce manufacturing costs by providing a lower cost alternative to the rising costs of energy and waste disposal. Depending on the waste and the fossil fuel it is replacing, it may even help reduce our carbon footprint.

The volume of waste we generate continues to grow each year. In the United States, MSW generation has increased from 88 million tons in 1960 to 250 million tons in 2008 [4]. The amount of industrial waste has grown as well. Some of this increase is due to the fact that we have 120 million more people in the United States today than we did 50 years ago, but much of it is a result of our changing lifestyles and consumption habits. Today, we use significantly more disposable items than we did 50 years ago, and we have developed thousands of new chemicals, plastics, paints, and adhesives; all of which generate their own production by-products that need to be disposed of. Figure 15.1 shows the growth in MSW generation rates from 1960 to 2008.

Utilizing waste in the production of energy is not a new concept. A number of industries initiated programs in the mid-1970s and early 1980s after the Arab oil embargo and the Iran-Iraq war drove up energy prices exponentially and stricter environmental laws caused waste disposal costs to escalate. To remain competitive, companies began searching for ways to reduce their energy costs and to manage their waste in a more environmentally friendly yet cost-effective manner. What they found was that by recovering the energy value of the by-products they generated from their manufacturing processes, they could produce their own heat and power on site. This not only lowered their energy and waste dis-

posal costs, but it also eliminated the environmental liabilities associated with land filling their waste.

There are numerous non-hazardous industrial by-products and post-consumer wastes that contain recoverable energy including black liquor from the pulp and paper industry, distillation bottoms from chemical purification processes, wood collected from construction and demolition (C&D) debris, scrap tires, used oil, and carpet waste. There are also certain hazardous wastes such as spent solvents, paints, inks, and other discarded organic chemicals that have energy value. And of the 250 million tons of household waste generated each year, over 80% of this is organic material consisting of paper and paperboard, plastics, wood waste, yard trimmings, and food scraps; all of which can be used as a source of energy [4].

In this chapter, we will explore some of the benefits and challenges associated with using these wastes and industrial by-products as a source of energy, and we will look at some of the current and potential opportunities for doing so. Using these materials as a source of energy is not without its challenges. Many of these streams, in their "as generated" form are of low energy quality, or they are heterogeneous and need to be processed to recover their energy value. If the material is to be sent off site for energy recovery, there must be an efficient mechanism in place to collect, process, and transport the material to the ultimate energy consumer. The entity using the waste or by-product will need to obtain permits and will most likely have to modify their existing fuel handling and combustion system to accommodate using the alternative material. All these things add costs. However, what makes a waste-to-energy project unique is that the economics are almost always driven by disposal cost avoidance or regulatory compliance. Therefore, they are not completely dependent on the price of fossil fuels or government support to make them viable. For this reason, the economics are generally more favorable for these types of projects.

15.2   REGULATORY OVERVIEW

15.2.1   Defining a Waste

Before undertaking a waste-to-energy project, it is important to have a clear understanding of the regulatory status of the material

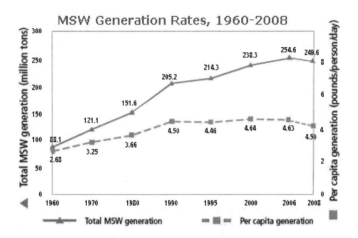

FIG. 15.1 MSW GENERATION RATES, 1960–2008 (Reproduced from EPA Non Hazardous Waste) [51]

you will be handling. Terms such as by-product, sludge, solid waste, hazardous waste, municipal waste, and spent material all have specific meanings within the regulatory world that define how a material is classified when it is discarded. These waste classifications determine the regulatory requirements as to how, when, and where the material can be disposed of or beneficially reused and can affect the economics of handling the material. Before spending any capital, you need to make sure there are no regulatory obstacles or added environmental costs that would make the project uneconomical due to the material's regulatory status.

In the United States, the governmental agency regulating the generation, treatment, and disposal of waste materials is the US Environmental Protection Agency (EPA). The primary goal of the EPA in regard to waste generation and disposal is to reduce, reuse, and recycle. Although they do not consider energy recovery, a form of reuse or recycling, they do consider it a viable option in the waste management hierarchy. Figure 15.2 depicts the EPA's preferred solid waste management hierarchy.

The Resource Conservation and Recovery Act (RCRA) is the primary law governing the disposal of solid and hazardous waste in the United States. It was enacted in 1976 to address the problems

associated with the growing volume of municipal and industrial waste. It was amended and strengthened in 1984 with the passing of the Hazardous and Solid Waste Amendments, which among other things required phasing out land disposal of hazardous waste. Nonhazardous industrial solid waste and MSW are regulated under Subtitle D of RCRA, and hazardous waste is regulated under Subtitle C [5].

The definition of a solid waste under RCRA serves as the starting point for determining the regulatory system under which a waste must be managed. For example, a material cannot be a hazardous waste unless it is first defined as a solid waste [6]. Under RCRA, the term "solid waste" means any garbage, refuse, sludge from a waste treatment plant, water supply treatment plant, or air pollution control facility and other discarded material, including solid, liquid, semisolid, or contained gaseous material resulting from industrial, commercial, mining, and agricultural operations, and from community activities. The term does not include solid or dissolved material in domestic sewage, or solid or dissolved materials in irrigation return flows or industrial discharges that are point sources subject to permits under the Federal Water Pollution Control Act. It also does not include source, special nuclear, or byproduct material as defined by the Atomic Energy Act [7].

In defining a solid waste, the EPA intended to capture the waste generated by all industries. With the recent emphasis on the use of biomass as a source of renewable energy, it is important to point that the definition of solid waste includes discarded materials from agricultural activities. Many sources of biomass come from agricultural activities, and you need to be aware that just because a material is considered renewable does not mean it is not a solid waste. Materials such as animal manure and food waste treated by anaero-

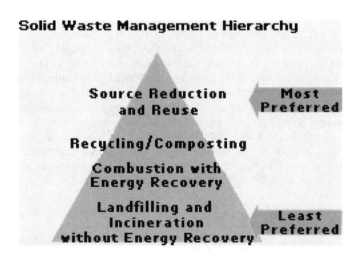

FIG. 15.2 SOLID WASTE MANAGEMENT HIERARCHY (Reproduced from EPA Non Hazardous Waste) [51]

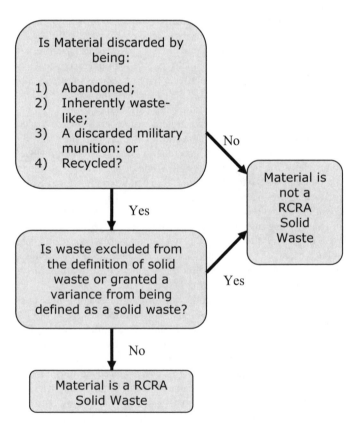

FIG. 15.3 IS A MATERIAL A SOLID WASTE? (Reproduced from EPA RCRA Orientation Manual) [52] [53]

bic digestion to produce biogas and crop residue burned for energy recovery could be subject to the solid waste rules. Non-hazardous solid waste is regulated at the state level, and different states have different rules. Thus, it is important to review the solid waste regulations for the state where the project will be implemented as this could affect the economics. Figure 15.3 shows a simple flowchart of some of the questions to be addressed in determining whether a material is an RCRA solid waste.

Hazardous waste is a subset of solid waste. The EPA defines hazardous waste very broadly as liquid, solid, contained gas, or sludge wastes that contain properties that are dangerous or potentially harmful to human health or the environment [8]. Because this definition is so broad the EPA has developed a set of criteria and definitions to help in making hazardous waste determinations. "Characteristic" hazardous wastes are wastes that are hazardous because of their ignitability, corrosivity, reactivity, or toxicity. "Listed" hazardous wastes are wastes generated from a specific list of industries and common manufacturing processes or from a list of over 500 commercial chemical products that are considered hazardous waste when discarded [9]. Figure 15.4 shows a simple flowchart of some of the questions to be addressed in determining whether a material is an RCRA hazardous waste.

To complicate matters the EPA has developed some exclusions and exemptions from the definition of solid waste and/or hazardous waste for certain wastes when they are reused or recycled (40 CFR 261.2(e) and 261.4) [6] [53]. Other recycled wastes may still be considered a solid waste or hazardous waste, but are subject to less stringent regulatory controls [6] [52] [53]. Although the purpose of this is to promote beneficial reuse and recycling, it can create confusion. Figure 15.5 shows a flowchart of some of the

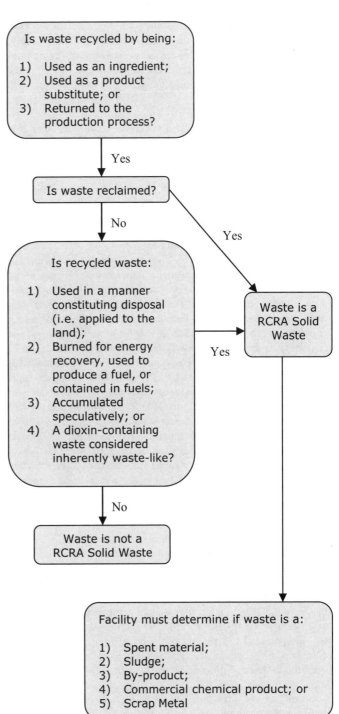

FIG. 15.5 **ARE ALL RECYCLED WASTES HAZARDOUS WASTES?** (Reproduced from EPA RCRA Orientation Manual) [52] [53]

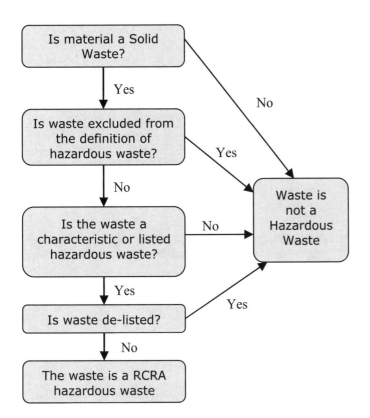

FIG. 15.4 HAZARDOUS WASTE IDENTIFICATION PROCESS (Reproduced from EPA RCRA Orientation Manual) [52] [53]

questions to be addressed in determining whether a recycled waste is an RCRA solid waste and, if so, if it is a hazardous waste.

Interpreting the rules can be tricky, so you should seek professional advice if you are not familiar with them. The flowcharts are only meant to be a guide and do not replace reviewing the regulations. In addition, the regulations change and get amended from time to time. Today, you might be utilizing a by-product as a source of energy that is excluded from the definition of a solid waste, and tomorrow, that exclusion could be taken away. Your economics could change significantly if you have to start complying with a new set of regulations. This is why most companies look to recover their initial capital fairly quickly as a change in the regulations could make the project no longer cost-effective.

15.2.2   Air Permits

If you are going to combust a material as a fuel whether it is a by-product, a waste, or a fossil fuel, you are going to need an air permit. Even if you are just switching from a fossil fuel to a by-product fuel, you will most likely be required to amend your permit. Depending on whether the material is classified as a by-product, a non-hazardous solid waste, or a hazardous waste will determine what kind of permits you will require. The type of air permit required will determine how much capital you will have to allocate to items like air emission controls.

In general, the combustion for energy recovery of non-hazardous industrial by-products and industrial solid wastes is regulated under section 112 of the Clean Air Act (CAA) [10]. This is better known as the "Boiler MACT" and affects any boiler or process heater that has the potential to emit more that 10 tons/yr of any single hazardous air pollutant (HAP) or more than 25 tons/yr of any combination of (HAPs). There are 188 compounds that are considered HAPs [11].

Facilities that combust materials classified as MSWs are regulated under section 129 of the CAA [10]. The air-permitting requirements and controls for this type of device are more stringent than under Section 112. In addition to meeting the MACT standards for HAPs, they also have emission limitations for nine additional pollutants, and there are requirements for operator training, preconstruction site assessments, and monitoring that are not included in Section 112.

Facilities that burn hazardous waste as a fuel or incinerate it are regulated under what is known as the Hazardous Waste Combustion MACT (40 CFR 63 — Subpart EEE) [48]. Needless to say, the regulatory standards for burning hazardous waste are much more restrictive than for burning by-products or MSW.

The CAA also regulates a number of common air pollutants such as particulate matter, ground-level ozone, carbon monoxide, sulfur oxides, nitrogen oxides, and lead, which are known as "criteria pollutants" [12]. The EPA sets specific allowable levels for these pollutants, and if a geographic region has not attained these standards, they are considered non-attainment areas. If you plan to locate your waste-to-energy project in a non-attainment area and have the potential to emit any of the criteria pollutants in excess of the specified regional thresholds, you will have more stringent permitting requirements and potentially require more expensive air emission controls.

As you can see, a waste-to-energy facility can be subject to different regulatory requirements based on the type of waste it will be handling, the potential it has to emit certain air pollutants, the amount of those pollutants it is going to emit, and whether it is located in a non-attainment area. Thus, a project may be viable if it is located in an attainment area, but the same project may not be economical if it is located in a non-attainment area requiring expensive emissions controls.

It is important to evaluate all the potential regulatory costs associated with an energy recovery project before you begin. In addition to air permits, you may be subject to other permitting and regulatory requirements depending on the state you are located in and the type of waste you are handling. You may be subject to local siting and zoning requirements, and the project may even change a facilities current zoning permit (i.e., from agricultural to industrial).

As you might expect as you go from handling a by-product to a non-hazardous solid waste to a hazardous waste, each permit gets more onerous and more expensive than the next. While you may be able to comply with the permit restrictions, the cost of all the regulatory controls and reporting requirements may become too expensive relative to the energy savings.

15.3   EVALUATING THE ENERGY VALUE OF A WASTE

In evaluating a waste-to-energy project, you need to determine the economic value of the waste you will be using. This value is based on energy content of the waste as compared with the fossil fuel it is replacing — less any costs associated with handling the waste. Differing types of materials have different energy values, require different handling equipment, and require different air pollution control measures. Thus, it is important to make sure that the energy savings realized from using the waste are sufficient to overcome any additional capital and operating costs required to handle the material.

In the United States, the energy value of a material is based on its BTU (British Thermal Unit) content. The most common fossil fuels used for energy production are coal at 11,000 to 13,000 BTU per pound, fuel oil at about 138,500/gal and natural gas at about 1030 BTUs/cu ft. If electricity is being used as a source of energy, which is expressed in kilowatts per hour, it is necessary to convert kilowatts to BTUs to make an energy comparison. One kilowatt is equal to 3412 BTUs.

Since the BTU content of different materials is expressed using different denominators, the economic value of a fuel is usually stated on a price per million BTU basis ($/MMBTU) to compare one fuel to another. Current pricing in the United States for some of the more common sources of energy as of the end of May 2010 is listed in Table 15.1. (Coal pricing is as of the end of 2009.)

Thus, if you were burning natural gas in your boiler at $5.02 per MMBTU and you had the opportunity to supplement this with a high energy content liquid by-product from your manufacturing process at minimal additional cost, it would certainly make economic sense to do so. If you were replacing fuel oil at $14.73 per MMBTU, you could probably afford to pay something for the liquid by-product fuel, assuming you had the permits to accept the material. On the other hand, if you are looking at substituting for the use of coal in your boiler at $2.59 per MMBTU with a solid material such as tires that required shredding and the steel belts removed to feed the material to the boiler, the economics may not warrant the investment unless you could receive a disposal fee for the tires.

TABLE 15.1   US ENERGY PRICES (See reference to 13, 14, 1, and 16 under "Energy Source")

Energy Source	Price	$/MMBTU
Coal (average delivered price - industrial) [13]	$64.87/ton	$2.59[1]
Heating oil (NY Harbor) [14]	$2.04/gal	$14.73
Natural gas (industrial) [15]	$5.02/MMBTU	$5.02
Electricity (industrial average) [16]	$0.0669/kwh	$19.60

Note:
(1) Assumes 12,500 BTUs/lb for coal.

A waste-to-energy program can generate savings in two ways: first by reducing your cost of fossil fuels and second from receiving a tipping fee for handling the material. If you are recovering energy from a by-product you generate on site, the tipping fee is not a necessary component as you have the benefit of eliminating the cost to disposal of the material. However, if you are operating a program that is based on bringing in outside waste as a source of energy many times, the energy savings is not enough to cover all the costs of handling and processing the waste, so you will need to charge a tipping fee. You also need to look at how efficient your system is at converting the material to useable energy. When dealing with a heterogeneous waste, do not assume what you calculate on paper is reality. In determining the true net energy value of a waste, you need to make sure you consider all the associated costs including environmental, processing and storage, transportation, and energy efficiency losses.

15.4 EXAMPLES OF WASTE MATERIALS AND BY-PRODUCTS THAT CAN BE USED AS A FUEL

There are a number of different business models that can be deployed in developing a waste-to-energy project. The model used depends on the waste generator's need for energy, the type of waste generated, the volume generated, and the logistics involved in transporting the material. For those manufacturing facilities that generate significant amounts of energy-bearing by-products, one method is to utilize the material on site to produce its own heat and power. Another method is for energy-intensive industries like utilities, cement plants, and asphalt plants to develop a program to accept energy-bearing wastes from third parties to procure a lower cost fuel source. In yet another model, a company might set themselves up as a merchant plant specifically to accept and process wastes generated by others and produce a fuel or generate power which they would sell, presumably for a profit.

In this section, we will look at some examples of industries utilizing their own by-products as a source of energy, and we will also look at some examples of post-consumer wastes that can or are being used for energy or power production. This will be a broad overview, as in the space provided not much detail can be given. However, a number of articles have been written on recovering the energy value of many of these materials, and you are encouraged to investigate them further if you would like more detailed information.

15.4.1 Industries Utilizing Energy Recovery

The industrial sector offers numerous opportunities to recover energy from the by-products generated off their manufacturing processes. Industries such as in the chemical, steel, pulp and paper, petroleum products, and aluminum industries all generate large quantities of energy-bearing by-products that need to be managed. They also consume a great deal of energy for process heating and cooling and power. By using the energy value of their by-products to fuel their boilers or produce power, they can

- reduce energy costs
- eliminate transportation and disposal costs
- reduce fossil fuel use
- eliminate the long-term environmental liabilities associated with off-site disposal
- reduce CO_2 emissions associated with transporting materials off site

The net effect of all of these is a reduction in operating costs, which will help them remain completive in the global market and a reduction in their carbon footprint.

15.4.1.1 Chemical Industry The chemical industry consumes about 6% of the total US domestic energy use and about 25% of all US manufacturing energy use. Half the energy consumed is used for process heating and cooling and for power, with the other half being used as raw materials feedstock. About one-fifth of electricity used by the chemical industry is produced onsite, primarily by means of cogeneration [17].

In most organic chemical production processes, chemical distillation is used to separated and purify the products. The distillation process generates by-products that are typically high-molecular-weight organic compounds. These compounds can have an energy content as high as 16,000 to 18,000 BTUs/lb. Typically, they are relatively clean streams with low moisture content and very little other contaminants such as toxic metals that might create air emission issues when combusted. Most are liquids so the cost to convert a boiler to burn these by-products is relatively small. In addition, the majority of the by-products are not considered a hazardous waste under RCRA, so the facility does not have to go through an expensive permitting process to burn the material. Thus, these materials offer an excellent opportunity to be used as an alternative fuel.

Many chemical companies have been using these by-product streams in their boilers for over 30 years. Faced with significant increases in energy costs beginning in the mid 1970s and higher disposal costs beginning in the early 1980s, the industry has been able to reduce operational costs through the use of these materials as a fuel. A facility with a 100 MMBTUs per hour boiler fueled with natural gas at \$5.02/MMBTU could save about \$2,200,000 per year by replacing 50% of their natural gas with a zero-cost by-product fuel. This would be in additional to the avoided cost to transport and dispose of the material.

15.4.1.2 Pulp and Paper Industry The US pulp and paper industry uses about 15% of the total energy consumed by all domestic manufacturing sectors [18]. In 2008, the industry generated 65% of its own energy [19]. The industry is one of the largest users of cogeneration generating 37% of the total energy produced by cogeneration within all manufacturing sectors in 2008 [19].

The manufacture of paper involves a pulping process where the lignin and hemicellulose in wood are broken down and separated from the cellulose. The most widely used process for this is the Kraft Process, which uses sodium hydroxide and sodium sulfide to break the bonds that link the lignin to the cellulose. This process generates a by-product stream called black liquor, which is about 85% water, but contains the lignin and hemicellulose along with the reaction by-products of sodium carbonate and sodium sulfate. Most of the water is evaporated through the use of multiple effect evaporators to concentrate the material to 65% to 80% solids. At this point, the energy value of the black liquor is about 6000 BTUs/lb [18]. The black liquor is then sent to a recovery boiler where the material is combusted to produce steam for the cogeneration of electricity and process heat. The recovery boiler also recovers sodium sulfide and sodium hydroxide, which is returned to the pulping process.

In the process to recovery, the energy value of the black liquor is energy intensive in itself and very complicated. The evaporation of the black liquor consumes around 3.78 million BTUs of steam per ton of pulp. The recovery boiler consumes another 1.13 million BTUs per ton of pulp in the form of supplemental fuel and electricity. However, the boiler produces 9 to15 times as much energy in the form of heat, which is used to produce steam and electricity through the use of cogeneration [18].

The effort the pulp and paper industry goes through to recover the energy value of the black liquor is a good example how complex the process can get. Unlike the distillation by-products from the chemical industry, which can be used directly as a fuel, the by-products of the pulp and paper industry require multiple processing steps to recover their energy value, and the process requires a significant capital investment on the part of the industry. However, the combination of energy cost savings and avoided disposal cost makes the effort well worth it.

15.4.1.3 Steel Industry Energy accounts for about 17% of the total manufacturing cost in the production of steel. About 10% of the electricity used by the steel industry is produced on site, primarily through cogeneration [20]. Two by-product streams are generated in the production process: coke oven gas and blast furnace gas. Both are used as an alternative source of energy in the coking and steel-making process.

Coke oven gas is a by-product of converting coal to coke. The coal is heated in coke ovens, and the volatiles in the coal are driven off at temperatures approaching 2000°F. This raw gas is made up of water vapor, tar vapor, aromatics, naphthalene, ammonia gas, hydrogen sulfide gas, and hydrogen cyanide gas. For the coke plant to use the raw gas for energy, it is sent to an on-site by-products plant to be cleaned and to recover useful chemicals. After the gas is cleaned and the chemicals recovered, the resulting gas is sent back to the coke oven and to the blast furnace to be used as a source of energy [21]. The energy content of the gas is low at 500 BTUs/cu ft, but the volume is significant. One ton of coal converted to coke generates 9500 to 11,500 cu ft of coke oven gas or the equivalent of about 5,000,000 BTUs [22]. This is more energy than the coke oven requires, so about half of the gas is used elsewhere in the plant or sold.

Blast furnace gas is a by-product of the blast furnace where iron ore is mixed with coke and reduced to pig iron, which goes on to the steel-making furnaces. The gas has a very low BTU content of only 90 BTU/cu ft, but the volume is so large (2.5–3.5 tons of gas per ton of iron produced) that a significant amount of energy can be recovered from the stream [23]. The gas is cleaned and used in the boilers and in the blast stoves, which pre-heat the blast furnace air.

Similar to the pulp and paper industry, the steel industry goes through a tremendous amount of effort to recover the energy content of their by-products. However, it provides a significant energy savings versus flaring the gas, which they did in the past.

15.4.2 Non-Hazardous Solid Waste

US industries generate 7.6 billion tons per year of non-hazardous industrial waste [1]. Although non-hazardous waste is not subject to the same stringent regulatory requirements as hazardous waste, handling and disposing of these materials are still a large expense for industry. Much of the non-hazardous industrial waste generated each year is not suitable for energy recovery. Ninety-seven percent of the volume is wastewaters, and a majority of the remaining waste consists of inorganic materials such as coal ash, foundry waste, slag, and cement kiln dust. However, there is still a significant amount of material that has recoverable energy content. The United States also generates a significant amount of municipal "post-consumer" waste such as scrap tires and used oil filters, which can be used for energy recovery. Examples of non-hazardous industrial waste and municipal post-consumer waste that have recoverable energy content include

- C&D debris
- scrap tires
- municipal wastewater treatment sludge
- used oil
- auto shedder residue
- rice hulls, oat hull and peanut shells
- beer yeast
- food processing wastes
- cooking oil grease from food establishments
- used oil filters
- rail road ties
- non-hazardous refinery wastes
- wire chop
- carpet waste
- used pallets
- non-hazardous spent solvents

15.4.2.1 Scrap Tires There are approximately 300,000,000 scrap tires generated each year in the United States [24]. For years, they were simply stockpiled on vacant land until a number of major fires caused states to begin regulating the collection and disposal of tires in the late 1980s. Most states now ban the land filling of whole tires, although many still allow land filling of shredded tires.

Tires have an energy content averaging 14,000 BTUs/lb [24]. As a solid fuel, they offer some advantages over coal in that they are higher in BTU content and lower in sulfur and nitrogen content, which helps reduce SOx and NOx emissions. The main consumers of tires as a fuel are cement plants, pulp and paper mills, and utilities. Since they already burn a solid fuel in the way of coal, the capital costs required to utilize tires as a supplemental fuel is relatively small. Most cement plants are able to burn whole tires, reducing the processing costs of the collectors and enabling the cement plants to charge a tipping fee for disposing of the tires. They can also handle the steel belts in the tires because one of the raw materials in making cement is iron. Utilities and pulp and paper mills require the tires to be chipped and, in some cases, require the steel belt to be removed so they generally pay a fee for the tire derived fuel. Currently, 54% of the scrap tires generated each year are used as a fuel. The use of the tires as a fuel has also helped reduce the number of stockpiled tires from 1,000,000,000 in 1990 to an estimated 128,000,000 by 2007 [24].

ILLEGALLY DUMPED TIRES ALONG A CREEK (Illegal Dumping Prevention Handbook — EPA)

The market for tire-derived fuels took time to develop, as initially there was no collection or processing system in place to manage the logistics of moving the scrap tires from the consumer to the energy user. As with many potential waste-to-energy streams, if there is no efficient method to move the material, it is difficult to establish a recycling market. Today, there is a well-developed network of collectors, processors, and energy consumers for scrap tires.

Scrap tires can be recycled to make products such as crumb rubber for use in asphalt or artificial turf, and these end uses do compete for the tires as a source of energy. However, even though these markets provide a more value-added end use for the scrap tires, the cost of shredding the tires to such a small size generally makes these options too expensive. If there is not a fuel outlet for the tires in a particular region, and they have to be shipped a long distance, then these other uses can become a viable outlet.

15.4.2.2 Construction and Demolition Debris
C&D debris is generated from the renovation and C&D of both residential and commercial properties. The amount of C&D material can change significantly from year to year based on the economy. In 2003, the EPA estimated that 164 million tons of C&D waste was generated in the United States [25].

C&D debris is made up of 22% to 45% combustibles such as wood (20%–30%), asphalt shingles (1%–10%), and plastics (1%–5%). The remaining 47% to 75% consists of non-combustibles such as concrete and rubble, drywall, metals, and bricks [25]. Most of the asphalt shingles are recycled by crushing and mixing with hot mix asphalt, which offers a higher economic return than use as a fuel. However, this still leaves about 25% of the stream available for energy recovery or about 42,500,000 tons/yr. There is not a lot of data on the energy value of wood from C&D debris, as it can vary based on moisture content, but it most likely averages between 7000 and 8000 BTUs/lb. Wood from demolition projects typically has lower moisture content and subsequently higher BTU content than other biomass as the material is aged.

The C&D industry is made up of a fairly sophisticated network of collectors and transfer stations where the materials are sorted and the useable materials are sent for recycling and the rest sent to landfill. Most of the wood is used as a fuel in industrial boilers; however, the clean wood has to be separated out from the painted wood or wood treated with wood preservatives. The EPA estimates that only about 50% of the wood is clean enough to be used for energy recovery because of potential air emission issues. This would equate to about 21,250,000 tons. Before the wood can be used as a fuel, it has to be chipped. The estimated market value for wood chips delivered to a utility power plant is $18 to $30 per ton, depending on a lot of factors including transportation costs, moisture content, and market demand [55]. Assuming an average of 7400 BTUs/lb for C&D wood waste, this would equate to between $1.21 and $2.03 per MMBTU or about 45% to 78% of the price of coal.

15.4.2.3 Municipal Wastewater Treatment Sludge
There is approximately 7,000,000 dry tons/yr of sewage sludge generated from the roughly 16,000 publicly owned wastewater treatment plants in the United States [26]. Sewage sludge is a by-product of the municipal wastewater treatment process. As generated, it is mostly water, and it is typically sent through a belt press or centrifuge to bring the solids content up to 20% to 30%. The resulting cake, commonly referred to as biosolids, is an odiferous substance that must be carefully handled and monitored to ensure public safety.

SEWER SLUDGE (30% Solids and 70% Moisture)

Sludge disposal represents a major cost of the wastewater treatment process. Prior to the end of 1991 when the practice was banned in the United States, much of the sludge was disposed of inexpensively by ocean dumping. In 1993, the EPA established management standards for the final use or disposal of sewage sludge including rules for land application, land filling, and incineration (Title 40 CFR Part 503) [49]. As municipalities began looking for cost-effective ways to dispose of their sludge in compliance with the new rules, a number of treatment technologies were developed to meet the new treatment standards including alkaline (lime) stabilization, composting, heat drying, anaerobic digestion, and incineration.

On a dry basis, sewer sludge typically has an energy value of between 6000 and 9000 BTUs/lb. The problem is that mechanical drying of the sludge can only get the moisture content down to 70% to 80%. Thus, the material is still too wet and too low in energy content to be used directly as a fuel. However, a number of methods have been explored to recover the energy in the wet sludge. Some municipalities that incinerate the sludge have added heat recovery systems to capture the waste heat off the incinerator to produce steam. The steam is used to produce power for the facility through cogeneration.

Heat drying the sludge on site has also become a popular alternative, particularly in densely populated areas, as it reduces the volume of the sludge and destroys the pathogens. In addition, the product produced off the dryer can be used as a fertilizer because of its nitrogen content or as a fuel because of its energy content. The problem with heat drying is that it requires a lot of fuel, typically natural gas, to dry the sludge as it is 75% water, and it requires a significant capital investment. Therefore, the industry has been looking at ways to capture the energy value of the sludge to assist in the drying process and reduce fuel costs.

In 2005, a group out of Chicago built a unique gasification and drying system at the Philadelphia wastewater treatment plant. The system would gasify the dried sludge and then use the energy produced from the gasification process, in the form of waste heat, in a dryer to partially dry the sludge prior to feeding it to the gasifier. The benefit of this was that it was a closed-loop system, and by using the energy content of the sludge to assist in the drying process, no fossil fuels would be required for the dryer. The facility had operational issues due to the quality of the sludge, but it did have limited success. The facility has since ceased operations due to some political factors, but other municipalities are currently evaluating or operating similar technologies.

There is also resurgence at municipal wastewater treatment plants in using anaerobic digestion to treat the sludge and produce biogas. The biogas, which usually contains 50% to 60% methane, can be used as a fuel for the sludge dryer or in a boiler to provide heat and hot water to the plant. If the biogas is cleaned up, it can also be used in a cogeneration system to produce on-site power. Municipalities have historically flared their biogas as they viewed anaerobic digestion as part of the wastewater treatment process rather than a source of renewable energy. This mind set has changed in recent years as energy costs have increased. Municipalities are now looking for ways to improve the efficiency of their digesters to produce more biogas, which can be used on site to reduce energy costs. This will also help reduce sludge disposal costs as improving the efficiency of the digester will reduce the volume of solids they generate.

The next few years will be an interesting period for the municipal wastewater treatment industry. They have the potential to transform themselves from a wastewater treatment industry that consumes a great deal of energy to an industry that produces its own power. The problem today is that there are so many new technologies being marketed to the industry, each touting itself as the most efficient in recovering the energy value of the sludge, that the industry does not know which method to focus on. Municipalities are also very conservative, and they take a long time to make changes so it is too early to predict if a single technology will rise to the top or if a number of different technologies will be used based on each facility's particular situation. Time will tell.

15.4.2.4 Used Oil Approximately 1.35 billion gallons of used oil are generated each year in the United States [27]. This includes spent automotive lubricating oils, hydraulic fluids, compressor oils, and metal working oils.

Used oil has been used as an alternative fuel for decades. However, because of the issues over improper management of used oil over the years, the EPA established a set of regulations in 1985 primarily directed at used oil burned for energy recovery. The main driver of these rules was to keep off-specification oil from being burned in non-industrial boilers such as those in apartment buildings. Then, in 1992, the EPA finalized standards establishing requirements for generators, transporters, transfer facilities, collections centers, processors, re-refiners, burners, and marketers of used oil (Title 40 CFR Part 279) [50]. Although not as stringent as the hazardous waste rules, it does require the generators, transporters, and burners to maintain records and ensure the proper management of the used oil [28].

Of the 1.35 billion gallons of used oil generated each year, the EPA estimates that 780 million gallons are used as a fuel, 165 million gallons are re-refined back into lube oil base stock, and 426 million gallons are land filled or disposed of improperly. Of the material used as a fuel, most is burned in asphalt plants (38%), with the rest being burned in small space heaters in garages and automotive centers (14%), industrial boilers (12%), utility boilers (10%), steel mills (10%), cement kiln (4%), and other (12%) [27].

Depending on the quality of the used oil, it can have up to 140,000 BTU/gal. It is mostly used a replacement for fuel oil in large industrial boilers or furnaces capable of burning the material with no additional air emission controls. As a secondary market, used oil will compete with natural gas. If natural gas is priced at $5.00 per million BTUs and it takes 7.14 gal of used oil to equal one million BTUs, then the equivalent value of the used oil would be $0.70/gal. Rarely, it is used as a replacement for low-cost coal as the price would have to be below $0.36/gal to be competitive. The market forces are dynamic in the used oil business. Depending on the price of fuel oil and the available

outlet for the used oil, the collection company may charge the generator, pick it up for free, or even pay for it.

Used oil is generated in small quantities from many different sources. Thus, there must be an efficient system in place to collect all this volume and move it to the used oil burner. Over the years, a well-established network of collection points (e.g., automotive repair shops, oil change shops, and garages) and collectors has developed to consolidate the used oil before shipping it to processing facilities and the end users. Being a liquid with high energy density, the used oil can be shipped fairly long distances and still be competitive as an alternative fuel, depending on the fuel oil market.

Used oil can also be re-refined back into a base stock for lubricating oil, and this outlet does compete against the fuel outlets for the supply of used oil. However, this market is still developing, and it has taken a long time to do so for two reasons. First, even though it creates a higher value-added end product, the economics have been marginal in the past versus using the material directly as a fuel. Second, there has always been a stigma attached to products made from a waste material that they are inferior. As such, someone driving a $40,000 car has generally not been willing to take the risk (perceived or not) of damaging his/her engine for a few dollars in savings from using a lube oil made from a re-refined base stock. This has been an issue with recycled anti-freeze as well. This perception is changing as re-refining technologies improve and the industry strives to educate the consumer, but it is still an issue. Thus, most used oil will likely continue to be used an alternative fuel.

15.4.2.5 Auto Shredder Residue Approximately 12 to 15 million automobiles are disposed of annually in the United States [29]. The cars are stripped of reusable parts and then sent to shredding operations, where they are shredded into smaller pieces. The metal is recovered and sold to scrap metal processors. The remaining material, called the auto shredder residue (ASR), comprises about 15% to 20% of the vehicle [30].

The United States generates approximately 5 million tons of ASR each year, and nearly all of it is land filled. ASR is composed of plastics, rubber, foam, residual metal pieces, paper, fabric, glass, sand, and dirt. Approximately 20% to 50% of the ASR, on a dry basis, is combustible, including the plastics, fabric, and rubber. To

TYPICAL AUTO SHREDDER RESIDUE (California DTSC — Evaluation of Shredder Residue as Cement Manufacturing Feedstock)

use the material as a fuel, you would first have to remove all of the incombustibles. Even then, the fuel quality of the material would be marginal because of the heterogeneous nature of ASR. It is estimated that the fuel value of the unprocessed material is roughly 5000 BTUs/lb. After removal of the incombustibles, the fuel value could be as high as 9000 to 10,500 BTUs/lb [29].

Over the years, the cement industry has looked at using ASR as an alternative fuel, but because of the contaminants and the cost to process the material, it has never been economically viable. Another option to recover the energy value of ASR is to use gasification. Gasification is more amendable to heterogeneous materials such as ASR, and it has a better emission profile than combustion, which would be an advantage due to the contaminants in the ASR. The biggest issue is that the material can still be land filled inexpensively. A recent report showed landfill costs averaged $43/ton in the United States in 2009 [31]. Given the capital costs and the costs to segregate out the non-combustibles, it would be difficult to make the economics work. However, because of regional differences in tipping fees, this might be a viable option in some parts of the county.

15.4.2.6 Carpet Waste An estimated 2,940,000 tons of post-consumer carpet was disposed of in 2009; most of it in landfills [32].

Recycling post-consumer carpet is a difficult task, not so much from a processing standpoint, but because a mechanism for collecting, sorting, and transporting used carpet has not established. There are also no regulatory drivers other than the push to reduce the amount of waste sent to landfills. Waste carpet does create waste management problems as takes up a large amount of landfill space due to its bulk. However, only 278,500 tons or 9.5% of the carpet waste stream were recycled or used for energy in 2009 [32].

There is gradually a network of collectors and processors being developed to manage the logistics of moving waste carpet from the consumer to the processor, but it is still in its infancy. Like any new waste recycling industry, these challenges are being taken on by individual entrepreneurs. The carpet industry has created an association, Carpet America Recovery Effort (CARE) to try and accelerate the recovery process. A key factor in trying to develop the fledging collection industry is to have a high-volume outlet for the collected material. It is a which-comes-first scenario. You cannot develop a successful recycling business if you cannot get the supply, and at the same time, companies are not going to set up the collection and transfer network if they do not have a viable outlet for the material. Thus, the industry is trying to develop fuel outlets where large volumes of waste carpet could be consumed in the short term while the recycling market is being developed.

Depending on the type of carpet (nylon versus polypropylene), its BTU value can range from 7400 BTUs/lb to 12,000 BTUs/lb (17.17 MJ/kg to 28.10 MJ/kg) [33]. The average is closer to 7500 BTUs/lb as most carpet is made of nylon. The carpet can have up to 25% calcium carbonate as this is used in the backing of the carpet. This increases the ash content, which would create a problem if it was used as a fuel in a traditional boiler. However, cement kilns use calcium carbonate as a raw material, so this would be less of a concern if they were to use the material as a fuel. There may be a potential emissions issue though as nylon carpet has higher nitrogen content than coal, and this could create a potential for increased NOx emissions.

CARE conducted a study on the cost of collection, shredding, and providing the waste to a cement plant as a fuel. Because of the low cost to landfill carpet (CARE estimated this at $35 per ton),

CARPET SCRAP AT LANDFILL (CARE web site: Paul Humber)

it was determined that producing a fuel product for cement plants was not cost-effective when you take into account the capital costs and operating costs of shredding the carpet. Cement plants burn coal, so there, fuel costs are low, and therefore, there would not be any economic advantage for them to use the material unless they could receive a tipping fee.

The carpet manufacturing industry does generate carpet scrap from their production process, and most of this material ends up in local landfills. However, one company, Shaw Floors, in an effort to reduce landfill and energy costs installed a gasification system in 2006 to manage their carpet waste. The syngas produced from the gasification process is combusted, and the waste heat is used in a waste heat boiler to produce steam for their manufacturing process [34]. Key to the economics of this program is that the material does not have to be collected and transported, which saves costs.

15.5 REGULATORY DRIVERS AND OBSTACLES

Regulations can be both a driver and an obstacle to the utilization of a waste or a by-product as an alternative source of energy. In most cases, waste disposal regulations have created opportunities by restricting how a particular waste stream can be disposed of, which usually drives up the costs to manage the waste.

Prior to the enactment of the Solid Waste Disposal Act in 1965 and more specifically RCRA in 1976, there was very little interest in capturing the energy value of a waste. The cost to landfill materials was inexpensive as there were no landfill standards and there were no restrictions on what could be land filled. Thus, most waste, toxic or not, was land filled, as this was the most cost-effective disposal option. As the regulations began to change, many of these materials were restricted from landfills or forced to go to a hazardous waste landfills so companies were faced with rising disposal costs. Driven by the need to find alternatives, many industries like the chemical industry and the pulp and paper industry began to look at using their by-products as an alternative source of energy.

The ban on land disposal of certain hazardous waste in the late 1980s and early 1990s is another example of the regulations driving the development of waste-to-energy programs. In 1984, Congress enacted the Hazardous and Solid Waste Amendments to RCRA, which imposed a gradual ban on the land disposal of just about all organic hazardous waste including refinery and chemical wastes. This caused a rapid increase in disposal prices, as commercial incineration was the only option for some of these wastes. This was an expensive option costing as much as $1000 per ton compared with the $100 to $200 per ton they had been paying to landfill their waste. Faced with these higher costs, companies began to look for ways to use the energy value of their wastes to produce a hazardous waste fuel for the cement industry. The refinery waste was a sludge, and at the time, the cement industry was set up only to burn hazardous waste liquids. Seeing an opportunity a group of enterprising companies began to develop methods to blend the refinery solids with slop oil, another by-product of the refineries, to produce a liquid fuel. The programs were successful, and the cement industry began accepting a significant volume of blended refinery wastes as a fuel. At one point in the early 1990s, this accounted for almost 30% of the hazardous waste fuels being burned by the cement industry. While the cement industry still charged a disposal fee, the cost to the refineries was much less than incineration as the cement plants benefited from using the energy value of the waste.

Regulating the disposal of scrap tires and the prohibition on ocean dumping of sewage sludge are other examples of regulations creating opportunities for waste-to-energy projects. In all these situations, the regulations drove up the cost to manage the particular waste stream, making energy recovery an attractive option.

The enactment of regulations does not always drive waste-to-energy projects from a command and control perspective. Over the years, there have been a number of federal programs initiated to encourage energy recovery as a method of waste disposal. During the late 1970s and 1980s, as a result of rising energy prices brought on by the Arab Oil embargo and the Iran-Iraq war, a number of government programs were implemented to help promote the production of electricity through MSW to energy projects. Initially, the burning of hazardous waste fuels was exempted from the RCRA regulations to help foster alternatives to landfills. Currently, the government is providing grants and low-interest loans to help promote new technologies that will convert wastes into transportation fuels. This level of support is critical for many of these wastes to energy technologies as they are not yet economically viable based on today's energy prices.

Regulations can also be a deterrent to a waste-to-energy project. When handling a material that is classified as a solid waste, you become subject to a myriad of rules and regulations that can significantly drive up your costs. Regulatory compliance is expensive. Depending on the waste, you will most likely need a solid waste permit or even a hazardous waste permit. If you have the potential to emit certain air pollutants, you will be required to obtain air permits, and you may be required to install additional air pollution control equipment. If you will be storing any type of oil or grease, you may be subject to regulation under the Spill Prevention, Control, and Countermeasures program of the Clean Water Act. Thus, you will be opening yourself up to a whole new set of rules and reporting requirements that may affect the economic viability of the project.

It is also important to be aware that regulations do change. You may be handling a material today as a by-product that is subject to one set of rules, and suddenly, it gets reclassified as a solid waste subject to a whole new set of rules. This could change the economics of the project such that it is no longer viable.

A good example of what can happen is currently taking place. On June 4, 2010, the EPA published a proposed rule under the CAA that would classify certain non-hazardous secondary materials such as scrap tires, distillation bottoms, and sewage sludge as solid wastes when burned in a combustion unit. Under the current regulations, when these materials are burned for energy recovery, they are not considered a solid waste, and the activity is regulated under Section 112 of the CAA, which regulates emissions from industrial and commercial boilers. If the new rule goes into effect, these materials, when combusted for energy recovery, may be considered a solid waste subject to the more stringent Section 129 rules of the CAA, which regulates the emissions from MSW incinerators. If the rule is promulgated, facilities currently using these materials as a fuel such as cement plants, utilities, and pulp and paper plants would have to comply with the more stringent Section 129 rules. This will most likely increase their costs of using these materials as a fuel and may cause them to discontinue the practice if the additional costs outweigh the fuel savings.

15.6 ECONOMIC AND ENVIRONMENTAL BENEFITS OF WASTE TO ENERGY

Rarely do you run across an opportunity that not only benefits the economics of the manufacturing sector, but also has a positive impact on the environment. Using wastes and industrial by-products as a source of energy in the manufacturing process rather than disposing of them in landfills can accomplish this goal.

Waste-to-energy projects can provide a cost benefit to both the facility using the waste and the company generating the waste. The facility using the energy value of the waste is able to reduce manufacturing costs, which helps them remain competitive in today's global economy. For some cement plants, implementing the hazardous waste fuels programs in the 1980s was a question of survival at a time when the industry was struggling with high energy costs, overcapacity, and low-cost imports. Some of these plants would not be operating today had it not been for these programs. Not only did it save jobs, but it actually created jobs because of the testing and processing required in operating a waste-to-energy program.

The generator of the waste benefits from the waste-to-energy programs by way of reduced disposal costs. While they will most likely still pay a disposal fee, it will usually be less than their other alternatives due to the fact that the disposal facility is recovering the energy value of the waste. In the case of a hazardous waste, which would otherwise have to be incinerated, the disposal cost savings can be significant. There can also be intangible cost savings to the generator by avoiding the future environmental liabilities associated with landfilling if that is their alternative. If the generator is recovering the energy value of their own waste on site, they will have the added benefit not having to pay to transport and dispose of their material.

There are a number of environment benefits to waste-to-energy projects as well. The waste is destroyed, which reduces the volume sent to landfills, and it conserves natural resources by reducing the amount of fossil fuels we use. If the waste is a biomass, and it is replacing a fossil fuel, this can help reduce our carbon footprint. Depending on the waste and the fuel it is replacing (i.e., coal), it can also help reduce emissions such as sulfur dioxide. If the waste would have otherwise been incinerated, using it as a fuel can create a net reduction in overall emissions. This is due to the fact that you are using it to replace an existing emission source (i.e., fossil fuel) rather than creating a new source by incinerating the material. If you are recovering the energy value of the material on site, you also eliminate off-site transportation, which reduces the use of transportation fuels and the corresponding emissions from burning the fuel.

15.7 GENERATING HEAT VERSUS POWER

There are a number of methods that can be used to recover the energy value of a waste or industrial by-product. This includes

- using it directly as fuel in an industrial furnace to provide heat to a process such as in a cement kiln or blast furnace
- using it as a fuel in a boiler to produce steam for process heating or to power a steam turbine to produce electricity (or both in the case of co-generation)
- using it directly in internal combustion engine to power a generator to produce electricity
- changing its physical form (i.e., gasification of a solid or anaerobic digestion of a liquid) to produce a useable fuel

The decision on which method to use depends a lot on whether you plan on producing heat or power. If you are paying a high cost for electricity or if you have no requirement for heat, you will probably want to use the energy value of the waste to produce electricity. If you have the option of producing combined heat and power (CHP), you may also want to consider producing electricity. However, for most facilities that have a heat requirement and operate a boiler or industrial furnace, the simplest and generally most economical method to recovery energy from a waste or by-product is to use it directly as a fuel. One of the biggest advantages in using it as a fuel is that there is less energy efficiency losses as you are combusting the material directly to produce heat. While other options may appear to provide more economic value, when you consider the additional capital and material processing costs coupled with the energy losses inherent in some of these processes, you may find that they are not the most cost-effective option.

As an example, the economic value of producing your own electricity from a by-product stream looks attractive on a BTU basis ($19.60/MMBTU) if you are paying $0.0669/kwh. However, once you take into account the additional capital and operating costs of producing electricity and consider the efficiency losses in converting the energy in the material into electricity, the economics may not look as attractive; particularly if you produce less than 1 MW.

In the case of a steam turbine, the electrical efficiently can range from a high of 37% to a low of 10%, depending on its size [35]. The rest of the energy is either lost internally or exhausted in the form of low-pressure steam. If you are using a boiler to produce the steam, there will also be a 10% to 15% energy loss in converting the fuel to steam. Thus, if you assume a 25% electrical efficiency for the turbine and an 85% efficiency for the boiler, the total amount of electricity produced will amount to only 21.25% of the original energy input. Therefore, the $19.60/MMBTU value for the electricity you are purchasing is really worth only $4.16. MMBTU if you are producing it yourself. In other words, this is what you could afford to pay for fuel and break even on making your own electricity versus buying it at $0.0669/kwh. This excludes any operating and maintenance costs and capital payback so the actual value would be lower. However, this also assumes you are not capturing and utilizing the low pressure steam off the turbine (CHP), which would significantly improve overall efficiencies and the economics of the project. If the fuel is essentially free as it is for the chemical industry in using their distillation by-products, the economics of producing electricity can be attractive, particularly if they are using the low pressure steam off the turbine for process heating. However, if you have to spend a lot of money to clean up the waste before you can use it to produce electricity, the economics can fade fairly quickly.

If the waste fuel is clean enough, you may be able to use it as a fuel in an internal combustion engine to power a generator. These systems typically have higher electrical efficiencies (30%–35%), but they are not as flexible in the types of fuels they can handle. If the material is not a liquid or a gas, you will have to process it into a form that can be used by the engine. If you use anaerobic digestion or gasification to convert the waste into a gaseous fuel, you will have to do some clean up on the gas before you can use it directly in the engine. Most of these engines are designed to run on clean fossil fuels, and while the industry is working to modify them to handle dirtier, lower-quality fuels, there are still issues to overcome. In the case of gas turbines, they have a very low tolerance for sulfur, and many gaseous streams like digester gas have a high hydrogen sulfide content. Reciprocating engines are more forgiving, but they may require more maintenance than normal due to contaminants in the fuel and they may lose some efficiency using lower-quality fuels. Digester gas from the anaerobic treatment of municipal sewer sludge is high in siloxanes, which will condense out on engines parts, so these need to be removed before using the gas in an internal combustion engine. It should be pointed out that these issues are not insurmountable, and there is standard equipment available to clean up the biogas. It just becomes a question of how expensive it is to clean up the gas versus the value of the electricity produced. It may be more economical to use the gas in a boiler to produce heat.

In the case of syngas produced from the gasification of a solid, this gas will also have to be cleaned as it will contain tars that will condense out on the engine parts. In addition, the fuel value of syngas, as produced, can be as low as 125 to 150 BTUs/cu ft, which is generally too low for most engines to run on. Thus, you may be better off combusting the syngas to produce heat and using a waste heat boiler to produce steam. You also need to consider the additional labor costs involved in processing and feeding the materials to the gasifier. As with the biogas, it becomes a question of the cost versus savings.

There is no simple answer to the question of which method is the most efficient in recovering the energy value of a waste. It is obviously less capital intensive if you can just use it direct in a boiler or a furnace to produce heat. However, there are many factors that can influence the decision, and no single technology will work in all situations. It can depend on the cost of electricity versus the cost of fossil fuel. It can depend on the physical form of the waste and quality of the waste. Environmental costs and capital costs can be a factor. Processing costs, including labor, need to be taken into account. These all have to be looked at on a project-by-project basis.

15.8 BUSINESS RISKS, LIABILITIES, AND RESPONSIBILITIES

In any new business venture, you need to evaluate the risks versus the reward. Implementing a waste-to-energy project is no different. You need to evaluate all the risks, liabilities, and responsibilities you will be taking on before you begin. You may understand the risks in your own business, but in implementing a waste-to-energy project, you may be opening yourself up to a whole new set of risks and liabilities you are not accustomed to dealing with. This includes environmental risks, operational risks, employee health and safety risks, community relations risks, and waste supply risks. If these are not managed properly, you could jeopardize the any savings you may have enjoyed from the project or, worse, adversely affect your entire business.

If the management of waste is not your primary business, you may want to employ the services of others who are versed in this area to help run the program. Your facility will most likely be subject to a new set of environmental rules due to the waste materials you will be handling, and if you are not familiar with the regulations governing these materials, you could find yourself subject to penalties and fines for not properly managing the wastes.

The waste may contain chemicals that you or your employees are not familiar with. Thus, you will need to train your employees in the proper handling of these materials and equip them with the proper personal protective gear. You will also need to make sure you have the proper handling and storage equipment in place, so you do not have an environmental release. You need to make sure you have all the proper permits and proper paperwork controls. From an operational standpoint, you need to make sure that handling and using the waste do not interfere with the ongoing operations of the plant or affect the quality of the products you are manufacturing.

You should never underestimate the importance of good community relations when it comes to implementing a waste-to-energy project. If you let the relationship with the community become adversarial, your entire project could be at risk. Do not make the mistake of assuming no one will object to the project because it is "green" or because it will reduce the use of fossil fuels and the use of landfills. Whether the concern is truck traffic, noise, dust control, or air emission, there is always a possibility that someone will object to some aspect of the project. Thus, it is critical to keep the community involved from the beginning and keep an open dialogue with them about the progress of the project.

Risk of losing your supply is an obvious but important factor to consider in the waste-to-energy business. The generator of the waste will always be looking to find a more cost-effective way to handle their waste. Thus, it is important that you do a thorough investigation of the market for the waste not only to ensure there is

enough supply, but also to understand the dynamics of the market. You may find that you have to reach out further to find enough supply, and the added transportation costs could make the project uneconomical. You need to be aware of other projects that may compete for the same material and what effect that may have on the availability of supply and costs.

Most risks can be managed and minimized, but you need to be conscious of them and develop a plan to address them before they become a problem.

15.9 STORAGE AND HANDLING OF WASTES

The storage, handling, and feeding of the waste material can be a critical bottleneck in operating any waste-to-energy project, particularly if it is in a form you are not used to handling. Many projects run into operational problems not with the energy recovery technology, but with the material handling system. If the material is a solid, it gets hung up or bridges in the storage tanks or the feed system because it is too wet or sticky. If it is a liquid, it gets too thick to pump, or it turns out to be corrosive to the materials of construction. Waste materials never seem to arrive in the condition you expect based on the initial sample you received. Thus, you need to try and find out as much as you can about how others handle the material and what problems they have run into to minimize down time.

You also need to take the necessary safety precautions based on the material you will be handling. If you will be handling flammable materials, make sure you have the proper explosion proof equipment. If you are going to be shredding materials, you need to be aware of the potential for dust explosions. If you will be producing a biogas, you may need to compress it and store it in pressurized tanks. You need to make sure you have the proper storage containment to prevent any environmental releases.

All these things may add costs on the front end of the project, but they will seem trivial compared with the lost production time if you have a material handling problem after the project has begun.

15.10 SOURCING WASTE MATERIALS: UNDERSTANDING THE SUPPLY CHAIN

Probably the most critical aspect of a waste-to-energy project is ensuring the supply of waste. For those companies who are recovering the energy value of their own on-site generated material, this is not a factor. However, for those who plan on accepting waste from outside sources to supply their waste-to-energy program, this becomes the most important piece of the puzzle in developing a project. Whether you plan on using scrap tires, hazardous waste, or agricultural biomass waste, if you do not have a plan to secure your feedstock, you run the risk of a failed project.

To understand how waste moves from the generator to the recycler, you need to understand the concept of reverse distribution. In a typical supply chain system, goods move from the producer to the wholesaler to the retailer and, ultimately, to the consumer. In the waste business, everything moves in reverse.

In every successful waste management program, whether it is municipal waste or industrial waste, there is an intricate network of haulers, processors, and transfer stations that collect, process, and deliver the waste from the generator to the end disposal site or a recycler. Each type and class of waste generally has its own network set up for the movement of the waste, and these networks take time to evolve and grow when a new material enters the waste recycling market. For example, there is a whole mechanism in place today for scrap tires and used oil whereby small haulers pick up the waste from the tire store or the oil change shop and move it to sites where the materials are consolidated and then on to the processor and, ultimately, to the fuel user. The hazardous waste industry has a network of specially licensed haulers that collect small quantities of waste solvents from body shops, printers, and dry cleaners and bring them to blenders who produce a hazardous waste fuel that gets shipped to cement plants to be used as an alternative fuel. Solid waste moves from the curb to a transfer station, where it is sorted and bulked for hauling to a landfill or a recycler. A whole cottage industry usually made up of small entrepreneurs, each providing a link in the chain, gets built up over time in every successful waste management system, which allows for the smooth flow of materials from the generator to the ultimate disposer or recycler.

Depending on the material you plan on using for your waste-to-energy project, you generally need to work within the collection network, if it is already set up, to be assured of a steady supply. Many times, the volume from an individual generator is not large enough to supply your entire requirement, or they do not have enough to ship full truckloads. The material might not be in the proper physical or chemical form to be used directly in the energy recovery system. Thus, the waste flows through a system of collectors and intermediary transfer points to be segregated, sorted, shredded, or blended to produce a material in the form that the energy recovery facility can handle. Once established, these systems are very efficient and cost-effective at taking small volumes and bulking them up to deliver full truckloads to the ultimate user. While you could go out and set up your own supply chain, this will add marketing and logistical costs to the project, which you may or may not wish to incur.

A key aspect of these networks of collectors and processors is that they act as a buffer between the generator of the waste and the consumer. The collectors and processors will typically have more than one outlet for the waste to ensure they can service the generator in case your facility is temporarily down and cannot receive material. It is important to understand that using waste for energy is not like buying fuel oil. You cannot just turn it off when you do not need it. Generators continue to produce waste whether you are taking it or not. The facility generating the waste may only have so much storage capacity, and if it gets full and they cannot move the material off site, they may have to shut down their production. The collectors and processors smooth out the distribution chain to ensure this does not happen. This is an important point to consider if you decide to enter into a 100% exclusive supply agreement with a waste generator. What happens if your facility goes down and you temporarily cannot take the waste? Similarly, what happens if the supplier of the waste has an operational problem and temporarily cannot supply the waste? Sole sourcing has its advantages, but it has its risks as well.

15.11 TRANSPORTATION LOGISTICS

Transportation is one of the biggest costs associated with waste management, and it can make or break a waste-to-energy project. As discussed in the last section, the logistics involved in moving small quantities of waste from the point of generation to the waste-to-energy facility can be a complex task. This is where transfer stations and processors can provide a valuable link in the reverse

distribution chain to help hold down transportation costs and make the waste collection and disposal system more economical.

Even with the use of transfer stations and collection networks, sometimes the transportation economics just won't work. If a material is not very dense like MSW or ASR, you cannot afford to ship it very far and be competitive with local landfill prices. This is especially true today as transportation costs keep increasing due to the rising cost of transportation fuels. If the waste is not generated near the location where it will be used, you will always run the risk that increased transportation costs could kill the project. When developing a waste-to-energy project, it is important to look at the available supply in the market area and determine what effect a large increase in transportation costs will do to the economics.

15.12 COMMUNITY RELATIONS

One of the biggest mistakes people make when they implement a waste-to-energy project is to not involve the public in the process from the beginning. This is particularly important when you are planning to accept waste from off site. Most of us get caught up believing that, because we are performing a valuable service by keeping wastes out of landfills and reducing fossil fuel consumption, there would be no reason for anyone to object to the project. Never underestimate what objections the community may have with the project, as doing so could be fatal to the project. Local residents could raise concerns about increased truck traffic, potential odor issues, dust control, noise, fire, or explosions. They will want you to assure them that you are not emitting pollutants that could create health issues. It takes only one concerned citizen with a loud-enough voice in the community to make your permit review and approval process a long, frustrating, and potentially futile experience.

It is important to build and maintain a positive relationship with the entire community, not just the city council. If there is a concern from a citizen about some aspect of the project, you need to listen and be empathetic to their concerns. Their perception of a situation may or may not be accurate, but the fact that they are worried or concerned should be acknowledged and respected. As technical individuals, we tend to assume that risk communication is simply an educational process. Explain the scientific facts, and once the community understands them, they will agree with your assessment of the risk [54]. Unfortunately, risk communication does not work that way. People want you to address their concerns, perceived or not, and they definitely do not what to be talked down to.

So before you get to far down the road on your project, it is a good idea as part of your feasibility study to identify what community relations issues you could have. Remember, while air permits and solid waste permits are handled at the state and federal level, siting and zoning permits are made by the community. Their decisions are made on trust and credibility, and no matter how much good science you present, they will not believe you if they do not trust you. While you may not need to hire a risk communications consultant, you need to be versed in how to handle any issues that may arise.

15.13 EFFECT OF WASTE MINIMIZATION AND THE ECONOMY OF CONTINUITY OF SUPPLY

One thing to remember when managing a waste-to-energy project is that you do not always control your own destiny. A change in the regulations, a change in your waste supplier's manufacturing process, or a downturn in the economy can all affect the volume and/or quality of waste you receive. While you cannot control these events, you need to be aware of them and be prepared to mitigate their effect.

15.13.1 Economic Conditions

This risk applies to both the generator of the waste and the entity operating the waste-to-energy program. On the supply side, the volume of a particular waste stream could be reduced if the generator's business slows down, or it could be eliminated completely if they stop manufacturing the product that generates the waste or they move their operations overseas. On the demand side, the waste-to-energy facility's requirements for the waste could decline if their energy demand is reduced due to a cut in production as a result of a slow economy.

This scenario is actually being played out right now because of the current state of the economy. Many cement plants have reduced production or even shut down because of the lack of cement sales. This cutback in production has affected their consumption of hazardous wasted fuels such that waste generators are now having difficulty finding an outlet for their material. That is why it is not always a good idea for either the generator or the user to be sole sourced, at least without a backup plan.

15.13.2 Waste Minimization

In managing a waste-to-energy program, you need to remember that companies are not trying to make waste as it generally costs them money to get rid of it. Plus, it is a cornerstone of the EPA's policy to reduce the amount of waste generated. Thus, companies are always looking for ways to reduce the volume of waste they generate, and this can affect the volume of waste available for energy recovery. A good example of this is the US cement industry that in 1995 was burning close to 325,000,000 gal/year of hazardous waste fuels in 23 cement plants. Today, due to waste minimization, the economic slowdown, a move away from organic solvents in paints and coatings, and manufacturing moving overseas, the volume has dropped to approximately 200,000,000 gal/year burned in 11 cement plants.

15.14 RECYCLING VERSUS ENERGY RECOVERY

When evaluating a waste-to-energy project, you need to be aware of the other value-added recycling alternatives that may compete for the supply of the waste. The EPA's preference is to reuse or recycle a waste if it cannot be reduced. While they accept energy recovery as a viable waste management option, it is not their first choice. If the waste can be reused or recycled, they will promote this over energy recovery.

Almost all waste materials have a recycling or reuse option. Scrap tires can be made into crumb rubber to be used in asphalt or artificial turf. Waste solvents can be recycled, and used oil can be re-refined back into lube oil base stock. The biogenic portion of MSW, municipal wastewater treatment sludge, and animal manure can all be composted to make fertilizer. It comes down to a question of economics. For some of these wastes, the current value of the recycled product does not justify the cost it takes to recycle the waste. Thus, using them as a source of energy is the most cost-effective option. This is not to say things could not change. If you can acquire a raw material at no cost or even get paid to take it as is

the case with solid waste, there will always be entrepreneurs trying to perfect a way to produce a value added product from the waste stream. Thus, it is important to stay current on any new value-added recycling technologies that are being developed so you are not wondering someday what happed to your waste supply.

15.15 USE OF ANAEROBIC DIGESTION AND GASIFICATION FOR WASTE

Many wastes streams have the potential to produce energy, but they are either too dilute (i.e., organic wastewaters and animal manure) or, due to their physical characteristics (i.e., carpet waste or auto fluff), are not suitable to be burned in a convention combustion device.

There are a number of technologies available today that can convert these materials into a usable fuel. Some are thermal, some are chemical, some are biological, and some are thermochemical. Two technologies that are currently getting a lot of attention are anaerobic digestion and gasification. Anaerobic digestion is a biological process that can convert the organics in a wastewater stream into a biogas, which is mostly methane and carbon dioxide. The biogas can be used as a fuel in a boiler or an internal combustion engine. Gasification is a thermochemical process that can convert solid carbon containing wastes like MSW, carpet waste, or ASR into a gaseous fuel, which is mostly carbon monoxide and hydrogen. This "syngas" can be combusted as a fuel or cleaned up and further refined to produce liquid chemicals.

The question is whether these technologies can be cost-effective. The technologies are not in question. Anaerobic digestion is a naturally accruing process that produces methane from the decomposition of organic matter in the absence of oxygen. The use of anaerobic digesters has been around for over 150 years. In developing countries, simple home-based anaerobic digestion systems are used to provide energy for cooking and lighting. Gasification was used in the 1800s to produce town gas from coal for heating and lighting, and it was used during WWII to power motor vehicles in Europe.

The potential issue with these technologies is that they can become too complex and too expensive for the particular application. The anaerobic digestion process can be simple or complex, depending on how efficient you want to make the process. More efficient means shorter hydraulic retention times, better conversion of the volatile solids to biogas, and higher methane concentrations in the biogas. One of the things you can do to optimize the system is to put more energy into it in the form of heat. A mesophilic digester, which is the most common, operates most efficiently at around 98°F. It can operate at 68°F, but you compromise efficiency. A thermophilic digester operates in the range of 122°F to 149°F. The higher temperature accelerates the anaerobic process and generates more biogas, which more than compensates for the extra energy input. This also translates into smaller tanks and less capital costs. However, thermophilic digestion is more sensitive to upsets, so it requires better monitoring. There is also an optimum carbon-to-nitrogen ratio for the anaerobic process to work efficiently. Most waste streams do not have the perfect ratio, so you either have to blend different waste streams to make up the deficiency of one with the other or accept the lower yields.

As you can see, there are a lot of variables to consider if you are going to use anaerobic digestion to recover energy from a waste stream. Depending on how complex the system becomes, you may need additional manpower to monitor the system, which will add operating costs. The biogas may also need to be upgraded, depending on your end use. The gas will be saturated with water vapor, and it will generally contain hydrogen sulfide. The moisture will have to be removed, and depending on what you will be doing with the biogas, the hydrogen sulfide may also have to be removed. If you are digesting municipal wastewater treatment sludge, the biogas will also contain siloxanes, which have to be removed if you plan on burning the gas in an internal combustion engine.

So why go through all this effort. If the organic concentration in the stream is high enough, there is the potential to produce significant amounts of biogas. This biogas can be used in an on-site boiler to provide heat or to produce process steam, or it can be used in an internal combustion engine to produce electricity. It can also be upgraded to natural gas pipeline specifications to be fed into the gas pipeline system or to produce compressed natural gas for vehicle fuel. The process can also reduce disposal costs. If you are an industrial facility discharging a high-strength organic wastewater stream to a municipal wastewater treatment plant, you will most likely be subject to significant surcharges. Anaerobic digestion will reduce the level of volatile organic compounds in the stream as well as producing biogas. The main use for biogas is as a replacement for natural gas. Unfortunately, with current natural gas prices in the United States being so low, it is difficult to make the economics work for many of these digester projects. However, if disposal cost avoidance is factored into the equation the project may make economic sense.

Gasification can also be a challenging operation, depending on how complex you want on make the system. The advantage of the gasification is that it can handle a wide variety of feedstock. However, the feedstock may still require pre-processing to be fed it into the gasifier. This could add significant costs to the project, making it uneconomical. Gasification does have lower emissions than combustion in a boiler due to the limited amount of oxygen and lower temperature profile. The syngas produced from a simple gasification process will be relatively low in BTU content. Thus, the easiest way to use the energy value of the syngas is to combust it and use the resulting heat for process heat or in a waste heat boiler to produce steam. If it you want to use it in an internal combustion engine, the syngas would need to be cleaned as it contains tars that could condense out in the engine. Cleaning the gas can be a very expensive option.

Gasification has its advantages and trade-offs just like any other energy recovery technology. Whether it becomes a significant player in the waste-to-energy market remains to be seen. The syngas does have the potential to create a number of different value-added chemicals and fuels through the Fischer-Tropsch process, so it will definitely play some role in the future.

Both anaerobic digestion and gasification offer promising alternatives for recovering the energy from waste. Numerous municipal wastewater treatment plants use anaerobic digestion, and a lot of work is being done to improve the efficiency of the process. It is increasingly be used on farms to treat animal manure. Gasification has applications with difficult to handle wastes like ASR and carpet waste. As an example, Shaw Floors, a carpet manufacturer, currently gasifies 14,000 tons/yr of carpet scrap generated from their production process and produces steam for their facility. There are also projects being explored to gasify MSW and produce ethanol, methanol, butanol, and other chemicals from the syngas through the Fischer-Tropsch process.

The current issue in the United States with all of these alternative technologies is that land disposal costs and energy costs are still relatively inexpensive. The economics of many of these ap-

plications are based on replacing natural gas as a fuel, and while oil prices have been increasing, natural gas has stayed relatively low over the last year, at $4.00 to $5.00 per MMBTU. Thus, when you take into account the capital costs and operating costs for some of these projects, the returns are marginal at best. The economics can make sense in certain situations particularly if there is a disposal cost issue. Refineries are currently looking at gasification for some of their hazardous oil-bearing waste to reduce disposal costs and produce energy for the refinery. Each project needs to be evaluated individually, as what works for one waste stream will not necessarily work for another.

15.16 UTILIZING HAZARDOUS WASTE FUELS IN THE CEMENT INDUSTRY: CASE STUDY

The cement industry's experience over the last 30 years using hazardous wastes as a fuel in the production of cement can provide a valuable insight into the benefits and challenges of implementing of a sustainable waste-to-energy program.

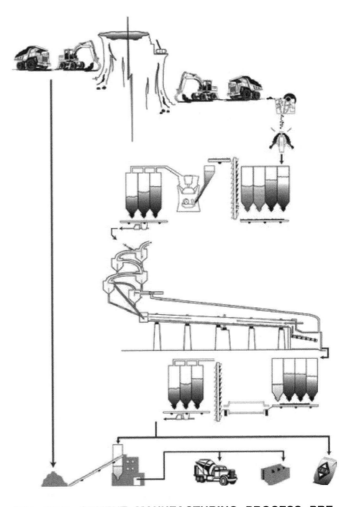

FIG. 15.6 CEMENT MANUFACTURING PROCESS PRE-HEATER/PRE-CALCINER (Reproduced from US Department of Energy: Energy and Emission Reduction Opportunities for the Cement Industry) [36]

Cement is produced in long rotary kilns, which can be 10 to 25 feet in diameter and 200 to 750 feet long. The principal raw materials are limestone (71%), clay, shale, silica sand, and iron ore. The raw materials are ground together and fed into the kiln, where the mixture is heated to 2700°F. The limestone is calcined driving off carbon dioxide, and the resulting calcium oxide reacts with the other ingredients to form golf ball–sized nodules called clinker. The clinker is ground with 5% gypsum to make cement [36]. This process requires a lot of energy in the way and heat, which is produced from burning fossil fuels. Fuel can account for as much as one-third of the cost of producing cement.

There are three basic types of cement plants: wet process, long dry, and pre-heater/pre-calciner. In the wet process, the raw feed is mixed with water and fed into the kiln as a slurry. In the long dry process, the raw materials are fed to the kiln as a dry powder. The pre-heater/pre-calciner technology is a dry process, but the feed is sent through a series of cyclones before it enters the kiln, and the waste heat off the kiln is used in the cyclones to pre-heat and pre-calcine the raw materials. Figure 15.6 shows the flow of materials through a pre-heater/pre-calciner plant. The energy required to produce 1 metric ton of clinker and the thermal efficiency for each process is shown in Table 15.2.

Thus, a wet process plant that produces 500,000 metric tons/yr of clinker would use about 3,125,000 million BTUs per year or about 125,000 tons of coal, while a pre-heater/pre-calciner plant of the same size would use only 2,155,000 million BTUs per year or about 86,200 tons of coal. At current coal prices, that would be a savings of about $2,517,000 per year.

In the early 1980s the cement industry in the United States was struggling. They had just recovered from the effects of the 1973 through 1975 recession as the economy was entering another downturn in 1981. At the same time, fuel costs were increasing due to the energy crisis created by the Arab oil embargo in 1973 to 1974 and the supply disruptions caused by the Iran-Iraq war, which began in 1980. Even though the cement industry used coal as their primary fuel source, the policies put in place by Congress in the late 1970s to encourage the use of alternate fuels including coal put pressure on coal supplies causing its price to increase as well. The dollar had also begun to strengthen against other currencies in the early 1980s, which made exporting cement to the United States an attractive option for other countries who were experiencing their own economic downturns. Thus, even as the US economy began to improve in 1982, low-cost imports caused cement prices to decline further in an already depressed market.

The older wet cement plants were much more vulnerable to the depressed pricing due to their higher energy use. Some of these plants had explored the use of waste solvents as an alternative fuel in the mid 1970s in an attempt to reduce costs, but their experiences were a dismal. There was no real collection network set up to supply the waste fuels, and since there were no regulations in place

yet governing the disposal of waste solvents, landfills were still an inexpensive option. Thus, the cement plants had difficulty finding enough material to supply their needs. The waste fuel they did get was of poor quality as the suppliers did not understand the fuel needs of the cement plant, and there was little, if any, quality control. Therefore, most of the early programs failed, and the cement industry abandoned the practice of using waste solvents as a fuel.

The enactment of RCRA in 1976 began to change of all this. Solvents and other organic chemicals could no longer be disposed of as a liquid in landfills, and thus, it became more expensive to dispose of these materials. A number of enterprising entrepreneurs began to see the value of these wastes as a source of energy, and gradually, a network of collectors and blenders began to emerge to collect the solvent wastes from the generator, blend it, and provide it as a fuel to the cement plants. Waste solvents typically contain about 9000 to 14,000 BTUs/lb, depending on moisture content and other contaminates, so if blended correctly, they could provide a fuel comparable in energy value to coal. From an environmental standpoint, the cement plants provided a sound method to destroy the waste due to their high temperatures and long gas residence times. In addition, since 71% of their raw feed is limestone, the process acts as a natural scrubber to neutralize any acid gases that may be formed. The plants also use bag houses or electrostatic precipitators to collect any particulate they generate. Thus, the cement industry already had the necessary controls in place to combust the waste effectively.

As the network of collectors and blenders began to develop, another group of enterprising companies began to emerge as on-site fuel managers for the cement plants. The cement industry did not understand the hazardous waste regulations, and they were not familiar with handling the flammable organic chemicals that were in the waste fuels they were receiving. There was also poor communications between the blenders and the cement plants regarding scheduling, acceptance criteria, and temporary plant outages, so both groups were becoming frustrated with the other. The on-site fuel managers became the bridge between the two. These fuel managers would take care of getting the proper permits for the cement plant, constructing the tanks and ancillary equipment to process the wastes, and setting up an on-site laboratory to test the incoming material. They also entered into long-term agreements to operate the fuel-blending facility on behalf of the cement plant and to source the waste fuels for them. This allowed the cement plant to go back to making cement, letting those knowledgeable in the waste business manage the waste fuel operations and deal with the suppliers.

Cement plants are huge energy consumers capable of using 50,000 to 100,000 gal/day of waste fuel. Much of the hazardous waste was generated in small quantities or even drums, which the cement plants were not set up to take. Thus, the on-site fuel managers worked closely with the network of collectors and blenders to provide an uninterrupted flow of waste from the generator to the cement plant. The collectors would pick up the drums from the small generators, bring them to a processor who would bulk and blend the materials, and then provide a fuel product to the on-site fuel managers. The on-site fuel manager would test the incoming loads and make the final fuel blend prior to providing it to the cement kiln. The fuel density of the waste solvents was high to allow the materials to be shipped long distances economically. Disposal prices were also high enough to cover any additional transportation costs. Gradually a national network of transporters, brokers, transfer stations, and fuel blenders was developed to efficiently move the material from the generator to the cement plants.

TABLE 15.2   APPROXIMATE ENERGY REQUIRED TO PRODUCE 1 METRIC TON OF CLINKER [36]

Type of Kiln	Energy Required (Million BTUs per metric ton)	Percent Thermal Efficiency
Wet process	6.25	27%
Long dry process	5.64	30%
Pre-heater/pre-calciner	4.31	40%

From 1981 through 1991, the industry grew from three cement plants burning hazardous waste fuels to over 23 plants burning more than 325,000,000 gal/year and saving over 1,000,000 tons/yr of coal. Regulations promulgated in 1984 by the EPA banning most organics including refinery waste from landfills fueled the growth of the industry. The hazardous waste fuel programs became a win/win for everyone as the generators paid a lower disposal fee than if they had to incinerate their waste, the cement plants were able to reduce their operating costs, and the programs had the support of the regulators as it offered a competitive alternative to landfills at a time when the EPA was trying to eliminate the land disposal of hazardous waste.

Unfortunately, the success of these programs brought some unexpected consequences. As the burning of hazardous waste in cement plants grew, it brought a lot of attention to itself. The hazardous waste incinerator industry who had spent millions of dollars permitting and building their facilities felt the cement plants had an unfair advantage since the burning of hazardous waste fuels was not regulated under RCRA. The storage was regulated under RCRA, but the burning was regulated by the individual states through air permits. Thus, they began to lobby against the waste fuel programs. In addition, environmental groups felt that because the burning was not regulated under RCRA, it was causing harm to the environment. Numerous battles were waged in Washington over a 4- to 5-year period on how to regulate the burning of hazardous waste fuels, and hundreds of thousands of dollars were spent by all parties.

This all came to a head in 1991 when the EPA finally enacted the Boiler and Industrial Furnace (BIF) regulations, which regulated the burning of hazardous waste as a fuel. The rule put standards in place for the types of fuels that could be burned and the levels of containments in the fuels. It also required that the cement plants to perform stack tests and maintain continuous emissions monitoring on their stacks. While the industry felt this rule would finally legitimize the waste fuels business and quiet the retractors, it had the opposite effect as it brought an even higher level of scrutiny to the industry. There were also some ambiguities in the new rules that left things open to interpretation, which, in turn, led to some EPA regions enforcing the rules differently than others. Cement plants began getting hit with significant fines for non-compliance, and suddenly, the cement companies who had always been viewed as a good neighbor in their community providing high-paying jobs were being looked upon as villains who were threatening the health of the residents. The cement plant managers were spending an inordinate amount their time going to public meetings and responding to the concerns over these programs.

As you might expect, the cement industry began to sour on the use of hazardous waste fuels. The industry was operating in a sold-out condition at the time, and there were some in the industry who claimed the burning of waste fuels reduced production due to the water content of the fuels. The cement industry was also on an expansion mode, building new and more energy efficient plants that did not need the waste fuel programs to be competitive. In addition, the costs to operate the hazardous waste fuel programs under the new BIF rules were increasing. Then, beginning in 1993, the supply of waste fuels started to decline as the effects of waste minimization, recycling, and other EPA initiatives began to take hold. As the volumes declined, so did the tipping fees as the industry started to fight over the reduced volume. Each cement company began to search for ways to protect their fuel supply. Some formed joint ventures with the large hazardous waste management companies; some purchased a network of fuel blenders, and others bought their on-site fuel managers in an effort to backward integrate. Unfortunately, tipping fees continued to decline, and eventually, many cement plants decided that the environmental risks, potential fines, lost production, and loss of focus on their core business were not worth the savings, and they discontinued their hazardous waste fuels programs. By 2006, there were only 14 cement plants left burning about 240,000,000 gal/year of hazardous waste fuels.

Today the hazardous waste fuel industry has developed into a mature business. The network of collectors and processors is fully developed, and the waste moves smoothly from the generator to the cement plant. The cement plants that are left burning understand the regulations and the waste materials that they are handling. Most have taken over the on-site operations from the fuel managers as a way to reduce costs. Gone are most of the entrepreneurs that helped create the business. Due in part to the down turn in the economy, only 11 cement plants are left burning hazardous waste fuels, and the volume burned is approximately 200,000,000 gal/year. Only three of these plants continue to use the older wet process cement technology.

It may be good for those in the biomass-based renewable energy business to take stock of the cement industry's experience with hazardous waste fuels as they strive to develop their own network of collectors, processors, and energy recovery facilities. There appears to be a lot of similarities.

15.17 MUNICIPAL SOLID WASTE AS A SOURCE OF ENERGY

Combustion of MSW for energy recovery is receiving renewed interest as a result of increasing energy prices over the last few years and the push to develop renewable energy alternatives. Although regulated as a solid waste, the biogenic portion of the waste stream is considered a renewable source of energy. These waste-to-energy facilities burn MSW in large furnaces and recover the waste heat to produce steam for heat and power, which is sold. Some of the existing plants in large metropolitan areas also support district heating and cooling systems by providing steam to supplement the city's heating, cooling and hot water system. In 2008, approximately 32 million tons of MSW was combusted for energy recovery, representing about 13% of the total volume of MSW generated [37].

The municipal waste-to-energy industry has had its challenges throughout the years. The plants have high capital costs with an average plant, costing between $110,000 and $140,000 per daily ton of capacity, and they have high operating costs requiring about 60 people to operate a 1000-tons-per-day plant [38]. An official from a major operator of municipal waste-to-energy plants testified before Congress in 2005 that a new facility capable of processing about 2250 tons of trash per day and generating 60 megawatts of electricity would cost about $350 million and have an operating cost of about $28 million a year [39].

These high capital and operating costs make it difficult for municipal waste-to-energy plants to compete against inexpensive landfill prices. In the example above, the operating costs alone for a 2250-tons-per-day plant equates to $34/ton. A recent report showed the average landfill price in the United States for 2009 at $43/ton [31]. This number is also skewed by the higher prices paid in the Northeast. Landfill prices in the Midwest and Southeast are closer to $36/ton. As a result, many municipalities that operate waste-to-energy plants require the MSW collected within the community to be sent to the facility. As you might expect, most waste-to-energy plants are concentrated in highly densely populated areas like the Northeast, where landfill costs and the costs to transport the waste to the nearest landfill are high.

Most of the growth in the US waste-to-energy business came between 1978 and 1995. As a result of the Arab oil embargo in 1973, a number of laws were enacted by Congress to aid in the development of alternative energy technologies including waste to energy. In 1978, the Public Utility Regulatory Policies Act (PURPA) was enacted, which required utilities to purchase electricity from small power plants like waste-to-energy plants. By guaranteeing a market for the electricity produced from waste-to-energy plants, PURPA enabled developers to find funding for these projects that would have otherwise been viewed as too risky by the financial community. Other laws were also enacted that provided insured loans, loan and price guarantees, and purchase agreements for biomass projects, including waste-to-energy projects using MSW [40]. To ensure these facilities received enough waste, municipalities also enacted flow control measures that required solid waste within the municipality to be sent to the waste-to-energy facility. Thus, the facilities were guaranteed a market for the electricity they produced, and they were guaranteed a supply of waste at the tipping fee they needed. As a result, by 1987, 110 waste-to-energy plants were in operation and an additional 220 in planning or under construction [40].

Everything began to change by the early 1990s. Most of the favorable tax laws were repealed or allowed to expire, and in 1994, the US Supreme Court upheld a challenge to flow control, ruling it violated interstate commerce. Landfills had lowered their tipping fees, making it difficult for waste-to-energy plants to compete due to their high capital and operating costs. The public also started to question the quality of air emissions from the plants and began opposing siting new facilities in their communities. Meanwhile, environmental groups were promoting recycling and warning that waste-to-energy would discourage recycling. A number of waste-to-energy projects were canceled in favor of waste recycling programs. Despite these obstacles, 37 additional waste-to-energy facilities came on line from 1990 to 1995, but by 1995, growth of the industry ground to a halt [40]. While there have been some capacity expansions of existing plants in the last few years and communities have found ways around the flow control ruling, not a single new plant has been build since 1995. In fact, the number of plants has decreased from 102 plants in 2000 to 87 plants in 2007 [41].

MSW waste-to-energy plants make their money in three ways. They charge a tipping fee for the waste they receive, they sell any scrap metal recovered, and they sell the energy they produce in the form of power or heat. The amount of energy they can produce is dependent on the energy value of the incoming material. Figure 15.7 shows the breakdown of materials that make up the MSW stream [37]. Over 80% is combustible, making the stream a good candidate for energy recovery.

To quantify the energy value of MSW, the US Energy Information Administration (EIA) has developed estimates of the energy content for each of the individual components that make up the MSW stream. The current values are given in the Table 15.3 [42].

Using these values, EIA has come up with an estimated energy value for MSW of 5865 BTUs/lb as of 2005. This compares to 5040 BTUs/lb in 1989 [43]. They have also started to track the energy value of the biogenic versus non-biogenic portions of MSW as there is an ongoing debate over what portion of the MSW stream should be considered renewable energy. While this method provides a good way to follow trends, the energy content of the MSW received at each individual facility will be different based on geo-

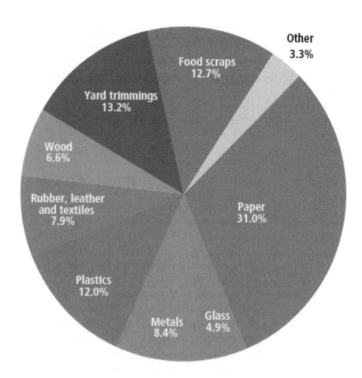

FIG. 15.7 BREAKDOWN OF MSW STREAM 2008 (Reproduced from EPA Municipal Solid Waste Generation, Recycling, and Disposal in the United States: Facts and Figures for 2008) [37]

graphic nuances and whether any materials were removed from the stream to be recycled. The trend does show that the energy value of MSW has been increasing over the years as plastics continue to make up a larger share of the mix.

Estimates of the energy value of MSW from the Waste to Energy Research and Technology Council (WTERT), an industry trade group, would seem to put the energy level somewhat lower. According to WTERT a typical waste-to-energy plant will produce a net 500 to 600 kWh of electricity per ton of MSW [38]. Using the earlier assumption of 21.25% for the electrical efficiency of a boiler/steam turbine system and using an average of 550 kWh per ton, the calculated gross energy input would be about 8.83 MM BTUs per ton of MSW or about 4417 BTUs/lb. This appears to be more in line with other estimates of industry data.

WTERT also suggests that the price a waste-to-energy plant receives for the electricity they sell is about $0.04/kwh [38]. For a 1000-tons-a-day plant generating 550 kWh per ton of MSW this would equate to $22,000 per day in revenue from the sale of electricity or about $22 per ton of MSW. With an estimated $140,000,000 capital cost for a plant this size, the sale of electricity would barely cover the cost of capital based on a 20-year amortization rate, let alone operating costs, overhead, and profit. Thus, the tipping fee makes up an important component of the revenue stream if the facility is going to cover its costs. And based on my experiences, if the facility was located in the Midwest, where most of the electricity is produced from coal, the price received for the electricity probably would not be much more than about $0.025/ kwh or about $13.75 per ton of MSW.

A number of other technologies including pyrolysis, gasification, and plasma arc gasification are currently being explored

TABLE 15.3 TYPICAL HEAT CONTENT OF MATERIALS IN MUNICIPAL SOLID WASTE (MSW) (MILLION BTUS PER TON)

Materials	Million BTUs per Ton
Plastics	
Polyethylene terephthalate[c, e] (PET)	20.5
High density polyethylene[e] (HDPE)	19.0
Polyvinyl chloride[c] (PVC)	16.5
Low density polyethylene/ Linear low density polyethylene[e] (LDPE/LLDPE)	24.1
Polypropylene[c] (PP)	38.0
Polystyrene[c] (PS)	35.6
Other[e]	20.5
Rubber[b]	26.9
Leather[d]	14.4
Textiles[c]	13.8
Wood[b]	10.0
Food[a, c]	5.2
Yard trimmings[b]	6.0
Newspaper[c]	16
Corrugated Cardboard[c, d]	16.5
Mixed paper[e]	6.7

[a]Includes recovery of other MSW organics for composting.

[b]Energy Information Administration, Renewable Energy Annual 2004, "Average Heat Content of Selected Biomass Fuels," Washington, DC, 2005.

[c]Penn State Agricultural College Agricultural and Biological Engineering and Council for Solid Waste Solutions, Garth, J. and Kowal, P. Resource Recovery, Turning Waste into Energy, University Park, PA, 1993.

[d]Bahillo, A. et al. Journal of Energy Resources Technology, "NOx and N2O Emissions During Fluidized Bed Combustion of Leather Wastes," Volume 128, Issue 2, June 2006, pp. 99–103.

[e]Utah State University Recycling Center Frequently Asked Questions. http://www.usu.edu/recycle/faq.htm.

(Reproduced from US Energy Information Administration, Methodology for Allocating Municipal Solid Waste to Biogenic/Non-Biogenic Energy, May 2007) [42]

as alternatives to combustion in converting MSW into energy. The technologies are not new, but they have not been applied to processing MSW until now. Pyrolysis and gasification are similar technologies. Both expose the waste to high temperatures with limited or no oxygen present. In pyrolysis, the MSW is heated by an external source to temperatures of 750°F to 1500°F with no oxygen present, which causes the materials to decompose and produce a gaseous fuel, a liquid fuel, and a solid fuel (char). In gasification, the MSW is exposed to just enough oxygen to allow a small amount of the material to combust creating an internal source of heat to drive the decomposition process and produce a gaseous fuel. Plasma arc gasification exposes the waste to temperatures in excess of 10,000°F using plasma torches to convert the waste material into gas and a slag by-product. In all these technologies, the gas or liquid produced can be used to produce electricity similarly to a typical waste-to-energy facility. The potential advantage of these technologies is that they produce fewer emissions than combustion, and they are more efficient at recovering the energy content of the waste. However, they have not yet been scaled up on MSW to determine their economic viability.

Capturing the methane gas off of landfills provides another opportunity to recover energy from MSW. Over time, the organics in MSW in the landfill start to decay, and the landfill becomes a huge anaerobic digester. Like any anaerobic digestion processes, this creates biogas, which is mainly methane and carbon dioxide. Prior to the last few years, the gas was either vented or flared. Rising energy cost and improved technology have created an opportunity to collect and clean the gas and burn it in an internal combustion engine to generate electricity or produce pipeline quality gas and feed it into the natural gas pipeline.

Capturing methane off landfills provides not only an economic benefit by lowering energy costs, but also a powerful environmental benefit as well. MSW landfills are the second largest anthropogenic source of methane emissions in the United States, accounting for 22% of the total and releasing an estimated 30 million metric tons of carbon equivalent to the atmosphere in 2008 alone [44]. Methane is 21 times more potent of a greenhouse gas than CO_2 so reducing its emissions while recovering its energy value provides a unique opportunity to help solve two issues [44].

Another area that is garnering some interest these days is to combine the gasification of MSW with the Fischer-Tropsch process to convert the syngas off the gasification process to ethanol, methanol, and other chemicals. The Fischer-Tropsch process uses catalysts to convert the carbon monoxide and hydrogen in the syngas to various chemicals. It has been used commercially in a number of industries for years, mostly using coal as the feedstock. There are a number of US and Canadian companies currently working to commercialize this in the MSW market. Another US-based company has developed a process to produce ethanol from the syngas using microorganisms instead of catalysts. Once again, time will tell whether these technologies will be economical in a full-scale operation.

15.18 WASTE HEAT RECOVERY

Just about every industry exhausts a certain amount heat from their industrial processes. For some industries such as petroleum, chemical, power, steel, and glass manufacturing, the amount can be significant. This exhausted heat is lost energy and lost profits. Recovering the energy value contained within the waste heat could reduce production costs, save non-renewable fossil fuels, and reduce greenhouse gas emissions. One such project was initialed in 2005 by Integral Power, LLC, a private company, at a petroleum coke pant in Port Arthur, Texas. The 1800°F to 2000°F flue gas off the coke calcining kilns, which had been exhausted out the stack in the past is now used to produce 450,000 lb per hour of high-pressure stream. The steam is used to make electricity, and the excess is sold to a neighboring refinery.

Unfortunately, many waste heat recovery projects cannot justify the capital expense because the streams are too low in temperature, or the volume of exhaust gas is too small, making energy recovery impractical. In other situations, the gas may be too dusty or corrosive, making energy recovery too expensive due to the frequent maintenance required on the recovery equipment. For some companies, the competition for capital within the organization may be such that the return on invested capital for these projects is not quick enough. This is why a lot of projects are taken on by third parties under a build-own-and-operate scenario.

For low- to medium-temperature waste heat, there are a number of recovery technologies such as absorption chillers (200–400°F) to provide air conditioning or refrigeration, organic rankine cycle, and kalina cycle systems (200–1000°F) to produce electrical power and heat pumps (<200°F) for heating and cooling. The use of these technologies on small projects has been limited in the past due to the capital cost relative to the amount of energy recovered [45].

Waste heat recovery offers an opportunity to produce tremendous amounts of energy that is currently being wasted. However, further work needs to be done to develop technologies to make low-temperature heat recovery more economical.

15.19 CONCLUSION

Waste-to-energy projects are usually driven by one of two factors: reducing energy costs or reducing disposal costs. Many times, both are a factor. This was the case in the 1980s when we saw incredible interest in alternative fuels spurred by high oil costs. A multitude of legislative incentives were initiated to spur investment in waste-to-energy. At the same time waste disposal, costs were rising due to the enactment of new regulations that put stricter controls on the management of solid waste. The combination of higher energy prices, higher waste disposal costs, and government support led to a number of waste-to-energy projects being implemented both in the industrial sector and in the MSW sector. The same effect is happening today. While disposal costs are not increasing as rapidly as they did in the 1980s, the increase in fossil fuel prices over the last few years has spurred a flurry of activity both in the private and public sectors to develop alternative sources of energy including waste-to-energy.

The question for any waste-to-energy project becomes an economic one: How much effort must be put in to collecting, transporting, and processing of the waste to recover its' energy value? If the amount of energy recovered compared with the cost to recover that energy is marginal, the project is almost always destined to fail. Energy markets are extremely volatile and unpredictable, so if the energy savings are marginal, a minor shift in energy prices or a loss of governmental support could quickly make the project uneconomical. Thus, it is important not to get caught up in the euphoria of waste-to-energy and implement a project just because it is the latest trend. It needs to be based on a sound, sustainable economic model. This model can take different forms. It could be based on solving a waste disposal problem, or it could be based on the fact that you are located in a high-cost electricity region, so producing power from waste makes sense. It could even be based on environmental sustainability. Whatever the reason, there needs to be a driver.

In the United States, if there is not a regulatory driver for a particular waste, it is difficult to make the economics of a waste-to-energy project work because of low landfill costs and relatively low energy costs. This was evidenced by the carpet industry's effort to try and develop a program to use scrap carpeting as a fuel within the cement industry. The exception is if you can use the energy value of your own wastes or by-products on site. Here you have the advantage of saving on the cost of transporting and disposing of the material. This avoided cost can offset the additional capital and processing costs associated with a waste-to-energy project.

Many European countries as well as Japan use waste to energy far more extensively than the United States. However, this has been driven by reduced landfill capacity and subsequent environmental laws restricting the use of landfills and governments promoting waste-to-energy. In addition, energy costs in these counties are generally much higher than they are in the United States. Denmark, a country with a population of about 5.5 million people, has 30 municipal waste-to-energy plants that supply of good portion of the district heating and electricity for cities and communities [46]. However, the country's average industrial cost of electricity in 2009 was about $0.17/kwh, and their natural gas cost was about

$14/MMBTU [47]. These higher energy costs make waste-to-energy projects a viable alternative.

Waste-to-energy will continue to play an increasing role in our future energy needs. As our population continues to grow, so will the volume of waste we generate, and it will need to be managed in some fashion. While the battle of energy recovery versus recycling will continue to be waged, most agree that there is a need for both. It will be important for policy makers to continue to support the private sector in their efforts to make energy recovery more efficient and affordable. While waste-to-energy is not going to solve the world's energy problem, it can certainly be a part of it.

15.20 REFERENCES

1. U.S. Environmental Protection Agency, Waste - Non Hazardous Waste. http://www.epa.gov/epawaste/nonhaz/index.htm

2. U.S. Environmental Protection Agency, The National Biennial RCRA Hazardous Waste Report (Based on 2007 Data). http://www.epa.gov/waste/inforesources/data/br07/national07.pdf

3. U.S. Environmental Protection Agency, Municipal Solid Waste Generation, Recycling and Disposal in the United States: Facts and Figures for 2008. http://www.epa.gov/epawaste/nonhaz/municipal/pubs/msw2008rpt.pdf

4. U.S. Environmental Protection Agency, Municipal Solid Waste in the United States. 2007 Facts and Figures. http://www.epa.gov/epawaste/nonhaz/municipal/pubs/msw07-rpt.pdf

5. U.S. Environmental Protection Agency, Wastes — Laws and Regulations. History of RCRA. http://www.epa.gov/waste/laws-regs/rcrahistory.htm

6. U.S. Environmental Protection Agency, Introduction to: Definition of Solid Waste and Hazardous Waste Recycling (40 CFR 261.2 and 261.9). http://www.epa.gov/epawaste/inforesources/pubs/hotline/training/defsw.pdf

7. U.S. Senate, Solid Waste Disposal Act (As Amended), Section 1004. Page 10. http://epw.senate.gov/rcra.pdf

8. U.S. Environmental Protection Agency, Wastes — Hazardous Waste. http://www.epa.gov/waste/hazard/index.htm

9. U.S. Environmental Protection Agency, Wastes — Hazardous Waste Types. http://www.epa.gov/waste/hazard/wastetypes/index.htm

10. U.S. Environmental Protection Agency, Clean Air Act — Title I — Air Pollution Prevention and Control. http://epa.gov/oar/caa/title1.html

11. U.S. Environmental Protection Agency, The original list of hazardous air pollutants. http://www.epa.gov/ttn/atw/188polls.html

12. U.S. Environmental Protection Agency, Six Common Air Pollutants. http://www.epa.gov/air/urbanair/

13. U.S. Energy Information Administration, Average Price of Coal Delivered to End Use Sector by Census Division and State. http://www.eia.doe.gov/cneaf/coal/page/acr/table34.html

14. U.S. Energy Information Administration, Spot Prices of Crude Oil, Motor Gasoline, and Heating Oils, January 2009 to Present. http://www.eia.gov/pub/oil_gas/petroleum/data_publications/weekly_petroleum_status_report/current/pdf/table11.pdf

15. U.S. Energy Information Administration, Natural Gas Prices. http://www.eia.gov/dnav/ng/ng_pri_sum_dcu_nus_m.htm

16. U.S. Energy Information Administration, Average Retail Price of Electricity to Ultimate Customers. http://www.eia.doe.gov/cneaf/electricity/epm/table5_3.html

17. U.S. Energy Information Administration, Chemicals Industry Analysis Brief. http://www.eia.doe.gov/emeu/mecs/iab98/chemicals/energy_use.html

18. U.S. Department of Energy, Energy Efficiency and Renewable Energy — Energy and Environmental Profile of the U.S. Pulp and Paper Industry. http://www1.eere.energy.gov/industry/forest/pdfs/pulppaper_profile.pdf

19. American Forest and Paper Association, 2010 AF&PA Sustainability Report.

20. U.S. Energy Information Administration, Steel Industry Analysis Brief. http://www.eia.doe.gov/emeu/mecs/iab98/steel/expenditures.html

21. American Iron and Steel Institute, The Coke Oven By-product Plant, Mike Platts, http://www.steel.org/AM/Template.cfm?Section=How_Steel_is_Made&TEMPLATE=/CM/ContentDisplay.cfm&CONTENTID=12307

22. Metals Advisor, Coking Process Description. http://www.energysolutionscenter.org/HeatTreat/MetalsAdvisor/iron_and_steel/process_descriptions/raw_metals_preparation/coking/coking_process_description.htm

23. U.S. Department of Energy Office of Industrial Technologies — Energy and Environmental Profile of the U.S. Iron and Steel Industry, pp. 46, 51. http://www1.eere.energy.gov/industry/steel/pdfs/steel_profile.pdf

24. Rubber Manufacturers Association, Scrap Tire Markets in the United States, 9th Biennial Report (May 2009). https://www.rma.org/publications/scrap_tires/index.cfm?PublicationID=11502&CFID=342464&CFTOKEN=87096242

25. U.S. Environmental Protection Agency, Proposed Rule Making: Identification of Non-hazardous Secondary Materials That Are Solid Waste — Construction and Demolition Materials — Building-Related C&D Materials (March 18, 2010). http://www.epa.gov/epawaste/nonhaz/define/pdfs/cd-building.pdf

26. U.S. Environmental Protection Agency, Biosolids Generation, Use and Disposal in the United States. http://www.epa.gov/osw/conserve/rrr/composting/pubs/biosolid.pdf

27. U.S. Environmental Protection Agency, Proposed Rule Making: Identification of Non-hazardous Secondary Materials That Are Solid Waste — Used Oil (March 18 2010). http://www.epa.gov/epawaste/nonhaz/define/pdfs/used-oil.pdf

28. U.S. Environmental Protection Agency, Introduction to: Used Oil (40 CFR 266 Subpart E and Part 279). http://www.epa.gov/solidwaste/inforesources/pubs/hotline/training/uoil.pdf

29. U.S. Environmental Protection Agency, Proposed Rule Making: Identification of Non-hazardous Secondary Materials That Are Solid Waste — Auto Shredder Residue (March 18 2010). http://www.epa.gov/epawaste/nonhaz/define/pdfs/auto-shred.pdf

30. California Environmental Protection Agency, Evaluation of Shredder Residue as Cement Manufacturing Feedstock (March 2006). http://www.dtsc.ca.gov/TechnologyDevelopment/upload/auto_shredder_report.pdf

31. The Envirobiz Group, Waste Business Update, Average US Municipal Solid Waste Gate Rates Increased by 2% in 2009 (May 20, 2010). http://www.envirobiz.com/envirobiz-waste-business-update.asp

32. Carpet America Recovery Effort, Annual Report 2009. http://www.carpetrecovery.org/pdf/annual_report/09_CARE-annual-rpt.pdf

33. U.S. Environmental Protection Agency, Homeland Security Research - Emissions from Combustion of Post-Consumer Carpet in a Cement Kiln (2005). http://www.epa.gov/nhsrc/pubs/reportConsumerCarpet050905.pdf

34. Shaw Floors Web site, Energy. http://www.shawfloors.com/Environmental/EnergyDetail

35. U.S. Environmental Protection Agency, Combined Heat and Power Partnership — Technology Characterization: Steam Turbines, (page 8). http://www.epa.gov/chp/documents/catalog_chptech_steam_turbines.pdf

36. U.S. Department of Energy, Energy Efficiency and Renewable Energy — Energy and Emission Reduction Opportunities for the Cement Industry. http://www1.eere.energy.gov/industry/imf/pdfs/eeroci_dec03a.pdf

37. U.S. Environmental Protection Agency, Wastes - Non-Hazardous Waste - Municipal Solid Waste. Municipal Solid Waste Generation, Recycling, and Disposal in the United States: Facts and Figures for 2008. http://www.epa.gov/epawaste/nonhaz/municipal/pubs/msw2008rpt.pdf

38. Waste-to-Energy Research and Technology Council Web site, Answers to FAQ. http://www.seas.columbia.edu/earth/wtert/faq.html

39. Statement of Michael Norris, Director of Business Development, American Ref-Fuel Company. Testimony Before the Subcommittee on Select Revenue Measures of the House Committee on Ways and Means. http://webharvest.gov/congress110th/20081126141752/http://waysandmeans.house.gov/hearings.asp?formmode=printfriendly&id=2697

40. U.S. Department of Energy, National Renewable Energy Laboratory. Managing America's Solid Waste. http://www.nrel.gov/docs/legosti/fy98/25035.pdf

41. U.S. Environmental Protection Agency, Wastes — Non-Hazardous Waste — Municipal Solid Waste. Municipal Solid Waste in the United States, 2007 Facts and Figures, p. 150. http://www.epa.gov/epawaste/nonhaz/municipal/pubs/msw07-rpt.pdf

42. U.S. Energy Information Administration, Methodology for Allocating Municipal Solid Waste to Biogenic/Non-Biogenic Energy (May 2007). http://www.eia.doe.gov/cneaf/solar.renewables/page/mswaste/msw.pdf

43. U.S. Energy Information Administration, Municipal Solid Waste (MSW) Heat Content and Biogenic/Non-Biogenic Shares. http://www.eia.doe.gov/cneaf/solar.renewables/page/mswaste/mswtable1.html

44. U.S. Environmental Protection Agency, Landfill Methane Outreach Program. http://www.epa.gov/lmop/basic-info/index.html

45. Northwest CHP Application Center. An Overview of Industrial Waste Heat Recovery Technologies for Moderate Temperatures Less Than 1000°F (September 2009). http://chpcenternw.org/NwChpDocs/AnOverviewOfIndustrialWasteHeatRecoveryTechForModerateTemps.pdf

46. Waste-to-Energy Research and Technology Council, Energy Recovery — European Countries. http://www.seas.columbia.edu/earth/wtert/globalwte_europe.html

47. Europe's Energy Portal. http://www.energy.eu/

48. Code of Federal Regulations, Title 40, Part 63, Subpart EEE — National Emission Standards for Hazardous Air Pollutants from Hazardous Waste Combustors. http://ecfr.gpoaccess.gov/cgi/t/text/text-idx?c=ecfr&sid=91c20f8e8507bf661f597fcc2cb29648&tpl=/ecfrbrowse/Title40/40cfr63b_main_02.tpl

49. Code of Federal Regulations, Title 40, Part 503 — Standards for the Use or Disposal of Sewage Sludge. http://ecfr.gpoaccess.gov/cgi/t/text/text-idx?c=ecfr&sid=bc8b9aa7babf9677b5fdde559a2f19b7&rgn=div5&view=text&node=40:29.0.1.2.41&idno=40

50. Code of Federal Regulations, Title 40, Part 279 — Standards for the Management of Used oil. http://ecfr.gpoaccess.gov/cgi/t/text/text-idx?c=ecfr&sid=bc8b9aa7babf9677b5fdde559a2f19b7&rgn=div5&view=text&node=40:26.0.1.1.9&idno=40

51. U.S. Environmental Protection Agency, Non-Hazardous Waste. http://www.epa.gov/osw/basic-solid.htm

52. U.S. Environmental Protection Agency, RCRA Orientation Manual. http://www.epa.gov/osw/inforesources/pubs/orientat/rom.pdf

53. U.S. Environmental Protection Agency, Definition of Solid Waste and Recycling Presentation. http://www.epa.gov/epawaste/hazard/dsw/downloads/soliddef.ppt

54. Katherine E. Brown. Purdue University, Department of Communication. *Why Rules for Risk Communication are not Enough: A Problem-Solving Approach to Risk Communication.*

55. Burlington Electric, Joseph C. McNeil Generating Station, Wood Fuel Facts, Wood Fuel Costs. https://www.burlingtonelectric.com/page.php?pid=75&name=mcneil.

GEOTHERMAL ENERGY AND POWER DEVELOPMENT

Lisa Shevenell and Curt Robinson

ABSTRACT

Geothermal energy constitutes an indigenous, sustainable, continuous, base load renewable resource available to power developers on most continents. This chapter discusses geothermal energy basics, resource exploration, and types of resources along with their utilization, sustainability, benefits, and the potential environmental consequences of resource development. The current state of knowledge and possible expansion of the resource base via Enhanced Geothermal Systems technologies are also discussed.

16.1 INTRODUCTION

Geothermal literally means "earth heat." Areas where energy can be tapped due to high heat flow in the top 5 km of the Earth's crust are considered geothermal. Geothermal is one of the least recognized renewable energy resources, and as such, is significantly underutilized in the US despite being the only base load renewable power source, exclusive of hydroelectric, available 24 hours per day. The Earth is a natural source of geothermal energy, which can be converted into electrical power. In the simplest and most understandable sense, geothermal energy can be extracted from a subsurface region of hot rock "the reservoir" via fluids that are heated as they flow through permeable rocks and fractures in the reservoir. Those fluids can then be brought to the surface under pressure via wells, where they can drive a turbine or heat exchanger to produce electricity. Sometimes these heated fluids reach the surface naturally to form hot springs or geysers. One of the best-known of such geothermal features in the world is the Old Faithful Geyser in Yellowstone National Park. Geothermal and hydrothermal systems are virtually synonymous, because a geothermal power system requires a working fluid (water or steam) to facilitate the transfer of the earth's heat to the surface and then into a turbine or heat exchanger.

"Although the Earth is an immense source of heat, most of this heat is either buried too deeply or is too diffuse to be exploited economically" [1]. However, technologies have advanced, resource assessment and surveillance tools have improved, and economics have changed in the last 30 years. The April 2010 meeting of the World Geothermal Congress resulted in The Bali Declaration [2], which proclaimed that geothermal energy can change the world.

This declaration recognized that "Geothermal energy is the only renewable energy source which is totally independent of daily, seasonal, and climatic variation, allowing it to provide power with a higher availability than any other energy source including fossil fuels and nuclear." While not strictly true in the case of being independent of seasonal fluctuations as efficiency declines in the summer months, geothermal is being recognized as a base load renewable power source in contrast to others such as wind and solar that operate intermittently.

Hot springs have been utilized for thousands of years for heating, cooking, bathing, and preparing hides. Some of the earliest traces of human use, and settlement near hot springs, can be found in Asia and Europe. Geothermal hot springs and their minerals were believed to have healthful benefits (balneology) and used for at least 5000 years. One of the best-known sites for healthful hot springs is Bath, England, which was developed by the Romans in 75 AD and was named Aquae Sulis (the waters of Sulis), after a Celtic deity associated with the site. Other sites were used by humans as many as 10,000 years ago in the US [3].

It is estimated that annually, geothermal energy is more than a $3 billion world-wide business and the prospects for geothermal power development increase each year. The recent global recession in 2009 and 2010 slowed the infusion of private equity, but several national governments continue to commit financial resources to geothermal development. In 2009, the US government granted $400 million to more than 120 geothermal research, development, and deployment efforts. This is a 20-fold increase in geothermal research dollars compared to the average annual amount invested in the previous two decades. This commitment of capital is opening the door to innovative research into drilling technologies, operations in high-temperature environments, improved resource assessments, enhanced geothermal systems, and more efficient geothermal heat pumps.

In 2010, geothermal energy generated 10,715 megawatts (MWe) of electric power worldwide and another 50,583 MWt (megawatts thermal) through direct use applications [4]. It is thought that geothermal power, if developed and managed properly, can provide sustained and continuous energy for the United States and much of the world for thousands of years. "Domestic resources are equivalent to a 30,000-year energy supply at our current rate of use for the United States!" [5]

Commercial production of geothermally generated electric power humbly began in 1904 in Larderello, Italy. By 1913,

Larderello was producing 12.5 MW of electricity and it continues to produce today. Other areas that have notably been developed over the years include New Zealand (1958), the United States (1960), and Japan (1961) [6].

The ten nations leading geothermal power production in 2010 (in descending order) are: US (3093 MWe), Philippines (1904 MWe), Indonesia (1197 MWe), Mexico (958 MWe), Italy (843 MWe), New Zealand (628 MWe), Iceland (575 MWe), Japan (536 MWe), El Salvador (204 MWe), and Kenya (167 MWe) [7]. Seven of these ten countries occur in the Pacific Ring of Fire (Fig. 16.1), named for its volcanic and seismic activity related to subduction of the tectonic plates. Additional countries developing or expanding existing geothermal resources are Turkey, Australia, Chile, China, Guatemala, Honduras, Nicaragua, New Guinea, and Peru. Turkey and Iceland have had the greatest relative geothermal development of the last 5 years. Bertani [7] observed that a total of 10,715 MWe of geothermal electric energy is being produced worldwide in 2010 and 18,500 MWe of production are projected by 2015.

At the time of this writing, the US is the leading producer of geothermal power. The most productive states in 2010 are: California (2553 MWe), Nevada (442 MWe), Utah (46 MWe), Hawaii (35 MWe), and Idaho (16 MWe) [7]. Bertani [7] predicts that by 2015, the US will have 5400 MWe in production. Other anticipated geothermal producing states are Alaska, Florida (from oil and gas wells), New Mexico, Oregon, Texas, and Wyoming.

Estimates on the amount of available geothermal energy in the US have varied widely, but a recent report by the US Geological Survey (USGS) estimates 39,000 MW from conventional hydrothermal systems and 517,000 MW from Enhanced (engineered) Geothermal Systems (EGS) [8]. The estimate for conventional hydrothermal systems may be conservative. For instance, the USGS estimate for conventional systems in Nevada is only 1391 MW, whereas an estimate using additional data not utilized by the USGS indicates there are likely over 5000 MW that could be developed in the near term [9]. The USGS EGS estimate is of the same order of magnitude as was obtained in an extensive study on the feasibility of exploiting EGS systems, which notes up to 100 GWe (100,000 MW) of generating capacity over the next 50 years [10].

It is obvious that renewable energy development is sensible, economical, and environmentally necessary. Geothermal power development has several advantages over other forms of renewable energy. First, geothermal energy systems provide continuous base load power. Second, geothermal energy power plants require a smaller footprint than other renewable power plants such as wind or photovoltaic. Third, geothermal energy power plants have comparatively small environmental impacts, if they are managed effectively.

The Earth was formed nearly five billion years ago and it can be said that it never quite cooled off. The earth is losing heat continuously to space. The Earth's core is molten and thought to be approximately 5500°C [11]. There are at least four sources of heat which create some form of geothermal energy: (1) convection and the migration of denser materials downward, (2) heat from radioactive decay of potassium, thorium, and uranium isotopes, (3) the remnant heat from gravitational collapse, and (4) impacts from planetesimals [12–14].

Geothermal areas occur in areas of high heat flow in the upper part of the crust where it can be utilized. Normal geothermal gradients are approximately 25°C/km, and geothermal areas occur in areas with greater gradients.

There are several technological approaches to developing geothermal power. Geothermal heat pumps (also known as ground source heat pumps) and direct use applications (building heating, greenhouses, spas, fish farms) are relatively simple technologically and inexpensive to develop. In contrast, the production of electricity can require considerable inputs of capital, technology, and labor, as will be discussed in subsequent sections.

There are two general categories of natural geothermal systems that have been exploited to date, those with an upper crustal source of magmatic heat (magmatic-heated systems), and those without that source of heat (amagmatic or extensional systems). Many actively exploited magmatic-heated systems are associated with magmatism along crustal plate boundaries. The convecting, viscous mantle is moving large, rigid portions of the crust in different directions at different speeds. Molten material moves up along spreading centers at mid-ocean ridges to create new crust while older pre-existing oceanic crust moves outward toward continental plates (Fig. 16.2). As oceanic crust encounters the continental plates, it subducts underneath the continental crust, and descends into hotter portions of the earth. Partial melting of the descending "slab" forms magma bodies, which, because of their buoyancy, begin to rise through the overlying continental crust. These magma bodies heat the upper crust, and produce volcanic activity when they reach the surface (Fig. 16.2). Figure 16.1 shows the locations of plate boundaries and the relative motion along them. Active magmatic-heated geothermal systems commonly occur along these plate boundaries in the vicinity of volcanoes and favorable fault structures.

The second type of geothermal system is the amagmatic or extensional-type. These systems do not typically occur along plate boundaries, nor do they occur in close proximity to young magmatism or volcanic activity, but instead they are commonly found in areas where the Earth's crust is pulling apart and thus thinning. Where thinning occurs (Fig. 16.3), depths to the hot mantle are less and temperature gradients are higher. Additionally, extension produces open fracture networks (faults) through which geothermal fluids (rain, snow) are able to circulate to relatively great depths (up to 4+ km) and attain temperatures of up to 200°C or greater. There are indications that the depth of penetration of meteoric fluids in fault systems is enhanced if there is a strike-slip component to crustal motion, or in other words, if the crustal tectonics can be characterized as transtensional (shear plus stress that pulls fault planes apart) instead of purely extensional [16, 17]. Examples of electricity-producing amagmatic systems can be found in the Great Basin of the western US, and western Turkey. Other types of hydrothermal systems are discussed in some detail in the section on Six Potential Geothermal Production Systems.

The ability of a liquid-dominated geothermal reservoir to produce electricity is correlated with the temperatures in fluids developed from a geothermal well, because the enthalpy or heat content of a liquid-dominated resource is largely determined by temperature. The USGS and other expert sources [8] consider liquid-dominated geothermal resources above 150°C to be high-temperature, 90°C to 150°C to be moderate-temperature (power generation), and resources below 90°C to be low-temperature (historically, directly used). Low-temperature resources (<90°C) were historically believed to not yield enough heat to generate power. That changed in 2006 when a power system developed by UTC (United Technologies Corp.) was installed at Chena Hot Springs, Alaska and began generating approximately 225 kWe of power using 74°C well water.

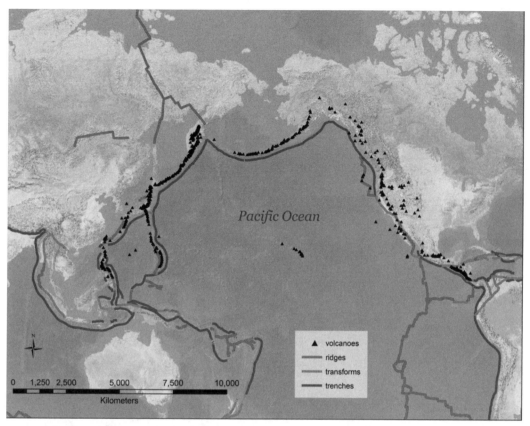

FIG. 16.1 MAP OF THE WORLD SHOWING LOCATIONS OF PLATE BOUNDARIES (RING OF FIRE) AROUND THE PACIFIC OCEAN (Source of data used to plot the map can be found at [15])

16.2 BRIEF SURVEY OF THE LITERATURE

Geothermal technologies are evolving and there are several key sources (some referenced in this chapter) that can provide the basis for future study of geothermal systems. Many geothermal practitioners regard Ronald DiPippo's first and second editions of *Geothermal Power Plants* [18] and [19] to be outstanding resources for geothermal power developers. Some of the chapters are: Geology of Geothermal Regions; Exploration Strategies and

Techniques; Geothermal Well Drilling; Reservoir Engineering; Single-Flash Steam Power Plants; Double-Flash Steam Power Plants; Dry-Steam Power Plants; Binary Cycle Power Plants; Advanced Geothermal Energy Conversion Systems; Energy Analysis Applied to Geothermal Power Systems; and several case studies; and the second edition includes a chapter on the environmental impact of geothermal power plants. The so-called MIT report, *The Future of Geothermal Energy* [10], is a useful reference and is available online, and includes chapters on subjects such as

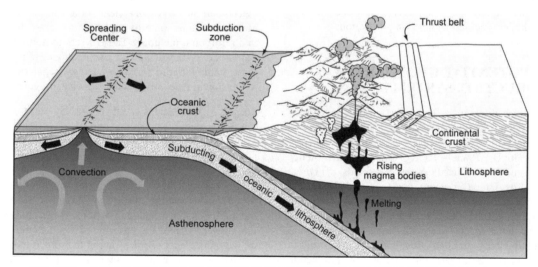

FIG. 16.2 CROSS SECTION OF THE EARTH SHOWING PLATE TECTONIC PROCESSES

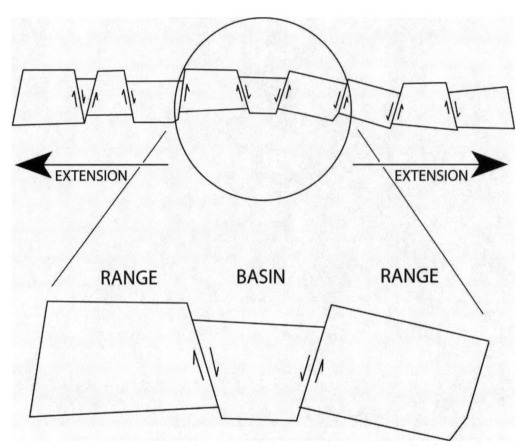

FIG. 16.3 IDEALIZED CROSS-SECTION ACROSS THE BASIN AND RANGE SHOWING THE ALTERNATING BASINS AND RANGES CHARACTERISTIC OF THE GREAT BASIN OF THE WESTERN U.S.

Geothermal Resource-Base Assessment, Recoverable EGS Resource Estimates, Review of EGS and Related Technology, Subsurface System Design Issues, Drilling Technology and Costs, Environmental Impacts, and Economic Analysis. Another good reference is *Geothermal Systems: Principles and Case Histories* [20]. Additionally, *Geothermal Energy: An Alternative Resource for the 21st Century* [6], has chapters that address heat transfer, geothermal systems, exploration, and assessment; it does not discuss some of the largest geothermal fields in production.

16.3 SIX POTENTIAL GEOTHERMAL PRODUCTION SYSTEMS

Most experts agree there are essentially six different types of geothermal production systems: (1) Natural Hydrothermal Systems, (2) Enhanced/Engineered Geothermal Systems - EGS (previously known as Hot Dry Rock — HDR), (3) Magmatic Systems, (4) Geopressured and Co-Produced Geothermal Fluids, (5) Direct Use Systems (or Low-Grade Systems), and (6) Geothermal Heat Pumps (Ground Source Heat Pumps or GeoExchange) [21, 5]. Each system requires different engineering and technological approaches for their development (see below). All systems require heat, permeability/porosity for circulation, and a working fluid to extract the heat. Strategies for sustaining (renewing) production over time are also important.

16.3.1 Natural Hydrothermal Systems

Natural hydrothermal systems represent the oldest and most conventional source of geothermal power production. These systems are subterranean reservoirs that transfer and circulate water that contains extractable heat toward the surface (the fluid is commonly called brine or working fluid). There are two types of natural hydrothermal systems; those that contain water (liquid) and others that contain steam (vapor). Steam-dominated systems are relatively rare but include some of the world's largest producing geothermal reservoirs, including The Geysers in northern California and Larderello in Italy. There are two geologic environments that host these natural hydrothermal systems: magmatic and amagmatic (extensional) (see previous discussion and figures).

16.3.2 Enhanced Geothermal Systems

Deep systems are the domain of EGS or HDR and are expected to reside from 3 km to 10 km or deeper and require considerable technology to develop, because sufficient permeability for fluid flow is usually not present under natural conditions. In such cases, a 3-dimensional network of permeable fractures needs to be created (or engineered) in the subsurface, through which fluid flow can be induced to enable production of fluids as in a conventional geothermal power system. Engineering efforts historically have focused on the inducement of fracturing in the subterranean region, and/or expanding or creating a fracture space (that increases permeability/porosity) to develop a geothermal aquifer. Research is ongoing to determine how to keep these fractures open to allow long term aquifer development. Other technological approaches may require injection of working fluids to create a reservoir or aquifer where no fluids are present. This practice would be impractical in arid climates where water supplies are scarce. It is believed that these systems could be developed on most continents if the technological challenges can be overcome.

While some EGS systems are expected to exploit high temperatures located at great depths, other EGS may be relatively shallow but are underproductive (or naturally non-productive) geothermal systems that are manipulated in an attempt to make energy extraction economical. Only a few systems of this nature have been created and none have been commercially successful to date. In the 1970s and 1980s, an experimental site was developed at Fenton Hill, New Mexico that produced up to 10 MW of power over a 30 day period [22]. More recently there have been developments in Australia [23] and in Europe at Basel, Switzerland, and Soultz, France [24, 25]. The Basel EGS project was put on hold following inducement of a magnitude 3.4 earthquake during fluid injection in 2006. This earthquake unnerved the local population and caused a small amount of non-structural damage in the town [26].

A study published in 2006 evaluated the quantity of potential energy available and development issues related to EGS. "Based on growing markets in the United States for clean, base load capacity, the panel thinks that with a combined public/private investment of about $800 million to $1 billion over a 15-year period, EGS technology could be deployed commercially on a timescale that would produce more than 100,000 MWe or 100 GWe of new capacity by 2050. This amount is approximately equivalent to the total R&D investment made in the past 30 years to EGS internationally, which is still less than the cost of a single, new generation, clean coal power plant" [10].

16.3.3 Magmatic Systems

These subterranean regions are systems that are molten or semi-molten and host temperatures greater than 650°C. It is believed that several technological breakthroughs will be required to drill and extract heat from these promising regions [21]. Natural sites occur adjacent to stratovolcanos (Cascades), hot spots (Hawaii), and divergent tectonic plate boundaries (Iceland, East African Rift Valley). In 2009, drilling experiments penetrated the supercritical temperature zone at the Iceland Deep Drilling Project (IDDP) and encountered magma [27], but significant fluid flow rates were not encountered. It is thought that these magmatic resources may be developed for geothermal power in the future [14].

16.3.4 Geopressured and Co-Produced Geothermal Fluid Systems

Found in deep sedimentary basins in various locations in the United States and other parts of the world, geopressured systems are known to "contain three forms of energy: methane, heat, and hydraulic pressure," [5]. Geothermal fluids, co-produced in conjunction with oil and gas, represent a resource that has long been ignored by oil and gas developers. It is thought that this resource could be harnessed to produce electrical power for the petroleum producer or for sale. "Like geopressured resources (areas of rapid sedimentation that are characterized by pressures and temperatures), co-produced water can deliver near-term energy savings, diminish greenhouse gas emissions, and extend the economical use of an oil or gas field" [5].

16.3.5 Direct Use System

As the name implies, Direct Use involves the direct use of hot water for providing heat energy for residential, commercial, or municipal purposes. Generally, source fluid temperatures are less than 100°C — temperatures not formerly considered sufficient for geothermal power production. Although typically underutilized in the US relative to many other parts of the world, direct use systems are employed in several cities in the United States, including Boise, Idaho, Klamath Falls, Oregon, and Reno, Nevada. Boise has an extensive district-heating system for the State Capitol region. Direct use can be employed for heating, drying, aquaculture, horticulture, and de-icing sidewalks and roads.

16.3.6 Geothermal Heat Pumps

Also referred to as GHPs, geothermal heat pumps use the constant temperature of the earth for heating and cooling applications for residential and commercial buildings. The principle behind this application is that temperatures remain fairly constant a few feet below the surface of the earth. These systems use heat pumps to extract heat from underground for space heating and to pump heat into the ground when cooling of buildings is required. It is interesting to note that Fort Knox, Kentucky installed the largest geothermal heat pump system in the world with more than 150 geothermal heat pumps in use in 2006, and has thereby reduced their fossil fuel energy use by 20% [28]. "Geothermal heat pumps provide an important example of how low-grade thermal energy, available at shallow depths from 2 to 200 m, can be used to provide substantial savings in the cost of heating and cooling of buildings" [5].

16.4 GEOTHERMAL RESOURCE DEVELOPMENT — EXPLORATION, DRILLING, AND RESERVOIR ENGINEERING

Considerable attention must be given to a number of details during the development of a potential geothermal power production site (well). The following variables must be considered: exploration and assessment of the resource site, depth required to develop the well, production temperature, fluid flow rate, power potential, connectivity to a grid, geology, and geomorphology, hydrochemistry, drilling resources, power generation systems, and human resources. In many cases, the human dimension (labor, government officials, regulatory agencies, etc.) can be just as challenging as the geotechnical problems an organization might encounter.

Exploration of a potential geothermal development site is a critical component of a project's success. There are many exploration techniques that can be employed to achieve an accurate assessment of the resource. Geologists, geophysicists, geographic information system (GIS) experts, hydrogeologists, remote sensing experts, and others can be called upon to assist in assessing the potential location, depth, and capacity of a resource. The developer must understand that the average geothermal gradient will yield 25°C to 30°C per kilometer. Steeper gradients will yield higher temperatures at shallower depths, and thus greater opportunities for economical extraction. Geological expressions on the surface, such as geysers, fumaroles, thermal springs, mud pots, faults, sinter, tufa, borates and other deposits can provide clues that the explorer can use to evaluate and develop the resource. Combs and Muffler [29] suggested six steps in assessing a potential resource: (1) literature search, (2) airborne survey, (3) geologic and hydrologic survey,

(4) geochemical survey, (5) geophysical survey, and (6) drilling. Although published in 1973, these steps are still relevant today. Methods of assessment have improved, yet two general techniques have proved critical for early stage geothermal exploration for decades: (1) chemical geothermometer calculations of spring and well waters, and (2) temperature gradient measurements, and heat flow calculations.

Chemical geothermometers can be used to estimate subsurface reservoir temperatures, if an accurate thermal spring or well water chemical analysis is available. Calculation of geothermometers from spring chemistry has been a key component of early stage geothermal exploration for decades. Although there are many different geothermometers (e.g., see [30] for a partial listing, and [31] for geothermometer and plotting spreadsheets), two general forms are most commonly used: SiO_2 geothermometers (the equation for which varies depending on the fluid temperature at depth; e.g., [30]), and the empirical Na-K-Ca [32] geothermometer, often employed with a Mg correction [33].

The second critical type of data collected and used for decades in geothermal exploration, is temperature gradient and related heat flow calculation. A comprehensive heat flow database can be found at the Southern Methodist University (SMU) geothermal laboratory web site [34]. As of fall 2010, SMU has been funded to update thermal data for North America. Maps and products (to be discussed in a later section) will be updated with new data and data types (e.g., those not necessarily obtained purely for geothermal exploration purposes).

Wright [35] suggested seven geophysical techniques in exploration: (1) shallow surveys of thermal gradients and heat flow, (2) electrical methods, including resistivity surveys, telluric currents, electromagnetic, magnetotelluric, and audio magnetotelluric, (3) gravity and microgravity surveys, (4) airborne and ground magnetic surveys, (5) seismic studies including microseismics, refraction surveys, and reflection surveys, (6) radiometric heat production surveys, and (7) borehole geophysical logging and seismic profiling. "The culmination of an exploration program is marked by exploratory drilling of borewells" [35]. The exploratory wells can be used to evaluate anomalous temperature conditions, assess permeability, and conduct production tests to assess the capacity for power production [6]. Many of the electrical methods mentioned by Wright [35] are no longer in use, but magnetotelluric (MT) and Controlled Source Audio Magnetotelluric (CSAMT) surveys have become very important tools for the geothermal exploration community.

In the fall of 2006, a geothermal exploration workshop was hosted by the Geothermal Resources Council in which both traditional and new methods of geothermal exploration were highlighted [36]. Many of the more traditional methods of exploration (as noted in the previous paragraph) are the more advanced and expensive forms of exploration that typically take place after geochemical evaluation of the thermal waters. However, one older, inexpensive, early stage exploration tool has recently been revitalized and significantly improved: the measurement of shallow subsurface temperatures. Measurements of temperatures at 2 m depths at many locations throughout a site over a short period of time can be used to locate thermal anomalies and define possible drilling targets [37, 38]. These surveys are helpful for locating blind geothermal systems, as documented by [39] which provide several examples where 2 m temperature surveys were used to identify geothermal anomalies as a follow-up investigation of borate occurrences initially located in the field with remote sensing techniques.

Figure 16.4 shows an example of one such shallow temperature survey at Columbus Marsh, Nevada. Historically, borate was known to have been mined at Columbus Marsh based on information in the literature. A search was conducted on a geochemical database of existing water analyses, and water sampling was also conducted near these borate crusts. Both sets of data indicated high subsurface temperatures. In 2006, Kratt [40] used satellite remote sensing data (ASTER satellite imagery) to identify borate-rich evaporite crusts on the playa. Subsequent field work included further characterization of these borates using a field spectrometer. Shallow temperature surveys were then conducted to delineate possible drilling targets (Fig. 16.4). Temperature surveys identified three anomalies on the margin of the playa, although anomaly A is an artifact removed after performing an albedo (a measure of how strongly an object reflects light) correction. However, the other two anomalies remain and can be used to delineate drilling targets. As a measure of corroboration, a 7-m-deep temperature probe at the southwest anomaly confirmed that temperatures are continuing to rise rapidly with depth.

After a potential geothermal resource has been assessed via exploration and the prospects for developing the geothermal resource are positive, a more comprehensive drilling program will begin. Geothermal developers have been the beneficiaries of technological breakthroughs in the oil and gas industries. There are many similarities between oil and gas development and geothermal development. The one key difference is that geothermalists are seeking heat and therefore require equipment that can be used in higher temperature environments. Metallurgical breakthroughs have provided more resistant drill bits that require less replacement and less downtime during the drilling operation.

Drilling a geothermal production well can be one of the most challenging and costly aspects of developing a geothermal resource, and production wells are estimated to represent 25% to 40% of the costs associated with developing a power plant, according to some experts (e.g., [41]). These costs can include road and drill pad construction, casing, cementing, labor, mud to avoid boiling of the water, etc. The depth of the well and the geological formations present will determine how long drilling will actually take. A good understanding of the subsurface geology can help the developer better manage drilling costs. Conventional geothermal wells can be shallow or exceed several thousand meters. In the future, it is expected that developers will embrace the development of EGS, that is, enhanced or engineered geothermal systems. Since temperatures increase with depth, improved drilling technology in the future will facilitate the economic extraction of heat from greater depths [6].

Many geothermal plants employ a system of wells for both extraction and injection of geothermal fluids. One of the keys to successfully managing and sustaining a geothermal reservoir is the development of an aquifer replenishing system, which means that the optimal siting of and development of injection wells can be just as important as the drilling of production wells.

16.5 GEOTHERMAL ELECTRICAL POWER GENERATION

The technology of geothermal power generation is very similar to other methods of electrical power generation. In the simplest of terms, steam is produced to turn the turbines that charge the generator to produce electrical power. Generally, there are three types of geothermal power plant systems: Dry Steam, Flash Steam, and Binary-Cycle. Power generation systems vary in size, ranging from kilowatts to multi-megawatts, and variations of design may result in hybrid systems such as flash-binary systems.

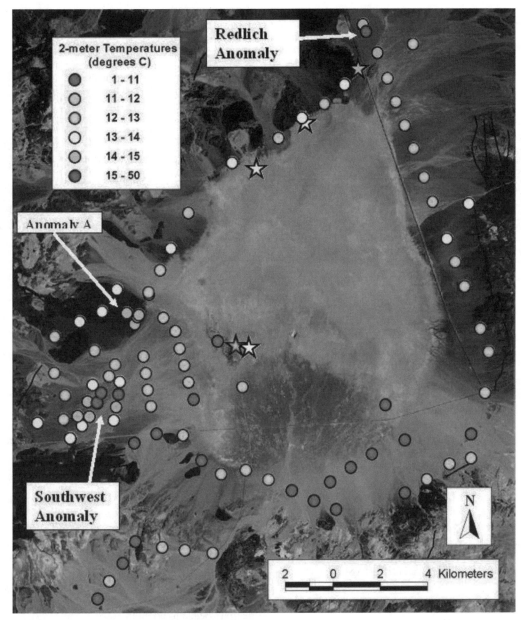

FIG. 16.4 ASTER SATELLITE IMAGE SHOWING RESULTS OF A SHALLOW TEMPERATURE SURVEY (UNCORRECTED) AT COLUMBUS MARSH, NEVADA (*Source*: [39])

16.5.1 Dry Steam

Dry Steam, the oldest and simplest type of power plant (Fig. 16.5) was first developed in the Larderello field in Italy in 1904. It has also been employed commercially at The Geysers (California) since 1960. Steam from a geothermal well drives the turbine directly. Larderello and The Geysers are the two main dry steam production areas in the world. Other dry steam areas are found in Matsukawa, Japan; Kamojang, Indonesia; Poihipi Road at Wairakei, New Zealand; and Cove Fort, Utah in the United States. "The general character of a dry-steam reservoir is that it comprises porous rock features fissures or fractures . . . that are filled with steam," [18].

16.5.2 Flash Steam

Superheated fluids are introduced into a tank and "flashed" (rapidly boiled) to vapor, the steam from this vaporization drives the

turbine (Fig. 16.6). According to DiPippo [18], "The single-flash steam plant is the mainstay of the geothermal power industry. It is often the first power plant installed at a newly-developed liquid-dominated geothermal field." It is most efficient for high-temperature systems, typically >180°C.

16.5.3 Binary-Cycle Power

These systems require a heat exchanger (Fig. 16.7), which allows the heated brine (water) to cycle through a heat exchanger and heat a secondary working fluid that has a lower boiling point than water. These working fluids are then vaporized, the steam from which turns the turbine. "The working fluid, chosen for its appropriate thermodynamic properties, receives heat from the geo-fluid, evaporates, expands through a prime-mover, condenses, and is returned to the evaporator by means of a feed pump" [18].

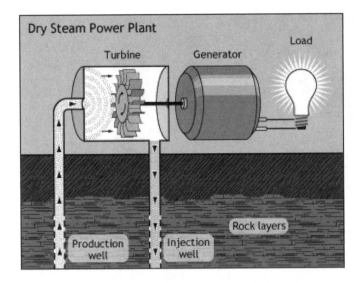

FIG. 16.5 DRY STEAM POWER PLANT (*Source*: [42])

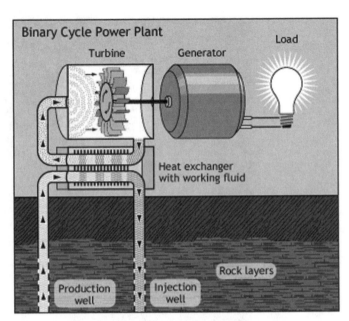

FIG. 16.7 BINARY CYCLE POWER PLANT (*Source*: [42])

Some of the benefits derived from a binary-cycle system include the ability to use a lower-temperature resource as well as a closed loop such that water is not lost and all produced water is re-injected. Binary cycle systems also reduce the likelihood of build-up of calcium carbonate or other mineral scaling in the wells and/or turbines. It should be noted that some binary systems use the Rankine cycle or organic Rankine cycle (ORC). Working fluids used are typically isopentane or isobutene. Experiments are currently being considered that would use CO_2, metal organic compounds, and other compounds as working fluids.

These examples clearly do not present an exhaustive discussion of power plant systems, but they do suggest that some power plant systems might be more applicable to a particular geothermal resource's temperature and geochemistry. However, some reservoir fluids can be used with both binary and flash, an example of which is the Bradys power plant in Nevada (Fig. 16.8).

16.6 CURRENT US GEOTHERMAL POWER PRODUCTION EFFORTS

Geothermal development in the US largely occurred in the late 1980s and early 1990s, with relatively little additional geothermal development until 2005 when the 30 MW Galena 1 power plant was installed at Steamboat, NV. New BLM leasing rules were written over a 2-year period and once completed, the result was the first totally competitive lease sale on August 14, 2007. Table 16.1 shows the results of leasing from 2007 to 2010 for Utah, Nevada, and California, the latter two in which geothermal development in the US has been most active as of 2010.

During 2005 and 2006 the Bureau of Land Management (BLM) was re-writing rules for leasing. One of the main changes was that all leases would go to competitive bid; they would only be leased non-competitively if they were not sold during auction. After parcels were available for lease following the two year rule-writing and decade long lull in the geothermal industry as a whole, leasing activity began to boom in 2007.

Geothermal activity is compared among western states in the next two figures. As can be seen in Fig. 16.9, California produces more geothermal electric power than all of the other states combined, but Nevada leads the country in active geothermal projects as of April 2010 (Fig. 16.10). More geothermal exploration and research has been conducted in Nevada both recently and historically, and the business climate and regulation in Nevada are more favorable than in some surrounding states, resulting in Nevada having the most geothermal exploration and development of any state in the country in recent years.

Geothermal development has been steadily increasing since 2005 (Figs. 16.11 and 16.12) following the resumption of leasing in 2007. Figure 16.12 shows the annual capacity in the western US installed between 2005 and 2010; a steady trend of increasing geothermal capacity is seen after 2005. A total of 7057 MW of capacity were under development as of April 2010 [49]. This is more than double the installed capacity in 2010, indicating that geothermal energy production had begun to accelerate by 2010, partly in response to renewable portfolio standards set for utilities and the public desire for clean, renewable, indigenous power

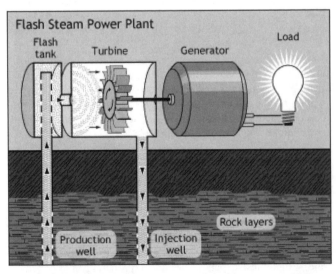

FIG. 16.6 FLASH STEAM POWER PLANT (*Source*: [42])

FIG. 16.8 VIEW OF BRADYS POWER PLANT AND FUMAROLES FROM A DISTANCE. BRADYS GEOTHERMAL PLANT IS OPERATED BY ORMAT, NV, AND HAS BOTH FLASH AND BINARY UNITS PRODUCING POWER (*Source:* [43], Photo by Mark Coolbaugh. Nov. 2, 2002)

sources. Of the traditional renewable power sources (wind, solar, biomass, and geothermal), geothermal is a base load power source with the highest capacity factor (discussed below) and is, thus, of greatest utility in displacing traditional, fossil fuel, base load power sources.

16.7 COMMENTS ON ENHANCED GEOTHERMAL SYSTEMS AND THE PROSPECTS FOR THE FUTURE

Enhanced or Engineered Geothermal Systems (EGS; Fig. 16.13) have received renewed and focused attention since the publication of the MIT Report [10]. Some of the earliest experiments for EGS were conducted by the Los Alamos National Laboratory in the 1970s at Fenton Hill, New Mexico. At the time, the experiments attempted to exploit so-called Hot Dry Rock (HDR) resources by drilling to a depth of about 3000 m, identifying a 195°C resource, penetrating and hydrofracturing the granite, injecting and extracting a working fluid, and ultimately producing about 4 to 10 MWe of power [6, 22, 50]. The Fenton Hill experiment was active between 1974 and 1979 and was eventually closed in 1995. Other experiments followed, including the Rosemanowes Quarry, in southwestern England. Several EGS experiments are taking place at Soultz-sous-Forets in France, Landau in Germany, Cooper Basin in Australia; and in the US.

Why are researchers interested in EGS and HDR? Duchane and Brown [50] aptly summarize its potential:

"The total amount of heat contained in HDR at accessible depths has been estimated to be on the order of 10 billion quads (a quad is the energy equivalent of about 180 million barrels of oil and 90 quads represents the total US energy consumption in 2001). This is about 800 times greater than the estimated energy content of all hydrothermal resources and 300 times greater than the fossil fuel resource base that includes all petroleum, natural gas, and coal [52]. Like hydrothermal energy resources already being commercially extracted, HDR holds the promise for being an environmentally clean energy resource, particularly with regard to carbon dioxide emissions, which can be expected to be practically zero" [50].

The MIT Report advanced the enormous potential that EGS might hold:

"By evaluating an extensive database of bottom hole temperature and regional geologic data (rock types, stress levels, surface temperatures, etc.), we have estimated the total EGS resource base to be more than 13 million exajoules (EJ, where one EJ is equivalent to 10^{18} joules). Using reasonable assumptions regarding how heat would be mined from stimulated EGS reservoirs, we also estimated the extractable portion to exceed 200,000 EJ or about 2000 times the annual consumption of primary energy in the United States in 2005. With technologic improvements, the economically extractable amount of useful energy could increase by a factor of 10 or more, thus making EGS sustainable for centuries" [10].

TABLE 16.1 RESULTS OF COMPETITIVE GEOTHERMAL LEASING IN THE WESTERN U.S. BETWEEN 2007 AND 2010

	Parcels Offered	Parcels Sold	Acres	Total Receipts	Average per Acre
Nevada BLM geothermal leases					
2007	43	43	122,849	$11,669,821	$94.99
2008	35	35	105,212	$28,207,806	$268.11
2009	108	82	323,223	$8,909,445	$27.56
2009N	26	12	91,987	$49,223	$0.54
2010	114	75	212,370	$2,762,292	$13.01
Total	326	247	855,640	$51,598,587	

Source: [44]

2009N — refers to non-competitive sales.

	Parcels Offered	Parcels Sold	Acres	Total Receipts	Average per Acre
California BLM geothermal leases					
2007	17	17	14,722	$53,761	$3.65
2008	10	10	5,352	$256,242	$47.87
2009	15	15	11,399	$131,126	$11.50
2010	9	9	4,887	$1,984,943	$406.16
Total	51	51	36,362	$2,426,073	

Source: [45] and [46]

	Parcels Offered	Parcels Sold	Acres	Total Receipts	Average per Acre
Utah BLM geothermal leases					
2007		3	6017	$3,675,696	$610.86
2008		44	144372	$5,400,738	$37.41
2009	1	1	228	$57,250	$251.05
2010	17	17	60,320	$335,393	$5.56
Total	18	65	210,938	$9,469,077	

Source: [47] and [48]

The 13 million EJ is the geothermal resource base (stored thermal energy) and not the power that can be generated. The MIT report [10] includes several maps of estimated temperature at depth. This data was used to estimate available energy, and one such map is reproduced here for 6.5 km depth (Fig. 16.14).

EGS continues to offer great potential, and experiments in developing geothermal power using this approach are ongoing. The U.S. Department of Energy committed $132.9 million specifically to EGS-related experiments in 2009 (EGS demonstrations and EGS components research) and another $98.1 million to innovative exploration and drilling. Experiments are taking place at Desert Peak, Nevada and at The Geysers in California. The EGS projects funded include sites at already productive hydrothermal systems (i.e., Desert Peak has a power plant, but one unproductive well can be stimulated) as well as at existing oil and gas wells that also produce hot water from considerable depths (4.6+ km).

However, experts are not yet in agreement on how geothermal resources will be developed. Some argue that conventional hydrothermal systems ought to be developed first, while others claim that more than $1 billion needs to be invested in deep EGS before it can be developed, and another group asserts that the appropriate technology for economical drilling in high-temperature regimes is not yet available. However, it is clear that some fairly significant hurdles need to be cleared before this technology will be economical. Research should include an improved knowledge of the rock mechanics of potential EGS reservoirs. How to create and predict where fractures will occur is difficult, so the practice is to drill one well, stimulate (hydrofracture) it, and try to map the fractures, in part by monitoring microearthquakes. A second well is then drilled to intercept the fractures. Because much of this EGS resource may be fairly deep, keeping fractures open at depth will be a challenge. Research is being conducted to improve the assessment of the systems, so as to help locate and drill into fractures.

Other areas that may provide prodigious amounts of energy are at the margins of volcanic activity. Experiments are continuing at the Icelandic Deep Drilling Project and Newberry Caldera in Oregon, as well as other sites for this type of resource, including Hawaii.

16.8   UTILITY IMPACT ON GEOTHERMAL DEVELOPMENT

Utilities are continuing on the path of integrated resource planning (IRP) to provide energy services to their customers. Utilities use IRP principles to assess the cost effectiveness of all supply and demand side options available to provide electric service. Based on the assessment, utilities build a renewable energy and energy efficiency portfolio to serve customer needs. Two geothermal technologies — power production and geothermal heat pumps (GHP) — can be included in the utilities' portfolios. As a result, utilities can play an important role in promoting more geothermal development.

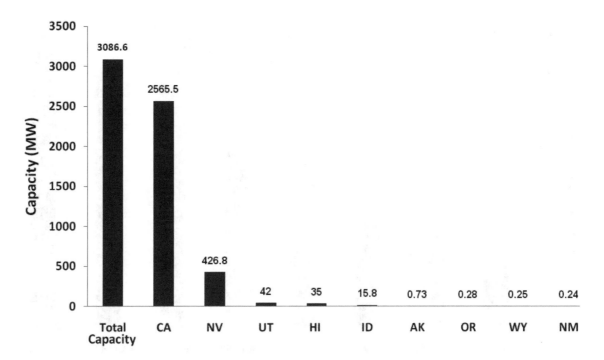

FIG. 16.9 INSTALLED CAPACITY IN THE U.S. AS OF APRIL 2010 (*Source:* [49])

In recognition of this role, a Utility Geothermal Working Group (UGWG) was formed at the 2005 Geothermal Resources Council Annual Meeting. This group provides annual update reports. The UGWG's mission is to accelerate the appropriate integration of geothermal technologies into mainstream applications. To help accomplish its mission, the group conducts periodic training events in the form of webcasts and workshops.

16.9 GEOTHERMAL POWER PRODUCTION FINDINGS

Based on information gained at various public events, geothermal power generation is of great interest to utilities; however, some utilities regard geothermal power plants as a risky proposition based on the possibility that the first wells drilled into the geothermal

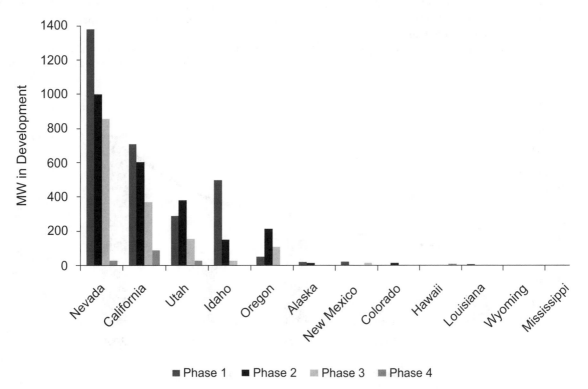

FIG. 16.10 TOTAL CAPACITY IN DEVELOPMENT IN THE US AS OF APRIL, 2010 (*Source:* [49])

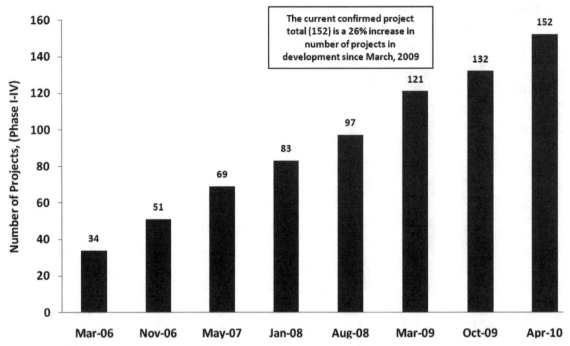

FIG. 16.11 THE NUMBER OF PROJECTS UNDER DEVELOPMENT HAS BEEN STEADILY INCREASING SINCE 2006 (*Source:* [49]).

reservoir will not be successful. Geothermal power plants are also capital-intensive, requiring significant funding up front before the project produces any revenue. Utilities are more confident with the plants and are more willing to negotiate a financeable power purchase agreement (PPA) with a developer, if the following five conditions are met:

- A delineated geothermal resource, with a bankable report that defines probable long term performance
- A defined permitting path without pitfalls
- A credible developer with a proven project management track record

- The control of the entire geothermal resource to preclude competing interests for the same fluid/steam supply
- The use of proven technologies

Utilities may be willing to enter into PPAs if the output compares favorably with the "default power plant," which currently is a gas-fired combined cycle plant. The utilities estimate purchasing power from the default choice in the range of $65 to $90/MWh. The price includes capital, operation and maintenance (O&M), and fuel costs.

The price that a geothermal power plant developer can offer to a utility in a PPA largely depends on (1) the exploration, drilling,

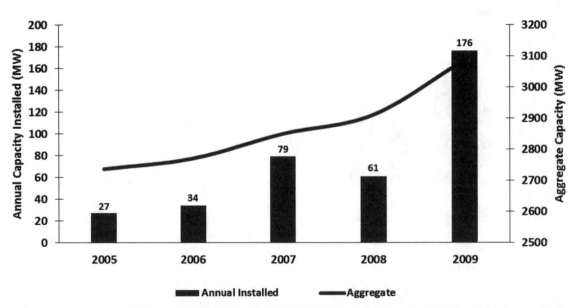

FIG. 16.12   TOTAL INSTALLED CAPACITY IN THE WESTERN U.S. FROM 2005 TO 2009 (*Source:* [49]).

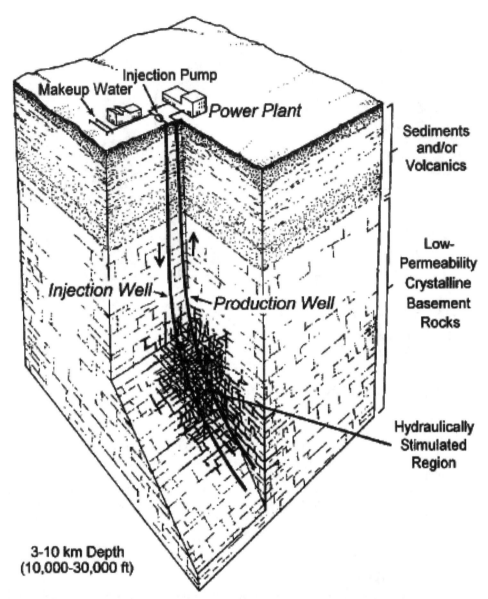

FIG. 16.13   SCHEMATIC OF A TWO-WELL EGS IN A LOW PERMEABILITY FORMATION SHOWING HYDRAULICALLY STIMU-LATED AREA WITH CREATED FRACTURES (*Source:* [10] and [51])

and development costs of getting the project on line and (2) the financing charges associated with the costs. The costs for a typical 20 MWe power plant can be seen in the Table 16.2.

A major impact to development cost is the local, regional, national, and global competition for commodities such as steel, cement, and construction equipment. Geothermal power is competing against other renewable and non-renewable power development, building construction, road and infrastructure improvements, and all other projects that use the same commodities and services. Until equipment and plant inventories meet the increase in demand for these commodities and services, project developers can expect the costs of them to rise.

Using the above costs as a basis, a typical geothermal power plant has a capital cost of $4000/kW (kW of generating capacity). This capital cost is translated to a MWh cost by applying an annual factor reflecting interests rates for financing the total capital cost. At an annual factor of 0.2, reflecting an interest rate of 18% to 20%, the capital financing costs are $104/MWh. The financing costs as-

sume that the plant is on-line 90% of each year. At an annual factor of 0.15, reflecting an interest rate of 13% to 15%, the capital financing costs are $76/MWh [55].

TABLE 16.2   TYPICAL COSTS FOR CONSTRUCTION OF A 20 MWE POWER PLANT IN THE US

Development Stage	Cost (Millions of $)
Exploration and resource assessment	8
Well field drilling and development	20
Power plant, surface facilities, and transmission	40
Other costs (fees, operating reserves, and contingencies)	12
Total Cost	80

(*Source:* [54])

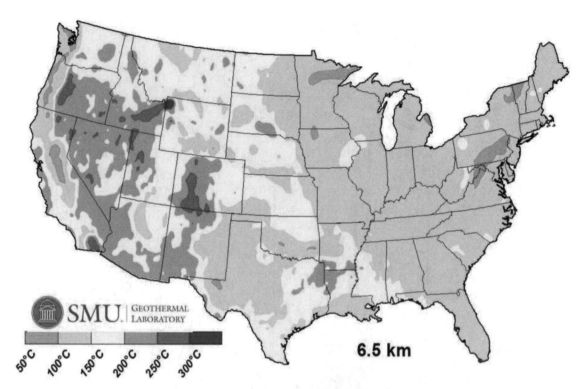

FIG. 16.14 CALCULATED TEMPERATURE AT A DEPTH OF 6.5 KM (*Source:* [53])

Typical O&M cost for a plant is about $15/MWh [55]. The O&M costs include reservoir management and assume that the power plant uses Organic Rankine Cycle (ORC) technology for energy conversion with air to air cooling towers. ORC technology uses a moderately high molecular mass organic fluid such as butane or pentane to absorb the heat from the geothermal fluid and drive the turbine. The technology has the benefits of high cycle and turbine efficiencies, low mechanical stress on the turbine, reduced turbine blade erosion, and the reduced need for full time operators to be present.

If the power plant uses a different technology or water to air cooling towers, the O&M costs are likely to be higher. Using these two annual factors and adding the O&M cost to the annualized capital costs, the developer may be able to offer a utility output in

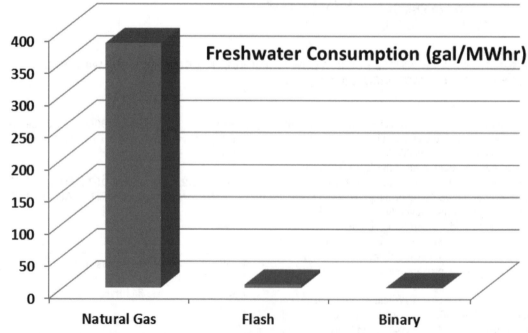

FIG. 16.15 COMPARISON OF FRESHWATER CONSUMPTION FOR POWER PRODUCTION SHOWING NATURAL GAS CONSUMES FAR MORE WATER THAN GEOTHERMAL. THIS IS AN IMPORTANT CONSIDERATION AS MANY GEOTHERMAL SYSTEMS ARE LOCATED IN ARID ENVIRONMENTS (*Source:* [56])

the range of $91 to $119/MWh. This price could be lowered if the utility were to finance the power plant construction.

16.10 ENVIRONMENTAL BENEFITS

Geothermal, along with other renewables, is well known to be a more environmentally friendly source of power than traditional fossil fuel power sources that produce considerable green house gases. Figures 16.15 to 16.17 illustrate this point graphically in comparison to fossil fuel sources. Figure 16.15 shows the significant difference in freshwater consumption between a state of the art design of a natural gas-fired power plant and a state of the art design of a geothermal plant in California [56]. In this comparison, the natural gas-fired plant consumes about 361 gal/MWhr in power production whereas the flash plant consumes approximately 5 gal/MWh and binary geothermal plants consume none [56]. Clearly, geothermal plants are more practical in many western states with geothermal potential because of the lack of un-appropriated water rights in the arid states. In fact, most geothermal plants under construction or development as of late 2010 were binary plants that would not impact local water resources at all, assuming all would be air cooled. Most will be air cooled due to the paucity of water in many western states.

Particulate emissions from geothermal plants are also considerably less than in fossil fuel plants. Figure 16.16 clearly illustrates that coal plants (on average) emit 2.23 lbs/MWh of particulates, whereas natural gas plants emit only 0.14 lbs/MWh. Geothermal power plants have no particulate emissions and pose no risk to human health with respect to this type of pollutant.

Geothermal plants also have relatively small emissions of greenhouse gases relative to fossil fuel plants. Figure 16.17 shows approximate CO_2 emissions from power generated from coal, oil, natural gas, and geothermal flash and binary. The geothermal plants clearly have negligible (or none in the case of binary) CO_2 emissions relative to fossil fuels, with coal emissions at approximately 2191 lbs/MWhr, and natural gas at approximately half that amount [56]. However, CO_2 emissions are relatively unknown for geothermal systems. For instance, emissions are not measured at all in Nevada geothermal power plants. California power plants only report CO_2 emissions above a threshold of 2,500 metric tons of CO_2 per year per facility [57]; thus values reported by the California Air Resource Board are biased toward higher numbers because the data does not include the relatively low emitting geothermal sources, and ignore binary sources all together. Also, these values do not take into consideration that CO_2 emissions will likely vary with time and older plants have lower emissions because water has cycled through earlier via re-injected water which had previously degassed the originally entrained CO_2. Hence, the values illustrated in Fig. 16.17 are to be thought of as rules of thumb more than hard and fast numbers. They illustrate that geothermal plants emit far less CO_2 than fossil fuel power sources, and are a more desirable source of power from the perspective of greenhouse gas emissions and the accompanying concerns over climate change.

Figure 16.18 illustrates the dramatic difference in environmental impact between geothermal power and fossil fuels. The figure shows the skyline of Reykjavik, Iceland in 1932 when fossil fuels were used as the power source in contrast with current day Reykjavik where the power is now from hydroelectric and geothermal power plants. Currently, Iceland as a whole produces approximately 18% of its electricity from geothermal power generation. A total of 88% of Iceland's housing is heated using geothermal district heating [58]. This combination results in a much cleaner environment

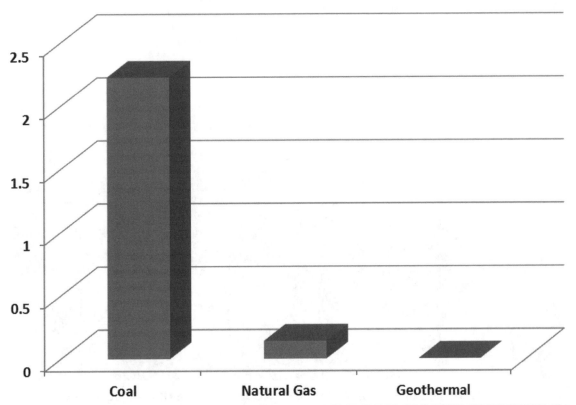

FIG. 16.16 COMPARISON OF PARTICULATE EMISSIONS AMONG THREE TYPES OF POWER PRODUCTION (*Source:* [56])

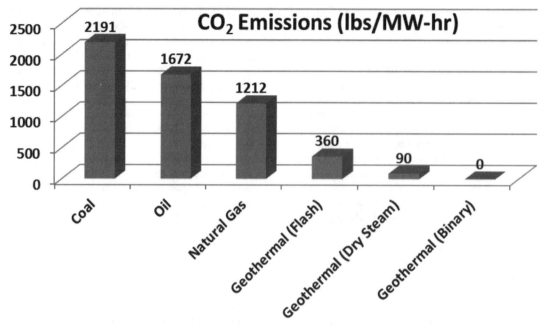

FIG. 16.17 COMPARISON OF CO$_2$ EMISSIONS AMONG SIX TYPES OF POWER PRODUCTION. (*Source:* [56])

and better air quality than was historically the case during fossil fuel power generation.

16.11 POTENTIAL ENVIRONMENTAL IMPACTS

Although geothermal power plants have had a strong record of safety and performance as many potential environmental impacts are routinely and successfully mitigated, there is some potential for adverse impacts if proper management strategies are not implemented.

Potential environmental impacts from geothermal power plants include:

- Gaseous emissions (i.e., H$_2$S pollution of atmosphere (routinely mitigated))
- Visual impacts, noise, construction, disturbance of wildlife habitat and vegetation
- Water pollution; (Brine pollution of environment (routinely mitigated))
- Reservoir drawdown
- Land subsidence
- Induced seismicity
- Water use

FIG. 16.18 REYKJAVIK, 1932 (ON FOSSIL FUELS) ON THE LEFT. MODERN DAY REYKJAVIK USING GEOTHERMAL ENERGY (*Source:* [59])

- Disturbance of natural hydrothermal manifestations (i.e., cessation of spring discharge)
- Landslides; catastrophic and creeping
- Hydrothermal explosions and induced boiling (rare)

16.11.1 Gaseous Emissions

Hydrogen Sulfide (H_2S) is a common gas in geothermal systems, but is routinely mitigated at power producing systems using scrubbers. In fact, this unwanted H_2S has been used to help produce scale inhibitor in countries around the world and has improved profitability at these sites. Scale (precipitates of minerals) can be inhibited from forming in pipes, flash vessels, and heat exchangers using waste from the production of geothermal fluids [60].

16.11.2 Visual Impacts, Noise, Construction, Disturbance Of Habitat And Vegetation

Aesthetic impacts can be expected from any type of development, some of which are transient, and many of which, such as visual impacts, are mitigated with vegetation, paint that blends with the surroundings, and smaller structures.

16.11.3 Water Pollution

Because geothermal fluids typically have total dissolved solids greater than potable water, there is the potential to impact potable water supplies if not managed properly. However, this issue is routinely mitigated by re-injecting the fluids back into the geothermal reservoir, which has the added, substantial advantage of maintaining reservoir pressures and increasing the lifetime of the resource relative to practices that do not include re-injection.

16.11.4 Reservoir Drawdown and Land Subsidence

Overproduction without sufficient re-injection can lead to fluid drawdown and subsidence. Such was the case at the Dixie Valley power plant in Nevada, which resulted in increased fumarole activity, new thermal features, a "dead zone" in vegetation from excess CO2 emissions, and ground subsidence in the late 1990s. Water from a shallower, cool well has been added to the re-injected fluid to maintain reservoir pressures. Insufficient re-injection is common at flash power plants where a portion of the produced fluid is lost to the atmosphere (Fig. 16.19).

16.11.5 Induced Seismicity

Induced seismicity has been reported in both EGS reservoir stimulation (e.g., Basel, Switzerland) and natural hydrothermal systems under production. The Geysers, California present one example of a natural hydrothermal system being associated with induced seismicity. An excerpt from the Berkeley Seismological Laboratory (BSL) web site quotes a 1992 bulletin:

"There were no earthquakes (ML ≥ 2.5) observed in the area of the geothermal field between 1949 and 1975. Starting in 1976 earthquakes were observed in the geothermal field and the rate steadily increased to a rate of at least 12 earthquake (ML ≥ 2.5) per year since 1984. The maximum annual rate was 26 earthquakes (ML ≥ 2.5) which occurred in 1988 (this is equivalent to 40 percent of the annual rate for the entire central coast ranges)."

The Berkeley Seismological Laboratory (BSL) has published yearly maps of seismicity for The Geysers, California area [62]. The year-to-date map for 2010 is reproduced below (Fig. 16.20) showing the concentration of seismic activity

FIG. 16.19 EXCESS PRODUCTION WITHOUT SUFFICIENT RE-INJECTION BETWEEN 8000 AND 10,000 FT AT DIXIE VALLEY, NEVADA CAUSED REDUCTION IN RESERVOIR PRESSURE AND SURFACE CRACKS FROM SUBSIDENCE (*Source:* [61])

at The Geysers relative to the surrounding area. The BSL has published these annually since 1998, and the pattern retains the same essential configuration each year, showing that production at the Geysers has consistently induced seismicity since at least 1998. As can be seen in Fig. 16.20, most of the earthquakes have been relatively small; however, some notably (felt) larger events took place in 2005: two earthquakes with magnitudes >4.0 on May 8 and 9, and three between magnitude 3.0 and 3.99.

Note, Fig. 16.20 shows a concentration and greater density of small earthquake events in part due to the higher density of microearthquake sensors deployed at The Geysers than is the case for the rest of the state. The sensors at The Geysers are a combination of surface and borehole instruments, including up to 49 different borehole monitoring sites; this is clearly not the typical sensor deployment density [63]. In contrast, the Northern California Seismic Network (NCSN) reports data from 794 operating stations [64]. Hence, the borehole sites alone account for over 6% of the sensors used by the NCSN to monitor all of northern California.

16.11.6 Water Use

Some parts of the world are relatively arid (e.g., western US where most natural hydrothermal systems occur in the country) and water consumption can be a serious impediment to geothermal development. For instance, in the western US, water rights are typically fully appropriated and not available for use in cooling towers for binary systems. Binary working fluids are commonly cooled in air-cooled condensers, resulting in a significant reduction in power production efficiency in the summer months (up to 25%). Flash plants lose some percentage (typically ≤ 20%) of fluid in the process of boiling to turn the turbines and subsequent evaporation of the condensate in cooling towers. In such cases, it may be necessary to locate a secondary source of water to re-inject into the reservoir to maintain pressures.

16.11.7 Disturbance of Natural Hydrothermal Manifestations

Overproduction of a resource can result in disturbance of natural hydrothermal features, or even cessation of spring discharge.

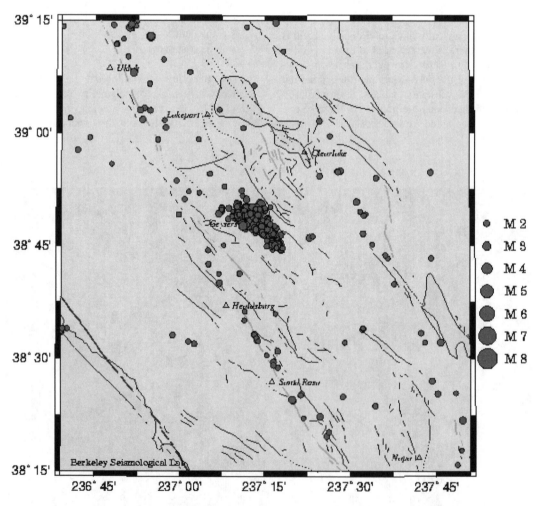

Geysers Area Seismicity - 01/01/2010-08/26/2010

FIG. 16.20 INDUCED SEISMICITY AT THE GEYSERS GEOTHERMAL SYSTEM FROM THE FIRST PART OF 2010 (*Source:* [63])

FIG. 16.21A OHAAKI POOL IN 1963 PRIOR TO GEOTHER-MAL DEVELOPMENT (Photograph provided by R. Glover. *Source:* [65])

FIG. 16.21B OHAAKI POOL IN 1969 FOLLOWING THE INITIATION OF GEOTHERMAL DEVELOPMENT (Photograph provided by R. Glover. *Source:* [65])

One illustrative example occurred at Ohaaki Pool in New Zealand. Prior to 1963, the pools were active (Fig. 16.21a–c). Following initiation of well testing during which the discharge flowed into the river in 1969, the spring levels dropped significantly. Subsequently, the produced geothermal waters were re-injected into the ground through wells, and also, in 1989, some led directly into the Ohaaki Pool; so that it overflowed again, although not without significant setbacks and displeasure to people of the Ngati Tahu people for which the Ohaaki pool is a taonga (treasure). One of the early attempts at filling the pool was with water from a well 0.5 km distant so that the water was cooled before it reached the pool, this resulted in the formation of an unsightly gray amorphous silica sludge in the pool. This problem was overcome by discharging hot separated water at 160°C directly into the pool [65]. Hence, production can have unintended consequences, but in many cases can be mitigated successfully with some effort.

16.11.8 Landslides

Geothermal systems tend to occur in altered rocks, and when they are on steep, poorly vegetated slopes, natural creep and landslides can ensue (Fig. 16.22). An example of a landslide associ-

FIG. 16.21C OHAAKI POOL FOLLOWING VARIOUS REME-DIATION MEASURES AT THE POOL TO MITIGATE IMPACTS OF GEOTHERMAL DEVELOPMENT (Photograph provided by R. Glover. *Source:* [65])

ated with a geothermal system occurred at the Zunil (Guatemala) geothermal field in 1991. Locals believed the slide was caused by a well explosion or volcanic activity, but later studies attribute the slide to natural causes. The landslide occurred at night and buried 23 people alive. It also decapitated a 260°C geothermal well, which required 14 months to repair [66].

FIG. 16.22 LANDSLIDE ON HYDROTHERMALLY ALTERED SLOPE AT ZUNIL, GUATEMALA (*Source:* [61], Photos courtesy of Fraser Goff)

16.11.9 Hydrothermal Explosions and Induced Boiling

These impacts are rare, but induced boiling can occur in over-produced systems. These effects might occur as a result of hot water withdrawal, subsequent pressure reduction, with or without inflow and contact of cold water with hot rock. This is another reason to have a sound reservoir management and re-injection strategy along with pre-production data gathering to understand steady state conditions.

16.12 COMPARISON WITH OTHER RENEWABLE ENERGY SOURCES

Capacity factor is the ratio of the net electricity generated, for the time considered, to the energy that could have been generated at continuous full-power operation during the same period. The following is a comparison of capacity factors for renewable energy sources; clearly, geothermal has the highest, and hence, is the best choice for production of base load power (24 hours per day, 7 days per week).

Technology	Capacity factor
Geothermal	97%
Biomass	80%
Hydro	40–70%
Wind	26–40%
Solar	22–32%

However, the efficiency in geothermal power production decreases in summer time, particularly in air cooled systems (the predominant binary systems in the western US). Efficiency can decrease by 25% in the hottest summer months as air cooling of the binary working fluid becomes less efficient.

16.13 COMPETING IN ENERGY MARKETS

Some factors that help geothermal power plants compete in energy markets include:

• Renewable energy portfolio standards (RPS)
• Federal tax credits and programs
• Increased fossil fuel costs
• Climate change concerns
• Security concerns

Renewable energy portfolio standards (RPS) vary from state to state but, essentially, they are regulations that require a certain percent of a utility's energy portfolio to contain renewable energy (geothermal, wind, solar, hydro, biomass, municipal solid waste, qualified hydropower, marine and hydrokinetic) by a particular date (Table 16.3 lists RPS for states that will be producing new geothermal energy in the near term). With RPSs, the electric utility companies are required to generate or purchase this percentage of renewable energy, subject to varying degrees of enforcement depending on the state, the rigor of which varies by state. Note that Texas reported their values as 5880 MW, which is approximately 10% of their current utilization (listed as ~10% in Table 16.3). Clearly, California has the most aggressive RPS at 33%, with the overall average power generated by renewable energy expected to be a minimum of 19.4% by 2025 for the western US (excluding Alaska and Wyoming). However, the adoption of tradable renewable energy credits (TRECs) by some states (e.g., California in 2010)

could impact how the RPS is implemented. Currently, geothermal is a preferred energy source to apply toward an RPS because it is base load power. However, if TRECs are an option, they could be applied in lieu of building additional geothermal capacity because the market would be focused on the value of the electricity energy rather than its reliability. The impact of TRECs on RPSs and potential impact on geothermal development are currently unclear.

Federal Production Tax Credits (PTC) are also in place, although many of the past PTCs in the US have not truly provided an incentive to geothermal power development because of the short duration of the PTC (i.e., a short time between enactment of the incentive and the required "placed in service date."). Often a PTC has been put into place with a two year sunset clause, and it can take 5 years for a power plant to be constructed and placed into initial operation. Hence, the PTC has been largely ineffective as an incentive for the geothermal industry, and most projects that have benefited from a PTC have been well underway before the PTC law was passed or extended. Based on a report from 2007, the PTC has had very little effect on base load renewables, such as geothermal [72] because of the short time between announcement of the PTC and the placed in service date.

In 2009, the PTC was extended through 2013, which provides a more realistic window over which to develop a project. As of 2010, the PTC for geothermal and other renewables was 2.2 cents per kilowatt-hour (kWh) for the first 10 years of operation. This 2.2 cents/kWh is inflation adjusted from 1.5 cents/kWh from the original bill passed in 1992 ([73]EPACT, 1992; 26 USC § 45). Since the PTC inception, it has expired and been reinstated three times, and has been extended two additional times [72], adding to the overall ineffectiveness of this tax credit for the geothermal industry because it did not allow for long term planning. Now that there is a longer time period to develop projects, the PTC should lead to additional geothermal development. Unfortunately, this longer PTC time frame took effect during the severe economic downturn of 2009 to 2010.

The PTC is a credit that is applied to the taxes due rather than adjusted income, and hence is more beneficial to the operator than deductions from income would be. However, if there is no profit in the first years of operation, there is no tax liability, and the tax credit is not helpful as it cannot be applied. This situation may give rise to the formation of alliances between the geothermal operator and a partner with tax liabilities toward which this credit could be applied (a more detailed discussion of PTCs can be found at [74]).

Because PTCs have not been as valuable to the geothermal industry as they could be, the American Recovery and Reinvestment Act of 2009 (H.R. 1) extends a 30% investment tax credit (ITC) to geothermal in lieu of the PTC. The ITC was added to help the industry during economically uncertain times as well as provide an alternative to the PTC, which are not enticing to early stage developers due to routinely short sunset dates. ITCs provide a pre-development tax credit so that developers need not wait until their operation is in production to realize the benefits. However, the ITC could potentially benefit unscrupulous would-be developers seeking investor capital. Because production revenues are not required to reap the benefits of ITC, there is less incentive for developers to complete projects than in the case of PTC, where production must occur for the tax benefits to be realized. The ITC could also encourage some developers to overestimate the resource in order to obtain 30% of a larger capital project cost.

As of fall 2010, in lieu of either of these tax credits, geothermal, wind, biomass, and solar projects could opt to receive a grant

TABLE 16.3 RENEWABLE PORTFOLIO STANDARDS FOR WESTERN U.S. STATES THAT HAVE THE ABILITY TO PRODUCE GEOTHERMAL POWER IN THE NEAR TERM, INCLUDING GULF STATES WITH EXISTING HOT OIL WELLS (TEXAS IS THE ONLY GULF STATE WITH A CURRENT RPS)

	2010	2015	2020	2025	Type
AK					None
AZ	2.5%	5%	10%	15%[1,2]	RPS
CA	20%		33%[1,2]		RPS
CO			20%[1]		RPS
HI			20%[1]		RPS
ID			10%		Target
MT		15%[1,2]			RPS
NM			20%[1,2]		RPS
NV		20%[1]		25%[2]	RPS
OR				25%[1,2]	RPS
TX		~10%			RPS
UT				20%[1,2]	RPS
WA		15%[1,2]			RPS
WY					None

The bulk of the data were obtained from:
[1] ____[67]
and
[2] [68]
Additional Sources:
Arizona: [69]
California: [70]
Idaho: [71]

of up to 30% of the capital cost of the project. Because financing can be difficult for geothermal operations in the early stages of development due to high risk/cost resource definition and resource characterization, this incentive could prove very useful to the advancement of the industry. A comprehensive list of federal and state incentives for renewable power generation can be found at Reference [75].

In addition to these various government programs (PTC, ITC, grants), volatile and increasing fossil fuel costs have made geothermal energy development more cost competitive and attractive over the years. However, fossil fuel costs fluctuate both up and down, thus creating the possibility that at any given time, fossil fuel power generation may be more or less expensive than geothermal power generation. One advantage to offset this uncertainty is the use of long term Power Purchase Agreements (PPA), which are routinely used by utilities to purchase from geothermal power producers. The advantage to these is that geothermal has no fluctuation in fuel costs and reasonably estimated operation and maintenance and debt service costs such that a fixed price over a fixed period of years (often 20 years) can be negotiated. In many of these PPAs, the utility is required to purchase the specified level of power and must adjust other sources of power as load values fluctuate throughout the year. This guarantees stable pricing for the utility as costs of other sources of power fluctuate, while providing the geothermal power producer a guaranteed income stream. Hence, after the high risk of exploration and development of a geothermal resource has been overcome, long term operations are relatively low risk and operating costs are fairly stable.

An additional factor that makes geothermal, as well as all other renewables, more attractive, is the prevalence in the media of discussions of climate change partially due to electricity generation using fossil fuels. Additionally, the US in particular would benefit from greater energy independence and ability to use its indigenous sources of power while increasing economic growth through development of geothermal resources. Geothermal generation can help in these matters and is also a good alternative for construction of distributed power sources that are less vulnerable to large scale power disruption through natural or human caused (terrorism) disasters than are large (e.g., 2000 MW) coal fired power plants. As an example, Nevada currently has 11 producing power plants, most in remote areas, with another approximately 60 projects in various stages of development. Another example is a small power station at Chena, Alaska where the geothermal waters are used to heat, cool, and power a small resort. These types of power plants are not as vulnerable to disruption as large megawatt, high value facilities. Similarly, as in the case of Nevada, it is highly unlikely that widespread wildfires or earthquakes would disrupt more than one power station at a time.

Hence, there are a variety of factors that are making geothermal power plants more attractive and economical including PTC, ITC, PPA and general public opinion. However, reduced risk in drilling is needed as this is a very expensive up-front cost that entails considerable risk. Also, additional research is needed to better identify productive faults. Reservoir management strategies have been evolving, but the geothermal systems are all very complex, so trial and error, along with sophisticated numerical modeling, is still part of reservoir management. Research is also needed to improve efficiencies in energy conversion as well as efficiencies in air cooling for working fluids in binary systems.

Additionally, the limited availability of electric transmission lines in areas of high geothermal potential could limit development. However, many states have done transmission studies to account for the most likely placement of renewable energy power plants. Renewables are different than fossil fuel generation because the resource is location specific, whereas fossil fuels can be transported for use at generating facilities near existing transmission. So, although geothermal is becoming increasingly competitive in the energy markets, there are yet many hurdles to overcome and manage.

16.14 SUSTAINABILITY

Sustainability of a resource is the ability to utilize it in such a way as to meet the needs for the present generation without compromising the needs of future generations. In the case of geothermal, this requires that the resource is not overproduced so that it can be used in the future. Natural, unexploited hydrothermal systems can have durations of 5000 to 1,000,000 years. However, it is difficult to anticipate how a resource will respond to development and reinjection, and the strategy will likely be altered throughout the life of the resource. It is unknown how long a geothermal resource will be sustainable because it is possible to overproduce or mismanage the resource. It also is unknown how sustainable a particular resource will be because there are limited data available worldwide (106 years at Larderello, Italy), and less in the US (a maximum of 50 years at one site, The Geysers). In theory, a geothermal resource should be sustainable for hundreds of years if properly managed, but the length of sustainability will depend on the initial quantity of hot fluid, rate of power generation, and consumption (water and heat loss). The age of waters in geothermal systems varies, but can be quite old (e.g., ~10,000 years in NV), so re-injection into these systems in particular is vital as there will not be any natural recharge in a reasonable time frame. Fluids taken from the reservoir must be returned to it after power generation to maintain the resource. Exploitation that exceeds natural recharge greatly reduces system lifetimes. Re-injection of water is the key. The longevity of a geothermal system will be a function of the power output, well density, injection strategy, initial reservoir temperature and pressure, and permeability.

16.15   DIRECT USE OF GEOTHERMAL

Geothermal water for direct use applications has typically been underutilized in the US. Other parts of the world (e.g., Japan, Iceland, 28 countries in Europe) [76] take significant advantage of lower temperature resources, which have many uses. Geothermal fluid can be used directly in many applications: aquaculture, bathing, biofuels production, biogas production, cement drying, concrete block curing, cooking, fabric dyeing, food processing, greenhouses, heat pumps, heating and cooling, hydrogen production, lumber drying, onion and garlic dehydration, paper processing, pasteurization, pulp drying, refrigeration, snow melting, soft drink carbonation, soil sterilization, soil warming, vegetable dehydration and water heating.

One use of geothermal water is space heating and this is often accomplished using heat exchangers (Figs. 16.23 and 16.24) to extract the heat and transfer it to a secondary fluid, typically potable water. This water is then circulated through the building, complex, or district in a closed loop to provide space heat. The geothermal water is re-injected into the reservoir, making this side of the heat exchanger a closed loop as well. One reason heat exchangers are used rather than using the geothermal water directly is that geothermal fluids typically have varying levels of dissolved constituents that may precipitate in pipes as it cools along a heating loop. Use of potable water minimizes mineral precipitation because the dissolved solids are typically less than what is found in geothermal fluids. Space heating results in no net consumption of either fluid.

FIG. 16.23   THIS IS A "PLATE TYPE" HEAT EXCHANGER, WHICH PASSES HOT GEOTHERMAL WATER PAST MANY LAYERS OF METAL PLATES, TRANSFERRING THE HEAT TO OTHER WATER PASSING THROUGH THE OTHER SIDE OF EACH PLATE WHERE THE HEATED FLUID IS USED FOR SPACE HEATING (*Source:* [59])

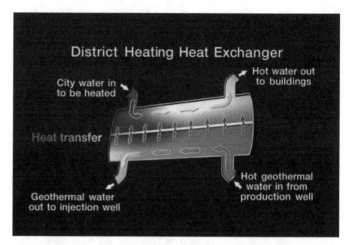

FIG. 16.24   SCHEMATIC OF HEAT EXCHANGER (*Source:* [59])

16.16   GEOTHERMAL HEAT PUMP FINDINGS

Geothermal heat pumps (GHP), also known as ground source heat pumps (Fig. 16.25), provide an energy efficient technology used in the U.S. and around the world [77]. Although this technology has been in existence since the 1940s, it still has not realized its full market potential, but it is gaining ground. A December 2008 Oak Ridge National Laboratory Report [78], described the barriers to GHP system adoption and methods to overcome them. The barriers include (1) high installation costs, (2) consumer's and regulator's lack of awareness of the technologies, (3) lack of business models that support long term adoption, (4) lack of infrastructure to install and maintain systems, and (5) lack of new technologies and methods of installation.

The report describes that utilities, individually and collectively, can push through the barriers by adopting large, pilot scale GHP installation programs for new and retrofit sites. The size of a system needed to deliver comfort (heat in the winter, cool in the summer) is measured in tons, where a one ton system can heat or cool 46.5 m^2 (500 ft^2). Model utility programs could start with a goal of several hundred tons of GHP systems installed in the first year, and then scale up to thousands of tons per year based on the results of the earlier years. An average house is approximately 186 m^2 (2000 ft^2), so 1000 installed tons would heat and cool approximately 250 homes.

The program can consist of four segments, some of which follow one another, while others can be done at the same time: (1) providing education that maintains and enhances customer, installer, and other stakeholder awareness and skill levels, (2) selecting GHP installation sites, (3) installing and commissioning GHP equipment, and (4) evaluating retrofit performance and revising project implementation.

The UGWG used the above model program concept and assisted the Oklahoma Municipal Power Authority (OMPA) in submitting a GHP system marketing and rebate proposal to the Oklahoma State Energy Office. The state accepted the proposal for the American Recovery and Reinvestment Act of 2009 funding. The state will offer a rebate of $1000 per ton for up to 2340 tons of GHP installed by OMPA members by March 31, 2012. This rebate is in addition to the $800 per ton that OMPA and its members offer their customers.

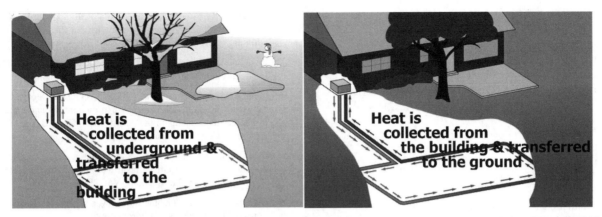

FIG. 16.25 FIGURE ILLUSTRATING THE CONCEPT OF GEOTHERMAL HEAT PUMPS. THE ONE ON THE LEFT IS IN THE WINTER TIME WHEN HEAT IS HARVESTED FROM THE GROUND, WHEREAS THE ONE ON THE RIGHT IS FOR SUMMER WHEN HEAT IN THE BUILDING IS TRANSFERRED TO THE GROUND (*Source:* [59])

OMPA members have conducted studies showing that GHP systems provide a 0.5 kW per ton reduction in summer peaks. Over a 25-year period and a 5% discount rate, using current capacity costs of $100 per kW/year, the savings represent a net present value of $1400 per ton. The GHP systems also reduce the building's carbon footprint by 17 million Btu/ton annually.

In an effort to take the OMPA model nationally, the UGWG has formed a "Geo Hero" Working Group. The Geo Hero Working Group has six objectives:

1. Enhance communications between and among utilities that have geothermal heat pump (GHP) programs or are considering such programs,
2. Use the combined group purchasing power to reduce the cost of GHP installations by taking advantage of marketing efficiencies,
3. Quantify the cost and benefits of GHP programs, including carbon footprints and the "non-energy" benefits such as jobs, comfort, safety, and extended equipment life,
4. Transform the market of installing, operating, maintaining, evaluating, and improving GHP systems in homes and businesses in the U.S.
5. Work with utility oversight organizations to describe the benefits of GHP in meeting the requirements of renewable energy and energy efficiency portfolios, and
6. Work with state utility regulatory commissions to gain approval for utility investments in GHP systems to be included in the rate base.

GHP systems appear to be promising because of the 2010 ARRA (American Recovery and Reinvestment Act) funding that is available to help them move to the market place.

16.17 SUMMARY AND CONCLUSIONS

This chapter reviewed and discussed potential geothermal production systems, resource development and engineering, electrical power production, enhanced geothermal systems, potential environmental benefits and impacts, energy market, sustainability, direct use, and geothermal heat pumps. It is our conclusion that the environmental benefits of geothermal energy far outweigh the potential impacts. Geothermal power production is particularly important to developing economies; not only can economies historically dependent on imported fossil fuels change their consumption practices,
but these emerging economies can also mitigate their production of greenhouse gases, and rely on indigenous power supplies.

There are a few impediments to geothermal power production. The most significant has been the lack of public and private capital, particularly in the wake of the global economic recession, which began in 2008. The global capital markets have been recovering, and it is the belief of the authors that within the next two to three years this obstacle will have been removed. Some projections on geothermal power development suggest that 18,000 MWe may be in production by 2015. There are many ambitious and sound development plans in the US, East Africa, Central America, South America, and Australia.

Geothermal energy is an attractive renewable energy source. The beauty of a geothermal resource is that it constitutes a ready and reliable source of power, easily developed in some regions, but certainly developable, as technology improves, in many parts of the planet. Geothermal power production promises to be a significant source of energy in the future. From its humble origins in 1904 in Larderello, Italy to now producing about 11,000 MWe globally, new geothermal power technologies will enable many countries to embrace and develop this vast and dependable source of base load energy.

16.18 ACKNOWLEDGMENTS

There are several individuals who greatly assisted in assembling this information. Many thanks go to Marcelo Lippmann, Ron DiPippo, Subir Sanyal, Sabodh Garg, and Guy Nelson for their kind and able advice and input. Many thanks to Mark Coolbaugh, Larry Garside, and Karl Gawell for their thoughtful reviews and input.

16.19 ACRONYMS

ARRA American Recovery and Reinvestment Act
ASME American Society of Mechanical Engineers
ASTER Advanced Spaceborn Thermal Emission and Reflection
BLM Bureau of Land Management
BSL Berkeley Seismological Laboratory
CPUC California Public Utilities Commission
CSAMT Controlled Source Audio Magnetotelluric
DOE Department of Energy
EGS Enhanced Geothermal Systems
EJ Exajoules

GHP   Geothermal Heat Pumps
GIS   Geographic Information System
HDR   Hot Dry Rock
IDDP   Iceland Deep Drilling Project
IRP   Integrated Resource Planning
ITC   Investment Tax Credit
MIT   Massachusetts Institute of Technology
MT   Magnetotelluric
MWh   Megawatt-hour
O&M   Operations and Maintenance
OMPA   Oklahoma Municipal Power Authority
ORC   Organic Rankine Cycle
PPA   Power Purchase Agreement
PTC   Production Tax Credits
R & D   Research and Development
RPS   Renewable Energy Portfolio Standards
TREC   Tradable Renewable Energy Credits
UGWG   Utility Geothermal Working Group
USGS   US Geological Survey
UTC   United Technologies Corporation

16.20   REFERENCES

1. Lumb, J.T., "Prospecting for Geothermal Resources," in Rybach, L. and Muffler, L.J.P. (eds). Geothermal Systems: Principles and Case Histories. Chichester: Wiley & Sons, 1981, pp. 77–108.

2. World Geothermal Congress. "The Bali Declaration," April 30, 2010.

3. Lund, J., "Balneological Use of Thermal Waters," International Geothermal Days, Bad Urach, Germany, 2001a.

4. Lund, J.W., and R. Bertani, 2010. "Worldwide Geothermal Utilization 2010". Transactions Geothermal Resources Council v. 34, pp. 195–198.

5. Green, B. D. and Nix, G. R., 2006. "Geothermal—The Energy Under Our Feet, Geothermal Resource Estimates for the United States," NREL Technical Report 840-40665.

6. Gupta, H. and Roy, S., 2007. "Geothermal Energy: An Alternative Resource for the 21st Century," Amsterdam: Elsevier.

7. Bertani, R, 2010. "Geothermal Power Generation in the World, 2005—2010 Update Report," Proceedings World Geothermal Congress 2010, Bali, Indonesia, April 25-29.

8. Williams, C. F., Reed, M. J., Mariner, R. H., DeAngelo, J., and Galanis, P. S., Jr., 2008, "Assessment of moderate- and high-temperature geothermal resources of the United States," U.S. Geological Survey Fact Sheet, 2008, 3082, p. 4.

9. Shevenell, L., and Blackwell, D., 2011. "Update on Near-Term Geothermal Potential in Nevada, Fall 2010" Bulletin Geothermal Resources Council (in review).

10. Tester, J. et al. (MIT-led Interdisciplinary Panel). "The Future of Geothermal Energy: Impact of Enhanced Geothermal Systems (EGS) on the United States in the 21st Century," Boston: MIT, 2006. <http://www1.eere.energy.gov/geothermal/future_geothermal.html>.

11. Tarbuck, E.J., and F.K. Lutgens. Earth, an Introduction to Physical Geology. 9th. Upper Saddle River, NJ: Pearson Prentice Hall, 2008. p. 334.

12. Glassley, W., 2008. "Geothermal Energy Systems," UC Davis Extension Class Notes.

13. Hancock, P. and Skinner, B. J., 2000. "Radioactive Heat Production in the Earth," The Oxford Companion to the Earth. Encyclopedia. com. June 15, 2010. <http://www.encyclopedia.com/doc/1O112-radioactivehtprdctnnthrth.html>.

14. Wohletz, K., and Heiken, G., "Volcanology and Geothermal Energy," Berkeley: University of California Press, 1992. <http://ark.cdlib.org/ark:/13030/ft6v19p151/>.

15. Plate boundaries data (trenches, ridges, and transforms): http://www.ig.utexas.edu/research/projects/plates/data.htm.

16. Faulds, J.E., Coolbaugh, M., Blewitt, G., and Henry, C.D., 2004, "Why is Nevada in hot water? Structural controls and tectonic model of geothermal systems in the northwestern Great Basin." Geothermal Resources Council Transactions, pp. 649–654.

17. Blewitt, G., Hammond, W.C., and Kreemer, C., 2005, "Relating geothermal resources to Great Basin tectonics using GPS" Geothermal Resources Council Transactions, Vol. 29, pp. 331–336.

18. DiPippo, R., 2005. "Geothermal Power Plants – Principles, Applications and Case Studies," Oxford: Elsevier.

19. DiPippo, R., 2008. "Geothermal Power Plants – Principles, Applications, Case Studies and Environmental Impact," Oxford: Elsevier Ltd., p. 493.

20. Rybach, L. and Muffler, L.J.P. (eds). "Geothermal Systems: Principles and Case Histories," Chichester: Wiley & Sons, 1981.

21. Tester, J. et al. 2005. "Sustainable Energy: Choosing Among Options," Boston: MIT Press.

22. Brown, D., 2009. "Hot Dry Rock Geothermal Energy: Important Lessons from Fenton Hill," Proceedings, 34th Workshop on Geothermal Reservoir Engineering, Palo Alto: Stanford University, 2009: np. Web. <http://pangea.stanford.edu/ERE/pdf/IGAstandard/SGW/2009/brown.pdf>.

23. Chen, D., 2010. "Concepts of a basic EGS model for the Cooper Basin, Australia," 2010 World Geothermal Congress (WGC 2010), Bali, Indonesia, April 25-29. Web. <http://www.wgc2010.org/index.php?option=com_content&view=article&id=102&Itemid=124>.

24. European deep geothermal energy programme. Retrieved from http://www.soultz.net/version-en.htm.

25. Genter, A., Goerke, X., Graff, J., Cuenot, N., Grall, G., Schindler, M., and Ravier, G., 2010 "Current status of EGS Soultz geothermal project (France)," Proceedings World Geothermal Congress 2010, Bali, Indonesia, April 25–29.

26. Kraft, T., Mai, P.M., Wiemer, S., Deichmann, N., Ripperger, J., Kaestli, P., Bachmann, C., Faeh, D., Woessner, J., Giardini, D., 2009. "Enhanced geothermal systems; mitigating risk in urban areas," EOS, Transactions American Geophysical Union, 90(32), pp. 273–274.

27. Elders, W. and Friðleifsson, G. Ó., 2009. "Iceland Deep Drilling Project Finds Magma," GRC Bulletin 38/4 (2009): 31–32.

28. Helman, C. "Geothermal Power: Fort Knox's Buried Treasure." Forbes, October 9, 2006.

29. Combs, J. and Muffler, L.P.J., 1973. "Exploration for Geothermal Resources," in: Kruger, P. and Otte, C., eds., Geothermal Energy, Palo Alto: Stanford University Press, pp. 95–128.

30. Verma, S.P., K. Pandarinath[a] and E. Santoyo,2008. "SolGeo: A new computer program for solute geothermometers and its application to Mexican geothermal fields." Geothermics, v. 37(6), pp. 597–621.

31. Powel, T. and Cumming, W., 2010. "Spreadsheets for geothermal water and gas geochemistry," PROCEEDINGS, Thirty-Fifth Workshop on Geothermal Reservoir Engineering, Stanford University, Stanford, California, February 1–3, 2010, p. 10.

32. Fournier, R.O., and Truesdell, A.H., 1973. An Empirical Na-K-Ca Geothermometer for Natural Waters. Geochimica et Cosmochimica Acta 37: 1255–1275.

33. Fournier, R.O., and Potter, II, R.W., 1979. Magnesium correction to the Na-K-Ca chemical geothermometer. Geochim. Cosmochim. Acta 43: 1543–1550.

34. Southern Methodist University Geothermal Laboratory. Retrieved from http://smu.edu/geothermal/.

35. Wright, P. M., Ross, S. H., and West, R.C., "State-of-the-art Geophysical Exploration for Geothermal Resources," Geophysics, 50, pp. 2666–2699. 1985.

36. Geothermal resources council. (n.d.). Retrieved from http://www.geothermal.org/powerpoint06_explor.html.

37. Sladek, C., M.F. Coolbaugh, and R.E. Zehner, 2007. "Development of 2-Meter Soil Temperature Probes and Results of Temperature Survey Conducted at Desert Peak, Nevada, USA." Transactions Geothermal Resources Council, v. 31, pp. 363–368.

38. Sladek, C. and Coolbaugh, M. F., 2009. "Improvements in Shallow (2-Meter) Temperature Measurements and Data Interpretation," Transactions Geothermal Resources Council, 33, pp. 535–541.

39. Kratt, C., Coolbaugh, M., Peppin, B., and Sladek, C., 2009. "Identification of a New Blind Geothermal System with Hyperspectral Remote Sensing and Shallow Temperature Measurements at Columbus Salt Marsh, Esmeralda County, Nevada," Transactions Geothermal Resources Council, 33, pp. 481–486.

40. Kratt, C., Coolbaugh, M. F., and Calvin, W. M., 2006a. "Remote detection of Quaternary borate deposits with ASTER satellite imagery as geothermal exploration too," Geothermal Resources Council Transactions, 30, pp. 435–439.

41. Capuano Jr., L., 2009. "Drilling presentation". GeoAmericas Conference, San Francisco, CA, February, 2009.

42. U.S. Department of Energy (DOE). "Hydrothermal Power Systems," 2010b. <http://www1.eere.energy.gov/geothermal/powerplants.html>.

43. Mark Coolbaugh, Reno, Nevada, supplied unpublished photos.

44. Bureau of land management. (2010, August 09). Retrieved from http://www.blm.gov/nv/st/en/prog/minerals/leasable_minerals/geothermal0/ggeothermal_leasing.html.

45. Quimby, F., and Krauss, J. (2009, July 14). Geothermal lease sale generates more than $9 million. Retrieved from http://www.blm.gov/wo/st/en/info/newsroom/2009/july/NR_0715_2009.html.

46. Oil and gas lease sale information by calendar year. Retrieved from http://www.blm.gov/ca/st/en/prog/energy/og/instructions/leasesale.html.

47. Geothermal leasing. Retrieved from http://www.blm.gov/ut/st/en/prog/energy/oil_and_gas/oil_and_gas_lease.html.

48. Geothermal leasing. Retrieved from http://www.blm.gov/ut/st/en/prog/energy/geothermal0.html.

49. Jennejohn, D., 2010. "US Geothermal Power Production and Development Update," Geothermal Energy Association, April 2010, 33 p. <geo-energy.org/pdf/reports/April_2010_US_Geothermal_Industry_Update_Final.pdf>.

50. Duchane, D. and Brown, D., 2002. "Hot Dry Rock (HDR) Geothermal Energy Research and Development at Fenton Hill, New Mexico," GHC Bulletin, December 2002:13-19. Web. <http://geoheat.oit.edu/bulletin/bull23-4/art4.pdf>.

51. Mock, J.E., Tester, J.W., and Wright, P.M., 1997. "Geothermal Energy from the Earth: Its Potential Impact as an Environmentally Sustainable Resource." Annual Review of Energy and the Environment, 22:305–356.

52. Tester, J. W., D.W. Brown, and Potter R.M., 1989. "HotDry Rock Geothermal Energy-A New Energy Agenda for the 21st Century," Los Alamos National Laboratory report LA-11514-MS.

53. Blackwell, D., Negraru, T.P., Richards, M.C., 2007. "Assessment of the Enhanced Geothermal System Resource Base of the United States," Natural Resources Research, 15(4), pp. 283–308.

54. U.S. Department of Energy (DOE), 2008. "Geothermal Tomorrow," September 2008, p. 32.

55. Nelson, G., 2009. "Utility Geothermal Working Group Update," Geothermal Resources Council Transactions, 33, pp. 1019–1022.

56. Kagel, A., Bates, D., and Gawell, K., 2007. "A Guide to Geothermal Energy and the Environment," Geothermal Energy Association. Apr. 2007. <http://geo-energy.org/reports/Environmental%20Guide.pdf>.

57. California Air Resource Board (CARB). "Mandatory Greenhouse Gas Emissions Reporting Tool." <http://www.arb.ca.gov/regact/2007/ghg2007/frofinoal.pdf>.

58. Yamaguchi, N. and Albertsson, A.L., 2006. "Reykjanes Geothermal Power Plant Utilization of Geothermal Energy in Iceland," Transactions Geothermal Resources Council, 30, pp. 773–776.

59. Geothermal Education Office (GEO), Tiburon, California, 2000. <http://geothermal.marin.org/GEOpresentation/>.

60. Gallup, D., and Kitz, K., 1997. "Low-Cost Silica, Calcite and Metal Sulfide Scale Control Through On-Site Production of Sulfurous Acid from H_2S or Elemental Sulfur," Transactions Geothermal Resources council, 21, pp. 399–403.

61. Fraser Goff, White Rock, New Mexico supplied unpublished photos.

62. Seismicity maps for The Geysers area. Retrieved from http://seismo.berkeley.edu/weekly/geyser.html.

63. Calpine/unocal geysers network. Retrieved from http://www.ncedc.org/geysers/.

64. Locations of seismic monitoring stations http://www.ncedc.org/ftp/pub/doc/NC.info/NC.stations.

65. Glover, R.B., Hunt, T.M., and Severne, C.M., 2008. "Impacts of development on a natural thermal feature and their mitigation — Ohaaki Pool, New Zealand," Geothermics 29 (4–5): 509–523.

66. Goff, S. and F. Goff, 1997. "Environmental impacts during geothermal development: some examples from Central America." Proceedings of the NEDO International Geothermal Symposium Sendai, Japan (1997), pp. 242–250.

67. States with renewable portfolio standards. Retrieved from http://apps1.eere.energy.gov/states/maps/renewable_portfolio_states.cfm#map.

68. Database of state incentives for renewables and efficiency. Retrieved from http://www.dsireusa.org/.

69. R14-2-1804. Annual Renewable Energy Requirement.

70. RPS from EXECUTIVE ORDER S-14-08: http://gov.ca.gov/executive-order/11072/.

71. Renewable portfolio standards. Retrieved from http://www.creativeenergies.biz/go.php?id=26

72. Sissine, F., 2007. Renewable Energy: Background and Issues for the 110th Congress. CRS (Congressional Research Service) Report for Congress, 37 p.

73. EPACT, 1992. Section 1914 of the Energy Policy Act of 1992 (EPACT92, P.L. 102-486).

74. Energy law alert: the production tax credit and wind power investments. (2004, September 1). Retrieved from http://www.stoel.com/showalert.aspx?show=2395.

75. Financial incentives for renewable energy. Retrieved from http://www.dsireusa.org/summarytables/finre.cfm.

76. Lund, J., 2001b. "Geothermal Use in Europe," Geo-Heat Center (GHC) Bulletin, June 2001, 21, No. 2.

77. Johnson, K., "Geothermal Heat Pump Guidebook," 3rd Edition May 2007.

78. 2008 Oak Ridge National Laboratory Report, (Geothermal (Ground Source) Heat Pumps, ORNL/TM 2008/232.

16.21 OTHER SOURCES FOR INFORMATION

Geothermal Resources Council, www.geothermal.org
Geothermal Energy Association, www.geo-energy.org/
Geothermal Education Office, geothermal.marin.org/
International Geothermal Association, www.geothermal-energy.org

US Department of Energy, Geothermal Technologies Office, www1.eere.energy.gov/geothermal/
Oregon Institute of Technology, Geo-Heat Center, geoheat.oit.edu/
Stanford University, pangea.stanford.edu/ERE/research/geoth/
Southern Methodist University, smu.edu/geothermal/
University of Utah, Energy and Geoscience Institute, www.egi.utah.edu/
University of Nevada Reno, Great Basin Center for Geothermal Energy, www.unr.edu/geothermal

DEVELOPMENT OF ADVANCED ULTRA SUPERCRITICAL COAL FIRED STEAM GENERATORS FOR OPERATION ABOVE 700°C

Paul S. Weitzel and James M. Tanzosh

17.1   INTRODUCTION

Advanced Ultra Supercritical (A-USC) is a term being used to describe a coal fired power plant design with the inlet steam temperature to the turbine at 700°C to 760°C (1292°F to 1400°F). Nickel alloy materials are required. The term Ultra Supercritical (USC) is a term for plants currently designed and operating at 600°C (1112°F) using available and suitable ferritic and stainless steels. Increasing efficiency of the Rankine regenerative-reheat steam cycle to improve the economics of electric power generation and achieve lower cost of electricity has been a long sought after goal. Efficiency has more recently been recognized as a means for reducing the emission of carbon dioxide and its capture costs, as well as a means to reduce fuel consumption costs. Programs have been established by nations, industry support associations and private companies to advance the technology in steam generator design and materials development of nickel based alloys needed for use above 700°C. The worldwide abundance of less expensive coal fuel has driven economic growth. The challenge is to continue to advance the improvement of efficiency for coal fired power generation technology, representing nearly 50% of the United States' (U.S.) production, while helping to maintain economic electric power costs with plants that have favorable electric grid system operational characteristics for turndown and rate of load change response.

The Newcomen steam engine operated at about 0.5% thermal efficiency in 1750 [1]. Major efficiency milestones such as James Watt's 1769 patented improvement to a Newcomen steam engine by adding a separate condenser is credited with achieving 2.7% thermal efficiency by 1775 and became a major propellant of the Industrial Revolution due to the economic benefit of the fuel savings attained. Watt's 1782 patent for expansive working, double acting cylinder engine, is credited with achieving 4.5% efficiency by 1792. James Watt, however, is also assailed for causing delay in economic advancement due to his reluctance to raise the working pressure of the steam engine. Lacking in good boiler instrumenta-

tion such as reliable water level and pressure gages, Watt's concerns were probably well founded in limiting the operating pressure to about 0.034 MPa (5 psig). After Watt's patents began to expire, Richard Trevithick is credited with engine improvements that permitted increasing steam pressure, to 1MPa (145 psig), to achieve 17% thermal efficiency by 1834.

Another milestone, American Electric Power's (AEP) Philo plant Unit 6 steam generator was the first commercial supercritical unit in service early in 1957, Figure 17.1. Philo was 120 MW, 85 kg/s, 31 MPa/621°C/565°C/538°C (675,000 lbs/hour, 4500 psi/1150°F/1050°F/1000°F) supplied by The Babcock and Wilcox Company. In 1959, Philadelphia Electric Company's Eddystone steam generator provided by Combustion Engineering, Inc., delivered 325 MW, 252 kg/s, 34.5 MPa/649°C/565°C/565°C (2,000,000 lb/hour, 5000 psi/1200°F/1050°F/1050°F and later operation at 4700 psi/1130°F/1030°F/1030°F) [2]. The net plant heat rate for Eddystone was 8534 Btu/kWh, 39.99% (HHV) net plant efficiency without environmental system auxiliary power. These two units led the world toward many supercritical boilers that followed and are the predominant type for large electric power steam thermal plant selection. Stainless steel materials were used for this service. Nickel alloys are being evaluated for ASME Code acceptance up to 760°C (1400°F) steam conditions.

Note that the reheat steam temperatures at Philo and Eddystone are stepped lower than the main steam temperature. Obtaining enough heat to reach the reheat temperature set points through a reasonable control range was a problem handled in this manner. Current USC and proposed A-USC turbine designs incrementally step up the reheat temperature and use a feedwater train with a HARP configuration (extraction to a heater above reheat point) so that the percentage of reheat flow to main steam flow is 75 to 80% instead of 85 to 92% as with the earlier vintage designs. This helps improve efficiency, meet the required heat absorption through the steam temperature control range and allows the reheat tubing to better utilize the stress capability of the materials where thickness is set by manufacturing rules and not as much by pressure stress limits.

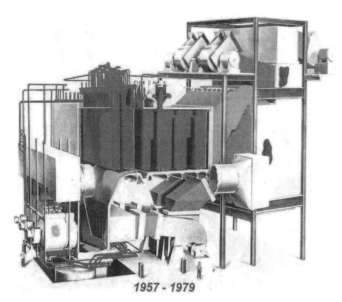

1957 - 1979

FIG. 17.1 AEP PHILO UNIVERSAL PRESSURE STEAM GENERATOR B&W CONTRACT UP-1 (*Source:* The Babcock & Wilcox Company)

Special research programs both in Europe (such as the THERMIE AD700 program) and in the US DOE Boiler Materials for Ultrasupercritical Coal Power Plants have set a goal to improve thermal efficiency and reduce carbon dioxide emission through application of materials with higher temperature capability to 760°C (1400°F) [3-5]. The project, managed by the Electric Power Research Institute (EPRI) and the consortium also includes the US domestic boiler manufacturers Alstom Power, Babcock Power, Babcock and Wilcox, Foster Wheeler. In addition there is a sponsored program for the development of A-USC steam turbine materials through meetings and collaboration between the boiler and turbine manufacturers which has included Alstom, General Electric and Siemens. It is recognized that increasing the steam temperature must be accompanied by the development of improved materials that are capable of surviving causes of failure and providing useful economic life. The effort of this materials development consortium is to address the competitive industry wide data needs such ASME Code allowable stress and other properties to qualify the new materials. Achieving the ASME Code acceptance of materials for high steam temperature is demonstrated in simulated laboratory experiments, in actual coal firing field service trials and in fabrication and construction practices has been a mission of the Ohio Coal Development Office (OCDO) and the US Department of Energy (DOE) NETL by sponsoring this research effort. As an example representing the work of the industry program, this chapter will describe steam generator design and materials development work of the Babcock and Wilcox Company's efforts, supported both B&W internally and externally by the OCDO / DOE program.

Advanced ultra supercritical coal fired generators (A-USC) have the potential for lower cost of electricity especially when combined with the requirements to capture carbon for sequestration (CCS). The plant production costs per megawatt-hour are the lowest for A-USC w/CCS based on plant economic studies for coal firing, see Chapter 18. Advanced cycles, with steam temperatures up to 760°C, will increase the efficiency of coal-fired boilers, before adding CCS, from an average of 36-39% efficiency (for the current domestic fleet) to about 47% (HHV). This efficiency increase will enable coal-fired power plants to generate electricity at com-

petitive rates while reducing CO_2 and other fuel-related emissions by as much as 17-22%. Steam temperatures and pressures up to 760°C/35 MPa (1400°F/5000 psi) are required [3]. Combining CCS with A-USC plants will provide lower cost of electricity generation with 90% carbon capture.

17.2 HIGHER NET PLANT EFFICIENCY

Net plant efficiency is a relative measure of how much fuel is used to generate a net output of electricity. The higher the efficiency of a plant, the less fuel input is required and fewer emissions are produced for the electrical output. Net plant efficiency is affected by three main components. These are net turbine heat rate (NTHR), boiler efficiency, and auxiliary power consumption. The NTHR is the ratio of steam turbine input from the net boiler heat flow, Btu/hour, divided by the gross generation, kW. The heat flows are the main and reheat steam energy flow minus feedwater and spray attemperation energy flow. Usually, for large units, the boiler feed pumps are steam turbine driven saving on the electrical conversion losses over electric driven pumps while lowering overall gross generation. The cycle efficiency is improved with use of the turbine driven boiler feed pump (TDBFP). Variable frequency or hydraulic coupling electric motor driven feedpumps may alternatively be used.

The Net Plant efficiency or its reciprocal term Net Plant Heat Rate (NPHR) is a key evaluation parameter for electric power plant economics and the wholesale market cost of electricity. Practice in the US is to define net plant efficiency as the ratio of net generated electric energy by the fuel energy, on a higher heating value (HHV) basis. The net plant heat rate (NPHR) is:

$$\text{NPHR} = \text{NTHR}/ \, ((\eta_{Blr}/100) \times (100 - \%AP)/100) \quad ((kJ/kWh \, (Btu/kWh)) \qquad (17.1)$$

Where:
NTHR = net turbine heat rate, Btu/kWh, input heat by steam divided by net generator output power

η_{Blr} = boiler fuel efficiency, %, this is the fuel higher heating value (HHV) energy input to steam

%AP = percent auxiliary power in % of gross power generation

The net plant efficiency is the reciprocal of the net plant heat rate multiplied by 3412.9 Btu/kWh times 100% and relates the fuel energy to the net power to the electric grid.

$$\eta_{Plant} = 100 \, \% \, (3412.9 / \text{NPHR}) \qquad (17.2)$$

Additional factors lower the net plant efficiency from the starting point of net steam turbine efficiency. These are the auxiliary power consumption and boiler fuel efficiency (defined by ASME Power Test Code, PTC 4). Boiler fuel efficiency is the percent of fuel input heat absorbed by the steam (working fluid), and is primarily dependent on ultimate fuel analysis (constituents C, H, S, N, O, H_2O, ash, etc.) and the operating conditions of the fuel and gas side of the boiler. The Heat Loss Method accounts for the dry gas loss (excess air and exit gas temperature), the loss due to moisture (air, fuel and the product from fuel hydrogen), ash residue sensible heat, unburned carbon loss, radiation and convection loss from the boiler setting and other unmeasured loss. Boiler efficiency is typically in a range from about 85 to 92%. If the allowance for auxiliary power is 6% of gross generation a result for the A-USC plant efficiency with 52.7% steam turbine efficiency is computed to be about 45%, on a HHV basis.

17.3 BOILER FUEL EFFICIENCY

The net plant heat rate discussed above was based on the US standard practice of boiler efficiency based on higher heating value (HHV). In the US the standard is HHV whereas in Europe the practice is to use lower heating value (LHV). The fuel HHV is obtained by laboratory analysis in an oxygen bomb calorimeter. The Lower Heating Value (LHV) of the fuel is computed by subtracting the latent heat of vaporization for water produced by fuel hydrogen combustion and fuel moisture content. The formula for calculation of lower heating value is:

$$LHV = HHV - H_{fg} (M + 8.94 \times H_2)/100 \qquad (17.3)$$

Where: M, fuel moisture % by weight
H_{fg}, water latent heat at reference temperature 25°C (77°F)
H_2, fuel hydrogen % by weight

Example is the difference for an Ohio coal (12540 Btu/lb HHV, 5.2% moisture, 4.83% hydrogen):

$$LHV = 12540 \text{ Btu/lb} - 1049.7 (5.2 + 8.94 \times 4.83)/100$$
(Eq. 17.3 with coal values)
= 12540–507.85
= 12032 Btu/lb LHV (27,998 kJ/kg)

In accounting for the boiler energy input, the LHV basis will not charge the boiler with as much energy input as the HHV basis so the efficiency will be higher. For the Ohio coal example this will be different by +4.1% in the NPHR (HHV basis) and the efficiency would be 45% net HHV and 46.9% net LHV.

Example difference for Pittsburgh Natural Gas (23170 Btu/lb HHV, 0% moisture, 23.53% hydrogen):

$$LHV = 23170 \text{ Btu/lb} - 1049.7 (0 + 8.94 \times 23.53)/100$$
(Eq.17.3 with gas values)
= 23170–2208.13
= 20899 Btu/lb LHV (48,545 kJ/kg)

The LHV basis efficiency will be higher than the HHV basis. For the Pittsburgh Natural Gas example this will be +9.53% in the NPHR (HHV basis). A natural gas power plant with this fuel achieving 55.43% (LHV) net efficiency would be 50.0 % (HHV) net plant efficiency.

Steam turbine Rankine cycle efficiency should be compared to the Carnot cycle heat engine efficiency. The Carnot cycle efficiency, with a reversible externally heated engine operating in a cycle where the working fluid undergoes isothermal heat addition, isentropic expansion, isothermal heat rejection, and isentropic compression represents the ideal cycle for maximum thermal efficiency. The formula for the Carnot Cycle thermal efficiency of a heat engine is provided in fundamental engineering thermodynamics textbooks, also in STEAM/Its Generation and Use, Edition 41 [6], and is:

$$\eta_c = 1 - T_c/T_h \qquad (17.4)$$

Where Carnot efficiency, η_c, is the ratio of useful work output divided by heat input, T_c is the colder absolute temperature of the energy rejected from the cycle, and T_h is the absolute temperature of the energy supplied to the working fluid.

Using the example where heat is supplied at 760°C (1033°K, 1400°F, 1860R) and the cycle rejects energy at 38°C (311°K, 100°F, 560R), the Carnot cycle attains an efficiency of 69.9%, (= (1 − 311/1033) × 100%). Note that pressure is not involved in the calculation of Carnot Cycle efficiency. The 760°C inlet working fluid Carnot cycle efficiency of 69.9% compares to an expected example A-USC steam turbine net efficiency of 52.7% (6825 kJ/kWh (6475 Btu/kWh) NTHR). This efficiency only considers the output generated for the heat input by the steam to the prime mover and not the efficiency losses in producing the heat input from the fuel to the steam.

How to best apply the capital funding available on a power plant project is a critical question for the plant designer. The cost basis of technological improvements must be known to make an economic evaluation in today's competitive marketplace. A good reference on this issue is an ABB paper written by H. Kotschenreuther, "Future High Efficiency Cycles."[7] This paper reported the ranking of several technology improvement steps for better plant efficiency. From least cost to highest cost per efficiency improvement, million German Marks DM / % net LHV efficiency (million $/% net LHV efficiency), these were:

1. Reducing condenser back pressure, 5.4 (4.6)
2. Increasing to 8th extraction point feedwater heater, raising feedwater temperature, 6.7 (5.7)
3. Raising live steam and reheat temperature, 14.5 (12.3)
4. Raising live steam temperature, 15.0 (12.7)
5. Using separate boiler feedpump turbine (BFPT) instead of main turbine driven pump, 16.8 (14.2)
6. Raising live steam pressure, 46.2 (39.1)
7. Changing from single to double reheat, 67.0 (56.7)
8. Using separate BFPT condenser, 71.7 (60.7)

This tabulation clearly shows that to optimize plant efficiency, raising steam temperature before raising steam pressure is better by a 3:1 cost/benefit ratio.

17.4 SELECTION OF TURBINE THROTTLE PRESSURE

In earlier studies proposing higher efficiency power plant design there was the belief that increasing the throttle pressure to 45 MPa (6500 psig) and greater was required in order to meet the plant efficiency goals [2]. Throttle pressure for the desired throttle temperature of 700°C to 760°C (1292°F to 1400°F) will ultimately be determined by the turbine manufacturers' recommendations. At the beginning of the present work of the US A-USC consortium, throttle pressure was discussed and determined most feasible in the range of 5000 to 5500 psi. Pressure vessel thickness for cycling operation and the practicality of the optimum or maximum available work per mass of the working fluid are major considerations.

According to the Second Law of Thermodynamics the maximum useful work possible per pound mass of working fluid in a process between two states is the difference of the available energy. The available energy, AE, for a process is defined between the two end states as:

$$AE = h_1 - h_2 - T_0 (s_1 - s_2) \quad \text{kJ/kg (Btu/lb)} \qquad (17.5)$$

Where:
h_1, enthalpy at state 1, from the heat supplied, kJ/kg (Btu/lb)
h_2, enthalpy at state 2, the heat rejected state, kJ/kg (Btu/lb)
T_0, absolute Temperature at the ambient condition for heat rejection, K (R)

s_1, entropy at state 1, kJ/kg K (Btu/lb R)
s_2, entropy at state 2, kJ/kg K (Btu/lb R)

A process cannot achieve more useful work than the available energy between the two end states. Figure 17.2 shows the available energy for steam at five operating throttle temperatures, 1000°F, 1100°F, 1200°F, 1300°F and 1400°F, and each exhausting to 1.5 in. Hg and 100°F. Considering a single reheat cycle, the optimum available energy peaks at about 2500 psia for 1000°F, about 4000 psia for 1200°F and at about 5000 psia for 1400°F. Because higher pressure results in higher component costs, the optimum available energy should be sought. Due to concerns that very thick pressure parts will require a very limited rate of load change and longer start up times, the steam generator design must optimize the operating pressure with the design temperature and select materials with optimum properties and cost.

Throttle pressure for the Rankine cycle prime mover, a steam turbine where work is produced by expansion of the steam, is fundamental to the optimum amount of available energy of the working fluid at the specified operating throttle temperature. Temperature is the more important factor regarding cycle efficiency. Selection of pressure is of secondary importance toward efficiency while of major importance in steam turbine design and operation. Setting the HP throttle pressure, the IP inlet pressure and the LP exhaust pressure is important for the optimization and meeting acceptable operating conditions for the prime mover. The turbine expansion line end states, throttle inlet and exhaust pressure must be set to achieve available energy utilization and not operate the exhaust too wet or too superheated.

Achieving the optimum available energy per mass of working fluid at the higher design temperature is the key to higher cycle efficiency. The increased efficiency means a smaller set of equipment is needed to produce the equivalent capacity compared to the plant designed for lower efficiency at lower temperature and pressure. Higher steam pressure does help reduce the flow path pipe size needed to deliver the energy flow. The more compact plant equipment will help with cost savings as long as pressure vessel thickness and material costs are at the optimum. The costs of the higher priced nickel alloys must be balanced with the savings in less fuel consumption, lower weight and size of equipment, and cost avoidance for emission allowance requirements.

17.5 DOES THE DOUBLE REHEAT CYCLE MAKE SENSE?

The double reheat cycle has normally been considered to provide 1.5% to 2.0% points of efficiency above single reheat for 538°C to 593°C (1000 to 1100°F) throttle temperature. A study [8] reported the double reheat cycle would only provide 0.7% advantage above single reheat turbine with 35 MPa 680°C/700°C (5075 psig 1256°F/1292°F) inlet conditions. The lower throttle temperature was due to reluctance to spin at 3600 RPM with 1292°F or higher main steam temperature. The cost of nickel alloy steam leads for the second reheat pressure at 300 psi requiring larger diameter piping is also a cost concern. The promotion to a double reheat steam cycle still needs careful study and advocacy by steam turbine vendors. Operation of a double reheat cycle was considered more difficult because of controlling the differential between the HP, IP 1 and IP 2 steam temperatures. The first A-USC plants will more likely be single reheat and double reheat may be adopted later.

17.6 ADVANCED USC STEAM TURBINE HEAT BALANCE AND TURBINE CYCLE DESCRIPTION

The A-USC steam generator for the development study is designed for 750 MW gross and designed to provide high efficiency using nickel alloy components for higher temperature operation. The steam turbine throttle conditions are 35 MPa/732°C/760°C with 67.7 mBar condenser pressure (5000 psi/1350°F/1400°F and 2 in. Hg). An additional new requirement of an A-USC steam generator is to deliver cooling steam from a source such as the primary superheater outlet at 1.5 % [8] of main steam flow rate to the HP outer casing. 1.5% cold reheat steam is retained at the turbine for IP turbine outer casing cooling. Feedwater temperature to the economizer for proposed steam turbine designs has ranged from 630°F to 649°F at MCR. An adapted single reheat steam turbine cycle is used in this analysis. The turbine stage isentropic efficiency used for the example in the study is HP — 89%, IP — 96%, LP — 92%.

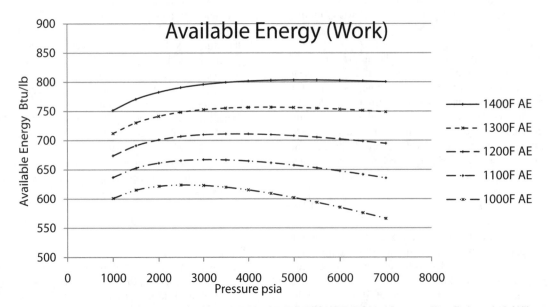

FIG. 17.2 AVAILABLE ENERGY BTU/LB AS A FUNCTION OF PRESSURE PSIA (*Source:* The Babcock & Wilcox Company)

A-USC steam turbine stage efficiencies are expected to lie in the following ranges depending on configuration offered by the various vendors; HP-89.2 to 93.3%, IP-90.5 to 96.6%, LP-90.6 to 95.8%. USC turbine performance is expected relate similarly for A-USC turbines [8,9]. The steam generator outlet conditions are 36.2 MPa/735.5°C/760°C (5250 psig, 1356°F, 1400°F). Inlet feedwater temperature is 332.8°C (631°F).

A-USC Plant Turbine Heat Balance is shown in Figure 17.3. Alstom originally provided the steam turbine and feedwater heater configuration to the materials development consortium for a single reheat machine. There are five low-pressure feedwater heaters. Heater number 2 has a drip pump to forward the drain flow into the condensate to heater 3. Heater 4 is an open type and supplies the electric booster pump suction. Heater 5 is after the booster pump and this is where one minor modification was made to the Alstom cycle, the heater 5 drip pump is removed and the drains cascade to heater 4. This sacrifices 3.9 Kcal/kWh (7 Btu/kWh) on the heat rate. Heater 6 is the deaerator operating at 1.2 MPa (170 psia). The turbine driven boiler feedpumps are before heater 7. Heater 8 is on cold reheat extraction. Heater 9 is a heater above reheat point (HARP). A desuperheater 10, on the third extraction point is the last feedwater heater before the economizer and exhausts to heater 7.

17.7   REDUCING CARBON DIOXIDE EMISSIONS

New plants designed with A-USC steam conditions for 45% net plant efficiency will produce about 20% less CO_2 than the average subcritical plants that are the majority of units in services, Figure 17.4. New high efficiency A-USC plants could be built carbon capture ready for later addition of carbon capture equipment. A-USC will lower the CO_2 per MW thus reducing the size of the CCS equipment and the auxiliary power consumption required [3]. Efficiency improvements and auxiliary power reduction are possible after the first results of pilot tests for both post-combustion and oxy-combustion capture methods and need to be demonstrated at larger scale. Heat integration with the water/steam cycle is important to produce the steam and meet the functional requirements for solvent regeneration on post-combustion. Flue gas cooling prior to CO_2 purification and compression heat recovery can be useful in the reduction of fuel input.

Carbon capture and sequestration (CCS) will affect plant net efficiency. Auxiliary power for the CPU, additional cooling tower, air separation unit (ASU), polishing scrubber, with some reduction of the standard plant auxiliary power, the oxy-combustion total auxiliary power is about 20.5% of gross power generation. Oxy-combustion A-USC plant efficiency with Ohio coal is estimated to be 38.1% (HHV) with 90% capture of carbon dioxide, comparing very well with current supercritical plants without carbon capture.

Reducing the economizer gas outlet temperature is a little more difficult because of the higher feedwater temperature of A-USC. The gas temperature of the SCR for NOx control needs to remain high enough through the load range and have the excessive gas temperature reduced for the structural design limitations of the flue and airheater. Rather than quenching the hotter flue gas in the FGD scrubber, integrated flue gas heat recovery is useful before and after the airheater, especially if it will reduce costs. Removal of flue gas moisture is needed for CCS equipment and the utilization of latent heat of the flue gas moisture should be considered in the overall plant net efficiency based on HHV of the fuel. Often water recovery from the flue gas moisture is also needed to help alleviate the plant water consumption.

Oxy-combustion methods for carbon capture incorporate heat integration for oxygen preheating and compressor cooling. Oxygen is supplied instead of air so that the nitrogen is low and the carbon dioxide concentration increased.

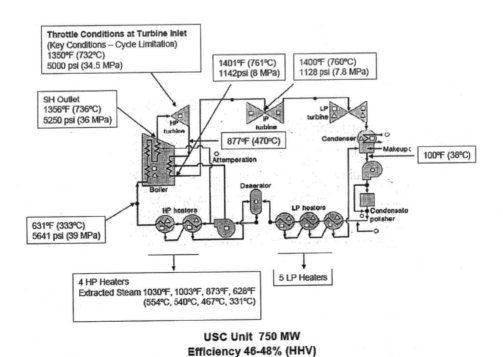

USC Unit 750 MW
Efficiency 46-48% (HHV)

FIG. 17.3   A-USC STEAM TURBINE HEAT BALANCE FOR MCR (*Source:* EPRI Report [12])

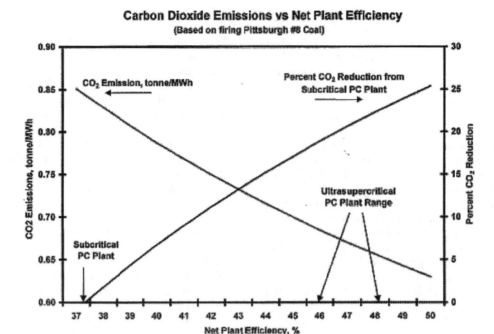

FIG. 17.4 CO₂ REDUCTIONS VS. NET PLANT EFFICIENCY — NATIONAL COAL COUNCIL [11]

Recycled flue gas is returned to the burner zone to dilute the flame temperature, control furnace exit gas temperature and maintain the level of convection heat transfer capability of the boiler. This protects the furnace tube metals, avoids refractory and allows better operational characteristics with load turndown. The amount of recycle gas is determined considering the furnace and convection pass protection and steam temperature control functions.

Three forms of flue gas recycle have been considered: hot, cold, and warm. Hot recycle of flue gas, also called gas recirculation (GR), was popular starting as early as the 1950s for furnace protection and reheat steam temperature control. GR has lost favor with coal firing primarily due to maintenance problems with GR fan erosion. Gas bypass control is now preferred. Hot gas at about 700°F and higher was taken after the economizer and before the airheater and then recirculated to the furnace. Hot recycle needs a high temperature fan and filtration system. Sorbent injection, depending on the fuel's sulfur content, may be needed. Otherwise, an increase in the sulfur concentration in the furnace will accelerate the rate of corrosion. Hot recycle is not preferred for advanced ultra supercritical designs. Cold recycle is used to return flue gas to the pulverizers meeting the primary air function of the coal milling system. The recycle gas is cooled and cleaned (SOx, particulates). The same gas is forwarded to the compression purification unit (CPU). The CPU does provide further cleaning of the carbon dioxide while meeting the pipeline requirements. An SCR is not required for oxy-combustion, as the CPU will have capability to remove the small amount of NOx. Warm recycle is flue gas pulled from the airheater gas outlet, cooled using a condensate heat exchanger, filtered using an electrostatic precipitator or a baghouse and sent by the secondary recycle stream back through the airheater to the burner windbox. Flue gas exiting the boiler is cooled in the regenerative airheater with two gas streams returning to the furnace (pulverizer primary and secondary gas). Refer to Chapter 18 to diagram of Gas Cycle Flow Diagram for Oxygen-Combustion.

17.8 EXAMPLES OF CURRENT ULTRA SUPERCRITICAL (USC) OPERATING PRACTICE FOR B&W STEAM GENERATOR DESIGNS

Millmerran 1 and 2 — Queensland, Australia, Figure 17.5. Two 420-MW units supplied for 354.36 kg/s 24.9 MPa/568°C/596°C 0.018 MPa (2,812,430 lb/hour 3610.5 psi/1054°F/1105°F/556°F 5.4 in. Hg) with heater above reheat point (HARP), firing 34.8% ash fuel. Tube materials used are 347H, T91, T22, and T12.

Weston 4 — Wisconsin Public Service, see Figure 17.6. Steam generator firing PRB coal supplies steam to the 590 MW turbine at a rate of 458.75 kg/s 26.0 MPa/585°C/585°C/291°C (3,641,000 lb/hour 3775 psi/1085°F/1085°F/556°F) with a HARP feedwater cycle. The operating mode is sliding pressure. The main and reheat steam leads are P92 grade [10].

The current state-of-the-art limits in the United States for USC units has main steam temperatures to about 605°C (1121°F), and hot reheat temperatures to about 613°C (1135°F). The highest main steam pressures of about 30.5 MPa (4423 psi) have been designed in Europe while the highest main steam and reheat temperatures have been commissioned in Japan. The most advanced cycle conditions for the current market in the U.S. have been developed for the John W. Turk, Jr. AEP Hempstead project, commercial in 2012. B&W is supplying the 600 MW net boiler design at 26.1 MPa/602°C/608°C (3785 psi / 1116°F/1126°F/570°F feedwater).

17.9 ADVANCED USC STEAM GENERATOR OPERATIONAL DESIGN

A supercritical plant uses a steam generator design that operates at full load above critical pressure 22.1 MPa (3208 psia). The surface tension property of water becomes zero at the critical point 373°C (704°F) and water does not boil to form two-phase bubbles

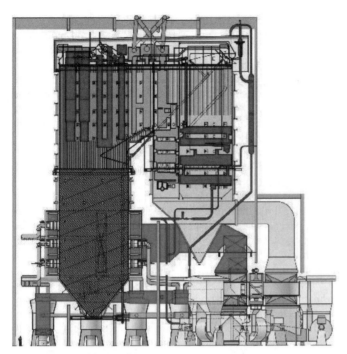

FIG. 17.5 MILLMERRAN 1 AND 2 (*Source:* The Babcock & Wilcox Company)

as in a subcritical pressure boiler. The term Benson boiler is a design that is capable of operating on a variable pressure ramp and is capable of start up from zero pressure at initial firing and up to supercritical pressure at higher load.

Traditionally, supercritical steam generators were controlled to maintain constant throttle pressure. Split pressure allows the furnace waterwalls to operate supercritical and the final superheater and steam turbine operate variable pressure to reduce steam temperature throttling and cyclic stress on the pressure part metals. Unit load demand is set to demand firing rate and feedwater flow. Steam temperature is controlled by the adjustment of the ratio of firing rate to feedwater flow. The furnace enclosure is kept supercritical to avoid subcritical two-phase flow and the high temperature excursions that occur with dryout or Departure from Nucleate Boiling (DNB). DNB causes high metal temperature tube failure due to the inadequate steam film heat transfer coefficient. The inside tube wall temperature is suddenly higher when the water film dries out so that the steam cooling mass flux is too low. A Benson boiler must be designed with furnace circuitry that with correct operating conditions and variable pressure mode will be capable of permitting appropriately located dryout to occur.

Supercritical plants require feedwater purity so that tube side deposition will not cause overheating damage. Condensate polishing with oxygenated water treatment (OWT) is required to achieve excellent water purity. Even many natural circulation (drum type) units now use OWT. The deposition has been greatly reduced so that the requirement for frequent chemical cleaning is almost eliminated. The steam generator pressure drop of the furnace enclosure stays very close to new unit values. This also stops the increase of boiler feedpump power over time between acid cleaning.

Controls and design of equipment of an A-USC plant must achieve cold start up, warm restarts, hot restarts, load cycling, and shutdown. The design is set so the Benson point, defined below, is estimated to be at 45% load. Final steam temperature control range

meets set point from 50 to 100% load. Reheat steam temperature control range meets set point from 60% to 100% load.

Control of the variable pressure Benson steam generator is accomplished with a method of logic that handles the transition from the recirculating mode using the boiler circulation pump to the 'once through mode' where all the water entering the economizer leaves from the superheater outlet. A minimum circulation flow rate is established with the boiler feed pump and the boiler circulation pump to permit initial firing. Water that is not evaporated is drained from the vertical separator and water collection tank system to the condenser and polishing system. The load and steam generator pressure increase with increased firing rate demand. The turbine valves are controlling the minimum pressure build up and the pressure ramp curve. Two-phase flow is accommodated until the critical point is reached at about 75% load where the transition to single-phase flow occurs at critical pressure. The load when the vertical steam separator runs dry because the entering fluid is converted completely to steam is called the Benson point. This load will usually be at about 30% to 40% At a point, the boiler circulation pump is shut off and the boiler feed pump will control the feedwater flow to meet the furnace enthalpy pickup demand function (from the economizer outlet to the superheater inlet). By this method, the proportion of steam generator surface serving as "evaporator" and as superheater is controlled as a function of load and stabilized. Steam temperature is controlled by multiple stages of spray attemperation and the temperature difference across the first stage attemperators is used to trim the relationship between spray and feedwater flow in the furnace. The steam temperature control for faster transients must account for the time delay of the water entering the economizer to leave the superheater outlet, which takes about 15 minutes at minimum circulation flow load and about 3 minutes at MCR load. If outlet steam temperature of a variable pressure steam generator is controlled by the feedwater to firing rate ratio, the tube metals will suffer damage due to the error in required feedwater flow caused by transit time delay.

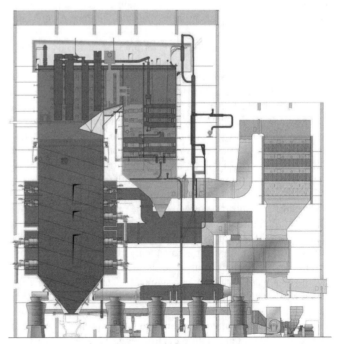

FIG. 17.6 WESTON 4 (*Source:* The Babcock & Wilcox Company)

The A-USC design study has focused on the boiler arrangement and pressure part details requiring specification to meet the high-temperature, high-pressure steam conditions of an advanced ultra supercritical steam plant with steam turbine conditions, 35 MPa / 732°C / 760°C 67.7 mBar (5014.7 psia / 1350°F / 1400°F / 2" Hg). The use of steam at 1350°F was considered because of the material allowable strength limits required much thicker pressure vessels and the need to permit cycling service was a major factor. Higher allowable stress values have been specified since earlier studies applying the 740 nickel alloy used for superheater tubes and headers and will permit considering temperatures to 760°C (1400°F). The limiting areas for the design are the secondary superheater outlet headers, the startup system steam separators and water collecting tank, and the furnace enclosure. The superheater outlet conditions are set with about 5 % main steam lead pressure drop. In order to minimize the thickness and size of these headers, a two header arrangement with double end outlets is selected. This also allows the header ligament spacing to be larger with tube sections alternating to the headers. The steam turbine would have two inlets to the HP casing, which means a forging is required to transition from the four main steam pipes from the steam generator into two.

The startup system steam separators (VS) and water collecting tank (WCT) are the other large diameter thick wall vessels, which are significantly affected by the selection of the turbine throttle conditions and rate of transient load change. These "water side" components must be fabricated from ferritic or nickel material to avoid internal corrosion. The steam separator design is based on an inside diameter mass flux (steam flow per cross sectional area). Therefore, for a required separator cross sectional area, the number of smaller diameter separators must be greater in order to maintain reasonable thicknesses for the much higher turbine throttle conditions. Water collecting tanks are sized based on required volume for transients. Use of two tanks is considered where diameter and thickness must be limited and the elevation above the boiler circulation pump needs to be high enough to protect the inlet suction head. Note: "water side" refers to components that will normally start up containing water, and may contain steam at higher load. "Steam side" refers to components that will contain air or vapor at start up.

Setting the design is needed to determine the material selection, size, quantity and the requirements necessary to provide an economic estimate for a steam generator that would meet the higher plant efficiency desired. A steam turbine cycle analysis was needed to evaluate the benefit of performance level compared to cycle configuration, complexity, and potential economic cost. The steam generator economic evaluation was needed to validate the premise that the gain in plant efficiency would be justified and support the higher capital cost of the equipment requiring special metal alloys while limiting the increase of the levelized cost of electricity (LCOE).

Variable Pressure operation, using a mode sometimes called pure sliding pressure, of the boiler and turbine provides a better heat rate at part load and reduces the thermal stress damage to the turbine and boiler very high temperature pressure parts. The criterion is to have small changes in the temperature of the turbine metals, which are very thick pressure parts. Turbine valves are operated wide open from about 25% load to MCR reducing the steam temperature quench from throttling. The turbine high-pressure section exhausts at a higher temperature to the cold reheat steam line. The reheat steam temperature control range is wider down to as low as 55% load. The throttle pressure rides on a ramp based on the steam flow pushed through the throttle area. Boiler feed pump

power is reduced at part load. The boiler can operate with zero pressure on start up due to the waterwall design and pumping a minimum circulation flow rate. There are no valves between the economizer inlet and the superheater outlet.

Modified sliding pressure operation uses throttle valve reserve so the valves are slightly closed at 90% to 95% load and some boiler stored energy can respond more quickly to rapid load change near full load. Temperature loss due throttling is milder and not a very severe problem on metals. Split pressure is used in some systems to operate the boiler waterwalls a full pressure and throttle at a location between the primary and final superheater. The turbine metal temperatures remain at set point values and the throttling loss in temperature can be made up in the final superheater. Many once through boilers both subcritical and supercritical operated at constant pressure where the turbine valves are modulated through the load range for turbine control. There will significant temperature throttling loss and metal temperature change with load modulation.

Partial load cases for 80%, 60% and 40% load are included normally to evaluate the design at steam temperature control load points and to help determine the Benson point load where the steam generator achieves once through operation producing entirely vapor leaving the evaporator and no longer produces water drain flow from the vertical steam separator. Variable pressure operation reduces throttling temperature loss by the turbine valves and maintains full steam temperature to the steam turbine over as wide a control range as possible, part load operation may be a constraining limit and may set material selection and tube thickness requirements at section transition points. Full load, MCR, normally will set the material alloy selection requirements. A wider steam temperature control range, even using variable pressure, will require increased tube thicknesses and higher pressure drop unless higher alloy materials are used in a greater portion of the heating surfaces. The steam temperature control range for main and reheat steam and the minimum circulation flow load requirements are parameters determined by the partial load analysis.

The steam generator needs to provide auxiliary steam for sootblowing, steam for emissions equipment and steam turbine HP cooling steam. The source for sootblowing steam is usually the primary superheater. The A-USC steam turbine HP and IP outer casings require cooling steam and special inlet nozzle designs to the inner casing connection otherwise the outer casing material would need to be a nickel alloy or would be unacceptably thicker, and a limit to faster load change rates. HP casing cooling will involve an additional steam lead from a source on the steam generator. The primary superheater outlet steam before attemperation has been selected. This will provide about 1.5% of the main steam flow rate at about 550°C (1025°F) flowing along the inner casing and be admitted to the first stage blades. The cold reheat steam will provide about 1.5% of reheat flow rate for IP casing cooling and admit the steam to the inlet blading [9]. The turbine cycle would provide steam to the air heater preheating steam coils as is normally required.

The steam generator heat absorption process results in an enthalpy increase along a high process pressure path through the series of components starting at the economizer inlet and leaving the final superheater outlet. This is shown in the Temperature versus Enthalpy (T-h) diagram, Figure 17.7. There is no high capacity and economic way to reach supercritical pressure steam conditions at lower pressure and lower boiling temperature in order to use lower alloy enclosure walls and then compress steam for superheating in the convection tube bank sections. The placement and adequate design of the heating surface to accomplish this thermodynamic process is the art and experience of steam generator design. The extension to higher steam temperatures requires careful evaluation and selection of materials for the furnace enclosure at these conditions. There is no flat region on the T-h diagram for a steam generator process at 34.5 MPa (5000 psi) as there is for a 241 MPa steam generator. Ferritic alloys T23 and T92 with coatings are some of the materials being considered for higher sulfur fuel and low NOx combustion.

The throttle conditions required for this high efficiency plant necessitate boiler and piping materials, which are stronger than the currently used ferritic and austenitic materials. Nickel based super alloys are required for major portions of the superheater and reheater. This is due to the need to maintain reasonable thin wall pressure components with the high design pressure and temperature requirements of cycling service. Economizer sections can be fabricated with currently available carbon steel materials. Enclosure walls can be fabricated from SA213T92 (and SA213T23material).

Several heating surfaces that function as preheating, "evaporator" or steam generating, superheater, and reheater handle the heat absorption duty of the steam generator. Preheating is the job of the economizer, which receives the entering feedwater. The amount of preheating is limited so that two-phase steam is not produced to enter the furnace wall enclosure circuits. Operating variable pressure must be handled like a drum boiler to avoid a steaming economizer. Evaporation is accomplished in the furnace wall enclosure. The lower enclosure will be spiral panels or may be vertical tube panels where the unit size is large enough to have the proper steam flow per foot of perimeter, lb/hour ft. The lower furnace enclosure will become superheating surface near full load and will increase temperature quickly with absorption. The reheating duty is treated the same as superheating except the location sequence in the gas path and the allowed steam side pressure drop must be carefully designed. Figure 17.8, h-P, Enthalpy versus Pressure diagram

shows the different heating surface banks and the steam conditions for several loads. The line originating at the critical point, 22.1 MPa (3208 psia) and running out to about 1055 Btu/lb enthalpy at 6000 psi is the "pseudo film boiling" line where the fluid below acts like water and above the line acts like steam. It can be seen that the A-USC steam generator is primarily in the greater proportion through the load range, a superheater.

It should be noted for the design case displayed on the h-P diagram, Figure 17.8, the steam temperatures of the components run nearly parallel to the isotherms so the metal temperature change in the upper load range has a very desirable minimal change.

The advanced ultra supercritical steam generator is fundamentally the same as current supercritical boiler design because fuel type rules govern arrangement of the "gas side." The working fluid "side" will be stretched out in the load range that operates above critical pressure. The overall heating surface that functions for superheating duty is of greater proportion. The sequence of placement for which type of heating surface is first, second, etc., may be different than is common in the current USC supercritical boiler. The heating surface for reheating duty is required to operate over a wider temperature range, inlet to outlet.

Fuel characteristics, burner placement and gas injection points still govern the size of the furnace required, the spacing distance to the first tube bank and the side to side spacing of the banks as a function of gas temperature. It is evident there is usually more than one way to address the requirements and criteria for a successful design arrangement. In addition to the requirements to use advanced material alloys, the major differences between a conventional supercritical unit and the A-USC unit are in the heating surface required for the superheater and reheater resulting from lower gas to fluid temperature difference and the much greater superheating absorption duties that are required of these components. The higher temperature feedwater from the turbine cycle affects the ability of the economizer bank to achieve lower gas outlet temperatures and

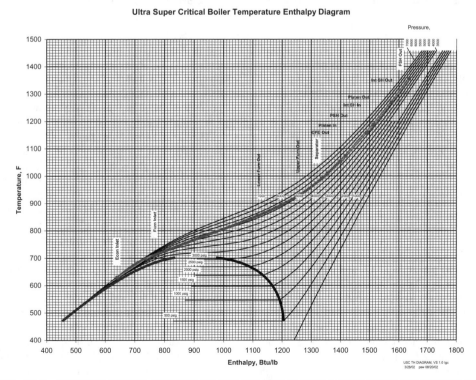

FIG. 17.7 TEMPERATURE–ENTHALPY T-H DIAGRAM FOR A-USC (5250 PSI 1350°F) [3]

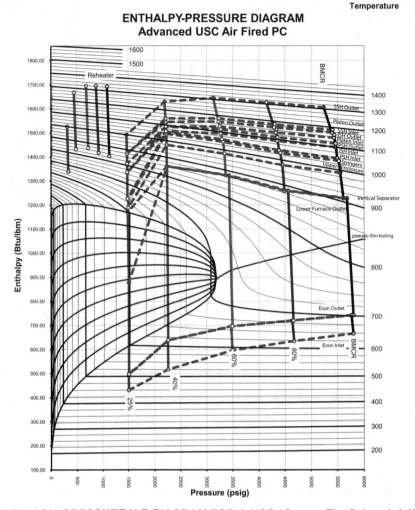

FIG. 17.8   ENTHALPY–PRESSURE H-P DIAGRAM FOR A-USC (*Source:* The Babcock & Wilcox Company)

meet furnace panel inlet temperature limitations through the load range. The proportion of heating duty is mostly superheating.

The steam generator arrangement for electric utility service has evolved with the changes to industry requirements. Unit size has become as large as up to 1400 MW, and sometimes as small as 20 MW. Supercritical plants are usually larger than 400 MW in the present market place. The economy of scale improves electricity production cost and yet the production capacity and operating flexibility of the unit must be a good fit with the electric grid service area requirements and the utility financial outlook.

Utility steam generator size currently seems to be in a range from 400 to 1000 MW, with about 750 to 850 MW being generally the most representative. The first A-USC demonstration unit planned in Germany at Wilhelmshaven by Eon was to be 400 MW. Commercial sizes would be planned at 1000 MW. There is a wide range of desired features sought by the marketplace, mostly due to customer preference in the need to utilize particular coals. A steam generator designed to use a wide variety of coals and maybe gas and oil too, would require several additional features to cope with wide differences in operational needs due to the shift in the heat absorption characteristic for the fuel type.

The selection of certain features and arrangement used for the A-USC steam generator may evolve due to the need to better suit

the 200 to 300°F higher steam temperature above current normal practice. The following "DNA chart" in Table 17.1, shows the various feature design options that may be incorporated in the steam generator. There has certainly been more than one way to achieve a functional and market successful design.

There are several arrangement styles. The two pass design versus tower design is often a subject of debate in the selection process. Selection preference should be well founded on experience for the fuel utilized. The two pass design is said to be the most predominant configuration worldwide. The tower design has also found success. There is much literature and third party opinion that can be found, particularly on the internet about the benefits of each design. There are also variants on design arrangements worthy of evaluation, the ranch style (like the Philo 6 design) and the modified tower design, which is a combination of two pass and tower features.

Some of the positive opinions for the two pass design are shorter steel structure, timesavings of parallel construction sequence, less complicated high temperature tube sections support, considered more economical to erect, less sootblowing required to clean the pendant surfaces than high temperature horizontal surface in the tower design. The examples of negative opinions are concern for ash concentration and erosion at the arch turn, non-drainable surface, and wall enclosure thermal differentials.

TABLE 17.1 ONCE THROUGH ADVANCED ULTRA SUPERCRITICAL PRODUCT STRUCTURE TREE "DNA CHART"

Arrangement	"Two Pass" (B&W Carolina Design)	
	Tower	
	Modified Tower	
	Ranch	
	New Configuration	
Reheat	Single Reheat	
	Double Reheat	
	Non-reheat	
Pressure Profile	Variable Pressure	
	Split Pressure	
	Fixed Or Constant Pressure	
Furnance Enclosure	Up-up First, Second, Third Pass	Partial Mix Header System
	Spiral Tube	Full Mix Header System
	Vertical Tube	Bifurcated Transition
		Transition Header
Convection Pass Surface	Platens	
	Horizontal In Furnace (Tower)	
	Pendants	
	Wingwalls	
	Curtain Wall	
	Division Wall	
	Horizontal Banks	
	Screen	
	Stringer Tubes	
Convection Enclosure Surface	Water Cooled	Economizer Enclosure
	Steamwater Colled	Boiler Enclosure
	Steam Colled	RH Enclosure
		Primary SH Enclosure
		"Birdcage" Enclosure
Backpass	Series Downpass	
	Parallel Downpass	
	Three Downpass	
Furnace Bottom	Flat	
	7.5 Degree Dihedral Floor	
	Hopper	Dry
		Wet
CCS (Carbon Capture Storage)	Air Fired, Scr, Post-combustion	LP Auxilliary Generation
	Oxy-combustion	GR (Gas Recirculation)
		GT (Gas Tampering)
Firing	Pulverized	Single Wall Firing
	Cyclone	Opposed Walls
	BFB	4 Walls
	CFB	Corner Tangential
Emissions	OFA	
	Low NOx Burners	
	SCR	

Some of the positive opinions for the tower design are better gas flow distribution resulting in lower tube metal upset temperatures, high ash fuels allow removal falling back through wider tube spacing to a single furnace hopper. Examples of negative opinions are taller steel structure and crane required, complication of tube section support for higher loads and thermal differentials with the enclosure.

Both the two pass and tower designs have found success and preferences in selected applications. Countermeasures for the neg-

ative aspects of both designs have been incorporated that keep the designs viable in the market place.

The furnace enclosure is one of the first considerations in the design process. The enclosure must contain the combustion gas products and lower the adiabatic gas temperature to a low enough value entering the first convection surfaces. This is the furnace exit gas temperature (FEGT), a design criteria set by the coal ash characteristics for slagging and fouling properties.

Predominantly, the furnace enclosure is a vertical updraft structure with burners firing either on a single wall, front and rear walls (opposed firing), all four walls, or from at or near the four corners. The furnace plan area and height to the FEGT exit plane are geometric controlling parameters for the enclosure that are set by thermal intensity limits and environmental emissions requirements.

An ash hopper configuration is normally used for "dry bottom" furnaces with front and rear walls inclined at angles greater than 45°. A chain conveyor in a water trough is the most used method to remove and dewater the slag and ash. Slagging or "wet bottom" furnaces are sometimes used for certain types of coal. The molten ash runs like a lava stream to a tank or chain conveyor. The furnace floor for a "wet bottom" may be nearly flat and retain the layer of molten slag that flows to the openings. Normally pin stud and refractory covering are needed to prevent erosion with wet bottom furnaces and the furnace absorption is reduced.

Combustion gas from the burners is cooled primarily by radiation to the furnace enclosure to a temperature (design FEGT for the coal) suitable to prevent unmanageable slagging before passage through the immersed tube rows for convection heating duty. Radiant platens are very wide spaced tube sections that can prevent bridging of ash. Platens serve to lower the FEGT in the same manner as the furnace enclosure. As the gas becomes cooler, convection banks with closer and closer spaced tube rows are allowed. This maintains the gas velocity while the gas density increases permitting lower free flow area and preventing excessive erosion and draft loss. The economic engineering design of the convection surface tends to be placed so the steam is flowing counter to the gas flow direction (counter-flow) thus requiring less heating surface area than needed if parallel flow (gas and steam flow in same direction) or in mixed flow. Steam flow in the same direction as the gas flow is sometimes needed and is used to moderate the gas and steam temperature upsets and unbalances. Final main steam and reheat steam outlet banks will usually be parallel flow design. Outlet tube rows are shielded behind rows in the tube bank interior to limit the direct radiation from the open cavities between banks. This combination of flow arrangements is usually the best economically considering the tradeoff in metals alloy requirement and quantity of heating surface.

The sequence of which type of heating duty surface is placed in the gas flow sequence is judged by the dependence on the heat available through the load range to meet the outlet temperature control set point. An intermediate superheater bank is most likely placed before the final superheater bank. This is to better handle the metal temperature upsets. Next is the final reheat superheater, and then the primary superheater, the primary reheater and economizer follow. With higher reheat outlet temperature, the placement of more final bank surface may be needed with caution to handling the wider upset temperatures possible. Intermediate reheat attemperation may be selected. Control of reheat outlet temperature without use of attemperation is desired because spray will reduce efficiency. Gas recycle from an oxy-combustion process would lend a hand in controlling reheat steam temperature. Coal fired steam generators for A-USC combined with carbon capture sequestration (CCS) appears to offer an advantage in lower cost of electricity.

17.10 FURNACE ENCLOSURE

The primary steam generator design case is a once-through steam generator or "boiler" for pulverized coal (PC) fired, dry hop-

per bottom, designed for A-USC steam conditions. The steam generator is referred to as a Benson™ boiler designed to operate in the variable pressure mode.

The lower furnace normally includes spiral tube circuitry up to about the arch (refer to Figures 17.5 and 17.6) where interconnected transition headers on each wall provide for the conversion to vertical upper furnace tubes. A more recent design improvement includes a vertical tube lower furnace enclosure with optimized multiple lead ribbed (OMLR) tubing and is expected to simplify fabrication and construction, improve operation and maintenance and lower costs (refer to Figure 17.9). The boiler convection pass configuration is typical of a B&W design that includes two parallel gas paths in the downpass. Control dampers, located at the bottom of the downpass, bias the flue gas flow to each of the gas paths to control reheat outlet steam temperature throughout the load range. Regenerative air heaters are located downstream of the economizer flue gas outlet to heat both the primary and secondary air.

A vertical tube enclosure is recommended for the lower furnace of the A-USC plant. This technology uses special ribbed tubing and operates at mass-flux conditions comparable to drum type boilers. When the vertical tube furnace becomes accepted it is expected to replace the spiral wound enclosure technology for units larger than 500 MW with high steam flow per foot of enclosure perimeter. Spiral wound furnace circuitry is normally utilized in the lower furnace for a sliding or variable pressure unit. Fewer slightly larger tubes

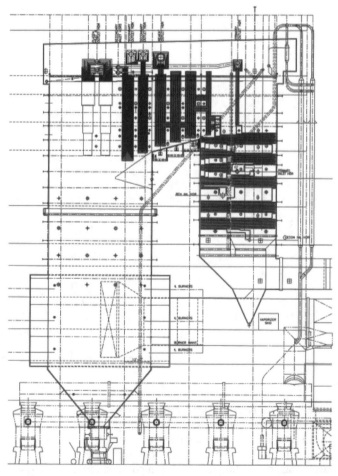

FIG. 17.9 A-USC STEAM GENERATOR B&W TWO PASS OR "CAROLINA" DESIGN (*Source:* The Babcock & Wilcox Company)

are used in the spiral, increasing the mass-flux and routing the tubes around the furnace periphery to pass through the varying heat flux zones providing more even fluid outlet temperatures. The spiral wound tubes with internal multi-lead ribbing eliminate problems with distribution, flow instability and uneven heat absorption by the furnace tubes. Near the elevation of the furnace arch work point, transition headers and piping are used for the conversion from spiral wound tubes to vertical smooth tubes in the upper furnace and arch.

The furnace walls utilize a gas-tight welded membrane construction, with wall tubes bent to accommodate openings for the low NO_X burners, OFA ports, observation ports, access doors, soot blowers and test connections. The furnace construction is such that the walls will be shipped in panels, with headers attached where feasible. Integral windboxes are attached to the furnace walls on the unit providing optimum air distribution controlled by the burner and OFA port assemblies.

The lower furnace uses vertical tubes with a particular internal configuration called optimized multiple-lead ribbed tube (OMLR). Using vertical tubes in the lower furnace would save on fabrication and erection cost, especially in the case where T92 material is needed. The pressure drop would also be lower saving pumping power and slightly improving unit efficiency. There will also be cost savings due to less pressure part design thickness requirements for the lower furnace, economizer and pre-boiler equipment.

Optimized multiple lead ribbed (OMLR) tubes permit operation with a lower design mass-flux that is comparable to drum type boilers, in sliding pressure once-through boilers thus achieving furnace circuitry with a Natural Circulation Characteristic (NCC). The Natural Circulation Characteristic is an effect where an increase in heat absorption from an upset results in an increase of tube mass-flux and less excursion of the outlet enthalpy and temperature. NCC results in lower tube-to-tube outlet differential temperatures during upset/ unbalances. Once- through boilers normally have tubes operating with a high mass-flux, equal or greater than 2000 kg/s m^2 (1,475,000 lb/hour sq ft.) and will have a Forced Circulation Characteristic (FCC). FCC produces an effect where increased heat absorption due to upsets results in reduced mass flux for tubes in a panel at fixed pressure drop. This causes tube outlet enthalpy and temperature to increase. The FCC design mass flux must be high enough so that when upsets reduce the mass flux, it is still high enough to moderate the increase in tube temperatures.

A-USC boilers may use spiral or vertical tube lower furnace designs depending on the ability to achieve Natural Circulation Characteristics, (NCC). Natural circulation drum boilers usually have tubes operating with mass-flux lower than 1150 kg/s m^2 to 850,000 lb/hour sq ft. A supercritical boiler tube circuit, with special ribbed tubes, can be designed to operate with a Natural Circulation Characteristic through nearly the complete load range and achieve the benefit of moderated excursions during heat upsets. There are special design considerations for applying vertical tube lower furnace walls. Unit size and minimum control range for steam temperature are particular criteria.

This new OMLR vertical tube development also requires coordination with the economizer outlet enthalpy conditions. The higher economizer outlet enthalpy from the advanced ultra supercritical conditions used in this study, most likely will require the furnace to be spiral wound circuitry using T-92 or T-23 material. The lower furnace area must absorb the same amount of heat whether spiral or vertical. While the spiral tube routing passes through all the horizontal zones of varying heat flux, certain vertical tubes will pass upward through only the higher heat flux zones. Reducing the economizer outlet enthalpy for application of T-23 in a vertical tube lower fur-

nace is required for the ultra supercritical steam generator design. T-91 or T-92 will be required otherwise. Vertical tube lower furnace may be utilized if the cooling flow per foot of perimeter is high enough. Larger units, greater than about 750 MW and with fuels that tend to less emissions design constraints, may accept this design.

17.11 FURNACE ROOF

The roof is made up of sections of membrane tube panels and sections of loose tubes. In areas where loose tubes are used, the tubes are flat studded where applicable to protect the backup refractory. Roof tubes, tie bars, tight roof casing and seal boxes are arranged to form a structural grid to contain furnace or penthouse pressure.

Seals around penetrating vertical tube legs are constructed with a layer of refractory, at the roofline, and shop applied seal plates welded to the tube legs which are seal welded together in the field to afford a completely metallic enclosure. The superheater and reheater tube legs are restrained at the roofline. Tube leg length and flexibility for excessive differential motion is provided in the header distance from the roof for all expected temperature differentials.

17.12 CONVECTION PASS ENCLOSURE

The convection pass enclosure is the wall surface enclosing the convection pass heating surface tube banks. The convection pass enclosure utilizes a welded membrane construction. It is partially shop assembled to reduce field erection. Tubes are bent to accommodate doors and other openings. The convection pass enclosure is supplied with steam from the vertical separators and could be cooled by reheat steam or the second reheat steam in the case of a double reheat cycle.

17.13 CONVECTION PASS HEATING SURFACE

The superheater, reheater and economizer heating surface arrangement are a combination of radiant and convective heat transfer surface. Pendant radiant platens are supported from the roof or by stringers. At the rear of the boiler, the gas turns down into the horizontal convection pass area where the tube surface is end supported at the enclosure walls or stringer supported. The longer length banks are stringer supported by economizer tube legs and are usually primary superheater and horizontal economizer surface. Double reheat cycles would have portions of the second (low-pressure) reheater in both down passes and some final outlet pendant surface.

Tube bank side spacing is determined based on the relative flue gas temperatures and the fouling characteristics of the coal ash. Maximum gas velocities based on the erosion potential of the fly-ash may also affect the required side spacing. Tube shielding and erosion barriers may be utilized selectively to provide local protection in areas of potentially high ash concentration. The surface is also arranged to promote effective use of sootblowing in maximizing heat transfer performance.

Even with the best designed burner and furnace system and the most cautious operating instructions and practices, there will be periods of operation where very rapid load changes will occur that cause flue gas and superheater flow unbalances with resultant gas and steam temperature upsets and unbalances. The convection tube metal requirements for materials and thicknesses are determined in accordance with ASME Code allowable stress and temperature oxidation limits by considering temperature and flow unbalances of the flue gas and steam paths. Superheater and reheater outlet tubes with the highest expected steam temperatures are typically placed in the interior of the tube bank to avoid high tube metal temperatures due to cavity radiation. Coal-ash corrosion in the high temperature areas is also considered in developing the tube metal requirements.

The 750 MW A-USC steam generator is smaller and narrower in width by 20% compared to a conventional supercritical unit because of the lower heat input requirements resulting from the higher efficiency. However, header and piping weight is less than the conventional supercritical unit. The overall suspended weight of the boiler is 10% more and the increase in tubing weight is 16% more than a conventional supercritical unit.

17.14   MATERIALS DEVELOPMENT

Major components, in-furnace tubing for the waterwalls, superheater/reheater and external piping, headers and other accessories require advancements in materials technology to allow outlet steam temperature increases of 300°F up to 1400°F. Experiences with projects such as the pioneering Eddystone supercritical plant and the problems with the stainless steel steam piping and superheater fireside corrosion provided a lesson for A-USC development [12]. Industry organizations recognized that a thorough program was required to assure adequate development of new and improved materials and protection methods to advance to the high temperature steam conditions.

The DOE/OCDO Materials Development Program for A-USC technology includes nine task categories [13,14].

Task 1. Conceptual Design and Economic Analysis. Design and estimating to provide component material selection, quantity, economic evaluation.
Task 2. Mechanical Properties of Advanced Alloys. Develop data on new advanced alloys.
Task 3. Steamside Oxidation and Resistance. Characterize rates, temperature limits, and protection methods for exfoliation.
Task 4. Fireside Corrosion Resistance. Characterize alloys and protection methods in laboratory, pilot and field tests for Eastern, Midwestern and Western coals on waterwall and superheater components.
Task 5. Weldability. Test alloys and dissimilar materials in component forms to develop processes and procedures.
Task 6. Fabricability. Test alloys using standard shop processes (machining, bending, swaging, etc.) to develop data and procedures.
Task 7. Coatings. Test protection methods, both existing and new material combinations and application methods.
Task 8. Design data and rules. Provide improved analysis techniques and compile data, rules and guidelines, and submit code cases.

The program tasks were led by a boiler supplier or the Electric Power Research Institute (EPRI), with participation in most cases by other boiler suppliers, U.S. laboratories and universities. EPRI and Energy Industries of Ohio (EIO) are providing technical and administrative program management.

17.14.1   Problems to Overcome
Material related failures in steam generators are often due to inadequate data on material properties, internal oxidation and corrosion, fireside corrosion and erosion, inadequate welding procedures and fabrication techniques [6]. Proponents of the advancement of steam plant efficiency believed that finding new materials, and by adapting those from other fields of applications, that steam conditions up to 1400°F were possible. Industry experience gained in the materials development program improved analysis methods, will help achieve lengthy service exposure time of the materials that is essential to the new product introduction.

17.15   FAILURE MECHANISMS

Three failure mechanisms appear to raise a higher level of concern of proponents and future owners of A-USC plant technology. Adequate mechanical properties, in particular creep rupture strength, steam oxidation and exfoliation, and coal ash corrosion, pose threats to the economic service life of components.

17.15.1   Mechanical Properties
Determining the thermal coefficients of expansion, conductivity, ductility and other mechanical properties is important to the design and fabrication of materials. In addition, welds, and weldments for both thick sections and tubes were tested. To achieve 1400°F (760°C) steam temperatures, longer creep rupture strength testing at higher temperatures is very important to the A-USC design. Creep rupture tests have now achieved in excess of 30,000 hours [14]. To gain benefits from the higher costs, high strength nickel-based materials must be used to the optimum level of their strength capability.

17.15.2   Steamside Oxidation
Laboratory testing of plain and coated specimens at 650°C, 750°C and 800°C with exposure times out to 10,000 hours, have produced some interesting results [15]. Oxidation rates and weight loss were lower for materials with chromium content of more than 12% with ferritic steels and 19% Cr for iron based austenitic materials. Shot peening or blasting has been effective for increasing oxidation resistance at lower test temperatures, but such surface cold work treatment of materials, used above 700°C does not produce effective results. Additional results can be found in these references [16,17].

17.15.3   Fireside Coal Ash Corrosion
Fireside corrosion is due to attack by molten coal ash containing elements such as sulfur forming alkali sulfates, etc., that attack the outside tube surfaces. The presence of chloride can aggravate the corrosion. In the lower furnace, corrosion by sulfidation can occur and is aggravated by reducing or alternating by reducing or alternating oxidizing and reducing conditions, the outside tube surfaces. Low NOx burners and unburned carbon may also attribute to corrosion of the waterwalls, superheater and reheater [18]. With a dependence on the fuel type, the corrosion rates typically increase up to a maximum, at about 690 to 730°C (1274 to 1350°F), and then decrease at still higher temperatures. A-USC at 700 to 760°C will result in outside tube metal temperatures greater than 815°C (1500°F) and it is thought that above certain tube metal operating temperatures, corrosion will be reduced. Chromium content of the base mate-

rial and protection measures with weld cladding should provide adequate service lifetimes.

Laboratory tests with coal ash and in situ testing programs have been performed to expose various materials and coatings/claddings [13,14,18]. Testing was conducted to determine rates in typical coal ash environments with three coal types: Eastern, Midwestern and Western. Western coal would be preferred as the fuel for the first launch of A-USC. Higher chromium content in the base material or with coatings, at a level of about 27%, will help to reduce the corrosion rates. New projects with conditions for oxy-combustion CCS are in progress at pilot testing facilities [19].

17.16 FABRICATION METHODS AND WELDING DEVELOPMENT

To avoid problems that introduce defects in the fabrication processes, various tests were performed to acquire knowledge on handling the new alloys in processes such as bending, machining, swaging and welding.

Shop process trials formed a basis for establishing fabrication procedures. Shop welding practices, particularly with dissimilar metal welds (DMW), were tested in many combinations of product forms and materials. Field welding procedures were also evaluated when installing test sections to determine procedural limitations [14,20].

17.17 DESIGN CODES, DESIGN BY RULE, DESIGN BY ANALYSIS

The design methods for ASME Section I have used formula that have undergone improvements in the history of the Code to meet high standards of safety and better calculation techniques for the product form. Unfired pressure vessels may be designed by formula or by analysis (Section VIII Div 1 or Div 2). Analysis will be required to supplement section I "design by rule" methodology for these advanced plants.

17.17.1 Materials Selection

Many candidate materials are under evaluation for A-USC. In the specific process of design, a primary selection list provides consideration of the increasingly capable metals for the progres-

sion of temperature in the steam generator components. Table 17.2 shows a possible listing of materials selected for consideration in component applications in the more recent A-USC design study.

17.17.2 Design by ASME Section I

Through sponsorship by the DOE/OCDO Materials Development Consortium, a new ASME formula for pipe and tube thickness design was submitted and accepted, Section I Appendix A-317, July 1, 2006. The new formula will be more accurate for thicker components at high temperature. ASME A-317 minimum thickness, t:

$$t = D\,(1 - e^{-P/SE})/2 + C + F \qquad (17.6)$$

Where:
D, outside diameter of component, in.
P, design pressure, psi
S, ASME Code allowable stress, ksi
E, efficiency, -
C, allowance for threading, in.
F, allowance for expanding, in.
For an A-USC boiler, C = F = 0.

The ASME Section I allowable stress values are provided in Figure 17.10. Alloy 740 and Haynes 282 data are being submitted for code cases to ASME.

17.18 MATERIAL SUPPLY CHAIN

One of the major questions regarding the readiness to construct an A-USC plant is the material supply chain ability to meet schedules starting as soon as 2015 for components and new materials in the quantities and sizes required. Several meetings and discussions have been ongoing with the worldwide materials suppliers to advise them on the planning and required materials. The production of large diameter nickel alloy for piping and headers is presently limited by the ingot size of the equipment for the current vacuum induction melting (VIM) process. Obtaining longer lengths of major piping will incur cost and delivery impacts to the A-USC technology projects. Suppliers have indicated that when production investments can be justified by the market, there should be suitable availability of material supplies.

TABLE 17.2 MATERIAL SELECTION

Alloy	Composition (Nominal)	Application
210 C, 106C	Carbon Steel	Economizer, Piping, Headers
T12	1Cr-.5Mo	Waterwalls
T22	2.25Cr-1Mo	Waterwalls, Reheater
T23	2.25Cr-1.6W-V-Nb	Waterwalls, Reheater
T91	9Cr-1Mo-V	Waterwalls, Reheater
T92	9Cr-2W	Waterwalls, Reheater, Piping
347 HFG	18Cr-10Ni-Nb	Superheater, Reheater
310 HCbN	25Cr-20Ni-Nb-N	Superheater, Reheater
Super 304H	18Cr-9Ni-3Cu-Nb-N	Superheater, Reheater, Piping, Headers
617	55Ni-22Cr-9Mo-12Co-Al-Ti	Superheater, Reheater, Piping, Headers
230	57Ni-22Cr-14W-2Mo-La	Superheater, Reheater, Piping, Headers
740	50Ni-25Cr-20Co-2Ti-2Nb-V-Al	Superheater, Reheater, Piping, Headers
282	58Ni-10Cr-8.5Mo-2.1Ti-1.5Al	Piping Headers

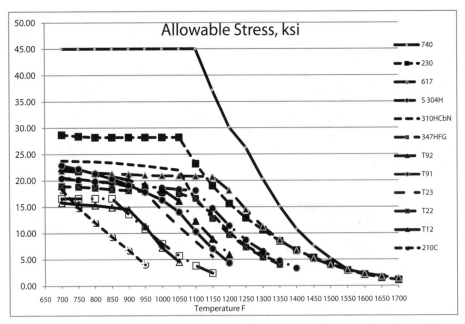

FIG. 17.10   MATERIAL ALLOWABLE STRESS (*Source:* The Babcock & Wilcox Company)

Relative cost ratio (to T22 = 1 on weight basis) of alloy tubes is generally expected to range as shown in Table 17.3.

The cost of nickel alloys is such that the full extent of their temperature stress capability should be utilized.

Estimated material weights and sizes are shown in Table 17.4 and Table 17.5 for new alloys in 800 MW A-USC steam generators. Boiler dimensions are approximately 100 ft depth, 65 ft width, and 150 ft height.

17.19   COST OF ELECTRICITY

Levelized cost of electricity (LCOE) is an evaluation method used to determine electric plant production costs of a kWh over the plant's economic life in terms of capital, operating (fuel and maintenance), and fixed costs. While there is a variance of costs between studies for regional conditions, the LCOE may be used to provide judgment criteria between the relative merits of different power generation technology options. Capital and fuel are two major categories with the highest expenses for a thermal plant.

The net plant heat rate (NPHR) or net plant efficiency is the factor with high impact to the cost of electricity. This will be especially true with the addition of CCS. The day-ahead pricing offer to the independent system operator (ISO) is made considering expense recovery primarily for the fuel. The competitive order for selection by the ISO is based on the unit's offer curve and the ability to compete will be the measure of success.

17.20   ECONOMICS

Evaluation of the economic cost of these new candidate materials necessary in the application to ultra supercritical steam power plants shows that the potential thermal efficiency improvement may be viable and is within the allowable margin expected for achieving equal or better cost of electricity and significant reductions in emissions per kilowatt-hour. The improved heat rate of the A-USC boiler and steam turbine would allow a 750 MW plant to have a capital cost of 13% more than a conventional PC subcritical plant. It has been determined that considering cost reductions due to smaller equipment in other areas of the plant along with fuel savings, the capital cost of the A-USC boiler itself could be 40% higher than a subcritical boiler.

TABLE 17.3   RELATIVE COST RATIO

Alloy	Cost Ratio
T22	1
T23	2
T92	3
Super 304 H	5
310 HCbN	9
Nickel Alloys	46

TABLE 17.4   WEIGHT OF BOILER TUBE

Total Tube Weight	7,070,000 Lbs
Carbon steel	420,000 lbs
T12	500,000 lbs
T23 to T92	2,600,000 lbs
Stainless Steel	1,600, 000 lbs
Alloy 230	1,100,000 lbs
1.75″ OD × 0.400″ MW	
2.00″ OD × 0.165″ / 0.355″ MW	
Alloy 740	850,000 lbs
1.75″ OD × 0.290″ / 0.400″ MW	
2.00″ OD × 0.280″ / 0.400″ MW	

TABLE 17.5 WEIGHT OF BOILER PIPE

Total Pipe/Header Weight	2,000,000 lbs
Carbon Steel	400,000 lbs
P91/92	1,000,000 lbs
Alloy 230	100,000 lbs
18.5″ OD × 2.5″ MW	
26.5″ OD × 4.0″ MW	
Alloy 740	375,000 lbs
1.75″ OD × 0.825″ MW	
20″ OD × 1.750″ / 2.250″ MW	
35″ OD × 3.0″ MW	

The material cost for the conceptual boiler design including the nickel based super alloys was developed based on current pricing for the commercially available materials and estimated pricing for the new super alloys [4]. This cost was compared to the cost of a subcritical PC boiler of approximately the same capacity output. The advanced ultra supercritical boiler was found to have a capital cost of approximately 32% more than a comparable subcritical boiler. This percentage increase is below the estimated 40% increase allowable for an A-USC plant to be economically viable on an equal cost of electricity basis. Figure 17.11 provides a graphic depiction of the equipment cost allocation.

The tubing weight increase is due to the heat transfer at a lower temperature difference between the hot gas and the colder fluid in the tubing. The ultra supercritical boiler has the same gas side temperature conditions as a conventional supercritical unit while the steam is at higher temperature. Accordingly, more heating surface is needed to achieve the required heating duty. However, the energy absorption from the boiler inlet to outlet is lower and the amount of steam flow to be heated is lower, making the steam path smaller for the same power output as compared to a lower efficiency steam cycle. The higher efficiency reduces the fuel input and the combustion products gas weight flow, which also means the size of equipment from pulverizers, furnace, flues, air heaters, and environmental systems are smaller.

Key materials identified that will have the greatest impact on the ability to meet the efficiency and LCOE targets, are Inconel® alloy 740 and Haynes® alloy 230. The development of these two alloys and their performance capability and price are critical in attaining the eventual market success of the A-USC technology. Providing an exercise in the total delivery process steps through an adequate period of operational performance testing and materials examination is planned in the COMTEST1400 program.

17.21 COMTEST1400

To answer the perceived questions and gain acceptance of A-USC, a concept to design, build and operate A-USC components at small scale delivered into the hands of an electric utility operator has been developed by the U.S. Materials R&D Consortium. The COMTEST1400 project would be similar in purpose to the AD700 project conducted over 5 years at Eon's Scholven F Plant, Gelsenkirchen, Germany. Potential owner/operator criteria and concerns are being addressed for a test facility to operate and *put into practice* the technology of an advanced ultra supercritical steam generator possibly at up to 350 bar (5000psi) 760°C (1400°F).

The COMTEST 1400 program is to show potential owners that the new materials and delivery process procedures of fabrication and construction are well defined and put through acceptance trial of first commercial practice. This will also help with bounding the costs of the A-USC steam generator as a deliverable product with an affordable and practical plant life cycle. The design, material supply, fabrication, quality procedures, construction, start up and operating procedures would be provided, put through first use so the product is better known to the utility owners and operators.

Four different concept approaches are being considered:

1. A-USC components placed inside the setting of an existing boiler,
2. Totally external components with separate firing system, or
3. A gas slipstream boiler outside of an existing boiler that can be isolated from the gas side and be better protected while

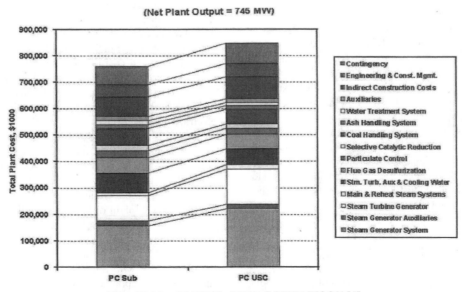

FIG. 17.11 CAPITAL COST COMPARISON [4]

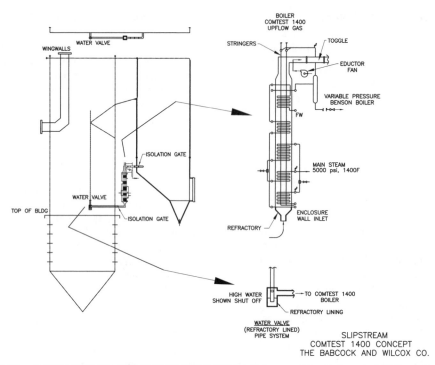

FIG. 17.12   COMTEST1400 SLIPSTREAM CONCEPT (*Source:* The Babcock & Wilcox Company)

also posing less chance of damage or loss of operations to the host boiler, or,

4. A hybrid combination of the slipstream and internal components.

The slipstream concept is shown in Figure 17.12. The slipstream boiler could be modular and operate at a different gas flow loading than the host boiler. It is expected to operate for about 15,000 hours, allow for cycling and subsequently, the components would be examined for evidence of deterioration. It would provide reassurance to potential electric utility A-USC plant owners that the materials and fabrication procedures developed for critical components will perform satisfactorily at these conditions. Extensive test instrumentation would be included to permit performance and material evaluations. The major U.S. OEM's participating in the materials R&D consortium would take part in the COMTEST1400 program. The facility would be seen as a U.S. national R&D facility for demonstrating features of a high efficiency steam generator. This program would undertake the complete exercise of all aspects of placing A-USC into practice, design specification, supply chain, quality assurance, fabrication procedures, field assembly, start up and test operation.

Control and operational experience would also be gained on the relative newer methods used for a variable pressure boiler that has some, but fewer plants introduced in the U.S. Operation and welding practices were difficult lessons to learn when once-through boilers were first introduced in the late 1950s and 1960s. Operation and maintenance training must be supported extensively when a new A-USC product is introduced.

17.22   FIRST COMMERCIAL PLANTS

The true customer of an operating electric generating unit is the transmission grid independent system operator (ISO) that is meeting the load demands at various locations by taking delivery and transmitting power produced by the many generating plants. This competitive market is served by units at plants which must bid a pricing offer, dollars per MW, for a range of load, one day ahead and that have various operating restrictions, cost structures and characteristics that may or not align well with the power grid's needs at each hour of the day. Low net plant heat rate, low fuel cost, wider load turndown range, and fast rate of response are characteristics that will enable the unit to achieve positive net income to pay the bills. The owner planning a new generating unit must factor these considerations into the design. The entry level for a new unit in this market is larger capacity and very capital intensive. A-USC plants require careful planning and development to gain market acceptance.

The tendency has been to select a larger unit size, greater than 450 MW and averaging about 750 MW. Economy of scale has normally resulted in the delivery of a lower cost per MW installed and a lower levelized cost of electricity (LCOE). The steam turbine/steam generator capacity for an A-USC plant must be carefully studied to address the market conditions of how well the capacity and operating characteristics will fit with the requirements of the electric grid operation and needs of the ISO.

Eon, Gelsenkirchen, Germany, had previously announced the planned development of an A-USC demonstration plant at Wilhelmshaven (about 500 MW with 700°C conditions). The plant would have started in 2014 and is now delayed or abandoned due to market and technical reasons. Components were tested at the Scholven F plant over 5 years in the AD700 program. Subsequent inspections found cracking with the alloy 617 material. The preferred commercial size unit was stated (for Europe) to be at 1000MW. China has been constructing 1000MW units. Going forward with the COMTEST1400 program is needed to further the acceptance of advanced ultra supercritical power plants.

The U.S. R&D consortium for A-USC is proposing that the timeline for COMTEST1400 is between 2010 and 2015. An air fired A-USC demonstration plant would be proposed to be built in the period from 2015 to 2020. The demonstration of oxy-combustion at scale also should occur from 2015 to 2020. Oxy-combustion A-USC should be demonstrated 2020 to 2025. Air fired A-USC pulverized coal plants could be commercial after 2020, with oxy-combustions A-USC plants commercial by 2025. This schedule keeps the introduction of the first technology practice separate so that two new features are not handled at the same time. National and industry support is needed to gain experience and acceptance of these first plants.

17.23   ADVANCED ULTRA SUPERCRITICAL POWER PLANT

The success of the development of Advanced Ultra Supercritical steam generators for thermal power plants will be measured in the perceptions of how well these criteria are met in the design:

- Steam generator and plant process with high (HHV) net efficiency
- Minimize the cost of CO_2 capture and sequestration
- Near zero emissions for NOx, SOx, PM, PM10, Hg
- Utilizing abundant lower cost coal
- Reduce auxiliary power consumption
- High availability
- Better maintainability
- Safe and simple operation
- Be the favored plant for base load operation and ability to dispatch with lower LCOE
- Compatible with grid requirements providing load cycling and sustained turndown capability
- Fast starting; cold, warm and hot restart; turn down to minimum load—all with low stress damage accumulation
- Achieve economies of scale
- Better by comparison in lower cost of ownership for the higher perceived value

The power generation units delivering the best value could very well be those operating with A-USC conditions and CCS technology. Pulverized coal-fired steam power plants offer potential advancement to even higher thermal efficiencies through the development and application of pressure part metallurgy capable of higher temperatures at higher pressures to optimize the available energy of the working fluid.

Design concepts of a 750 MW advanced ultra supercritical steam generator have been developed starting with the current arrangements typical of present day application. New and significantly different steam generator arrangements are very likely to be developed.

Customer training for operation and maintenance becomes very important. The introduction of the once-through boiler had some bad experiences with problems besides the design for subcritical operation, also with bad operation and control and with lack of good maintenance practice such as welder qualification. One such example was the problem with the fluid transit time delay causing the slow action to control events created 15 minutes earlier. Careful design of the advanced ultra supercritical plant, simulation, training and qualification will be of prime importance.

17.24   ACKNOWLEDGEMENT

This development effort is supported by several organizations and individuals besides the authors, who would like to thank the many contributions. The B&W participants in the Materials Development Consortium project — Al Bennett, Walt Mohn, Jeff Sarver, Steve Kung, Ed Robitz, John Sanders, and John Siefert. The project management and sponsors, Robert Romanosky, Patricia Rawls, Fred Glaser of the U.S. Department of Energy, Robert Purgert, Energy Industries of Ohio (EIO), Marrio Marrocco, Ohio Coal Development Office (OCDO), Ramnath Viswanathan and John Shingledecker of Electric Power Research Institute (EPRI). The DOE, OCDO, and The Babcock & Wilcox Company have provided funding. Other organizations and personnel from Alstom, Babcock Power, and Foster Wheeler are contributing in their project tasks and internal company development efforts. Keeping the sound economic foundation for lower cost electric power has been the goal of these efforts by advancing the state of the art for higher efficiency power generation using abundant low cost coal and reducing carbon emissions of the major energy source of power.

17.25   REFERENCES

1. Burstall, AF., "A History of Mechanical Engineering", MIT Press, The Massachusetts Institute of Technology, Cambridge, MA, 1965.

2. Silvestri, GJ., et.al., "Optimization of Advanced Steam Condition Power Plants", Diaz-Tous, IA., (ed.), Steam Turbines in Power Generation — PWR-Vol. 3, Book No. H00442, ASME, 1992.

3. Bennett, AJ, Weitzel PS, Boiler Materials for Ultrasupercritical Coal Power Plants — Task 1B, Conceptual Design, Babcock and Wilcox Approach, USC T-3, Topical Report, DOE DE-FG26-01NT41175 & OCDO D-0020, February 2003.

4. Booras, G. "Task 1 C, Economic Analysis", Boiler Materials for Ultra-supercritical Coal Power Plants, DOE Grant DE-FG26-01NT41175, OCDO Grant D-00-20, Topical Report USC T-1, February 2003.

5. Viswanathan, R., Shingledecker, J., Phillips, J., In Pursuit of Efficiency in Coal Power Plants, (ed. Sakrested, BA) 35th International Technical Conference on Clean Coal and Fuel Systems 2010, Clearwater, FL, June 2010.

6. Kitto, JB, Stultz, SC., Steam/its generation and use, Edition 41, The Babcock & Wilcox Company, Barberton, OH, 2005.

7. Kotschenreuther, H., "Future High Efficiency Cycles", Coutsouradis, D., et al. (eds.), Materials for Advanced Power Engineering, Part I, 31-46, 1994 Kluwer Academic Publishers, Printed in the Netherlands.

8. Wheeldon, J., Engineering and Economic Evaluation of 1300°F Series Ultra-Supercritical Pulverized CoalPower Plants: Phase 1. EPRI, Palo Alto, CA: 2008. 1015699.

9. Zachary, J., Kochis, P., Narula, R., "Steam Turbine Design Considerations for Supercritical Cycles", Coal Gen 2007, August 2007.

10. Bennett, AJ., Progress of the Weston Unit 4 Supercritical Project in Wisconsin, (BR-1790), Power-Gen International, Orlando, Fl, November 2006.

11. National Coal Council, "Opportunities to Expedite the Construction of New Coal-Based Power Plants", Library of Congress Catalog # 2005920127.

12. Silvestri, GJ., "Eddystone Station, 325 MW Generating Unit 1-A Brief History, ASME, March 2003.

13. Viswanathan, R., "U.S. Program on Materials Technology for Ultrasupercritical Coal Power Plants", Electric Power Research Institute, Palo Alto, CA, March 2006.

14. Viswanathan, R., et al., "U.S. Program on Materials Technology for Ultrasupercritical Coal-Fired Boilers", in Proc. of the 5th International Conference on Advances in Materials Technology for Fossil Power Plants, ASM International, 2008.

15. Sarver, JM., Tanzosh, JM., "Characterization of Steam-Formed Oxides on Candidate Materials for USC Boilers", Sixth International Conference on Advanced Materials for Fossil Power Plants, Sante Fe, NM, September 2010.

16. Unocic, KA., Pint, BA., Wright, IG., "Characterization of Reaction Products from Field Exposed Tubes", Sixth International Conference on Advanced Materials for Fossil Power Plants, Sante Fe, NM, September 2010.

17. Totemeier, TC., Goodstine, SL., "Oxidation of Candidate Alloys and Coatings for A-USC Applications", Sixth International Conference on Advanced Materials for Fossil Power Plants, Sante Fe, NM, September 2010.

18. Gagliano, MS., Hack, H., Stanko, G., "Fireside Corrosion resistance of Proposed USC Superheater and Reheater Materials: Laboratory and Field Test Results", 33rd International Technical Conference on Coal Utilization and Fuel Systems, Clearwater, Fl, June 2008.

19. Kung, SC., "On Line Measurements of Gaseous Species in Pilot Scale Combustion Facility for Fireside Corrosion Study", Sixth International Conference on Advanced Materials for Fossil Power Plants, Sante Fe, NM, September 2010.

20. Mohn, WR., Tanzosh, JM., "Considerations in Fabricating USC Boiler Components from Advanced High Temperature Materials", in Proc. of the 4th International Conference on Advances in Materials Technology for Fossil Power Plants, Hilton Head Island, SC, ASM International, October 2004.

17.26 LIST OF ACRONYMS AND ABBREVIATIONS

ASME PTC	ASME Power Test Code
ASTM	American Society for Testing and Materials
A-USC	Advanced Ultra Supercritical
B&W	The Babcock & Wilcox Company
CCS	Carbon Capture Sequestration
DOE	United States Department of Energy
EIO	Energy Industries of Ohio
EPRI	Electric Power Research Institute
FEGT	Furnace Exit Gas Temperature
FCC	Forced Circulation Characteristics
FGD	Flue Gas Desulphurization
HHV	Higher Heating Value
ISO	Independent System Operator
LCOE	Levelized Cost of Electricity
LHV	Lower Heating Value
MCR	Maximum Continuous Rating
NCC	Natural Circulation Characteristics
NPHR	Net Plant Heat Rate
NTHR	Net Turbine Heat Rate
OFA	Over Fire Air
OMLR	Optimized Multi Lead Ribbed Tube
OWT	Oxygenated Water Treatment
PC	Pulverized Coal
PRB	Powder River Basin
SCR	Selective Catalytic Reduction
TDBFP	Turbine Driven Boiler Feed Pump
USC	Ultra Supercritical
VS	Vertical Separator
WCT	Water Collection Tank

CARBON CAPTURE FOR COAL-FIRED UTILITY POWER GENERATION: B&W'S PERSPECTIVE

D. K. McDonald and C. W. Poling

18.1   INTRODUCTION

Changing climate and rising carbon dioxide (CO_2) concentration in the atmosphere have driven global concern about the role of CO_2 in the greenhouse effect and its contribution to global warming. Since it has become widely accepted as the primary anthropogenic contributor, most countries are seeking ways to reduce CO_2 emissions in an attempt to limit its effect. This effort has shifted interest from fossil fuels, which have energized the economies of the world for over a century, to non–carbon-emitting or renewable technologies.

For electricity generation, the non-carbon technologies include wind, solar, hydropower, and nuclear, whereas low carbon technologies use various forms of biomass. Unfortunately, all current options are significantly more expensive than current commercial fossil-fueled technologies. Although use of wind and solar for power generation is increasing, they are incapable of supplying base load needs without energy storage capacity which is currently impractical at the scale required. The only technologies capable of sustaining base load capacity and potential growth are coal, natural gas, and nuclear. Nuclear has a long lead time to commercial operation, and concerns about long-term disposal of waste have not been resolved. Natural gas, although lower in carbon emissions than coal, still produces CO_2, and its availability and price have proven to be volatile. Although recent discoveries have significantly increased availability of natural gas reserves, the pricing impact of higher cost extraction methods is not yet known.

In many countries, including the United States, coal is an abundant and low-cost source of fuel for generating electricity. Figure 18.1 shows the International Energy Agency's (IEA's) global electricity-generating capacity by fuel in 2007 and that projected for 2030. It should be noted that the current IEA forecast indicates more than a 70% increase in coal-based power generation by 2030. In the United States, coal currently fuels about 50% of power generation and represents an enormous infrastructure investment. Consequently, it will take considerable investment to move away from this base load fuel.

Figure 18.2 shows the EIA's estimate of total recoverable coal resources, which shows the United States having the largest supply in the world. Since energy security is a major issue in the United

States, it is very difficult to ignore the abundance and low cost of domestic coal. Coal is the most abundant and lowest-cost fuel for power generation in the United States that preserves long-term energy security. In addition, developing nations such as China and India are relying heavily on coal to fuel their growing economies, and most of the increased coal-fired electricity-generating capacity projected in 2030 will be in those countries.

As the world grapples with CO_2 management, it is becoming increasingly clear that for the sake of the infrastructure in many developed countries, and to provide a low-carbon emissions option for developing countries relying heavily on coal, some form of carbon capture and storage (CCS) will be necessary. In a carbon-constrained world, the long-term viability of coal depends on the technical and economic success of emerging technologies for capture and storage. This may first be in the form of retrofit for the existing fleet, but considering the growth rate in other countries, new plants will also be necessary as soon as practical.

Deployment of CCS is currently hindered by regulatory and social issues related primarily to long-term CO_2 storage. Capture technologies are in the demonstration phase and are expected to be commercially available in the next decade, but storage remains a hindrance to deployment globally. Europe has been in the lead regarding carbon controls, but the value of CO_2 remains too low to support CCS projects and unresolved public concerns about long-term storage reliability, and liabilities are blocking demonstration projects. In North America, the Canadian province of Alberta has taken bold steps to define a reasonable value for CO_2 and to address storage issues, but in general, Canada is relying on the United States to determine the best methods and timing of carbon controls. As a result, CCS has stalled in many countries until the storage issues are resolved, and a reasonable value is set for CO_2.

The two primary storage options are enhanced oil recovery (EOR) and deep saline reservoirs. The deep saline reservoirs have more than sufficient capacity, but the impact of CO_2 and co-sequestered constituents on the reservoirs are not clear. This option has not been widely practiced, and demonstration projects to study the effects are in progress. The near-term hope for deployment in the United States hinges primarily on EOR. There is significant capacity, but it is limited, and although millions of tons of CO_2 have

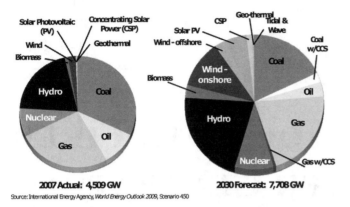

Source: International Energy Agency, *World Energy Outlook 2009*, Scenario 450

FIG. 18.1 WORLDWIDE INSTALLED ELECTRIC GENERATING CAPACITIES BY FUEL, 2007 AND 2030

already been used for EOR, concerns about use of EOR for long-term storage are mounting, which may either significantly slow or even block this avenue.

Despite these challenges, carbon capture technologies are being developed by several companies for use with fossil fuels, especially coal. There are several potential carbon capture technologies for coal. The three coal-based technologies closest to deployment are pre-combustion, including Integrated Gasification Combined Cycle (IGCC), oxy-combustion, and post-combustion including amine- and ammonia-based systems. In addition, chemical looping, membrane, and biological carbon capture methods are in the early development stages.

The Babcock & Wilcox Company (B&W) is aggressively pursuing several approaches for generating electricity that address the carbon issue including nuclear, concentrated solar receivers for steam generation, and biomass firing in addition to capture from existing and new coal- and natural gas–fired steam-generating plants. This chapter provides an overview of oxy-combustion and post-combustion technologies for coal-fired steam (and electricity) generation, which are being developed for nearer-term deployment.

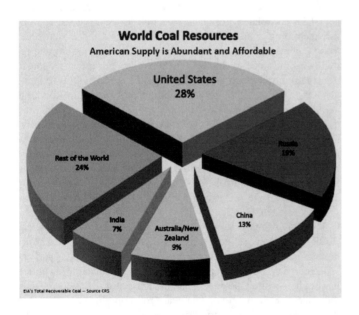

FIG. 18.2 WORLDWIDE COAL SUPPLIES (2008)

18.2 OXY-COMBUSTION

Oxy-combustion is a means of removing nitrogen from the combustion process to produce a gas with a high concentration of CO_2, which can then be easily purified, compressed, and used or stored. Typical combustion of coal with air produces a flue gas that is primarily nitrogen with some water, about 14% CO_2, and some minor constituents. Capturing the CO_2 at this low concentration is challenging and requires reactive regenerable solvents as described in Section 18.3. By concentrating the CO_2 to 70% or greater by volume, it can be further purified to 99% purity, compressed, and either stored or used. To concentrate the CO_2, the coal is burned with nearly pure oxygen producing combustion byproduct, which contains CO_2 at about 85% by volume dry.

Combustion with gases having a higher concentration of oxygen than air has been used for decades in the glass and metal-melting industries, and experimentation with coal is not new. However, adapting oxy-combustion to modern power plant scale, understanding its impact on coal combustion byproducts and the performance of constituent removal systems, plant performance, economics, and reliability have been seriously addressed only in the past decade. This section will describe the oxy-combustion process, the development path B&W and Air Liquide (AL) have jointly pursued and the current state-of-the art.

18.2.1 Process Description

An oxy-combustion plant consists of a steam generator or boiler, coal preparation equipment, and systems to remove various undesirable constituents such as particulate, SO_2, SO_3, Hg, and water from the combustion byproducts. These components are generally of conventional, proven design and materials. In addition, oxygen is supplied by an air separation unit (ASU), and the product gas, which is primarily CO_2, is treated and compressed in a compression and purification unit (CPU).

Combustion of coal with nearly pure oxygen results in a byproduct that is primarily CO_2, some water, and minor constituents such as SOx, NOx, Hg, CO, oxygen, nitrogen, and argon. Because of the small amount of nitrogen, which composes 78% of air, the quantity of combustion byproduct is only about one-fourth of what would result if air were used. Figure 18.3 shows the relative flows and compositions for the oxidant (air or nearly pure oxygen mixed with recycled flue gas) and the resulting combustion byproducts. Removal of nitrogen from the combustion process not only significantly reduces the mass flow, but it also increases the concentration

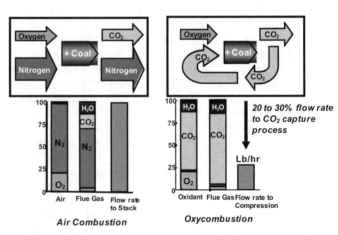

FIG. 18.3 OXIDANT AND COMBUSTION BYPRODUCTS

of the constituents in the combustion byproduct by at least a factor of four and, unless tempered by recycling flue gas, significantly increases the flame temperature.

In most oxy-combustion processes, flue gas is recycled to maintain an acceptable flame temperature, dilute corrosive constituents, and provide mass flow for heat transfer. Some or all of this recycle gas is cleaned of undesirable constituents such as particulate, SO_2, SO_3, and water prior to recycling so it acts as a diluent for corrosive constituents. In the absence of recycle, high temperatures and high concentrations of corrosive constituents would increase boiler tube corrosion rates exponentially requiring much more expensive high-grade materials to resist this combination. Opinions vary concerning the location in the process from which the recycle is extracted and the quantity of recycle that would result in the lowest cost and highest plant efficiency. Most studies by those in the industry would agree that for pulverized coal combustion, sufficient recycle to approximate the combustion conditions with air firing is optimum. The reasons will be explained further in sections 18.2.3.1 and 18.2.3.2.

18.2.2   State of Development

Oxy-combustion was first explored by B&W for EOR in 1979, but low oil prices proved it to be uneconomical. As interest in CO_2 emissions was increasing in the mid-1990s, it was given further consideration as a means of concentrating CO_2 for storage or use. Since then, oxy-combustion has been aggressively developed, and in 2001, B&W teamed with AL to bring the ASU and CPU into the process and improve the overall efficiency and economics. Several engineering studies and cost estimates for both retrofit and new unit applications were carried out. In addition, proposals to build a 25-MWe retrofit and a 350-MWe commercial greenfield plant were made to potential customers. Small pilot testing co-sponsored by the US Department of Energy (DOE) and the State of Illinois was carried out from 2001 to 2004 in B&W's 1.5-MWth (5 MBtu/hr)

small boiler simulator (SBS) burning bituminous and sub-bituminous coals. The results of this testing using the cold recycle process proved that the process could be safely and easily operated and demonstrated NOx reductions up to about 70% compared with air firing, volumetric concentrations of CO_2 of 80% and higher, and indicated heat transfer and constituent removal processes performed similar to air firing.

These encouraging results prompted modification of B&W's 30-MWth Clean Environment Development Facility (CEDF) for oxy-firing and testing with bituminous, sub-bituminous, and lignite coals in both cold and warm recycle configurations in 2007-2008 [4, 7]. As shown in Figure 18.4, the CEDF is small-scale power plant using a single 100-MBtu/hr (essentially full-scale) burner firing into a specially designed boiler followed by a full complement of backend gas cleanup systems. Although CO_2 was "caught and released," all of the processes were tested. At the time, it was the first oxy-combustion plant to operate at this scale in the world. Coal was successfully fed both indirectly for bituminous and directly from the pulverizer using recycled flue gas to dry and convey the coal. The results of these tests confirmed control philosophies for transitioning and all major trips, as well as showing NOx reduction as high as 70% and SO_2, SO_3, Hg, and particulate removal performance similar to air firing.

In 2008, a comprehensive effort was made to determine the best processes including optimization and integration options for the power block, ASU, and CPU and to further refine individual component designs. Twenty-seven cases were analyzed using Aspen's HYSYS software modeling the entire plant including steam and gas cycles as well as the ASU and CPU processes. Performance and costs for the various equipment and integration options were analyzed using the same fuel and site conditions, and the same time frame and financial assumptions used in the DOE study "Pulverized Coal Oxy-combustion Power Plants" DOE/NETL-1291 Rev. August 2008 [8] so it could be used as the base case for

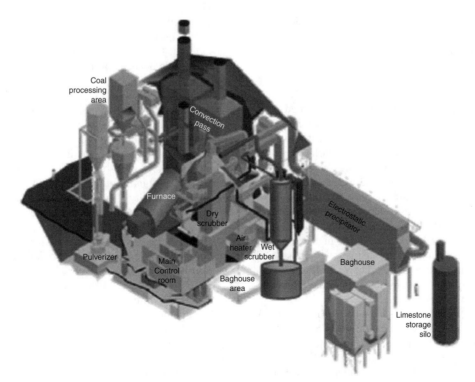

FIG. 18.4   CLEAN ENVIRONMENT DEVELOPMENT FACILITY (30 MWTH) OXY PILOT

comparisons. All three recycle location options and the impact of reducing the recycle quantity were considered. The benefit of conventional super-critical and ultra-super-critical steam conditions was also evaluated. As a result, the warm recycle process was selected for low-sulfur fuels, and the best approach for heat integration was identified.

This work led to development of a demonstration plant design based on 150 MWe of gross electrical generation and a commercial plant design for 700-MWe gross.

18.2.3   Design Considerations

18.2.3.1 Recycle Quantity As mentioned, the quantity of recycle for an oxy-combustion pulverized coal-fired boilers is generally about the same as would be present with air firing on a mass basis primarily due to material considerations in the boiler. It has been shown that some reduction in recycle and cost may be possible with relatively low steam temperatures, but such a unit would not have the same air firing capability. However, for the purpose of an oxy-combustion plant, low or no recycle is not practical and would deliver minimal cost savings.

Current designs recycle about 70% of the flue gas back to the boiler to control flame temperature, dilute corrosive constituents, and provide mass for convective heat transfer. The expectation is that with low or no recycle, the equipment will be proportionately smaller and less expensive. This misconception is based on the idea that the cost of the equipment is closely related to the gas volume flow, and since oxy-combustion produces 25% to 30% of the flow from an air-fired plant (see Figure 18.3), substantial savings can be gained. But low or no recycle is not practical for steam generation, and it would not deliver the savings that might be expected.

Recycle provides the gas mass flow to balance radiant and convective heat transfer in the boiler, so steam temperatures can be achieved economically. In a boiler, the heat is transferred to the water and steam contained in metal tubes by radiation and convection. Radiation dominates within the furnace at high gas temperatures, but as the gas temperature decays, convection tends to dominate (see Figure 18.5). To withstand the much higher heat fluxes from radiation in the furnace, which are driven by flame temperature, water is used to cool the tubes, and boiling occurs in the upper furnace. Steam is generally one-fifth as effective as water at cooling the tubes, so super-heating is done in the convection pass where heat fluxes are lower and cost-effective materials can be used.

The amount of furnace absorption is set so the furnace exit gas temperature is not too high. For coal, this temperature is limited to prevent slagging and fouling of the heat transfer surfaces, which would inhibit steam generation. Reduction of the recycle quantity significantly increases flame temperature and burner zone heat release, also increasing furnace slagging concerns and radiation heat fluxes. This results in much higher gas and tube metal temperatures and requires higher alloys in the furnace because radiation is a function of temperature to the fourth power. The higher gas temperature and resulting greater log mean temperature difference will require less furnace heat transfer surface, but that savings in material quantity is quickly overcome by the higher alloy material, fabrication, and installation costs.

Modern furnaces are made of carbon and low-alloy steels. A significant metal temperature increase would require higher alloys, many of which costing at least five times that of current furnace materials. So even if furnace surface could be reduced by 50%, the material cost alone would almost triple, and the welding costs would be much higher due to pre- and post-weld heat treating requirements.

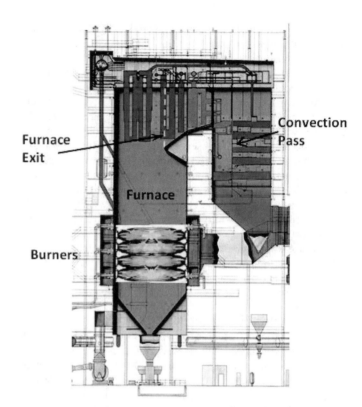

FIG. 18.5   TYPICAL BOILER

The total heat absorption in the boiler is defined by the desired steam flow, temperature, and pressure to the steam turbine. The heat absorption necessary to limit the furnace exit gas temperature to the desired value, which is a function of the slagging properties of the coal, sets the furnace heat transfer area. The remainder of the heat is absorbed in the convection pass.

The lower gas flow will reduce the cross-section proportionally assuming that gas velocity, which drives heat transfer, is maintained, but surface reduction will be driven by the lower required absorption. Tube rows will be much shorter requiring more bends.

As recycle decreases, furnace absorption increases, convection pass absorption decreases, and convection pass volumetric gas flow decreases. The lower gas flow leads to a smaller convection pass cross-section to maintain gas velocities that drive convective heat transfer. However, the absorption required does not decrease as rapidly as the volumetric flow so the amount of surface required is not proportional to the gas flow. This means a longer convection pass with shorter straight tubes and more tube bends per unit area, making significant cost reduction unlikely. As a result, even though the quantity of surface decreases, the cost will not decrease as drastically and may actually increase.

As a result of these factors, the overall cost of the boiler may actually increase despite its smaller size, that is, if suitable materials even exist that will provide properties that preserve current reliability. Recent work has also shown that decreasing convection increases difficulty in achieving higher advanced steam temperatures and steam cycle efficiencies. In fact, to achieve super-critical steam temperatures of 1300°F or higher, it may be necessary to increase rather than decrease recycle (relative to air firing mass flow) owing to material limitations for even the latest materials like Inconel 740. Increasing recycle reduces flame temperature and radiation and, more importantly, increases the flow available for convective heat transfer, which is essential in achieving high steam temperatures. Therefore, significantly decreasing recycle would not be expected to reduce the boiler cost, and it makes design for higher steam temperatures and efficiencies more difficult or impossible.

There is also less cost savings than might be expected in the equipment downstream of the boiler. The expectation of savings proportional to gas flow reduction for the downstream equipment overlooks three important facts: (1) the cost of these systems is not primarily in the vessels, flues, and casings; (2) the performance and sizing of the gas cleaning systems are linked to the relationship between pollutant concentration and the gas volume flow, not the volume flow alone; and (3) perimeter is proportional to the square root of the area; meaning if the gas flow is 30% compared with air firing, the enclosure material will be reduced on the order of 45%.

Although the gas volume decreases as recycle is decreased, the same mass of pollutant must be treated. This increases the concentration impacting performance. For example, since the quantity of ash does not change, the ash removal systems must have the same capacity and will be about the same cost. For an electrostatic precipitator, the same power is required for the same loading, and sparking may tend to occur at a lower voltage. The plate area is set by these factors, not gas flow alone. In a fabric filter, the lower gas flow decreases the equipment size by the square root of the volume, but higher loading requires a lower filtration velocity to maintain a practical back-pulsing frequency increasing cost.

When the complete systems are considered, cost reductions in the equipment downstream of the boiler on the order of 30% to 40% are realistic if it were possible to have no flue gas recycle at all. In a typical power plant, the equipment downstream of the boiler can represent 20% to 25% of the total boiler island cost, and the boiler island in an oxy-combustion plant is about 33% of the total plant cost. So assuming the boiler cost does not increase, the net savings would be less than 10% of the boiler island and less than 3% of the total plant cost total, while introducing significant performance and reliability risks, since such a boiler design, if even possible, would be very challenging and a first of a kind. Material limitations, the need to drive to higher steam temperatures for efficiency, and marginal cost benefit at higher risk make reduction of recycle unattractive as a design improvement.

Design of an oxy-combustion pulverized coal plant with low or no recycle is questionably feasible using known materials, fabrication, and construction methods. It is also doubtful whether, if achieved, it would be economical. Since many components would operate under unprecedented conditions, performance and economic risks would also be greatly increased until proven, and it would not be capable of air firing.

18.2.3.2 Recycle Location There are four main options recycle options with oxy-combustion: "cold, cool, warm, and hot recycle". Generally, recycled flue gas is returned to the boiler in two streams. Primary recycle is used to dry and convey the coal, so it must be cooled and dried for all three options. The secondary recycle is returned to the furnace to provide mass for the boiler. Oxygen is added at any of three primary locations: in the burner, in the primary recycle, and in the secondary recycle.

The cold recycle process for very high sulfur coals, takes the secondary recycle after flue gas drying as shown in figure 18.6, cleans and cools the entire flue gas flow from the boiler removing particulate, sulfur compounds, Hg, and moisture before sending the primary and secondary recycle back to the boiler. The remainder of the flue gas is sent to the CPU and then to storage. Cool

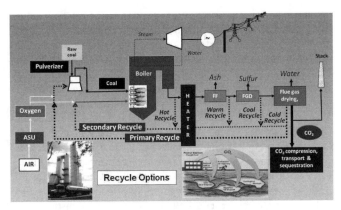

FIG. 18.6 FLUE GAS RECYCLE OPTIONS

recycle, for high sulfur coals, takes the secondary recycle after the FGD but before flue gas drying.

Warm recycle takes the secondary recycle immediately after the gas heater at about 300°F to 400°F and particulate removal. There is no other treatment of this stream before entering the boiler. The hot recycle process takes the secondary recycle immediately after the boiler at about 700°F to 750°F. Hot particulate removal (not shown) is necessary to protect the fan needed to blow it into the boiler, but there is no other treatment.

Each of these options has advantages and disadvantages. Cold recycle cleans the recycled gas the most and reduces the constituent concentrations in the boiler the most. Both warm and hot recycle return hotter gas to the boiler, improving fuel efficiency, but the gas is untreated so the concentrations of constituents like SO_2 and moisture within the boiler are higher. Warm and hot recycle are reasonable with low-sulfur coals, but not when coal sulfur is high. In addition, particulate removal for hot recycle is challenging, and it has been shown that the increased fan power, owing to the higher gas temperature, offsets the fuel efficiency gain, so warm recycle is more practical. Testing has also shown that the impact of higher moisture levels on combustion is manageable. As a result, warm recycle is preferred by B&W and AL for low-sulfur coals. For high-sulfur coals sufficient sulfur removal in the secondary recycle stream is necessary, or cold or cool recycle should be used.

18.2.3.3 Solid Particulate, SO_2, SO_3, and Moisture Removal
If no constituents are removed in the recycled flue gas, the concentrations in the boiler will be the same as the products of combustion of the coal with oxygen, which can be four or more times as high as with air firing. If a constituent is completely removed from the flue gas before recycling, the recycled flue gas will dilute the constituent to the same level as experienced with air firing. To design the boiler for reasonable conditions using conventional materials and equipment, solid particulate, sulfur, and moisture must be removed to achieve acceptable levels throughout the process.

As in an air-fired unit, particulate is removed downstream of the air-heater to protect the fans handling recycled flue gas and the CPU compressor. This can be accomplished using conventional equipment such as electrostatic precipitators or fabric filters in the same configuration relative to the primary scrubber for a conventional plant. A fabric filter follows a dry scrubber and precedes a wet scrubber. The induced draft fan and recycle fans are downstream of the primary scrubber and particulate removal equipment.

The recycle configuration and design of the power block equipment are dependent on the sulfur, chlorine, and moisture content in the fuel. For low-sulfur coals, warm recycle can be used, removing only solid particulate in the secondary recycle stream, but the primary recycle stream must be scrubbed, and moisture must be removed if it is used for drying and conveying coal in the pulverizers. For medium sulfur coals, some sulfur removal in the secondary stream will be necessary, and for high-sulfur coals, all of the flue gas leaving the boiler must be scrubbed. The removal must be sufficient to prevent unacceptable levels of SO_2 in the furnace, which will result in more rapid and aggressive corrosion of the high-temperature portions of the secondary and reheat super-heaters, especially if significant chlorine is also present. In addition, higher SO_2 will produce higher SO_3 levels, which can be deleterious to the downstream equipment owing to acid formation.

Because of the higher concentrations of SO_3 and moisture in the flue gas, the acid dew point is elevated compared with air firing. To prevent corrosion in the equipment downstream of the boiler as much SO_3 must be removed as practical, and the gas temperature should be kept above the acid dew point sufficiently to prevent acid condensation in cooler areas of flues and equipment. SO_3 removal can be accomplished with a dry scrubber and fabric filter combination, or if a wet scrubber is used, dry injection of a sorbent such as Trona ($Na_3(HCO_3)(CO_3) \cdot 2H_2O$) upstream of the fabric filter can be effective.

Since the flue gas will leave the primary scrubber saturated with moisture, moisture must be removed prior to the compression process and, as mentioned previously, if recycled flue gas is used in the pulverizers. This can be accomplished by several means such as direct contact or quench coolers. Saturated flue gas at relatively low temperature leaving quench-type coolers must be sufficiently reheated before fans or compressors to prevent damage from formation of water droplets.

18.2.3.4 NOx During oxy-coal combustion, NOx production is much lower than with air firing. This is partially due to the absence of nitrogen, since nearly pure oxygen is being used instead of air, and partially due to a re-burning effect from the recycled flue gas. Together, these have resulted in 50% to 70% lower NOx production. In addition, NOx is oxidized to NO in the compression and purification step.

However, since oxy-combustion plants currently start with air firing to a minimum stable load and then transition the oxidant from air to oxygen, NOx emissions during startup can be an issue if there is a limit on short-term emissions, and there is no removal equipment such as selective catalytic reduction.

18.2.3.5 Air Infiltration Careful design and installation are necessary to minimize infiltration of air into the process during oxy-combustion. Ingress of air has two key detriments: (1) it adds mass and dilutes the CO_2 concentration, which must be handled in the CPU; and (2) it adds nitrogen, which increases NOx formation in the combustion process. Only those portions of the process that operate below atmospheric pressure are susceptible to air infiltration. This is generally from the boiler to the ID and secondary recycle fan inlets and potentially just before the inlet of the primary recycle fan.

Location of fans, including booster fans, to minimize negative pressure differences relative to atmosphere, reduces the driving force. Design of the flues and equipment in this portion of the process to minimize potential for infiltration and careful quality control of welding during fabrication and installation will minimize leakage. Replacement of air with clean, dry CO_2 for uses such as flame scanner cooling, sealing of openings in the boiler, and back-

pulsing in the fabric filter will also significantly reduce the amount of air ingress. Provisions for temporary events such as on-line maintenance activities must also be considered to avoid exceeding the inlet capacity of the compression and purification equipment and reducing delivery of purified CO_2.

18.2.4 Power Block Equipment Options

In arranging the power block equipment, the options are essentially the same as with air firing. The combustion process can be pulverized coal, cyclone, or a fluidized bed. The boiler is followed by a gas-to-gas heat exchanger that reduces the exiting flue gas temperature and transfers the heat to the recycle flue gas returning to the boiler in the same manner as an air-heater for air firing. Internal leakage is less of an issue since gas is leaking to gas and the impact is a tradeoff in power between the fans. However, if oxygen is introduced into the recycle gas prior to the air-heater, care must be taken to minimize loss to the flue gas exiting the boiler since it will be lost to the compression process, resulting in a double loss: the power to produce the oxygen and the power to process it in the CPU.

For low- or medium-sulfur coals, a dry scrubber followed by a fabric filter can be used as the primary removal system. For high-sulfur coals, a wet scrubber may be required. The choice is determined by (1) the process being used ("warm, cool, or cold" recycle) and (2) the resulting SO_2 concentration in the furnace and corrosion constraints imposed by materials. The higher moisture level in the flue gas from oxy-combustion increases the saturation temperature. For a dry scrubber, an approach margin leaving the dry scrubber must be maintained to prevent condensation in the fabric filter or downstream equipment. These constraints may limit the amount of slurry that can be introduced and the corresponding removal efficiency. For a wet scrubber, the outlet temperature will be naturally higher by 20°F to 30°F owing to the higher saturation temperature.

As a result, moisture is removed after the primary scrubber and before sending to the compression equipment. If the primary recycle is used to dry the coal, as with pulverized coal combustion, both streams must be dehumidified. In addition, a very low SO_2 concentration entering the compression system may be required. To accomplish these requirements, a secondary scrubber and flue gas cooler may be

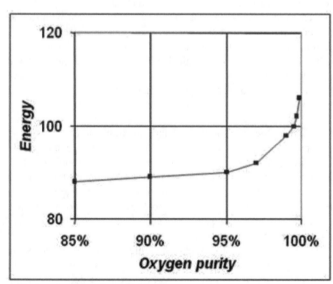

Use by permission from Air Liquide

FIG. 18.7 ASU POWER VERSUS PURITY

necessary. Babcock & Wilcox Company has developed a polishing scrubber combined with direct contact cooling for this purpose.

18.2.5 Air Separation and CO_2 Purification and Compression

In oxy-combustion, the nearly pure oxygen is produced by an ASU, and the flue gas produced is prepared for transportation and storage by a CPU. Oxygen purity impacts the CO_2 purity entering the CPU and affects the cost and power required. The lower the CO_2 concentration, the more energy it takes to purify and compress the product gas, and the larger the equipment must be to accommodate the greater flow. At the same time, producing oxygen with higher purity requires more energy, so an optimum is desired. Studies have shown that above 95% purity, the cost of producing oxygen increases significantly, whereas the benefit to the CPU is minimal (see Figure 18.7). As a result, 95% purity is generally accepted, with the remaining 5% being about two-thirds argon and one-third nitrogen.

Three primary technologies for separating oxygen from air are available: cryogenic separation, adsorption (Vacuum Swing Adsorption) and permeation (hollow fiber membranes). Cryogenic separation is the only one that allows production of oxygen in large quantities (up to 7000 metric tons of oxygen per day in a single train). Two new technologies are under development: chemical looping and ion transport membranes. Chemical looping is the use of an oxygen carrier, such as iron, oxidized in air, conveyed to the combustion process where the carrier is depleted of oxygen through participation in the combustion process. The carrier is then recovered, re-oxidized, and the cycle is repeated. Various processes and carriers are currently being tested at the laboratory scale but are not yet economical or practical for large power plant application. Ion transport membranes are made of solid inorganic ceramic materials, which permit a diffusion of oxygen ions (O^{2-}) through the ceramic crystal structure under the influence of oxygen partial pressure. This kind of membrane operates at high temperature between 700°C and 950°C. It means that natural gas is required to heat the air going to the membrane, making this solution less attractive for a pulverized coal boiler.

The conventional method of air separation is based on cryogenics operating on the principle that various gases liquefy at different pressures and temperatures. Air is 78% nitrogen, 21% oxygen, and 1% minor constituents, mostly argon. The condition where oxygen becomes liquid differs from nitrogen or argon, although argon is very close. The moisture and CO_2 in air would freeze earlier than oxygen becomes liquid and therefore are removed after compression by adsorption on fixed beds. The air flows through a first vessel filled with adsorbent that stops water, CO_2, and some hydrocarbons and N_2O. When the adsorbent is saturated, the air is directed to another vessel when the first is regenerated with dry and warm nitrogen flowing counter to the air. This temperature swing adsorption process is a very efficient and well-proven technique to achieve the very low levels of freezing components required by cryogenic applications. The dry air is cooled to the point where oxygen becomes liquid, leaving most of the nitrogen and argon in gaseous form. The degree of purity depends on how precisely the process is controlled to the specific temperature and pressure, and as previously shown, more purity requires more energy.

In recognition of the unique and different requirements for the oxygen in a power plant versus conventional air separation for commercial oxygen production, B&W teamed with AL in 2001. In addition, AL was interested in developing the CPU. These two

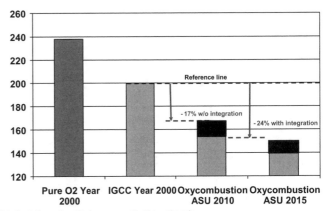

Note: the darker portion of the bars represent heat integration gain
Used by permission of Air Liquide

FIG. 18.8   REDUCTIONS IN ASU SPECIFIC ENERGY

systems contribute all but a small portion of the incremental cost and parasitic power relative to a conventional air-fired plant. Air Liquide has made significant investment in addressing cost and power consumption and has achieved impressive gains.

Air separation is a proven technology. But, oxy-combustion requires much different oxygen conditions than typical cryogenic air separation applications, including IGCC, and provides opportunity for heat integration with the steam cycle. Through our collaboration, AL has gained a better understanding of boiler requirements and has optimized the ASU design (see Figure 18.8). First the ASU was adapted to oxy-combustion conditions, resulting in a 17% reduction in specific energy. Further improvement produced an additional 7% when the heat of compression is integrated into the steam cycle. This represents a 24% reduction in specific energy compared with a state-of-the-art IGCC ASU. This design is commercially available, and AL has established a target for another 10% reduction in the next 5 years to reach 140 kWh/metric ton of pure O_2 (contained in an oxygen stream at 95% O_2, atmospheric pressure and under ISO conditions). Primarily as a result of this significant power reduction, oxy-combustion can have efficiency and cost advantages over the other options [1].

Air Liquide has also demonstrated the ability to change load at a rate as high as 5% per minute at industrial scale (>2000 metric tons per day), and it has developed operating methods to maximize revenue and/or allow partial capture. To maximize revenue, design for "peak-shaving" uses a highly flexible ASU incorporating cryogenic liquid storage. During the hours when the electricity price is high, full output is achieved by running the ASU at reduced load and making up the oxygen shortfall from the stored LOX. This maximizes revenue by minimizing parasitic power and maximizing net power output. When electricity prices are low and the lost generation has minimal economic impact, the ASU is operated at a higher load to allow for oxygen storage. One other advantage of this design is that it allows constant and stable operation of the ASU at its design load, which increases its efficiency.

To achieve partial CO_2 capture the plant is designed for both air and oxy-combustion capability, and a partial sized ASU is installed with LOX storage. When CO_2 capture is desired, the plant is operating in the oxy-mode using the ASU and LOX, and when it is air fired, the remainder of the time, only the ASU is operated to store oxygen. This approach minimizes capital investment

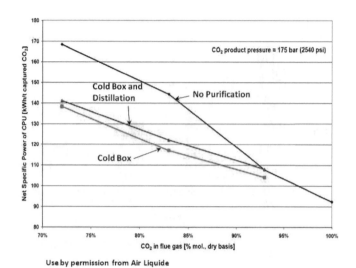

FIG. 18.9 CO₂ PURIFICATION METHOD AND SPECIFIC ENERGY

and parasitic power for oxy-operation. The amount of time for air or oxy-firing depends on the amount of CO_2 capture necessary. As CO_2 value increases over time, additional ASU capacity can be added to achieve a full capture plant. This approach has been estimated to reduce the initial capital for the ASU and CPU and increase the CO_2 avoided cost. The impact on CO_2 cost avoided could be less than 5% compared with a full (90%) capture plant when coupled with "peak shaving" operation, operating air fired during peak hours and oxy-firing off-peak.

Purification and compression of the flue gases are achieved in the CPU. Three levels of purification are possible, depending on the product gas purity requirements: no purification, cryogenic separation (cold box), and cryogenic and distillation. Each of these requires a certain amount of energy to achieve a desired purity and discharge pressure. Figure 18.9 shows that the lowest energy requirement for purification of CO_2 concentrations between 70% and about 93% involves refrigeration for partial condensation or refrigeration and distillation, depending on the degree of purity desired. The higher the purity, the higher the capital cost.

The CPU is designed to achieve a low specific energy, on the order of 120 kWh/metric ton of CO_2 captured. In an oxy-combustion plant, the non-condensable gases, such as oxygen, nitrogen, and argon, are vented from the CPU as the only source of air emissions. As a result, the emissions from an oxy-combustion plant are extremely low, essentially no particulate, Hg, heavy metals, HF, HCl, or SO_x. Only small amounts of NO, CO, O_2, Ar, and CO_2 are emitted to the atmosphere. CO is the only constituent with no removal in the process, but methods to promote oxidation and increase the CO_2 captured exist, although at added cost.

The ASU and CPU together currently represent about 30% of the capital cost of a new plant and about 70% of the total parasitic power. However, AL supports ongoing efforts to further reduce power and cost.

18.2.6 Demonstration Plant Description and Features

Figure 18.10 shows a model of the demonstration plant design indicating all major equipment including the boiler, dry scrubber (SDA), fabric filters, moisture removal, and sulfur polishing sys-

tem (WFGD/DCC), as well as the ASU and CPU. The warm recycle process is used, and the secondary recycle with only particulate removal is shown with green flues. The remainder of the flue gas shown with blue flues has SO_x, particulate, and moisture removed. The smaller yellow flue is the primary recycle, and the gas pipe to the CPU is shown in lighter blue. Because it is only 150-MWe gross, the steam cycle is sub-critical at 2400 psi and 1050°F main and reheat steam temperatures. It was designed for a western location where water is scarce so it uses dry cooling with very low fresh water makeup and is zero liquid discharge [2, 5, 6].

The only air emissions in the oxy-mode come from the CPU vent and are extremely low. In fact, during oxy-combustion the NO_x, SO_x, Hg, and particulate are predicted to be below current power plant measurement accuracy. Since CO is not removed in the process, unless oxidation to CO_2 is incorporated, it would be about the same as an air-fired plant. The plant would capture over 90% of the CO_2 produced, which is more than 1 million tons per year. That leaves about 105,500 tons/yr of CO_2 that would also be emitted from the CPU vent.

18.2.7 Commercial Plant Considerations

Oxy-combustion has yet to be demonstrated at utility scale. At the date of publication, the largest plant that has been successfully operated is 40 MWth. A 30-MWe electric plant is being designed for retrofit in the Callide power plant for CS Energy in Australia. However, it is ready for commercial-scale demonstration. Following a demonstration at a meaningful scale, B&W and AL expect to be ready to provide full-scale commercial plants. In preparation, a 700-MWe gross (518-MWe net) reference design has been developed to more thoroughly understand and resolve design and cost issues. It is based on sub-bituminous coal with a modern super-critical boiler with steam conditions of 3500 psi and 1100°F main and reheat steam temperatures. The net plant efficiency is expected to be about 31.5%, which is approaching the current coal-fired fleet average without CCS (note this process achieves 33.6% with bituminous coal at sea level as shown in Figure 18.12). Like the demonstration plant, the air emissions in the oxy-mode are very low with NOx, SO_x, Hg, and particulate below current power plant measurement accuracy and CO about the same as with air firing. This size plant would capture about 4.5 million tons of CO_2 per year, which equates to more than 90% of the CO_2 produced [3].

18.2.8 Current Performance and Cost Comparison With Other CCS Technologies

As part of the integration study work mentioned in Section 18.2.2, performance and cost data were taken from DOE

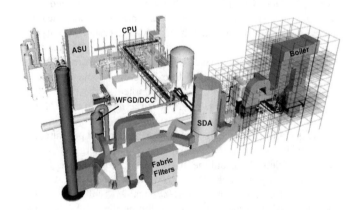

FIG. 18.10 150-MWE GROSS DEMONSTRATION PLANT

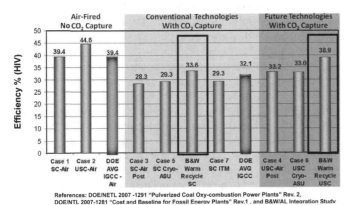

References: DOE/NETL 2007-1291 "Pulverized Coal Oxy-combustion Power Plants" Rev. 2, DOE/NTL 2007-1281 "Cost and Baseline for Fossil Energy Plants" Rev.1 . and B&W/AL Integration Study

FIG. 18.11   COMPARISON OF NET PLANT EFFICIENCY FOR CCS OPTIONS

reports for comparison with the B&W/AL warm recycle design. Figures 18.11 and 18.12 show the net plant efficiency and levelized cost of electricity for several options based on data from DOE reports noted at the bottom of the charts [8]. Air fired technologies are shown with the white shaded background, conventional carbon capture technologies with a light gray shaded background, and future carbon capture technology predictions with a darker gray shaded background. The technologies are identified along the bottom; Cases 3 and 4 are post-combustion with super and ultra supercritical steam conditions, Cases 5 and 6 are oxycombustion with super and ultra supercritical steam conditions, and the two options encircled in black are the B &W-AL warm recycle process with super and ultra supercritical steam conditions. Supercritical steam conditions are 3500 psi, 1110/1150°F, and ultra-super-critical are 4000 psi, 1350°F/1400°F. Case 7 uses ion transport membrane technology for oxygen separation. The average IGCC performance from the referenced DOE/NETL 1281 study on the same basis is also shown.

Figure 18.11 compares the net plant efficiencies. The warm recycle design, encircled in black, promises noticeably higher efficiency compared with the other carbon capture technologies. When ultra-super-critical steam conditions become available in a decade or so, the efficiency is expected to be comparable to a today's current modern air-fired super-critical power plant design.

Figure 18.12 compares the levelized cost of electricity and its components for the same options using the same shading scheme. All are based on the same financial assumptions burning the same bituminous coal and estimated in 2007 dollars, not including owner's costs. Again, warm recycle promises much lower costs, and future high-temperature steam boilers will improve costs even further. These comparisons indicate that oxy-combustion promises to be very competitive with both IGCC and post-combustion technologies.

18.2.9   Partial Capture and Swing Operation

Although oxy-combustion requires conversion of the entire process, a plant can be operated to achieve partial capture and benefit from both lower capital cost and higher revenue. This is accomplished by designing the plant to be capable of air firing and providing liquid oxygen storage and an ASU sized for part load. The plant would operate air fired for part of the time and oxy-fired for part of the time, depending on what percentage of CO_2 capture is desired. The smaller ASU would require less capital and consume

less power. It would be operated when in the oxy-mode and during all or a portion of the air firing period to store LOX. If more capture is desired later, the ASU can be retrofit to provide more capacity.

Revenue can also be optimized by "swing operation" in which the plant is operated in the air-fired mode during peak periods when net output is higher (due to smaller ASU), power is in high demand, and electricity selling prices are high, and operated in the oxy-mode when the selling price is low and net output is lower.

If the operating profile for the plant is well defined, swing operation can also improve revenue for an oxy-plant. The ASU would be designed for part load as defined by the operating profile and LOX storage provided. When full load output is not required, the boiler load is reduced, but the ASU is operated at full output to produce more oxygen than required for combustion. The additional oxygen is stored as LOX, and the net electricity generation will be reduced according to the higher power consumption by the ASU. When full load is needed, LOX is used to make up the difference between ASU capacity and operating load. Obviously, a plant designed in this manner must have a well-understood operating profile. By generating more net power when electricity prices are higher and storing LOX when electricity prices are lower, profitability can be maximized.

18.2.10   Retrofit Potential for Oxy-combustion

Initially oxy-coal combustion was favored as a retrofit technology. However, early studies using small (<400 MWe), old units with no scrubber or SCR required a high investment because (1) basic SOx and NOx equipment was included in the retrofit cost, (2) small unit costs are negatively impacted significant by economy of scale, (3) the older units suffered from relatively high air infiltration for a variety of reasons, and (4) the subcritical steam cycles were of low efficiency by comparison to current capability. After accounting for the significant increase in parasitic power, the net heat rates were high, and along with the high capital costs, high costs of electricity resulted. Because of these high costs, many began to view oxy-combustion only as a greenfield technology. In addition, partial capture with oxy-combustion is not as flexible as with post-combustion carbon capture (PCCC) and may require more capital investment.

It was also believed by some that oxy-combustion is not as easy to retrofit as PCCC. Actually, oxy-combustion and PCCC require very similar site characteristics, and oxy-combustion has the advantage of not imposing on the steam turbine. Both require particulate

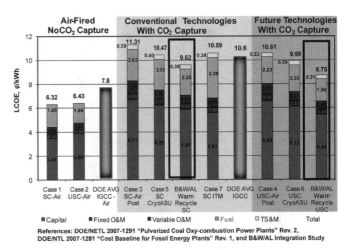

References: DOE/NETL 2007-1291 "Pulverized Coal Oxy-combustion Power Plants" Rev. 2, DOE/NTL 2007-1281 "Cost Baseline for Fossil Energy Plants" Rev. 1, and B&W/AL Integration Study

FIG. 18.12   COMPARISON OF LEVELIZED COST OF ELECTRICITY FOR CCS OPTIONS

removal and SO_2 scrubbing and space for compression equipment. Post-combustion carbon capture requires space for a polishing scrubber, the absorption and regeneration towers, re-boiler, piping, pumps, and so on, and it requires either major modifications to the existing low-pressure steam turbine or a separate boiler to produce the significant amount of low-pressure regeneration steam. Oxy-combustion requires space for a dehumidifier, which also polishes some SO_2, and the ASU. Overall, the total footprint of this equipment is essentially the same. Thus, the base plant requirements for oxy-combustion compared with PCCC are essentially the same, but PCCC has the added complication of steam turbine modifications or another boiler for low-pressure regeneration steam.

For 90% carbon capture, both technologies have about the same retrofit requirements and similar cost. The greatest advantage PCCC has for retrofit applications is the ability to install a smaller, lower-cost system and remove only part of the CO_2. But it has been shown by DOE and others that the cost of CO_2 removed or avoided for partial capture is higher than 90% capture.

18.3 POST-COMBUSTION CARBON CAPTURE

18.3.1 Process Description

In the remainder of this section, PCCC technology shall simply be referred to as "PCCC." It is called "post-combustion" because it captures CO_2 from the flue gas downstream of the boiler. The technology lends itself well to retrofits because a PCCC system is a stand-alone system that does not require boiler modifications to accommodate its inclusion in the flue gas train. Babcock & Wilcox Company's PCCC system works by absorbing CO_2 directly from flue gas in an absorber unit using a regenerable solvent. The CO_2-laden solvent from the absorber is then sent to a solvent regenerator unit where the solvent is heated, and the CO_2 is released in a concentrated stream. This CO_2 stream can then be compressed and permanently stored in appropriate geological strata. The solvent, after being relieved of CO_2 in the solvent regenerator, is then recycled to the absorber unit for additional CO_2 capture. Figure 18.13 shows an overview of the major chemical/physical processes within the basic PCCC system design.

18.3.1.1 Flue Gas System Flue gas enters the CO_2 absorption system after pre-treatment to reduce particulate matter, NOx, and SOx to very low levels. The reduction of particulate matter is necessary to help decrease maintenance costs and because impurities in particulate matter (coal fly ash, for example) can contribute to corrosion and solvent degradation. The reduction of NOx and SOx is necessary because these compounds are reactive toward many solvents and because of purity specifications in the transport lines. The absorber design pressure is generally higher than a typical FGD system, so the flue gas must be conveyed through the CO_2 absorption system using a blower or compressor.

The optimal scrubber inlet temperature for the process described herein is somewhat lower than the conditions at which the FGD system operates. The flue gas must therefore be cooled before entering the absorption section of the CO_2 absorption system. The process utilizes a direct contact cooler to accomplish this.

The direct contact cooler is an open spray tower in which the flue gas is cooled by contact with water before entering the absorber. Water is pumped to an array of spray nozzles in the top of the tower, and the flue gas enters from the bottom. The flue gas is cooled as it passes upward through the tower counter-currently to the liquid. The water is then collected in the vessel sump, where it is recycled to the spray nozzle array after cooling.

The direct contact cooler has dedicated re-circulation pumps to circulate cooling water through the tower. A dedicated heat exchanger is used to provide the required cooling to the re-circulated water. Process water is used as makeup to the circulation system as required, and a bleed stream is purged from the system to remove trace amounts of acid gases and other contaminants captured during the cooling process in the spray column.

18.3.1.2 CO_2 Absorber System

18.3.1.2.1 Packed Bed CO_2 Absorber Column The flue gas leaving the direct contact cooler then enters the CO_2 absorber. The absorber is usually a packed column-type unit utilizing metal packing to increase the effective area for mass transfer occurring within the tower. Structured packing generally facilitates the desired absorption performance at a relatively low pressure drop, compared with random packing or trays. This can become a significant advantage in terms of energy savings, given the large volumes of flue gas generated by electric utilities.

The flue gas entering the CO_2 absorber is contacted counter-currently with a solvent solution. The solvent used is typically an amine solution or an amine solution with additives. The regenerated (or lean) solvent solution is introduced at the top of the column, using low-pressure weir/trough-type liquid distributors to feed the solution over the packing via gravity overflow. The solvent solution subsequently absorbs CO_2 from the flue gas in the packed bed section(s). At the bottom of the absorber column, the CO_2-laden (or rich) solvent solution exits the absorber via a CO_2 absorber bottoms pump and is then pumped to the solvent regeneration system.

The flue gas that exits the top of the final absorption section of packing often contains some residual amount of solvent, which is either entrained or present in the gaseous phase of the flue gas stream. Since this solvent solution is costly and must be conserved as much as possible, a water wash packed section is typically located at the top of the absorber between the final absorption packed section and the mist eliminator, so that the solvent is recovered (by either physical impaction or by gaseous absorption) before the flue gas exits the absorber.

After passing through the water wash section of the absorber, the flue gas then passes through an entrainment separator (which may be either a chevron-type or a mesh pad) before exiting the absorber.

18.3.1.2.2 Rich Solvent Solution Heating/Lean Solvent Solution Cooling CO_2 absorption by amine solvents is a temperature-swing

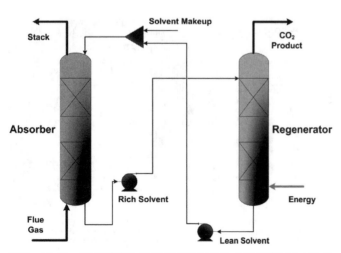

FIG. 18.13 GENERIC REGENERABLE SOLVENT-BASED PCCC FLOW SHEET

process. CO_2 is absorbed from flue gas with relatively cool solvent and is then stripped from the solvent solution by heating the solution in a regenerator tower. Before entering the regenerator, it is advantageous to pre-heat the rich solvent stream by exchanging heat with the hot, lean solvent stream. This conserves energy that would otherwise be used to heat the rich solvent solution to the regenerator operating temperature. This serves a dual purpose, since the lean solvent stream leaving the bottom of the absorber is too hot for effective CO_2 absorption. Cooling the lean solvent stream by exchanging heat with the rich solvent solution saves a significant quantity of cooling water. The heat required in the regenerator is provided by a re-boiler heat exchanger, which uses steam to boil the solvent stream, thereby releasing the absorbed CO_2.

18.3.1.3 Solvent Regenerator System

18.3.1.3.1 Solvent Regenerator Tower The heated, rich solvent stream from the CO_2 absorber enters the top of the regenerator tower where it is heated as it flows downward by gravity through the tower. The regenerator tower may utilize trays or beds of packing to provide the required mass transfer surface area for the regeneration of the solvent. As the rich amine solution is heated, CO_2 is released from the liquid stream and passes upward through the tower as a vapor, along with steam. The CO_2/steam mixture is progressively purified as it passes upward in the tower and ultimately leaves through the top of the tower where it passes to the condenser. The released CO_2 passes through a short rectification section at the top of the column to remove any traces of solvent, and then leaves the solvent regenerator from the top of the tower and passes to the regenerator condenser.

The re-boiler heat exchanger for the regenerator column uses steam to heat the solvent solution to the required temperature for effective stripping. The reboiler units create a vapor stream, which passes upward from the bottom of the regenerator tower and through the tray section. Condensation of a portion of the stripping vapor as it contacts the solvent, which cascades downward through the tray section, heats the solvent solution, and drives off the absorbed CO_2.

The regenerator unit may also have an additional reboiler, which is referred to as the reclaimer. The reclaimer is similar in size to the reboiler, but is used to purify the solvent. As the solvent runs through successive absorption/desorption cycles over time, there is a small amount of degradation that occurs due to side reactions. The byproducts created by these side reactions must be periodically removed from the solvent. The reclaimer unit performs this function. A small slipstream of solvent from the lean solvent stream is periodically directed to the reclaimer unit where it is vaporized into the regenerator unit. The byproducts are left behind in this process and are recovered from the reclaimer as a "heavy heel."

18.3.1.3.2 CO₂ Stream Conditioning From the top of the regenerator column, the gaseous CO_2 stream leaves the unit and passes through a condenser, which cools the gas stream enough to condense steam and excess solvent solution. After passing through the condenser, the nearly pure CO_2 stream passes through a regenerator reflux drum to further remove condensed water and solvent solution entrained in the gas stream.

The purified CO_2 stream from the top of the regenerator reflux drum then proceeds to the CO_2 dehydration/purification/compression system. The liquid stream from the bottom of the regenerator reflux drum is returned to the regenerator column.

18.3.1.3.3 Lean Solvent Stream Conditioning The lean solvent solution from the bottom of the regenerator column passes through a heat exchanger, as described in Section 18.3.1.2.2, and is then returned to the CO_2 absorber by the regenerator bottoms pump. The lean solvent is cooled further by a trim cooler before entering the absorber. This ensures positive control of the lean solvent inlet temperature, which is a critical parameter.

Some solvents may require additional purification other than the reclaimer as previously described. In this case, a small portion of the lean solvent stream is directed to additional processes such as filtration units, ion-exchange units or activated carbon beds, to remove impurities (particulate and dissolved) from the solvent stream.

18.3.2 State of Development

B&W is conducting a major effort to develop a PCCC process suitable for the capture of CO_2 at coal-fired utility plants. If CO_2 can be economically captured from these large point sources and subsequently stored outside the atmosphere, emissions of greenhouse gases, and their potential impacts on the climate system, could be reduced significantly. This effort is being conducted in parallel with a variety of other B&W studies aimed at the reduction or capture of CO_2, including major programs in ultra-super-critical boilers and oxy-coal combustion. Processes such as regenerable solvent-based PCCC have several advantages for near-term application at coal-fired power plants: (1) they are commercially available (for other industrial applications), (2) they are applicable to both new and existing coal-fired plants (unlike IGCC, for example), and (3) they can be applied to either the entire flue gas stream, or only a portion of it (unlike both IGCC and oxy-coal combustion).

Regenerable solvent scrubbing using simple amines, as deployed by the petrochemical industry since 1934, is currently the only commercially available PCCC technology. These conventional systems are capital intensive and have a high operating cost. It is B&W's intent to find better solvents, improve the process flow schemes, make better use of heat sources in the power plant, and better manage solvent life. Babcock & Wilcox Company believes that the development of a superior PCCC process is a key development target to ensure the continued use of coal in a carbon-constrained world.

In a regenerable solvent-based PCCC process, CO_2 is removed using a wet scrubber, or absorber, similar to the wet scrubbers used for removing SO_2 from flue gases. There are some important differences between CO_2 and SO_2 scrubbers, however. First, CO_2 is present in much greater quantities than SO_2, meaning that the solvent must be regenerable, or reusable. After absorbing CO_2, the rich solvent is therefore directed to a regenerator wherein the CO_2

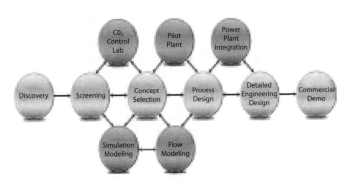

FIG. 18.14 B&W'S PCCC PROCESS DEVELOPMENT PROGRAM

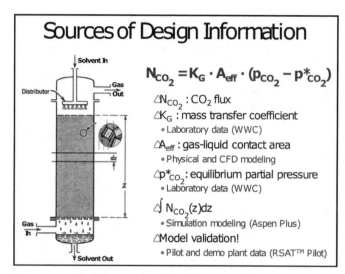

FIG. 18.15   SOURCES OF PROCESS DESIGN INFORMATION

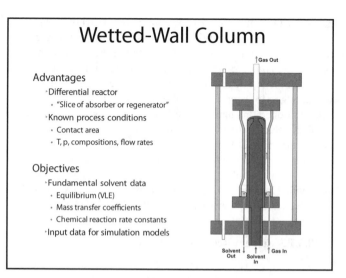

FIG. 18.16   WETTED-WALL COLUMN

is removed from the solvent and directed to compression and storage, while the lean solvent is returned to the absorber. Second, CO_2 is less soluble in water and therefore more difficult to remove from flue gases than SO_2. This means that the solvent must have a strong affinity for CO_2 to allow for its capture in a reasonably sized absorber. Unfortunately, solvents with a strong affinity for CO_2 can be undesirable for the regeneration process because they tend to require significant energy to release the captured CO_2. The net result of these factors means that if conventional amine scrubbing were to be used for CO_2 capture at a coal-fired power plant, the electrical output of the plant would be reduced by about one-third and the cost of electricity would increase by 50% to 100%.

B&W's PCCC process development approach, illustrated in Figure 18.14, is aimed at finding better solvents, improving process design, making better use of heat sources in the power plant, and better managing solvent life and corrosion issues. This work is being conducted using a stage-gate approach to carry ideas from discovery through initial screening, selection of an improved process concept, preparation of process and detailed engineering designs, and finally demonstration of the process at commercial scale.

Figure 18.14 also illustrates a variety of tools that have been put into place by B&W to support the PCCC development activities. These tools are invaluable for the support of a commercial PCCC product. Significant resources and funding have been devoted to the creation of these tools. The rationale for selecting these particular tools can be illustrated by considering the kinds of information needed to develop a conceptual design for an improved regenerable solvent-based CO_2 absorber. This information is illustrated in Figure 18.15. The regenerator design requires essentially the same information, since it simply runs the absorption process in reverse.

The rate of CO_2 absorption, or flux N_{CO2}, can be expressed as the product of an overall mass transfer coefficient, KG, the effective gas-liquid contact area, A_{eff}, and the driving force for mass transfer expressed as the difference between the actual partial pressure of CO_2 and the partial pressure of CO_2 that would be required to bring the phases into equilibrium, $(p_{CO2} - p^*_{CO2})$. Key information on the fundamental characteristics of a solvent can be generated in laboratory-scale equipment. This information is generated for conditions at a particular point in the absorber or regenerator. By generating information over a wide range of conditions, a computer simulation

model can be used to predict the overall performance of the process. The model predictions must be verified at each step.

18.3.2.1 CO_2 Control Laboratory This laboratory is used to evaluate potential CO_2 solvents to reduce the regeneration energy required while maintaining a reasonable rate of absorption. It contains two major test facilities: a wetted-wall column (WWC) for precise measurements of fundamental mass transfer and chemical kinetics data and a fully integrated bench-scale process for studies of process design concepts. These laboratory-scale tools facilitate the characterization and selection of solvents, as well as the evaluation of novel process flow sheets and heat integration approaches. These tools have been carefully characterized and validated against data from other laboratories.

The WWC, shown in Figure 18.16, is a gas-liquid contactor wherein CO_2 absorption or desorption can be studied under precisely controlled conditions. Its primary advantage is that, due to its simple geometry, the area of contact between the gas and liquid (solvent) is accurately known. The solvent flows verti-

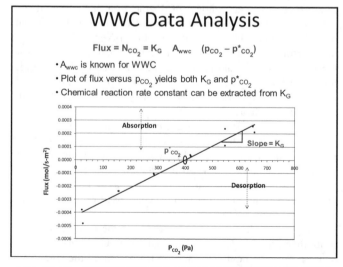

FIG. 18.17   FUNDAMENTAL DATA PRODUCED BY WWC

cally upward through the tube at the center, exits at the top, and flows downward over the outside surface of the tube in a thin film. This solvent is contacted with the CO_2-containing gas, which flows upward through the annular space around the tube. Careful control of temperature, pressure, and concentrations permits the generation of high-quality fundamental data on the mass transfer, chemical reaction kinetics, and thermodynamic properties of the solvent. This information is then used in computer simulation models to predict process performance.

Figure 18.17 illustrates how WWC data is used to characterize a solvent. If the CO_2 flux — the rate at which CO_2 is either absorbed or desorbed — is plotted as a function of the CO_2 partial pressure in the gas, a straight line is obtained. Since the area of gas-liquid contact is known, the overall mass transfer coefficient can be calculated from the slope of the line. The point at which the line crosses the x-axis defines the thermodynamic equilibrium value of CO_2 vapor pressure in the liquid phase for the given conditions.

Through careful design of the experiments, the chemical reaction rate constant for the reaction of the solvent with CO_2 can be extracted from the overall mass transfer coefficient. All of this information is crucial for proper prediction of solvent performance in the process.

The bench-scale PCCC unit, illustrated in Figure 18.18, is a fully functional process testing facility, designed to operate at a very small scale. The unit contains most of the equipment that would be

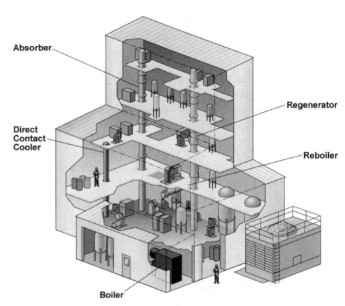

FIG. 18.19   PCCC PILOT PLANT

used at larger scale, including the absorber column at the left, the regenerator column at the right, and the electrically heated reboiler at the lower right in the photograph. In a full-scale process, heat to the reboiler would normally be provided by steam. The unit is designed to capture approximately one kilogram of CO_2 per hour.

The columns are of modular design, and the process can be operated in a variety of modes, providing flexibility for development work. The unit provides the opportunity to examine the performance of a new solvent in a fully integrated process, facilitates parametric studies of the important independent process variables, and provides data for computer simulation model validation. It can also be used to evaluate solvent management issues such as degradation and corrosion.

18.3.2.2 PCCC Pilot Plant The PCCC pilot plant, which is shown in Figure 18.19, is used to evaluate potential CO_2 solvents and process flow schemes aimed at reducing the regeneration energy required while maintaining a reasonable rate of absorption. Relative to the bench-scale data provided by the CO_2 control laboratory, the primary advantage of the PCCC pilot plant is that it can provide high-quality, quantitative data representative of full-scale equipment. The pilot plant is installed in a building adjacent to B&W's small boiler simulator (SBS-II), which is a facility that simulates a coal-fired power plant from the coal pile to the stack. The pilot plant operates on approximately half of the flue gas flow from the SBS-II, which is approximately 3100 lb/hr of flue gas, or 7 tons/day of captured CO_2. It can also be operated in a re-circulation mode using synthetic flue gas, which is a mixture of nitrogen and manufactured CO_2.

Conditioned flue gas exiting the SBS-II system first flows to a blower, which provides the motive force required to move the gas through the direct contact cooler and absorber tower. The flue gas is then cooled to the desired absorber operating temperature in the direct contact cooler, wherein the flue gas is directly contacted with cooling water as it flows through a bed of packing. The flue gas then enters the base of the absorber tower (2 ft in diameter and approximately 55 ft high) where it is contacted by the solvent, and CO_2 is removed. The CO_2 removal portion of the absorber consists of two beds of packing. A third bed of packing at the upper end of

Bench-Scale PCCC Process

Advantages
- Integrated process
- Modular column design
- Multiple modes of operation

Objectives
- First look at new solvents
- Impacts of operating conditions
- L/G, T, P, CO_2 loading
- Mass and energy balances
- Solvent management studies

0.9 kg/hr CO_2 capture

FIG. 18.18   BENCH-SCALE PCCC UNIT

the absorber tower is a water-wash section to reduce solvent losses to the flue gas stream. Clean flue gas is exhausted to the atmosphere, first through a carbon filter, and then through the SBS-II stack.

Rich solvent from the base of the absorber is pumped through a carbon bed and particulate filter, the cross heat exchanger, and a trim heater before it enters the upper portion of the regenerator (also 2 ft in diameter and approximately 55 ft high). In the regenerator, CO_2 is thermally driven from the solvent solution as it passes down through beds of packing. A natural gas-fired boiler provides steam to the reboiler, which drives the regeneration process.

The captured CO_2 is exhausted through the carbon filter and SBS-II stack. The lean solvent leaving the bottom of the regenerator is cooled as it flows, first through the cross heat exchanger and then a trim cooler before it flows into the solvent cycle tank. From the solvent cycle tank, the lean solvent is pumped back to the top of the absorber.

Nearly all pilot plant operations are controlled from the control room using four workstations. Personnel and equipment are protected by a safety interlock system. Over 400 instruments provide extensive process information to the control system as well as to a data acquisition system. Data collected allows for the completion of both mass and energy balances. The building is well ventilated, and an ambient air monitoring system continuously samples the air at 10 locations throughout the building for the presence of solvent vapor. All waste water is sequestered in the waste water storage tank, and the water is analyzed to determine the proper disposal method.

Construction of the pilot plant began in June 2008. First full operation on an amine solvent was achieved in June 2009. Baseline tests to fully characterize pilot plant performance on a conventional 30 wt% monoethanolamine solvent have been completed. The results of these tests will serve as a basis of comparison for other solvents, and as validation data for computer-based process simulation models.

After completion of the baseline tests, development testing of advanced solvents and processes has begun. The most promising process concepts identified through bench-scale CO_2 control laboratory testing, and supported by computer simulation modeling, are selected and fully characterized and optimized through testing in the pilot plant.

18.3.3 Reference Plant Description and Features

At the time of writing, a commercial-scale demonstration of regenerable solvent-based PCCC has not been performed at a coal-fired utility power plant. Thus, the size of the first demonstration plant is not yet known, so a complete engineering design of a commercial plant is not required at this time. However, much work can be done in advance of the first commercial plant design to gain understanding in key areas such as chemical and physical process simulation, equipment sizing and selection, supply chain development, capital and operating cost evaluation, understanding the intellectual property landscape, and integration of the PCCC system with the power plant.

Thus, B&W launched the PCCC Reference Plant design project. Using design tools such as Aspen Plus® and ProTreat®, the project team performed numerous process simulations to calculate equipment sizes for a theoretical plant, which is designed to remove 90% of CO_2 from a flue gas stream containing 1500 metric tons per day of CO_2, based on burning a typical eastern bituminous (Illinois #6) coal and using a typical amine-based solvent (30 wt% aqueous monoethanolamine, or MEA) as the scrubbing solution.

One of the primary goals of this project was to create a scalable model, which could be used as a basis for comparison for techno-economic evaluations, as well as to enable B&W to respond to requests for preliminary pricing and equipment descriptions for PCCC systems. During the design phase of the project, most of the key departments within B&W, which would typically be involved in a contract to engineer, procure, and construct the PCCC system, were engaged in the process. This methodology led to further benefits such as the establishment of preliminary design teams for the future product, as well as the development of a preliminary supply chain of vendors and fabricators for the various materials, equipment, and services required.

The PCCC Reference Plant design includes contributions from nearly all departments in the company, including engineering disciplines such as structural mechanics, technical design, instrumentation and controls engineering, and process engineering, as well as non-engineering disciplines such as purchasing, cost estimating, scheduling, and transportation. Babcock & Wilcox Company Construction Co., Inc., was also involved in the process, not only providing estimated pricing for erection of the proposed equipment, but also assisting in designing the plant for ease of constructability.

The PCCC Reference Plant project has led to the development of a full package of engineering documentation, which includes a complete set of piping and instrumentation drawings, a 3-D plant layout model using PDMS (Figure 18.20), various equipment lists and schematics, mechanical designs and fabrication sketches for major process vessels, foundation and structural steel designs, RFQ packages and equipment specifications for all major pieces of equipment, various construction estimates, man-hour estimates for all technical and non-technical disciplines, process flow diagrams with complete material and energy balances, and more.

18.3.4 Commercial Plant Considerations

Babcock & Wilcox Company's regenerable solvent-based PCCC system is based on technology, which has been used for over 75 years in the oil and gas industry. However, many challenges

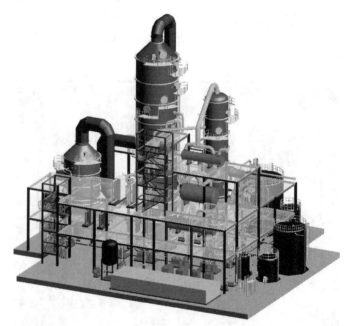

FIG. 18.20 A 1500-METRIC TONS PER DAY PCCC REFERENCE PLANT

that have been addressed in existing applications have entirely new ramifications and impacts when this technology is used in the power generation industry. Moreover, a host of new issues arises when the existing technology is applied to coal flue gas, which involves a different process stream in terms of volume, temperature, pressure, and chemistry. New contaminants are present in coal flue gases, which have never been addressed before by the existing technology. Operating pressure is much higher, and gas volume is much lower for oil and gas applications. The sheer size of the equipment required for power plant applications may reach the upper design limits for some of the important sub-systems within the PCCC system.

This section includes general discussions of some of the key issues that must be considered when applying the technology described herein to PCCC for coal-fired flue gas at commercial scale for the power generation industry.

Flue Gas Pre-Scrubbing. Depending on the concentration of NOx and SO_2 in the flue gas entering the PCCC system, a pre-scrubber may be required upstream of the regenerable solvent absorption train. Residual NOx and SOx in the flue gas are absorbed in the solvent, and at significant levels, this can lead to formation of byproducts, which can be detrimental to the performance of the solvent (this is described in more detail later).

Pre-scrubbing of the flue gas upstream of the PCCC system can take several forms. If the flue gas has already been treated by traditional flue gas desulfurization (FGD) systems, such as dry or wet scrubbers, then the performance of these processes could be enhanced by either chemical additives or physical modifications to the existing FGD systems, where only nominal performance enhancements are required. However, if a significant amount of additional NOx/SOx removal is required, then a separate pre-scrubber unit may be necessary. Depending on the design of the PCCC system, it may be possible to achieve the required level of pre-scrubbing by using chemical additives such as sodium bicarbonate, lime, or other alkaline reagents to the direct contact cooler.

Partial Boiler Load Operation. Many larger power generation plants are considered to be base loaded and as such operate at or near full load — or at least at steady state — for extended periods. However, many of these plants still do adjust their operation based on changes in electrical demand. For other plants, adjusting load is a constant process, which takes place weekly, daily, or even hourly. Thus, the PCCC system must be designed with this important process consideration in mind.

The amount of flue gas generated is one of the primary sizing criteria for the direct contact cooler and CO_2 absorber, whereas the amount of CO_2 removed from the flue gas is one of the primary sizing criteria for the regenerator. These vessels are generally designed according to the maximum values of these parameters, but the internals for these vessels, and the ancillary equipment that supports their operation, must be designed for maximum operational flexibility. Equipment such as pumps, piping systems, heat exchangers, and tanks must all be designed to perform within acceptable limits throughout the expected range of the operation of the power plant.

In addition to designing specific equipment for maximum flexibility, the overall operation of the PCCC system must be considered as well. Especially in cases where the amount of flexibility required cannot be achieved simply by changing flow rates, temperatures, and/or pressures, for example, options such as re-circulation loops, bypasses, or semi-batch operation can also be considered to provide reliable operation during low boiler load.

Emissions. Depending on the solvent properties, as well as the type(s) of solvent reclamation equipment used in the PCCC system, the potential exists for new emissions in the various gas, liquid, and solid effluent streams. For example, depending on the volatility of the solvent, additional treatment beyond the rectification/wash section at the top of the absorber (as described previously) may be required. Also, depending on whether the system is designed to operate in a water-generating or water-consuming mode — both of which may be possible, depending on the solvent and the final system design — certain compounds may be present in any liquid effluent streams that are generated, which may require treatment by the power plant's waste water treatment system. Finally, depending on the solvent used and the type of solvent reclamation system used, the effluent streams from such equipment would be a solid waste, which could require off-site disposal. All of these effluent streams must be considered not only when evaluating various PCCC technologies, but also during the PCCC system design phase once a technology is selected.

Solvent Degradation and System Corrosion. Two of the most influential and least understood parameters in designing a PCCC system for coal flue gas applications are solvent degradation and system corrosion. These concepts are highly interrelated and interdependent, and a change in one area often directly impacts the other.

Solvent degradation is the primary mechanism for the loss of the available solvent loading capacity in the absorber of a PCCC system. As the concentration of degraded solvent increases in the circulating solution, there is less free (un-reacted) solvent to react with the CO_2 in the inlet flue gas to the absorber. This directly impacts the removal capacity of the circulating solution, thus causing the absorber CO_2 removal efficiency to decrease. This can be offset by increasing the circulation rate of solvent; however, this also increases the reboiler duty, which increases steam (energy) consumption. Another option is to add solvent reclamation equipment to lower the concentration of degradation species to acceptable levels. Several different equipment options are discussed below.

Solvent degradation occurs through several different mechanisms: (1) oxidative, (2) thermal, and (3) chemical.

Oxidative degradation occurs when the re-circulating solvent comes into contact with oxygen, which is contained in the entering flue gas from the power plant. Another possible location for oxidative degradation, which is more often overlooked and less significant, is in the regenerator. Oxygen can be regenerated as a byproduct of thermal degradation reactions, released from oxidative degradation species as they are processed in the regenerator, or carried to the regenerator as dissolved oxygen. This oxygen can then react with a solvent molecule to form further degradation species.

A second degradation mechanism is thermal degradation, which is the formation of heat-stable salts at elevated temperatures. These compounds cannot be regenerated to the original solvent in the base PCCC system regenerator. These reactions are significantly enhanced by elevated temperatures and are most prevalent at the rich solvent exit of the cross-heat exchanger, the reboiler, regenerator bottoms, and the lean solvent entrance to the cross heat exchanger. In addition, the solvent reaction to form thermal degradation species often produces NH_3 as a breakdown byproduct.

A third solvent degradation mechanism is chemical degradation, wherein the circulating solvent reacts with SO_2 and NO_2 in the inlet flue gas to the PCCC system. Both of these contaminants will quickly react with the solvent, forming heat-stable salts.

Other conditions that increase the solvent degradation rate in an existing system are as follows: (1) increased temperatures in the cross heat exchanger, reboiler, or regenerator (due to fouling, for example); (2) higher corrosion rates leading to higher concentrations of metals in the circulating solution; (3) excursions of ash (metals) to the PCCC system; (4) higher CO_2 loading in the reboiler (which leads to increased temperature and therefore increased thermal degradation); and (5) lower CO_2 loading in the lean solvent feed stream at the top of the absorber (which increases the driving force of and therefore the rate of various oxidative and chemical degradation reactions).

There are many methods to decrease the concentration of degradation species in the re-circulating solution including ion exchange, thermal re-claimers (which are discussed above), activated carbon beds, filtration, electrodialysis, vacuum distillation, and degradation inhibitors. Ion exchange and electrodialysis are most effective at removing ionic degradation species such as oxidative species. Activated carbon beds are generally used to remove degradation precursors. Filtration is used to remove particulate (e.g., ash) and pipe scale resulting from corrosion. Thermal reclaimers and vacuum distillation are used to regenerate and remove all types of degradation species.

Solvent degradation is further compounded by corrosion products. As corrosion occurs in the PCCC system, the metals concentration in the system increases catalyzing degradation reactions and forming higher concentrations of degradation species. Conversely, as the concentrations of degradation products increase for many solvents, the corrosivity of the solution increases. This, in turn, accelerates the corrosion rate within the PCCC system.

Corrosion takes many forms and occurs in several locations in the PCCC system. One of the most significant locations for corrosion is in the regenerator bottom, where aqueous (liquid-phase) water and gaseous CO_2 are present, creating the possibility of CO_2 pitting, which forms small pinpoint holes that can be difficult to detect. This type of corrosion is also often found at galvanic cells, which are created where two different metals are welded together. Both corrosion and erosion are a concern at the rich solvent inlet of the regenerator, where the feed stream is present in both liquid and gaseous phases owing to flashing. Both erosion and corrosion in this area can cause significant vibration and, ultimately, mechanical failure if the system is not designed properly. Two other areas where corrosion is a concern, because of both high temperatures and the ability for gaseous CO_2 to evolve from solution, are the rich outlet and the lean inlet of the cross heat exchanger. Corrosion in both of these areas can be addressed by proper material selection during the design phase of the PCCC system.

General Process Optimization. In addition to the options discussed above, many other opportunities exist both in the design phase and during actual operation of the PCCC system. Power consumption can be minimized by using heat integration strategies as discussed in the next section, as well as by selecting the proper equipment for the intended application, and by tuning the system during startup. Split feed arrangements, wherein a semi-lean solvent stream is taken from the middle of the regenerator and fed to the middle of the absorber, can help improve absorber performance in some cases. Inter-cooling can be of benefit in moving the so-called absorber temperature bulge (an area of maximum temperature in the absorber tower as the gradient is mapped out from top to bottom) to decrease the amount of regeneration steam required. Vapor recompression in the regenerator system can also help to decrease the amount of regeneration steam required in some cases. In addition, the operating temperature and pressure, as well as the lean and rich solvent loading, can all be optimized based on the specific solvent being used.

Another area where the PCCC system can be optimized is to the footprint/layout. Since PCCC technologies are generally well suited for retrofit applications, the proposed equipment will likely be required to fit into an existing space at a power plant, and many plant locations have moderate to severe space limitations. Certain types of equipment can be configured to require minimal footprint, where real estate is at a premium. For example, vertical thermosiphon re-boilers could be used in lieu of horizontal shell and tube designs, or absorber/regenerator tower diameters could be minimized based on the type of internals used therein. Pumps, heat exchangers, and instrumentation can be located in areas, which not only provide the functionality required, but also allow for ease of maintenance while being located such that the amount of footprint required is minimized.

Power Plant Steam Usage/Heat Integration. As discussions continue regarding the concept of installing PCCC to help retain coal as a viable fuel source for power generation in the future, perhaps no single aspect of PCCC has garnered more discussion than the amount of energy required to operate such systems. Because the solvents used to absorb CO_2 are too expensive for once-through operation, they must be regenerated and recycled. However, the so-called "energy penalty" for doing so is significant.

Energy requirements for PCCC can be minimized in a number of ways. Electrical power consumption (i.e., from motors, heaters, and instrumentation) can be minimized as indicated above. Heat recovery can be accomplished by integrating systems such as boiler feed water heaters, heat exchangers within the PCCC system, and inter-coolers within the CO_2 compression system. Perhaps the most significant impact can be made by integrating or modifying the existing boiler systems to generate some or all of the steam required for solvent regeneration in ways that reduce the total amount of steam required (or maximize the efficiency of energy recovery from this steam).

In addition to using novel approaches to maximizing energy consumption in the steam cycle, one must also consider the source of the solvent regeneration steam. Typically, the assumption in many PCCC engineering studies to date has been that the steam required for solvent regeneration will be withdrawn from the cross-over header located between the intermediate- and low-pressure steam turbines. However, the amount of steam generally required for most current PCCC system designs may require a quantity of steam that would negatively affect the operation of the turbine, to the extent that such schemes are not practical.

In cases such as this, alternative sources of regenerator steam must be considered. Such sources could include over-firing the boiler to create additional steam, placing additional heating surface in the existing boiler, or perhaps a separate steam generator altogether, such as a circulating fluid bed fired with biomass or a gas turbine with a heat recovery steam generator. Of course, such sources may also create additional amounts of CO_2 as well as other regulated contaminants, which may trigger major permit changes and require additional environmental controls. Ultimately, the optimum approach for a given power plant will depend on factors such as the amount of land and type of fuel available, age and condition of the existing boiler(s), permit limitations, and federal, state, and local regulations.

18.4   ACKNOWLEDGEMENTS

Figure 18.1 was developed by B&W based on data from IEA, *World Energy Outlook 2009*. The 2007 pie chart on the left is based on Annex A — Tables for Reference Scenario Projections, Table

Reference Scenario: World, Page 623. The 2030 pie chart on the left is based on Table 6.2, Capacity additions by fuel and region in the 450 Scenario (GW), Page 234 and for retirements, Figure 6.7, World installed coal capacity and retirements/mothballing in the 450 Scenario, Page 235 was used. The results were cross-checked that with Figure 9.3, World power-generation capacity in the 450 Scenario (GW), Page 323.

Figure 18.2 was generated from the EIA data Table 8.2 World Estimated Recoverable Coal, December 31, 2008 (public domain).

Figures 18.7, 18.8, and 18.9 are used by permission from AL.

Figures 18.11 and 18.12 were developed by B&W using data generated by B&W and taken from the DOE reports referenced on the figures (public domain).

All other figures and information were developed by B&W Power Generation Group.

18.5   REFERENCES

1. Tranier J. P., et al, "Air Separation Unit and CO_2 Compression and Purification Unit for Oxy-coal Combustion Systems," *35th International Technical Conference on Coal Utilization & Fuel Systems Conference, Clearwater, FL, June 2010.*

2. McDonald (B&W), Tranier (Air Liquide), "Oxy-coal is Ready for Demonstration," *35th International Technical Conference on Coal Utilization & Fuel Systems Conference, Clearwater, FL., June 2010.*

3. McDonald, D. K., et al, "Status of B&W PGG's Oxy-coal Combustion Commercialization Program" *Electric Power Conference 2010, Baltimore, MD, May 2010.*

4. Farzan, McDonald (B&W), Varagani, Docquier, Perrin (Air Liquide), "Evaluation of Oxy-coal Combustion at a $30MW_{th}$ Pilot," *1st International Oxyfuel Combustion Conference, Cottbus, Germany, September 2009.*

5. McDonald, D. K., et al, "Oxy-fuel Process Developments Leading to Demonstration Plant Design," *1st International Oxyfuel Combustion Conference, September 2009, Cottbus, Germany.*

6. McDonald, D. K., et al, "B&W and Air Liquide's 100 MWe Oxy-fuel Demonstration Program," *34th International Coal Utilization and Fuel Systems Conference, June 2009.*

7. McDonald, D. K., et al, "30 MW_t Clean Environment Development Oxy-Coal Combustion Test Program," *33rd International Coal Utilization and Fuel Systems Conference, June 2008.*

8. Ciferno, J., et al, "Pulverized Coal Oxy-combustion Power Plants," *Final Report, August 2007, DOE-NETL-2007/1291*, Revision 2 August 2008.

PETROLEUM DEPENDENCE, BIOFUELS — ECONOMIES OF SCOPE AND SCALE; US AND GLOBAL PERSPECTIVE

James L. Williams and Mark Jenner

The economic behavior of all fuels share many common characteristics and it is important for the reader to apply principles concerning one fuel to others as the need arises. In many cases, we use the United States as an example. This is because there is better long-term data available than for other countries. In addition, it is a reference point to compare countries especially when evaluating the impact of policies. When comparing fuels we will focus on the cost per Btu and Btu/lb and Btu/ft^3. Efficiency in converting Btu's to mechanical or electrical forms of energy is important as well. The scale of the "conversion plant" may affect efficiency and fuel use for a particular purpose.

19A.1 INTRODUCTION

Our purpose is to examine petroleum dependence, identify risks of dependence, methods to mitigate dependence and risks and provide a methodology for comparison of alternative sources of energy. We will not emphasize carbon emissions, but will provide tools for examining the economic impact of various methods of reducing emissions. We will compare petroleum with selected biofuels. Our goal is to present a disciplined approach to the analysis of fuels and to distinguish between the fuel and its carrier or storage device. As terminology is often not well defined, we will define terms that could lead to ambiguity. This chapter has two (2) Parts: Part A dealing with Measures of Petroleum Dependence, and Part B addressing Bioenergy Economics.

In the following discussion, the term "petroleum" includes crude oil, natural gas liquids and petroleum products, unless mentioned otherwise.

19A.2 MEASURES OF PETROLEUM DEPENDENCE

A nation dependent upon petroleum imports faces two major risks: a cessation of imports and a dramatic increase in price. For importing nations, a spike in prices can cause extreme economic damage. Figure 19A.1 shows oil prices in 2009$ and nominal dollars. The

bars show U.S. recessions. There is a clear economic risk of high prices for nations that import a significant volume of their oil.

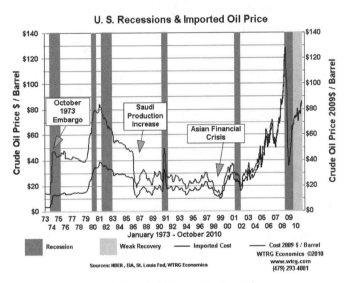

FIG. 19A.1 *Source* [1, 3, 4, 5]

We can use the graph to argue that high oil prices cause recession. However, risks to oil exporters can be seen in the same graph. If the graph demonstrates that high prices cause recessions, it can also be viewed from the producer's perspective. Recessions cause low oil prices. Dramatic dips in prices are also evident in economic weakness outside the U.S. as in the case of the Asian Financial Crisis. A dramatic increase in production by one or more OPEC members can have a similar effect and large decreases in production or exports often lead to spikes in price. The October 1973 Embargo is an example. The embargo was a response to importing countries supporting Israel in the Yom Kippur War. While it is often called the OPEC or Arab Oil Embargo, neither is correct. OPEC member Venezuela did not participate; Iran, which is Persian, did join the embargo. Oil price fluctuations pose economic risks to producers and consumers alike.

We say a nation is dependent if it is a net importer of petroleum (crude oil and/or petroleum products). There are several commonly used measures of dependence: (1) petroleum imports as a percentage of total petroleum consumption, (2) the number of days total petroleum stocks cover petroleum imports, (3) the number of days total stocks cover consumption, (4) the percentage of petroleum in total energy consumption, and (5) the percentage of petroleum imports from the top 5 suppliers [6, 7]. In the long run, only imports as a percentage of consumption matter. The others help understand how much risk an importing country might have to mitigate in the short-term.

19A.2.1 Import Percent of Consumption

Net petroleum imports as a percent of consumption is the most general and intuitive measure of import dependence. U.S. dependence on imports of crude oil and petroleum products was about 35% at the time of the 1973 Oil Embargo [4]. By the late 1970s, it approached 50% before declining to below 30% in the early 1980s. It rose steadily to 60% in 2006 before falling below 50% during the 2008 to 2009 recession. Import dependence generally declines during periods of high prices and recession. The causes will become clear later (Fig. 19A.2).

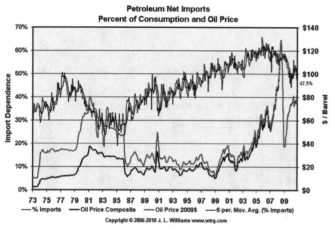

FIG. 19A.2 *Source* [3–5]

19A.2.2 Days Coverage — Stocks

Days coverage (Fig. 19A.3) is the number of days petroleum stocks (crude oil and petroleum products) can supply the market. In the graph below days coverage (stocks/consumption) is presented with and without the stocks in the U.S. Strategic Petroleum Reserve (SPR). It is worth noting that as the stocks in the Strategic Petroleum Reserve increased in the early 1980s there was a corresponding

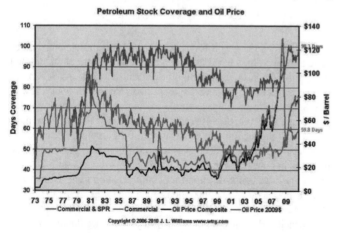

FIG. 19A.3 *Source* [3–5]

decrease in the number of days that commercial stocks could cover consumption. Before the SPR, the petroleum industry maintained levels it thought appropriate to cover normal business risk as well as interruption due to hurricanes and geopolitical risks of embargos, wars and revolutions. With the advent of the SPR there was no need for the industry to maintain stocks to cover geopolitical and hurricane risk and commercial inventories declined to a level commensurate with the normal business risk.

19A.2.3 Days Coverage — Imports

Days import coverage is the number of days petroleum stocks or inventories can replace imports (stocks/imports). If we view imports as a percent of consumption as a strategic measure of dependence, then days of import coverage is tactical in the sense that it is a relatively short-term measure. It is more focused on the threat of import interruption than days coverage. This graph shows the number of days petroleum inventories can replace imports for commercial stocks and total stocks, which includes the Strategic Petroleum Reserve. Including the SPR, current import coverage is only marginally better than it was at the time of the 1973 embargo. Most of the increase since 2008 is due to the impact of the recession (Fig. 19A.4).

Every U.S. President since Richard Nixon has called for energy

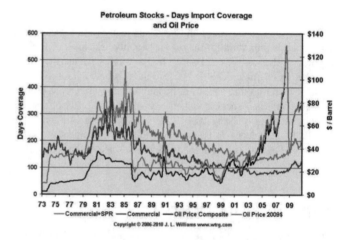

FIG. 19A.4 *Source* [3–5]

independence by which they mean petroleum independence. It is patently obvious that by this measure the U.S. is not significantly better off than it was at the beginning of the 1973 Oil Embargo or in 1979 at the time of the Iranian Revolution and Iraq-Iran War which followed. The problem is not unique to the United States. However, it is only one side of the dependence issue.

19A.2.4 Export Dependence

Oil exporters are generally more dependent upon oil and its price than those dependent upon imported oil. Figure 19A.5 shows how wildly OPEC export revenue fluctuates over time. Half of the time since 1975 the annual fluctuation in inflation adjusted OPEC revenue exceeded 20%. In five of the 35 years, the annual change exceeded 40%.

OPEC net export revenue shows the organization is in a far better position than it was in most of the 1980s and 1990s, but a better measure might be per capita export revenues (Fig. 19A.6). While there is improvement from the period of 1980s and 1990s, we see in Fig. 19A.6 that on a per capita basis, even in 2008 at the time of the highest oil price in real or nominal dollars, OPEC's inflation adjusted per capita exports were far below the peak in 1980. The large fluctuations in revenue from oil make planning and budgeting very difficult for exporters, and place heavy stress on the political establishment.

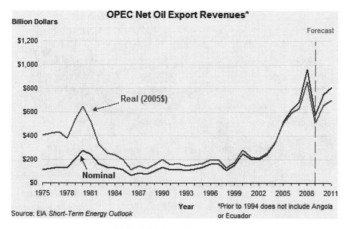

FIG. 19A.5 *Source* [2]

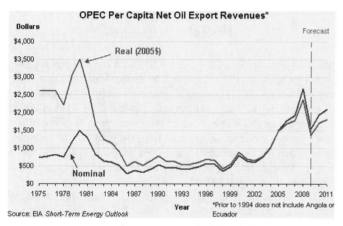

FIG. 19A.6 *Source* [2]

Most exporters maintain sovereign wealth funds where they invest money during periods of high oil prices to draw on when prices are low. However, as in the U.S. most, if not all, governments can reverse actions of their predecessors. The most recent example among the exporting nations is Venezuela. Venezuelan president Hugo Chavez has spent most of his country's funds and leveraged future earnings with loans from foreign investors to fund military expansion and social programs, which have been less than effective. The mere existence of large funds in exporting nations is attractive for those who hope to assume power in the future.

With OPEC's heavy dependence on export income, it is little wonder that in late 2008 in the face of declining demand in OECD member countries and falling prices, OPEC was asking for guarantees of consumption before investing in additional production capacity.

19A.3 CONSUMPTION

We return to the case of importing nations with the U.S. as an example. If petroleum independence can be achieved it is most likely through a combination of an increase in production and a decrease in consumption. Lower consumption comes from improving efficiency, fuel substitution, conservation or a weak economy. An examination of the history of U.S. petroleum consumption gives insight into where to place the current emphasis (Fig. 19A.7).

Until the end of World War II, U.S. production was roughly equal to consumption and the U.S. was not petroleum dependent. For most of the post-WWII era consumption increased faster than

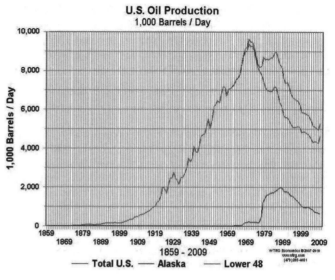

FIG. 19A.7 *Source* [4]

production. In Fig. 19A.8, we show both oil production and total petroleum production, which includes petroleum liquids from natural gas wells compared to consumption.

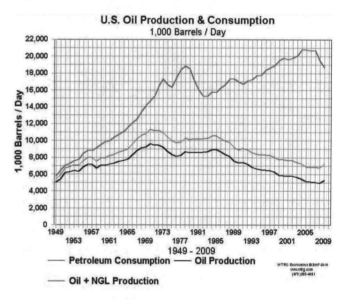

FIG. 19A.8 *Source* [4]

There are only three periods where U.S. petroleum consumption fell for more than 2 years in a row. The two largest declines started in 1978 and 2005.

Consumption fell 19% from a peak of 18.8 million barrels per day in 1978 to 15.2 million b/d in 1983. The second period started in 2005 at the all time high of 20.8 million barrels per day for U.S. consumption. By 2009, it was down 10% at 18.7 million b/d. Both periods shared two characteristics: (1) A dramatic spike in crude oil prices and (2) an exceptionally severe recession. The reasons the most recent period showed only half the decline of the period starting in 1978 may provide clues toward an energy policy that could significantly reduce petroleum consumption and with it dependence on imports.

Oil consumption declined following the 1973 oil embargo and ensuing spike in price, but quickly resumed an upward trend. The big change came in 1979 when oil prices spiked again. That spike was the result of lost production from the Iranian Revolution and the Iraq-Iran War that followed. With that spike came back-to-back recessions in the United States and a rapid decline in petroleum consumption. There were multiple factors involved. Higher prices led to conservation and some improvements in efficiency, but there was also considerable fuel switching.

19A.4 SECTORS

Changes in consumption from the standpoint of conservation, policy and from fuel switching vary by sector. We will look at each of the five sectors: residential, commercial, industrial, electric power and transportation. This will help us later when we examine the potential for alterative fuels and the characteristics required. We look at not only petroleum but also how its consumption compares with other fuels. We use 1949 and 2009 data because they are the first and most recent complete years of data available. We use 1978 because it was the year of the first peak in total U.S. petroleum consumption (Fig. 19A.9).

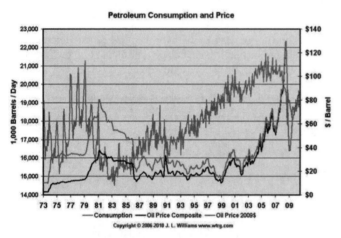

FIG. 19A.9 *Source* [3–5]

19A.4.1 Residential

Residential petroleum consumption peaked in 1972 at 1.523 million barrels per day drifting down to 1.377 million b/d in 1978. By 1981, it was down to 800,000 b/d. Relatively stable since the early

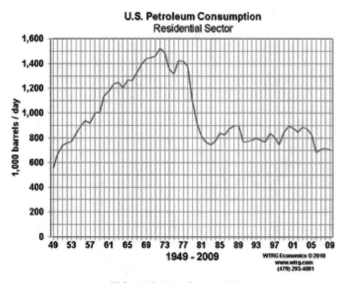

FIG. 19A.10 *Source* [4]

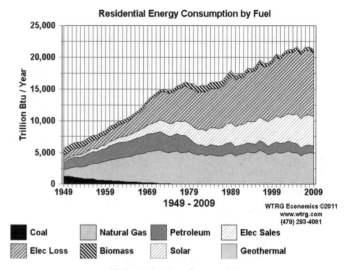

FIG. 19A.11 *Source* [4]

1980s, it stands at 706,000 b/d for 2009. With some liberal rounding of percentages, in 1949 at the end of World War II residential energy consumption equally split with 20% each for biomass, coal, petroleum, natural gas and electricity. At that time, 41% of the energy consumed at home required the homeowner to bring in logs from the woodpile or shovel coal from the coal bin into the coal-fried furnace. In some cases, the owner would do both, heating the home with coal and cooking with an old-fashioned wood stove (see Figs. 19A.10 through 19A.15).

By 1978, at the time of the first peak in U.S. petroleum consumption, direct consumption of coal represented less than one percent of residential energy consumption and biomass was down to four percent. Electricity including system losses to generate and distribute power constituted 49% of total residential energy use. Petroleum (16.2%), natural gas (30.8%) and electricity (48.8%) replaced coal (0.3%) and most of the wood (3.8%) used for heating. Coal and wood shared two characteristics. They were labor intensive for the consumer and particulate matter in the smoke was visible at the point of consumption i.e. the residence.

In 2009 we find electricity the dominant residential fuel (68.8%) followed by natural gas (23.0%), petroleum (5.5%), biofuels (2.0%), solar (0.5%) and geothermal (0.2%).

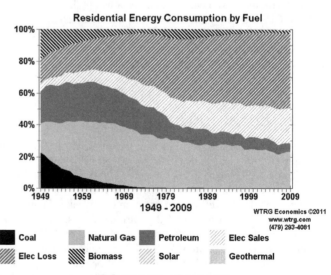

FIG. 19A.12 *Source* [4]

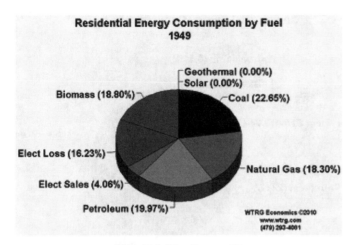

FIG. 19A.13 *Source* [4]

In the very short-term, residential energy consumption is more or less fixed. Consumers can adjust temperatures lower in the winter and higher in the summer, but switching fuels or increasing efficiency requires capital investment in a different heating or cooling system. These changes take place over years or decades. The switch to an electric furnace in a home can happen quickly, but switching to natural gas

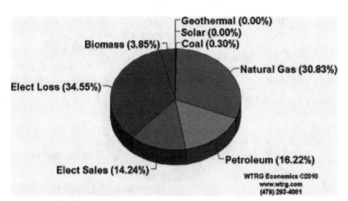

FIG. 19A.14 *Source* [4]

requires a new furnace and at a minimum a pipe from the residence to the main line. At the extreme, it takes the development of more pipeline infrastructure at the local, state and/or national level. Better insulation, doors, and windows that are more efficient can obviously reduce residential consumption of energy for heating and cooling.

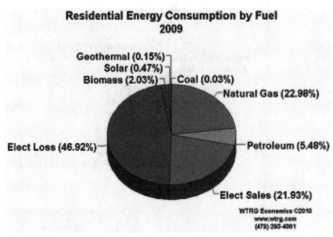

FIG. 19A.15 *Source* [4]

Changes in residential mix of petroleum fuels over time give additional insight into fuel switching. In 1949, 58.9% of residential petroleum consumption was distillate (heating oil), 25.0% was kerosene (heat, light & cooking) and 16.1% propane (heat, light and cooking). In the 1950s, it was common to heat with distillate (heating oil) and cook with propane. By 1972 when residential petroleum consumption peaked heating oil was 61.5% of the residential petroleum market, kerosene declined to 8.6% and propane was up to 29.3%. Between 1978 and 1983, residential heating oil use dropped 53% and propane went down by 30%. Heating oil consumption never recovered, but propane returned to the 1972 level of consumption and is now 58.8% of the total residential market for petroleum. Propane is "ideal" for residences where natural gas is not available because of its flexibility for cooking or heating and the advantage that it is stored on site. Many rural areas are subject to extended power outages during severe winter storms giving propane an advantage over electricity. It is commonly used to fuel backup generators when the power is out (Fig. 19A.16).

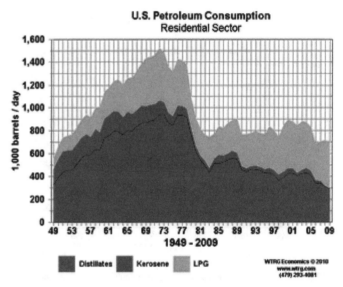

FIG. 19A.16 *Source* [4]

Petroleum use for residential energy has been flat since the early 1980s despite a 33% increase in population. There is relatively little opportunity for additional switching from petroleum to natural gas or electric power. Biofuels, solar and geothermal sources may make some inroads into the residential market, but with only 700,000 b/d of residential petroleum consumption (5.5%) it cannot repeat the 800,000 b/d decline in the 1970s and early 1980s. The residential segment has little to contribute to future energy security.

19A.4.2 Commercial

The commercial sector's consumption of petroleum peaked at 746,000 b/d in 1973 a year later than residential use and by 1978 when total petroleum consumption peaked commercial consumption was down to 685,000 b/d. Its decline was less abrupt not reaching a low point of 332,000 b/d until 1998. Since then it has been relatively stable near 350,000 b/d (see Fig. 19A.17 through 19A.23).

In 1949, commercial energy consumption was dominated by coal (37%) followed by petroleum (24%) and natural gas (21%). Electricity was only 14% of commercial consumption. By 1978, coal was virtually eliminated as a commercial fuel (0.3%). Electricity (49%) was dominant followed by natural gas (31%) and petroleum (16%). The introduction of air conditioning in commercial build-

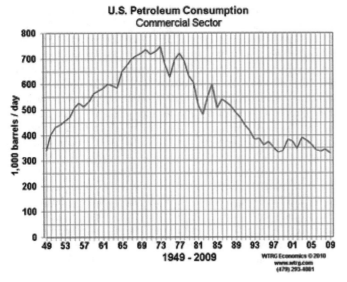

FIG. 19A.17 *Source* [4]

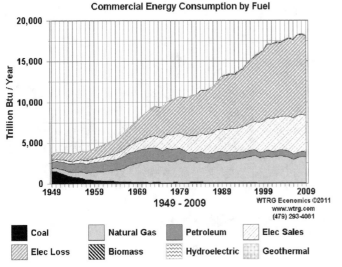

FIG. 19A.18 *Source* [4]

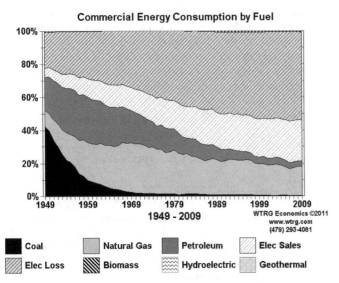

FIG. 19A.19 *Source* [4]

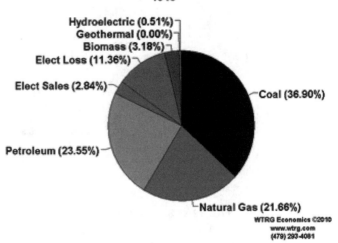

FIG. 19A.20 *Source* [4]

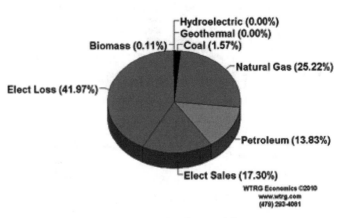

FIG. 19A.21 *Source* [4]

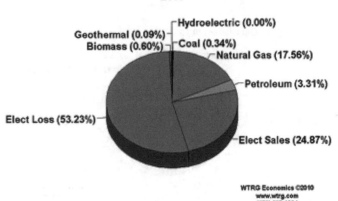

FIG. 19A.22 *Source* [4]

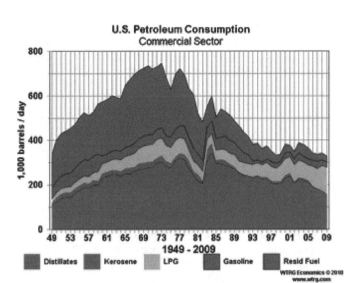

FIG. 19A.23   *Source* [4]

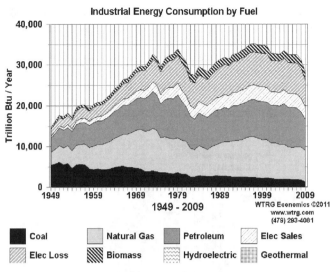

FIG. 19A.25   *Source* [4]

ings fuels exceptionally rapid growth in electricity use from the mid-1960s to mid-1970s. Supermarkets and other businesses even advertised air-conditioned shopping. In little more than a decade, air-conditioning became the norm in the commercial sector. In the last 30 years, all of the increased energy consumption in this sector has been electric power. By 2009, electricity (69%) was the norm followed by natural gas (23%) and petroleum (5%).

The use of residual fuel oil and distillates dominated commercial petroleum consumption in the first half of the post WWII. The heavy residual fuel oil was used primarily as a boiler fuel for heating and in some cases for electric power generation. Residual fuel oil generally has a relatively high sulfur content and greater pollution than other petroleum products making it less desirable in urban settings. Similar to the residential sector, the commercial sector uses relatively little petroleum and little prospect for improving future energy security through lower consumption.

19A.4.3 Industrial

While the residential and commercial sectors use petroleum primarily for heating the industrial sector also uses it as process

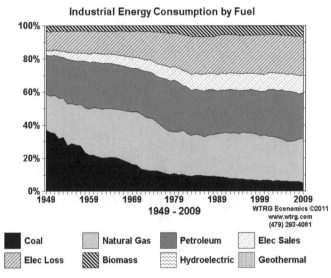

FIG. 19A.26   *Source* [4]

FIG. 19A.24   *Source* [4]

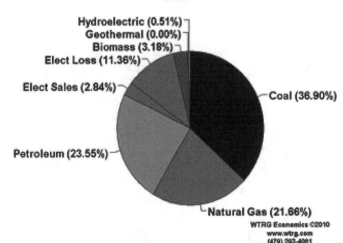

FIG. 19A.27   *Source* [4]

fuel and a feedstock to make products. The products are primarily what we refer to as plastics but also include products such as rubber, tar and asphalt. While some biofeedstocks can substitute for petroleum, we anticipate relatively little substitution and that petroleum will maintain something close to its current market share of 29% for the foreseeable future (see Fig. 19A.24 through 19A.30).

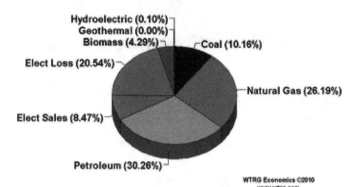

Industrial Energy Consumption by Fuel 1978

FIG. 19A.28   *Source* [4]

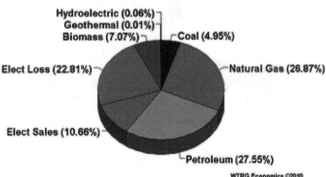

Industrial Energy Consumption by Fuel 2009

FIG. 19A.29   *Source* [4]

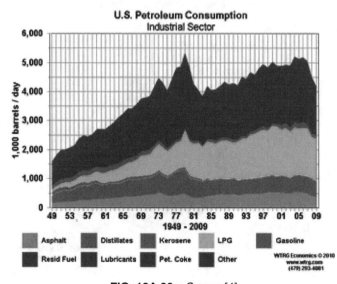

FIG. 19A.30   *Source* [4]

Liquid petroleum gases dominate petroleum products used by the industrial sector. Since liquids striped from natural gas well are a major source significant source of LPG this has less impact on petroleum dependence than is indicated by Fig. 19A.30.

19A.4.4   Electric Power Generation

Petroleum consumption for electric power generation peaked in 1978. It then plummeted from 1.747 million b/d to 478,000 in 1985. In addition to a recession, it is clear was a structural change between 1978 and 1985. Use fell dramatically between 2006 and 2009 to 175,000 b/d. Since 1949, the only fuel to see declines in the power sector has been petroleum. The advent of nuclear power slowed the rate of growth in the use of other fuels, but for decades, coal use grew at a slightly rater rate than total fuel use in this sector. In 1949, it was 43% and it now represents 48% of all fuel used on a Btu basis. Because increased concern about CO_2 and other emissions cleaner burning natural gas is making inroads into coal's market share (see Figs. 19A.31 through 19A.38).

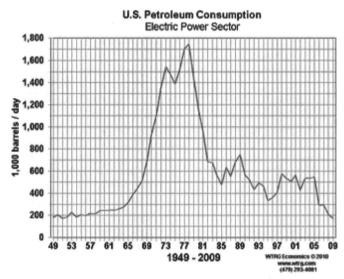

FIG. 19A.31   *Source* [4]

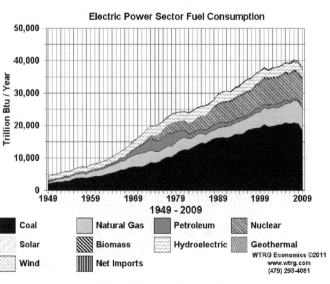

FIG. 19A.32   *Source* [4]

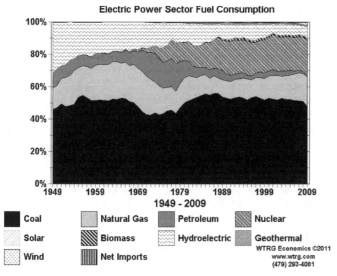

FIG. 19A.33 *Source* [4]

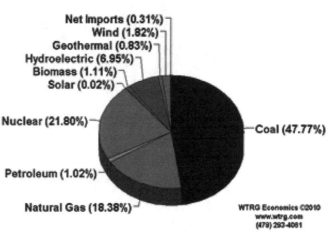

FIG. 19A.36 *Source* [4]

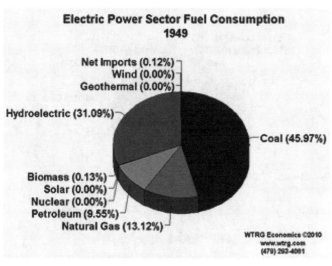

FIG. 19A.34 *Source* [4]

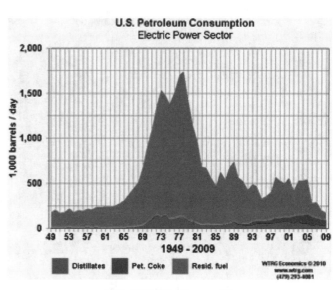

FIG. 19A.37 *Source* [4]

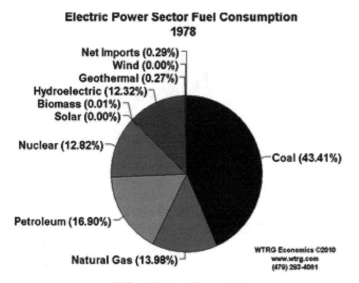

FIG. 19A.35 *Source* [4]

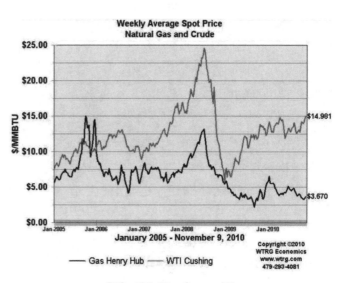

FIG. 19A.38 *Source* [3]

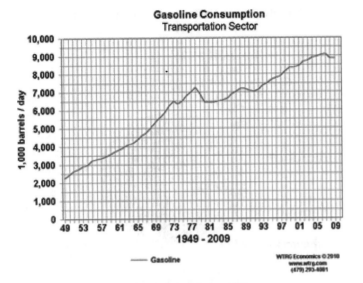

FIG. 19A.39 *Source* [4]

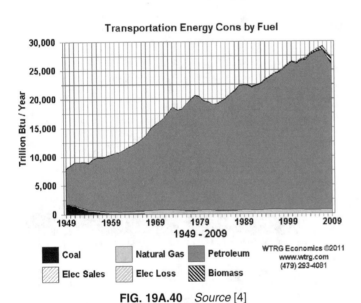

FIG. 19A.40 *Source* [4]

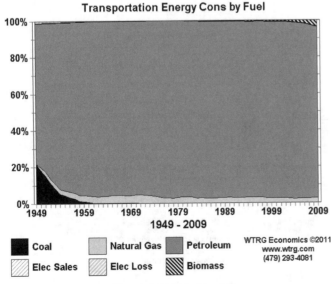

FIG. 19A.41 *Source* [4]

When we look at fuel use in the petroleum for power generation, we see the now expected pattern of a strong decline in the use of residual fuel since the late 1970s. While pollution is certainly one of the major factors price is another. Coal is less expensive but does not explain the most recent drop since 2005.

While some of the latest decline could be attributed to the recession, the primary reason is the recent disparity between the price of petroleum and the price of natural gas on a Btu basis. With the development of technology to produce natural gas from shale formations and the prolific production from these wells the linkage between oil and gas prices collapsed. Natural gas (Fig. 19A.38) is now far less expensive than petroleum based on energy content and absent a collapse in oil price will continue to be the fuel of choice for power generation in facilities that have the ability to switch between oil and natural gas. Natural gas and petroleum are used in peaking plants that can be quickly brought online to meet changes in demand. This is particularly important in the summer for air-conditioning. However, the use

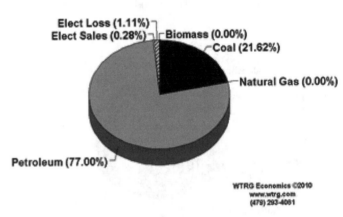

FIG. 19A.42 *Source* [4]

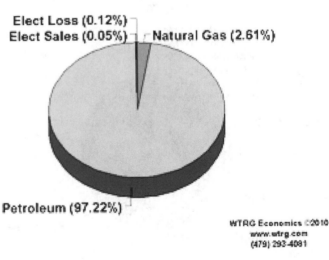

FIG. 19A.43 *Source* [4]

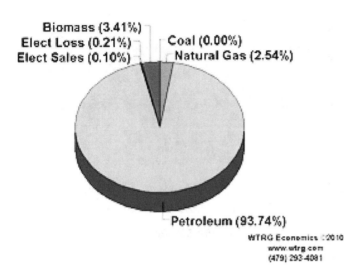

FIG. 19A.44   *Source* [4]

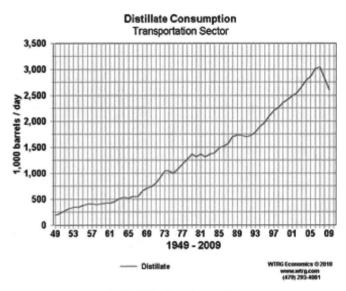

FIG. 19A.46   *Source* [4]

of natural gas as a base load fuel is on the rise (Figs. 19A.39 to 19A.44).

19A.4.5 Transportation

The consumption of petroleum for transportation has exhibited consistent growth through out the last six decades. In 1949, petroleum represented 77% of fuel use for transportation with almost all of the remainder coming from coal to power steam locomotives. By 1978, 97% of transportation fuels came from petroleum and 3% from natural gas. In 2009, petroleum was down slightly to 94% with natural gas and biofuels at 3% each. The recent increase in biofuels is due primarily to ethanol as a gasoline additive. It is used as an oxygenate to reduce pollution and is heavily subsidized.

Transportation shows consistent growth in petroleum consumption. It is the only sector besides industry to have significantly higher levels of consumption than 1949. Transportation is unique. Unlike other sectors that use fuel at a fixed location, vehicles whether they are automobiles, trucks, trains of airplanes, must carry their fuel with them. The volume of the fuel, its weight and the weight of its fuel tank are major considerations in vehicle efficiency. For this reason, unlike other sectors fuel substitution is difficult because you have to carry your Btus with you. While this does not preclude the use of natural gas the weight of containers to carry the gas under high pressure to achieve a similar range to gasoline or diesel is a major impediment (Figs. 19A.45 and 19A.46).

19A.5   NIMBY (NOT IN MY BACK YARD)

Perception and reality in assessing pollution impact of fuels are often quite different. Advertisements for electric powered vehicles often give the impression that they are nonpolluting. Clean electric vehicles are not "clean" and neither are electric furnaces or air conditioning. They appear to be clean at the point of consumption (my back yard), but at primary level of fuels to generate power (somebody else's back yard) they are not. It is important to note that 47.8 percent of the electricity generated in the U.S. in 2009 came from coal vs. 31.4 percent combined from nuclear (21.8%), hydroelectric (7.0%), wind (1.8%), geothermal (0.8%), and solar (0.02%). The remainder comes from natural gas (18.4%), biomass (1.1%) and petroleum (1.0%) (Figs. 19A.47 to 19A.49).

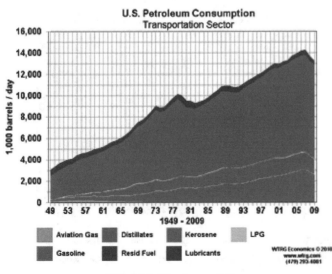

FIG. 19A.45   *Source* [4]

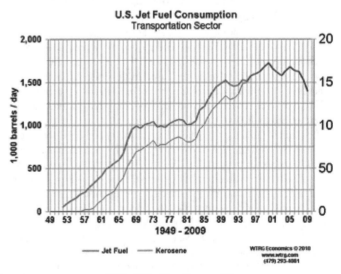

FIG. 19A.47   *Source* [4]

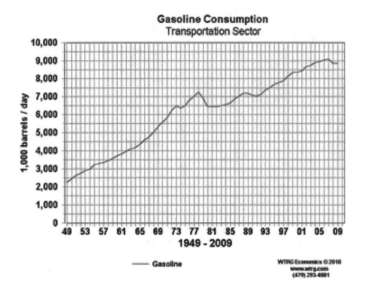

FIG. 19A.48   *Source* [4]

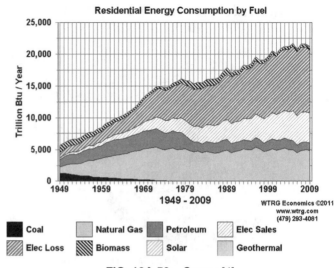

FIG. 19A.50   *Source* [4]

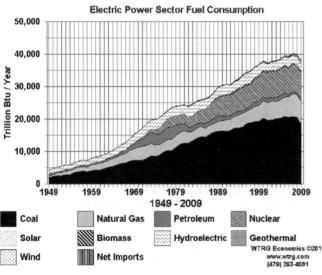

FIG. 19A.49   *Source* [4]

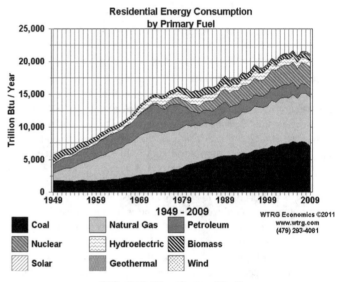

FIG. 19A.51   *Source* [3, 4]

19A.5.1 End Use or Consumer Energy versus Primary Energy Consumption

To illustrate the point we will look at the residential sector. Figures 19A.50 and 19A.52 show energy use by source at the point of consumption. Figures 19A.51 and 19A.53 replace the electric power use with the primary fuel used to generate the power. The contrast is stark. Instead of disappearing, use of coal increases 240% over our six-decade time horizon. Natural gas increases to 12 times the 1940 level.

Looking at the share of each fuel on a primary fuel basis shows that coal use, which was 32% of the energy used by the residential sector in 1949, is about the same in 2009 at 33%. Petroleum use declines from 22% to 6%, hydroelectric slips from 6% to 5%, natural gas grows from 21% to 37% and by 2009 nuclear power is the source of 15% of residential energy. Despite the recent interest in renewable energy, biofuels have fallen from 19% of residential energy consumption in 1949 to 3% in 2009. Renewable energy from hydroelectric, biofuels, solar, geothermal, and wind power has decline from 25% of total residential use in 1949 to 10% today. Over that period residential energy consumption grew 278% while renewable energy consumption only

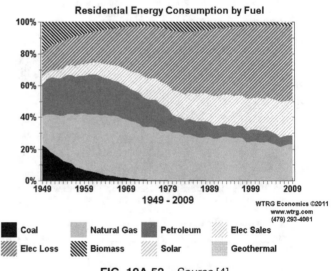

FIG. 19A.52   *Source* [4]

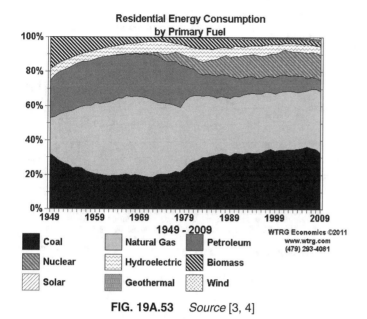

FIG. 19A.53   *Source* [3, 4]

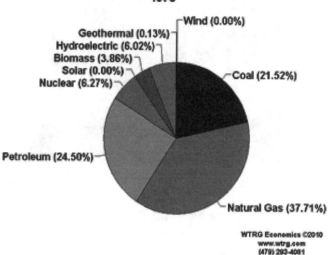

FIG. 19A.55   *Source* [4]

increased 51%. If nuclear is lumped in with renewables the market share at 25% is the same as it was in 1949 (Figs. 19A.54 to 19A.58).

For brevity, we will avoid primary fuel graphs for commercial and industrial consumption, which tell a story similar tends to the residential sector. So little electricity is used in the transportation sector (Fig. 19A.40) that use at the point of consumption is virtually the same as primary fuel consumption.

19A.5.2   Ultimate Fuel Use

If we role transportation in with other sectors we see that petroleum has been the fuel with the greatest growth over the last three generations. As it is the only fuel imported in significant quantity it must necessary be the focus on the consumption side of energy independence.

Section 19A.4.5 demonstrated that growth in petroleum consumption is primarily due to demand for transportation fuels. Substitution is difficult because unlike the other sectors transportation requires that consumers carry their energy with them. To substitute electricity or natural gas for petroleum effectively, it will be necessary to have an

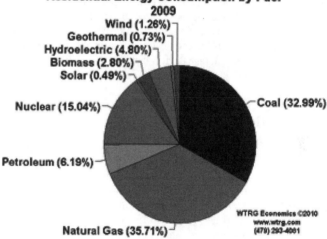

FIG. 19A.56   *Source* [4]

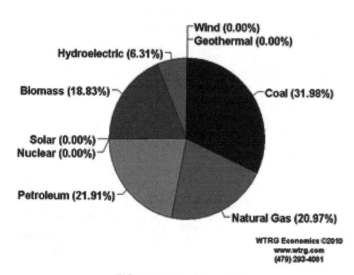

FIG. 19A.54   *Source* [4]

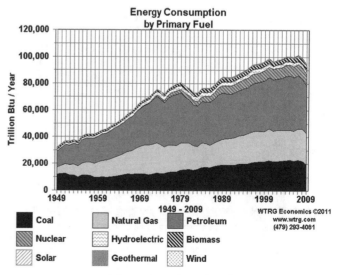

FIG. 19A.57   *Source* [4]

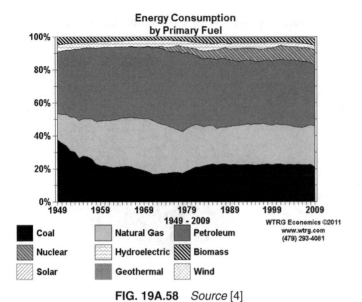

FIG. 19A.58 *Source* [4]

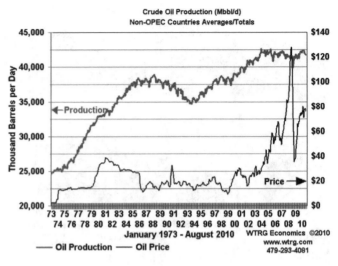

FIG. 19A.60 *Source* [3, 4]

energy storage device with similar capabilities to the gasoline tank in an automobile. It must store enough energy to allow a 300+ mile range, weigh no more than a full tank of gasoline and occupy a similar volume. Once an easily rechargeable battery can meet those conditions, any primary fuel can substitute of petroleum. While automobiles that are more efficient can help, significant changes in energy dependence will come from better storage devices that allow for fuel substitution.

19A.6 PRICE, PRODUCTION, AND GEOPOLITICS

Any discussion of the reaction of oil production to prices must be separated between OPEC and non-OPEC nations. Presented with a graph of world oil production and oil prices it would be difficult to explain the influence of prices on production. Rising prices in the late 1970s were met with lower production when higher production would be expected (Figs. 19A.59 to 19A.64).

If production is separated into OPEC and non-OPEC the picture becomes clearer. An examination of non-OPEC production shows there are long lags between a change in crude oil price and changes

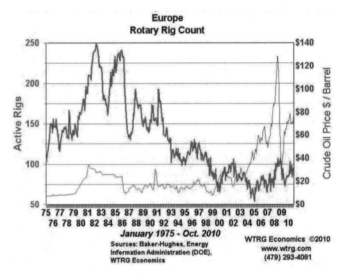

FIG. 19A.61 *Source* [3, 4, 8]

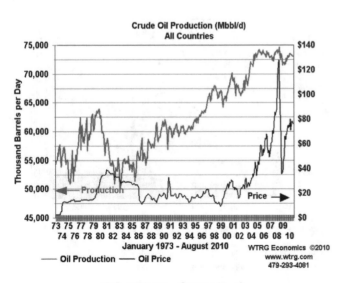

FIG. 19A.59 *Source* [3, 4]

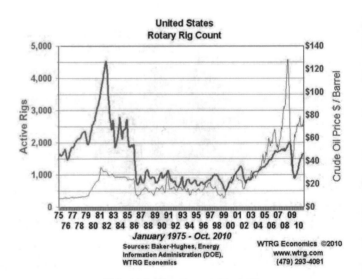

FIG. 19A.62 *Source* [3, 4, 8]

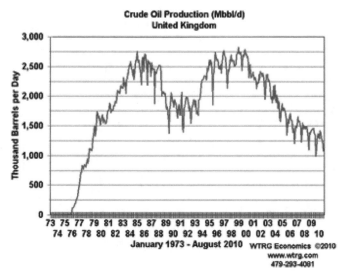

FIG. 19A.63 *Source* [3, 4, 8]

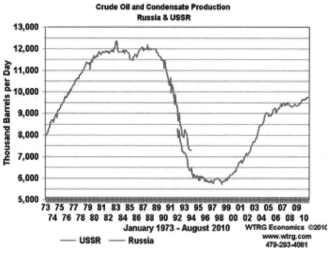

FIG. 19A.65 *Source* [3, 4, 8]

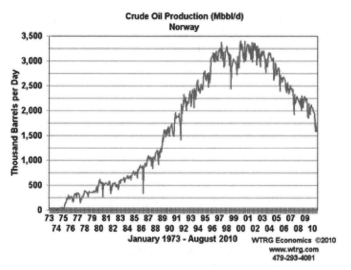

FIG. 19A.64 *Source* [3, 4, 8]

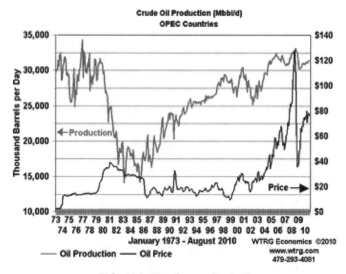

FIG. 19A.66 *Source* [3, 4, 8]

in production. A higher price tends to lead to higher production, but it comes with a long delay.

Oil prices surged in the late 1970s. In response, drilling activity in the U.S. and Europe skyrocketed to historic highs.

However, it was several years before the investment paid off in higher production. Drilling for oil is like most other capital investments. It requires a major initial investment and may take years to complete a drilling program. In new areas for production pipeline infrastructure must be put in place to bring the oil to market. If the reservoir is offshore, the costs are higher and infrastructure takes longer to build.

In general, non-OPEC production is not subject to much political risk. However, the breakup of the Soviet Union had a major impact on crude oil production. Production fell over 5 million barrels per day. At least half of the FSU decline was made up by higher production in other non-OPEC countries.

Most oil producing countries left to their own devices produce near capacity to maximize short-term revenue. If all countries behaved this way, oil production would be higher and prices lower. Outside OPEC, production is almost never shut-in with low prices.

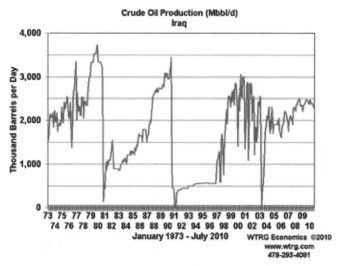

FIG. 19A.67 *Source* [4]

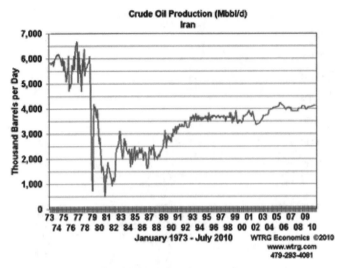

FIG. 19A.68 *Source* [4]

19A.6.1 OPEC

OPEC formed 50 years ago to represent the interests of exporting nations. Member countries share the common characteristic that they are net exporters of crude oil and those exports are a significant portion of their international trade. While dependable sources of oil and oil prices are of great importance to importing counties like the U.S. and China, on a relative scale, they are far less important than reliable customers, access to shipping lanes or pipelines, and prices are to exporters. For example, the latest CIA World Factbook shows Venezuela dependent on petroleum for 90% of exports, 50% of government revenue and 30% of GDP. For Saudi Arabia oil was 90% of exports, 80% of the government budget and 45% of GDP. For either country, a 20% change in oil price is the difference between a growth economy and recession (Figs. 19A.65 to 19A.70).

19A.6.2 Price Relation Not as Clear with OPEC

OPEC production changes do not follow same pattern as non-OPEC production. In time of low oil prices OPEC will often cut back on production to shore up prices, but unlike most non-OPEC producers, OPEC attempts to maintain a cushion of spare capacity to guarantee supply in the case of a supply interruption or greater than expected demand for crude oil. OPEC also increases production if they deem the price too high.

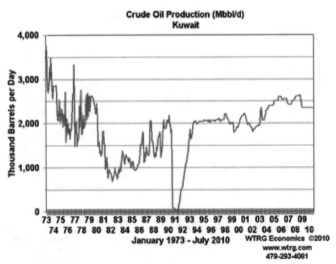

FIG. 19A.69 *Source* [4]

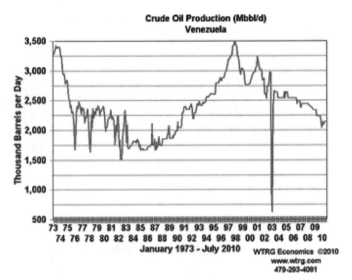

FIG. 19A.70 *Source* [4]

Many of the major changes in OPEC production and international oil prices were not the result of deliberation within the organization. Historically, OPEC production has been heavily influenced by wars & revolutions. The following is an abbreviated list:

- 1973 Embargo (not just Arab, Persian)
- Iranian Revolution
- Iraq-Iran War
- Gulf Wars I & II
- PDVSA strike in Venezuela
- Attacks by revolutionary groups in Nigeria

Each of the events above led to higher oil prices. Often other members of OPEC made up for supply interruptions.

A more detailed discussion is available on the "Oil Price History" page at www.wtg.com.

19A.7   SUMMARY OF PETROLEUM DEPENDENCE

A nation that depends on imports for a significant portion of its consumption is dependent on petroleum. An international shortage of oil, which causes a price spike, can cause as much damage to an oil dependent economy as an interruption in imports. Dependence can be lowered by either increasing domestic production or decreasing domestic consumption of petroleum. In countries with non-petroleum energy resources, other primary sources of energy can be substituted over time. Substitution is most difficult in the transportation sector because the high Btu content of petroleum relative to its weight and volume makes it a relatively efficient fuel for the purpose. Significant substitution will occur when batteries or onboard natural gas storage comes close to matching the energy per pound and energy per cubic foot of petroleum fuels.

PART B

19B.1   THE ECONOMICS OF BIOENERGY

The economics of bioenergy is much the same as the energy economics of fossil fuels. The frontier nature of the fledgling bioenergy industry adds enormous complexity. Biomass energy feedstocks can be wet, with different levels of moisture, dry, little to no water, or liquid without any water. Fuels can be produced specifically for energy, or they can be a byproduct residual from the production of a higher-valued product. Waste-derived fuels bring with them many layers of environmental regulation, further adding to the complexity.

Conversion technologies are nearly as variable from thermal gasification and pyrolysis technologies to biological anaerobic digestion technologies. Access to oxygen changes the emissions and conversion efficiency for all systems. Technology innovations are also occurring in biomass production, harvest, transportation, and storage. Densification technologies, such as baling and pellet mills, can have a large economic impact on the storage and handling of biomass.

The bottom line is that the *supply* of bioenergy inputs and outputs must be balanced with the everyday *demand* for energy, food, fiber, building materials, and all the other products and services that rely on plant-based materials to succeed. The lack of market infrastructure for bioenergy projects means that input prices, technology efficiency, output yields and prices must be created for each project. This all adds considerable costs.

Like fossil fuel supplies, biomass feedstock quantities can be increased, but ultimately are limited by what is accessible in the short-run. Bioenergy supplies and reserves will increase as we learn more. Biomass fuels have emissions also. Some pollutants of concern are a greater liability than fossil fuels, although many are lower.

This chapter is written to provide the most objective perspective possible. Again, like any industry, and certainly like any of the fossil fuel industries, the foundation for economic success is rooted in technology. That is why this manual focuses almost completely on technology. The basic outline of this part is to provide a broad overview of infrastructure, technology, and policy before outlining bioenergy demand factors, supply factors, and policy influences in bioenergy.

19B.2   CURRENT US BIOMASS UTILIZATION IS POLITICAL

Bioenergy and biomass today are mired down in the politics of food, energy, forestry, solid waste remediation, air quality, water quality, and climate policy. A convergence of rising oil prices, demand for ethanol to replace the fuel oxygenate, heighten awareness of climate issues, and global conflict placed biomass energy and bio-based products firmly in the political arena.

In the political vernacular of bioenergy, *biomass* has the legal qualifier that it composed of is carbon of recent origin as opposed to prehistoric fossil carbon. Fossil coal and crude oil can be described as ancient biomass. Ecologically speaking biomass is anything made from plant material. Until the politics of climate change and greenhouse gas emissions imposed time constraints on biomass, fossil fuels like coal, crude oil, and natural gas were generally considered ancient biomass. The current political vernacular biomass now implies "recently created," therefore ancient biomass (fossil carbon) is not included as biomass in this discussion.

The basic elements of various biomass materials are: carbon (C), hydrogen (H), oxygen (O), nitrogen (N), sulfur (S), chlorine (Cl), and ash. These are the fundamental building blocks of life: carbohydrates, fats, oils, water, proteins, oxygen and carbon dioxide. They are also the components of environmental pollutants like greenhouse gases (carbon dioxide, methane gas, and nitrous oxide), VOCs (Volatile Organic Compounds), ammonia gas, as well as treated organic wastes in wastewater and municipal solid waste (MSW). The basic chemical footprint of biomass is increasingly held hostage by the policies that have evolved over many decades.

Since 2005, billions of public and private money have been invested in the development of bioenergy projects. Much of this in-

vestment in a new bioeconomic infrastructure has been halted by the downturn in the US economy.

Sadly, competition for scarce biomass resources has turned industry counterparts solidly against each other. Many livestock farmers and industry groups have become anti-ethanol because demand for corn has been strained with the expansion of the corn ethanol industry. Similar battles are brewing over forest residuals. Should the natural forest debris be kept in the forest creating a threat to fire or removed and used for fuel? The biggest and meanest political battles are occurring across conflicting regulatory boundaries. Is waste paper fodder for recycled paper, a source of fuel, or a regulated waste?

This is the political chaos in which this narrative on bioenergy begins. Navigation of this political minefield is not as difficult as it appears on the surface. There are no easy steps to follow to economic success. There are no assurances that the policy makers will get it together while there is still time. However, if one stays rooted in technology and efficiency, the economics of bioenergy follow closely behind.

In the Energy Information Administration (EIA) analysis of US energy consumption in 2009, renewable energy accounted for over 8 percent of energy consumption [20]. Half of this renewable energy consumption in the US in 2009 was derived from biomass fuels. This includes 10.7 billion gallons of corn-based ethanol, 0.7 billion gallons biodiesel, 1,800 megawatts (MW) generation capacity from landfill gas power plants and several million tons of pellet fuel production.

The 2008 recession caused transportation fuel consumption to plummet, but expansion in ethanol, wind and solar fuels have also increased. EIA also reported that during the years 2008 and 2009, carbon dioxide (CO_2) emissions declined by 10 percent (Fig. 19B.1) [9]. Most of the decline in CO_2 emissions is due to the decline in liquid petroleum products, but other reductions came from fuel switching away from costly coal to natural gas and increased utilization of existing nuclear power capacity to save money on the increased cost of coal. On much smaller levels the increases in renewable fuels, ethanol, biogas, wind, and solar, have played a role.

The interrelationship between the economy, technology and policy is complex. As the economy was sputtering in 2008 and into 2009, it was a race to identify the fundamental cause of the economic chaos. There were plenty of explanations of what was happening, but there were as many different explanations as there were economists to be asked. In March of 2009, the USDA, Economic Research Service (ERS) identified 17 factors that have influenced the long and short run supply and demand for commodities [10]. The analysis compares underlying factors of economic shocks in three time periods: 1970s, 1990s, and 2006 to 2008.

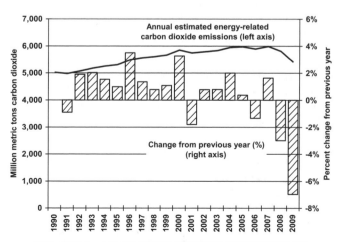

FIG. 19B.1   US CARBON DIOXIDE EMISSIONS [9]

The 17 economic factors are laid out in Table 19B.1. A very instructive part of the presentation is that supply and demand effects are identified individually. The decisions that are made in an economy depend on who is buying (demand) and who is selling (supply). Economic supply decisions are different than economic demand decisions. It is highlighted here because too often prices and quantities get discussed as they only have one implication.

The US economy is not driven by a single factor. In this case, a serious economic shock was caused by 17 factors. We tend to look for binary, yes/no, 0/1, solutions to any of our economic and political crises. The reality is nearly always a heterogeneous, interdependent cluster of many issues. Leaders and law makers tend to do a knee-jerk, rapid response when often times that response only aggravates the problem. This is brought up here because this is much like bioenergy economics. There are multiple long and short-run supply and demand effects that ultimately determine if a bioenergy industry or project will enjoy economic success or failure.

Even on the surface, the influence of policy on economics becomes clear. Rapidly increasing prices, particularly in agriculture and energy markets, were the precursor to this significant contraction in economic activity during the last 2 years. Success in understanding the complex biomass economics requires a fundamental understanding of the interrelationship of technology, policy and economics, Fig. 19B.2 [11].

Policies that influence economic behavior and choice will be demand-related policies. Policies that influence science and technology will be supply-related policies. When the laws and policies serve as a guide for human choice without becoming irrelevant, they operate in the intersection of the three regions identified with a

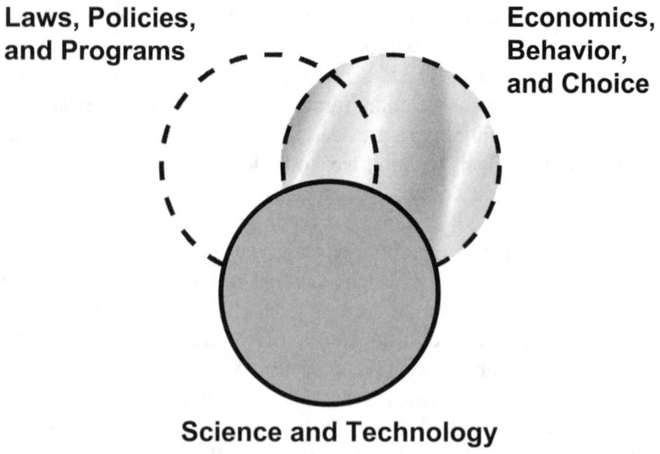

FIG. 19B.2   RELATIONSHIP OF POLICY, ECONOMICS, AND TECHNOLOGY [11]

star. The most compelling region however is the intersection of the human choice and technology — without the influence of policy.

While there are many factors motivating the adoption of bioenergy like energy independence and local economic development, a principal driver is greenhouse gas (GHG) emission control. This is driven almost entirely by the definition of carbon dioxide (CO_2) as the standard of measure in greenhouse gas (GHG) degradation. With the Kyoto Treaty (1997) [12], CO_2 switched from being a quality-of-life enhancer, to that of a quality-of-life destroyer. As this is chapter is being written the US public and private sectors simultaneously love carbon and hate carbon. The subsequent morass of public policies creates a harsh environment with which to commercially develop biomass energy and products.

19B.3   TECHNICAL INNOVATION DRIVES ECONOMIC GROWTH

In an economy without political and administrative costs, economic growth is driven by technical change and efficiency. This is important because it is technical change and innovation that drives economic growth. The laws and policies can speed technical change up or slow it down, but without technical change there is no economic growth. Successful industries are able to operate with regulatory overheads that are much lower than the economic returns of doing business. Technical innovation can offset increased regulatory administrative costs.

One of the great examples of increasing technical and economic efficiency in agriculture is the dairy industry. In the last 60 years, pounds of milk produced per cow each year has grown from an average of about 5,000 pounds/cow/year in 1949 to over 20,000 pounds/cow/year in 2009. Over the same time period, the number of cows in the US dairy herd has dropped from 24,000,000 in 1949 to less than 10,000,000 in 2009. Remarkably, total milk produced has also increased by 50 percent. This means that sixty percent of the 1949 dairy herd is now producing 50 percent more milk than the 1949 dairy herd. To get that amount of production in 1949 it would have required 3.5 times more cows than were being milked in 2009. That is some significant technical innovation.

These increases in productivity are fairly constant across all of US agriculture. USDA, Economic Research Service (ERS) keeps track of changes in total factor productivity for the industry (Fig. 19B.3) [13]. Total factor productivity (TFP) is essentially a

TABLE 19B.1   FACTORS AFFECTING THE US ECONOMY DURING ECONOMIC SHOCKS [10]

Contributing factor	1970s	1990s	2006–2008
Long run			
Demand			
Export demand growth	X	X	X
Due to food demand growth		X	X
Due to population growth			X
New use/innovation: biofuels			X
Supply			
Slow production growth	X	X	X
Declining R&D investment		X	X
Land retirement	X	X	
Short run			
Demand			
Government food policies	X	X	X
Supply			
Government food policies	X	X	X
Weather-induced crop losses/failure	X	X	X
Macroeconomic			
Economic growth		X	X
Depreciation of U.S. dollar	X	X	X
Rising oil prices	X		X
Accumulation of petrodollars/foreign reserves	X		X
Futures market/speculation	X		X
Inflation	X		
Financial crisis		X	X

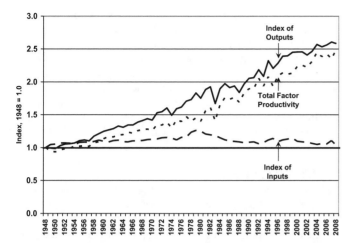

FIG. 19B.3 INCREASES IN THE TOTAL FACTOR PRODUCTIVITY OF US AGRICULTURE, 1948 TO 2008 [13]

measure of the output from all the inputs. By indexing annual values, ERS tracks changes in productivity. In Fig. 19B.3, the inputs are indexed relative to 1948 being equivalent to 1.0. Since 1948, TFP in US agriculture has increased 2.5 times. These gains in efficiency imply fewer residuals (wastes).

Gains in agricultural efficiency and productivity can be extended to reduced environmental stress and even reduced energy consumption. In 2009, the Keystone Center put out a report on environmental indicators for principle crops in US agriculture [14]. In this Field to Market report, improvements in US soybean production from 1987 to 2007 were shown to decrease energy use, soil loss, climate impact based on sustainability indicators. Land-use and irrigation water use also increased in efficiency.

As commercial production systems expand beyond single processes, multiple, integrated technologies are better able to utilize all the inputs and byproducts more completely. When the proper combination of technologies is aligned, it is possible to move to a production system with negligible or zero wastes. This is particularly true with agricultural, forestry, and biomass energy systems. Leftovers must simply be processed to enter the market with as high a value as possible.

With increasing popularity corporate America is establishing departments of sustainability. These new units are charged with conserving power, fuel, and water. They represent market-based efficiency gains. Just like the examples above with agriculture, conservation of energy and resources is a win for the environment, energy, and economics. Sadly, even when environmental agencies have units that focus on conservation and value-added programs, nearly always they are subjugated to more politically popular punitive compliance programs.

19B.4 BUILDING A BIOENERGY INFRASTRUCTURE

The thermodynamic efficiency of energy utilization is not enough to provide economic success. It also takes a market demand, which may not be driven by mass-balance energy utilization.

Biomass energy does not have well-developed supply systems and markets. Plan A in the biofuels commercial experiment was to develop a fuel product that dovetailed seamlessly into the existing liquid transportation fuel sector. The corn-based ethanol is the result of that commercial experiment.

The US ethanol industry really took off after about a third of the states legislatively banned, or phased-out, methyl tertiary butyl

ether (MTBE) from fuel about 7 years ago [15]. It was replaced relatively quickly with ethanol. This created a very large demand for ethanol and spawned the construction of ethanol plants across the country. Then that ethanol like "gold-rush" came to an abrupt halt with the oxygenate demand having been met, just as the US economy went into a recession.

The economic roller-coaster of the US biofuels industry expansion has left the nation a bit confused about whether the homegrown ethanol industry has been a worthwhile investment or not. It is important to note that the addition of MTBE was to reduce fuel emissions. That was a healthy step forward. The replacement of MTBE with ethanol was to further reduce water quality issues from underground gasoline storage tanks. That is an environmental benefit that we still enjoy. Significant environmental progress has been made even as the economy wrestles with continued investment in biofuels.

In addition to having little infrastructure in place for primary bioenergy markets, there is an additional challenge in adapting significant existing biomass processing infrastructure. Management of liquid and solid wastes treat undervalued carbon as an economic liability. There are actually multiple infrastructure issues here.

- A new infrastructure must be developed for new bioenergy products.
- The existing waste treatment infrastructure must be transformed to maximize resource value rather than minimize liabilities.
- Biomass materials are light and bulky, so local cost reduction favors locally fueled projects. The central processing/distribution networks of existing fossil fuels will be supplemented/offset by distributed generation of biomass feedstocks.

A quick look at the 80-year history of US soybean production provides a nice illustration on the impact of a viable, mature economic infrastructure (Fig. 19B.4) [16]. Back in the 1920s and 1930s, soybeans were grown as a forage crop for cattle. It has taken a full 80 years to reach the production levels of over 3 billion bushels. It was not a management decision, but the development of production technologies, new processing technologies, new uses and markets, and the development of ocean going vessels that could move large bulky protein products around the globe economically. It has even taken a few wars and famines to have developed the current levels of US soybean production.

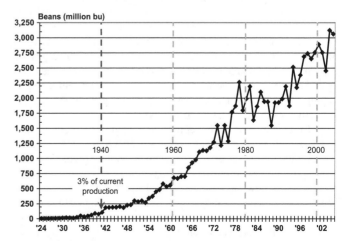

FIG. 19B.4 EIGHTY YEARS OF US SOYBEAN PRODUCTION [16]

In many ways this is not unlike the development of the US biofuels industry. At the 1940 levels of production, soybeans were grown at only 3 percent of 2005 levels. Based on the data in this chart, 1940 was about 16 years into the development of the mature soybean marketing infrastructure.

On the other hand, developing an economic infrastructure on the reuse of residuals requires a realization that carbon emissions are biomass resources that are in the wrong place. We should all be working together to pull leftover carbon (emissions) back into the economy as more biomass for energy, compost, organic nutrients, and biobased fibers. The goal is a higher value of life using fewer resources. By intensively managing our resources there are fewer emissions. The result (not the goal) is zero waste.

Adding value to waste organics is about getting more from less. Solid waste recycling and composting pulls solid waste "emissions" back into the economy. Now the same thing needs to be done with greenhouse gas emissions. Our new carbon policies must make recycling emissions easy and relatively costless.

Unfortunately, that is not the direction we are going in at the moment. The Energy Independence and Security Act of 2007 (EISA) is the enabling statute for the revised Renewable Fuel Standards (RFS2) regulations. EISA was legislatively written to create an economic demand for biofuels. The motivation was to increase the supply and demand for a locally grown, cleaner liquid fuel. Between 2007 when the statute was written by the legislators and 2010, EPA switched the priority from increasing demand to restricting greenhouse gas emissions.

As promulgated in 2010, the rule makes it difficult for biofuels to qualify for the RFS economic benefits. While EPA can legally do this, proceeding as the agency has proposed will halt economic growth in biofuels production. The additional emission criteria EPA has added to biofuels make other non-RFS uses like making electricity from biomass more compelling.

Substantive and sustainable change is achievable, but the risks are real. Policies can be written that promote enhanced quality of life and more efficient use of resources. The biomass industries need to guide the policies.

19B.4.1 Intergenerational Costs of Increasing Government Administrative Costs

A newer infrastructure based on small, locally grown biomass and renewable energy systems will cost trillions of dollars [35]. In addition to these large physical system capital costs, we must add in interest on the trillions of dollars of federal stimulus money that must eventually be repaid across multiple generations.

If interest on this new production and distribution network is 7 percent, and the cost of living is 3 percent, then we effectively pay 10 percent interest on the new infrastructure until the trillions of dollars of borrowed stimulus capital are paid off. This can work if the new bioeconomic infrastructure is returning more. If the average of all renewable energy projects is a 20 percent return before paying the debt and half of that (10 percent) goes to service the collective debt, the economy still grows.

Unfortunately neither the legislative (Congress) nor administrative (agency) lawmakers are very good at writing costless laws. Policy collisions between poorly crafted oversight regulations and funding programs are costly to industry, government and consumers. For years now the USDA and the Department of Energy have been funding biomass energy projects and research in which industry and government have invested billions of dollars. When the biomass utilization/carbon emission laws are written to exclude those investments, the false-start dollars get added to the debt without any hope of a return.

Arbitrary economic friction threatens the success of this risky, bold approach. For example, if the policy implementation cost increases the debt servicing fees described above by 5 percent, then the new bioeconomy must pay an effective debt service cost of 15 percent before we have economic growth. If the policy implementation cost increases the effective interest by 10 percent, the breakeven economic bar now requires more than a 20 percent return. In this last example, the additional implementation costs have consumed all economic growth — which is not economically sustainable.

19B.4.2 Uncertainty and Prices

The US economy in general works like it does because there is a lot of historical information on production, purchases, and prices [17]. If something changes quickly, then all the analysts get nervous and there are "adjustments" to compensate. In the bioproduction ecosystem, there is insufficient historical data. This makes investing much more risky. Investors do not like to lose money so when they loan money for an anaerobic digester or a gasifier, they charge higher interest rates to cover additional risks. Startup costs are higher for new technologies, because it is difficult to estimate budgets without years of experience. Also data collection is expensive and sometimes data on test burns or on the feedstock quality will need to be measured for the first time.

19B.4.3 All Prices are Relative

This is always true in economics. It is even more apparent in emerging market structures that do not have robust prices [17]. Most biomass energy economic analyses have been conducted by modeling stand-alone enterprises. The economics tend to reflect the retrofitting of an existing facility with new equipment, using product prices based on a similar existing market. These economic studies are excellent first steps into an industry that barely exists.

All studies are most relevant the minute they are complete. The more time that passes for most price-based economic studies, the less relevant they become. The anaerobic digestion studies done 20 years ago have little economic value today. The livestock we are producing today are more efficient at converting feed to meat than they have ever been and we produce less manure per animal than we did 20 years ago. And in the case of anaerobic digesters, the newer digesters being installed operate more consistently than the earlier digesters.

In addition, biomass energy studies conducted on an energy price assumption tied to $40 a barrel crude oil, will have a significantly different result when crude oil sells for $130 a barrel.

All biomass energy markets must be contractually established or used internally at the production site. There are no terminal markets (grain elevators or stockyards) for a producer to deliver a load of biomass energy to sell. This is a profoundly different marketing strategy than existed for grain and livestock 50 years ago. If a facility intends to develop a biomass energy or alternative market, it must first find a buyer/client — and sign a contract.

19B.4.4 Data Collection

On the bioenergy frontier, the risks of failure can be capitalized into a proven, turn-key system or each valuable lesson-learned can be pieced together from scratch. Data collection is very expensive, but one way to lower the risks of failure is to collect the best data available on the frontend. If a project fails for lack of good preliminary data, then the cost of not collecting that real-time data becomes very high [17].

Kinds of data that need to be collected fall into three basic categories: input quality, technology emissions, and product quality (relative to the intended market). Technology providers that offer turn-key conversion technology systems may charge more for their services than it appear to merit, but they are trying to recapture their costs of collecting all the data to make their technology work. It is possible to independently develop a commercial biomass conversion technology from scratch, but then the independent developers must collect the input quality, emissions and product quality data themselves.

One caution about using historically tabled data. It is possible to find data tables on manure and biomass quality and energy content. These are excellent starting points for preliminary planning. They are used extensively throughout this document. However, once money begins to be invested in any project, it is time to begin collecting project-specific data. Unfortunately, there are millions of dollars that get spent on investments that are made on the average data that has been created for planning, and too often the site-specific inputs do not have the same quality characteristics as the average tabled data. When the project advances to the point where money is being spent on the project, it is time to collect project-specific data.

19B.4.5   Waste Treatment versus Biomass Processing

The evolution of the valuable environmental benefits and the increasing value of producing energy have created some confusion about whether it is better to save the environment, or have economic growth. The right answer is both, and that is possible with well planned bioenergy enterprises. Just as the specialized, ethanol plants are learning the benefits of diversification, businesses built on making a profit are learning the economic value of environmental remediation. Many waste management businesses are watching the economic excitement associated with biofuels and are adjusting their cost-minimizing business models to operate more like profit centers [18].

Cost recovery is not the same as a profit center. Waste treatment functions minimize liabilities and costs. The highest value of a cost-minimizing project is zero cost. Zero costs (without profits) do not excite investors and bankers. Even when costs are zero, a profit only occurs if there is also revenue (profit = revenue costs).

Waste remediation projects can generate a profit, but that profit is based on remediation services rather than commodity production. Biomass methane energy projects (landfills, wastewater treatment digesters and manure digesters) have evolved as energy (commodity) enterprises with in a waste treatment facility. The facility's primary function is to remediate the waste and stabilize the carbon. Then they attempt to recover some costs by developing methane gas utilization. The first goal is to remediate waste and the cost recovery benefit from energy is an afterthought.

Waste treatment industries have a legal, health and environmental mandates that impose enormous costs. They are offsetting some of those expenses through bioenergy production. Bioenergy profit centers in turn are learning to more efficiently utilize their resources similar to the cost recovery management of waste treatment facilities. The two different business models are merging into a single, very efficient low-cost profit center.

As waste remediation moves toward energy production, a second benefit emerges. Waste remediation shifts from a service-oriented industrial sector to a product-producing industrial sector. Economic wealth is generated from what was initially only a service industry.

19B.4.6   The Price Impact of Recycling

The concept of reuse and recycling enhances efficiency, but it is not without its impacts [17]. In the case of paper recycling, increased efficiencies gained by salvaging used paper created a greater supply of paper from the original supply of pulp wood. The excess supplies of waste biomass feedstocks exist because these waste materials are under developed and so they have a negative value. As the supply of paper increased from recycling, the price also dropped.

Inputs will have more competition. The feedstock price will move upward as competition for undervalued feedstocks increases. Corn for ethanol, for instance became more expensive when corn for animal feed became less available. As more industries compete for the same materials they will put pressure on keeping their own input costs down. As bioenergy, chemical and fiber products become more reliant on biomass feedstocks, each firm will compete for the least cost input into the process, driving biomass feedstock prices down. Even though there are contradictory forces on feedstock prices, this is part of the challenge of establishing a bioeconomy.

19B.5   BIOENERGY DEMAND/ CONSUMPTION

The demand component of the supply-demand economic relationship focuses on markets and prices. Demand is about "How much are consumers willing to pay for this?" And, "Where can this product be sold?" Demand can be thought of as managing markets and prices.

There are two important things to understand about demand. First, demand choices are different than supply decisions. Second, it is important to recognize that there are specific factors that influence demand choices [19]. These are: increases in income, a change in preferences, changes in the prices of related goods, number of buyers, and expectations of future price. These factors become most important in the policy discussion, since policy seeks to influence either supply or demand decision choices.

Economics is ultimately about balancing the supply with demand. In the emerging bio-production systems, pollutants that are in surplus quantities are being aligned with legitimate energy needs. The beauty is that the environment is enhanced, alternative energy forms are developed and the economy grows. While that is the vision, reality requires a few more steps.

The Department of Energy reports that the consumption of energy in 2009 has been growing. The US energy consumption has been hovering at about 100 quadrillion btu since 2003. In 2009, that level of energy consumption fell a healthy 5 percent to 94.8 quadrillion btu [20]. US energy consumption has not been this low since 1997. Fossil fuel consumption fell 5 percent to 78.6 quadrillion btu. Most of this reduction in energy consumption has been driven by the harsh economic reality that the recession created for energy consumers. As illustrated in Fig. 19B.1 though, GHG emissions have dropped by 10%, even without explicit CO_2-reduction policies.

Renewable energy consumption increased to 7.75 quadrillion btu in 2009. This is 8.2 percent of US energy consumption. While renewable energy consumption that is less than 10 percent this is not trivial. Written out in long hand, 7.75 quadrillion btu is 7,745,000,000,000,000, or 7.745×10^{15}.

Half of the renewable energy (3.884 quadrillion btu) came from biomass energy sources. Wood and derived fuels made up 24 percent of renewable energy, which the EIA defines this as wood and wood pellet fuels. Sometimes this EIA category also includes black liquor and other wood waste liquids.

Biofuels (ethanol and biodiesel) made up 20 percent of the renewable energy consumed last year. This is derived from 10.7 billion gallons of ethanol 0.7 billion gallons of biodiesel in that were produced in 2009. This 11.4 billion gallons of biofuels accounted for over 4 percent of the combined US petroleum and biofuels consumed. That small, but growing share of biofuels in US energy consumption represents a home grown fuel with environmental benefits.

The remaining 6 percent of bio-based renewable energy, 0.447 quadrillion btu, comes from biogenic MSW, landfill gas projects, and digester projects. The only category that has been increasing the last few years has been energy production from landfill gas projects. This is primarily because digester projects must include the high cost of biomass liquid storage containers, while landfill biomass container costs are included in the landfill operation costs.

Other non-bio renewable energy consumption sources include hydroelectric (35 percent), wind (9 percent), geothermal (5 percent), and solar (1 percent). Hydroelectric has remained fairly constant, but hydroelectric energy availability has more to do with annual droughts and floods than with other energy policies. Wind continues to grow rapidly and solar energy consumption has grown at a slower rate.

The other non-renewable energy sources consumed in 2009, include petroleum (37 percent) and coal (21 percent), both of which declined. Nuclear (9 percent) and natural gas (25 percent) consumption remained about the same. The shifts in nuclear and natural gas consumption resulted in lower CO_2 emissions because these fuels emit less CO_2 per btu than petroleum and coal. These shifts have been due more to economic concerns than an interest in emission reductions. Only time will tell if it is a temporary adjustment or one that is longer-term.

19B.5.1 The Price of Carbon, or Carbon in Demand — $/lb

The big-money question in bioenergy development is, "What is the price of carbon?" This is asked in reference to the cost of

biomass feedstocks and also about the market price of the final product. It is part of the same question, "Can bioenergy production and use be profitable?" All biomass is chemically composed of carbon. Green plants take in CO_2 and respire oxygen, O_2, through photosynthesis. The life-giving benefits of living biomass are that it constantly replenishes our supply of breathable O_2 and stores solar energy biochemically in carbon-based plant products.

A common emerging carbon policy theme is to establish a price of carbon through an emissions market. The underlying principle is based on capping an upper level of emissions and then establishing an emission trading market based on emission credits and debits. This has worked fairly well for some criteria air pollutants and has also been explored for water quality permit trading.

19B.5.1.1 Positive Carbon Value
Carbon is already valued in many markets. Foods and feeds that are bought for sugar, starches, fiber, and oils are being purchased for a value-added carbon market. Fuels that contain hydrocarbons are purchased for a carbon market. The implicit value of our carbon-based, food and energy markets already represent coarse approximations of the value of carbon. These food and energy carbon prices are also traded on the basis of their inherent positive value rather than on the cost of mitigating a negative liability (emission market).

An easy example of a common commodity that is traded on its carbon value is corn grain or shelled corn. Corn grain has a primary market in the US for livestock feed. While there are multiple components within corn that make it an excellent ingredient in livestock feed, the main component is the energy in starch. Starch is a basic polysaccharide carbohydrate $(C_6H_{10}O_5)_n$ [21].

Figure 19B.5 contains general food, fuel, and livestock feed "demand-side" prices. These prices represent prices that consumers of these goods actually pay for these products. These values should be considered as illustrative rather than definitive. The prices in dollars per ton are presented on a log scale because the price differences between different markets differ by orders of magnitude. The conventional food prices include meat from $2 to $5 per pound, butter, cheese, and even cold breakfast cereal at $3 per pound (as-is basis).

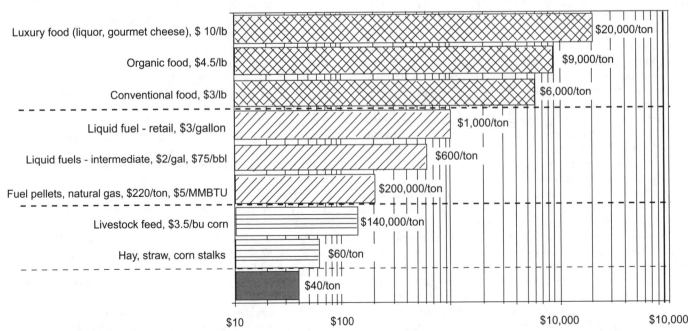

FIG. 19B.5 COMMON CARBON PRODUCTS AND THEIR VALUE IN DOLLARS PER TON ($/TON) [22]

Each horizontal bar in Fig. 19B.5 represents a different class of products. Each of these classes can be divided into dozens of more narrowly defined products. These classes and categories represent separate economic demand functions. There is no single demand function for biomass, there are literally thousands demand functions for products that are derived from plant products. The commodities listed in Fig. 19B.1 are not bought and sold on carbon content, but on their inherent end use value which is a function of carbon content. The higher valued products are generally not sold in the units of $/ton either, but doing so at least puts the prices on the same unit.

Water is a principle component, particularly for the food and feed products. Monetizing the price of the product as carbon is very conservative. If carbon represents only 5 percent of the actual product for instance, the value of that product in terms of carbon alone would increase 20 fold. In a similar way, the greenhouse gases are standardized on carbon-equivalence. Greenhouse Gas (GHG) Global Warming Potential (GWP) values, which are established to estimate climate change, are not based on a chemical equivalent of carbon, but an estimated risk of radiation of carbon dioxide (CO_2) [23].

Finally, it is noteworthy that the value in Fig. 19B.5 ($/ton) is on a log scale. If presented on a linear scale, the lesser carbon end-use values do not show up. A case can be made that food, fuel and feed values drop by nearly a magnitude for each market. Only coal is the outlier in this hierarchy. The price of coal, in this case the wholesale price of IL coal, is well below the other fuels. Because it is such a low priced fuel, the inefficiencies associated with power generation and costs of transportation are less concerning.

19B.5.1.2 The Price of Emission or the Negative Value of Carbon

It is fundamental to understand that emissions, pollution, and wastes are the production of materials that are produced in quantities that overwhelm the demand for those products. While there may be academic value in identifying an output that is external to an analytical system as an externality, it does not facilitate the utilization of the isolated materials. One way of thinking about emissions is to consider them as unused surplus inputs that can not be assimilated back into the current system as it has been designed. The key is to redesign the system to assimilate the surplus materials.

The price of carbon as defined by the CO_2 emissions market has a story of its own. Figure 19B.6 represents the vintage 2003 spot carbon credit prices from the Chicago Climate Exchange (CCX) from January 2006 until mid-2010 [24]. The CCX, Carbon Financial Instrument (CFI) contract is just one product that is handled by the CCX, and the spot prices generally are much lower than the futures prices and other instruments. However, the spot prices work best for illustrative purposes. The term "vintage" refers to the origination year of the traded CFI. Each CFI contract represents one ton of CO_2 equivalent. The log scale prices in Fig. 19B.6 are below the prices indicated in Fig. 19B.5 ($10/ton).

This value of carbon (CCX credit traded prices) has a rather unique character. The carbon emission value in this market can be either a positive value (revenue) or a negative value (cost), depending on who is buying and selling. If an energy user is emitting ancient, fossil carbon, this carbon emission price is a cost that must be mitigated. If the energy end-user is using a recent carbon, biomass fuel, this carbon emission price is a revenue.

The price of the CFI, carbon credits in Fig. 19B.6 rose through June of 2008 for several reasons. First, the European Union has implemented mandatory carbon emission limits. This has increased the value of CO_2 emission credits globally. Second, there was general anticipation that the US legislature was on the verge of establishing its own federal carbon legislation. The more confident early carbon traders and emitters became that legislative action was on the horizon, the more stable the price became. Oddly when those legislative discussions did begin in 2009, the carbon emission price began to fall. Other carbon trading instruments are trading at a higher value than the spot prices presented here, but the CFI contracts represent a ton of CO_2 equivalent.

Remediation costs are commonly paid on emissions. Homeowners and industries pay for treatment and remediation of water, air and solid waste emissions. A Municipal Solid Waste (MSW) landfill may collect a $40/ton tipping or gate fee to take a ton of MSW that would be considered an emission of sorts. Like the categories in Fig. 19B.5, a ton of MSW is not equal to a ton of carbon. Generally a ton of MSW is about 60 percent biogenic material [25]. As the solid was remediation costs move from service provider back to the consumer these costs increase. A residential solid waste fee of $10 per month for pick up of 200 pounds per month (50 lbs/week) becomes $100 per ton to the homeowner.

The issue of whether biomass is a resource or a liability will continue to be developed in subsequent sections.

19B.5.2 BTUs and the Best Value

If energy is the currency by which value is to be assigned, it is fairly straight-forward to convert nearly any carbonaceous material into an energy value (dollars per million btu, $/MMBTU). These energy values take historical price data and add a tabled higher heating value, (HHV) value [26]. The resulting information contained in the new constructed values is diminished in quality by addition of the tabled values. The $/MMBTU energy value is not a perfect measure, but it allows comparisons that are not possible between barrels and bushels. It also uses the information of interest, energy, as the denominator.

Prices in terms of $/MMBTU is not a conventional energy value [27]. These new energy values work like prices in that if one is buying energy, a low $/MMBTU is the goal. If someone is selling energy, a high $/MMBTU is not bad.

There are trade-offs in using this new metric.

- Care must be exercised in comparing inputs with outputs. Crude oil or shelled corn are intermediate fuels, not final products. Neither fuel product can be put into a gas tank, they both

FIG. 19B.6 VINTAGE 2003 CARBON CREDIT CLOSING VALUES — CHICAGO CLIMATE EXCHANGE [24]

must be refined and delivered to a retail market. The refining and processing costs are not captured in this simple energy value metric.

- Some historical price data is wholesale, some is retail. Some moves into industrial markets, some through residential markets. These are all distinctly different markets — that are not captured in the energy values.
- Conversion efficiencies, which are discussed frequently here, are not included in these energy values.

19B.5.3 Energy Values and Ethanol

With crude oil, gasoline, shelled corn, and ethanol all in the same $/MMBTU units, they can be graphed on the same chart (Fig. 19B.7). Crude oil prices serve as a good benchmark for energy prices. The

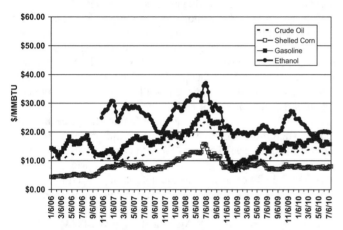

FIG. 19B.7 WHOLESALE ENERGY VALUES OF ETHANOL, CORN, GASOLINE, AND CRUDE OIL [27]

y-axis scale has been set to the same scale of $/MMBTU for each chart so the charts can be compared. The fuels of most interest are gasoline and ethanol since they are both final products rather than intermediate fuels like crude oil and corn.

This chart indicates that on an energy value-basis that gasoline is generally a better value than ethanol as a transportation fuel. If the goal is to purchase energy, then lower energy values are preferred. This is really all the chart indicates. It does not show that ethanol is not a viable transportation fuel. Ethanol has had more commercial success as a fuel additive rather than as a primary fuel. Ethanol has become an industry in part because of its environmental qualities and probably even more so because of its energy independence and local rural development benefits. These other parameters indicate that benefits other than energy content need to be part of the energy value metric.

Like crude oil refineries, ethanol plants produce multiple products: ethanol, distillers grains, corn (vegetable) oil, heat, and CO_2. As this industry matures, the biofuel facilities will become even more efficient. This chart also shows that on a dollar/energy unit basis, that corn is a lower-valued feedstock than crude oil. It is grown locally, which feeds into the energy independence and local economic development opportunities.

19B.5.4 Energy Values and Biodiesel

In July of 2010 as this chapter is being written the US biodiesel industry has gone through a severe contraction. In addition to moving through the same economic recession that the rest of the country went through, the biodiesel facilities lost their

$1 per gallon production tax credit. This is really a production (supply-side) economic issue. These energy values are based on the purchase price of these materials, so they reflect a demand-side event.

Biodiesel fuel can be made biologically with many different plant and animal oils. It is possible using thermal technologies like pyrolysis to convert solid fuels like wood and grass into biodiesel. Some of the thermal conversion technologies used for biomass are not too different from the chemical and thermal technologies already established in the petrochemical industry.

The initial feedstock in use before the economic down turn in 2008 was nearly always virgin vegetable oil, or soybean oil. Since the economy turned south, there has been great interest in converting existing facilities to handle more of the used and waste oils. But demand for biodiesel has declined with the supply, so the market price of biodiesel fuel has followed the diesel fuel prices.

Figure 19B.8 shows crude oil, rendered, number 2, yellow grease, soybean oil, diesel fuel and biodiesel fuel. The diesel fuel prices

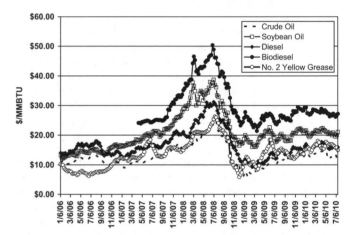

FIG. 19B.8 WHOLESALE ENERGY VALUES BIODIESEL, SOYBEAN OIL, YELLOW GREASE, DIESEL FUEL, AND CRUDE OIL. (*Source* [27])

come from the retail prices reported by the EIA, which reflect the fuel taxes that consumers pay. The biodiesel prices are whole sale prices for B-100 (100 percent biodiesel). These do not include the blender credit, or the other fuel taxes. This means that the spread between the energy value of diesel fuel and biodiesel fuel is much greater than what is shown here.

This is the entire point of Fig. 19B.8. For biodiesel to be competitive with diesel fuel, it must have a similar energy value. Currently, it does not. In fact for much of the time since January of 2006 the principal biodiesel feedstock, soybean oil, has had the same energy value as retail diesel fuel.

Soybean oil is far too valuable to be used as a fuel in its first use. Soybean oil is generally used as a food ingredient. The demand for food-grade vegetable oil is too strong for the biodiesel industry to compete. This is part of the reason the food vs. fuel debate is not the perceived threat that some have implied. The food uses of biomass will always out-compete the fuel and animal feed uses. Edible biomass can be used for other uses, but when there is a food scarcity, the fuel uses will lose.

Figure 19B.9 shows the spread between the wholesale, untaxed biodiesel energy value and the taxed, retail price of diesel fuel. It could be argued that this additional cost per energy unit of biodiesel fuel is the non-monetary benefit of energy independence, local economic development and environmental benefits of using biodiesel

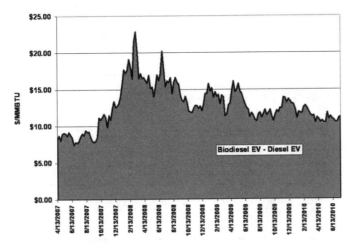

FIG. 19B.9 ENERGY VALUE MARGIN BETWEEN WHOLE-SALE BIODIESEL AND DIESEL FUEL [27]

over fossil diesel fuel. The bottom line is that there has to be a viable market for this more costly fuel alternative. The development of more efficient and cost-effective production technologies for biodiesel production will be the fastest way to reap the other non-monetary benefits using biodiesel fuel. Currently the additional costs are not sustainable.

19B.5.5 Energy Values and Residential Heat

The residential heat energy values tell a different story than the transportation fuels. The residential appliances are stationary, so solid fuels play a role. Wood has been burned as long as man has known how to make fire. As cord wood, it is produced locally and consumed locally. It is imported into urban areas.

There is no estimation of energy conversion losses. Since all the values in this chart are being discussed as heating fuel, this chart would be more accurate if it included the conversion efficiencies that Dennis Buffington, Pennsylvania State University, uses [28]. The value of efficiency will continue to be developed in other sections.

Figure 19B.10 compares the wholesale value of various heating fuels. The two most common for heat are natural gas and heating oil. Coal is used extensively in power generation. The three fossil fuels presented here maintain a fairly equidistant margin between each other. The one significant shift in the fuels presented here is

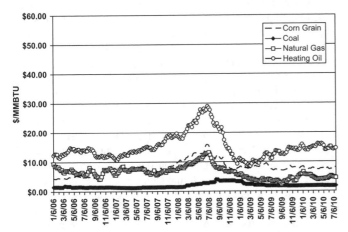

FIG. 19B.10 WHOLESALE ENERGY VALUES OF COAL, HEATING OIL, NATURAL GAS, AND CORN [27]

that during 2009 natural gas has been running very close to coal on a $/MMBTU basis. The coal price series used here is Illinois basin coal. The coal in Appalachia can be 50 percent higher in price. This might push the energy value of coal above the energy value of natural gas. This is significant since natural gas emits about half the CO_2 as coal. Corn is presented for perspective. It is not used as a significant commercial fuel.

Figure 19B.11 presents the retail prices for residential heating fuels. The EIA only presents heating season prices for propane and heating oil. This is one market where biomass fuels are very competitive with the conventional fuels. Particularly during the heating seasons, fuel pellets have been more affordable that propane and heating oil. Depending on the region, the same is likely true for electricity. Only corn grain as a fuel is a better value. Which is interesting because at current prices, corn is an expensive ingredient into both biofuels and livestock feed, but it still a good value as a heating source for residential heat.

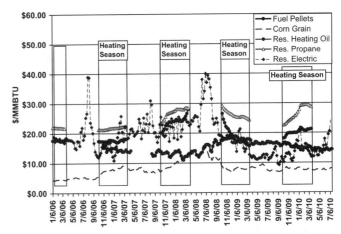

FIG. 19B.11 RETAIL RESIDENTIAL HEATING ENERGY VALUES OF HEATING OIL, PROPANE, ELECTRIC, FUEL PELLETS, AND CORN [27]

19B.5.5.1 More Benefits than Energy As discussed energy is no longer the only component of carbon-based fuels that play a role. In the case of fossil fuels and biofuels, one significant characteristic is density. In general bio-based fuels tend to be less dense than the more conventional fossil fuels. This has significant economic implications. Figure 19B.12 shows a progression of charts that begin with fuel data graphed with energy value plotted against fuel density.

The first chart of Fig. 19B.12a shows a scatter chart of liquid, solid and gaseous fuels in terms of energy value and their density. Coal for instance has the lowest energy value and the highest density. Natural gas has a low energy value and a low density. Natural gas is popular though because it piped directly to the end user, so the density issue does not interfere with delivery. For the liquids and solid fuels though, the more dense the fuel the cheaper the delivery and utilization costs. Figure 19B.12b simply identifies the fuels by various phases (gas, liquid, or solid).

In Fig. 19B.12c, the groups of fuels by phase are bisected by arbitrary transects to illustrate that within each phase, the lower energy value with a high density has greater value. This is primarily to lower the transportation and handling costs of both the feedstocks and the fuel end products. Mixed grass hay for instance has a low density and energy value, but to replace the absolute energy value of a ton of coal, it may require 10 to 12 tons of grass hay.

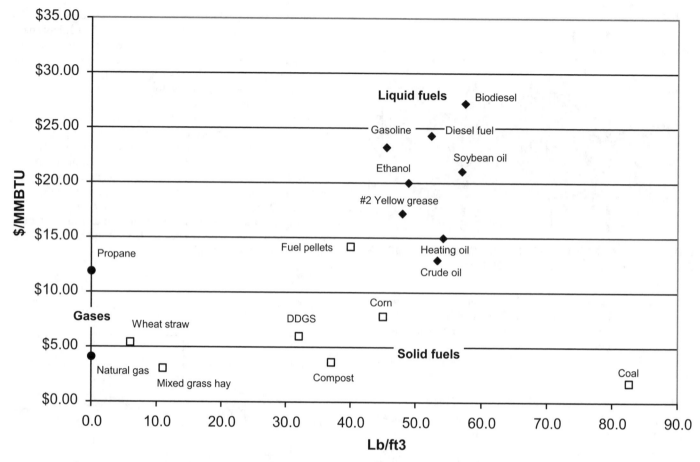

FIG. 19B.12A   ENERGY VALUE BY DENSITY FOR VARIOUS FUELS [29]

So far in this discussion of non-energy characteristics that have economic value, the only one that has been discussed is density. One other non-energy characteristic that merits discussion is the impact of carbon dioxide emissions on value.

When a million btu of natural gas is burned, about half the CO_2 is emitted as with the combustion of a million btu of coal. Assigning CO_2 emission values collected from EIA to the same collection of fossil and biofuels and plotting energy, value, density, and CO_2 emissions on the same chart produces and effect like that presented in Fig. 19B.13. Here the least cost value (energy and density) is one that also emits the least CO_2. These are only relative relationships presented to illustrate some of the pressures that various fuel characteristics bring to the economic success of a project.

19B.6   BIOMASS SUPPLY AND PRODUCTION ECONOMICS

The "supply" side of the economic supply-demand relationship focuses on quantities available for sale. This discussion provides information on where biomass fuel is going to come from to power and fuel this nation. Economists tend to focus on prices and markets (demand), while engineers look more intently at inputs, technology and output quantity — which supply economics (Fig. 19B.14). Supply economics describe from where and how much biomass it ill take to fuel an energy project.

There is more to supply economics than may appear on the surface. Factors that typically influence supply are: the price of inputs,

technology, number of producers, expectations of future prices, taxes, subsidies, and regulations [31]. These factors make the entire supply curve, or relationship between quantity and price, increase or decrease (shift). Only when looking at a specific commodity, like the price of the current sugarcane crop, does a price-point move up or down a static supply curve. Unfortunately very specific supply relationships get lumped together in ways that add confusion and blur the lines of biomass economic opportunities.

19B.6.1   Defining Biomass Supplies

Supply and demand curve charts are generally used by economist to make their point to other economists. It is important in understanding these conceptual tools that a supply curve for corn grain will look different than one for corn stalks. There is a tendency to discuss biomass opportunities as though biomass is a single commodity. It is not. Supply curves for switchgrass, hybrid poplars, forest slash, sawdust, manure, food waste, sugarcane, and any other source of biomass for energy, will have a different price/quantity relationship.

Corn grain, for instance, can be sold for food, feed, ethanol biofuels, or as solid-fuel for residential heat. The dry, left-over corn plant after grain harvest can be harvested for fuel in advance cellulosic biofuels or it can be left on the field to provide residual organic matter and nutrients to the soil. Or it can be harvested before it matures as corn silage, wet, and this can be used as animal feed or digested anaerobically for the production of biogas fuels. The popular discussion tends to focus on the end-use or markets available. These are demand issues. In this example, supply-side component is just corn.

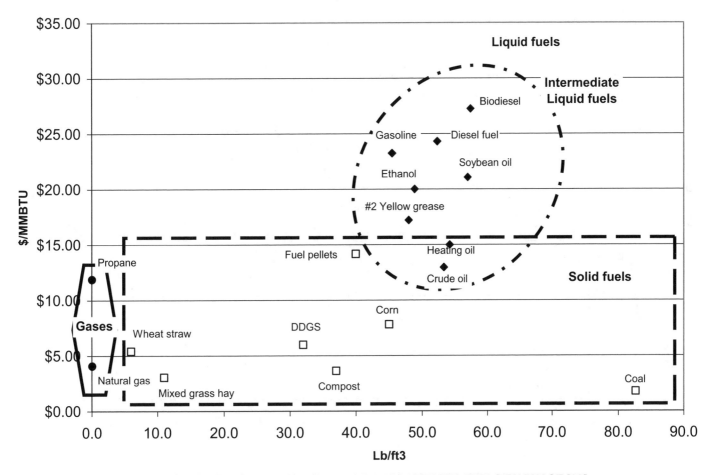

FIG. 19B.12B   ENERGY VALUE BY DENSITY FOR VARIOUS FUELS BY PHASE [29]

The bioenergy markets are only just developing, so there is limited fuel-specific demand information. It is not uncommon for supply quantities and qualities to be discussed in terms of demand, or the conditions of marketability. This is a reasonable first step, but it also creates allocation challenges. If a biomass supply has been determined for a specific use, it is dangerous to allocate that single source for multiple applications. While supply contracts can aid in dedicating specific supplies for specific uses, they are not without limitations. On the one hand, until a project is far enough along that biomass suppliers are willing to sign supply contracts, a given biomass supply is available to the highest bidder.

On another hand, if a dedicated, 10 megawatt (MW) biomass power plant is being built in the shadow of a 800 MW commercial coal-fired power plant even supply contracts may be challenged. It may be 5 years after the biomass power plant is operational that the commercial plant begins to co-fire biomass and coal. Even at a fuel mix of only 2 percent biomass at the 800 MW power plant will require an equivalent of biomass to operate 16 MW of power. Whether the co-fired plant is looking at fuel volumes or power production the new biomass demand at competing power plants will test the strength of the initial biomass supply contracts.

This co-firing example brings up the issue of whether it is best to blend biomass with other fuels, or co-fire, or whether it is best to utilize biomass alone in a conversion technology. The answer is that there are some situations where it makes the best economic sense to focus purely on biomass alone. There will be many more opportunities to co-fire or blend a small amount of biomass with existing conventional fuels. Unfortunately the politics and policies

of biomass often play a greater role in this decision than the available feedstocks, technologies and market do.

19B.6.2   Conversion Technology Project Scale and Scope

Project Size or Scale Production data on emerging conversion technologies is difficult to find. Often times the data collection requires access to the technology of interest. This is accomplished by beginning on a very small scale, or size, and gradually work up to a commercial scale project. The basic economic sizes are laboratory scale, pilot scale, commercial scale or industrial scale. These scales are set by the intended goal or purpose rather than by specific size. For instance, a commercial biodiesel plant might be as small as 3 million gallons per year, while a pilot cellulosic ethanol plant might be more than 3 million gallons of production capacity per year. It is more about whether the biodiesel plant can make money or the pilot cellulosic ethanol plant is being used to gather data for a larger plant.

The scale categories can be defined as: [32]

• A laboratory-scale project is small enough to fit on a lab table-top. The purpose of this size is to perfecting the process; there are no commercial economic considerations.
• Pilot-scale projects are the intermediate step between laboratory and commercial. Technology developers need confidence in their technologies — a proof of concept — to attract investors and clients. The scale of these pilots is large enough to test the equipment at a rigorous level, but small enough to minimize economic risks to the developer.

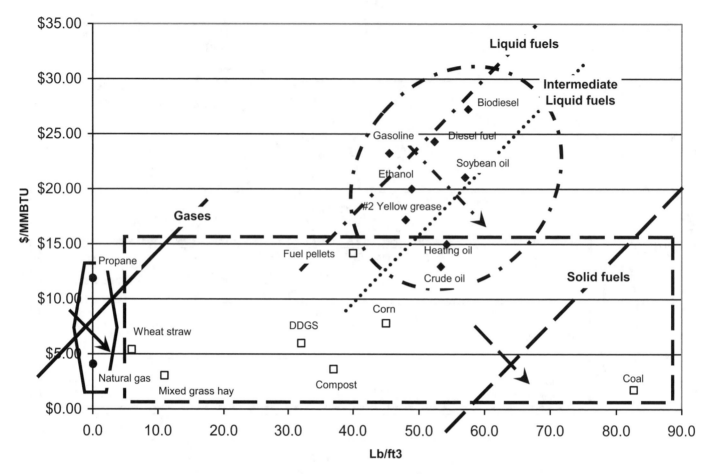

FIG. 19B.12C   ENERGY VALUES BY DENSITY FOR VARIOUS FUELS [29]

- The purpose of a commercial-scale plant is to be economically viable. These projects are economic experiments. Hundreds of millions of dollars are on the line. Public grants and loans do not guarantee success, but provide some risk protection. Once the "first of its kind," commercial-scale project risks are removed, subsequent projects should operate without public assistance.

Finally, industrial-scale projects are replications of the successful "first" commercial-scale projects. A key factor in the rapid expansion of the dry-mill corn ethanol industry was that investors knew the established technologies were replicable.

Should a biomass energy project be "Large and Specialized (Industry)" or "Small and Diversified (Sustainable Agriculture)?" Both business models work, but the "large"/"small" designations are political terminology. For biomass energy projects to succeed over the long-run, they will need to avoid playing political games for as long as possible.

In economics, specialization is referred to as economies of scale. By specializing, firms spread capital investments over more units of production output. This means they get larger and focus on a single output. The result is a lower per-unit of production and more competitive cost structure. In agriculture we talk about farms getting larger and fewer in numbers. This is due to economies of scale.

Diversification of an asset, means that assets have multiple uses and produce multiple outputs. This is referred to as economies of scope. A "diversified" farm produces multiple commodities. The

"conventional" corn-based, dry mill ethanol plant was designed to produce ethanol and manage the byproducts of distillers grains (DDGS) and carbon dioxide (CO_2). Since energy costs continue to rise, this fledgling industry continues to innovate combining biomass gasification or anaerobic digestion for fuel sources. They are cultivating research in new markets for the DDGS, reusing municipal wastewater, developing CO_2 product enterprises, and finding new ways to utilize the waste thin stillage. Highly specialized biofuels facilities are diversifying.

Combining economies of scale (specialization) and scope (diversification) are creating exciting new realities. This is happening across the nation. Landfills are pumping their methane to manufacturing facilities for a fuel source. Large livestock facilities are diversifying into energy projection from methane digesters. Ethanol, biodiesel and fuel pellet mills are all continuing to innovate to be large enough to capture economies of scale, but diversify into multiple product lines to reduce waste and increase revenue. Large efficient, facilities are nesting lesser enterprises within the umbrella of the specialized enterprise to more completely utilize company resources.

Economies of scope imply that the sum of the total system is greater than the sum of the parts (individual enterprises).

19B.6.3 Biomass Supplies

In 2005, the US Department of Energy (DOE) and the US Department of Agriculture (USDA) published their landmark report on biomass available for use in US petroleum-derived markets

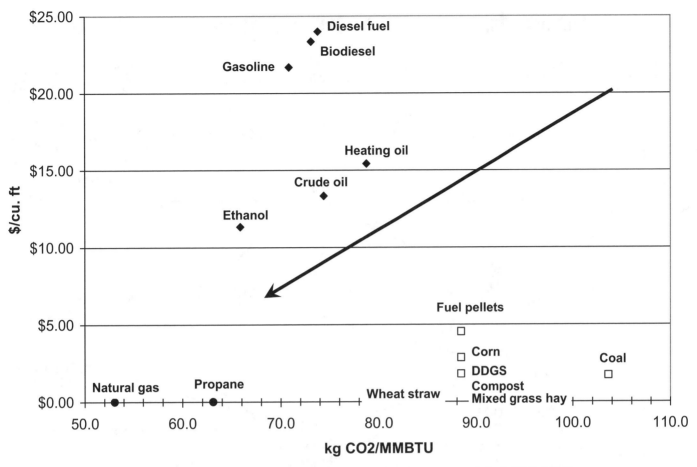

FIG. 19B.13   DENSITY VALUE BY ENERGY EMISSION FOR VARIOUS FUELS [29]

[33]. The goal was to see if the biomass land resources could sustain a feedstock supply of 1 billion tons of biomass to replace 30 percent of the US petroleum consumption. They determined that 1.3 billion tons: 998 billion tons from agriculture and 368 billion tons from residual forest biomass; had the potential to offset one third of the US petroleum consumption — without taking away from the current food and finished wood uses.

This study "fueled" the race to a biofuel platform. It occurred when oil supplies were challenged with pipeline repairs and hurricanes. Greenhouse Gas (GHG) mitigation was on the horizon. The race was on. The Billion Ton Study, as it has become known as, is a seminal, thoughtful estimate of biomass resources. Over time, cases have been made that sources of biomass that were considered in the study will not be allowed or available [34]. Other cases have been made that the Billion-Ton Study is very conservative and in fact biomass can do more in time. Gains from technology efficiency, reductions in energy consumption, and the inclusion of sources of biomass not considered, seem to indicate that on a technical supply basis, biomass can play a significant role as a fuel.

The most notable biofuel production in the US is the growth of the starch-based ethanol industry. It has grown from a vision of agriculture in the 1980s to a significant industry today. In 2009, the EIA reports that US fuel consumption relied on 4 percent biofuels in 2009 [20]. The Renewable Fuels Association reports that in 2009, 10.6 billion gallons of ethanol were produced in the US (Fig. 19B.15) [36].

Ethanol demand grew faster than supplies as states began switching from using MTBE as a fuel oxygenate to ethanol as an oxygenate in 2005 and 2006. The industry has been trying to move from a fuel additive market to a fuel market through the development of cars and engines that run well on E-85 fuels (85 percent ethanol and 15 percent gasoline), but progress has been slow. This change is requiring both the production of more cars that can run on either gasoline or E-85 (flex-fuel vehicles) and the installation of E-85 pumps to provide easy access to commuters.

One of the challenges that ethanol has to overcome is that it is not as energy dense as gasoline. Ethanol has about 76,000 million btu (MMBTU)/gallon while gasoline has about 125,000 MMBTU/gallon. Any economic value from using E-85 instead of gasoline has to also overcome the lower energy density.

Advanced biofuel production (from cellulose) is a new great challenge. On a technical basis, turning fiber into fuel is quite feasible. The primary constraint is in doing it economically, such that the revenues are greater than the costs. The yields for converting cellulose to ethanol have been in the 80 to 100 gallon per ton range. Some yields have been reported higher, but to really understand the commercial yield of conversion of fiber to fuel, there needs to be a commercial sector producing it.

The two longest-running leaders in producing ethanol from biomass are Verenium Corp in Jennings, Louisiana (1.5 million gallons/year) and KL Energy in Upton, Wyoming (1.0 million gallons/year). Most of the other projects are fairly fluid. Biomass Rules, LLC maintained a database of proposed projects that totaled 700

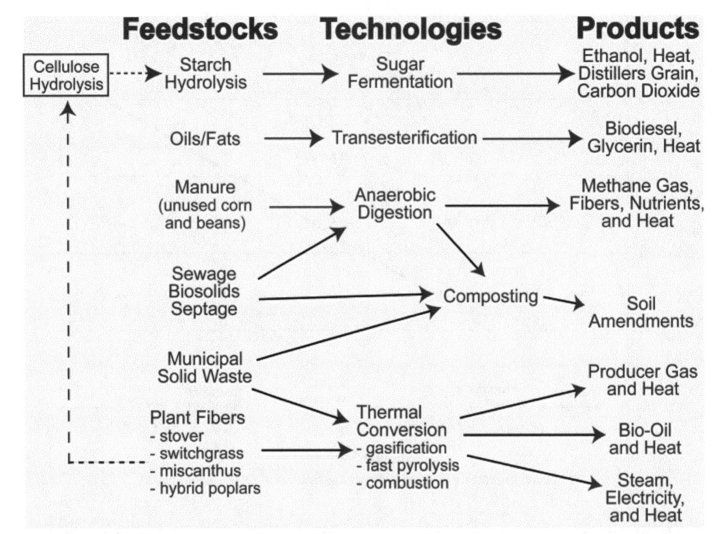

FIG. 19B.14 MATRIX OF BIOMASS FEEDSTOCKS, TECHNOLOGIES AND OUTPUTS, BIOTOWN, USA SOURCEBOOK, 2006 [30]

million gallons of annual capacity through 2009 (Fig. 19B.16) [37]. These 55 projects are presented on the basis of feedstock source. Unfortunately the cellulosic biofuels policies such as the 2010 Revised Renewable Fuels Standards (RFS2) may not include some of these proposed projects.

The risks associated with starting up a new biofuel project cannot be understated. When the "normal" risks are combined with the economic shocks of the last 2 years, maintaining an advanced bio-

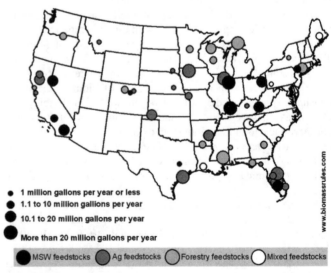

FIG. 19B.15 US ETHANOL PRODUCTION IN MILLIONS OF GALLONS PER YEAR [36]

FIG. 19B.16 PROPOSED US CELLULOSIC BIOFUELS PROJECTS [37]

fuels project list also became more difficult. An excellent alternative advanced biofuel project list that is available on the internet is maintained by the Biofuels Digest on their industry data page [38].

Algae projects have a life of their own. Algae have many potential markets available to algae production including: human food and nutrition, livestock feed, water quality remediation, CO_2 remediation, and energy. These markets can also serve multiple markets at the same time. As a result the investment risk can be spread out more easily for algae markets. Two algae industry associations have been formed in the US with scores of projects under development. Algal research has been conducted in all the market areas for decades, so there is a wealth of knowledge already available in the scientific literature and community. The challenge here again is to develop a commercial production system that provides greater revenues than costs (Fig. 19B.17).

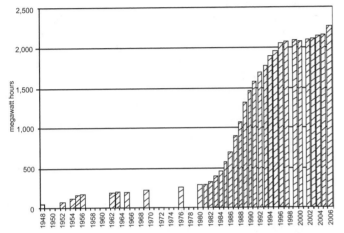

FIG. 19B.17 US BIOMASS POWER PLANT CAPACITY BY YEAR, [39]

Biomass power projects are one of the least documented uses of biomass energy. There are datasets around that track various components of biomass power, but it seems that each dataset track just a little different data than each other. An excellent resource on biomass energy data is the Department of Energy, Biomass Energy Data Book [39].

Using biomass for heat is even more difficult to quantify. The paper and forest products industry have become very good at providing process heat for their industries by converting production residuals into fuel. The 2005, DOE, Billion-Ton Study referenced above identifies 140 million tons of biomass that are already used in existing wood and paper industry facilities.

The electricity generating capacity from solid waste landfills is impressive. The US EPA Landfill Methane Outreach Program (LMOP) reports roughly 500 landfill gas projects have a generation capacity of 1,800 MW, not including the significant amount of landfill gas that provides fuel for direct use in local industries [40]. Those 500 projects have over 6 billion tons of waste in place.

There are more resource-efficient ways to utilize the carbon stored in solid waste than entombing it in landfills. Composting, recycling, and energy production are some of the ways to add value to these underutilized materials. Utilizing municipal solid waste (MSW) as a fuel is a popular feedstock because MSW is already delivered to the existing facilities. Biomass fuel transportation and handling costs are cost-prohibitive. Since biomass tends to be less dense than fossil sources, transportation and storage costs for biomass are proportionally higher on a per ton basis. The installation of a conversion technology project at or near a landfill can capitalize on the delivered biogenic MSW materials.

Currently it is more cost effective to install a generator on a large landfill than it is to install a generator on a large anaerobic digester, because the landfill costs the "container" out as a landfill operation cost. A digester project, however, must include the "container" as well as the electrical generator or end-use technology in the project budget.

Industrial digesters are experiencing new-found interest. Anaerobic digesters provide some recoverable energy in addition to organic stabilization. Sources of industrial organic wastes can have higher biological oxygen demand (BOD)/energy contents than more common feedstocks such as manure. In fact, many on farm manure digesters seek out higher strength feedstocks from industry to add to their farm digester projects.

On-farm manure, anaerobic digesters for energy production, have been underdevelopment in the US for 40 years. There was a steep learning curve in the 1970s and 1980s with many failures, but they have reached a certain level of technological maturity. There are two US digester manufacturers that have each installed over 30 digesters each. Manure digesters that are built by the commercial venders today operate pretty close to their specifications. US EPA AgSTAR program keeps track of operating digester projects in the US [41]. Currently they report over 130 manure digesters with a generation capacity of about 50 MW.

Finally the U.S. fuel pellet industry is growing at an unprecedented rate and is estimated to have a production capacity of over 6 million tons of fuel pellets per year. This is up from about one million tons of production capacity 5 years ago. Statistics from the Pellet Fuels Institute show that in the 2000 to 2001 heating season North America produced 730,000 tons of wood biomass pellets [42]. That annual production capacity from 10 years ago is about the size of pellet plants that are being constructed currently (Fig. 19B.18).

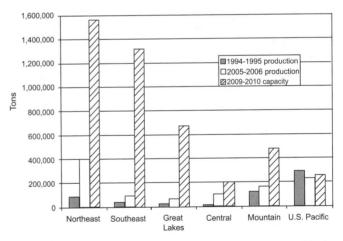

FIG. 19B.18 CHANGE IN THE US FUEL PELLET INFRASTRUCTURE OVER THE LAST 15 YEARS [42]

Before the turn of the 21st century, an average sized pellet mill was as small as 10,000 tons/year capacity. That increased to 30,000 tons/year, and now is over 100,000 tons/year. Among this class of mills was the biomass crop fuel pellet cooperative, Show Me Energy, Missouri. In 2008, the Green Circle Bio Energy, Inc. opened a 500,000 tons/year facility in Cottondale, Florida.

Not all the new pellet mills are super-sized. Many mills using local materials with domestic markets are still being built in the 30,000 tons per year capacity range. Some are driven by more than foreign markets. Colorado has two "mid-sized" pellet mills in the 100,000 to 200,000 ton/year capacity range. They are making fuel out of the dead trees left behind by the devastating pine beetle epidemic.

The fuel pellet industry is not immune to a challenged economy. Dixie Pellets in Selma, Alabama, which had an annual production capacity of 400,000 tons, closed its doors last September. Another older mill in northwest Montana reportedly also closed its doors in January 2010. The Montana mill had an annual capacity of 30,000 tons and is nestled in the Rocky Mountains.

19B.7   NON-ENERGY MATERIALS

Finally, regarding biomass supply economic issues, is managing the non-energy considerations. Since biomass is composed of plant parts, there are always nutrients and other residuals either going into the process or coming out of a process that may not have direct energy value, but have other values. In fact, unless biomass energy projects optimize all the energy and non-energy components, economic success will be limited. Biomass energy production almost always has low profit margins — if any. When biomass projects become profitable, it is because all the parts have found the highest-valued market in which to contribute to project revenues.

Once again algae makes a great example. One algal technology development driver is the opportunity to use algae to capture waste CO_2 from power plant exhaust. CO_2 is a limiting nutrient in green plant production. When algae are exposed to enriched concentrations of CO_2, they grow more biomass. Co-locating algal CO_2 sequestration projects with power plants to absorb the surplus (waste) CO_2 is intuitive.

Eqn. 19B-1: Photosynthesis
$$6CO_2 + 6H_2O + sunlight \Rightarrow C_6H_{12}O_6 + 6O_2$$

Several commercial carbon dioxide manufacturers are co-locating with corn ethanol plants because approximately one third of the bushel of corn leaves the ethanol process as CO_2. Needless to say, CO_2 is a non-energy resource with significant value.

Other common non-energy resources that need to be included in the economic evaluation are: nitrogen, phosphorus, water, and even salt remediation. Consider if a common wastewater treatment technology of the future included secondary treatments that aerated algal tank to burn off the energy of the wastewater. Instead of relatively valueless biosolids, wastewater treatment plants may produce a value-added algae for bioenergy or biopower production.

19B.8   BIOENERGY POLICY BENEFITS AND COSTS

Biomass policies have many benefits as well as costs. Statutory legislation is written by legislators. Laws are also created by the courts through tort law. These are the two primary vehicles to establishing new laws. A third vehicle for establishing law is through promulgation of regulations, which are the implementation language for statutory law. Regulatory law is written by Administrative agencies as opposed to the legislators or the courts.

Biomass shows up as regulated wastes in liquid, gaseous, and solid forms. Biomass can be food, animal feed, clothing, forestry products, and energy products. It is grown on private and public lands for wildlife preservation, erosion control, human recreation, and for waste remediation. Consequently, production and harvest of all of these forms of biomass are subject to the laws that oversee and govern the use of these products. There is nothing simple or straight forward about biomass policy.

If biomass is regulated as a pollutant then the chances of developing an industry become increasing limited. As mentioned above, the difference between waste treatment and biomass processing is often one of semantics. Publicly supported wastewater treatment technology aerates wastewater to grow biota to consume the carbon energy in the wastewater and then removed as biosolids. These stabilized biosolids are hauled off to be land applied generally at a cost greater than their value. Alternatively, wastewater could grow beneficial species of algae, consuming the carbon energy in the wastewater and then processed as value-added biobased products and energy.

This is not an engineering problem, but a policy problem. Statute and regulation-specific, legal responsibilities by public agents entrusted with public health and the environment, have no incentive to adapt policies that achieve multiple objectives. This biomass policy barrier is looming-large for the carbon-regulated industries.

Biomass residuals in many ways form the foundation for the environmental pollutants that are emitted everyday. In the air, they are greenhouse gas, carbon dioxide equivalents of CO_2, CH_4, N_2O, and fluorinated gases; and other air pollutants such as Volatile Organic Compounds (VOC); Criteria Pollutants like carbon monoxide and Nitrogen Oxides (NO_2) as well as Hazardous Air Pollutants. In the water they are Biological Oxygen Demand (BOD), ammonia, nitrite, phosphorus, and total organic carbons. In solid waste they are wood, paper, and cardboard, but also include the liquid and air components also. These compounds are made up of the basic building blocks of life.

These are residuals, surpluses, and leftovers from using and producing life-giving food, fiber, and fuels. The residuals are not inherently bad. In fact, it was only after they began to accumulate in large enough concentration that we became aware of the environmental impact of these organically-derived pollutants. The process of regulating surplus biomass became formalized in the 1970s with the landmark environmental statutes like the Clean Air Act and the Clean Water Act.

19B.8.1   A Brief History of US Environmental Regulation

In the early 1970s, the US was facing serious environmental challenges. Rivers were on fire because of combustible chemicals that were discharged from industry. Acid rain was a serious problem. Even litter, a pollutant without a discharge source (non-point), was everywhere. A suite of federal statues including the Clean Air Act and the Clean Water Act opened a new era of environmental oversight and regulation.

There are many legitimate reasons for the current federal and state regulatory structure. The regulation of pollution in air, water, and solid waste are defined in each enabling environmental statute. A starting place is an abbreviated listing of the 40-year history of the U.S. EPA [43].

1970   President Nixon forms the EPA
1970   The Clean Air Act (CAA) is passed.
1972   Federal Water Pollution Control Act Amendments of 1972 are passed.
1976   Resource Conservation and Recovery Act (RCRA, solid/hazardous waste) is passed.
1977   Clean Water Act (CWA) of 1977 is passed.
1980   Comprehensive Environmental Response Compensation and Liability Act (CERCLA, superfund) was passed.

In the 1990s a new environmental imbalance began to surface, greenhouse gases (GHG). In 1992, the United Nations Framework

Convention on Climate Change (UNFCCC) was adopted by the UN and later made available for member signatures in Rio de Janeiro, Brazil in June of 1992 [44]. The Framework Convention continues to evolve. In December of 1997 the Kyoto Protocol for the UNFC-CC was adopted in Kyoto, Japan. This Kyoto protocol sets binding targets for 37 industrialized countries to reduce GHG emissions back to 1990 levels from 2008 to 2012 [45]. While the US has not ratified the Kyoto Protocol, its inception has created a benchmark reference to carbon emissions as a driving force in policy.

Today, references to air pollutants (GHG, CO_2, CH_4), water pollutants (Biological or Chemical Oxygen Demand — BOD, COD), and solid waste (MSW, yard waste, manures, biosolids); are managed as costly social and economic liabilities. The policies isolate and segregate surplus and residual forms of carbon, while the solutions to recycling these resources back into the economy depends on using them — not isolating them. Technically biomass and bioenergy holds great potential to convert all waste treatment into resource and economic development. However, carbon-phobic law makers, regulators, and community groups stand against it.

There is confusion in the laws at every level about whether carbon and biomass are wastes or resources. Under federal law, biomass is legally defined as, "any organic material that is available on a renewable or recurring basis." [46] The Agriculture Title of the US Code of Regulation explicitly defines the term "biomass" to include:

. . . agricultural crops; trees grown for energy production; wood waste and wood residues; plants (including aquatic plants and grasses); residues; fibers; animal wastes and other waste materials; and fats, oils, and greases (including recycled fats, oils, and greases).

It also excludes paper that is commonly recycled, and unsegregated solid waste. This is a regulatory definition established by an administrative agency like the US Department of Agriculture as implementation language to a statute, which is established by legislators.

Statutory language carries more weight than regulatory language. There are numerous statutory waste remediation definitions that take precedence over the USDA regulatory definition. For instance, the statutory definition of a solid waste is not based on the physical form of the material, (i.e., whether or not it is a solid as opposed to a liquid or gas), but on the fact that the material is a waste. RCRA §1004(27) defines solid waste as [47]:

Any garbage, refuse, sludge from a wastewater treatment plant, water supply treatment plant, or air pollution control facility, and other discarded material, including solid, liquid, semisolid, or contained gaseous material, resulting from industrial, commercial, mining, and agricultural operations and from community activities.

These differences between these technical and legal definitions are barriers to biomass energy development. The greater the differences are, the more costly the adoption. Within the context of solid waste, the above legal definitions are still workable. However, as the legal priorities shift to carbon mitigation, new statutes will take a precedence over the current legal prioritization.

The USDA and DOE continue to fund biomass energy projects, and EPA has begun to regulate carbon emissions under the Clean Air Act (CAA) through the "Title V GHG Tailoring Rule," which is different than greenhouse gas legislation (GHG) [48]. All of these agencies focus on the management of carbon, hydrogen, oxygen, nitrogen, phosphorus, and sulfur. Again, these elements are the basic building blocks of life.

Environmental compliance does not need to cost society. If the policies are structured correctly they can be neutral or contribute to economic growth (Table 19B.2). Keeping in mind the factors that influence demand (prices) and those that influence supply (costs), it is pretty easy to see the role various federal statutes play on the economy.

TABLE 19B.2 HISTORICAL ECONOMIC IMPLICATIONS OF THE LEGISLATIVE ECONOMIC IMPACT [48]

	Economic Impact		
	Positive	Neutral	Negative
Energy Independence and Security Act (EISA)	X		??
2008 Farm Bill	X		
Generic Cap and Trade Legislation	??	X	??
Clean Air Act (CAA)			X
Clean Water Act (CWA)			X
Resource Conservation and Recovery Act (RCRA)			X

Basically the Energy Independence and Security Act (EISA) is statutorily directed at creating economic growth by reducing US dependence on foreign energy, though EPA did not interpret the implementation regulations (Revised Renewable Fuel Standards) in this context. The 2008 Farm Bill is intended to strengthen food, feed, and fuel markets for US agriculture. It also has some environmental directives. Cap and Trade policy, in general, simulates a market based on reducing costs and liabilities. Like waste remediation, this is policy is driven by minimizing costs. Participants enter the carbon-sequestration market to reduce their carbon emission compliance costs. While it simulates a market, it does not create wealth. Carbon has a lot of value, so this can have either positive or negative economic impacts so some of the policy outcomes are unclear.

19B.9 PAYMENTS OR SUBSIDIES

Commercial agriculture, whether organic or conventional, large or small, is a small business. Farmers make significant financial and quality of life sacrifices to keep from having their facilities foreclosed upon when the weather or the markets move in ways that cost them. Generally, because the Department of Agriculture (USDA) and the Department of Energy (DOE) understand the very real economic benefit of bioenergy industries, they tend to provide economic incentives for farmers and energy entrepreneurs to get into the business of biomass production and bioenergy conversion.

Both the USDA and DOE have given away billions of dollars in funding for these purposes. It is classical infrastructure development policy. Grants and loans help defray the multitude of financial and technical risks associated with developing start-up industries. Providing financial incentives to keep the cost of the raw products low is paid back many times as value is added along the processing chain. Honest wealth creation provides taxable income and economic growth. Payments to producers and processors lower the cost on the supply-side of economics.

Payments made from the government to the private sector to promote industry development and commerce are also subsidies. Investing public money in startup biobased product industries is generally viewed as a worthwhile public investment. Subsidizing an industry is not.

A tremendous resource to all involved in the bioenergy industry development is the Database of State Incentives for Renewables & Efficiency [49]. This is a repository of nearly all state and federal incentive programs for renewable energy and energy efficiency. It is easy to navigate and it is populated with more information continuously.

19B.10 BIOBASED EVERYTHING

USDA is administering a demand-side program for the federal government. It is referred to as the Bio-preferred program authorized through the last two Farm Bills. Through this authorizing legislation the federal government is required to buy biobased products when they are available and competitive.

Essentially the USDA is developing market standards for thousands of biobased product the government uses [50]. This benefit is tremendous in itself as it increased the demand for products made from biomass, including energy. A secondary benefit is that the cost of developing market standards for new products is borne by the federal government. Once established, the private sector also has access to these standards. The BioPreferred program has established a list of 40 biobased product categories and their minimum biobased content [51]. This is a great biomass policy!

19B.11 FOOD VERSUS FUEL

The politics surrounding food and fuel are about as serious as two political issues can be. There were some food shortages in Asia (rice) in 2008, which had nothing to do with US ethanol policies. Ethanol has been made primarily of corn in the US and there is a perception that corn is a critical component of human diets in this country. According to the latest USDA, World Agricultural Outlook Board Report, in most recent market year of 2009/2010, 41 percent of US corn was used for livestock feed, 34 percent was used for ethanol (some of which also is used as livestock feed) [52]. About 15 percent of the US corn supply last year was exported.

Food prices went up with oil prices in 2007/2008. They are coming back down. No one likes the price of food and fuel to go up. Therefore the US government has a long-standing precedent of working at keeping food and fuel prices very affordable. The US farm program payments and many other programs directed toward food and fuel programs are the frequent topic of political change. They get labeled as subsidies and heralded as the bane of the misuse of government spending. These programs, as with all government programs, should be transparent and openly debated. What is often poorly communicated though is that by keeping the prices of raw food and fuel commodities low, the input costs of development of the end uses is also kept low.

According to the US Department of Commerce, Bureau of Economic Analysis (BEA) statistics the share of income is about 12 percent of disposable income while the share of disposable income spent of energy is about 6 percent [53]. Figure 19B.19 shows that less

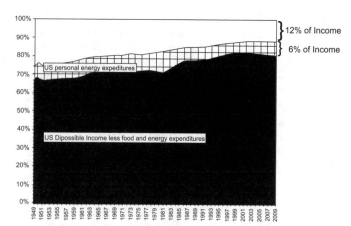

FIG. 19B.19   US DISPOSABLE INCOME SPENT ON FOOD AND ENERGY (1949–2009) [53]

than 20 percent of personal consumption in the US is spent on food and energy leaving over 80 percent of disposable income to spend on other things. The benefit is the US food and fuel component of the per capita disposable income is some of the lowest in the world.

Regarding the food and energy price inflation, the prices of food and energy are volatile. Figure 19B.20 illustrates the last 5 years of food and energy price changes relative to the rest of consumer prices [54]. The food and energy price indices add enough distortion that the US Department of Labor, Bureau of Labor and Statistics (BLS) keeps track consumer prices with food and energy and without food and energy. The indices in Fig. 19B.20 represent changes from the previous 12-month period.

Food and energy are considered inelastic in demand. This means that consumer buying habits do not change as rapidly with a change in price as they change with regard to other purchases [55].

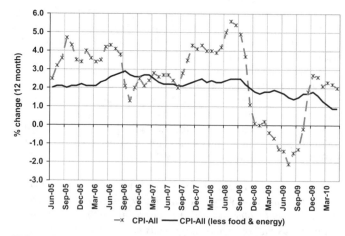

FIG. 19B.20   CONSUMER PRICE INDICES FOR FOOD AND ENERGY RELATIVE TO ALL OTHER CONSUMER PRICES [54]

19B.12 CO_2 EMISSIONS FROM BIOMASS — OR NOT

Over the past year, the Department of Energy (DOE) and the Environmental Protection Agency have been battling it out over whether biomass emissions add to GHG accumulation. In December 2009, EPA declared carbon dioxide a human health hazard. This announcement came as a climate summit was beginning in

Copenhagen. It also was necessary for EPA to regulate it as they do with other Clean Air Act emissions. Since then EPA has been going about establishing federal regulations on restricting the emissions of carbon dioxide from industry.

The Department of Energy, also part of the federal Administration, has been following the IPCC protocols on accounting for greenhouse gas emissions (including carbon dioxide). The central difference between the EPA intent to regulate CO_2 and the DOE/IPCC efforts is that the latter group does not count emissions from biomass. The DOE/IPCC methodology considers biomass carbon to be resident in the biosphere. With about 4 percent of our energy consumption coming from biomass fuels, there is a large difference between counting biomass carbon dioxide emissions and not counting them.

In May of 2010, the DOE, Energy Information Administration (EIA) released its Annual Energy Outlook on the state of energy on out into the future [56]. The EIA has been following the United Nations International Panel on Climate Change (IPCC) for more than 10 years [57]. In the Annual Energy Outlook 2010 (AEO) report there is a section entitled, "Accounting for carbon dioxide emissions from biomass energy combustion," in which the DOE discusses the IPCC history [58]. The last footnote in this AEO section (Footnote #78), presents the emission estimates for various biofuels and technologies if carbon dioxide emissions were counted.

In June 2010, the EPA issued its Greenhouse Gas (GHG) tailoring rule, which contradicts the IPCC characterization of biomass GHG emissions [59]. The EPA recently solicited comments on how to bridge this gap [60]. Maintaining the existing path of EPA to regulate all CO_2 is an economic barrier to biomass energy adoption. There seems to be no concern for the current sagging economic growth by EPA. Without a recognized differential between biomass CO_2 emissions and fossil CO_2 emissions, there is little reason to develop a bioenergy industry. This US policy "Waterloo" will require one of the two biomass policies to emerge as a clear leader. The DOE policy is more conducive to wealth creation and will promote a healthier environment.

19B.13 CHALLENGES WITH THE CURRENT RFS2 POLICY

The revised Renewable Fuels Standard regulations (RFS2) of the Energy Independence and Security (EISA) Act of 2007 became effective on July 1, 2010 [61]. This law is intended to benefit the nation by developing energy independence, renewable fuels, a cleaner environment and economic growth. Two and a half years after becoming law, the clarity expressed in the language of the EISA statute is less apparent in the RFS2 implementing regulation overseen by the U.S. Environmental Protection Agency. EPA is responsible for developing and implementing regulations to ensure that transportation fuel sold in the United States contains a minimum volume of renewable fuel.

The objective of the new law is to establish the utilization of 36 billion gallons of biofuels by 2022. The minimum limits are 15 billion gallons for "renewable fuels" and 21 billion gallons of "advanced biofuels." The difference is that renewable fuels (corn-based ethanol) must reduce greenhouse gas (GHG) emissions by a minimum of 20 percent lower than gasoline or diesel fuel. Advance biofuels must reduce GHG emissions by a minimum of 50 percent.

Within the advanced biofuels category are two other subsets: "biomass-based diesel" and "cellulosic biofuel." Biomass-based diesel must account for one billion gallons of the 21 billion gallons

of advance biofuels, while cellulosic biofuels must account for 16 billion gallons. The renewable fuels and the biomass-based diesel are already within reach. The less certain standards are for the other 20 billion gallons of advanced and cellulosic biofuels.

In the great toolbox of government policies (commercial, environmental, health, age, etc.), some of the more efficient contain outcome or performance criteria. The less efficient policies focus on managing the inputs and technologies to try to shape the outcomes. Renewable Fuel Standards are performance criteria, but the accounting system that was developed for corn-based ethanol in RFS1 is too complicated to expand to all biomass feedstocks, technologies and fuels that could possibly be used for transportation fuels. That, however, did not stop EPA from developing very restrictive layers of compliance and record keeping in the RFS2 rules.

While the premise is still on performance criteria of utilizing 36 billion gallons of biofuels, most of the regulation deals with qualifying feedstocks, technologies, combinations of feedstocks and technologies and keeping records of billions of gallons of fuel produced. The cost of noncompliance and recordkeeping falls on commercial fuel producers.

The RFS2 rules may work in the technical sense — especially for those fuels that closely resemble ethanol. For other less traditional transportation fuels like bio-oils from thermal conversion, natural gas from biogas, hydrogen and electricity, the RFS2 rule serves as a barrier to entry in the transportation fuel market.

Adhering to an EPA-specific mandate that causes the RFS2 standards to fail economically is a barrier to the Congressional intent. This does not mean renewable fuels standards cannot be created that promote energy independence and reduce carbon emissions. It does mean that the EISA can not be implemented as written.

19B.14 MATH IN THE NAME OF THE LAW

Because essentially every biomass conversion technology is experimental, or existing commercial technologies are nearly obsolete, there is limited data to put into the economic planning models. If actual data exists it is typically proprietary or in too small of quantity to be useful. Models can be populated with the latest commercial technology data, but the new technologies are so much more efficient that the answers that come out of the models based on older technologies have no value.

This does not stop models from generating numbers. Reputable policy institutions have taken funding to model scenarios for EPA, and then later publicly declared that the method in which the numbers are being employed is without meaning. Estimates are being inserted into sophisticated process models as placeholders for better data in the future. Outputs from models which were generated as a terminal use of the data and model assumptions are being used as input data into further processing. Government agencies and institutes are required by law to produce a number, but it appears to matter less whether the output value has any mathematical or statistical value. Excellent planning tools like Life Cycle Analyses (LCA) are being codified into laws that cannot be adequately supported by neither the data nor the models. It is math in the name of the law.

19B.15 POLICY CHALLENGES FOR THE BIOENERGY INDUSTRY

Unfortunately in this depressed economy there is little hope for a reversal in the ever-increasing costs of the increasing number of regulations required to startup a bioenergy project. The two forces working here generally are the technical efficiency gains from innovations by engineers which are beginning to be overwhelmed by the economic cost of poorly conceived, politically influenced costly regulations.

For economic success in bioenergy, there must be economic growth. If the cost of regulatory compliance overwhelms the chance for profit, economic success will not happen. Because many of the environmental components are made from the elements of biomass, policies that promote the complete utilization of these materials will also keep them out of the environment. Technical efficiency leads to environmental quality. This is a tough sell to the legislators and regulators though.

19B.16 BIOENERGY SUMMARY

There are many reasons to be excited about the potential expansion of a biomass energy industry: energy security, reduced carbon dioxide emissions, local rural economic development, and distributed generation (local production and local use) to name a few. Unfortunately there is a great deal of confusion about whether these benefits outweigh the social trade-off and costs of doing so.

New markets are being developed (economic demand) as well as new biomass feedstocks (economic supply). In addition a new market and distribution infrastructure is under development. These new economic structures can ease traditional energy supplies as well as remediate traditional environmental stresses. It is possible for the energy and environmental efficiencies to translate into new wealth creation and economic growth. This can not happen without rapid and severe assistance with public policy.

Public policy can add incentives to reduce the costs and risks of supplying biomass fuels. It can also add incentives to enhance the demand and market prices for biomass-derived energy. Existing air, water, and solid waste laws and regulations dealing with organic wastes (biomass) are also part of the solution. There must be public acknowledgement that existing human health and environmental benefits that are currently being overseen by respective agencies actually can be substituted through the biological and thermal energy conversion of biomass to energy. So from a carbon or biomass policy perspective new incentives must be implemented to enhance supply and demand, and the old single-function environmental regulations must be adapted and transformed into multi-function energy and bioproduct development laws.

These are not small requisites, particularly as the US economy struggles to recover from a significant recession. There is no room for missteps. The most likely risk is that the cumulative administrative costs of complying with multiple new regulations and investment in new capital will slowly overwhelm the economic benefits from wealth creation. Economic success of the biomass energy industries will be determined by the efficiency and speed at which leaders and the general public can find confidence in a biomass energy paradigm. It is now more of a political and public policy question than one of technology. Because even if the technical advantage today is marginal, these frontier technologies will continue to become more efficient as they develop. The technical aspect in the near term will take care of itself.

19.1   CONCLUDING REMARKS

Fossil fuel economics are profoundly global in nature and bioenergy economics are local. However, the two meet and compete directly at the consumer level. For example, biodiesel must be price competitive with petroleum diesel at the pump. The policy questions about whether it is worth subsidizing biofuels to reduce dependence on imported petroleum are central to the issue. It could be argued that subsidizing biofuels or other alternatives would allow importing nations to reduce military budgets as a nation that is not dependent on petroleum imports has less need to field a navy to protect petroleum shipping lanes. Those savings could be used for subsidies or development of better technologies.

Wealth creation is what drives economic growth. This originates from gains in technological efficiency. Public policies influence the economics positively and negatively, but the foundation of economic growth is derived from technological innovation. With or without government intervention innovation will shape the energy future.

19.2   ACRONYMS

AEO	Annual Energy Outlook, Energy Information Administration, Department of Energy.
BEA	Department of Commerce, Bureau of Economic Analysis
BLS	Department of Labor, Bureau of Labor Statistics
BOD	Biological Oxygen Demand
Btu	British Thermal Unit
$C_6H_{12}O_6$	Molecular formula of glucose/sugar, a common carbohydrate.
$(C_6H_{10}O_5)_n$	The molecular formula of Starch, which is a basic polysaccharide carbohydrate.
CAA	Clean Air Act, a US federal statute
CCX	Chicago Climate Exchange, a US carbon credit market place.
CERCLA	US federal statute, Comprehensive Environmental Response Compensation and Liability Act
CFI	The CCX, Carbon Financial Instrument, a financial contract representing an avoided carbon dioxide emission.
CH_4	Molecular formula of methane. Methane is a hydrocarbon and the principle component of natural gas and biogas. It has also been designated as a greenhouse gas and determined to be harmful to human health by the EPA.
CO_2	Molecular formula of carbon dioxide.
COD	Chemical Oxygen Demand
DDGS	Dried Distillers Grains and Solubles. The corn mash remaining from dry mill ethanol production is distillers grains. DDGS refers to distillers grains that have been dried for transportation and preservations. The solubles refer to solids from other processes that are combined with the distillers grains. This is a
EIA	The US Department of Energy, Energy Information Administration
EPA	The US Environmental Protection Agency
ERS	USDA, Economic Research Service. The agency within the United States Department of Agriculture that conducts economic analyses on the agricultural industry.
GHG	Greenhouse Gases http://www.epa.gov/climate change/emissions/index.html
GWP	Global warming potential is a constructed measure based on a compounds influence on radiative forcing. For further reading see: http://www.ipcc.ch/publications_and_data/ar4/wg1/en/ch2s2-10.html

H₂O	Molecular formula of water
HHV	Higher heating value includes the condensation of combustion products and for biomass appears to by 6% to 7% greater than the lower heating value (LHV, without condensation included) [25].
HAP	Hazardous Air Pollutants, a component of the Clean Air Act http://www.epa.gov/ttn/atw/pollsour.html
LCA	Life Cycle Analysis
LMOP	The US EPA Landfill Methane Outreach Program http://www.epa.gov/lmop/
MMBTU	Million btu (1 btu x 1,000,000)
MSW	Municipal solid waste
MTBE	Acronym for methyl tertiary butyl ether http://www.eia.doe.gov/emeu/steo/pub/special/mtbeban.html
MW	megawatts
NBER	National Bureau of Economic Research
NIMBY	Not in my back yard
NO₂	Nitrogen Oxides, a criteria pollutant of the Clean Air Act, National Ambient Air Quality Standards
OPEC	Organization of Petroleum Exporting Countries
RCRA	Resource Conservation and Recovery Act, a US federal statute
RFS2	Acronym for the 2010 revision of the renewable fuels standards. These are the implementation regulations the 2007.
SPR	Strategic Petroleum Reserve
TFP	Total Factor Productivity [13]
UNFCCC	United Nations Framework Convention on Climate Change [43, 44]
USDA	United States Department of Agriculture
VOC	Volatile Organic Compounds
WTRG	WTRG Economics

19.3   REFERENCES

1. NBER National Bureau of Economic Research www.nber.org.

2. EIA Short-Term Energy Outlook www.eia.doe.gov.

3. WTRG Economics (J. L. Williams www.wrtg.com 479-293-4081).

4. Energy Information Administration, Department of Energy www.eia.doe.gov.

5. FRED® Federal Reserve Economic Data, Federal Reserve Bank of St. Louis research.stlouisfed.org.

6. "The Coming Energy Crisis?" by James L. Williams and A. F. Alhajji, February 2003. WTRG Economics www.wrtg.com.

7. "Measures of Petroleum Dependence and Vulnerability in OECD Countries" by A. F. Alhajji and James L. Williams, April 21, 2003. WTRG Economics www.wrtg.com.

8. Baker Hughes rig count page www.bakerhughes.com.

9. "U.S. Carbon Dioxide Emissions in 2009: A Retrospective Review" EIA. May 5, 2010. http://www.eia.doe.gov/oiaf/environment/emissions/carbon/index.html.

10. This article, "Agricultural Commodity Price Spikes in the 1970s and 1990s: Valuable Lessons for Today," by May Peters, Suchada Langley and Paul Westcott, in the March 2009 issue of *Amber Waves* is just a great piece of economics. http://www.ers.usda.gov/AmberWaves/March09/Features/AgCommodityPrices.htm.

11. Ineffective, Minus-5 Regulations. Mark Jenner, PhD. 2004. http://biomassrules.com/which.html.

12. The Kyoto Protocol is an international agreement linked to the United Nations Framework Convention on Climate Change. The Kyoto Protocol was adopted in Kyoto, Japan, on 11 December 1997. http://unfccc.int/kyoto_protocol/items/2830.php.

13. Agricultural Productivity in the United States. USDA, Economic Research Service. http://www.ers.usda.gov/Data/AgProductivity/.

14. "Environmental Resource Indicators for Measuring Outcomes of On-Farm Production in the United States." Field to Market: The Keystone Alliance for Sustainable Agriculture. First Report, January 2009. http://www.keystone.org/spp/environment/sustainability/field-to-market.

15. "Motor Gasoline Outlook and State MTBE Bans." Tancred Lidderdale. US Department of Energy, Energy Information Administration. http://www.eia.doe.gov/emeu/steo/pub/special/mtbeban.html.

16. "Lessons learned from 80 years of US soybean production." Mark Jenner, PhD. Biomass Rules, LLC. 2005. http://biomassrules.com/biomass_economics.html.

17. "San Benito County Sourcebook of Biomass Energy." Mark Jenner, PhD, Biomass Rules, LLC. Central Coast Recycling Market Development Zone. October 2008. http://www.recycleloan.org/San%20Benito%20County%20Biomass%20SourceBook.pdf.

18. "New Bioenergy — Profit Meets Waste Remediation." Mark Jenner. BioCycle. Biomass Energy Outlook. Volume 48, Number 8. page 50. August 2007. http://www.jgpress.com/archives/_free/001400.html.

19. Economics, Roger A. Arnold, South-Western, Thomson. Seventh Edition. Chapter 3.

20. "Renewable Energy Consumption and Electricity Preliminary Statistics 2009. US. EIA, DOE. August 2010." http://www.eia.doe.gov/cneaf/alternate/page/renew_energy_consump/rea_prereport.html.

21. Definition of "starch," Webster's Online Dictionary http://www.websters-online-dictionary.org.

22. Mark Jenner, PhD. Biomass Rules, LLC. 2010. www.biomassrules.com.

23. It is convenient for illustration to standardize a potential effect or outcome of a compound, but there is no substitute for direct relationships. The data regarding CO₂ increases in the atmosphere is pretty straight-forward. The data regarding the GWP of GHG is less so. The GWP is not covered in this chapter but more information can be found on the Intergovernmental Panel on Climate Change (IPCC) website at: http://www.ipcc.ch/publications_and_data/ar4/wg1/en/ch2s2-10.html.

24. Chicago Climate Exchange (CCX), Market Data http://www.chicagoclimatex.com/market/data/summary.jsf Accessed 08/08/2010.

25. "Methodology for Allocating Municipal Solid Waste to Biogenic and Non-Biogenic Energy." Energy Information Administration, Office of Coal, Nuclear, Electric and Alternative Fuels, US Department of Energy. May 2007. http://www.eia.doe.gov/cneaf/solar.renewables/page/mswaste/msw.pdf.

26. Oak Ridge National Laboratories, Bioenergy Feedstock Development Program, Feedstock Conversion Factors. http://bioenergy.ornl.gov/papers/misc/energy_conv.html.

27. Biomass Energy Values & Prices, Biomass Rules, LLC. http://www.biomassrules.com/price_energy_value_home.html.

28. Performing Energy Calculations, Dennis E. Buffington, Professor, Agricultural and Biological Engineering, Pennsylvania State University, Cooperative Extension. H83. First Ed. 06/08 http://www.abe.psu.edu/extension/factsheets/h/H83.pdf.

29. Interaction of Energy Value, Density, and CO₂ Charts. Biomass Rules, LLC, www.biomassrules.com, and WTRG Economics, www.wrtg.com. 2010.

30. "BioTown, USA Sourcebook of Biomass Energy." Indiana State Department of Agriculture & Reynolds, Indiana. Mark Jenner, Biomass Rules, LLC. 2006. http://www.in.gov/oed/files/Biotown_Sourcebook_040306.pdf.

31. Economics, Roger A. Arnold, South-Western, Thomson. Seventh Edition. Chapter 3. 2005.

32. "Scaling the biomass energy Mountain." Mark Jenner. BioCycle Magazine. Biomass Energy Outlook. July 2007. Vol. 48. No. 7. p. 60. http://www.jgpress.com/archives/_free/001380.html.

33. "Biomass as feedstock for a bioenergy and bioproducts industry: The Technical Feasibility of a Billion-ton Annual Supply." Perlack, Robert, D, Lynn L. Wright, Anthony F. Turhollow, Robin Graham, Bryce J. Stokes and Donald C. Erbach. Oak Ridge National Laboratory, US Forest Service, and Agricultural Research Service. Department of Energy and Department of Agriculture. April 2006. http://feedstockreview.ornl.gov/pdf/billion_ton_vision.pdf.

34. The 2010 EPA Revised Renewable Fuel Standards (RFS2) do not include woody biomass from federal lands.

35. "Building An Infrastructure From Scratch," Mark Jenner, PhD. BioCycle Biomass Energy Outlook Column. Vol. 50, No. 7. July 2009.

36. Renewable Fuels Association. Industry Statistics. Historical US. Fuel Ethanol Production. http://www.ethanolrfa.org/pages/statistics/#A [Referenced 9/18/2010].

37. Burning Bio News. Biomass Rules, LLC, Greenville, IL. Vol. 3. No. 5 http://biomassrules.com/eNews/BBNv3n5.pdf.

38. Advanced Biofuels Tracking Database release 1.1, Biofuels Digest Data Page, Biofuels Digest. http://biofuelsdigest.com/bdigest/free-industry-data/.

39. "Biomass Energy Data Book, Energy Efficiency and Renewable Energy. Department of Energy." Biopower. Table 3.2. http://cta.ornl.gov/bedb/biopower.shtml [accessed 9/18/2010].

40. US EPA Landfill Methane Outreach Program Website. [Access 9/18/2010] http://www.epa.gov/lmop/projects-candidates/operational.html Other excellent sources of landfill biomass data are the State of Garbage in America Survey produced by BioCycle Magazine http://www.jgpress.com/archives/_free/001782.html#more and the Methodology for Allocating Municipal Solid Waste to Biogenic/Non-Biogenic Energy http://www.eia.doe.gov/cneaf/solar.renewables/page/mswaste/msw_report.html.

41. US EPA AgSTAR Program. Guide to Anaerobic Digesters. http://www.epa.gov/agstar/operational.html.

42. Data combined from historical files of the Pellet Fuel Institute and Biomass Rules, LLC. www.biomassrules.com.

43. EPA Website, Timeline of Accomplishments. http://www.epa.gov/history/timeline/70.htm.

44. United Nations Framework Convention on Climate Change. UNFCCC. Status of Ratification. http://unfccc.int/essential_background/convention/status_of_ratification/items/2631.php.

45. Kyoto Protocol. United Nations Framework Convention on Climate Change. http://unfccc.int/kyoto_protocol/items/2830.php.

46. USC Title 7, Chapter 107—Renewable Energy Research and Development, Section 8101. http://caselaw.lp.findlaw.com/uscodes/7/chapters/107/sections/section_8101.html.

47. EPA. RCRA, Superfund & EPCRA Call Center Training Module, Introduction of Solid Waste and Hazardous Waste Recycling (40 CFR section 261.2 and 261.9). October 2001. http://www.epa.gov/wastes/inforesources/pubs/hotline/training/defsw.pdf.

48. Mark Jenner, Sorting Priorities of Economic Growth, Energy, and the Environment. Burning Bio News. Vol. 3, No. 5. http://biomassrules.com/eNews/BBNv3n5.pdf. Also "Prevention of Significant Deterioration and Title V Greenhouse Gas Tailoring Rule" EPA. 40 CFR Parts 51, 52, 70, and 71. Federal Register, Vol. 75, No. 106, June 3, 2010. http://www.gpo.gov/fdsys/pkg/FR-2010-06-03/pdf/2010-11974.pdf#page=1.

49. Database of State Incentives for Renewables & Efficiency, www.dsireusa.org.

50. USDA BioPreferred Program to increase the purchase and use of renewable, environmentally friendly biobased products. www.biopreferred.gov.

51. The list of 40 biobased product categories and their minimum biobased content can be found at: http://www.biopreferred.gov/files/BioPreferred_product_categories_June_2010_FINAL.pdf.

52. USDA, "World Agricultural Supply and Demand Estimates," World Agricultural Outlook Board. WASDE-486 http://www.usda.gov/oce/commodity/wasde/latest.pdf.

53. US Department of Commerce, Bureau of Economic Analysis. National Income and Product Accounts. Table 2.5.5 Personal consumption expenditures by function. http://www.bea.gov/national/nipaweb/TableView.asp?SelectedTable=74&Freq=Year&FirstYear=2008&LastYear=2009. The USDA calculates a similar figure at less than 10 percent of disposable income. The values presented here are more conservative and still quite low.

54. US Department of Labor and Statistics, Bureau of Labor and Statistics (BLS) Consumer Price Index (CPI) Food and energy price indices for last five years. http://www.bls.gov/cpi/.

55. "Elasticity of Demand," Don Hofstrand. Co-Director, Ag Marketing Resource Center. Iowa State University. March 2007. http://www.agmrc.org/business_development/getting_prepared/business_and_economic_concepts_and_principles/elasticity_of_demand.cfm#.

56 Annual Energy Outlook 2010. US Department of Energy, US Energy Information Administration. May 11, 2010. http://www.eia.doe.gov/oiaf/aeo/index.html.

57. The earliest reference I could find was the IPPC methodologies on the EIA website was a 1994 manual created in Paris, France. http://www.eia.doe.gov/oiaf/1605/archive/87-92rpt/chap2.html#Estimating_Emissions.

58. http://www.eia.doe.gov/oiaf/aeo/index.html.

59. US EPA Title V Greenhouse Gas Tailoring Rule, Fact Sheet. http://www.epa.gov/nsr/documents/20100413fs.pdf.

60. Call for information on Greenhouse Gas Emissions Associated with Bioenergy and Other Biogenic Sources. http://www.epa.gov/climatechange/emissions/biogenic_emissions.html.

61. "RFS2 Takes Effect," Mark Jenner, PhD. Biomass Energy Outlook. BioCycle Magazine, July 2010. Volume 51. Number 7.

COAL GASIFICATION

Ravi K. Agrawal

20.1   INTRODUCTION

Fossil fuels supply almost all of the world's energy and feedstock demand. Among the fossil fuels, coal is the oldest, most abundant, and widely available form of fossil fuel. Coal constitutes over 75% of the world's fossil fuel. Since the beginning of the industrial revolution, coal has been the backbone of the world energy system. Before the discovery of oil and gas, coal was used to generate town gas via gasification for lighting and heating purposes. During World War II, coal gasification technology was extensively used in Germany to produce oil substitutes. After World War II, oil replaced coal as the major source of energy, and the interest in coal gasification remained dormant until the recent rise in energy prices. Since 2000, several gasification plants have become operational, most of these plants are for the production of chemicals and only a handful of plants for power generation.

Coal is among the cheapest fuel available. Figure 20.1 compares the historical and projected price of major energy sources in the United States [1]. Projections of the Energy Information Administration (EIA) indicate that coal is likely to remain the cheapest fuel in the foreseeable future and will likely become cheaper as the price of other preferred energy sources rise. Based on the energy content in terms of price per Btu, electricity commands the highest premium, followed by oil and gas. This price differential is a key driver that determines the interest in coal and its conversion into other energy substitutes.

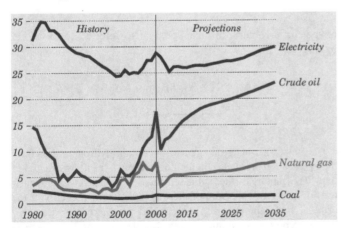

FIG. 20.1   UNITED STATES ENERGY PRICES AND PROJECTIONS (2008 DOLLARS PER MILLION BTU) [1]

Since electricity trades at the highest premium, the cheapest source of energy, coal, has been extensively used to generate elec-

tricity. About 60% of global electric power is generated from coal, and about two-thirds of coal produced is used for power generation. Therefore, generation of electricity and the consumption of coal are intricately linked. Unfortunately, coal has the largest carbon footprint when compared with conventional fuels derived from oil and natural gas. Historically, most coal-based power plants are considered to be "dirty" as only a few had controls to lower the emissions of SO_2, NOx, and particulates, and even to date, virtually none have controls on mercury and CO_2 emissions. Tightening of environmental regulations in countries such as United States, Europe, and Japan has prompted installation of emission control devices. Table 20.1 summarizes the current US New Source Performance Standards (NSPS) on SO_2, NOx, particulates, and Hg. Also shown in Table 20.1 are emissions from a typical coal-fired power generation unit and a coal-based integrated gasification combined cycle (IGCC) unit.

TABLE 20.1   COMPARISON OF US NSPS EMISSION STANDARDS VERSUS EMISSIONS FROM AN IGCC UNIT AND COAL-FIRED POWER PLANT [2,3]

Emissions	Typical Coal Unit	NSPS	Projected IGCC
SO_2, lb/MMBTU	0.17	1.2	0.03
NOx, lb/MMBTU	0.15	0.15	0.02
Particulates, lb/MMBTU	0.02	0.03	< 0.01
Mercury Removal	–	–	90+%

The control of contaminants and pollutants from an IGCC unit is generally dictated by the more stringent downstream requirements, especially those of the gas turbine and/or the downstream catalyst requirements and not necessarily the environmental regulations. As a result of these requirements, the emissions from a coal-based IGCC units are significantly lower than the current NSPS requirements. For IGCC applications, additional power can be generated by diluting the syngas with hot water, steam, or nitrogen from the air separation unit (ASU) unit, while simultaneously reducing NOx. Therefore, the design requirements for IGCC ensure that the emissions from coal gasification plants are significantly lower than those from a traditional coal-fired steam generation units.

Future regulations on greenhouse gas emissions will have a major impact on traditional coal-fired power plants. This will lead to accelerated implementation of coal gasification–based power plants, as gasification provides for an efficient method to generate power from coal while minimizing the environmental impact.

Another important application of coal gasification is for the production of chemicals and oil substitutes. Clearly, gasification is one technology that converts "dirty fuels" such as coal, petroleum coke, and other carbonaceous materials, into alternates for cleaner and high-value energy sources such as oil and natural gas. This conversion is possible because all fossil fuels come from the same source and have very similar chemical composition. All fossil fuels are made up of carbon and hydrogen, with a small amount of oxygen. Coal is very heterogeneous and varies widely in chemical composition. Carbon is the dominant ingredient of coal and forms the basis for classifying coal. In broad terms, coal is classified into four ranks: anthracite, bituminous, sub-bituminous, and lignite. Anthracite and bituminous coals are referred to as high-rank coals and have the highest carbon content. Sub-bituminous and lignite are considered low-rank coals as they contain significant amount of moisture. Bituminous and sub-bituminous are the most common coals used. Table 20.2 summarizes the properties of selected fuels [4].

duced from gasification is rich in CO and H_2 and can be used for the production of electricity, chemicals, and liquid fuels. Typical chemicals that can be produced from syngas include: hydrogen, methanol, ammonia, acetic acid, and oxygenates. Liquid fuels, such as naphtha, diesel, ethanol, dimethyl ether (DME), and methyl tertiary butyl ether, can also be produced from syngas. In recent years, interest in using syngas to reduce iron ore has picked up, especially in India. Thus, coal gasification can be effectively used to provide syngas as a feedstock or an agent for a variety of applications.

20.2 THEORY

Gasification can be considered to be the conversion of a carbonaceous material into a gas that has a chemical value. For example, consider the reaction of carbon with half the amount of oxygen required for combustion. This reaction may be referred to as gasification, as it results in an intermediate gaseous product, CO that has approximately 70% of the fuel value of carbon. Since

TABLE 20.2 CHEMICAL AND FUEL PROPERTIES OF SELECTED FUELS [4]

Elemental Analysis (wt %)	Bituminous	Sub-Bituminous	No. 2 Fuel Oil	Nat. Gas
Carbon	75.5	70.8	86.1	75.0
Hydrogen	5	5.0	13.2	25.0
Oxygen	4.9	15.3	0.8	
Nitrogen	1.2	1.0	0.005	
Sulfur	3.1	0.4	0.7	
Ash	10.3	7.5		
Higher Heating Value, dry basis, BTU/lb	13,100	12,100	19,700	23,800
Atomic H/C Ratio	0.795	0.855	1.840	4.000
Atomic O/C Ratio	0.049	0.162	0.007	0.000

Methane is the main ingredient of natural gas, and its properties are shown in Table 20.2 to represent natural gas. Table 20.2 shows that coal has similar constituents as oil and natural gas, but also contain undesirable components such as oxygen and inorganic mineral matter (ash). Ash is an oxidized form of inorganic mineral matter that is inherently present in coal. The presence of undesirable components lowers the heating value of the fuel and/or increases material handling requirements. The fuel characteristics defined by the atomic H/C ratio indicate that the normalized hydrogen content of coal is about half that of oil, and only one-fifth that of natural gas. This ratio indicates that coal is lower in hydrogen and higher in carbon, when compared with oil or natural gas. Therefore, coal can be converted to oil or natural gas by either adding hydrogen or rejecting carbon. Gasifying coal with steam to generate hydrogen and rejecting excess carbon as carbon dioxide can make up the hydrogen deficiency in coal. The excess O/C ratio of coal is adjusted by rejecting the oxygen with hydrogen and carbon. Gasification also helps in removal of inorganic matter from carbon and hydrogen as ash. Syngas pro-

oxygen is involved, this reaction is also referred to as partial oxidation.

$$C + \tfrac{1}{2} O_2 \xrightarrow[\text{Gasification}]{} CO \qquad (20.1)$$

$$C + O_2 \xrightarrow[\text{Combustion}]{} CO_2 \qquad (20.2)$$

Carbonaceous materials generally contain hydrogen; therefore, the gasification products will include hydrogen resulting in a higher than 70% capture of fuel value in syngas. The efficiency of a gasification process is the sum of thermal and chemical energy of the original fuel captured in syngas, minus any auxiliary energy required to process coal into syngas. The thermal energy represents the amount of energy that can be recovered in form of steam while cooling the hot syngas. The chemical energy is represented by the heating value of the syngas.

Reactions occurring during coal gasification are quite complex. As coal is heated to gasification temperature, coal devolatilizes into volatile products rich in hydrocarbons, leaving behind a carbon-rich residue referred to as char [4]. The devolatilization products react with oxygen and steam to yield syngas rich in carbon monoxide, carbon dioxide, hydrogen, steam, and methane. Modern gasifiers operate at high temperatures, where most devolatilized products do not survive or are not favored by equilibrium. Therefore, modeling of gasification reactions may be simplified by considering only the principal reactions involving carbon, carbon monoxide, carbon dioxide, hydrogen, steam, and methane. The most important of the reactions are:

Partial oxidation: $\quad C + \frac{1}{2}O_2 \longrightarrow CO$ $\qquad$ (20.3)

Water gas shift: $\quad CO + H_2O \longleftrightarrow CO_2 + H_2$ $\qquad$ (20.4)

Methanation: $\quad CO + 3H_2 \longleftrightarrow CH_4 + H_2O$ $\qquad$ (20.5)

Steam-carbon reaction: $\quad C + H_2O \longrightarrow CO + H_2$ $\qquad$ (20.6)

Boudouard reaction: $\quad C + CO_2 \longrightarrow 2CO$ $\qquad$ (20.7)

The partial oxidation and other gas phase reactions are generally fast and approach equilibrium at high gasification temperatures. The methanation reaction is favored at lower temperatures and high pressures. Steam-carbon and the Boudouard reactions are the slow reactions and are the rate-limiting step in some gasifiers. The syngas composition depends on a number of things including the type of coal, gasifier residence time, oxidant, and gasification temperature and pressure.

Historically, gasification was conducted at low temperatures (<1300°F), where some of the products of devolatilization survived gasification conditions. Syngas generated by low temperature gasifiers is rich in hydrocarbons, including methane and tars (condensable volatiles). Early in the 1900s, prior to the development of air separation plants for producing O_2, gasification reactions were conducted using air to generate heat for the endothermic gasification reactions. The dominating reaction in these gasifiers was the slow steam carbon reaction, which required multiple reactors.

The development of large air separation plants in the mid-1900s aided the development of modern gasifiers that use oxygen. These modern gasifiers are truly partial oxidizers as steam is used only for moderation purposes and not necessarily for the steam carbon reaction. These gasifiers typically operate at pressures above 300 psia and temperatures above 2500°F. Fast partial oxidation reactions make these modern gasifiers compact. Partial oxidation–based gasifiers now dominate the present gasification arena.

20.3 GASIFICATION METHODS AND TECHNOLOGIES

Gasification is an old technology that was initially developed for production of town gas for heating and lighting purposes. During World War II, the German interest in gasification was for production of syngas for motor fuels and chemicals. Interest in gasification waned after World War II owing to the discovery of cheaper energy sources, except in South Africa, where due to an oil embargo imposed by the United Nations, South Africa continued to rely on its coal reserves to meet their fuel needs. The Arab oil crisis in mid-1970s renewed interest in gasification, only to be lost again to the oil price crash in mid-1980s. With each decline in oil price, several companies abruptly abandoned their gasification efforts. This, along with the mergers, acquisitions, and spin-offs, has frag-

mented the gasification technology. Of over 60 gasification technologies developed in the 1970s [5-9], only a fraction appear to have survived the crash of the 1980s. With each boom, technology providers attempt to resurrect previously abandoned technologies. Additional discussion on these recent coal gasification activities has been documented elsewhere [2,10-15].

The classification of methods for gasification has predominantly been based on the type of reactor used to contact coal with gasifying agents. Air, oxygen, and steam are the most common gasification agents. Gasifiers are lumped into three broad categories: moving bed gasifiers, fluidized bed gasifiers, and entrained bed gasifiers. All of these technologies were invented in Germany prior to World War II. Lurgi invented the moving bed gasifier, Winkler invented the fluidized bed gasifier, and Kopper-Totzek (K-T) invented the entrained bed gasifier.

In a moving bed gasifier, coal is fed from the top, and the packed reacting bed of coal moves slowly downwards by gravity. The gasification agents introduced from the bottom of the gasifier gasify the coal as it moves up the gasifier. In a fluidized bed gasifier, the gasification agent lifts coal particles as it bubbles through the bed and gasifies the coal. In an entrained bed gasifier, the coal particles are entrained with the gasification agent.

Other gasification techniques such as molten salts and underground gasification are still in their early stages of demonstration and are not expected to contribute significantly in the near future. Discussion on these technologies can be found elsewhere [10-15].

20.3.1 Moving Bed Gasifier

The moving bed technology is the oldest of the gasification technologies used for gas production. In a moving bed gasifier, coal is introduced from the top and gasified counter currently by gasification agents from below (Figure 20.2).

The gaseous products are removed from the top of the gasifier, and the ash is removed from the bottom. Initial gasifiers had air mixed with steam that was passed below a rotating grid at the bottom. The outgoing ash heats the incoming air to gasification temperatures. As the gasification agents move up the gasifier, they oxidize the remaining char in the ash to carbon monoxide and carbon dioxide. The oxidation reactions produce heat required for the endothermic gasification reactions. As shown in Figure 20.2, this is the hottest zone in the gasifier. As the remaining gasification agents

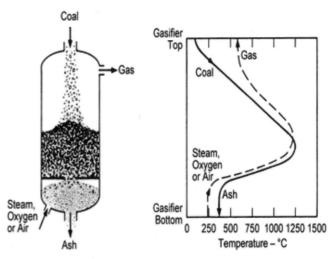

FIG. 20.2 TYPICAL DEPICTION OF A MOVING BED GASIFIER AND RELEVANT TEMPERATURE PROFILE [7,16]

and oxidation products move up, they gasify the downward moving char produced in the zone above. When the hot product gas moves up, it heats the raw coal. As the coal is heated, moisture is initially driven off followed by devolatization of coal where gaseous products rich in hydrocarbons and char are formed. Some of the hydrocarbons escape the gasifier and remain in product gas. The char product continues to react as it moves down the gasifier. The thermal efficiency of these counter-current gasifiers is good as the outgoing ash heats the incoming gasification agents, and the outgoing product gases heat the incoming coal. This can be seen from the temperature profiles shown in Figure 20.2. In later versions, air is replaced with oxygen to improve the quality of syngas.

Ash can either be rejected as a solid or as a liquid slag. In the dry ash mode of operation, excess steam is injected with air/oxygen to maintain the temperature below the ash-slagging temperature, the temperature at which ash melts. In the slagging version, the steam flow is limited to maintain gasification temperature above ash-slagging temperature, and molten slag is removed from a pool in the base of the gasifier.

The Lurgi gasifier is one of the oldest and most well-established moving bed technologies. Through the years, the Lurgi gasifier has undergone improvements, and several versions are now available. Figure 20.3 shows two versions of a Lurgi gasifier as differentiated by the ash rejection technique [5,15]. The pressurized version of Lurgi developed in 1930s uses top-mounted lock hoppers to feed the coal. The Lurgi dry ash gasifier uses a pressurized oxygen-blown counter-current gasifier and has been used widely in South Africa, the United States, Germany, the Czech Republic, and China. British gas modified a Lurgi style gasifier to operate at high temperatures, causing the ash to be rejected as slag and improved the throughput and carbon conversion efficiency. This slagging version is currently marketed as a British Gas and Lurgi (BGL) gasifier.

In a Lurgi gasifier, coal is spread evenly on the bed by means of a motorized distributor. A stirrer is included in some designs to handle caking coals. As the coal moves down, it is heated by the upward-flowing syngas that leaves the gasifier. The heat causes the coal to dry followed by devolatilization. Some of the devolatilized products escape gasification and leave the gasifier with syngas. As the devolatilized coal moves down, it is gasified with combustion products from the combustion zone below. In the bottom combustion zone, char is oxidized with oxygen to burn off the remaining carbon from ash and to provide energy for the gasification reactions. In the dry ash mode of operation, excess steam is injected with oxygen to keep the temperature below the ash fusion temperature. A motor driven rotating ash grate is used to remove ash in a "dry" state and also to support the coal bed. In order to improve throughput, carbon conversion, and decrease tar production, British Gas modified their Lurgi gasifier by increasing the combustion temperature to melt the ash. In the BGL gasifier, water-cooled nozzles replace the ash grate at the bottom. The water-cooled nozzles are used to introduce steam and oxygen into the gasifier. The molten slag is cooled in the slag trap by quenching, and the solidified slag is removed through a lock-hopper arrangement.

Despite the popularity of the Lurgi gasifiers, there are several limitations that increase the capital and operation costs. Some of the limitations of Lurgi are moving internal parts (that require maintenance), long residence time due to low reaction rates (which limits throughput), inability to feed coal fines, and the requirement to include additional gas cleanup systems to handle tars, ammonia, and phenols. Because of these reasons, recent activity in commercial application of moving bed technology has been limited. A summary of other "historical" moving bed gasifiers can be found elsewhere [6-9].

20.3.2   Fluidized Bed Gasifiers

A generic depiction of a fluidized bed gasifier and the relevant temperature profile are shown in Figure 20.4 [7,16]. In a fluidized bed gasifier, a gasification agent is passed through a bed of coal fines at a high-enough velocity to suspend the solid and cause the

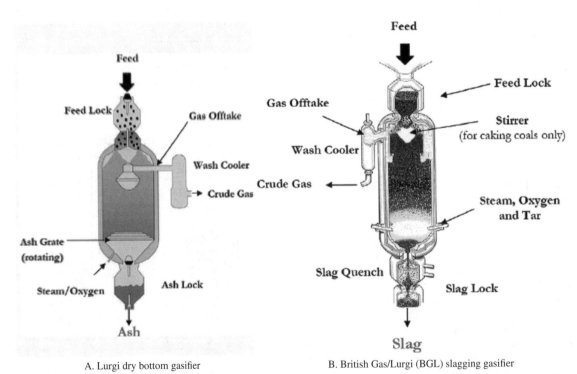

A. Lurgi dry bottom gasifier   B. British Gas/Lurgi (BGL) slagging gasifier

FIG. 20.3   DEPICTION OF POPULAR MOVING BED GASIFIERS [5,15]

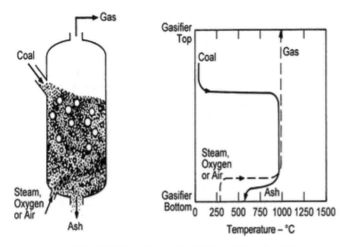

FIG. 20.4 DEPICTION OF FLUIDIZED BED GASIFIER AND THE RELEVANT TEMPERATURE PROFILE [7,16]

bed to behave as a fluid. The dry coal feed introduced into the hot fluidized bed mixes rapidly with the bed material and almost instantaneously attains the gasification temperature. This leads to higher throughputs in fluidized beds, as compared with moving bed gasifiers. Because of rapid mixing of raw coal with the bed material, the product gas composition is uniform, and the formation of tars and phenols are minimized. The maximum bed temperature of a fluidized bed gasifier is limited by the ash softening temperature, where ash begins to stick to other particles and solid surfaces. Fluidized bed gasifiers have the flexibility to handle all types of coal and are especially efficient in gasifying high-ash coals.

The Winkler gasifier developed in 1920s was the first commercial application of a fluidized bed process [17]. The Winkler process was used extensively by the Germans during World War II to generate syngas required for the Fischer-Tropsch process. The Winkler process operates at atmospheric pressure and has limited application due to its high capital and operating cost. The low throughput and high cost of syngas recompression led to the development of pressurized version of the Winkler technology. Rheinbraum, the developer of the high-pressure version of the Winkler technology, refers to the process as high-temperature Winkler (HTW) process, a misnomer as it is the high-pressure feature that distinguishes it from the original Winkler process. Pressurized operations were achieved by use of a lock-hopper arrangement followed by a screw feeder to feed dry coal into the gasifier. In 1986, Rheinbraum built a 600-t/d oxygen-blown HTW plant in Germany [13]. The plant operated at 10 bar, and the syngas was used for the production of methanol.

Some developers are also working on modifying the HTW to better adapt to local coal properties, such as high ash or high moisture [10,13]. Bharat Heavy Electrical Limited (BHEL) in India is developing an air-blown fluidized bed gasifier to handle high-ash Indian coals. Similarly, Herman Research Laboratories (HRL) in Australia is developing another version of an air-blown fluidized bed gasification process that is more suitable to high-moisture coals. The uniqueness of the HRL technology is the use of hot syngas to dry high-moisture coals in an up-flow entrained dryer. Coal is pressurized in lock hoppers and screw fed into the hot syngas leaving the gasifier. Dried coal is separated in a cyclone and then screw fed to the gasifier. Herman Research Laboratories refers to this technology as Integrated Drying & Gasification Combined Cycle.

Both the BHEL and Drying & Gasification Combined Cycle. Technologies have been planned for demonstration in the future.

Another version of the pressurized fluidized technology has been developed by the Gas Technology Institute. Gas Technology Institute's U-Gas pressurized fluid bed gasifier is one of the few survivors from the late 1970 era and is shown in Figure 20.5 [15,18]. In the U-Gas process, dried and milled coal is fed into the gasifier via a lock-hopper. The gasification takes place in a fluid bed, which is kept fluidized by the gasification agents steam and air or oxygen. The sloping grid at the bottom serves as a distributor of the gasification agents and to route the agglomerated ash out of the gasifier. The key to the operation is ash agglomeration and separation of low carbon ash from the fluidized bed. This is accomplished by keeping the temperature near the sloping grid close to the ash softening temperature. Because of relatively low operating temperature, the predominant reaction in the gasifier is the slow steam carbon reaction, which results in limited throughput. In addition, lack of ability to preferentially oxidize refractory-like char from coal in the fluidized bed results in low carbon conversion [19].

In order to overcome these drawbacks, KBR and Southern Company Services have jointly developed a transport reactor technology to obtain higher circulation rates, velocities, and riser densities than in conventional circulating beds, resulting in a more uniform temperature profiles, higher throughputs and higher carbon conversion [20,21]. The KBR transport reactor integrated gasification (TRIG) is presented in Figure 20.6. The TRIG gasifier is refractory-lined pipe with no internals. In TRIG technology, dry coal is fed through lock-hoppers into the upper section of the mixing zone, where it undergoes rapid devolatilization and gasification reactions. The char along with recirculating ash is separated from the syngas in a cyclone and is recycled back to the gasifier via a

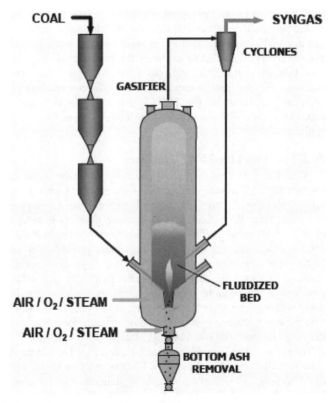

FIG. 20.5 DEPICTION OF U-GAS PRESSURIZED FLUIDIZED BED GASIFIER [15,18]

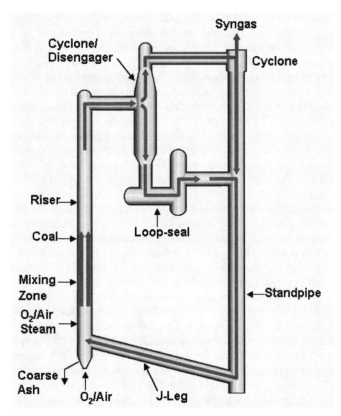

FIG. 20.6 DEPICTION OF KBR TRIG TECHNOLOGY [20,21]

standpipe and J-leg. The gasification agent, air or oxygen, along with a relatively small amount of steam, is introduced into the bottom section of the mixing zone where it is rapidly consumed by the re-circulating char from the J-Leg. The rapid mixing of bed material results in a uniform temperature and minimizes the amount of steam needed to moderate the partial oxidation reaction. This makes the TRIG technology more of a partial oxidizer, and therefore, it is more compact than fluidized bed gasifiers. An additional advantage of the TRIG is that it can be designed in either air-blown mode or oxygen-blown mode, depending on the application.

20.3.3 Entrained Flow Gasifiers

In entrained flow gasifiers, coal and the gasification agents are co-currently fed, and the coal particles are entrained with the reacting gases (see Figure 20.7). Since high velocities are required to entrain coal, the residence time in these gasifiers is short to maintain reasonable gasifier size. High-carbon conversion at these conditions can be achieved by using finely ground coal and high gasification temperatures (> 2500° F). Typically, coal and gasification agents are introduced into the gasifier at high velocities through one or more burners. The burners can be oriented in many different ways: tangentially, radially opposed, or axially. Consequently, there are several variations of entrained flow gasifiers available in today's market [10,13,22-34].

The entrained bed gasifier performance is affected to a large degree by the mixing efficiency of the burners and flow characteristics. Gasifiers with multiple burners have higher carbon conversion because of high turbulence and high gasification temperatures. The entrained bed gasifier is operated at temperatures higher than the ash melting temperature to enable removal of ash as liquid slag. Since

ash is rejected as a liquid slag, the efficiency of slagging gasifiers with high-ash coal is low, owing to the rejection of the sensible heat of the molten ash. High temperature used in the entrained bed gasifiers reduces the residence time requirement and results in syngas rich in CO, CO_2, H_2, H_2O, and very low amounts of hydrocarbons. Both slurry and dry feed entrained bed gasifiers are available. Compared with slurry feed gasifiers, the dry feed gasifiers operate at higher temperatures. Entrained bed gasifiers have many advantages that include high capacity per unit volume, high carbon conversion, and a product gas that is essentially free of heavy hydrocarbons. On the other hand, high gasification temperatures result in a number of challenges: burner and gasifier reliability due to corrosive and sticky slag, and the requirement to cool extremely hot syngas. Oxygen consumption in entrained bed gasifiers is typically the highest among the gasifier types.

The original entrained bed gasifier is the Koppers-Totzek (K-T) gasifier. The K-T gasifier was invented and commercialized in Germany in the 1940s. The K-T process has been extensively used to produce ammonia. In 1950, Heinrich Koppers GmbH installed three gasifiers for Tippi Oy at Oulu, Finland, for the production of ammonia. Since then, K-T gasifiers have been installed in Greece, Turkey, India, South Africa, Zambia, and elsewhere, mostly for ammonia manufacture [6].

Many modern entrained bed gasifiers have features that resemble aspects of the original K-T gasifier. The traditional atmospheric pressure K-T gasifier is horizontal and ellipsoidal with two burner heads mounted on the ends. Some plants installed in India use four burner gasifier designs [6]. In the K-T process, pulverized coal is premixed and entrained with steam and oxygen and injected into the gasifier burners. In addition to the high gasification temperatures of 3300°F to 3500°F, the multiple burner assembly combination improves turbulence, resulting in high carbon conversion and liquefaction of ash. Most of the ash flows down the gasifier walls as molten slag and drains into a slag quench tank. A waste heat boiler is mounted on the top of the gasifier to recover heat from the hot product gases and to generate steam.

In the late 1970s, Shell and Koppers started a joint program to develop a pressurized version of the K-T gasifier. After the takeover of Koppers by Krupp, the partnership was dissolved and each

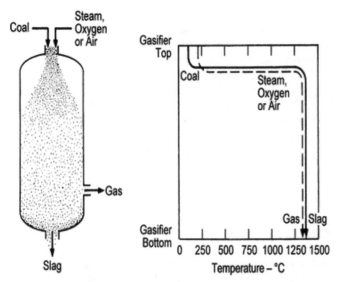

FIG. 20.7 DEPICTION OF GENERIC ENTRAINED BED GASIFIER AND RELEVANT TEMPERATURE PROFILES [7,16]

started working independently on developing the gasifier technology. Shell built a demonstration facility at their refinery in Deer Park, Texas that operated from 1987 to 1992. Krupp-Koppers built a pilot plant at Furstenhausen, Germany that operated from 1988 to 1992. The Krupp-Koppers process was called Prenflo (Pressurized Entrained Flow). Both Shell and Prenflo gasifiers share many common features (see Figure 20.8).

In 1990s, Udhe joined Krupp for the development and marketing of the Prenflo technology. Early in 2002, Shell and Krupp-Udhe announced that the Shell and Prenflo technologies would be merged and offered as a single technology. More recently, this partnership has been dissolved and Shell and Prenflo technologies are once again being offered separately [10].

Many features of the Shell Coal Gasification Process (SCGP) and Prenflo processes are very similar to the original KT process. All three processes have two horizontally opposed burners through which dry pulverized coal is introduced into the gasifier. The hot product gases flow upward through a vertical membrane cylindrical wall. This is similar to the waterwalls used in boilers to keep the refractory cool by using water in tubes. Molten ash entrained with the upward-flowing syngas is deposited on the waterwalls and flows downwards. The slag coats the waterwalls and offers a protective insulating layer augmenting the thin layer of refractory on the membrane wall. The slag is removed through the bottom of the gasifier floor and quenched in a water bath. The sudden drop in temperature of the slag causes it to solidify and break up into fine, inert, glassy, black grit. The cooled and solidified slag is removed through a lock-hopper arrangement. The top of the gasifier is quenched with a portion of the cooled product gas to ensure solidification of fly slag before it enters the syngas coolers. The solidified fly ash carried with the product gas is removed in a cyclone downstream of the syngas cooler.

Since the breaking of the Shell-Prenflo consortium, each has added additional features to include minor variations such as addition of the quench zone to cool the syngas. A variation of the Prenflo design is also being offered as Prenflo Direct Quench (PDQ) process. The PDQ process is a down flow design where dry coal is fed through four opposed burner at the top into a gasifier equipped with a membrane wall. The raw gas is quenched directly with water at the bottom exit. The quenched Shell design maintains the up-flow gasifier design, but incorporates a water quench system downstream of the gasifier and upstream of the syngas cooler [33].

The origins of the GE Coal gasification process lie in Texaco's gasification technology gasifying heavy petroleum refinery streams. Texaco's gasification technology has been widely used since the early 1950s for production of chemicals [9]. In the 1980s, Texaco designed the Kingsport Gasification facility in Tennessee for production of chemicals and the first coal-based IGCC plant at Southern California Edison's Cool Water Station in California.

GE Energy acquired the Texaco Coal Gasification process from Chevron in 2004 after Texaco was purchased by Chevron [10]. The GE process uses slurry feed down-flow entrained flow gasifier (Figure 20.9). Coal is wet milled, slurried with water, and charged into the gasifier with a piston pump. Slurry feed is introduced into the gasifier with oxygen through a single burner that is located centrally at the top of the gasifier. The reactor shell is a refractory-lined vessel. The slag is quenched and removed from the bottom of the gasifier via a lock-hopper arrangement. The water leaving the lock-hopper is separated from the slag and recycled for slurry preparation. GE offers three variations to cool the syngas: a radiant boiler, a total water quench, and a combination of the two. The selection among these three alternatives is a choice of cost and application. In the total water quench, less heat is recoverable as high-pressure steam, whereas with a radiant boiler, high-pressure

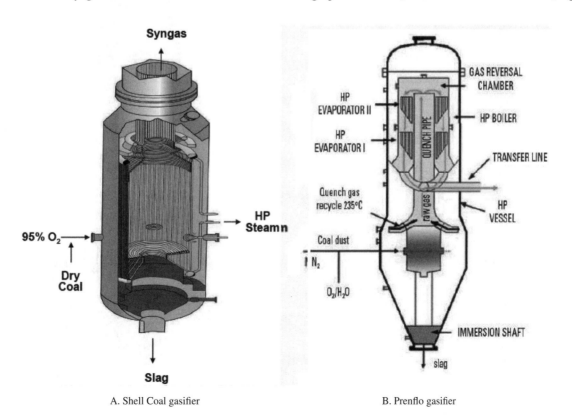

A. Shell Coal gasifier   B. Prenflo gasifier

FIG. 20.8 DEPICTION OF SHELL AND PRENFLO GASIFIERS [25,26]

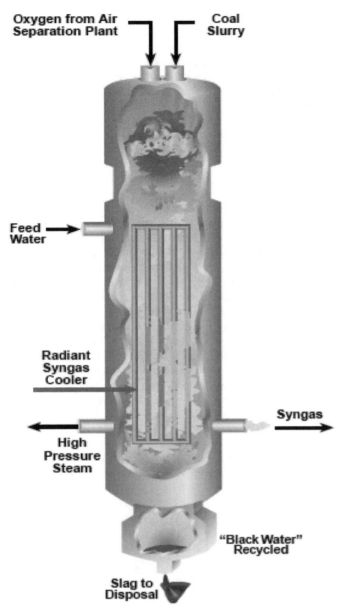

FIG. 20.9 DEPICTION OF GE GASIFIER [29-31]

The ash is removed as molten slag through a tap hole at the bottom of the gasifier. The molten slag is quenched in water and crushed and is continuously removed via a pressure let down system. The E-Gas process is one of few technologies that do not use a lock hopper arrangement to remove slag.

The Siemens Gasification process also has had a long history of ownership. In 1975, Deutsches Brennstoffinstitut Freiberg developed this process for gasification of German coals. In the past, the technology has been marketed under the name GSP, Noell, and Future Energy. Siemens acquired Future Energy gasification technology in 2006 and is marketing it as Siemens Fuel Gasification (SFG) technology [10].

Siemens technology features a top-fired burner through which coal and gasification agents are introduced and is shown in Figure 20.11. This feature somewhat resembles that of GE technology. But unlike the GE process, the Siemens process is based on a dry feed system, and the gasifier wall is refractory lined and cooled with water. The spirally wound cooling screen is covered with SiC castible refractory and layers of solid and molten slag. To provide for flexibility, Siemens offers both total quench and partial quench design.

The Mitsubishi Heavy Industries (MHI) process is an air-blown, two-stage, dry feed, up-flow gasifier with a membrane wall and no refractory [23]. The two-stage concept, as shown in Figure 20.12, is similar to the E-Gas technology; the differences being an

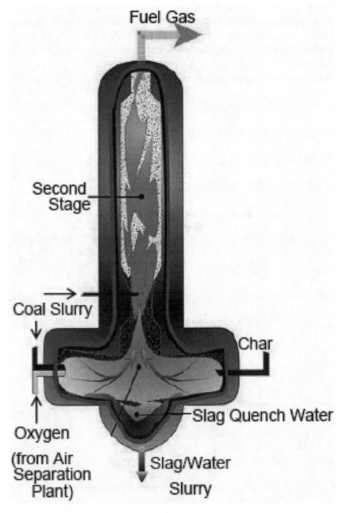

FIG. 20.10 DEPICTION OF E-GAS GASIFIER [10,28]

steam is generated. The radiant boiler option is more efficient but a more capital intensive option because of the exotic alloys needed for the syngas cooler.

Dow originally developed the E-Gas Technology in 1970s. The E-Gas technology is now being marketed by ConocoPhillips [10,28]. Similar to the GE technology, E-gas uses a slurry coal feed system that is fired into the bottom of a horizontal cylinder with two opposed burners (see Figure 20.10). The burner section is very similar to the KT technology. All the oxygen required for gasification is fed through these two burners. In addition, a portion of the coal is fed just above the first-stage burner zone to cool the syngas. In theory, this behaves as a staged fuel burner. Since there is no oxygen fed to the second stage, some products of devolatilization remain in the syngas product as it flows upward. The syngas is cooled in fire tube syngas coolers to generate steam. Char produced in the second stage is removed in a hot, dry filter downstream of the syngas coolers and recycled back to the first stage. The two-stage operation improves the efficiency and reduces oxygen requirements, but increases capital cost.

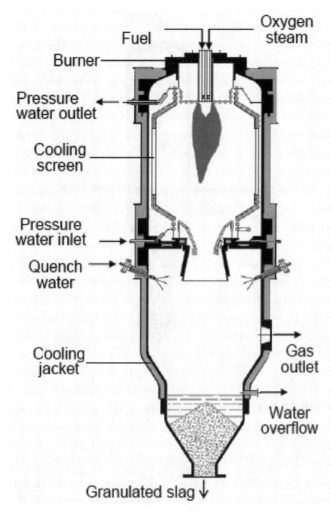

FIG. 20.11 DEPICTION OF SIEMENS GASIFIER [10,15]

The TPRI gasifier is a dry coal–fed, oxygen-blown, two-stage up-flow entrained bed reactor with a membrane wall. The two-stage concept is similar to the slurry fed, oxygen-blown, E-gas and the dry-feed, air-blown, MHI processes. In the TPRI process, coal along with steam and oxygen is fed to the first stage. In the second stage, only coal and steam are fed. The TPRI gasifier is also offered in two variations to cool the syngas: steam generation mode and quench mode. The TPRI gasifier has been proposed for a 250-MW IGCC GreenGen project in China, similar to the shelved US FutureGen project, to demonstrate IGCC with carbon capture and sequestration. The TPRI is jointly owned by major Chinese power companies including China Huaneng Group, China Huadian Corp, China Power Investment Corp., China Datang Corp, and China Guodian Corp. Therefore, TPRI is well positioned to be a major technology provider both inside China and internationally. For the US market, TPRI has formed an alliance with Houston-based Future Fuels and is working on a 270-MW Good Spring IGCC project. This project will be the first-of-a-kind IGCC project that applies non-Western technology to a US project.

The Electric Power Development Corporation in Japan has been working on developing the EAGLE gasifier, which is similar in concept to TPRI. The EAGLE gasifier is a dry coal–fed, oxygen-blown, up-flow entrained bed reactor with a membrane wall. Dry coal is fed at two levels into the gasifier. The first level operates

air-blown and dry feed system. The first stage is operated in combustion mode at high enough temperatures to allow the ash to be removed as slag. In the second stage, coal is introduced to cool the syngas by the devolatilization reactions. The char is separated from the gas in a cyclone and candle filter and recycled back to the first stage. This technology was recently demonstrated at Nakoso Power Plant in Japan using a modified M701DA gas turbine [23].

Among several technologies being developed in China, two technologies, opposed multi-Bruner (OMB) and Thermal Power Research Institute (TPRI), have recently appeared in the market place and are threatening to break the monopoly of Western technologies (see Figure 20.13). The OMB technology developed by East China University of Science and Technology is a slurry fed down-flow–entrained gasifier. Four top-mounted, opposed fired burners introduce the slurry with oxygen into the gasifier. The reactor is a quench design with special internals to reduce water carryover. The slag is removed via a lock-hopper arrangement. A unique feature of the OMB technology is the particulate removal system, where the particles are removed from the syngas with a combination of jet mixer, cyclone, and a water scrubber. The OMB technology is also being developed using a dry coal feed system with a membrane wall. In principle, the dry feed OMB gasifier design feature will make the gasifier design very similar to the Pren-flo Direct Quench.

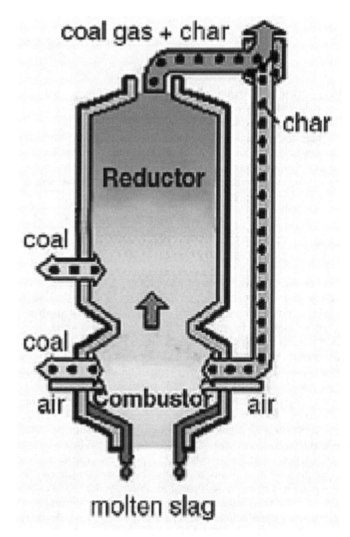

FIG. 20.12 DEPICTION OF MHI GASIFIER [15]

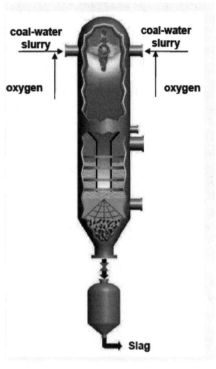

A. OMB gasifier

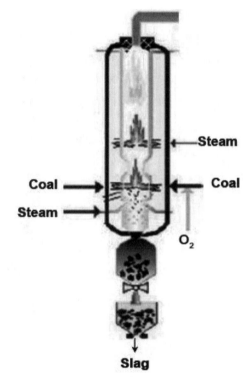

B. TPRI gasifier

FIG. 20.13 DEPICTION OF OMB GASIFIER AND TPRI GASIFIER [15]

oxygen rich, whereas the second level is starved of oxygen. The hot syngas leaving the burner zone is quenched with cooled syngas, very similar to the art practiced in the Shell technology.

There are several other gasification technologies currently being developed by various organizations in several countries. Most of these technologies are in various stages of development and further away from commercialization. Additional information on these developing technologies may be found elsewhere [10,13,15].

20.4 GASIFICATION ISLAND DESIGN ISSUES AND COST IMPACTS

Even for a known feedstock and a known end application, deciding on a type of gasifier and the design of the gasification island is a complex issue. The most cost-effective path is not always straightforward. Often there are conflicting issues between capital cost, operating cost, and efficiency. High efficiency is desirable as it reduces the plant size and the amount of emissions from the plant. This is the key reason why most of the gasification technology providers are now offering various options to address these issues. For syngas cooling, some technology providers are providing at least two options: syngas cooler or quench. The use of a syngas cooler to generate high-pressure steam increases both efficiency and capital cost. The cost of the syngas coolers can be a very significant portion of the overall IGCC capital, and in several of the currently operating plants with large water tube syngas coolers, it represented as much as 20 % of the total Gasifier Island capital cost. On the other hand, quench systems result in lowering both capital cost and efficiency, but increase operating cost. Dry feeding coal improves efficiency but increases both capital and operating cost due to the need for lock hoppers and the pressurization

requirement. Although it may be more economical to slurry feed the coal, there is a significant energy penalty associated with it because valuable energy is consumed in vaporizing water used for slurring the coal.

During the initial stages of planning, care should be given to the development of a block flow diagram, which clearly defines the processing steps required to meet the project objectives. This is a challenging task, since a wide variety of products including: power, chemicals, fuels, and metals can be produced with syngas. In addition, the design provides the ability to generate multiple products. Therefore, project developers like to have flexibility built into the plant, making the design a complex task.

Irrespective of the end application, the unit processes and unit operations in the gasification island are similar. However, the design criterion may not necessarily be the same. Figure 20.14 is a typical block flow diagram that describes various unit operations in a typical gasification island. Further definition of the block depends on the specific application. This flow sheet is typically used to define the project basis and establish a cost estimate of the gasification island.

Several studies on cost indicate that for a given gasification island scheme, the capital cost of various gasification technologies are within 10% of each other, with the dry feed systems being on the high side [14,35]. Air blown gasifiers would fall on the lower side of the cost spectrum. Although these studies focus on the type of gasifier as a differentiating factor in determining cost, packaging other unit operations within the gasification island are critical for optimizing cost and efficiency.

Relative costs of various gasification blocks identified in Figure 20.14 are shown in Table 20.3. These cost numbers are based on several studies available in the open literature [14,35-38]. The wide range in relative cost reflects the various options available to

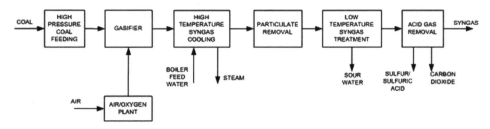

FIG. 20.14   BLOCK FLOW DIAGRAM OF A TYPICAL COAL GASIFICATION ISLAND

perform the same unit operation. For coal preparation and feed, the low cost option reflects slurry feed systems and the high cost option is for dry feed systems. During the initial stages, the high-pressure coal feeding block would need to be defined for the type of feeder required: slurry feed or dry feed. Although most of the gasifier developers promote a particular type of feed, GE's pursuance of dry feeding coal clearly indicates that the developers are not closed to improving their technology to improve capital cost and/or efficiency.

Table 20.3 indicates that the ASU is one of the costliest blocks in oxygen-blown gasifiers. The compression cost for an air-blown system would be about half that of ASU. Entrained bed gasifiers generally promote the use of oxygen; however, MHI's recent success with an air-blown entrained bed gasifier suggests that air may also be an option in the future, at least for power generation. In general, air-blown gasifiers are considered to be more efficient and economical for power generation purposes. In fact such flexibility is already built into the fluid bed systems that can be designed in either oxygen to air, depending on the production mode.

For IGCC applications, the higher efficiency gasifiers will likely be based on a dry feed and air-blown system. Combustion turbines designed to fire natural gas offer the option to extract air while firing in syngas mode. The extracted air can be used to supplement all or part of the air requirement to the ASU for oxygen-blown systems. In the air-blown mode, the extracted air can be recompressed to supplement the main air compressor. The use of extraction air reduces auxiliary power requirements.

The gasifier is another area of the gasifier island that contributes significantly to the capital cost. The high end numbers reflect entrained flow membrane lined gasifiers and fluidized bed gasifiers. Low reaction rates in fluidized bed gasifier will require multiple gasifiers. The low number reflects cost of a transport gasifier that operates as a partial oxidizer and at lower temperatures [38]. The mode of operation, gasification versus partial oxidation has significant impact on gasifier size and oxidant requirements. The low

temperature moving bed gasifiers are probably not going to be economical going forward as they require large reactor volumes.

The ash removal system is generally considered to be an integral part of the gasifier and its cost is reflected in the cost of the gasifier block.

The high cost of syngas cooling is one area where a number of technology providers are already providing options to balance cost and efficiency. Well known options for cooling syngas include: quench, syngas coolers, and a combination of both. The syngas cooler provides an option to raise steam while cooling the raw syngas. The steam can be either saturated or superheated. Production of superheated steam requires the use of exotic alloys, making it an expensive option. The efficiency improvements due to raising superheated steam will need to be evaluated during the initial stages of the gasifier island design. The low cost range in Table 20.3 reflects the direct quench option, where the energy from the hot raw syngas in not recovered, thereby lowering the plant efficiency.

Selection of a particulate filtration device is another area where options such as filters, cyclones, and/or wet scrubbing are available. In many gasification applications, a barrier filter is used. Removal of the entrained particulate matter from the raw gas at temperatures above the dew point avoids condensation problems and keeps the fly ash dry to facilitate its removal. Dry fly ash recovery significantly reduces the buildup of salts in the recycle process water and the cost of wastewater clean-up. Fluid bed gasifiers use cyclones at the gasifier exit to recover the bulk of the particulates and char material which are then recycled back to the gasifier. The quench systems used for cooling syngas also remove fly ash, chlorides, metals, and some sulfur from the raw syngas. Therefore, in a quench system the treatment and disposal of water becomes challenging. The syngas cooling and particulate filtration are two operations whose costs are comparable to that of the gasifier. Therefore, careful evaluation of these blocks is essential to optimize cost.

The definition of the low temperature syngas treatment block is mainly dependent on the end use application. For example, if the application is for IGCC, then removal of condensate from the syngas prior to mercury and sulfur removal is desirable. Removal of condensate facilitates the removal of ammonia and chlorides from the syngas. For chemical production or IGCC carbon capture, the sour syngas has to be shifted, and heat integration and gas humidification prior to shifting for systems using syngas coolers (rather than a quench system) become important. The number of stages required for shift, either one or multiple stages, depends on the application. For example, for hydrogen production either two or three stages would be preferred to maximize hydrogen production. For IGCC carbon capture, a single stage may be sufficient to meet the project requirements. Depending on the shift catalyst requirement, the syngas will need to be conditioned to meet the required H_2O/CO ratio. In schemes using upstream quench systems the conditioning step may be eliminated. The conditioning can be performed in a saturator

TABLE 20.3   RELATIVE COST OF VARIOUS BLOCKS IN THE GASIFICATION ISLAND

Gasification Island Block	Low	High
Coal Prep and Feed	5	15
ASU	20	25
Gasifier	15	25
Syngas Cooling	5	20
Particulate Removal	5	15
LT Syngas Treatment	5	10
Acid Gas Removal Unit	10	20
Utilities and Offsites	15	20

or a scrubber that not only humidifies the sour syngas, but also facilitates removal of mercury, ammonia, and chlorides.

The Acid Gas Removal (AGR) processes are very effective at removal of H_2S down to very low levels but are less effective at COS removal. Some of the shift catalyst hydrolyzes COS to H_2S and HCN to NH_3. Depending on the hydrolyzing capability of the shift catalyst and the application, an additional hydrolysis reactor may be required to remove COS and HCN. Downstream of the shift and hydrolysis reactors the syngas is further cooled to remove excess condensate prior to treating the sour syngas for mercury removal.

The syngas treatment area also includes the removal of trace components such as mercury. In existing IGCC plants, mercury removal is not practiced. This will change in the future, at least in United States, Europe, and Japan. In most IGCC projects being explored in these regions, mercury removal is being seriously considered. For chemical applications, mercury removal is generally practiced to avoid coating and poisoning downstream catalysts. Typically, sulfided carbon or impregnated metal are used as bed material to remove mercury.

Technologies for acid gases removal such as H_2S, CO_2 or combination of the two have been used extensively in the natural gas processing industry. Several types of AGR technologies are available to remove acid gases. Most popular technologies use solvents to absorb acid gases. The AGR technology is generally categorized by nature of the solvent bonding with the acid gas, which is either physical or chemical. In a physical solvent process, the bonds between the solvent and acid gas are weak, and therefore, reducing pressure regenerates the solvent. In a chemical solvent process, heat is required to break the gas-solvent bonds and regenerate the solvent. Therefore, the selection of the solvent type will need to be based on the comparison of capital cost and efficiency for a given application.

In a typical AGR process the sour syngas is contacted countercurrently with a solvent in an absorber to remove the acid gases. For carbon capture case, two absorbers will be required — one for H_2S removal and the other for CO_2 removal. The treated syngas from the absorbers then leaves the gasification island for its applications, power generation or chemical synthesis.

The solvent leaving the absorber contains the acid gases and is regenerated. The regeneration is performed by either heat and/or by pressure reduction, depending on the nature of the solvent. Both physical and chemical solvent processes are well suited to remove both H_2S, as well as, CO_2. In the carbon capture mode, the solvent is regenerated in a two-step process; one for generating CO_2, and the second for liberating H_2S. After regeneration, the acid-gas free lean solvent from the bottom of the regenerator is cooled and pumped back to the absorber. The H_2S containing acid gas leaving the regenerator is further processed to recover sulfur. H_2S produced during solvent regeneration is typically converted to sulfur or sulfuric acid. Conversion to sulfur is more expensive as sulfur plants are more expensive than sulfuric acid plants. However, if there is no market for sulfuric acid, conversion of H_2S to sulfur may be the only choice.

For low sulfur coals, direct oxidation technologies are available. In direct oxidation technology, H_2S is directly oxidized to sulfur in an absorber. Typically, the oxidation agents include SO_2 or oxygen that is dissolved in an AGR solvent. The sulfur crystals are removed by filtration. The solvent along with make-up oxidant is recycled back to the absorber. Direct oxidation technologies have an advantage as they oxidize H_2S to sulfur in one step and do not require a Claus unit. Direct oxidation technology is not applicable for the removal of CO_2.

Clearly, the definition of each block within the block flow diagram (Figure 20.14) will need careful evaluation to optimize the cost and efficiency. Search for a "cookie cutter" design (one design fit all) is desirable but will be challenging as the optimum design will depend on a number of items including: feedstock type, location, permits, end application, developer preferences, etc. Therefore, options to define the block flow diagram during the initial planning stage should be carefully evaluated to minimize changes to the project schedules and cost.

20.5 APPLICATIONS OF COAL GASIFICATION

Syngas generated from coal can be used for a wide variety of products including: power, chemicals, fuels, and metals. In addition, the plant can be designed to produce multiple slates of products. In this sub-section, application to individual products is highlighted. Co-production can be achieved by integrating the desired applications.

20.5.1 Power Generation

In the United States and throughout the world, most of the electricity is generated from coal. In most of the power plants, coal is burned in a boiler to generate steam. Steam is then expanded in a turbine to generate power. The turbine exhaust is condensed and pumped back to the boiler to generate steam. Figure 20.15 represents a typical coal-based power plant that is based on Rankine cycle.

The cycle of converting heat to work, better known as the Rankine Cycle, is about 25 to 30% efficient. In many power plants the exhaust (flue) gas from the boiler is generally not treated and emitted to the atmosphere. This is the main reason why coal-based power plants are considered "dirty". Tightening of the emission regulations globally has resulted in installation of scrubbers for the removal of sulfur dioxide and selective catalytic reduction for the removal of NOx.

In order to improve efficiency, newer power plants use gas turbines to generate power. The gas turbine is based on Brayton cycle that uses the fuel in a combustion turbine to generate electricity by expanding hot gases over rotating turbine blades. Typical efficiency of this cycle is 35% to 45%. In addition, heat can be recovered from the hot turbine exhaust to generate steam. The steam is expanded in a steam turbine to generate additional electricity. The combination of combustion turbine and steam turbine is shown in Figure 20.16. The "combined cycle" arrangement increases the overall energy efficiency to about 60%.

Typically, natural gas is the fuel of choice that is used to generate hot combustion gases to be expanded in a combustion turbine

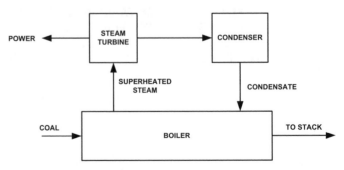

FIG. 20.15 BLOCK FLOW DIAGRAM OF A GENERIC COAL-FIRED POWER PLANT BASED ON RANKINE CYCLE

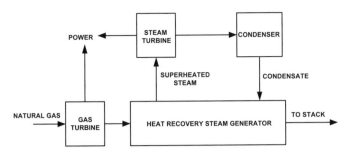

FIG. 20.16 BLOCK FLOW DIAGRAM OF A GAS FIRED COMBINED CYCLE UNIT BASED ON COMBINED BRAYTON AND RANKINE CYCLES

to generate power. In some regions of the world due to high natural gas price or limited natural gas supply, alternate fuels such as coal derived syngas is being pursued for power generation. In IGCC applications, natural gas is replaced with syngas as fuel in the gas turbine (see Figure 20.17). The gasification process is more complex and involves significantly more capital (see Figure 20.14 and Table 20.3). In places such as Europe, India, China, and Africa where natural gas is not available in abundance, coal gasification is being seriously considered. Where natural gas commands a high price, there may be enough incentive to process coal into a gaseous fuel for power generation purposes.

For IGCC applications, syngas needs to free of particulates and sulfur. In recent projects, removal of trace metals such as mercury and arsenic is also being explored. The IGCC application facilitates the integration between the steam and air/oxygen systems to improve efficiency. While firing syngas in a combustion turbine designed with flexibility of firing natural gas, some of the air from the combustion turbine compressor can be extracted for use in other parts of the plant, as discussed earlier.

The extracted air can be used in the air separation plant to generate oxygen (for oxygen-blown gasification systems), or further compressed to provide air for air-blown systems. For the steam side integration, steam generated in the gasification island during syngas cooling is used in the steam turbine to generate additional power, thereby increasing power output. A typical integration of the gasification island with the power island is depicted in Figure 20.18.

The combined cycle efficiency of a coal-based power plant is less than that of natural gas fired units due to the additional energy requirements to process coal and syngas. In several countries, regulators and environmental lobbyists are calling for the capture and sequestration of CO_2, especially from coal-based power plants. Removal of CO_2 from traditional power plants will be challenging as the exhaust from these plants are close to atmospheric pressure

leading to a large volume of gas to be treated. Removal of CO_2 from high-pressure syngas provides an advantage for the IGCC application as compared with traditional coal-fired power plants. For carbon capture based IGCC application, CO in syngas will need to be shifted to CO_2 to facilitate the capture. Additionally, the acid gas removal system will need to be designed to remove both H_2S and CO_2, and the gas turbine will need to be modified to use a fuel rich in hydrogen. Search for higher efficiency and tightening environmental regulations will make coal-based IGCC plants an attractive option to generate electric power in the future.

20.5.2 Liquid Fuels

One of the more attractive applications of gasification is the conversion of coal into liquid fuels. Fischer-Tropsch (F-T) synthesis was used in World War II to convert coal to liquid fuels [39-43]. Since then this technology has been used extensively in South Africa where due to an oil embargo, South Africa had to rely on its extensive coal resources to provide liquid fuels, as mentioned earlier.

The process to generate liquid fuels was originally invented by Badische Anilin und Soda Fabrik (BASF) in 1913. BASF used a high-pressure catalytic process to produce long chain hydrocarbons and oxygenates from CO and H_2. In the mid-1920s, Fischer and Tropsch used coal derived syngas to generate straight chain hydrocarbons (popularly referred to as wax) and light hydrocarbons. They used an active iron based catalyst for their synthesis. The conversion of CO and H_2 can be represented by the equation:

$$CO + 2H_2 \longrightarrow -[CH_2]- + H_2O \qquad (20.8)$$

Although BASF originally invented the process, the reaction is better known as the F-T reaction. The F-T reaction requires H_2/CO ratio of 2. The H_2/CO ratio of syngas produced from coal varies from about 0.6 to 0.8, depending on the type of gasifier and the amount of steam or water added to the gasifier. Therefore, the syngas has to be shifted to produce additional hydrogen by the water gas shift reaction:

$$CO + H_2O \longleftrightarrow CO_2 + H_2 \qquad (20.9)$$

The overall F-T reaction is represented by combining these two reactions to yield:

$$2CO + H_2 \longrightarrow -[CH_2]- + CO_2 \qquad (20.10)$$

The use of iron based catalyst promotes the water-gas shift reaction which makes them ideally suited for coal-based syngas. In recent years, cobalt based catalyst has become popular for F-T synthesis. Cobalt does not promote the water-gas shift reaction and the coal derived syngas has to be shifted prior to the F-T synthesis, thereby requiring additional processing. Cobalt based catalyst are more suited to process syngas derived from natural gas as it inherently contains more hydrogen. The H_2/CO ratio of natural gas based syngas is typically about 2 and therefore does not require shifting. This process is popularly referred to as Gas-to-Liquids (GTL) technology. Similarly, conversion of coal derived syngas to liquid fuels is popularly referred to as Coal-to-Liquids (CTL).

The wax products from the F-T synthesis reactions need to be hydro-treated, hydro-cracked, and fractionated to produce naphtha and diesel. Figure 20.19 summarizes major processes and unit operations that are typically required to convert syngas into liquid fuels, such as naphtha and diesel. Liquid fuels from the F-T process are essentially sulfur free and therefore command a higher premium in the market.

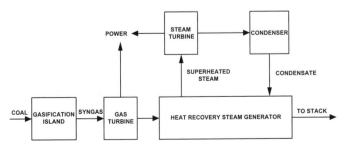

FIG. 20.17 GENERIC BLOCK FLOW DIAGRAM DEPICTING THE SUBSTITUTION OF NATURAL GAS WITH COAL IN A COMBINED CYCLE PLANT

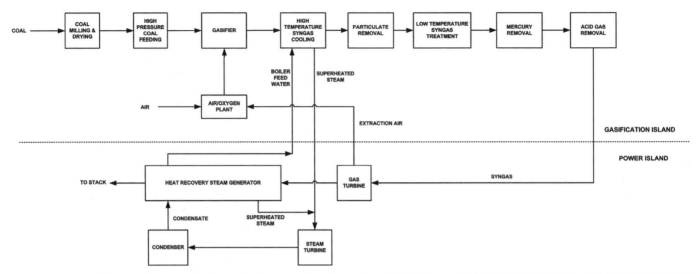

FIG. 20.18 BLOCK FLOW DIAGRAM DEPICTING A TYPICAL INTEGRATION OF THE GASIFICATION ISLAND WITH THE POWER ISLAND

Methanol, one of the top ten chemicals produced globally, can also be produced from syngas. Methanol can be used directly or blended with other petroleum products to produce a clean burning transportation fuel. Similar to the F-T reactions, methanol can also be produced by CO and H_2 by using copper and zinc based catalyst according to the reaction:

$$CO + 2H_2 \longrightarrow CH_3OH \qquad (20.11)$$

The methanol process also requires H_2/CO molar ratio of 2. The CO/H_2 ratio from a typical coal gasifier is less than 0.8. Therefore, CO has to be shifted to prior to conversion of syngas to methanol. Both F-T catalysts and the methanol synthesis catalyst are expensive and susceptible to poisons such as chlorides and sulfur. Therefore, prior to the synthesis, syngas must be treated to remove these contaminants. Ammonia and halides are typically removed by water scrubbing. During the water wash step, the ratio of H_2O/CO can be adjusted to meet the shift catalyst requirements. H_2S can be removed with any preferred AGR system that will satisfy the specifications required to prevent poisons from reaching the catalyst. Typically, a COS hydrolysis unit may be required to convert COS to H_2S as most AGR systems do not remove COS. The acid gas removal will need to remove H_2S to match the syngas specifications for liquid fuel synthesis, as discussed earlier.

In recent years, China has been converting methanol to DME for use as a substitute for liquefied petroleum gas (LPG) and diesel. The dehydration reaction of methanol takes place over an acid catalyst by the following reaction:

$$2 CH_3OH \longrightarrow CH_3OCH_3 + H_2O \qquad (20.12)$$

In addition to its usage as a liquid fuel substitute, methanol is an important chemical intermediate to produce formaldehyde, methyl tertiary butyl ether (MTBE), acetic acid, methyl amines, and methyl halides.

20.5.3 Synthetic Natural Gas

Conversion of coal into synthetic natural gas (SNG) is sometimes desirable to facilitate the transport of fuels with higher energy density to the market place. SNG is primarily methane, and is synthesized over a nickel based catalyst by the following reaction:

$$CO + 3H_2 \longrightarrow CH_4 + H_2O \qquad (20.13)$$

The methanation reactions are exothermic and the heat can be used to generate steam as shown in Figure 20.20. Capability of producing superheated high-pressure steam has created interest in the development of a catalyst that can operate at higher temperatures.

This reaction requires the H_2/CO ratio of 3, which indicates that CO will need to be shifted to a higher extent than required for the

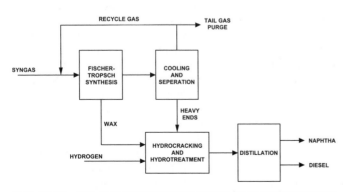

FIG. 20.19 MAJOR PROCESS BLOCKS FOR CONVERTING SYNGAS TO LIQUID FUELS

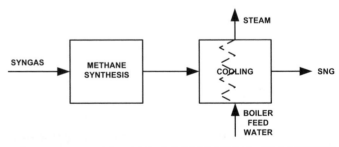

FIG. 20.20 MAJOR PROCESS OPERATIONS TO CONVERT SYNGAS TO SNG

liquid fuel processes discussed earlier. To eliminate methanation catalyst poisons, the syngas will need to be scrubbed with water to remove ammonia and halogens. During the water wash step, the ratio of H_2O/CO can be adjusted to meet the shift catalyst requirements. Sulfur compounds will need removal by the AGR system to eliminate catalyst poisons. Some CO_2 may also need to be removed by the AGR system to meet the natural gas specifications.

20.5.4 Hydrogen and Ammonia

For hydrogen synthesis from syngas, all of the CO will need to be shifted. The sour shift catalyst can accomplish most of the desired shifting. However, to bring the CO concentration down to low levels, low temperature sweet shift may be required. To eliminate catalyst poisons, the syngas will need to be scrubbed with water to remove ammonia and halogens. Both sulfur compounds and CO_2 will need removal by a dual AGR system. Since the entire CO needs to be shifted in multiple stages, the syngas will have to be adjusted to meet the required H_2O/CO ratio. Humidifying with steam or quenching with water can achieve the desired adjustment. If the hydrogen is for ammonia production purposes, then the gasifier can be either enriched air blown or oxygen blown. For all other purposes the gasifier must be operated in oxygen-blown mode to minimize diluting the hydrogen with nitrogen.

Ammonia catalyst is poisoned with CO and CO_2. Typically after the three stages of shift and AGR, about 0.1 to 0.5 % CO and CO_2 still remain. They are removed by using nickel based catalyzed reaction with hydrogen to produce methane, which is relatively inert to the catalyst. Figure 20.21 shows the schematics for a typical gas conditioning in an ammonia plant.

The gases after methanation steps contain residual water vapor. Water vapor acts as a temporary poison to the ammonia synthesis catalyst and must be removed. The gases are cooled and dried with molecular sieves to remove trace amounts of water. The molecular sieves also remove CO_2, H_2S, COS, H_2O and organic solvents. A liquid nitrogen wash column removes any remaining CO, CH_4, and Ar. The cryogenic effect is generated by injecting liquid nitrogen into the purified hydrogen stream. The amount of nitrogen added in the nitrogen wash is adjusted to obtain the desired ratio of 3:1 between nitrogen and hydrogen for the ammonia synthesis.

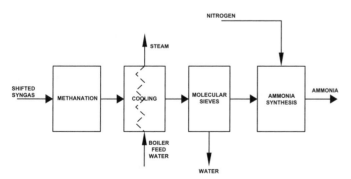

FIG. 20.21 SCHEMATICS FOR CONVERTING SHIFTED SYNGAS TO AMMONIA

20.6   OUTLOOK

Coal plays a pivotal role in the global energy market. Coal is widely available with reserves estimated to last over 140 years and is among the lowest cost fuel. In addition, coal can be gasified and converted to other energy resources such as liquid hydrocarbons

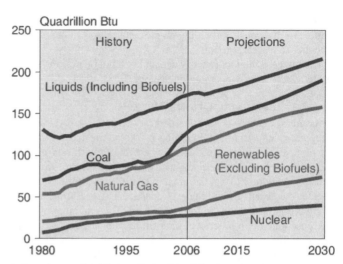

FIG. 20.22 GLOBAL ENERGY CONSUMPTION BY FUEL TYPE [44]

and SNG. As discussed earlier, for some applications, syngas generated from coal can be substituted for oil or natural gas. Two of the fastest growing economies, China and India, have limited oil and gas resources, but are rich in coal reserves. Therefore, incentive for coal usage via gasification as an alternative to supplement oil and natural gas will be strong. Gasification processes are capital intensive, and therefore, its future is closely linked to the projected energy prices. The interest in coal gasification is intimately linked to the global energy market, but not vice-versa. The impacts of global energy consumption pattern and consumption growth areas on coal utilization are briefly reviewed in the following sections.

20.6.1   Global Energy Consumption Pattern

The global energy and the industrial feedstock markets are dominated by fossil fuels. In 2006, fossil fuels accounted for about 450 Quadrillion (10^{15}) Btu of energy consumed globally (see Figure 20.22) [44]. Among the fossil fuels, liquid fuels dominate the market at about 38%, followed by coal at about 28% and natural gas at about 24%.

All of these fossil fuels satisfy special niches in the marketplace. Over 74% of liquid fuel is consumed by the transportation sector, and the industrial sector consumes the balance. Industrial sector represents those that use the fuel as a feedstock to produce end products such as: chemicals, plastics, metals, glass, cement, paper, etc. About 62% of coal is used for electricity generation, while about 34% is consumed by the industrial sector. Major consumers of natural gas are the industrial sector, which consumes about 43%, followed by electricity production at about 35%.

Figure 20.22 shows that fossil fuels will continue to dominate the energy markets, and its usage will continue to grow in the foreseeable future. The oil consumption trend shows a slow down during 2009 to 2011 due to global recessions. As the world economy picks back up, the consumption is expected to return to the normal consumption levels. Natural gas consumption is expected to follow traditional usage pattern. Since the new millennium (2000 A.D.), coal usage has accelerated, and according to EIA, its usage in the near future is expected to grow at a much faster rate than all of the other fossil fuels. However, the type of technology used to utilize the coal will be influenced by future regulations implemented to address concerns regarding greenhouse gas emissions. Increase in energy demand and regulations on greenhouse gas emissions will likely promote coal usage via gasification. This is because gasification provides for a cleaner and more efficient method to utilize coal while simultaneously providing an efficient option to capture the greenhouse gas carbon dioxide.

The increase in global energy demand will not be uniform and is expected to come from emerging markets.

20.6.2   Identifying Growth Markets

Based on the growth rate the markets can be classified into two categories: industrial and emerging markets. The industrial market represents a mature economy where the growth rate is slow and predictable. EIA defines the industrial market as the Organization for Economic Cooperation and Development (OECD) members that represent United States, Canada, Mexico, Europe, Japan, South Korea, Australia, and New Zealand. The emerging markets are those that are in process of industrialization and represent rapidly growing economies. The emerging market is represented by countries not belonging to OECD and includes Russia, China, India, Brazil, and countries in Middle East and Africa. This group is referred to as the non-OECD nations. The energy consumption of these two groups is shown in Figure 20.23 [44].

Figure 20.23 shows that before the new millennium, the non-OECD nations consumed less energy and at a slower rate than OECD nations. Since the beginning of the new millennium, the energy demand in non-OECD nations has grown at a much faster rate than OECD nations. And, in 2007, the energy consumption in non-OECD nations surpassed the total energy consumed by the OECD nations. According to EIA, by 2030 non-OECD nations are estimated to consume about 60% of the total world energy. This accelerated consumption of energy resources in non-OECD nations is primarily due to demand from China and India.

According to the EIA [44], the combined energy usage in China and India has grown from 10% of the total world's energy consumption in 1990 to 19% in 2006 and is projected to grow to 28% in 2030. Other non-OECD economies in Middle East, South America, and Africa are also expected to grow at a robust rate. Therefore, in the foreseeable future the growth in energy demand will be in non-OECD nations.

20.6.3   Energy to Fuel the Growth Markets

The coal consumption data shown in Figures 20.22 and 20.24 suggests that most of the energy demand in non-OECD nations has been met by coal; and coal will continue to play a dominant role

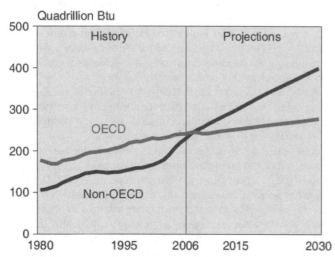

FIG. 20.23   COMPARISON OF ENERGY CONSUMPTION BY OECD AND NON-OECD NATIONS [44]

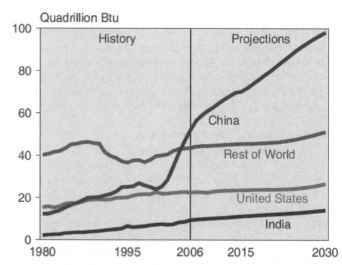

FIG. 20.24   COAL CONSUMPTION IN SELECTED WORLD REGIONS [44]

in meeting the future energy demand. As mentioned earlier, coal reserves are abundant and amount to about 930 billion short tons, or about 140 years of reserves at the present consumption rate [44]. Over 90% of coal reserves are in countries such as United States, Russia, China, India, Australia, South Africa, and Eastern Europe; the nations where most of the growth in energy demand is projected by EIA to occur. Therefore, most of the growth in energy demand will most likely be met by coal.

Coal consumption in selected world regions is shown in Figure 20.24 [44]. According to EIA, United States, China, and India account for over 90% of net increase in coal consumption. Figure 20.24 also shows that China's coal usage has grown exponentially since the beginning of the new millennium and is projected to grow rapidly in the foreseeable future. China's coal usage is expected to quadruple from 2000 to 2030 and will account for about half of the world coal consumption [44]. Figure 20.24 also indicates that India consumes about one-fifth as much coal as China. Also, India's coal usage growth is faster than that of the United States.

Over two-thirds of the coal consumed by China, United States, and India is predominantly for power generation. Over half of the electric power in the United States is generated from coal, compared with 80% in China and 70% in India. Furthermore, coal accounts for two-thirds of electricity generation in non-OECD. Therefore, in the foreseeable future coal is likely to continue dominating the power market in these countries. Although most of the coal usage will follow the traditional usage, there exists a good opportunity for coal gasification to capture part of the market.

20.6.4   Outlook for Coal-to-Liquids

In 1970s oil shortage caused by the Arab oil embargo led to a steep increase in oil prices (See Figure 20.25).

To minimize impact of imported oil, United States and Europe led conservation efforts to minimize oil usage by replacing oil with natural gas, especially in the existing electricity generation sector. They also explored development of alternate energy sources. Clearly, high oil prices have had the most impact on the interest in coal gasification. Several books and reviews documenting the results of these developments have been discussed earlier in this chapter and elsewhere [5-15]. The collapse of oil prices in the 1980s led to a steep decline in gasification activities. Most of the projects were

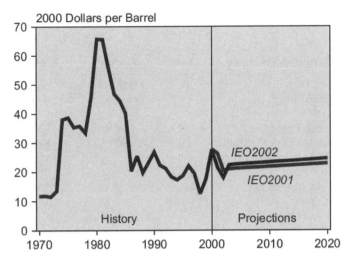

FIG. 20.25 OIL PRICES — WHERE IT WAS AND WHERE IT IS HEADED [45]

terminated by the mid-1980s, and gasification remained mostly dormant through the later part of the century. The future in gasification exists as long as fortunes can be made in the energy markets.

Figure 20.25 also shows the decline in the world oil prices in 2000 due to global economic slowdown led by a mild recession in the United States caused mainly by the bursting of the dotcom bubble and the September 11th terrorist attacks on the United States. To stop the decline in oil price, the Organization of the Petroleum Exporting Countries (OPEC) implemented production quota cuts. Even after three production quota cuts, OPEC was unable to stabilize the oil price. This led the EIA in 2002 to forecast a mere rise of 1.3% in the price of oil between 1999 and 2020 [45]. Low energy prices basically implied that there would be little interest in coal gasification.

In the new millennium, most energy forecasters, including China's leaders had expected the energy growth in China to follow the past growth pattern at 3 to 4% [46]. However, in reality the growth in China's economy during the early part of the millennium caused the energy consumption to increase more than four fold of the estimate and surprised the world markets. This rapid growth in China caused realignment in world's energy market in which China established itself as the fastest growing market worldwide. The rapid growth in China led to increased oil demand and was one of the reasons that caused oil price to soar to $150 per barrel in 2008. The rapid increase in the oil price curbed oil consumption, and the world's economy slowed down once more. This caused oil price to plummet to about $30 per barrel, before recovering to the $60 to 70 per barrel range. The slow-down in economic growth and the bursting of the financial bubble in early 2009, will likely be short-lived, and the oil price and consumption are expected to recover in 2010 [44]. With the recovery in global markets, EIA sees a robust growth in oil consumption leading to high oil price as the likely outcome. The high oil price theory can also be supported by the fact that the per capita energy consumption in China and India is still below world average, even after the rapid growth in energy consumption seen in recent years.

In order to bracket an estimate of future oil price, EIA presented three oil-price cases representing different scenarios (see Figure 20.26) [44]. The reference case reflects the assumption that OPEC members will maintain their share of world oil supply with a quota system. The high-price case represents a scenario where the non–OPEC oil–producing nations restrict world access to their

oil supplies, and OPEC members do not increase their production quota to meet the oil demand. The low-price case represents a scenario where both OPEC and non-OPEC nations increase access and production limits. This low-price case scenario played out in 1980s when the OPEC nations flooded the oil market causing the oil prices to drop precipitously. The low oil prices caused high cost producers and unconventional sources to decrease, if not stop their production. The low-price scenario is the most unlikely of all of three scenarios mainly because the energy demand in non-OECD nations is unfulfilled [44]. In addition the sources for cheap new oil appear to have been depleted, and oil companies are exploring options to dig deeper and deeper to produce oil. Therefore, oil price is expected to increase in the foreseeable future.

As the price of oil rises, liquid fuels from alternate sources become increasingly attractive. In 2006, only 3.1 million barrels per day of alternate liquid fuels were produced worldwide [44]. Alternate liquid fuels currently being produced include biofuels, Fischer-Tropsch liquids and oxygenates. Biofuels are derived from biomass, typically food sources such as corn and sugar cane. Ethanol and "bio-diesel" (esterified vegetable oil or animal fat) constitute the bulk of the biofuels being produced today. Biofuels produced from food sources are sometimes referred to as the first Generation Biofuels. The second-generation Biofuels technologies use non-food sources such as cellulosics to produce mainly ethanol (or "cellulosic ethanol"). Cellulosic ethanol can be produced by gasifying non-food biomass or by breaking down the cellulosic structure with enzymes. Technologies are also being developed to pyrolytically convert biomass to liquid fuels. In recent years, due to environmental concerns and government subsidies, there has been a significant increase in the second-generation biofuels activity [47]. However, the contribution to the alternate liquid fuel pool due to cellulosic biofuels is expected to be small.

Oxygenates are defined as oxygen containing organics such as alcohols and ethers. Methanol is the most popular alcohol produced. Methanol is typically not used as a fuel; instead it is converted to MTBE or DME to be used as a fuel substitute or an additive. DME is also used as a substitute for LPG and diesel. Oxygenates can be produced via gasification, as discussed earlier.

Production of alternate liquid fuels from non-food biomass, coal, natural gas, etc. typically involves the use of gasification to produce syngas for synthesis. Table 20.4 summarizes coal gasification–based liquid fuels plants that have recently become opera-

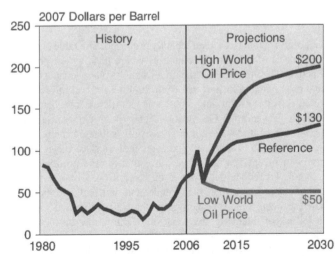

FIG. 20.26 FORECAST OF GLOBAL OIL PRICES BASED ON THREE PRICE CASES [44]

TABLE 20.4 RECENT ACTIVITY IN COAL-TO-LIQUIDS PROJECTS

Plant/Project Name	Year	Country	Product
Arckaringa	2014	Australia	CTL
Haolianghe, Heilongjiang	2005	China	Methanol
Jinling, Nanjing	2005	China	Methanol
Shaanxi Shenmu Chemical Plant	2005	China	Methanol
Yankuang Cathy	2005	China	Methanol
Yulin	2006	China	Methanol
Dahua Chemicals, Dalian	2007	China	Methanol
Shenmu	2007	China	Methanol
Wison	2007	China	Methanol
Yongcheng Chemicals	2007	China	Methanol
Kaixiang Chemical Plant	2008	China	Methanol
Puyang Plant	2008	China	Methanol
Shenhua, Majiata	2008	China	H2/CTL
Jiangsu Sopo Group/ Praxair	2009	China	Methanol
Shenhua Baotou	2009	China	Methanol
Shenhua Ningxia	2009	China	Methanol
Yankuang Luan Fertilizer	2009	China	Methanol
Anhui Huayi	2010	China	Methanol
China Blue Chemical	2010	China	Methanol
Shandong Jiutai	2010	China	Methanol
Shenhua Ningxia	2010	China	DME
Shilin	2010	China	Methanol
Inner Mongolia Chemical Plant	2011	China	Methanol
Hulunbeir	2013	China	Methanol
Tata/Sasol	2014	India	CTL
Medicine Bow		United States	CTL
Rentech Strategic Fuels and Chemicals Center	2012	United States	CTL

TABLE 20.5 RECENT ACTIVITY IN COAL-TO-SNG PROJECTS

Plant/Project Name	Year	Country	Product
POSCO	2013	South Korea	SNG
Secure Energy Systems SNG	2010	United States	SNG
Hunton Energy	2013	United States	SNG
Peabody/ConocoPhillips SNG Plant	2013	United States	SNG
South Heart	2013	United States	SNG
Taylorville Energy Center	2015	United States	SNG
Indiana Gasification SNG Project		United States	SNG
Mississippi Gasification SNG Project		United States	SNG

the EIA numbers indicates that over 800 new plants will be built by 2030 worldwide! In the US, EIA [44] estimates that the coal usage for CTL purposes will increase by 0.7%, and CTL plants may consume about 68 million tons of coal in 2030. Coal-based liquid fuel production by 2030 is estimated to be 257,000 barrels per day [44] or about 20 plants over the next 20 years.

At an average capital cost of about $50,000 per barrel per day capacity, CTL plants will demand a major commitment of capital investments. High capital cost and lack of demand for oil are perhaps key reasons as to why many of these projects have not yet materialized, at least in the OECD countries.

20.6.5 Outlook for Coal-to-SNG

High natural gas prices in early 1980 led to the interest in production of SNG from coal. As discussed earlier, syngas generated from coal gasification process can be converted to methane. The use of existing pipelines to transport SNG created interest in several mine mouth coal gasification projects. The collapse in energy prices in the late 1980s caused most of these projects to be put on hold. In recent years the concept of coal-based non-IGCC unit attracted some attention [48]. The non-IGCC concept reflects a combination of SNG unit and a combined cycle unit, but not necessarily at the same site. Typically, SNG is transported via pipelines to an off-site combined cycle unit. This concept appears attractive in the United States where the regulatory permitting process for a SNG plant or a chemical plant is less cumbersome than permitting a power plant. In Europe, the threat of being cut-off from natural gas supplies originating in Russia has increased interest in coal-based SNG units to enhance energy security. In China and India, SNG projects appear attractive due to lack of natural gas resources. Table 20.5 summarizes recent activities for SNG applications. The data has been compiled from several sources, as mentioned earlier. The interest in coal gasification–based SNG appears low, even in China and Europe, where one would have assumed that national security interest would have prompted a few SNG projects. In recent years no SNG plants have been built, but there are some projects in the planning phases. In the United States, competition from the liquefied natural gas (LNG) and shale gas has limited interested in SNG.

20.6.6 Outlook for Coal-to-Electricity

Currently the global consumption of electricity is about 20 trillion kilowatt-hours. By 2030, EIA estimates electricity consump-

tional or those that have been recently proposed. The table is sorted by country to highlight the countries where gasification projects are active. The year denotes the time when the plant became operational or is planned to start-up. Table 20.4 is based on information available from several sources and news articles. The databases of NETL [11] and the Gasification Technology Council [12] are probably two of the more comprehensive lists that are freely available on the web. Table 20.4 indicates that most of the reported activity in recent years has been in China for the production of methanol. To reduce dependence on foreign oil, China has been using methanol as a fuel, fuel supplement and fuel intermediate. There has been some CTL activity in the United States. However, due to lack of financing and weak oil demand, these projects are either being delayed or being shelved.

By 2030, EIA [44] estimates that the worldwide production of alternate liquid fuels will increase by another 10 million barrels per day. Assuming an average plant size of 12,000 barrels per day,

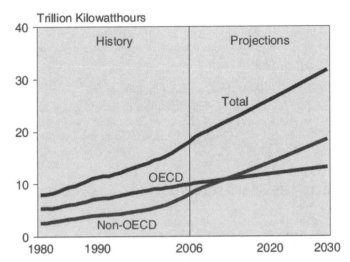

FIG. 20.27 GLOBAL NET ELECTRIC POWER PRODUCTION [44]

tion to increase by 50% (see Figure 20.27) [44]. This implies that the global generating capacity has to increase by at least 50%, to maintain the present availability ratio.

In the OECD countries the electricity markets are well established and mature, and therefore, the demand in electricity requirements is predictably slow as seen in Figure 20.27. In the non-OECD, a large amount of electricity demand goes unmet at present. In addition, due to the strong economic growth in the non-OECD countries, the growth rate of electric generation capacity in the non-OECD is projected by EIA to be about three times that of OECD countries. In China and India, the electricity demand is even higher, and generating capacity is estimated to increase at an annual rate of 4.4%. The per capita energy consumption in China and India is a fraction of the average per capita consumption in OECD nations and indicates a strong continuing demand for power. Strong continuing demand for power reflects a strong continuing demand for coal as discussed earlier in section 20.6.3.

To meet the growing power demand shown in Figure 20.27, electricity generation capacity is estimated to grow by about 1800 gigawatts (GW) by 2030, effectively doubling the present generation capacity [44]. China, United States, and India account for about 60% of the estimated generation capacity increase. Over two-thirds of this new demand in these three countries is estimated to be met by coal. By 2030, EIA anticipates China to add about 600 GW of coal-based capacity, doubling its coal usage for power generation. India's coal-based power capacity is project to increase by 64 GW in 2030, representing a growth of over 80%. In the United States, about 46 GW of coal-fired generating capacity is projected to be added by 2030 [44]. Assuming an average power plant size to be 400 MW, the EIA estimate indicates that over 1700 coal-based power plants will be built over next two decades in these three countries, most of them in China.

Table 20.6 summarizes recent coal gasification–based power projects that have recently become operational or those that have been planned. The table represents data from several sources, as mentioned earlier. Several coal gasification projects have been

TABLE 20.6 RECENT ACTIVITY IN COAL-TO-ELECTRIC POWER PROJECTS

Plant/Project Name	Year	Country	Product
Mulgrave IDGCC	2013	Australia	Power
ZeroGen	2015	Australia	Power
Huadian Banshan	2010	China	Power
China-Tai Yangzhou Power	2011	China	Power
Dongguan	2011	China	Power
GreenGen	2012	China	Power
Thermoselece Vresova	2007	Czech Republic	Power
RWE	2015	Germany	Power
Sanghi IGCC Plant	2002	India	Power
APGENCO/ BHEL/NTPC	2011	India	Power
Sulcis IGCC Project	2005	Italy	Power
Nakoso IGCC	2007	Japan	Power
Clean Power CCPR&D	2008	Japan	Power
J-Power/ Chugoku-Eagle	2016	Japan	Power
Kowepo Taean IGCC # 1	2013	South Korea	Power
Powerfuel	2013	UK	Power
Delaware Clean Energy Cogen Project	2002	United States	Power
Edwardsport IGCC	2011	United States	Power
Sweeny IGCC	2013	United States	Power
Kemper County IGCC Project	2014	United States	Power
Goodspring IGCC	2015	United States	Power
Summit Texas Clean Energy	2015	United States	Power
Cash Creek	2015	United States	Power
Hydrogen Energy California	2015	United States	Power

proposed worldwide. Projects listed in Table 20.6 account for less than 10 GW, or a fraction of projected power demand. One reason for this low number could be that the developers are choosing the more economical traditional direct coal combustion for steam cycle based power plants over the gasification route. In order to compete with these traditional units, the gasification technology has to be better packaged to be financially attractive. Perhaps co-production of a high-value product such as a liquid fuel substitute or a high-value chemical along with power is

one option to generate additional revenues. Concerns regarding greenhouse gas emissions will make the gasification units financially more attractive than traditional steam cycle based power plants in the future.

20.6.7   Outlook for Coal-to-Chemicals

As discussed earlier, a major end user of natural gas is the industrial sector. Because of the strong economic growth in non OECD nations led by China and India, and the demand for "green"

TABLE 20.7   RECENT ACTIVITY IN COAL-TO-CHEMICALS PROJECTS

Plant/Project Name	Year	Country	Product
Australian Energy Co	2015	Australia	Ammonia
Hefei City Ammonia Plant	2000	China	Ammonia
Puyang Ammonia Plant	2000	China	Ammonia
Zhong Yuan Dahua Group Ltd.	2000	China	Chemicals
Haolianghe Ammonia Plant	2004	China	Ammonia
Huala Hengsheng	2005	China	Ammonia
Jinling	2005	China	Ammonia
Liuzhou Chemicals	2005	China	Ammonia
Sinopec, Zhijiang	2005	China	Chemicals
Dong Ting Ammonia Plant	2006	China	Ammonia
Hubei Ammonia Plant	2006	China	Ammonia
Shuanghuan Chemical	2006	China	Ammonia
Sinopec Wuhan, Hubei	2006	China	Ammonia
Sinopec, Anqing	2006	China	Ammonia
Sinopec, Yueyang	2006	China	Ammonia
Sinopec/Nanjing	2006	China	Ammonia
Weihe Chemical	2006	China	Chemicals
Yunnan Tianan	2006	China	Ammonia
Sinopec Qilu	2007	China	OxoChemicals
Synthesis Energy /Hai Hua	2007	China	Chemicals
Yunnan Zhanhua	2007	China	Ammonia
Yuntianhua Chemicals, Anning	2007	China	Ammonia
Yunzhanhua Chemicals, Huashan	2007	China	Ammonia
Dalian	2008	China	Ammonia
Shanxi Jincheng Anthra	2008	China	Chemicals
Shenhua, Majiata	2008	China	Hydrogen
Tengzhou, Fenhuang	2008	China	Ammonia/Methanol
Yongcheng Shell Plant	2008	China	Chemicals
Jiangsu Linggu	2009	China	Ammonia
NCPP	2009	China	Chemicals
Guiozhou Kaijang	2010	China	Ammonia
Guizhou Chemical Plant	2010	China	Ammonia
Ningbo	2010	China	Ammonia/Methanol
Shanxi Chem Yangkuang Coal, JinCheng	2010	China	Ammonia
Tianjin Chemical Plant	2010	China	Ammonia
Datong, Shenyang	2018	China	Chemicals
DEWA/Sino Global	2015	Dubai	Chemicals
Paradip Gasification H2/Power Plant	2010	India	Hydrogen
Jindal Steel	2011	India	Chemicals
Gail India RCF		India	Chemicals
Elnusa		Indonesia	Chemicals
Coffeyville Syngas Plant	2000	United States	Ammonia
Faustina Hydrogen Products LLC	2010	United States	Ammonia

ethanol in the United States, there has been a growing demand for chemicals, especially ammonia, for production of fertilizers. In the United States, the demand for ammonia has been met with natural gas. Recent finds of abundant shale gas in the United States have kept natural gas prices low and limited the use of coal for chemicals. In Asian countries, China, India, and Japan, limited supply of cheap natural gas has increased the deployment of coal gasification projects for production of chemicals. Table 20.7 summarizes coal gasification–based chemical plants. Not included in this table are the projects for methanol that have been listed earlier in Table 20.4. During the recent years, the rapid rise in the chemical production rate in China has caused production surpluses [49]. Oversupply in the market results in suppression of profit margins, and perhaps, this is one reason why fewer plants have been proposed beyond 2010.

In summary, recent activity in coal gasification is due to the desire to have diversity in energy sources due to political conflicts, national security, and terrorism. The Chinese government has continued to encourage development of coal-based fuels and chemicals as a strategic issue due to their reluctance to become more reliant on foreign oil. In order to encourage coal usage, the Chinese government has eliminated requirements for special permission for these units; as a result many projects were simultaneously implemented causing a surplus in the market [46]. In the long run, the growth and success of coal gasification for power, chemicals, fuels, or metals will be based on capital cost–related issues and the price of oil/natural gas. Therefore, as noted in Section 20.4, careful consideration must be paid while selecting and defining the process blocks within the gasification island. Finding ways to protect against the relative pricing volatility between coal and conventional fuels will be a key commercial issue. Since there are numerous applications for syngas, gasification offers the advantage to coproduce chemicals, fuels, or metals along with power. These polygeneration facilities provide for flexibility, lower exposure to uncertainties, and lower production cost.

20.7   REFERENCES

1. US Energy Administration Information. Annual Energy Outlook 2010 Early Release Overview. http://www.eia.doe.gov/oiaf/aeo/overview.html.

2. Ratafia-Brown, J., L. Manfredo, J. Hoffmann, and M. Ramezan, 2002, Major Environmental Aspects of Gasification Based Power Generation Technologies, Final Report, US Dept. of Energy.

3. Bechtel, Global Energy and Nexant, 2003, Gasification Plant Cost and Performance Optimization, Final Report to US Dept. of Energy, DE-AC26-99FT40342.

4. Agrawal, R. K., 1984, "Kinetics of Biomass and Coal Pyrolysis," Ph.D. thesis, Clarkson University.

5. Reimert, R., G. Schaub, et al., 1989., Gas Production, Ullmann's *Encyclopedia of Industrial Chemistry*, 5th ed., Vol A12, VCH Verlagsgesellschaft.

6. Encyclopedia of Chemical Processing and Design, J. J. McKetta and W. A. Cunningham, eds., Vol 9, 1979.

7. Hebden, D., and H. J. F. Stroud, 1981, Coal Gasification Processes, in Chemistry of Coal Utilization, 2nd Suppl. Vol, M. A. Elliott, ed., Wiley-Interscience, pp. 1599-1752.

8. *Proceedings of 10th Synthetic Pipeline Gas Symposium*, Chicago, IL, Oct 1978.

9. Nowacki, P., ed., 1981, Coal Gasification Process, Noyes Data Corp.

10. Gasification Technology Council. http://www.gasification.org/

11. NETL, Gasification World Database 2007, http://www.netl.doe.gov/technologies/coalpower/gasification/database/Gasification2007_web.pdf.

12. Gasification Technologies Council Database, http://www.gasification.org/database1/search.aspx.

13. Higman, C., and M. van der Burgt, 2008, Gasification, 2nd ed., Elsevier.

14. Holt, N., 2001, Integrated gasification combined cycle power plants, In *Encyclopedia of Physical Science and Technology*, 3rd ed., Academic Press.

15. Stiegel, G., http://www.netl.doe.gov/technologies/coalpower/gasification/pubs/pdf/DOE%20Gasification%20Program%20Overview%202009%2009-03%20v1s.pdf.

16. Alpert, S. B., M. J. Gluckman, 1986, Coal Gasification Systems for Power Generation, *Annual Review of Energy*, **11**, pp. 315-355.

17. Squires, A. M., 1982, Proceedings of Joint Meeting of Chemical Industry & Eng Society of China and AIChE, Beijing, Sept 19-22, pp. 322-353.

18. Lau, F., 2009, Commercial Development of the SES U-Gas Gasification Technology, 2009 Gasification Technologies Conference, Colorado Springs, CO.

19. Agrawal, R. K., 1991, Effect of Pyrolysis Conditions on Gasification Reactivity of Chars, Presented at the 8th Annual International Pittsburgh Coal Conference, Pittsburgh, PA.

20. Smith, P. V., KBR Transport Gasifier, 2005 Gasification Technology Conference, San Francisco, CA.

21. Challand, T., KBR Transport Gasifier — The Path Forward, 2006 Gasification Technology Conference, Washington, DC.

22. Trapp, B., N. Moock, and D. Denton, 2004, Coal Gasification: ready for prime time, *Power*, **148**(2), pp. 42-50.

23. Peltier, R., 2007, IGCC demonstration plant at Nakoso Power Station, Japan, Power, pp. 32-36.

24. van Holthoon, E., 2009, Shell Coal Gasification — Leading Technology across Multiple Applications, 2009 Gasification Technologies Conference, Colorado Springs, CO.

25. Radtke, K. R., D. Battensby, and C. Marsico, 2005, Renaissance of Gasification based on Cutting Edge Technologies, *VGB Power Tech*, **9**, pp. 106-115.

26. Elcogas, IGCC Puertollano, http://212.170.221.11/elcogas_body/images/, IMAGEN/TECNOLOGIAGICC/thermie.pdf.

27. US Dept of Energy, Wabash River Coal Gasification Repowering Project: A DOE Assessment, DOE/NETL-2002/1165.

28. Wabash River Energy Ltd., Wabash River Coal Gasification Repowering Project, US Dept. of Energy Report No. DE-FC21-92MC29310.

29. McDaniel, J., 2002, Tampa Electric Polk Power Station IGCC Project, US Dept. of Energy Report No. DE-FC-21-91MC27363.

30. Tampa Electric Company and US Dept. of Energy, The Tampa Electric IGCC Project, Topical Report No. 19, July 2000.

31. Tampa Electric IGCC Project, Topical Report No. 19, DOE, July 2000.

32. Zhou, Z. Project and Technology Updates — Power, Chemicals, and Fuels, GTC 2009 Conference.

33. van den Berg, R. E., et al., Method and System for Producing Synthesis Gas, Gasification Reactor and Gasification Systems, US Patent, Appl. No. US2006/0260191A\2006.

34. Wabash River Coal Gasification Repowering Project, 2002, DOE/FE-0448.

35. DOE/NETL-2007/1281 Cost and Performance Baseline for Fossil Energy Plants, May 2007.

36. van der Burgt, M. J., 1998, How to reduce capital cost of IGCC power plants, 17th EPRI Conference on Gas-Fired Power Plants, San Francisco.

37. Holt, N., 1998, IGCC Power Plants — EPRI Design and Cost Studies, EPRI/GTC Gasification Technologies Conference, San Francisco, CA.

38. DeLallo, M. R., J. S. White, T. L. Buchanan, 1998, Economic Evaluation of Advanced Coal Gasification Technologies for Power Generation, EPRI/GTC Gasification Technologies Conference, San Francisco, CA.

39. Storch, H. H., R. B. Anderson, L. J. E. Hofer, C. O. Hawk, and N. Golumbic, 1946, Synthetic Liquid Fuels from Hydrogenation of Carbon Monoxide, Bureau of Mines, US Government Printing Office, Washington, DC.

40. Stranges, A. N., Germany's Synthetic Fuel Industry 1927-45, AIChE 2003 Spring National Meeting, New Orleans, LA.

41. Agrawal, R. K., 1991, Design of Fischer Tropsch Synthesis Facility, US Department of Energy, Pittsburgh Energy Technology Center, Contract No. DE-AC22-91PC89854.

42. Falbe, J., ed., 1982, Chemical Feedstocks From Coal, John Wiley.

43. Steynberg A., and M. Dry, 2004, Fischer-Tropsch Technology, Studies in Surface Science and Catalysis, Volume 152, Elsevier.

44. Energy Information Administration, 2009, International Energy Outlook, DOE/EIA-0484.

45. Energy Information Administration, 2002, International Energy Outlook, DOE/EIA-0484.

46. Rosen, D. H., and T. Houser, 2007, China Energy, Joint Project by the Center for Strategic and International Studies and Peterson Institute for International Economics.

47. Ethanol Producer Magazine, Proposed Ethanol Plant List, 2009, United States and Canada, http://www.ethanolproducer.com/article.jsp?article_id=5461.

48. Agrawal, R. K., and T. Joyner, "Integrated Gasification Combined Cycle Growth Sector Projections," ASME Power 2008 Conference, Lake Buena Vista, FL, July 22-24, 2008.

49. UNESCO/Shell Chair in Coal Gasification, UNESCO 2007.

CONSTRUCTION OF NEW NUCLEAR POWER PLANTS: LESSONS TO BE LEARNED FROM THE PAST

Roger F. Reedy

21.1 BACKGROUND

During the 1960s and 1970s, when most nuclear plants in the United States were being designed and constructed, costs escalated in an uncontrolled manner. This was due to many factors, and all interested parties had an influence on the results. This chapter will address these issues, not to accuse or place blame, but to analyze and discuss in order to learn from the past.

The first commercial nuclear power plants were designed and constructed in the mid 1950s and early 1960s. Because these plants were based on the new technology of using nuclear reactions to heat water, thus producing steam to turn the turbines, there were no Codes or Standards that specifically addressed nuclear power plant components. Therefore, these early plants used the most applicable Codes and Standards that were available at the time [1–3]. That was the proper course of action.

At the time, the question was, what was available in terms of Codes and Standards? What Codes and Standards were really appropriate for pressure vessels and piping in a nuclear power plant? The designers of some nuclear power plant systems considered that question and determined that the nuclear reactor that contained the nuclear fuel was equivalent to a boiler in a fossil fuel power plant. Other designers felt that the nuclear reactor was an unfired pressure vessel. The available Code for the design and construction of the boiler were the American Society of Mechanical Engineer's Boiler and Pressure Vessel Code, Section I, "Power Boilers" and Section VIII, "Unfired Pressure Vessels." [1, 2]. Both approaches were correct because the differences between the two ASME Codes were minimal for the design and construction of the pressure vessel. The appropriate Piping Code used was ASME B31.1, "Power Piping" [3]. Both Section I and Section VIII were used for the first nuclear power plant reactor pressure vessels, and B31.1 was used for the associated pressure piping, regardless of whether Section I or Section VIII was used for the nuclear reactor pressure vessel.

The designers of the first nuclear power plants felt that a leaktight pressure vessel should be built around the nuclear reactor and critical piping so that any leakage from the reactor vessel or primary coolant piping would not allow radioactive steam to escape to the atmosphere. There were several important technical issues associated with the construction of a large pressure vessel that could contain any leakage from the reactor pressure vessel or the associated piping. Because this large pressure vessel was designed to contain any leakage, it was called a containment vessel.

Although the military of the United States had the first nuclear reactors constructed in the early 1950s, the first commercial nuclear power plant, Shippingport, went into operation on December 2, 1957 [4]. Although the Shippingport power plant produced commercial power, it was financed and jointly owned by the AEC and Duquesne Light Co. The first privately owned commercial nuclear power plants, Dresden, owned by Commonwealth Edison Co., and Rowe, owned by Yankee Atomic Electric Co. were placed in operation in 1960 and 1961, respectively [5, 6].

The Shippingport plant had a cylindrical containment vessel, but because they were much larger, the Dresden and Rowe nuclear plants used spherical containment vessels which were more than 300 ft in diameter. The primary reason for using spherical containment vessels was to prevent the required thickness less than 1 1/2 in. thick, because pressure vessels greater than 1 1/2 in. thick would have to be stress relieved (post-weld heat treated) in the field. At the time, this was thought to be an impossible task. However, in the mid-1970s, two different containment vessels, each 2 in. thick, were easily post-weld heat treated in the field by Chicago Bridge & Iron Co. [7]. For these three plants, the containment pressure vessels were designed and constructed to the rules of the ASME Section VIII Code.

Today, many engineers believe that the American Society of Mechanical Engineers (ASME) published the Section III Code for Nuclear Components in order to make nuclear power plants safer. That is not the case, as will be identified in the short history outlining the development of ASME Code for Nuclear Components.

21.2 HISTORY OF THE SECTION III CODE

21.2.1 Nuclear Components

As stated earlier, the first commercial nuclear power plants were designed, constructed, and placed in service in the mid-1950s and early 1960s. The success of these commercial nuclear power plants were an incentive for system designers to design and construct larger plants, which meant the nuclear reactors would also be larger and thicker.

At the same time, oil refineries and chemical plants were being constructed for new processes. These new processes required higher pressure and temperatures which also caused the pressure vessels to be much thicker than those designed and constructed previously. In order to reduce costs associated with the thicker steel, the increased amount of welding, the cost of new equipment for forming operations, and the requirements for volumetric examination of welds, there was a need for pressure vessel Codes to reduce required thickness.

That means that the nuclear industry and other industries were looking for ways to reduce costs and schedules. Companies fabricating pressure vessels were seeking a new approach for the design and fabrication of heavy wall pressure vessels. In 1955, the ASME Code Committee responded to this need by forming the Special Committee to Review Stress Basis. The purpose of the Committee was to evaluate how pressure vessels could be made thinner [8]. The Special Committee was not investigating how pressure vessels could be made safer. The Criteria document clearly states that the investigation centered on how the allowable stresses for pressure vessels could be increased, thus permitting thinner pressure vessels at less cost.

The original assignment for the Special Committee was to develop a new ASME Code for heavy wall Section VIII pressure vessels [9]. By 1958, the Committee was making good progress on developing a new Code. However, the nuclear industry was also growing and needed some means to permit the use of thinner nuclear reactors. The nuclear industry was able to convince the Committee that it would be easier to develop a Code for nuclear components than to develop a general Code that would be available for all industries, regardless of service requirements. After all, in the nuclear industry, the pressure vessel is only required to contain steam or water. In other industries, the service products are far more complex. The fluids contained in pressure vessels for other industries were sometimes corrosive, toxic, poisonous, or explosive. Based on this, the task group changed courses and developed a Pressure Vessel Code for nuclear components, with the idea that a new Section VIII Code would be easy to develop after the original concepts for making thinner vessels were incorporated into the Nuclear Pressure Vessel Code [10].

The only way to reduce the required thickness was to reduce the safety factor (design factor) and to compensate for that reduction by other means. The concept to be followed was that the new ASME Code could be used to construct pressure vessels that would be as safe as those constructed to the rules of Section I or Section VIII. In order to achieve the proper compensation, each part of the Code had to be critically evaluated.

The approach taken was that new rules would be required for permitted materials, design methodologies, fabrication details, non-destructive examination, and testing. Also, new rules would be required for overpressure protection of the pressure vessels.

With regard to materials, the new Code would require limitations on melting practice and provide rules for Charpy impact testing. Many fabrication details permitted in Section VIII would be prohibited. The new Code would require that all main seams in the reactor vessel be examined by radiography. The rules for testing and overpressure protection were also to be modified.

At the time the new Nuclear Code was published in 1963, the design factor or safety factor for Section VIII was 4 [11]. That is, the allowable stresses were based on the minimum tensile strength of the material to be used divided by 4. For the new Code, the allowable stresses were to be based on a design factor of 3.

The rules for design of the reactor pressure vessel were drastically different from the rules of Section VIII. Formulas were given to determine the minimum thickness of the shell, but a complete stress analysis considering all operating loads and conditions was required. In addition, the stress theory was changed from maximum tensile stress theory to a shear stress theory. Stress limits were provided for primary and general membrane stress, primary local membrane stress, and primary bending stress. In addition, secondary stresses and peak stresses were required to be evaluated. The new allowable stresses were based on using three-dimensional stresses identified as "stress intensities." Rules were also added for the evaluation of fatigue or cyclic service.

It was felt by the ASME Committee that the additional restrictions and requirements would compensate for the reduction in the safety factor or design factor. When ASME contacted the Atomic Energy Commission (AEC) to obtain acceptance of the new Nuclear Code (Section III), the Commission was skeptical and concerned that Section III would not be as safe as Section VIII [12]. It took several years of investigation before the AEC was able to accept the new Section III Code. The AEC never stated that Section III was a better or safer Code than Section VIII. Records of pressure vessel and boiler explosions from the mid-1960s demonstrate that no explosions can be attributed to faulty rules in Sections I or VIII [13].

Although Section III was first published in 1963 the new Section VIII Pressure Vessel Code, based on the technical requirements of Section III was first published in 1968 as Section VIII, Division 2 [14]. The publication of the new Section VIII, Division 2 all along with Section III was the culmination of the effort to publish ASME Codes that would provide for the use of thinner and less costly pressure vessels. There is no doubt that Section III is more appropriate for the design and construction of pressure vessels in nuclear service, but it cannot be proven that Section III or Section VIII, Division 2 provides for the design and construction of safer pressure vessels.

21.3 HISTORY OF RULES FOR NUCLEAR PRESSURE VESSELS: PRIOR TO SECTION III

Having identified the development of ASME Section III above, it is important to also review how the ASME Code addressed the issue of design and construction of nuclear reactor and containment pressure vessels prior to the publication of ASME Section III.

In general, ASME publishes Code Cases to the Boiler and Pressure Vessel Code in order to provide alternative rules to those already in the Code. Because there were no Code rules that addressed the design and construction of the pressure vessels needed for the new nuclear power plants, Code Cases were published to modify the rules of Section I and Section VIII [15–21].

The first ASME Code Cases that addressed design and construction of nuclear pressure vessels appeared in late 1959. Note that this is a number of years after the first commercial nuclear power plants were placed into operation. It should also be noted that these new Code Cases were Section VIII Code Cases.

Case 1270 N — "General Requirements for Nuclear Vessels" [15]
Case 1271 N — "Safety Devices" [16]
Case 1272 N — "Containment and Intermediate Containment Vessels" [17]

Case 1273 N — "Nuclear Reactor Vessels and Primary Vessels" [18]

Case 1274 N — "Special Material Requirements" [19]
Case 1275 N — "Inspection Requirements" [20]
Case 1276 N — "Special Equipment Requirements" [21]

21.3.1 Analysis of Nuclear Cases

21.3.1.1 Case 1270 N

Case 1270 N states, "All vessels that are an integral part of nuclear installations shall be constructed in accordance either with the requirements of Section I or else with the requirements of Section VIII" [15].

This Code Case provides the General Requirements for the design and construction of pressure vessels to be used at nuclear power plant sites.

The nuclear reactor vessel is identified as, "any vessel, any tube, or an assemblage of tubes, regardless size, in which nuclear fuel is present and in which the nuclear chain reaction takes place."

Primary vessels are defined as, "those vessels regardless of size, other than the reactor vessels, which are an integral part of the primary coolant system during normal operation and which contain or may contain coolant at the operating conditions of radioactivity level, pressure and temperature."

Containment vessels are, "outer vessels which enclose the reactor vessel or portions of the primary coolant circuit or both. The containment vessels are not normally pressurized and are built to contain the radioactive substances that may be released in case of an accident or failure of the reactor vessel or the primary coolant circuit or both."

Intermediate containment vessels are defined as, "those vessels within the containment vessel that encloses a portion or all of the primary reactor vessel. The intermediate containment vessels may or may not be pressurized during normal operation but they are intended to contain the primary coolant that may be released in case of an accident or failure of the vessel which they enclose."

Secondary vessels are, "all other vessels which do not contain reactor coolant or are not otherwise subject to irradiation."

21.3.1.2 Case 1271 N

Case 1271 N provided the rules for safety devices on pressurized water reactors and boiling water reactors as well as containment vessels. This Case was important because it was essential that radioactive effluent never be released to the public [16].

21.3.1.3 Case 1272 N

Case 1272 N identified the fabrication rules addressing welded joints, stress relieving, and Charpy impact testing for containment vessels. It also provided design rules requiring that the formulas of Section VIII be followed with the stress values increased by a factor of 10%. It further provided that both primary and secondary stresses be calculated and that the combination of the two was not to exceed three times the allowable stress values without the 10% increase [17].

21.3.1.4 Case 1273 N

Case 1273 N addressed the design and construction of nuclear reactor vessels and primary vessels. It provided limitations on weld joints and attachment welds and identified the types of nozzles that were to be used to facilitate volumetric examination of the attachment welds. The case required the use of the design formulas as well as a complete stress analysis of the pressure vessel. The stress limitations were the same as those for containment vessels except that the 10% increase in allowable stress was not permitted. In other words the design of the reactor vessel was more conservative than the design of the containment vessel [18].

21.3.1.5 Case 1274 N

Case 1274 N provided rules for the materials to be used in the design and construction of nuclear reactors and containment vessels [19].

21.3.1.6 Case 1275 N

Case 1275 N provided rules for non-destructive examination of the pressure vessels. Primarily the Case addressed the use of ultrasonic examination of the welds [20].

21.3.1.7 Case 1276 N

Case 1276 N provided rules for the construction and use of bellows expansion joints [21].

When these seven Code Cases are reviewed in detail, it can be seen that they are the true forerunner of Section III. The only significant difference between the seven nuclear Code Cases and Section III is that the design factor or safety factor in Section III has been reduced from 4 to 3 and the stress theory has been changed from maximum stress to shear stress using the Tresca theory of failure. The Cases did not provide rules for evaluating fatigue.

21.4 NRC ACCEPTANCE OF SECTION III

The basic Code for all these nuclear Code Cases was Section VIII (Division 1). Outside of the fact that Section III had a reduced design factor (safety factor) and required increased stress analysis, including the valuation for fatigue, Section III offered almost nothing that was not already required by Section I for Section VIII plus the seven nuclear Code Cases. The prominent new rule in Section III that was not included in the seven nuclear Code Cases was provisions for fatigue analysis [22].

Many people have argued that Section III provided construction rules for a safer pressure vessel than one constructed to the rules of Section I or Section VIII with the seven nuclear Code Cases. Unless a reduction in design factor (sometimes erroneously referred to as the safety factor) with some increased stress analysis can be shown to make a pressure vessel safer, the argument is completely without merit. A truer statement and a better argument is that Section III is just as safe as Section I or Section VIII (Division 1), even though the design factor or safety factor had been very significantly reduced from 4 to 3 [11, 22].

From the above background, it is easy to understand why the AEC was reluctant to accept Section III with its lower the design factor. Much of the compensation for the reduced design factor was already required by the rules in the Code Cases for the reactor vessel. The Code Cases depended on the use of Section I or Section VIII with a design factor of 4. The AEC was really questioning how Section III with a safety factor of 3 was as good as the Code Cases that were based on rules of section VIII along with the safety factor of 4. In fact, at a public meeting in Chicago, the AEC representatives stated they could not accept the new Section III Nuclear Code until they were convinced that pressure vessels constructed to that Code were as safe as pressure vessels constructed to the rules of Section VIII. It took a number of years for that to happen [12].

The items used to justify that Section III was as safe as Section VIII were steam receivers for jet aircraft catapults on board the aircraft carrier USS Constellation (CV-64). The steam receivers replaced old Section VIII pressure vessels that only lasted about one year before they had to be replaced. The new steam receivers were designed and fabricated by Chicago Bridge & Iron Co. to the rules for nuclear reactors in the draft Section III Code before the final version of the document was published [23]. The new steam receivers never had to be replaced. When the Atomic Energy Commission (AEC) reviewed the data, they were impressed and accepted Section III with the reduced design factor and the new fatigue design rules as being as safe as Section VIII.

Today, many people including regulators and Code users are convinced that Section III is required to be met because it produces safer pressure vessels. The facts are completely contrary to that assumption. Section III came into existence as a means to reduce thickness of pressure vessels used in the nuclear industry. The driving factor behind the new Section III was really an effort to the reduce costs of heavy wall pressure vessels.

21.5 ASME CODE PHILOSOPHY

In order to effectively use the lessons learned from the past mistakes it is necessary for engineers and others that work with the ASME Section III Code to understand ASME Code philosophy. This misunderstanding of Code philosophy cost hundreds of millions of dollars in the construction of the old nuclear plants. Some key philosophical points in the ASME code are identified below.

21.5.1 Compliance With Code

Once a pressure vessel or other component has been completed and is CodeStamped, Section III requirements have been met. The Code no longer applies to that item and the only way the Code Stamp can be removed is by an action initiated by ASME. Inspectors, regulators, Owners, or other interested parties cannot, under any circumstances, remove the Code Stamp. If there are any questions concerning the Code Stamp being applied to the pressure vessel or other component someone must take the case to ASME and request an evaluation and hearing. This hearing is a legal process and lawyers are expected to participate.

21.5.2 Codes are Never Retroactive

New Code provisions cannot be made a requirement for any pressure vessel or other component that has been completed and Code stamped. In other words, the Code is not a retroactive requirement.

21.5.3 Engineering Judgment

For situations that are not addressed by the Code, the Code requires the responsible engineers to use engineering judgment based on the principles and philosophy of the Code. In general, these engineering judgments should be documented.

21.5.4 Role of the Engineer

The engineers responsible for the Code design are professionals in terms of education and experience. Their judgments may be questioned by others, but in no case may the engineer be overruled by someone not in the chain of responsibility. Quality assurance personnel are never allowed to overrule engineers. It must be remembered that the ASME Code is an engineering document.

21.6 LESSONS LEARNED — MOSTLY FORGOTTEN

It has been more than 25 years since the last nuclear power plant was constructed in the United States. From 1970 to 1985, many costly mistakes were made during the construction of nuclear power plants. Many experts determined that nuclear plant construction costs were not properly controlled. There is a lot of truth in that statement.

Many of the mistakes that were made during construction were administrative in nature and were due to misunderstandings and lack of knowledge regarding Codes, Standards, and Regulations. The big problem is that these mistakes were expensive and time-consuming, and resolution of the issues was often entrusted to the wrong people.

Now that construction of new plants is starting again, and in some cases, restart of the construction of partially completed plants is proceeding, many of the issues being raised are the same issues that were a significant factor in the unwarranted costs of construction many years ago. In the past 25 years, most of the experienced managers from the previous construction efforts have either died, retired, or left the nuclear industry. Some companies have worked on new nuclear plants overseas, but administrative and regulatory issues overseas are different from those in the United States. If we don't use our knowledge of the lessons learned from the last generation of power plants, the cost of nuclear power construction will unnecessarily skyrocket again.

This chapter is intended to highlight the Lessons Learned from the past, provide background for the avoidance of some of the pitfalls, and suggest innovative solutions for the future. There is no doubt that a comprehensive program to avoid the mistakes from the past will significantly reduce costs and schedules. However, to do so will require a firm commitment by all levels of management.

21.7 BAD WELDING — BEFORE VWAC

The issue of bad welding and nuclear power plant sites was highlighted by the television program, "60 Minutes" [24]. The program showed inspectors and others proclaiming that the welds on the nuclear structures were inadequate and undersized. The primary allegation was against the fillet welds on piping supports. In order to address this criticism, many of the plants under construction at the time underwent a review of all the fillet welds on piping supports [25]. One nuclear power plant went so far as to re-inspect all the support welds three times. This was a considerable effort because there were 17,000 supports in the plant that were re-examined three times.

One of the basic problems was that inspectors were measuring the size of fillet welds using fillet weld gages distributed by companies that sold welding materials. The inspections were performed by placing the gauge over the fillet welds and shining a flashlight from behind the gauge. If the inspector could see light the between the gauge and the weld, the weld was documented as an acceptable. This is the origin of all the "bad welding." It is no coincidence that the use of these gauges required the use of more welding materials. There is no doubt that these gauges were effective sales tools.

Obviously fillet welds on a large structure will vary in profile. These welds are not part of a Swiss watch. Something had to be done to correct the problem.

Because this welding issue was a significant concern to the whole industry, Reedy Associates (now known as Reedy Engineering) contacted a group of nuclear power plant utilities and helped form the Nuclear Construction Issues Group (NCIG) to resolve the welding issue [25]. Participants included nuclear power plant utility companies, Architect/Engineers, the American Institute of Steel Construction (AISC), the American Welding Society (AWS), and Reedy Associates.

One of the problems to be addressed was the lack of uniform acceptance criteria for fillet welds. While working on the correction of this problem, NCIG continually informed the US Nuclear Regulatory Commission (NRC) of progress toward resolution of the issue, which was a concern for all parties involved.

The following is a quote from the American Welding Society (AWS) D1.1 Structural Welding Code-Steel [26].

"1.1.1.1 The fundamental premise of the Code is to provide general stipulations adequate to cover any situation. Acceptance criteria for production welds different from those specified in the Code may be used for a particular application provided they are suitably documented by the proposer and approved by the Engineer. These alternate acceptance criteria can be based upon evaluation of suitability for service using past experience, experimental evidence or engineering analysis considering material type, service load effects, and environmental factors."

In addition, the D1.1 Commentary provides further clarification.

"C1.1.1.1 The workmanship criteria provided in Section 3 of the Code are based upon knowledgeable judgment of what is achievable by a qualified welder. The criteria in Section 3 should not be considered as a boundary of suitability for service. Suitability for service analysis would lead to widely varying workmanship criteria unsuitable for a standard code. Furthermore, in some cases, the criteria would be more liberal than what is desirable and producible by a qualified welder. In general, the appropriate quality acceptance criteria and whether or not a deviation is harmful to the end use of the product should be the Engineer's decision. When modifications are approved, evaluation of suitability for service using modern fracture mechanics techniques, a history of satisfactory service, or experimental evidence is recognized as a suitable basis for alternate acceptance criteria for welds." [26].

These statements helped form the basis for the uniform acceptance criteria identified in the welding report produced by NCIG.

Although many people stated that the issue could not be resolved using the methodology and approach used by NCIG, in less than 12 months, NCIG published an acceptance criteria for inspection of fillet welds, a sampling plan for determining whether or not an issue of undersize welds really existed, and a training program for inspectors. The NRC accepted the NCIG Visual Weld Acceptance Criteria (VWAC) in June 1985, 11 months after the task was started [27]. Although this project was initiated and managed by Reedy Associates, EPRI agreed to publish the document. The full document can now be downloaded from the EPRI website at http://myepri.com. On the website, the document can be found by searching for "NP-5380." The document consists of three volumes which can be downloaded. The first volume is the acceptance criteria, the second volume is the sampling plan, and the third volume is a training program for inspectors [28–30].

In summary, it can be stated that there never was any significant undersized welds (bad welds) at nuclear power plant construction sites. The real problem was that the inspectors were not properly trained to inspect structural welding. Also, inspection from plant site to plant site was not uniform. For that reason a training program was included as part of the final NCIG report on structural welding.

The training program and specific inspection guidelines addresses cracks, weld size, incomplete fusion, weld overlap, craters, weld profiles, undercut, ferocity, arc strikes, length and location of welds, slag, and spatter. The training program for inspectors was specifically accepted by the NRC.

Another significant part of the report is a sampling plan. The sampling plan provides a common sense approach to determining whether or not there is a significant issue regarding undersize welds. An important aspect of the sampling plan is that it could be used for sampling almost anything and it has a 95/95 criteria. That is, the plan provides 95% assurance that 95% of the sample meets the required criteria.

The sampling plan was accepted by the NRC in April 1987 [31]. The training program for inspectors was accepted in June 1985 [32]. The use of the VWAC documents in the 1980s resulted in savings greater than $100 million [33]. If the inspectors have been properly trained at the start of construction, the cost of nuclear power plant construction before 1985 could have been reduced by many hundreds of millions of dollars [33].

21.8 NRC AND ASME INTERPRETATIONS

ASME Interpretations provide answers to questions raised by users of the ASME Code [34]. These Interpretations have been published since 1976 because of a ruling of the US the Supreme Court [35]. The ASME Interpretations are not new rules, but are mandatory because they help clarify what is written in the ASME Code. Many ASME Interpretations are written primarily because code users are not familiar with nor understand the rules and philosophy of the ASME Code.

During the 1980s, the NRC published some documents that made statements that were completely contrary to requirements in the Code. As a result questions were sent to ASME that asked for clarification of the Code requirements. The replies to the questions asked were not taken seriously by the NRC. As a result, ASME and the NRC staff held meetings to resolve the issue of acceptance of the ASME Interpretations. The result of the meeting was that the NRC agreed to accept ASME Interpretations of the Code.

The resolution of this issue was very significant. Hopefully the issue will not arise again. Code users should be aware that ASME will answer most questions sent to them. The Foreword of the ASME Code states:

"Only the Boiler and Pressure Vessel Committee has the authority to provide official interpretations of this Code. Request for revisions, new rules, Code Cases, or interpretations shall be addressed to the Secretary in writing and shall give full particulars in order to receive consideration and action

(see Mandatory Appendix XX covering preparation of technical inquiries)" [36].

Mandatory Appendix XX states,

"Code Interpretations provide clarification of the meaning of existing rules in the Code, and are also presented in question and reply format. Interpretations do not introduce new requirements. In cases where existing Code text does not fully convey the meaning that was intended, and revision of the rules is required to support an interpretation, an Intent Interpretation will be issued and the Code will be revised."

Obviously, the clarifications given in Code Interpretations must be followed.

21.9   CONTENT OF RECORDS

A significant problem in the nuclear industry is the huge number of documents to be maintained for equipment. A common joke was that a valve could not be shipped until the weight of the records for that single valve was equal to the weight of the valve itself. Both the ASME Code and NRC documents identify records to be maintained at a nuclear power plant.

As stated in the NCIG report entitled, "Guidelines for the Content of Records to Support Nuclear Power Plant Operation, Maintenance and Modification (NCIG-08)":

"Utilities are committed to meet the provisions of the NRC regulatory requirements which require that the Owner keep records which would be of "significant value" in demonstrating capability for safe operation;

• Maintaining, reworking, repairing, replacing, or modifying an item;
• Determining the cause of an accident or malfunction of an item; and
• Providing required baseline data for inservice inspection.

"The intent of the regulatory requirements is to ensure that the necessary data is available in the records to accomplish these four purposes. These Guidelines provide a means to define this data and provide a method for rationally selecting the best source of data which is necessary for support of operation, maintenance, and modification of the plant."

The problem is that the records are identified by the name of the record, and not by content. The result has been maintenance of a huge set of documents for each item in the nuclear plant. NCIG took on the task of identifying the essential data required for each item, and suggested practical means to condense the records.

For example, the ASME Code requires post-weld heat treatment (PWHT) records to be kept. The result is that for one item, that record maybe 6 or 8 in. thick if all data is maintained. The task of resolving this issue involved members of the team to review the records for post-weld heat treatment and report what was found. As might be expected, the data in the records were different at each plant. There was no understanding of which data was essential for the maintenance and operation of the equipment. However,

the records contained readings from thermocouples, furnace temperature, and miscellaneous data that was superfluous in every respect.

Based on that, the team was asked to identify the essential data that was in each record for PWHT. The result was that the essential PWHT data in the records was a very small portion of the data being kept. All that was required to be maintained was a record of the number of hours and temperature of post-weld heat treatment for each item.

After reviewing the PWHT records the team looked at all the essential records using the same approach. The result was a tabulation of the essential data for each of the essential records. The change in record-keeping was phenomenal. One utility conservatively estimated that it had saved more than $50 million each year in record-keeping costs.

The full NCIG report, "Guidelines for the Content of Records to Support Nuclear Power Plant Operation, Maintenance and Modification (NCIG-08)" can be downloaded from the EPRI website, http://myepri.com [37], and searching for "NCIG-08." This task was also initiated and managed by Reedy Associates (Reedy Engineering) [38].

21.10   PIPING SYSTEM TOLERANCES

In 1987 the NRC found some instances were piping system supports were not placed as shown in the drawings. The result was that the NRC issued IE Bulletin No. 79-14. All utilities were requested to have a walk down of the piping systems to assure that plants were constructed in accordance with the drawings. The piping system supports were the primary concern.

The inspectors examining the supports found deviations in support locations that varied from a fraction of an inch to 1 ft or more. The physical relocation of the supports to correct the situation was a very significant problem. Correction was extremely costly and time-consuming. The problem was that engineers did not specify tolerances on the dimensions for locating supports. It was often impossible, because of other constraints, to place supports at the exact locations shown on the drawings.

Placing tolerances on dimensions has always been a task that engineers like to avoid. Until IE Bulletin No. 79-14 was published, many engineers felt that it really was not necessary to specify tolerances for piping and the related supports. IE Bulletin 79-14 changed that assumption, but it was a very costly lesson learned.

The Foreword to the ASME Code states,

"The Code does not fully address tolerances. When dimensions, sizes, or other parameters are not specified with tolerances, the values of these parameters are considered nominal and allowable tolerances or local variances may be considered acceptable when based on engineering judgment and standard practice as determined by the engineer."

NCIG decided to address the issue of tolerances. The first thing the NCIG team did was to ask each of the architect/engineering firms to identify the largest tolerances they had ever used for location of supports in any plant. That is, tolerances at any plant, whether it was a refinery, a chemical plant, or a fossil fuel power plant were to be tabulated. Each of the architect/engineers helping in the task group gathered the data.

Based on the data gathered, the largest of all the tolerances identified were judged to be acceptable. The NCIG report for this task is, "Guidelines for Piping System Reconciliation (NCIG-05)" [39]. The Executive Summary in that document states:

"The Nuclear Regulatory Commission and the ASME Boiler and Pressure Vessel Nuclear Code require a reconciliation of the as-built installation of piping systems with the as-analyzed piping system. The need for a consistent approach by the industry in performing these reconciliations led to the development of a position paper prepared by the Pressure Vessel Research Committee (PVRC) and entitled 'Technical Position on Piping System Installation Tolerances.' The paper, published as Welding Research Council (WRC) Bulletin 316, provides recommendations for performing piping system reconciliations and includes tolerances on piping configuration and locations of piping supports."

This document provides guidelines for implementing the PVRC Technical Position. Specific tolerances are provided for piping dimensions, component weights, and support locations and orientations. Piping systems that meet the specified tolerances are acceptable without need for further reconciliation or reanalysis. When the specified tolerances are exceeded, the out-of-tolerance item must be evaluated by the Designer. Recommendations are provided for performing these design evaluations.

The NRC staff has reviewed this document and issued a favorable review letter(Appendix B). The NRC concludes that this document represents a technically acceptable approach for performing reconciliations of the as-built installation of piping systems with the as-analyzed piping system."

Further, the Introduction states:

"Section III of the ASME Boiler and Pressure Vessel Code contains reconciliation requirements in NCA-3554, 'Modification of Documents and Reconciliations with Design Report.' NCA-3554 requires that significant differences between the as-built piping system and the construction drawings must be reconciled with the Design Report.

The implementation of IE Bulletin No. 79-14 [39] and NCA-3554 [40] has proven to be both costly and time-consuming because, at present there is no agreement on what are significant differences or the tolerances to be applied, the definition of detail required for as-built documentation, and the detail of analysis required for the evaluation process. The approach provided in these Guidelines represents collective industry judgment and is based on good engineering practice."

In spite of all the efforts to resolve the issue of piping system tolerances in the 1980's, some utilities today (2010) are making new efforts to walk down the piping systems for the purpose of re-analyzing the stresses in the piping due to the location of the supports. In many cases, the piping systems were designed and constructed to the 1967 B31.1 Power Piping Code. In that regard, the Introduction to B31.1 states:

"This Code shall not be retroactive, or construed as applying to piping systems erected before or under construction at the time of its approval by the United States of America Standards Institute."

The important point is that once an ASME Code piping system has been installed, it is completely inappropriate and unwise to re-analyze the piping system to the provisions in later Codes and Addenda. To do so is a complete waste of time and money. This is the type of problem that engineers with no design and construction experience encounter. These engineers must be trained to understand Code philosophy before they undertake any reanalysis projects. The reanalysis of the piping systems will cost millions of dollars and nothing will be achieved. There is no Code, Standard, or a NRC regulation that requires any of this type of reanalysis effort.

It should be understood that most consultants and design engineers practicing today have never designed any new ASME Code pressure vessels, piping systems (including supports), or other components in nuclear power plants. Further, most did not understand ASME Code philosophy. Without that experience and understanding, proper design evaluations cannot be made. This is a lesson that must be learned from past experience.

The full NCIG report, "Guidelines for Piping System Reconciliation (NCIG-05)" can be downloaded from the EPRI website, http://myepri.com, and searching for "NCIG-05" [37].

This task was also initiated and managed by Reedy Associates (Reedy Engineering) [41].

21.11 USE OF COMPUTERS AND FINITE ELEMENT PROGRAMS

A common problem in the nuclear industry is that many engineers and consultants are performing stress analysis of components that have already been designed. This is a futile task that has no value. These engineers and consultants often use finite element analysis (FEA) to evaluate the design of ASME Code components [42]. The use of FEA as a stress analysis tool to evaluate components previously designed, demonstrates that the engineer or consultant have no comprehension of the ASME Code philosophy.

The ASME Code is not compatible with finite element programs. Finite element programs are not even consistent. For example, different answers will be obtained if a different mesh is used. Even if the same program is used by two different engineers, there is little chance that the stress results will be consistent. Without consistent results, the work performed is meaningless.

It must further be recognized that finite element programs will identify maximum stress at some points, but it cannot break down these maximum stresses into general membrane stresses, general bending stresses, or secondary stresses. There are no post-processors available to perform this task. Yet without knowing the category of stress, it is impossible to evaluate the design in accordance with ASME Code requirements. However many consultants make a lot of money running these futile programs.

There is an article in May 2010 issue of Mechanical Engineering Magazine [43] that addresses the use of finite element computer programs. It points out the significant problems and a lack of understanding that most engineers have about the use and application of these FEA programs. The article is written by Jack Thornton.

For the most part, ASME Code formulas and stress analysis are based on the theory of shells of revolution and beams on elastic foundation. To evaluate a stress analysis based on Code formulas and stress analysis, with a finite element program is a completely futile task and waste of money.

21.12 CYCLIC LOADINGS — IS COUNTING CYCLES VALID?

Some consulting firms sell hardware and software to be used to count the number of cycles on some pieces of equipment in a nuclear power plant. It is extremely difficult to understand the value of counting the number of cycles. The design life of any piece of equipment is not based on a hypothetical number of stress cycles. Further the ASME Code methodology for evaluating cyclic loadings is an approximate guess at best.

The authors of the fatigue rules in Section III stated in a commentary that the rules are extremely conservative and may be incorrect by a factor of 5. That is, the number of allowable stress cycles may be 5 times the allowable number obtained using the Code rules. In a nuclear plant, welds are inspected on a periodic basis. Before any weld would fail, a crack would show. There would be plenty of time to repair the weld before any failure would occur. Counting stress cycles for any piece of equipment is another futile task that is time-consuming and costly.

Before cyclic loading can cause failures, cracks will be developed that can be found by die penetrant examination. Cyclic design rules are theoretical and cannot be used to accurately predict failure.

21.13 NEW APPROACHES — THINK OUTSIDE THE BOX (THE CASE FOR INNOVATIVE PIPING SUPPORT SYSTEMS)

During the last phase of nuclear power plant construction, piping supports consisted of large welded structures. Typically, these individuals supports weight hundreds of pounds and were difficult to install. Often the installed location was not as shown in the drawings because of different types of interferences. This caused problems with tolerances and interfaces with other equipment.

It was difficult, time-consuming, and very costly to move for modify the large welded supports. A new approach is required for the construction of the new nuclear power plants.

Some organizations have adopted a modular construction approach to reduce time, control costs, and shorten construction schedule. Large welded supports for piping, cable trays, and HVAC are not really ideal for modular construction. However there is a practical solution to the problem.

With regard to the ASME Code, Section III, the solution is in the use of Code Case N-500-2, "Alternative Rules for Standard Supports for Classes 1, 2, 3, & MC, Section III, Division 1." This Case allows for the use of MSS SP-58, "PIPE HANGERS AND SUPPORTS — MATERIALS, DESIGN AND MANUFACTURE."

There is at least one company that provides standard supports for piping, HVAC, and cable trays. The supports have been used around the world and refineries, chemical plants, power plants, and other similar facilities. The same standard supports are now available for use in nuclear power plants by implementing Case N-500-2.

These MSS SP-58 [44] pipe supports as permitted in the Case are not welded, are movable, light weight, and easily modified. They consist of standard parts, with special bolts that can be tightened with a 1/4 turn. These simple supports can be installed in less than 5 to 10 min. whereas a large welded (still simple) support often takes more than one man-day to fabricate and install. (This does not include moving or modifying the welded support.)

These SP-58 supports can be compared to a child's ERECTOR SET, but the MSS SP-58 supports are easier to be installed and can be installed more quickly. The use of standard supports should save time for installation, and should save money and material and welding costs. Standard supports seem to be a logical next step; modular construction for nuclear power plants.

21.14 QUALITY ASSURANCE AND MANAGEMENT

In the past, the role of quality assurance was not fully understood by management. Often management would assign all responsibility for body control or on the assurance issues to the Quality Assurance Manager. This was the wrong thing to do for many reasons.

One reason for not assigning quality assurance issues to be resolved by the Body Assurance Manager is that the resolution of the issues must be directed by the manager responsible for the condition being evaluated. It is up to the manager involved to provide the resolution because he is the only person in authority who has the technical understanding and know how to determine the resolution. Further, the Quality Assurance Manager has to be independent of production. If the Quality Assurance Manager is making technical decisions, he is violating the basic principle of Quality Assurance [45]. The NRC should be notified any time this type of violation occurs.

It was common years ago for the Quality Assurance Manager to determine whether or not a provision of the Code was violated. The Quality Assurance Manager is absolutely the wrong person to make decisions regarding violations of the Code. The Quality Assurance Manager does not have the knowledge and background that would be required to make these decisions and even if properly qualified, the quality assurance manager must never make production decisions.

Unfortunately, the situation is still occurring at some utilities. The result is very costly and time-consuming as well is being worthless. This is a lesson that must be learned for the new plants that will be constructed.

21.15 GUIDELINES FOR EFFECTIVE QUALITY ASSURANCE PROGRAMS

21.15.1 Lessons Learned

During the last phase of nuclear power plant construction, one of the largest factors, if not the largest factor, in cost over-run was the inappropriate application of quality assurance. For the most part, senior management assigned "quality assurance" to the Quality Assurance Manager and the quality assurance team. Management had the attitude that quality assurance was "their" job.

Because the Quality Assurance Manager and his team were not (and could not) be fully cognizant of all aspects of plant construction, they adopted the concept of verifying that all procedures were fully complied with. In other words, the quality of an item was judged on the basis of compliance with procedures when it should have been judged on whether or not it complied with the technical specifications.

21.15.2 General Principles

(a) The purpose of a Quality Assurance Program for fabricating items is to ensure that the items meets the requirements of the Design Basis Documents, drawings, and specified Codes and Standards. Quality of an item is assured when it is verified that the item meets the specified technical acceptance criteria.

(b) Because documentation by itself is not necessarily sufficient to establish the quality of the item, it may be necessary to review items during the appropriate stages of fabrication.

(c) It is the responsibility of senior company management to determine the most effective way to implement the production process. Management must ensure that production activities are modified when necessary to correct problems.

(d) Inspectors may work at any level in the organization, as long as the Quality Assurance Program requires them to report significant non-conforming conditions to senior management. These personnel may offer suggestions for change, but must remain completely independent of all engineering or production decisions. Inspectors and quality assurance personnel must never be responsible for the final determination of whether or not equipment complies with the technical specifications.

21.15.3 Quality Assurance Program [46]

(a) An effective Quality Assurance Program must be practical and performance-based.

(b) Responsible management must determine which work activities affect the quality of an item or service.

(c) Responsible management must determine the extent of training required for personnel. Personnel need not be trained for activities they are not required to perform.

(d) Responsible management must define requirements of the Quality Assurance Program and resolve issues associated with program implementation.

(e) The responsible engineering personnel must identify the requirements to be met from Codes and Standards used for design and construction.

21.15.3.1 Comments

(1) An effective Quality Assurance Program will improve quality and safety and also reduce costs. Quality Assurance Programs that do not achieve this result are inadequate and counter-productive.

(2) Company management must take complete possession of and must fully control the Quality Assurance Program and its implementation. The Quality Assurance Manager must not control the Program or its implementation. Because this principle was not fully understood in the past, the implementation of the program became extremely expensive without improving quality in any way.

(3) Quality Assurance Programs must be simple, straight forward, and concentrate on the achievement of quality by doing things correctly the first time. It is possible to eliminate non-conformance reports. When items are non-conforming, the work effort is triple. (The work is performed, then the non-conformance is removed, and finally the work is correctly performed. One way to eliminate some non-conformances is perform examinations of the work while it is still in progress.)

(4) Management must instill the "team concept" and eliminate any "we/they" attitudes. The Quality Assurance Program must then reflect the value of and management's support for working together to resolve all issues in a cooperative fashion. Managers and supervisors must continually assure that "we/they" attitudes are eliminated immediately. "We/they" attitudes lead to fights which no one can win.

(5) Inspectors and auditors must fully understand the technical activities they are monitoring. When these personnel do not understand the technical activities, the "we/they" attitudes immediately come to the forefront. Preferably,

the examiners must be capable of performing the work activity they monitor.

(6) The goal of the Program must be quality products — not merely compliance with procedures. Anyone can verify compliance with the procedure, but only skilled technical personnel can verify that items meet the technical requirements of the acceptance criteria.

(7) In practice, compliance-based Quality Programs tend to concentrate on process, activities, tasks, procedures, and paperwork without necessarily relating them to end results. In the past, some Programs required all activities to be performed to procedures and documented, but this approach causes increased costs and reduced product quality. (Often inspectors felt that if there were no procedures and the activity was not documented, it must be assumed the activity was not performed.)

(8) Performance-based Quality Programs tend to be more focused on the end results that directly contribute to safe and reliable plant operation and de-emphasize the processes used to obtain the desired results. If the primary focus of the program is other than product quality, the result will be that unnecessary documentation will be generated, and other costly, but non-productive, activities will needlessly increase the cost of the products or services.

(9) 10 CFR 50, Appendix B allows a "graded approach" for Quality Assurance Programs. Management must take full advantage of this in order to have an effective Program [47].

(10) In the past, Quality Assurance Programs concentrated on documentation and procedures and the assurance of quality for items was a secondary goal. This approach was often completely ineffective and costly in time and material. Many problems were caused by the fact that verifiers often did not understand the production processes used and the acceptance criteria established for these items. Often items were judged to be non-conforming because of minor inconsequential details in the documentation. For example, if the procedure required signatures in the documentation to be in black ink and the actual signature was in blue ink or red ink, the equipment was identified as "non-conforming" because the documentation did not comply with the procedure.

21.15.4 Organization

(a) Senior management must be responsible for establishing the provisions of the program, verifying its implementation, and assessing its effectiveness. Senior management must ensure that production and engineering management are continually involved in the implementation of the Program.

(b) Questionable conditions must be reported to the level of management responsible for production or engineering of the item. Management must then evaluate the condition to determine whether or not it is a concern.

(c) Items and services must be accepted or rejected on the basis of established acceptance criteria. These criteria may be modified, amplified, or eliminated, as determined by responsible production or engineering management. With proper justification, these criteria may be changed by the organization that developed the criteria, for the purpose of accepting a non-conforming item.

(d) Qualification examination of lead auditors must include technical questions regarding the work activities they are expected to monitor and the proper use of acceptance criteria. These questions must be developed by engineering and production personnel, not the Quality Assurance Team.

(e) Auditors may require training to understand the technical aspects of the work they audit. The qualification documentation of all auditors must include the specific work activities they are qualified to audit and their specific technical qualifications for this work. The auditors' technical qualifications must be updated yearly.

(f) Management responsible for design or construction must identify equipment attributes that are required to be verified by independent inspectors.

(g) Management responsible for design or construction must specify hardware items that are subject to the requirements of the Program. Management may determine that some items are outside the scope of the Program.

(h) Management must determine acceptance criteria and program scope, requirements, and goals.

(i) Quality must be achieved and maintained by those who are assigned responsibility for performing the work [line organizations such as design, fabrication, construction, installation, testing operations, maintenance, etc.].

21.15.4.1 Comments

(1) Inspectors must report questionable conditions to responsible production or engineering management for the appropriate managers to resolve the condition.

(2) Products and services must be accepted/rejected on the basis of established acceptance criteria. These criteria may be modified, amplified, or eliminated as determined by responsible production or engineering management.

(3) An item must not be considered "non-conforming" until it is so classified by the responsible production or engineering management.

(4) Engineering is responsible for determining the appropriate acceptance criteria for equipment.

(5) If the quality program is not being implemented as documented, responsible line management may change the system to meet required objectives.

(6) Quality Control personnel must be part of the production organization, but have the organizational freedom to express any concerns regarding product quality to the appropriate level of management.

(7) Responsible line management must identify the characteristics to be verified. Not all activities or characteristics require verification by independent personnel.

(8) The QA group may report to any level in the organization as long as they are able to keep the appropriate level of responsible production or engineering management informed regarding the effectiveness of the program.

21.15.5 Design Control

(a) Engineering is the art of making practical application of the knowledge of pure sciences, in the economical design and construction of engines, bridges, buildings, mines, ships, and industrial plants.

(b) Engineering design and analysis is a highly professional discipline and must be conducted using qualified engineers, peer reviews, and good supervision (use of step-by-step

procedures for design and analysis without comprehensive understanding by the engineer has led to faulty designs).

(c) Registered Professional Engineers cannot be overruled on technical issues by anyone not a Registered Professional Engineer without raising serious issues regarding safety. Quality Assurance personnel must never be allowed to overrule the responsible Professional Engineer.

(d) Responsible engineering management must establish requirements for the extent of, and methods for, design control.

(e) The criteria to determine the adequacy of design inputs, level of detail, and extent of verification must be identified by the manager responsible for the design of the item. The methods and extent of verification of design activities must be determined by the responsible engineering manager.

(f) Engineering standards (industry or company-originated) may be used rather than develop new written procedures for designing items or performing stress analyses. For design and analysis activities, emphasis must be placed on the knowledge, experience, and understanding by designers. If procedures are used to control design activities, they must provide administrative details only and must allow the design engineer to determine how detailed design work is to be accomplished.

(g) Engineering management is responsible for assuring that engineers are qualified for the engineering work they perform.

(h) Engineering judgment must be verified and accepted by engineering management.

(i) Designs must be checked by qualified engineers or reviewed by engineers knowledgeable in the design requirements for the items being fabricated. It is desirable that the verification of the design work must be on a continuing basis with the involvement of the responsible designer during the design process.

21.15.5.1 Comments

(1) Responsible engineering management must establish requirements for the extent of, and methods for, design control.

(2) Established engineering standards (industry or company-originated) must be used rather than written procedures on how to design or analyze items. Emphasis must be placed on knowledge, experience, and understanding and supervision, not procedures and checklists.

(3) Ensure that engineers are qualified and, when possible, use Registered Professional Engineers.

(4) Design control procedures must be limited to administrative details. In other words, technical information and data must not be included in the design control procedures.

(5) Engineering management must set the criteria for any limits on the use of engineering judgment.

(6) Designs must be checked by peer engineers or reviewed by engineers knowledgeable of design requirements. It is desirable that design reviewers or checkers work on a continuing basis with the responsible designer.

(7) Responsible production or engineering management must determine how computer programs are to be veri-

fied. Benchmarking computer programs is not a valid means of program software verification.

(8) Engineers must be aware that they are responsible for all technical assumptions in every computer program they use.

(9) Responsible engineering management must ensure that engineers using computer programs fully understand the computer program, all built-in assumptions, and the limits of its scope. Further, the responsible engineers must know how to verify their engineering designs and analyses by other means.

(10) All designers must understand that when computer programs are used to design or perform stress analysis, the designer is fully responsible for all assumptions built into the computer program. The engineer/designer must assure that the computer program is compatible with the theory of failure inherent in the type of stress analysis and design requirements of the applicable construction code used. (Finite Element Analysis computer programs are not compatible with the stress analysis requirements of the ASME Boiler and Pressure Vessel Code, Section III, which requires the determination of primary membrane, primary bending, secondary, and peak stresses.)

21.15.6 Procurement Document Control

(a) Management is responsible for defining the requirements necessary to ensure the quality of hardware furnished by suppliers.

(b) The responsible manager must determine the detail necessary to be specified in procurement documents for equipment and services to be procured.

(c) Emphasis must be on items and services, rather than programmatic activities.

(d) The responsible production or engineering management must provide input to the quality assurance provisions to be included in procurement documents.

21.15.6.1 Comment

(1) The manager must determine the level of quality assurance to be consistent with the importance and complexity of the item or service to be procured.

21.15.7 Instructions, Procedures, and Drawings

(a) The responsible manager must determine the level of detail to be contained in instructions, procedures, and drawings. It is intended that these documents be clear and concise, and minimized to allow flexibility and judgment in routine matters. The department manager must ensure that procedures are in place to ensure that applicable Codes and Standards are met.

(b) Responsible production or engineering management must identify those technical activities that specifically require written instructions, procedures, and drawings, because not all activities require instructions, procedures, or drawings.

(c) Personnel may perform activities routine to their occupation without written instructions or procedures. Written procedures or instructions are only required when management determines they are not necessary to properly control the work activities.

21.15.7.1 Comments

(1) It is intended that instructions, procedures, and drawings be clear and concise, and minimized to allow flexibility and judgment in routine matters.

(2) The department manager must ensure that procedures are in place to ensure that applicable Codes and Standards are met.

21.15.8 Document Control

(a) Responsible production or engineering management must define which documents contain or define acceptance criteria and the associated controls to ensure that items and services are properly accepted.

(b) The appropriate manager responsible for the work being documented must determine the level of documentation required, and the level of reviews and approval for changes.

(c) Documents that define processes or procedures within the scope of the Program must be controlled to ensure that work activities are being performed as required.

21.15.8.1 Comments

(1) There is often too much documentation being controlled because responsible management has not clearly identified which documents must be controlled.

(2) If a document provides technical criteria for acceptability of products or services, that technical data is required to be retained. However, it is not necessary to control all related documentation.

(3) Responsible production or engineering management must define which documents contain or define acceptability criteria and the associated controls to assure products and services are properly accepted.

(4) Documents which define processes or procedures necessary to attain quality must be controlled to assure work activities are being performed as required by line management.

(5) Only retain essential technical data.

(6) Only Engineering can define essential technical data.

21.15.9 Control of Purchased Items and Services

(a) The manager responsible for specifying use of purchased items or services must determine whether source evaluation, evaluation of objective evidence of quality, or inspection and examination upon delivery is to be used for acceptance.

(b) The manager is responsible for acceptance or rejection of purchased items and services, as well as resolution of non-conformances.

(c) Procured items must be evaluated by review of objective evidence.

(d) Some, but not all, documentation can be considered "objective evidence." "Objective evidence" means evidence that does not contain opinions or judgments.

(e) Documented test results are objective, but any document which relies on judgments is "subjective evidence."

(f) Most documentation is "subjective evidence." If the documented evidence cannot be directly verified as "firsthand" evidence, it is not "objective." Subjective evidence is based on judgments, not direct facts.

(g) Responsible management is responsible for establishing necessary controls regarding traceability.

21.15.9.1 Comments

(1) Source evaluation may include surveys, audits, interviews, and review of supplier history and non-conformances, as appropriate to the scope of work to be performed.

(2) The primary source of objective evidence is the product itself and its characteristics, not documentation.

(3) Anyone reviewing objective evidence of quality must be technically knowledgeable of the items and services being evaluated.

(4) Responsible production or engineering management must identify when source receipt inspections are adequate to assure conformance with specified requirements.

(5) Audits of quality programs, documentation, and procedures are a completely inadequate method to determine the quality of items.

21.15.10 Identification and Control of Items

(a) Responsible engineering management must specify the identification and traceability requirements for items to ensure that only correct and accepted items are used or installed.

21.15.10.1 Comment

(1) Management must assure that the Program provides for actions to be taken once traceability of an item is lost. In most cases, traceability can be easily reestablished using common sense approaches.

21.15.11 Control of Processes

(a) The responsible manager must determine the controls necessary for special processes, such as welding, nondestructive examination, and heat treatment of items.

(b) The program must define acceptable tolerances for qualification time limits.

21.15.11.1 Comment

(1) The responsible manager must establish the type of documentation required and the control of personnel qualifications.

21.15.12 Inspection

(a) Responsible production or engineering management must identify which attributes are required to be inspected, and when and how results are to be documented.

(b) The responsible manager must ensure that inspectors have the necessary technical expertise to understand and accept the work they inspect.

(c) "Activities affecting quality" must be determined by the appropriate management, not by the Quality Assurance Team.

21.15.12.1 Comments

(1) Responsible production or engineering management must identify which characteristics are required to be inspected, and when and how results are to be documented.

(2) If quality control is part of the production group, it will help eliminate the "we/they" syndrome, which creates hard feelings and conflicts.

(3) Management must emphasize that production workers or supervisors are permitted to perform in-process verifications.

(4) Workers and their supervisors are responsible for product conformance with specified requirements.

(5) When a technical operation is reviewed, it must be by personnel who fully understand the work being performed.

(6) Program auditors must never be allowed to require extra provisions to be added to a supplier's program without concurrence of responsible production or engineering management of the auditing company after it is determined these added measures are absolutely necessary.

(7) The quality control organization must be part of production.

(8) Many inspectors improperly consider in-process verifications to be inspections. In-process verifications may be performed by any personnel and are never to be considered inspections or examinations. These verifications are for informational purposes only.

21.15.13 Test Control

(a) The responsible manager must determine which items are to be tested, the type of test required, and the appropriate acceptance criteria.

21.15.13.1 Comment

(1) If tests are performed in accordance with Codes or Standards, the engineering manager must assure that the provisions of those documents are met.

21.15.14 Control of Measuring and Test Equipment

(a) Management must identify the attributes, of measuring and test equipment used to verify conformance to specified requirements that affect the quality of the item or process.

(b) Management must determine calibration periods, because it is a risk decision.

(c) Equipment required to be controlled and calibrated must be identified by the responsible manager, who must also determine the type, range, accuracy, and tolerances required for calibrated devices.

(d) An organization may perform periodic checks on measuring and test equipment to determine that calibration is maintained.

21.15.14.1 Comments

(1) More frequent calibrations reduce risks. If measurement with a calibrated instrument is required, equipment measured with instruments found to be out of calibration can

be cause for rejection. If the out-of-calibration condition is found after shipment, the item may be still be acceptable after re-verification.

(2) The manager must determine corrective actions when non-conforming test equipment is found.

(3) When periodic checking is used, discrepancies need only be resolved to the prior check, provided that the discrepancy is discovered by the periodic check and there is documentation of the work performed since the previous check.

(4) Management must determine calibration periods, because it is a risk decision. More frequent calibrations reduce risks. If measurement with a calibrated instrument is required, equipment measured with instrument out of calibration can be cause for rejection. If the out of calibration condition is found after shipment, the item may be rejectable.

(5) Not all measuring equipment must be calibrated. Only calibrate equipment when close, critical tolerances must be verified. Often machined surfaces (gasket and flange surfaces) can be visually verified.

(6) Whether or not it is believable, in the past, a number of prominent organizations calibrated even rulers and measuring tapes. Such activities clearly illustrate quality assurance programs that are completely out of control.

21.15.15 Handling, Storage, and Shipping

(a) The responsible manager must identify the necessary controls for handling, storage, and shipping of equipment.

21.15.15.1 Comment

(1) The necessary controls must be practical, yet anticipate possible accidents, and methods for correction.

21.15.16 Inspection, Test, and Operating Status

(a) The responsible manager must identify the inspection, test, and operating status to be identified and controlled.

21.15.16.1 Comment

(1) The procedures for controlling inspections, tests, and operating status must be clear and understood by all those working in the vicinity.

21.15.17 Control of Non-Conforming Items

(a) The quality of an item is rendered unacceptable (non-conforming) only by identification of a deficiency in an essential hardware characteristic.

(b) A deficiency in documentation or procedure may render the quality of an item indeterminate, but it does not mean that the item is non-conforming. Further evaluation may be used to determine the acceptability of the item.

21.15.17.1 Comments

(1) The emphasis must be placed on control of hardware, not on documentation which has no impact on hardware quality.

(2) Example of an appropriate definition:

A non-conformance is a deficiency in characteristic that renders the quality of an item unacceptable.

(3) A deficiency in documentation or procedure may make an item or activity questionable, but by itself will never make an item non-conforming.

(4) A deficiency in procedure could result in an item being unacceptable, but a non-conformance of an item can be found only by examination or evaluation of the item, not review of a procedure.

(5) The quality of an item is rendered unacceptable (non-conforming) only by identification of a deficiency in a hardware characteristic.

(6) A deficiency in documentation or procedure may render the quality of an item indeterminate and require an evaluation and disposition of the item by the responsible engineering organization.

(7) Reports of deficiencies in documentation or procedures that identify the questionable condition of an item may be documented on a questionable condition report and resolved by verification of the acceptable condition of the hardware.

(8) An example of a completely inappropriate non-conformance report is an 800 page NCR that was written because some inspectors (illogically, to say the least) measured the outside diameter of 1/2 in. diameter tubes at 90° and 180° to determine the ovality of the tubes. They determined that the differences in ovality exceeded the 8% limited identified in Section III of the ASME Code. Each measurement was identified as a non-conforming condition. In reality, there were no non-conforming conditions because the tubes were many times thicker than required. The inspectors never should have been allowed to identify the conditions found as being non-conforming.

(9) At one nuclear power plant construction site, 2000 NCRs were evaluated to determine how many NCRs actually described equipment that was non-conforming. The reason the results of the evaluation was that only one non-conformance report correctly identified a non-conforming condition for the hardware. Four of the non-conformance reports were questionable, but further evaluation showed that the conditions were not of concern. All of the other 1995 reports identified nothing that was non-conforming.

At the time, the cost of processing one non-conformance report was estimated by 16 different nuclear utilities as somewhere between $3500 and $7000. If $3500 is conservatively taken as the true cost, almost $7 million was spent on nothing. However that cost does not include the cost of writing up and identifying the conditions in the first place.

21.15.18 Corrective Action

(a) When a condition adverse to quality is identified, the responsible manager must determine the cause and identify the corrective action. When similar non-conformances are identified, the root cause must be reevaluated to determine the proper course of action to preclude repetition.

(b) Responsible production or engineering management must establish appropriate corrective action to ensure the final acceptance of the hardware.

(c) The Quality Assurance Program must distinguish between "significant conditions adverse to quality" of the hardware or service and issues concerned with administrative procedures and documentation.

(d) The responsible production or engineering manager must define conditions that are "significant."

(e) Production organizations responsible for the adverse quality condition of the hardware are responsible for determining and taking both corrective action and preventative action.

21.15.18.1 Comments

(1) Whether or not the condition is, or could be, adverse to the quality of an item, and its significance, must be determined by the responsible engineering group.

(2) The corrective action program must focus on correction of significant conditions adverse to the quality of hardware.

(3) The quality program must distinguish between significant conditions adverse to quality of the hardware and corrections to administrative procedures and documentation.

21.15.19 Quality Assurance Records

(a) Responsible production and engineering management must determine which data is essential and provide means for collection, storage, and retrieval of that essential data.

(b) Required records must be indexed.

(c) Radiographs may be reproduced.

(d) Quality assurance records are to be maintained with the goal of saving the necessary technical information that may be required during plant operation.

(e) If data is required to be kept, it is permissible to verify and summarize the required data in a new format, and keep the summary, rather than the original record.

(f) Records may be generated and stored electronically. Retention that are unique to electronic records.

21.15.19.1 Comments

(1) Non-essential data must not be maintained as a permanent record.

(2) Documentation must be maintained on the basis of technical information that may be required later, not on the basis of the title of the document. It is desirable to eliminate collection of unnecessary information that has no usefulness after the plant is in operation. Senior management is responsible for ensuring that this goal is met.

(3) The Program must address how electronic records are generated, identified, authenticated, stored, and maintained for as long as required. If records are generated electronically, the Program must describe the means of identifying the individual who generated the record and the means of verifying that the individual is authorized to generate the record. If records are converted from one media to another, (e.g., paper to magnetic or magnetic to optical disc) the Program must describe the process for verification of the contained information.

(4) Responsible production and engineering management must determine which data is essential and provide means for collection, storage, and retrieval of that essential data. Non-essential data must not be maintained as a permanent record.

21.15.20 Audits

(a) When a technical operation is audited, it must be by personnel who understand the work being performed.

(b) The audit frequency must be specified in the Quality Assurance Manual. The frequency must be commensurate with the schedule of activities.

(c) Senior management must ensure that all personnel performing audits have the proper technical experience and skills to understand the work they are auditing.

(d) Auditors of suppliers may recommend to responsible management that extra provisions be added to a supplier's Quality Assurance Program, but the final decision to do so belongs to the responsible line manager in the auditor's organization.

21.15.20.1 Comments

(1) Only qualified engineers should be used to audit design control.

(2) Auditors may recommend changes to a supplier's quality assurance program, but must never be allowed to force suppliers to change their program.

(3) Auditors can only accept or reject a supplier's quality assurance program. However, the auditors manager may require the supplier to make changes before an order is placed.

21.16 SUMMARY

New nuclear power plants are now in the early phases of design and construction. In order to assure that costs and schedules do not get out of control, it will be necessary to evaluate and learn the lessons from the past. Although new people are involved, everyone should be aware and learn from the mistakes of the past.

The ASME Code has been looked upon by many as a hindrance. People should recognize that the Code changes all the time and that if new changes are required people should inform ASME. The ASME Code Committee is anxious to address and resolve new issues and to modify the Code and in the third as necessary to meet new ideas for design and construction.

21.17 ACRONYMS

ASME	American Society of Mechanical Engineers
AEC	Atomic Energy Commission
VWAC	Visual Weld Acceptance Criteria
AISC	American Institute of Steel Construction
AWS	American Welding Society
NCIG	Nuclear Construction Issues Group
NRC	US Nuclear Regulatory Commission
EPRI	Electric Power Research Institutes
PWHT	Post-weld heat treatment
PVRC	Pressure Vessel Research Committee

HVAC	Heating, Ventilation, and Air Conditioning
BPVC	Boiler and Pressure Vessel Code
ASA	American Standards Association
BPVC	Boiler and Pressure Vessel Code

21.18   REFERENCES

1. ASME BPVC Section I, Power Boilers—1952

2. ASME BPVC Section VIII, Unfired Pressure Vessels—1958

3. ASA B31.1 Code for Pressure Piping—1955

4. Duquesne Light Co.

5. Commonwealth Edison Co.

6. Yankee Atomic Light Co.

7. Chicago Bridge & Iron Co., Author

8. Criteria of Section III of the ASME Boiler and Pressure Vessel Code for Nuclear Vessels

9. Author's discussions with Committee members

10. Author's attendance at Special Committee meetings

11. ASME BPVC Section VIII, Unfired Pressure Vessels—1962

12. Author's discussion with Mr. James Mershon of the AEC

13. National Board of Boiler and Pressure Vessel Inspectors

14. ASME BPVC Section VIII, Division 2 Alternative Rules for Pressure Vessels—1968

15. Case 1270 N—"General Requirements for Nuclear Vessels

16. Case 1271 N—"Safety Devices"

17. Case 1272 N—"Containment and Intermediate Containment Vessels"

18. Case 1273 N—"Nuclear Reactors and Primary Vessels"

19. Case 1274 N—"Special Material Requirements"

20. Case 1275 N—"Inspection Requirements"

21. Case 1276 N—"Special Equipment Requirements"

22. ASME BPVC Section III, Nuclear Vessels—1963

23. Author participated in the design

24. CBS Telecast of "60 Minutes" early 1984

25. Author's personal knowledge

26. AWS D1.1 Structural Welding Code-Steel—1985

27. NRC letter from Mr. James P. Knight to Mr. Douglas E. Dutton, dated June 26, 1985

28. Visual Weld Acceptance Criteria, Volume 1: Visual Weld Acceptance Criteria for Structural Welding at Nuclear Power Plants (NCIG-01, Revision 2) (EPRI NP-5380)

29. Visual Weld Acceptance Criteria, Volume 2: Sampling Plan for Visual Reinspection of Welds (NCIG-02, Revision 2) (EPRI NP-5380)

30. Visual Weld Acceptance Criteria, Volume 3: Training Manual for Inspectors of Structural Welds at Nuclear Power Plants Using the Acceptance Criteria of NCIG-01 (NCIG-03, Revision 1) (EPRI NP-5380)

31. NRC letter from Themis P. Speis to Mr. Walter H. Weber, dated April 9, 1987

32. NRC letter from Mr. James P. Knight to Mr. Douglas E. Dutton, dated June 26, 1985

33. Information from nuclear utility members of NCIG to author

34. Mandatory Appendix XX of Section III of the ASME Code, 2010 Edition

35. ASME v. Hydrolevel, 456 U.S. 556 (1982) No. 80-1765

36. Foreword to Section III of the ASME Code, 2010 Edition

37. EPRI website—Not necessary to sign in as Member

38. "Guidelines for the Content of Records to Support Nuclear Power Plant Operation, Maintenance and Modification (NCIG-08)," Volumes 1 & 2 (EPRI NP-5653, November 1988)

39. NRC Bulletin published in 1979.

40. NCA-3554, "Modification of Documents and Reconciliation With Design Report, Section III of the ASME Code

41. "Guidelines for Piping System Reconciliation (NCIG-05)" (EPRI NP-5639, May 1988

42. Author's review of stress analysis and discussions with inexperienced engineers

43. Mechanical Engineering Magazine, May 2010, "The Question of Credibility" by Jack Thornton

44. MSS SP-58, "PIPE HANGERS AND SUPPORTS—MATERIALS, DESIGN AND MANUFACTURE." 1993 Edition

45. Author's discussions with NRC officials

46. These guidelines meet NRC Regulations 10CFR50, Appendix B, and Section III of the ASME Code

47. Minor items should not be considered the equivalent to major safety related items

NUCLEAR POWER INDUSTRY RESPONSE TO MATERIALS DEGRADATION – A CRITICAL REVIEW

Peter Riccardella and Dennis Weakland

22.1 INTRODUCTION

In the 1960s, nuclear power for electricity production was one of the greatest future economic promises, both in the United States and internationally. Electricity "too cheap to meter" was the claim, and US utilities and supplier corporations jumped in whole-heartedly. Yet 50 years later, the industry is just pulling out of decades-long doldrums. No new nuclear plants have gone on line in the United States for over 30 years. All US Nuclear Steam System Suppliers (NSSSs) are either foreign owned or have major overseas partners, and we no longer have the domestic infrastructure and capacity to produce the components needed for new plants. The current nuclear fleet, although operating safely and economically, generates only about 20% of US electricity. Given shortages of domestic fossil fuels and concerns about greenhouse gases and global warming, not taking greater advantage of this United States–developed energy resource defies logic.

What happened? And what can the industry do to keep this from reoccurring during the current reemergence of nuclear power as a major clean energy source of the future? Many of the problems were no doubt political and beyond the scope of this technical article. Yet most were self-inflicted, many of them related to component mechanical and structural integrity that can and should be avoided in the future, if we can learn from the lessons of the past.

The purpose of this article is to highlight the issues that the authors have personally seen and been involved in, as engineers and technical managers in the nuclear industry for over 40 years. The authors' backgrounds combine employment at major nuclear vendors during the formative days of the industry, plus utility engineering perspective provided by a co-author who was a utility participant and ultimately chairman of the industry's generic Materials Reliability Program. The article looks retrospectively at the root causes of these problems, itemizes the lessons learned, and recommends an approach going forward that will anticipate and hopefully prevent the industry from repeating these mistakes in the future.

22.2 THE ISSUES

22.2.1 Cast Stainless Steel Piping Inspection Issues

For economic reasons, one PWR vendor switched to centrifugally cast stainless steel (SS) for primary coolant loop piping. A few years after this decision was made, it was discovered that the piping could not be ultrasonically inspected as required by the in-service inspection (ISI) code [1]. The inspection problem still exists today, some 40 years later, and although the cast piping has not yet shown any propensity for degradation, there are risks involved because the lack of inspection has not allowed for the identification of precursors to serious material degradation problems that might occur.

22.2.2 Flow Accelerated Corrosion of Carbon Steel Piping

With the public and regulatory focus on the integrity of the reactor coolant boundary, the industry got a wake-up call in late 1986 when catastrophic failure of a large feedwater line resulted in the deaths of several workers at a nuclear power plant. Although the failure of the line was not a nuclear safety concern and was not the first such failure from flow accelerated corrosion (FAC) in the power industry, it was the first in the nuclear industry to result in a fatality. This event demonstrated the need to look beyond the nuclear safety related boundaries. This failure resulted in Bulletin 87-01 [2] and Generic Letter 89-08 [3] being issued by the NRC mandating inspections of carbon steel piping that may be affected by FAC. Prior to this event, there were approximately eight events associated with FAC on the secondary side of nuclear power plants between 1981 and 1986 [4], and although this failure mechanism was known in other industries, the transfer to nuclear power plants was not made because of the focus of resources and inspections on the systems within the nuclear systems. As with nuclear safety-related piping systems, there was a general reliance on the robust engineering and materials of construction resulting from the use of the ASME Code. However, that code does not address corrosion, stating that preventing it is the responsibility of the plant owner. The Figure 22.1 is an example of the type of failure that can occur

FIG. 22.1 RUPTURE OF A 6-IN. (152.4 MM) FEEDWATER PIPE IN AN AREA THINNED BY INTERNAL CORROSION [5]

from internal erosion of the pipe wall of moderate energy piping in a power plant. All power plants are now routinely inspecting and monitoring over long periods the areas where flow-induced erosion/corrosion may occur.

22.2.3 Nickel Base Alloy Primary Water Stress Corrosion Cracking [6]

In the early 1970s, Alloy 600, and its weld metals, Alloys 82 and 182, came into favor for numerous PWR locations because of its good mechanical properties, weldability, the fact that it is a single-phase alloy (requiring no post-weld heat treatment), good general corrosion, and pitting resistance, and it was thought to be a good transitional material (i.e., its thermal expansion coefficient is intermediate between those of SS piping and low-alloy steel components). However, it has proven to be susceptible to primary water stress corrosion cracking (PWSCC) in the PWR primary coolant environment, resulting in cracking problems in numerous components, including:

o steam generator tubing
o pressurizer nozzles and heater sleeves
o reactor pressure vessel (RPV) top head penetrations
o RPV bottom-mounted instrument penetrations
o butt welds between component nozzles and piping, often referred to as "safe-ends" (see Figure 22.2)

PWSCC and associated leakage have cost the PWR industry hundreds of millions of dollars in mitigation and repair activities, not to mention the dollars associated with extended outage times, increased regulatory burden, and damage to its reputation.

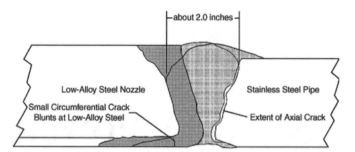

FIG. 22.2 THROUGH-WALL CRACK AND PART-DEPTH CIRCUMFERENTIAL CRACK IN V. C. SUMMER REACTOR VESSEL HOT-LEG OUTLET NOZZLE [7]

22.2.4 IGSCC of Sensitized SS in BWR Piping [8]

Even though sensitization of SS was a relatively well-known phenomenon, unstabilized SS was used extensively in BWR components and piping, which were subjected to sensitization-inducing fabrication practices (heat treatment and welding). The concern first revealed itself in nozzle safe-ends, which were sensitized owing to their inclusion in the vessel post-weld heat treatment process. These "furnace-sensitized" components cracked relatively early in BWR plant lifetimes. Higher dissolved oxygen levels in BWR coolant makes this form of intergranular stress corrosion cracking (IGSCC) almost exclusively a BWR concern. Somewhat ironically, however, it was one of the reasons cited for the use of Inconel and cast SS piping in PWRs.

The sensitized safe-ends were replaced relatively quickly, and ceased to be used in newer plants, but a more insidious, far-reaching problem lurked — weld sensitization of essentially all welds in SS piping systems (typically 300~500 welds per plant). In 1974, there was a massive outbreak of IGSCC in small diameter (25 cm [<10 in.]) SS piping. In typical fashion, a great deal of effort was expended by the BWR vendor and utilities to justify the belief that it would not affect larger diameter piping, until IGSCC was discovered in larger diameter lines. (1978 — first incidence of IGSCC in large diameter, 24 in. [61 cm] piping at KRB in Germany; 1982 — first domestic incidence of IGSCC in large diameter, 28 in. [71 cm] piping at Nine Mile point 1.)

The impact of this one issue on capacity factor losses has been massive (see Figure 22.3), and it shows that a generic issue, if left unmanaged for too long, has the potential to affect the entire fleet of plants. Some BWR utilities have spent on the order of $50 to $100 million, replacing entire piping systems in their plants. All BWRs eventually implemented some form(s) of remedy, including

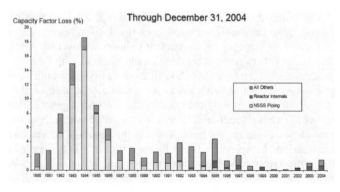

FIG. 22.3 CAPACITY FACTOR LOSSES DUE TO IGSCC IN BWRS [17]

piping replacement, water chemistry improvement, residual stress improvement, weld overlays, or combinations thereof.

22.2.5 Reactor Vessel Embrittlement [9]

Another problem that has proven to be life limiting for some plants (e.g., the early shutdown of Yankee Rowe) and has led to expensive analyses and operating restrictions for others is embrittlement of the reactor vessels due to neutron irradiation. The issue arose from the effects of neutron damage to low-alloy steel material not being understood prior to the manufacturing. As data was obtained from material surveillance programs, it was found that small amounts of some alloying materials in base metal and welds could significantly impact the toughness of the materials when subjected to neutron irradiation damage. For example, the presence of copper in reactor vessel welds was found to accelerate neutron embrittlement. This copper had been added as a coating to the weld rods to prevent them from rusting while in storage. In most low-alloy steels, these elements are non-contributors to the material properties in the un-irradiated state and therefore not essential. The industry and the manufacturers quickly limited the levels of elements found to be detrimental during neutron exposure in power reactors. Plants continue to monitor the effects on the materials of construction through the use of test specimens installed in the reactor, which are periodically removed for testing. Although this issue has been managed well over the past two decades, it continues to demand significant resources for research and continued vigilance. Most of the issues concerning neutron damage of reactor vessel steels are long-term issues (1 to 2 decades in the future), making it hard to dedicate current resources to them when other plant issues have nearer term consequences. However, the safety implications of a brittle fracture of a RPV make this one of the top priorities for the industry to monitor and manage successfully.

22.2.6 Buried Piping

At the time of construction, it was recognized that, like water and sewage lines in municipalities, corrosion of buried carbon steel and cast iron piping systems was inevitable without some form of corrosion control measure. This is why these systems were designed with protective coatings applied to the exterior of the piping to isolate the pipe material from the potentially corrosive environment or had sacrificial or impressed current cathodic protection systems applied. So with all the forethought, why are these systems failing? The coating systems commonly applied were hydrocarbon-based materials, and over time, they lose adhesion or become brittle and crack. The coating systems performance can be further reduced when, during installation, a nick in the coating occurred and provided a location for corrosion to take hold. This occurred more often than one would assume since the installers did not know or have an appreciation of the long-term impact of a small nick or "holiday" in the coating. When we turn to systems protected by various cathodic protection means, we find that lack of understanding and the costs associated with maintenance have taken a toll on the systems. Maintenance of an impressed or sacrificial cathodic protection system requires maintenance of the voltage sources, connections, control mechanisms, or anodic consumables, which became expensive for plants on limited budgets. And with a failure mechanism that is long term (5 to 10 years) in nature having no near-term failure manifestation or impact on system operation or system integrity, these protection systems began to fall into disrepair, eventually resulting in leakage failures of the piping systems. As is typical with buried systems (regardless of industry), they are out of sight and out of mind. To be fair,

more than just costs impacted the system failures of buried piping components. Some of the integrity issues for buried piping were related to changing environmental conditions on both the inside and outside walls of the piping. For example, many fire protection systems were installed using cast iron cement–lined piping due to its good history of resistance to OD corrosion damage. However, if soil around the buried gray cast iron becomes acidic due to run-off or other sources, it is subject to corrosive damage. The damage mechanism called "graphitization" occurs. Unlike the typical wastage observed during the corrosion of steels, there may be no apparent wall loss when cast iron pipe corrodes. The iron simply "rusts" in place, leaving behind only brittle graphite flakes, and the remaining matrix reaches a level that cannot withstand the pressures or mechanical loads imposed on the pipe. Since gray cast iron is brittle by nature, when a crack occurs, it tends to be much larger than would occur in ductile materials. The rusted item keeps its net shape but lacks strength as shown in the Figure 22.4.

However, a more widespread failure mechanism was coming into play for many of the inland waterway power plants, which affected both buried and above-ground piping. The intake of raw water from rivers and water sources for bulk cooling at the plants brought with it active organic life. These organisms adhered to the inside of the piping and bred in stagnant or low-flow areas, corroding the piping materials by a mechanism known as MIC or microbiologically influenced corrosion. As industrial sites along the rivers and water sources reduced their discharges, the once polluted rivers became even more alive with active organic life (mussels, clams, and bacteria), thus aggravating the condition. Although MIC damage itself is very localized, there are thousands of localized sites along the length of a piping system, creating small areas of leakage. Figure 22.5 shows a typical example of the degree of wall thickness loss in piping in localized areas.

The industry developed programs to address the concerns with the failure of safety related buried piping in the 1990s (Nuclear Energy Institute [NEI] 07-07 Groundwater Initiative). These programs provided sound guidance, but each plant addressed the concerns on a plant-specific basis, and since the failures lacked significant nuclear safety consequences, most buried piping failures were treated as a repair upon self-identification (leakage) basis. Recently, some of the plants have taken on the very expensive option to replace the piping with high-density polyethylene piping that is unaffected by the failure mechanisms currently impacting

FIG. 22.4 AN EXAMPLE OF A FAILURE IN CAST IRON PIPING [18]

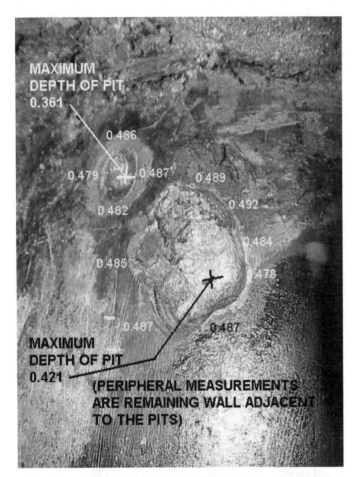

FIG. 22.5 EXAMPLE OF LOCALIZED WALL THICKNESS LOSS DUE TO MICROBIOLOGICALLY INFLUENCED CORROSION (MIC) [18]

metal buried piping systems. In addition, the industry has recently developed an initiative to address the maintenance, repair, and replacement of these systems through NEI Initiatives APC 09-53 [10] and NEI-09-14 [14].

22.3 DISCUSSION

The above paragraphs highlight some key materials degradation issues that have affected nuclear power plants. The balance of this chapter will present the authors' view of how the industry management approach has contributed to the material issues listed. Some common threads run through all of these issues, namely,

- plant design and construction decisions focused on reducing cost without a full understanding of their impact
- short-term perspective versus an entire life cycle perspective of managing plant issues
- outage/maintenance cost focus
- cost- and compliance-based management of technical/regulatory based issues
- failure to learn from foreign and other plant operating experience (rather than ask the question, "Why does it not apply?" ask the question: "Could this be me?")

The overall approach to issue management is complex and individual aspects of it cannot be addressed without impacting others.

The Nuclear Power Industry is an industry that is constantly under the microscope of public opinion and perception. The importance of managing materials degradation in a proactive manner so as to preclude failures and public reaction cannot be overstated. No executive wants to read in the morning newspaper that his plant is the subject of a piping or component failure due to any reason.

To understand the industry's historical approach to materials degradation issues, one must look to the initial design and licensing process as the foundation upon which the approach was built. The materials of construction drew heavily from the Nuclear Navy experience, which in large part has provided robust materials for use within the power generation industry. Although the materials had proven to be robust in the operating environment of a ship, there was little true academic understanding of how the materials would perform when scaled up from small and exceptionally well-maintained Navy power plants to larger power-generating plants, with little fundamental research into the applicability of such scaling. Access to detailed materials degradation information developed or observed by the military was generally unavailable to NSSSs and the utilities. Nonetheless, there was a fundamental belief and assumption that the materials of construction would not be subject to significant degradation during the 40-year design operating life of the power plants.

Acceptance of the hypothesis that robust materials of construction would not be subject to degradation leads to an assumption that any material failure must be caused by factors other than basic material degradation. As a consequence, significant engineering, quality control, and manufacturing resources were spent and focused on addressing factors such as fatigue, fabrication defects, and stress loading. It should be noted that this effort has proven to be very successful, since relatively few industry failures have resulted from these factors. However, few if any resources were focused on the potential for material degradation due the interaction of system operating environment with the materials of construction.

The lack of focus on materials degradation issues early in the design and construction time frame was understandable since the available experience base indicated that no degradation had been observed, material testing in the power reactor environment was very expensive, and the time horizon for the materials testing was extremely long before one could see any results. Coupling these thoughts with the competitive economic considerations that existed for building a plant in the 1960s and 1970s (i.e., four vendors competing for an expanding but finite market), it is understandable that one would not spend significant resources looking for unobserved and unexpected degradation of materials that have been demonstrated to be robust. Other important factors at the time were the high interest rates and long construction time frames for the plants, causing the abandonment of some construction projects due to public fears of high electric rates from a technology that was purported to be "too cheap to meter."

The utility perspective on how to address the materials integrity issues of nuclear power plants is rooted in two areas: (1) an expectation of minimum of maintenance costs and (2) a fix upon identification of issues. As noted in the opening paragraph, the expectation of the nuclear power proponents was that the costs of operation were going to be small and that the materials of construction were so robust that maintenance problems would be virtually non-existent. Second, much of the utility management experience was rooted in the fossil fuel power generation arena, where a "repair and continue operating" philosophy was the standard approach. The impact of managing in a highly regulated environment (from a technical rather than a rate base perspective) was not fully appreciated. The management approach to fossil generation plants was to fix what failed when it self-identified or replace/repair known wear locations at each scheduled outage. However, nuclear power plants brought significant demands for mandated periodic inspections (ISIs [1]), extensive documentation of all activities, and determination of the root cause for any form of degradation. These costs and the costs associated with the loss of production and the associated repair/replacement activities led many in the utility industry to adopt an analytical approach to plant materials degradation issues, which allowed the plant to continue to operate while working toward regulatory resolution.

As a result, an initial mind set of trying to analyze an issue into acceptance was widely applied. Since the plants were designed and manufactured with very conservative assumptions of loading and material properties (don't forget that the experience base still showed very few failures), from an engineering perspective, when material degradation was identified, one could typically show that all safety margins, regulatory requirements, and code requirements are met, even though material degradation was identified. It was manageable by analysis, and therefore, the power plant could continue to operate and generate revenue. The industry became adept at defining why a particular degradation mechanism was applicable only to specific components or specific operational histories at a small number of plants, essentially putting our problems into small manageable boxes rather than integrating our operating experience over broader and more generic horizons.

The nuclear industry was considered so 'technical' and robust through the application of NRC regulation and ASME code requirements that there was a general feeling/consensus that, if a utility was meeting these requirement there was no need to go any further. There was little or no benefit for a utility to go beyond the regulatory or Code requirements because these additional actions would have to be justified from a cost perspective to the rate regulators or at a minimum to the shareholders. Additionally, the utility would have to address why a deviation from the standard or an enhancement action was necessary or desirable to industry oversight organizations and other utilities. Generally, the nuclear power community would view this "extra" action as becoming mandated or highly recommended for use at other plants, thus increasing the cost of current operation. With all the baggage that would come along with going beyond the NRC or ASME requirements, there is little wonder that the management approach was to meet the requirements, period.

This led to a period of what we will call "management by regulation" by the NRC in the late 1970s through the 1990s. This was a natural evolution from the compliance-driven mind set of construction and licensing used by personnel making the leap from a plant under construction to a power production facility. Couple this with a large influx of retired Navy personnel into the utility business following the TMI event, who were comfortable with the rule/procedure-based approach to issues that was very successful in the US Navy. Both the utility industry and regulator were comfortable with this approach; however, the regulations and codes were developed and based on a robust material philosophy and not geared to the detection of degradation in robust materials, such as Alloy 600. The Codes and regulations used a sampling approach to the identification of materials degradation that focused on areas of high stress and potential for construction/fabrication defects. Again, it was assumed by the Code committees and the regulator that the likely locations for a piping system integrity issue would be a weld location since the materials in general would not be subject to degradation. So little emphasis was placed on the need for guidance

in the area of repair and replacement of power plant components that, in the 1972 Edition of the ASME Inservice Inspection Code [1], it was relegated to just a few pages. There were essentially no formal written rules for addressing generic material degradation mechanisms; all the regulation and Codes were focused on local defect identification and repair. Again, this is a natural extension from the assumption of robust materials of construction, not subject to generic degradation, so defect identification was considered appropriate. The NRC regulation focused on the maintenance of pressure boundary integrity and identified defects needed to be repaired and managed to conform to the safety and regulatory margins identified in the plant licensing commitments. Therefore, the industry's approach was to determine what was allowable under their licensing basis and meet the Code of construction. When a defect or condition was identified by the industry, the regulator would generally issue generic guidance to require all plants to assess the potential for the defect/condition. Each plant would ensure that they met their regulatory commitments and determine if the condition was applicable to their plant. As one can imagine, it was in each plant's best interest to first ensure that all regulation and Codes were met and then to define the issue in a manageable fashion, which meant to technically define the problem/issue addressing the specific attributes of the condition and determine if those attributes are applicable to the individual plant (i.e., How can I justify that this is not an issue at my plant?).

Since construction of PWR plants happened over a time frame of 20+ years, materials degradation mechanisms did not affect all plants at the same time or to the same degree. With the exception of IGSCC in the BWR fleet (which were constructed over a shorter ~10-year period), degradation identified at a power plant was more or less considered to be a plant-specific issue in the late 1970s and early 1980s. The treatment of each issue or set of issues as plant-specific led to a wide array of solutions. One of the most prevalent was to use an engineering approach to show how your plant was different from the plant that had the problem, thus eliminating your plant from the affected population and thereby minimizing the economic impact on your plant or utility.

The first significant materials degradation issue for the PWR fleet was the cracking of Steam Generator Alloy 600 tubing. When the issue was first identified, the immediate reaction was that this cracking affected only those plants that either were on phosphate chemistry or changed from phosphate to AVT (all volatile chemistry), then it was only those plants with tight U-bends that were over-bent, then it was the expansion methods used at the tubesheets, then the thermal treatment of the alloy 600. As the issue became more widespread, all the plants came to the understanding that the issue was all Alloy 600 tubing and that the stresses and other conditions were simply drivers for the degradation process. The industry finally decided that component replacement with Alloy 600, thermally treated tubing was the answer, only to find that although it lasted longer, it also degraded. Now the industry has moved to Alloy 690 tubing, which will most likely have a longer operating life than Alloy 600TT, but given enough time, it will likely also suffer degradation events. Hopefully, this will be longer than the operating lifetime of the power plants.

22.4　FLEET-WIDE RECOGNITION

The first segment of the US Nuclear fleet to recognize that a fleet-wide approach to materials degradation issues was required was the BWR plant owners. To address the issue of IGSCC in the

SS piping and components within the reactor coolant system, a consortium of all BWR owners was formed under the auspices of EPRI. The BWRVIP (Boiling Water Reactor Vessel and Internals Program) was born in part because all these plants were of relatively similar material design and had a common NSSS supplier, which made addressing proprietary design issues easier than in the PWR fleet, which had three NSSS vendors at the time. The BWR-VIP executive leadership was determined to manage the inspection and repair approaches from a utility perspective and not wait for the NRC to drive actions. The approach was to proactively address the degradation and prepare repair and mitigation strategies as a fleet, then meet and discuss with the regulator the technical basis for the strategies, gain their insights, and address their concerns in a non-reactive manner. The BWRVIP issued guidance documents requiring each BWR owner to perform certain actions, and the executives of each BWR plant held each other accountable. Thus, the BWRVIP guidance became a method of self-governance.

The self-governance approach allowed the BWR owners to plan and pool their resources to address potential issues. The early identification and investigation of the degradation mechanism allowed for the development of repair and mitigation strategies that could be accomplished during non-outage or crisis periods. The BWR-VIP developed inspection, repair, and mitigation approaches that were required to be performed by all BWR owners, hand in hand. Taking a fleet approach to standardize the inspection and remediation approach, the BWRVIP actively engaged the regulator to keep them informed of the intended actions and technical basis for applying the measures. The regulator watched the implementation of the guidance closely to determine if all plants were indeed implementing the inspection and repair strategies specified. The net result of this proactive self-governance by the BWR fleet led to a significant decrease in the NRC mandated regulation in the 1990s and 2000s, allowing the utility resources to be focused on addressing and resolving the issue rather than responding to regulatory mandates.

Although this fleet-based approach was working for the BWR owners in the late 1980s and 1990s, the PWR fleet was still in a mode of engineering their problems away through differences in design or defining failure mechanisms as unique to particular operating approaches or transients. As a result, the NRC staff issued nearly a dozen generic communications to the PWR fleet and only three to the BWR fleet between 1985 and 2000. Despite the experience gained by the PWR fleet in addressing the degradation of Alloy 600 materials in steam generators, the PWR owners continued to hold fast to technical beliefs developed for the tubing that it was environment and stress related and pertained to thin wall materials only. It was felt that the stress distributions and environments were so different that Alloy 600 piping in the reactor coolant loop would not be affected. In fact, the PWR arguments were strong enough that the regulator and ISI Code Committee allowed a risk-based approach to inservice inspections to actually reduce the frequency of inspection of Alloy 600 locations based on inspection history to date and the assumed robustness of the material.

The French indicated in the 1990s that they had observed degradation of the Alloy 600 penetrations in their reactor vessel closure heads and other Alloy 600 weld locations. What was the US PWR reaction to this operational experience in France? Initially, it was that the cracking was caused by the French operating approach of load-following with their reactors and that this additional thermal cycling was the cause. Since US plants do not

load-follow, it was argued that the US fleet was unaffected. Additionally, it was also determined that there are differences in the Alloy 600 chemistries and that only certain manufacturing heats may be affected. When one US plant identified leakage from their Alloy 600 head penetrations, the immediate reaction was that this was a vendor-specific (B&W) design issue and affected only those heats of materials. The approach of finding reasons why it did not affect me was alive and well. Why did we assume that the French and B&W Alloy 600 experience was not applicable to the entire PWR fleet?

We believe the answer lies in the emphasis on outage and cost impact. To assume that all Alloy 600 welds and materials would crack would require the development of new inspection technologies, increase outage durations, and risk the potential extension of an outage for repair and treatment of any findings. The PWR fleets waited until head inspections were mandated by the regulator to begin inspections. It should be noted that research dollars were not being spent to determine the degradation mechanism drivers at the time. In the early 2000s, another PWR experienced through-wall leakage of a reactor coolant loop dissimilar metal weld [7], Figure 22.2 and in a B&W plant, the reactor vessel head was found to have a nearly through-wall cavity the size of a football at the base of an Alloy 600 reactor vessel head penetration [6]. These events prompted the regulator to issue inspection orders for all reactor heads [11].

This served as a wake-up call to the PWR fleet. The change in the PWR approach came after the reactor vessel head degradation event in 2002. The PWR fleet executives established a NEI initiative to develop a similar proactive self-governance approach for the PWR fleet as the previously discussed BWRVIP. The regulatory order for PWR reactor head inspections [11] was difficult and a potential harbinger of things to come for other Alloy 600 weld locations. The industry formed a task force to develop an Industry Materials Degradation Initiative that would provide guidance to utilities that would go beyond the current regulation and ensure a proactive approach to potential materials degradation issues that affect reactor coolant pressure boundary components. NEI 03-08 [12] was issued in May 2003 for implementation by January 2004.

This initiative identified all the current industry guidance of the time into three categories: mandatory, needed, and good practice. Each utility was required to establish a program in compliance with the requirements of the NEI 03-08 Initiative, reviewing all industry documents identified in the initiative to ensure that each plant operated by the utility was in compliance with all aspects of the referenced guidance documents. Each utility was required to implement the referenced document guidance as written if it was identified as either mandatory or needed. This prompt action forestalled further regulatory action addressing Alloy 600 inspections. The industry guidance for the inspection of reactor vessel heads and the NRC Order 03-09 were used to develop an ASME Code Case N-729 [15] addressing RPV head inspections, and subsequently, the requirements of the order were rescinded. The industry has subsequently developed a number of similar inspection and repair documents for Alloy 600 concerns, including EPRI Materials Reliability Program Document MRP 139 [16] and several ASME Code Cases.

One would think that, after living through the Steam Generator tubing, BWR piping, and PWR Alloy 600 experiences, the industry would take a more enlightened perspective to all materials degradation issues, namely, if it happened somewhere else, why should I not think it will happen here. Buried piping degradation has taken a long time for plants to address. The system is largely out of mind (being buried), and failures self-identify and can be fixed by a patch or replacement of a short segment of piping. These failures are of minor safety significance, but are a recurring maintenance headache and public embarrassment for both utilities and the regulator. When a failure occurs at a plant, how far does one go to resolve the issue? The industry has recently adopted NEI Initiatives APC 09-53 [9] and NEI-09-14 [14] for buried piping. These documents will require aggressive assessment and inspection and reasonable assurance of structural/leak integrity of all buried piping, with an emphasis on pipes containing radioactive materials.

22.5   NEW PLANTS

Can the US nuclear industry move successfully forward in the future? There is certainly hope in that the industry, in general, has been adopting a more generic view of materials degradation events and that it has demonstrated a significant resolve to proactively address materials issues. The development of a technical understanding of the physics of degradation processes and longer-term testing of materials will certainly help the industry manage the issues in a responsible manner. Also, as a result of the materials degradation problems experienced, new plants will most certainly implement the remedies and new materials that have been developed as solutions to these problems in the operating fleet. However, the industry will need to maintain its vigilance in the identification of degradation throughout the plant systems. One cannot assume that a failure is always a one-time event. This evolution from a repair/crisis response approach is evident in the recent industry initiatives. Adopting the industry-wide approach to address material failures and mitigation activities should be integrated into the plant performance metrics to provide a gauge on how well we, the industry, are addressing the material condition of the plant on a holistic level rather than a component-specific level.

With all the positive activities going on throughout the industry, it would be remiss not to mention one ongoing concern: the loss of corporate technical knowledge as many 30+ year utility and vendor employees are retiring. The loss of the technical staff that lived through the problems of the 1980s, 1990s, and 2000s without orderly succession planning and knowledge turnover can lead to the improper response when new forms of degradation are identified at plants (as they most certainly will). There are fewer materials personnel in the industry following the retirement of many "old timers," as most utilities are not replacing this discipline in their engineering ranks and those that remain have less available time to ensure that industry organizations and researchers are addressing both the current plant concerns and looking to the future to anticipate the next issue(s) on the horizon.

The NSSSs and NRC have staffed up to be prepared for a surge of new plant licensing and construction, bringing much needed new blood into the industry. With proper succession planning and technology transfer, it is hopeful that we can stem the loss of corporate technical knowledge described above. But if the surge loses steam and takes longer to realize than currently planned, it is not clear how long the economics will support this increased staffing. It is also unclear how well our understanding of the materials and issues in 30- and 40-year-old plants will translate into, "new plant" knowledge. Like all phases of life, there is definitely a generation gap in knowledge between the existing plants and the new passive design plants.

22.6   SUMMARY

In summary, a review of material degradation issues that have affected the current fleet of nuclear plants, and the industry's approach to addressing them, has led the authors to observe several trends that have been pervasive in many of them. These include the following:

- Plant design and construction decisions focused on reducing cost without a full understanding of their impact
- Short-term perspective verses an entire life cycle perspective
- Outage/maintenance cost focus
- Cost- and compliance-based management of technical/regulatory-based issues
- Failure to learn from foreign and other plant operating experience. (i.e., rather than "Why it does not apply?" — ask the question: "Could this be me?")

The authors have identified a number of steps that can and are being adopted to avoid such problems in the future, or at least to minimize their financial and political impact. These include the following:

- Adopt a generic perspective to materials degradation issues, such as embodied in NEI 03-08 (Materials Initiative) [12], NEI 97-06 (Steam Generator Guidelines) [13], and NEI 09-14 (Buried Piping) [14].
- Utilities need to take a long-term approach to plant ownership rather than focus on short-term goals such as outage length and quarterly profitability.
- We also encourage a nuclear fleet approach rather than an individual utility approach to materials degradation issue resolution.

The industry must recognize that it is managing an academic/technical/political regulated industry and go beyond an analysis and design basis approach to material issue resolution. And finally, they must understand the need for basic physical understanding of degradation mechanisms, accepting a holistic approach to issue resolution.

22.7   REFERENCES

1. ASME Boiler and Pressure Vessel Code, Section XI, Rules for Inservice Inspection of Nuclear Power Plant Components. In: ASME Boiler and Pressure Vessel Code. New York: American Society of Mechanical Engineers, first published in 1968, various issues and addenda through 2007.

2. NRC Bulletin No. 87-01: Thinning of Pipe Walls in Nuclear Power Plants, issued July 9, 1987.

3. NRC Generic Letter 89-08, Erosion/Corrosion-Induced Pipe Wall Thinning, issued May 2, 1989.

4. NRC IEN 82-22, Failure in Turbine Exhaust Lines, issued July 9, 1982.

5. PSB-54A from Babcock & Wilcox Power Generation Inc.

6. Gorman, J., Hunt, S., Riccardella, P., and White, G. A. PWR Reactor Vessel Alloy 600 Issues, Chapter 44, ASME B&PV Code Companion Guide.

7. US NRC Information Notice 2000-017, 2000, Crack in Weld Area of Reactor Coolant System Hot Leg Piping at V. C. Summer (Supplement 1, 2000; Supplement 2, 2001). Washington, DC: US Nuclear Regulatory Commission.

8. Ford, F. P., Gordon, B. M., and Horn, R. M., 2006, Corrosion in boiling water reactors, *ASM Handbook*, Volume 13C, Corrosion: Environments and Industries, Cramer, A. D., and Covino, B. S., Jr., eds., ASM, Metals Park, OH, p. 341.

9. NRC Generic Letter 92-01: Reactor Vessel Structural Integrity, issued March 2, 1992.

10. NEI Initiative APC 09-53: Buried Piping Integrity Initiative, issued December 9, 2009.

11. NRC Order US NRC Issuance of Order Establishing Interim Inspection Requirements for Reactor Pressure Vessel Heads at Pressurized Water Reactors (EA-03-009). Washington, DC: US Nuclear Regulatory Commission; 2003.

12. NEI 03-08: Guideline for the Management of Materials Issues.

13. NEI 97-06: Steam Generator Program Guidelines.

14. NEI 09-14: Buried Piping Integrity Initiative.

15. ASME Code Case N-729, Alternative Examinations Requirements for PWR Reactor Vessel Upper Heads with Nozzles Having Pressure-Retaining Partial-Penetration Welds.

16. EPRI Materials Reliability Program, MRP-139, Primary System Piping Butt Weld Inspection and Evaluation Guidelines.

17. Griesbach, T. J., and Gordon, B. M., Materials Aging Management Programs at Nuclear Power Plants in the United States, IAEA 2nd International Symposium on Nuclear Power Plant Life Management, October, 2007, Shanghai, China.

18. Weakland, D. P., Personal Files.

NEW GENERATION REACTORS

Wolfgang Hoffelner, Robert Bratton,[1] Hardayal Mehta,
Kunio Hasegawa and D. Keith Morton[1]

ABSTRACT

In 1999, an international collaborative initiative for the development of advanced (Generation IV) reactors was started. The idea behind this effort was to bring nuclear energy closer to the needs of sustainability, to increase proliferation resistance, and to support concepts able to produce energy (both electricity and process heat) at competitive costs. Six reactor concepts were chosen for further development: the sodium fast reactor (SFR), the very-high-temperature gas-cooled reactor (VHTR), the lead or lead-bismuth cooled liquid metal reactor, the helium gas-cooled fast reactor, the molten salt reactor (MSR), and the super critical water reactor. In view of sustainability, the Generation IV reactors should not only have superior fuel cycles to minimize nuclear waste, but they should also be able to produce process heat or steam for hydrogen production, synthetic fuels, refinery processes, and other commercial uses. These reactor types were described in the 2002 Generation IV roadmap. Different projects around the world have been started since that time. The most advanced efforts are with reactors where production experience already existed. These reactors include the SFR and the VHTR. The other reactor types are still more in a design concept phase. This chapter briefly describes the six Generation IV concepts and then provides additional details, focusing on the two near-term viable Generation IV concepts. The current status of the applicable international projects is then summarized. These new technologies have also created remarkable demands on materials compared with light water reactors (LWRs). Higher temperatures, higher neutron doses, environments very different from water, and design lives of 60 years present a real engineering challenge. These new demands have led to many exciting research activities and to new Codes and Standards developments, which are summarized in the final sections of this chapter.

23.1   INTRODUCTION

The history of mankind repeatedly provides examples where a need is recognized and creative thinking is able to determine appropriate solutions. This human trait continues in the field of energy,

[1]This manuscript has been co-authored by Battelle Energy Alliance, LLC, under Contract No. DE-AC07-05ID14517 with the US Department of Energy. The US Government retains, and the publisher, by accepting the article for publication, acknowledges that the US Government retains a non-exclusive, paid-up, irrevocable, worldwide license to publish or reproduce the published form of this manuscript, or allow others to do so, for US Government purposes.

especially in the nuclear energy sector, where advanced reactor designs are being refined and updated to achieve increased efficiencies, increased safety, greater security through better proliferation control of nuclear material, and increased use of new metallic and non-metallic materials for construction. With minimal greenhouse gas emissions, nuclear energy can safely provide the world with not only electrical energy production but also process-heat energy production. Examples of the benefits that can be derived from process heat generation include the generation of hydrogen, the production of steam for extraction of oil-in-oil sand deposits, and the production of process heat for other industries so that natural gas or oil does not have to be used. Nuclear energy can advance and better the lives of mankind while helping to preserve our natural resources. These are noble and lofty goals. This chapter provides the reader greater understanding on how advanced reactors are not a "pie-in-the-sky" idea but are actually operational on a test scale or are near term. In fact, efforts are currently under way in many nations to build full-scale advanced reactors. The reader is encouraged to enjoy this chapter, for the future is just around the corner.

23.2   GENERATION IV INITIATIVE AND RELATED INTERNATIONAL PROGRAMS

The challenge of global warming and the resulting necessity to promote energy technologies with low carbon dioxide emissions led to an increased interest in nuclear power toward the end of the last century.

To support this interest on an international scale, a worldwide initiative was launched with the aim of overcoming certain limitations of current nuclear power plants (LWRs) with respect to sustainability, economics, safety and reliability, and proliferation resistance.

These developments led to the Generation IV International Forum (GIF). The goals of this initiative are given in the Generation IV roadmap [1], and they are summarized in Table 23.1.

The development of nuclear power can be divided into several plant-generations as shown in Figure 23.1. Advanced reactors, based on current nuclear power plant technology (EPR, AP1000, ESBWR, advanced Canadian deuterium uranium reactor [CANDU], APWR, etc.), are called Generation III+. Generation IV reactors go beyond LWR technology. They are intended to be commercially available along the guidelines given in Table 23.1 by about 2030. It was also recognized that joint international research and development (R&D) would be necessary to meet this ambitious

TABLE 23.1   GOALS FOR GENERATION IV NUCLEAR POWER PLANTS AS DEFINED BY THE GENERATION IV INTERNATIONAL FORUM) [1]

Goals for Generation IV Nuclear Systems	
Sustainability	Generation IV nuclear energy systems will provide sustainable energy generation that meets clean air objectives and promotes long-term availability of systems and effective fuel utilization for worldwide energy production.
	Generation IV nuclear energy systems will minimize and manage their nuclear waste and notably reduce the long-term stewardship burden, thereby improving protection for the public health and the environment.
Economics	Generation IV nuclear energy systems will have a clear life-cycle cost advantage over other energy sources.
	Generation IV nuclear energy systems will have a level of financial risk comparable to other energy projects.
Safety and reliability	Generation IV nuclear energy systems operations will excel in safety and reliability.
	Generation IV nuclear energy systems will have a very low likelihood and degree of reactor core damage.
	Generation IV nuclear energy systems will eliminate the need for offsite emergency response.
Proliferation resistance and physical protection	Generation IV nuclear energy systems will increase the assurance that they are a very unattractive and the least desirable route for diversion or theft of weapons-usable materials, and provide increased physical protection against acts of terrorism.

goal. The following six concepts were chosen for further R&D consideration:

- (Very) High-temperature reactor (VHTR)
- Sodium fast reactor (SFR)
- Lead fast reactor (LFR)
- Super critical water reactor (SCWR)
- Gas-cooled fast reactor (GFR)
- Molten salt reactor (MSR).

Fusion reactors are sometimes called Generation V nuclear power plants.

According to [2], the R&D performed within GIF focuses on both the viability and performance phases of system development. The former phase examines the feasibility of key technologies, such as suitable or novel structural materials or advanced fuel concepts. The latter phase focuses on the development of performance data and optimization of the system. However, the scope of GIF

activities does not extend to a demonstration phase, which involves the detailed design, licensing, construction, and operation of a prototype or demonstration system in partnership with industry.

Other international collaborations in the field of advanced reactors exist. They are, however, complementary to GIF rather than competing with GIF. Only the three most important projects should be mentioned here: GNEP, INPRO, and SET-plan.

According to a very recent announcement [3], the Global Nuclear Energy Partnership (GNEP) is to change its name to the International Framework for Nuclear Energy Cooperation (IFNEC) and establish a new mission statement as it aims to broaden its scope with "wider international participation to more effectively explore the most important issues underlying the use and expansion of nuclear energy worldwide."

Historically, the US Department of Energy (DOE) has outlined four overarching goals for the original GNEP Program: (1) to decrease US reliance on foreign energy sources without impeding

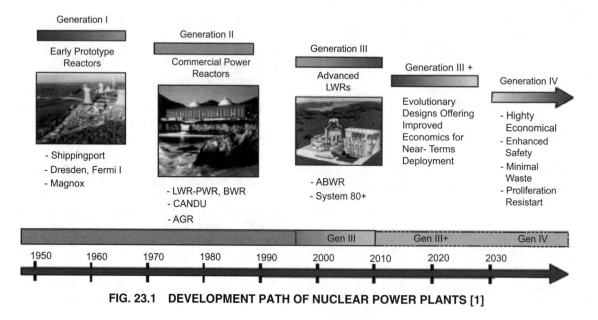

FIG. 23.1   DEVELOPMENT PATH OF NUCLEAR POWER PLANTS [1]

US economic growth, (2) to use improved technologies to recover more energy and reduce waste when recycling spent nuclear fuel, (3) to encourage the use of energy sources that emit the least atmospheric greenhouse gases, and (4) to reduce the threat of nuclear proliferation.

The partnership has a three-tiered organization structure. The Executive Committee comprised of Ministerial-level officials, provides the high-level direction. The Steering Group, whose members are designated by the Executive Committee, carries out actions on behalf of GNEP at the direction of the Executive Committee. At a September 2007 meeting of the Executive Committee, two working groups were established to address matters concerning "reliable nuclear fuel services" and "infrastructure development."

Currently, the Nuclear Fuel Service Working Group is addressing how to design and implement an effective nuclear energy infrastructure using fuel leasing and other economically viable and proliferation-secure arrangements. The Infrastructure Development Working Group is addressing the financial, technical, and human resource issues involved in creating an international nuclear energy architecture based on GNEP's Statement of Principles.

In October 2007, the DOE announced the first set of technical and conceptual design development awards — over $16.3 million to four multinational industry consortia led by Areva, Energy Solutions, GE-Hitachi Nuclear Americas, and General Atomics. In announcing the decision, the Assistant Secretary of Nuclear Energy indicated that the grants "enable DOE to benefit from the vast technological and business experience of the private sector as we move toward the goal of closing the nuclear fuel cycle."

In an April 2009 statement, the DOE announced that the Department cancelled the US domestic component of the GNEP [4]. It further said, "The long-term fuel cycle R&D program will continue but not the near-term deployment of recycling facilities or fast reactors. The international component of GNEP is under interagency review."

The International Atomic Energy Agency (IAEA) project IN-PRO was established in 2001 by bringing together technology holders, users, and potential users to consider jointly the international and national actions required for achieving desired innovations in nuclear reactors and fuel cycles [5, 6]. Since the early part of 2009, it has been determined to structure the project's task into the following four areas, with a forum for dialogue by members as a crosscutting vehicle for communication:

- methodology development and its use by members,
- future nuclear energy vision and scenario,
- innovative technologies,
- innovation in institutional arrangement.

The first results of the INPRO activity are listed in reference [7] for the assessment of innovative nuclear reactors and fuel cycles.

The visions for a sustainable energy supply in Europe are described in the European Strategic Energy Technology (SET)-plan [8]. With respect to high temperature reactors, thermal technologies for the use of heat and/or steam from high-temperature reactors (oil, chemical, and metal industry; synfuels and hydrogen production; seawater desalination; etc.) shall be proposed. Two reactor concepts are included: a prototype sodium-cooled fast reactor coupled to the electricity grid and a demonstrator reactor (either lead or gas cooled), not coupled to the grid. The operation of the prototype and demonstrator reactors from 2020 will allow a return of experience that, coupled with further R&D, will enable commercial deployment starting around 2040. At the same time, a coordinated program of crosscutting research will be conducted in all aspects of nuclear reactor safety, performance, lifetime management, waste handling, and radiation protection to serve both the development of future Generation IV reactors but also the continued safe and competitive operation of current nuclear plants that are providing 30% of the European Union (EU) electricity.

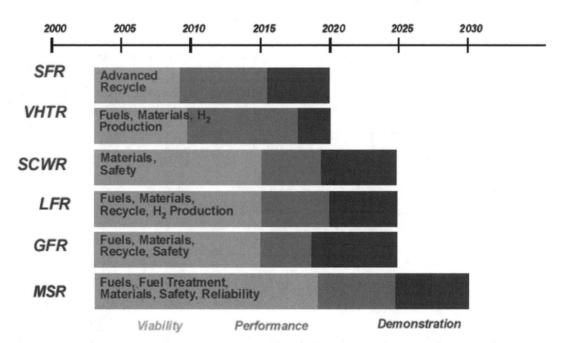

FIG. 23.2 THE DEPLOYMENT PERSPECTIVES OF ADVANCED NUCLEAR PLANTS, AFTER [13]. MOST IMPORTANT R&D ACTIVITIES ARE ALSO SHOWN

23.3 SHORT DESCRIPTION OF THE GENERATION IV SYSTEMS

The six nuclear technologies proposed within GIF are not entirely new plants. They are based on some experience gained with experimental reactors or even with large-scale pilot plants like the sodium-cooled French Superphenix [9] or the gas-cooled German HTR [10]. The SCWR is basically a pressurized LWR running a super critical steam cycle that has an impact on pressure and temperature. Most plant experience exists with SFRs and HTRs. Hence, this chapter will emphasize these two types of reactors. Lists of SFR and HTR plants can be found in the literature (e.g., [11, 12]). An assessment of the timeline for deployment of several Generation IV nuclear systems is shown in Figure 23.2 [13]. The current status of the main operation and design parameters of the different systems is shown in Table 23.2.

The next group behind the most advanced concepts of the SFR and VHTR are as follows: SCWR, LFR, and GFR with expected demonstration plant availability later than the SFR and the VHTR. The least developed is the MSR with the longest expected time for a demonstration plant. Even if the time scale is questionable, Figure 23.2 gives quite a good picture regarding the maturity of the different systems.

Besides fuel and fuel cycle issues, structural materials are also considered a significant development goal for several of the concepts. Performance of components under service conditions that are significantly different than the current LWRs is also a considerable challenge for design and design codes.

23.3.1 Liquid Metal-Cooled Reactors

Liquid metal reactors (LMRs) are designed for high-power density cores, taking advantage of the high heat removal and high heat transport capability of the coolant. Coolants can be mercury, sodium or lead/lead-bismuth. The very first liquid metal cooled nuclear reactor (Clementine [15]) used mercury coolant, which is liquid at room temperature. However, because of disadvantages including high toxicity, low boiling point, a high neutron cross section and other disadvantages, it has fallen out of favor. The sodium reactor technology is rather mature but remains to be commercialized successfully. Lead-cooled systems are less mature but they show some advantages due to the high boiling point of the coolant.

23.3.1.1 Sodium Fast Reactors
The sodium-cooled fast reactor (SFR) system features a fast-spectrum reactor and closed fuel recycle system. The primary mission for the SFR is management of high-level wastes and, in particular, management of plutonium and other actinides. With innovations to reduce capital cost, this mission can be extended to include electricity production, given the proven capability of sodium reactors to utilize almost all of the energy in the natural

TABLE 23.2 SUMMARY OF IMPORTANT OPERATION AND DESIGN DATA FOR THE GENERATION IV REACTORS [14]

Data	PWR	SCWR	VHTR	SFR	LFR	GFR	MSR
Coolant inlet temperature (°C)	290	280	400-600	370	400	450	656
Coolant outlet temperature (°C)	320	510	750-950	550	550	850	700-850
Pressure (MPa)	16	25	7	0.1	0.1	7	0.1
Maximum radiation dose (dpa)	100	10-70	1-10	200	200	200	200
Coolant	water	water	helium	liquid sodium	liquid Pb/PbBi	He/CO_2 super-cooled	molten salt
Critical components	RPV, internals, cladding	RPV, internals, cladding	RPV, internals, cladding	cladding	cladding	RPV, fuel/core	core
Metals	ferritic steels, austenitic steels, zircaloy	ferritic-martensitic steel, austenitic steels, Ni-base, ODS	ferritic-martensitic steels, Ni-base, ODS	ferritic-martensitic steels, austenitic steels, ODS	ferritic-martensitic steels, austenitic steels, ODS	ferritic-martensitic steels (RPV)	Ni-base
Ceramics			graphite, C/C, SiC/SiC, SiC			SiC, TiC, other ceramics	graphite
Main damage mechanisms	corrosion, embrittlement, low cycle fatigue	corrosion, embrittlement, low cycle fatigue	high temperature corrosion, creep, low cycle fatigue	corrosion, creep (thermal/irrad.), low cycle fatigue, irrad.	corrosion, creep (thermal/irrad.), low cycle fatigue, irrad.	corrosion, creep (thermal/irrad.), low cycle fatigue, irrad.	corrosion, creep (thermal/irrad.), low cycle fatigue, irrad.
Design Codes	ASME Div 1, RCC-M	ASME Div. 5, RCC-MRx	ASME Div. 5, RCC-MRx	ASME Div. 5, RCC-MRx	ASME Div. 5, RCC-MRx	ASME Div. 5, RCC-MRx	ASME Div. 5, RCC-MRx

uranium versus the 1% utilized in thermal spectrum systems. An important safety feature is that the system operates at atmospheric pressure. However, sodium reacts chemically with water and with air, which has to be taken into consideration by reliable design measures.

Three concepts are currently studied for the SFR:

- A medium- to large-size (600-1500 MWe) loop-type SFR with minor actinide (MA) bearing mixed uranium-plutonium oxide (MOX) fuel, supported by a fuel cycle based upon advanced aqueous processing at a central location serving a number of reactors.
- A medium size pool-type SFR with uranium-plutonium-minor-actinide zirconium metal alloy fuel, supported by a fuel cycle based on pyro metallurgical processing in facilities co-located with the reactor.
- A small size (50-150 MWe) modular pool-type SFR with similar metal alloy fuel, supported by a fuel cycle based on pyro processing at a central or regional location.

Details about these three concepts including expected operational parameters are described in [16].

23.3.1.2 Lead-Cooled Fast Reactor (LFR)

The LFR uses a fast-neutron spectrum and a closed fuel cycle for efficient conversion of fertile uranium. It can also be used as a burner of minor actinides from reprocessed spent fuel. Lead is relatively chemically inert compared with sodium, which is an important safety aspect. However, corrosion issues with structural materials still have to be resolved. In the GIF Roadmap, the LFR was primarily envisioned for missions in electricity and hydrogen production, and actinide management. The following two designs are currently considered in the GIF-framework:

- The Small Secure Transportable Autonomous Reactor (SSTAR) with mixed uranium-plutonium nitride (MN) fuel.
- The European Lead-cooled System (ELSY) with MOX fuel.

Details about the current status of the LFR developments are given in [17].

23.3.2 Gas-Cooled Reactors

Gas-cooled reactors use gas (carbon dioxide or helium) as coolant. Using a thermal neutron spectrum, these reactors have been used in the past, and some of them are running today, or they are in engineering stage. Projects on gas-cooled fast reactors also existed as an alternative to sodium-cooled fast reactors from the 1960s through the 1980s, but no prototype was ever built. However, the concept was considered in the GEN IV roadmap, and development work for GFR is ongoing [18]. The attractive characteristic of a gas-cooled reactor is the high coolant temperature that could in principle be reached (about 1000°C). This would enable using nuclear energy not only for electricity generation but also for process heat or even combined processes. One main drawback of the GFR is its low thermal inertia, which could have detrimental temperature excursions in case of a loss of coolant accident.

23.3.2.1 Very-High-Temperature Reactor

The unique capability of very-high-temperature reactors (VHTRs) to produce process heat above 600°C makes them a strategic reactor type that can efficiently produce hydrogen through steam electrolysis, or supply both hydrogen and high-temperature heat for producing synthetic fuels from coal or biomass, or supply high-temperature process heat and hydrogen or synthetic fuels as chemical reactants to varied industrial plants including petro-chemistry and steel making. Today, the VHTR system shows the most active R&D cooperation in the framework of the GIF and the greatest number of national projects of prototypes in the next two decades.

The VHTR operates with thermal neutrons currently with a once-through fuel cycle. It uses helium as the cooling medium and graphite as the moderator and core structural material. The VHTR can be designed either with a pebble bed core or with a prismatic core. For both designs, tri structural-isotropic (TRISO) fuel is used. It is a type of micro-fuel particle. It consists of a fuel kernel composed of UO_X (sometimes UC or UCO) in the center, coated with four layers of three isotropic materials. The four layers are a porous buffer layer made of carbon, followed by a dense inner layer of pyrolytic carbon (PyC), followed by a ceramic layer of SiC to retain fission products at elevated temperatures and to give the TRISO particle more structural integrity, followed by a dense outer layer of PyC. The fuel kernels remain the same for both core concepts.

The VHTR can be considered as a good example of how the availability of suitable structural materials can influence the technical concepts. Table 23.3 shows the changes of proposed concepts

TABLE 23.3 SHORT-TERM CHANGES OF THE VHTR DESIGN AND OPERATION DATA AS A RESULT OF MATERIALS LIMITATIONS AND VIABLE DESIGN OPTIONS (IN617, 800H, 230 ARE DESIGNATIONS FOR NICKEL-BASED SUPERALLOYS, T91 IS AN ADVANCED 9CR1MO MARTENSITIC STEEL) [ORIGINAL WORK]

	2002	2009	Comments
Gas outlet temperature	at least 1000°C	probably 850°C, eventually 920°C	no structural ceramic components required
Reactor pressure vessel temperature	600°C	eventually < 450°C, probably < 350°C	use of T91 questionable, probably PWR RPV steel
Intermediate heat exchanger	He-He	He-CO₂, He-steam, He-He	IN617/800H/230 sufficient, steam generator required
Electric energy conversion	direct cycle He-gas turbine (vertical)	indirect cycle, steam turbine	no high temperature He turbine with vertical shaft
Thermal plant	I-S hydrogen	synfuel, hydrogen (not limited to I-S), chemical industry, steel industry	new requirements

for gas-cooled reactors from 2002 (original GEN IV roadmap) to 2009. Both the gas outlet temperature and the gas inlet temperature were considerably reduced, which was mainly a result of current structural material limitations and minimal design data.

However, the variety of processes that can make use of heat or steam coming from a VHTR increased considerably even with decreasing operation temperature. Hydrogen production, petroleum refining, oil recovery, coal and natural gas derivatives, petrochemicals, and ammonia and nitrate production are currently considered as priority candidates for use of nuclear heat from a high-temperature reactor. A detailed discussion about envisaged combined processes can be found in the literature [19].

23.3.2.2 Gas-Cooled Fast Reactor

The GFR is a fast neutron spectrum system using helium at about 7 MPa pressure as the cooling medium [18]. The GFR can be considered as a complement to the SFR deployment, which benefits from a more mature technology The GFR combines the advantages of a fast neutron spectrum with those of high temperatures. It can be deployed for closed fuel cycles for the minimization of wastes, when the minor actinides are recycled. The high outlet temperatures potentially provide improved economy in the power conversion units and also permit process heat applications, similar to the VHTR. In the VHTR, the use of graphite increases the thermal inertia of the core, thereby limiting the maximum temperature during transients. On the other hand, GFR cores have relatively low thermal inertia that would lead to the development of very high temperatures in case of a loss of coolant accident. Current efforts are investigating design features to overcome this problem.

23.3.3 Other Gen IV Systems

23.3.3.1 Super Critical Water Reactor

The SCWR is basically a LWR using water in a super critical status. It is a high-temperature, high-pressure water-cooled reactor that operates above the thermodynamic critical point of water (above 374°C, 22.1 MPa). The main advantage of the SCWR is improved economics because of the higher thermodynamic efficiency and plant simplifications. The various design options include fast and thermal neutron spectra as well as opportunities to utilize conventional or advanced fuel and fuel cycles. Two SCWR concepts have evolved from existing LWR and pressurized heavy water reactor designs: (1) a large reactor pressure vessel containing a reactor core analogous to conventional LWR designs and (2) designs with distributed pressure tubes or channels containing fuel bundles, analogous to conventional CANDU and Reactor Bolshoy Moschchnosty Kanalny (RBMK) nuclear reactors. The following concepts are considered within GIF:

University of Tokyo Thermal and Fast Spectrum Designs: These are pressure vessel design concepts that have been under development at the University of Tokyo since 1989 (thermal version) and 2005 (fast version). The thermal version is called "Super LWR," and the fast reactor version is called "Super Fast LWR."

High-Performance Light Water Reactor (HPLWR): This is a pressure vessel design that is under development in Europe and is partially funded by the European Commission.

CANDU-SCWR: This is a pressure-tube reactor that is being developed by the Atomic Energy of Canada Ltd that uses a thorium fuel cycle and a separate heavy water moderator with enhanced safety functions.

SCWR-SM: This is a pressure vessel design under development in the Republic of Korea that utilizes a solid ZrH_2 moderator.

Mixed Core Design: This is a pressure vessel concept that is being evaluated at Shanghai Jiao Tong University. The core consists of a fast spectrum inner region and a thermal spectrum outer region.

More details about current GIF-SCWR projects can be seen in [20].

23.3.3.2 Molten Salt Reactor

An MSR uses fuel that is dissolved in a fluoride salt coolant. The technology was partly developed in the 1950s and 1960s in the United States at the Oak Ridge National Laboratory (ORNL), where two prototype reactors were built. Compared with solid-fuelled reactors, MSR systems have lower fissile inventories, provide the possibility of continuous fission-product removal, avoid the expense of fabricating fuel elements, and employ a homogeneous isotopic composition of fuel in the reactor. These characteristics, among others, enable MSRs to have potentially unique capabilities and competitive economics for actinide burning and extending fuel resources. Earlier MSRs were mainly considered as thermal neutron spectrum graphite-moderated reactor concepts. Since 2005, R&D has focused on the development of fast spectrum MSR concepts (molten salt fast reactor [MSFR]). This would allow extended resource utilization and waste minimization. Additionally, advantages of the molten salt concept like high boiling point and liquid fuel could be used.

Apart from MSR systems, other advanced reactor concepts are being studied within Generation IV using liquid salt technology. Liquid salt is considered a primary coolant for the advanced high-temperature reactor (AHTR). It is also considered for intermediate heat exchange use as an alternative to secondary sodium in SFRs or to secondary helium in VHTRs.

Two baseline concepts are currently considered:

- The MSFR as a long-term alternative to solid fuelled fast neutron reactors. The potential of MSFR has been assessed but specific technological challenges must be addressed and the safety approach has to be established.
- The AHTR as a high-temperature reactor with higher power density than the VHTR and passive safety potential from small to very high unit power (>2400 MWt).

Both concepts have large commonalities in basic R&D areas, particularly for liquid salt technology and materials behavior (mechanical integrity and corrosion).

More details about current GIF-MSR projects can be seen in [21].

23.4 ADVANCED REACTOR PROJECTS

For the near future, traditional light or heavy water reactors will be the choice of technology. This group of reactors underwent and continues to undergo significant improvements in safety and in performance. Besides traditional large nuclear power stations, small reactors for local energy supply (electric thermal) are being studied in different countries. Therefore, it may be useful to briefly consider these reactor developments in this chapter.

23.4.1 Concepts and Innovation Technologies for Next-Generation Light Water Reactors

Light water reactors are still considered the best choice for future electrical energy supply. The design concepts for the next-generation LWR system are (1) best safety and economy in the

2030 time frame, (2) simplifying operation and maintenance, (3) dramatically shortening the time for construction, (4) dramatic reduction of the quantity of spent fuel produced, reduction of the consumption of uranium, radioactive waste, and exposure to radiation, and (5) improvement of performance for plant life (approximately 80 years).

23.4.1.1 Advanced Light Water Reactors This section summarizes a more detailed description given in [22]:

Advanced boiling water reactor (ABWR) derived from a General Electric design.

System 80+ is an advanced pressurized water reactor (PWR), which was ready for commercialization but is not currently being promoted for sale.

The Westinghouse AP1000, scaled up from the AP600, received final design certification from the NRC in December 2005, the first Generation III+ type to do so. It represents the culmination of a 1300 man-year and $440 million design and testing program.

GE Hitachi Nuclear Energy's ESBWR is a Generation III+ technology that utilizes passive safety features and natural circulation principles and is essentially an evolution from its predecessor design, the SBWR at 670 MWe.

Mitsubishi's large APWR (1538 MWe) — advanced PWR — was developed in collaboration with four utilities (Westinghouse was involved in the early stages).

Areva NP (formerly Framatome ANP) has developed a large (1600 and up to 1750 MWe) European pressurized water reactor (EPR), which is currently under construction in Finland.

Together with German utilities and safety authorities, Areva NP (Framatome ANP), is also developing another evolutionary design, the SWR 1000, a 1250-MWe BWR with 60-year design life now known as Kerena.

Toshiba has been developing its evolutionary advanced BWR (1500 MWe) design, originally BWR 90+ from ABB then Westinghouse, working with Scandinavian utilities to meet European Utilities Requirements (EUR).

A third-generation standardized VVER-1200 reactor of 1150 to 1200 MWe is, among others, an evolutionary development of the well-proven VVER-1000 reactor in Russia.

TABLE 23.4 TYPICAL NEXT-GENERATION BWR PLANT CONCEPT [23]

Electric output	1700 to 1800 MWe
Fuel	large bundle
Safety system	hybrid (optimized passive and active system)
Primary containment vessel	double containments outer: steel containment vessel inner: steel plate reinforced concrete containment vessel
Countermeasures of external events: earthquake airplane crash	seismic isolation systems reinforced building

TABLE 23.5 TYPICAL NEXT-GENERATION PWR PLANT CONCEPT [23]

Electric output	1780 Mwe (plant efficiency 40%)
Fuel average burn-up	70 gigawatt days/ton
Primary coolant temperature hot leg	330°C
Steam generator surface area	8500 m³ high efficiency type
Primary coolant flow rate	29000 m³/hour/loop
Safety system	4 train direct-air-cooling hybrid system
Ultimate heat sink	air and sea water

23.4.1.2 Heavy Water Reactors The CANDU-9 (925-1300 MWe) was developed as a single-unit plant. It has flexible fuel requirements ranging from natural uranium through slightly enriched uranium, recovered uranium from reprocessing spent PWR fuel, mixed oxide (U & Pu) fuel, direct use of spent PWR fuel, to thorium.

India is developing the advanced heavy water reactor (AHWR) as the third stage in its plan to utilize thorium to fuel its overall nuclear power program. The AHWR is a 300-MWe reactor moderated by heavy water at low pressure.

Projects in many other countries are based on these types of plants. Table 23.4 tabulates major specifications for the BWR. Major specifications for the next-generation PWR are shown in Table 23.5. The values in the tables are typical values based on Japanese expectations. Several innovative technologies are needed to develop these designs of next-generation LWRs.

There are some very important technological steps involved with the new generation of LWRs that are shown in Figure 23.3.

23.4.2 Small Modular Reactors

As nuclear power generation has become more established since the 1960s, the size of the reactor units has grown from 60 MWe to more than 1600 MWe, with corresponding economies of scale in operation. At the same time, there have been many hundreds of smaller reactors built both for naval use (up to 190 MWt) and as neutron sources, yielding enormous expertise in the engineering of small units. The IAEA defines "small" as under 300 MWe. The contents of this subsection are based on [24, 25].

Designs for small modular reactors (SMRs) are being developed in several countries, often through cooperation between government and industry. Countries involved include Argentina, China, Japan, Korea, Russia, South Africa, and the United States. Small modular reactor designs encompass a range of technologies, some being variants of the six Generation IV systems selected by GIF, whereas others are based on established LWR technology.

Such reactors could be deployed as single or double units in remote areas without strong grid systems, or to provide small-capacity increments on multi-unit sites in larger grids. They feature simplified designs and would be mainly factory-fabricated, potentially offering lower costs for serial production. Their much lower capital cost and faster construction than large nuclear units should make financing easier. Other advantages could be in the area of proliferation resistance, as some designs would require no on-site

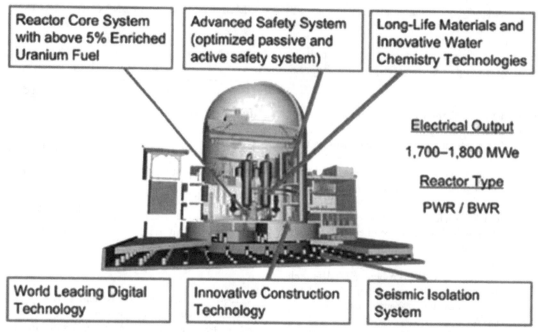

FIG. 23.3 KEY TECHNOLOGIES FOR DEVELOPING NEXT-GENERATION LWR [23]

re fueling, whereas others would require re fueling only after several years. Some could be used with advanced fuel cycles, burning recycled materials.

Numerous concepts exist for SMRs based on LWR technology. Several such designs are being promoted by nuclear industry companies, including AREVA, Babcock & Wilcox (mPower), General Atomics, NuScale, and Westinghouse (IRIS). Others are being developed by national research institutes in Argentina, China, Japan, Korea, and Russia. Two small units designed to supply electricity and heat are under construction in Russia, based on existing icebreaker propulsion reactors. These will be barge-mounted for deployment to a remote coastal settlement on the Kamchatka peninsula. Some other designs are well advanced with initial licensing activities under way.

Several SMR designs are HTGRs (e.g., the Pebble Bed Modular Reactor [PBMR]). These designs are well suited to heat or co-generation applications. There are also several other concepts for advanced SMR designs, including liquid metal-cooled fast reactors. These are generally at an earlier stage of development, with some the subject of GIF collaborative efforts. One example in this category is the 4S design from Toshiba of Japan, a sodium-cooled "nuclear battery" system capable of operating for 30 years with no re fueling. It has been proposed to build the first such plant to provide 10 MW of electricity to a remote settlement in Alaska, and initial licensing procedures have begun. Another example in this category is the Hyperion Power Module, a lead-bismuth cooled LMR, developed by Hyperion Power.

A recent SMR candidate is the travelling-wave reactor, which is currently promoted by TerraPower [26]. According to [27], a traveling-wave reactor requires very little enriched uranium, reducing the risk of weapons proliferation. The reactor uses depleted-uranium fuel packed inside hundreds of hexagonal pillars. In a "wave" that moves through the core at only a centimeter per year, this fuel is transformed (or bred) into plutonium, which then undergoes fission. The reaction requires a small amount of enriched uranium (not shown) to get started and could run for decades without re fueling. The reactor uses liquid sodium as a coolant; core temperatures are rather hot, about 550°C, versus the 330°C typical of conventional reactors.

Other concepts for advanced SMRs have been proposed by commercial and research organizations in several countries, and some aim to commence licensing activities in the next few years. As an example of the licensing activities, on the regulatory side, the US Nuclear Regulatory Commission held a workshop in October 2009 on small- and medium-size nuclear reactors covering issues such as licensing, design basis, staffing, etc. However, no firm plans to construct demonstration plants have yet been announced.

If multiple modular units on a single site were to become a competitive alternative to building one or two large units, then SMRs could eventually form a significant component of nuclear capacity. They could also enable the use of nuclear energy in locations unsuitable for large units, and some designs could extend the intended use for non-electricity applications. Whether SMR designs can be successfully commercialized, with an overall cost per unit of electricity produced that is competitive with larger nuclear power plants and other generating options, remains to be seen.

23.5 STATUS OF GENERATION IV PLANTS: SODIUM FAST REACTORS

Sodium fast reactors belong together with the high-temperature reactors in the group of Generation IV reactors where industrial experience from the past already exists. Table 23.6 lists the reactors according to history and future expectations. It can be seen that there is a strong current activity in Asia.

23.5.1 The Japan

In Japan, the Fast Reactor Cycle Technology Development (FaCT) project is carried out by the Japan Atomic Energy Agency (JAEA) in corporation with Japanese utilities under the sponsorship of the Japanese government. The FaCT project was planned based on a feasibility study of the commercialized fast reactor

TABLE 23.6   CURRENT AND PAST SFR PROJECTS [ORIGINAL WORK]

	U.S.	Europe	Russia	Asia
Past	Clementine, EBR I/II, SEFOR, FFTF	Dounreay, Rhapsody, Superphenix	BN-350	
Cancelled	Clinch River, IFR	SNR-300		
Operating		Phenix	BN-600	Joyo, FBTR Monju,
Under Construction			BN-800	PBFR, CEFR
Planned	S4, PRISM	ASTRID	BN-1800	S4, JSFR, KALIMER

cycle system in 1999 to 2005 [28]. From the feasibility study, an advanced loop-type SFR was selected for future nuclear energy systems. This is because the SFR meets sustainability, economy, nuclear proliferation resistance, safety, etc. Japan sodium-cooled fast reactor (JSFR) was launched in the FaCT project to develop for commercialization through innovation technologies.

The development of innovative technologies on advanced component structures is expected to be completed by about 2015. System and component designs and trial manufacturing of components will start after 2015. A demonstration fast reactor is scheduled for operation around 2025. A commercialized fast reactor cycle system will be constructed around 2050.

Innovative concepts and technologies for SFRs are shown in Figure 23.4 taking JSFR as an example. Table 23.7 shows the design specification for the JSFR. The size of a reactor vessel for the JSFR of an advanced loop-type SFR will be minimized, and the reactor core internals will be simplified.

The diameter and wall thickness of the reactor vessel are considered to be 10.7 m and 50 to 60 mm, respectively. A shortened piping, two-loop cooling system, and integrated intermediate heat exchanger (IHX) with a primary pump are introduced into the design from the view point of reduction of cost, safety, maintainability, and manufacturability. A containment vessel (CV) would be rectangular in shape, because the pressure load to the CV is not high compared with that to LWRs. A double-wall structure of steel plate reinforced concrete is applied to all parts of the building. The volume of the reactor building is about 150,000 m^3, which is less than one-half of a current advanced PWR.

Regarding demonstration and commercialization of the JSFR, there are several innovative technologies for design study. The current status of innovative technologies under development includes the two-loop cooling system, increased reliability of the reactor system, a simplified fuel-handling system, a passive reactor shutdown system, mitigation measures against core disrup-

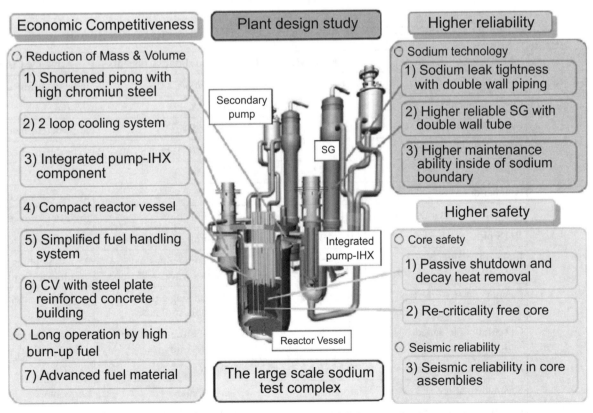

FIG. 23.4   INNOVATIVE TECHNOLOGIES FOR THE JSFR [29]

TABLE 23.7 SPECIFICATION FOR JSFR [29]

Item	Specifications
Electricity output	1500 MWe
Thermal output	3570 MWt
Number of loops	2
Primary sodium temperature and flow rate	550/395°C 3.24×10^7 kg h^{-1} / loop
Secondary sodium temperature and flow rate	520/335°C 2.7×10^7 kg h^{-1} / loop
Main steam temperature and pressure	497°C 19.2 MPa
Feedwater temperature and flow rate	240°C 5.77×10^6 kg h^{-1} / loop
Plant efficiency	≈42%
Fuel type	TRU-MOX
Burn-up (average)	≈150 gigawatt days / ton
Breeding ratio	breakeven (1.03), 1.1, 1.2
Cycle length	26 months or less; four batches

tive accidents, and a minor actinide–bearing MOX (U/Pu mixed oxide) fuel core.

A method to evaluate the economic competitiveness was developed by the Demonstration Fast Breeder Reactor (DFBR) project and was modified and improved for the JSFR at the present currency value. Based on the reduction of reactor building volume and structural weight, adopting a simplified configuration, adopting the twin-plant concept, and pursuing scale merit by enlargement of the power output, the construction cost per unit of electricity for the JSFR would be competitive with that of future LWRs.

23.5.2 Korea

In Korea, there are 20 commercial nuclear power reactors under operation, producing about 36% of the total electricity in Korea. Energy resources of crude oil, natural gas, and coal are insufficient, and hydropower is also limited. Nuclear energy is a key energy resource, especially with the aspect of large reductions of greenhouse gas emissions. About 97% of energy resources are imported from abroad, and it is expected that the share of nuclear power should increase to 48% after 12 years. The Korean Atomic Energy Research Institute (KAERI) is currently designing Generation IV fast reactors, aiming for construction completion and operation in 2028 [30].

Technology developments for a SFR are categorized as (1) advanced design studies of the core design, (2) development of basic technologies, and (3) development of advanced technologies. The advanced design studies consist of core design, heat transport system design, and associated mechanical and structural design. The core designs are being performed for 600 to 1200 MWe transuranium (TU) burner core and a 1200 MWe break even core. System analysis codes and sodium technology studies are performed as the development of basic technologies. Passive decay heat removal circuit (PDRC) experiments, S-CO$_2$ Brayton cycle system, Na-CO$_2$ chemical reaction test, and metal fuel technologies are conducted in the development of advanced technologies.

Specifications of an advanced reactor system concept for Generation IV are tabulated in Table 23.8. Many items for the advanced concept are to be determined. The conceptual design for KALIMER-600 was finished in 2006, and the advanced concept

TABLE 23.8 SPECIFICATION OF ADVANCED CONCEPT FOR GENERATION IV [30]

		KALIMER-600	Candidate Concepts	Advance Concepts
Reactor	Power, MWe	600	600/900/1200	to be determined
	Conversion ratio	1.0	0.5-0.8, 1.0	0.5-0.8, 1.0
	Core exit temperature °C	545	510-550	to be determined
	Cladding material	Mod. HT9	Mod. HT9 / FMS	to be determined
	Fuel type	U-TRU-Zr	U-TRU-Zr	U-TRU-Zr
NSSS	Number of loops	2	2, 3	to be determined
	Reactor vessel diameter, m	11.4	minimization	to be determined
	IHTS pipe length, m	118	minimization	to be determined
	In-vessel rotating plug	2 rotating plugs	2 rotating plugs with multi wave-guide tubes	2 rotating plugs with multi wave-guide tubes
	Steam generator tube type	helical single tube	helical single tube / double wall tube	to be determined
	GDC-4: double-ended guillotine break	-	leak-before-break for RV, IHTS piping	leak-before-break for RV, IHTS piping
	Residual heat removal system	PDRC	PDRC	PDRC
	Seismic isolation	horizontal	horizontal	horizontal
BOP	Energy conversion system	Rankine	Rankine / S-CO$_2$ Brayton	to be determined

is currently being developed. After testing the passive decay heat removal circuit, an integral testing loop will be constructed. A draft action plan was prepared by the Korean government in 2007, and a standard safety analysis report and final safety analysis report will be approved by the Korean government. A demonstration reactor will be constructed and operated by 2028.

23.5.3 India

23.5.3.1 Historical Perspective India adopted a nuclear energy strategy taking into account indigenous resources for long-term energy security and sustainability. The currently known Indian nuclear energy resources comprise 61,000 tons of uranium and about 800,000 tons of thorium. Considering low inventory of uranium and for long-term sustainability of nuclear energy, India selected a closed fuel cycle–based three-stage nuclear power program in the 1950s. At that time, the country had just become independent and had no large-scale industrial infrastructure. The necessary pressure vessel technology for LWRs did not exist at that time. Moreover, LWRs would have also needed a technologically complex uranium enrichment facility. Therefore, using indigenous LWRs was not a feasible option at that time. Also, India had a smaller grid necessitating installation of small power reactor systems. Considering these aspects, the country adopted the 220 MWe PHWR technology. The details of these PHWRs and the newer indigenously designed 540 and 700 MWe reactor systems are provided in [31–33]. Pressurized heavy water reactors were also chosen in the first stage so as to produce maximum plutonium essential for the fast breeder reactor (FBR)–based second stage of the program.

The second stage comprises FBRs, utilizing plutonium-based fuel. The plutonium and uranium reprocessed from the first stage would be effectively utilized in the initial part of the second stage. This initial part will use well-proven oxide fuel-based reactors (FBRs), and subsequently, at an appropriate stage, when all the new necessary technologies have been developed and demonstrated, metallic fuel-based FBRs will be introduced.

India has envisaged robust thorium reactor technologies as a promising sustainable future energy resource for the country. Studies indicate that once the FBR capacity reaches about 200 GWe, thorium-based fuel can be introduced progressively in the FBRs to initiate the third stage of the program, where the U-233 bred in these reactors is to be used in the thorium-based reactors. The proposed roadmap for the third stage therefore comprises thorium-based reactor technologies, incorporating the (Th-U-233) cycle. India is one of the leading countries in the world in thorium research and has gained that experience through thorium irradiation and the operation of U-233 fuelled research reactors.

23.5.3.2 Future Nuclear Cycle Program Nuclear power supplied 2.5% of India's electricity in 2007 (coal, hydro, and gas provided 68%, 15%, and 8%, respectively). This share is forecasted to reach 25% by 2050 [34]. With about six times more thorium than uranium, India has made utilization of thorium for large-scale energy production a major goal in its nuclear power program, utilizing a three-stage concept [35,36]:

- Pressurized heavy water reactors (PHWRs) fueled by natural uranium, plus LWRs, producing plutonium-239 as a by-product.
- Fast breeder reactors using plutonium-based fuel to breed uranium-233 from thorium-232. The blanket around the core will have uranium as well as thorium, so that further plutonium (particularly plutonium-239) is produced as well as uranium-233.
- Advanced heavy water reactors (AHWRs) burn the uranium-233 and plutonium-239 with thorium-232, getting about 75% of their power from thorium. The used fuel will then be reprocessed to recover fissile materials for recycling.

23.5.3.3 Fast Breeder Test Reactor The second stage of the Indian nuclear program is based on plutonium fuelled FBRs. A 40 MWt fast breeder test reactor (FBTR) is in operation in India since 1985 [35,37]. Reference [37] provides a description of the reactor and summarizes the operating history of the reactor. It is a loop-type SFR located at the Indira Gandhi Center for Atomic Research (IGCAR), Kalpakkam. The reactor design is based on the French reactor Rapsodie, with several modifications, which include the provision of a steam-water circuit and turbine-generator in place of a sodium-air heat exchanger in Rapsodie. Heat generated in the reactor is removed by two primary sodium loops and transferred to the corresponding secondary sodium loops. Each secondary sodium loop is provided with two once-through steam generator modules. Steam from the four modules is fed to a common steam-water circuit comprising a turbine generator and a 100% dump condenser. The reactor uses a high plutonium mono-carbide as the driver fuel. Being a unique fuel of its kind without any irradiation data, it was decided to use the reactor itself as the test bed for this driver fuel. The FBTR was synchronized with the grid in July 1997. The operating experience of this FBTR has provided sufficient feedback and confidence for India to launch upon the construction of a 500 MWe fast reactor.

23.5.3.4 Prototype Fast Breeder Reactor (PFBR) The prototype fast breeder reactor (PFBR), designed by the IGCAR, is a 500 MWe, sodium-cooled, pool-type, mixed-oxide (MOX) fuelled reactor having two secondary loops. Reference [38] describes the salient design features including the design of the reactor core, reactor assembly, main heat transport systems, component handling, steam water system, electrical power systems, instrumentation and control, plant layout, safety, and R&D.

Figure 23.5 shows the flow diagram of the main heat transport system. The primary objective of the PFBR is to demonstrate techno-economic viability of FBRs on an industrial scale. The reactor power is chosen to enable adoption of a standard turbine, as used in fossil power stations, to have a standardized design of reactor components resulting in further reduction of capital cost and construction time in the future and compatibility with regional grids.

Better safety features of a pool-type reactor, i.e., the main vessel with no nozzles leading to high integrity of the vessel, relatively large thermal inertia leading to ease in design of decay heat removal with lower heat capacity requirements and availability of more time for the operator to act, and large diameter of the main vessel with internals leading to significantly lower strain in the main vessel in case of core disruptive accident, led to a selection of the pool-type design for primary circuit configuration. The pool-type concept also enables further extension of the design to larger power reactors in the future.

The main vessel is made of highly ductile AISI 316LN material, and it satisfies leak-before-break criteria. The reactor is designed to meet the regulatory requirements of Atomic Energy Regulatory Board (AERB). The responsibility for construction, commissioning, operation, and maintenance of the reactor is with Bhartiya Vidyut Nigam (BHAVINI), a part of India's Department of Atomic Energy.

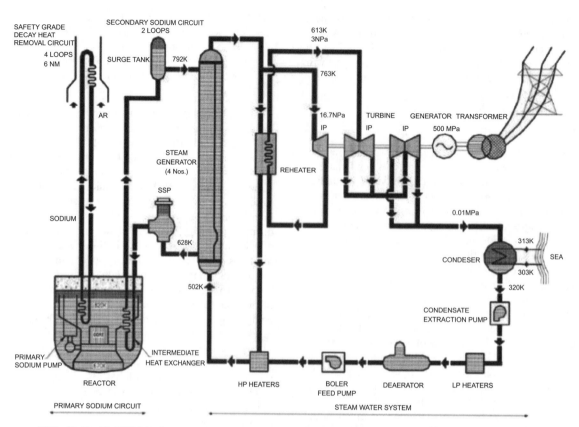

FIG. 23.5   FLOW DIAGRAM OF THE MAIN HEAT TRANSPORT SYSTEM OF THE PFBR [38]

23.5.3.5 Advanced Heavy Water Reactor The AHWR is a 300 MWe, vertical, pressure tube-type, heavy water moderated, boiling light water-cooled natural circulation reactor [39]. The fuel consists of Th-Pu oxide and Th-U233 oxide pins. The fuel cluster is designed to generate nearly 65% of energy out of U-233, which is bred in situ from thorium. In AHWR, minor actinides produced in U-233 pins are much smaller in quantity as compared with those produced in Th-Pu pins. This shows the advantage of a thorium fuel cycle in this regard.

Advanced heavy water reactor has adopted the well-proven pressure tube technology. There are several passive safety systems for normal reactor operation, decay heat removal, emergency core cooling, confinement of reactivity, etc. A number of major experimental facilities have been built and planned to validate the new concepts used in AHWR. The core damage frequency targeted is 1×10^{-7} per year or less. Even though the AHWR is primarily designed to work with U-233/Pu-Th MOX fuel, the design has the flexibility to accept other fuel types. Figure 23.5a shows the schematic arrangement of different systems in AHWR.

India has also acquired considerable experience in thorium irradiation in research reactors and thorium also has been introduced in PHWRs for initial flux flattening. With the sustained efforts over years, India has acquired experience over all aspects of the thorium fuel cycle. An example is the KAMINI (Kalpakkam Mini) reactor at IGCAR, perhaps the only reactor in the world that uses U-233 as fuel. This fuel was bred, processed, and fabricated at Bhabha Atomic Research Center, Mumbai.

23.5.3.6 International Collaborations and Availability of Materials and Technical Knowledge India has a flourishing and largely indigenous nuclear power program. Because India is

outside the Nuclear Non-Proliferation Treaty due to its weapons program, it has been for over three decades largely excluded from trade in nuclear plants or materials. This forced it into nuclear energy self-sufficiency extending from uranium exploration and mining through fuel fabrication, heavy water production, reactor design and construction, to reprocessing and waste management. Now, foreign technology and fuel are expected to boost India's nuclear power plans considerably. All plants will have high indigenous engineering content. In the context of India's nuclear trade isolation over three decades, Larsen & Tubros (L&T), India's largest engineering group, has produced heavy components for PHWRs and has also secured contracts for 80% of the components for the FBR at Kalpakkam. It is qualified by the American Society of Mechanical Engineers to fabricate nuclear-grade pressure vessels and core support structures, achieving this internationally recognized quality standard in 2007. L&T plans to produce 600-ton ingots in its steel melt shop and has a very large press to supply finished forgings for nuclear reactors, pressurizers, and steam generators. Other nuclear equipment manufacturers in India include BHEL, Bharat Forge, HCC, etc.

The AERB, formed in 1983, is responsible for the regulation and licensing of all nuclear facilities and their safety. To fulfill its mission of stipulating and enforcing rules and regulations concerned with nuclear and radiological safety, AERB is in the process of developing codes and guides for various aspects of nuclear reactor design. Reference [33] provides some details of the design and construction codes followed in the Indian nuclear program. Design of Indian PHWRs is based on ASME B&PV Code, Section III. In those cases where the code rules could not be met in letter, they have been met in spirit by carrying out the required material characterization and component qualification tests. The in service inspection

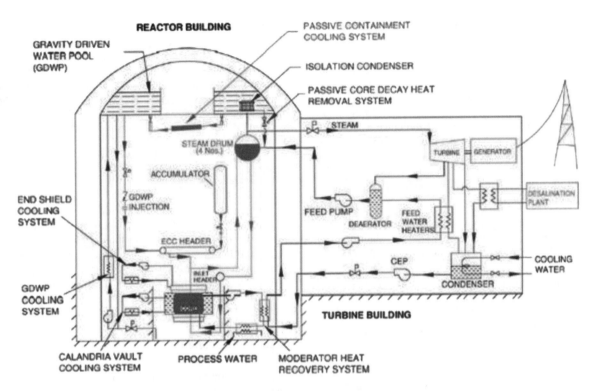

FIG. 23.5A   SIMPLIFIED SCHEMATIC ARRANGEMENT OF THE AHWR [39]

program follows a combined philosophy of ASME Code Section XI and the Canadian Code CSA/N281.5-05. Design of the concrete containment is based on the French Code RCC-G. To avoid a mixture of codes, AERB is in the process of writing its own codes. It has already published five Safety Codes, 45 Safety Guides covering various aspects of nuclear power plants, and 51 other safety support documents. A few others are in various stages of preparation.

23.5.4   Russia

Following recent information from the literature [40], Russia has significant experience with sodium-cooled reactors. The BN-350 prototype FBR generated power in Kazakhstan for 27 years (to 1999), and about half of its 1000 MWt output was used for water desalination. It used uranium enriched to 17% to 26%. Its design life was 20 years, and after 1993, it operated on the basis of an annual license renewal. Russia's BOR-60 was the demonstration model preceding it.

The construction of the first BN-800 reactor is well under way. It has improved features including fuel flexibility — U+Pu nitride, MOX, or metal, and with breeding ratios up to 1.3. However, during the plutonium disposition campaign, it will be operated with a breeding ratio of less than one. It has enhanced safety and improved economy — operating cost is expected to be only 15% more than VVER. It is capable of burning up to 2 tons of plutonium per year from dismantled weapons and will test the recycling of minor actinides in the fuel. In 2009, two BN-800 reactors were sold to China, with construction due to start in 2011.

The BN-1800 is next in this development chain. This power generating unit is designed to meet the requirements of the strategy for developing atomic energy in Russia in the first half of the 21st century. The development time is the next 15 years, and construction could start after 2020. The design includes the development of advanced technical solutions as compared with the

BN-800 reactor, which is now under construction. The new technical solutions are based on the substantial positive experience in operating fast reactors in Russia (~125 reactor years), specifically the BN-600 reactor. The innovations make it possible not only to solve strategic problems, such as increasing safety, improving ecology (including burning actinides), and non-proliferation but also to make large improvements in economic performance.

The development of BN-1800 is based on the maximum possible use of tested solutions, implemented in BN-350, -600, and -800 reactors and the use of new technical solutions that increase safety and cost effectiveness. The following technical solutions have been tested:

- three-loop scheme for the power generating unit, sodium in the first and second loops, working body water/steam;
- integrated arrangement of the first (radioactive) loop with the main and backup vessels.

23.5.4.1 Economic Performance Economic performance is improved by the following:

- increasing the power;
- increasing the efficiency of the steam-power cycle up to 45.5% to 47% by increasing the coolant temperature in the three loops, using the working body in the third loop with trans-critical pressure, using schemes with sodium intermediate superheating of steam, and optimizing the construction and layout of the turbine system;
- increasing the rated service life of the power-generating unit up to 60 year, increasing the service life of the replaceable equipment by a factor of 1.5–2 compared with that achieved in BN-600;
- increasing the time interval between re-loadings to 1.5 to 2 years and reaching an installed capacity utilization factor of 0.9.

23.5.4.2 Safety Safety is improved by using safety systems based on a passive principle of action: implementing a system for emergency cool down with natural circulation in all loops, developing passively actuating (based on a change in flow rate and temperature at the exit from the core) systems for influencing the reactivity, passively actuating systems for protection from an increase in pressure in the first and second loops, optimizing safety systems that localize the consequences of the chemical activity of the coolant (systems for automatic protection of the steam generator, quenching sodium fires).

23.5.4.3 Proliferation Resistance Proliferation resistance is increased by the following processes:

- using nitride fuel, which permits satisfying the conditions that are important for safety (core breeding ratio — about 1);
- sodium void effect over the core as a whole (about zero), achieving nitride-fuel burnup 13% h.a. and higher;
- implementing a cost-effective fuel cycle, which permits burning actinides, and technological support for the non-proliferation regimen — elimination of the fuel-cycle stage where plutonium is separated in a pure form.

23.5.5 China

According to recent information [41], basic research on fast reactor technology in China was started in the mid-1960s and continued for approximately 20 years. During this period, there was no clear civilian nuclear power program in the country; thus, little manpower and few resources were devoted to this research area. In 1986, the National High-Tech Program was launched, and fast reactor technology development was brought into line with this program. An applied basic research project was executed until 1993 with an engineering target of a 60 MW experimental fast reactor. It was considered to fully meet the requirements of the reactor design and safety analysis. The conceptual design of the China Experimental Fast Reactor (CEFR) was a 65 MW reactor matched with a 20 MW turbine generator and was finished in 1993, as a result of international cooperation with the Russian FBR Association for the technical design of the CEFR main systems and with France

CEA for the R&D. It was followed by the preliminary design and then a detailed design by a Chinese design team. The third phase (1993–2006) of the R&D activities was concentrated on the CEFR design demonstration. Nearly 50 subjects were proposed from the conceptual, preliminary and even the detailed design. For the recent and near-term R&D activities of fast reactor technology, the emphasis of their efforts is as follows:

(1) Use the CEFR as a tool to verify the computer codes used for the CEFR design and to be used for the prototype and demonstration of a fast reactor, which have the same main technical selections.
(2) The CEFR is a model for studying CEFR safety properties.
(3) Research to support the safe operation of the CEFR.
(4) Applied research for the following fast reactor prototype or demonstration, including the establishment of rigs and facilities for models that test key components and systems, as well as research related to the simplification of the systems.
(5) R&D for an advanced SFR system mainly including an innovative SFR concept, reactor design and safety features, advanced fuel cycle strategy, and new technology and new materials.

23.5.6 Europe

In Europe, particularly in France, strong interest in SFRs already exists. With respect to industrial application, the Superphenix was the most important plant. Also in Germany, a fast sodium breeder reactor existed (SNR 300 in Table 23.6). However, this plant never went into operation. Currently, several fast reactor concepts are being considered in Europe as shown in Figure 23.6. The SFR is considered as the reference technology.

Sodium as a coolant has several advantages: high conductivity, liquid from 98 up to 883°C at a pressure of 1 bar, low viscosity, compatible with steels, industrial fluid, and low cost. The main disadvantages of sodium are its reactivity with air and water and the fact that it is opaque. Lead is basically a good coolant showing no reactivity with air and water, but it is corrosive, it is toxic, has a very high density and it can become solid. Helium is considered as an inert and transparent alternative having no temperature constraints. However, helium has low density and a high pressure reactor concept is needed. Current ongoing research, design, and

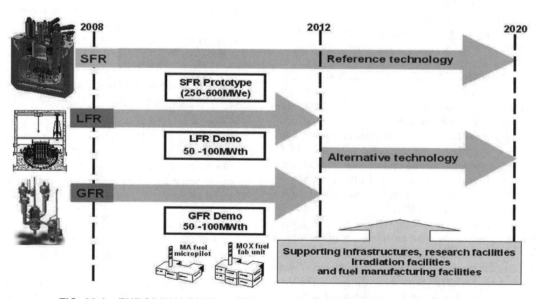

FIG. 23.6 EUROPEAN PERSPECTIVE CONCERNING FAST REACTORS [42]

prototype projects shall provide the basis for the future fast reactor plant concept chosen.

In France, the SFR is the candidate prototype of a Generation IV system to be built as early as 2020. This project is called ASTRID and mixed oxide fuel (U, Pu) O₂ is considered the reference fuel for the core of this reactor [43]. The core design of French advanced sodium-cooled fast reactors is mainly driven by safety, competitiveness, and flexibility margins compared with previous SFR projects. Performance objectives include improvement of safety features, flexible management of plutonium (optimization of uranium resources) and transmutation of minor actinides (environmental burden decrease), high burnup rate, high operating availability, and proliferation resistance enhancement with integrated fuel cycle. The ASTRID prototype reactor will provide valuable information about SFR-MOX fuel reprocessing that will help determine an industrial treatment process that can meet the Generation IV objectives of economics, proliferation, waste minimization, and safety. Metallic fuel and carbide fuel are being considered for long-term use.

ASTRID is called a "self-generating" fast reactor rather than a breeder in order to demonstrate low net plutonium production. ASTRID is designed to meet the stringent criteria of the GIF in terms of safety, economy, and proliferation resistance. CEA plans to build this plant at Marcoule.

The ASTRID program includes development of the reactor itself and associated fuel cycle facilities: a dedicated MOX fuel fabrication line (possibly in Japan) and a pilot reprocessing plant for used ASTRID fuel. The program also includes a fabrication of fuel rods containing actinides for transmutation (called Alfa), scheduled to operate in 2023, though fuel containing minor actinides would not be loaded for transmutation in ASTRID before 2025 [44].

23.5.7 United States

In 1983, DOE began the Advanced Liquid Metal Reactor (ALMR) program with a goal to increase the efficiency of uranium usage by breeding plutonium and create the condition wherein transuranic isotopes would never leave the site. The ALMR pro-

gram was funded from 1984 to 1994. One of the products of this program was the PRISM reactor design [45,46].

PRISM is an advanced fast neutron spectrum reactor plant design with passive reactor shutdown, passive shutdown heat removal, and passive reactor cavity cooling. The reactor supports a sustainable and flexible fuel cycle to consume transuranic elements within the fuel as it generates electricity. The essence of the reactor technology is a reactor core housed within a 316 stainless steel reactor vessel. Liquid sodium is circulated within the reactor vessel and through the reactor core by four electromagnetic pumps suspended from the reactor closure head. Two intermediate heat exchangers inside the reactor vessel remove heat for electrical generation.

The PRISM technology is deployed as a power block with two reactors side by side supporting a single steam turbine generator set. The plant is divided into two areas: the nuclear island (reactors through steam generators) and balance of plant (steam turbine to generate electricity). The nuclear island is two reactors in separate containments, plus steam generators, and shared services, in a single, seismically isolated, partially buried building as depicted in the cutaway view of a PRISM reactor nuclear island shown in Figure 23.7. Each reactor heats an intermediate coolant loop, sending heat to a steam generator. Steam from the steam generators is combined and sent to the balance of plant, where a single turbine generator produces electricity.

A draft Licensing Plan and preliminary Design Control Document (DCD) for the PRISM reactor were delivered to the US NRC in May 2010.

23.6 VERY-HIGH-TEMPERATURE GAS-COOLED REACTOR

The other type of advanced reactor for which former industrial experience exists is the (very) high-temperature gas cooled reactor. Past and current projects are listed in Table 23.9.

The main advantage of a VHTR is its capability to provide high-temperature heat for industrial processes. It is therefore primarily considered as a technology for the production of CO₂-free process heat.

In the Generation IV roadmap, the VHTR was still assumed to be operated with a direct cycle helium gas turbine and/or a heat source for hydrogen production with thermochemical water splitting (iodine- sulfur process). Lacking structural materials and economic considerations led to a considerable change in the original project parameters as discussed already in an earlier section. The VHTR has a graphite core and it can be designed as a pebble bed or with prismatic blocks. Fuel for either core option utilizes the TRISO coated fuel particles (Figure 23.8). The performance of both concepts is comparable and therefore both designs are currently being pursued.

23.6.1 United States

The most significant Generation IV effort in the United States is the Next Generation Nuclear Plant (NGNP) effort. This demonstration plant is expected to be operational in 2021.

In May 2004, the US DOE released a Request for Information and Expressions of Interest on the NGNP, the objective of which was "… to conduct research, development, and demonstration of a next-generation nuclear power reactor in order to establish advanced technology for the future production of safe, efficient, and environmentally acceptable power and to demonstrate the economic and technical feasibility of such facilities to the US electric power industry." President George W. Bush signed the Energy Policy

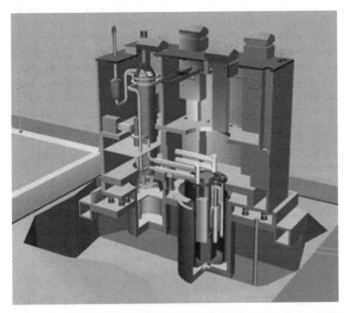

FIG. 23.7 CUTAWAY VIEW OF A PRISM NUCLEAR ISLAND [45]

TABLE 23.9   PAST AND PRESENT HIGH-TEMPERATURE GAS-COOLED REACTORS [ORIGINAL WORK]

	U.S.	Europe	Africa	Asia
Past	Peach Bottom and Fort St. Vrain both prismatic (P)	AVR and THTR-300, Germany, both pebble bed (PB)		
Cancelled			PBMR (PB), South Africa	
Operating				HTR-10 (PB), China, HTTR (P), Japan
Under construction				HTR-PN (PB), China
Planned	NGNP			

Act of 2005 on August 8, 2005 (Public Law 109-58 2005), creating the Next Generation Nuclear Plant project. The law instructed that DOE shall establish the Next Generation Nuclear Plant project, with a prototype to be sited at Idaho National Laboratory. The centerpiece is to be the development of reactor, fuel, and associated technology for the production of hydrogen as well as electricity. The DOE and the NRC submitted a joint NGNP licensing strategy to the US Congress in 2008.

R&D specific to NGNP mentioned in the Energy Policy Act (2005) and conducted to date is based on the gas-cooled VHTR concept promulgated in the Generation IV technology roadmap. The VHTR system uses a thermal neutron spectrum and a once-through uranium cycle. The VHTR system is primarily aimed at relatively faster deployment of a system for high-temperature process heat applications, such as coal gasification and thermochemical hydrogen production, with superior efficiency. The reference reactor concept has a 600-MWt helium-cooled core based on either the prismatic block fuel of the Gas Turbine–Modular Helium Reactor (GT-MHR) or the pebble fuel of the PBMR. The primary circuit is connected to a steam reformer/steam generator to deliver process heat. The VHTR system has coolant outlet temperatures at 1000°C. It is intended to be a high-efficiency system that can supply process heat to a broad spectrum of high-temperature and energy-intensive, non-electric processes. The system may incorporate electricity generation equipment to meet cogeneration needs.

NGNP is envisioned to extend the application of nuclear energy into the broader industrial and transportation sectors, reducing fuel use and pollution and improving on the safety of existing com-

mercial LWR technology. Next Generation Nuclear Plants will use new, high-temperature, gas-cooled reactor technologies to produce energy that is able to run both a primary and a secondary industrial application — for example, generating electricity while supporting petroleum refining or bio- and synthetic-fuel production through the provision of hydrogen and/or the provision of process heat.

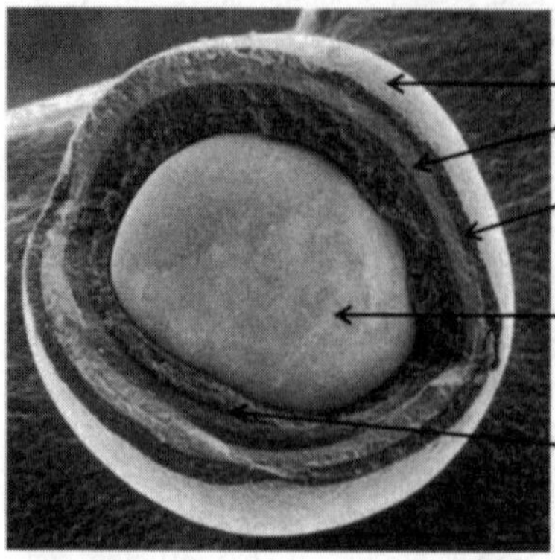

About 40% of the nation's greenhouse gas emissions come from industrial processes in high-energy consuming sectors. With NGNP systems, the process heat or steam generated by the high-temperature nuclear reactors will be used to power applications such as power generation using advanced highly efficient turbines; plastics manufacturing; petroleum refining and fuels production; and producing ammonia for fertilizer. By integrating energy generation and production operations, NGNP technologies will allow high energy consuming industries and sectors to reduce carbon dioxide emissions, limit their need for fossil fuels, and become more competitive. The basic technology for the NGNP has been established in former high-temperature gas-cooled reactor plants shown in Table 23.9.

After studying a variety of options, DOE determined that a high-temperature gas-cooled nuclear reactor (HTGR) with a 750°C to 800°C outlet would best meet the operating parameters associated with these objectives. DOE, through the Idaho National Laboratory, has conducted design and trade studies with input from the commercial industry to define the gross operating parameters of the NGNP so that it will be (1) capable of generating process heat for electricity and/ or hydrogen production, or for other uses and (2) configured for low technical and safety risk with highly reliable operations. The target core outlet temperature of 750°C to 800°C associated with the NGNP design would be capable of meeting the needs of many industrial-process-heat end users. Studies conducted to date can be accessed via the NGNP Web site [48].

The NGNP project includes R&D, design, licensing, and construction activities conducted in two phases, leading to operation of a NRC-licensed prototype Generation IV reactor and an associated energy delivery system. Phase 1 is the phase that covers selecting and validating the appropriate technology; carrying out enabling research, development, and demonstration activities, determination of whether it is appropriate to combine electricity generation and hydrogen production in a single prototype nuclear reactor and plant and to carry out initial design activities for a prototype reactor and plant, including development of design methods and safety analytical methods and studies. Phase 2 is the phase that covers development of a final design for the prototype nuclear reactor and plant through a competitive process; application of licenses

FIG. 23.8   TRISO COATED FUEL PARTICLE [47]

Pyrolytic Carbon

Silicon Carbide (SiC)

Kernel

Porous Carbon Buffer

to construct and operate the prototype nuclear reactor from the US NRC and construction and start up operations for the prototype nuclear reactor and its associated hydrogen or electricity production facilities. Both phases include R&D and licensing activities. The prototype reactor should produce hydrogen and/or electricity and could also demonstrate other uses for the high-temperature process heat generated by the reactor. The prototype reactor design would be generic enough that plants can be replicated at multiple sites within the United States.

23.6.1.1 Pebble Bed Version One HTGR technology being examined by the NGNP is the pebble bed design based on a fundamental fuel element, called a pebble, that is a graphite sphere (6 cm in diameter- size of a tennis ball) containing about 15,000 uranium oxide particles with a diameter of 1 mm (Figure 23.8). The uranium oxide kernel is encased in a pressure vessel made of several high-density ceramic coatings. The strongest layer is a tough silicon carbide ceramic. This layer serves as a pressure vessel to retain the products of nuclear fission during reactor operation or accidental temperature excursions. Figure 23.9 illustrates the construction of a pebble. About 330,000 of these spherical fuel pebbles are placed into a graphite core built from graphite blocks. In addition, as many as 100,000 unfueled graphite pebbles are used with the 330,000 fuel pebbles to shape the core's power and temperature distribution by spacing out the hot fuel pebbles. The graphite core is constructed from graphite blocks forming an open cylindrical volume. An optional center graphite column can be placed at the center of the void forming an annular core for the pebbles. The graphite acts:

- as a structure forming the core,
- as a neutron moderator and reflector, and
- as a solid heat absorber and conduction path to an ultimate heat sink in case of an accident.

The graphite core is restrained by lateral restraint straps to keep the graphite blocks compressed in a cylindrical structure. On the outside of the graphite core is a metallic core barrel that restrains the core during an earthquake and acts as a thermal shield to the reactor vessel. The core barrel and graphite core are located in a large pressure vessel.

Helium gas enters the vessel near the bottom and flows up in the outer risers in the permanent graphite reflector blocks reaching the plenum above the core where the gas is forced down through the pebbles and out the vessel to the secondary side of the plant. A small portion of the gas in the top plenum flows down the openings for the control rods, cooling them during operation. The helium coolant is an inert noble gas that does not react with materials in the core at high temperatures nor changes phase with temperature increase. Further, because the pebbles and reactor core are made of refractory materials, they cannot melt and will degrade only at the extremely high temperatures encountered in accidents (more than 1600 degrees C), a characteristic that affords a considerable margin of operating safety. The graphite core structure represents a large thermal capacitance combined with the low core power density results in slow thermal transients. Because the pebbles form a packed bed, the helium is distributed evenly through without the need of flow channeling.

To re fuel the pebble core, pebbles pass through the bottom of the graphite core and new pebbles are added at the top of the core. This operation is performed continuously during reactor operations allowing the reactor to stay on line. During operation, one pebble is removed from the bottom of the core about once a minute as a replacement is placed on top. In this way, all the pebbles gradually move down through the core like gumballs in a dispensing machine, taking about six months to do so. This feature maintains the optimum amount of fuel for operation without requiring excess activity. It eliminates an entire class of excess-reactivity accidents that can occur in current water-cooled reactors. Each expended pebble is measured to determine the remaining fuel and is stored. The stored pebbles are recycled through the core until the remaining nuclear fuel is below a minimum quantity. Also, the steady movement of pebbles through regions of high and low power production means that each experiences less extreme operating conditions on average than do fixed fuel configurations, again adding to the unit's safety margin. After use, the spent pebbles must be placed in long-term storage repositories, the same way that spent fuel rods are handled today.

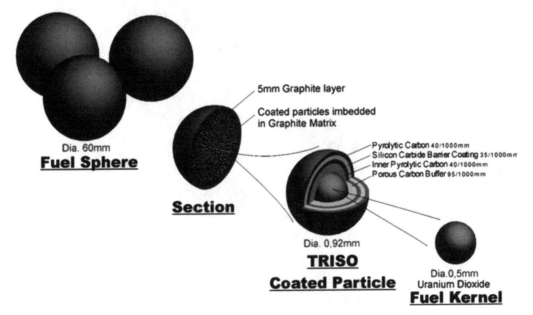

FIG. 23.9 PEBBLE CONSTRUCTION [49]

23.6.1.2 Prismatic Core Version The basic fuel element in a prismatic high-temperature gas-cooled reactor is a ceramic fuel particle approximately 1 mm in diameter. The spherical fuel particle is a ceramic pressure vessel containing a uranium oxy-carbide kernel. The ceramic pressure vessel retains the products of nuclear fission during operation or accidental temperature excursions. The particles are placed in a fuel compact typically containing 4000 to 7000 particles. The fuel compacts are typically 12.7 mm in diameter by 50 mm in length. The fuel compacts are pressed into channels drilled into graphite blocks. There are 14 to 15 compacts in each channel. Graphite fuel blocks have 210 channels; thus, each fuel block contains approximately 3126 compacts.

The reactor core consists of an assembly of hexagonal prismatic graphite blocks in annular configuration consisting of three annular rings (Figure 23.10). The center and outer portions of the core are made from unfuelled graphite reflector blocks. The center ring contains the active ring of graphite fuel element blocks. The outer reflector blocks have full core height channels for control rods. Some of the fuel blocks also contain full height vertical channels for control rods and the reserve shut down system. The reserve shut down system uses ceramic-coated boron carbide pellets using gravity to fill the channels upon activation. Inherent in the design of high-temperature gas reactors (both prismatic and pebble bed core designs) is the ability to shut down the reactor during an accident. As the core heats during an accident, the inherent large negative temperature coefficient stops the chain reaction in the active core effectively shutting down the reactor.

The active core is 10 blocks high with 102 fuel columns. With the inner and outer reflector blocks, the physical graphite reactor structure is 6.83 meters in diameter and 13.59 meters high. Graphite pedestals or columns support each graphite column. The area between the columns is the lower plenum. A metallic core barrel restrains the graphite structure during seismic events and acts as a thermal heat shield for the reactor vessel. The graphite reactor structure is the solid neutron moderator and reflector in the core. Graphite remains solid at temperatures well above those experienced during accidents. The graphite has a high heat capacity creating a large heat sink for the core in case of the accident. Further, the high heat capacity and low power density of the reactor core results in very slow and predictable temperature transients.

The reactor vessel contains the reactor core structure and shutdown cooling system used for re fueling. Helium coolant enters near the bottom of the reactor vessel and flows up the outside of the core barrel to the plenum above the graphite core structure. Helium flows out of the plenum down through the coolant holes in the fuel blocks to the lower plenum and out the vessel. The outer and inner reflector has no helium flow with all convection cooling occurring in the active core. A considerable margin of safety is gained by the use of the inert noble gas helium as the coolant. The gas does not react with the reactor core materials at high temperatures encountered in accidents (>1600°C). The helium coolant does not moderate neutron; its use does not add or subtract reactivity.

Refueling the core is handled remotely using a re fueling machine located above the reactor vessel. A lever arm is attached to an extendable shaft lowered through an opening in the reactor vessel into the core. The grapple on the end of the lever arm interfaces with the graphite block. Each block is then transferred to a lift station (another extendable shaft into reactor vessel) where it is pulled up into the shielded re fueling machine. The shielded re fueling machine then takes the block to adjacent dry storage. The remaining fuel blocks

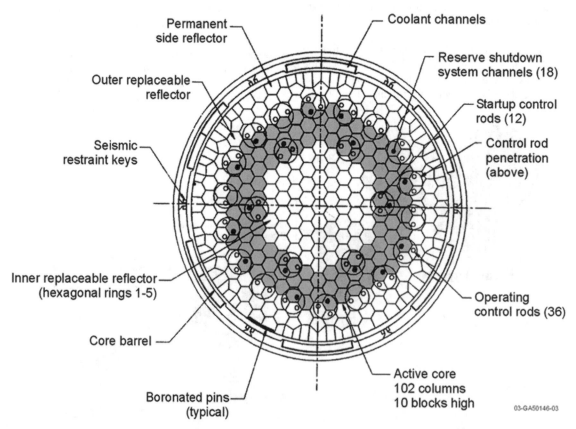

FIG. 23.10 CORE DESIGN OF A PRISMATIC HIGH-TEMPERATURE REACTOR [50]

are distributed in the core to control power peaking and flux profile in the core. The fuel cycle is a once through, three-year cycle with one-half of the active core re fueled every 20 months.

23.6.1.3 Secondary Side of Prismatic or Pebble Bed Reactors

The helium coolant leaving the reactor vessel can be used for process heat as well as electricity generation. A direct Brayton cycle can use the reactor coolant in a high-temperature gas turbine. An indirect Rankine cycle requires a steam generator to transfer heat from the helium coolant to produce steam. Typical efficiencies of the two cycles depend on the outlet temperature of reactor. At a 700°C reactor outlet temperature, the Rankine cycle can achieve approximately 40% efficiency. At a higher reactor outlet temperature of 900°C, the Brayton cycle efficiency is approximately 47%. Transferring heat for industrial applications requires unique and custom design of heat exchangers to interface with the industrial application.

23.6.2   South Africa

The South African PBMR started in 1999 with the development of a direct Brayton cycle plant for electricity generation and low temperature cogeneration applications such as desalination. The plan was to build a demonstration plant called the DPP400 at Eskom's Koeberg site and the RSA national utility Eskom was the targeted customer. This plant was designed to generate 165 MW of electricity using a 400-MWt annular core pebble bed reactor coupled to a direct Brayton cycle power conversion unit.

During the last few years, growing interest in HTRs for high-temperature process heat or cogeneration applications became visible. Particularly, the US NGNP could become the first customer for a pebble bed plant of this type. As a result of these developments and also of national funding problems, the board of PBMR decided to change to an indirect steam plant that could be used for electricity generation and/or process heat. The most recent plant design is based on a 2 × 250-MWt reactor layout where each reactor has its own primary cooling circuit and steam generator. On the secondary side, the steam generators are connected to a common steam header. Although the project advanced quite far, the South African government, in September of 2010, decided to stop funding the effort.

23.6.3   Japan

A demonstration plant for a prismatic core is in operation in Japan (HTTR) [51]. This system was originally designed as a heat source for hydrogen production with the thermochemical iodine-sulfur process [52]. The main parameters of the plant are summarized in Table 23.10.

TABLE 23.10   CHARACTERISTICS OF THE JAPANESE HTTR [51]

Thermal power	30 MWt
Fuel	coated fuel particle/prismatic block type
Core material	graphite
Coolant	helium
Inlet temperature	395°C
Outlet temperature	950°C (max.)
Pressure	4 MPa

TABLE 23.11   PERFORMANCE DATA OF THE CHINESE HTR-PM [54]

Reactor module numbers	2
Thermal power/module	250 MWt
Lifetime	40 years
Core diameter/height	3.0 / 11m
Primary system pressure	7.0 MPa
Helium inlet/outlet temperature	250°C / 750°C
Helium mass flow rate	96 kg / second
Fresh steam temperature/pressure	566 °C / 13.2 MPa
Electric power	210 MWe

On March 13, 2010, long-term (50 days) full power operation of HTTR at a reactor outlet coolant temperature of about 950°C was successfully completed, and various performance data was obtained. Main future demonstration activities will go toward industrialization of the I-S hydrogen process and a HTGR cascade energy plant for 79% efficient production of hydrogen, electricity and freshwater. A nuclear commercial hydrogen production plant is envisaged by 2030.

23.6.4   South Korea

The South Korean NHDD project intends to build a VHTR for hydrogen production. No decision has been taken with respect to core design (prismatic block or pebble bed). The gas outlet temperature is expected to be 950°C, and the reactor power should be 200 MWt. A cold reactor vessel option is being considered. Hydrogen shall be produced in a five-train sulfur-iodine thermo-chemical plant. Technology selection should be finished by 2012, and starting of operation of the demonstration plant is scheduled for 2026.

23.6.5   China

China built a pebble bed type of demonstrator (HTR-10) that is based on the former German experience. The HTR-10 experience shall be used for the new HTR-PM demonstration plant [53]. The HTR-PM plant will consist of two nuclear steam supply systems. Each of these modules consists of a single zone 250 MWt pebble-bed modular reactor and a steam generator. The two modules feed one steam turbine and will generate 210 MW of electric power. A pilot fuel production line will be built to fabricate 300,000 pebble fuel elements per year. This line is closely based on the technology of the HTR-10 fuel.

The main performance data of the HTR-PM is listed in Table 23.11.

23.7   OTHER GENERATION IV SYSTEMS

In contrast to the SFR and VHTR systems, only little to no experience exists with industrial plants for the other Generation IV systems. Projects for these reactor types are currently in a conceptual or preliminary design phase.

23.7.1   Lead-Cooled Fast Reactor (LFR)

Designs of LFRs are closely related to the SFR. The experience with reactors having lead or lead bismuth as coolants is by far, less established than with SFRs. Russia has experimented with several

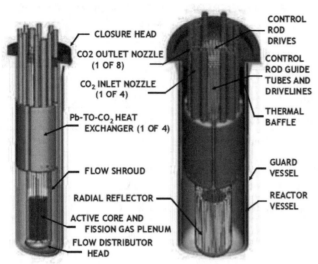

Figure 4: Small Secure Transportable Autonomous
Reactor (SSTAR).

FIG. 23.11 SMALL SECURE TRANSPORTABLE AUTONOMOUS REACTOR (SSTAR) [17]

lead-cooled reactor designs and has used lead-bismuth cooling for 40 years in reactors for its Alfa class submarines. A significant new Russian design is the BREST fast neutron reactor, of 300 MWe or more with lead as the primary coolant, at 540°C, and super-critical steam generators. A pilot unit is planned at Beloyarsk and 1200 MWe units are proposed. A smaller and newer Russian design is the Lead-Bismuth Fast Reactor (SVBR) of 75–100 MWe. This is an integral design, with the steam generators sitting in the same Pb-Bi pool at 400°C to 495°C as the reactor core, which could use a wide variety of fuels.

Rosatom, the Russian Nuclear Energy State Corporation, has put forward two fast reactor implementation options for a government decision in relation to the Advanced Nuclear Technologies Federal Program 2010–2020. The first focuses on a lead-cooled fast reactor such as BREST with its fuel cycle. The second scenario assumes parallel development of fast reactors with lead, sodium and lead-bismuth coolants and their associated fuel cycles. The second scenario is viewed as the most favored, since it is believed to involve lower risks than the first one.

If implemented it would result in technical designs of the Generation IV reactor and associated closed fuel cycles technologies by 2013, and a technological basis of the future innovative nuclear energy system featuring the Generation IV reactors working in closed fuel cycles by 2020.

Interest in the LMR technology exists also in other countries like the EU, as indicated already earlier in the SFR-section. One design project is the small secure transportable autonomous reactor (SSTAR) [55] shown in Figure 23.11 with mixed uranium-plutonium nitride fuel [17]. The other design project is the European lead-cooled system (ELSY) [56] with MOX fuel (design data see Table 23.12) [17].

23.7.2 Gas-Cooled Fast Reactor (GFR)

An experimental demonstration reactor (ALLEGRO) [57] is currently planned in the EU as already indicated above. It should be built in the coming decade. It will have a thermal power of around 80 MW and it will not produce any electric energy. Its main purpose is the demonstration of the technical feasibility of

TABLE 23.12 KEY DESIGN DATA OF GIF LFR CONCEPTS [17]

Parameter / system	SSTAR	ELSY
Power (MWe)	19.8	600
Conversion ratio	≈1	≈1
Thermal efficiency (%)	44	42
Primary coolant	lead	lead
Primary coolant circulation (at power)	natural	natural
Primary coolant circulation for decay heat removal	natural	natural
Core inlet temperature (°C)	420	400
Core outlet temperature (°C)	567	480
Fuel	nitrides	MOX, (nitrides)
Fuel cladding material	Si-enhanced F/M stainless steel	T91 (aluminized)
Peak cladding temperature (°C)	650	550
Fuel pin diameter (mm)	25	10.5
Active core height/ equivalent diameter (m)	0.976 / 1.22	0.9 / 4.32
Primary pumps	–	Eight pumps, mechanical, integrated in the steam generator
Working fluid	supercritical CO_2 at 20 MPa, 552°C	water-superheated steam at 18 MPa, 450°C
Primary /secondary heat transfer system	Four Pb-to-CO_2 heat exchangers	Eight Pb-to-H_2O steam generators
Safety grade decay heat removal	reactor vessel air cooling system plus multiple direct reactor cooling systems	reactor vessel air cooling system plus four direct reactor cooling systems plus four secondary loops systems

this system, which has never been built before. ALLEGRO shall incorporate (on a reduced scale) all of the architecture of a GFR, excluding the power conversion system. It shall contain main materials and components foreseen later for an industrial GFR. The safety principles are those proposed for GFRs: core cooling through gas circulation in all situations and ensuring a minimum pressure level in case of leaks. This is done with a specific guard

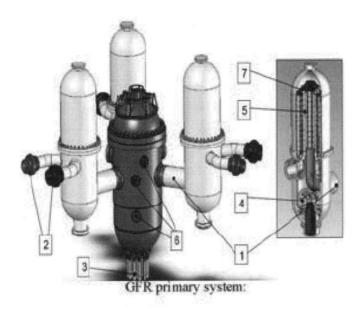

GFR primary system:

1. *Primary cross-duct*
2. *Secondary pipes with isolating valves*
3. *Control Rod Drive Mechanisms*
4. *Primary blower and associated motor*
5. *Compact Heat Exchanger modules*
6. *Pipe connections for Decay Heat Removal systems*
7. *Primary isolation valve*

FIG. 23.12   PROPOSED GFR ARRANGEMENT [18]

containment. The ALLEGRO demonstrator (Figure 23.12) will also support the development and qualification of an innovative refractory fuel element that can operate at the envisioned high reactor temperatures.

The reactor pressure vessel is a large metallic structure (inner diameter 7.3 m, overall height 20 m, weight about 1000 tons, and thickness of 20 cm in the belt line region). The material selected, a martensitic 9Cr1Mo steel (industrial grade T91, containing 9% by mass chromium, and 1% by mass molybdenum) undergoes negligible creep at the operating temperature (400°C). The reference material for the internals is either 9Cr stainless steel or SS316LN. The primary arrangement is based on three main loops (3 × 800 MWt), each fitted with one IHX–blower unit, enclosed in a single vessel.

This single component limits the consequence of a simultaneous rupture of the first and second safety barrier (fuel cladding and the primary system).

Specific loops for decay heat removal in case of emergency are directly connected to the primary circuit using a cross duct piping in extension of the pressure vessel. They are additionally equipped with heat exchangers and forced convection cooling.

This system arrangement allows for the extraction of the residual power in any accident condition. Additionally, a passive natural gas circulation system can be used in most cases due to the low pressure drop of the core design [18].

23.7.3   Super Critical Water Reactor

The SCWR is a water-cooled reactor that operates above the thermodynamic critical point. Although the SCWR builds on current

PWRs, plans for a demonstration plant are not very far developed. Several design options using pressure vessel and pressure tube technologies are currently under consideration with the aim of providing a spectrum of possibilities for consideration for the next generation of water-cooled reactor technology. These design options are being used to define high priority R&D areas and will contribute to the definition of a future design that will improve and optimize all GIF metrics.

The SCWR system is primarily designed for efficient electricity production, with an option for actinide management based on two options in the core design: the SCWR may have a thermal or fast-spectrum reactor and a closed cycle with a fast-spectrum reactor and full actinide recycle option based on advanced aqueous processing at a central location. Figure 23.13 shows a schematic of a "super fast LWR" that has been developed by the University of Tokyo.

23.7.4   Molten Salt Reactor

The MSR projects are also primarily in a conceptual design phase. Two options of a MSR are currently being considered [21]: the thermal version of the AHTR and the MSFR.

The defining aspects of an AAHTR are the use of coated particle fuel embedded within a graphitic matrix cooled by liquid fluoride salt [58]. A Pebble Bed Advanced high-temperature Reactor (PB-AHTR) operating at ~900 MWt is the most actively developing commercial scale plant design [59]. The plant design is currently transitioning from a conceptual to an initial engineering scoping phase.

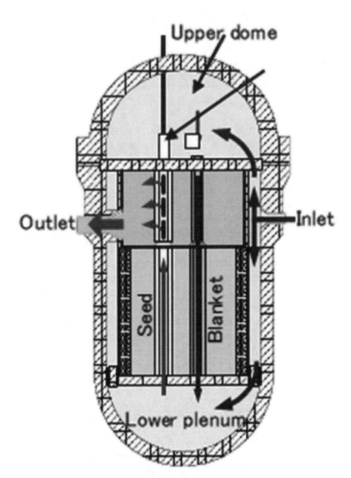

FIG. 23.13   SUPER FAST LWR (SCHEMATIC) [20]

TABLE 23.13   REFERENCE DESIGN CHARACTERISTICS OF THE MSFR [21]

Thermal power (MWt)	3000				
Fuel molten salt composition (mo1%)	LiF-ThF$_4$–^{233}UF$_4$ or LiF-ThF$_4$–^{233}UF$_4$– (Pu-MA)F$_3$ with LiF = 77.5 mol%				
Fertile blanket molten salt composition (mol%)	LiF-ThF$_4$ (77.5–22.5)				
Melting point (°C)	550				
Operating temperature (°C)	700–800				
Initial inventory (kg)	^{233}U-started MSFR		TRU-started MSFR		
	Th	^{233}U	Th	Actinide	
	38300	5060	30600	Pu	11200
				Np	800
				Am	680
				Cm	115
Density (g/cm^3)	4.1				
Dilatation coefficient (/°C)	10^{-3}				
Core dimensions (m)	Radius: 1.15 Height: 2.30				
Fuel salt volume (m^3)	18 9 out of the core 9 in the core				
Blanket salt volume (m^3)	8				
Thorium consumption (ton/year)	1.112				
^{233}U production (kg/year)	93(^{233}U-started MSFR) 188 during 20 years then 93 (TRU-started MSFR)				
Breeding ratio (^{233}U-started MSFR)	1.085				

Starting from the ORNL Molten Salt Breeder Reactor project (MSBR), an innovative concept has been proposed, resulting from extensive parametric studies in which various core arrangements, reprocessing performances and salt compositions were investigated. The primary feature of the MSFR concept is the removal of the graphite moderator from the core (graphite-free core).

In the USA, a PB-AHTR (900 MWt) is being actively developed. A research, development and demonstration roadmap is under study for component testing to support a PB-AHTR prototype scale plant and a development path for the structural materials is being established. In Europe, since 2005, R&D on MSR has been focused on fast spectrum concepts (MSFR) that have been recognized as long-term alternatives to solid-fuelled fast neutron reactors with attractive features (very negative feedback coefficients, smaller fissile inventory, easy in service inspection, and simplified fuel cycle). MSFR designs are available for breeding and for MA burning. The MSFR design characteristics are shown in Table 23.13.

23.8   TECHNICAL CHALLENGES

Main technical challenges for advanced reactors are fuel, fuel cycle (high level waste), and structural materials able to operate safely over a long period of time (60 years and more) under very demanding exposure conditions.

23.8.1   Fuel and Fuel Cycle

In a once-through fuel cycle, the spent fuel consisting of plutonium, uranium, neptunium, minor actinides (americium and curium), and fission products is disposed of in a final repository. In case of fuel reprocessing, uranium and plutonium are separated. Only the still usable portion of uranium is recycled and the rest are disposed of. Separation can be done either chemically (liquid extraction) or electro-metallurgically. High amounts of uranium are lost this way and plutonium together with the minor actinides are the long-living elements in the nuclear waste. Additionally, plutonium bears a proliferation risk. Fast reactors can operate with mixed fuel containing uranium, plutonium, and minor actinides that allows fuel cycles where only the fission products remain as waste. This means that the uranium resources last for a very long period of time and nuclear waste would no longer contain long-living products as shown in Figure 23.14.

Basically, there are two routes for fuel treatment being considered. Separate U and Pu (as already done) or separate U and Pu but also separate the minor actinides and produce mixed fuel. Weapon grade plutonium remains separate in this process chain until mixing, which is considered a proliferation risk. Therefore, concepts are under development to separate uranium, plutonium, neptunium, and the minor actinides in one-step, where plutonium does not appear as a separate fraction. The two concepts are summarized in Figure 23.15.

The fuel and fuel cycle options for the different plants are comprehensively described for the different reactor types in [61] in the

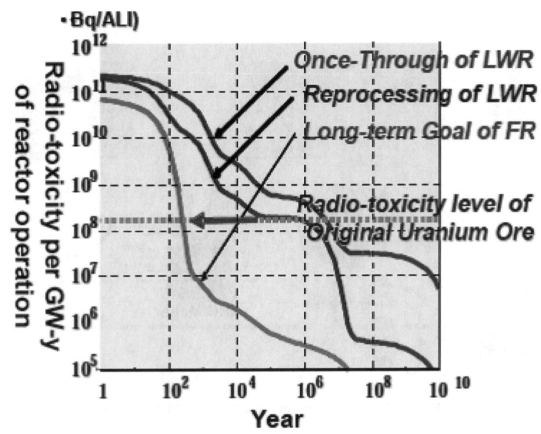

FIG. 23.14 INFLUENCE OF ADVANCED FUEL CYCLE ON LIFE-TIME AND RADIO-TOXICITY OF HIGH LEVEL WASTE (ALI : ANNUAL LIMIT ON INTAKE) [60]

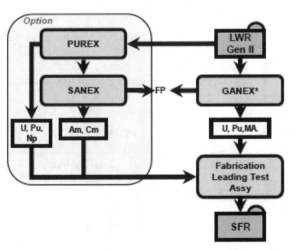

GANEX : Group ActiNides EXtraction

FIG. 23.15 CONCEPTS FOR ADVANCED FUEL RECYCLING. OPTION 1 CONSISTS OF TWO AQUEOUS SEPARATION STEPS WHERE U, Pu, AND Np ARE EXTRACTED IN ONE STAGE AND THE MINOR ACTINIDES ARE EXTRACTED IN ANOTHER STAGE. THE GANEX PROCESS RELEASES U, Pu, AND THE MINOR ACTINIDES IN ONE PROCESS STEP. FOR BOTH OPTIONS, ONLY THE FISSION PRODUCTS (FP) HAVE TO BE DISPOSED [60]

respective sections: VHTR, SFR, LFR, SCWR, GFR, and MSR. The Global Actinide Cycle International Demonstration (GACID) [60] project shall demonstrate that the SFR can manage effectively all actinide elements in the fuel cycle, including uranium, plutonium, and minor actinides (neptunium, americium and curium).

23.8.1.1 Advanced Recycling Center (ARC) The ARC starts with the separations of spent nuclear fuel into three components: (1) uranium that can be used in CANDU reactors or re-enriched for use in LWRs; (2) fission products (with a shorter half-life) that are stabilized in glass or metallic form for geologic disposal; and (3) actinides [the long-lived radioactive material in spent nuclear fuel (SNF)] that are used as fuel in the Advanced Recycling Reactor (ARR).

The electrometallurgical process is proposed to perform separations. This process uses electric current passing through a salt bath to separate the components of SNF. A major advantage of this process is that it is a dry process (the processing materials are solids at room temperature). This significantly reduces the risk of inadvertent environmental releases. Additionally, unlike traditional aqueous MOX separations technology, electrometallurgical separations does not generate separated pure plutonium making electrometallurgical separations more proliferation resistant.

The actinide fuel (including elements such as plutonium, americium, neptunium, and curium) manufactured from the separations step is then used in PRISM to produce electricity in a conventional steam turbine. Figure 23.16 shows a schematic of the ARC. The sodium coolant in the PRISM or 'burner' reactor allows the neutrons to have a higher energy, converting them into shorter-lived fission products. An ARC is proposed to consist of an electrometallurgical

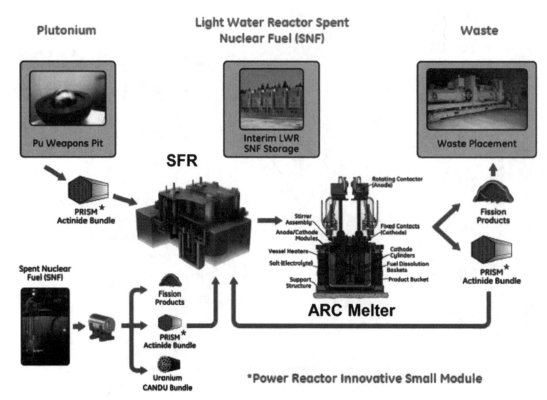

FIG. 23.16 SCHEMATIC OF THE ADVANCED RECYCLING CENTER ARC [62]

separations plant and three power blocks of 622 MWe each for a total of 1866 MWe [62].

Besides the well-known oxide fuel, other types like carbides, nitrides, or metallic fuel options are currently considered as options.

23.8.1.2 Other Fuel Cycles

23.8.1.2.1 Thorium Cycle As alternatives to uranium/plutonium based fuels, thorium fuel cycles are being explored to become independent from the uranium supply. India has envisioned robust thorium reactor technologies as a promising sustainable future energy resource for that country. Studies indicate that once the FBR capacity reaches about 200 GWe, thorium-based fuel can be introduced progressively in the FBRs to initiate the third stage of the program, where the U-233 bred in these reactors is to be used in the thorium-based reactors [35]. The proposed roadmap for the third stage therefore comprises thorium-based reactor technologies, incorporating the (Th-U-233) cycle. India is one of the leading countries in the world in thorium research and has gained that experience through thorium irradiation and the operation of U-233 fuelled research reactors.

23.8.1.2.2 Molten Salt A totally different type of fuel is used in MSRs. Here fuel and cooling are the same medium. Table 23.14 lists the different salts that are currently being considered.

Included are also molten salt options as primary coolant of the AHTR and as an intermediate coolant for the SFR.

TABLE 23.14 FUEL AND COOLANT SALTS FOR DIFFERENT APPLICATIONS (AN MEANS ACTINIDES) [21]

Reactor type	Neutron Spectrum	Application	Carrier Salt	Fuel System
MSR breeder	thermal	fuel	$^7LiF\text{-}BeF_2$	$^7Lif\text{-}BeF_2\text{-}ThF_4\text{-}UF_4$
	non-moderated	fuel	$^7LiF\text{-}ThF_4$	$^7LiF\text{-}ThF_4\text{-}UF_4$
				$^7LiF\text{-}ThF_4\text{-}PuF_3$
MSR breeder	thermal / non-moderated	secondary coolant	$NaF\text{-}NaBF_4$	
MSR burner	fast	fuel	$LiF\text{-}Naf$	$LiF\text{-}(NaF)\text{-}AnF_4\text{-}AnF_3$
			$LiF\text{-}(NaF)\text{-}BeF_2$	$LiF\text{-}(NaF)\text{-}BeF_2\text{-}AnF_4\text{-}AnF_3$
			$LiF\text{-}NaF\text{-}ThF_4$	
AHTR	thermal	primary coolant	$^7LiF\text{-}BeF_2$	
SFR		intermediate coolant	$NaNO_3\text{-}KNO_3\text{-}(NaNO_2)$	

TABLE 23.15 LIST OF SEVERAL STRUCTURAL MATERIALS CONSIDERED FOR ADVANCED NUCLEAR REACTOR SYSTEMS [63]

Alloy Class	Description
Ferritic-martensitic steels (e.g., HT9,T91, NF616, HCM12A, ODS steels)	Iron-based bcc steels, typically containing 2.25-12% wt% chromium for those used in various nuclear applications. These steels have high thermal conductivity, and good resistance to radiation induced void swelling, and low thermal expansion coefficients. They are limited by creep strength and maximum service temperature. Ferritic alloys with higher chromium contents have higher strength at high temperatures. Oxide-dispersion-strengthened (ODS) variants are designed to add strength while retaining other good properties.
Austenitic stainless steels (e.g., PNC316, Inconel 800H)	Iron-based fcc steels containing 16-25wt% chromium. These steels have higher strength and better corrosion resistance than ferritic-martensitic steels but are typically limited by void swelling resistance under irradiation and have some susceptibility to stress-corrosion cracking in water-cooled nuclear systems.
Nickel-based alloys (e.g., IN617, HA230, X-750, IN600, IN718, IN792LC)	Nickel-based alloys that have the greatest high-temperature strength, creep strength and corrosion resistance for alloys in the Fe-Cr-Ni system. These alloys have not typically been considered for in-core applications with exposure to significant neutron damage because of radiation embrittlement. These alloys have some susceptibility to stress-corrosion cracking in water-cooled nuclear systems.
Ceramics and ceramic composites (e.g., C, SiC, SiC/C, SiC/SiC)	Ceramics and ceramic composites are considered for applications at temperatures above those where metals can be used. Ceramic composites have improved ductility under tension as compared to pure ceramics.

23.8.2 Structural Materials

This subsection will concentrate on structural materials, including fuel cladding. The main structural components besides cladding are the reactor pressure vessel, core and core support materials, intermediate heat exchanger, steam generator and piping. There are many requirements for all of these nuclear reactor structural materials:

- The material must be available, affordable, and it must have good fabrication and joining properties.
- Cladding and ducts/piping must possess low neutron absorption.
- The materials must have good elevated temperature mechanical properties, including creep resistance, long-term stability, and compatibility with the reactor coolant.
- Materials used in a high-energy and high-intensity neutron field must be resistant to irradiation-induced property changes (radiation hardening and embrittlement, swelling, phase instabilities, creep, and helium-induced embrittlement).

A summary of the materials considered for Generation IV applications is given in Table 23.15 [63].

23.8.2.1 Pressure Boundaries Reactor pressure vessel steels and other pressure bearing components for LWRs are low alloy steels with good weldability and high fracture toughness. A disadvantage of this well-established class of materials is the inability to develop a creep resistant microstructure. For hot vessel options, alternative materials have to be considered. They must have more elements for solid solution strengthening and precipitate forming.

Figure 23.17 shows a comparison of the allowable stresses for 1000 hours [64]. This means that for higher temperatures, the 1000-hour stress rupture data was used. The steep drop of grade A 508 (which is the conventional low alloy RPV steel) that is a result of missing creep properties can be seen. For more than 1000 hours, the effect becomes much more pronounced. The mod 9Cr-1Mo

steel is certainly the best-suited material (as far as high-temperature strength is concerned) but material uniformity through thick sections and welding issues are concerns. This steel belongs to the class of martensitic 9-12% Cr. The development and the properties of these steels are described in [65]. Generally, the microstructures of the 9% to and 12% Cr steels are designed by balancing austenite and ferrite stabilizers to produce 100% austenite during austenitization and 100% martensite during a normalizing (air-cooling) or quenching treatment following austenitization.

The superior high temperature and corrosion behavior together with the relatively moderate costs made this class of materials attractive for several high-temperature applications. For martensitic steels developed in 1970–1985 (to which grade 91 steel belongs), carbon, niobium, and vanadium were optimized, nitrogen (0.03%-

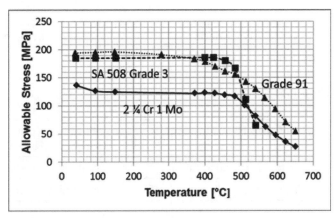

FIG. 23.17 ALLOWABLE STRESS VALUES FOR REPRESENTATIVE RPV-STEELS (1000 HRS STRESS RUPTURE DATA) [64]

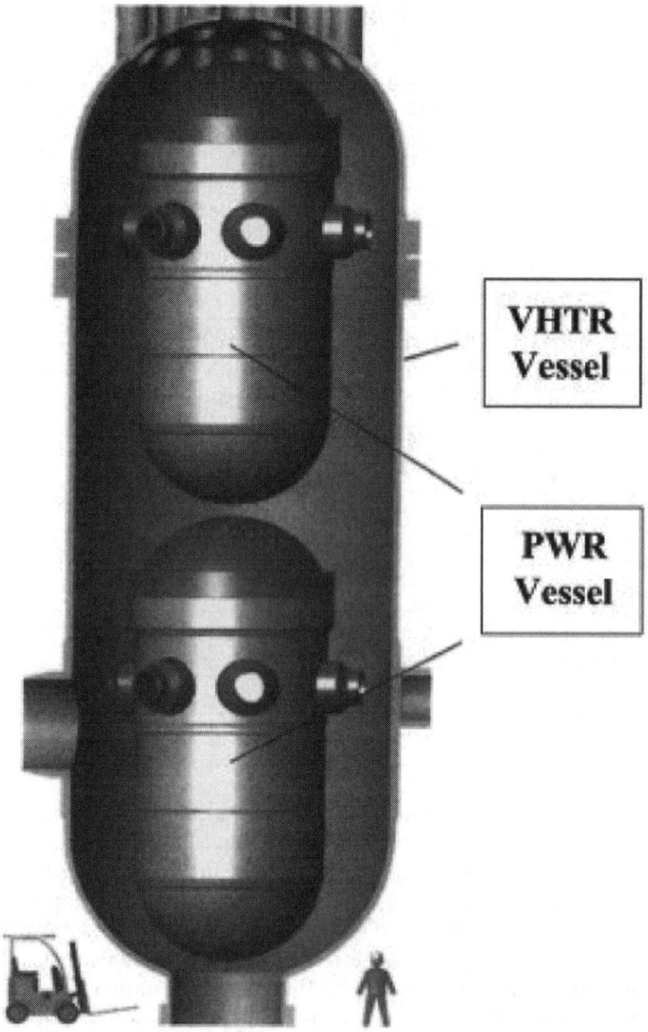

FIG. 23.18 DIMENSIONS OF A VHTR PRESSURE VESSEL COMPARED WITH A PWR VESSEL [66]

vessel applications. Negligible creep, creep-fatigue interactions, cyclic softening, and the necessity to produce large welded vessels for more than 60 years design life need still more information about this materials than currently available. The less creep resistant 2¼ Cr-1Mo is sometimes considered as an alternative. Figure 23.18 gives an impression about the dimensions of a RPV for a VHTR [66] resulting from the low power density of the core, which is a major contributor to the passive safety feature of these reactors. Castability, forgability, and weldability are the most important material properties with respect to manufacturing.

Creep resistant materials with good corrosion resistance and superior high-temperature strength are also required for other advanced nuclear plant components.

Austenitic steels and super alloys (iron-based and nickel-based) belong to this class. They have been significantly improved over the last few decades similar to the ferritic-martensitic materials. They are of high importance for several components that operate in corrosive environments at elevated temperatures. They have some problems to accommodate higher irradiation levels as shown in the next section. Nevertheless, they are prime candidates for high-temperature applications outside the irradiation environment, like the intermediate heat exchanger for a VHTR, steam generator, or turbo-machinery.

23.8.2.2 Cladding Materials Cladding materials in advanced reactors are exposed to an extremely challenging environment including neutron irradiation, corrosion, and fuel interaction combined with significant stresses. The specific conditions in advanced fast reactors can subject cladding and duct/piping to doses of greater than 200 dpa within 5-7 years of operation. Such doses combined with other environmental factors cause degradation of material properties. Increase of ductile to brittle transition temperature (DBTT), helium embrittlement, swelling, thermal creep, irradiation creep and environmental damage by coolant-clad or fuel-clad interactions are predominant damage mechanisms posing very challenging demands on cladding materials. A summary of irradiation-induced damage mechanisms is shown in Table 23.16 [67].

Austenitic materials (iron-based and nickel-based) would be candidates for high corrosion and high thermal creep resistance. However, they are extremely susceptible to irradiation-induced damage [68]. Figure 23.19 [69] shows a comparison of the swelling behavior of materials with austenitic matrix and materials with ferritic-martensitic matrix. Results from ferritic-martensitic oxide

0.05%) was added, and the maximum operating temperature was increased to 593°C. Although these steels have been extensively in use in the power-generation industry throughout the world, they are still considered only as potential candidates for Generation IV

TABLE 23.16   IRRADIATION-INDUCED TYPES OF DAMAGE [67]

Effect	Consequence in Material	Type of Degradation in Component
Displacement damage	formation of point defect clusters and dislocation loops	hardening, embrittlement
Radiation-induced segregation	diffusion of detrimental elements to grain boundaries	embrittlement, grain boundary cracking
Radiation-induced phase transitions	formation of phases not expected according to phase diagram, phase dissolution	embrittlement, softening
Helium formation and diffusion	void formation (inter- and intra-crystalline)	embrittlement, creep type damage
Irradiation creep	irreversible deformation	deformation, reduction of creep life
Swelling	volume increase due to defect clusters and voids	local deformation, eventually residual stresses
Irradiation-induced stress corrosion cracking	grain boundary effects	enhanced stress corrosion cracking

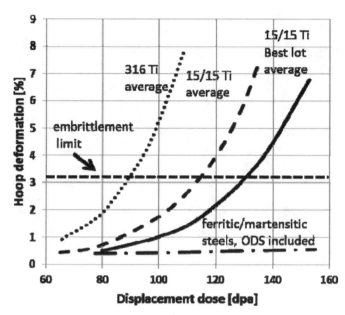

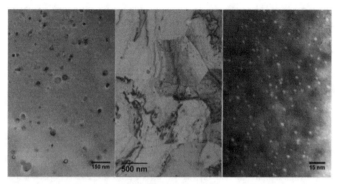

FIGURE 23.20 MICROSTRUCTURE OF DIFFERENT FERRITIC ODS MATERIALS [14]. A) COMMERCIAL ALLOY PM2000. B) COMMERCIAL ALLOY PM2000 AFTER SEVERE PLASTIC DEFORMATION TO PRODUCE NANO-GRAINS (MATERIAL FROM G. KORB [72]). C) ADVANCED FERRITIC 19% CR ODS ALLOY (JAPANESE DEVELOPMENT [73])

FIG. 23.19 SWELLING AND EMBRITTLEMENT BEHAVIOR OF DIFFERENT CLADDING MATERIALS RE-PLOTTED [69]

dispersion strengthened materials (ODS) are also included. This class will be described later.

Besides irradiation-induced damage, helium related damage can occur for fast spectra and particularly in fusion plants. Helium is a result of trans-mutant reactions. Helium diffuses to all kinds of sinks: point defects, dislocations, and grain boundaries. Depending on its concentration, it can form intra granular as well as inter granular clusters and bubbles. Helium bubbles at the grain boundaries considerably reduce toughness of a material.

Increasing the creep strength and decreasing the tendency for formation of helium bubbles at the grain boundaries is a challenge for material design. One possible solution is the introduction of very fine non-metallic inclusions, so called dispersoids, in the metallic matrix. This class of materials is called oxide dispersion strengthened (ODS). Dispersoids can be introduced into austenitic, ferritic, and martensitic matrices. The idea to improve the creep properties of alloys by the introduction of ceramic particles is not new. Dispersion strengthened nickel-base alloys (e.g., MA6000) were seriously considered and already researched as material for un-cooled gas-turbine vanes in the 1980s [70]. At Asea Brown Boveri, a second stage vane was even put into operation in an experimental land-based gas turbine. It consisted of an ODS-blade that was brazed into a precision cast top and root pieces [71]. Shaping of the ODS-blade, and finally costs, were the major reasons why this development never went into production. At the same time, interest in ferritic and ferritic-martensitic ODS alloys emerged from nuclear fusion research, since the ferritic matrix showed both swelling and irradiation creep properties superior to that found in an austenitic matrix. Nickel-based alloys could also not be used in a nuclear fusion environment due to their tendency for helium formation and embrittlement. Satisfactory qualities of ODS alloys could only be achieved by powder metallurgical techniques. In a first step, an alloy powder and dispersoids (usually yttrium oxide Y_2O_3) are mechanically alloyed. This step leads to finely distributed ceramic dispersoids in the powder mass. The homogenous milled powder product is consolidated by hot isostatic pressing and/or hot extrusion. Finally, the material undergoes a heat treatment to get opti-

mum properties. A typical representative for a ferritic ODS alloy is the commercial alloy PM2000 (Plansee), which is, however, no longer produced (due to weak market demand). It contains Y_2O_3 particles with an average diameter of about 25 nanometer (nm) [Figure 23.20 (a)]. Within the European Project EXTREMAT [72], nano-grained PM2000 was produced by severe plastic deformation (SPD). This process had, however, no influence on the dispersoid size.

A typical microstructure resulting from the SPD-process is shown in Figure 23.20 (b). In recent years, advances in understanding the mechanical alloying process have resulted in the development of the advanced ODS ferritic alloy, known as 14YWT nano-structured ferritic alloy [74], which contains a high number density of O-, Ti-, and Y-enriched clusters, or nano-clusters with sizes of ~2–5 nm. These nano-clusters possess an unusually high degree of thermal stability and are primarily responsible for the excellent combination of mechanical properties of the nanostructured ferritic alloys at room and elevated temperatures.

Helium is also attracted by dispersoids (Figure 23.21), which would be a key advantage of ODS steels in nuclear applications. The combination of a high number density of nano-clusters and nano-size grains typical of the 14YWT NFA may improve its tolerance to neutron irradiation damage by providing efficient sinks for trapping point defects and transmutation products such as helium. This represents a promising direction for developing materials for applications in advanced nuclear energy systems. Ferritic steels with 19% Cr could also be produced with nano-sized dispersoids. Figure 23.20 (c) shows the TEM-micrograph of the Japanese development 19 Cr ODS [73,76]. The dispersoid size is about one order of magnitude lower than that of alloy PM2000.

Manufacturing of fuel pins made of advanced ODS alloys has already reached a stage where experimental claddings were produced for reactor irradiation or fuel pins were produced for actual reactor use [77]. Figure 23.22 shows pins fabricated to be tested in the BOR-60 reactor. First results were reported in [78] and they can be summarized as follows. The fuel pin irradiation test of the ODS claddings was carried out at high cladding temperatures of 700°C and 650°C in BOR-60. Vibro-packed MOX fuels with low oxide to metal (O/M) ratio were used. The integrity of the ODS cladding fuel pins was maintained up to the burnup of 5.1 at% and the neutron dose of 21 dpa, and the fuel cladding chemical

FIG. 23.21 MIGRATION OF HELIUM TO OXIDE DISPERSOIDS IN THE FERRITIC ODS ALLOY PM2000. LARGE BUBBLES ARE FORMED AROUND DISPERSOIDS (3), INTERMEDIATE SIZE BUBBLES ARE EITHER IN THE MATRIX OR ALONG DISLOCATIONS (2), AND SMALL BUBBLES ARE LOCATED AT LOOPS (1). DISLOCATIONS AND LOOPS ARE NOT VISIBLE UNDER THESE CONTRAST CONDITIONS. (PLOTTED FROM CHEN ET AL [75])

FIG. 23.23 GRAPHITE CORE ELEMENTS FOR A PEBBLE BED REACTOR [80]

interaction (FCCI) of the ODS claddings was very small. These results suggest that the ODS claddings and low O/M ratio fuels are the most appropriate fuel pin systems for high-burnup fuel. The irradiation of ODS cladding fuel pins continues in BOR-60 to a burnup of 15 at%, and post-irradiation data will be acquired.

ODS materials are basically under consideration also for other structural applications like high-temperature reactor internals.

23.8.2.3 Non-metallic Materials The most important non-metallic component is the graphite used for core construction of the high-temperature reactor. This includes the permanent inside and outside reflectors, the core blocks, and the core supports. Such cores have been built for earlier HTRs, but insufficient information about the old grades of graphite hamper usage today (and the old grades may no longer be available) and improvements were found to be necessary. New graphite grades that are anticipated to show improved performance under VHTR in service conditions are

being procured. New fine-grained isotropic graphite types with high strength and low irradiation damage are required to achieve high outlet-gas temperature, long life, and continuity of supply. Dimensional changes and creep under irradiation are the most important design parameters that are currently being established for different grades of graphite. For a more detailed discussion of graphite in a VHTR environment, refer to the literature [79]. Graphite components for a pebble bed core and for a prismatic core are shown in Figures 23.23 and 23.24.

Other components where the temperature requirements of gas-cooled reactors (VHTR and GFR) cannot be accommodated with metallic materials are GFR cladding and parts of a VHTR control rod.

Another non-metallic material of interest is ceramics. However, the main problem with solid ceramics is the very low ductility. To overcome this problem, ceramics are usually "toughened" by making compounds of ceramic fibers and the ceramic matrix. The production procedure consists in principle in a woven (fibers) preform, which is infiltrated by the precursor of the matrix (often a resin). This assembly is then treated at very high temperatures for graphitization or carbonization.

There are different qualities of such materials available that were produced with different methods. Sometimes a porous matrix remains, which is not necessarily an obstacle for application. Ma-

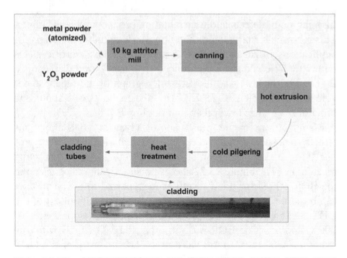

FIG. 23.22 PROTOTYPES OF ODS FUEL PINS FOR SFR REACTORS [14]

FIG. 23.24 GRAPHITE FUEL ELEMENT BLOCK FOR A PRISMATIC VHTR CORE [81]

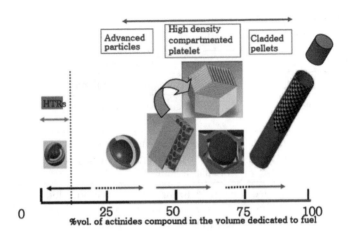

FIG. 23.25 FUEL ELEMENT OPTIONS FOR A GFR. THE CLADDING MATERIAL PROPOSED IS A SICF/SIC COMPOUND [82]

terials of this type exist already (SiCf/SiC, Cf/SiC, Cf/C where the f stands for fiber) in aerospace applications. They are very expensive and they can undergo similar damage under irradiation as described for graphite. Such composites were proposed as cladding base-material for the GFR (Figure 23.25) [82]. Particular attention must be paid to gas-tightness of these materials in case of GFR cladding where the fission gas that has to remain contained in the cladding provides an internal pressure. It is very difficult to guarantee the necessary tightness without long-term experience. This is the reason why currently for claddings, more complex assemblies consisting of fiber reinforced base material with a liner are being studied for the GFR.

23.8.2.4 Influence of Environment The cooling media used in the different advanced reactors provide an additional challenge for the materials used. Under operating conditions, they can deteriorate the surface of components, which can lead to a loss of the load carrying cross section or to local damage like grain boundary attacks or pitting. From these defects, cracks can grow that finally leads to component failure. It is not possible to cover corrosion issues in detail in this chapter. This discussion will confine the considerations to only a few key aspects.

Liquid metals may selectively deplete constituents from alloys, and impurities may chemically react with them. Chemistry control will be required for all coolants. The primary issue with sodium is its chemical reactivity in air and water. Experience with existing SFRs have shown that there is only a moderate corrosion problem expected. In the case of lead, the temperature becomes important. Between 400°C and 500°C, a compact stable oxide is formed on austenitic and ferritic-martensitic steels. The metal-oxide layer becomes unstable above 550°C. Therefore, protective coatings (e.g., FeAl) are envisioned. For an assessment of the long-term behavior, the interactions of irradiation, corrosion, and erosion must be better understood. The effects of embrittlement and cyclic thermal/mechanical loading on the design life of components must be studied [83].

In case of gas-cooled reactors (particularly HTRs), the impurities in the chemically inert helium determine the corrosion behavior. Oxidation, carburization, and de-carburization can occur depending on the temperature.

The key variables affecting corrosion identified for SCWRs are temperature, water density (pressure), dissolved oxygen con-

centration, water conductivity, and surface finish. Corrosion data under SCWR conditions shows that the oxidation rate of steels, especially ferritic-martensitic (F/M) steels, is rather high, increasing rapidly above 500°C. In addition to general corrosion, stress corrosion cracking (SCC) (inter-granular and trans-granular) is expected to be a critical degradation mode in an SCWR [84].

Molten salt can also affect the surface of (unprotected) structural materials. Nickel-alloys (Ni-W-Cr) are expected to have the required stability. However, details of chemical composition and long-term behavior still need to be investigated [85].

23.8.2.5 Materials Modeling After the introduction of quantitative descriptions of creep and creep-damage mechanisms in metals in the middle of the last century (e.g., collective work by Norton, Monkman-Grant, etc. [86,87]), it took about 25 years to develop a working engineering understanding of creep-fatigue interactions (e.g., collective work by Manson, Coffin, Mowbray, etc. [88]). The introduction of damage mechanics in terms of subcritical crack growth and the introduction of constitutive laws for creep-fatigue interactions (e.g., Chaboche) was a further improvement in lifetime assessments of structures. With the current availability of large computer clusters operating in parallel mode, numerical solutions of equations for atomistic behavior became very attractive. Although it is well accepted that damage starts at atomistic levels, it is not easy to bridge the gap between atomic and structure levels and requires an understanding of the related physical phenomena on a range of scales from the microscopic level all the way up to macroscopic effects.

Determination of the lifetime of components exposed to severe environments such as in VHTRs is very demanding, particularly when damage interactions (like creep-irradiation or strength-microstructure, toughness-irradiation induced phases) must be considered. The simulation of materials behavior under such extreme conditions needs to encompass broad time (nanoseconds to 60 years) and length (nanometers to meters) scales from atomistic descriptions of primary damage formation to a description of bulk property behavior at the continuum limit. This requires a multi-scale, multi-code modeling approach that begins at the atomistic level with ab initio and molecular dynamics techniques, moving through the meso-scale using reaction rate theory models, lattice kinetic Monte-Carlo and dislocation dynamics, and ends with the macro-scale using finite element methods and continuum models. Experimental validation of the modeling results is mandatory. Materials modeling is currently part of many national and international research projects. For further details see [89,90].

23.9 DESIGN CONSIDERATIONS

Finally, components operating safe and reliably for the anticipated design life (60 years for some components) have to be manufactured. These components undergo different types of damage during their lifetime, as listed in Table 23.17.

It can be seen that damage occurs on different scales in a material. The micro-scale where micro- and nano-structure play the most important role. Accumulation of these effects on the micro-scale leads to response on the macro-scale, which finally manifests itself in component failure. One of the big challenges is to predict or to monitor the development of damage under the very demanding and complex conditions in an advanced nuclear plant. Currently many attempts to improve lifetime prediction and lifetime assessments are made. Besides programs for the establishment

**TABLE 23.17 DAMAGE EVENTS OCCURRING ON
DIFFERENT SCALES [91]**

Exposure	Microscale	Macroscale
Temperature	phase reactions, segregations	hardening/softening, embrittlement
Irradiation	displacement damage, phase reactions, segregations, helium damage	hardening, embrittlement, swelling
Environment	surface layer, local attack (pitting), grain boundary attack, formation of local stress raisers	reduction of carrying cross section, subcritical crack growth, unexpected premature failure
Impact and static load	dislocation movement, diffusion controlled dislocation and grain boundary processes,	plastic deformation, creep deformation, buckling, plastic collapse, sub-critical and critical crack growth, unexpected premature (catastrophic) failure
Cyclic Load	dislocation movement, local micro-crack formation, intrusions/extrusions	cyclic softening, ratcheting, subcritical crack growth, premature failure
Combined exposures: creep-fatigue, irradiation creep, corrosion fatigue, stress corrosion cracking	(synergistic) damage accumulation	(synergistic) damage accumulation, unexpected damage, premature failure

of materials data for the relevant materials under relevant conditions, many activities exist to use advanced materials modeling tools like ab initio, molecular dynamics, kinetic Monte Carlo, dislocation dynamics together with constitutive models, fracture mechanics and other methods to get an improved understanding of materials damage as mentioned before (see [91]). Although very good progress has been achieved, it is clear that enormous efforts are still necessary to extend these techniques to real-life component assessments.

Structural components have the additional requirement of passing licensing requirements. As load-bearing elements that maintain the integrity of the reactor core, structural materials face considerable scrutiny. While creep, fatigue, corrosion, and irradiation all present a challenge for nuclear materials, they also pose a significant challenge for designers. A key requirement in the design process of any power system is the use of proven, conservative design criteria. high-temperature structural design methodology uses inputs of material properties (tensile, creep, fatigue, compression, toughness, etc.), but also mechanical models covering multi-axiality, creep-fatigue interactions or negligible creep criteria are required.

Design is limited by design rules given in design codes like the ASME B&PV code. These codes are challenged by the requirements of advanced nuclear plants. New materials have to be considered, viable temperature limits established, and the design procedure as a whole need to be reconsidered. It has to be mentioned that irradiation and corrosion are typically not part of most design codes and discussions about this issue are ongoing. The following tasks were performed to improve ASME B&PV Code, Section III rules with respect to the needs of constructing a VHTR. These tasks were funded by the US DOE with the exception of Task 12, which was funded by the US NRC.

- Task 1: A Review of Structural Material Properties for Both Base Metal and Weldments for Alloy 800H and Grade 91 Steel

- Task 2: Regulatory Safety Issues in the Structural Design Criteria of ASME Section III Subsection NH and for Very High Temperatures for VHTR & GEN IV
- Task 3: Improvement of ASME NH Rules for Grade 91 (negligible creep and creep-fatigue)
- Task 4: Updating of ASME Nuclear Code Case N-201 to Accommodate the Needs of Metallic Core Support Structures for High-Temperature Gas Cooled Reactors Currently in Development
- Task 5: Collect Available Creep-Fatigue Data and Study Existing Creep-Fatigue Evaluation Procedures for Grade 91 and Hastelloy XR
- Task 6: Review and Assessment of Operating Condition Allowable Stress Values and Recommended Corrective Actions
- Task 7: Review of Current Intermediate Heat Exchanger Experience and Recommended Code Approach for Intermediate Heat Exchanger
- Task 8: Creep and Creep-Fatigue Crack Growth at Structural Discontinuities and Welds
- Task 9: Update and Improve Subsection NH — Simplified Elastic and Inelastic Design Analysis Methods
- Task 10: Update and Improve Subsection NH — Alternative Simplified Creep-Fatigue Design Methods
- Task 11: Selection of Candidate Materials for Incorporation Into ASME Subsection NH
- Task 12: NDE and ISI Technologies for HTRs

The main portion of these tasks is centered around creep and creep-fatigue. This task list describes work that has been performed between 2006 and 2010. This work is going on with the aim to deepen findings of the first period and to convert these results into new code rules. This has led to the situation that new code rules for high-temperature gas reactors and for liquid metal reactors are currently under development and approval. It is the final aim to develop a separate ASME Section III, Division 5 code containing rules for high-temperature gas-cooled reactors (HTGRs) and for liquid metal reactors (LMRs). This code should also contain guidance with respect to corrosion and irradiation damage. Advanced bonding techniques and surface coatings should also be considered.

23.10 NDE (NON-DESTRUCTIVE EXAMINATION)

Tasks 8 and 12 point directly to the need for advanced condition monitoring and non destructive evaluation. Compared with current LWRs, monitoring of the condition of an advanced plant is much more demanding. Coarse-grained materials, different types of damage, missing long-term experience, and visibility are only some key issues to be addressed.

Reliability is an essential factor for the SFR. It is important to design and construct plants with highly reliable reactor components, taking into account manageable maintenance, easy repair, and inspection capability. Simple configurations of components such as a single-piece forged core support structure for compact reactor vessels, shortening lengths of piping, etc., are developed from the viewpoint of inspection accessibility and reducing inspection locations. Reduction of high stress regions and welding lines are also considered for component structures to reduce inspection workloads.

One of the innovative technologies for SFRs is to develop advanced non destructive examination (NDE) technology. Inspection for the core support structures is considered a key point for reliability. As sodium is optically opaque, conventional inspection techniques are not easy to apply for core structures immersed in or under a sodium environment. An inspection technique using an under-sodium viewer (USV) is being developed to visualize the components using ultrasonic waves [92]. The USV system consists of a matrix arrayed transducer, which has a large number of small piezoelectric elements to transmit and receive ultrasonic echoes without a mechanical scanning device. The transducer has a signal processing device using the synthetic aperture focusing technique to synthesize images of the inspection targets with high resolution. Tests under sodium confirmed that the target could be visualized clearly within 2 mm resolution [93].

Development of an under-sodium area monitor (USAM) is also in progress to reduce the size and weight of the transducer with higher resolution. The monitor system is based on the same principles of the above USV system. The USAM uses optical diaphragms as the ultrasonic wave receiver instead of piezoelectric elements. A simulated flaw width of 0.2 mm was successfully detected in an underwater imaging test.

To apply the USAM system to core support structures, a transport system is required to deliver the USAM sensor to a target location under a sodium environment. An under-sodium vehicle driven by electromagnetic pumps is also being developed. Experimental fabrication of a prototype vehicle and a performance test in a static water pool have confirmed that the vehicle has enough speed and stability.

Further developments of non destructive examination technologies under a sodium environment are ongoing to confirm the applicability of the systems, utilizing underwater and under-sodium testing facilities.

The fact that other types of damage are expected to occur in advanced reactors led to a re-consideration of surveillance samples.

Modern advanced analysis, testing, and modeling techniques would also allow taking samples directly from highly exposed components for further examination and residual life assessments. Such techniques are: advanced transmission electron microscopes, X-ray and neutron beamline techniques, nano-indentation or mechanical testing of extremely small (typically 1-2 micrometers) samples produced by focused ion beam techniques [94]. Modeling for deeper understanding of materials behavior can be done on different scales ranging from atomistic scale to macro-scale. These methods are currently under consideration and under development in different research projects.

23.11 CROSSCUTTING GIF ISSUES

A few crosscutting tasks were defined within GIF that are relevant for several of the reactors. Only two of them will be briefly mentioned: the working groups on proliferation risk and physical protection (PR&PP) and the risk and safety working group (RSWG).

PR&PP has been defined as follows:

- Proliferation resistance is that characteristic of a nuclear energy system that impedes the diversion or undeclared production of nuclear material or misuse of technology by the Host State seeking to acquire nuclear weapons or other nuclear explosive devices.
- Physical protection (robustness) is that characteristic of a nuclear energy system that impedes the theft of materials suitable for nuclear explosives or radiation dispersal devices (RDDs) and the sabotage of facilities and transportation by sub-national entities and other non-Host State adversaries.

In addition to the establishment of the PR&PP Working Group, the GIF has recognized the need for a Risk and Safety Working Group to address the approach to be adopted for the safety of future nuclear energy systems. The GIF also recognized that an interface with the activities of the PR&PP Working Group would be needed, and thus noted:

- A need for integrated consideration of safety, reliability, proliferation resistance, and physical protection approaches in order to optimize their effects and minimize potential conflicts between approaches.
- A need for mutual understanding of safety priorities and their implementation in PR&PP and RSWG evaluation methodologies.

The efforts of these two groups continue to be carefully coordinated. This has been largely accomplished so far via the close working relations between the leaderships of the two groups. Advances by either group have relevance to the other and are mutually beneficial to both. It also continues to be important to assess and understand the impact of all specific design features in relation to objectives of safety performance, physical protection, and proliferation resistance.

23.12 INTERNATIONAL R&D

The GIF provides a forum for multi-lateral R&D cooperation. Its member countries meet regularly to discuss the research required to support the development of next-generation reactors. According to Bouchard [2], it has resulted in a tremendous brainstorming effort, from R&D teams from over twelve countries and EURATOM, on a scale rarely matched in history, which, in turn, produced numerous collaborative projects in reactor and fuel technologies. Even if the six nuclear systems studied within the GIF correspond to concepts already known, their development within the GIF has benefited from the exchanges between technical experts representing most of the main world nuclear actors, originating from both academia as well as industry sectors.

Though technical exchanges started earlier, sometimes on a bi-lateral level, multi-lateral cooperation emerged to principles, accepted by all, which duly recognize background property information and dealt satisfactorily with all property rights (intellectual, commercial, etc.). The GIF thus appears as the only existing large scale international structure enabling multi-lateral cooperation within a sound legal basis that ensures that its R&D activities are carried out in an equitable manner between partners.

The R&D performed in GIF focuses on both the viability and performance phases of system development. The former phase examines the feasibility of key technologies, such as adequate corrosion resistance in lead alloys or super critical water, fission product retention at high temperatures for particle fuel in the very-high-temperature gas-cooled reactor. The latter phase focuses on the development of performance data and optimization of the system.

The scope of GIF activities does not extend to any demonstration phase, which involves the detailed design, licensing, construction, and operation of a prototype or demonstration system in partnership with industry. To help prepare for future commercialization of Generation IV systems, a Senior Industry Advisory Panel (SIAP) provides advice on GIF R&D priorities and strategies. Specifically, this panel contributes to discussion on strategic review of R&D progress and plans for the GIF systems from the industry perspective, by addressing issues such as industrial interest, technical viability, economics, licensing, risk management, project management, and industrial infrastructure. The SIAP contributes valuable views on system deployment, future nuclear fuel cycles, and international frameworks for nuclear safety standards and regulations.

But it is not only the GIF that promotes and performs international R&D collaborations. Much work is also done within IAEA or within the EU or other countries. At the inception of the NGNP Project, experts from DOE national laboratories, gas reactor vendors, and universities collaborated to establish technology R&D plans to help guide NGNP R&D. These plans [95 thru 98] outlined the testing and computational development activities needed to qualify the materials and to validate the modeling and simulation tools to be used in the design and operation of the NGNP. The technology R&D plans drew on worldwide experience gained from the six demonstrations and/or prototype HTGRs that were built and operated over the past 60 years. The plans included detailed descriptions of the required technical activities with associated schedules and budgets for project completion, which formed the baseline for execution of the R&D needed for the NGNP project. The R&D activities are organized into the following four major technical areas: (1) fuel development and qualification, (2) graphite qualification, (3) high-temperature materials qualification, and (4) design and safety methods validation. To accomplish these objectives, the R&D program draws upon expertise at DOE national laboratories (including Idaho National Laboratory and ORNL) and a broad array of universities, along with international facilities and expertise accessible to DOE via the GIF.

Countries like India, with strong nuclear R&D programs, are also part of international R&D networks. India has a flourishing and largely indigenous nuclear power program. Because India is outside the Nuclear Non-Proliferation Treaty due to its weapons program, it has been for over three decades largely excluded from trade in nuclear plant or materials. This forced it into nuclear energy self-sufficiency extending from uranium exploration and mining through fuel fabrication, heavy water production, reactor design and construction, to reprocessing and waste management. Now, foreign technology and fuel are expected to boost India's nuclear power plans considerably. All plants will have high indigenous engineering content. In the context of India's nuclear trade isolation over three decades, L&T, India's largest engineering group, has produced heavy components for PHWRs and has also secured contracts for 80% of the components for the FBR at Kalpakkam. It is qualified by the American Society of Mechanical Engineers to fabricate nuclear-grade pressure vessels and core support structures, achieving this internationally recognized quality standard in 2007. L&T plans to produce 600-ton ingots in its steel melt shop and have a very large press to supply finished forgings for nuclear reactors, pressurizers and steam generators. Other nuclear equipment manufacturers in India include BHEL, Bharat Forge, HCC, etc.

The AERB of India, formed in 1983, is responsible for the regulation and licensing of all nuclear facilities and their safety. To fulfill its mission of stipulating and enforcing rules and regulations concerned with nuclear and radiological safety, AERB is in the process of developing codes and guides for various aspects of nuclear reactor design. Reference 6 provides some details of the design and construction codes followed in the Indian nuclear program. Design of Indian PHWRs is based on the ASME B&PV Code, Section III. In those cases where the code rules could not be precisely met, they have met the intent by carrying out the required material characterization and component qualification tests. The in service inspection program follows a combined philosophy of the ASME B&PV Code, Section XI and the Canadian Code CSA/N281.5-05. Design of concrete containments are based on the French Code RCC-G. To avoid a mixture of codes, AERB is in the process of writing its own codes. It has already published 5 Safety Codes, 45 Safety Guides covering various aspects of nuclear power plants, and 51 other safety support documents. Additional safety documents are in various stages of preparation.

23.13 OUTLOOK

An overview of advanced options for the generation of nuclear power was given. The considerations were focused on power plants expected to become commercially available in 10 to 30 years from now. Concerns about uranium resources and long living constituents in LWR waste (plutonium, minor actinides) support fast reactor concepts. In addition, nuclear power is also considered as a low CO_2 emission, thermal heat source for industrial processes like hydrogen production or refinery processes. Currently, the most mature advanced reactors are the SFR and the high-temperature gas-cooled reactor (VHTR). The high-temperature gas-cooled reactor has the advantage that it produces high-temperature heat that can not only be used for electricity production but also for process heat in applications like synthetic fuel, hydrogen production, steel and other industries. Better fuel economy can certainly be reached with fast reactors, like the SFR. With respect to combined cycles, they are less flexible. There are also economic challenges still to be overcome. Projects for both technologies exist in different countries and the technical obstacles appear to be manageable for currently envisioned operating conditions. Challenges still exist in the field of fuel and fuel cycle optimization as well of structural materials. Fuel claddings are of particular importance in order to reach the ambitious goals of thermal efficiency and burnups. Increasing the gas outlet temperature for the next-generation VHTRs to 950°C will also require more advanced materials. Another big challenge is the transfer of existing concepts like the GFR, MSR, and SCWR into prototypes.

23.14 CONCLUSION

After reading this chapter, one cannot be immune from the excitement that all of the people involved in this great adventure are experiencing around this planet of ours as advanced Generation IV reactors begin their journey from the design boards to reality. This chapter reveals just how rapid the next generation of nuclear reactors are proceeding. To support this effort, ASME is currently developing new Codes and Standards for these types of reactors, as are other societies and other nations. People worldwide are diligently working now for the future benefit of all.

23.15 ABBREVIATIONS

ABWR: advanced boiling water reactor
AECL: atomic Energy of Canada Ltd
AERB: atomic Energy Regulatory Board
AHTR: advanced high-temperature reactor
AHWR: advanced heavy water reactor
AISI: American Iron and Steel Institute
ALLEGRO: prototype French gas-cooled reactor (GFR)
ALMR: advanced liquid metal reactor
APWR: advanced pressurized water reactor
ARC: Advanced Recycle Center
ASME: American Society of Mechanical Engineers
ASTRID: French SFR prototype reactor
at%: atomic percent
BHAVINI: Bhartiya Vidyut Nigam
BHEL: Bharat Heavy Electricals Ltd./India
BREST: Russian lead fast reactor
CANDU: Canadian deuterium uranium reactor
CEA: French Atomic Energy Commission
CEFR: China experimental fast reactor
CV: containment vessel
DBTT: ductile to brittle transition temperature
DCD: design control document
DFBR: demonstration fast breeder reactor
DOE: US Department of Energy
ELSY: European lead-cooled system
EPR: European pressurized water reactor
ESBWR: Economic simplified boiling water reactor
EURATOM: European Atomic Energy Community
EXTREMAT: materials for extreme conditions (EU-FW6-project)
FaCT: fast reactor cycle technology development
FBTR: fast breeder test reactor
FBR: fast breeder reactor
FCCI: fuel-cladding chemical interaction
F/M: ferritic-martensitic
FP: fission product
GACID: global actinide cycle international demonstration
GANEX: group actinides extraction
GEN IV: Generation IV

GFR: helium gas-cooled reactor, gas-cooled fast reactor
GIF: Generation IV International Forum
GNEP: Global Nuclear Energy Partnership
GT-MHR: gas turbine modular helium reactor
GWe: electric power in gigawatts
h.a.: fraction of heavy metals being fissioned
HCC: Indian nuclear power engineering company
HPLWR: high-performance light water reactor
HTGR: high-temperature gas-cooled nuclear reactor
HTR: high-temperature reactor (gas-cooled)
HTR-PM: Temperature Gas-Cooled Reactor–Pebble Bed Module (China)
HTTR: high-temperature Engineering Test Reactor
IHTS: intermediate heat exchanger system
IHX : intermediate heat exchanger
IAEA: International Atomic Energy Agency
IFNEC: International Framework for Nuclear Energy Cooperation (has replaced designation "GNEP" since 2010)
IGCAR: Indira Gandhi Center for Atomic Research
INPRO: International Project on Innovative Nuclear Reactors and Fuel Cycles of IAEA
IRIS: International Reactor Innovative and Secure
I-S: iodine sulfur process for hydrogen production
ISI: in service inspection
JAEA: Japan Atomic Energy Agency
JSFR: Japanese sodium cooled fast reactor
KAERI: Korean Atomic Energy Research Institute
LFR: lead fast reactor
LMR: lead or lead-bismuth cooled liquid reactor, liquid metal reactor
L&T: Larsen & Tubros
LWR: light water reactor
MA: minor actinide
MN: mixed uranium-plutonium nitride
MOX: mixed uranium-plutonium oxide
MSBR: molten salt breeder reactor
MSFR: molten salt fast neutron reactor
MSR: molten salt reactor
NDE: non destructive examination
NGNP: Next Generation Nuclear Plant
NHDD: Nuclear Hydrogen Demonstration Project of KAERI
NRC: US Nuclear Regulatory Commission
ODS: oxide dispersion strengthened
O/M: oxide to metal
ORNL: Oak Ridge National Laboratory
PB-AHTR: pebble bed advanced high-temperature reactor
PBMR: pebble bed modular reactor
PDRC: passive decay heat removal circuit
PFBR: prototype fast breeder reactor
PRISM: Power Reactor Innovative Small Module
PR&PP: proliferation risk and physical protection
PWR: pressurized water reactor
PyC: pyrolytic carbon
RBMK: Reactor Bolshoy Moschchnosty Kanalny
R&D: research and development
RDD: radiation dispersal device
RPV: reactor pressure vessel
RSWG: risk and safety working group
SCC: stress corrosion cracking
SCWR: super critical water reactor
SET: European strategic energy technology
SFR: sodium (cooled) fast reactor

SIAP: Senior Industry Advisory Panel
SMR: small modular reactors
SNF: spent nuclear fuel
SPD: severe plastic deformation
SSTAR: small secure transporable autonomous reactor
SVBR: Russian modular lead–bismuth fast reactors
TRISO: tri-structural-isotropic
TU: trans-uranium
USAM: under-sodium area monitor
USV: under-sodium viewer
VHTR: very-high-temperature (gas-cooled) reactor
VVER: Vodo-Vodyanoi Energetichesky Reactor; Water-Water Energetic Reactor

23.16   REFERENCES

1. GEN IV Roadmap, http://gif.inel.gov/roadmap (2002).

2. J. Bouchard, "The Global View," GIF Symposium, Paris (France), 9-10 September, 2009, www.gen-4.org/GIF/About/documents/GIFProceedingsWEB.pdf (2009), pp. 11-13.

3. http://www.world-nuclear-news.org/NP-New_name_and_mission_for_GNEP-2106108.html, 21 June (2010).

4. "US GNEP Program Dead, DOE Confirms," *Nuclear Engineering International*, April 15 (2009).

5. http://www.iaea.org/INPRO/

6. A. Omoto, "International Project on Innovative Nuclear Reactors and Fuel Cycles (INPRO) and its Potential Synergy with GIF," GIF Symposium, Paris (France), 9-10 September, 2009, www.gen-4.org/GIF/About/documents/GIFProceedingsWEB.pdf (2009), pp. 263-268.

7. International Atomic Energy Agency, "Methodology for the Assessment of Innovative Nuclear Reactors and Fuel Cycles," Report of the Phase 1B of the International Project on Innovative Nuclear Reactors and Fuel cycle (INPRO), IAEA-TECDOC-1434, IAEA, Vienna (2004).

8. European SET-plan http://ec.[7europa.eu/energy/technology/set_plan/doc/2009_comm_investing_development_low_carbon_technologies_roadmap.pdf (2009).

9. A. Camplani, A. Zambelli, "Advanced Nuclear Power Stations: Superphenix and Fast-breeder Reactors," Endeavour, Volume 10, Issue 3 (1986), pp. 132-138.

10. H. Nickel, K. Hofmann, W. Wachholz, I. Weisbrodt, "The Helium-cooled High-temperature Reactor in the Federal Republic of Germany — Safety Features, Integrity Concept, Outlook for Design Codes and Licensing Procedures," *Nuclear Engineering and Design*, 127 (1991), pp. 181-190.

11. Fast Breeder Reactor, http://en.wikipedia.org/wiki/Fast_breeder_reactor, visited June (2010).

12. High Temperature Reactor, http://en.wikipedia.org/wiki/Very_high_temperature_reactor, visited June (2010).

13. J. Bouchard, "The Global View," GIF Symposium-Paris (France)-9-10 September (2009), slides only.

14. W. Hoffelner, "Development and Application of Nanostructure Materials", in *Understanding and Mitigating Aging in Nuclear Power Plants: Plant Lifetime Management (PLIM) for Safe, Long-Term Operation*, Ph. Tipping, Ed., Woodhead Publishers/UK, 2010, to be published.

15. Clementine fast reactor, http://en.wikipedia.org/wiki/Clementine_(nuclear_reactor), visited June (2010).

16. M. Ichimiya, B.P. Singh, J. Rouault, D. Hahn, J.P. Glatz, H. Yang, "Overview of R&D Activities for the Development of a Genera-tion IV Sodium-cooled Fast Reactor System," GIF Symposium, Paris (France), 9-10 September, 2009, www.gen-4.org/GIF/About/documents/GIFProceedingsWEB.pdf (2009), pp. 213-221.

17. L. Cinotti, C.F. Smith, H. Sekimoto, "Lead-cooled Fast Reactor (LFR) Overview and Perspectives," GIF Symposium, Paris (France), 9-10 September, 2009, www.gen-4.org/GIF/About/documents/GIFProceedingsWEB.pdf (2009), pp. 173-178.

18. P. Anzieu, R. Stainsby, K. Mikityuk, "Gas-cooled Fast Reactors (GFR): Overview and Perspectives," GIF Symposium, Paris (France), 9-10 September, 2009, www.gen-4.org/GIF/About/documents/GIFProceedingsWEB.pdf (2009), pp. 127-133.

19. T. J. O'Connor, "Gas Reactors — A Review of the past, an Overview of the Present and a View of the Future", GIF Symposium, Paris (France), 9-10 September, 2009, www.gen4.org/GIF/About/documents/GIFProceedingsWEB.pdf (2009), pp. 77-91.

20. H. Khartabil, "SCWR: Overview," GIF Symposium, Paris (France), 9-10 September, 2009, www.gen-4.org/GIF/About/documents/GIFProceedingsWEB.pdf (2009), pp. 143-147.

21. C. Renault, M. Hron, R. Konings, D.E. Holcomb, "The Molten Salt Reactor (MSR) in Generation IV: Overview and Perspective," GIF Symposium, Paris (France), 9-10 September, 2009, www.gen-4.org/GIF/About/documents/GIFProceedingsWEB.pdf (2009), pp. 191-200.

22. Advanced Nuclear Power Reactors, http://www.world-nuclear.org/info/inf08.html, visited June (2010).

23. Tsuzuki, K., Shiotani, T., Ohno, I., and Kasai, S., "Development of Next Generation Light Water Reactors in Japan," International Congress on Advances in Nuclear Power Plants (ICAPP'09), Tokyo, May 12 (2009).

24. "Small Nuclear Power Reactors," World Nuclear Association Document, February 2010.

25. "Technology Roadmap — Nuclear Energy Agency (NEA) and International Energy Agency (IEA), 2010.

26. TerraPower, http://www.intellectualventures.com/OurInventions/TerraPower.aspx (last visit August 2010).

27. M. L. Wald, TR10: Traveling-Wave Reactor, http://www.technologyreview.com/biomedicine/22114/ (last visit August 2010).

28. Sagayama, Y., "Feasibility Study on Commercialized Fast Reactor Cycle Systems," Proc. Global 2005, Tsukuba, Japan (2005).

29. Kotake, S., Sakamoto, Y., Mihara, T., Kubo, S., Uto, N., Kamishima, Y., Aoto, K., and Toda, M., "Development of Advanced Loop-type Fast Reactor in Japan," *Nuclear Technology*, Vol. 170, April (2010).

30. Koo, G. H., "Overview of LMR Program and Code Rule Needs in Korea," ASME Codes & Standards, Working Group on Liquid Metal Reactors, San Diego, USA, May 12 (2009).

31. Bajaj, S.S. and Gore, A.R., "The Indian PHWR," *Nuclear Engineering and Design*, Vol. 236 (2006), pp. 701-722.

32. Bhardwaj, S.A., "The Future 700 MWe Pressurized Heavy Water Reactor," *Nuclear Engineering and Design*, Vol. 236 (2006), pp. 861-871.

33. Kushwaha, H.S., Vaze, K.K. and Dixit, K.B., "Design of Indian Pressurized Heavy Water Reactor Components," Chapter 68, *Companion Guide to the ASME Boiler & Pressure Vessel Code*, ASME Press, 2009.

34. "Nuclear Power in India," World Nuclear Association Document, updated March 30 (2010).

35. Kakodkar, A., "Technology Options for Long Term Nuclear Power Deployment," *Nu-Power*, Volume 23 (1-4) (2009) pp. 22-28.

36. "Thorium," World Nuclear Association Document, October (2009).

37. Srinivasan, G., Suresh Kumar, K.V., Rajendrann, B. and Ramalingam, P.V., "The Fast Breeder Test Reactor — The Design and Operating Experiences," *Nuclear Engineering and Design*, 236 (2006), pp. 796-811.

38. Chetal, S.C., et al., "The Design of the Prototype Fast Breeder Reactor," *Nuclear Engineering and Design*, 236 (2006), pp. 852-860.

39. Sinha, R.K. and Kakodkar, A., "Design and Development of the AHWR — the Indian Thorium Fuelled Innovative Nuclear Reactor," *Nuclear Engineering and Design*, Vol. 236 (2006), pp. 683-700.

40. V. M. Poplavskii, A. M. Tsibulya, A. A. Kamaev, Yu E. Bagdasarov, I. Yu. Krivitskii, V. I. Matveev, B. A. Vasil'ev, A. D. Budyl'skii, Yu L. Kamanin, N. G. Kuzavkov, A. V. Timofeev, V. I. Shkarin, K. L. Suknev, V. N. Ershov, S. V. Popov, S. G. Znamenskii, V. V. Denisov, V. I. Karsonov, "Prospects for the BN-1800 Sodium Cooled Fast Reactor Satisfying 21st Century Nuclear Power Requirements," *Atomic Energy*, Vol. 96, No. 5 (2004), pp. 308-314.

41. Xu Mi, "Fast Reactor Technology R&D — Activities in China," *Nuclear Engineering and Technology*, Vol. 39, No. 3 (2007), pp. 187-192.

42. Sustainable Nuclear Energy Platform (SNETP), *Strategic Research Agenda*, May 2009, http://www.snetp.eu/www/snetp/index.php?option=com_content&view=article&id=63&Itemid=36.

43. F. Varaine, N. Stauff, M. Masson, M. Pelletier, G. Mignot, G. Rimpault, A. Zaetta, J. Rouault, "Comparative Review on Different Fuels for Gen IV Sodium Fast Reactors: Merits and Drawbacks," International Conference on Fast Reactors and Related Fuel Cycles (FR09), December 7-11, 2009 — Kyoto, Japan (2009).

44. Fast Neutron Reactors, http://www.world-nuclear.org/info/inf98.html, visited June (2010).

45. US Nuclear Regulatory Commission, "Preapplication Safety Evaluation Report for the Power Innovative Small Module (PRISM)," NUREG-1368, February 1994.

46. US Nuclear Regulatory Commission, "Preapplication Safety Evaluation Report for Sodium Advanced Fast Reactor (SAFR) Liquid-Metal Reactor," NUREG-1369, December 1991.

47. "Advanced Gas Reactor Fuel Program's TRISO Particle Fuel Sets A New Record For Irradiation Performance," http://www.ne.doe.gov/geniv/neGenIV9.html, November 16, 2009.

48. Next Generation Nuclear Plant, http://www.nextgenerationnuclearplant.com/, visited June (2010).

49. NGNP Project, "NGNP Fuel Qualification White Paper," INL/EXT-10-17686, July 2010.

50. M. Richards, et al., "The H2-MHR: Nuclear Hydrogen Production Using the Modular Helium Reactor," ICAPP '05, Seoul, Korea, May 15-19, 2005, Paper 5355.

51. HTTR top page, http://httr.jaea.go.jp/eng/index_top_eng.html, visited June (2010).

52. K. R. Schultz, "Use of the Modular Helium Reactor for Hydrogen Production," World Nuclear Association Annual Symposium, 3-5 September 2003 — London (2003).

53. Zuoyi Zhang, Zongxin Wu, Yuanhui Xu, Yuliang Sun, Fu Li, "Design of Chinese Modular High-Temperature Gas-Cooled Reactor HTR-PM," 2nd International Topical Meeting on High Temperature Reactor Technology, Beijing, China, September 22-24, 2004 (2004), Paper 15.

54. Z. Zhang, Z. Wu, D. Wang, Y. Xu, Y. Sun, Fu Li, Y. Dong, "Current Status and Technical Description of Chinese 2×250 MWt HTR-PM Demonstration Plant," *Nuclear Engineering and Design*, Vol. 239 (2009), 1212–1219.

55. Smith, C., W. Halsey, N. Brown, J. Sienicki, A. Moisseytsev, D. Wade, "SSTAR: The US Lead-cooled Fast Reactor (LFR)," *Journal of Nuclear Materials*, Volume 376, Issue 3, 15 June 2008 (2008), pp. 255-259.

56. Cinotti, L., G. Locatelli, H. Aït Abderrahim, S. Monti, G. Benamati, K. Tucek, D. Struwe, A. Orden, G. Corsini, D. Le Carpentier, "The ELSY Project", Paper 377, Proceedings of the International Conference on the Physics of Reactors (PHYSOR), Interlaken, Switzerland, 14-19 September (2008).

57. R. Stainsby, K. Peers, C. Mitchell,1 Ch. Poette, K. Mikityuk, J. Somers, "Gas Cooled Fast Reactor Research and Development in the European Union," (2009), http://www.hindawi.com/journals/stni/2009/238624.html, visited June 2010.

58. AHTR-reactor, http://www.ornl.gov/sci/ees/nstd/research_ahtr.shtml, visited June 2010.

59. PB-AHTR http://www.nuc.berkeley.edu/pb-ahtr, visited June 2010.

60. F. Nakashima, T. Mizuno, H. Nishi, L. Brunel, S. Pillon, K. Pasamehmetoglu, J. Carmack, "Current Status of Global Actinide Cycle International Demonstration Project," GIF Symposium, Paris (France), 9-10 September, 2009, www.gen-4.org/GIF/About/documents/GIFProceedingsWEB.pdf (2009), pp. 239-246.

61. GIF Symposium, Paris (France), 9-10 September, 2009, www.gen-4.org/GIF/About/documents/GIFProceedingsWEB.pdf (2009).

62. "GE Hitachi Advanced Recycling Center — Solving the Spent Nuclear Fuel Dilemma," GE Hitachi Nuclear Energy Press Release, 2010.

63. T. Allen, H. Burlet, R. K. Nanstad, M. Samaras, S. Ukai, "Advanced Structural Materials and Cladding", in Advanced Nuclear Energy Systems, special edition, MRS Bulletin 34, 20 (2009).

64. S. Fazluddin, K. Smit, J. Slabber, "The Use of Advanced Materials in VHTRs," 2nd International Topical Meeting on High Temperature Reactor Technology, Beijing, China, September 22-24 (2004), Paper E06.

65. R. L. Klueh, D. R. Harries, ASTM, "High-Chromium Ferritic and Martensitic Steels for Nuclear Applications," Eds.: R. L. Klueh, D. R. Harries ASTM Monograph 3, ISBN: 0-8031-2090-7, ISBN13: 978-0-8031-2090-7 ASTM (2001).

66. I. Charit, K. L. Murty, "Structural Materials for Next Generation Nuclear Reactors," Second ACE Workshop, May 9, 2007 Boise, Idaho (2007), http://coen.boisestate.edu/iucace/Presentations/I.%20Charit%20and%20K.%20Murty%20-%20Structural%20Materials%20for%20Nuclear%20Reactors.pdf , visited June 2010.

67. M.A. Pouchon, J. Chen, R. Ghisleni, J. Michler, W. Hoffelner, "Characterization of Irradiation Damage of Ferritic ODS Alloys with Advanced Micro-Sample Methods," *Experimental Mechanics*, 50[1] (2010), 79-84.

68. T.R. Allen, J.T. Busby, R.L. Klueh, S.A. Maloy, M.B. Toloczko, "Cladding and Duct Materials for Advanced Nuclear Recycle Reactors," *Journal of Materials*, Vol. 60, No. 01, January (2008), pp. 15.

69. P. Yvon, F. Carré, "Structural Materials Challenges for Advanced Reactor Systems," *Journal of Nuclear Materials*, 385 (2009), pp. 217–222.

70. "High Temperature Alloys for Gas Turbines and Other Applications," 1986, W. Betz et al Eds., D. Reidel Publ. Comp. (1986).

71. C. Verpoort, US Patent 4817858: "Method of manufacturing a workpiece of any given cross-sectional dimensions from an oxide-dispersion-hardened nickel-based superalloy with directional coarse columnar crystals".

72. http://www.functional-materials.at/rd/rd_ptc_extremat_de.html (currently only German Version available), visited June 2010.

73. Material 19Cr ODS was provided by A. Kimura, University of Tokyo.

74. M.K. Miller, K.F. Russell, D.T. Hoelzer, "Characterization of Precipitates in MA/ODS Ferritic Alloys," *Journal of Nuclear Materials*, Volume 351, Issues 1-3 (2006), pp. 261-268.

75. J. Chen, P. Jung, W. Hoffelner, H. Ullmaier, "Dislocation Loops and Bubbles in Oxide Dispersion Strengthened Ferritic Steel After Helium Implantation Under Stress," *Acta Materialia*, Volume 56, Issue 2 (2008), pp. 250-258.

76. A. Kimura, et al., "Super ODS Steels R&D," OECD NEA NSC in co–operation with the IAEA, Workshop on Structural Materials for Innovative Nuclear Systems Structural (SMINS), Karlsruhe, June 4-6, 2007, http://www.nea.fr/html/science/struct_mater/Presentations/KIMURA.pdf.

77. S. Ukai, T. Kaito, M. Seki, A. A. Mayorshin, O. V. Shishalo, "Oxide Dispersion Strengthened (ODS) Fuel Pins Fabrication for BOR-60 Irradiation Test," *Journal of Nuclear Science and Technology*, Vol. 42, No. 1, p. 109–122 (January 2005).

78. T. Kaito, S. Ukai, A. V. Povstyanko, V. N. Efimov, "Fuel Pin Irradiation Test at up to 5 at% Burnup in BOR-60 for Oxide-Dispersion-Strengthened Ferritic Steel Claddings," *Journal of Nuclear Science and Technology*, Vol. 46, No. 6 (2009), pp. 529–533.

79. STP-NU-009, "Graphite for High Temperature Gas-Cooled Nuclear Reactors," ASME (2008).

80. T. Burchell, R. Bratton, and W. Windes, "NGNP Graphite Selection and Acquisition Strategy," ORNL/TM-2007/153, September 2007.

81. J. W. Sterbentz, et al., "Reactor Physics Parametric and Depletion Studies in Support of TRISO Particle Fuel Specification for the Next Generation Nuclear Plant," INEEL/EXT-04-02331, September 2004.

82. P. Yvon, "Impact of Coolant on Design, Materials and Technological Developments for the Gas-cooled Systems," First topical seminar on coolants and innovative reactor technologies, Aix en Provence, November 27-28, 2006, GEDEPEON (2006).

83. C.F. Smith, L. Cinotti, H. Sekimoto, "Lead-cooled Fast Reactor (LFR) Ongoing R&D and Key Issues," GIF Symposium, Paris (France), 9-10 September, 2009, www.gen-4.org/GIF/About/documents/GIFProceedingsWEB.pdf (2009), 181-189.

84. D. Guzonas, "SCWR Materials and Chemistry Status of Ongoing Research," GIF Symposium, Paris (France), 9-10 September, 2009, www.gen-4.org/GIF/About/documents/GIFProceedingsWEB.pdf (2009) 163-171.

85. S. Delpech, E. Merle-Lucotte, T. Auger, X. Doligez, D. Heuer, G. Picard, "MSFR: Material Issues and the Effect of Chemistry Control," GIF Symposium, Paris (France), 9-10 September, 2009,

www.gen-4.org/GIF/About/documents/GIFProceedingsWEB.pdf (2009), 201-208.

86. F.H. Norton, *The Creep of Steels at High Temperatures*, McGraw-Hill, New York, NY, 1929.

87. F.C. Monkman and N.J. Grant: Proc. ASTM., 1956, Vol. 56, pp. 593-620.

88. A. E. Carden, et al. Eds., *Fatigue at Elevated Temperatures*, ASTM STP 520, ASTM (1973).

89. M. Samaras, W. Hoffelner, M. Victoria, "Modelling of Advanced Structural Materials for GEN IV Reactors", *Journal of Nuclear Material* (ISSN 0022-3115), 371 (2007), 28-36.

90. M. Samaras, M. Victoria, W. Hoffelner, "Nuclear Energy Materials Prediction: Application of the Multi-scale Modelling Paradigm," *Nuclear Engineering and Technology*, Volume 41, Issue 1 (2009), 1-10.

91. W. Hoffelner, "Damage Assessment in Structural Metallic Materials for Advanced Nuclear Plants," *Journal of Material Science* (2010), 45:2247–2257.

92. Kotake, S., Sakamoto, Y., Mihara, T., Kubo, S., Uto, N., Kamishima, Y., Aoto, K., and Toda, M., "Development of Advanced Loop-type Fast Reactor in Japan," *Nuclear Technology*, Vol. 170, April (2010).

93. Karasawa, H., Izumi, M., Suzuki, T., Nagai, S., Tamura, M, "Development of Under-Sodium Three-Dimensional Visual Inspection Technique Using Matrix-Arrayed Ultrasonic Transducer," *Journal of Nuclear Science Technology*, Vol. 37, 9 (2000).

94. W. Hoffelner, M.A. Pouchon, M. Samaras, J. Chen, A. Froideval, "Condition Monitoring of High Temperature Components with Sub-sized Samples," Proceedings of the 4th International Topical Meeting on HTR Technology, HTR 2008, September 28-October 1, Washington, DC, USA (2008), HTR2008-58195.

95. W. Windes, T. Burchell, R. Bratton, "Graphite Technology Plan," INL/EXT-07-13165, September, 2007.

96. R. N. Wright, J. K. Wright, T. L. Sham, R. Nanstad, and W. Ren, "Next Generation Nuclear Plant Reactor Pressure Vessel Materials Research and Development Plan," INL/EXT-08-14108, PLN-2803, Revision 0, April, 2008.

97. R. N. Wright, J. K. Wright, T. L. Sham, R. Nanstad, and W. Ren, "Next Generation Nuclear Plant Intermediate Heat Exchanger Materials Research and Development Plan," INL/EXT-08-14107, PLN-2804, Revision 0, April, 2008.

98. R. Schultz, "Next Generation Nuclear Plant Methods Research and Development Technical Program Plan," INL/EXT-06-11804, PLN-2498, Revision 1, September, 2008.

PRESERVING NUCLEAR POWER'S PLACE IN A BALANCED POWER GENERATION POLICY

Owen Hedden

24.1 END OF THE NUCLEAR ROAD?

In considering the future of nuclear power for electricity production in the U.S., it is necessary to consider the present public perception of nuclear power. It is also necessary to consider public perceptions of the various competing sources of electricity production. These include coal, natural gas, and the several "green" or "renewable" sources, including hydro, wind, and solar. Note that petroleum is no longer part of this discussion; its use has diminished to barely 1% of electricity production.

Chapter 24 makes extensive use of referenced publications of the Nuclear Energy Institute, to take advantage of their expertise on specific subjects. This chapter also makes extensive use of referenced quotations from the works of established authors in the field to minimize misinterpretation of their work.

In 1994, about the time that the last nuclear power plant was completed, USA Today announced [1]: "Essentially, the nation decided that nuclear power wasn't worth the price." . . ."Nuclear energy provides only 21% of our electrical needs, 4% of our total energy consumption. Is it worth the trouble? Not until the problems of waste, safety and cost are completely resolved."

No new nuclear power plants have gone on line since then, but increases in plant efficiency mean that nearly 20% of our electric power is still provided by nuclear power. While there has been an increase of over 17% in our electric power consumption since 1994, nuclear power has been able to maintain its share without building new plants, by increasing availability and up-rating power output. Those increases are remarkable, but we cannot plan on that continuing.

Now, in 2010, the media view is changing [2]: "The newest nuclear power plant in the U.S. opened nearly 15 years ago, but many more may be coming soon. A new reactor in Tennessee is scheduled to come on line by 2013, two more are under construction in Georgia, and the NRC is reviewing applications for another 21 reactors (27 reactors, NRC reported in August) at 17 sites."

24.2 NUCLEAR POWER'S PLACE: MAINTAINING 20% OF TOTAL ELECTRIC POWER GENERATION

Table 24.1 shows the share of U.S. electricity production from each of the usually considered sources for the years 1989 and 2008.

TABLE 24.1   U.S. ELECTRICITY GENERATION FUEL SHARES, PERCENT

	1989	2008
Coal	53.4	48.2
Natural gas	11.9	21.4
Nuclear	17.8	19.6
Hydro	9.2	6.0
Petroleum	5.5	1.1
Wind	0.1	1.3
Wood	0.9	0.9
Waste	0.3	0.4
Geothermal	0.5	0.4
Solar	0.0	0.0
Other	0.1	0.7

(Derived from EIA/doe.gov *Monthly Energy Review*)

Year-by-year trends in the 20-year period spanned by this Table were remarkably consistent. Coal and hydro gradually diminishing, natural gas rapidly increasing, petroleum almost disappearing, wind starting to pick up, and everything else little changed.

The nuclear power plants we have keep running better and better, and new ones are now on their way, despite vocal opposition. These should insure that nuclear power's place will be maintained, or even expanded. Since opposition is expected to continue, nuclear power's place will also need to be continually defended.

24.2.1   Increasing Use of Renewables - Going "Green"

When you really want to go "green," by reducing electric power plant emission of carbon dioxide, sulfur dioxide, nitrous oxides, or mercury, remember that nuclear produces none of them. Hydropower has similar advantages, and is a significant contributor in Europe. Wind and solar have similar emission omission, but are not available continuously; and must have equivalent backup power available on short notice. Through links to Norway and Sweden, which have extensive hydro facilities, Denmark is able to take advantage of its wind power generation by sending and receiving electricity through underwater cables as needed. You may recall that Denmark and Sweden are separated by as little as 2 mi of water. In the U.S., hydro is perceived to have environmental disadvantages and is actually seeing a decrease in national output.

24.2.1.1 Denmark's Example Denmark is often cited as a prime example of use of wind energy, producing 20% of its electricity and no longer needing to import oil. That is somewhat misleading, since Denmark now has its own (North Sea) oil and exports its surplus. It still has to import as much coal as it did before it developed wind power. Denmark also has by far the highest electricity rates in Europe; the residential rate in 2008 was 38 cents per kilowatt-hour [3a]. The average U.S. residential rate in 2008 was 11.26 cents per kilowatt-hour [4].

24.2.1.2 Alternative Energy Solutions — Renewable Hydro, Wind, Solar Development of alternatives to coal and petroleum, and more recently, natural gas, for generation of electricity, was an on-going goal long before the recent concerns of global warming and carbon dioxide. Think Boulder/Hoover Dam on the Colorado River, built 75 years ago to supply electric power to Los Angeles, still going strong. Thirty-five years ago, hydro-electric provided 15.8% of U.S. electricity generation, about the same as petroleum and natural gas, with coal providing 44.4%, nuclear already up to 9%, and no recordable generation from solar, wind, or even wood or waste. Table 24-1 shows more recent distribution of electric power sources. Hydro-electric has become less popular. Hydro's share has fallen by 35% since 1989, and kilowatt-hour generation has fallen 10%. Evidently some old, small dams have been retired.

Solar and wind now get most of the attention, and subsidies. Individual use of both solar panels and windmills has been common in some regions for many years, so proposals for large-scale applications were no surprise. Little is said about their liabilities. First are availability and reliability, 24 hours a day, 7 days a week. Another source with the same generating capacity must always be available. The American public demands no less. Batteries would be ideal, but are ruled out by cost. Hydro would be a good green source, but in most parts of the country hydro is already fully utilized. Existing coal, natural gas, or nuclear-driven steam generators take time to warm up and bring on-line, unless they are running on stand-by, which is very inefficient/expensive. New natural-gas fired turbines (like jet engines) or combined-cycle units are used in some areas; these can come on-line much more quickly than steam generators.

Conservation and environmental advocacy groups say the problem can be easily addressed by a "smart" electricity distribution system, but that is more easily said than done. It requires electricity distribution infrastructure we do not have.

Solar and wind are also require a large amount of land in relation to their power output. Wind farms make poor neighbors, because of rumble from the turbine blades; at least, that is what the neighbors within a half mile say. Bird-kill, ranging from eagles to bats, is an additional disadvantage. Another aspect is the requirement for high-voltage power lines from the wind or solar panel farms located where the conditions are right and the land is cheap, to the urban power distribution centers. The same conservation groups that advocate the "energy farms" fight the construction of new power lines to serve them. Not-in-my-back-yard (NIMBY) is alive and well.

Renewable energy sources are discussed much more thoroughly and objectively in earlier chapters of this handbook, solar in Chapters 1 through 5, wind in Chapters 7 through 10 and hydro in Chapters 11 and 12. Chapters 13 through 16 discuss other renewables that altogether presently provide less than 4% of U.S. electric power output [5].

24.2.2 Change in Public Perception

The primary factor affecting nuclear power has been the accident at Three Mile Island in 1979. Fortunately, there is recent evidence of a change in media/public automatic reaction to fear of nuclear power proposals:

Quoting **The Economist**, [6] on the subject **Lessons to be Learned:** "After the accident on Three Mile Island in 1979, Americans grew scared of nuclear power and stopped building new reactors, even though no one died in that accident. Had the nation not panicked, it would now have many more nuclear reactors, making the shift to a low-carbon economy significantly easier."

Similarly, in **The New Yorker**, [7] in the context of changing attitudes toward nuclear power as a source of electricity, from its initial acceptance until that acceptance "was gone, obliterated by the accidents at Three Mile Island, Pennsylvania, in 1979 (where no one was killed) and at Chernobyl, Ukraine, in 1986 (which caused thousands of deaths). But the giant anti-nuclear demonstrations of the time, in Europe and America, were fuelled at least as much by the fear of nuclear war as by fear of nuclear reactors." "Then, inconveniently for the world but conveniently for the nuclear power industry, a new truth intruded. Nuclear winter was over, but global warming was here, and the ideological categories got thoroughly scrambled.". . . ."And nuclear power plants have one great advantage over the fossil-fuel kind: they do not emit carbon dioxide, a green-house gas that is hastening the world toward climatic disruption and disaster."

24.3 FIVE PRIMARY ARGUMENTS OF GROUPS OPPOSING NUCLEAR POWER

This chapter includes discussion of nuclear power opposition's five primary arguments;

1. Fear and Three Mile Island,
2. Nuclear proliferation,
3. Waste/spent fuel disposal,
4. New construction costs, and
5. Vulnerability to terrorist attack.

The Three Mile Island and Chernobyl accidents are considered, as well as comparison of risks of the competing generation technologies. Also, comparisons are made with wastes and waste disposal associated with those technologies - coal is an easy target.

24.3.1 Three Mile Island vs. Chernobyl — It Can't Happen Here

Fundamental differences in reactor design and operation say "no." All U.S. commercial power reactors are "light water" reactors designed to "fail-safe" — shut down automatically in case of an accident without operator intervention. The reactor vessel must be designed so that its structural failure would be an incredible event. In addition, a containment structure is required that will contain all the products of the failure of a major component (steam generator, pump, etc.) other than the reactor vessel. Even the U.S. graphite-moderated pressurized water reactors designed using the same principles as the RBMK reactors are designed to fail-safe; shut down automatically in case of an accident. Not the RBMK reactors at Chernobyl.

24.3.1.1 Three Mile Island (TMI) On March 29, 1979, on TMI Unit 2, main feedwater flow was inadvertently blocked. With a relief valve stuck open things quickly went to hell. Everything else that could go wrong did. The operators did not know quite what to do. Not realizing the relief valve was stuck open, they thought the reactor had too much water, and shut down some emergency pumps. As pressure in the reactor dropped, the water flashed into steam, which prevented fuel rods from cooling. About a third of the core eventually melted. This was all abetted by instrument problems, equipment malfunctions, and computer-system breakdowns.

From the point of view of the owner, the plant was lost. But from the point of view of the public safety, nothing happened. Radioactive water was confined to the plant. Radioactive gas releases were never significant and were quickly diluted and released. Physicists from my employer at the time were all over the area downwind of the plant immediately afterward and never found any evidence of radioactive fall-out.

A recent Fact Sheet from the Nuclear Energy Institute [7a] reported:

"In 1990, the National Cancer Institute of the National Institutes of Health released the results of a two year study of cancer data in 107 U.S. counties that contained, or were adjacent to, major nuclear facilities that had begun operations before 1982. Among the counties were York, Lancaster and Dauphin near the TMI plant in Pennsylvania. The study, which compared cancer mortality rates in the 107 counties with rates in counties with no nuclear facilities, found no increased cancer mortality for people living near the nuclear installations. The study also found no evidence that leukemia for any age was linked to routine operations at the TMI reactors or to the accident at TMI 2."

The accident did have beneficial consequences for U.S. nuclear reactor safety due to improvements in regulation, safety systems, and operator training and supervision. Radiation physicist Bernard Cohen [8] of the University of Pittsburgh, wrote "the average number of operating hours per year for (U.S.) plants has increased by 12%. Unplanned shutdowns have been reduced by 70% accident rates for workers have declined more that three-fold, the volume of radioactive waste has declined by 72%, and radiation exposure to workers has been cut in half." Associated costs averaged $20 million per plant.

24.3.1.2 Chernobyl Then, in April 1986, there was Chernobyl. All the four Chernobyl reactors, and many other Russian reactors, are of a design they call RBMK. That design was known here to be fundamentally faulty, with a built-in instability. On loss of coolant, unlike U.S. reactors which then shut themselves down, the RBMK can run away unless shut down by the operators. But the electrical engineers, not the operators, were running an experiment on Unit 4. They had by-passed or disabled the safety and emergency systems, including the backup diesel generators. They chose to continue the experiment when things started to go wrong. Unfortunately, the core meltdown, explosion, and radioactivity release followed.

U.S. commercial reactor design vs. RBMK, summarized by Cohen [8]:

1. A reactor which is unstable against loss of water could not be licensed in the United States.
2. A reactor which is unstable against a temperature increase could not be licensed here.
3. A large power reactor without a containment [structure] could not be licensed here.

Cohen also states, regarding Chernobyl, "Post-accident analyses indicate that had there been a U.S.-style containment, none of the radioactivity would have escaped, and there would have been no injuries or deaths."

24.3.2 Public Tolerance for Loss of Life in Energy Production

If those dead-set against nuclear power are concerned because of loss-of-life, they should be aware of the risks of other energy sources. If they are genuinely concerned with preventing deaths in energy production, they should be directing their energies elsewhere, toward natural gas production, pipelines, oil drilling and refining, coal mining and burning, and various related emission sources.

All energy sources cause health risks. We just pay more attention to some than to others. Particularly severe accidents capture our immediate attention, but as they accumulate over time they become routine and we accept them as part of the cost of their use. Long term accumulation of lower level risks is even more readily accepted. A report on severe accidents worldwide, 1969 to 1987 [8a] summarizes annual deaths related to energy production worldwide, but it also provides a reminder of annual deaths related to electricity production in the United States. From coal, primarily mine disasters, more than 200 deaths annually; from dam failures, more than 200; from oil, from sources ranging from transport to refineries to tank farms to oil platforms, more than 135 deaths annually; from nuclear radiation, 31, once, at Chernobyl.

While none of these directly involve U.S. electricity generating facilities (except perhaps dam failures) it does not take much thought to recall loss of lives due to coal mine cave-ins and refinery fires, vs. zero losses in U.S. nuclear power plants.

24.3.2.1 Chernobyl Deaths — Radiation Exposure There were 31 immediate deaths at Chernobyl, most due to radiation. Additional deaths due to the accident have been the subject of considerable speculation. Four months later, that total remained the same. About 200 had been treated for radiation sickness and survived. Nobody beyond the site was reported to have symptoms of direct radiation sickness.

Five years later, a Russian physicist claimed on British television that 7000 of the 600,000 cleanup workers had died, of acute radiation sickness. It was promptly pointed out that that was about the expected figure of deaths from all causes among 600,000 people of the age spectrum involved.

Ten years later the OECD Nuclear Energy Agency prepared a factual report. From the beginning, the effect of radioactive iodine on the thyroid, particularly in children, had been a primary concern. Nearly 700 cases of childhood thyroid cancer had been reported, with three deaths; many more cases were expected. The other major effect was psychological, fear of becoming a cancer victim. There had been no increase in overall cancer rates in adults in the affected regions.

Almost 20 years later [9], the Chernobyl Forum said the accident would lead to far fewer deaths from cancer than had at first been predicted. "The Chernobyl Forum comprises a number of United Nations agencies, including the International Atomic Energy Agency and the World Health Organization, and also the governments of Belarus, Russia, and Ukraine." The Chernobyl Forum estimates 9,300 will die from cancer as a result of exposure to radiation from the plant. Greenpeace says 100,000. "However, this figure also includes deaths from causes other than cancer . . ." What is the pre-accident number of expected deaths over the time period and population considered? That is the type of information that must be included to put such claims into context.

Regarding Chernobyl, Cohen [8] also states that the effect of Chernobyl radiation exposures worldwide, after about 50 years, will be enough to cause about 16,000 deaths. For comparison, he also points out that that is still less than the number of premature deaths caused *every year* by air pollution from coal-burning power plants in the United States.

24.3.2.2 Comparative Risk — Nuclear versus Coal-Burning Power Plants Consideration of comparative risk and associated costs can pose controversial conclusions. As an example, Prof. Cohen [8] notes that NRC's 1975 Reactor Safety Study estimates that a nuclear power plant will cause an average of 0.8 deaths across its lifetime (the total of various risks including accidents and radiation exposure). He then notes that the NRC-mandated safety improvements following the Study increased plant costs by an average of $2 billion. This equates to $2.5 billion per life saved, which amounts to $4.16 billion in 2010 dollars [10]. Cohen then draws two conclusions; first, that what we spend to save a single life in making nuclear power safer could save thousands of lives if spent on causes such as cancer screening or transportation safety. Second, that the addition to the cost of nuclear plants made them more expensive than coal-burning plants, which the utilities then ordered. But a coal-burning plant causes an estimated 1000 to 3000 deaths over its operating lifetime, primarily from air pollution. So all those premature deaths from a coal-burning plant are caused by saving one life caused by an equivalent nuclear power plant. As is noted elsewhere in this chapter, there are about 600 coal-burning power plants and 100 nuclear power plants in the United States.

When there is no alternative to the need for a new source of electric power available 24/7, these are powerful conclusions to present to groups opposing new nuclear power plants. The disparity is still conclusive, even using the 125 deaths per plant estimated following the Reactor Safety Study by the Union of Concerned Scientists, a leading nuclear power opposition group. Add to this the fact that many new power plants will be needed in the coming decades, and this "safety" factor becomes more powerful.

Much needed advances in coal-burning technology are presented in Chapters 17 through 21.

24.3.2.3 Risks of Uranium Mining This section is included because of recent environmentalist protests about re-establishment of a uranium milling operation at a site in southwestern Colorado that has been dormant since 1979. A recent New Yorker article [11] reviewed local attitudes in an area where many miners have died of lung cancer. The first large mill to process ore was built in 1912, to extract radium, which is a decay product of uranium. Eventually, radium was replaced by less poisonous and more effective substances, and vanadium became the primary product, Vanadium mills discarded uranium in their tailings. Then, in 1943, a new mill was built to process vanadium tailings into uranium oxide, for the Manhattan Project, the national project to build the atomic bomb. After the war, with the nuclear arms race, around 900-uranium mines were opened across the Colorado Plateau, with essentially no regulation. Mill tailings, which contain radium and other radon emitters, were a major risk for public contamination. Many miners died of lung cancer in an epidemic, first

documented in 1956, but locally it was an accepted risk because mining paid well. Mining was discontinued in the 1980s, because of public concern about radioactive sites. The State of Colorado forced a Superfund cleanup that required the complete destruction of the town of Uravan, site of a major mill. It took 20 years to remove virtually all traces of radioactive contamination from the site. It is now fenced in with warning signs "Radioactive, Do Not Enter." There are no homes within nine miles. Now, with the promise of a new uranium mill and resumption of mining in the area, some of the assumptions forcing the shutdown and cleanup are being questioned.

Everyone assumed there was or would be an increase in leukemia cancer rates, birth defects, and other serious conditions. However, sources cannot be found for the "documented increases." The "yellowcake" uranium oxide as produced by the mills, before being enriched elsewhere, has a negligible radioactivity level. The locals are actually very well informed technically regarding radiation levels, unlike the environmental activists. One of locals who worked in the Uravan mill noted that the radiation levels he received there were about the same as someone who worked in Grand Central Station in New York, constructed of radon-emitting granite. This has been documented; a worker at Grand Central receives a larger dose than the Nuclear Regulatory Commission allows a uranium mill to emit at the site boundary. A recent study of health records of Uravan from 1936 to 2004 showed a significant increase in lung cancer, but in men only, in the underground miners, many of whom were heavy smokers. The rates for those who worked in the mill, or the women in town, who received environmental exposure, were essentially normal. The overall Uravan mortality rate was 10% lower than the national average. There was less heart disease, probably reflecting the life-style of people who liked outdoor activities. Now, mines are required to be ventilated, and have severe limits on radon exposure.

The World Health Organization does not classify uranium as a human carcinogen. There is no compelling evidence that low amounts of radiation cause health problems. The U.S. Nuclear Regulatory Commission regards amounts below about 10 rems (100 mSv) to be in that category. It sets the annual nuclear worker dose limit at 5 rems, and notes that the actual worker dose is 1 rem. The average U.S. annual dose is 0.62 rem. For 100 people receiving a 10 rem dose, one cancer could be expected; however, 42 would get cancer for other reasons.

24.3.3 Nuclear Proliferation

"There is no connection between reprocessing of used nuclear fuel in the U.S. with the proliferation of nuclear weapons" [3b]. Major nuclear power electricity generating countries France, Russia, England, Canada, and Japan continue used fuel reprocessing, recycling about 95% of the material in the fuel rods to produce new fuel rods, while reducing the volume of high-level nuclear waste by one-half or one-third. France, Israel, North Korea, and Pakistan all developed nuclear weapons before they developed nuclear power, and of them, only France has developed significant electricity generating capability.

24.3.4 Radiation Exposure, Radioactive Waste, Waste Disposal

Waste accumulation in dumps seldom gets our attention, and air-borne waste gets even less attention. We know it is there but there is little we can do about it. What we do not know, which can hurt us, is that nuclear power plants are not the only source of radioactive or toxic emissions.

Here is an interesting statement in **Nuclear News** [12], quoting Lord Marshall, Chairman of Britain's Central Electricity Generating Board, in 1986:

"Earlier this year, British Nuclear Fuels released into the Irish Sea some 400 kilograms of uranium, with the full knowledge of the regulators. This attracted considerable media attention, and, I believe, some 14 parliamentary questions.

"I have to inform you that yesterday the CEGB released about 300 kilograms of radioactive uranium, together with all its radioactive decay products, into the environment. Furthermore, we released some 300 kilograms of uranium the day before that. We shall be releasing the same amount of uranium today, and we plan to do the same tomorrow. In fact we do it every day of every year so long as we burn coal in our power stations. And we do not call that 'radioactive waste;' we call it coal ash."

Nuclear News then continued "A similar situation exists for radium and thorium in fly ash and bottom ash collected as a result of burning coal."

Recently, in **The New Yorker** magazine [7] . . ."In the United States, coal plants (there are 600 of them, as against 100 nuclear ones) . . . dispose of their most dangerous waste by sending it up the chimney free of charge. And these plants — which, according to one estimate, cause 24,000 premature deaths every year — could help trigger a worldwide "accident" far bigger than a Three Mile Island or even a Chernobyl: catastrophic, irreversible climate change." (Cohen [8] reaches a similar conclusion; over 16,000 deaths.)

The toxic elements in these annual air-borne emissions from coal-fired power plants in the United States have been quantified [3c]: 48 tons of mercury, 88 tons of lead, 80 tons of chromium, and 50 tons of arsenic, total 266 tons. The Environmental Protection Agency says coal-fired power plants account "for 40% of all domestic human-caused mercury emissions." Solid waste from coal, which includes both ash and scrubber sludge, totals 130 million tons, which is three times as much volume as all U.S. municipal garbage. Coal ash is usually contaminated with heavy metals uranium, radium, and thorium.

The Environmental Protection Agency is presently holding hearings on a proposal to reclassify waste from coal-burning power plants from special waste to hazardous waste, increasing the amount of regulation.

Annually U.S. nuclear power plants produce 2000 tons of used fuel. In *50 years* of operation, the U.S. nuclear power industry has produced about 60,000 tons of high-level waste. Not to minimize the problem, but compare this with the *annual* U.S. production of 130 million tons of coal ash, much of it contaminated with heavy metals.

Political will is a major factor obstructing efficient reduction of high-level nuclear waste. In the 1970s the Carter Administration halted recycling of spent fuel rods from nuclear power plants claiming that recycling the fuel would make plutonium available for nuclear weapons. Subsequently, in 1982, Congress passed the Nuclear Waste Policy Act. It required the Federal Government to take possession of the high-level waste produced by the nuclear power plants by 1998. The utilities have paid ($17.4 billion to March 31, 2010) for the development and operation of a waste repository which was built at Yucca Mountain, Nevada. Its use has been blocked politically, and the utilities have been forced to build "interim" spent fuel storage facilities on their own plant sites. Meanwhile, their payments to the Nuclear Waste Fund continue.

24.3.4.1 Used Fuel Disposal Strategy The Nuclear Energy Institute (NEI) has provided a detailed description of the present status of used nuclear fuel disposal in *Industry Supports Integrated Used Fuel Management Strategy, April 2010* [13]:

"The nuclear energy industry supports a three-pronged, integrated used fuel management strategy:

1. Interim storage of used fuel at centralized, volunteer locations.
2. Research, development and demonstration of advanced fuel utilization and recycling technologies.
3. Development of a permanent disposal facility.

Used fuel storage at nuclear plant sites is safe and secure. However, interim storage sites at centralized volunteer locations will enable the movement of used fuel from both decommissioned and operating plants before recycling facilities or a repository begin operating.

A research and development program should be implemented for advanced nuclear fuel utilization and recycling technologies, including a commercial demonstration plant. The objectives of reprocessing and recycling uranium fuel are to reclaim a significant amount of energy that remains in the fuel and to reduce the volume, heat and toxicity of byproducts placed in the repository.

An integrated used fuel management program includes key elements phased in during the short, medium and long terms.

Short-term goals include:

- Continuing the U.S. Nuclear Regulatory Commission's endorsement of waste confidence.
- Signing of standard contracts between the U.S. Department of Energy and energy companies for managing used fuel at new nuclear plants, which was accomplished in 2008.
- Adequately funding the repository licensing process, including the NRC's review of DOE's Yucca Mountain repository construction application. The Obama administration has announced its intent to terminate this project and withdraw DOE's license application with prejudice, signaling that it does not intend to resubmit the application. DOE has established the Blue Ribbon Commission on America's Nuclear Future to recommend strategies for used fuel management. The industry believes the Yucca Mountain licensing process should continue. Ultimately, a geologic repository will be needed somewhere. Even if a facility is not built at Yucca Mountain, completion of the licensing process will yield vital lessons that will inform the commission's deliberations and facilitate completion of a facility when a new site is selected. However, if the administration halts the licensing process, the industry believes it should be done in a manner that would facilitate resuming the process at a later date should that be warranted.
- Establishing a research and development program for advanced fuel utilization and recycling technologies, including government partnerships with industry.
- Identifying and developing volunteer sites for interim storage and advanced fuel utilization and recycling development facilities

Medium-term goals include:

- Moving used fuel to interim storage sites, ideally at advanced fuel utilization and recycling development sites.
- Continuing research, development and demonstration of advanced fuel utilization, recycling and fuel fabrication technologies to make them more cost effective and efficient.
- Licensing a repository

Long-term goals include:

- Commercial advanced fuel utilization and/or recycling.
- Operating the repository"

24.3.4.2 Established Record of Safe Transportation Integrity of the containers used to ship and subsequently store used nuclear fuel at interim storage sites and recycling facilities or the final geologic repository is essential. In the Nuclear Energy Institute's Fact Sheet *Experience, Testing Confirm Transportation Of Used Nuclear Fuel Is Safe, Reliable, February 2009,* [14] NEI provides the criteria for, and experience with, used fuel shipping containers:

"Over the past 40 years, the U.S. nuclear energy industry has safely transported more than 3,000 shipments of used nuclear fuel over 1.7 million miles. It has completed these shipments with no injuries, fatalities or environmental damage resulting from the radioactivity of the cargo. Since 1971, nine accidents involving commercial used nuclear fuel containers have occurred — four on highways and five during rail transport. Four of these accidents involved empty containers, and none resulted in a breach of the container or any release of its radioactive cargo."

"The containers that transport the used nuclear fuel are extremely robust, with multiple layers of steel, lead and other materials to confine radiation from the used fuel. These specially designed containers weigh between 25 and 40 tons for truck transport and between 75 and 125 tons for rail shipments, including the weight of the used fuel. Typically, for every ton of used fuel, there are about 4 tons of protective shielding."

24.3.4.3 The NRC-Required Tests of Used Fuel Containers "The NRC must approve containers that transport used nuclear fuel. Before the agency certifies container designs, they must meet rigorous engineering and safety criteria. In addition, the container designs must be shown, by test or analysis, to survive a sequence of four hypothetical accident conditions simulating the cumulative effects of impact, puncture, fire and submersion. The test sequence involves:

- a 30-ft free fall onto an unyielding surface, which would be equivalent to the cask being struck by a train traveling 60 mi/hour
- a puncture test allowing the container to fall 40 inches onto a steel rod 8 inches long and 6 inches in diameter
- a 30-minute exposure to fire at 1,475 degrees Fahrenheit that engulfs the entire container submerging the same container under 3 ft of water."
- Separate, undamaged containers are also subjected to immersion in 50 feet of water. Furthermore, casks must survive greater than 600 feet of water pressure for one hour without collapse, buckling or inleakage of water."

"In combination with actual testing, transportation container manufacturers use computer programs and scale models to evaluate the containers' protective capabilities and verify — with a substantial margin of safety — that the containers meet NRC and international requirements. For example, drop testing of full-scale and partial-scale transportation containers in Germany and Japan have validated previous simulations."

24.3.4.4 ASME Requirements for Construction of Used Fuel Containers The American Society of Mechanical Engineers has provided rules for construction of these containers in *Boiler and Pressure Vessel Code Section III Rules for Construction of Nuclear Facility Components, Division 3, Containments for Transportation of Spent Nuclear Fuel and High Level Radioactive Material and Waste* [15].

24.3.4.5 NRC Requires a Shipment Security Plan Continuing the quotation of the Nuclear Energy Institute's Fact Sheet: [14] "NRC regulations also require the establishment of a security plan to ship used nuclear fuel safely to the used fuel repository at Yucca Mountain, Nev., and implementation of this plan before shipments begin. The shipper must track and monitor these shipments carefully over the entire route. The agency must review and approve in advance the plan and procedures to protect against radiological sabotage or theft."

24.3.4.6 Low Level Radioactive Waste Disposal of low level radioactive waste also follows an orderly Regulatory process. Most of the following is taken from a fact sheet provided by NEI: *Disposal of Commercial Low-Level Radioactive Waste,* [16]

What is Low-Level Waste? "Low-level waste is solid material contaminated with low levels of radioactivity. It includes such items as gloves and other personal protective clothing, glass and plastic laboratory supplies, machine parts and tools, filters, wiping rags, and medical syringes that have come in contact with radioactive materials. Low-level waste from nuclear power plants typically includes water purification filters and resins, tools, protective clothing and plant hardware. Low-level waste does not include used fuel from nuclear power plants or waste from U.S. defense programs."

"The low-level waste from nuclear power plants accounts for half the volume and most of the radioactivity in low-level waste produced in the United States. The remaining low-level waste is produced by several thousand other industrial facilities and institutions that use radioactive materials. They include medical research laboratories, hospitals, clinics, pharmaceutical companies, government and industrial research and development facilities, universities and manufacturing facilities." This waste is produced from use of radioactive isotopes or X-ray machines in a number of beneficial activities. These are as varied as diagnosis and treatment of disease, medical research, non-destructive testing of pipelines, airplanes and bridges, eradication of insect pests, food preservation, and smoke detectors.

The radioactivity in approximately 95% of all low level waste fades to background level within 100 years.

The radioactive isotopes for these beneficial applications are typically produced in small research or teaching reactors. An interesting development a few years ago was that groups opposed to nuclear power forced an end to production of these isotopes in the United States. Hospitals, researchers, and inspection agencies had to import them from Canada.

24.3.4.7 Safe Disposal of Low-Level Waste — Three Sites "The Nuclear Regulatory Commission established technical requirements for low-level waste disposal sites. They require, among other things, that natural resources in the area, such as wildlife preserves, be avoided. The site also must be sufficiently isolated from groundwater and surface water, and the site must not be in an area of geological activity (such as volcanoes or earthquakes). Regardless of design, all low-level waste disposal sites use a series of natural and engineered barriers to prevent radioactivity from reaching the biosphere."

"The Transportation Department and the NRC regulate the shipment of radioactive materials, including low-level waste."

There are three disposal facilities currently accepting low-level radioactive waste:

- "Barnwell, S.C. Barnwell is licensed by South Carolina to receive wastes in Classes A, B and C. The facility accepts waste from Connecticut, New Jersey and South Carolina.
- Richland, Wash. The facility is licensed by the state of Washington to receive wastes in Classes A, B and C. It accepts waste from states that belong to the Northwest Compact (Washington, Alaska, Hawaii, Idaho, Montana, Oregon and Wyoming) and the Rocky Mountain Compact (Colorado, Nevada and New Mexico).
- Clive, Utah. Clive is licensed by the state of Utah to accept Class A waste only. The facility accepts waste from all regions of the United States.

One new LLW disposal facility has been licensed but is not yet operating:

- Andrews County, Texas. In September 2009, the Texas Commission on Environmental Quality issued a license for Waste Control Specialists LLC to build and operate a facility at its site in Texas. The facility, expected to begin operating in 2010, will accept Classes A, B and C low-level radioactive waste from Texas and Vermont, as well as the federal government."

24.3.5 Cost of New Construction

Citing government support of a particular energy source as resulting in competitive advantage is pointless - all benefit from substantial research and development incentives (support) or nonregulation. Some benefits just have been in place longer than others.

Bryce [3d], cites 2009 estimates of construction cost per kilowatt of capacity and capacity factors for

(a) nuclear, $4000 to $6700, with 90% capacity factor
(b) offshore wind, $5000, with 30-40% capacity factor
(c) solar, $6000, but with only 23% capacity factor
(d) coal, $2300, with 72% capacity factor [4b]
(e) natural gas, $850, with 41% capacity factor [4b]

These are cited to again make the point that construction costs for the most popular renewables are comparable to nuclear construction but that efficiency and availability are much less. While their fuel costs may be nil, nuclear fuel cost is less than that of coal, which is the only competing high-availability generating plant. And again, nuclear power generation produces no carbon dioxide, and, compared to coal, infinitesimal amounts of toxic waste.

24.3.5.1 Excessive Nuclear Plant Construction Cost Rhodes [18], p.4, cites a report in Forbes magazine in 1985 blaming mismanagement (described in great detail) for the disparity in construction costs for the 35 plants not on line in 1984. These costs ranged from $932/kW for McGuire 2 (1100 MW PWR, commercial operation, 1984) to $5192/kW for Shoreham ([800 MW BWR, closed 1989). This range of 5.6/1 indicates clearly that some utilities were much better able to manage design and construction costs. These figures are assumed to include added cost from engineering and Regulatory changes following both the Reactor Safety Study, noted in 24.3.2.2 and TMI accident discussed in 24.3.1.1.

One factor seldom mentioned in discussion of the costly delays in construction is the cost of money at that time of high inflation. For instance, utility bonds were paying as much as 16%. At that rate compounded annually, cost doubles in five years. Presently, utility bonds pay around 4%; in five years those will have paid a little over 20%. Now is a good time for a utility to borrow.

Roger Reedy discusses construction costs and "lessons learned" much more informedly and completely in Chapter 21.

24.3.6 Vulnerability to Attack

These discussions of the five primary objections to nuclear power plants will conclude with this section addressing security and resistance to terrorist attack. Since September 11, 2001, these threats have been prominent in public concerns. Physical and administrative barriers to plant access have been established. Before, plant tours would include a visit to the control room and the spent fuel pool. Not anymore.

24.3.6.1 Nuclear Plant Safety Features Now, in addition to the property perimeter fencing, a double chain-link fence surrounds the key components of the plant, including the cooling towers. There is single-point road access, and a perimeter intrusion detection system. Within that, additional authorization is required for access to "vital areas."

The Nuclear Energy Institute [19] provides the particulars on the containment structures: "The same features that safeguard the public and the environment from a radiation release also defend the reactor from outside interference. The reactor is typically protected by about 4 ft of steel-reinforced concrete with a thick steel liner, and the reactor vessel is made of steel about 6 in. thick. Steel-reinforced concrete containment structures are designed to withstand the impact of many natural disasters, including hurricanes, tornadoes, earthquakes and floods, as well as airborne objects with a substantial force."

"An independent study confirms that the primary structures of a nuclear plant would withstand the impact of a wide-bodied commercial airliner. The Electric Power Research Institute conducted a state-of-the-art computer modeling study on the impact of a Boeing 767 crash. EPRI concluded that typical nuclear plant containment structures — as well as used fuel storage pools and steel and concrete fuel storage containers — would withstand the impact forces and shield the fuel"

Nuclear Energy Institute [20] provided a practical example of implementation of their emergency preparedness planning when a strong Gulf Coast hurricane struck three nuclear power plants. While this did not test the plant's ability to resist an attack by stealth, deceit, or force, it demonstrated the effectiveness of their programs in response to a forced shutdown, subsequent power restoration, and their interaction with the surrounding communities.

Nuclear power plants are among the few power sources unaffected by destruction of surrounding infrastructure, interruptions in the transportation of fuel supply and other factors. Because nuclear power plants refuel every 18 to 24 months, they can operate despite prolonged interruptions that impact fuel supplies for other energy sources. This capability underscores the need for a diverse energy portfolio.

24.3.6.2 Cyber Security Cyber security is another potential form of attack that is now being addressed by the U.S. Department of Homeland Security (DHS). The goal of the DHS National Cyber Security Division's Control Systems Security Program is to reduce industrial control system risks. The Control Systems Security Program coordinates activities to reduce the likelihood of success and severity of impact of a cyber attack against critical infrastructure control systems through risk-mitigation activities. The Control Systems Security Program activities include:

1. Research for the rapid development and deployment of hardened control systems with built-in security
2. Conduct of rigorous tests to reveal exploitable system vulnerabilities and then develop system fixes
3. Development of integrated risk analysis to assess the ability of power companies to mitigate potential risks.

US-CERT, United States Computer Emergency Readiness Team, is the operational arm of the National Cyber Security Division at the Department of Homeland Security. US-CERT interacts with federal agencies, industry, the research community, state and local governments, and others to disseminate reasoned and actionable cyber security information to the public. Its role is to provide information updates, warnings, training and assistance in avoiding internet attacks that could cripple or commandeer plant computer systems.

24.3.6.3 Cyber Security Implementation Implementing a cyber security solution includes people, processes and technology to mitigate identified security risks [21]. Control systems need to incorporate authorization, authentication, encryption, intrusion detection and filtering of network traffic and communication. The National Electricity Reliability Council works with Department of Homeland Security to protect the North American electric system from cyber and physical attacks. The National Electricity Reliability Council has created a collection of practices to help power companies protect critical facilities against a range of cyber and physical threats.

24.4   PRESENT CONCERNS

24.4.1   Maintaining Supply Diversity

"Fuel diversity is one of the great strengths of the U.S. electric supply system. Each source of electricity has unique advantages and disadvantages, and each has its place in a balanced electricity supply portfolio" [22]. Coal-fired power plants still generate nearly half of our electricity.

24.4.1.1 Natural Gas It can be concluded, as indicated in Table 24.1 and Section 24.3.5, that natural gas is presently the leading competitor for increased share of the electrical generation market. "Natural gas-fired electricity generation has more than doubled since 1990. Nearly all power plants built over the past 15 years are fueled by natural gas. However, natural gas is subject to significant price fluctuations because it also is used as a heating fuel and in industrial processes" [22]. In 2008 the average price of natural gas increased 27% to 5.00 cents per kilowatt-hour (peaking at 7.36 cents per kilowatt-hour on July 2). Those prices triggered

increased development. Natural gas production has been revolutionized by commercialization of horizontal drilling and fracturing of gas-bearing shale deposits. Just within the city of Fort Worth, Texas, there are now 1200 productive gas wells tapping the Barnett shale. With the jump in gas production, its price dropped. The 2010 price is now around 2.2 cents per kilowatt-hour. With new field development in Texas and Pennsylvania, predictions are that prices will stay down for a few years, and utilities are wondering if they should delay new nuclear plant construction.

Environmentally, while natural gas is more polluting than nuclear or renewables, it is less polluting than coal. Gas-fuelled electric generating units, even the combined cycle units, are relatively cheap and quick to build. Cross-country gas transmission lines are a serious liability, however. Pipelines blow up only occasionally but when they do they destroy lives and property. This is a problem that nuclear power avoids.

24.4.1.2 Renewables Given our inability to store energy from wind and solar, they will remain bit players. But technological disruptions such as occurred with natural gas drilling into shale deposits keep occurring. A breakthrough in energy storage technology could do the same for electricity production from wind and solar.

24.4.1.3 Nuclear Nuclear electricity production cost is 2.03 cents per kilowatt-hour. "The uranium fuel for U.S. nuclear plants is abundant, readily available from reliable allies, such as Canada and Australia, and low in cost. Coupled with industry success over the past 20 years in reducing operating costs, the low fuel cost makes America's 104 nuclear energy plants among the lowest-cost sources of electricity available. Once built, new nuclear power plants would provide the same degree of price stability for consumers." [22]. Note: availability of cheap uranium is one of the justifications for accumulating and storing used fuel rods rather than reprocessing them to recover usable fuel and reduce the volume of waste. A recent study [23] has reported that the availability of natural uranium will not constrain any reasonably expected growth of nuclear power in this century. It further recommends, strongly, interim storage of spent nuclear fuel for a century or so in regionally consolidated sites.

24.4.2   License Renewal Activities

The current fleet of nuclear power plants was licensed for either 30 or 40 years. Most have developed programs to extend their licenses for another 20 years. Four units have already begun operating beyond 40 years, and three more are scheduled to begin in 2010. An additional 59 units have received their license extensions and 27 are under review. These total 93 units; a few more applications are expected as the other units approach the end of their licenses. The point is, even with license renewal, additional plants will be needed to satisfy demand for more clean electric generation.

24.4.3   New Plant Designs - New Licensing Process

The licensing process imposed by the U.S. Nuclear Regulatory Commission (NRC) on the present fleet of nuclear power plants evolved as a series of steps that contributed to delays in construction with each design change. A new, more efficient process was addressed as part of the 1992 Energy Policy Act. The new process provides for certification of standardized designs, early site approval and combined construction and operating licenses.

Status as of July 2010:

NRC had issued four new design certifications and was currently reviewing four advanced new designs. There are two advanced designs for pressurized water reactors of 1600 to 1700 MW and two for boiling water reactors of 1350 to 1600 MW capacity.

NRC had issued four early site permits and had two additional applications under review. It is currently reviewing applications for 17 combined construction and operating licenses; with several two-unit proposals, they include 27 new reactor units.

There is concern in the industry that the increased workload for NRC's review staff for new reactors as well as license renewal will result in licensing delays, and that this will be exacerbated by politically-driven budgetary demands for reduction of expense/staff for all Federal agencies.

There is also concern regarding recruitment and training of new nuclear power plant operators. The existing generation of plant operators is maturing. With renewal of licenses for existing plants and licensing of new plant designs and construction, there is going to be a significant need for new licensed plant operators.

24.4.4 Training for Plant Operation

The TMI accident, in which operator error was a major factor, brought about major changes in reactor operator training, plant operating experience sharing, emergency response capability, radiation protection, and many other areas of nuclear plant design, operations and maintenance. The Institute of Nuclear Power Operations in Atlanta was formed nine months after the accident to drive operational excellence, open communication and continuous improvement among all U.S. nuclear plant operators.

Operators receive rigorous training and must hold valid federal licenses. Nuclear power plant operators spend one week out of every six in training. Proficiency in the operation and maintenance of plant components and systems is essential. All nuclear power plant staff are subject to background and criminal history checks before they are granted access to the plant. The NRC requires companies that operate nuclear power plants to have a fitness-for-duty program for all personnel with unescorted access to vital areas of the plant. The NRC requires companies to conduct random drug and alcohol testing on their employees. At least half of all employees are tested annually.

24.4.4.1 ASME Nuclear Training Seminars
ASME Nuclear Training Seminars offer a number of training courses that would be of value for plant engineers, planning and maintenance personnel. One course that might be of use to plant operators is Nuclear Power Plant Startup Operations and Maintenance. It is generally during plant startup when plant operating staff learns to address problems such as piping and equipment vibration, thermal expansion and loose parts monitoring. The course provides background and lessons learned regarding the operation and maintenance of plant components and systems.

24.5 FUTURE CHALLENGES, OPPORTUNITIES

24.5.1 Longer-Term Advanced Nuclear Plant Designs

The Nuclear Energy Institute has prepared a listing [24] of highly advanced new reactors based on new technologies. Some could be ready for commercial use in the United States by the end of this decade, while others are expected to be available after 2020:

AREVA Antares

AREVA based the design for the Antares on the concept of a gas-cooled (helium) reactor. The company is developing the 250 MW design in the context of the Generation IV International Forum.

Babcock & Wilcox Co. mPowerTM Reactor

The mPowerTM reactor design is a passively safe, modular 125-MW advanced light water reactor with a below-ground containment. B&W intends to apply for NRC design certification.

GE Hitachi (GEH) Nuclear Energy Power Reactor Innovative Small Module (PRISM)

The PRISM is a 311-MW advanced reactor cooled by liquid sodium. GEH plans to apply for a combined license for a prototype.

General Atomics Gas Turbine Modular Helium Reactor (GT-MHR)

The GT-MHR is a high-temperature gas reactor with advanced gas turbine technology.

Hyperion Power Generation Hyperion Power Module (HPG)

The HPG is a 25-MW liquid metal-cooled reactor about the size of a hot tub. The company plans to apply for NRC design certification.

NuScale Power Inc. NuScale Reactor

The NuScale is a 45-MW light water reactor. The company plans to apply for NRC design certification.

Pebble Bed Modular Reactor Ltd. PBMR

The PBMR is a high-temperature 165 MW reactor with a closed-cycle, gas turbine power conversion system. PBMR Ltd plans to apply for NRC design certification.

Toshiba 4S (Super-Safe, Small and Simple)

The 4S is a 10-MW sodium-cooled reactor. Toshiba plans to apply for NRC design approval.

Westinghouse-Led Consortium International Partnership International Reactor Innovative and Secure

The IRIS is a modular, light-water reactor design, with each module capable of producing between 100 and 300 MW. The consortium plans to apply for NRC design certification.

As it happens, three designs employ light-water reactors, three use liquid-metal reactors, and three use high-temperature gas reactors.

Further information on each is found on the nrc.gov site for nuclear reactors; advanced reactors.

As the U.S. demand for clean electric power increases, the availability of these small modular reactor designs should be attractive for applications or locations that do not need or want the 1000+ MW size.

24.5.2 U.S. Fabrication Capability-Limitations or Opportunities

The cessation of new plant construction in the U.S. was accompanied by the abandonment of much of the U.S. heavy component manufacturing capability. When a market for replacement of steam generators and reactor vessel heads finally developed, these had to be fabricated in Canada or overseas. France, Germany and Italy in Europe, and Japan and South Korea in Asia, have all developed capabilities for major component fabrication. China's participation is inevitable.

The only U.S. commercial nuclear power plant supplier maintaining a manufacturing facility is Babcock and Wilcox; they have been manufacturing major components for the U.S. (nuclear) Navy. These are in the same size range as most of the modular designs discussed above, indicating that market is available for U.S. participation.

24.5.3 Unanticipated Operating Problems

The existing nuclear power plants were designed with material degradation due to fatigue and irradiation as the primary concerns. Neither has turned out to be a concern, but unanticipated forms of degradation have appeared and been dealt with. It has been noted that these have occurred at about ten year intervals, so more may be anticipated in the present plants, and also in the new plants after they come on line and develop their own operating experience. The new participants in the U.S. nuclear power industry must not be complacent.

24.5.4 The Nuclear Fuel Cycle

A recent study [23] focuses on the "nuclear fuel cycle," a concept that includes both the kind of fuel used to power a reactor and what happens to the fuel after it has been used. It observed that there has been very little research on the fuel cycle for the last 30 years and has looked at the underlying assumptions. Then the expectation was that the uranium supply was limited and that fuel reprocessing would be necessary. Now, it is expected that the supply of natural uranium will not be a concern for the rest of this century, and that the present once-through cycle will continue to be the most economic. This will provide time for research to determine the best fuel cycle for the coming generation of nuclear power plants. The report concludes that significant changes are needed in the planning and implementation of used fuel storage and disposal options, and that these must be integrated with studies of the optimal fuel cycle.

24.6 PROTECTING THE ENVIRONMENT — UNFINISHED BUSINESS

24.6.1 An Inconvenient Truth

Inconveniently for the world but conveniently for the nuclear power industry, global warming is here. Nuclear technology is the only large-scale baseload, energy generation technology with a near-zero carbon footprint. Several prominent opponents have come to appreciate this fact. "Such founding fathers of the envi-

ronmental movement as Stewart Brand, the creator of the Whole Earth Catalog, and Patrick Moore, an early stalwart of Greenpeace, now support nuclear" [7]. For example, as compiled by the Nuclear Energy Institute [25]:

- **Stewart Brand**: "Now we come to the most profound environmental problem of all . . . global climate change. Its effect on natural systems and on civilization will be a universal permanent disaster. . . . So everything must be done to increase energy efficiency and decarbonize energy production. Kyoto accords, radical conservation in energy transmission and use, wind energy, solar energy, passive solar, hydroelectric energy, biomass, the whole gamut. But add them all up and it is still only a fraction of enough. . . . The only technology ready to fill the gap and stop the carbon dioxide loading of the atmosphere is nuclear power. . . . It also has advantages besides the overwhelming one of being atmospherically clean. The industry is mature, with a half-century of experience and ever improved engineering behind it. . . . Nuclear power plants are very high yield, with low-cost fuel. Finally, they offer the best avenue to a 'hydrogen economy,' combining high energy and high heat in one place for optimal hydrogen generation. "If all the electricity you used in your lifetime was nuclear, the amount of waste that would be added would fit in a Coke can."
- **Patrick Moore:** "A more diverse mix of voices are taking a positive second look at nuclear energy—environmentalists, scientists, the media, prominent Republicans and Democrats and progressive think tanks. They are all coming to a similar conclusion: If we are to meet the growing electricity needs in this country and also address global climate change, nuclear energy has a crucial role to play."
- Anglican bishop **Rev. Hugh Montefiore** Former chairman and trustee for Friends of the Earth: "I have been a committed environmentalist for many years. It is because of this commitment and the graveness of the consequences of global warming for the planet that I have now come to the conclusion that the solution is to make more use of nuclear energy."

24.6.2 Integrated Energy Parks

Nuclear power plant waste disposal, as discussed in Section 24.3.4, clearly needs further action for resolution. There have been discussions within the U.S. Department of Energy (DOE) to utilize their nuclear-focused national laboratories, Los Alamos, Idaho, Sandia, Savannah River, and Oak Ridge as "integrated energy parks." These laboratories have decades of experience with nuclear materials and technologies, and their communities are familiar with nuclear issues and interested in keeping the jobs. Their use for interim nuclear waste storage would allow the federal government to meet its obligations under the Nuclear Waste Policy Act, which requires DOE to take possession of the nuclear waste. This would alleviate security concerns regarding present storage in about one hundred twenty locations across the country; these regional national laboratories have well-established security. It is also consistent with the NEI strategy [13] and a recent study [23] recommending regional consolidation of used fuel.

These DOE sites would also appear to be appropriate sites for the research needed to develop an optimal nuclear fuel cycle and the eventual optimal reprocessing and geologic repository needs for present and future energy resource or waste management benefits.

24.6.2.1 Feedback on Problems The possibility exists that an additional benefit of the merging of commercial nuclear activities with the military nuclear activities at the DOE national laboratories would result in feedback on U.S. Navy's operating experience with the pressurized water reactors in over 100 nuclear-powered submarines and 11 aircraft carriers, information that could be important to the continued safe and reliable operation of commercial nuclear power reactors. This has never been permitted because of security concerns that originated before the global spread of nuclear power plant technology.

Only when problems have come up and the commercial nuclear power industry discovers that a technology exists to address it do we get a clue that the Navy must have had the same problem. An example is steam generator tube corrosion. U.S. heat exchanger manufacturers were building vertical U-tube steam generators having stainless steel tubes for the nuclear Navy (In 1957, I worked in a shop that had forty steam generators under construction for the nuclear Navy). The manufacturers followed that practice in steam generators for their first nuclear power plants. In the early 1960s, the Navy switched to Ni-based tubing for steam generators for a new design prototype (I was a project engineer for the D1G/D2G), and the manufacturers soon followed suit on their new commercial units. Also at this time, the manufacturers discovered that there was a small company in Richland, WA that had developed a technology for inserting a flexible probe through the U-tubes to quickly detect location and measure depth of corrosion. Very convenient. Coincidentally, Richland WA was also the location of the DOE Hanford Laboratory, and not much else.

24.6.3 Recycling Used Fuel

Used fuel recycling has been mentioned previously, in 24.3.3 in the context of nuclear proliferation and in 24.3.4.1 as an important element in a used fuel disposal strategy. The present situation is summarized very concisely by the Nuclear Energy Institute [26]: "For economic and national security reasons, the United States does not currently recycle used nuclear fuel. After its use once in the reactor, companies remove it for ultimate disposal in a repository. This 'once-through' fuel use is called an 'open' fuel cycle. The recycling and reuse of nuclear fuel is called a 'closed' fuel cycle. This approach would capture the vast amount of energy still remaining in used nuclear fuel."

"The federal government plans to evaluate both the open and closed fuel cycles, including the benefits and availability of advanced recycling technologies. The nuclear industry endorses this plan, which could result in long-term environmental and energy security benefits for America."

24.6.3.1 Converting Used Fuel into New Fuel
After one cycle in a reactor, about 95% of the material in the fuel can be reused. France, Great Britain, and Japan all have had fuel reprocessing operations as part of their nuclear power programs for many years. The reprocessed fuel, which contains plutonium as well as uranium, is called MOX (mixed oxide). Reprocessing also reduces the volume of high-level nuclear waste by one-half or one-third. Reactors must be converted to use MOX fuels. The French have been converting some of their reactors to operate with MOX fuel of 95% depleted uranium oxide and 5% plutonium oxide.

24.6.3.2 Converting Weapons into New Fuel
While the U.S. does not permit reprocessing of U.S. used reactor fuel, it does dismantle nuclear weapons. Since the mid-1990s, Russia has been shipping bomb-grade nuclear material from dismantled warheads to the U.S. for conversion into fuel for commercial reactors. That program will continue for years to come [3e].

24.6.4 Permanent Geologic Repository

Whatever the outcome on once-through fuel cycle versus fuel recycling, a permanent repository for high-level nuclear power plant waste will be required ultimately. Section 24.3.4.1 quotes a discussion of this by the Nuclear Energy Institute in *Industry Supports Integrated Used Fuel Management Strategy* [13]. The Nuclear Energy Institute has also issued a Key Issue [27] - Repository Development, quoted extensively herein:

24.6.4.1 Government Mandates Used Nuclear Fuel Repository
Congress passed legislation in 1982 directing the U.S. Department of Energy (DOE) to build and operate a deep geologic repository for used nuclear fuel and other high-level radioactive waste. Under this legislation—the Nuclear Waste Policy Act—Congress set a deadline of 1998 for DOE to begin moving used nuclear fuel from nuclear power plants. Because of delays, however, the 1998 deadline is long past due.

In 1987, Congress adopted an amendment to the Nuclear Waste Policy Act that directed DOE to study Yucca Mountain, Nev. — a remote desert location — as the site for a potential repository for geologic disposal of used nuclear fuel. DOE's study of the site was delayed until 1992, partly because Nevada refused to issue the environmental permits needed for surface-disturbing work. After several court cases, the state issued the permits, and DOE began its studies.

In 1994, DOE started building a system of tunnels at the site. Scientists conducted extensive volcanic, seismic, geological, hydrological and geochemical studies in these tunnels to assess how a repository would perform over tens of thousands of years. DOE published the results of these scientific and technical analyses in a comprehensive evaluation of the site that demonstrated a Yucca Mountain repository is capable of protecting public health and safety.

Based on this comprehensive evaluation, in 2002, Congress and President George W. Bush approved Yucca Mountain, Nev., as the site of the repository.

To fund the federal program, the 1982 legislation established the Nuclear Waste Fund. Beginning in 1983, consumers of electricity produced at nuclear power plants have paid a fee into the fund of one-tenth of a cent for every kilowatt-hour of electricity produced. Commitments to the Nuclear Waste Fund, including interest, now total over $34 billion."

24.6.4.2 DOE Submits Yucca Mountain License Application
In June 2008, the U.S. Department of Energy submitted to the U.S. Nuclear Regulatory Commission a license application to build a deep geologic repository for used nuclear fuel and other high-level radioactive waste at Yucca Mountain, Nev., a remote desert location.

24.6.4.3 Blue Ribbon Commission Formed to Evaluate Alternatives
The Obama administration announced plans in 2009 to terminate the Yucca Mountain program and empanel a blue-ribbon commission of experts to study alternatives. On Jan. 29, 2010, the Energy Secretary announced the formation of a Blue Ribbon Commission on America's Nuclear Future to provide recommendations for developing a safe, long-term solution to managing the nation's used nuclear fuel and high-level radioactive waste from defense programs. The commission will produce an interim report within 18 months and a final report within 24 months."

24.7 CONCLUSION

Nuclear power has had an important role in U.S. electric power production for over 20 years. The other primary competitors over this period have been coal and natural gas. Coal use has been gradually diminishing; while cheapest, it has many environmental disadvantages. Natural gas use has been increasing rapidly because on increasing supply and decreasing cost. It has few disadvantages apart from producing carbon dioxide, a "greenhouse" gas. Renewables such as wind and solar may be popular currently, but are presently not capable of baseload operation and thus require baseload-capable full time backup, and are expensive. Hydro-electric power is the only renewable baseload (available 24/7) electricity generation source. However, hydro-electric percent of total electric generation has actually been decreasing; evidently all the best sites have already been utilized.

Nuclear power is relatively economical and is the only remaining baseload electricity generation source that produces no carbon dioxide and thus does not contribute to global warming. The primary barrier to increasing nuclear power generation evidently is long-term waste disposal, which, while not a technical problem, has become a political barrier.

24.8 REFERENCES

1. **USA Today**, Dec. 13, 1994, p. 14A.

2. **Parade**, June 6, 2010: *The Rise of Nuclear Power. Intelligence Report:* (**Parade** is a nationally distributed Sunday newspaper supplement.)

3. Robert Bryce, *Power Hungry, The Myths of "Green" Energy and the Real Fuels of the Future*, PublicAffairs, 2010

3a. Bryce, p. 107.

3b. Bryce, p. 271.

3c. Bryce, pp. 58–59.

3d. Bryce, p. 262.

3e. Bryce, p. 273.

4. U.S. Energy Information Administration, Electric Power Annual 2008 Table 7.4.

5. U.S. Energy Information Administration, Electric Power Annual 2008 Table 8.4b.

6. The Economist, May 8, 2010, p. 36.

7. The New Yorker, March 22, 2010, p.19–20.

7a. Nuclear Energy Fact Sheet *The TMI 2 Accident: Its Impact, Its Lessons, August 2010.*

8. Bernard L. Cohen, *The Nuclear Energy Option: An alternative for the 90s,* Plenum Press, 1990.

8a. Nuclear News, April 1991.

9. The Economist, April 22, 2006, p. 77.

10. Bureau of Labor Statistics, CPI inflation data, bls.gov/data/inflation.

11. The New Yorker, *The Uranium Widows*, Peter Hessler, September 13, 2010.

12. **Nuclear News**, October 1986.

13. Nuclear Energy Institute Policy Brief *Industry Supports Integrated Used Fuel Management Strategy, April 2010.*

14. Nuclear Energy Institute Fact sheet *Experience, Testing Confirm Transportation Of Used Nuclear Fuel Is Safe, Reliable, February 2009.*

15. ASME *Boiler and Pressure Vessel Code Section III Rules for Construction of Nuclear Facility Components, Division 3, Containments for Transportation of Spent Nuclear Fuel and High Level Radioactive Material and Waste,* The American Society of Mechanical Engineers, 2010 Edition.

16. Nuclear Energy Institute Fact sheet *Disposal of Commercial Low-Level Radioactive Waste, September 2009.*

17. U.S. Energy Information Administration, Electric Power Annual 2008 Table 5.2.

18. Richard Rhodes, *Nuclear Renewal, Common Sense about Energy,* Whittle Books, 1993.

19. Nuclear Energy Institute Key Issues, Plant Security, *Nuclear Plant Safety Features.*

20. Nuclear Energy Institute Key Issues, Safety and security *Nuclear Plants' Structural Strength, Emergency Plans Perform Well Through Hurricane Katrina,* 2005.

21. ENERGY-TECH magazine, August 2010, *Wireless Internet plant security,* K.S. Raj.

22. Nuclear Energy Institute policy brief, *U.S. Needs New Nuclear Plants to Meet Energy Demand, Maintain Supply Diversity.*

23. R&D magazine, *The future of the nuclear fuel cycle,* September 17, 2010.

24. Nuclear Energy Institute Key Issues, New Reactor Designs, *Longer-Term Advanced Nuclear Plant Designs.*

25. Nuclear Energy Institute News and Events - Environmentalists.

26. Nuclear Energy Institute Key Issue, Recyclng Used Nuclear Fuel.

27. Nuclear Energy Institute Key Issue, Repository Development.

24.9 OTHER RESOURCES

The Necessity of Fission Power, Hans Bethe, Scientific American, January 1976, pages 21-31, and a book by the same title, 1991, Touchstone.

Blaming Technology, Samuel C. Florman, 1981, St. Martin's Press.

International Energy Agency, 2009 World Energy Outlook.

STEAM TURBINE AND GENERATOR INSPECTION AND CONDITION ASSESSMENT

Lawrence D. Nottingham

CREDITS

The material presented in this chapter includes certain text owned by Structural Integrity Associates, Inc. (SI), reproduced with the permission of SI. All figures, photographs, and diagrams presented are owned by SI.

25.1   INTRODUCTION

The fundamental design concepts embodied in steam turbines and generators are fairly simple—allow high pressure steam to expand through a series of blades mounted to a rotating member to extract thermal energy from pressurized steam and convert it to mechanical energy in the form of torque. Then utilize this torque to rotate a large electromagnet past a series of conductors to produce current flow in the conductors—pretty simple. However, as simple as this seems, these are very complicated machines that involve essentially all aspects of engineering—statics and dynamics, heat transfer, fluid mechanics, thermodynamics, ferrous and non-ferrous metallurgy, organic and inorganic chemistry, steam cycle chemistry, materials behavior (stress/strain), fracture mechanics, corrosion, erosion, electrical (power) engineering, electric circuits and circuit models, electromagnetics, control theory and controls, power system analysis, dielectrics and electrical insulation, tribology, and various forms of materials joining from glues and resins to soldering, brazing, and welding.

Looking back at the evolution of steam turbines and generators, it quickly becomes clear that the single-most limiting factor that has controlled machine output capacity has almost universally been the ability to dissipate heat from the generator. While numerous, significant advances have been made in turbine efficiency and unit rating, were it not for the generator output limitations, the additional mechanical work required to power the generator could easily be provided by simply adding more turbines. In fact, that has been one means of powering larger generators—adding turbines in tandem driving a single generator. Early generator designs featured air cooling, at ambient pressure and temperature, and conventional cooling schemes—convective cooling of the rotor steel and stator core and indirect cooling of the electrical coils

by conduction through the insulation to the rotor and stator core where the convective cooling occurred. Step increases in unit output were achieved coincidental with major ventilation improvements—sealed machines with integral gas coolers; increases in gas pressure from 15 to 30 to 45, 60, and 75 psi, and therefore increased heat dissipation capacity; the use of lower density and higher heat capacity cooling gases as the cooling medium; direct gas cooling of the stator and rotor coils; liquid (water) cooling of the stator coils; and even water cooling of the rotor coils in some non-domestic designs. Additionally, advances in insulation materials enabled operation at higher temperature without degradation of the insulating properties of the materials. At one point in the mid-1970s through mid-1980s, cryogenically cooled superconducting generators were considered the next design generation, and at least one prototype machine was built domestically.

As these advances were made, machine complexity increased dramatically. Hydrogen gas was found to be a very effective cooling medium, and it quickly became the standard. Where it was once necessary only to pass ambient air through the machine, it became necessary to create sealed systems and maintain high Hydrogen purity because of the explosive nature of Hydrogen when mixed with air. Elaborate labyrinth and gland sealing systems for the rotors, integral hydrogen coolers, hydrogen dryers, and auxiliary support systems resulted. Where the inherent rigidity of components in small machines generally avoided vibration issues, elaborate coil bracing systems were required as the machines grew is size and the coils became longer and more flexible and therefore more prone to vibration and fatigue. Rotor bearing systems evolved from simple sleeve bearings with pooled lubricant to relatively complex tilting pad bearings with elaborate seals, pumped lubricant, and complex auxiliary support systems to remove particulates, cool the oil, and safely remove hydrogen gas.

In parallel, advances in materials were required to keep pace with the increased size of major components, particularly rotor components. Stress in a rotating component is directly related to size (diameter) and rotational speed—larger and faster both equate to increased stress. For integral pole rotors, i.e., those having magnetic poles integral to a single piece rotor as found exclusively in steam turbine generators, a maximum of four magnetic poles was established as the practical limit for number of poles. To produce electrical current at a specific frequency, which in the US is 60 Hz, the magnetic poles must pass the stator coils at that frequency; consequently, 3600 and 1800 RPM are the operating speeds for 2-pole and 4-pole rotors, respectively. With the speed defined, stresses can be determined based on the size, and limits can be established based on the stress capacity of the material. So, to build larger machines, materials had to evolve accordingly. In addition to the load carrying capabilities of the materials, for generators, specific alloys evolved to provide specific electrical characteristics.

Turning to the turbines, many of the advances came via materials improvements, and alloy evolution and selection were based on operation at high temperature in a steam environment. Material improvements have allowed designers to take turbines to higher temperatures and pressures, with increased output as a result. Similar to the generator rotors, at a given speed, turbine rotor size effectively establishes the stress and therefore the material properties required for the rotor. Additionally, and again in a similar fashion as for the generator rotor, specific alloys evolved for the various turbines based on certain design conditions. For higher temperature rotors, which tend to be the smaller, typically high pressure (HP) and some intermediate pressure (IP) units, high temperature creep and temper embrittlement may be issues. Consequently, specific alloys evolved to provide enhanced resistance to these damage mechanisms. For the larger low pressure (LP) units, on the other hand, the lower operating temperatures mean that creep and embrittlement likely are not issues. However, the larger size dictates higher strength materials because of the generally higher stress associated with larger size. So, different materials evolved to meet these requirements. Other considerations involved in material evolution and selection include corrosion resistance and possibly steam, water droplet, and hard particle erosion resistance. And since blade velocity is directly related to diameter, different considerations come into play in these regards. Design of the smaller HP and IP rotors with the shorter blades will involve different considerations than the large LP rotors with long blades running at blade tip velocities near or exceeding Mach speeds.

Vibration and vibration analysis also was an integral and important part of the expansion of turbine size and capacity. Analogous to the small generators, in which the components were relatively small and stiff, growth in size typically equated to increased flexibility and therefore lower natural frequencies. As turbine blades grew in size, so did the need for dynamic analysis to assure that natural frequencies remained well away from any operational frequencies.

Operating hand in hand with all of these design and material considerations and evolution, improved design tools and computerization have enabled improved precision and reduction of conservatism, resulting directly in increased turbine and generator output. And this leads directly into the subject of this document—turbine and generator inspection and condition assessment. In the older machines, there was greater uncertainty associated with certain aspects of the designs, particularly materials and design calculations. Coincident with the evolution of improvements in materials and reduction in the uncertainty associated with certain behavioral characteristic was an increased ability to take advantage of these; that is, to work closer to the limits. And so, condition assessment is an issue for old and new alike.

Inspection and condition assessment of steam turbine and generator components is a complex process involving a significant number of components, an array of different materials, numerous potential damage mechanisms, and the full range of available non-destructive evaluation (NDE) methods and techniques. Add to this data analysis and then the stress and fracture evaluations, and it is easy to understand that many, many combinations are possible.

Insofar as the scope of this document is concerned, there are many, many tests routinely performed on various turbine and generator components and typically classified as maintenance tests. For the generator, these may include electrical tests to assess the viability of ongoing operation of the existing generator insulations, tests to confirm that the stator end turn bracing continues to maintain the coil end turns properly detuned from natural frequencies, tests for stator core tightness, wedge tightness tests, and so on. For the turbine, maintenance inspections include inspection of the stationary and rotating blades for cracking and other signs of distress including droplet erosion, water cutting, tenon cracking, tie wire cracking, and many others; they include dimensional measurements to confirm clearances at seals, and so on. These maintenance inspections generally detect conditions that can and are remedied at that time. Each test or measurement typically has an established limit, and once the limit is exceeded, the parts are either repaired or replaced. This might involve, for example, tightening of the generator stator end turn bracing, re-wedging the winding slots to tighten the compression on the coils, removing and replacing damaged hard metal erosion shields on the turbine blades, and an array of similar actions for the various other components. This class of inspection and assessment is not subject to further discussion within the scope of this document. Treatment of all of these tests, inspections, and assessments would require many, many volumes of material, even if covered only at the most rudimentary level.

Rather, the body of this document deals for the most part with the larger rotating components. These components are not necessarily subject to regular inspection at each turbine-generator overhaul, but are subject to critical analysis and re-inspection based on projections of damage progression into the future, or at intervals prescribed by the manufacturer based on historical information. They are included because the inspections are much more involved, time consuming, and costly, and so are performed on an as-needed basis rather than as part of the routine maintenance. Additionally, failure is typically much more catastrophic than the components covered by the maintenance tests, and immediate replacement is typically not an option because of long lead times and high costs that preclude the possibility in most instances of maintaining spares. Even repairs are typically very long, costly operations.

25.2 NON-DESTRUCTIVE INSPECTION METHODS

A number of inspection methods are used to assess various turbine and generator components for flaw conditions that could lead to failure or other operational issues if left unattended. Without going into great detail, some brief description of the relevant inspection methods is necessary as a background for the component-specific information to follow. Most of the basic inspection methods recognized by the American Society for Nondestructive Testing (ASNT) are employed in some form for one or more of the various components that are subjected to periodic inspection. For those unfamiliar with non-destructive evaluation (NDE), there are a number of basic NDE methods: ultrasonic, eddy current, radiography, and a number of others. Each method has very specific requirements that must be met to become a certified practitioner. In general, there are three levels of qualification for each method, Levels I, II, and III, listed in order of increasing responsibility. For example, a Level I must work under the supervision of a Level II or III, can implement an inspection, but cannot interpret results. A Level II can work independently and perform all stages of an inspection, including interpretation. A Level III can additionally develop and write procedures and administer a qualification program, as examples. Each qualification level has specific requirements in terms of education, experience (typically while certified at the next lower level), and training, and each require periodic requalification by examination.

Visual Testing (VT) is exactly as it sounds—visually assessing components or assemblies for signs of distress. While this might sound simple and even trivial insofar as recognizing it as a discipline, quality VT is actually one of the more demanding of the various NDE methods because it requires extensive knowledge of operative damage mechanisms and the visual signs that are associated with each. VT obviously requires access to the surface(s) under examination, although direct access is not mandatory. Video probes, mirrors and other devices are often used to view surfaces and assemblies that are otherwise inaccessible.

Ultrasonic Testing (UT)—UT is the material testing analogy to active sonar—ping and listen. UT involves the introduction of high frequency sound waves into a component using a transducer that converts electrical voltage into deformation and the opposite. Certain materials possess this piezoelectric property. The voltage is applied as a very short pulse, which causes the material to resonate briefly. By coupling the transducer to the surface of a component, the mechanical oscillation of the transducer can be coupled into the component as a short duration sound wave. Similarly, when a sound wave strikes the piezoelectric element of the transducer, a voltage is created. The basic concept of UT is to introduce the sound into the part, where it will reflect from a flaw back to the transducer and therefore produces a measurable voltage. The propagation mode and direction are controlled via the use of a refracting wedge between the transducer and the component. The sound field propagation direction is used to define the direction in the part from which the reflection occurs. Because propagation velocity for a specific propagation mode in a given material is constant, the propagation time provides a measure of reflector location along the beam, and the voltage of the return signal gives some indication of the significance of the reflector—larger flaws cover more of the field and therefore in general produce more significant reflection amplitudes. UT can often be effectively used to detect and characterize flaws at locations completely otherwise inaccessible, including flaws located at hidden internal surfaces of components and flaws internal to the material itself, i.e., subsurface flaws. There are numerous variations, or techniques, employed in optimizing an inspection for a specific application. These will be discussed to the degree necessary along with the various applications.

Magnetic Particle Testing (MT)—This is a method for detecting surface connected and near-surface flaws at accessible surfaces of ferromagnetic materials. When the material is magnetized, any flaw lying in the magnetized material disrupts the field, forcing the magnetic flux around the flaw. This establishes magnetic poles at the flaw. The magnetic field can be developed directly by passing current through the part or indirectly by passing current through a conductor located in close proximity to the part. By then distributing fine particles of ferromagnetic material over the part, the particles are attracted to the field established by the flaw. Color contrast particles are available in different colors to provide the best contrast to the background color of the component. Fluorescent particles are available, typically in liquid suspensions, for use with ultraviolet light to make them more visible. Wet fluorescent MT (WFMT) is typically considered to be more sensitive than dry, color-contrast MT. For subsurface flaws, because the leakage field associated with a flaw is relatively weak and localized, detection does not typically extend to a significant depth into the part. However, MT procedures can be effectively applied for the detection of flaws under thin coatings, for example, paint.

Liquid Penetrant Testing (PT)—This method is limited to the detection of flaws that are open to an accessible surface of the component. The principle upon which PT relies is that of capillary action of a liquid. A penetrant liquid is applied to the surface to be inspected and is drawn into surface flaws by capillary action. The surface is then cleaned of all remnant penetrant, and a developer is applied to draw the penetrant from the flaw to the surface to make it visible. Similar to MT inspection, the liquid penetrants come in color contrast and fluorescent, the latter requiring the use of ultraviolet light and also the more sensitive. The penetrant material itself is specifically designed via controlled color, surface tension, and viscosity to enhance the process. Cleanliness and surface preparation are very important considerations in PT inspection. Any contaminant that prevents the introduction of the penetrant into the flaws can reduce the effectiveness of the inspection or even render it completely useless. Similarly, contaminants that hold penetrant can lead to false calls. Surface cleaning methods that upset the surface, such as abrasive grit blasting and grinding, are generally considered incompatible with PT inspection as the flaws can be smeared over or peened closed.

Eddy Current Testing (ET)—ET inspection is an electromagnetic inspection method that is effective in detecting and characterizing surface and near-surface flaws in conductive materials. When current is passed through a coil that is held in close proximity to the surface of a conductive material, the magnetic field developed by the coil induces current flow, i.e., eddy currents, in the conductive material. So long as all things remain constant–material conductivity, magnetic permeability, proximity, material condition–the characteristic impedance of the coil remains constant. However, any change, including the presence of a discontinuity within the eddy current field, changes the impedance. Consequently, the impedance can be monitored as a means of assessing the surface for the presence of flaws. Alternatively, a secondary pickup coil can be used to measure the eddy currents directly, with the same result. Eddy current inspection provides advantages over other surface inspection methods in that the output can easily be recorded and positionally correlated. It therefore lends itself to automation. ET inspection can also be very sensitive, even to very small, tight cracks that likely might go undetected by other surface inspection methods.

There are other inspection methods such as Radiographic Testing (RT), Infrared Thermography (IRT), and others that are not generally used for any of the covered turbine and generator applications, so additional detail on these is not provided.

25.3 TURBINE DESIGN AND COMPONENT OVERVIEW

While it is well beyond the scope of this work to provide significant detail on turbine and generator design, certain information is required in order to put into the right context the information that is provided on inspection and condition assessment. As is typically the case, certain damage and failure mechanisms, and therefore the means of dealing with them, are related directly to the design or to specific design details. There are many different types of steam turbines, the differences founded in the underlying principles involved with the conversion of thermal energy to work. There are condensing and non-condensing turbines, reheat turbines, extraction turbines, induction turbines, and so on. In the power plant, the most commonly found is the condensing turbine, which exhausts steam that is already partially condensed and at a pressure below atmospheric into a condenser. The basic concept is to extract work by allowing the steam to expand through a series of blades that turn the rotor onto which the blades are attached. The blades through which the steam initially passes are very small. Small blades maintain acceptably low stress when exposed to the high pressure steam; yet extract significant work because of the high pressure even though acting on relatively small areas. To allow the expansion and make most efficient use of each successive row of blades, the blades increase in size along the flow direction. As the pressure drops, greater area is required to extract meaningful work, and the blades can be larger and yet maintain acceptable stress levels because of the lower pressure. Typically, turbines for power generation use have been manufactured in one of three basic types: high pressure (HP) turbines, intermediate pressure (IP) turbines, and low pressure (LP) turbines. In any of these designs, the blades are arranged around the rotor in rows, where each row has duplicate blades within it and the blades increase in size along the steam flow direction through the turbine. In addition to these basic types, there are also hybrids, for example, HP/IP and IP/LP turbines that combine the two into a single unit. In such case, the two are still separated internally in sections, e.g., the HP section and the IP section, but are combined into a single rotor in a single casing.

Regardless, the HP turbines will typically be the smallest machines having relatively small blades, IP will be larger machines with larger blades, and the LPs are the largest machines with the largest blades. In terms of pressures and temperatures, the HP sees the highest pressures and temperatures, followed by the IPs which might see equivalent temperature but not equivalent pressure, followed by the LP. The last few blade stages in the LP typically see some level of condensation before the steam exhausts into the condenser. In addition to the various pressure designations for the turbines, they are also characterized by flow direction, single flow or double flow. Whereas HP and IP machines are generally single flow, i.e., having steam flow in only one axial direction, it is more common for LP rotors to have the steam enter at the axial center of the machine and flow equally in both outward directions, i.e., double flow. In the latter double flow (DF) turbines, there is no net axial thrust and therefore no requirement for thrust support provisions in the design. A typical DFLP rotor is shown in Figure 25.1

FIG. 25.1 DFLP ROTOR SITTING IN THE BOTTOM HALF-SHELL

sitting in the lower half of the casing with the upper half removed. In most cases, the two sections of the hybrid HP/IP and IP/LP machines are opposing flow, but this is not always the case.

In terms of the turbine combinations, any combination that can be imagined has probably been built at some time or other. The predominant arrangement in more modern turbines is the tandem compound arrangement in which the machines are all aligned along a single shaft axis and coupled directly one to the next. Such an arrangement might have an HP turbine, followed by an IP turbine, followed by one or more LP turbines, and then the generator. As mentioned above, there are many, many variations on the number of turbines involved in driving the generator and on how they are arranged. There are single turbines driving the generator, up through and including five turbines (most that could be found) including an HP, IP, and 3 LP turbines. In these arrangements, the same steam flows from the HP, then to the IP, and then to the LP(s). In a reheat unit, the steam returns to the boiler after passing through the HP, where it is reheated before passing through the IP, which for this reason is also commonly called a reheat turbine. Cross compound is another option in which there are two generators driven by separate turbines. A cross compound unit typically has a 3600 RPM, 2-pole generator driven by an HP or an HP/IP, or an HP and an IP, plus an 1800 RPM, 4-pole generator driven by an one or more LP rotors. These still use the same steam and can involve reheat steam to the IP turbines or IP sections.

Output for steam driven generators in the US, of which there are on the order of 2500 and 3000 operating units devoted to commercial power generation, ranges for the most part between 50 and 1200 megawatts (MW), although a significant number of smaller units also can be found, even as small as 2 MW, operating typically at municipal facilities. Regarding the age of the units, not too many steam power plants have been built in the US since the mid- to late 1970s, so the fleet is aging, with the majority near or past the half century mark.

So, just from these generalized descriptions some of the important considerations that could impact the approach to inspection and life assessment might include:

- Types of turbines comprising the unit—HP, IP, HP/IP, LP, etc.;
- Unit rating and therefore the sizes of the machines—also with second order affects such as the materials used as impacted by imposed stress, which is related to size;
- Age of the unit as this dictates the materials that would have been used and certainly the accumulated service hours;
- Rotational speed of the unit;
- Pressure and temperature of the steam;
- Temperatures of the various components; and,
- State of the steam along the flow, i.e., condensing or not.

For the generator, it is a bit simpler. A generator is a generator in terms of the overall design concept. The rotor has sections of the body containing a number of axial slots running the full length of the body and solid sections that form the magnetic poles. Coils are wound into the axially machined slots and around the magnetic poles that are integral to the rotor forging. The coils are held in place by slot wedges, and where the coils exit the slots and wrap around to the opposite side of the pole, large rings are assembled over the end turns to support them. A DC voltage applied to the rotor windings, i.e., the excitation voltage, creates the magnetic field associated with each magnetic pole.

The stator is essentially a large core of ferromagnetic, electrical grade material surrounding the rotor and built up by many thousands of very thin laminations stacked axially and having integral slots running the full length to contain and support the stator coils. Similar to the rotor windings, the stator coils are held in the slots by wedges and connected in the end turn area to form the 3-phase winding.

As opposed to the turbines, of which there are many different variety, as described above, all generators take on these same common features. The distinguishing features in the generator are generally linked directly to output, heat generation, heat capacity, and heat dissipation:

- Unit rating;
- Stator cooling method, e.g., conventional, gas inner cooled, direct water cooled, etc.;
- Ventilation gas, typically air or Hydrogen;
- Ventilation gas pressure (even if having water cooled coils, the rotor and the remainder of the stator are still gas cooled); and,
- Temperature rating of the insulation materials.

Having now established some basic design fundamentals, we can delve deeper into the issues associated with inspection, condition assessment, and remaining life analysis. We will start with turbines and move to generators.

25.4 TURBINE COMPONENTS REQUIRING PERIODIC LIFE ASSESSMENT

Figure 25.2 provides a schematic of a turbine rotor cross section. Remember that the stationary components were defined as routine inspection and maintenance components, at least for the purposes of this document, and are therefore not included.

Regardless the type of rotor, i.e., HP, IP, or LP, there are many common features rotor to rotor. Generally, there is a central shaft and most often (at least for older vintage rotors) the rotor shaft has a central bore hole. The disks, positioned appropriately along the shaft to provide spaces between for the stationary blades that adjust the steam flow direction at the admission side of the rotating stages for optimum performance, form the support members for the rotating blades. The disks can be non-integral to the central shaft, as shown, partially integral meaning that the smaller disks are integral to the shaft and the larger disks are separate, assembled components, or fully integral to the shaft. Assembled disks typically involve an interference (shrink) fit. Historically, the smaller HP and IP rotors had integral disks and LP rotors had assembled disks. The evolution from fully non-integral, to partially integral, to fully integral disks essentially followed directly the evolution and availability of larger, quality forgings. Some HP and IP designs feature blades that are directly mounted to the shaft without the presence of disks, per se. Other major components include the bearings that support the rotors and couplings that form the connection interface between adjacent rotors.

25.5 ROTOR FORGING ASSESSMENT

Considerable research and development activities have been expended over the years to develop inspection and analysis tools that reduce outage and analysis times and provide realistic assessments of remaining life. For the central shafts, forging production standards have provided the appropriate surface from which to launch relevant inspection protocols. The process that results in a useful rotor forging begins with one of a number of rotor forging specifications which have evolved over the years to provide optimum performance under a number of operational conditions. The classic material breakdown for applications in fossil plant rotors is:

- HP and IP rotors having operating temps up to 1050° F—Cr-Mo-V (typically 1Cr-1Mo-1/4V) forgings;
- LP rotors and discs—3.5Ni-Cr-Mo-V;
- Generators—Ni-Cr-Mo-V; and,
- Some very old units may be just Cr-Mo or Ni-Mo-V (1940s) but those would be the exceptions.

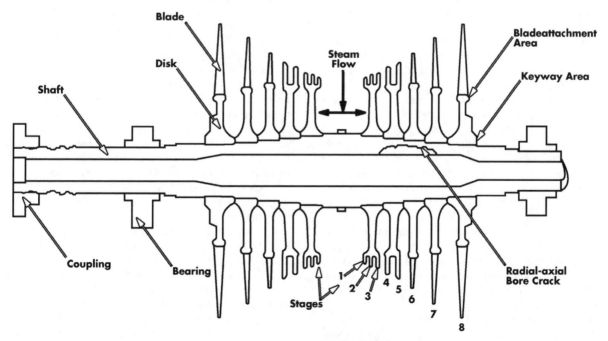

FIG. 25.2 TURBINE ROTOR (DFLP) CROSS SECTION

Specifications include:

- A-469 Vacuum-Treated Steel Forgings for Generator Rotors;
- A-470 Vacuum-Treated Carbon and Alloy Steel Forgings for Turbine Rotors and Shafts (this is a Ni-Cr-Mo-V materials); and,
- A-471 Vacuum-Treated Alloy Steel Forgings for Turbine Rotor Disks and Wheels (also Ni-Cr-Mo-V materials).

One designation that is referenced occasionally is that of C-grade rotor material. C-Grade is a designation that was applied to 1950s and earlier vintage 1Cr-1Mo-1/4V forgings that were austenitized at 1850°F. These forgings were found to exhibit low ductility and notch sensitivity (low fracture toughness). The austenitizing temperature was lowered in the mid-1950s to 1750°F which improved ductility and toughness, designated as D-Grade material. Much of the early development activities associated with forging in-service inspection (ISI) was centered on inspection of C-Grade rotors, although the inspections now are recommended for most, if not all, rotors.

Older forgings, particularly those made prior to the advent of vacuum degassing practices, were prone to concentrations of impurities and flaws near the forging centerline. In the ingot, as the solidification of the molten steel progressed from the outside inward, impurities tended to concentrate within the last to solidify, so toward the ingot center. Then, as the solidified ingot was forged to form the basic shaft shape, these impurities were further condensed to the center section, concentrating them along and around the shaft axis. A simple solution for removal of these impurities and flaws was to bore the rotor along the axis.

In their final configurations, turbine and generator rotors are subjected to any number of significant stress conditions. However, the single most significant concern in terms of consequences of a failure is a rotor burst. High speed rotation creates significant hoop stress that is highest at the shaft center and falls off with increasing distance from the center. The central bore hole, in fact, acts as a stress concentration and thereby increases the stress at the bore surface even further. Even so, until recently, when steelmaking practice evolved to the point of making suitably clean large ingots and forgings, the benefits associated with flaw removal via the presence of a bore overrode the negative impact on stress. Consequently, in the US nearly all large rotors have central bore holes. There are some exceptions, for example some rotors made during World War II, when boring machines were otherwise occupied, and more recently when forging quality improvements have allowed for the elimination of the bore, but, in general, the presence of a bore is relatively standard for the vast majority of the US fleet.

The raw forging, typically in the form of a series of contiguous cylindrical sections made to encompass the final rotor shape, is fairly easy to inspect from the outer periphery, and specifications evolved accordingly for acceptance inspection of the forging. For the central material, ASTM A-418 is one such inspection standard. In the early days of ultrasonic inspection and still in many cases to this day, ultrasonic inspection instruments were (are) calibrated using a series of calibration blocks containing reflectors of the size of interest and located at various test distances representative of the volume of interest in component being inspected. For large rotors, this means very large blocks. ASTM A-418 was developed for the specific purpose of eliminating the calibration blocks by using either the bore in the case of a hollow cylinder or the far wall in the case of the solid cylinder as the calibration reflector. Theoretically derived relationships between the reflector of interest and the reflection from a bore or backwall are used to establish the correct calibration sensitivity based on the bore or backwall reflection.

Considering the complex configuration of the final machined rotor, however, ISI is not so easily accomplished from the outer periphery. This is true regardless whether the subject rotor is a turbine rotor, which has the disk configurations at least and possibly assembled disks covering the OD, or a generator rotor that has the full-length integral winding slots cut into the body. However, the presence of the central bore hole provides a convenient, typically uniform surface from which to inspect the central material for potentially detriment flaws. Failures that initiate at the rotor bore surface, where in-service stresses are highest, propagate by low cycle fatigue or high temperature creep. Failures that initiate subsurface typically initiate by link-up and growth of inherent discontinuities remnant from the steelmaking process, but then propagate by low cycle fatigue. Because the predominant stress is a hoop stress, i.e., tangential, the flaw of main concern is in a radial–axial orientation, normal to the primary stress direction. The study of linear elastic fracture mechanics tells us that a surface flaw of axial length (L) by radial depth (a) has a crack tip stress intensity that is the same as a subsurface flaw that is the same length but twice the radial dimension, i.e., $L \times 2a$, if acted on by the same stress. This factor, when combined with the fact that the stress is highest at the bore surface, means that the bore surface flaw is of major concern. However, the subsurface flaws, particularly those in close proximity to each other such that linkup is possible, cannot be discounted.

Ultrasonic inspection of the rotor material from the bore surface, a practice that has become known generically as boresonic inspection, has become standard throughout the industry. Typically, the inspection involves the introduction of ultrasonic waves in different directions relative to the bore surface to look for flaws at different orientations. Because of the criticality of surface and near-surface flaws, located where stresses are highest, the inspection is typically designed to concentrate on the first few inches of material radially outward from the bore. In most applications of ultrasound, there is a limitation on the ability to detect and assess flaws located at and very near the test surface, i.e., the surface from which the ultrasonic waves are launched, in this case the bore surface. This is typically a very shallow distance, measured in hundredths of an inch, but in this case it is, nonetheless, the most important material in terms of flaw criticality. Consequently, boresonic inspection is supplemented using reliable surface inspection methods, generally MT or ET inspection.

Boresonic systems currently in use are all fully automated, featuring motorized, external scanning mechanisms that transport the transducers systematically through the bore via a probe head and series of drive rods to accomplish complete and reliable coverage of the material. Figure 25.3 presents a photograph of a turbine rotor inspection underway. The scan is accomplished either in a raster mode in which the transducers undergo alternating clockwise and counterclockwise 360° rotations with an axial index between, or in a continuous helical scan. The raster mode allows the transducers to be hard wired to the external instrumentation and data acquisition system, while the helical scan mode requires the use of slip rings in the circuit to prevent continuous twisting of the wires. The automated systems contain motion pickups to record probe head position along with the inspection data. This enables the generation of positionally-correlated data images for analysis purposes. More information on data analysis is presented later.

Boresonic inspection is supplemented with a surface inspection to cover the zone in which the UT is ineffective. The surface inspection is typically either MT or ET. Bore MT is conducted

FIG. 25.3 BORESONIC DRIVE ATTACHED TO AN HP/IP ROTOR

using a central conductor and therefore indirect magnetization. The inspection relies on residual magnetization, i.e., the magnetizing field is applied and then released while the magnetic particle suspension is applied. The inspection utilizes color-contrast particles in a wet suspension. Viewing of the surface is performed using an optical borescope or miniature camera in combination with a high intensity light source. Bore ET inspection is conducted using a more or less conventional eddy current probe that is scanned over the entirety of the bore surface.

For rotor bore surface inspection, MT and ET each provide certain advantages and disadvantages. By virtue of the fact that ET is an electronic inspection method, the data can be digitally recorded for archival retention purposes. The inspection method requires that the ET probe is scanned over the surface, thereby making the inspection compatible with the boresonic scanning. In fact, the probe is typically included in the boresonic probe head such that the data can be acquired concurrently with the boresonic inspection data, which results in a real time savings. This advantage notwithstanding, the main advantage of ET owes directly to the fact that the data is acquired along with the digitally stored positional information and so the coordinate reference system is consistent with that used for the boresonic inspection. In the analysis phase of a bore evaluation, the data is combined based on proximity of indications and the local stress level. Closely proximate indications in regions of high stress must be considered as a single flaw, the boundary of which encompasses the entire volume containing the individual flaws. The logic is that, if the ligament stress is high enough, the individual flaws will soon link. Even small errors in flaw location therefore can lead to significant errors in combining the individual indications into an equivalent flaw size. And the error can go either way—overly conservative by combining individual flaws that have wider separation or treating individually flaws that should be combined. The primary deterrent to bore MT inspection is that the inspection must be conducted using some visual device, such as an optical borescope or camera, to view the surface. Unfortunately, when using such devices, there is no effective way to provide the same positional precision as can be attained when using precision optical encoders in conjunction with automated positional systems. And the fact that the ET inspection can be conducted concurrently with the UT means that both sets of data have exactly the same position reference datum. However, the advantage of the bore MT is that it provides the ability to "see" the indication, a capability

that is still very attractive to many. A typical MT indication associated with an axial bore crack is shown in Figure 25.4. Likely the best of both worlds is the use of ET as the primary inspection, with full digital data acquisition and storage concurrently with the UT inspections, followed by bore MT if/as necessary to corroborate ET indications.

Bore surface preparation typically involves honing the bore to provide a smooth, uniform surface condition, void of any conditions that would upset the transducers or otherwise interfere with effective coupling of the ultrasound into and out of the rotor material. In manual UT, the operator has full control of the transducer and can compensate for minor surface irregularities by over-scanning and feel. But in automated UT, the transducers traverse the surface one time and with limited capability to compensate for irregularities. Consequently, the surface must be smooth and regular. Most bore inspection specifications require that all oxide and contaminants are removed and that a 63-μ-inch finish is provided for the inspection. This surface finish is critical for the proper contact of the UT probes and equally important for reliable detection and characterization of surface MT indications. Rotors typically have shrink-fit assembled end plugs that must be removed to access the bore, and this operation is typically done by machining the plugs. For bores that have a uniform bore diameter, typically only a single plug is removed and the inspection is conducted from one end. Often, however, rotors may have steps, transitions, and even bottle bores. These features likely were put in at the time of manufacture to remove material containing detected flaws. Depending upon the details of the bore configuration, access from both ends of the bore for honing and delivery of the inspection systems may be required, in which case both end plugs must be removed.

Data analysis, for the most part, is now a largely automated process that converts the ultrasonic data into flaw tables that include the location of each flaw centroid, expressed in most cases in terms of radial depth, R (inches), axial position, Z (inches), and circumferential position, θ (degrees), as well as dimension in each direction as well. At one time, sizing of indications was based solely on the amplitude of the ultrasonic response using area/amplitude correlations based on the response from flat bottom holes (FBHs) in calibration blocks. However, this sizing method has long been recognized as inaccurate because response amplitude is impacted by

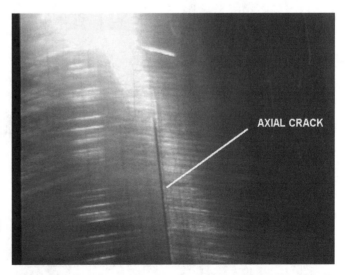

FIG. 25.4 IMAGE OF A BORE AXIAL CRACK IN AN HP TURBINE ROTOR

many variables in addition to flaw size. Flaw orientation relative to the beam direction, flaw shape, morphology, and composition can all impact the reflection amplitude, not to mention material structure and its affect in terms of attenuating the sound. Consequently, without a priori knowledge on all of these other variables, size analysis based on amplitude alone is prone to significant error. And any such errors are always non-conservative; that is, all of the variables that impact amplitude tend to lower the amplitude and therefore make the flaws appear smaller than actual.

This is not meant to imply that early practitioners did not recognize and account for these errors. To the contrary, significant correction and safety factors were applied to account for the potential sizing error. Unfortunately, the flaw sizing processes and algorithms based on response amplitude simply resulted in significant uncertainty. As a direct result, extremely conservative approaches were utilized to make it absolutely certain that failures would not occur. For many years, beginning in the early 1980s, related R&D efforts concentrated on the removal of conservatism without assuming unacceptable risk of failure.

More recently, flaw sizing algorithms based on dynamic echo response, hit envelope, tip diffractions, and amplitude drop have evolved and have improved the accuracy of the flaw sizing results, and therefore reduction in the safety factors that must be applied. Ultrasonic images like those shown in Figure 25.5 are also reviewed as part of the data analysis process and to perform spot checks on selected flaws from the flaw tables.

As stated earlier, the primary concern for the central material in a rotor forging is low cycle fatigue. The stress cycle is related to start/stop operation and to a much lesser degree to load swings. When a unit is brought up to speed and then put on line, the turbine rotors heat up and the rotational (centrifugal) stress increases with increasing rotational speed. The startup procedure can have a significant impact on peak stress, as the peak thermal stress occurs during the startup transient conditions and then is significantly reduced once temperatures equilibrate and stead state operating conditions have been reached. It is therefore good practice to control the startup rate to assure that the peak thermal stress is not present concurrent with the peak mechanical stress. This typically means slow ramp up to full speed or even hold points at certain speeds to allow the rotors to soak at the temperatures to equilibrate there before continuing. Material properties are also impacted by temperature, and in general, higher temperature means more ductility and less susceptibility to brittle fracture. These warming practices also allow for the materials to pass the brittle-to-ductile transition temperature range before reaching peak stress. For the generator, temperatures are much lower and thermal stresses are typically not even a consideration. Once at speed and fully on line and at operational temperature, the hoop stresses that predominate in the shaft central material remain effectively constant until the unit shuts down. Minor stress cycles accumulate with major load, and therefore thermal swings, but these account for only a small fraction of the accumulated damage and life consumption. By far, cold starts account for the majority of life consumption, followed by hot starts for which the thermal stresses are not additive, followed by major load swings.

Engineering analysis of indications detected via the rotor bore inspections is performed using component-specific analysis programs. While there are other analysis programs used by various analysis providers, the SAFER (Stress And Fracture Evaluation of Rotors) computer analysis program that was developed by the Electric Power Research Institute (EPRI) sets the standard and will be presented here in that regard. SAFER provides two distinct analysis options, deterministic and probabilistic. In a deterministic analysis, single-value input parameters are used. These are typically worst case values to provide the most conservative solution, which is in the form of a go–no-go answer. That is, the future duty cycle is input, along with conservative parameters in the form of flaw sizes, material properties, and stresses, and a conservative estimate of remaining life is determined. Alternatively, the SAFER program provides the means to perform probabilistic analyses. Under this analysis option, all input parameters are input as statistical distributions, and a Monte Carlo simulation is exercised to deter-

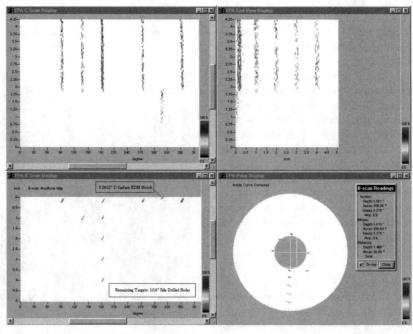

FIG. 25.5 ULTRASONIC IMAGES (SECTIONAL VIEWS) USED IN DATA ANALYSIS, IN THIS CASE REPRESENTATIVE OF A CALIBRATION BLOCK CONTAINING MACHINED REFLECTORS

mine the probability of failure in a specified operational interval. This analysis removes a great deal of conservatism and therefore is more realistic. However, in order to take advantage of the lowered conservatism, the owner must be willing to deal with the output in the form of a probability of failure in a specified operational life. The owner must decide the probability of failure that represents an acceptable risk. Typically, operators consider a 0.1% probability of failure over the intended operational period until the next scheduled inspection to be acceptable, and for Nuclear units the acceptable risk is more typically 0.01% (1 in 10,000). But the comfort level is something that only the owner/operator can define based on their tolerance for risk. In reality, there is risk associated with operation of a unit regardless the assessment that is conducted. The probabilistic approach merely quantifies this risk.

The analysis process involves three primary tasks. The first is a transient thermal-elastic finite element stress analysis to determine the distribution of stress and temperature in the near-bore region of the rotor. It is important to evaluate the typical cold start-up condition(s) for that specific unit to ensure that these factors are included in the analysis. Information required to perform the transient stress analysis of a rotor is listed below.

- Description of the rotor periphery as a function of axial location;
- Description of the rotor bore geometry as a function of axial location;
- Blade weights and mass center location or rim loads for each stage;
- Inlet and outlet steam temperature and pressure versus time during a 'typical' startup;
- RPM versus time during a typical startup;
- Heat balance diagram for the unit at full load;
- Service history of the unit (online time and starts); and,
- Rotor material data—fracture appearance transition temperature (brittle/ductile temperature), composition, yield strength, fracture toughness.

This is followed by the linkup analyses of the NDE data based upon the proximity of indications and local stress and temperature. This linkup analysis utilizes the spatial variation of stress and temperature in the rotor to predict whether or not neighboring indications should be combined and considered as larger indications.

The third task is to perform the critical crack size evaluation. This is either done deterministically using specific values for the various input parameters or probabilistically using statistical ranges derived from historical databases. Where the deterministic approach is exercised here as the third step, it can be followed by the probabilistic analysis if additional rigor is considered justifiable, or the analysis can proceed directly to the probabilistic analysis, bypassing the deterministic. In this probabilistic treatment, the values of stress, initial crack size, fatigue crack growth rate, creep crack growth rate, and fracture toughness are treated as random variables. If available, rotor-specific material properties and their associated distributions are used. In those instances where rotor-specific materials property data is not available, literature values available for the same class of material in similar rotors are used. It is recognized, however, that the distributions for the various database properties are relatively broad, i.e., having relatively large standard deviations. This factor alone can have a dramatic impact on the outcome. If this level of analysis falls short of indicating the desired remaining operating life with acceptable risk, then the next level of analysis would require rotor-specific material properties, with the hope that the properties are determined to be toward the

better side of the distribution rather than toward the worse side. This means that material samples must be removed from the rotor for testing purposes. Sample removal and testing is discussed later.

Another consideration in assessing certain rotors is that of damage by high temperature creep. As with other topics mentioned at various points within this document, the purpose is not to describe creep in great detail, but simply to provide sufficient information to enable a reasonable understanding of its importance. Creep occurs essentially by atomic level diffusion of defects in the atomic structure. When subjected to elevated temperature above about 850°F and sufficient stress, vacancies and dislocations migrate within the atomic structure until they reach a grain boundary, at which point they are pinned. As the process proceeds, additional "flaws' continue to accumulate at the grain boundaries, eventually forming microscopic voids or cavities. At first, these will be distributed relatively randomly, but eventually, as the process continues, they become "oriented", i.e., more prevalent along grain boundaries that are normal or near-normal to the principal stress direction. Ultimately, the size and density of cavities becomes sufficient that they begin to link to form micro-cracks. These eventually coalesce to form macro-cracks. Once cracks are formed, they can propagate by creep or fatigue, or a combination creep-fatigue mechanism. In its initial stages, creep is consider more a materials properties degradation mechanism in that the properties begin to degrade before any measurable flaw can be detected. However, once cracks are formed, creep is a double edged sword because not only is a fracture mechanics flaw present, but it is growing into material around the crack that is likely also degraded via creep cavities and/or micro-cracks. Once creep reaches this point of progression, failure is typically not far behind.

Temper embrittlement is yet another damage mechanism that can be experienced in higher temperature rotors. Whereas creep involves diffusion of atomic level vacancies and dislocations, temper embrittlement involves diffusion of certain impurity elements (i.e., phosphorus, tin, lead, sulfur, arsenic, etc.) to prior austenite grain boundaries. Here they provide easy fracture paths, lowering fracture toughness and increasing the fracture appearance transition temperature (FATT) accordingly. Temper embrittlement typically occurs in the temperature range of 600°F - 1000°F, although degree of embrittlement is a function of exposure temperature within this range, with the greatest effects occurring in the range of 600°F to 750°F, targeting the back ends of the HP and middle of the IP sections. It seems more than logical that increased content of tramp elements would increase susceptibility to temper embrittlement and quantitative chemical analysis can be used to determine the level of tramp elements in the material. However, no correlation to composition with which to predict susceptibility is known to exist.

Certain HP and IP rotors, or at least regions within these rotors, operate within the temperature ranges at which creep and/or temper embrittlement are potential considerations. In the analysis of the rotor forgings, in fact, even in the absence of detectable flaws, these mechanisms may become life-limiting. Every NDE method and technique, as implemented for a specific application, has an inherent detection threshold beneath which flaws will not be detected. As the properties degrade, at some point the flaw size needed to grow to failure within some defined operating interval can fall below the detection threshold of the inspection, at which point the detection threshold flaw must be considered as the life-limiting case. Said another way, if a flaw is present just below the detection threshold, and if that flaw is larger than that needed to grow to failure within the next duty cycle, then the properties in combination with a near-detectable flaw become life-limiting and the reinspection interval must be shortened to preclude risk of a failure.

25.6 TURBINE DISKS

For the purposes of this discussion, the disk is divided into three parts, the hub, which is the relatively large, heavy body, central part including the bore and keyway, the thinner web section that extends outward from the hub, and the rim section where the blades are attached. See Figure 25.6.

Turbine disks are generally susceptible to intergranular stress corrosion cracking (IGSCC) and are particularly sensitive when operating in unsuitable steam chemistry. Additionally, and potentially as important as or even more important than operating conditions are the conditions present during shutdown. While cracking can occur essentially anywhere on the disks, certain areas are more prone to damage than are other areas. Highly susceptible areas include those having higher stresses combined with either high temperature exposure or geometric considerations that tend to trap contaminants that contribute to the stress corrosion damage progression. This typically means that the blade attachment regions are prone regardless. All operate at elevate stress because

of the stress concentrations associated with attachment geometries. Some operate at temperatures sufficient to cause long term thermal damage, and others operate in regions where steam conditions are suitable for the formation of stress corrosion damage.

The other region that is considered highly susceptible to cracking is the bore/keyway region of shrink-assembled disks. The designs of shrunk-on disks typically involve some form of an axial key at the shrink-fit interface to prevent the disk from rotating relative to the shaft under upset loading conditions. While the entire disk is susceptible to IGSCC, the bore is more prone than most other surfaces because the nominal rotational stress is highest at the bore. Additionally, because of the stress concentration effect of the keyway, the keyway surfaces are even more prone to cracking. Another significant factor that can impact IGSCC in the bore and keyway regions owes to the fact that there is no steam flow in these regions. While other surfaces exposed to steam flow can be cleaned of contaminants by the flow, the bore and keyway surfaces are not flushed in any way. Consequently, once contaminants are introduced into these regions, the IGSCC process continues unabated.

Disks operating at and beyond the phase transformation zone (PTZ), i.e., where the steam is transforming from dry to wet and beyond, are particularly prone to IGSCC. For many years, IGSCC was known as an operative damage mechanism and was considered to be a function only of steam chemistry and the local operating conditions, i.e., those conditions present during operation in and beyond the PTZ. More recently, the process has been better defined and it now appears that the initiation occurs during unprotected shutdowns [1]. The droplets and liquid films which form on the blade/disk surfaces in and beyond the PTZ during operation do not initially create damage because neither contains any oxygen, even in units with hundreds of ppb of oxygen in the steam. However, in addition to the liquids, deposits of contaminants such as chlorides can form during operation. During shutdown these deposits are still present, and if there is no protection the deposits absorb moisture. Unlike the moisture present during operation, however, this moisture does contain oxygen, and when present in combination with the deposits of contaminants remaining from operation, becomes highly acidic. This environment first breaks down the passivity on the PTZ surfaces, which leads to pitting. Repetition of the unprotected shutdown situation eventually leads to a critical pit size which, during operation in the concentrated liquid film, will grow into a micro-crack. This micro-crack can then grow (propagate) as a corrosion-fatigue or stress corrosion crack only during operation because this is the only time that the stress exists. However, initiation stems directly from unprotected periods of shutdown. If the turbine and the PTZ are protected with dehumidified air, starting day one and continuing during subsequent period of shutdown, the deposits remain dry and therefore do not cause passivity breakdown and pitting. As a consequence there are no initiating centers for corrosion fatigue and stress corrosion cracking.

Pulling all of these factors together, it means that the most susceptible disks are the last downstream in the steam flow, typically the last couple stages in the LP rotors. These disks are also the largest on the rotors and therefore the ones that can do the most damage in the event of a catastrophic failure, which can occur if bore/keyway cracks propagate to a critical size. A failed disk is shown in the photograph of Figure 25.7.

Turning now to the inspection of shrunk-on disk bores and keyways because the cracks occur at inaccessible surfaces, UT is the only available inspection option unless the rotor is de-stacked, i.e., the disks removed to enable implementation of surface inspec-

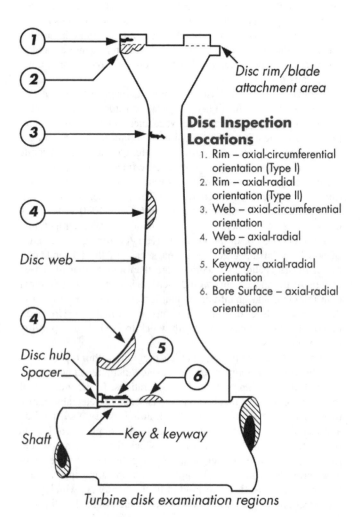

Disc Inspection Locations

1. Rim – axial-circumferential orientation (Type I)
2. Rim – axial-radial orientation (Type II)
3. Web – axial-circumferential orientation
4. Web – axial-radial orientation
5. Keyway – axial-radial orientation
6. Bore Surface – axial-radial orientation

Disc rim/blade attachment area

Disc web

Disc hub

Spacer

Shaft

Key & keyway

Turbine disk examination regions

FIG. 25.6   SHRUNK-ON DISK AND INSPECTION LOCATIONS

FIG. 25.7   FAILED TURBINE DISK

tion methods. Because of the cost associated with de-stack and re-stack, which may even necessitate high speed balancing of the rotor, inspection with the disks remaining on the shaft is by far the preferred approach. However, in many cases, the geometries of the surfaces available for transducer placement often detract from the effectiveness of the inspection.

Stresses at the bore and keyway are tangential; consequently, cracks tend to grow in a radial/axial plane. In UT, optimum detection for cracks at the opposite surface of a component occurs when the beam is in a plane normal to the corner created by the intersection of the crack with the component surface. Additionally, the best approach angle within this plane is within the range of 30° to 60° and best at 45° relative to the corner. For this application, the beam is therefore optimum when maintained in a radial/circumferential plane and at 45° relative to the ID surface. At the outer extremes of the hub, some designs have relatively short cylindrical sections that make it easy to accomplish this optimum beam propagation direction, but only over the lengths of the cylindrical sections. Looking back to Figure 25.6, it certainly goes without saying that a beam cannot be maintained in a radial/circumferential plane under the web section, and other geometry can impact the effectiveness of the inspection as well. This is not to say that adequate coverage cannot be achieved. However, going back several years when ultrasonic data was presented as a time/amplitude trace on a CRT (or more recently, digital) screen and when the ultrasonic beam angle was fixed to a single value for a particular transducer/wedge combination, achieving high reliability was very difficult and tedious. On one particular manufacturer's designs, the web sections are continually changing contoured radii; consequently, to best cover the entirety of the bore, a series of compound curved wedges had to be used, and each applied only to a single location on the web.

The advent of computer based data acquisition and analysis systems and the use of linear phased array ultrasonic technology have resulted in improved flaw detection and sizing in general, and particularly so for this difficult application. Since this is the first mention of phased array UT, some brief description of the technology will be provided before proceeding on the disk inspection application.

An ultrasonic phased array system includes an array transducer, which contains multiple, precisely positioned transducer elements

(the array), an electronic pulser/receiver unit that provides independent pulsers and receivers for each of the transducer elements, and timing circuitry that controls the sequence and the timing intervals (phasing) at which the elements are pulsed. This, as opposed to the single element of conventional pulse-echo transducers or the transmit/receive pair of conventional pitch–catch search units. If some or all of the elements are pulsed sequentially with a small, yet precise, timing delays imposed one to the next, certain beam characteristics can be varied, and in fact controlled. A linear array consists of a number of linear elements, arranged either in a single row or in a two-dimensional array pattern, while an annular array is an arrangement of concentric ring elements.

The linear phased array (LPA) technology provides the abilities to steer the ultrasonic beam to angles other than straight in front of and normal to the transducer element and to focus the beam at some point along the beam axis. The easiest way to understand how this works is to view it from a point in front of the transducer but not on the axis normal to the transducer face. The propagation time required for each beam to get to that point from each of the transducer elements can then be calculated based purely on geometric considerations since the propagation velocity and the precisely defined element locations within the array transducer are known. So, if the elements of the array transducer are pulsed at very precise timing intervals based on the different propagation times such that all of the beams reach the defined point at the same time, the beam has effectively been steered in that direction and focused at the defined position along the beam direction. All of the beams are in phase and additive at this defined point, but out of phase and destructively interfering elsewhere. The timing sequence and delays required to do this are collectively called the phasing or focal law. By picking another beam direction and/or another point along beam direction, a similar calculation can be made, thereby creating a new focal law. Because these operations are all performed very quickly and efficiently in the control computer, a series of focal laws can be implemented to sweep the beam through a series of angles at a fixed increment, for example, 30° through 60° at 1° increment, very rapidly.

As described above, the linear array is used primarily to influence beam direction, electronically focus the beam, and/or a combination of the two. A true spatial representation of the linear array data requires that the data be presented in polar coordinates rather than the Cartesian plots that are used to present x/y raster scanned data. The amplitudes at all digitization points along each of the waveforms are typically presented in colors, thereby creating a 2-dimensional view in the plane of the swept beam. These plots have become known as sectorial, or S-scans, because they represent sectors of the cross section of the component in the plane of the beam. A typical S-scan image is provided in Figure 25.8. In this case, the scan is of a calibration block containing a stack of side drilled holes, with a cartoon of the block shown at the left and the S-scan image to the right to show the relationship between the component and the resulting S-scan image.

This is the same technology that has been used in the medical field for many years, baby's first fetal picture being a prime example. However, because of the extreme amounts of data needed to create such an image, instruments required very high speed processors, which made them very expensive and limited their use to medical applications. Advances in processor speed and the corresponding reduction in processor cost have now permitted the entry of array systems into industrial applications over the past ten years or so.

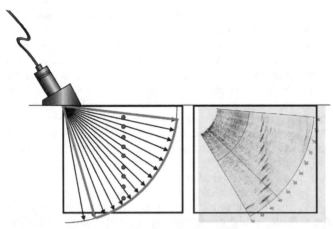

FIG. 25.8 ULTRASONIC LINEAR ARRAY S-SCAN IMAGE OF A CALIBRATION BLOCK CONTAINING A SERIES OF SIDE-DRILLED HOLES

Now that the inspection technology has been described, attention can return to the application at hand, inspection of turbine disk bore and keyway surfaces. Quality inspections are performed using some form of scanning device to assure complete and accurate coverage of disk. Some scanners are more-or-less crawling devices that somehow traverse the OD of the disk to transport the transducer(s) over the various surfaces. More often than not, however, the scan is accomplished using a fixed transducer positioning device and by then rolling the rotor to scan the transducer around the periphery. A scanner of this type is shown in Figure 25.9. Positioning devices used in this application can be relatively simple, more-or-less a transducer on a stick. However, such devices are often not very effective in the tight confines between blade stages

and proper adjustment of the transducer can be difficult to achieve. Consequently, the more prevalent design incorporates motor driven motion axes to properly position the transducer on the surface. The scanner shown in Figure 25.9 has motor driven raise and lower of the scan arm, extension along the scan arm, and two rotational axes at the transducer. Once the transducer is properly positioned, the rotor rotation is typically provided by power rolls or a large lathe. Rotational position is acquired along with the ultrasonic data such that the data can be positionally correlated via an optical encoder or similar device.

Full automation, digitally recorded data, and full imaging capabilities have significantly enhanced the inspection of turbine disks, reduced inspection and analysis time, and improved the accuracy and reliability of the inspections. And the introduction of the LPA technology described above has further enhanced the process. Figure 25.10 provides an S-scan LPA image of cracked disk superimposed over a dimensionally accurate drawing of the disk. In this case, the crack extends along the top of the keyway and axially along the bore of the disk beyond the end of the keyway. The particular S-scan image was acquired with the transducer on one face of the disk web. The image clearly shows the nominal responses from the disk bore and the top of the keyway, plus the cracks extending from the keyway. Because the data provides an accurate dimensional representation of the various reflectors, it is possible not only to detect the presence of a crack, but also to measure its depth, a very important consideration in the stress and fracture remaining life evaluation to follow.

For the remaining life assessment, the geometries of the hub and web sections of shrunk-on turbine disks vary too radically one to the next to enable development of any generic life assessment computer program that would cover any appreciable portion of the fleet. Consequently, stress and fracture evaluations are typically performed on a case-by-case basis using general purpose finite element analysis programs such as ANSYS.

FIG. 25.9 TYPICAL TRANSDUCER POSITIONING DEVICE FOR TURBINE DISK INSPECTION

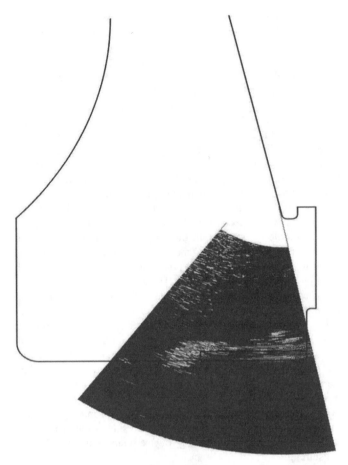

FIG. 25.10 LINEAR PHASED ARRAY S-SCAN IMAGE OF CRACKED DISK KEYWAY

25.7 DISK RIM BLADE ATTACHMENT DOVETAILS

Turbine blades can be attached directly to the rotor shaft or to the rim of the disks, whether integral or shrunk-fit. The photograph of Figure 25.11 shows an example of one blade attachment configuration referred to as a fir tree or Christmas tree dovetail for obvious reasons. Blades attached at the rim of a disk of the fir tree dovetail attachment design are typically one of two basic designs, an axial entry blade attachment as shown in Figure 25.11, or a straddle-mount or tangential entry blade attachment. In the axial entry design, the fir tree configuration is machines in a repeating pattern axially through the rim of the disk, thereby forming a series of female slots around the periphery of the rim to accept the blades. In this configuration, the slots and mating blades can be straight through the disk rim, straight but at an angle, or curved in passing through. The general configuration of the fir tree is fully exposed on both faces of the disk rim, as in Figure 25.11.

In the straddle-mount or tangential-entry design, the fir tree geometry is machine fully around the periphery of the disk, the disk forming the male side of the attachment dovetail. The blades have the mating fir tree dovetail configuration machined up and into each blade base such that when full engaged on the disk rim, the blades straddle the rim of the disk. To permit loading of the blades onto the rim, the dovetail hooks are removed from the rim for a short segment of the rim sufficient to allow one blade to be dropped over the loading slot and then moved circumferentially to engage the dovetail. In this design, the blades are actually referred to as buckets, and the loading slot on the rim is called the loading notch. The buckets are loaded sequentially by dropping them over the notch and then sliding them tangentially until the stage is full and only the notch remains. A notch block is then pinned between the buckets located on either side of the notch to keep the buckets in place and prevent tangential motion during service. Figure 25.12 provides a photograph of a 3-hook straddle-mount dovetail attachment. Loaded blades can be seen to the extreme left and right of the photograph, with the blades between removed to expose the fir tree dovetail geometry and the loading notch.

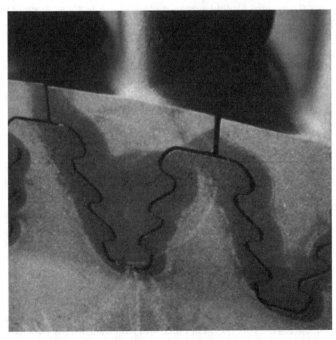

FIG. 25.11 TYPICAL FIR-TREE BLADE ATTACHMENT

FIG. 25.12 STRADDLE-MOUNT, TANGENTIAL ENTRY BUCKETS AND RIM WITH LOADING NOTCH SHOWING

There are other designs as well, for example a tangential entry design in which the dovetail on the disk rim forms the female side of the attachment and the male side is on the blades. However, these are found more prevalently on relatively small industrial turbines and infrequently in the turbines built for commercial power generation. The vast majority of the US fleet involves one of the two designs presented above at lease on most, if not all stages.

Within the majority of the US fleet, there is one exception found frequently on LP rotors. That is, on the last stage and sometimes last two stages of certain LP rotors, a pinned blade attachment configuration has been used extensively and is still in use. In this design, a series of side-by-side, parallel slots are machined fully around the disk rim. The slots are initially narrow at the slot bottoms with transitions at different radial positions to form wider slots and narrower plates toward the OD of the arrangement. The buckets have matching fingers extending from beneath the bucket foil and into the rim slots. Holes are then drilled axially through the disk rim plate fingers and the bucket fingers, and pins are inserted to hold the buckets in place.

Exceptions also typically exist on the first few stages of HP sections, all of the HP stages in some designs, and typically, some of the entry stages in reheat IP sections. This encompasses the smaller blades, located at much smaller diameters, where disk, per se, is therefore not needed. In such case, the blades are typically attached directly to the shaft proper without the radial extension of a disk. In such case, the shaft typically provides the female geometry of the attachment and the blade has the male attachment geometry. Designs encountered routinely included T-slots and fir tree slots directly into the shaft OD.

To this point, the only damage mechanisms presented for the disks have been related to stress corrosion cracking, and it is clear that this mechanism is important only for the last few stages of the LP that operate in or past the PTZ. This mechanism was discussed earlier primarily in the context of the bore/keyway cracking. However, the same considerations apply equally for the blade attachments for the stages that operate in or past the PTZ. Cracking typically occurs in the inside radii at the corners of the fir tree dovetail hooks on the disk side of the attachment and propagate either across the fir tree or more radially across the hook. Such cracking occurs in both the axial-entry and straddle-mount designs in the appropriate wet stages. Figure 25.13 shows a cracked dovetail in a typical tangential-entry, straddle-mount blade attachment dovetail.

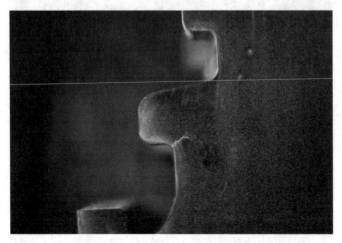

FIG. 25.13 WET FLUORESCENT MT INDICATION OF A CRACKED DOVETAIL HOOK

Cracking also occurs in the finger-pin design, in which cracks can occur at the bottoms of the slots in the disks, around the pin holes, and at the geometry transitions along the sides of the slots.

But corrosion-related cracking mechanisms are not the only ones possible in the blade attachment dovetails. Unfortunately, creep and temper embrittlement are not only possible in the rotor body but also in the high temperature blade attachments. Consequently, inspection and condition assessment are not limited to just the last few stages in the LP rotor blade attachments. In the first few stages of the HP and also the first few stages of reheat IP turbines, temperatures are adequate to develop high temperature creep and/or temper embrittlement, particularly given the accumulated service hours that the majority of the fleet has now experienced.

Each of these variations on blade attachment configuration presents its own inspection challenges. Starting with the tangential-entry, straddle-mount design, this inspection is typically not too difficult if implemented using the LPA UT inspection technology. Because each hook represents a potential crack initiation site, each must be inspected independently. The inspection is performed by introducing a UT beam from the side of the disk immediately beneath the blade. The beam is directed up into and across the dovetail to the opposite side at the appropriate beam angle to address a given hook directly. Prior to the availability of linear phased array UT, this meant that an assessment had to be performed to define the geometry of the underlying dovetail configuration and then carefully selecting the appropriate fixed beam angle and transducer position on the side of the disk for optimum interrogation of each hook. From there, implementation was merely a matter of implementing each of the inspections from each side of the disk—not a terribly challenging undertaking, but certainly tedious and time-consuming.

With the introduction of the LPA technology, this inspection can now be performed very quickly and effectively, and with much greater reliability and sizing precision. Transducer placement is essentially the same as for the fixed angle inspections—on the side of the disk beneath the blade attachment, with the beams directed up and into the dovetail hooks on the opposite side of the attachment. However, the ability to sweep the beam makes it much easier to infer the geometry and to establish the most effective transducer position. The real time imaging capability of the LPA technology makes it possible to perform some manual scanning to define the most optimum transducer location along the face of the available test surface, and the resulting image provides instant recognition of the associated dovetail geometry to the trained operator. Additionally, inspection of each side of each disk dovetail can be accomplished in a single scan pass around the disk owing to the beam sweeping capabilities of LPA, which results in significant time savings. Figure 25.14 provides an LPA S-scan image of a typical tangential-entry, straddle-mount dovetail design having significant cracking. The image is superimposed over a sketch of the geometry to assist with recognition of the various reflectors. In this image, the more significant reflections are from the tapered geometry of the hooks, while the numbered smaller reflections are from the cracks, which are located on geometric features of the dovetail that produce no reflections unless cracks are present.

Blade attachments that enter the blades directly into contoured slots in the shaft, i.e., smaller blades stages in the HP and IP sections, are inspected using the same protocol except that the ultrasound enters from available surfaces on the shaft between the blade stages and is directed inward toward the dovetails. This inspection is made difficult only by the limited space for transducer placement and manipulation between the stages. Often, special transducers

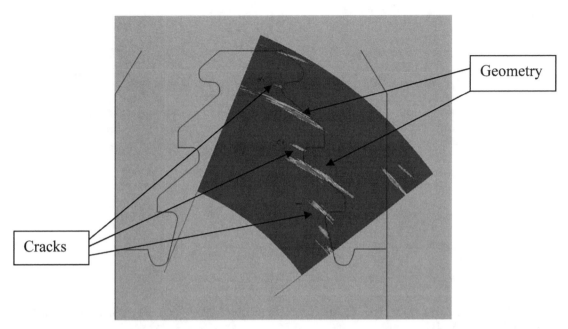

FIG. 25.14   TYPICAL LPA S-SCAN IMAGE OF CRACKED DOVETAILS

and transducer wedges are required to enable inspection from very limited and often complex configured surfaces. Higher frequencies are typically used to increase resolution, keeping in mind that the entirety of the geometry of the attachment is downsized significantly relative to the typical straddle-mount designs, and the ultrasonic propagation distances are reduced, as well.

Inspection of the exposed faces of the axial-entry design can be accomplished using appropriate surface inspection methods, primarily MT or ET. However, cracks are not always present only at the exposed faces, but can initiate away from these surfaces where detection is much more difficult and requires subsurface inspection methods and techniques. Ultrasonic inspection has become the standard inspection approach for this application. However, axial-entry blade attachment dovetails present a significant inspection challenge, even when utilizing the LPA technology. The inspection is complicated by the fact that the geometries of available test surfaces for transducer placement can and do vary significantly from rotor to rotor and from one disk to the next on the same rotor. In general, for this design the disk web is relatively thin and the disk rim section typically is much wider. From a design viewpoint, the idea is to reduce the mass acting on the hub to the degree possible to minimize bore/keyway stresses as possible, yet to provide sufficient web to adequately support the rim and blade loads. Additionally, where stages involve relatively small, low mass blades, a single disk can support multiple blade stages, in fact up to three per disk. Going all the way back to Figure 25.6, which depicts a disk associated with axial-entry blade attachments, this particular disk would support two stages of blades and has tapered sides beneath the disk rim section. Other configurations might involve radii on the underside of the rim or an elevated rim with straight, parallel sides for some distance beneath the rim before the contour to the relatively thin web starts.

The best ultrasonic approach for these attachments involves introduction of the beam from the underside and/or sides of the rim section, with the beam looking up through the rim at the dovetail. This involves any number of contoured wedges needed to match the many different surface contours that might be encountered during

an inspection. And the inspection may involve variable axial angulation of the beam or circumferential angulation, or some of each, depending on the exact disk configuration and the crack location and orientation.

Moreover, even after having addressed the various test surface configurations and optimum beam angulation requirements, the intermittent nature of the blade dovetails as they pass the transducer during a rotational scan means that the data analysis must be performed on a pattern recognition basis. And, there is a question related to the number of beam angles required to cover the geometry properly. Going back to the days of fixed angle UT, the possible combinations of beam angle and transducer location were almost endless, meaning many transducer/wedge combinations and many scans resulting directly in significant inspection duration. By comparison, for the tangential-entry, straddle-mount design, the geometry is constant within the circumferential scan and limited to two or three fixed locations to interrogate, i.e., two or three hooks. In the axial-entry design, however, the geometry changes in a direction essentially normal to the circumferential scan direction, so it is difficult to define the number of beam angles required to cover the geometry properly. So, more is better. Then, there is the question of whether to introduce the beam along the geometry (i.e., axially), or into the geometry (i.e., circumferentially), or both.

The LPA technology simplifies this considerably, and the optimized inspection typically utilizes a combination of the two: beams angulated axially and beams angulated circumferentially, both introduced during a circumferential scan. For the latter, a variety of fixed axial interrogation angles should be performed to cover the full axial length of the attachment dovetail. Alternatively, two dimensional arrays can be used to steer the beam both axially and circumferentially. Regardless, because the scan is circumferential and the geometry is machined axially across the rim, the resulting data represents intermittent, yet sequential passage by the dovetails. Consequently, data analysis is primarily performed on the basis of pattern recognition, i.e., looking for response(s) that do not fit the typical pattern created as the dovetails pass. Because so many beam angles and so many crack locations are possible,

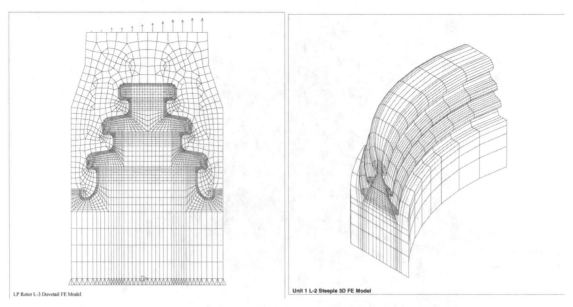

FIG. 25.15 LPRIMLIFE FINITE ELEMENT MODELS OF TWO DIFFERENT DOVETAIL TYPES, TANGENTIAL-ENTRY (LEFT) AND CURVED AXIAL-ENTRY (RIGHT)

data analysis can be a very time consuming, tedious undertaking. Because of the generation of a large number of beam angles, the LPA technology can produce a huge amount of data very rapidly and very reliably. However, sorting through all of the data looking for relevant indications can be a very detailed and time-consuming process.

Remaining life assessment for disk rim dovetails can be as demanding an undertaking as the inspections, owing to the large number of different geometries that may be encountered. The state-of-the-art for the LP disk rim analysis of SCC in LP blade stages at and beyond the PTZ is the EPRI LPRimLife computer code. This program includes geometry modeling leading to finite element analysis for any of the geometric configurations

comprising the vast majority of those currently in service, including: the tangential-entry, straddle-mount; the straight axial-entry; the curved axial-entry; and, the dovetail finger plates. Figure 25.15 shows finite element meshes generated within LPRimLife for typical tangential-entry, straddle-mount and the curved axial-entry blade attachment designs, while Figure 25.16 shows a model of the pinned plate dovetails.

On the various fir tree dovetail designs, one of the prime considerations that must be factored into the analysis, and which is integral to the LPRimLife program, is stress redistribution. As a crack grows, the cracked hook loses capacity to carry load; consequently, the load must be redistributed to other hooks on the fir tree dovetail. The LPRimLife program includes consideration for

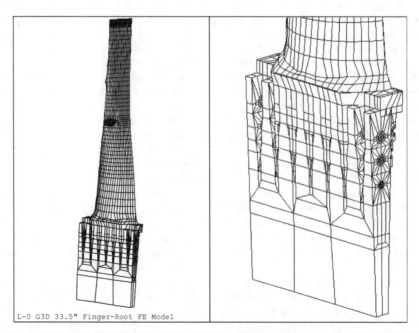

FIG. 25.16 LPRIMLIFE FINITE ELEMENT MODEL OF THE PINNED PLATE ATTACHMENTS

both initiation and crack growth, and it is extremely important to consider the impact of redistributed stresses on initiation times and on subsequent crack growth rates at other hooks, and upon diminished growth rate because of lowered stress at the initial crack. This analysis is extremely complex in execution, requiring an iterative process at very fine crack extension intervals to properly assess this primary consideration as it impacts remaining life.

One outcome of an analysis can be that the worst (i.e., deepest or biggest) crack in a dovetail does not represent the worst case condition. Consider a dovetail design having three hooks, as an example. If only one hook is cracked, that means that the other hooks are still in the initiation stage and can continue to support additional load redistributed from the cracked hook. This might take significant time, allowing continued operation over an extended period. In another case, if all three hooks have cracks, even small cracks, at the same position along the dovetail (i.e., stacked), then all have passed initiation and are in the growth mode. In such case, the remaining life may well be less than for the large, single-hook crack. So, the analyst must understand these interactions and carefully choose the crack conditions to analyze such that the worst case condition is defined during the analysis. The natural tendency is to pick the largest crack and analyze that, but in this case, the largest crack may not be the most life limiting case.

Another prime consideration in the analysis of tangential-entry, straddle-mount dovetails is the stress analysis around the notch block. The hooks immediately adjacent to the notch block are prime locations for crack initiation, typically ahead of other regions around the periphery, due to elevated stress associated with the stress singularity created by the notch in addition to the added load of the notch block being supported by the blades on either side. This is not always the worst case crack location, as local machining variations and localized exposure history may cause cracking elsewhere ahead of the notch block region. Consequently, comprehensive inspection should not be waived and just the area around the notch block tested. However, the tendency is for the notch block area to initiate first; consequently, this is a good location to concentrate initial inspection efforts.

25.8   SOLID (UNBORED) TURBINE ROTORS

As alluded to earlier, some older turbine rotors were made without bore holes. Additionally, modern steelmaking and forging practices have now made it possible to make forgings for which the benefits of lowering the central stress via elimination of the stress concentrating bore typically outweigh the potential for having an undetected critical flaw in the forging. The compound effects of lowering the stress and increasing the tolerable flaw size simple based on the difference between a surface and subsurface flaw increase the critical flaw size to the point that it is simply not likely to occur by modern practices.

For the older unbored rotors, the need for periodic inspection is obvious. They can contain significant flaws and clusters that can link up to eventually form critical flaws. Unfortunately, failure of modern solid forgings is not totally unheard of; consequently, owners seek to have these rotors inspected as well. Inspection of these rotors is not a trivial undertaking, and even under the best of circumstances and when using the most comprehensive and effective inspection approaches available, the best that can be accomplished is a partial inspection.

Access and the ability to direct beams into the material at the appropriate angles for detection of the more critical radial–axial flaws impose the most restrictive limitations. Consider, for example a built up LP turbine rotor having a central shaft and assembled disks such as shown in Figure 25.2. This certainly would not be a prevalent design, and such a rotor may not even exist–solid shaft with assembled disks–but it is useful to make a point. In such case, there would be very few, if any, locations along the body of the rotor at which an ultrasonic beam could be introduced into the shaft material. In the more likely cases in which the involved rotors have integral disks, access to OD surfaces appropriate for launching the ultrasound is still very limited, typically to the small lands between the disks and the sides of the disks themselves. When using the surfaces between disks, the inspection is often further complicated by the presence of relatively narrow lands and grooves placed there for sealing purposes. When using the sides of the disks themselves, the geometries may well complicate the inspection because the transducer wedges must be contoured to fit. Consequently, for any but flat sides on the disks, special wedges may be required.

Regardless, even for locations where appropriate transducer placement can be accomplished, an optimum inspection for radial–axial flaws is achieved only in the cross sectional plane of the transducer—that is, directly under the transducer. For locations axially removed from this plane, for example, under the disk center, some level of coverage can be achieved using beam spread. However, beam intensity decreases with increasing distance from the beam axis, so sensitivity diminishes accordingly. Additionally, for the flaw of interest, i.e., the radial–axial flaw, the reflectivity of a given flaw in the direction of the transducer decreases due to its orientation relative to the beam direction. Together, these factors mean that reflection amplitude for a given flaw decreases away from the cross sectional plane of the transducer, and the wider the disk, the more significant the sensitivity reduction. In some cases involving relatively wide disks, inspection beneath the disk center simply cannot be accomplished using single transducer, pulse-echo techniques. Even if the beam angle is appropriately modified such that beam intensity is maintained, the directionality of the reflection makes detection of the radial–axial flaw less than optimum. In such case, the only option is to use a transmitter on one side of the disk and a receiver appropriately positioned on the opposite side to receive the reflected signal. This then introduces the necessity to coordinate the positions and motion of two independently held transducers.

For the narrower disks, there are ways to compensate for the reduced sensitivity effects, for example by simply increasing gain. However, there are other complications that cannot be accounted. The way to best assess the position of a reflector is to scan it until the peak response is obtained, which generally means that the flaw is then located on the peak intensity point within the beam, i.e., the beam axis. On these rotors, the ability to scan axially is typically limited severely by the geometry; consequently, for any fixed beam angle the exact axial position of the reflector remains unknown and the appropriate gain increase needed to compensate for its off-axis position therefore cannot be defined.

There is also the possibility that the reflector can be offset radially, i.e., away from the rotor centerline. The same considerations apply for these. Looking first only within the cross sectional plane of the transducer and considering only a straight, radial beam, a radial–axial flaw that is offset significantly from the rotor centerline is more aligned with the beam than normal to it as the transducer passes over it; consequently, sensitivity to this flaw is significantly diminished because of its orientation. By comparison, the same reflector located at or near the rotor centerline comes into a plane normal or near normal to the beam at two positions within the scan around the rotor. To inspect for off-axis, radial–axial flaws, angle beams are required, and to cover any appreciable radial distance, multiple angles are required as shown in Figure 25.17. The further the zone of coverage from the rotor axis, the higher the refracted angle required for optimum detection.

And then there is the combination of these limitations when attempting to address off-axis, radial–axial flaws that are additionally under the disk and away from the cross-sectional plane of transducer placement. Compound refraction of the beam away from the rotor axis and pitched to direct it under the disk is required to detect flaws that are located in the rotor.

So, from all of these, it can be seen that a good inspection will require multiple angles radially and axially from all access locations and typically implemented in all combinations, plus possibly pitch–catch arrangements for some areas. It is easy to see that the inspection can become very complicated and very time-consuming to implement. Fortunately, the LPA technology discussed previously as implemented for turbine disks can also be used here to simplify and streamline this inspection as well. Certainly, it is feasible and relatively straight forward to implement the various radial angle beam inspections in the plane of the transducer, i.e., those shown in Figure 25.17, during a single scan by utilizing the beam steering capabilities of a single dimensional linear array. By adding a refracting wedge that pitches the beam appropriately in the axial direction under the disk, and then using the same steering process to steer the beam radially, the inspection under the disk can be accomplished. As the axial length of the disk increases, multiple axial angles may be required. Alternatively, a 2-dimensional array can be used to steer both axially and radially.

An optional or supplemental technique that can be used in certain situations involves introduction of the beam from the sides of the disks. This is relatively easy to implement on disks having flat sides and particularly effective when they are tapered, getting narrower toward the OD of the disk. In this case, the beam is the first introduced radially toward the rotor axis and steered axially using the array, as shown in Figure 25.18. Then, to cover for

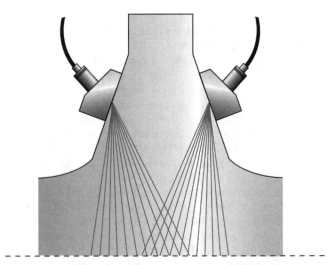

FIG. 25.18 INSPECTION FROM THE DISK SIDE SURFACES

off-axis flaws, the transducer can be skewed somewhat and then the axial steering repeated during a second scan around the disk. Several skew angles may be required to cover fully adequate radial depth from the rotor centerline, and the inspection should be repeated from both faces as shown and in both skew directions. As with the other inspections defined earlier, the axial sweep and beam skew +/– can all be accomplished concurrently using a 2-D array.

Once the inspection data have been acquired and processed and the flaws defined as best possible, the rotor analysis follows the same procedure using the same computer program as defined for bored rotors. Typically this is done using the EPRI SAFER program or similar.

25.9   GENERATOR ROTORS

Turning now from the turbines to the generators, those having central bore holes are inspected with the same boresonic equipment and following the same inspection protocols and procedures as those used for turbine boresonic inspection. Disassembly and reassembly of the rotor is a bit more involved than for the turbine rotors, which merely require bore plug removal to access the bore surface. The generator rotor bore contains insulated conductors that carry the DC current to the rotor windings. There must be some means of getting the DC to the rotor windings, and the bore provides a convenient path to get the excitation from the exterior, past the seals, and past the bearing journals. Once beyond the seals and journals, conductive radial studs bring the DC to the shaft OD where additional conductors contained typically in wedged slots running along the shaft OD connect from the studs to the rotor windings. These conductors are insulated from each other and from the ground potential of the rotor itself, and are typically installed as a completed assembly. Once the bore plug has been removed from the collector or exciter end of the rotor, the insulated and sealed radial stud components must be removed, and then the insulated axial conductor assembly can be removed.

The axial conductor assembly extends into the rotor only to a point somewhat inside the bearing journal and not to the rotor body. This rotor section is typically much smaller in diameter and

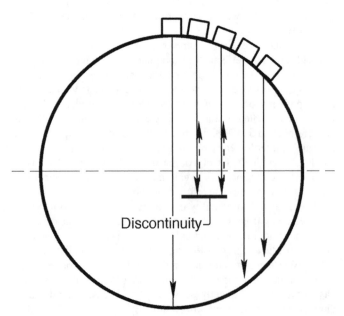

FIG. 25.17   MULTIPLE ANGLE BEAM INSPECTION FOR OFF-AXIS PLANAR REFLECTORS

Discontinuity

therefore operates at much lower stress and, because the sections are typically uninterrupted cylindrical shapes, these areas can alternatively be inspected conveniently from the OD. Consequently, a plug can be installed in the bore from the turbine end of the rotor, and a bore inspection of the full body length can be conducted without disassembly of the leads. This saves the expense of removing the additional rotor end plug, the radial stud assemblies, and the axial conductors. It eliminates the need for certain replacement parts and for a new replacement bore plug. And an effective inspection can be conducted from the OD toward the shaft end where the leads are located. But the plug must be 100% reliable in containing the honing oil and honing debris and containing the couplant fluids used during the ultrasonic inspection; otherwise, the leads will have to be removed and reinsulated after the fact, potentially extending the outage.

Additionally, some manufacturers install steel filler bars in the generator rotor bores over the length of the body section to replace the magnetic material removed by placement of the bore. Because the magnetic flux lines pass through the rotor body, the theory is that the use of the flux bars improves the magnetic efficiency of the rotor. Some manufacturers adhere to this practice, while others do not. Where flux bars are present, obviously these must be removed to provide access to the bore for honing and inspection.

Otherwise, inspection of a generator rotor from the bore is identical to inspection of a turbine rotor. Typically, the generator is longer than the turbines, somewhat longer than the LP rotors, and significantly longer than the HPs, IPs, and even HP/IP combined rotors, so inspections take a bit more time. Additionally, there is typically a step in the bore near the end of the axial conductors, the nominal bore being of a diameter selected to best clear indications and the conductor bore being of a fixed diameter set by standard conductor sizes. Any steps and transitions take special bore preparation processes, and the different sizes may require that some of the fixtures be changed to accommodate and/or recalibration of the test system. So, additional time must be provided accordingly.

Inspection of unbored generator rotors, however, is an altogether different application than described for turbines. Winding slots run from one end of the body to the other over more than half of the rotor periphery—likely closer to 65% - 70%, leaving some 30% or so comprising the poles, which in most cases do not have slots. Even in the pole sections, some designs include ammortisseur windings, axial conductors that are there to carry surface currents cause by electrical fault conditions, including motoring. While the ammortisseur slots, if present, are shallow and leave sufficient space between for spot inspections, the winding slots run on the order of six inches or so deep, leaving narrow teeth between that are typically on the order of ½-inch or so wide at the narrowest point. So, inspection of the rotor central material from the slotted sections of the rotor is effectively not possible. That is, the thin, long rotor teeth do not lend themselves to reliable inspection of the rotor central material. Additionally, the rotor typically has other machining that interferes with implementation of any meaningful inspection from the OD, even from the poles. For example, some manufacturers machine some form of land and grove configuration along the rotor to increase the surface area for cooling purposes. In some designs, the lands and grooves as part of the ventilation delivery design. Even here, though, where the machining is useful only in the slotted region, because the pattern is produced by turning the rotor, it is present at the poles as well as the slotted portions. Figure 25.19 shows an air gap pickup ventilated rotor and a typical rotor OD machining configuration.

FIG. 25.19 TYPICAL AIR GAP PICKUP VENTILATED GENERATOR ROTOR

For the most part, inspection of a generator rotor from the OD is very difficult at best, and provides little value because of the limitations imposed by the slots and the surface machining configurations. There are exceptions, but most cases are very difficult.

Limited inspections can be performed on unbored generator rotors during rewinds, when the slots are stripped of their windings. Special probe wedges can be contoured to fit the slot bottoms, thus enabling inspection of the body length over much of the volume. If including beam spread in the equation, and if performed using LPA swept beams, this approach attains approximately 75%-80% coverage of the rotor volume. If inspection can additionally be conducted from the OD surface of the rotor at the magnetic poles, then the remainder of the rotor has good coverage as well. Otherwise, the limited inspection from the slot bottoms is the best that can be achieved.

Analysis of a generator rotor is typically much simpler than that of a turbine for several reasons. First the geometries are relatively uniform, so fewer sections need to be modeled. For example, the rotor body cross section is the same from one end to the other and can be modeled as a single section. Turbines, on the other hand, have numerous sections, disk sizes, blade loads, etc., each requiring geometry-specific analysis. Second, because generator rotors operate much cooler than the turbines, thermal stresses and transient thermal events are negligible. These typically are not even considered in the analysis unless specifically analyzing the impact of significant transient events or fault conditions. And finally, because the generator rotors operate much cooler, thermal degradation mechanisms, such as creep and temper embrittlement, do not pose any form of a threat. Aside from these factors that make the generator rotor analyses much simpler, the treatment of the flaws is performed exactly the same as for the turbine rotors.

25.10 GENERATOR ROTOR RETAINING RINGS

Retaining rings are high strength ring forgings that are assembled onto each end of AC generator rotors to support the rotor end turn windings. The rotor windings are contained over the length of

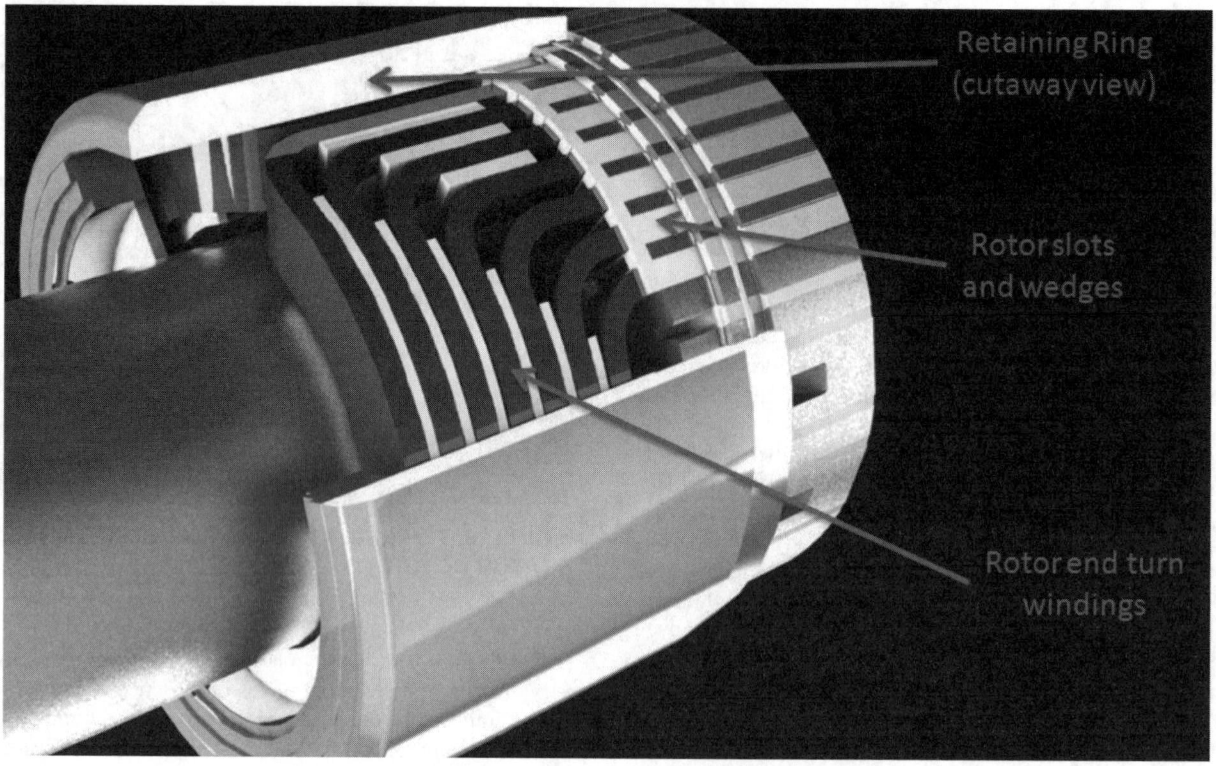

FIG. 25.20   GENERATOR ROTOR RETAINING RING AND SUPPORTED END TURN WINDINGS

the rotor body in axial slots machined into the rotor body. Provisions are included in the configuration of the rotor teeth, i.e., the material remaining between the slots, for wedges to contain and support the windings in the slots. At the ends of the rotor body, the winding coils exit the slots and wrap around the solid pole sections of the rotor, re-entering a symmetrical slot on the opposite side of the pole. The retaining rings support the windings in the end turn region where they are otherwise unsupported and cannot support their own centrifugal load in the cantilevered configuration. Figure 25.20 shows the coil/slot/ring arrangement. The retaining ring is assembled in all cases using an interference fit which places the ring under tensile stress when at standstill. Most often, the shrink fit is between the end of the rotor body and the inboard end of the retaining ring, although some limited designs place the shrink fit at the outboard end of the retaining ring, typically onto a fluted section of the rotor. Typical designs also include a circumferential key assembly at or near the shrink fit region to prevent axial movement of the ring. This can be a separate key in a machined keyway, scalloped bayonet fit, mating step, or others. The shrink fit at the nose of the retaining ring and the circumferential key can be seen in Figure 25.20.

The shrink fit is designed such that the ring maintains intimate contact with the rotor even under maximum overspeed and electrical fault heating conditions such that full and continuous support of the winding coils is maintained at all times, even under the most extreme fault conditions. The shrink fit interface also provides a current path for surface currents circulating under certain abnormal operating conditions, and must maintain contact to minimize arcing and subsequent related damage. Some designs include specific provisions, i.e., specific additional current conductors under the rings to carry these surface currents, while others rely solely on the retaining ring and conduction through the fit.

The rings used on modern generator rotors are typically made from a high strength, non-magnetic, austenitic steel alloy which is one of several possibly compositions. These rings operate in the magnetic field created by the generator rotor and therefore are heated magnetically and thus producing heating losses. The use of non-magnetic material minimizes the heating losses and so increases generator efficiency. Domestic generator manufacturers first used non-magnetic materials during the 1940s, and continue to use non-magnetic materials to the present. Over the years, catastrophic retaining ring failures have occurred occasionally, and for some non-magnetic ring failures, the investigations that ensued found the failures to be attributable to intergranular stress corrosion cracking (IGSCC). In addition, vast amounts of related research have shown that IGSCC can occur even in the presence of relatively benign environments, including moisture. Following the discovery that IGSCC is an operative damage mechanism under certain exposure conditions, this damage mechanism quickly became a major concern. One alloy in particular, containing nominally 18% Manganese and 5% Chromium (generically known as 18-5), received a great deal of attention, in fact the majority, because of the large number of rings made from this alloy. This alloy was used extensively by most generator manufacturers world-wide for many years. In fact, it was the industry standard; consequently, thousands of the 18-5 rings have been placed into service.

The history of retaining ring failures shows that there have been very few domestic failures involving the 18-5 material and attributable to IGSCC, although failures of this alloy have been more pervasive world-wide. In a study conducted by EPRI in the early 1980s, a total of 39 failures were found to have occurred on retaining rings in general world-wide [2]. Of these, nine were of the 18-5 alloy. Of the nine 18-5 ring failures, two were not really retaining rings but zone cooling rings that additionally had pre-existing machining tears, two occurred in Europe in water-cooled rotors that had water leaking directly onto the rings (water cooled rotors are not offered by domestic manufacturers), one was caused by condensation during prolonged, improper storage, one was attributable to an overspeed incident and not IGSCC, one was attributed to low ductility material and not IGSCC, and one involved a crack that was discovered during an inspection but not associated with a failure in the normal definition of failure. This accounts for eight of the defined failures, leaving a single 18-5 retaining ring failure that is germane to a discussion of IGSCC in rings of designs typically implemented by domestic manufacturers. And, even this failure had extenuating circumstances including low gas temperature, operation of the ring in the cold gas (not typical), high moisture content, and lower bound fracture toughness [3]. Other failures may have occurred since, but the noted EPRI study is the last known comprehensive study that has been published, so information on more recent failures is sketchy at best.

So, a reasonable question might be—with thousands of rings in services in the US alone for many, many years, why only one IGSCC failure? The simple answer is that moisture exposure is typically not that usual an event. If we look at the generator as a whole and specifically at the retaining rings, in general the rings in most typical designs operate at elevated temperature. Typical operating temperatures are, in fact, well above the point at which moisture can condense on the rings, even if the cooling gas contains significant moisture. One could then speculate that exposure must occur when the machines are shut down, whether remaining closed or open to the environment. All available literature on this topic suggests that this is most likely the case. In one possible exposure scenario, consider the operating generator that has moisture in the cooling gas when it is shut down and goes on standby. This could definitely lead to exposure. Or consider the rotor sitting unprotected on an outdoor turbine deck. An earlier study, for example, showed that most of plants where damage in 18-5 rings has been found are outdoor units located in high humidity regions of the country—along the southeastern seaboard and gulf coast [4]. So that at least implies that the exposure occurred when the rotors were sitting on the turbine decks.

But laboratory experiments and published data indicate rapid growth rates, so even under the limited time exposures experienced during standby, why have we found many damaged rings, but generally having relatively minor damage, and only one failure? One published report addresses a potential re-incubation time [4]. The laboratory data shows that crack growth stops when the moisture is removed. However, when reintroduced to an existing crack, several hundred hours are required before the crack begins to grow again. Perhaps, this provides the explanation—perhaps the rings simply do not see sufficient exposure, over sufficient durations, and at sufficient frequency to grow crack to failure.

None of this is meant to imply that this problem is trivial or that it should be ignored. This is a very demanding application for any material, requiring extremely high strength of up to 180 ksi yield strength or possibly even a bit higher, good fracture toughness on the order of 150 ksi root inch, high corrosion and stress corrosion resistance, and appropriately low magnetic permeability. In the wake of the discovery that these rings are susceptible to IGSCC, a new alloy having 18 Manganese, 18 Chromium composition (18-18) was developed as a suitable replacement. Ring suppliers and generator manufacturers quickly adopted recommendations for wholesale ring replacement, an expensive proposition considered by many to be unsupported by the failure history. However,

for those not able or inclined to implement appropriate mitigation measures with which to enable continued operation of 18-5 rings, replacement likely is the most reliable and least involved solution. Still, many owners have not been inclined to replace rings, particularly given the relatively exemplary operating history. As a result, through the mid- and late-1980s, EPRI invested heavily in related research directed toward justifiable avoidance of ring replacement based on valid condition assessment and effective moisture mitigation. The EPRI program was directed toward all aspects of an effective asset management approach. It addressed inspection, moisture monitoring, moisture mitigation, and analytical assessment protocols, with the overall objective of providing viable options to the utility owner-operator to allow continued safe, reliable operation of 18-5 rings. This research led to the following advancements:

- effective non-destructive evaluation procedures with which to detect extremely small flaws, well smaller than those required to cause sudden failure;
- effective equipment and procedures to ensure that rings are maintained at all times in appropriately dry environments;
- monitoring systems with which to assess ongoing environmental conditions, make the appropriate adjustments if/as necessary, and discontinue operation when unable to control properly;
- quantification of the actual crack depths necessary for sudden failure and demonstration that they are relatively large—for example, even the low fracture toughness of the single IGSCC failure described above withstood a crack 0.6-inch deep × 1.1-inch long before it failed; and,
- development of a probabilistic life assessment computer code (RRing-Life) to statistically account for all of the factors potentially impacting crack initiation, crack growth, and failure.

Ring replacement, which is very expensive to the owner, can be avoided in many cases except those in which flaws have already grown to excessive sizes or where other specific conditions cannot be maintained within acceptable limits. Effective non-destructive evaluation (NDE) provides the means to assess current condition in terms of existing cracks and/or less dramatic evidence of prior moisture exposure. Moisture mitigation procedures can be implemented to minimize future exposure to hostile environments. And, probabilistic condition assessment provides a realistic means to assess the potential for crack initiation and growth and for ring failure.

From an owner's viewpoint, inspections are ideally conducted with the rings assembled to the rotors in order to avoid ring removal and reassembly costs. From an NDE standpoint, it is also advantageous to perform the inspections with the rings assembled to the rotor because of the shrink-fit that tends to hold cracks open and make them more readily detectable. The materials used for retaining rings acquire their properties via work hardening, typically mandrel expansion. This process leaves the resulting forging in a state of relatively high compressive residual stress, likely in the tens of thousands of pounds per square inch (psi). On the other hand, when assembled to the rotor, shrink fit stresses are highly tensile, on the order of 80, 90, 100 or even greater psi tensile, depending on the rotor size. The change in stress from highly tensile to moderately high compressive when the ring is removed from the rotor has a crack closure affect that profoundly impacts the ability to detect the cracks using ultrasonic techniques. When an interface (gap) that provides the acoustic impedance mismatch required to cause a reflection is reduced to some value that is a fraction of the UT wavelength, the sound merely passes through and beyond the gap. Squeezing the cracks closed has this affect. Consequently, ultrasonic inspection (UT) approaches, performed with the rings on the rotor and under tensile shrink fit stress, provide the most effective means of inspecting inaccessible surfaces.

EPRI research [3] clearly demonstrated that the reliability of detecting any specific crack in retaining rings is appreciably enhanced through the application of multiple inspection methods and modes. Clearly, the detectability of a crack is affected by many variables—size, shape, morphology, orientation, location, and so on. One of the most likely locations for cracking to occur is along the edges of the rotor teeth in the shrink fit region. This is because the moisture, when present, tends to wick along the coil/slot-wall interface, where it then contacts the ring at the tooth edges. Tooth edges, unfortunately, also constitute ultrasonic reflectors in and of themselves; consequently, it can be difficult at times to distinguish between pure geometry and geometry with a co-located crack. This is true of conventional, contact shear wave (S-wave) UT techniques, which are typically used for the basic detection scans on the rings. However, other techniques provide for improved discrimination between the two reflectors and are typically used to supplement the conventional inspection for this exact reason.

Time-of-flight diffraction (TOFD) UT is a dual transducer ultrasonic technique that places the two transducers on opposite sides of area of interest, directed toward each other. One is the transmitter and the other is the receiver, and reflector depth below the test surface is calculated based on the time of flight using a simple geometric triangulation algorithm. TOFD relies on forward-reflected and diffracted responses and therefore does not respond at tooth edges in the same way as the conventional back-reflection approaches. And TOFD can discriminate very positively and reliably between a surface indication (geometry) and a subsurface indication (crack tip diffraction) via the propagation time depth calculation. However, TOFD relies on tip diffraction responses to make this distinction, and these diffraction responses are generally low in amplitude relative to reflections. Because cracks are intergranular, they do not always produce sufficiently coherent tip responses for reliable detection. Consequently, this technique can miss cracks. However, even so, TOFD still can play a significant role. TOFD is extremely effective for discrimination and sizing purposes when it does detect, and can easily be performed as a supplemental inspection at the shrink-fit regions, at all other geometric discontinuities such as steps and transitions where cracks are most likely to occur, and as needed in other locations where indications have been detected during the basic pulse-echo detection scans to further characterize the indication source. Other techniques can also be utilized as needed to aid in flaw discrimination and proper classification including a mode-converted L-wave (MCLW) UT inspection technique, an inner surface creeping wave inspection, and focused beam approaches. Collectively, when implemented properly, these techniques result in a reliable inspection. The appropriately prepared inspector will have a number of tools in the toolbox in the way of multiple inspection approaches with which to optimize detection reliability and characterization accuracy.

Regardless which inspections are performed, inspection effectiveness is dramatically dependent upon full automation. The use of robotic scanners, effective position manipulation and position acquisition, digital UT data acquisition, and state-of-the-art data imaging capabilities significantly enhance the inspections.

Manual inspections are very difficult and reliability suffers appreciably.

As compared to UT, eddy current (ET) inspection does not suffer loss of sensitivity due to crack closure. When the surface cracks are squeezed shut under high compressive stress, the oxide that has formed on the faces of the crack serves as an electrical insulator such that, even when the faces of the crack are pressed together, there is still an electrical discontinuity which ET picks up very readily. PT inspection is about the only other option, but because PT relies on capillary action to pull the penetrant into the crack, PT does suffer from crack closure. Therefore, for removed rings, ET provides a superior inspection option.

For outer surface inspection of assembled rings, PT and ET provide the two options. Even when the rings are assembled, damage forms on retaining rings tend to be tight and not particularly conducive to rudimentary PT approaches. The application of a visible PT inspection using the typical procedures appropriate for these PT systems is simply not adequate for reliable detection. For optimum results, the PT should be lipophilic and non-hygroscopic, such that the PT in and of itself does not constitute moisture exposure potential. It should be high resolution, fluorescent, and it should be implemented using lengthy penetrant application and developer dwell times. Eddy current inspection is also effective on the outer surface for the inspection of assembled rings. ET can be implemented either manually, typically only to corroborate PT indications locally, or using the same basic data acquisition and imaging system when testing large areas, in which case, automation is considered necessary. Eddy current arrays are now also available for rapid surface inspection on retaining rings.

For the analysis of retaining rings, the only available computer analysis program specifically used to perform remaining life assessments of IGSCC in non-magnetic retaining rings is the EPRI RRing-Life program. This is a fully integrated program that includes built-in finite element modeling capabilities with which to perform stress analyses for specific, generic ring designs. It is a probabilistic code that provides exposure probabilities based on detailed assessment of a number of exposure sources. For example, the code includes a link to National Weather Service statistics such that the local environmental conditions can be applied for the location of the unit during machine-open intervals. This is not exact data, for example, humidity conditions on exact dates, but statistical representations that provide probabilities of exposure when open to the environment in general for the local climatic conditions. It provides the means to bias this information based on implementation of specific moisture mitigation provisions applied at the plant or where known exposure events have occurred. It provides statistical exposure probabilities from specific events such as cooler leaks and leaks in water-cooled windings. The program divides the analysis into three basic generator operational modes—running, stand-by, and open. Consequently, it provides the ability to define when the rings are most likely to see exposure such that maintenance and mitigation dollars can be spent most effectively. One of the most effective uses of the program, once all of the modeling is complete and the basic exposure scenarios have been quantified, is to perform sensitivity studies. The variables can be systematically altered to determine which are most sensitive to change. Again, by defining those variables that are most sensitive to change, the owner can define the most effective mitigation program and thereby maximize return on investment.

A class of magnetic materials was also utilized on older machines and is still used in some instances, even on new generators. For the most part, the primary damage mechanism in these rings is fatigue—primarily low cycle fatigue due to start/stop operation and possibly high cycle fatigue under limited, abnormal conditions. However, when discussing fatigue issues in rings, they apply equally to the non-magnetic alloys as to the magnetic rings. The magnetic rings just don't appear to suffer IGSCC as the non-magnetic rings do. One possible high cycle fatigue driver involves torsional oscillations of the rotor due to electric system unbalance, both steady state and those related to transient events. A disturbance on the electrical system puts a sudden torque on the rotor. Because of the very limited torsional damping in the rotor system, the torsional oscillations decay slowly and therefore accumulate many, many cycles on the rotors from a single torsional event. These can lead to torsional cracking of shafts, excitation of higher order turbine blade frequencies that can fatigue blades, and torsional relative motion of shrink assembled components. Years ago, a shrunk-on turbine disk failed in a unit operating near a steel mill that was operating an electric furnace. The cracking initiated by fretting fatigue, but once initiated grew by torsional fatigue. The transients introduced by the electric furnace were blamed for the failure. Similarly, a retaining ring failure that occurred more recently was attributed to exactly the same situation. It too was located near a steel mill operating an electric furnace. The torsional events caused fretting fatigue initiation of cracks at the shrink fit of the retaining ring, and the cracks then were driven to failure by the continuing torsional oscillations produced frequently by the transient torque events.

Otherwise, operative damage mechanisms are not generic to the fleet, but limited to specific designs, specific alloys or subsets of alloys, and possibly to specific operating conditions. An issue known to exist for certain magnetic rings is that of hydrogen embrittlement, but it has only been defined for rings of a specific alloy and, even for this alloy, for rings beyond a certain yield strength within the particular alloy. Programs for these rings include hardness tests to predict yield strength, coupled with inspections for the detection of existing cracks. Another non-magnetic alloy known as Gannaloy, a Nickel–Titanium–Chromium–Aluminum alloy, was used by one manufacturer for a number of years on certain generators. Several failures have occurred and the manufacturer has recommended replacement of all Gannaloy rings. Failure investigations have not provided adequate insight relative to the failure mechanisms nor any means to predict or prevent failure; consequently, the manufacturer's recommendations appear to be appropriate and well-founded, without options as those available for well-maintained and periodically inspected 18-5 rings.

Inspections for retaining rings other than the 18-5 rings do not vary appreciably form the 18-5 inspections. The same procedures are valid and the same implementation means, i.e., full automation, digital acquisition, and quality imaging capabilities, are still important. Crack closure is still an issue for free standing rings, although MT inspection provides and additional option for free standing magnetic rings. Additionally, fatigue cracks and similar as found in magnetic rings do not have the branched, intergranular nature of IGSCC cracks. Consequently a more coherent crack tip is present, increasing the detection reliability when using ultrasonic tip diffraction techniques. As a result, the application of redundant inspection techniques is not as much an issue. Having said that, however, geometric reflectors still present a discrimination challenge when using conventional, pulse-echo techniques; consequently; some form of positive tip diffraction UT, such as TOFD, adds significantly to the inspection, and application of redundant inspection approaches still improves reliability of the inspection.

For an analytical assessment of remaining life of retaining rings other than 18-5 and/or damage mechanisms other than IGSCC, typically these would be conducted using general purpose finite element stress analysis programs, such as ANSYS or similar, and general fracture analysis procedures. In general, however, cracks propagating by any mechanism other than IGSCC are not so readily disrupted. Whereas IGSCC growth in 18-5 rings can be arrested by eliminating the corrosive environment, fatigue cracks for example will continue to grow so long as the stress cycling continues. Additionally, initiation time for fatigue cracks typically represents the vast majority of life, leaving little for growth. As a result, once detected, cracks typically must be dealt with at the time, either by local removal or ring replacement.

25.11 GENERATOR ROTOR TOOTH-TOP CRACKING

Two specific cracking mechanisms exist on typical domestic generator rotors. Each is specific to a particular design approach and therefore is generic to a particular manufacturer or to manufacturers producing similar designs. The first occurs in the retaining ring shrink-fit area on rotors having Tee top teeth and in designs that leave some of the Tee-top in the shrink fit region. The mechanism is fairly straight forward. When the ring is assembled to the rotor, the shrink fit imposes a compressive load on the rotor teeth. This further imposes an inward bending stress on the cantilevered sides of the Tee-top. The function of the Tee-top is to mate with a matching wedge that supports the outward winding load developed by rotation of the rotor. When at speed, therefore, the winding load bears outward on the cantilevered sides of the Tee-top. The stress, in this condition, is a concentrated tensile stress at the inside radius where the Tee-top intersects the radial side of the tooth. The general configuration and loading is shown in Figure 25.21. In this diagram, the downward load occurs at standstill from the retaining ring shrink fit, and the upward load occurs during operation due to the winding and wedge load.

The stress cycle is start/stop operation, and the stresses are high enough that low cycle fatigue cracks can initiate in the radius, as shown, in as few as several hundred start/stop cycles. The end result of this cracking is unknown as no failures are known to have occurred. The proposed end result is that the teeth lose their ability to support the shrink fit load at some point, so the retaining ring bears inward on the windings, thereby crushing the insulation around and between the winding coils. The failure would then be an electrical failure involving turn to turn shorts or coil to shaft grounds, the latter of which can be very destructive, melting copper and rotor steel and resulting directly in retirement of the rotor.

Inspection of this region requires removal of the retaining rings and the end wedges to expose the radius where the cracks initiate. Inspection can then be by any surface technique inclusive of PT, MT, or ET, although because of the proximity to the winding components, the introduction of magnetic particles and/or liquid penetrants certainly introduces significant risk of contaminating the windings and the winding insulation. Therefore ET inspection provides the better option.

A second generator rotor tooth-top issue exists for many rotors that have steel wedges having hardness that closely matches the rotor hardness and involves a fretting fatigue initiation mechanism.

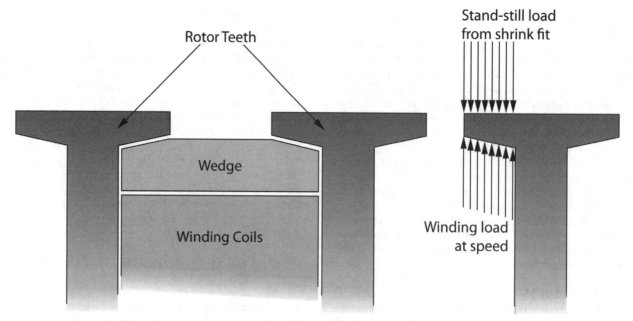

FIG. 25.21   TOOTH TEE-TOP CONFIGURATION (LEFT) AND LOADING (RIGHT)

Fretting occurs more readily where the materials are of similar hardness, i.e., where there are steel wedges, and therefore is worse where harder ferritic steel wedges are used. Relative motion is created by the once-per-revolution bending of the rotor. If a location on the rotor surface is selected and then followed through a full rotation, the location will be at maximum compressive bending stress when at 12:00, at neutral bending stress when at 3:00, at maximum tensile bending stress when at 6:00, back to neutral at 9:00, and finally back to maximum compressive when the location returns to 12:00. Put another way, the rotor surface length changes as the rotor rotates, shorter at 12:00 and longer at 6:00.

The wedges, however, are not necessarily locked to the rotor, at least at slow speed, so they are not forced to undergo equivalent strain. Consequently, there is a relative motion between the wedges and the rotor. The relative motion experienced by any given wedge is equivalent to the strain accumulated in the rotor over a length equivalent to the wedge length minus any length that is locked to the rotor. If viewed from the wedge, there is symmetry over the length of the wedge; the axial center of the wedge therefore sees no relative motion at any speed. During startup as the rotor speed increases, the bearing load of the windings and wedge increases and, as the normal force increases, the friction force increases accordingly. This locks more and more of the wedge to the rotor, working from the wedge center toward the ends as the rotational speed increases. At some point the entire wedge locks to the rotor and no relative motion occurs from this point on until speed is reduced during shutdown. Fretting cannot occur at slow speed because the pressure load on the bearing interface is not sufficient to cause damage, even though the relative motion is high. At full speed and at some point before reaching full speed, the wedges lock to the rotor over their full lengths; consequently, the wedges see the same strain as the rotor, but without relative motion. At some intermediate speed, however, the bearing load and the relative motion occur concurrently at levels sufficient to cause fretting fatigue damage. Because the critical motion is limited to the material around the wedge ends, damage is limited to these same localized regions on the rotor, i.e., at and around the rotor wedge ends.

Fretting damage can lead to transverse (radial–circumferential) cracks which, under certain circumstances, can grow under pure bending fatigue. If a crack reaches the critical depth at which the bending stress is sufficient to grow the crack, stress cycles accumulate very rapidly, once per revolution, so the cracks typically grow very rapidly to failure. However, it also appears that there are some mitigating circumstances that reduce the likelihood of a fretting crack reaching the critical depth necessary for growth by pure bending. First of all, fretting in and of itself is a deflection controlled process, so as the fretting crack grows deeper, it is growing out of the surface deformation; consequently, it arrests at some point. Additionally, because the cracks are relatively small, the presence of a crack is not sufficient to impact the deflection. A second influencing factor is that associated with Poisson effects in both the rotor and the wedges. As the rotor comes up to speed and expands radially, Poisson dictates that it must grow shorter as well. Additionally, the radial compressive load on the wedges causes a Poisson lengthening of the wedges. The axial contraction of the rotor and axial expansion of the wedges, coupled with the fact that the wedges are locked to the rotor at the wedge centers, introduces a compressive component of stress on the surface of the rotor teeth in the rotor material that has not yet locked to the wedge. As the speed increases and the Poisson effects increase, the compressive loads accumulate toward the ends of the wedges. A second potential source of axial compressive stress in the rotor derives from the forging process and residual stresses that can remain in the rotor. When these compressive stresses are combined with the axial bending stresses, the resultant axial tensile stress is reduced; consequently, the fretting crack must reach a greater depth before the axial bending stress alone can take over and grow the crack. Overall, a conclusion that must be drawn is that the stress situation is relatively complex and includes a number of contributory factors, many of which reduce the severity or the end result of this crack initiation mechanism.

Susceptibility to fretting damage is associated with a number of rotor design considerations. First is the bending flexibility of the rotor, as related to the length/diameter ratio. Longer, leaner rotors are more susceptible than are short, stout rotors. Susceptibility is further a function of the number of accumulated start/stop cycles, given that the combination of relative motion and compressive bearing load occur only at intermediate speeds, which are experienced only during startup and shutdown. Location along the length of the rotor is also important. Because maximum bending stress occurs at the axial center of the rotor, the shaft axial center is more susceptible than the ends. And susceptibility is a function of wedge length. The amount of strain that accumulates over the portion of the wedge that is not locked down at any given speed equates directly to the amount of relative motion at that speed.

Another issue that involves a totally different initiation mechanism can ultimately lead to similar transverse cracking in the rotor in the immediate vicinity of the wedge ends on the underside of the tooth-top dovetails where the wedges contact the teeth. In this case, damage initiates from electrical arcing between the rotor teeth and winding wedges. Certain transient fault and steady state conditions can cause unbalance between the electrical phases, which reflects over to the rotor as motor currents. Events that can cause these transient events include motoring incidents, line faults such as closely proximate lightning strikes, and so on. Steady state conditions can result, for example, if the unit is operated with unbalanced transformers on the three phases. The currents in the rotor conduct at the rotor surface, traveling down the rotor and typically around it at the retaining rings and them back along the opposite side of the rotor. In the rotor winding slot region, the currents tend to conduct in the wedges, arcing to the rotor teeth at the ends of the wedges and then back into the next wedge. Unlike the fretting damage, which tends to occur only where the rotor wedges are steel and therefore relatively closely matched in hardness to the rotor material, these surface currents tend to cause arcing damage at the ends of any of the slot wedges. In fact, where other materials are used for the wedges, for example, aluminum or other lower electrical resistance material, the arcing is more prevalent because of the lower resistance of the wedges, and, therefore, the increased potential for the current to travel in the wedges and arc to the rotor at the wedge ends.

Where arcing occurs, localized heating results and the severity is a function of the electrical current level. If the arcing is sufficient, the heating can actually be sufficient to cause phase transformation, i.e., re-austenitization and even melting. Damage can run the gamut, from very minor heating, localized spark erosion, re-austenitization, to melting, in which case, the melted zone is also going to be surrounded by re-austenitized material, and so on. When the heat source is removed, the re-austenitized material is quenched, resulting directly in a nugget of untempered martensite, which is very brittle and almost certainly cracks immediately under the imposed stress. Because the local stress is primarily an axial bending stress, any resulting cracks tend to form in a radial–circumferential plane, just like the fretting cracks. In this case, however, the depth of the resulting crack is not limited like it is for fretting cracks. Arcing damage is related to current levels and can immediately generate relatively deep cracks and otherwise damaged material. It is certainly possible to cause damage sufficient to retire the rotor. Other areas on the rotor where damage may also occur is at the retaining ring shrink fit and at the ends of transverse flexibility slots that are cut across the poles to equalize the moments of inertia about the pole and quadrature axes and therefore to minimize twice per revolution vibration.

Inspection for tooth-top cracking of this type can take any of a number of forms and include any of a number of NDE methods. For moderate to severe electrical arcing damage, there is nothing

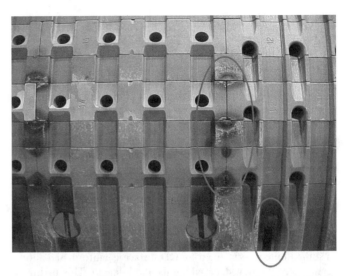

FIG. 25.22 GENERATOR ROTOR SHOWING OVERHEATING DAMAGE ASSOCIATED WITH CIRCULATING CURRENTS AT THE WEDGE ENDS AND THE END OF THE TRANSVERSE FLEXIBILITY SLOTS

like a good visual inspection of the rotor, concentrating on the areas defined above as being most prone to damage: the teeth surfaces adjacent to the wedge ends and the ends of the wedges themselves, the teeth and wedges and nose of the retaining ring adjacent to the shrink fit, and the rotor material around the ends of the transverse flexibility slots. Figure 25.22 provides a photograph of a rotor displaying definite signs of overheating damage on the teeth and wedges in the immediate vicinity of the wedge ends. In this photograph, the slot run horizontally, and one of the poles is shown covering the bottom ~1/3 of the photo. The end of one of the transverse flexibility cross slots can be seen at the bottom right corner of the photo. This also exhibits very definite signs of overheating associated with circulating surface currents.

For less extensive electrical damage and for any level of fretting damage, however, the damage does not manifest itself in any way that produces visible symptoms on the accessible surfaces. To detect these damage forms requires either ultrasonic inspection techniques or disassembly of the rotor–at least removal of the retaining rings and the rotor slot wedges–to access the crack initiation surfaces and permit inspection by some surface inspection method. For wound rotors, even when the rings and wedges are removed, the use of PT and MT methods for the detection of cracks carries the same risks as defined earlier for the axial cracking of the tooth-tops. The inspections must be conducted in very close proximity to the winding components, including the slot insulation, and the risk of contamination is significant. This can be done, but it requires extreme caution and extensive steps to protect the windings and insulation. For rotors that are being rewound, i.e., those from which the winding coils and insulation have been removed, either MT or PT are perfectly acceptable test methods, although viewing the underside of the tooth-top can be challenging, particularly if using the more sensitive fluorescent MT or PT, which requires viewing under UV illumination.

An alternative is to inspect the disassembled rotor using the ET inspection method. This, of course, still requires full disassembly to expose the initiation surfaces, and special ET probes, configured to match the geometry. This inspection benefits immensely from automation, i.e., automated or semi-automated scanning, and full digital data acquisition and surface imaging. This requires a scanner that matches the slot geometry in some way and that holds the ET probe at the proper attitude in close proximity to the examination surface. The use of an ET array, i.e., an array of probes that collectively cover the entirety of the test surface width, speeds up the inspection significantly in that the entirety of a tooth top surface can be inspected in a single scan pass along the length of the rotor. One manufacturer, in fact, the primary manufacturer in the US for which fretting fatigue of generator rotor tooth-tops is an issue, recommends removal of the retaining rings and slot wedges and implementation of an ET inspection protocol. Unfortunately, the disassembly and reassembly costs far exceed the inspection costs, likely by at least an order of magnitude.

An alternative for the detection of underside tooth-top cracking without rotor disassembly involves UT from the outer surface and, in most cases, is implemented using the linear phased array. In some limited cases, the rotor OD surface is uniform and allows scanning of the transducer along the tooth length. This permits a fairly simple inspection protocol that utilizes either an LPA or conventional UT approach. However, in most cases, the linear array is required because of the rotor OD surface conditions. The complex surface configurations that are typical in these rotor designs can be observed by looking back at Figures 25.19 and 25.22. The machined lands and grooves effectively prevent axial scanning of the tooth top. The LPA technology provides the means to set the transducer in one stationary position along the tooth and to then sweep the beam through a series of angles to cover some length of the underside surface of the tooth-top. By utilizing multiple of the lands and grooves, the entirety of the wedge end regions can be covered adequately.

Analysis of rotors for transverse cracking due to either fretting fatigue or arcing damage relies on finite element analysis using general purpose finite element programs such as ANSYS. That is, there is no custom, special purpose program with which to analyze specific tooth-top dovetail cracking situations on specific rotors. The analysis of fretting fatigue indications involves determination of the crack depth required to grow by once-per-revolution bending fatigue. If detected indications are smaller than this critical flaw growth size, with appropriate safety factors applied, then further operation can be justified with limited risk of failure. For damage caused by electrical arcing, the analysis is not this simple. Arcing damage includes not only the resulting cracks, but also the damaged material surrounding the cracks. Consequently, the only way to assure future safe operability involves removal of all damaged material and metallographic verification of complete removal. Once this has been accomplished, then the modified rotor configuration can be modeled and analyzed to determine the impact of the new configuration on remaining life.

25.12 MATERIAL PROPERTIES CHARACTERIZATION

When stress and fracture analysis indicates less that desired remaining life for a primary component, the uncertainty associated with material properties most often is a significant contributor. Consider, for example, that many of these components–rotor forgings, shrunk-on disks, retaining rings–were forged prior to the development of modern fracture mechanics approaches to crack propagation and ultimately to failure of the component. Consequently, even when the original material properties data is available, they may well not include fracture toughness or even FATT.

And then there is the issue of the availability of the data to the owner/operator. Often this information is simply not readily available. So, in all likelihood, the analysis will be conducted using public domain database properties, of which significant quantities are typically available through various sources.

Unfortunately, the spread in data for the key material property input variables is typically significant; consequently, one is faced with making a decision on which value to use for the analysis. If performing a deterministic analysis, this means deciding the specific value that will be used, and for the sake of conservatism in the analysis, a worst case assumption is typically used. While this is the conservative approach, it means that the real value for the variable will, in nearly all cases, be better than the value used for the evaluation. In most cases, this analysis results in significant overly conservative outcomes as a direct result. And, when one considers that this may apply for multiple variables, conservatism stacks on conservatism to a point at which it is not even reasonable. Even if performing a probabilistic analysis, the spread in the data still results in overly conservative assessments in the form of higher than actual probability of failure because of the inclusion of the worst case values even on a statistically applied basis.

The solution is in the ability to more closely define the variables, in this case, the specific material properties, and the most effective means of better defining the material properties uses samples extracted from the most critical region of the component being assessed. For a turbine rotor forging, this most often means the near-bore material. For a turbine disk, it could mean the material near the rim for blade attachment studies, and so on. A miniature sample removal tool is available for extraction of samples from the surface of a rotor bore, the side of a turbine disk, or other locations as needed and as appropriate. This specialized machining tool extracts a small wafer, about the size of a quarter dollar, leaving a smooth dimple in the surface. The tool is shown in a rotor bore in Figure 25.23.

Analysis of the dimple has shown that it produces no meaningful increase in local stress, so the dimple can remain with no further conditioning. Miniature samples can be analyzed in the lab to quantify material chemical composition, measure hardness (thereby estimating yield strength), and define material microstructure. Additionally, mechanical properties can be better characterized using automated ball indenture testing. Properties that can be quantified using miniature samples and the automated ball indenture test

FIG. 25.23 MINIATURE SAMPLER INSIDE A ROTOR BORE

include yield strength, ultimate tensile strength, fracture appearance transition temperature, and fracture toughness.

Additionally, the sampler can be used to extract samples from the high temperature regions of HP and IP rotors to assess the material for creep damage and for temper embrittlement.

25.13 INSPECTION VALIDATION

The Electric Power Research Institute (EPRI) has supported an extensive turbine and generator condition assessment program over the years. This program has effectively dealt with the primary aspects of T/G condition assessment, i.e., inspection validation and remaining life assessment. EPRI maintains NDE validation specimens for rotor bore inspection and turbine disk inspection. These test beds comprise samples with known flaws (known only by EPRI, not the participating inspection vendor) that are used in blind tests to assess the capabilities of various inspection approaches. Some of the samples contain artificial flaws in the form of electron discharge machined (EDM) notches, drilled holes, etc. Some of the mockups contain more realistic, yet still man-made, flaws in samples made by a hot isostatic press (HIP) process to metallurgically join segments with artificial flaws inserted at the interfaces. Still other samples have been removed from service-exposed components and so contain naturally occurring flaws.

The EPRI T/G program has additionally led the way on the remaining life assessment aspect of condition assessment via production of the SAFER stress and fracture program for rotor bore assessment, the LPRimLife computer code for assessing turbine disk rim blade attachment dovetails, and RRing-Life for probabilistic assessment of generator retaining rings.

These resources have contributed immensely to the advancement of the state-of-the-art for T/G assessment and are available to interested parties for further use as the need arises.

25.14 REPAIRS

The first and most obvious repair that can be implemented for any number of the described flaw conditions, at least those at or very near a free surface, is to remove them by local excavation. This is typically done by grinding, carefully removing the flaw, and ultimately putting the geometry at the site of the local material removal in the best condition possible to minimize the impact on local stresses. Confirmatory inspections, most often MT or PT, are implemented along the way to track the flaw for minimum material removal while confirming complete flaw removal. Because these inspection methods have inherent detection thresholds, i.e., a flaw size that cannot be detected, final confirmation by metallographic replication often follows the MT or PT. Machining for flaw removal is another option. Over-bores and bottle bores are often implemented to remove flaws detected at or near the rotor central bore surface. The generator rotor tooth-top cracking issues each have repairs that involve crack removal by machining, and in these cases, the repairs actually do improve the stress situation, so there is dual benefit. More is presented on these later.

Unfortunately, except for very rare situations, as for the generator rotor tooth-top repairs, repair by local material removal can only increase the local stress. Consequently, each repair has a maximum depth beyond which the benefits of removing the flaws are overridden by the negative impact of increased stress, and some other repair or replacement must be considered. And each repair typically requires extensive analysis, most often involving 3-D finite element stress analysis, to define the appropriate maximum allowable repair depth and to define the optimum final repair geometry.

Repair of assembled components can often become very involved owing to the necessity of disassembling the components from the rotors. Generator rotor tooth-top repairs, regardless the design and the elected repair method, require fan, retaining ring, and wedge removal and reassembly upon completion of the repairs. Because this involves removal and reassembly of relatively massive components, it is not unusual for the rotor to require rebalancing before being returned to service. Similarly, repair of shrunk-on turbine disks and repairs for generator rotor coupling keyway cracking involve removal and reassembly of shrunk-on components, with similar processes involved.

For the generator rotor tooth-top axial cracking associated with the retaining ring shrink fit, there are a number of fixes available. One, identified as a short ring fix, involves removal of any existing cracks by local grinding or machining of the radius, increasing the radius at the initiation site in the process. Then the original or duplicate retaining rings are reassembled, and the crack initiation and growth clock is essentially reset. This fix can be implemented only up to certain crack depths; otherwise the reduction in area associated with crack removal in and of itself structurally compromises the tooth integrity and ability to perform its intended function.

In a long ring fix, a longer retaining ring is used such that the circumferential key can be removed from the original shrink fit area, thereby permitting full removal of the Tee-top over the length of the existing shrink fit. The key is moved inboard on the rotor body relative to the shrink fit, which necessitates the longer retaining ring. The machining dimensions at the two fits, the new shrink area where the Tee tops have been completely removed and the extended nose where the key has been moved, are designed such that the former carries the shrink fit load. The latter sees no appreciable shrink fit load, if any, but carries the outward load at speed. This fix requires new retaining rings, and additionally a rotor rewind to accommodate insulation modifications associated with the new ring design. However, this is a more permanent fix, providing increased fatigue life, whereas the short ring fix only resets the fatigue clock. Other repairs involve removal of the Tee-tops by machining the sides off rather than machining the OD, or combinations of the two.

A number of repairs are available for the transverse cracking at the wedge ends that occurs either by fretting or negative sequence arc damage. While these repairs vary in detail, they generally improve the situation by removing any pre-existing damage by machining, and in the process, relieve the contact at the ends of the wedges such that the accumulated strain is distributed over greater length of the rotor tooth. This lowers the axial stress that drives the process.

In addition to these very specific repairs, most if not all domestic manufacturers and any number of independent repair organizations now have established repair welding procedures and processes in place. In the case of turbine disk rim blade attachment cracking, it is not at all uncommon to simply machine away the entire outer periphery of the disk and then replace the material with a weld build-up of material. The attachment geometry is than machined into the new material and the blades are reassembled. Another approach involves similarly machining the entire dovetail region from the disk and then replacing it with a ring forging that is welded to the periphery. Weld repairs are also routinely used, for example, to replace cracked shaft ends and to build up material to enable restoration to original configurations where rubs and other events have forced material removal in the journal and seal areas.

Much has been presented on the state-of-the-art for inspecting major turbine and generator rotating components. Advances in modern inspection technology provide the ability to characterize flaws as never before. LPA ultrasonic inspection technology is a prime example. A great deal of effort and expense has been expended to develop very specific, computerize remaining life assessment programs for specific turbine and generator components, and the accuracy of the analyses has improved as a direct result. Additionally, probabilistic approaches are available and are typically used to remove some of the conservatism inherent to the deterministic analysis process. We have the means available through miniature sample removal and testing to better define material properties. And, we have any number of repair options available, some relatively generic and others applying to specific components. And so, yes, the state-of-the-art for turbine and generator condition assessment has evolved, and continues to do so, with an ever present objective of extracting all possible useful life from these very expensive machines and components, while maintaining acceptable exposure to risk.

25.15 REFERENCES

1. Engelhardt G, Macdonald DD, Zhang Y, and Dooley B. "Deterministic Prediction of Corrosion Damage in Low Pressure Steam Turbines." 14th ICPWS, Kyoto, Japan. August 2004. Also published in PowerPlant Chemistry, 2004, 6(11), pp 647–661.

2. Viswanathan R. "Retaining Ring Failures." *Workshop Proceedings: Retaining Rings for Electric Generators*. Palo Alto, CA: Electric Power Research Institute, August 1983. EPRI EL-3209.

3. Nottingham LD, Ammirato FV, MacDonald DE, Zayicek PA, and Elmo PM. "Evaluation of Nonmagnetic Generator Retaining Rings" Palo Alto, CA: Electric Power Research Institute, October 1994. TR-104209.

4. Kilpatrick NL, Schneider M. "Update on Experience with In-Service Examination of Nonmagnetic Rings on Generator Rotors" *Workshop Proceedings: Generator Retaining Ring Workshop*. Palo Alto, CA: Electric Power Research Institute, May 1988. EL-5825.

STEAM TURBINES FOR POWER GENERATION

Harry F. Martin

26.1   INTRODUCTION

Steam turbines have historically been the prime source of power for electric power generation. Turbines come in a variety of types with regard to inlet and exhaust steam conditions, casing and shaft arrangements and flow directions. This chapter will focus on steam turbines currently being applied to power generation. The steam conditions will include those currently applied to fossil fired power plants, combined cycle and nuclear power units.

Currently, fossil fired plants, whether coal fired or gas fired, typically supply steam at 1800 to 3500 psig steam pressures with 950 to 1050°F main steam and reheat temperatures. Double reheat units are few and will not be discussed in this chapter in much detail. Combined cycle plants (CC) typically have multiple drum heat recovery steam generator (HRSG) with inlet temperatures of 1050°F for main steam and reheat steam (see Chapter 27). While no new nuclear units have been installed in recent years, plans are underway to apply the AP1000 nuclear systems (see Chapters 23 and 24). Past steam conditions were typically saturated steam (0.2% moisture) and reheat temperature of 500°F. All current applications are condensing designs with regenerative extraction, except for CC units. Current application can be used up to 8 to 10 inHgA exhaust pressure. The application of air cooled condensers is more prevalent in CC applications.

Current designs are axial flow turbines. However, in some applications, radial flow stages are used in the inlet stages. Typical arrangements are tandem compound. This means more than one turbine casing's rotors are coupled together on the same shaft. For example a turbine train consisting of a high-pressure turbine (HP), an intermediate-pressure turbine (IP) and two low-pressure turbines (LP) would have all of these in line and coupled to an electric generator. Cross compound units (two or more shafts) are currently in use. Some arrangements have the HP and IP on one shaft along with a generator and the LP turbines on a separate shaft along with another generator. In many instances the low-pressure turbine shaft is running at half speed and has larger annulus areas to reduce LP turbine exhaust velocity and associated leaving losses. However, due to increased blade height and associated exit annulus area for full speed units and the higher costs, cross compound units are not currently in favor. Casing configurations for fossil units are typically high-pressure unit, intermediate-pressure unit, and one or more low-pressure units exhausting to the condenser. The high-pressure and intermediate-pressure units are frequently combined into one casing (HP–IP).

Steam turbines use both reaction and impulse designs. The fundamentals of these blade design types will be discussed as they are currently applied. The selection of the type blading has a significant impact on the turbine design.

Steam turbines employ digital control systems that can provide automatic turbine control to avoid mal-operation. These systems can be supplied by the turbine supplier or by others. In some instances these interface directly with the overall plant and boiler control systems.

Equipment supplied for the last 20 years have greatly increased inspection intervals per the recommendations of the turbine suppliers. The inspection interval for major inspections can be 8 to 10 years. Sometimes the inspection interval is defined in terms of equivalent operating hours (EOH) which includes the number of startups in the determination.

26.2   GENERAL INFORMATION

Any discussion of steam turbines should include some thermodynamic concepts. The first law leads to the writing of the general energy equation for a steady flow processes.

$$(PE)_1 + (KE)_1 + u_1 + (pv)_1 + Q = (PE)_2 + (KE)_2 + u_2 + (pv)_2 + W \tag{26.1}$$

Where:
  PE = the gravitational or stored mechanical energy of the fluid
  KE = the mechanical kinetic energy of the fluid
   u = the internal energy of the system
  pv = the pressure volume product or flow work
   Q = heat added
   W = shaft work

Since enthalpy (h) is equal to u + pv, the work for a flow system with no heat added, or negligible changes of potential and kinetic energy can be written as:

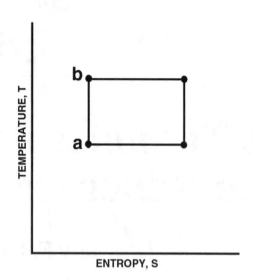

FIG. 26.1   CARNOT CYCLE

$$W = \Delta h \qquad (26.2)$$

The second law of thermodynamics dictates that all cycles must reject heat. The efficiency of a cycle can be defined as:

$$\eta = work \,/\, heat \; added$$

or

$$\eta = (heat\;added - heat\;rejected) \,/\, heat\;added \qquad (26.3)$$

A reversible cycle is one that is reversible in all aspects and could produce the same thermodynamic state points run in either direction. This of course is an idealization. However this concept is useful. For example the Carnot cycle is defined as one in which heat is added and rejected at constant temperature and the expansion and compression is accomplished with no losses. This cycle is shown on a temperature entropy diagram in Figure 26.1.

The resulting efficiency of such a cycle is defined as:

$$\eta = (T_a - T_b) \,/\, T_a \qquad (26.4)$$

This is the maximum possible efficiency for a cycle operating between temperatures Ta and Tb.

The steam turbine cycle is a Rankine cycle. This cycle consists of compression of water, heating and evaporation of steam in a boiler, expansion in turbine and heat rejection in a condenser. A theoretical version of this cycle is shown as Figure 26.2. Most units use a reheat cycle with regenerative heating though the use of feedwater heaters using steam extracted at various locations throughout the turbine (see Chapter 29). Regenerative heating reduces the heat added and the heat rejected by the cycle. While the output power is also reduced, the net effect is an improved cycle efficiency.

There are a number of gas dynamic concepts that have significant impact on turbine design and application. These will be discussed in the following paragraphs.

The speed of sound can be written as:

$$V^* = (dp \,/\, d\rho)^{1/2} \qquad (26.5)$$

Where:
V^* = velocity at the critical point or the sound velocity
p = pressure
ρ = density

Mach No. (M) is defined as the ratio of the local velocity to the local sound velocity.

The relationship between area and pressure for an isentropic flow process can be written as:

$$\frac{dA}{dp} = \frac{A}{\rho V^2}[1 - M^2] \qquad (26.6)$$

Where:
A = area
V = Velocity

From this relationship we can deduce the flow passages for different flow velocities. These are summarized in Figure 26.3. The significance of this figure is the differences in the behavior of a diffusing passage between subsonic and supersonic diffusers and the development of supersonic velocity in turbines.

From the definition of an isentropic process the following useful relationships can be developed:

$$h_1 - h_2 = \left(\frac{V_2^2 - V_1^2}{2gJ} \right) \qquad (26.7)$$

The condition when the velocity is zero is referred to as the stagnation state or total condition:

$$h_t - h_2 = \frac{V_2^2}{2gJ} \qquad (26.8)$$

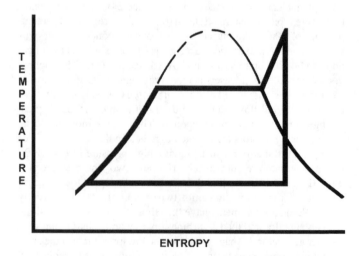

FIG. 26.2   THEORETICAL RANKINE CYCLE

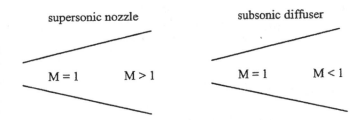

FIG. 26.3 SUBSONIC AND SUPERSONIC EXPANSION

Where in the English system:
J = 778 ft-lbf/Btu
g = 32.2 lbm ft/lbf sec^2

From perfect gas relationships and the continuity equation the following non-dimensional flow term can be developed:

$$\frac{mR\sqrt{T_t}}{A_2 p_t} = \sqrt{2gJCp\left[\left(\frac{p_2}{p_t}\right)^{\frac{2}{k}} - \left(\frac{p_2}{p_t}\right)^{\frac{k+1}{k}}\right]} \qquad (26.9)$$

Where:
m = mass flow
T = temperature in degrees Rankine
R = gas constant
Cp = specific heat at constant pressure
k = specific heat ratio

A plot of non-dimensional flow and pressures ratio is shown as Figure 26.4. Note the dashed portion of this curve illustrates the double value of this equation. The flow reaches a maximum at the critical pressure ratio (p*/Pt). For real gases such as steam, the values of critical pressure ratio and acoustic velocity are found in steam tables. However, there are many instances where perfect gas analysis is very useful in steam turbine analysis.

The purpose of blading in a turbine is to transform the kinetic energy of the incoming fluid into useful work. Newton's second law of motion can be applied to develop a momentum theorem.

$$F_\theta = \frac{w}{g}\left(V_{\theta 1} - V_{\theta 2}\right) \qquad (26.10)$$

Where:
F = force in the tangential direction
V = velocity in the tangential direction

We can derive m the following equations:

$$\overline{W} = \frac{w}{g}\left(U_1 V_{\theta 1} - U_2 V_{\theta 2}\right) \qquad (26.11)$$

Where:
$\overline{W}$ = power
U = turbine wheel speed
V_o = tangential velocity
w = mass flow rate

For a constant diameter the relationship may be written as:

$$\overline{W} = \frac{wU}{g}\left(V_{\theta 1} - V_{\theta 2}\right) \qquad (26.12)$$

For steady flow with uniform conditions and no leakage we may write:

$$\overline{W} = \frac{w}{g}\varpi\left(r_1 V_{\theta 1} - r_2 V_{\theta 2}\right) \qquad (26.13)$$

Where:
ϖ = rotational speed in radians/second
r = radius
w = flow rate

Writing this equation in terms of enthalpy produces:

$$\overline{W} = w\left(h_1 - h_2 + \frac{V_1^2 - V_2^2}{2gj}\right) \qquad (26.14)$$

A turbine stage consists typically of a stationary row of blades, and a rotating row of blades. The stationary blades are often referred to as nozzles. The stationary row and the rotating blades act together to allow the steam flow to produce work on the rotor. Many stages are included in a turbine casing to produce the shaft output to drive the generator.

Turbine stages, with the exception of last stages of the low-pressure turbine, are typically classified as either an impulse stage or a reaction stage. In a pure impulse stage the entire stage pressure drop is taken across the stationary row of blades. However, in practice this is not practical and about 5% to 20% of the stage drop is taken across the rotating blade. The addition of another stationary and rotating row to this stage forms a "Curtis" stage. However, this type of design is not in use in current designs for large steam turbines.

The velocity triangles and general stage arrangement are shown in Figure 26.5. Referring to this figure, the power developed by an impulse turbine can be written as:

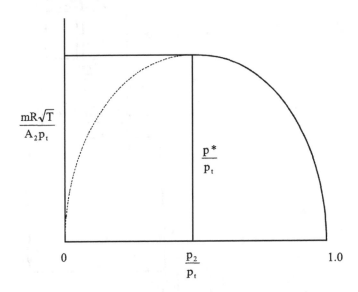

FIG. 26.4 FLOW VS. PRESSURE RATIO

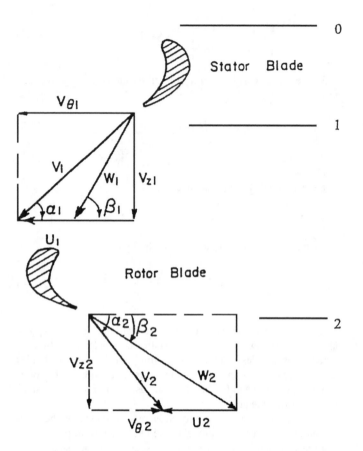

FIG. 26.5 VELOCITY DIAGRAM FOR SIMPLE STAGE

$$\overline{W} = \frac{wU^2}{g}\left[\frac{\cos\alpha_1}{v} - 1\right]\left[1 + \frac{\cos\beta_2}{\cos\beta_1}\right] \qquad (26.15)$$

Where
v = velocity ratio

Reviewing Equation 26.15, the advantages of small values of α_1 and β_2 are evident. This is a characteristic of impulse blading.

In a reaction turbine pressure drop occurs in both the stationary and rotating blades. In a reaction turbine there is an increase in the relative velocity leaving the passage. In a symmetrical stage, the enthalpy drops are equal for the fixed and rotating rows. This is a 50% reaction stage. For a 50% reaction turbine the power can be written as:

$$\overline{W} = \frac{wU}{g}(2V_1\cos\alpha_1 - U) \qquad (26.16)$$

The typical constructions of these two types of turbine stages are shown as Figure 26.6. Due to the higher pressure drops across the stationary row in an impulse stage, the sealing for the stationary row is at a lower diameter. However, a flow system design involving the leakage along the shaft, the sizing of the pressure balance holes in the rotor and the possible inclusion of a seal at the inner diameter between the rotating and stationary row is required in an impulse or Rateau turbine. This is discussed in Reference [1]. In an impulse stage the design should set the reaction at the hub to be positive to ensure good stage performance. Therefore, some degree of reaction is required. Considering the radial variation of station-

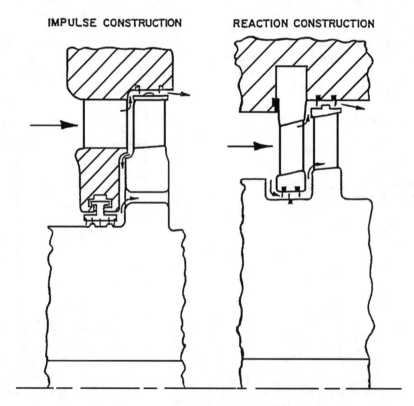

FIG. 26.6 IMPULSE AND REACTION STAGE

ary exit pressures, the reaction is usually increased as the stage heights get larger as you proceed through the turbine. Generally these stages are designed such that there is a radial in flow into the cavity between the stationary row and the rotor.

Important stage characteristics of a turbine stage application are velocity ratio, and stage loading coefficient. The velocity ratio is defined as the ratio of the blade wheel speed to the velocity that would be obtained by the isentropic expansion through the stage stagnation pressure drop.

$$v = U / Cis \qquad (26.17)$$

Where:
  v = velocity ratio
  U = wheel speed
  Cis = isentropic stage velocity

The stage loading coefficient is defined as the work done in the stage divided by the wheel speed squared.

$$\psi = \frac{\Delta H}{U^2} \qquad (26.18)$$

Impulse turbines can have higher stage loadings and therefore, fewer stages than can be used efficiently for a reaction turbine for the same expansion. This generally means higher losses, as will be discussed later.

Figure 26.7 shows the variation of efficiency for nominal reaction and impulse blading with velocity ratio. The figure illustrates the advantages of impulse blading for lower stage velocity ratios and higher blade loadings. In addition, the benefit of reaction blading when lower stage loadings can be utilized is illustrated.

In the first stage of a high-pressure turbine, an impulse or Rateau stage is often utilized. This stage is commonly called the Control Stage. Currently, single stage control stages are utilized. However, in the past Curtis stages were used. The Curtis stage was popular when large temperature drops were required due to rotor and blade mate-

rial limitations for high temperature inlets. The Rateau type control stage has better performance than the Curtis type control stage.

The control stage combined with inlet valves serves as the flow control for the turbine. As load is reduced the control valves can be operated in a sequential valve mode with constant pressure operation meaning that a 4 valve unit could operate with between 1 and 4 valves open and the control stage operating at admissions of 25% to 100%. Therefore, the control stage blade loading at 25% arc of admission is significantly higher and has a higher pressure ratio than at 100% admission. This is true since the pressure downstream of the control stage is essentially proportional to flow and the nozzle inlet pressure is still at 100% pressure. The nozzle is typically choked at this condition but the absolute velocity entering the rotating blade is supersonic. At partial admission, the losses are higher, however, the reduced throttling loss and the higher inlet pressures provide overall part load cycle efficiency improvements. There are many operating strategies that exist with today's plants and control systems. These will be discussed in more detail in 26.6. However, a partial arc design when operating at part load will generally produce a better part load heat rate than a typical full arc design.

In the low-pressure turbine, the steam is expanded from inlet pressures of 60 to 230 psia to typical condenser pressures of 1 to 3 inHgA. This increase in specific volume requires rapid increases in flow area especially in the last few stages. In these stages the design is significantly different than typical impulse or reaction blade design mentioned in the preceding paragraphs. In these stages the work done is quite large. The LP turbine can contribute up to 60% of the unit output and the last stage alone can produce 10% of the total output. Therefore, the efficient design of these stages is important.

The stage design is driven by consequences of the radial equilibrium equation in the Meridional (axial–radial) plane.

$$\frac{1}{\rho}\frac{\partial P}{\partial r} = \frac{V_\theta^2}{r} - V_z \frac{\partial}{\partial Z}\frac{V_r}{} - V_r \frac{\partial}{\partial r}\frac{V_r}{} \qquad (26.19)$$

The first term on the right hand side shows the effect of a swirling flow such that the static pressure increases with radius. This

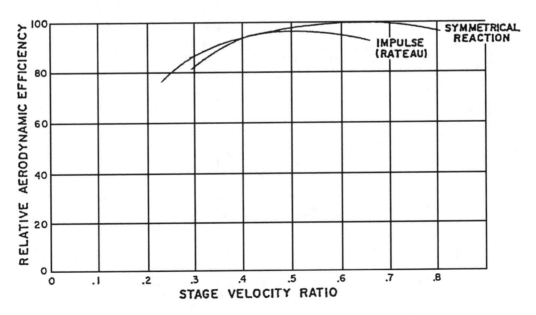

FIG. 26.7 EFFECT OF STAGE TYPE ON AERODYNAMIC EFFICIENCY

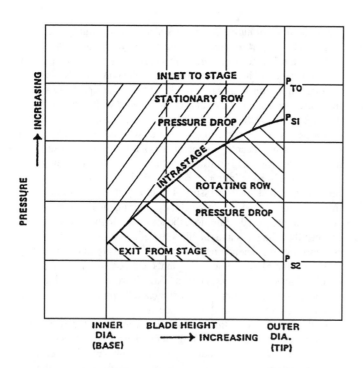

FIG. 26.8 PRESSURE DISTRIBUTION LAST STAGE

effect is significant at the exit of stationary rows where high tangential velocities occur. The second and third terms are referred to as the streamline curvature terms. The effect of the streamline curvature terms become significant at end wall regions of the blade path where wall tapers and abrupt changes occur.

Figure 26.8 shows the pressure drop across the stationary blade row is greater at the base than at the tip. Consequently, the steam velocity leaving the base or inner diameter is greater than at the tip as shown in Figure 26.9. The space between the stationary and rotating blades (see Figure 26.10) is the region with significant change in relative inlet angle to the rotating blade. There is a radial variation of reaction. Therefore, the base sections approach an impulse design with high turning. Therefore, blades with a large amount of twist are required.

Computerized flow field programs both axi-symmetric and three-dimensional are used in the design of low-pressure end blade paths. The process usually starts with the axi-symmetric approach. Very early versions of these methods are discussed in References [2] and [3].

Annulus areas have increased with larger blades and high tip speed. Tip speeds exceed Mach 2 and exit relative velocities are supersonic. Hub-to-tip ratios have been reduced from the standard 0.5 of the past to values near 0.4. New LP flow field and stationary blade design concepts have permitted this change, while maintaining hub reactions to acceptable levels. Many designs use

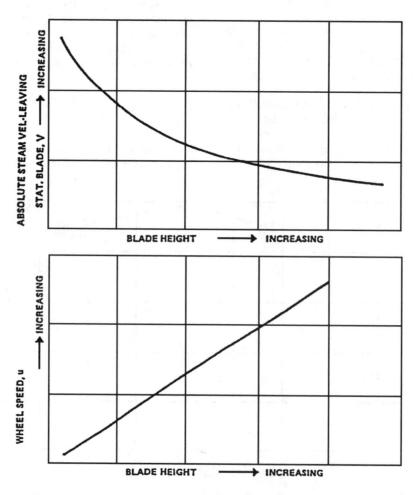

FIG. 26.9 VELOCITY DISTRIBUTION BETWEEN STATIONARY AND ROTATING BLADES

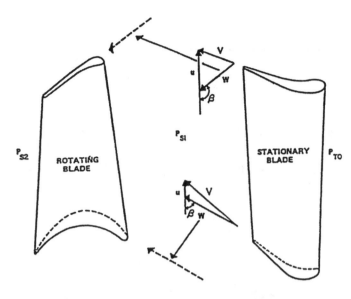

FIG. 26.10   BLADING VELOCITY

converging diverging sections to limit losses. Stationary blade designs utilize blade lean and sweep [4, 5] to control the flow field and limit losses.

The performance of the LP turbine is significantly affected by the performance of the diffuser and the flow path directing the steam to the condenser. Diffuser design is currently done using computational fluid dynamics (CFD) computer programs. Reference [6] discusses effective use of such computer codes. The inclusion of the effect of the last row blade flow distribution is emphasized in the reference. Modeling, using uniform flow distributions or model test data using uniform distributions, has shown to be ineffective in predicting exhaust diffuser performance.

Figure 26.11 shows the variation in total exhaust loss as a function of volumetric flow for three LP turbine designs using a variety of last row blade sizes and exhaust annulus. This factor is a significant influence in selecting which design should be applied to a given cycle. However, one must also consider the variations in condenser pressure and load demand through the year to select the correct configuration.

Leaving loss is the kinetic energy of the absolute velocity leaving the last row (L-0R) of the LP turbine. This is generally defined in terms of Btu's. Actually there are leaving losses for the exit of each turbine. However, the leaving loss of the last stage of the LP turbine is the most significant. The leaving loss for the L-0R for an LP turbine is obtained through integration of the exit flow properties obtained from a CFD calculation. The leaving loss is a part of the total exhaust loss. Figure 26.12 shows a diagram representing the expansion of the last stage of an LP turbine on an enthalpy-entropy diagram (HS diagram). Note that the stage exit static pressure, Ps-b, is less than the downstream condenser pressure. Therefore, the exhaust diffuser has recovered some of the exit kinetic energy from the blade path. The total exhaust loss (T.E.L.) is the value shown in Figure 26.11 at various volumetric flows. At low volumetric flows, the flow at the base of the blade will be recirculating and high hood losses are obtained.

Transonic turbine blades behave similarly to converging-diverging (CD) nozzles. This is true even if the blade is not a CD section per se. The throat area of the blade at this point is sonic. As the exit pressure decreases the passage shock wave will move down-

stream towards the exit plane and the blade loading will increase. At a certain pressure, the flow in the passage is fully expanded and the blade loading reaches its maximum value. Further decreases in back pressure will not affect the flow pattern inside the blade passage downstream of the throat of the blade. Therefore, the blade loading will not change with lowering of the exhaust pressure. This condition is referred to as "limit load." This is the reason that the load–vacuum correction curves provided by turbine suppliers reach a point where the power of the unit will not increase with further decreases in back pressure. Figure 26.13 shows isobars for a blade passage at the transonic point and after limit load has been reached.

LP turbines expand into the wet region, and they usually employ moisture removal devices to reduce moisture content. This moisture removal has a twofold benefit. First, the performance is improved through the reduction in moisture losses in the blade path and the reheating effect caused by the reduction of moisture. Second, the less moisture available the less blade tip erosion will occur.

The causes of moisture losses are:

- Condensation shock losses
- Braking losses
- Drag losses
- Miscellaneous losses

Condensation shock losses occur when the rapidly expanding steam, after crossing the saturation line, fails to reach equilibrium. In other words although the moisture level has theoretically reached a level as high as 3.5%, fog formation has not taken place. The longer the delay in reaching equilibrium the greater the buildup of super-saturation and the larger the loss.

Braking losses occur when moisture strikes the rotating blades and causes a negative torque.

Drag losses are of two types. Drag losses due to fog drops and drag loses due to large drops torn off trailing edges of stationary rows. The loss occurs due to dragging and accelerating of the droplets by the steam.

Miscellaneous losses include centrifugal losses of water being centrifuged out by the rotating blades, boundary layer losses caused by the waviness of the deposited water film, the losses generated by the kinetic energy of the blades being transformed into heat and continuous under cooling losses. The under cooling is the

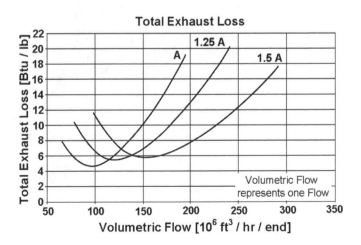

FIG. 26.11   TOTAL EXHAUST LOSS

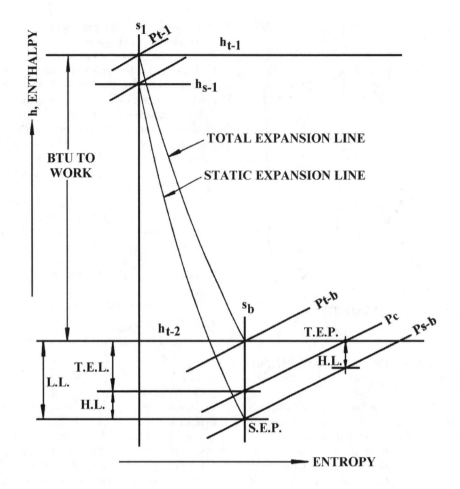

FIG. 26.12   LP TURBINE LAST STAGE EXPANSION

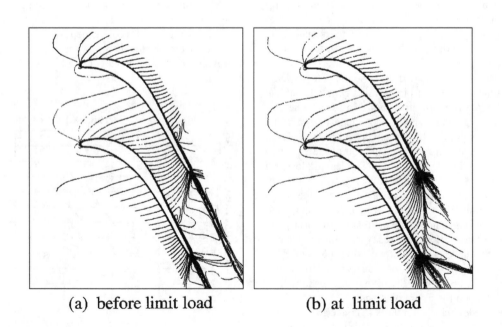

(a) before limit load   (b) at limit load

FIG. 26.13   LIMIT LOAD

loss associated with the rapidly expanding steam not being able to catch up with the equilibrium moisture level.

Not only is there a loss associated with non-equilibrium but the flow passing capability of the stage can be affected. Depending on the wetness, the flow passing capability could be increase by 5% to 6%. Reference [7] provides information on both efficiency and flow passing effects of moisture in turbines.

Addressing last row erosion has become more important with the larger tip speed designs in use today. Different blade designs use a variety of protection. Stellite tips have been brazed onto the upper portions of the blades. Blades have been flame and laser hardened. The protection strategy is a function of the blade material used. Reference [8] provides both a description of the erosion process and relative erosion resistances of various materials. The primary source of erosion are the droplets that come from water collected on the upstream stator blade. The water droplets are torn off the blade trailing edge and accelerated to a speed less than steam speed. Therefore, the droplets have a relatively low absolute velocity but a high relative velocity to the rotating blade. See Figure 26.14 for the velocity diagram of this process. The water droplets will vary in size up to the largest droplet size that can be supported in the flow stream.

There are many types of sealing currently used in steam turbines depending on the application. A good general equation for seal flow is the "Martin's equation." A form of this equation can be written as:

$$W = \frac{1700KAP_0}{\sqrt{P_0V_0}}\sqrt{\frac{1-\left(\frac{1}{\rho}\right)^2}{N+\ln\rho}} \qquad (26.20)$$

Where,
W = leakage flow, lb/hour
A = area, sq. in.
P_0 = Absolute inlet total pressure, psia.
V_0 = inlet specific volume, ft³/lbm
ρ = ratio of inlet total pressure to exit static pressure

N = number of throttling
K = kinetic energy annihilation coefficient (usually used as 0.8 for step seals

Equation 26.20 was shown to produce high flow rates for straight through seals in Reference [9]. This reference provides useful coefficients for certain commonly used seal designs.

Finite element analysis computer codes and individual supplier special computer codes are the dominant mechanical design tools in use today. However, it is useful to verify design concepts with an appropriate equation. Reference [10] provides explanations and equations that have been used for this purpose.

26.3 TURBINE CONFIGURATIONS

The primary applications of steam turbines utilized in power generation are used in conventional fossil fired (coal or oil) power plants, nuclear units and combined cycle power plants. Most new applications in the past 15 years have been combined cycle units.

In fossil plants the sizes range from 100 to 1500 MW. Typical steam conditions are 2400 psig and 1000°F main stream and 1000°F reheat. However, higher pressure and temperature units are operating. New fossil units have been few but significant efforts have been made in retrofitting existing plants to improve reliability and efficiency.

Figure 26.15 is a longitudinal section of a fossil steam turbine rated at greater than 800 MW The main steam inlet pressure is supercritical and main steam and reheat temperatures are 1050°F. This is single reheat cycle. The unit is a tandem compound design with a shaft speed of 3600 rpm. There are eight feedwater heaters in the cycle. The design utilizes shared bearings. This means that each elements rotor does not have two bearings.

Flow enters the single flow HP turbine using side inlets through control valves (not shown) and flow through the blade path to the HP exhaust. The HP cylinder exhaust is in the base and steam flows to the reheater. The HP turbine is a single flow design therefore, requiring a dummy piston for thrust balancing. Feedwater heating

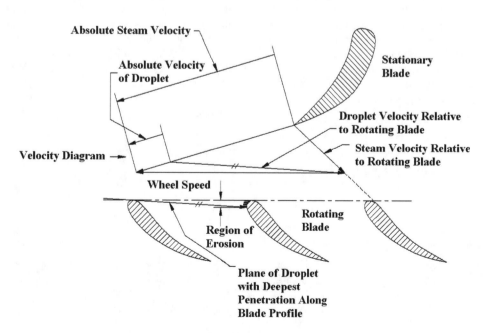

FIG. 26.14 DROPLET IMPACT VELOCITY

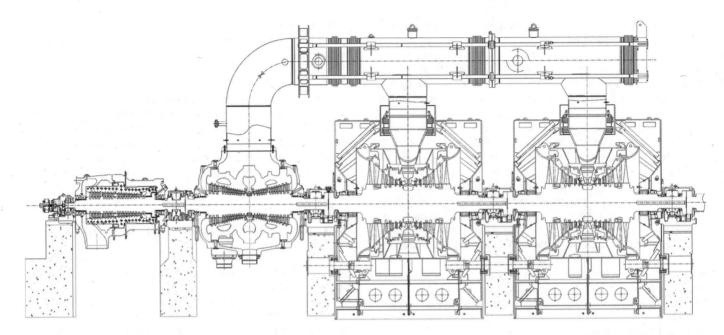

FIG. 26.15   LONGITUDINAL SECTION OF SUPERCRITICAL PRESSURE TURBINE 4 FLOW LP FOR 3600 RPM

steam is extracted from the HP section blade path and the HP exhaust. The turbine is a full arc design and inlet pressure exists between the inner and outer cylinder for a portion of the HP section. The HP extraction chamber is formed between the inner and outer cylinder. All of the rotating rows use single tee root attachment to the rotor. The rotor does not have a bore except for a test specimen bore located away from the main body of the rotor. The stationary blades are attached to the inner cylinder.

Flow enters the IP turbine through side mounted reheat valves (not shown). The IP turbine is a double flow design without the need for a dummy piston. The IP first row is a diagonal low reaction design. This design reduces the axial space requirements of the blade path and provides a lower relative total temperature to the first rotating row for mechanical strength considerations. The blade attachments are tee root. The IP flow exits to the LP turbines through the cover using a large diameter large turning radius elbow. Extraction flow is supplied from two internal locations and the IP exhaust. The internal extractions are unbalanced with the higher pressure feedwater heater supplied from the generator end and the lower pressure heater from the turbine end. The extraction zones are contained inside the inner casing. The IP rotor is no bore except for the test section.

The LP is a four-flow 2 cylinder design. The LP inlet zone uses a flow guide to reduce losses. The blade attachments are tee root in the front end and side entry at the exhaust end. The last row blade is a free-standing blade. The exhaust diffuser outer diameter section is attached to the inner casing and the inner diameter the diffuser section is part of the outer cylinder. The diffuser performance is somewhat negatively affected by the LP outer cylinder bracing. The unit is down flow exhaust to the condenser. The rotor again is a no bore rotor except for test bore locations. The extraction zones are contained in the inner cylinders. The extractions in each turbine are symmetrical. However, the lowest pressure and the next to the lowest pressure heaters are supplied from different LP turbines.

Most fossil units being manufactured today use no bore rotors. However, at least one manufacturer supplies welded rotors. These welds are usually between similar materials but welding between dissimilar materials is also done depending on the application.

The unit in Figure 26.16 is for 50 Hz application and therefore rotates at 3000 rpm. It is very similar to the unit in Figure 26.15 except that it is a six flow LP end (3 double flow turbines). The unit is also for supercritical pressures. Some components are scaled from 50 to 60 Hz or from 60 to 50 Hz depending on the original design. LP end rotating blades are typically scaled.

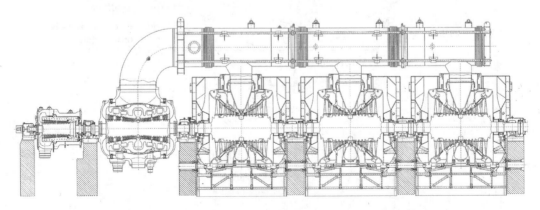

FIG. 26.16   LONGITUDINAL SECTION OF SUPERCRITICAL PRESSURE TURBINE 6 FLOW LP FOR 3000 LP

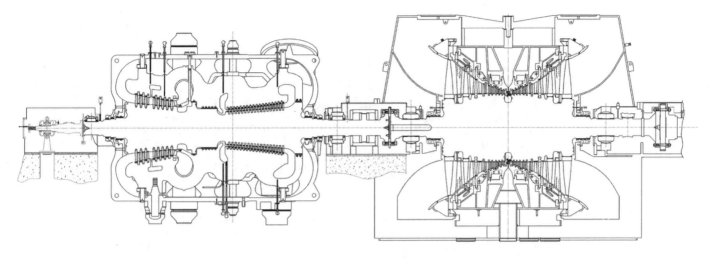

FIG. 26.17 LONGITUDINAL SECTION OF HP–IP AND DOUBLE FLOW LP

Figure 26.17 shows a retrofitted unit. The HP is a full arc admission design. The high- and intermediate-pressure blade paths are combined into one casing (HP–IP). The retrofitted components are the inner cylinder and rotor. The rotor is a no bore design. The HP–IP inner cylinder appears much thicker than in reality. This is due to the fact that longitudinal section includes a large support rib. The rib controls cylinder deformation. The HP uses integrally shrouded reaction blades with tee root attachments. The IP blade path utilizes a diagonal low reaction stage to limit axial space required for the blade path and to reduce relative total temperature on the first rotating row for mechanical reasons. The IP rotor utilizes reaction stages with double tee root blade attachments for the integral shrouded blading. The IP blade path exhaust flow turns and flows between the inner and outer cylinders. This provides cooling for the inner and outer cylinders. There are three dummy pistons in this design (HP, IP, and LP) to balance thrust. The HP and the IP blade paths are thrust balanced independently. There is an equilibrium pipe connecting the HP exhaust to downstream side of the HP dummy. Spring back seals are utilized over the rotating row of blades in the HP and IP. The dummies use spring back seals and in some cases retractable seals (see section 26.5). Another supplier's HP–IP turbine design is show as Figure 3-1 of Reference [11]. This design uses impulse type blading and has a disc diaphragm type construction. This requires essentially no dummy piston.

The LP rotor and inner cylinder were retrofitted. The LP cylinder utilizes a single inner cylinder design, and the LP extractions are symmetrical and separated by the partitioning of the inner cylinder. There is a no bore rotor. The blades use tee root and side entry blade attachments. The last stationary blade (L-0C) uses slots on the surface of the blade to remove moisture for improved performance and reduced erosion. In addition, a low diameter seal is used for the last stationary row (L-0C). Figure 3.2 of Reference [11] shows the LP design of the same supplier as Figure 3.1 of this reference. Note all low diameter sealing is used and the front end stages are impulse design.

Figure 26.18 shows essentially the same design HP–IP as Figure 26.17, except that the HP uses a partial arc design with a control stage. The space between the exit of the control stage and the inlet row of the reaction blade path facilitates uniform flow into the reaction blade path during partial admission for improved performance and reduced non-uniform blade loading effects. The nozzle block of the control stage is a slide in design compared to previously used bolted on nozzle blocks. The control stage rotating blade is mounted to the rotor using a pinned root design. Note that the control stage blade is quite wide relative to the width of the other HP stages. This is due to high loading requirements during partial arc admission.

Current combined cycle units do not have feedwater heaters but instead have inductions (reference Chapter 27). Typically, there

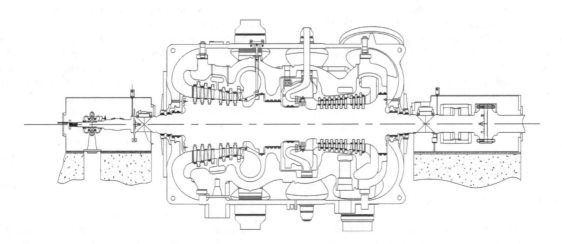

FIG. 26.18 LONGITUDINAL SECTION OF HP–IP TURBINE WITH CONTROL STAGE

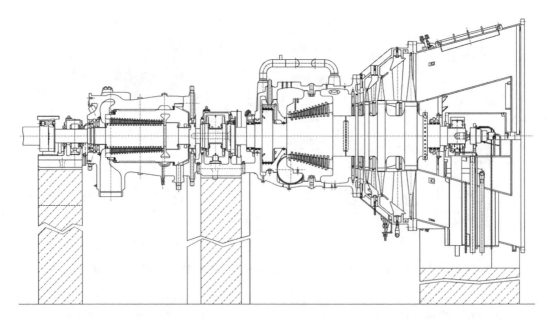

FIG. 26.19   LONGITUDINAL SECTION OF AN HP AND IP/LP TURBINE FOR COMBINED CYCLE APPLICATION

is an induction from the intermediate-pressure drum downstream of the HP turbine and a low-pressure drum induction at the LP inlet or inside the LP turbine. Therefore, the flow increases from the HP inlet to the exhaust of the LP turbine, as opposed to flow decreasing in a conventional fossil steam plant with extractions for feedwater heating. For this reason, the HP turbine typically has smaller blade heights and sealing is more important for improved efficiency. However, due to the increasing flow downstream of the HP turbine, the incentive for very high HP efficiencies is less in CC power plant.

Figure 26.19 shows a combined cycle stream turbine with 1800 psig inlet stream pressure and main steam and reheat temperatures of 1050°F. The design is an HP–IP/LP design using shared bearings. The HP is a separate single flow turbine with a barrel construction. These types of turbines do not have a horizontal joint on the HP outer cylinder. The inlet is volute design and inlet pressure exists for almost the entire section of the space between the inner and outer cylinders. The unit has a bottom HP exhaust. These applications have non-return valves in the piping going to the reheater since they are operated with bypass systems. The HP rotor is a no bore design uses tee root integral shrouded blades.

The IP/LP design is a straight through design from the IP inlet to the axial flow exhaust. The inlet is a bottom side entry. A very large dummy is required at the inlet end for thrust balancing. Due to the large exhaust area of the IP/LP design the unit is not independently thrust balanced. The region between the inlet to the first stage and the dummy is cooled with HP exhaust steam to provide long creep life to the rotor in this area. The rotor is welded between the IP and LP parts of the shaft to permit the use of different materials to satisfy high temperature and high strength requirements of the different ends of the turbine (see Section 26.4). The rotor uses integral shrouded blades in IP and front stages of the LP sections. Interlocked blades are used in the LP end. The last stage of this particular turbine uses a titanium blade due the very large blade height.

There is an induction between the IP and LP blade paths. This induction may not be operative below 20% load. The exhaust uses an axial diffuser. The design of this diffuser is improved through

the use of airfoil shaped supports in the base of the turbine. These supports are also used for oil flow to the bearing and instrumentation connections. The outer cylinder has two vertical joints. These are; the connection of the IP and LP cylinders and the connection of the LP cylinder and the exhaust diffuser section. After initial assembly, the joints need not be broken and the cylinder can be lifted as one outer cylinder. A large dummy leak off connection is shown connecting the low-pressure side of the dummy to the induction zone.

Figure 26.20 shows the HP turbine for a CC plant using an impulse blading design. Note the smaller dummy piston. The rotor is machined to a disc diaphragm construction with axial pressure balancing holes to control leakage flow and axial thrust. The sealing mounted in the stationary blades is on a low diameter with small radial clearance spring back seals.

Figure 26.21 shows the longitudinal section of an HP–IP, LP design for combined cycle application. This design is used for higher megawatt applications than the arrangement shown in Figure 26.19. The design uses an HP–IP design similar to what is used in typical fossil power plants. The LP is a double flow unit using a titanium last stage and has side exhausts since the application uses side mounted condensers.

The HP–IP turbine uses a large inner cylinder that is also the carrier of the stationary blades and HP dummy seals. The design utilizes dummy pistons to balance the thrust. The HP thrust is balanced with the HP dummy while the IP thrust is balanced with a unique dummy piston arrangement at the turbine end of the shaft. The HP is a full arc of admission design.

The latest designs applied to nuclear units have been in the retrofit market. New units will most likely reflect this approach for design. Especially if a verified design concept is required. Nuclear units typically run at half speed relative to fossil applications. Therefore, in the countries using 60 Hz electrical systems the rotational speed would be 1800 rpm.

Figure 26.22 shows a retrofit application for nuclear HP turbine that formerly had a control stage design. Nuclear units operate primarily at full power and the application of full admission units has some advantages. In all applications the ability to pass the licensed

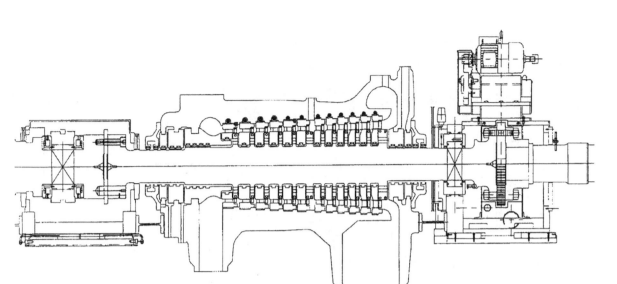

FIG. 26.20 IMPULSE HP TURBINE FOR COMBINED CYCLE APPLICATIONS

power flow is a critical criterion. Non-equilibrium two phase flow has been shown to be a significant issue in some applications [7]. This unit uses a diagonal design first stage. The unit has unbalanced extractions in the HP. One extraction supplies the high-pressure feedwater heater and the other supplies the first stage of a two stage reheater for superheating the flow to the low-pressure turbine. The inlet shows a diagonal stage with a flow guide to reduce losses. The design features a small inner cylinder and blade rings for car-

rying the stationary blades and providing extraction chambers. The blades are tee root construction.

In the nuclear cycle, there is only an HP and LP turbine due to the relatively low inlet pressures of nuclear cycles. Between the HP exhaust and the LP inlet, steam usually flows through a moisture separator reheater. The separator is required since HP exhaust steam is wet. The reheater uses either HP inlet steam or HP inlet and HP extraction steam to reheat the HP exhaust steam after the

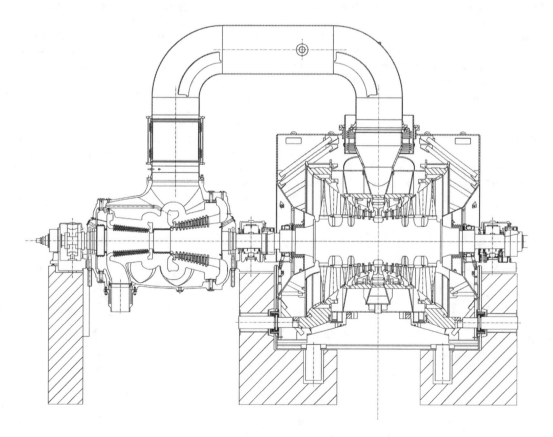

FIG. 26.21 HP–IP, LP TURBINE FOR COMBINED CYCLE APPLICATION

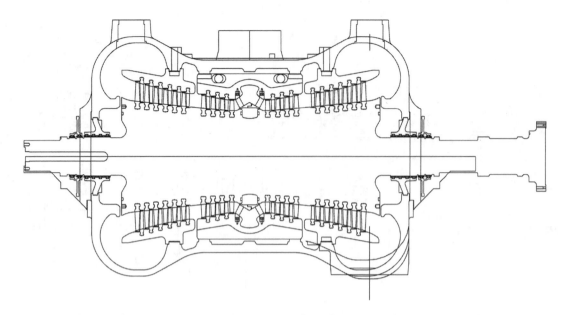

FIG. 26.22   HP NUCLEAR TURBINE

moisture is removed. Moisture removal effectiveness of today's separators is quite high and essentially all of the moisture is removed before reheating.

Figure 26.23 shows a nuclear LP turbine. The unit is a double flow turbine using tee root and side entry blade attachments. This design utilizes a shrunk on discs design. The first stage disc is keyed, as well as, shrunk on to the rotor. Many nuclear LP turbines today use integral rotors to mitigate high stress and stress corrosion. However, the disc design of Figure 26.22 has been verified with service and analysis and is accepted in the industry.

The extraction zones are separated by sections of the inner cylinder. These extractions in the wet regions of the blade path also provide moisture removal. In addition, moisture removal slots are located at the outer boundary of the flow passage and slotted hollow stationary blades are employed at the L-0 location.

26.4   DESIGN

Turbine design technology has developed to analysis based methodology as opposed to a scaling or experienced based methodology. The application of finite element analysis (FEA) and

computational fluid dynamics (CFD) has made complicated analysis routine. This has improved the performance and reliability of today's steam turbines. The use of advanced analysis tools has also improved costs through more effective material utilization.

Steam turbine materials for the current applications have not changed significantly over the years but the processing of the materials has improved to reduce impurities. Blading materials are typically 12%Cr stainless steels. However, Cr content can be as high as 16%. Currently, there a number of titanium designs for the very high tip speed blades. These are mostly being operated in combined cycle units. For moisture protection of low-pressure blades, either flame hardening or stellite shields brazed to the blade are used. Some alloys, that cannot be flame hardened, can be laser hardened. Titanium blades, in a combined cycle application, do not require shielding. Reference [8] provides useful information on droplet erosion and the relative resistance of materials to erosion.

Most HP and IP rotor materials are made of low alloy steel. These steels are nominally 1%Cr. However, the rotor should have high strength and toughness at the high operating temperatures. Most rotors are made from one forging per element. However, some designs utilize smaller forgings welded together at the OD to

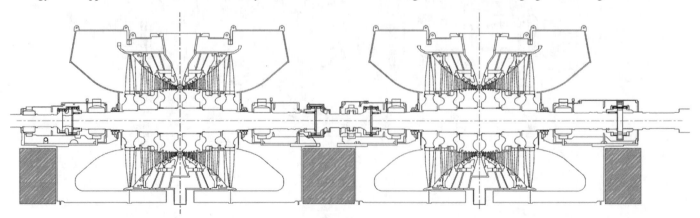

FIG. 26.23   LP NUCLEAR TURBINE

form the rotor body. The design of rotors for future higher temperatures will most likely utilize more welded construction. Reference [12] discusses design changes for higher temperature applications. In this reference the welding of 12%Cr and 3.5NiCrMoV is discussed. Figure 26.19 illustrates a current application where the LP and IP rotors are joined by welding. This is discussed in Reference [13]. Some rotors of this type have used solid rotors with different heat treatment to obtain the appropriate high strength for the LP and the higher temperature properties for the IP end.

For HP and IP rotors, fracture mechanics evaluations are made with the objective that the any flaw that cannot be detected in the rotor does not propagate to critical crack size during the life of the rotor in order to preclude a rotor burst. This evaluation includes the material toughness and the effect of operation during the target life of the unit.

In many applications the HP and IP cylinders are designed to carry the stationary blades as opposed to have a special blade ring. Figures 26.15, 26.16, 26.17, 26.18, 26.19 and 26.21 show this type of construction. In this arrangement, the inner cylinder must be designed to maintain the required degree of roundness to provide proper orientation during operation. These cylinders typically use an axial rib to aid in this aspect. Figure 26.19 illustrates the concept of a blade ring construction in the IP portion of the IP/LP turbine. Figure 26.22 shows the blade ring construction for a nuclear HP turbine. Blade rings provide some advantages for control of deformation to control seal clearances but can increase cost and outage time at inspections and other outages requiring significant disassembly.

HP and IP casings, rotors and valves are designed to a particular life. The life factors in operating time and cycles. The cyclic life will vary depending on the application. For example, a nuclear unit does not see many startup and shutdown cycles or load changes. Therefore, duty cycles are required, as well as, startup and shutdown profiles for cyclic duty evaluation.

HP and IP rotors generally are the limiting items in low cycle fatigue, with HP rotor usually the most critical. The rotors are designed with the objective that the limiting location meets the duty cycle requirements for the application. The allowable stresses must include the effect of high temperature on the fatigue strength. Typical limiting locations are in the blade attachment areas or in some other fillet location at the intersection of a large and small radius on the shaft, e.g., HP dummy piston. If the limiting blade attachment location is not visible, the geometry can be altered to make this location visible during inspections. Stress-relieving grooves are generally used to both relieve the stress of the blade attachments and to put the limiting stress location in a more inspectable location.

The operating temperatures and transient temperature distributions are required for this type of analysis. The temperature throughout the turbine needs to be specified since most analysis is done using the entire component, such as: rotor, cylinder or valve body. A heat transfer analysis is done using appropriate flow information and convective heat transfer coefficients. In some designs radiation heat transfer is important, since steam is an absorbing re-emitting medium [14]. References [15] and [16] describe in great deal heat transfer analysis that can be applied to steam turbines. Some of these correlations have been verified from actual steam turbine data. Reference [15] even includes data verified from actual turbine rotor bore temperature measurements. This reference is specifically for a reaction turbine design. Reference [16] is primarily for an impulse design. However, many of the methods are transferable.

For heat transfer the appropriate temperature to be used is the relative adiabatic wall temperature. However, since the Prandtl number in steam is generally near one the use of relative total temperature is appropriate.

$$Trt = Ts + \frac{Vr^2}{2gJCp} \qquad (26.21)$$

Where:
Trt = the relative total temperature
Vr = relative velocity,
g = gravitational constant
J = Joules constant
Cp = specific heat at constant pressure

The control of relative velocity is often used to reduce the operating temperature of the components such as blades and rotors. This concept for cooling IP rotors is discussed in Reference [17].

For LP rotors, solid rotors, welded rotors and shrunk on disc rotors are used in the industry. For shrunk on disc rotors the required disc shrink fit is evaluated with the objective to preclude discs coming loose during startup transients and overspeeds. Some discs are keyed to the shaft intending to prevent the disc from overspeeding relative to the shaft in the event that the disc would come loose. Fracture mechanics is also applied in the design of LP rotors.

Rotors are evaluated for high cycle fatigue. The calculations need to address all anticipated loading on the shaft. In addition, the effect of torsional stress for short circuits and the stress due to misalignment are included.

Thrust balance of the rotor is usually done such that HP, IP and LP turbines are balanced independently. This limits the number of design cases needed to check thrust and the size of the dummy pistons. Large dummy pistons are not required with impulse blading. This leads to lower dummy leakage rates. In some designs independently balancing of thrust is not practical such as the HP, IP/LP design shown in Figure 26.19. In this case, the axial thrust produced in the LP end of the IP/LP is too large to contain. Therefore, the thrust can be offset through a built in unbalance in the HP or the thrust bearing is selected to accommodate the maximum loading.

In the past, double inner cylinder construction was typical for LP turbines. However, with the use of FEA, the single inner cylinder design can be designed for the required operating life. However, the blade carriers may be made of a higher alloy. This is particularly true when moist steam is present in the blade path at that location. The increase in carbon content can help reduce the potential for flow assisted corrosion. The casings need to provide sufficient flow area for the inlets and extraction areas to prevent large pressure losses in these areas. The peak stresses of these designs usually occur at full temperature. In this case, the cyclic duty of the casing can be assessed just knowing the times and temperature ranges during operation.

Since the outer pressure boundary of the LP inner cylinder is the condenser, thermal shields are generally used over most of the cylinder outer diameter. These help to limit the thermal stresses in the casings. The large axial gradients seen in LP turbines and the stiffness variation in the casing, lead to the potential for ovality at the exhaust end of the machines, prudent design changes and appropriate setting of radial clearances are used to address this situation.

HP and IP blade paths are typically designed on a unit specific basis. This is done, in order to better match the individual cycle requirement, while achieving high levels of efficiency. Typically, the process is computerized to achieve optimization. The process

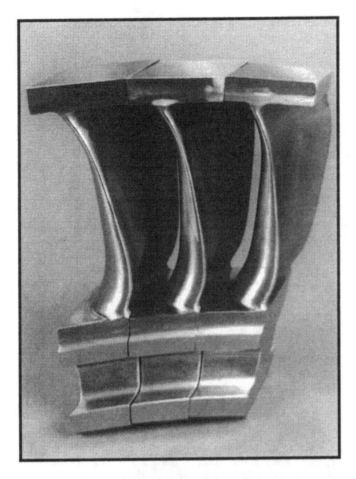

FIG. 26.24 BOWED BLADE

would have prescribed limits that might include a set of basic sections, blade root attachments, etc. The design process would select blade section, stage loading and reaction in order to achieve the best efficiency while meeting cycle parameters such as: inlet temperature, inlet and exit pressures and flow. The blade could be bowed or just taper twisted. Typically, in the HP turbine, the blades would be bowed due to the low aspect ratio, blade height to blade chord. Figure 26.24 depicts a bowed blade arrangement for an HP turbine. Reference [18] discusses the concept of bowed blading. At low aspect ratios the reduction of secondary flow losses is more significant. Bowed blades are also applied in IP turbines and the front stages of LP turbines. See Reference [19] for a description of one approach to automated design and optimization. This process can be utilized in HP, IP turbines and for the un-tuned blades of LP turbines.

The HP, IP, and front stages of LP blades are typically integrally shrouded, or one piece construction as compared to older designs that may have used shrouds that were riveted to the blade. The shrouds typically are made tight at assembly for added mechanical strength. Some integrally shrouded blades are designed to run with small gapes between blades. The concept is to reduce tip deflection and add mechanical damping. Typically, these blades are un-tuned which means they do not have speed or frequency limitations during operation to address harmonic effects.

The blade attachments are typically tee root, double tee root (for larger blades), and side entry roots. The designs need to address aerodynamic and centrifugal loadings. Flow disturbances, windage

heating (bypass operation in HP), and transient overloads are usually addressed in blade mechanical design margins.

The LP end blading design is usually the result of an iterative process in which axi-symmetric and three-dimensional flow field methods are used to produce a blade path that meets aerodynamic efficiency and mechanical limitations. CFD codes are even capable of some degree of unsteady analysis. For example in Reference [4] unsteady analysis shows that the stationary blade wake effect is more pronounced at the base of L-0R blade than at the tip.

The blade profiles are generally Mach number dependent such that sections with subsonic, transonic and supersonic relative exit Mach numbers would have different shapes and velocity distributions. A converging diverging section is often applied at exit Mach number is greater than 1.4. However, LP turbines are subject to large variations in flow conditions due the large variation in volumetric flows. Therefore, each design should address the specifics of the intended application.

In addition to improved airfoil aerodynamics, flow field effects to improve performance have been aided by improved CFD capability. For example, the use of lean and sweep in LP stationary blades increases performance. Figure 26.25 shows an illustration of a stationary blade design using improved CFD analysis. This concept is discussed in References [4] and [5].

The LP end blades are tuned blades. This means that all modes of vibration below some harmonic of running speed are tuned. For example one supplier may use the 7th harmonic. Free-standing and interlocked blade designs are currently being used. The interlocked designs have much greater stiffness after the blades are "locked up." Lock up occurs when the blade interlocking features come into contact. This is typically between 50% and 70% of running

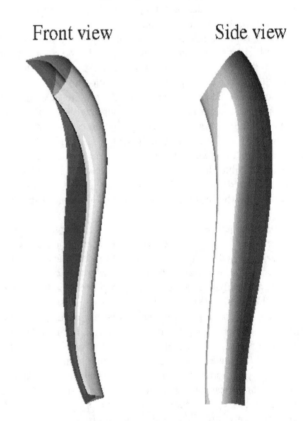

FIG. 26.25 LAST STAGE STATIONARY ROW WITH SWEEP AND LEAN

FIG. 26.26 INTERLOCKED LP TURBINE LAST ROW ROTATING BLADE

speed due to the centrifugal untwist of the blade. Tuned blades are verified on prototype designs by use of a rotating test. The blade frequency control for subsequent units typically uses some type of stationary frequency test. The fact that these blades require tuning for reliable operation leads to operating speed restrictions at no load and off frequency limits on line. However, not all modes require a speed restriction. Figure 26.26 illustrates an interlocked last row rotating blade design that includes a mid-span snubber and a shroud interlocking feature. This figure also illustrates a four lug side entry root "fir tree root."

The blade attachment for these large blades is typically a side entry fir tree root or a pinned root. The root steeple stresses will limit the number of speed cycles a design can tolerate and thus could affect the cyclic duty potential for some designs. Blade roots are sometimes rolled or shot peened to improve their HCF, LCF, and stress corrosion capability.

For LP end blades, the start of stall flutter can be assessed and consideration for this built into the operating guidelines (see Section 26.6). However, un-stalled flutter does not occur at off design conditions and therefore, the protection is in the design itself. In the past a variety of methods were used to predict un-stalled flutter such as:

- Dynamic cascade tests and analytical predictions based on these results
- Model turbine tests
- Using experienced based criteria such as Strouhal number or reduced frequency.

These methods have produced moderate success.

Mix-tuning is a process of slightly modifying a nominal blade to achieve some prescribed natural frequency variation. In un-stalled flutter, the blades vibrate at the same frequency. If there is suffi-

cient variation in natural frequency blade to blade, this vibration is suppressed. In free-standing blades mix–tuning has been quite successful in eliminating un-stalled flutter. However, in interlocked blades mix-tuning is not possible due to the interlocking. However, mis-tuning requires a more significant change to the blade and attempts to change the aerodynamic characteristic of the blade to blade interaction. This is quite difficult as explained in Reference [20]. However, a computer code (TRACE) that solves the Reynolds-Averaged Navier Stokes equations in the relative frame has been applied to analyzing un-stalled flutter. Figure 26.27 (taken from Reference [20]) shows the computation of aerodynamic damping for three different Mach numbers. The aerodynamic damping is expressed in terms of logarithmic decrement. ND represents the nodal diameter pattern of the vibrating coupled blade structure. Negative ND indicates a backward traveling wave while a positive ND indicates a forward traveling wave. Un-stalled flutter is always seen in the backward traveling wave. Negative log decrement means the blade is unstable. Combined with knowledge for the mechanical damping of the system, this approach can be used to determine the susceptibility to un-stalled flutter.

Moisture erosion of rotating blades in the wet steam region results from liquid droplets impinging on or near the leading edge. See Reference [8] for a detailed description of this process. Damaging droplets are not the moisture in the primary blade flow. The primary source of the water that erodes the blades comes from moisture deposited on the upstream stationary row. This deposition is caused by inertial and diffusion processes. The moisture flows to the trailing edge of the blade and drops are formed in the slow moving wake behind the trailing edge of the stator. The droplets are broken up into smaller droplets by the drag force of the steam. This process creates a distribution of coarse moisture that can range up to 500 microns or more. These droplets are slowly accelerated and carried to the rotating blade. Figure 26.14 illustrates the water droplet velocity in this process.

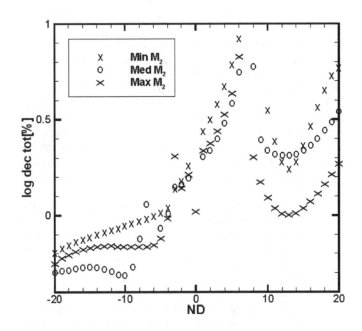

FIG. 26.27 AERODYNAMIC DAMPING FOR THREE DIFFERENT MACH NUMBERS AS A FUNCTION OF NODAL DIAMETER

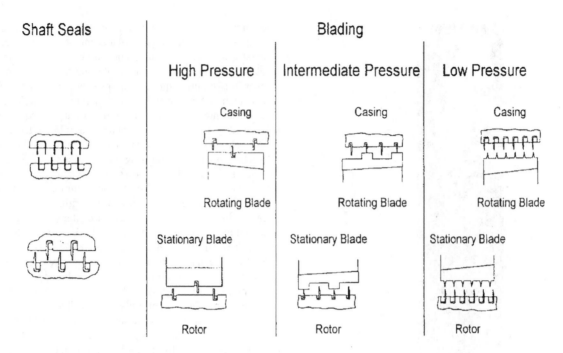

FIG. 26.28   SEAL CONFIGURATIONS

Erosion protection is in general twofold: blade tip protection through flame hardening, or adding stellite strips, etc., and moisture removal. Moisture removal can employ slotted end walls, slotted stators and even heating of the stator blade to evaporate the moisture. Generally, the slotted stator uses two locations for removal of stator blade surface water. One is located on the pressure surface and one on the suction surface. If a hollow blade is used for evaporating surface moisture, the steam is extracted from another location, generally a LP extraction zone, and piped into the stator with appropriate drainage provisions. This water can be sent either to the condenser or to a low-pressure feedwater heater.

Turbine exhaust hood design can have a significant impact on turbine performance. Exhaust hoods can be either axial (see Figure 26.19), radial or as is most common radial/axial, see Figure 26.21). Axial diffusers in general provide the best performance.

The ideal situation is to have a diffusing exhaust hood where the turbine blade exit pressure is less than the condenser pressure. However, this cannot exist over the entire operating range. Therefore, the design point of the diffuser is quite important. As discussed in Reference [6], CFD can be used quite effectively in design. However, the blade path hood interaction needs to be included.

For units that have significant condenser pressure variations seasonally, it may be prudent to design the diffuser to a slightly off-design condition. For example, if a unit that needs to generate peak power in the summer time, it may be prudent to sacrifice some of the performance in the winter months in order to maximize the performance when the power demands are the highest.

Sealing has made a significant improvement in turbine performance. In addition, some of the sealing concepts being applied reduce the potential for seal rubs and therefore, reduce the efficiency degradation. Turbine efficiency degradation is inevitable but the degree of degradation between overhauls can be affected by seal selection as well as prudent turbine operation (see section 26.6).

Figure 26.28 shows one supplier's seal application philosophy. The seals in the HP and IP turbines are stepped design for reduced leakage. Because of the reduced blade height, sealing is most important in the HP turbine and becomes less of a factor as the blade height increases. For this reason, impulse blading with its disc diaphragm construction with low diameter sealing (see Figure 26.6) is sometimes applied for low flow small blade height turbine designs. Seals in the LP turbine in this application use tip to tip sealing that employ many seals but can accommodate large differential expansions.

One new seal concept is the use of abradable seal material on the seal carrier ring. The concept of this approach is shown in Figure 26.29. The clearance is reduced. However, in the event of a rub the abradable material wears as opposed to seal. The leakage flow reduction with this approach has been approximately 20%.

Currently brush seals are being applied in steam turbines (see References [21, 22]). The concerns of large pressure ratio and high fence height requirements have been addressed and brush seals have been successfully operated in steam turbines. Figure 26.30 shows a diagram of a brush seal application. The brush seal is installed with a clearance and the "bristle blow down effect" resulted in as much as a 50% reduction in leakage. Brush seals can be installed in a variety of arrangements such as a spring back or retractable seals.

Retractable seals have been used in certain locations in HP and IP turbines. To date, applications have been in the glands and dummy pistons. The concept is to use spring force to hold the seal ring segments at a larger diameter until pressure forces overcome the spring load and move the ring segments to a lower diameter. This approach permits smaller radial clearances during operation while having more rub protection during startup, thus reducing the concern for seal rubs affecting seal performance. This concept is discussed in Reference [23].

Honeycomb seals have been applied in LP turbines. These have mostly been applied over free-standing blades (Figure 26.31). They provide for better sealing and greater rub protection than would be

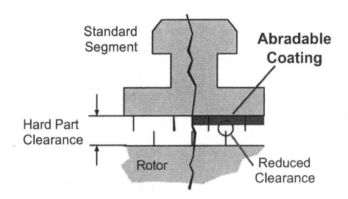

FIG. 26.29 ABRADABLE COATING

achieved with a hard liner. Therefore, clearances are smaller with this seal compared to a hard liner.

Solid particle erosion (SPE) damage to turbine blade paths is the result of iron oxide particles (magnetite) entering the steam turbine. These particles can result from exfoliation from the boiler and steam piping. There have been many papers discussing the generation of these particles and ways to remedy the situation external to the turbine. Reference [24] discusses this topic in great detail. Minimizing exfoliation is a most effective way to reduce SPE in steam turbines.

SPE is found in the regions of the high-pressure turbine (HP) and intermediate-pressure turbine (IP) near inlets to the turbine. Damage has also been seen in valves, cylinders, blades and rotors in the dummy pistons. The HP damage can be in the blade path for units with or without a control stage. If hard particle damage is an issue, the control stage blades would be significantly impacted primarily due to the high velocities at partial arc operation.

In the blade path, the larger particles cannot follow the steam as it accelerates and turns in the blade passage. This causes the particles to impact blade pressure surface of the stationary row. Reference [24] provides calculations that show that damaging particles do not travel at steam speed. However, designs with higher nozzle exit velocities will produce higher particle impact velocities. The higher velocity and greater steam turning make impulse designs more susceptible to SPE. Particles that leave the nozzle will impact the rotating blade with more negative incidence than the steam due to the slower particle velocity. Therefore, particles will impact on the inlet suction or convex side of the blade. This effect cannot only cause leading edge erosion of the rotating blade but can under

certain instance produce particle rebound that can cause damage to upstream nozzles suction surface.

While the impact of transient operation (e.g., Startup) is debated in the literature, Reference [25] supports the impact of transient operation and concludes that bypass systems have a significant impact on reducing hard particle erosion.

In general, reducing velocity and flow turning will reduce SPE. Typically the impulse nozzle will have greater turning and higher velocity than a reaction stationary blade design. Erosion would be expected to vary with the kinetic energy of the particle, thus producing a variation with velocity squared. Reference [26] shows a power law dependence of velocity on solid particle erosion with the exponent varying between 1.91 and 2.52. Even in a reaction blade path the reduction of stage pressure ratio would have a significant impact on SPE reduction.

The impact of solid particle erosion can be reduced through the application of coatings. For control stages, coatings have been in service for more than 26 years. The boride diffusion coating can be applied to nozzle vanes and rotating blades as needed. The success of this coating application has been documented to extend the period of maintenance due to solid particle erosion [27]. The time interval for SPE damage to be addressed could double when coatings are used. The nozzle design can also have a significant impact on erosion as discussed in Reference [28]. In addition, the increased nozzle pitch reduces SPE. For many units with control stages, valve management features are available that will permit switch-

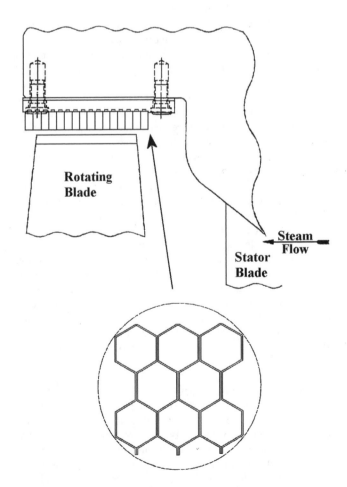

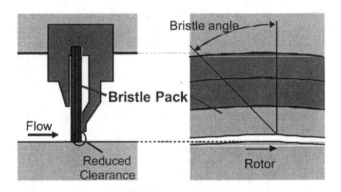

FIG. 26.30 BRUSH SEAL

FIG. 26.31 HONEYCOMB SEAL

ing from partial arc to full arc during startup. This can significantly reduce the stage velocity and therefore reduce solid particle erosion at a time when hard particles are being generated. This valve mode transfer feature permits the return to partial arc operation to improve performance at part loads.

Future designs will need to address significantly higher pressure and temperature. Designs are currently being developed for 5000 psi, 1300°F steam conditions. Therefore, there will be more use of materials that utilize higher chromium (10-12%) for rotors and casings. In addition, the need to address significant temperature variation in the same casing will lead to greater use of welding of dissimilar materials. In addition, there will be a need for more Nickel based materials for casings. In a study reported in Reference [29] in 1992, it was concluded that the next step in future optimized steam conditions for a double reheat pulverized coal cycle would be between 4000 psi, 1100°F/1100°F/1100°F and 5000 psi, 1200°F/1100°F/1100°F.

It is interesting to note that a 325 MW rated turbine generator with 5000 psi inlet pressure and 1150°F steam inlet temperature and double reheat of 1050°F/1050°F went into service at The Eddystone Generating Station in 1957. The unit operated until 1975 and is now an ASME Historical Mechanical Engineering Site.

26.5 PERFORMANCE

The common method of defining performance for the overall turbine is heat rate. This is defined as the heat added to the cycle to develop one kW of output. This is really the reciprocal of efficiency. Typically heat rate is expressed in Btu/kW-h or kJ/kW-h. The turbine heat rate only includes the heat that is transferred to the steam and not the heat released by the fuel. The plant heat rate will always be higher than turbine heat rate. Heat rates are calculated using the ASME steam tables.

The blading and sealing currently utilized in steam turbines have reached such high levels that significant improvements in heat rate will require higher steam pressures and temperature. The direction of these efforts was discussed in Section 26.4.

The performance of the steam turbine is obtained by computer models that are proprietary to the specific turbine supplier. However, most models have similar methodologies in that the essential losses are included and that some simplification is used to reduce the complexity of the blade path to a computer model.

Typically, losses addressed are blade profile losses, secondary flow losses, leakage losses, and moisture losses. The blade profile losses are typically corrected for surface roughness and Reynolds number, Mach No., trailing edge thickness, and incidence angle. Secondary flow losses are associated with the turning of the boundary layer on the end walls of the blade passage. The losses can be evaluated from cascade testing, CFD modeling, model turbine testing, etc. These losses are typically correlated with ratio of blade pitch to height. It is this loss that a bowed blade will affect.

The application of multi-axis numerically controlled machines to manufacture machined blades has eliminated the use of parallel sided blades. In HP, IP, and the front stages of LP turbines using reaction blading, shapes such as "bowed" blades are typically used for performance improvements. Bowed blades typically shaped as shown in Figure 26.24. The improvement comes from forcing flow toward the hub and tip of the blade passage thus suppressing the vortex that is developed in the passage. References [18] and [19] discuss this concept in detail. In impulse designs bowed blades are

not typically used and flow control concepts are used to improve performance.

In a typical reaction steam turbine the distribution of the average losses are as a percentage of the total loss are shown as follows:

	HP turbine	IP turbine
Profile loss	54	68
Secondary flow loss	32	24
Leakage loss	14	8

The efficiencies of today's HP and IP turbines are quite good relative to the use of parallel sided blading and straight seals. Currently efficiencies can be as high as 90% for the HP turbine element including valve losses, and 93.5% for the IP turbine including valve losses. The performance of the LP turbine will vary significantly with the level of moisture, the condenser pressure and the exhaust arrangement.

Performance verification is an important subject. Overall heat rate testing has been addressed by ASME PTC 6 committee. Reference [30] documents the code for standard fossil and nuclear units with regenerative feedwater heaters and Reference [31] is for combined cycles.

Reference [30] now combines the full-scale and the alternate test into one report. The PTC 6 committee recommends the use of the code for conducting acceptance tests of steam turbines and any other performance test where performance levels are to be obtained with a minimum of uncertainty. In fact, the code recommends the use of the full-scale test with a condensate flow measurement for fossil unit steam turbines. However, the alternative test uses fewer measurements and makes greater use of correction curves for cycle adjustments. The uncertainty of the full test is reported to be ±0.25% on unit heat rate, while the alternative test reports an uncertainty of 0.33%. For nuclear units the where most of the cycle is wet the accuracy is between ±0.375% and 0.5%. For combined cycles the test procedure of Reference [31] reports an uncertainty of ±0.50%.

These tests are quite expensive due to the instrumentation requirements including a flow section for flow measurement. In fact very few "pure" code tests are conducted since other arrangements are frequently made during the negotiation, installation or test phases. Therefore, unit by unit specific test definitions are frequently used for unit verification. When tests are conducted, the condenser pressure is of particular importance. The vacuum corrections are developed by the turbine supplier by a variety of methods, such as CFD modeling, model testing, and experience with similar units.

For measurement of HP, IP efficiencies a standard efficiency test can be conducted to by measuring the pressures and temperatures (enthalpy drop test). Tests on HP turbines can be conducted at valves wide open to eliminate valve losses in the data. In some cases pressures downstream of the valves are measured and the efficiency definition uses this point as the inlet condition. However, it can be difficult at times to measure pressures far enough downstream of bends, and other flow disturbances to get accurate pressure readings.

High- and intermediate-pressure turbine enthalpy drop tests are typically within ± 0.6% uncertainty. In many designs, the performance levels are affected by leakages from items, such as HP, IP and LP dummy piston leakages, and inlet seal leakages. To determine the level in some instances tests referred to "influence factor"

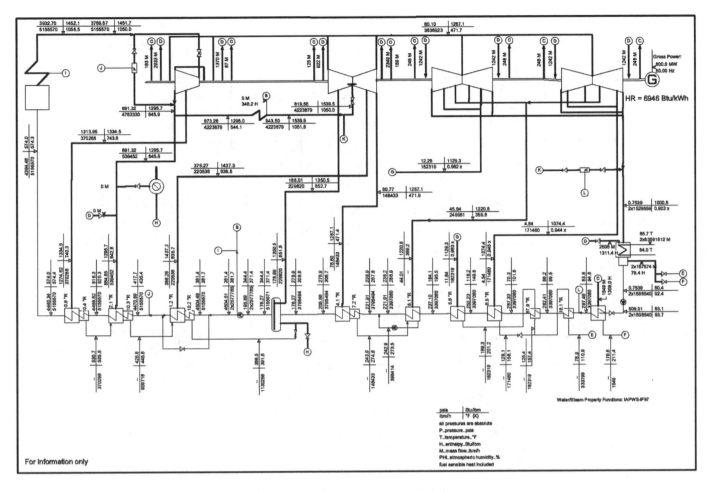

FIG. 26.32 HEAT BALANCE DIAGRAM

tests (see Reference [11]) are conducted to evaluate these effects. The need for this type of testing is certainly design dependent.

Figure 26.32 shows a turbine heat balance diagram for a supercritical pressure inlet HP turbine with single reheat. The inlet temperatures are 1050°F/1050°F. The unit has separate HP, IP and 2 double flow LP turbines. The unit has 8 feedwater heaters, which are supplied by extractions from the turbine. This unit has one heater above the reheat pressure. This means that an extraction comes from the blade path of the HP turbine before the HP exhaust. This cycle makes a big improvement in heat rate since in this instance the feedwater temperature was increased by almost 50°F. However, most plants use the HP exhaust (cold reheat) as the highest pressure extraction. Reasons for the lower pressure extractions are initial cost and potential reliability issues with high-pressure heaters.

The other extraction arrangements are somewhat typical. There are three extractions from the IP and three from the LP casing. The extraction flows varying from 5% to 11% of the local blade path flow. The LP inlet pressure for this design is relatively low compared to most US power plants for this size. This means larger piping from the IP to the LP turbine. However, with longer blades in the IP back stages relative to the shorter blades in the front stages of the LP turbine, there is a potential for performance improvement. The rated heat rate for this unit is quite good at 6946 Btu/kWh.

HP turbines can be either full arc or partial arc of admission units. If the unit is planned to be operated at part load for a significant period of time, a partial arc design ought to be considered. For

subcritical units sliding pressure operation is an option. Therefore, justification can be made for either full or partial arc designs with a control stage. Performance over the load range and load reserve requirements need to be considered. Reference [32] discusses many options for operation with full arc, partial arc and full arc designs with overload valves. Reference [33] discusses comparisons between full arc and partial arc designs. These two references illustrate that the design of the unit makes a significant difference in the off design performance. For a partial arc design, it has been shown that operating with full pressure partial arc admission until 50% admission and then sliding the inlet pressure for further load reduction is an attractive operating strategy for heat rate improvement and it can also minimize cyclic stresses in the rotor relative to pure constant pressure operation. One important consideration is the prediction, years in advance, as to how the unit will be operated. Therefore, operating flexibility ought to be considered.

26.6 OPERATION AND MAINTENANCE

Operational concerns for steam turbines encompass a variety of issues. Some of these are low cycle and high cycle fatigue, material creep, vibration, steam chemistry, and water induction. Steam turbines currently in use have a large variety in the types of operating instructions and control systems to address operation issues. However, many units today employ a digital hydraulic control system

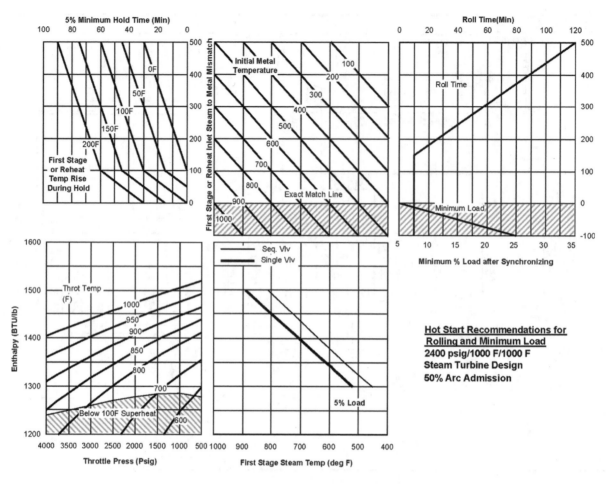

FIG. 26.33   STARTUP RECOMMENDATION

that permits starting and loading the steam turbine in a controlled fashion as to adhere to the limits recommended by the supplier.

There are a number of limits that must be adhered to prior to admitting steam to the turbine. The steam temperature must be high enough to limit too large of a mismatch between steam and metal temperatures for items such as steam chests, cylinders, and rotors. In addition, enough superheat is required to limit internal condensation. The inlet pressure and temperature recommendations are obtained from the controller or the instructions. Figure 26.33 shows one such recommendation in graphical form to limit rotor stress during initial admission of steam and rolling time to synchronous speed. For units that have bored HP–IP long heat soak times at less than full speed are required to heat the bore to temperature near or above the Fracture Appearance Transition Temperature (FATT). This is done to have sufficient ductility in the rotor to prevent crack growth of any cracks in the rotor of a size below or at the ability of inspection techniques to identify. Some reduced acceleration to full speed is required for a cold rotor even with today's rotors.

The startup of a steam turbine is a complicated process of checks if manually done. That is why controllers are now available that perform the following functions while controlling turbine component stress, avoiding blade and rotor resonance ranges:

- Starting the turbine from turning gear speed to nominal speed
- Synchronization
- Loading and unloading between zero and rated load

Since more units are cycled today, automated control is more critical since these units start and shutdown more frequently. In addition, with automatic control, there are varying operating strategies that can be implanted to reduce stress. Reference [34] discusses this topic for a combined cycle plant. If the operation is not controlled by a controller, the operator would need to follow a set of instructions for loading and unloading, as shown in Figure 26.34.

Turbine control systems usually use some type of on line computation of rotor stress to control low cycle fatigue damage during startup, shutdowns and load changing. These calculations are generally addressing the low cycle fatigue limiting location on the rotor. These locations are either in the blade attachments areas of the HP and IP turbines. The calculations are numerical but in some instances use mathematical relationships to determine the thermal stress. This approach can factor in the actual rates of temperature, speed and/or load change, as opposed to the linearization of the assumed event while using a chart. The control system output can also advise the operator regarding desired steam conditions for reducing stresses at startup. In addition, the amount of margin remaining or an extrapolated peak stress can be provided for either manual or controller action. These can be provided for a number of components such as HP and IP valves, and HP and IP rotors.

The boundary temperatures for the models can be obtained using steam temperatures and applying appropriate heat transfer rates to the surfaces. In some instances appropriate metal temperatures

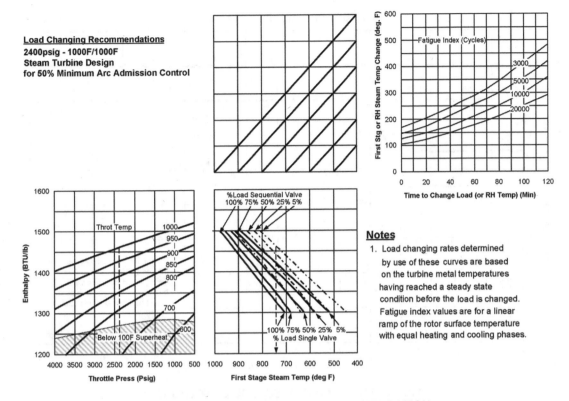

FIG. 26.34 LOAD CHANGING RECOMMENDATION

measured on casings are applied as the surface temperature for the rotor. Guidelines for allowable increases in temperature are provided to the operator. In general, the rotor limits the low cycle fatigue of the turbine. However, if a stationary part, such as a steam chest, needs additional limits, these need not be modeled. Thermocouples installed directly into the part can be used to monitor component stress. For example, in a steam chest shallow and deep thermocouples are routinely used to provide either a controller or operator information relative to the stress limits.

When starting a partial arc design unit that has the capability to transfer from full arc to partial arc operation, it could be beneficial to the transfer to governor valve control with the unit in single valve mode. The reason is that once the unit is loaded, the inlet temperature and the control stage exit temperature will increase rapidly and the rate of change may be more than desired. If the unit is in single valve mode, transferring to partial arc would reduce the control stage exit temperature, thus reducing the rotor stress. Data showing this is situation is shown in Reference [33].

Control systems can help protect the turbine from a variety of other issues that could affect turbine reliability. The rotor train acceleration rate can be set to provide a more rapid speed increase and to limit operation in regions of rotor critical speeds or tuned blade resonances for the tuned LP end blades (usually the last three stages of the LP turbine).

Condenser pressure limits are in intended to prevent blade damage through high cycle fatigue and to avoid excessive windage heating. One such limit curve is shown as Figure 26.35. This particular chart is for a combined cycle unit. The LP induction pressure is used as an indicator of last stage flow since there are no extractions or inductions beyond this location. The horizontal lines at low flows are the low flow limit line. One limit is for alarm and the other is for automatic trip. The angled lines that rise from about

18 psia to about 26 psia represent the stall flutter limit for the blade. Un-stalled flutter is not addressed by operation but by the design of the blade. These lines are determined by calculation and testing.

Exhaust hood sprays provide some protection for windage heating. This cooling of the blade by an exhaust hood spray is aided by the recirculating flow that would be present at the base of the large last row blade during windage heating. Thermocouples in the exhaust hood provide input to alarm messages or input to activate hood spray directly.

There are times when feedwater heaters need to be removed from service. In general, rules are applied on what heaters can be removed without reducing turbine load. As heaters are removed from service, flow is increased in the following stages of the turbine. This produces overload potential for the blades which is to be avoided. Reheater sprays increased above the design can also produce overloading situations throughout the turbine among other undesirable issues.

The control of steam purity is an important aspect of turbine operation. The presence of corrosive impurities in steam can cause damage to turbine components. The damage is caused by corrosion, stress corrosion, corrosion fatigue, and erosion-corrosion. Caustics, salts, and acids must be controlled. Deposition of impurities can also cause thermodynamic losses and distress by lowering the efficiency of blades and upsetting the pressure distribution. Deposits can reduce the flow passing capability of the blading due to deposits especially in HP turbines with their relatively small blade passages. Stress corrosion has been a significant issue in some rotor and blade attachment designs in LP turbines.

Limits for steam chemistry are sometimes provided with different allowables for different stages of operation. For example different limits for normal operation at load and startup. The normal operating limits are more restrictive. When operating with poorer

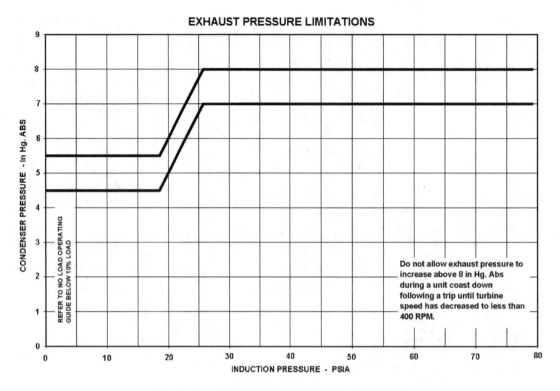

FIG. 26.35 EXHAUST PRESSURE LIMITATIONS

chemistry than ideal, the limit may have time restrictions. When operating with steam from an auxiliary boiler such as supplying gland steam during startup, more latitude is generally given for chemistry limits.

Generally, continuous or routine monitoring of samples taken at the HP inlet and the Hot Reheat inlet is recommended. When steam enters, the cycle from multiple pressures from separate steam sources such as in a combined cycle unit, each source should be monitored. References [35], [36], [37] and [38] provide more information regarding the issue of steam chemistry in power plants.

Some retrofit designs and new unit applications since the 1990s have recommended 8- to 10-year inspection intervals. Some of these inspection intervals may be based on equivalent operating hours. One definition in a published standard is defined as.

$$Te = Ta + AN \qquad (26.22)$$

Where:
Te = equivalent operating hours
Ta = actual operating hours
 N = number of total starts (cold, warm and hot)
 A = multiplier on the number of starts

Reference [39] discusses inspection and overhaul of large steam turbines. This reference recommends a value of A in Equation 26.22 of between 20 and 30.

Intermediate inspection intervals may also be recommended with varying degrees of component inspections. For example if minor, medium and major inspections are defined with only a major inspection requiring total opening of inner and outer casings. Over a 200,000 hour equivalent operating time frame, inspections could be recommended as follows

Minor	Medium	Minor	Major
25,000	50,000	75,000	100,000
125,000	150,000	175,000	200,000

Recommendations are provided for inspections to be made on line in varying time intervals to identify items such as instrumentation errors, drain malfunctions, feedwater heater drain valve and non-return valve operation.

Special attention to the oil lubricating system is warranted. Some units use large supplemental cleaning systems that vary from partial to 100% oil flow capability.

The inspection recommendations for individual units and, if applicable, the calculation of equivalent operating hours may be different than the example given above.

The collection and tracking of data on shaft vibration levels and phase angles is recommended. This is helpful information to review especially going into a long turbine outage, since it may highlight the need for additional work that is currently unplanned.

The potential for water damage is present. Therefore, it is important to review plant design, operating practices and instrumentation against the requirement of Reference [40]. New units should address these concerns in the initial design. Systems to detect the potential for water damage are sold as retrofits to existing plants.

26.7 SUMMARY

The state of technology available in the current steam turbines designs is quite advanced. Advances in Finite Element Analysis,

Computational Fluid Dynamics and Computer Aided Design have made this possible. Some of the design methods and concepts have been validated through the application to retrofit designs, as well as, new units. Very large annulus areas are available on full speed units. Significant heat rate improvements will require higher pressure and temperature steam conditions.

Extended service intervals are becoming more common for new and retrofit units due to the increased reliability of the current designs. The use of equivalent operating hours is currently being applied to steam turbines. Automated control systems have aided in improving reliability and operating flexibility.

This chapter attempts to provide information for use in understanding the current state of technology in steam turbines. Due to space limitations a great deal of material could not be included. However, the use of references has attempted to provide resources for additional information. In addition to references cited in the text, References [41] and [42] will also provide useful general information.

26.8   REFERENCES

1. Moroz, L., and Tarasov, A. "Coupled CFD and Thermal Steady State Analysis of Steam Turbine Secondary Flow Path," ASME Paper JPGC2003-40048, Presented at the 2003 International Joint Power Generation Conference, June 16-19, Atlanta, GA.

2. Steltz, W.G., Evans, D.H., and Stahl, W.F., "The Aerodynamic Design of High Performance Low Pressure Steam Turbine," Third Scientific Conference on Steam Turbines of Great Output, Gdansk, Poland, Sept. 1974.

3. Marsh, H. "A Computer Program for Through Flow Fluid Mechanics in an Arbitrary Turbomachine in Radial, Axial and Mixed Flow Types," Aeronautical Research Council R. and M, No. 3509.

4. Stueer, H., Truckenmueller, F., Borthwick, D., and Denton, J., "Aerodynamic Concept for Very Large Steam Turbine Last Stage" Paper GT2005-68746, Proceedings of ASME Turbo Expo 2005, Reno-Tahoe, Nevada, USA, June 6–9, 2005.

5. Lampart, P., "Numerical Optimization of Stator Blade Sweep and Lean in an LP Turbine Stage," ASME Paper IJPGC2002-26161, Proceedings of IJPGC, 2002 International Joint Power Generation Conference, Phoenix, AZ, USA, June 24–26, 2002.

6. Gray, L., Hofer, D.C., and Takenaga H., "Recent Advances in the Prediction of Exhaust Losses for Low Pressure Steam Turbines with Downward Exhaust Hoods," Advances in Steam Turbine Technology for Power Generation, PWR Vol. 26. ASME 1994.

7. Leyzerovich, A., "Wet-Steam Turbines for Nuclear Power Plants" Published by PennWell.

8. Heyman, F.J., "Liquid Impingement Erosion," ASM Handbook, Volume 18, "Friction, Lubrication and Wear Technology" ASM International, 1992.

9. Meyer, C.A. and Lowrie, J.A., "The Leakage Through Straight and Slant Labyrinths and Honey Comb Seals," ASME Journal of Engineering for Power, Vol. 97, pp. 495–502.

10. Young, W.C., "Roark's Formulas for Stress and Strain," 6th Edition, Published by McGraw Hill, Inc.

11. "Evaluating and Improving Steam Turbine Performance", Second Edition, by Cotton, K.C.

12. Magoshi, R., Nakano, T., Konishi, T., Shige, T., and Kondo, Y. "Development and Operating Experience of Welded Rotors for High Temperature Steam Turbines", ASME paper IJPGC200-15007, Presented at the 2000 International Joint Power Generation Conference, Miami Beach, Florida, July 23–26.

13. Zabrecky, J.S., Bezugly, J.A., Brown, M.K., and Martin, H.F., "High Power Density 60 Hz Single Flow Steam Turbine with 42 Inch Titanium Last Row Blade for Advanced Combined Cycle Application", PWR-VOL. 34, 1999 ASME Joint Power Generation Conference.

14. Hottel, H.C., and Egbert, R.S., "Radiant Heat Transmission from Water Vapor", AIChE Transactions, Vol. 38, 1942.

15. "Cyclic—Duty Turbine Boiler Operating Practices", Electric Power Research Institute Report EPRI CS-3800, Project 911-1 December 1984.

16. Brilliant, M., and Tolpadi, A. K., "An Improved Analytical Approach to Steam Turbine Heat Transfer", Proceedings of ASME Power 2004, March 30-April 1, 2004, Baltimore, Maryland.

17. Oeynhausen, H., Drosdziok, A., Ulm, W., and Termuehlen, H., "Advanced 1000 MW Tandem Compound Reheat Steam Turbine", Presented at 1996 ASME Joint Power Generation Conference.

18. Chen, S., and Martin, H.F., "Blading Design to Improve Performance of HP and IP Steam Turbines" PWR-Vol. 30, Proceedings of the International Joint Power Generation Conference, 1996.

19. Simon, V., and Oeynhuasen, H., "3DV Three-Dimensional Blades – A New Generation of Steam Turbine Blading" PWR-Vol. 33, 1998 ASME International Joint Power Conference Volume 2.

20. Stueer, H., Schmitt, S., and Ashcroft, G., "Aerodynamic Mistuning of Structurally Coupled Blades" GT2008-50204, Proceedings of ASME Turbo Expo 2008: Land, Sea and Air, June 9-13, Berlin, Germany.

21. Neef, M., Sulda, E., Suerken, N., and Walkenhorst, J., "Design Features and Performance Details of Brush Seals for Turbine Applications", GT2006-90404, Proceedings of ASME Turbo Expo 2006: Land, Sea and Air, May 8-11-13, Barcelona, Spain.

22. Stephen, D., and Hogg, S. J., "Development of Brush Seal Technology for Steam Turbine Retrofit Applications." IJGPC2003-40103, Proceeding of the IJGPC2003 International Joint Power, June 16-19, Atlanta, Georgia.

23. Little, N., Sulda, R., and Terezakis, T. "Efficiency Improvement on Large Mechanical Drive Steam Turbines", Proceedings of the 30th Turbomachinery Symposium.

24. Sumer, W.J., Vogan, J.H. and Lindinger R.J., "Reducing Solid Particle Erosion Damage in Large Steam Turbines", Proceedings of American Power Conference, Vol 47, 1985 pp 196–212.

25. Reinhard, K.G., "Turbine Damage by Solid Particle Erosion" ASME Publication 76-JPGC-PWR-15, 1976.

26. Sapate, S.G., and Roma Rao, A.V., "Effect of Erodent Particle Hardness on Velocity Exponent in Erosion of Steels and Cast Irons", Materials and Manufacturing Processes, Volume 18, Issue 5, January 2003, pages 783–802.

27. Kramer, L.D., Quereshi, J.I., Rousseau, R.A., and Ortolano, R.J., "Improvement of Steam Turbine Hard Particle Eroded Nozzles Using Metallurgical Coatings", ASME Publication 83-JPGC-Pwr-29, 1983.

28. Tran, M.H., "Aerodynamic Design to Minimize Solid Particle Erosion on Control Stage Blading", EPRI Conference on Solid Particle Erosion, March 1989.

29. Silvestri, G.J., Bannister, R.L., Fujikawa, T., and Hizume, A. "Optimization of Advanced Steam Condition Power Plants", Transaction of the ASME Vol. 144, October 1992, pg. 612–620.

30. ASME PTC 6 - 2004 (Revision of ASME PTC6-1996)

31. ASME PTC 6.2 – 2004

32. Termuehlen, H., "Variable Pressure Operation and External Turbine Bypass Systems to Improve Plant Cycling Performance", Presented at the Joint ASME/IEEE/ASCE Power Conference, Charlotte, N. C. October 9-10, 1979.

33. Silvestri, G.J., and Martin, H.F., "An Update On Partial-Arc Admission Turbines For Cycling Applications", Presented at the Electric Power Research Institute 1985 Fossil Plant Cycling Workshop, November 5–7, 1985, Miami Beach, FL.

34. Ulbrich, A., Gobrecht, E., Siegel, M. R., Schmid, E., and Armitage, P. K., "High Steam Turbine Operating Flexibility Coupled With Service Interval Optimization", ASME Paper IJPGC2003-40072, Presented at the International Joint Power Generation Conference June 16–19, 2003, Atlanta, Georgia.

35. "The ASME Handbook on Water Technology for Thermal Power Systems", Paul Cohen, ed., ASME, 1989.

36. "Proceedings: Eighth International Conference on Cycle Chemistry in Fossil and Combined Cycle Plants with Heat Recovery Steam Generators", June 20-22, 2006, Calgary, Alberta Canada. EPRI, Palo Alto, CA: 2007. 1014831.

37. "Proceedings: Sixth International Conference on Fossil Plant Cycle Chemistry", EPRI, Palo Alto, CA. 2001. 1001363.

38. "Proceedings: Fifth International Conference on Fossil Plant Cycle Chemistry", EPRI, Palo Alto, CA. 1997. TR-108459.

39. "Recommendations for Inspection and Overhaul of Steam Turbines", VGB-R115Me, Published by VGB Technical Association of Large Power Plant Operators"

40. "Recommended Practices for the Prevention of Water Damage to Steam Turbines Used for Electric Power Generation", ASME TDP-1-2006 (Revision of ASME TDP-1-1998).

41. "Marks Standard Handbook for Mechanical Engineers", Eighth Edition, McGraw Hill Book Company.

42. Sanders, W. P., "Turbine Steam Path", Published by PennWell.

COMBINED CYCLE POWER PLANT

Meherwan P. Boyce

27.1   INTRODUCTION

27.1.1 Thermal Power Plants are power plants operating on carbon-based fuel such as coal, natural gas, and petroleum products. There are five major types of thermal power plants in use on the power grid today:

27.1.1.1 Steam Turbine Power Plants produce electric power by creating steam at high pressure and temperature (superheated steam at 2000 psia/138 bars and 1500°F/815°C) in boilers, which is then expanded through a steam turbine causing the turbine to drive a generator, which then produces electric power. The steam leaving the turbine is usually sent to a condenser, which maintains a vacuum and where the steam is condensed back to a liquid condensate (water). The steam turbine follows the Rankine Cycle. Steam turbine power plants can produce up to 2000 MW of power and are most widely used plant types in the world. These plants have a thermal efficiency between 28% and 35%.

27.1.1.2 Simple Cycle Gas Turbine Power Plants are plants that follow the Brayton Cycle; in gas turbines, the air is compressed in the compressor section of the turbine to a high pressure (580 psia/40 bars) and temperature (1300°F/704°C), and in the combustor the air is further heated to a higher temperature (2600°F/1426°C) at constant pressure. The gas leaving the combustor is at a high pressure and high temperature and is then expanded through the turbine section. The gases leaving the turbine are at a pressure of about (15.0 psia/1.03 bars) and a temperature of about (1200°F/649°C). The turbine drives the gas turbine's compressor and the generator, which produces electric power. These gas turbines can produce up to 300 MW of power. The gas turbine power plant can be installed in 12-18 months, and thus have been used widely in developing nations where energy requirements change rapidly. These plants have a thermal efficiency between 25% and 45% depending on the size of the plant.

27.1.1.3 Diesel Engine-Based Power Plants are plants that follow the Diesel Cycle. These plants for the most part are based on the medium speed (700-1500 rpm) and low speed (100-300 rpm) diesel engines. High-speed diesel engines (1500 rpm and above) are used mostly for vehicular drives, and low output generator sets (below 1 MW). Most of the medium-speed diesel engines produce up to 7 MW, and some of the low-speed diesels produce as much as 100 MW. These plants are used in countries where there is a very low power requirement. They are widely used on islands or in remote locations where there is no major network grid. They are also used widely as standby power. The thermal efficiency of these plants vary between 28% and 42%.

27.1.2 Combined Cycle Power Plants are associated with electrical power plants, which uses the waste heat from the prime mover for the production of steam, and consequently, the steam is used in a steam turbine for the production of additional power. This is usually a combination of the Brayton Cycle (gas turbine) as the topping cycle and the Rankine Cycle (steam turbine) as the bottoming cycle. However, technically the term "Combined Cycle" can be used for any combination of two or more cycles. Many small plants use the Diesel Cycle as the topping cycle, with the Rankine Cycle as the bottoming cycle; there are plants that have used the Brayton Cycle as both the topping and the bottoming cycles. Figure 27.1 shows a large combined Cycle Power Plant. The Combined Cycle Power Plant is not new in concept, since some have been in operation since the mid-1950s. These plants came into their own with the new high-capacity and efficiency gas turbines in the 1990s.

Combined cycle power plants are the most efficient power plants operating on the power grids through out the world with an efficiency ranging between 45% and 57%. These power plants come in all sizes. In this chapter, we will be emphasizing the larger plants ranging in size from 60 to 1500 MW. These combined cycle plants have as their core the gas turbine, which acts as the Topping Cycle, and the steam turbine, which acts as the Bottoming Cycle. In between the gas turbine and the steam turbine is a waste heat recovery steam generator (HRSG), which takes the heat from the exhaust of the gas turbine and generates high pressure steam for the steam turbine.

The Fossil Power Plants of the 1990s and into the early part of the new millennium will be the Combined Cycle Power Plants, with the gas turbine as being the center piece of the plant. It is estimated that from the year 1997-2010, an addition of 147.7 GW of combined cycle power has been built. These plants have replaced the large Steam Turbine Plants, which were the main fossil power plants through the eighties. The Combined Cycle Power Plant is not new in concept, since some have been in operation since the mid-1950s. These plants came into their own with the new high-capacity and efficiency gas turbines.

The new market place of energy conversion will have many new and novel concepts in combined cycle power plants. Figure 27.2 shows the heat rates of these plants, present and future, and Figure 27.3 shows the efficiencies of the same plants. The plants referenced are the Simple Cycle Gas Turbine (SCGT) with firing temperatures of 2400°F (1315°C), Recuperative Gas Turbine (RGT), the Steam Turbine Plant (ST), the Combined Cycle Power Plant

FIG. 27.1 A TYPICAL COMBINED CYCLE FACILITY SHOWING TWO PLANTS OF 4 GAS TURBINES EACH

(CCPP), and the Advanced Combined Cycle Power Plants (ACCP) such as combined cycle power plants using Advanced Gas Turbine Cycles, and finally the Hybrid Power Plants (HPP).

In the area of performance, the steam turbine power plants have an efficiency of about 35%, as compared to combined cycle power plants, which have an efficiency of about 55%. Newer gas turbine technology will make combined cycle efficiencies range between 60% and 65%. As a rule of thumb, a 1% increase in efficiency could mean that 3.3% more capital can be invested. However, one must be careful that the increase in efficiency does not lead to a decrease in availability. From 1996 to 2000, we have seen a growth in efficiency of about 10% and a loss in availability of about 10%. This trend must be turned around since many analysis show that a 1% drop in the availability needs about 2%-3% increase in efficiency to offset that loss.

The time taken to install a steam plant from conception to production is about 42-60 months as compared to 22-36 months for combined cycle power plants. The actual construction time is about 18 months, while environmental permits in many cases take 12 months

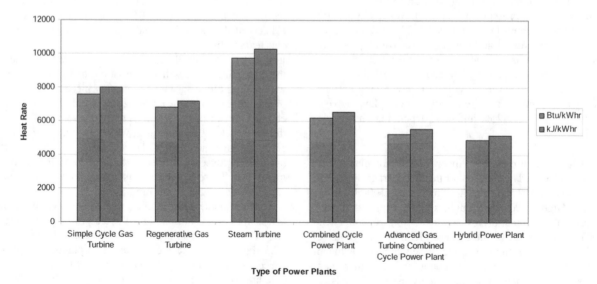

FIG. 27.2 TYPICAL HEAT RATES OF VARIOUS TYPES OF PLANTS

EFFICIENCY

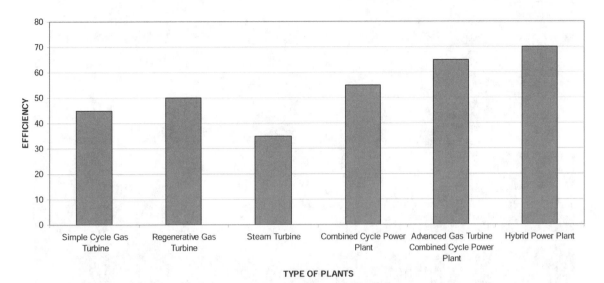

FIG. 27.3 TYPICAL EFFICIENCIES OF VARIOUS TYPES OF PLANTS

and engineering 6-12 months. The time taken for bringing the plant on line affects the economics of the plant, the longer the capital employed without return, accumulates interest, insurance and taxes.

27.1.3 Cogeneration is the production of two or more forms of energy from a single plant. The most common application of the term is for the production of electrical power and steam for use in process applications. This does not mean that other types of cogeneration plants are not being designed and used. Cogeneration plants are used to produce power and use the direct exhaust gases from the prime movers for preheating air in furnaces, or for the use in absorption cooling systems, or for heating various types of fluids in different process applications. Between 1996 and 2006, cogeneration plants accounted for 3% of the new power being generated worldwide, amounting to 19.6 GW of power.

A typical cogeneration plant uses the waste gases from the gas turbine to produce steam in a HRSG, or waste heat boilers (WHB); these terms are often used interchangeably, for use in various chemical processes. The steam could be used directly in an absorption chiller for producing refrigerated air or in a steam turbine to drive a cooling system, or to produce more power. Water is heated to provide hot water for all types of usage.

Cogeneration systems are also used in petrochemical plants where the prime mover drives are used to drive compressors to compress process gasses and then the heat used to produce steam for use direct in processes, or to operate an extraction, condensing or back-pressure steam turbine to drive a compressor or pump and the extracted steam used for a process application. Thermodynamically, the cogeneration plant is very similar to a combined cycle power plant.

A typical cogeneration plant in a refinery or a chemical plant generates high-pressure steam, which then is used in an extracting condensing steam turbine. It usually extracts part of the steam at a lower pressure for use in various chemical processes; the rest of the steam goes through the second part of the steam turbine and then to the condenser. The steam turbines usually drive separate generators; however, the system can be designed in a way that both the gas turbine and the steam turbine drive the same generator.

In most chemical plant applications, the gas turbine is a single-shaft unit if it is used for power generation, and a twin shaft is used

for mechanical drives such as driving a compressor or pump. The gases from the turbine exhaust at a temperature between 900°F (482°C) and 1100°F (593°C), depending on the turbine efficiency and the turbine firing temperature. The hot gases (approximately 90 lb/sec (40.8 kg/sec) for a 15-MW turbine to about 1400 lb/sec (636 kg/sec) for a 200-MW turbine) are piped into a boiler where steam is generated for use as process heat or for use in an extraction or back-pressure steam turbine.

Cogeneration plants are also known as Combined Heat Power (CHP) plants and are widely used in Europe, where they provide steam heating as well as power to large city centers. Figure 27.4 is a photograph of such a plant where the steam is bled after intermediate pressure (IP) stage of a three-stage steam turbine, which consists of a high pressure (HP) stage, and intermediate pressure (IP) stage and a low pressure (LP) stage.

27.2   TYPICAL CYCLES

The combined cycle power plants discussed in this chapter is using the Brayton Cycle as the topping cycle (gas turbine) and the Rankine Cycle (steam turbine) as the bottoming cycle. This by far is the most widely used configuration of the combined cycle power plant used in the power industry.

27.2.1   The Brayton Cycle

The gas turbine is a power plant that produces a great amount of energy for its size and weight. The gas turbine has found increasing service in the past 40 years in the power industry both among utilities and merchant plants as well as the petrochemical industry, and utilities throughout the world. Its compactness, low weight, and multiple fuel application make it a natural power plant for offshore platforms. Today, there are gas turbines that run on natural gas, diesel fuel, naphtha, methane, crude, low-Btu gases, vaporized fuel oils, and biomass gases.

The last 20 years has seen a large growth in gas turbine technology. The growth is spearheaded by the growth of new high-temperature materials technology, new coatings, and new cooling schemes for the hot section of the gas turbine. This, with the

FIG. 27.4 STEAM BLED AFTER IP STAGE OF STEAM TURBINE FOR TOWN CENTER HEATING PURPOSES

conjunction of increase in compressor pressure ratio, has increased the gas turbine thermal efficiency from about 15% in the 1950s to over 45% in the mid-1990s.

The Brayton Cycle, which is the base cycle of the gas turbine, in its ideal form consists of two isobaric processes and two isentropic processes. The two isobaric processes consist of the combustor system of the gas turbine and the gas side of the HRSG. The two isentropic processes represent the compression (compressor) and the expansion (turbine expander) processes in the gas turbine.

A simplified application of the first law of thermodynamics to the air-standard Brayton Cycle in Figure 27.5 (assuming no changes in kinetic and potential energy) has the following relationships:

Work of compressor

$$W_c = m_a(h_2 - h_1) \tag{27.1}$$

Work of turbine

$$W_t = (m_a + m_f)(h_3 - h_4) \tag{27.2}$$

Total output work

$$W_{cyc} = W_t - W_c \tag{27.3}$$

Heat added to system

$$Q_{23} = m_f \text{LHV (fuel)} = (m_a + m_f)(h_3) - m_a h_2 \tag{27.4}$$

Thus, the overall cycle efficiency is

$$\eta_{cyc} = W_{cyc}/Q_{23} \tag{27.5}$$

Increasing the pressure ratio and the turbine firing temperature increases the Brayton Cycle efficiency as shown in Eq. (27.5). This relationship of overall cycle efficiency is based on certain simplification assumptions such as the following: (1) $m_f \ll m_a$ (between 1.5% and 2.5% of the mass of air); (2) the gas is calorically and thermally perfect, which means that the specific heat at constant pressure (c_p) and the specific heat at constant volume (c_v) are constant; thus, the specific heat ratio γ remains constant throughout the cycle; (3) the pressure ratio in both the compressor and the turbine are the same, and (4) all components operate at 100% efficiency. With these assumptions the effect on the ideal cycle efficiency as a function of pressure ratio for the ideal Brayton Cycle operating between the ambient temperature and the firing temperature is given by the following relationship:

$$\eta_{ideal} = \left(1 - \frac{1}{\text{Pr}^{\frac{\gamma-1}{\gamma}}} \right) \tag{27.6}$$

where Pr is the pressure ratio, and γ is the ratio of the specific heats. The above equation tends to go to very high numbers as the pressure ratio is increased. In the case of the actual cycle, the effect of

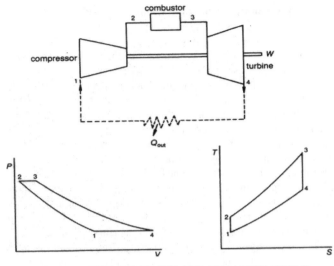

FIG. 27.5   THE AIR-STANDARD BRAYTON CYCLE

Figure 27.6 shows the effect on the overall cycle efficiency of the increasing Pressure Ratio and the Firing Temperature. The increase in the pressure ratio increases the overall efficiency at a given firing temperature; however, increasing the pressure ratio beyond a certain value at any given firing temperature can actually result in lowering the overall cycle efficiency. It should also be noted that the very high-pressure ratios tend to reduce the operating range of the turbine compressor. This causes the turbine compressor to be much more intolerant to dirt build up in the inlet air filter and on the compressor blades and creates large drops in cycle efficiency and performance. In some cases, it can lead to compressor surge. Surge is the reversal of flow in a compressor and can lead to a flameout, or even serious damage and failure of the compressor blades and the radial and thrust bearings of the gas turbine.

27.2.1.1 Inlet Cooling Effect The power in a gas turbine is greatly reduced by the increase in ambient temperature. A 10°F/5.6°C rise in temperature decreases the power output by about 4%. There are several techniques that are used for cooling the turbine compressor inlet from the simple evaporative cooling to the more complex and costly refrigerated inlet cooling.

- **Evaporative methods**: either conventional evaporative coolers or direct water fogging
- **Refrigerated inlet cooling systems**: utilizing absorption or mechanical refrigeration
- **Combination of evaporative and refrigerated inlet systems**: the use of evaporative cooler to assist the chiller system to attain lower temperatures of the inlet air
- **Thermal energy storage systems**: these are intermittent use systems where the cold is produced off-peak and then used to chill the inlet air during the hot hours of the day.

27.2.1.2 Evaporative and Fogging Systems Traditional evaporative coolers that use media for evaporation of the water have been

the turbine compressor (η_c) and expander (η_t) efficiencies must also be taken into account, to obtain the overall cycle efficiency between the firing temperature T_f and the ambient temperature T_{amb} of the turbine. This relationship is given in the following equation:

$$\eta_{cycle} = \left(\frac{\eta_t T_f - \dfrac{T_{amb}\, \mathrm{Pr}^{\left(\frac{\gamma-1}{\gamma}\right)}}{\eta_c}}{T_f - T_{amb} - T_{amb}\left(\dfrac{\mathrm{Pr}^{\left(\frac{\gamma-1}{\gamma}\right)}-1}{\eta_c} \right)} \right) \left(1 - \frac{1}{\mathrm{Pr}^{\left(\frac{\gamma-1}{\gamma}\right)}} \right) \qquad (27.7)$$

Tamb= 15 C EFF. COMP =87% EFF. TURB. = 92%

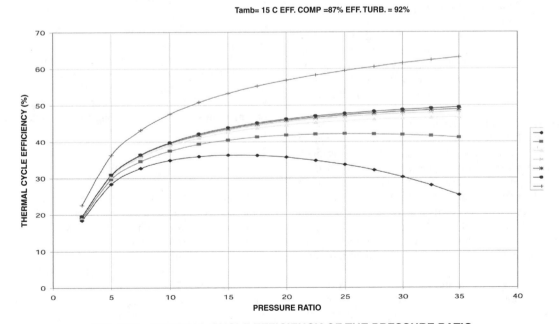

FIG. 27.6   OVERALL CYCLE EFFICIENCY OF THE PRESSURE RATIO

widely used in the gas turbine industry over the years, especially in hot climates with low humidity areas. The low capital cost, installation, and operating costs make it attractive for many turbine-operating scenarios. Evaporation coolers consist of de-mineralized water being sprayed over the media blocks, which are made of fibrous corrugated material. The airflow through these media blocks evaporates the water; as water evaporates, it consumes about 1059 BTU (1117 kJ) (latent heat of vaporization) at 60°F (15°C). This results in the reduction of the air temperature entering the compressor from that of the ambient air temperature. This technique is very effective in low-humidity regions. The work required to drive the turbine compressor is reduced by lowering the compressor inlet temperature, thus increasing the output work of the turbine. The inlet temperature is lowered by about 18°F (10°C), if the outside temperature is around 90°F (32°C). The cost of an evaporative cooling system runs around $50/kw.

Direct inlet fogging is a type of evaporative cooling method, where de-mineralized water is converted into a fog by means of high-pressure nozzles operating at 1000 to 3000 psi. (67-200 bars). This fog then provides cooling when it evaporates in the air inlet duct of the gas turbine. The air can attain 100% relative humidity at the compressor inlet and thereby gives the lowest temperature possible without refrigeration (the web bulb temperature). Direct high-pressure inlet fogging can also be used to create a compressor intercooling effect by allowing excess fog into the compressor, thus lowering temperatures in the first few stages of compression and thus further boosting the power output.

27.2.1.3 Refrigeration Systems

The refrigerated inlets are more effective than the previous evaporative cooling systems as they can lower the temperatures by about 45°F-55°F (25°C-30°C). Two techniques for refrigerating the inlet of a gas turbine are vapor compression (mechanical refrigeration) and absorption refrigeration. In a mechanical refrigeration system, the refrigerant vapor is compressed by means of a centrifugal, screw, or reciprocating compressor. Centrifugal compressors are typically used for large systems in excess of 1000 tons (12.4×10^6 BTU/13.082×10^6 kJ) and would be driven by an electric motor. Mechanical refrigeration has significantly high auxiliary power consumption for the compressor driver and pumps required for the cooling water circuit. After compression, the vapor passes through a condenser where it gets condensed. The condensed vapor is then expanded in an expansion valve and provides a cooling effect. The evaporator chills cooling water that is circulated to the gas turbine inlet chilling coils in the air stream.

Absorption systems typically employ lithium bromide (LiBr) and water, with the LiBr being the absorber and the water acting as the refrigerant. Such systems can cool the inlet air to 50°F (10°C). The heat for the absorption chiller can be provided by gas, steam, or gas turbine exhaust. Absorption systems can be designed to be either single or double effect. A single-effect system will have a coefficient of performance (COP) of 0.7 to 0.9, and a double-effect unit, a COP of 1.15. Part load performance of an absorption system is relatively good, and efficiency does not drop off at part load like it does with mechanical refrigeration systems.

The costs of these systems are much higher than the evaporative cooling system; however, refrigerated inlet cooling systems in hot humid climates are more effective due to the limitation of the evaporative cooling system in high humidity climates.

27.2.1.4 Combination of Evaporative and Refrigerated Inlet Systems

Depending on the specifics of the project, location, climatic conditions, engine type, and economic factors, a hybrid system utilizing a combination of the above technologies may be the best as seen in Figure 27.7. The possibility of using an evaporative system ahead of the mechanical inlet refrigeration system should be considered. This may not always be intuitive, since evaporative cooling is an adiabatic process that occurs at constant enthalpy. When water is evaporated into an air stream, any reduction in sensible heat is accompanied by an increase in the latent heat of the air stream (the heat in the air stream being used to effect a phase change in the water from liquid to the vapor phase). If fog is applied in front of a chilling coil, the temperature will be decreased when the fog evaporates, but since the chiller coil will have to work harder to remove the evaporated water from the air stream, the result would yield no thermodynamic advantage.

To maximize the effect, the chiller must be designed in such a manner that in combination with evaporative cooling the maximum reduction in temperature is achieved. This can be accomplished by designing a slightly undersized chiller, which is not capable of bringing the air temperature down to the ambient dew point temperature, but in conjunction with evaporative cooling the same effect can be achieved. In this manner, the system is taking the advantage of evaporative cooling to reduce the load of refrigeration.

27.2.1.5 Intercooling and Reheat Effects

The intercooling of the air in the compressor of the gas turbine by adding an intercooler between compressor stages and the reheating of the gases in the turbine section of the gas turbine will increase the work that the turbine will produce. The net work of a gas turbine cycle is given by:

$$W_{cyc} = W_t - W_c \qquad (27.8)$$

And can be increased either by decreasing the compressor work (intercooling) or by increasing the turbine work (reheating). Multi-staging of compressors is sometimes used to allow for cooling between the stages to reduce the total work input. Figure 27.8 shows a polytropic compression process 1-a on the P-V plane. If there is no change in the kinetic energy, the work done is represented by the area 1-a-j-k-1. A constant temperature line is shown as 1-x. If the polytropic compression from states 1 to 2 is divided into two parts, 1-c and d-e, with constant pressure cooling to $T_d = T_1$ between them, the work done is represented by area 1-c-d-e-j-k-1. The area c-a-e-d-c represents the work saved by means of the two-stage compression with intercooling to the initial temperature. The optimum pressure for intercooling for specified values P_1 and P_2 is

$$P_{OPT} = \sqrt{P_1 P_2} \qquad (27.9)$$

Therefore, if a simple gas turbine cycle is modified with the compression accomplished in two or more adiabatic processes with intercooling between them, the net work of the cycle is increased with no change in the turbine work.

The thermal efficiency of an ideal simple cycle is decreased by the addition of an intercooler. Figure 27.9 shows the schematic of such a cycle. The ideal simple gas turbine cycle is 1-2-3-4-1, and the cycle with the Intercooling added is 1-a-b-c-2-3-4-1. Both cycles in their ideal form are reversible and can be simulated by a number of Carnot Cycles. The Carnot Cycle is the most efficient

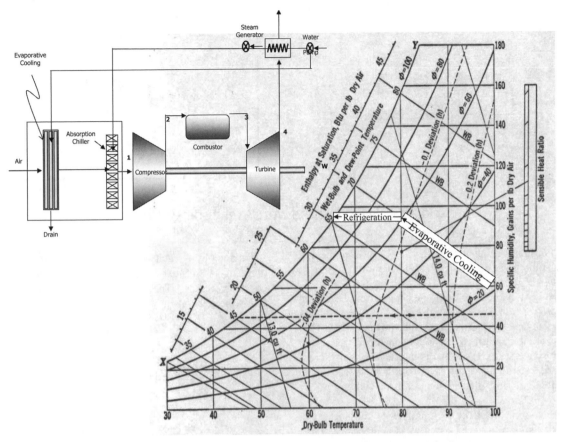

FIG. 27.7 EVAPORATIVE AND REFRIGERATED INLET SYSTEMS

cycle between two thermal reservoirs. Thus, if the simple gas turbine cycle 1-2-3-4-1 is divided into a number of cycles like *m-n-o-p-m*, these little cycles approach the Carnot Cycle as their number increases. The efficiency of such a Carnot Cycle is given by the relationship

$$\eta_{CARNOT} = 1 - \frac{T_m}{T_p} \qquad (27.10)$$

Notice that if the specific heats are constant, then

$$\frac{T_3}{T_4} = \frac{T_m}{T_p} = \left(\frac{P_2}{P_1}\right)^{\frac{\gamma-1}{\gamma}} \qquad (27.11)$$

All the Carnot cycles making up the simple gas turbine cycle have the same efficiency. Likewise, all of the Carnot cycles into which the cycle *a-b-c-2-a* might similarly be divided have a common value of efficiency lower than the Carnot cycles, which comprise cycle 1-2-3-4-1. Thus, the addition of an intercooler, which adds *a-b-c-2-a* to the simple cycle, lowers the efficiency of the cycle.

The intercooling of the compressed air has been very successfully applied to high-pressure engines. This system can be combined with any of the previously described systems. The addition of an intercooler to a regenerative gas turbine cycle increases the cycle's thermal efficiency and output work because a larger portion of the heat required for the process *c*-2i in Figure 27.9 can be obtained from the hot turbine exhaust gas passing through the regenerator instead of from burning additional fuel.

27.2.1.6 Mid-Compressor Flashing of Water Another manner of intercooling is the injection of demineralized water into the mid-stages of the compressor to cool the air and approach an isothermal compression process, as shown in Figure 27.10. The water injected is usually mechanically atomized so that very fine droplets are entered into the air. The water is evaporated as it comes in contact with the high pressure and temperature air stream. As water

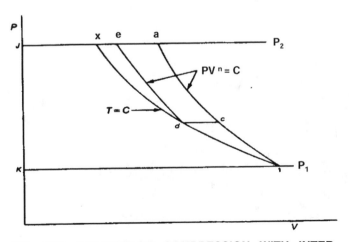

FIG. 27.8 MULTISTAGE COMPRESSION WITH INTERCOOLING

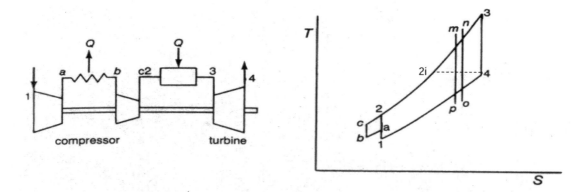

FIG. 27.9   AIR-STANDARD WITH INTERCOOLING CYCLE

evaporates, it consumes about 1058 BTU (1117 kJ) (latent heat of vaporization) at the higher pressure and temperature resulting in lowering the temperature of the air stream entering the next stage. This lowers the work required to drive the compressor. The steam or water injected for cooling purposes also increases the mass flow through the system and therefore increases the power output of the turbine. The steam to be injected, can be obtained from the use of a low-pressure single-stage HRSG and is injected at the first compressor section discharge and/or injection in the combustor if steam is being used for controlling the NO_x output in the combustor.

27.2.1.7 Regeneration Effect In a simple gas turbine cycle, the turbine exit temperature is nearly always appreciably higher than the temperature of the air leaving the compressor. Obviously, the fuel requirement can be reduced by the use of a regenerator in which the hot turbine exhaust gas preheats the air between the compressor and the combustion chamber. Figure 27.11 shows a schematic of the regenerative cycle and its performance in the T-S diagram. In an ideal case, the flow through the regenerator is at constant pressure. The regenerator effectiveness is given by the following relationship:

$$\eta_{reg} = \frac{T_3 - T_2}{T_5 - T_2} \tag{27.12}$$

Thus, the overall efficiency for this system's cycle can be written as

$$\eta_{RCYC} = \frac{(T_4 - T_5) - (T_2 - T_1)}{(T_4 - T_3)} \tag{27.13}$$

27.2.1.8 Reheat Effect The Reheat Cycle increases the turbine work, and consequently, the net work of the cycle can be increased without changing the compressor work or the turbine inlet tempera-

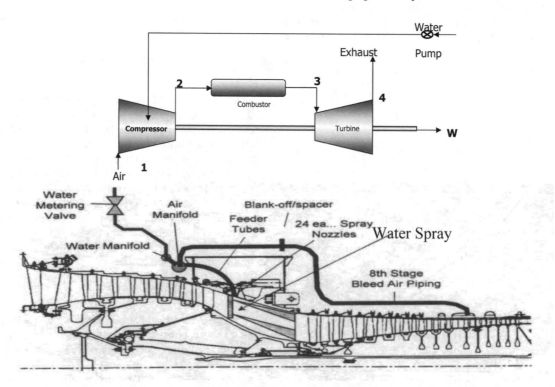

FIG. 27.10   MID-COMPRESSOR COOLING SHOWING A SCHEMATIC AS WELL AS AN ACTUAL APPLICATION IN A GE LM 6000 ENGINE (Courtesy of GE Corporation)

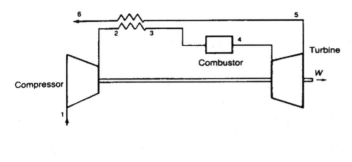

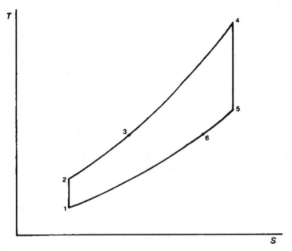

FIG. 27.11 THE REGENERATIVE GAS TURBINE CYCLE

ture by dividing the turbine expansion into two or more parts with constant pressure heating before each expansion as seen in Figure 27.12. This cycle modification is known as reheating. By reasoning similar to that used in connection with intercooling, it can be seen that the thermal efficiency of a simple cycle is lowered by the addition of reheating, while the work output is increased. However, a combination of regenerator and reheater can increase the thermal efficiency.

27.2.1.9 The Intercooled Regenerative Reheat Cycle The Carnot Cycle is the optimum cycle and all cycles attempt to reach toward this optimum. Maximum thermal efficiency is achieved by approaching the isothermal compression and expansion of the Carnot Cycle, or by intercooling in compression and reheating in the

expansion process. Figure 27.13 shows the intercooled regenerative reheat cycle, which approaches this optimum cycle in a practical fashion. This cycle achieves the maximum efficiency and work output of any of the modified Brayton Cycles. With the insertion of an intercooler in the compressor, the pressure ratio for maximum efficiency moves to a much higher ratio.

27.2.2 The Rankine Cycle

The Rankine Cycle employing water-steam as the working fluid is the most common thermodynamic cycle utilized in the production of electrical power. It is the cycle utilized by a steam turbine. A schematic of a steam power plant is shown in Figure 27.14; water enters the boiler feed water pump at point 1 and is pumped isentropically into the boiler. The compressed liquid at 2 is heated until it becomes saturated at 2a, after which it is evaporated to steam at 2b and then superheated to 3. The steam leaves the boiler at 3, expands isentropically in the ideal engine to 4, and passes to the condenser. Circulating water condenses the steam to a saturated liquid at 1, from which state the cycle repeats itself.

The thermodynamic diagrams corresponding to the steam power plant in Figure 27.14, showing the thermodynamic states, are shown in the pressure-volume (*P-V*) diagram in Figure 27.15, and the temperature entropy (*T-S*) diagram in Figure 27.16. The work done by the steam turbine (W_{st}) based on the *T-S* diagram in Figure 27.16 is given by:

$$W_{st} = \dot{m}_{st}(h_3 - h_4) \qquad (27.14)$$

where $\dot{m}_{st}$ is the mass flow of the steam, and h_3 and h_4 are steam enthalpies (Btu/lb; kJ/kg) at points 3 and 4.

The network produced by the system, W_{net}, is the turbine work less the pump work, W_P, required to raise the water to the desired pressure and is expressed below:

$$W_{net} = W_{st} - W_P \qquad (27.15)$$

$$W_{net} = \dot{m}_{st}(h_3 - h_4) - W_P \qquad (27.16)$$

The above analysis assumes an ideal isentropic expansion from points 3 to 4. In a steam turbine the actual process is not isentropic, and some loss does occur. The actual expansion is from points 3 to 4a. The pump work is much smaller than the turbine work and can be neglected when estimating the overall performance and efficiency of steam plants.

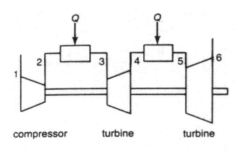

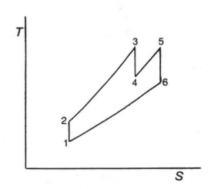

FIG. 27.12 REHEAT CYCLE AND T-S DIAGRAM

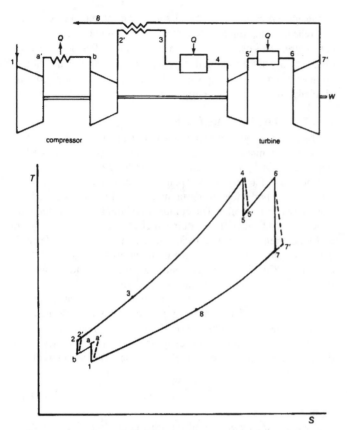

FIG. 27.13 THE INTERCOOLED REGENERATIVE REHEAT SPLIT-SHAFT GAS TURBINE CYCLE

The energy input requirement to the system, Q_{in}, is given by

$$Q_{in} = \dot{m}_{st}(h_3 - h_{2f}) \qquad (27.17)$$

The thermal efficiency of the system, η, is then given by:

$$\eta = \frac{W_{net}}{Q_{in}} \qquad (27.18)$$

$$\eta = \frac{\dot{m}_{st}(h_3 - h_4) - W_p}{\dot{m}_{st}(h_3 - h_{2f})} \qquad (27.19)$$

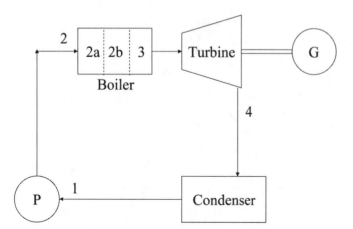

FIG. 27.14 SCHEMATIC OF A STEAM TURBINE POWER PLANT

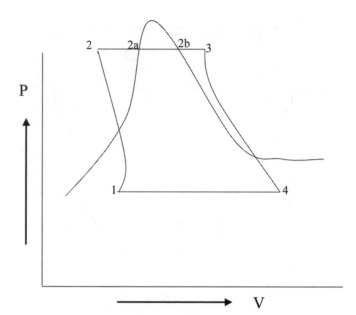

FIG. 27.15 PRESSURE-VOLUME DIAGRAM OF A TYPICAL STEAM TURBINE POWER PLANT

The overall thermal cycle efficiency of the system can be increased by preheating boiler feed water and the incoming combustion air by using hot exhaust from the boiler or gas turbine if available.

27.2.2.1 Heat Rate and Steam Rate The heat rate is a modified reciprocal of the thermal efficiency and is in much wider use among steam-power and turbine engineers. The heat rate for a turbine is defined as the heat chargeable in Btu per kilowatt hour or horse-power-hour turbine output. Again the basis upon which the turbine output is taken should be specified. Turbine heat rate should not be confused with the heat rate of the steam-power plant known as the station heat rate. The station heat rate, like station thermal efficiency, takes into account all the losses from fuel to switchboard.

The heat rate (HR) for a straight condensing or non-condensing turbine is

$$HR = \frac{\dot{m}_{st}(h_3 - h_{2f})3415}{W_{net}} \text{ Btu/kW-hr} \qquad (27.20)$$

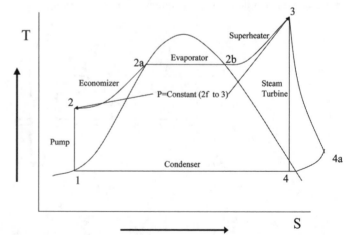

FIG. 27.16 TEMPERATURE-ENTROPY DIAGRAM OF A TYPICAL STEAM TURBINE POWER PLANT

or

$$HR = \frac{\dot{m}_{st}(h_3 - h_{2f})2544.4}{W_{net}} \text{ Btu/Hp-hr} \qquad (27.21)$$

where h_1 = enthalpy of throttle steam

h_{2f} = enthalpy of liquid water at exhaust pressure

W_{net} = output (kW or Hp)

Steam rate is defined as the mass rate of steam flow in pounds per hour divided by the power or rate of work development of the turbine in kilo-power-hour. The steam rate, therefore, is the steam supplied per kilo-watt-hour or horsepower-hour unit of output. The heat rate may be obtained by multiplying the steam rate by the heat chargeable.

27.2.2.2 Turbine Component Efficiency
The turbine may consist of three turbine sections the high-pressure (HP), intermediate-pressure (IP), and the low-pressure (LP) sections. Each of these sections has their own efficiencies. These are also known as isentropic efficiency and are given as:

$$\eta_{isentropic} = \frac{h_3 - h_{4a}}{h_3 - h_4} \qquad (27.22)$$

The turbine efficiency is the ratio of the real output of the turbine to the ideal output. The engine efficiency is, primarily, of interest to the designer as a means of comparing the real turbine with the ideal.

Mechanical Efficiency

The mechanical efficiency of a turbine takes into account the bearing and gear losses of the turbine, the power factor of the generator, and mechanical losses of the generator bearings.

27.2.2.3 The Regenerative-Reheat Rankine Cycle
It is evident from the Rankine Cycle shown in Figure 27.16 that considerable amount of heat is required to raise the temperature of the water from 2 to 2a. The Rankine Cycle has the disadvantage that the fluid temperature at the pump discharge is much lower than the fluid temperature at the turbine inlet. One way of overcoming this disadvantage is to use the internal system heat rather than the external heat to minimize this difference in the temperatures. This concept is called the *regenerative heating*. In the gas turbines, the regenerative heating is accomplished by using the high-temperature exhaust gases. In the steam turbines, intermediate-pressure steam rather than exhaust steam is used for heating the feed water.

As the high-pressure steam expands through the steam turbine, the steam gets very wet at low pressures. This wet steam is detrimental for a turbine; it results in reduction of efficiency and also nozzle and blade erosion. The reheat cycle involves heating of the steam withdrawn after partial expansion. This idea, combined with regenerative heating for improved thermal efficiency, is common practice in central power plants.

A simplified concept of the regenerative reheat steam cycle is depicted in Figures 27.17, and the thermodynamic cycle of the same is shown in Figure 27.18. The water enters the first pump at point 1 from where it enters the feed-water heater at point 2. In the LP economizer/ feed-water heater, the pressurized con-

densate is heated by part of the steam extracted from the high-pressure turbine at an intermediate pressure, point 6. The rest of the extracted steam is reheated in the reheater and enters the turbine at point 7. The heated water enters the second pump at point 3 from which it enters the boiler at point 4. The compressed liquid at 4 is heated until it becomes saturated at 4a, after which it is evaporated to steam at 4b, and then superheated to 5. The steam leaves the boiler at 5, expands isentropically in the ideal engine to 6, and in the real case to 6a where it is extracted for regeneration and reheat. The superheated steam now leaves the reheater at 7, expands isentropically in the ideal engine to 8 and in the real case to 8a where it passes to the condenser. Circulating water condenses the steam to a saturated liquid at 1, from which state the cycle repeats itself.

27.2.3 The Brayton-Rankine Cycle

The combination of the gas turbine with the steam turbine is an attractive proposal, especially for electric utilities and process industries where steam is being used. The schematic of this cycle is shown in Figure 27.19; the hot gases from the gas turbine exhaust are used in a steam recovery steam generator, which may be supplementary fired to produce superheated steam at high temperatures for a steam turbine. Figure 27.20 shows the distribution of the energy in a Brayton-Rankine combined cycle. About 40% of the energy is converted to power by the gas turbine; the remainder 60% of the energy is collected in the HRSG and is used to power a steam turbine, which produces about 20% of the energy as power.

The computations of the combined Brayton and Rankine cycle are divided up into three parts:

1. The work of the Brayton cycle (gas turbine) is the same as already outlined.
2. The work of the Rankine Cycle (steam turbine) is also the same as already shown for that cycle.
3. The computation of the exchange of the heat from the gas turbine to the steam turbine needs to be further examined.

The heat exchanged in the HRSG also known as the waste heat boiler is produced by the exchange of the heat from the gas turbine exhaust, to the steam condensate.

$$_4Q_1 = m_g(h_{4a} - h_{exhaust}) = m_s(h_{1S} - h_{4S}) \qquad (27.23)$$

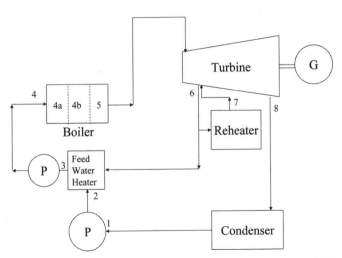

FIG. 27.17 SCHEMATIC OF A REGENERATIVE-REHEAT STEAM TURBINE POWER PLANT

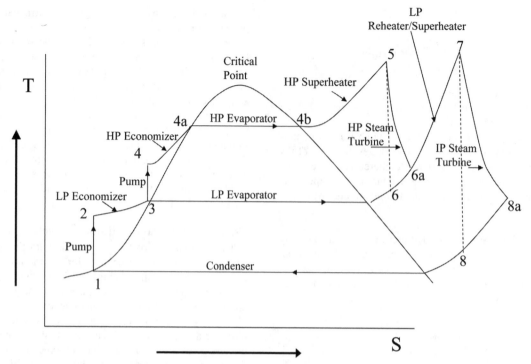

FIG. 27.18 TEMPERATURE-ENTROPY DIAGRAM OF A REGENERATIVE-REHEAT STEAM TURBINE POWER PLAN

where:

m_g = mass flow of the gas turbine exhaust

h_{4a} = enthalpy of gas turbine exhaust

$h_{exhaust}$ = enthalpy of the gas in the exhaust stack

m_s = mass flow of the condensate and steam

h_{1s} = enthalpy of the superheated steam

h_{4s} = enthalpy of the steam condensate (water)

Turbine work

$$W_{tS} = m_S(h_{1S} - h_{2S}) \qquad (27.24)$$

Pump work

$$W_p = m_S(h_{4S} - h_{3S})/\eta_p \qquad (27.25)$$

The combined cycle work is equal to the sum of the net gas turbine work and the steam turbine work. About 50%-60% of the design output of the gas turbine is available as energy in the exhaust gases. The exhaust gas from the turbine is used to provide heat to the recovery boiler. Thus, this heat must be credited to the overall cycle. The following equations show the overall cycle work and thermal efficiency:

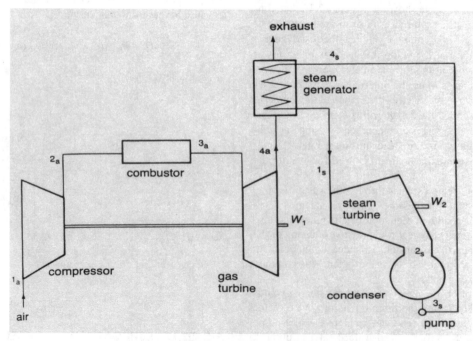

FIG. 27.19 SCHEMATIC OF THE COMBINED CYCLE

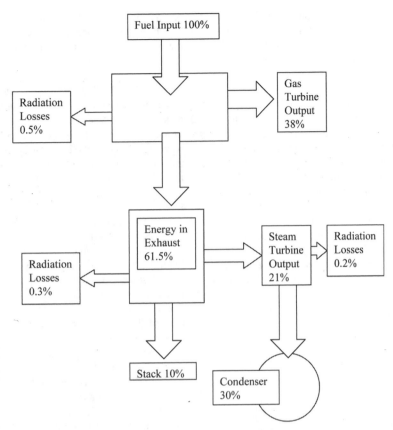

FIG. 27.20 ENERGY FLOW DIAGRAM OF COMBINED CYCLE POWER PLANT

Overall cycle work

$$W_{cyc} = W_{gt} + W_{st} - W_c - W_p \qquad (27.26)$$

Overall cycle efficiency

$$\eta_{cyc} = \frac{W_{cyc}}{m_{fgt}(\text{LHV})} \qquad (27.27)$$

The combination of the Brayton and Rankine cycles is one of the most efficient cycles in operation for practical power generation systems. The Brayton Cycle (gas turbine) and the Rankine Cycle (steam turbine) is combined as such to have the gas turbine as the topping cycle and the steam turbine as the bottoming cycle. Thermal efficiencies of the combined cycles can reach as high as 60%. In the typical combination, the gas turbine produces about 60% of the power and the steam turbine about 40% of the total power produced. Individual unit thermal efficiencies of gas turbines are about (28%-45%) and the steam turbine is about (30%-35%). The steam turbine utilizes the energy in the exhaust gas of the gas turbine as its input energy. The energy transferred to the HRSG by the gas turbine is usually equivalent to about the rated output of the gas turbine at design conditions. At off design conditions, the inlet guide vanes (IGV) are used to regulate the air so as to maintain a high temperature to the HRSG.

27.2.4 Components of a Combined Cycle Power Plant

In the traditional combined cycle plant, air enters the gas turbine where it is initially compressed and then enters the combustor where it undergoes a very high rapid increase in temperature at constant pressure. The high-temperature and high-pressure air then enters the expander section where it is expanded to nearly atmospheric conditions. This expansion creates a large amount of energy, which is used to drive the compressor used in compressing the air, plus the generator where power is produced. The compressor in the gas turbine uses about 50%-60% of the power generated by the expander.

The air upon leaving the gas turbine is essentially at atmospheric pressure conditions, and at a temperature between 950°F and 1200°F (510°C-650°C). This air enters the HRSG, where the energy is transferred to the water to produce steam. There are many different HRSG units. Most HRSG units are divided into the same amount of sections as the steam turbine. In most cases, each section of the HRSG has a pre-heater, an economizer, and feed-water, and then a superheater. The steam entering the steam turbine is superheated.

In most large plants, the steam turbine consists of three sections: a high-pressure (HP) turbine stage with pressures between 1500 and 1700 psia (100.7-114.2 bars), an intermediate-pressure (IP) turbine stage with pressures between 350 and 550 psia (23.51-36.9 bars), and a low-pressure (LP) turbine stage with pressures between 90 and 60 psia (6-4 bars). The steam exiting from the HP stage is usually reheated to about the same temperature as the steam entering the HP stage before it enters the intermediate stage. The steam exiting the IP stage enters directly into the LP stage after it is mixed with steam coming from the LP superheater.

The steam from the LP turbine enters the condenser. The condenser is maintained at a vacuum of between 2 and 0.5 psia (0.13-0.033 bars). The increase in back-pressure in the condenser will reduce the power produced. Care is taken to ensure that the steam leaving the LP stage blades has not a high content of liquid in the steam to avoid erosion of the LP blading.

The new gas turbines also utilize low NO_x combustors to reduce the NO_x emissions, which otherwise would be high due to the high firing temperature of about 2300°F (1260°C). These low NO_x combustors require careful calibration to ensure an even firing temperature in each combustor. New type of instrumentation such as dynamic pressure transducers have found to be effective in ensuring steady combustion in each of the combustors.

Combined cycle plants have several advantages. These include: (1) high thermal efficiencies (50% to 65%), (2) rapid startup (2-hour cold start), and (3) low first installed costs ($600 to $900 per kilowatt). Maintenance costs for combined cycles range from US $0.003 to US $0.007 per kilowatt hour (similar to cogeneration plants).

This section discusses the three major components of a combined cycle power plant:

1. Gas turbine
   a. Compressor
   b. Combustor
   c. Turbine expander
2. Heat recovery steam generator
3. Steam turbine

27.3 GAS TURBINE

The gas turbine consists of three basic components: the gas compressor, the combustor, and the gas expander (turbine). The air enters the compressor section of the gas turbine where the air is compressed to a very high pressure. Compressor pressure ratios of 17:1 to 35:1 in gas turbines used in a commercial combined cycle are common. The compressed air then goes into a combustor where the compressed air is injected with about 1.5% to 3.0% of fuel gas by weight to air and combusted to very high temperatures up to 2500°F (1371°C). This hot and pressurized gas is expanded through the turbine. The turbine produces energy to drive the air compressor of the turbine and drive the generator to produce power. To drive the turbine, air compressor requires about 50%-55% of the power generated by the turbine section of the gas turbine. The gas leaves the turbine at temperatures of up to 1200°F (648°C). The gas has energy to produce enough power to drive steam turbine rated at about 60% of the turbine rated power.

The industrial heavy-duty gas turbines employ axial-flow compressors and turbines. The industrial turbine consists of a multistage axial-flow compressor; and turbine. Figure 27.21 is a cross-sectional representation of the GE Industrial Type Gas Turbine, with can annular combustors, and Figure 27.22 is a cross-sectional representation of the Siemens annular combustor gas turbine.

27.3.1 Compressor Section

The industrial heavy-duty gas turbines employ axial-flow compressors and turbines. The industrial turbine consists of a 15-17 stage axial-flow compressor; in the industrial gas turbines, the loading per stage is considerably less and varies between 1.05 and 1.3 per stage. The adiabatic efficiency of the compressors has also increased, and efficiencies in the high 1980s have been achieved. Compressor efficiency is very important in the overall performance of the gas turbine as it consumes 55% to 60% of the power generated by the gas turbine.

An axial-flow compressor compresses its working fluid by first accelerating the fluid and then diffusing it to obtain a pressure increase. The fluid is accelerated by a row of rotating airfoils or blades (the rotor) and diffused by a row of stationary blades (the stator). The diffusion in the stator converts the velocity increase gained in the rotor to a pressure increase. One rotor and one stator make up a stage in a compressor. A compressor usually consists of multiple stages. One additional row of fixed blades (IGV) is frequently used at the compressor inlet to ensure that air enters the first-stage rotors at the desired angle. In addition to the stators, additional diffuser at the exit of the compressor further diffuses the fluid and controls its velocity when entering the combustors.

In an axial-flow compressor, air passes from one stage to the next with each stage raising the pressure slightly. By producing low-pressure increases on the order of 1.1:1-1.4:1, very high efficiencies can be obtained. The use of multiple stages permits overall pressure increases up to 40:1. The rule of thumb for a multiple

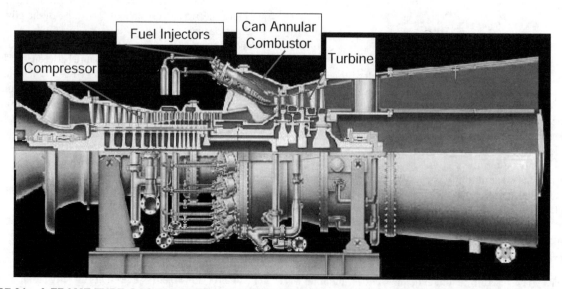

FIG. 27.21   A FRAME TYPE GAS TURBINE WITH CAN-ANNULAR COMBUSTORS (Courtesy of GE Power Systems)

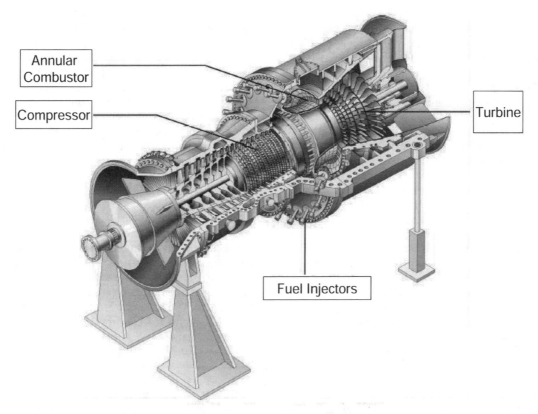

FIG. 27.22   A FRAME TYPE GAS TURBINE WITH AN ANNULAR COMBUSTORS (Courtesy of Siemens Power Systems)

stage gas turbine compressor would be that the energy rise per stage would be constant rather than the pressure rise per stage.

Figure 27.23 shows multistage high-pressure axial-flow compressor-turbine rotor. The turbine rotor depicted in this figure has a low-pressure compressor followed by a high-pressure compressor. There are also two turbine sections; the reason there is a large space between the two turbine sections is that this is a reheat turbine and the second set of combustors are located between the high-pressure

FIG. 27.23   A HIGH-PRESSURE RATIO COMPRESSOR-TURBINE ROTOR (Courtesy of ALSTOM)

and the low-pressure turbine sections. The compressor produces 30:1 pressure in 22 stages. The low-pressure increase per stage also simplifies calculations in the design of the compressor by justifying the air as incompressible in its flow through an individual stage.

27.3.2 Combustor

All gas turbine combustors perform the same function: they increase the temperature of the high-pressure gas. The gas turbine combustor uses very little of its air (10%) in the combustion process. The rest of the air is used for cooling and mixing. New combustors are also circulating steam for cooling purpose. The air from the compressor must be diffused before it enters the combustor. The velocity of the air leaving the compressor is about 400-600 ft/sec (122-183 m/sec), and the velocity in the combustor must be maintained below 50 ft/sec (15.2 m/sec). Even at these low velocities, care must be taken to avoid the flame to be carried on downstream.

The combustor is a direct-fired air heater in which fuel is burned almost stoichiometrically with one-third or less of the compressor discharge air. Combustion products are then mixed with the remaining air to arrive at a suitable turbine inlet temperature. Despite the many design differences in combustors, all gas turbine combustion chambers have three features: (1) a recirculation zone, (2) a burning zone (with a recirculation zone that extends to the dilution region), and (3) a dilution zone, as seen in Figure 27.24. The air entering a combustor is divided so that the flow is distributed between three major regions: (1) primary zone, (2) dilution zone, (3) annular space between the liner and casing.

The combustion in a combustor takes place in the primary zone. Combustion of natural gas is a chemical reaction that occurs between carbon, or hydrogen, and oxygen. Heat is given off as the reaction takes place. The products of combustion are carbon dioxide and water. The reaction is stoichiometric, which means that the proportions of the reactants are such that there are exactly enough oxidizer molecules to bring about a complete reaction to stable molecular forms in the products. The air enters the combustor in a straight through flow, or reverse flow. Most aero-engines have straight through flow type combustors. Most of the large frame type units have reverse flow. The function of the recirculation zone is to evaporate, partly burn, and prepare the fuel for rapid combustion within the remainder of the burning zone. Ideally, at the end of the burning zone, all fuel should be burnt so that the function of the dilution zone is solely to mix the hot gas with the dilution air. The mixture leaving the chamber should have a temperature and velocity distribution acceptable to the guide vanes and turbine. Generally, the addition of dilution air is so abrupt that if combustion is not complete at the end of the burning zone, chilling occurs, which prevents completion. However, there is evidence with some chambers that if the burning zone is run over-rich, some combustion does occur within the dilution region. Figure 27.25 shows the distribution of the air in the various regions of a typical reverse flow, can annular combustor. The theoretical or reference velocity is the flow of combustor-inlet air through an area equal to the maximum cross section of the combustor casing. The flow velocity is 25 fps (7.6 mps) in a reverse-flow combustor and between 80 fps (24.4 mps) and 135 fps (41.1 mps) in a straight-through flow turbojet combustor.

27.3.3 Air Pollution Problems

27.3.3.1 Smoke In general, it has been found that much visible smoke is formed in small, local fuel-rich regions. The general approach to eliminating smoke is to develop leaner primary zones.

27.3.3.2 Unburnt Hydrocarbons and Carbon Monoxide Unburnt hydrocarbon (UHC) and carbon monoxide (CO) are only produced in incomplete combustion typical of idle conditions.

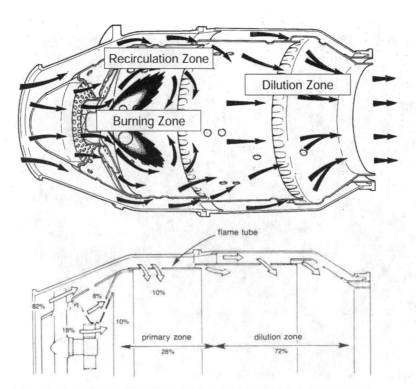

FIG. 27.24 CROSS SECTION OF A CAN ANNULAR COMBUSTOR

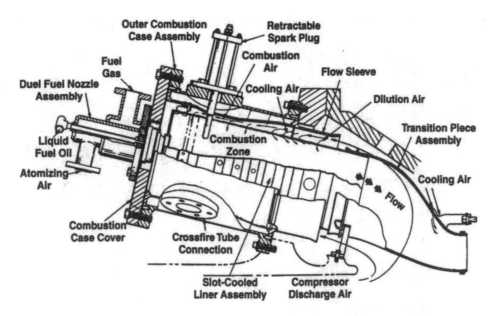

FIG. 27.25 A TYPICAL REVERSE FLOW CAN ANNULAR COMBUSTOR USED IN LARGE GAS TURBINES

27.3.3.3 Oxides of Nitrogen The main oxides of nitrogen produced in combustion are NO, with the remaining 10% as NO_2. These products are of great concern because of their poisonous character and abundance, especially at full-load conditions.

In 1977, NO_x control was required to meet the EPA standards of 75 ppm; it was recognized that there were a number of ways to control oxides of nitrogen:

1. Use of a rich primary zone in which little NO formed, followed by rapid dilution in the secondary zone
2. Use of a very lean primary zone to minimize peak flame temperature by dilution
3. Use of water or steam admitted with the fuel for cooling the small zone downstream from the fuel nozzle
4. Use of inert exhaust gas recirculated into the reaction zone
5. Catalytic exhaust cleanup

"Wet" control became the preferred method in the 1980s and most of 1990s since "dry" controls and catalytic cleanup were both at a very early stages of development. The catalytic converters were used in the 1980s and are still being widely used; however, the cost of rejuvenating the catalyst is very high.

There has been a gradual tightening of the NO_x limits over the years from 75 ppm down to 25 ppm, and now the new turbine goals are 9 ppm.

Advances in combustion technology now make it possible to control the levels of NO_x production at source, removing the need for wet controls. This of course opened up the market for the gas turbine to operate in areas with limited supplies of suitable quality water, e.g, deserts or marine platforms.

Although water injection is still used, dry control combustion technology has become the preferred method for the major players in the industrial power generation market. Dry low NO_x was the first acronym to be coined, but with the requirement to control NO_x without increasing carbon monoxide and unburned hydrocarbons; this has now become dry low emissions (DLE).

The majority of the NO_x produced in the combustion chamber is called "thermal NO_x." It is produced by a series of chemical reactions between the nitrogen (N_2) and the oxygen (O_2) in the air that occur at the elevated temperatures and pressures in gas turbine combustors.

27.3.3.4 Dry Low NO_x Combustor The gas turbine combustors have seen considerable change in their design as most new turbines have progressed to dry low NO_x Combustors from the wet combustors, which were injected by steam in the primary zone of the combustor. The DLE approach is to burn most (at least 75%) of the fuel at cool, fuel-lean conditions to avoid any significant production of NO_x. In the DLE combustor, a small proportion of the fuel is always burned richer to provide a stable "piloting" zone, while the remainder is burned lean.

The DLE combustors often experience the following major problems:

- Auto-ignition and flash-back
- Combustion instability

These problems can result in sudden loss of power because a fault is sensed by the engine control system and the engine is shutdown. Auto-ignition is the spontaneous self-ignition of a combustible mixture. For a given fuel mixture at a particular temperature and pressure, there is a finite time before self-ignition will occur. Diesel engines (knocking) rely on it to work, but spark-ignition engines must avoid it.

DLE combustors have premix modules on the head of the combustor to mix the fuel uniformly with air. To avoid auto-ignition, the residence time of the fuel in the premix tube must be less than the auto-ignition delay time of the fuel. If auto-ignition does occur in the premix module, then it is probable that the resulting damage will require repair and/or replacement of parts before the engine is run again at full load.

27.3.4 Turbine Expander Section

The axial-flow turbine is used in more than 95% of all gas turbine applications. The axial-flow turbine, like its counterpart the axial-flow compressor, has flow, which enters and leaves in the

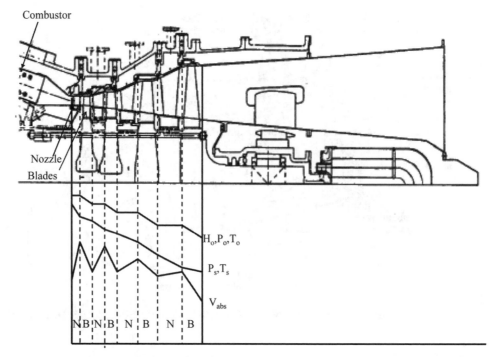

FIG. 27.26 SCHEMATIC OF AN AXIAL-FLOW TURBINE WITH THE DISTRIBUTION OF THE PRESSURE, TEMPERATURE, AND THE ABSOLUTE VELOCITY IN THE TURBINE SECTION

axial direction. There are two types of axial turbines: (1) impulse type and (2) reaction type. The impulse turbine has its entire enthalpy drop in the nozzle; therefore, it has a very high velocity entering the rotor. The reaction turbine divides the enthalpy drop in the nozzle and the rotor. Figure 27.26 is a schematic of an axial-flow turbine, also depicting the distribution of the pressure, temperature and the absolute velocity.

Most axial-flow turbines consist of more than one stage; the front stages are usually impulse (zero reaction) and the later stages

have about 50% reaction. The impulse stages produce about twice the output of a comparable 50% reaction stage, while the efficiency of an impulse stage is less than that of a 50% reaction stage.

Since 1950, turbine bucket material temperature capability has advanced approximately 850°F (472°C), approximately 20°F/10°C per year. Figure 27.27 shows the development of cooling schemes and blade materials as a function of firing temperatures. The importance of this increase can be appreciated by noting that an increase of 100°F (56°C) in turbine firing temperature can provide

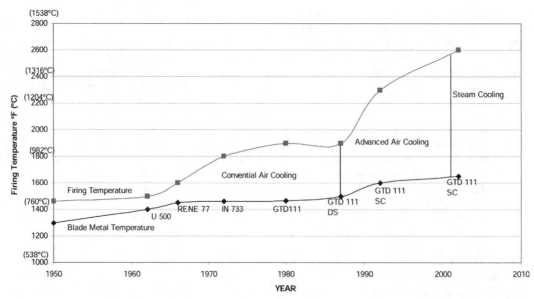

FIG. 27.27 FIRING TEMPERATURE INCREASE WITH BLADE MATERIAL AND ADVANCED COOLING SCHEMES

a corresponding increase of 8% to 13% in output and 2% to 4% improvement in simple-cycle efficiency.

The new advanced gas turbine blades have seen improvement in three major areas:

1. Blade material
2. Blade cooling schemes
3. Blade coating

27.3.5   Blade Material

The high temperatures that are now available in the turbine sections (turbine inlet temperature 2650°F/1454°C) are due to improvements of the metallurgy of the blades in the turbines. The stage 1 blade must withstand the most severe combination of temperature, stress, and environment; it is generally the limiting component in the machine.

In the 1980s, IN-738 blades were widely used. IN-738 was the acknowledged corrosion standard for the industry. Directionally solidified (DS) blades, first used in aircraft engines more than 25 years ago, were adapted for use in large airfoils in the early 1990s and were introduced in the large industrial turbines to produce advanced technology nozzles and blades. The directionally solidified blade has a grain structure that runs parallel to the major axis of the part and contains no transverse grain boundaries, as in ordinary blades. The elimination of these transverse grain boundaries confers additional creep and rupture strength on the alloy, and the orientation of the grain structure provides a favorable modulus of elasticity in the longitudinal direction to enhance fatigue life. The use of directionally solidified blades results in a substantial increase in the creep life, or substantial increase in tolerable stress for a fixed life. This advantage is due to the elimination of transverse grain boundaries from the blades, the traditional weak link in the microstructure. In addition to improved creep life, the directionally solidified blades possess more than 10 times the strain control or thermal fatigue compared to equiaxed blades. The impact strength of the directionally solidified blades is also superior to that of equiaxed, showing an advantage of more than 33%.

In the late 1990s, single-crystal blades were introduced in gas turbines. These blades offer additional creep and fatigue benefits through the elimination of grain boundaries. In single-crystal material, all grain boundaries are eliminated from the material structure, and a single crystal with controlled orientation is produced in an airfoil shape. By eliminating all grain boundaries and the associated grain boundary strengthening additives, a substantial increase in the melting point of the alloy can be achieved, thus providing a corresponding increase in high-temperature strength. The transverse creep and fatigue strength are increased, compared to equiaxed or DS structures. The advantage of single-crystal alloys compared to equiaxed and DS alloys in low-cycle fatigue life is increased by about 10%.

Development of directionally solidified blades as well as the new single-crystal blades, with the new thermal barrier coatings, and the new cooling schemes, is responsible, for the increase in firing temperatures. The high-pressure ratio in the compressor also causes the cooling air used in the first stages of the turbine to be very hot. The temperatures leaving the gas turbine compressor can reach as high as 1200°F (649°C), requiring the compressor cooling air to be cooled, and the cooling passages require to be coated.

27.3.6   Cooling Schemes

The cooling schemes are limited in the amount of air they can use, before there is a negating an effort in overall thermal effi-

ciency due to an increase in the amount of air used in cooling. The rule of thumb in this area is that if you need more than 8% of the air for cooling, you are loosing the advantage from the increase in the firing temperature.

In high-temperature gas turbines, cooling systems need to be designed for turbine blades, vanes, endwalls, shroud, and other components to meet metal temperature limits. Blade platforms also need to be cooled as well as transition pieces, which connect the combustor to the first stage turbine nozzles. The following five basic air-cooling schemes are the most commonly used, and in many cases, all five cooling schemes are used in cooling blades and turbine nozzles.

1. Convection cooling
2. Impingement cooling
3. Film cooling
4. Modified transpiration cooling
5. Water/Steam cooling

27.3.7   Coatings

There are three basic types of coatings: thermal barrier coatings, diffusion coatings, and plasma-sprayed coatings. The advancements in coating have also been essential in ensuring that the blade base metal is protected at these high temperatures. Coatings ensure that the life of the blades is extended and in many cases are used as sacrificial layer, which can be stripped and recoated. Life of coatings depends on composition, thickness, and the standard of evenness to which it has been deposited. The general type of coatings is very little different from the coatings used 10-15 years ago. These include various types of diffusion coatings such as aluminide coatings originally developed nearly 40 years ago. The thickness required is between 25 and 75 μm thick. The new aluminide coatings with platinum increase the oxidation resistance and also the corrosion resistance. The thermal barrier coatings have an insulation layer of 100-300 μm thick and are based on ZrO_2-Y_2O_3 and can reduce metal temperatures by 50°C-150°C. This type of coating is used in combustion cans, transition pieces, nozzle guide vanes, and also blade platforms.

The interesting point to note is that some of the major manufacturers are switching away from corrosion protection-based coatings toward coatings, which are not only oxidation resistant, but also oxidation resistant at higher metal temperatures. Thermal barrier coatings are being used on the first few stages in all the advanced technology units, as well as on turbine combustors, transition pieces and in some cases even on turbine compressor blades. The use of internal coatings is getting popular due to the high temperature of the compressor discharge, which results in oxidation of the internal surfaces. Most of these coatings are aluminide-type coatings. The choice is restricted due to access problems to slurry based, or gas phase/chemical vapor deposition. Care must be taken in production; otherwise, internal passages may be blocked. The use of pyrometer technology on some of the advanced turbines has located blades with internal passages blocked causing that blade to operate at temperatures of 35°C-70°C.

27.4   GAS TURBINE HRSG SYSTEMS

The waste heat recovery system is a critically important subsystem of a cogeneration system. In the past, it was viewed as a separate "add-on" item. This view is being changed with the realization that good performance, both thermodynamically and in terms of reliability, grows out of designing the heat recovery system as an integral part of the overall system.

The gas turbine exhaust gases enter the HRSG, where the energy is transferred to the water to produce steam. There are many different configurations of the HRSG units. Most HRSG units are divided into the same amount of sections as the steam turbine. In most cases, each section of the HRSG has a pre-heater, an economizer and feed-water, and then a superheater. The steam entering the steam turbine is superheated.

The most common type of an HRSG in a large combined cycle power is the drum-type HRSG with forced circulation. These types of HRSGs are vertical; the exhaust gas flow is vertical with horizontal tube bundles suspended in the steel structure. The steel structure of the HRSG supports the drums. In a forced circulation HRSG, the steam water mixture is circulated through evaporator tubes using a pump. These pumps increase the parasitic load and thus detract from the cycle efficiency. In this type of HRSG, the heat transfer tubes are horizontal, suspended from un-cooled tube supports located in the hot gas path. Some vertical HRSGs are designed with evaporators, which operate without the use of circulation pumps.

27.4.1 Once Through Steam Generators (OTSG) are finding quick acceptance due to the fact that they have smaller foot print and can be installed in a much shorter time and lower price. The OTSG unlike other HRSGs does not have a defined economizer, evaporator, or superheater sections. Figure 27.28 is the schematic of an OTSG system and a drum-type HRSG. The OTSG is basically one tube; water enters at one end and steam leaves at the other end, eliminating the drum and circulation pumps. The location of the water to steam interface is free to move, depending on the total heat input from the gas turbine, and flow rates and pressures of the feed-water, in the tube bank. Unlike other HRSGs, the once-through units have no steam drums.

Some important points and observations relating to gas turbine waste heat recovery are:

27.4.2 Multi-pressure Steam Generators: these are becoming increasingly popular. With a single pressure boiler there is a limit to the heat recovery because the exhaust gas temperature cannot be reduced below the steam saturation temperature. This problem is avoided by the use of multi-pressure levels.

27.4.3 Pinch Point This is defined as the difference between the exhaust gas temperature leaving the evaporator section and the saturation temperature of the steam. Ideally, the lower the pinch point, the More heat recovered, but this calls for more surface area and, consequently, increases the back-pressure and cost. Also, excessively low pinch points can mean inadequate steam production if the exhaust gas is low in energy (low mass flow or low exhaust gas temperature). General guidelines call for a pinch point of 15°F to 40°F (8°C to 22°C). The final choice is obviously based on economic considerations.

27.4.4 Approach Temperature- This is defined as the difference between the saturation temperatures of the steam and the inlet water. Lowering the approach temperature can result in increased steam production, but at increased cost. Conservatively high approach temperatures ensure that no steam generation takes place in the economizer. Typically, approach temperatures are in the 10°F to 20°F (5.5°C to 11°C) range. Figure 27.29 is the temperature energy diagram for a system and also indicates the approach and pinch points in the system.

Off-Design Performance-This is an important consideration for waste heat recovery boilers. Gas turbine performance is affected by load, ambient conditions and gas turbine health (fouling, etc.). This can affect the exhaust gas temperature and the air flow rate. Adequate considerations must be given to how steam flows (low pressure and high pressure) and superheat temperatures vary with changes in the gas turbine operation.

The HRSG consists of various pressure levels, which compare with the type of Steam Turbine used, the efficiency desired and the

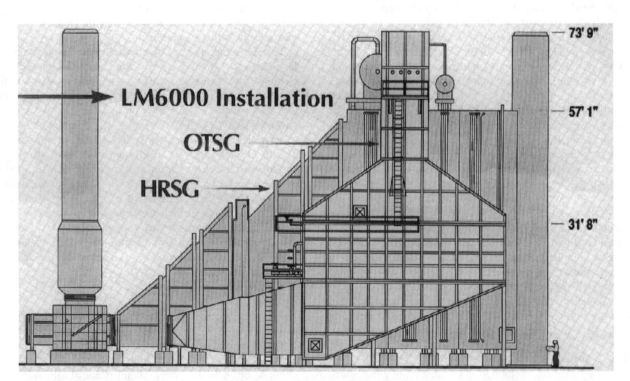

FIG. 27.28 COMPARISON OF A DRUM-TYPE HRSG TO A OTSG (Courtesy of Innovative Steam Technologies)

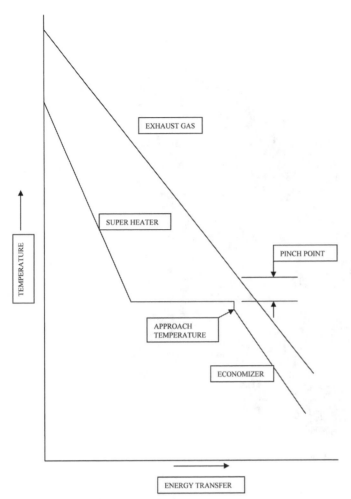

FIG. 27.29 ENERGY TRANSFER DIAGRAM IN AN HRSG OF A COMBINED CYCLE POWER PLANT

costs. The most common types of HRSGs used in Combined Cycle applications have two or three level of pressures, these pressure levels match the sections of the Steam Turbine. The major components of the HRSG as shown in Figure 27.30, consists of for each pressure level the following components:

- Preheater and deaerators
- Economizers
- Evaporators
  o Steam drums
- Superheaters and reheaters
  o Attemperators
  o Desuperheater

In addition to these components, the HRSG also houses the following components:

- Selective catalytic reduction system
- Ammonia injection grid system

27.4.5 Deaerator

A deaerator is a device that is widely used for the removal of air and other dissolved gases from the feed-water to steam-generating systems such as an HRSG. In particular, dissolved oxygen in

HRSG feed-waters will cause serious corrosion damage in steam systems by attaching to the walls of metal piping and other metallic equipment and forming oxides (rust). It also combines with any dissolved carbon dioxide to form carbonic acid that causes further corrosion. Most deaerators are designed to remove oxygen down to levels of 7 ppb by weight ($0.0005 cm^3/L$) or less. A deaerator is provided with each HRSG to remove the dissolved gases and add heat to the HRSG feed-water system.

27.4.6 Economizers

An HRSG economizer is a heat exchanger device that captures and transfers heat to the HRSG's feed-water. The feed-water or return water is pre-heated by the economizer from the exchange of heat between the exhaust gas from the gas turbine at that point, after most of its heat has already been transferred to the superheaters, reheaters, and evaporators. This results in the main heating circuit in the HRSG do not need to provide as much heat to produce a given output quantity of steam.

27.4.7 Evaporators

In the evaporator, water is converted to saturated Steam. The evaporator section is usually of a natural circulation design without the need for assisted circulation devices at any load. The entire evaporator section, including the steam drum, is designed to generate dry saturated steam. The evaporator shall be designed fully drainable and of all welded construction.

27.4.8 Attemperators

Attemperators can be classified into two types: direct contact and surface. The direct contact design uses a spray where the steam and the cooling medium (water and saturated steam) are mixed. In the surface design, the steam is isolated from the cooling medium by the heat exchanger surface. Spray attemperators are usually preferred on all units that have attemperation requirements; however, some surface type attemperators are occasionally used in some industrial units.

27.4.9 Desuperheaters

Desuperheaters (DSH) are used to deliver steam to auxiliaries, which may require the steam at lower temperatures and are often used with a pressure reducing station. These desuperheaters may be a drum type or an external spray type.

27.4.10 Selective Catalytic Reduction (SCR) System

Most of the HRSGs are furnished and installed with a complete selective Catalytic Reduction (SCR) system to control concentrations of NO_x generated by the gas turbine. SCR is a means of converting nitrogen oxides, also referred to as NO_x, with the aid of a catalyst to diatomic N2, and water, H2O. A gaseous reductant typically anhydrous or aqueous ammonia is added to the stream of exhaust gas and is absorbed onto a catalyst.

27.4.11 Design Considerations

27.4.11.1 Forced Circulation System Using forced circulation in a waste heat recovery system allows the use of smaller tube sizes with inherent increased heat transfer coefficients. Flow stability considerations must be addressed. The recirculating pump is a critical component from a reliability standpoint and standby (redundant) pumps must be considered. In any event, great care must go into preparing specifications for this pump.

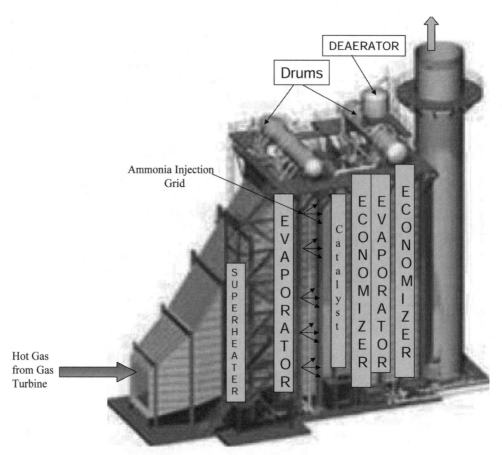

FIG. 27.30   A SCHEMATIC OF A TYPICAL HRSG

27.4.11.2 Back-pressure Considerations (Gas Side) These are important, as excessively high back-pressure s create performance drops in gas turbines. Very low-pressure drops would require a very large heat exchanger and more expense. Typical pressure drops are 8 to 10 inches of water.

27.4.11.3 Supplementary Firing of HRSG Systems There are several reasons for supplementary firing an HRSG unit. Probably the most common is to enable the system to track demand (i.e., produce more steam when the load swings upward, than the unfired unit can produce). This may enable the gas turbine to be sized to meet

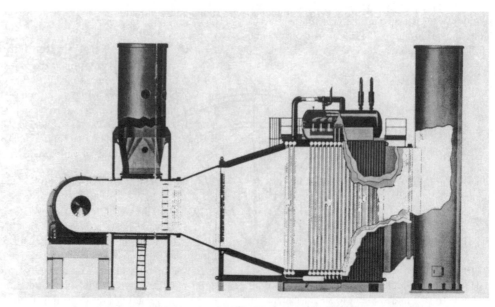

FIG. 27.31   SUPPLEMENTARY FIRED EXHAUST GAS STEAM GENERATOR

the base load demand with supplemental firing taking care of higher load swings. Figure 27.31 shows a schematic of a supplementary fired exhaust gas steam generator. Raising the inlet temperature at the waste heat boiler allows a significant reduction in the heat transfer area and, consequently, the cost. Typically, as the gas turbine exhaust has ample oxygen, duct burners can be conveniently used.

An advantage of supplemental firing is the increase in heat recovery capability (recovery ratio). A 50% increase in heat input to the system increases the output 94%, with the recovery ratio increasing by 59%. Some important design guidelines to ensure success include:

- Special alloys may be needed in the superheater and evaporator to withstand the elevated temperatures.
- The inlet duct must be of sufficient length to ensure complete combustion and avoid direct flame contact on the heat transfer surfaces.
- If natural circulation is utilized, an adequate number of risers and feeders must be provided as the heat flux at entry is increased.
- Insulation thickness on the duct section must be increased.

27.5   STEAM TURBINE

The usual steam turbine consists of four basic parts: the rotor, which carries the blades or buckets; the stator consisting of cylinder and casing, which are often combined, and within which the rotor turns; the nozzles or flow passages for the steam, which are generally fixed to the inside of the cylinder; and the frame or base for supporting the stator and rotor. In small steam turbines, the cylinder casing and frame are often combined. Several other systems such as the lubrication systems, steam piping systems, and a condensing system make up the rest of the turbine.

There are two major types of flow characteristics in steam turbines, the impulse turbine and the reaction turbine. The steam volume increases whenever the pressure decreases, but the resulting velocity changes depend on the type of turbine. These velocity changes are distinguishing characteristics of the different types of turbines.

The degree of reaction (R) in an axial-flow turbine is defined as the ratio of the change of enthalpy drop in the rotor to the change in total enthalpy drop across the stage:

$$R = \frac{H_{\text{rotor}}}{H_{\text{stage}}} \qquad (27.28)$$

By definition, the impulse turbine has a degree of reaction equal to zero. This degree of reaction means that the entire enthalpy drop is taken in the nozzle, and the exit velocity from the nozzle is very high. In practice, there must be a pressure drop across the rotating blades to generate flow. Since there is no change in enthalpy in the rotor, the relative velocity entering the rotor equals the relative velocity exiting from the rotor blade. Most steam turbine HP stages are typically impulse stages by design but average a reaction of 5% reaction at full load.

For a symmetric flow (50% reaction), the enthalpy drop in the rotor is equal to the drop in the stationary part of the turbine. This also leads to equal pressure drop across the stationary and rotating parts. Due to the difference in the turbine blade, diameters at the tip and the root of the blade the reaction percentages are different to counteract the centrifugal forces acting on the steam flow. If this was not done, too much flow would migrate to the blade tips. In

IP and LP turbines, the basic design is reaction; however; in these turbines, there is about 10% reaction at the root and at the tip going from about 60% reaction to about 70% reaction at the tips in the LP turbine. The impulse turbine produces about two times the power output as the 50% reaction type turbine; the reaction turbine on the other hand is more efficient. Thus, the combination of the high pressure stages being impulse and the later stages being about 50% reaction produces a high-power and efficiency turbine.

In most large plants, the axial-flow steam turbine consists of three sections: a high-pressure turbine stage (HP) with pressures between 1450 and 4500 psia (100-310 bars), with temperatures as high as 1212°F (656°C) an intermediate-pressure turbine stage (IP) with pressures between 300 and 600 psia (20.6-41.7 bars), and a low-pressure turbine stage (LP) with pressures between 90 and 60 psia (6.1-3.1 bars). The steam exiting from the HP stage is usually reheated to about the same temperature as the steam entering the HP stage before it enters the IP stage. The steam exiting the IP stage enters directly into the LP stage after it is mixed with steam coming from the LP superheater. The steam from the LP turbine enters the condenser. The condenser is maintained at a vacuum of between 0.13 and 0.033 bars. The increase in back-pressure in the condenser will reduce the power produced. Care must be taken to ensure that the steam leaving the LP stage blades has not a high content of liquid in the steam to avoid erosion of the LP blading reheat system.

Steam turbines used in modern combined cycle power plants are simple machines. The following are the important requirements for a modern combined cycle steam turbine:

1. Ability to operate over a wide range of steam flows
2. High efficiency over a large operating range
3. Reheat possibilities
4. Fast startup
5. Short installation time
6. Floor mounted installations

The plants operate over a wide range of steam flows, as the plants are now often cycled between base load and 50% of the base load in a 24-hour period. Thus, this requires a high efficiency over a wide operating range. Combined cycle power plants operate at many pressure levels. It is not uncommon that the same manufacturer's plant using the same gas turbine operates at two or three pressure levels. The combined cycle steam turbine has fewer or even no bleed points as compared to any where between four and eight bleed points for feed water heaters.

Rapid startup is very often important since many of these plants are started up on a daily basis. Great care in design must be exercised due to rapid increase in temperatures. This does not allow for rotor wheels, which use a shrink fit on to the shaft. There have been cases during rapid startup the rotor wheel "walking" on the shaft due to the different growth rates between the shaft and the rotor wheel.

Reheat of the steam is being used in many of today's steam turbine. Reheat improves overall combined cycle efficiency.

There are five basic types of shaft and casing arrangements of steam turbines: single casing, tandem-compound, cross-compound, double flow, and extraction steam turbines. Figure 27.32 shows schematics of various steam turbine arrangements.

27.5.1   Single-Flow Single-Casing Extraction Steam Turbines

In a single-flow turbine, the steam enters at one end, flows once through the blading in a direction approximately parallel to the axis, emerges at the other end, and enters the condenser. In turbines with a single casing, all sections are contained within one casing

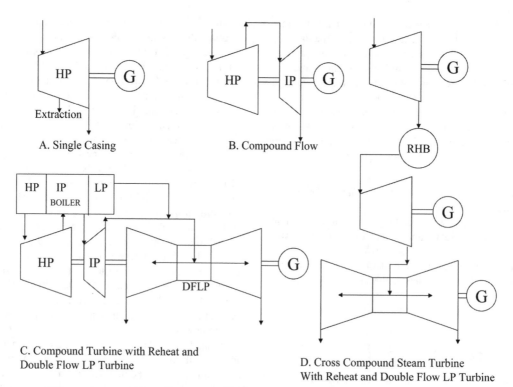

A. Single Casing

B. Compound Flow

C. Compound Turbine with Reheat and
Double Flow LP Turbine

D. Cross Compound Steam Turbine
With Reheat and Double Flow LP Turbine

FIG. 27.32   SCHEMATIC ARRANGEMENTS OF VARIOUS STEAM TURBINES

and the steam path flows from throttle to exhaust through that single casing. Figure 27.32A shows a simple path where steam enters a turbine and is exhausted to the atmosphere or a condenser; it also shows an extraction for cogeneration purposes. This is the most common arrangement in small and moderately large turbines.

The extraction flow stream turbine is the term applied to a turbine where part of the flow is extracted for various reasons such as steam for the plant or for absorption type chillers or for any other plant process. These turbines maybe back-pressure or condensing depending on the application and are used most commonly in a cogeneration application.

27.5.2   Compound-Flow or Tandem Compound Turbine

Compound-flow or tandem compound turbine is the term applied to a machine in which the steam passes in sequence through two or more separate units, expanding in each. The two units arranged in a tandem compound design have both casings on a single shaft and driving the same electrical unit. In most cases, the HP exhaust is returned for reheating before entering the IP turbine. Most often, the high-pressure and the intermediate-pressure portions are in one casing and the low-pressure portion in another. The IP exhausts into crossover piping and on to one or more low-pressure turbines. In Figure 27.32B, two sections are used, an HP and an LP section. Figure 27.32C indicates a condition where split steam is used in a double-flow, low-pressure section.

LP turbines are typically characterized by the number of parallel paths available to the steam. The steam path through the LP turbines is split into parallel flows because of steam conditions and practical limitations on blade length. Typically several LP flows in parallel are required to handle the large volume of flow rates. The steam enters at the center and divides the two portions passing axially away from each other through separate sets of blading on the

same rotor. This type of unit is completely balanced against end thrust and gives large area of flow through two sets of low-pressure blading.

27.5.3   Cross Compound Turbine

A cross compound design typically has two or more casings, coupled in series on two shafts, with each shaft connected to a generator. In cross compound arrangements, the rotors can rotate at different speeds but cannot operate independently as they are aerodynamically coupled. The cross compound design is inherently more expensive than the tandem—compound design, but has a better heat rate, so that the choice between the two is one of economics. Figure 27.32D shows a turbine setup where there are three casings each with their own generator and a reheat between the HP and the IP Section. Figure 27.33 is a schematic of a three section steam turbine in two casings. The HP and IP turbine section is in one casing, and the LP turbine is a double flow turbine in another casing. Figure 27.34 is a photograph of the lower half of a steam turbine casing housing the HP and IP turbines. Figure 27.35 is a close up of the regulating (first stage) nozzle and the regulating stage rotor followed by the other stages nozzles and rotors.

27.5.4   Steam Turbine Characteristics

Understanding the effect of the steam operating conditions on efficiency and load is very important in operating steam turbines at their optimum operating conditions. The two types of steam turbines are condensing steam turbines and the back-pressure steam turbines. Steam inlet temperature and pressure, and turbine exhaust pressure and vacuum, are the significant operating parameters of a steam turbine. The variations in these parameters affect steam consumption and efficiency. In a 100-MW steam turbine at a pressure of about 600 psia (41.4 bars) and 660°F (350°C), a 1% reduction in steam consumption can cost $500,000/year and about $900,000/

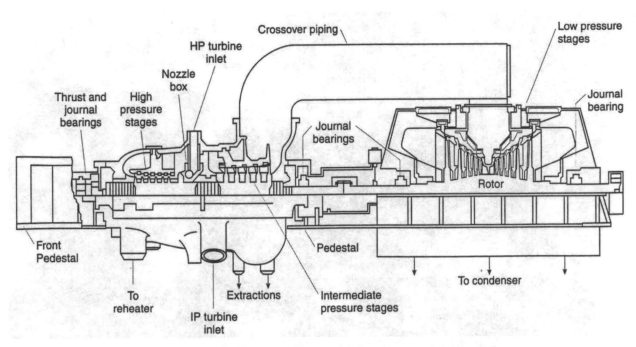

FIG. 27.33   SCHEMATIC OF A TYPICAL COMPOUND TURBINE

FIG. 27.34   LOWER HALF OF AN HP-IP TURBINE CASING WITH THE ROTOR IN THE CASING

year. This is based on a boiler efficiency of about 85% and LHV of fuel at 18,900 BTU/lb (10,500 kcal/kg).

Turbine steam inlet pressure is a major parameter, which affects turbine performance. To obtain the design efficiency, the steam inlet pressure should be maintained. Lowering steam inlet pressure reduces turbine efficiency and increases steam consumption. A 10% increase in steam pressure will reduce the steam consumption by about 1% in a condensing steam turbine and will reduce the steam consumption by about 4% in a back-pressure steam turbine. The effect on efficiency for 10% increase in pressure for a condensing steam turbine is about 1.5% and 0.45% for a back-pressure steam turbine.

Turbine steam inlet temperature is another major parameter that affects turbine performance. Reducing steam inlet temperature reduces the enthalpy, which is a function of both the inlet temperature and pressure. At higher steam inlet temperatures, the heat extraction by the turbine will also be increased. An increase of about 100°F (55°C) will reduce the steam consumption by about 6.6% in a condensing steam turbine, and 8.8% in a back-pressure turbine. The effect on efficiency for a 100°F (55°C) will be an increase of 0.6% in efficiency for a condensing steam turbine, and 0.65% in efficiency for a back-pressure turbine. It should be noted that the overall efficiency in most cases for a condensing steam turbine (30%-35%) is about twice that of a back-pressure turbine (18%-20%).

In condensing or exhaust back-pressure steam turbines, the increase of this back-pressure will reduce the efficiency and increase the steam consumption, keeping all other operating parameters. In condensing steam turbines, the condenser vacuum temperature will also increase if the removal of heat from the condenser is reduced. Thus, in a water-cooled condenser, if the temperature of the inlet water is increased, the power produced by the turbine is decreased because the back-pressure will be increased.

In summary, the condensing steam turbines are more efficient and produce more power than a back-pressure steam turbines. The

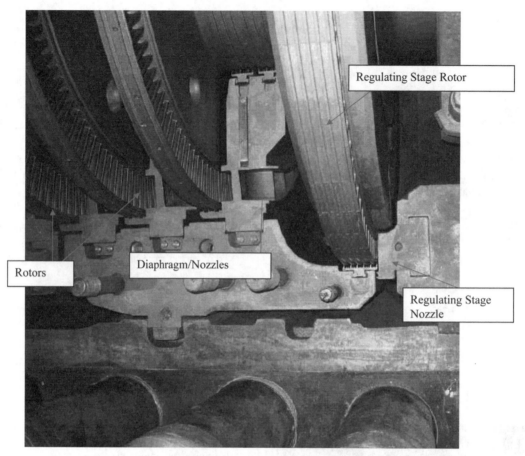

FIG. 27.35   HP STAGE NOZZLES/DIAPHRAGMS AND ROTORS

condensing steam turbine is also more efficient. The cost of a condensing steam turbine is about $25/kW more than a back-pressure turbine.

27.6   COMBINED CYCLE PLANTS

Combined cycle plants have several advantages. These include: (1) high thermal efficiencies (50%-65%), (2) rapid startup (2-hour cold start), and (3) low first-installed costs ($600-$900 per kilowatt. Maintenance costs for combined cycles range from US $0.003 to US $0.007 per kilowatt hour (similar to cogeneration plants).

As several gas turbine cogeneration plants utilize steam turbines (both back-pressure and condensing), cogeneration and combined cycles are similar in several aspects—design and operation lessons learned in one can be applied to the other.

The new combined cycle power plants placed into operation in the late 1990s have reached as high as 2800 MW. In this specific plant, each module consisted of the gas turbine, HRSG, and steam turbine on a single shaft producing 350 MW through a single generator; there were a total of 8 such modules in the plant. Other combined cycle configurations commonly used are two (commonly known as 2:1 configuration) or three (commonly known as 3:1 configuration) gas turbines each with their own generators and HRSG units, producing steam for one steam turbine with its own generator.

Another combined cycle gas turbine is a reheat gas turbine operating at a 30:1 pressure ratio with two combustors; each combustors

exit temperature is 2400°F (1315°C). The concept of reheat applied here involves the use of two combustors. The second combustor is used to reheat the air between the IP and LP gas turbines. The exit of the gas from the LP turbine is sent to an HRSG and the steam produced to a steam turbine. In many of these plants, there are two reheat gas turbines, each with their own HRSGs and one steam turbine.

Combined cycle power plants and cogeneration projects are capital intensive, with installed first costs ranging from $800 to $1000 per kilowatt. The choice of plants depends on a host of factors including:

- Location of the plant
- Annual utilization
- Power and heat demand ratios
- Utility electric rates
- Net fuel rate
- Type of fuel
- Fuel cost
- Rate of return criteria
- Plant first cost ($/kW)
- Operations and maintenance costs

Table 27.1 is an analysis of the competitive standing of the various types of power plants, their capital cost, heat rate, operation and maintenance costs, availability and reliability, and time for planning. In examining the capital cost and installation time of these new power plants, it is obvious that the gas turbine is the best choice for peaking power. Steam turbine plants are about 50% higher in initial costs than combined cycle plants, which are about

TABLE 27.1 ECONOMIC AND OPERATION CHARACTERISTICS OF PLANT

Type of Plant	Capital Cost ($/kW)	Heat Rate (Btu/kWh;kJ/kWh)	Net Efficiency	Variable Operation and Maintenance ($/MWh)	Fixed Operation and Maintenance ($/MWh)	Availability	Reliability	Time from Planning to Completion (mo)
Simple Cycle Gas Turbine (2500°F/1371°C) Natural Gas Fired	300–350	7582; 8000	45	5.8	.23	88%–95%	97%–99%	10–12
Simple Cycle Gas Turbine Oil Fired	400–500	8322; 8229	41	6.2	.25	85%–90%	95%–97%	12–16
Simple Cycle Gas Turbine Crude Fired	500–600	10,662; 11,250	32	13.5	.25	75%–80%	90%–95%	12–16
Regenerative Gas Turbine Natural Gas Fired	375–575	6824; 7200	50	6.0	.25	86%–93%	96%–98%	12–16
Combined Cycle Gas Turbine	600–900	6203; 6545	55	4.0	.35	86%–.93%	95%–98%	22–24
Advanced Gas Turbine Combined Cycle Power Plant	800–1000	5249; 5538	65	4.5	.4	84%–90%	94%–96%	28–30
Combined Cycle Coal Gasification	1200–1400	6950; 7332	49	7.0	1.45	75%–85%	90%–95%	30–36
Combined Cycle Fluidized Bed	1200–1400	7300; 7701	47	7.0	1.45	75%–85%	90%–95%	30–36
Nuclear Power	1800–200	10,000; 10,550	34	8	2.28	80%–89	92%–98%	48–60
Steam Plant Coal Fired	800–1000	9749; 10,285	35	3	1.43	82%–89%	94%–97%	36–42
Diesel Generator-Diesel Fired	400–500	7582; 8000	45	6.2	4.7	90%–95%	96%–98%	12–16
Diesel Generator-Power Plant Oil Fired	600–700	8124; 8570	42	7.2	4.7	85%–90%	92%–95%	16–18
Gas Engine Generator Power Plant	650–750	7300; 7701	47	5.2	4.7	92%–96%	96%–98%	12–16

$600-$900/kW. Nuclear power plants are the most expensive. The high initial costs and the long time in construction make such a plant unrealistic for a deregulated utility.

In the area of performance, the steam turbine power plants have an efficiency of about 35%, as compared to combined cycle power plants, which have an efficiency of about 55%. Newer gas turbine technology will make combined cycle efficiencies range between 60% and 65%. As a rule of thumb, a 1% increase in efficiency could mean that 3.3% more capital can be invested.

The time taken to install a steam plant from conception to production is about 36-42 months as compared to 22-24 months for combined cycle power plants. The time taken for construction affects the economics of a unit, the longer the capital employed without return, accumulates interest, insurance and taxes.

Natural gas is the best choice for a combined cycle power plant, both from initial cost as well as maintenance costs. Diesel fuel would be the second best option and would run about 15%-20% higher in the total life cycle cost of the plant.

The initial cost of a combined cycle plant, as shown in Figure 27.36, is made up of various components such as the Gas Turbine (30%), Steam Turbine (10%), the HRSG (10%), Mechanical Systems (15%), Electrical Systems (12%), Controls (3%) and the Civil Structures and Infrastructures (20%). Figure 27.37 shows the cost distribution over the life cycle of combined cycle power plant. It is interesting to note that the initial cost runs about 8% of the total life cycle cost, and the operational and maintenance cost is about 17%, and the fuel cost is abut 75%.

Worldwide experiences in Combined Cycles Power Plants indicate high availability, reliability, and thermal efficiency in most of these plants.

27.7 AVAILABILITY AND RELIABILITY

The Availability of a power plant is the percent of time the plant is available to generate power in any given period at its acceptance load. The acceptance load or the net established capacity would be the net electric power generating capacity of the power Plant at design or reference conditions established as result of the performance tests conducted for acceptance of the plant. The actual power produced by the plant would be corrected to the design or reference conditions and is the actual net available capacity of the power plant. Thus, it is necessary to calculate the effective forced outage hours, which are based on the maximum load the plant can

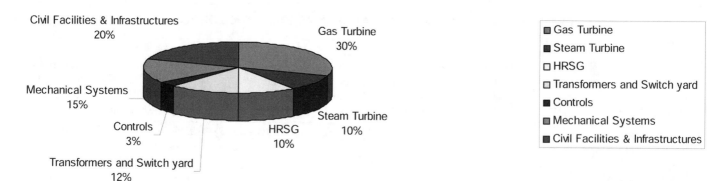

FIG. 27.36 COST COMPOSITION OF DIFFERENT PLANT AREAS IN A COMBINED CYCLE POWER PLANT

produce in a given time interval when the plant is unable to produce the power required of it. The effective forced outage hours is based on the following relationship:

$$EFH = HO \times \frac{(MW_d - MW_a)}{MW_d} \qquad (27.29)$$

where:

- MW_d = desired output corrected to the design or reference conditions. This must be equal to or less than the plant load measured and corrected to the design or reference conditions at the acceptance test.
- MW_a = actual maximum acceptance test produced and corrected to the design or reference conditions.
- HO = hours of operation at reduced load.

The availability of a plant can now be calculated by the following relationship, which takes into account the stoppage due to both forced and planned outages, as well as the forced effective outage hours:

$$A = \frac{(PT - PM - FO - EFH)}{PT} \qquad (27.30)$$

where:

- PT = time period (8760 hrs/yr)
- PM = planned maintenance hours
- FO = forced outage hours
- EFH = equivalent forced outage hours

Reliability of a plant depends on many parameters, such as the type of fuel, the preventive maintenance programs, the operating mode, the control systems, and the firing temperatures. The reliability of the plant is the percentage of time between planned overhauls and is defined as:

$$R = \frac{(PT - FO - EFH)}{PT} \qquad (27.31)$$

Availability and reliability have a major impact on the plant economy. Reliability is essential in that when the power is needed it must be there. When the power is not available, it must be generated or purchased and can be very costly in the operation of a plant. Planned outages are scheduled for non-peak periods. Peak periods are when the majority of the income is generated; as usual, there are various tiers of pricing depending on the demand. Many power purchase agreements have clauses, which contain capacity payments, thus making plant availability critical in the economics of the plant. A 1% reduction in plant availability could cost $500,000 in income on a 100-MW plant.

Starting reliability is another very important factor in a plant. This reliability is a clear understanding of the successful starts that have taken place and is given by the following relationship:

$$SR = \frac{\text{number of starting successes}}{(\text{number of starting successes} + \text{number of starting failures})} \qquad (27.32)$$

The insurance industry concerns itself with the risks of equipment failure. For advanced gas turbine combined cycle power plants, the

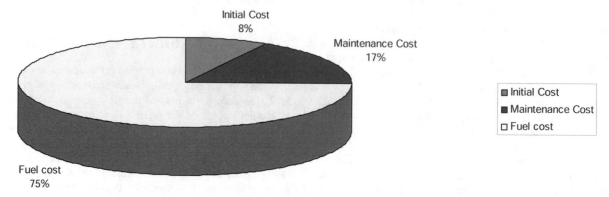

FIG. 27.37 COMBINED CYCLE POWER PLANT LIFE CYCLE COST

frequencies of failures and the severity of failures are major concerns. In engineering terms, however, risk is better defined as:

$$\text{Risk} = \text{probability of failure} \times \text{consequences of failure} \quad (27.33)$$

where the consequences of failure include the repair/replacement costs and the lost revenue from the downtime to correct the failure.

Actions taken, which reduce the probability and/or consequences of failure, tend to reduce risk and generally enhance insurability. Because of the high risks associated with insuring advanced gas turbines, demonstrating successful operation is important to the underwriting process.

Operating and maintenance costs are usually broken up into two categories: variable costs, and fixed costs. Variable operating costs include consumables and spare parts, while fixed operating costs include staff, insurance, taxes, and interest. The variable cost of a combined cycle is lower than the gas turbine, which is due to the fact that these costs are driven by the gas turbine spares, which can be distributed over a larger output of a combined cycle.

As these combined cycle power plants move toward coal-based fuels and the move toward higher turbine inlet temperatures and pressures continue, combined cycles will become more important. There is also a significant interest in converting existing gas turbines in simple cycle operation to combined cycle operation, and existing steam plants to modified combined cycle power plants. Results indicate that for a base-loaded plant running 5000 full-load equivalent hours per year, the production costs are between 20% and 25% less for a combined cycle plant at a fuel cost of $30/bbl. At fuel costs of $120/bbl, these costs are about 60%-70% less. A commonly used rule of thumb is that 50% to 60% of the gas turbine output can be added by using the Rankine Cycle (steam turbine) as the bottoming without any further fuel usage.

27.8   REFERENCES[1]

1. Boyce, M. P., 2006, *Gas Turbine Engineering Handbook*, 3rd ed., Elsevier, New York.

2. Boyce, M. P., 2010, *Cogeneration & Combined Cycle Power Plants*, 2nd ed, ASME Press, New York.

---

[1] Sources of figures and tables cited in this chapter are from the following publications of the author.

HYDRO TASMANIA — KING ISLAND CASE STUDY

Simon Gamble, Marian Piekutowski and Ryan Willems

ABSTRACT

Hydro Tasmania has developed a remote island power system in the Bass Strait, Australia, that achieves a high level of renewable energy penetration through the integration of wind and solar generation with new and innovative storage and enabling technologies. The ongoing development of the power system is focused on reducing or replacing the use of diesel fuel while maintaining power quality and system security in a low inertia system. In recent years Hydro Tasmania has undertaken several renewable energy developments on King Island with the aim to reduce dependence on diesel, reduce operating cost and greenhouse gas emissions, and demonstrate the potential for renewable energy penetration in power systems. This has been achieved through the substitution of diesel based generation with renewables such as wind and solar and the integration of enabling technologies such as storage and a dynamic frequency control resistor. The projects completed to date include:

- Wind farm developments completed in 1997 and expanded to 2.25 MW in 2003;
- Installation of a 200 kW, 800 kWh Vanadium Redox Battery (2003);
- Installation of a two-axis tracking 100 kW solar photovoltaic array (2008), and
- Development of a 1.5 MW dynamic frequency control resistor bank, that operates during excessive wind generation (2010).

The results achieved to date include 85% instantaneous renewable energy penetration and an annual contribution of over 35%, forecast to increase to 45% post commissioning of the resistor. Hydro Tasmania has designed a further innovative program of renewable energy and enabling technology projects. The proposed King Island Renewable Energy Integration Project, which recently received funding support from the Australian Federal Government, is currently under assessment to be rolled out by Hydro Tasmania (including elements with our partners CBD Energy) by 2012. These include:

- Installation of short term energy storage (flywheels) to improve system security during periods of high wind;
- Reinstatement or replacement of the Vanadium Redox Battery (VRB), that is currently out of service due to an operational event that led to damage to the system's cell stacks. It is envisioned that a proportion of the wind spill currently consumed

in the dynamic resistor could be recovered with the VRB and used to cover periods of low wind generation.
- Wind expansion — includes increasing the existing farm capacity by up to 4 MW;
- Graphite energy storage — installation of graphite block thermal storage units for storing and recovering spilt wind energy.
- Biodiesel project — conversion of fuel systems and generation units to operate on B100 (100% biodiesel); and
- Smart Grid Development—Demand Side Management — establishing the ability to control demand side response through the use of smart metering throughout the Island community.

This program of activities aims to achieve a greater than 65% long term contribution from renewable energy sources (excluding biodiesel contribution), with 100% instantaneous renewable energy penetration. The use of biodiesel will see a 95% reduction in greenhouse gas emissions. The projects will address the following issues of relevance to small and large scale power systems aiming to achieve high levels of renewable energy penetration:

- Management of low inertia and low fault level operation;
- Effectiveness of short term storage in managing system security;
- Testing alternative system frequency control strategies; and
- Impact of demand side management on stabilizing wind energy variability.

28.1 INTRODUCTION

28.1.1 Hydro Tasmania

Hydro Tasmania is Australia's leading renewable energy business contributing 50 per cent of Australia's electricity produced from renewable energy sources. Hydro Tasmania is a Government Business Enterprise, owned by the State of Tasmania. Hydro Tasmania owns and operates a portfolio of renewable generation assets, consisting of ~2,250 MW of hydropower and ~280 MW of wind generation. Hydro Tasmania retails electricity via Momentum, our wholly owned retail business. We offer consultancy services internationally to clients in the water and energy sectors via our engineering consultancy business, Entura.

Hydro Tasmania is responsible for the generation, distribution and retail of electricity on the Bass Strait Islands (BSI), specifically King and Flinders Islands. The cost to supply electricity on these islands is much greater than the revenue derived from selling the electricity to customers, who pay a tariff slightly higher than that available in mainland Tasmania but still well below the marginal cost of supply. The difference in cost and revenue is reimbursed to Hydro Tasmania by the Tasmanian State Government; this payment is known as the Community Service Obligation (CSO) [1].

To bridge the gap between the cost of production of energy and the revenue collected from customers Hydro Tasmania has implemented over the last ten years a number of improvements to the power system, focused to this date on King Island. The developments have been constructed in stages; an approach that has allowed a progressive reduction in diesel fuel usage (and cost associated with running hours and maintenance), building on lessons learned and utilizing available technologies at the time. By reducing the cost of electricity supply on the island, the level of CSO contribution from the Tasmanian Government is reduced. Each project has demonstrated a return on investment, where the cost to implement the changes has been less than the total reduction to CSO they generate.

With current rising trends and uncertainty in diesel fuel prices coupled with concerns of fuel availability and uncertain annual growth of the customer load on the islands, the cost of generating electricity on King Island is very likely to grow. To alleviate this increase in operational cost and mitigate the ongoing exposure to diesel price volatility, Hydro Tasmania is increasing the installed capacity of renewable generation on the island and improving control measures to maximize use of renewable energy.

This chapter will outline Hydro Tasmania's achievements to date and paint a picture of the way forward towards a sustainable energy future for King Island.

28.1.2 King Island

King Island is one of the larger islands located in Bass Strait, a body of water between the north coast of Tasmania and the south coast of Victoria on mainland Australia (Figure 28.1). The King

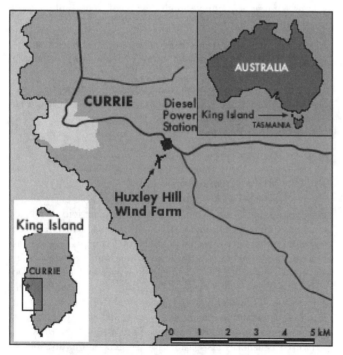

FIG. 28.1 KING ISLAND: LOCATED IN BASS STRAIT
(*Source:* Hydro Tasmania)

Island community consists of a population of around 1600 people, and is noted for its beef and dairy industries, as well as kelp (*seaweed*) farming and tourism. Being a remote island community, King Island is not connected to either mainland Tasmania or mainland Australia for its electricity supply. Until recent times, electricity generated on the island was derived entirely from diesel fuel; however, wind power now contributes to a significant portion of the island's annual energy demand. Wind power is recognized as being a cost effective supplement to diesel fuelled generation and is an important electricity source on King Island due to the excellent wind resource present in Bass Strait.

It is noted that King Island development includes substitution of diesel fuel by bio-fuels however this substitution while achieving significant reduction in GHG emission needs to be considered with all its limitations. There is limited supply of feed stock to produce the biofuel and the price of the feed stock will tend to shadow oil prices [2]. Consequently in further discussion the distinction is made between renewable energy options (wind, solar, etc.) and the supplementary role of fuel substitution.

28.1.3 Bass Strait Island Energy Vision

Hydro Tasmania's long term vision for the Bass Strait Islands is to develop innovative renewable energy power systems with the following objectives:

- Reduce cost of subsidy paid by the Tasmania Government;
- Achieve total sustainability in electricity supply — eliminate fuel price exposure (including minimal use of bio-fuels);
- Develop a world leading Tasmanian Renewable Energy demonstration project; and
- Implement an approach that can be utilized as an island power system solution around the world.

Achievement of the vision would see indigenous renewable resources, such as wind, wave and solar utilized in conjunction with sustainably harvested biomass resources to fuel power production. Enabling technologies such as storage and dynamic resistors would be utilized to maximize the contribution of renewable resources, with integration coordinated via an advanced system controller. Active management of demand side is achieved through the deployment of smart grid technology, with customers directly participating in interrupting and deferring load. The vision has been expressed diagrammatically in Figure 28.2.

Achieving this vision will result in significant improvements in emissions intensity on the islands, power sector and the local sustainable harvesting and production of energy will greatly improve environmental outcomes. Positive social impact will be achieved through job creation and skills development, and via enhancing tourism and the brand value of the region.

Hydro Tasmania's current BSI renewable energy development program will see significant progress towards this long term vision. Attainment of the vision will require broad and ongoing engagement and participation of the stakeholders, including the State Government, island communities and industry participants.

28.2 THE KING ISLAND RENEWABLE ENERGY INTEGRATION PROJECT

A key project for Hydro Tasmania under the BSI program is the King Island Renewable Energy Integration Project (KIREIP). This is a portfolio of innovative projects utilizing new and existing technologies to increase the use of renewable energy in an elec-

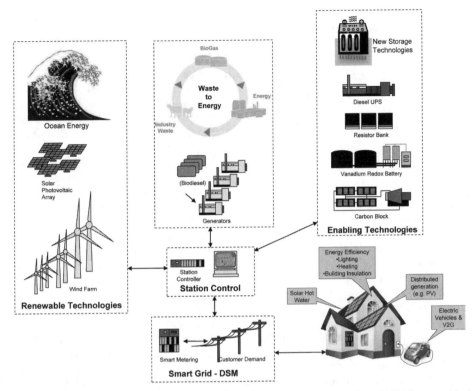

FIG. 28.2 LONG-TERM VISION FOR BSI — SUSTAINABLE ELECTCTRICITY SUPPLY (*Source:* Hydro Tasmania)

tricity network, reducing emissions and improving the quality of power supply on King Island. The project will build on previous accomplishments to demonstrate the potential of renewable energy working in conjunction with enabling technologies and establish a showcase for sustainable energy development.

KIREIP is partly funded by the Australian Government via a Renewable Energy Demonstration (REDP) grant [3]. Hydro Tasmania will receive up to \$15.28 m in grant funds to assist in the demonstration of a number of innovative technologies implemented in a power system for the first time.

Initial experience in small scale remote systems with integration of intermittent (variable) renewable energy sources to the power system, has revealed a number of issues which might be perceived as potential obstacles in large scale development of renewable generation in wider markets. The perception of renewable energy technologies being unsuitable to supply base load generation or to meet high priority loads has also pervaded the traditional players in the power industry [4]. One of the aims of KIREIP is to assist in dispelling these myths, and to demonstrate the portfolio of technologies that will be required to support a much more significant contribution from renewable energy in our future power systems.

The King Island system has shown some promising results with increased utilization of renewable energy beyond the norm for a system of this scale, successfully operating with wind generation supplying up to 80% (typically limited by the minimum output of a single diesel generator) of instantaneous customer demand and over a third of the annual energy demand. The proposed project aims to further improve the integration of wind generation with the system, by use of selected combinations of new technologies and operational strategies. The specific projects will address issues of direct relevance to both small and large scale power systems, aiming to achieve moderate to high levels of renewable energy penetration on an annual basis.

It is envisioned that the lessons learned in the development and operation of this program of projects will be applicable to other off-grid systems throughout the world and in the larger Australian electricity system to help achieve Australia's renewable energy target of 20% renewable generation by 2020. The advantages of using a system on the scale of the BSI to develop this understanding include:

- Generally a smaller system will experience problems with increasing variable generation much earlier than a large system;
- Wind variations in smaller systems (with a much higher installed capacity ratio and no locational diversity) are much greater and faster than in a large system;
- Cost effectiveness of pilot projects: some of the enabling technologies are not considered commercially viable, however, testing them in small scale system with much higher fuel costs makes pilot projects more attractive — so this can be achieved in a cost effective manner; and
- Establishment of test conditions with very high penetration of wind generation is not possible in a large interconnected system — having control of an entire power system is required.

The KIREIP represents a unique opportunity to develop and deploy renewable and enabling technology in a real world system at a scale significantly larger than lab scale test facilities.

28.3 OFF-GRID POWER SYSTEM DEVELOPMENT

The term Remote Area Power Supply (RAPS) generally refers to power systems servicing load in remote locations, or locations where a connection to a main grid is either not available or undesirable. There are many thousands of RAPS systems in use worldwide, accounting for thousands of MW of installed capacity, and cover a range of sizes and compositions, but share a few common characteristics:

- Consist of one or more reciprocating thermal generators, typically diesel fuelled;
- Usually supplied from a single power stations (no locational diversity of sources),
- Usually accepting lower supply standards than in large systems, and
- Are not normally associated with extensive distribution networks.

RAPS systems can range in size from several kW (for example a stand-alone power supply for a farm) to tens of MW, as would be typical for a RAPS system servicing load in Pacific Island [5]. They may be automated, requiring little in the way of servicing or operator attendance or may be completely manual in nature, requiring the intervention of onsite personnel to start and synchronize generators as the load varies during the day.

Traditionally in remote systems the energy is supplied by reciprocating generators burning fossil fuels such as diesel. This type of generation is cheap to establish, reliable, flexible to operate and has robust dynamic characteristics. However, the cost of purchasing and transporting diesel fuel to remote areas results in very high short term marginal costs. These costs are further affected by small volume transactions. High fuel costs result in higher electricity costs than those experienced in larger interconnected systems and may be borne by the end user, resulting in stunted economic development, or may be subsidized in part or full by a relevant government. In either case, there generally exists a motivation to reduce the marginal cost of electricity supply in these systems.

In recent years, this has been accomplished by integrating renewable energy generation with the aim of reducing the amount of fossil fuel required to service the customer demand. Depending on the abundance of resource, wind power and more recently solar photovoltaic (PV) technologies can provide electricity at less than the marginal cost of fossil fuel derived power. While the integration of intermittent renewable generation presents some operational challenges and generally increases overall system complexity, real savings in the cost of electricity supply can be realized, resulting in an overall improvement in local economic development while maintaining comparable level of power supply security.

28.3.1 Phases of Development

Typically the development of a remote power supply, such as the type on King Island will be developed in various stages. If one considers the stages of development targeting the achievement of certain renewable penetration targets (or diesel fuel reduction targets) then the levels may look like those outlined below in Table 28.1.

Each level will have differing incremental benefits depending on factors such as load, resource and complexity. The levels can be combined together into any given project development step. Figure 28.3 is a representation of the relationship between the level of renewable energy penetration achieved and the resulting technical complexity (and commercial immaturity) of the system. As renewable penetration is increased, the system becomes more complex, may require more exotic technologies and generally requires more effort to successfully integrate all components.

There are two key trade-offs present in the decision making process. One is a risk based assessment of the level of renewable penetration desired versus the risk associated with both the technology deployed (many are in early stages of commercialization) and the skills of the system operator and maintenance staff. The second relates to the degree of system security required versus the cost burden of achieving that level of performance, either in additional thermal plant operation or in enabling technology, such as power storage. Expectations on system security need to balance the objective of operational cost reduction through renewable technology deployment, and the provision of levels of power quality akin to main interconnected networks.

Each of these "levels" of RAPS development will now be discussed in turn in greater detail.

28.3.2 Level 1 RAPS Development

Historically the initial phase (Level 1) has been to deploy fossil fuel based energy systems. Typically this is in the form of diesel engines but could also be other form of gas generation operated on Liquefied Petroleum Gas (LPG). This level of development generally includes all the auxiliary equipment and systems required to operate the power station and associated distribution system, and will include protection devices and some level of automation. Level 1 systems are generally built with an emphasis on reliability and may not include provision for future integration of renewable technologies such as control system scheduling different generation technologies, separate from diesel generator reactive power sources.

Level 1 in the Bass Strait Island context refers to the original construction and commissioning of King and Flinders Island Power Stations. It is unlikely that integration of future renewable generation was a major influencing factor in the design of these original power systems. In today's more rapidly evolving power systems

TABLE 28.1 SUMMARY OF DEVELOPMENT LEVELS (*Source:* Hydro Tasmania)

Level	Description	% RE (Typical)
1	Initial fossil fuel based power supply with low capital costs and high fuel and O&M prices.	0%
2	Incorporate renewable energy at a proportion that needs minimal consideration for additional support equipment	30%
3	Increase the level of renewables penetration over that of Level 2 through the use of renewable energy enablers and other system support technologies. Integration of RE limited by the minimum output of diesel generation,	45%
4	Build upon the enabling technologies used in Level 2, to achieve the aim of 100% renewable penetration at times when conditions permit. That is, allowing the diesel engines to be temporarily shut down when RE output exceeds the demand. May also install further RE.	>60%
5	Substitution of remaining mineral diesel with biodiesel. Biodiesel still has a GHG footprint, so minimizing the total fuel use will minimize the GHG emissions.	>60%

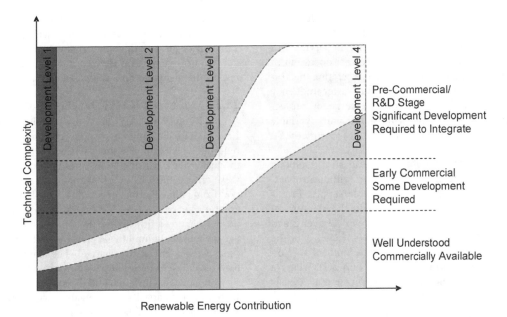

FIG. 28.3   LEVELS OF RAPS DEVELOPMENT (*Source:* Hydro Tasmania)

the selection of plant to allow greater flexibility under conditions of high renewable energy contribution is more critical.

28.3.3   Level 2 RAPS Development

The next phase (Level 2) would typically be to incorporate renewable energy at a proportion that needs minimal consideration for additional support equipment. The amount of renewable generation that can be integrated into the system will depend on the technology used; simple technology with low levels of control will generally be harder to integrate without support equipment, resulting in lower penetration levels without impacting on system reliability. The maximum amount of integrated renewable energy also reflects acceptable supply risks and associated costs of providing reserves. Generation with a higher level of control is more flexible in operation and thus easier to integrate, resulting in higher penetration potential. An example would be the comparison between a simple stall regulated wind turbine generator (WTG) and a more expensive pitch regulated machine. The power output from a stall regulated machine is not controlled, and as such tends to impact on the reliability of a power system at lower penetrations than would a controlled output machine.

In Level 2 systems, the key considerations are:

- Selection of appropriate renewable technologies;
  - o Availability of resource;
  - o Environmental factors such as climate;
  - o Ease of maintenance/accessibility;
    Operational impacts (capability diagram, availability of fault ride through, need to supply inertia, controllability, impact on quality of power supply);
  - o Cost of technology; and
  - o Renewable penetration requirements/expectations.
- Appropriate sizing of the capacity of renewable generators — higher instantaneous penetration possible with greater control capabilities. This affects the size of the largest contingency and required reserves; and
- Compatibility with existing plant.

Technology complexity can also influence selection, especially for sites where access and availability of maintenance personnel can be problematic. Power supplies in exceptionally remote areas typically utilize renewable technologies with little or no scheduled maintenance required. A prime example is the use of fixed tilt solar PV/battery systems in remote telecommunication repeater stations.

Finally the cost of a specific technology is a major influence in its selection. Whilst wind turbines such as simple stall regulated induction generators will usually cost less than an equivalent pitch regulated Doubly Fed Induction Generator (DFIG) turbine, the compromise is that generally a lower level of penetration can be achieved without the use of additional support technology. There is danger that the development using initially lower specification equipment may critically limit the potential amount of renewable energy integration. A greater level of control typically results in easier integration and higher levels of potential renewable penetration whilst maintaining an appropriate level of system reliability.

Typically, sizing of renewable generators will be dependent on several factors including the size and number of existing diesel generators, operational philosophy and required level of system security, expected load growth and the specific technologies selected.

As discussed above the level of technology used combined with the existing generation and load profile will have the greatest impact on the appropriate generator sizing. The primary consideration for selection of an appropriate generator size is the impact on the scheduling of thermal generation. The main constraint is generally the minimum level of thermal generator contribution required to maintain system security (provision of reserves, regulation, and adequate fault level current contribution), whilst maintaining the generators output within normal operating ranges. This is particularly relevant in retrofit systems where the original power system was not designed to incorporate renewable generation technologies. In these cases the existing generators may not be sized appropriately to make the most efficient use of the new renewable capacity. Ideally at this stage renewable generation may allow to reduce scheduled generation to a single diesel operating at or close to its minimum output. The system needs to be designed in such a manner that the generator in service will have sufficiently high fault current to allow correct operation of protective relays. With wind displacing diesel generation the system inertia becomes lower and this will result in faster frequency changes allowing less time for operation of emergency protections (under frequency). Operation with significantly variable system inertia and fault current makes it difficult to correctly set protective relays.

Level 2 projects will have renewable generators sized such that a meaningful contribution to the load is made without impacting on the stations ability to maintain a reliable supply. Provision of auxiliary equipment to support the thermal generators in maintaining system security is beyond the scope of Level 2 and falls under the scope of Level 3 developments. Level 2 in the King Island context is equivalent the initial development at Huxley Hill of 3 × N29 Wind Turbine Generators (WTG), where a nominal 15% utilization of renewable energy on an annual basis was achieved.

28.3.4   Level 3 RAPS Development

Level 3 projects increase the level of renewables penetration over that of Level 2 through the use of renewable energy enabling, and other system support technologies. Typically, Level 3 projects will seek to minimize the use of thermal generation by having an installed renewables capacity in excess of the load for some percentage of the load duration curve. Additionally, the use of support technology to minimize the role of the thermal generation in system security is required; including systems to support both regulation and reserve levels. Key problems/considerations include:

- Maintaining raise and lower reserves with high levels of renewable penetration;
- Providing fault level current for operation of conventional distribution network protection systems;
- Provision of frequency regulation;
- Provision of voltage regulation and reactive power; and
- Availability of secondary (additional supplementary) renewable resources.

It is noted that high renewable energy penetration requires a significant reduction in diesel output and preferably only diesel in operation so the maximum supply to load is equal to the demand minus minimum diesel generation. However this limits the raise reserve provided by the diesel min service to about 50% to 60% of rated diesel capacity. This is not sufficient to cover some large swings in wind generation output without activation of under frequency load shedding. In some systems an option called low load diesel is implemented [6]. Typically this option is supported by one diesel manufacturer and it is for small size diesel only (approximately 300 kW). Larger diesel will have minimum output defined at about 30% of rated value. It is noted that the approach adopted here is based on non-firm power supply. This means that under single diesel operation a trip of the diesel generator will result in system black out and the restoration of power supply in a diesel based system takes typically 5 to 30 minutes depending on the availability of the station operator. This approach may not be acceptable in all RAPS systems.

All of the above measures will contribute to maximizing the effectiveness of renewable generation in the system, thus reducing the reliance on fossil fuel based generation and the associated high cost. Generally, instantaneous renewable penetration achievable

by Level 3 systems will be governed by minimum loading requirements on the firm generation used, hence can be dictated by Level 1 plant decisions. At this stage of development it is useful to ensure that the firm generation is sized appropriately to take full advantage of the increased renewable generation capacity.

28.3.5   Level 4 RAPS Development

Level 4 systems build upon the enabling technologies used in Level 3, with the addition of support technologies that allow 100% renewable penetration at times when load conditions permit. Whist the technology considerations presented in Level 3 apply, all of these criteria must be met by supporting technology without support from online thermal generation. To allow for "zero-diesel" operation whenever renewable energy conditions allow reserve must be provided by a firm source. Reserve provided by an energy storage device would limit the achievement of zero-diesel operation to times when storage was charged. The length of time (hours/year) that zero-diesel operation is achieved will depend on the capacity of renewables installed and the available indigenous resources. Increasing the capacity of renewables will increase the duration of zero diesel operation; however, it is likely that there will be diminishing returns as with increasing installed capacity the instantaneous renewable energy generation will exceed the demand more frequently. Economically the contribution of the project may remain positive, but return on investment could decrease. This is a case of optimization such that a limit is set to stop "chasing the tail."

28.3.6   Level 5 RAPS Development

Although the previous development levels can achieve high levels of renewable energy penetration on utility scale RAPS systems there is currently a technical and economic limit to this resulting in a portion fossil fuel consumption remaining. Level 5 development aims to substitute this for a cleaner more sustainable alternative like biodiesel substitution for mineral diesel. Although the substitution of biodiesel could be undertaken at any stage of the system development it still must be sourced at a cost either directly as a processed fuel or through raw materials if the processing is internal to the system. In our Australian RAPS experience biodiesel costs are not significantly more cost effective than mineral diesel and are more costly in marginal cost terms to renewable alternatives. So undertaking the earlier levels of development to minimize diesel uses is more cost effective and will minimize the volume of biodiesel required leading to more sustainable outcomes. Additionally there is still a non-zero carbon footprint impact from the use of biodiesel, typically around 70% to 85% reduction [7], so minimizing the fuel use will also have a greater impact on the total carbon footprint.

At this level of development considerations such as the ability to locally produce fuel are a concern, with care to ensure the production is local if possible, feedstock are sustainably harvested and there are no adverse flow on effects, such as providing competition for food production. Having significantly reduced the volume of biofuel required management of these issues is more readily achievable. Also as mentioned before, the price of biodiesel will tend to shadow the oil prices.

28.4   KING ISLAND POWER SYSTEM OVERVIEW

28.4.1   Currie Power Station

Electricity on King Island is supplied by Currie Power Station an autonomous power system, not connected to mainland Australia or to Tasmania, located in the Bass Strait. The power system is owned and managed by Hydro Tasmania.

Power for the island is largely sourced from diesel and wind generation at the power station near the town of Currie. Diesel must be imported to the island, the costs of which make generation from wind energy an attractive alternative. The Currie power station incorporates the diesel, wind and photovoltaic generation panels on site. Key power system details are provided in Figure 28.4.

Annual consumption on the power system is 16 GWh and peak demand can reach 3.3 MW. Forty-five percent of annual demand is generated from wind energy. Under the current arrangement, 3-4 GWh of generated wind energy is spilled (not used) every year. This equates to 25% of the island's electricity consumption.

28.4.2 Distribution System

Radiating from the power station at Currie are four 11 kV feeders that supply power over 450 km of distribution network as shown in Figure 28.5. Two feeders supply the town of Currie and the abattoir and two longer, lightly loaded feeders supply beef and dairy farms around the island. There is some meshing of the network within the two longer feeders with manually operated switching points. The distribution system has most of its load within a short distance of the power station, although the system is generally lightly loaded and must cover substantial distances for an 11-kV distribution system.

28.4.3 Load Profile

The load on King Island is characterized by a typical residential load profile, with daily peak demands occurring in the morning and the evening (Figure 28.6). Due to the moderate climate on the island, there is little or no air-conditioning load, with annual peak loads experienced during the cooler winter months. Typical daily

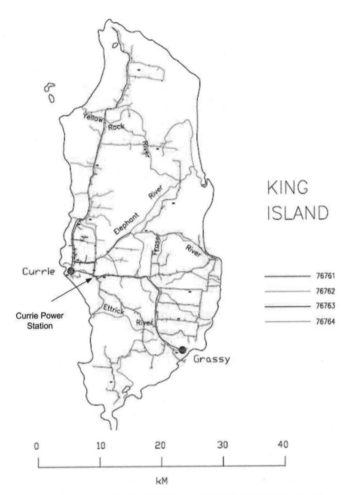

FIG. 28.5 KING ISLAND DISTRIBUTION SYSTEM (*Source:* Hydro Tasmania)

minimum load is around 1,200 kW overnight, with a peak load of around 2,800 kW occurring later in the morning. Due to the presence of some industrial loads such as the dairy and abattoir, greater loads are experienced during the week, with lower daily demand on weekends. These industrial loads account for roughly half of the King Island's annual electricity demand, and depending on maintenance schedules, can push the maximum daily load as high as 3,300 kW.

28.4.4 Power Station Development

In recent years Hydro Tasmania has undertaken several renewable energy developments on King Island with the aim to reduce the use of diesel fuel for the supply of electricity and therefore, reduce the associated cost and greenhouse gas emissions. This has been achieved primarily through the substitution of diesel based generation with renewable generation such as wind and solar. In addition, Hydro Tasmania has implemented a number of renewable energy enablers; these technologies do not directly generate renewable energy but rather enable the existing renewable generation to be utilized more effectively. To date, the projects implemented on King Island include:

- Huxley Hill wind farm development completed in 1997;
- King Island Renewable Energy Expansion (KIREX) project completed in 2004 and included:
  - An expansion of the existing Huxley Hill wind farm;

Diesel Generation	3 1,600 kW generators
	1 1,200 kW generator
	Total Capacity: 6MW
Wind Generation	3 250 kW Nordex wind turbines
	2 850 kW Vest wind turbines
	Total Capacity: 2.45 MW
Vanadium Redox Battery	800 kWh
	200 kW rating
Control System	Allen Bradley Programmable Logic Controllers
	Used to minimise diesel generation
Dynamic Resistor	1500 kW dynamic resistor for frequency control
Solar Generation	100 kW of solar generation

FIG. 28.4 CURRIE POWER KEY POWER SYSTEM DETAILS (*Source:* Hydro Tasmania)

Load Profiles

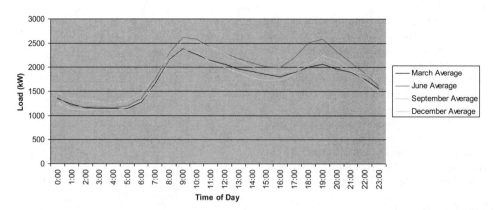

FIG. 28.6   2005 DAILY LOAD PROFILES [MONTHLY AVERAGE] (*Source:* Hydro Tasmania)

o Installation of a Vanadium Redox Battery (VRB);
o Upgrade to supervisory control system;
- Two-axis tracking 100 kW solar photovoltaic array installed by KI Solar Pty Ltd in 2008; and
- Dynamic resistive frequency controller completed in 2009.

28.4.5   Current Asset Configuration

The King Island power system consists of the following components:

- A Programmable Logic Controller (PLC) based system controller;
- Four Diesel Generators with a total installed capacity of 6,000 kW;
- Five Wind Turbine Generators (WTGs) with a total installed capacity of 2,450 kW;
- Vanadium Redox Battery (VRB), rated at 200 kW both charge and discharge with an energy rating of 800 kWh;
- Two-axis tracking solar array rated at approximately 100 kW; and
- A dynamic resistor capable of controlling frequency with a total rating of 1,500 kW. These assets are all connected to an 11-kV bus which supplies power to the 11 kV distribution network. The distribution network is split over four separate

feeders and covers the entire island; comprising of approximately 450 km of overhead lines. The current power system is shown schematically in Figure 28.7:

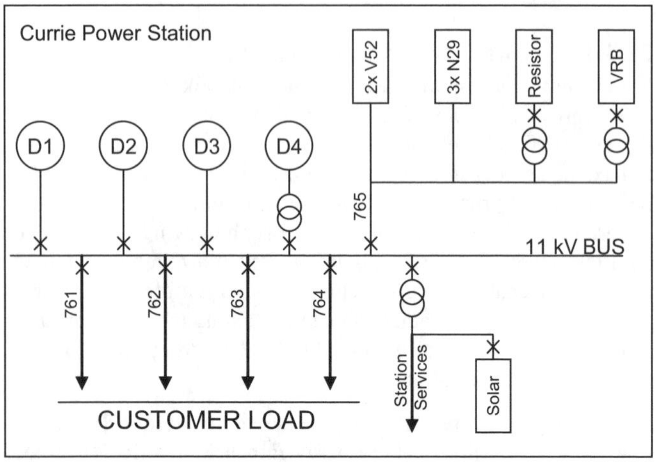

The system has been designed for unattended operation and employs a central station controller to automatically schedule generation assets according to system and renewable energy generation conditions. Known as the Sequencer, this PLC based control system is the overall controller of the power station.

28.4.5.1 Power Station System Controller The Sequencer schedules individual generators according to system loads and station operational constraints, and provides for fully automatic and unattended station operation. It also handles black start events, emergency load shedding and station alarming.

While the Sequencer is the overall controller, each of the power station elements (Diesels, WTGs, VRB, Solar, and Resistor Bank) are essentially self-regulating. That is, the sequencer only provide start/stop signals and/or provides an operating setpoint, with the exception of the solar which is self-dispatching and not controlled by the sequencer at all. The Sequencer does not manage functions such as voltage, frequency or load control, or low level fault handling.

The general philosophy of the system operation is to maximize the use of renewable energy sources to supply the customer load, whilst retaining sufficient "firm" generation capacity to ensure reliable operation. In general terms, the Sequencer issues a power set-point to the wind farm, and based on the actual renewable output will schedule one or more diesel generators to pick up the rest of the load. The Sequencer performs a high level supervisory and scheduling role and is essentially responsible for implementing operational constraints such as ensuring sufficient spinning reserve is carried by the system appropriate to the conditions, ensuring availability of reactive power and adjusting operational margins based actual variability of renewable energy resources.

28.4.5.2 Diesel Generation King Island Power Station was originally constructed to accommodate $4 \times 1,200$ kW diesel generator sets with provisions for adding a fifth machine if needed in the future. The initial installation comprised $2 \times 1,200$ kW sets and 1×800 kW set giving a firm capacity of 2,000 kW. Station capacity has since been increased by adding a fourth set (1,200 kW) and has recently been further upgraded with replacement of three

FIG. 28.7   ONE-LINE DIAGRAM OUTLINING THE OVERALL STATION TOPOLOGY (*Source:* Hydro Tasmania)

1,200 kW sets to 1600 kW machines, giving the station an installed capacity of 6,000 kW. This allows operation according to an N-1 redundancy regime for firm capacity.

The generating sets are arranged for automatic operation and can operate for periods of time without supervision. Essential maintenance and daytime supervised operation of the station is carried out under contract by skilled operators. Output from all the thermal generators are fed into the distribution system via 11 kV metal-enclosed switchgear and power cables to the overhead feeder lines.

There are four diesel generators, three of which are rated at 1,600 kW and produce electricity at 11 kV, and one 1,200 kW machine that produces power at 415 V. Each Diesel generator has its own PLC control system, which communicates with the Sequencer PLC. The real and reactive power load is shared between the running generators proportionally to machine rating; the kW load is shared through a load-share line setup and kVAr sharing handled by the Sequencer PLC, which determines the total kVAR load on the Diesel generators and sends a setpoint to each Diesel PLC.

There are a number of operational constraints in the diesel system that the Sequencer considers when scheduling diesels and wind generation:

- Maximum load: each diesel has a specified real and reactive power capacity that should not be exceeded. The Sequencer works to ensure a sufficient number of diesel generators are online at any given time to prevent overloading;
- Minimum Load: this is the minimum load that a Diesel generator should run at in order to ensure efficient operation and avoid problems associated with extended operation at low power levels; and
- Maximum Pickup: this is the maximum load that a Diesel generator can instantly pick up in the event of another generator failure. This parameter is used to ensure that there is sufficient spinning reserve available to cover loss of either diesel or wind generation without interrupting supply to the customer.

The diesel generators form the "backbone" of the power system. Whilst they are the most expensive power source to operate they are also the most reliable, and are considered to be the only "firm" generation capacity in the system as they can be relied upon to supply the customer load regardless of the time of day, season or environmental conditions.

28.4.5.3 Frequency Control Traditionally, the diesel generators provide both isochronous frequency control and voltage control/reactive power support. With the installation of the Dynamic Resistive Frequency Controller (DRFC) described below, the diesel system can at times be run in droop mode with the DRFC controlling frequency. This occurs when there is excess renewable energy generation available than is required to supply the load with the diesel generator(s) at minimum load. The advantage of this is the ability to hold diesel generation at the minimum load as well as conversion of excess renewable energy into spinning reserve. This concept is covered in more detail in section 28.8. In the current system the diesels are still required to provide voltage control and fault level current to ensure correct operation of the distribution system protection devices.

28.4.5.4 Huxley Hill Wind Farm In 1998 three Nordex N29 (250 kW) WTGs were installed. These are stall regulated simple induction machines. Their output is not regulated. Synchronizing N29 with the system during high wind conditions results in a step change in the output of the generator. If this happens in light load conditions a large overfrequency event is observed. Under some conditions this overfrequency may upset operation of other wind turbines if they have limited operational frequency range (e.g., 48-51 Hz). Also this type of wind turbine tends to produce greater fluctuations in wind turbine output possibly resulting in flicker problems.

In 2000 a proposal was developed to enhance the existing Huxley Hill wind farm to increase the penetration of wind power, with the aim of further reducing cost of electricity supply through savings in diesel fuel use. The project was known as the King Island Renewable Energy Expansion (KIREX) project.

KIREX supplemented the existing Huxley Hill wind farm with two additional WTGs, specifically Vestas V52s capable of producing a maximum of 850 kW of power each, increasing the total rated capacity of the wind farm to 2.45 MW. These WTGs were commissioned in 2003 and are pitch regulated doubly-fed induction machines.

A Vanadium Redox Battery (VRB) energy storage system was also installed as part of KIREX (see below). The objective of the storage system was to increase the recoverable portion of renewable energy and to smooth the variable output of the wind farm to enhance the use of wind power to displace diesel generation.

Finally, a substantial upgrade to the existing control system was needed in order to optimize the operation of system components to deliver power with adequate system security at the least cost.

28.4.5.5 Vanadium Redox Battery The Vanadium Redox Battery energy storage system was installed with the assistance of an Australian Federal Government Renewable Energy Commercialization Grant. The VRB technology was not widely available at the time, and this installation represented the first such project in Australia. The installation represented an early demonstration of the technology for the suppliers, and the commercial risk and demonstration value was recognized in the awarding of the Government grant.

Overall characteristics of VRB technology can be very attractive for RAPS and wind applications however the technology is not yet mature and its availability is limited. Very few VRB battery systems are currently in operation, these are predominately located in Japan, US, Austria and Germany.

Vanadium Redox Battery (VRB) technology was developed by Maria Kazacos at the University of New South Wales in 1985 [8]. This is a flow battery which stores separately positive and negative electrolytes. In a chemically neutral state the electrolyte is a mixture of vanadium pentoxide and diluted sulfuric acid. The energy is stored chemically in different ionic forms of vanadium. When the battery operates the electrolyte is pumped from storage tanks into cell stacks where one form of electrolyte is electrochemically oxidized and the other is electrochemically reduced. This reaction creates a current that is collected by electrodes and made available to an external circuit. The reaction is reversible allowing the battery to be charged, discharged and recharged. The reaction formula is as follows:

$$\text{Negative Reaction: } V2+-e- \rightarrow V3+ \qquad (28.1)$$

$$\text{Positive Reaction: } VO_2++2H++e- \rightarrow VO_2++H_2O \qquad (28.2)$$

The electrolytes are circulated through the cell stack by pumps. During the chemical reactions the imbalance in volumes build up and there is a need to rebalance electrolyte on regular basis.

FIG. 28.8   VRB LAYOUT (*Source:* Hydro Tasmania)

The VRB system consists of two major components — the battery (also known as the Wet Side) and the inverter. The VRB PLC controls the wet side, and interfaces with the inverter controller to load the battery. In order to use the VRB, the Wet Side must be running and the system pressurized. Figure 28.8 shows the wet side of the VRB in the foreground (electrolyte tanks, insulated pipe work and pumps) and the cell stack in the rear.

The VRB is able to both store energy and return the stored energy to the power system, and as such can act as a load or a generator. When storing energy the VRB is said to be charging and when returning energy to the power system the VRB is said to be discharging. Charging has priority over discharging, such that the battery is kept fully charged if wind is available.

VRB is a very attractive storage system as the energy storage (tanks) is separated from the power source (cell stack). This supports scalability of both power and storage elements. The electrolyte is very robust and difficult to damage provided a specified operating range is maintained. The electrolyte can be charged and discharged very frequently with more than a 10,000 charge cycles during the lifespan being quoted. The round cycle efficiency is reported about 70%. The battery has very good short term overload properties however to utilize them the converter needs to be sized appropriately. The battery does not have memory, it can be fully discharged without damaging the battery so the depth of discharge cycle in much greater than in other batteries. The maintenance requirements are minimal.

The King Island installation incorporated a 200 kW cell stack, with sufficient electrolyte storage for 800 kWh of energy storage.

28.4.5.6 Solar In 2008, six dual axis tracking solar photovoltaic systems (Figure 28.9) were installed by CBD Energy Ltd, adding approximately 100 kW of renewable energy generation to the power station. The units selected by CBD Energy were Solon XL movers. The use of solar power on King Island was pursued to demonstrate how a variety of renewable energy technologies may operate in such an integrated system.

28.4.5.7 Dynamic Resistive Frequency Controller Hydro Tasmania developed the dynamic resistive frequency control technology in response to a gap in the availability of existing control

equipment to utilize excess renewable generation to provide reserves and control frequency without the expense of additional energy storage. Similar systems have been trialed at much smaller scale in Alaska [9] and Antarctica [10]. This recently completed project was associated with a substantial upgrade to the power station control system and is currently undergoing extensive testing. The new control system employs the DRFC enabling the load to be varied rapidly in order to absorb excess wind generation rather than spill it through shutting down or throttling back wind turbines. Instead of reducing output, the wind turbines are be allowed to produce as much power as possible, with the excess generation absorbed by the DRFC. As the resistor loading can be adjusted rapidly, this effectively converts previously spilled wind into spinning reserve that is used to supplement diesel generation. Maintaining the power balance between generation and demand in this way allows the DRFC to maintain system frequency.

Shifting system reserve requirements away from the diesels enables the control system to operate diesel at minimum output level, reducing fuel use, whilst without the DRFC the diesel would need to do all the regulation. Whilst this can occur only during periods of wind spill, a further increase in renewable energy generation increases the amount of time that the system experiences spill, and thus increases the time that diesels can be run at minimum loading. This in turn reduces the amount of overall energy contribution from the diesel generators.

Figure 28.10 shows a simulation of the effect of the new DRFC and control method on diesel output and wind utilization over a half hour time period. Note that under the new control method the diesel output remains at minimum output of 300 kW for the majority of the time, saving an average of around 120 kW. Over a year this equates to a substantial amount of diesel fuel saved. The wind spill component of the chart is treated as system reserve.

The DRFC enables the running diesel generators to remain on minimum loading whenever the island's load is less than the combined solar and wind output (solar is self dispatched and included as a load offset). Modeling of the King Island energy demand compared with historic wind data and expected solar output indicates that this will occur for about half of the year — resulting in significant diesel fuel and thus GHG emission savings.

FIG. 28.9   SOLAR PV INSTALLATION, SOLON XL MOVERS (*Source:* Hydro Tasmania)

Difference in Diesel Generation (18/05/08)

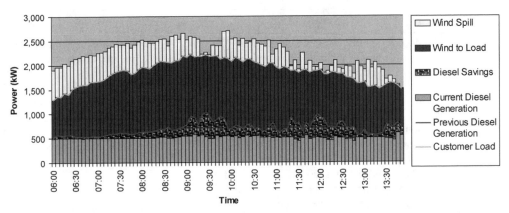

FIG. 28.10 DYNAMIC RESISTIVE FREQUENCY CONTROLLER — BENEFITS (*Source:* Hydro Tasmania)

The resistor is designed to maximize the level of wind generation that can contribute directly to the load thus maximizing wind utilization. The remaining portion of wind generation that cannot be used by the system to supply the customer load is currently converted into waste heat in the resistor. It should be noted that:

- The resistor heating (wasted energy) in short term provides fast frequency regulation capability and provides both rise and lower regulation if the operating point is within the active regulation range;
- In some cases the operating point can operate at the lower or upper limits providing reserve in one direction only;
- The air resistor units could be substituted by other forms of storage including thermal storage (water heating) or chemical storage with adequately specified properties matching variability of wind generation. It is noted that a portion of approximately 15% of currently wasted energy could be stored and then be recovered when required. The DRFC consists of 3 × 500 kW load banks, which provides a total installed capacity of 1.5 MW. Each bank is subdivided into two 250 kW sub-banks. The loading of each sub-bank is controlled by a phase-angle controller (PAC) and coordination of operation of individual PACs is carried out by the dedicated PLC. The DRFC has two operation modes:

1. Droop Mode. The DRFC receives a load (kW) setpoint from the Sequencer PLC then controls the resistive load to achieve the load setpoint.
2. Isochronous Mode. The DRFC controls system frequency autonomously.

A description of when and how the Resistor Bank changes between Droop mode and Isochronous mode is included in Table 28.2. The basic operating modes of the resistor are summarized in Table 28.2.

28.5 PERFORMANCE OF RENEWABLE ENERGY DEVELOPMENTS

28.5.1 System Performance

With the exception of the VRB, the renewable energy development program has been a resounding success; as a direct result of these projects an annual reduction of 1.39 million liters of diesel,

or approximately 35% has been achieved — this equates to approximately 4,000 tons of greenhouse gas emissions. The impact of the incremental integration of renewable energy technologies into the power system can be seen in Figure 28.11. Each phase of development has resulted in an increase in the utilization of renewable energy in system, and an associated reduction in emissions.

The integration of renewable energy generation and enabling technology in the King Island power system has demonstrated a significant benefit in terms of diesel fuel and Greenhouse Gas (GHG) reduction. The King Island power system is being recognized internationally as being world leading in terms of renewable energy integration in Remote Area Power Supply systems. The individual contribution and performance of each of these developments will now be discussed briefly in turn.

28.5.2 Huxley Hill Wind Farm

The first wind development on King Island, the three Nordex N29 (250 kW) WTG represented Australia's second commercial wind farm. The project received no external funding support and resulted in an immediate saving in fuel consumption.

Penetration of renewable energy into the system reached an annual average of 18% in the years following commissioning. This resulted in a 16% reduction in annual diesel fuel use or approximately 590 kiloliter reduction in diesel use per annum. In terms of the development level philosophy discussed above, Huxley Hill wind farm represented a Level 2 development project.

28.5.3 Vanadium Redox Battery

Commissioning and operation of the VRB was challenging as this occurred very early on in the development of the technology with no past operational experience available in the public domain. The VRB has suffered downtime as a result of cell stack failure in 2004, as well as a number of inverter failures. Following the failure of several cell stacks in 2004, the cell stacks were repaired by the manufacturer and returned to service. Significant downtime following the cell stack repairs has been caused by repeated failures of the VRB inverter. Inverter problems have also been experienced in a similar installation in Moab (Utah, US) [11]. At Moab it was decided to replace the inverter while Hydro Tasmania adopted the temporary solution of limiting VRB output to 150 kW.

Following an operational event a strategy decision was made to put the VRB in an extended "out of service" state while new cell technologies were assessed. A detailed investigation into the state

TABLE 28.2   OPERATING MODES OF DRFC (SECTION 28.9.3: VANADIUM REDOX BATTERY)
(*Source:* Hydro Tasmania)

Wind conditions	System	DRFC
Low winds	All energy produced by the wind turbines is used to supply the load. The diesel generator(s) operate in isochronous mode to provide frequency control	Resistor is in stand by mode to provide lower reserve to respond to over frequency events
Medium winds or high wind and high loads	Amount of wind energy used to supply load is limited by the firm reserves required for secure operation. Under some conditions part of wind generation will be absorbed by the resistor. During times of high load all wind energy is used to supply load with the remaining power balance supported by diesel generation. Diesel operates in isochronous mode	The resistor may be either on standby to provide over frequency control such as the case above, or it may absorb some excess wind under a setpoint issued by the sequencer.
Very high wind and low loads	At light loads a single diesel operates at minimum output and the rest of the load is supplied by the wind turbines. Spill of wind generation is absorbed by the resistor, providing effective reserves. The amount of wind energy used to supply the load is limited by the level of reserve available to support secure operation. Resistor operates in isochronous mode and diesel in the droop mode.	The resistor is partially loaded providing both rise and lower regulation.

of the VRB was conducted in July 2007. The subsequent report outlined the following main conclusions:

- Cell stack damage was extensive and stacks would need to be replaced in order to return the VRB to service;
- Failure was determined to be a result of a combination of overcharging and subsequent high temperature operation; and
- Temperature interlocks to limit operation at high temperatures were inadequate.

The operational regime adopted by Hydro Tasmania for the VRB reflected best practice as based on the knowledge of the technology at the time. Subsequent development of the VRB technology internationally led to the development of a different set of operational parameters and practices.

Hydro Tasmania believes that knowledge gained through investigations and experience would allow for successful operation of a restored VRB system.

The VRB battery offers excellent overload capability which has been specified as 100% overload for 100 seconds and 50% overload for five minutes. Unfortunately performance of the inverter has restricted the operation of the battery in this way. It is noted that VRB characteristics were matching the variability of wind generation very well with very frequent changes over between charge and discharge cycles.

The AC to AC efficiency of the VRB on King Island to date is 55%. Following an upgrade the similar plant at Moab has achieved an efficiency of 68%. The main reason for lower efficiency of King Island VRB was operation with too high electrolyte temperature due to a lack of coolers.

Hydro Tasmania is currently evaluating the role and function of the VRB battery with an aim to update the design and control system to improve availability, efficiency, and the effectiveness of the VRB to reduce diesel use. Subject to replacement or successful repair of the inverter, the VRB may be included under the resistor

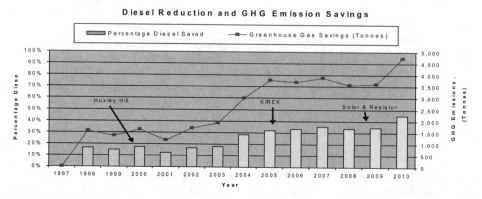

FIG. 28.11   HISTORICAL CONTRIBUTION OF RENEWABLE ENERGY ON KING ISLAND (*Source:* Hydro Tasmania)

control (DRFC) as the high priority load. The VRB will be charged as the first (highest priority) switchable load subject to its state of charge (SOC). If the state of charge is 90 to 100%, VRB charging load will be replaced by a resistor.

28.5.4 Vestas V52

The two Vestas V52 turbines have performed almost entirely without incident since commissioning in 2003, and have provided a significant contribution to the energy balance of King Island. The selection of variable speed turbines has proven to be a success and has resulted in much more stable output of wind generation compared to the unregulated Nordex machines. Between them, the Vestas WTGs have supplied the majority of the renewable energy contribution, and produced an average of around 3,900 MWh per annum, or around 25% of the average annual energy demand for FYE 2006. Operational experience with these turbines has been exceptional, with effective regulation of power output facilitated through the Vestas Online (VoL) control system providing stable wind power output even in challenging wind conditions. This regulation has allowed the sequencer control system to maintain high levels of wind penetration even in the absence of the VRB.

Performance of these turbines exceeded expectations — on the whole, the KIREX project achieved business case projections regardless of the VRB performance.

28.5.5 King Island Solar

The solar array is expected to deliver around 200 MWh a year to the Currie Power Station, resulting in CO_2 savings of around 180 tons per annum. To date the two-axis tracking array has demonstrated the advantage of diversifying resources, as there are periods during the year that are seasonally calm (little or no WTG output), yet there is sufficient sunshine to allow a reduction in diesel use.

28.5.6 Dynamic resistive frequency controller

Commissioning of the resistor presented a unique challenge, as this was the first installation of this size in the world. During testing and commissioning, the power system was operated with higher levels of renewable penetration than had been attempted previously. During this phase, a number of power station operations were improved, allowing the system to maintain reliability in the face of increased renewable energy penetration.

The resistor is currently undergoing a commercial test period, which will verify the benefit the resistor is providing to the system. The results so far have been very promising; even under constrained operation the resistor has allowed for the saving of over 850 MWh since May 2008. Annually the DRFC is projected to reduce diesel consumption by a further 8-10% without the addition of any further renewable generation or storage.

Technical performance of the resistor can be seen in Figure 28.12. Diesel regulation is reduced and system frequency is improved when the DRFC is in isochronous mode.

The DRFC will form an integral part of the King Island Power system when combined with the proposed KIREIP, resulting in one of the most advanced RAPS systems in the world.

28.6 ASSESSMENT OF BENEFITS

28.6.1 Introduction

Hydro Tasmania has developed an Energy Simulation Model (ESM) as a tool to simulate various energy production scenarios for the King Island power system. Specifically, the model is used to estimate future annual diesel fuel use based on a given system topology (e.g., number and type of wind turbines, storage technologies, etc.) and some assumptions such as annual load growth. The model has been designed to replicate the specific scheduling methodology/logic that is used now in the power station control system, as well as proposed future methodologies that will be enabled with proposed projects coming online. This tool provides the forecast performance data that underpins the investment decisions in new assets on King Island.

Inputs into the ESM include time series data as well as parameters that characterize the operation and performance of the simulated system components. The time series data consists of

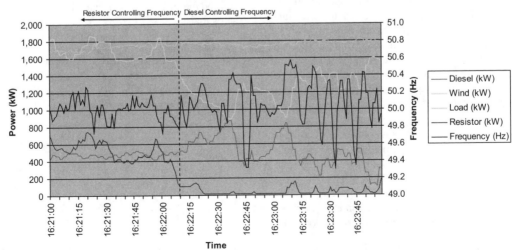

FIG. 28.12 DYNAMIC RESISTIVE FREQUENCY CONTROL — PERFORMANCE (*Source:* Hydro Tasmania)

10 minutely average values for customer load, wind speed and solar irradiation. The load and wind speed data are actual values recorded on the Island during the 2005/06 financial year, and the solar resource data is based on 25+ year ANZSES data for Melbourne and Launceston.

There are three system component categories that can be included in a scenario: "firm" generation such as diesel or gas engine generators, renewable generation including wind and solar and storage technologies such as VRB and Graphite Energy Storage, though any storage technology can be modeled if the round trip efficiency is known. In the King Island context, the system topology is set up to reflect the generating assets that are currently used in the system.

Scenarios that have been modeled include additional generating and power system enabling assets, such as:

- Vanadium Redox Battery;
- Graphite Energy Storage;
- Additional wind turbine generators;
- Diesel UPS; and
- Dynamic Resistive Frequency Controller.

The model is based on Visual Basic code as an Excel macro and works through several distinct "modules" of code. The first module is concerned with reading in the simulation inputs, such as the number and type of wind turbine generators to be modeled, number and size of diesel generators, the load scheduling of diesels, etc.

The next module is the annual energy simulation itself, which works on a 10 minute time step analysis of the energy flows within the power system. All calculations are based on time step averages, i.e., average customer load, average wind speed, etc. The model steps through each time step of the year, reading in customer load, wind speed, and solar resource from a time series of data stored within the model. The availability of storage devices is then determined, based on the previous state of charge. The amount of diesel generation required as per the selected control methodology is then calculated, based on the wind/solar/storage available. Wind/solar spill is then calculated, and distributed between available storage devices subject to the particular limits as set out in the system setup page. Each stage of the simulation process is discussed in more detail below.

28.6.2   Simulation Modules

28.6.2.1 Wind   Wind turbine output is determined according to wind speed as recorded in a time series. Wind turbine output for each simulated wind turbine is based on power production curves specific to each turbine. The simulation includes a number of turbine power curves as specified by manufacturers.

For each time step of the simulation the power output of each "available" wind turbine is calculated, and is separated between individual machines in order to differentiate between them when scheduling output in the control element of the simulation, as well as when distributing wind spill between storage devices.

28.6.2.2 Solar   Solar output is based on a number of different parameters including array size and tracking mode, as well as a time series of solar radiation resource — similar to the wind speed time series. The solar resource is given in terms of average global, direct and diffuse irradiation impinging on a horizontal surface Diffuse and direct solar values are used in order to determine the increase in solar radiation per unit area for a surface that is not horizontal, i.e., is tilted at some angle.

For each time step of the year, the position of the sun is calculated relative to the orientation of the solar array. For a fixed tilt array

this means that the sun's position relative to the surface of the array will change for each hour of the year — for a two axis tracking array the surface will always be pointed towards the sun. In addition, the angle of a tracking (or non-tracking) surface is calculated; this allows the calculation of the angle of incidence — the difference between the direction of the beam radiation and the direction that the array surface is pointed. The angle of incidence is also calculated for a horizontal plane.

The ratio between the angles of incidence for the horizontal plane and the tilted (PV surface) plane is a geometric factor which describes the increase in direct irradiance compared to the horizontal surface by tilting the plane towards the sun, as described in Figure 28.13.

This ratio between the horizontal and tilted surface radiation is calculated for each time step and applied to the direct beam radiation resource. The diffuse radiation component is then added to give the global solar resource for the PV array for that hour, in kWh/m².

Having calculated the global radiation impinging on the tracking surface in kWh/m², the total array output is calculated as the product of the calculated solar radiation and the rated output of the solar array. Note that no shading is modeled in the simulation.

28.6.2.3 Storage   Storage availability in the simulation is based on two parameters: the ability of the storage device to accept energy from renewable spill, and the ability to supply power to make up for shortfalls in renewable output. At this point in the simulation, availability of each storage device modeled is assessed according to the previous time step state of charge, as well as parameters particular to each device. Such parameters may include operating hours — it is possible to define a certain period of the day where operation is allowed, outside of which the storage device will be deemed unavailable. Availability is based on the ability of the storage device to produce minimum output for the time step (or hours if a minimum run time is set). Should a particular storage device become available for power production, a flag is set, and the subsequent control code determines required power output if the storage device is needed (i.e., if there is a shortfall in renewable energy production).

28.6.2.4 Diesel   Diesel generator output is simulated in the model. Diesels are configured by the user in setting up the simulation, including limits such as minimum and maximum output, fuel efficiency curves, and "load table" scheduling of machines. Given the calculated wind, solar and storage output available for the hour, the required diesel generation is calculated through one of two control philosophies.

28.6.3   Modeling Control Philosophy

Three discrete control philosophies can be modeled within the ESM tool, namely the Instantaneous Wind Penetration (IWP) method, the Resistor method, and the Diesel UPS method.

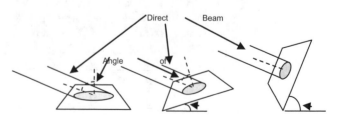

FIG. 28.13   EFFECT OF TILTING A SURFACE TOWARDS THE SUN

28.6.3.1 Instantaneous Wind Penetration Method The Instantaneous Wind Penetration method is a simulation of the station control philosophy which was introduced with the KIREX upgrade. It is has since been superseded by the resistive frequency control method, however has been modeled to provide a post-KIREX benchmark to determine the relative value of the new control method. This methodology has also been used to calibrate the model against years of historical data to ensure that the model provides a reasonably accurate estimation of diesel fuel use.

The IWP method works by limiting the penetration of renewables (exclusively wind in the past) to a factor known as the Instantaneous Wind Penetration. Allowable wind power is calculated by multiplying the customer load by the IWP, typically 70% in good wind conditions. Thus at any given time, at least 30% of the load must be supplied by diesel generators. In the actual control system, the IWP can be changed by the station operators, or modified by the sequencer itself to adjust to wind conditions. When wind conditions are such that wind turbine power output is highly variable (such as periods of frequent wind gusts) the IWP is typically lowered; this limits the effect of fluctuating wind output on the system, and allows more diesel generation to supply the load. In the past this has been effective in limiting system outages caused by erratic wind turbine output.

Historically, the average annual IWP for King Island was around 49%. Thus an average figure of 49% is assumed in the simulation for the entire year. In calculating the hourly diesel load under this control philosophy, the customer load is multiplied by 0.49 to determine the maximum renewables output for that hour.

$$R_a = IWP * L \qquad (28.3)$$

where:

- R_a is the maximum renewable contribution to the load (combination of wind and solar);
- IWP is the instantaneous Wind Penetration (0.49); and
- L is customer load.

If the combined wind and solar output is less than this figure, then the model examines the availability of storage devices to supplement the output. Should a storage device such as the carbon block or VRB be available for power production as assessed in the previous storage module then the required diesel output for that hour is equal to:

$$D = L - R_t - S \qquad (28.4)$$

where:

- D is the Diesel output;
- L is the customer load;
- R_t is the total renewables contribution to the load; and
- S is the storage output (0 if no storage available or required).

Should the total renewable output exceed the renewable contribution to load, the excess renewable component is spilled, and no output from energy storage is allowed. Renewables to customer is limited by the IWP, and the diesel generation required is simply the balance of customer load less renewables to customer. The resulting spill is then distributed between available storage devices.

Figure 28.14 indicates the energy balance under IWP control methodology (no storage is included). The minimum diesel requirement rises with the customer load and assumes a 70% IWP. Note that the diesel used in this example is higher when compared to the diesel generation required using the resistor based control method discussed below (purple line).

28.6.3.2 Resistor Method The resistive frequency control method is a new control methodology that has been implemented on the Island with the commissioning of the DRFC. Where the IWP method sought to limit wind generation to ensure a percentage of diesel generation was online to maintain system security, the resistor control method utilizes wind spill and a resistor in order to provide both raise and lower reserve, and regulates the power balance between supply and demand to maintain frequency. This allows the online diesel(s) to remain at minimum loading rather than above minimum in a regulating range. As a result, in periods of excess renewables generation, the total renewables supplying the customer load is not limited by an IWP factor, but rather by minimum diesel loading constraints.

To calculate the diesel generation requirement, it is first necessary to determine how many diesels are required due to reactive power requirements. Occasionally there exists a situation where the customer power load can be satisfied by renewables and a single diesel — however the reactive power demand requires that two machines be online. The number of machines is determined, and as such a minimum diesel requirement is calculated based on the minimum load limits on the machines. Maximum renewable output to supply customer load is limited by this minimum diesel requirement. Thus renewables allowed to supply the customer load is calculated:

$$R_a = L - (N_D \times D_{min}) \qquad (28.5)$$

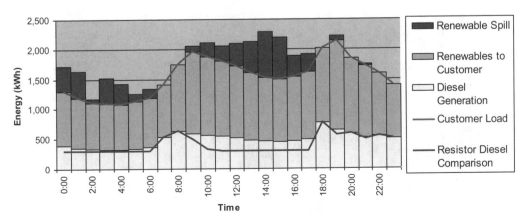

FIGURE 28.14   ENERGY BALANCE FOR IWP CONTROL LOGIC (*Source:* Hydro Tasmania)

where:

- R_a is the maximum renewable contribution to the load (combination of wind and solar);
- L is customer load;
- N_D is the number of diesels required (as calculated from reactive demand); and
- D_{min} is the minimum loading allowed on a diesel in kW.

Should the combined renewables output be less than R_a, then as in the IWP case, storage devices are examined to determine if any additional output can be used to supplement the renewables output. Thus Equation 28.4 is used to determine diesel output. As can be seen in Figure 28.15 the diesel output is kept to a minimum level during periods of wind spill under the resistor control method.

28.6.3.3 Diesel-UPS Method
An extension of the resistive frequency control methodology will occur with the installation of the proposed Diesel Uninterruptible Power Supply (D-UPS) project. The intention of the D-UPS is to provide ancillary services normally provided by an online diesel generator (such as voltage control, inertia and fault level contribution) through the use of a flywheel and diesel generator (normally offline) separated by a clutch arrangement. The D-UPS also provides an instant start capability, which will protect the system from generation shortfalls caused by fluctuations in renewable energy production. As discussed above, the intended outcome of this installation is to allow the power station to run without a diesel generator providing power at times when the renewable energy generation is well in excess of the total customer demand. While this operational methodology is not yet available at the station (as the D-UPS project is yet to commence), the modeling methodology is representative of what is intended to be implemented once the project is commissioned. This operating philosophy is required to examine projects that are scheduled to be implemented post D-UPS commissioning.

Note that a D-UPS is not entered into the firm generation table as a schedulable diesel. The purpose of the D-UPS is not to provide energy in the same way that the conventional diesels do, but rather allow the station to operate under a different logic, i.e., the ability to run the station without diesel generators online when there is sufficient renewable energy generation. Results of the energy balance using the D-UPS method are shown in Figure 28.15. Note that under D-UPS control philosophy, some hours require no diesel generation as there is sufficient excess renewable energy generation to cover the entire load.

28.6.4   Diesel Fuel Use
Once the loading on each online diesel is determined, a fuel efficiency curve is applied to calculate the fuel used for the time step. The fuel efficiency curve can be specific to each individual diesel (as would be the case for machines of different capacities and types). The fuel curves used in the King Island model are derived from manufacturer specifications for the models installed at the station.

The model can be set up to include a number of different fuels such as biodiesel and takes into account the effect on fuel efficiency for each fuel type.

28.6.5   Scheduling Order of Preference
There is a scheduling order of preference between renewable technologies in supplying the renewable contribution to the customer load. In the simulation, the default order of preference is:

1. Wind power generated by existing assets;
2. Solar power generated; and
3. Wind power generated by future assets.

This "merit order" reflects the philosophy adopted by Hydro Tasmania where the assets that are installed earlier get dispatched first; on a "first come first serve" basis. The merit order can be changed to suit the scenario and the solar power can be modeled as a direct load offset, which is useful when modeling distributed generation that is not controlled by the power station sequencer.

Thus the customer load is first addressed by existing wind output, followed by solar output, followed in turn by future wind output. For example, if the total allowable renewable contribution to the load can be supplied entirely by the existing wind assets, only existing wind output is used; solar and future wind output is spilled in the simulation, along with any excess existing wind generation.

The example in Figure 28.16 shows a case where all the existing wind is used, as well as some solar generation. The sequence of calculations in the model follows:

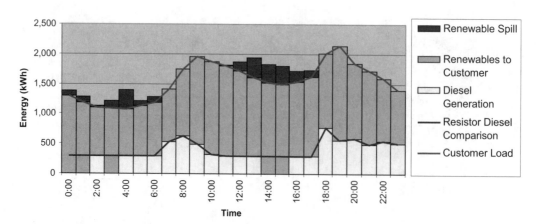

FIG. 28.15   ENERGY BALANCE FOR D-UPS BASED METHOD (*Source:* Hydro Tasmania)

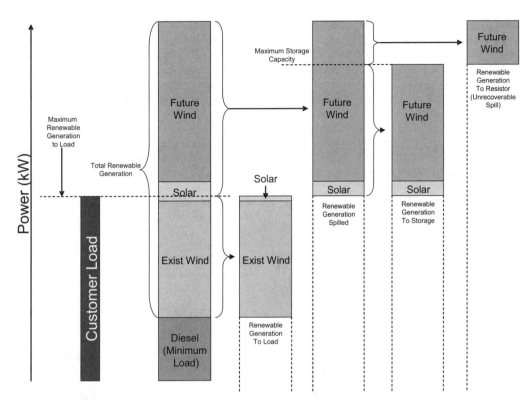

FIG. 28.16 RENEWABLE CONTRIBUTION BASED ON SCHEDULING ORDER OF PREFERENCE (*Source:* Hydro Tasmania)

A. Load balance calculation, determining the maximum renewables contribution to load based on customer load and minimum diesel requirements;

B. A percentage of the total renewable energy generation directly contributes to the load;

C. The remainder renewable energy generation that cannot be used to supply the load directly is "spilled" either to available storage or through unrecoverable means — resistor or through throttling of wind turbine blades (this is explained below);

D. A proportion of the spilled renewable energy is sent to any available storage. Storage availability may be limited by state of charge, i.e,. if the storage is full; and

E. Any remaining generation must be spilled through other means, such as a resistor, reduction in wind or solar output through set point control or switching off of generators.

28.6.6 Spill Distribution and Storage Balance

Should the system experience spill during the time step (indicated in column C in Figure 28.16), this spill is distributed between the available storage devices (D). Again, there is an order of preference between renewable sources when allocating spill, as well as an order of storage devices if more than one is present (e.g., VRB and Graphite Energy Storage). Any spill over the amount able to be absorbed by the storage device(s) is assumed to be spilled through other means (E) which could include a lower priority storage device, resistor or through reduction in output. In the case of spill where no storage is present it is assumed that spill is absorbed through a resistor or through reduction in output. It should be noted that an actual resistor is not modeled in the simulation, as any spill in excess of a theoretical resistive capacity will be achieved through reduction in output. Note that reduction in output can be

achieved practically through active pitch control on wind turbines as well as shutting down turbines and solar modules.

Spill absorbed by the storage device(s) is added to the state of charge, less any inefficiency assumed for each particular storage device. Alternatively, any output for the hour from storage devices is subtracted from the state of charge. Additional standing losses (such as Graphite Energy Storage cooling) are subtracted from the state of charge. The resulting state of charge is known as the "End of Time Step SOC", and is used to determine storage availability in the next time step.

Limitations on the amount of energy that can be accepted by a given storage device is governed by two factors: the state of charge of the device, and the maximum power input the device can handle. For example, a VRB storage device may be sized such that the inverter can produce or accept 200 kW of power (short term overload capabilities are not modeled in the simulation), and store 800 kWh of energy. If the VRB SOC was 500 kWh, then the ability to accept spill would be limited by the inverter rated power of 200 kW. If the SOC was 700 kWh, then the SOC is the limiting factor, and the spill acceptable is 100 kWh. Note that whilst in practice the VRB could accept 200 kW of power for half an hour to store 100 kWh, the simulation assumes an average power over the hour, and thus would assume 100 kW of input into the VRB for the hour.

28.6.7 Modeling Results

The above set of routines is repeated for each model time step for an entire year, with the resulting parameter values stored in an Excel spreadsheet. At the end of each year examined, the total energy figures for each generator and storage device modeled are calculated and copied into an annualized results page. For simulations requiring multiple years of simulation with load growth, the above is repeated for each year, with the customer load time series data multiplied by the specified load growth factor.

The end result of a simulation run is a 25 year forecast of annual energy production from each generation type (i.e., renewable, diesel and storage). The output of a particular scenario can then be compared to the modeling output of a base or reference case (typically the base case is a model run of the current system topology) which allows a comparison of diesel use between the two scenarios. For example, a model run of the current system can be compared with a model run of the current system with additional wind installed. The comparison of diesel use between each case will form the basis of diesel savings afforded by the additional wind project, which becomes the basis for the financial modeling The combination of energy and financial modeling underpin the business cases for each proposed project.

This scenario comparison approach allows for the assessment of individual project benefits under a number of different circumstances such as load growth, asset capacity, etc. This in turn enables the ability to optimize asset sizing to provide the greatest net benefit.

28.6.7.1 Performance of System Modeling The model has been calibrated against historic operating data to confirm the validity of the modeling results. Two years of the simulation were compared against historic operational data in order to calibrate the model to closely follow the operation of the pre-KIREX plant. The subsequent three years of simulation were generated using historical data, including both hourly load and wind speed data. Diesel savings have been calculated both against historic and KIREX simulated data. The monthly comparison between both simulated cases and actual historic data is shown in Figure 28.17.

Note that the red line indicates the combined output of the three Nordex machines under the original pre-KIREX operating philosophy. This line follows the actual historic WTG performance as measured prior to the implementation of KIREX. This serves to demonstrate the level of accuracy that the simulation tool models the system based on historic wind data. The blue line represents the simulated KIREX wind output. The discrepancies between the simulated output and the historic data are due to scheduled and unscheduled turbine down-time, variance between recorded wind speeds and actual wind speeds encountered by individual turbines, data resolution and shifts in KIREX operating philosophy as control

system parameters are "tuned" by operators over months of the year. The later is more prominent during the earlier months of KIREX operation, as parameters were initially set to conservative values to achieve system stability prior to tuning the system for maximum renewable penetration. The simulation model does not consider turbine availability or changes to operating parameters on a month to month basis. This leads to discrepancies in WTG output between months, however annual WTG energy production is quite comparable between the simulations and historic data.

The model has also been calibrated to simulate pre-KIREX operating conditions. Whilst only two years of pre-KIREX data is available to calibrate the model, this is countered by the fact that the system was not subject to constant tuning during this period as was the case for the beginning of the KIREX data. A comparison between the historic and simulated diesel fuel use is detailed in Figure 28.18

The ESM tool has provided the analytical capability to assess the potential benefits of the proposed suite of future developments on King Island.

28.7 KING ISLAND PROPOSED DEVELOPMENTS

Hydro Tasmania proposes to further develop the King Island system, to achieve greater levels of contribution from renewable resources and to demonstrate further innovative enabling technologies. The scope of work for the King Island Renewable Energy Integration Project (KIREIP) is comprised of the following discrete sub-project components:

- **Biodiesel**— Augmenting the existing diesel tanks and associated infrastructure to allow the use of B100 in the diesel generators;
- **D-UPS**— The installation of an auxiliary Diesel-UPS to allow 100% renewable energy penetration at times of high renewable energy production;
- **VRB**— Reinstatement or replacement of the existing Vanadium Redox Battery (VRB) energy storage system;
- **Wind**— Expansion of the existing wind farm by up to 4 MW for a total wind farm capacity of up to 6.45 MW;
- **Thermal Energy Storage** — Installation of a Thermal Energy Storage system with integrated energy recovery system, and

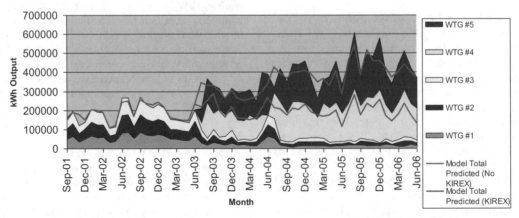

Comparison between historic WTG Output and Modelled Output

FIG. 28.17 RENEWABLE CONTRIBUTION BASED ON SCHEDULING ORDER OF PREFERENCE (*Source:* Hydro Tasmania)

Total Monthly Diesel Fuel Use (L): Comparison between Simulated and Historic Data

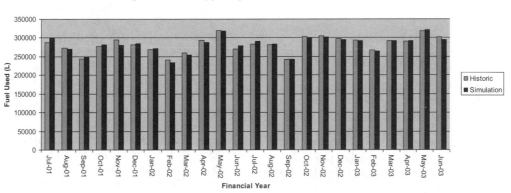

FIG. 28.18 COMPARISON BETWEEN HISTORIC AND SIMULATED DIESEL FUEL USE ON KING ISLAND (*Source:* Hydro Tasmania)

- **Smart Grid** — Deployment of Demand Side Management infrastructure and Smart Grid technology, in the way of an interconnected network of centrally controlled smart meters.

Schematically the project elements are described by Figure 28.19. The items shaded in blue represent the new elements proposed under KIREIP. The KIREIP aims to achieve a >65% utilization of renewable energy, including 100% instantaneous penetration when conditions allow, and a reduction in GHG of >95% through the use of biofuels. This project will be implemented over the next 3 years through the assistance of an Australian Government REDP grant of up to $15.28 m.

The following is a brief description of each of the projects proposed for King Island under KIREIP.

28.7.1 Biodiesel

Hydro Tasmania has been monitoring the development of biodiesel as a potential mineral diesel substitute for the BSI. By implementing biodiesel, Green House Gas (GHG) emissions can be substantially reduced. There are, however, differences in the properties between biodiesel and mineral diesel. It is desirable to test the performance of biodiesel in controlled conditions prior to implementation.

A trial will be undertaken at the King Island Power Station in order to prove the biodiesel in one of the existing engines, test fuel supply logistics, test the suitability of existing fuel infrastructure and trial the reliability of the biodiesel supplier.

The trial is proposed to run for 6 months. At the end of the trial, a decision will be made on whether to fully integrate biodiesel into

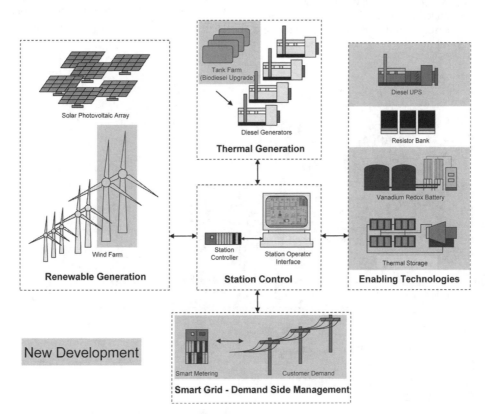

FIG. 28.19 PROPOSED KING ISLAND RENEWABLE ENERGY INTEGRATION PROJECT (*Source:* Hydro Tasmania)

the power station and to similarly implement biodiesel at the power station on Flinders Island.

28.7.2 Diesel Uninterruptible Power Supply

The Diesel Uninterruptible Power Support (D-UPS) project seeks to enhance the value of existing renewable energy generation on King Island by allowing all diesel units to remain offline during periods of time where renewable energy generation exceeds the customer demand, resulting in 100% renewable energy penetration.

The project involves the installation of uninterruptible Diesel Power Supply at the King Island Power Station. This is a commercial technology that is typically used as a backup power supply in industrial processes that are very sensitive to interruptions in the power supply, such as data centers, semiconductor manufacturing facilities and airports [12].

Although common in other industries (with thousands of units installed world-wide), diesel UPS systems are novel in a RAPS application. The proposed UPS consists of a diesel engine, generator/motor connected to a flywheel (rotating mass) and a clutch separating diesel engine and the generator. A conceptual arrangement is shown in Figure 28.20. Many different types of DUPS and UPS systems have been considered but this solution was selected due to its simplicity and suitability for operation at remote areas.

The purpose of the DUPS will be enhanced from its traditional use of providing supply in case of loss of mains to additionally provide a frequency stabilization function. As a frequency stabilizer it will provide fast reserves allowing minimization of reserves carried out by diesel generators, and supporting conditions under which the system could be supplied exclusively by wind generation with diesels on stand by.

It is intended to operate the King Island system with no diesel generators in service whenever possible, supported by diesel UPS operating in synchronous condenser mode with the engine disengaged by the clutch. During a frequency disturbance the high inertia of the flywheel will tend to resist the changes in frequency by releasing inertial energy. The flywheel is dimensioned to provide about 15 second support at the rated output. If the frequency was to fall below the threshold (49-49.5 Hz) or if the reserve levels on DRFC are very low the diesel engines starts and it achieves full speed in about 6 seconds. When the speed of diesel is the same as the system frequency the clutch recloses and diesel engine starts normal operation as controlled by the governor. It is observed that inertia of the flywheel is at least 10 times larger than inertia of standard diesel generator and this slows down significantly system response thus allowing more time for correct operation of control systems. D-UPS does not need to start the diesel engine in case of all frequency disturbances as some will be smoothed by inertial response. The big advantage of D-UPS is that it avoids synchronization of a new diesel engine at a time when the frequency is not stable. It is noted that D-UPS is usually equipped in the oversized generator/synchronous motor allowing significant supply of reactive power to the system and also injecting significant fault current when required.

After startup of diesel UPS and stabilization of the system frequency another diesel generator is started and the diesel UPS returns to its synchronous condenser (standby mode). The benefits of this application include reduction of diesel consumption, reduction in running hours of diesel engines and improved security of the supply. In the case of machine outages the diesel UPS can be used as a primary generator. With wind generation operating, and the DRFC is operating in isochronous mode the frequency is controlled by the DRFC and voltage, reactive power and fault level is supplied by the Diesel UPS in synchronous condenser mode. The diesel UPS will start providing load when available reserve is low or when the frequency falls below the threshold (49-49.5 Hz).

The D-UPS technology is mature and the changes required by this project will be limited to control algorithms, which are able to be implemented by the supplier. The plan is for the D-UPS to operate in parallel with the load; in a typical installation the D-UPS

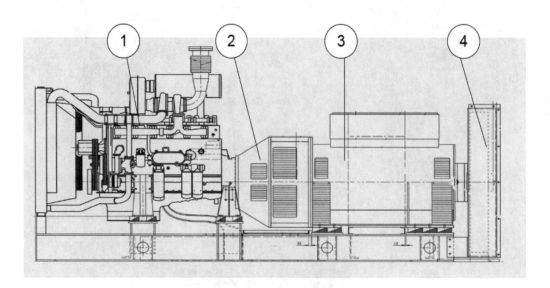

1. Caterpillar 3508BTA diesel engine
2. Electromagnetic Clutch
3. Synchronous Motor/Clutch
4. Flywheel

FIG. 28.20 DIESEL UPS — HITZINGER UNIT AS EXAMPLE (*Source:* Hitzinger)

is used in series with the main electricity supply. It is this change in system topology that entails the main integration risk of this project.

To address this risk the operation of the King Island system with a D-UPS has been investigated through dynamic system modeling (EMTDC). This model will also be used to refine the control logic associated with the D-UPS and will confirm the performance requirements of the D-UPS for the project to be technically feasible. Hydro Tasmania will then work with the D-UPS manufacturer to confirm the suitability of the standard product.

28.7.3 Vanadium Redox Battery Repair or Replacement

As discussed above the VRB unit has not been operational since March 2007. Much development in VRB cell stack technology has taken place since the original VRB project and it is envisioned that the existing VRB can be restored by replacing the cell stacks with those of a new design. The new VRB cell stacks incorporate improved materials, improving reliability. In addition, inverter technology has become commonplace since the original unit was installed; what was once a custom designed unit can now be purchased as a standard product with much greater confidence of fit for purpose and performance.

It is expected that by using the existing VRB infrastructure a refurbish option will be significantly less costly than a complete VRB development project. In addition to the cell stacks, the inverter will also be replaced in order to improve the utility of the battery, and allow it to fulfill additional roles. In addition to the review of VRB rectification options, Hydro Tasmania will be assessing the option for the replacement of the VRB against replacement with an alternate energy storage technology. Other chemical storage technologies, such as zinc bromine (ZBr) and sodium sulphur (NaS) will be assessed in this process.

28.7.4 Wind Farm Expansion

An increase in the wind farm capacity on King Island by up to 4 MW will increase the level of wind power available allowing for a greater percentage of time running in 100% renewable mode. This project derives much of its benefit due to the enabling nature of the D-UPS project. As such, this project cannot proceed without successful performance of the D-UPS. The wind farm will be built with well proven wind turbine generator technology. Development approvals are already in place for an additional two wind turbines on Hydro Tasmania land at the Huxley Hill wind farm site.

28.7.5 Graphite Energy Storage

The function of the graphite energy storage (thermal storage) technology as developed by Australian company CBD Energy Limited is to store large amounts of thermal energy for later use. The process takes advantage of the patented process in which graphite is heated to high temperatures, and the heat extracted using imbedded heat exchangers. This heat is then available for process heat requirements or to be converted back to electricity using a steam or Organic Rankine cycle turbine generator. The construction process for these units is outlined in Figure 28.21.

The Graphite Energy Storage technology is currently being commercialized by CBD Energy and represents a leading thermal storage option in the proposed KIREIP. A carbon block prototype module was constructed by CBD Energy for commissioning and testing in the final quarter of 2007. Independent verification of the performance of the prototype storage module has verified that:

- The Energy Storage Module accept electrical energy and successfully converts into thermal (heat) energy;
- The thermal energy was able to be stored within the module for a period of time with the measured loss of temperature ranging between 4% and 5% per day;
- The stored thermal energy was able to be extracted in the form of steam; and
- The steam produced was of temperature and pressure consistent with the requirements of a conventional steam turbine generator for electricity generation.

It is thought that the use of graphite energy storage technology within the King Island project will demonstrate both the potential

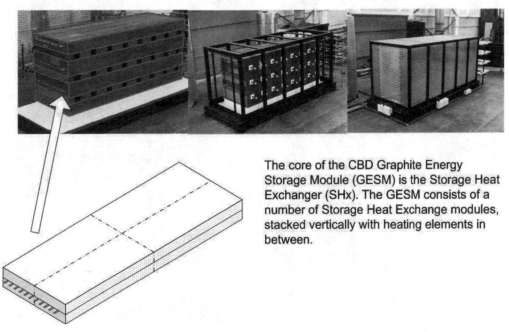

The core of the CBD Graphite Energy Storage Module (GESM) is the Storage Heat Exchanger (SHx). The GESM consists of a number of Storage Heat Exchange modules, stacked vertically with heating elements in between.

FIG. 28.21 GRAPHITE ENERGY STORAGE MODULE-CONSTRUCTION TECHNIQUE (*Source:* CBD Energy)

for the technology in RAPS and as a means of storage for utility scale applications, particularly associated with solar thermal generation, and with smoothing and firming of wind farm output.

28.7.6   Smart Grid

The Smart Grid is the next phase in the evolution of electrical power generation, transmission and distribution.

The King Island smart grid project is being designed to cater for a large variety of potential applications. The initial focus will be on facilitating demand response to improve system reliability and minimize generating costs. Later phases will be implemented in order of priority and as budget allows, improving customer retail services, addressing distributed generation, electric vehicles, increasing system reliability, improving power quality and reducing operating expenses.

The design will accommodate the same infrastructure as in other island power systems and the National Electricity Market (NEM). A key requirement is to design a smart grid that will remain relevant long into the future. Use of international standards wherever practical, including a standards based smart grid design and development process, aims to achieve this future flexibility.

High level use cases have been developed focusing initially on the smart grid coordinating demand response to improve system reliability and minimize generating costs. Use cases have also been developed for improving customer retail services, integrating distributed generation, electric vehicles, increasing system reliability, improving power quality and reducing operating expenses.

The cost and performance of the communication system between system control at Currie Power Station and customer meters, equipment and appliances is expected to significantly shape the smart grid solution. The technical requirements are currently being developed through system modeling of the demand response opportunity.

The plan proposed for implementing the King Island smart grid begins with community engagement prior to rolling out the necessary infrastructure, establishing baseline data and then deploying wind driven demand response.

In the BSI context the Smart Grid can provide matching of the demand to the available supply as much as possible through innovative tariffs such as real time pricing or using communication technologies including wireless and power line carrier technologies to shed and restore controllable loads. The objectives of the Smart Grid project are:

- Enable demand side response to improve reliability and minimize generating costs
- Engage customers in smarter energy consumption through innovative products
- Facilitate distributed generation, storage and electric vehicles
- Improve the performance and cost effectiveness of the power system and network.

The immediate benefits from a Smart Grid project on King Island will be the ability to:

- Switch off diesels at a greater frequency;
- Delaying diesel start up;
- Delay second diesel start up;
- Switch off second diesel at a greater frequency, and
- Provide the infrastructure to allow introduction of time of day tariffs and in particular a wind spill tariff to customers.

At this time the above smart grid benefits have yet to be modeled in terms of their positive impact on the state subsidy.

28.7.7   King Island Centre for Renewable Energy Excellence — KICREE

Hydro Tasmania has proposed a Renewable Energy Centre of Excellence for King Island. If developed this will be a unique and specialized Centre for renewable energy learning and research. It is envisioned that the Centre will:

FIG. 28.22   CONCEPT FOR KING ISLAND CENTRE FOR RENEWABLE ENERGY EXCELLENCE (*Source:* Hydro Tasmania)

- Be a showcase of integrated renewable energy systems to provide renewable energy case studies for conferences, workshops and courses such as a renewable energy implementation course;
- Have the potential to provide a live testing facility for pilot projects concerning new renewable energy products; and
- Provide the chance to utilize Hydro Tasmania's working relationship with UTAS more closely and provide a live testing facility for PhD/MSc research.

The center would leverage off the existing assets and those to be implemented under KIREIP. With future investment new connection infrastructure and research facilities could offer a world class working power system to test and confirm the performance of early development stage renewable technologies. Visually the concept is expressed in Figure 28.22

28.8   CONCLUSIONS

On King Island Hydro Tasmania has achieved some promising results in the increased utilization of renewable energy, successfully operating the power system with wind generation supplying up to 80% of instantaneous demand.

Off-grid systems are invariably more expensive to operate due to high fuel costs. Diesel fuelled power generation is the preferred choice for these systems, as evidenced by the prevalence of such systems internationally. These systems are excellent targets for the displacement of diesel fuel with renewable energy.

Hydro Tasmania's experience indicates that there is a natural limit of the penetration of renewable energy that can be achieved without specifically reinforcing the off-grid power system. In our experience this limit is typically of the order of 30% on an annual basis. Beyond this level we have experienced issues with frequency control, and high spinning reserve requirements to cover the variability of renewable energy sources such as wind turbines.

Hydro Tasmania has developed a dynamic resistive frequency controller to address the issue of spinning reserve under high wind conditions. Early results indicated that this approach will enable the displacement of a further 5-10% of diesel fuel on an annual basis.

The next major limitation on renewable energy penetration is the elimination of minimum diesel operation.

The proposed King Island Renewable Energy Integration Project is a portfolio of innovative projects utilizing new and existing technologies to increase the use of renewable energy in a grid connected network, reducing emissions and improving the quality of power supply. The project will demonstrate the future potential of renewable energy working in conjunction with energy storage and other enabling technologies, to contribute significantly to the development of advanced off-grid systems, and a lower carbon intensity Australian Electricity Market.

The proposed project aims to achieve 100% instantaneous renewable energy penetration, and a 65% annual contribution from renewable generation. Switching off of the last diesel requires the introduction of alternative sources to provide for voltage control and fault current to enable correct protection operation and system inertia. At this point in the project design the D-UPS is believed to be the most appropriate technology for the provision of these services.

The benefit of using a system of the scale of the BSI for this project in demonstrating these scenarios includes:

- In general a smaller system will experience problems much earlier than a large system so it is a very good test bed and will demonstrate the development path for enabling technologies to allow significant adoption of renewable energy in larger systems over a 10-year to 20-year timeframe;
- Wind variations in smaller systems (with a much higher installed capacity ratio) are much greater and faster than in large systems, allowing the identification and resolution of future issues likely to be encountered in large systems.
- Cost effectiveness of pilot projects, in general some of the enabling technologies are not considered commercially viable, however, testing them in small scale system with much higher base electricity cost (particularly systems using an expensive fuel such as diesel) makes pilot projects more attractive — so this can be achieved in a cost effective manner;
- Establishment of test conditions with very high penetration of wind generation is generally not possible in a large interconnected system — having control of an entire power system is a truly unique opportunity and would prove to be an excellent facility for international research and to demonstrate the world class innovation achieved in the Australian renewable energy sector.

It is envisioned that the lessons learned in the development and operation of the King Island power system will be applicable to other off-grid systems throughout the world.

28.9   REFERENCES

1. Community Service Obligation (CSO), http://www.treasury.tas.gov.au/, Accessed: 13/10/2010.

2. Schmidhuber, *Biofuels: An emerging threat to Europe's Food Security? Impact of an Increased biomass use on agricultural markets, prices and food security: A longer-term perspective*, 2006, http://www.notre-europe.eu/uploads/tx_publication/Policypaper-Schmidhuber-EN.pdf, Accessed: 13/10/2010.

3. Renewable Energy Demonstration Program, http://www.ret.gov.au/energy/energy%20programs/cei/acre/redp/Pages/default.aspx.

4. Diesendorf, *The base load fallacy*, 2007, http://www.sustainabilitycentre.com.au/BaseloadFallacy.pdf.

5. Pacific Island Renewable Energy Project, Pacific regional Energy Assessment, 2004, SPREP, UNDP, GEF, PIREP.

6. Diesel & Wind Systems, Denham, http://www.daws.com.au/projects/Denham.html, Accessed: 12/10/2010.

7. Beer, et al. "Final Report (EV45A/2/F3C) to the Australian Greenhouse Office on the Stage 2 Study of Life-cycle Emissions Analysis of Alternative Fuels for Heavy Vehicle", 2002.

8. Skyllas-Kazacos, M. An Historical Overview of the Vanadium Redox Flow Battery Development at the University of New South Wales, Australia, http://www.ceic.unsw.edu.au/centers/vrb/overview.htm, Accessed: 13/10/2010.

9. NREL, Power Flow Management in a High Penetration Wind-Diesel Hybrid Power System with Short-Term Energy Storage, 1999, http://www.nrel.gov/docs/fy99osti/26827.pdf. Accessed: 13/10/200.

10. PowerCorp, Mawson Wind-Diesel http://www.pcorp.com.au/index.php?option=com_content&task=view&id=12&Itemid=73. Accessed: 13/10/2010.

11. VRB Energy Storage for Voltage Stabilization Testing and Evaluation of the PacifiCorp Vanadium Redox Battery Energy Storage System at Castle Valley, Utah 1008434, , http://mydocs.epri.com/docs/public/000000000001008434.pdf. Accessed: 13/10/2010.

12. Hitzinger, *Dynamic Diesel UPS Pure sine wave 24/7/365*, http://www.hitzinger.at/assets/pdf/hitzinger_usv_e.pdf, Accessed: 13/10/2010.

HEAT EXCHANGERS IN POWER GENERATION

Stanley Yokell and Carl F. Andreone

29.1   INTRODUCTION

This chapter describes shell-and-tube and plate and frame types of power plant heat exchangers and tubular closed feedwater heaters and the language that applies to them. The chapter briefly discusses Header Type Feedwater Heaters and their application and use. It defines the design point used to establish exchanger surface and suggests suitable exchanger configurations for various design-point conditions and the criteria used to measure performance. It does not cover power plant main and auxiliary steam surface condensers because of the differences in how they are designed and operated The chapter briefly discusses the effects on the exchanger of normal and abnormal deviations from design point during operation.

29.2   SHELL-AND-TUBE HEAT EXCHANGERS

Shell-and-tube heat exchangers consist of tubes manifolded together in such a way that one fluid (the tube-side fluid) flows through inside of the tubes and another (the shell-side fluid) flows outside the tubes. In U-tube configurations, the tube ends are manifolded together in a single plate called a tubesheet or tubeplate. In straight tube configurations, there are two tubesheets. U-tube configurations are always an even number of passes. Straight tube configurations may be single or multipass.

The tubes are supported and flow of shell-side fluid is guided along the tube length by perforated plates or other devices called tube supports or baffles. The assembly of tubesheet, tubes, and tube supports or baffles is called a bundle or nest.

To guide the tube-side fluid into and conduct it from the tubes, a chamber is fastened to each tubesheet at its tube-end face. The seal of the chamber to the tubesheet is hydraulically tight. Depending upon its mechanical configuration, the chamber is called a channel, bonnet, or return cover.

The rear or inner face of the tubesheet(s) and the external surfaces of the tubes and the tube supports or baffles that make up the exterior of the bundle or nest are contained in a chamber called the shell. Its containment of the bundle of tubes is hydraulically tight.

If both tubesheets of a straight-tube unit are integrally fastened to the shell, the exchanger is called a fixed-tubesheet exchanger. If one tubesheet is fastened immovably to one end and the other tubesheet is free to move axially, the exchanger is called a floating-head exchanger. The tubesheet that does not move is called the stationary tubesheet. The end of the unit to which the stationary tubesheet is fastened is called the stationary end. The tubesheet that is free to move is called the floating tubesheet. The containment attached to the floating tubesheet is called the floating head.

U-tube and floating-head bundles may be welded to the shell at the stationary end or they may be held in place by flanged, gasketed joints. The latter are designated removable bundle heat exchangers. The closure of the shell at the U-bend or floating-head end is called a shell cover.

Shell-side fluid may also be arranged for single-pass, two-pass or multi-pass axial flow. It may also be arranged to flow in one of these ways: (1) across the tubes from a centrally located inlet to an outlet on a common centerline (cross-flow); (2) across the tubes split into two equal masses by a plate at the shell center line (split flow); (3) into the tube bundle through two equal-sized inlet connections equispaced along the shell, with each incoming stream split by a plate at the shell center line (double split flow); and (4) into a centrally located inlet and out through two separate, equal-sized outlets near the shell ends (divided flow). Figure 29.1 is picture of a typical shell-and-tube heat exchanger.

Closed Feedwater Heaters are a special type of shell-and-tube heat exchanger in which the inlet steam may be segregated in an enclosure surrounding a number of tubes whose function is to desuperheat steam extracted from various turbine stages optimized to reheat the feedwater, thereby optimizing the fuel needed too increase the plant efficiency and the fuel needed to generate the required steam for the turbine. Steam exiting the desuperheater enters the next zone called the condensing zone. A third group of tubes may be enclosed in a subcooling zone to subcool condensate. Feedwater heaters in boiling water and pressurized water nuclear power plants do not receive superheated steam, therefore they do not have desuperheating zones.

The Tubular Manufacturers Association's TEMA Standards [1] and the various Heat Exchange Institute Standards, such as HEI Standards for Closed Feedwater Heaters [2], HEI Standards for Power Plant Heat Exchangers [3] have systems widely used to designate the types of tubular exchangers used in power plants and schemes of designating exchanger sizes. Figure 29.2 shows the TEMA type designations. The HEI Standards have a figure that shows the various type designations. The HEI does not permit reproducing the figure. Readers can view the figure in the original standards.

FIG. 29.1 SHELL-AND-TUBE HEAT EXCHANGER [ATLAS Industrial Manufacturing, Clifton, New Jersey]

29.3 PLATE HEAT EXCHANGERS

A plate heat exchanger is a type of heat exchanger that uses metal plates to transfer heat between two fluids. It consists of a series of thin (0.005mm to 0.007mm), corrugated plates gasketed, welded or brazed together depending on the application of the heat exchanger. The plates which are available in sizes up to 4 ft 0 in. × 8 ft 0 in. (~1220 mm × ~2,500 mm) are compressed together with long bolts around the periphery in a rigid frame to form an arrangement of parallel flow channels with alternating hot and cold fluids. Figure 29.3 illustrates a typical plate heat exchanger.

Its major advantage over shell-and-tube exchangers is that, because the fluids spread over thin corrugated metal plates, they encounter more heat transfer surface in a given amount of space than do shell-and-tube units. The greater surface greatly increases the transmission of energy and rapidity with which the temperature of the exiting fluids changes.

Their major disadvantage is that the plates are assembled with gaskets that have to be made leak tight by means of long rods threaded on the ends. In recent years, some manufacturers have produced plate heat exchangers in which the gasketed joints are eliminated by welding the plates together. Such construction is acceptable only when the fluids will not foul the surfaces. The gap between adjacent plates is very small, usually less than 2.5 mm (0.10 –in.). Filters are required in many applications with river water or brackish sea water.

Plate heat exchangers (PHEs) were introduced to industry in 1923. The plate heat exchanger design restricts its use to medium- and low-pressure liquids. Welded and brazed units can accept higher pressures.

The plates are manufactured in hydraulic presses that press the thin sheets over dies to produce the corrugations. The corrugations in adjacent plates are at right angles to each other to provide 1-1/4-mm to 1-1/2-mm channels through which the fluid flows from plate pair to plate pair. Most plate heat exchangers used in power plants have stainless steel plates sealed with rubber or rubberized gaskets cemented to gasket surfaces on the plate edges.

The plates produce an extremely large surface area, which allows for the fastest possible transfer. Making each chamber thin ensures that the majority of the volume of the liquid contacts the plate, again aiding exchange. The troughs also create and maintain a turbulent flow in the liquid to maximize heat transfer in the exchanger. A high degree of turbulence can be obtained at low flow rates and high-heat transfer coefficient can then be achieved.

As compared to shell and tube heat exchangers, the temperature approach in a plate heat exchangers may be as low as 1°C whereas shell and tube heat exchangers require an approach that gives 5°C or more. For the same amount of heat exchanged, the size of the plate heat exchanger is smaller because of the large heat transfer area afforded by the plates (the large area through which heat can travel). Expansion and reduction of the heat transfer area is possible in a plate heat exchanger.

Another advantage of the plate heat exchanger is that it is easily dismantled for inspection and cleaning. The plates are also easily replaceable due to the fact that plates can be removed and replaced individually. The main weakness of the plate and frame heat exchanger is the necessity for the long bolts which hold the plates together. Although these bolts are seen as a weakness towards this type of heat exchanger, it has been successfully run at high temperatures and pressures.

All plate heat exchangers look similar on the outside. The difference lies on the inside, in the details of the plate design and the sealing technologies used. Hence, when evaluating a plate heat exchanger, it is very important not only to explore the details of the product being supplied, but also to analyze the level of research and development carried out by the manufacturer and the post-commissioning service and spare parts availability.

29.4 HEAT EXCHANGERS USED IN POWER PLANTS

Except for Closed Feedwater Heaters, which have their own set of HEI Standards, heat exchangers used in power plants are usually built in accordance with the requirements of the HEI Standard for Power Plant Heat Exchangers [3]. The standards specify acceptable materials of construction. Users of the standards use their data sheet templates to specify design point conditions, off design point conditions, and possible excursions that could affect the life of the exchangers. Virtually all heat exchangers used in power plants are built to the rules of Section VIII Division 1 of the ASME Boiler and Pressure Vessel Code (Code) [4]. Those exchangers installed in areas that are inaccessible because of radiation may be built to Section III of the Code. Some miscellaneous exchangers in the Reactor Building may be built to Section III Class C.

The following lists most of the exchangers used for various services, the heat transfer mode and the TEMA types used for the service.

1. Turbine building closed cooling water heat exchangers
2. Service water exchangers
3. Shutdown cooling water exchangers

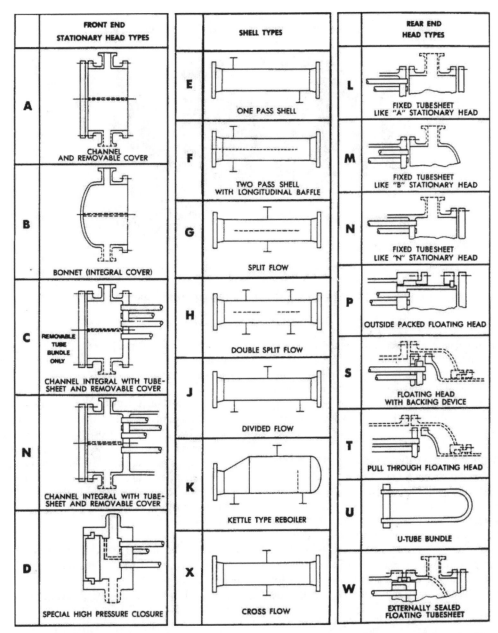

FIG. 29.2 TEMA TYPE DESIGNATION SYSTEM [Tubular Exchanger Manufacturers Association Tarrytown, New York Standards of the Tubular Exchanger Manufacturers Association, 9th Ed., 2007, The Tubular Manufacturer's Association, Tarrytown, New York.]

4. Turbine lube oil coolers
5. Gland steam condensers
6. Closed feedwater heaters
7. Header type feedwater heaters
8. External drains coolers
9. Steam generator blowdown coolers
10. Bearing oil coolers
11. Emergency diesel engine jacket water coolers
12. Sample coolers.

With the exception of closed feedwater heaters, the operating pressures on both shell and tube sides are moderate. The tube sides of the highest feedwater heaters in the cycle operate at the feedwater pump pressures. The following briefly discusses these exchangers.

29.4.1 Turbine Building Closed Cooling Water Heat Exchangers

The heat transfer mode of these exchangers is sensible heat transfer. The TEMA types used are AEL, AEM. BEM, AFL, AFM, BFM, AGL, AGM, BGM usually arranged horizontally. The units have large diameters. The shells may be E type (single pass), F type (two pass with longitudinal baffle) or G type (split flow) depending upon the flows and temperatures. The tube side is usually multipass. Channel construction depends upon accessibility requirements. There are few operating problems with these heat exchangers.

29.4.2 Service Water Exchangers

The heat transfer mode of these exchangers is sensible heat transfer. The TEMA types used are AEL, AEM. BEM, AFL, AFM,

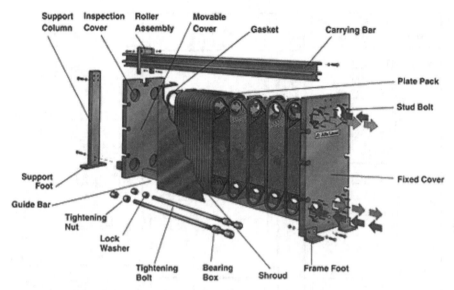

FIG. 29.3 ALL WELDED PLATE TYPE BEARING OIL COOLER [Alpha Laval, Richmond, VA]

BFM, AGL, AGM, BGM usually arranged horizontally. The units have large diameters. The shells may be E type (single pass), F type (two pass with longitudinal baffle) or G type (split flow) depending upon the flows and temperatures. The tube side is usually multipass. Channel construction depends upon accessibility requirements. There are few operating problems with these heat exchangers.

29.4.3 Shutdown Cooling Water Exchangers

Shutdown cooling water exchangers are provided in pairs. They are used only during shutdowns. Consequently the practice is periodically to test the integrity of the exchangers. The heat transfer mode of these exchangers is sensible heat transfer. The TEMA types used are AEM and NEN. Channel configuration depends upon accessibility requirements. The tube side is usually single or two pass. Tube-to-tubesheet joints are expanded. Some units are built with seal welded and expanded tube-to-tubesheet joints for maximum leak tightness. The terminal temperatures are sufficiently

close so shell expansion joints are not required. There are few operating problems. Units that lie idle between shutdowns should be cleaned regularly.

29.4.4 Turbine Lube Oil Coolers

The heat transfer mode of these exchangers is sensible heat transfer. The TEMA types used are BEU and AEU. Channel configuration depends upon accessibility requirements. Large vertical channel up. Tube joints are expanded. Some units have the tubes seal welded to the tubesheets to avoid the possibility of water entering the lube oil which flows in the shell. The shell is baffled to the extent that the allowable pressure drop permits. There are few operating problems.

29.4.5 Gland Steam Condensers

The heat transfer mode of these exchangers is transfer of latent heat of condensation transferred to the cooling stream. The TEMA types used are BEU and AEU, AEL, AEM, NEN, BEM. The

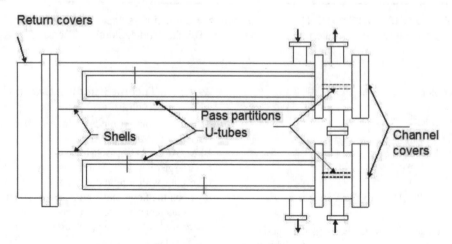

Schematic of Multitube Heat Exchanger Used as
Boiler Blowdown Cooler

FIG. 29.4 TYPICAL STEAM GENERATOR BLOWDOWN COOLER [Schematic drawing by MGT Inc., Boulder, CO]

channel type depends upon accessibility requirements. The fixed tubesheet types are single pass with shell expansion joints. This exchanger takes the full condensate flow which requires the unit to be short and fat. Tube joints are expanded but may be seal welded. There are few operating problems.

29.4.6   Closed Feedwater Heaters

Closed feedwater heaters in fossil stations may have one, two, or three heat transfer zones. Those in nuclear plants may have one or two. The distinction is that steam from the boiler in fossil plants is superheated, whereas steam from the steam generator in nuclear ones is saturated or wet saturated steam. The mode of heat transfer in desuperheating and subcooling zones is sensible. In condensing zones, latent heat of condensation is transferred to the feedwater.

Most modern feedwater heaters are U-tube designs, constructed with either full access or manway access channels (see Fig. 29.5). Most U-tube construction has the channel welded to the tubesheet, the tubesheet welded to a skirt and the skirt welded to the shell. To extract the bundle requires cutting the shell at a designated cut line. However, some few heaters have the channel and shell flanged enabling bundle extraction. At one time, some feedwater heaters used floating head designs such as TEMA types AES and DES. For complete descriptions and nomenclature, the reader should refer to the HEI Closed Feedwater Heater Standards.

Subcooler designs may be full pass, partial length or partial pass full length, with full pass, partial length being the most common. Feedwater heaters may be mounted horizontally or vertically, with some heaters mounted in the condenser neck.

Properly designed and operated feedwater heaters present few problems providing that they are not in cycling service. Those in cycling service may be subject to corrosion because of incomplete removal of non-condensibles during startups or having air fill the shell during shut down with tubes that will likely corrode. Tube bundles having corrosion-prone materials, such as carbon steel or copper alloys, must be protected by suitable blanketing arrangements that will prevent the air coming in contact with the wet tube surface. Proper operation includes careful monitoring of the liquid level, making sure operating vents are properly functioning and startup vents are used when bringing the heaters on line.

Feedwater heaters are installed in strings with the highest point heater receiving the highest pressure extraction steam from the turbine and the lowest point heater the lowest pressure extraction steam. The operation of the string is affected if one or more heaters is bypassed. Feedwater heaters may also be installed in parallel strings with one string carrying most of the load when the second string is bypassed.

Procurement of feedwater heaters requires preparing specific for-the-purpose procurement specifications. These include feedwater data sheets that describe design point conditions, overload conditions and low load conditions. Typically the User specifies the anticipated conditions and the Manufacturer fills in the data that reflects the Manufacturer's prediction of steam and condensate properties, heat transfer rates, outlet temperatures, and pressure drops.

It is important for Users to examine Manufacturers' designs for conformity to procurement specification including the manufacturer's procedures and standards, ASME Code calculations, vibration calculations and other calculations that determine the longevity of the heaters.

29.4.6.1 Duplex Closed Feedwater Heaters Duplex feedwater heaters are arranged horizontally and are normally inserted

into the condenser neck. A duplex heater consists of two heat exchanger modules, for example Low Pressure Heater 1 and Low Pressure Heater two in a common shell. The heater modules are either pure condensing or have a condensing zone with an integral drain cooler. They are separated by a partition wall in the shell. Turbine extraction steam of different pressure and temperature is fed through separate steam inlet nozzles. The feedwater flows from the channel through the U-tubes of the first module while extraction steam the lower pressure condenses on the outer surface of the tubes. The feedwater heated in module 1 flows through the tubes of the second module and heated to a higher temperature by the higher pressure extraction steam entering the second module steam inlet (Fig. 29.6).

The condensate is discharged at the shell bottom through nozzles with the flow from the second heater controlled by means of a valve that controls the levels in the shell. If a module has an integral drain cooler zone, part of the bundle is flooded.

Venting channels are set in the bundle lane at the lowest pressure. Flows in the channel are separated by dividing it into three spaces using internal shrouds or angular plates. The channel inlet nozzles is connected to the first shroud and the water outlet nozzle to the second shroud. Between the shrouds, the feedwater flows from heater 1 to heater 2.

29.4.6.2 Header Type Feedwater Heaters Header type, high-pressure feedwater heaters consist of a tube bundle with multiple bend tube coils enclosed in a shell. The tubes are individually welded to separate inlet and outlet header pipe nipples. They are designed as single-, two-, and three-zone heaters. Single-zone ones are straight condensing; two-zone ones have a condensing and subcooling zone and three-zone ones have desuperheating, condensing and subcooling zones. They may be installed horizontally or vertically.

The literature on Header Type Feedwater Heaters states that they have lower maximum stresses during transient operation condition and therefore fewer potential failure mechanisms than tubesheet heaters [5]. This is because the design minimizes thermal stresses by isolating the cold feedwater from the hot water stream. Header type feedwater heaters are also known as "snake" heater because of the sinuous passage of the feedwater tubes. It has traditionally been used in German high-efficiency, coal-fired plants. But it is seldom used in the United States. The reluctance in the United States has to do with the perception that no domestic heater manufacturer is sufficiently skilled to produce reliable header type heaters; the construction requires welding the tubes to nipples, which is one cause for concern.

The world's largest header-type feedwater heaters weigh more than 270 tons when empty. Three header type heaters and a separate desuperheater usually form a complete high-pressure train. The heaters are usually installed upright and, depending upon the piping system, have three or four water passes.

The principal cost savings for header type feedwater heaters is that the cylindrical headers require only relatively thin walls compared with conventional U-tube heaters designed for the same operating conditions. Reference [5] states that the wall thickness is only 10% to 20% of the wall thickness of conventional high-pressure hemi-head channel design U-tube heaters.

29.4.7   External Drain Coolers

The heat transfer mode of these exchangers is sensible heat transfer. The TEMA types used are AEL, AEM. BEM, NEN. Temperature levels are low on both sides with terminal temperatures relatively close,

FIG. 29.5 THREE-ZONE, HIGH-PRESSURE, MANWAY ACCESS FEEDWATER HEATER UNDERGOING HYDROSTATIC TESTING [Photograph by MGT Inc., Boulder, CO During Shop Inspection]

obviating the need for expansion joints. There are few operating problems. These coolers are also short and fat. They take all the feedwater matching up the same number of parallel strings as the heaters.

29.4.8   Steam Generator Blowdown Coolers

Blowdown is a hot mixture of feedwater and flashed steam. The heat transfer mode is a combinations of sensible heat transfer and condensation of flashed steam. Blowdown coolers typically are multitube U-tube types. See Fig. 29.4 for a typical blowdown cooler design. See the accompany figure for typical construction. There may be differential expansion problems between the passes of horizontal units with shell side condensation if the subcooling range is long.

29.4.9   Bearing Oil Cooler

The heat transfer mode of these exchangers is sensible heat transfer. Many utilities employ older, radiators with motor driven blowers, which have ineffective heat transfer rates, and require a large footprint, and frequent and involved cleanup and maintenance. Current technology is to use plate and frame units which have very close temperature approach capability. Construction is all welded with built in cleaning connections and self-draining capability. Depending on the source of the cooling water, unit may require frequent flushing and cleaning.

29.4.10   Emergency Diesel Engine Jacket Water Coolers

The heat transfer mode of these exchangers is sensible heat transfer. The TEMA types used are TEMA types AEL, AEM. BEM, NEN, and Plate heat exchangers. Shell and tube types do not usually require shell expansion joints. Tubes are expanded into grooved tubesheet holes. Plate exchanger types may be gasketed or all welded. There are few operating problems.

29.4.11   Sample Coolers

The heat transfer mode of these exchangers is sensible heat transfer. TEMA type does not apply. Sample coolers are small (1 to 2 sq. ft surface) catalog items with a tube(s) coiled in a tank. There are few operating problems.

29.5   DESIGN POINT

The design point is a set of specific steady state operating conditions, which fixes the heat content and state of the fluids entering and leaving an exchanger. The design point determines exchanger size. Exchanger configuration and construction are affected by design point conditions.

29.5.1   Design Point Conditions

The following conditions establish the thermal and mechanical design of a heat exchanger:

A. Inlet and outlet temperature of each stream
B. Operating pressure of each stream
C. Flow quantities of each stream
D. Composition of each stream
E. Condition of state of each stream
F. Permissible pressure drop in each stream.

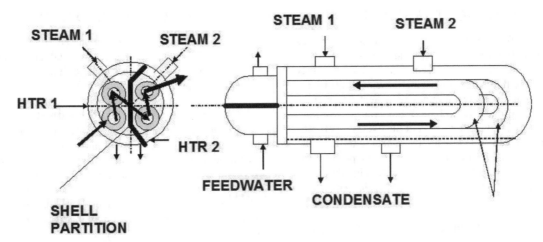

FIG. 29.6   SCHEMATIC REPRESENTATION OF DUAL FEEDWATER HEATER [Schematic drawing by MGT Inc., Boulder, CO]

29.5.2   Effect of Design Point on Configuration

For various fluid states, the design point affects choice of configuration, mechanical design and construction. The heat transfer mode is related to types of exchangers and design and construction details. The use to which the exchanger will be put also may affect the configuration. For example, a fluid that must remain contamination free should circulate at a higher pressure than the other fluid. This may affect the configuration. In addition, such a situation might call for double tubesheet construction.

For various fluid states, the following paragraphs show how design point affects choice of configuration, mechanical design, and construction. The heat transfer mode is related to types of exchangers and design and construction details. The reader of these notes should be aware that not all possibilities are covered and that the use to which the exchanger will be put also may affect the configuration.

In assessing the suitable configuration, designers consider the following:

1. Heat Transfer Mode
2. Type to Use Suitable for the Heat Transfer Mode.

A partial list of heat transfer modes and types of exchangers used is as follows:

TEMA types AEL, AEM, BEM, AFL, AFM, BFM, AGL, AGM, BGM, NEN are usually arranged horizontally.

  a) Gases heated or cooled leaving as gases.

Some feedwater heaters are arranged vertically to save space.

  b) Pure vapor condensing isothermally
  c) Mixed vapors of insoluble liquids condensing
  d) Mixed vapors of soluble liquids condensing
  e) Liquids heated or cooled
  f) Gases desuperheated, vapors condensed
  g) Vapors condensed and condensate subcooled (Feedwater heaters in fossil and nuclear stations
  h) Gases desuperheated, vapors condensed and condensate subcooled (Feedwater heaters in fossil stations)
  i) Vapors with large amounts of non-condensibles entering; condensate and saturated non-condensibles leaving

  j) Non-condensibles with vapor entering; non-condensibles and condensate leaving
  k) Superheated liquid in flashed and heated; two-phase vapor-liquid out
  l) Gas–liquid mixture in and out
  m) Saturated liquid in; two-phase liquid-vapor out

3. Construction
4. Possible Operating Problems

29.5.3   Normal Deviations from Design Point

Some normal operating deviations from design point occur when there is a difference in cooling water temperature between winter and summer and changes in demand elsewhere in the system.

Cooling water changes are dealt with by using bypasses to maintain constant service water velocity in the cooling stream. Changes in demand are dealt with by bypassing the process streams and recirculating to keep flow rates through the exchanger constant.

In feedwater heaters, changes in demand cause changes in the flow rates and pressures of the steam and feedwater to the heater. Ordinarily, when demand is moderately below capacity, the reduction in feedwater demand is balanced by generating less steam resulting in greater pressure drops between stages with correspondingly lower temperatures. In this circumstance, lower kW output is achieved.

When demand is intermittent and it is economical to take a boiler off the line thereby reducing feedwater demand, heaters may be bypassed or shut down. Putting heaters into and taking them out of operation frequently can result in more rapid erosion, corrosion, and mechanical damage than what would occur for continuous operation in the design point range.

29.5.4   Abnormal Deviations from Design Point

Abnormal deviations from Design Point conditions result from the following:

1. Insufficient design surface
2. Excess design surface
3. Inadequate venting
4. Inadequate level control in vaporizers, condenser-subcoolers and feedwater heaters.

29.6 PERFORMANCE PARAMETERS

To assess the performance of a heat exchanger, it is necessary to measure the flows through each side, the inlet and outlet temperatures and pressures on each side and to know the thermal properties of the fluids flowing. Using these data, the performance engineer can calculate the amount of heat being transferred (duty), the log mean or weighted temperature difference between the shell- and tubeside flows and the overall coefficient of heat transfer achieved and compare these data with the manufacturer's heat exchanger data sheet. By plotting performance versus time, the performance engineer can determine the rate of fouling and estimate the appropriate interval for cleaning the exchanger.

29.6.1 Performance Parameters for Closed Feedwater Heaters

Two parameters are used to describe the performance of Closed Feedwater Heaters-Terminal Temperature Difference (TTD) and Drains Subcooler Approach (DCA). The HEI Standard for Closed Feedwater Heaters defines TTD as the difference between saturation temperature corresponding to the entering extraction steam and the outlet feedwater temperature. The TTD may be positive or negative. The standard defines DCA as the temperature difference between the drains leaving the shellside of the heater and the entering feedwater on the tube side.

29.7 OVERLOADS AND LOW LOADS

An overload raises the duty demanded of a heat exchanger. The balance is upset, causing changes in metal temperatures. Shell expansion joints may not be adequate in fixed tubesheet equipment. Tube-to-tubesheet joints may fail.

29.7.1 Overloads in Closed Feedwater Heaters

If the turbine operates between its design load and its overload limit, more steam is demanded than at the design point. To provide the steam, the steam generator or boiler requires more feedwater. To maintain the specified terminal temperature difference (TTD) and drains cooling approach (DCA), more bleed steam is required. Therefore, the flows on both sides of the exchanger are increased. Steam flow is an uncontrolled flow that is a function of the operating conditions. At overload conditions, the heater will most likely maintain the specified TTD but the outlet temperature will be lower than the design basis.

On the shell side of the heater, the total pressure loss will be increased. If the steam and drain nozzles are not sized to handle the overload flows, the tubes and nozzles will be eroded because of the high steam velocity. Steam distribution may be impaired if distribution dome area is not large enough for the overload flow.

If the zonal cross flow baffle and tube support systems are not designed for the overflow, there will be vibration damage to the

FIG. 29.7 TYPICAL BOURDON TUBE TEST GAGE USED IN TESTING HIGH PRESSURE FEEDWATER HEATERS [Photograph by MGT Inc., Boulder, CO During Shop Inspection]

tubes and support plates. Vibration damage to the tubes will be located near the back face of the tubesheet, at the penetrations of the tube supports by the tubes, approximately midway between tube supports and in the U bend region. The baffles and tube supports will also be damaged by being rubbed and pounded by the tubes.

In desuperheating feedwater heaters, the desuperheater may not be able to handle the increased flow unless provision has been made in advance in sizing the unit or to bypass the excess extraction steam to the condensing zone.

In addition to the extra steam demand, the quantity of drains flow to the heater will be increased. The amount of steam flashed from drains cascaded from higher pressure heaters will, therefore, increase leading to surges and possible difficulties in controlling liquid level. If the shell contains a flash chamber for flashing the cascaded drains, and the chamber is only large enough to handle design point flow, it may flood and shift flashing to the condensing zone.

On the tube side, the higher feedwater flow can erode nozzles, pass partitions and pass partition covers and the tubesheet and tube ends. In addition, the pressure drop through the tubes will be increased.

29.7.2 Low Loads in Closed Feedwater Heaters

In fossil stations low load conditions in feedwater heaters with desuperheating zones result in the tube metal temperature being too low at the desuperheater exit. Consequently the steam will start to condense while the steam velocity is high and droplets of condensate will be hurled at the adjacent tube support and the tubes eventually eroding the support or tubes causing the tubes to erode and leak or the unsupported tube span to become longer than required to resist destructive tube vibration.

29.8 TESTING POWER PLANT HEAT EXCHANGERS FOR STRUCTURAL INTEGRITY AND TIGHTNESS

The purpose of the ASME-required hydrostatic testing ASME Code-stamped heat exchangers is to test the ability of the structure to withstand the test pressures. For this purpose, Manufacturer's use calibrated Bourdon tube type test gages as shown in Fig. 29.7. The ASME Code does not permit visible leakage during hydrostatic testing.

However, in constructions in which the back face of the tubesheet is not visible during hydrostatic testing, Users should be aware that hydrostatic testing does not disclose minute leaks (weeping) through the tube-to-tubesheet joints because the graduations of the test gages are too coarse to indicate minute pressure drops [5]. Depending upon the service of the exchangers such leaks may be insignificant or they may be unacceptable. In high-pressure feedwater heaters, minute leaks of feedwater through the tube-to-tubesheet joints can cause wire drawing, which erodes the tubesheet metal giving the eroded part the appearance of worm holes.

Most high-pressure and some intermediate-pressure feedwater heaters have the channel side faced with weld metal cladding. Worm holing can create cavities in the softer tubesheet metal leading to loss of capacity and sometimes to end of life of the heater.

To avoid such outcomes, prudent purchasers require Manufacturers to perform non-destructive testing in addition to the ASME Code required hydrostatic testing. These consist of liquid penetrant examination of tube-to-tubesheet joints, gas-bubble testing of tube-to-tubesheet joints, and mass spectrometer testing the heater with helium–air mixture in the shell and sniffer examination of the tube-to-tubesheet joints. The requirements for the procedures used in such testing are given in Section V, Non-destructive Testing of the ASME Code.

29.9 REFERENCES

[1] Standards of the Tubular Exchanger Manufacturers Association, 9th ed., 2007 The Tubular Exchanger Manufacturer's Association, Tarrytown, New York.

[2] Standards for Closed Feedwater Heaters, 8th Ed, 2008, The Heat Exchange Institute, Cleveland, Ohio.

[3] Standards for Power Plant Heat Exchangers, 4th Ed., 2004, The Heat Exchange Institute, Cleveland, Ohio.

[4] ASME Boiler & Pressure Vessel Code, American Society of Mechanical Engineers, New York, New York. A new edition is published at three-year intervals with Addenda published biennially along with Code Cases and Interpretations.

[5] Youssef, M., "Header type feedwater heaters as retrofits for cycling units." Power-Gen Europe 94: May 25-27, 1993 Paris.

[6] Yokell, Stanley, Pressure Testing Feedwater Heaters And Power Plant Auxiliary Heat Exchangers, Paper No. 2010-27106, presented at the ASME PowerGen Conference, July 2010, Chicago.

WATER COOLED STEAM SURFACE CONDENSERS

K. P. (Kris) Singh

30.1 INTRODUCTION

The subject of this chapter is the capital equipment in a steam power plant that is used to condense the exhaust steam from the lowest pressure turbine, by using water as the cooling medium. In the applied heat transfer literature, such heat transfer equipment is often simply referred to as the "Surface Condenser." A surface condenser is necessarily a large piece of equipment because more than 60% of the thermal energy produced by a power plant ends up as low enthalpy (waste) heat. This is because of the inherent thermodynamic limitation of the Rankine Cycle, which must be rejected by the condenser to the environment. The heat transfer area in a power plant's surface condenser easily dwarfs that in any other heat exchanger in the plant.

Classical thermodynamics holds that the lower the temperature of the heat sink, the higher the efficiency of the Carnot cycle. Therefore, attaining the lowest possible condensing temperature in the heat sink of the Rankine Cycle—the surface condenser—is a primary goal in surface condenser design. Since the saturation temperature and pressure of steam vary in a proportional manner at low pressures, the objective of low condensing temperature translates to that of a low condenser operating pressure. Accordingly, surface condensers are operated at as high a level of vacuum (typically 1.0 to 2.5 in. of mercury absolute) as the quantity and temperature of the cooling water would allow. The subatmospheric condition in the condenser and portions of turbine assembly and auxiliaries promotes leakage of air into the system. In boiling water type of nuclear plants the motive steam acquires additional non-condensibles due to radiological disassociation of water into hydrogen and oxygen. Whatever their origin, these so-called non-condensibles tend to collect in the lowest pressure region in the power cycle which is the steam space in the surface condenser. Unless removed continuously and efficiently, they may sharply interfere with the heat transfer process in the condenser. Kern [8] quotes observed data of Othmer which indicates that even 1% volumetric concentration of air in steam reduces the condensing film coefficient by approximately 45%.

The oxygen in the non-condensibles is another source of concern. Oxygen is known to actuate corrosion of condenser internals. Austenitic stainless steel tubing can become highly susceptible to stress corrosion in the presence of even small concentrations of oxygen. Therefore, efficient collection and removal of non-condensibles is of paramount importance in surface condenser design.

A reliable design for collection and expulsion of the non-condensibles is particularly important in the surface condensers used in geothermal plants where the steam extracted from the ground has a large fraction of associated gases.

Toward that end, the designer should ensure that the condensed steam — the condensate — does not have the opportunity to come in contact with the non-condensibles, lest it reabsorb some of the non-condensibles. Theoretically, condensate at its saturation temperature cannot absorb any non-condensibles. However, as we will discuss later, the mechanism of condensation implies a certain amount of subcooling no matter how artful the design. A faulty design can further accentuate the subcooling and increase the non-condensible absorption. Moreover, a subcooled condensate also brings down the plant cycle efficiency since the heat lost to the cooling water (outside environment) in the process of subcooling must be supplied by the boiler.

In addition to serving as the "heat sink" for the power plant, the surface condenser also acts as the dump for vents and drains and turbine bypass steam at various levels of pressure and quantity. While some of these so-called dumps are continuous, most are intermittent or sporadic.

Finally, the surface condenser performs a secondary, yet quite important, function of cooling the condensate so it can be "polished" (removed of solids and other dissolved impurities) in filters and demineralizers. These multi-faceted functions of the surface condenser accord it a pivotal place in the power cycle.

The above summary of condenser design considerations, however, only dwells on the system aspects. Many more considerations pertain to the condenser as a heat transfer hardware. Of particular importance in this category is steam jet erosion of the tubes. The steam leaving the last row of blades in a typical power cycle is usually very slightly wet. Consideration of economy in power plant design frequently leads to too small a distance between the low-pressure turbine exhaust flange and the top row of tubes in the condenser tube bank. The wet steam issuing from the turbine does not have adequate distance to sufficiently decelerate before hitting the tubes. Numerous condensers have suffered widespread tube failures due to this reason. This problem has become much more common with the increase in the size of the power plants. Large turbine wheels in bigger plants mean that the steam coming out of the turbine will be more maldistributed, resulting in tube wastage even in units where the *average* steam inlet velocity into the condenser is quite modest. It is common in coal fired and nuclear power plants to place low-pressure feed-water heaters, and other pipings in the path of steam inlet, which in many cases may further aggravate the steam velocity maldistribution. Finally, as mentioned above, the condenser is the recipient of vents and drains from the scores of equipment in the power plant. It is also the ultimate dump of "high energy" steam in the event of an abnormal condition. These connections sometimes contribute to the aforementioned problems of tube erosion and non-condensible blanketing. Of course, conventional heat exchanger operational problems, namely tube surface fouling, flow-induced vibration, structural integrity under thermal and pressure transients, etc., and perils of off-design operation, must be contended with.

The off-design operating condition pertaining to colder cooling water (in winter months) is particularly nettlesome. Colder cooling water means lower condensing pressures in the condenser and a higher rate of condensation. Since the specific volume of steam rises inversely with pressure, the twin conditions of lower pressures and higher steam flow rate can cause a marked increase in the steam inlet velocity and threaten the specter of rapid tube erosion. It must also be noted that the surface condenser produces the maximum reduction in the volume of the heating medium (steam) at only a small fraction of the pressure loss available in typical shell and tube heat exchangers. The incoming steam pressure is only 1 or 2 in. of mercury above the absolute vacuum. Ideally, the designer would wish that no pressure loss occur due to flow of steam across the bank of tubes. Loss of pressure would imply corresponding drop in the condensing temperature difference between steam and cooling water, causing a decreased temperature and a concomitant drop in the heat duty. Pressure loss due to flow of steam across tube banks cannot be eliminated; it can, however, be minimized. Indeed, minimization of the pressure loss, along with efforts to provide a steam distribution in natural accord with the steam demand pattern dictated by heat transfer laws, and efficient accumulation of non-condensibles, are three key considerations in surface condenser design. In the following sections, we will review the theoretical premise and state-of-the-art design practice in surface condenser technology with a brief discussion of industrial terms in design and operation.

30.2 SURFACE CONDENSER CONSTRUCTION

The foregoing narrative of the design objectives in a power plant condenser makes it abundantly clear that it cannot be a direct contact type unit. The importance of the purity of the condensate for reliable operation of the power plant was realized early on by James Watt who in 1765 is known to have preferred a shell-and-tube type condenser over a direct contact cooler. Modern surface condensers are housed in circular shells when the heat transfer surfaces are less than roughly 50,000 sq. ft. and in rectangular shells for larger units. In some instances, elliptical and obound shells have also been used. A rectangular shell provides maximum use of plant space, but is poorly shaped to withstand vacuum (see Photograph A). A circular shell is an excellent structural shape for vacuum conditions, but it does not lend too well to a tube layout leading to the most desired steam distribution. A circular shell is also the most rugged construction requiring minimum amount of bracing to withstand vacuum load. Circular cross-section condensers are typically used in small power plants (see Photograph B).

PHOTOGRAPH A: A TWO-TUBE PASS RECTANGULAR 900MW SUPERCRITICAL COAL-FIRED PLANT (Courtesy Holtec International)

PHOTOGRAPH B: CONDENSER FOR 80MW COMBINED CYCLE PLANT (Courtesy Holtec International)

No matter what the shape of the shell, due to the large volumetric flows involved, the designer always arranges the steam flow to be transverse to the tube bundle in the so-called cross flow arrangement. In large steam surface condenses the steam inlet — referred to as the "condenser neck" — is usually in the form of a tapered dome to provide a well distributed steam flow on to the tube banks (Fig. 30.1). The condenser neck is attached to the turbine exhaust by an expansion joint, which facilitates alignment during installation, and absorbs differential movements due to settling of the foundation and due to thermal expansions. This arrangement, however, gives rise to the potential of condenser uplift due to axial contraction of the expansion joint under vacuum conditions. In Europe, the common practice has been to eliminate the expansion joint, and "hang" the condenser from the turbine exhaust instead. The condenser is mounted on flexible foundations to provide thermal expansion absorption ability to the system. In either of the two schemes of connections, the designer must exercise extreme care while designing the condenser supports.

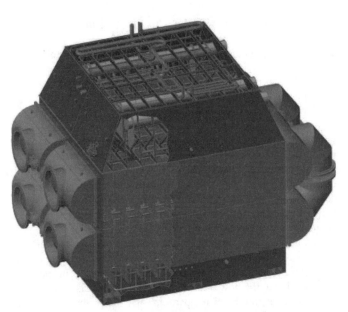

FIG. 30.1 TWO-PASS RECTANGULAR SURFACE CONDENSER

The bottom part of the condenser shell — the "hotwell" — serves to store the condensed steam for pumping up through the feedwater heater trains, and finally into the boiler.

Surface condensers are generally of the straight tube, fixed tubesheet type [1, Chapter 11]. An expansion joint is usually provided in the shell to reduce tube-to-tubesheet joint loads due to differential thermal growth between the tubes and the shell. Other concepts to alleviate start-up and operating conditions have been proposed and tried. One approach entails use of "compression bolts" to pull up the two tubesheets together, and thus place the tube bundle under a compressive axial pre-stress, before making the last shell-to-tubesheet joint. This pre-stress reduces the operating condition tube axial stress, which is assumed to be tensile owing to the circulating cold water inside the tubes.

Much of the condenser structural design effort is devoted to establishing the tubesheet thickness, which cannot be ascertained using standard formulas due to its shape (frequently rectangular), the presence of large untubed regions, and occasionally large mechanical loads transmitted from the cooling water nozzle connections. HEI surface condenser standards [2] recognize the wide variations in the tubesheet drilled hole pattern and other design parameters, and recommends the use of stress analysis techniques of the type described in reference [1] for determining the tubesheet thickness. Prompted by the overriding concern of condensate purity, the condenser suppliers have devised several schemes to ensure that the cooling water does not leak into the steam space. Using closely spaced double tubesheets [1, Chapter 10] is a classical recourse. The space between the two tubesheets is filled with pure condensate at a pressure higher than the waterbox pressure. This arrangement ensures that any leak path developed across the inner or outer tubesheet would cause the pure condensate to leak into the steam space or the waterbox, which would be a non-contaminative event, and which is quickly detected by monitoring the pressure in the inter-tubesheet space.

The double tubesheet construction also affords the opportunity to use a corrosion resistant exotic alloy, such as Muntz metal, in the waterbox closure tubesheet, and carbon steel for the shell closure tubesheet (inner tubesheet).

Another method is to employ an "integral double tubesheet." Such a tubesheet is made by drilling deep serrations in the tube holes at mid-tubesheet thickness. The outer boundaries of these cylindrical serrations intersect producing an interconnected set of cylindrical voids around each tube at mid-tubesheet thickness (see Fig. 30.2). The pressure in this catacomb of annuli can be monitored by an outside connection.

Most condensers are either single tube pass or two-tube pass. The number of tube passes is dependent on the available cooling water flow rate. The overall heat transfer coefficient in a surface condenser is strongly dependent on the tube velocity. Therefore, the condenser designer attempts to maximize the tubeside flow velocity within the constraints of the available pumping head and in-tube velocity limits for various tube materials derived from past operating experience. These considerations sometimes lead to a two-pass design. The heat transfer implications of one- and two-pass designs on the shellside steam flow will be discussed in a later section.

Unlike a feedwater heater, a power plant cannot be operated with the condenser out-of-service. For this reason, surface condensers are often made with the so-called divided waterbox, wherein the tube bundle is separated into two parallel banks, each equipped with its own circulating water inlet and outlet connections. This arrangement permits the tubes in one bank to be serviced (cleaned, plugged, etc.) by opening the waterbox while the other bank is operating. An obvious demerit of this set-up is the possibility of excessive loads on tube-to-tubesheet joints due to large metal temperature differences between the operating bank and out-of-service bank of tubes.

The long span of tubes between the two tubesheets requires lateral support to prevent excessive sagging and failure from flow induced vibrations. The plates used to support tubes are also used to support the rectangular shell panels in rectangular shells [1, Chapter 1], and therefore must be sized as load-bearing members. The tubes are, of course, the heart of any tubular heat exchanger. The selection of the correct tube material is particularly important in surface condenser design, because of the sheer quantity of heat transfer surface involved. Copper bearing materials, such as Admiralty, Copper Nickel, Aluminum Bronze etc., were commonly used for tube materials in the 1970s and 1980s. Problems associated with the copper contamination of condensate as well as the circulating water have prompted the use of stainless steel and titanium. Copper carryover has been named as a prime agent behind the so-called tube denting failure in steam generators of the pressurized water reactor nuclear plants. Boiling water reactor suppliers have also expressed concerns regarding the detrimental

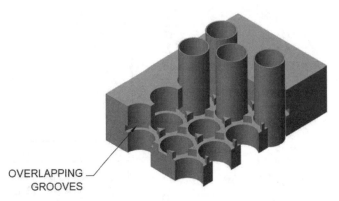

OVERLAPPING GROOVES

FIG. 30.2 INTEGRAL DOUBLE TUBESHEET

TABLE 30.1 COMMONLY USED TUBING MATERIALS

ASME Code designation	Comment
SA-249-304	Most common in the U.S.
SA-249-304L	
SA-249-316	Common the U.S.
SA-268-S44660	Sold under the trade name "Seacure"
SA-268 S43035	Sold under the trade name AL-6XN
SB-338 Grade 2	Titanium

effect of copper species in the BWR reactor internals. These developments have practically banished copper-based alloys from nuclear fueled power plants. As a result, titanium, despite its high cost, has emerged as a tubing material of choice of many power plants.

Table 30.1 provides a listing of the commonly used condenser tubing material at the present time.

30.3 DEFINITIONS

Certain terms are unique to the parlance of surface condenser technology. They are explained below to familiarize the reader with the established terminology.

Heat duty: Like all heat exchangers, the heat duty of a surface condenser is equal to the net heat transfer rate to cooling water. However, at the risk of small errors, the industry [2] has also adopted the simple definition of heat duty as λW_s, where W_s is the quantity of steam entering the condenser, and $\lambda = 950$ Btu/lb for condensers in turbine service, $\lambda = 1000$ Btu/lb for condensers in engine service.

Condenser pressure: The static pressure in the condenser varies in the shell due to pressure loss from steam flow across the tubes, and also due to changes in the local steam velocity. The industry defines the condenser pressure as the average static pressure in the plane transverse to the crossflow direction at a distance not greater than 12 in. from the first tube or tubes in the steam path. ASME

PTC 12.2 provides the guidelines for locating the measurement points.

Condensing steam temperature: It is the saturation temperature corresponding to the condensing pressure as defined above.

Initial temperature difference: $\theta_1 = T_c - t_i$ (in Fig. 30.3)

Temperature rise: $\Delta t = t_o - t_i$

Terminal temperature difference: $\theta_o = T_c - t_o$

Multi-pressure condenser: In some large power plants equipped with parallel turbines, the condensers for the different exhausts may utilize the same cooling water progressively heated in successive condensers. The condenser receiving the coldest water would, of course, maintain the lowest steam pressure. The successive condenser would operate at increasing higher steam pressures consistent with the entering cooling water temperature. Such a group of condensers is referred to as "multi-pressure condenser."

Condensate depression: As will be discussed later, the temperature of the condensate is always a few degrees lower than the coincident condensing steam temperature. The drop in the local pressure due to the flow of steam across the tubes also reduces the corresponding saturation temperature and hence the temperature of the condensate is reduced. These effects, among others, result in the temperature of hotwell condensates being a few degrees below the condensing steam temperature. This temperature difference is known as "condensate depression" or "condensate subcooling."

Multi-pressure condensers are usually arranged such that the coldest condensate is cascaded to the next coldest condensate and so on. This results in even greater condensate depression in the last (highest pressure condenser) hotwell. Excessive condensate depression is undesirable since it increases the likelihood of noncondensible absorption by the subcooled water.

Heat head: It is the old term for "initial temperature difference," defined earlier.

30.4 THERMAL CENTERLINE

It is of some interest to determine the plane transverse to the tube bank axis which divides the heat transfer area in such a manner that the steam consumption on both sides of it is equal. This

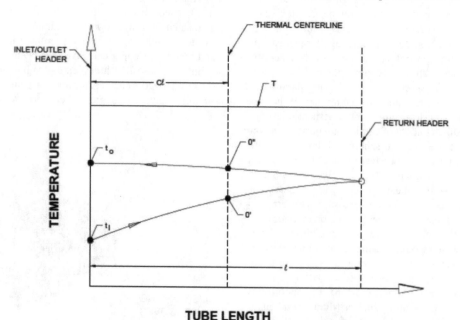

FIG. 30.3 THERMAL CENTERLINE IN A TWO-PASS CONDENSER

plane is known as the "thermal centerline." If the incoming steam were to have a flow front of uniform velocity, then locating the condenser steam inlet opening symmetrically with respect to the thermal centerline would be the ideal arrangement for steam flow. Velocity maldistribution in the steam, pointed out earlier in this chapter, tends to muddy up the significance of the thermal centerline; nevertheless, its notion is important in locating the condenser steam inlet section.

The amount of heat transferred to the cooling water over a differential surface area dA is given by the elementary heat exchange formula:

$$dQ = U \, dA \, (T_c - t) \qquad (30.1)$$

where U is the overall coefficient of heat transfer. Assuming U and the condensing temperature T_c to be constant all over, the heat transferred over a surface area A is given by integrating the above equation [3, Chapter 5] for the derivation of this equation).

$$Q = W_t \, C_t \, (T_c - t_i) \, (1\text{-}e^{-\eta}) \qquad (30.2)$$

where:
$W_t =$ Coolant flow rate
$C_t =$ Coolant specific heat (also implied to be constant over the range of cooling water temperatures in the condenser)
$t_i =$ Inlet temperature of cooling water and η is defined as

$$\eta = \frac{U \, A}{W_t \, C_t} \qquad (30.3)$$

The surface area A is measured from the location where cooling water enters the tubes. It is convenient to define A as a fraction of the total effective surface area in the condenser, A_T.

Let

$$A = x \, A_T \qquad (30.4)$$

then

$$\eta = x \, \eta_T \qquad (30.5)$$

where η_T is recognized as the number-of-transfer units (NTU) of the condenser in the heat exchanger literature [3, Chapter 6].

We will now proceed to derive the expressions for locating the thermal centerline in "one-tube pass" and "two-tube pass" condensers.

(i) Thermal centerline in one-tube pass condenser: let $A = cA_T$ define the location of the thermal centerline. Using Eq. (30.2), we have

$$Q_T = W_t \, C_t \, (T_c - t_i) \, (1\text{-}e^{\eta_T}) \qquad (30.6a)$$

$$0.5 \, Q_T = W_t \, C_t \, (T_c - t_i) \, (1\text{-}e^{-c\eta T}) \qquad (30.6b)$$

which yields

$$1\text{-}e^{-\eta}T = 2 \, (1\text{-}e^{-c\eta}T) \qquad (30.7)$$

Equation (30.6) can be solved for c to give:

$$c = \frac{1}{\eta_T} \, ln\left[\frac{2}{1+e^{-\eta_T}}\right] \qquad (30.8)$$

Thus, for a one-tube pass condenser of length ℓ, the thermal centerline is located at distance c from the inlet waterbox tubesheet, where $c\ell$ is defined by Eq. (30.8).

(ii) Two tubepass condenser: Let A_T denote the net effective surface area equally divided in the two-tube passes. ℓ denotes the net tube length. The surface area in the inlet tube pass to the left of the thermal centerline (point 0' in Fig. 30.3) is given by

$$\frac{c\ell}{2\ell} A_T \quad \frac{c \, A_T}{2}$$

Similarly, the net heat transfer surface in the inlet pass plus the tubes to the right of the thermal centerline in Fig. 30.3 is given by

$$0.5 \, A_T + \frac{(1-c) \, \ell}{2 \, \ell} = A_T \, (1-0.5c)$$

The heat duty in the tube surface in the first pass to the left of the thermal centerline is given by substituting $0.5c \, \eta_T$ for η in Eq. (30.2).

$$Q_1 = W_t \, c_t \, (T_c - t_i) \, (1 - e^{-0.5c\,\eta_T}) \qquad (30.9)$$

The heat duty provided by the outlet pass tube surface to the left of the thermal centerline is given by

$$Q_2 = Q_T - Q_{@\eta} = \eta_T \, (1-0.5c)$$

where $Q_T = W_t c_t \, (T_c - t_i) \, [1-e^{-\eta}{}_t]$
hence

$$Q_2 = W_t c_t \, (T_c - t_i) \, [(1\text{-}e^{-\eta}{}_T) - \{1\text{-}e^{-\eta}{}_T{}^{(1-0.5c)}\}] \qquad (30.10)$$

The requirement for the thermal centerline is

$$Q_1 + Q_2 = 0.5 \, Q_T$$

Substituting for Q_1, Q_2 and Q_T from the foregoing, we have

$$1 - e^{-0.5c\,\eta}{}_T + e^{-\eta}{}_T{}^{(1-0.5c)} - e^{-\eta}{}_T = 0.5 \, (1 - e^{\eta}{}_T)$$

This equation can be solved for c, resulting in:

$$c = \frac{2}{\eta_T} \, ln\left[\frac{-m + (m^2 + 4n)^{1/2}}{2n}\right] \qquad (30.11)$$

where $n = e^{-\eta_T}$
and $m = 0.5 \, (1 - n)$

The above expressions for the thermal centerline have been derived in the context of the surface condenser. They are, however, equally valid for any process condenser wherein the assumption of spatially constant U, c_T can be justified.

Table 30.2 shows the computed values of c for typical values of NTU, η_T, for one- and two-tube pass condensers.

We observe from Table 30.2 that for a given size condenser (measured by its NTU), the thermal centerline moves closer to the geometric centerline of the tube bundle in a two-tube pass unit in comparison to a one-tube pass unit. Therefore, we conclude that

TABLE 30.2   C FOR DIFFERENT VALUES OF η_T

Value of NTU; η_T	Value of C	
	One-tube pass tube	Two-tube pass tube
1	0.38	0.49
2	0.28	0.44
3	0.21	0.38
10	0.07	0.27

the consumption of steam in a two-tube pass condenser is more evenly distributed relative to a one-tube pass design.

It is also noted that the relative location of the thermal centerline as defined by c depends only on the NTU of the condenser. In particular, the location of the thermal centerline is *independent* of individual inlet and outlet temperatures.

30.5   CONDENSATE DEPRESSION

As stated before, subcooling of the condensate is undesirable on two counts: (i) it lowers the thermodynamic efficiency of the power cycle, and (ii) it enhances the propensity of the condensate to reabsorb non-condensibles. Although a multitude of effects are responsible for condensate subcooling (or condensate depression), three basic mechanisms have been identified which are innate to the process of condensing itself. These are:

i. Reduction of the partial pressure of steam around the condensing surface due to increased localized concentration of non-condensibles.
ii. Temperature difference between condensing vapor, and condensate film surface; and
iii. Temperature drop across the condensate film.

i. Reduction of partial pressure of steam: Silver [4] has shown, using partial pressure laws and Clapeyron's equation, that the drop in the local steam saturation temperature due to the presence of air is given by

$$\Delta T_a = \frac{RT_c^2}{\lambda} \, \ell n \left[\frac{1}{1-c_a} \right] \qquad (30.12)$$

where c_a is the fractional air concentration around the condensing surface. RT_c^2/λ for surface condenser conditions is of the order of 30°F. Table 30.3 shows ΔT_a as a function of local air concentration, c_a.

We observe from Table 30.3 that the depression of the saturation temperature can become significant in the air cooler region of surface condensers and in geothermal condensers, where air concentration of the order of 30% is possible. This effect is, however, rather unimportant in the condensing zone in well-designed units.

ii. Temperature drop between the condensing vapor and the surface of the liquid film:

Silver [5] pointed out that when liquid and vapor phases are not in equilibrium, but condensation is occurring, the escaping pressure of molecules must be less than the incident pressure of molecules on the liquid surface. Therefore, the vapor pressure of the vapor must exceed that of the liquid. Silver [4] used kinetic theory of

TABLE 30.3   SATURATION TEMPERATURE DEPRESSION DUE TO THE PRESENCE OF AIR ($RT^2/\lambda = 30°F$ ASSUMED)

C_a (%)	ΔT_a (°F)
0.05	0.015
0.1	0.03
5	1.54
30	10.70

gases to derive the effective film coefficient h_i corresponding to this vapor pressure difference. According to Silver:

$$h_i = f \left(\frac{g}{2\pi RT_c} \right)^{1/2} \frac{\lambda^2}{T_c V} \qquad (30.13)$$

where:

g = acceleration due to gravity
R = Gas constant for water per unit weight
λ = Latent heat of vapor at saturation temperature T_c
V = Specific volume of vapor
f = Molecular exchange condensation fraction: surmised to be equal to 0.036 by Silver & Simpson [6].

Table 30.4 reproduced from Silver [5] gives the interface coefficient h_i as a function of condensing temperature T_c.

We note from Table 30.4 that interface coefficient h_i drops rapidly as the condensing temperature is reduced, indicating that the temperature drop across the vapor–liquid film interface increases as the operating pressure is reduced. If m denotes the rate of condensate formation per unit heat transfer surface area, then continuity of heat transfer yields:

$$h_i \, \Delta T_i = m \, \lambda$$

or

$$\Delta T_i = \frac{m\lambda}{h_i}$$

or

$$\Delta t_i = \frac{m T_c V}{f \left(\dfrac{g}{2\pi RT_c} \right)^{1/2} \lambda}$$

TABLE 30.4 H_I AS A FUNCTION OF CONDENSER PRESSURE

Condenser temperature, T_c (°F)	Condenser temperature (°R)	Pressure, lb/in² abs.	h_i, Btu/ft²h °F
80	540	0.5069	3420
82	542	0.5410	3600
84	544	0.5771	3800
86	546	0.6152	4000
88	548	0.6556	4220
90	550	0.6982	4450
92	552	0.7432	4680
94	554	0.7906	4920
96	556	0.8407	5180

or

$$\Delta T_i = \frac{mT_c^{3/2}}{f\lambda}\left(\frac{2\pi R}{g}\right)^{1/2}$$ (30.14)

iii.   Temperature drop across the condensate film: In his landmark work on film condensation in 1916, Nusselt [7] derived the expression for the equivalent film coefficient under quiescent condensing conditions on a horizontal tube. Nusselt's result is usually expressed in the form:

$$h_f = 0.725\left[\frac{k_f^3 \rho_f^3 g\ \lambda^{1/4}}{\mu_f d_o \Delta T_f}\right]$$ (30.15)

where k_f, ρ_f and μ_f, respectively, denote the conductivity, density and dynamic viscosity of the liquid film. ΔT_f is the temperature drop across the condensate film. If m is the condensate drainage rate per unit of heat transfer surface area per unit time, then for steady state conditions

$$h_f\,\Delta T_f = m\,\lambda$$

or

$$\Delta T_f = \frac{m\lambda}{h_f}$$ (30.16)

If the average condensate film thickness is denoted by δ_f then elementary one dimension conduction relationship yields:

$$h_f = \frac{k_f}{h_f}$$ (30.17)

Equations (30.13) to (30.15) can be combined to furnish the condensate film thickness as a function of condensate drainage rate:

$$\delta_f = 1.54\left[\frac{\mu_f d_o m}{\rho_f^2 g}\right]^{1/3}$$ (30.18)

Equation (30.15) gives the corresponding film coefficient h_f. The temperature drop across the condensate film is given by rearranging Eq. (30.13).

$$\Delta T_f = 0.28\left[\frac{k_f^3 p_f^3 g^3 \lambda^3}{\mu_f d_o h_f^4}\right]$$ (30.19)

The total temperature drop due to the three aforementioned mechanisms is

$$\Delta T_d = \Delta T_f + \Delta T_i + \Delta T_a$$ (30.20)

The theoretically calculated value of the condensate temperature using the above equation is, however, is subject to much uncertainty for the obvious reason that the assumptions made en route deriving the equations do not hold up well in the practical hardware. The effect of vapor velocity, splashing of condensates on tubes from tubes above, velocity recovery, tube surface inundation, etc. tend to distance the practical conditions from the theoretical analysis. Nonetheless, the theoretical results provide valuable insight into the physical phenomena, and provide a rough quantitative measure of the relative importance of various mechanisms involved.

The concerns of condensate depression and excessive streamside pressure drop have led to the development of special surface condenser bundle designs. We will review them in the next section, before taking up the subject of thermal/hydraulic design of surface condensers.

30.6   BUNDLE DESIGNS

Numerous considerations govern the layout of the tube bundle. They can be summarized by the following seven, somewhat interrelated points.

i.   Pressure drop: The drop of steam pressure should be minimized to keep the saturation temperature as high as possible. This means providing generous steam penetration lanes throughout the bundle or shallow bundle depth.

ii.   Spatially Uniform Pressure Drop: The air cooler region is the pressure sink — the region of lowest pressure — in the condenser. The design should provide for equal pressure drop from the bundle periphery to the air cooler from all locations around the bundle. This would inhibit localized high velocities and result in maximum utilization of all available heat transfer surface. Emerson's rendering of an "ideal" condenser (Fig. 30.4) has this attribute. The so-called multi-folded bundle concept (Fig. 30.5), common in some European power stations, portrays this feature.

iii.   No stagnant regions: The bundle layout should provide for a complete sweep of all heat transfer surface in the condensing zone. Pockets of air accumulation can form in seemingly wide open bundles, due to steam lane arrangements.

iv.   Minimize velocity fluctuations: In a well-designed condenser, the steam continuously condenses as it penetrates the tube nest, causing a reduction in the volumetric flow rate. Designers provide for *gradually* narrowing steam lanes for this purpose. The so-called church window layout (Fig. 30.6) pictorially illustrates this concept.

v.   Optimize velocity distribution: High velocity in the primary condenser region would produce excessive (undesirable) pressure loss. On the other hand, the velocities in the air cooler portion must be high enough to maintain good heat transfer rate. Designers accomplish this by using a "tighter" layout pitch in and around the air cooler zone.

vi.   Optimize bundle cross-section: As the steam condenses on the upper row of tubes, it sends condensate "raining" down on the tubes below. A certain amount of the "rain" is desirable as it breaks the condensate film boundary layer and increases the heat transfer rate. But it may also cause tube surface "inundation" and reduce the heat transfer rate. The aspect ratio of the bundle (width-to-height ratio), and condensate loading on the tubes determines whether the condensate cascading through the bundle would aid or inhibit heat transfer. Intermediate condensate collection trays have also been used to mitigate tube inundation. The plant architect/engineers should consider the bundle aspect ratio in deciding upon the turbine foundation layout to enable optimal condenser performance.

vii.   Condensate reheat: As described in the preceding section, a certain amount of subcooling of the condensate is unavoidable. The condensate can, however, be reheated by providing for intimate contact with fresh steam as it falls toward the hotwell. The design must provide for adequate lateral clearances between bundles and the bundles and the condenser shell sidewall such that steam can reach

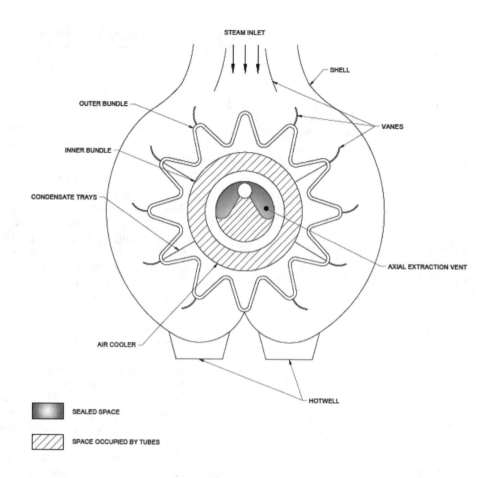

FIG. 30.4 EMERSON'S "IDEAL" BUNDLE

the space above the hotwell without sustaining excessive pressure drop.

Figures 30.7 and 30.8 show the tube lane layout for rectangular and cylindrical condensers used by Holtec International.

Bow [9] cites the standard U.S. Navy specification in use since the 1940s, which we quote below:

"The total area of the tube holes in the tubesheet shall not exceed 22% (for condensers designed for 1.25 psi abs. pressures or less at the steam inlet) or 24% (for condensers designed for higher pressure than 1.25 psi abs. at the steam inlet) of the total tubesheet area exposed to circulating waterflow, the tubesheet area being determined prior to drilling. In condensers in which the design includes space for steam flow between the shell and tube bundle, the tubesheet being smaller than the shell, this space shall be included as part of the tubesheet area exposed to circulating waterflow."

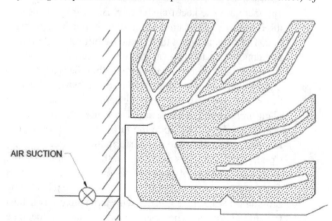

FIG. 30.5 MULTIFOLDED TUBE BUNDLE

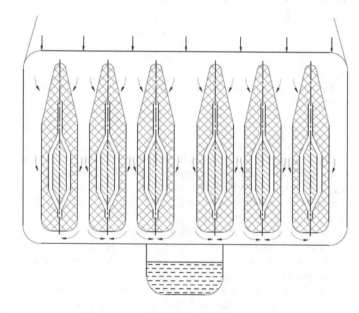

FIG. 30.6 "CHURCH WINDOW" LAYOUT

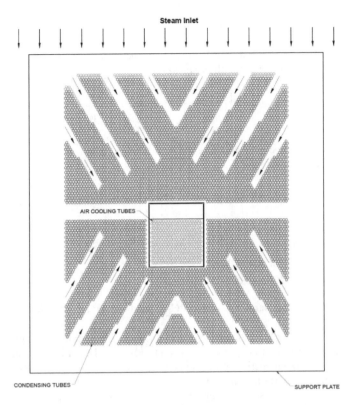

FIG. 30.7 TUBE BUNDLE & SUPPORT FOR A RECTAN-GULAR TYPE CONDENSER
(Courtesy Holtec International)

Bow [9] delineates several empirical dimensional requirements on tube bundle/shell design.

a. The cross-sectional shell area between the top and bottom row of tubes should have a tube-occupied area of not greater than

26% for $P_s > 3$ in. Hg
23% for 2.5 $P_s < 3$ in. Hg
20% for $p_s < 2.5$ in. Hg

b. Free fall height between the bottom tube row and top of the high water level (H.W.L.) in the hotwell:

Condenser HEAT TRANSFER SURFACE AREA, A_T (sq. ft.)	Minimum free fall height (in.)
$A_T < 50,000$	18
$50,000 \leq A_T < 150,000$	21
$150,000 \leq A_T < 300,000$	24
$300,000 \leq A_T$	27

c. Minimum height between the bottom of the condenser neck heater (for units with a feedwater heater installed in the condenser neck) and top of the tube bundle: 48 in.

d. Minimum hotwell condensate inventory: 4 minutes worth of condensation.

Bow's recommendations are rather onerous for commercial steam plants. They are, however, quite appropriate for small condensers used in ocean-going vessels.

30.7 SURFACE CONDENSER SIZING

The discussion on condensate depression in a preceding section indicated the intractable nature of the condensing phenomena in surface condensers. Faced with the enormity of the mathematical problem, the industry has chosen the pragmatic course of devising design formulas that exclude the shellside conditions. Instead, the HEI design rules [2] in the U.S. and BEAMA [10] in the U.K. (both widely used industry standards), rely on simple formulas combined with good design practice, with heavy reliance on the latter, to produce satisfactory designs. For example, the HEI design method defines the heat duty of the condenser Q_T as

$$Q_T = U A_T \theta_m \tag{30.21}$$

where θ_m is the nominal Logarithmic mean temperature difference based on the "steam condensing temperature" T_c,

$$\theta_m = \frac{t_o - t_i}{\ln \dfrac{T_c - t_i}{T_c - t_o}} \tag{30.22}$$

The overall heat transfer coefficient U is given by

$$U = F_1 F_2 F_3 C^* v_t^{1/2} \tag{30.23}$$

C^* depends on the nominal tube size. The factors F_1, F_2, and F_3 are correction factors for fouling, tube material and wall thickness, and cooling water inlet temperature, respectively. This correlation is largely based on Orrok's work in 1910 [12]. The approach of using multipliers for tube material resistance and cooling water temperature effects has been criticized in the literature [12,13].

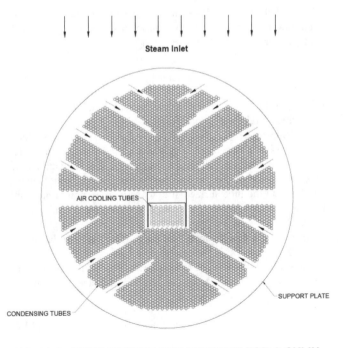

FIG. 30.8 TUBE BUNDLE AND SUPPORT FOR A CYLINDRICAL TYPE CONDENSER (Courtesy Holtec International)

Wenzel [14] reports that field data correlates rather poorly with the HEI formula, and errors of almost 100% are possible. These discrepancies arise from a host of factors such as air blanketing, steam maldistribution, high velocity effects, condensates splashing, and the like. There is little doubt that computer-based finite element modeling of fluid dynamics will eventually supplant the present day cookbook approaches, so far, there has been little movement away from empiricism in most industrial condenser design work. We herein summarize the design equations which may offer an alternate means to the HEI/BEAMA type procedures.

The overall heat transfer coefficient is defined by the well-known formula [3, Chapter 2].

$$\frac{1}{U} = \frac{1}{\frac{1}{h_o} + \frac{d_o}{d_i}\frac{1}{h_t} + \frac{d_o}{d_i}r_{fi} + r_{fo} + r_m} \tag{30.24}$$

The terms r_{fi} and r_{fo} represent tube fouling layer resistances on tube inside and outside surfaces, respectively. If these resistances cannot be estimated, then the designer may set them equal to zero, and apply a penalty factor on the overall clean coefficient thus calculated. The penalty factor is also known as the cleanliness factor. In condenser design practice it varies from 0.7 to 0.9, depending on the quality of the cooling water. Formulas to estimates h_i, r_m, and h_o are given below.

i.   Tubeside film coefficient, h_t: McNaught [15] is valid for Reynolds number > 10,000, can be used to estimate the in-tube film coefficient

$$\frac{h_t d_i}{k} = 0.015 Re^{0.835} Pr^{0.462} \tag{30.25}$$

where:
k = cooling water conductivity at its average bulk temperature
d_i = tube inside diameter
Pr = Prandtl number at average bulk temperature conditions
Re = Reynolds number at average bulk temperature conditions

i.   <u>Tube metal wall resistance</u>, r_m: Since condenser tubes are essentially of thin wall, the tube metal resistance is given sufficiently accurate by the formula:

$$r_m = \frac{t_t}{k_t} \tag{30.26}$$

where t_t and k_t are tube wall thickness and metal conductivity at mean tube metal temperature, respectively.

<u>Shellside condensing coefficient</u>, h_s: Estimation of the shellside coefficient is the core problem in surface condenser design. Reference [11] provides excellent information on this subject. The interaction of effects, such as condensate inundation, vapor velocity, splashing on the tube surface, droplet formation, non-condensible accumulation, etc., make the goal of an accurate evaluation all but impossible.

For practical design work, one may utilize the correlation presented by McNaught [15].

The shellside coefficient h_o is defined as the square-root-of-the-sum of squares of the shear uncontrolled coefficient h_{sh}, and the gravity controlled coefficient h_{gr}.

$$H_o = [h_{sh}^2 + h_{gr}^2]^{1/2} \tag{30.27}$$

where h_{sh} and h_{gr} are computed using the following correlations:
Shear controlled coefficient, h_{sh}:

$$H_{sh} = 1.26\, h_L\, (X_{tt})^{-0.78} \tag{30.28}$$

where h_L is the liquid phase forced convection coefficient, and X_{tt} is the Lockert–Martinelli parameter:

$$X_{tt} = \left[\left(\frac{1-x_L}{x_L}\right)^{1.8} \left(\frac{\rho_\upsilon}{\rho_L}\right) \left(\frac{\mu_L}{\mu_\upsilon}\right)^{0.2} \right]^{1/2} \tag{30.29}$$

h_L can be evaluated using the well-known liquid crossflow correlation [3, Chapter 4].

Gravity controlled condensation coefficient, h_{gr} (30.30)

$$h_{gr} = h_N \left[\frac{(\Sigma G_n)^{-s}}{} \right]$$

$s = 0.22$ for in-line bundles;
$s = 0.13$ for staggered bundles

h_N is the laminar single tube solution given by Nusselt (Eq. 30.15).

The above formula does not account for the heat transfer reduction due to non-condensibles. Therefore, it should not be used in the air cooler section of the condenser, wherein large non-condensible concentrations may occur.

<u>Air Cooler Section Design</u>: The most widely used procedure to calculate shellside heat transfer in this region is the Colburn & Hougen [16] method. It involves equating the heat transferred locally through the condensate film, tube wall, and cooling water film to the sum of the sensible cooling of the non-condensible and latent heat given up by the vapor condensed on the tube. The method is obviously an iterative one, which lends itself to convenient computerization.

Two equations for assessing non-condensible film coefficient are given.

a.   Standiford [17] proposed a conservative and highly approximate correlation which states that the non-condensible gas resistance is proportional to the volume fraction of air.

$$R_{nc} \approx 0.00037_w \tag{30.31}$$

where w is local volumetric concentration of air.

b.   Chisholm [18]: The heat transfer coefficient in the presence of non-condensibles is given by Chisholm by the following equation:

$$h_{nc} = \frac{aD}{d_o} Re_\upsilon^{1/2} \left(\frac{p_m}{P_m - p_\upsilon}\right)^b P_m^{1/3} \tag{30.32}$$

$$\left(\frac{p_\upsilon \lambda}{T_\upsilon}\right)^{2/3} (\Delta T_i)^{-1/3}$$

where for

$$Re_\upsilon > 350$$

$b = 0.6$; $a = 0.52$ first tube row; $a = 0.67$ second tube row; $a =$ third and later tube rows and for $Re < 350$; b: 0.7; $a = 0.52$ for all rows.

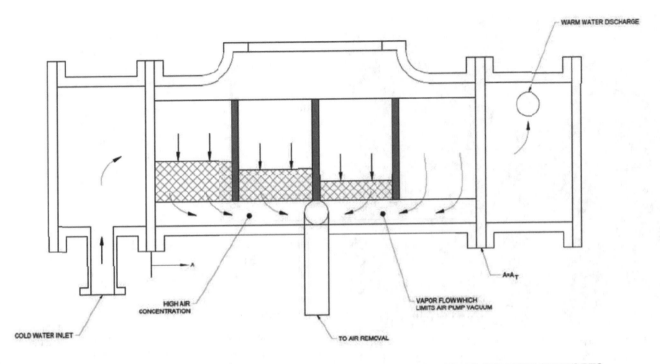

FIG. 30.9   NON-CONDENSIBLE ACCUMULATION IN A CONDENSER WITH PLATE TYPE SUPPORTS

30.8   TUBE SUPPORTS AND NON-CONDENSIBLES ACCUMULATION

Tube supports are a necessary evil in surface condensers. Unlike standard heat exchangers which handle single phase fluids, the tube support plates in a condenser serve no shellside film coefficient enhancement function: Their sole purpose is to protect the tubes from flow induced vibration, to inhibit the occurrence of acoustic resonance, and in rectangular units, provide structural support to the shell. However, if improperly designed, they may derate the thermal performance of the condenser. The mechanism for performance derating can be explained by considering the heat transfer characteristics of a surface condenser.

Let us consider a single pass surface condenser as shown in Fig. 30.9.

The heat duty as a function of the surface area coordinate for such an exchanger is given in Reference [3] as:

$$Q = W_t C_t (T_c - t_i) \left[1 - e - \frac{UA}{W_t C_t} \right] \qquad (30.33)$$

We note that at $A = 0$ (at the left end tubesheet location where the cooling water enters), $Q \equiv 0$.

The rate of heat transfer as a function of the heat transfer surface is obtained by taking the derivative of Q with respect to A. Accordingly

$$\frac{dQ}{dA} = U(T_c - t_i) e^{\frac{-UA}{W_t C_t}} \qquad (30.34)$$

Let us consider the ratio of the heat transfer rate at $A = 0$ to that at $A = A_T$; i.e., the ratio of the heat transfer per unit of tube surface at the two extremities of the condenser. Denoting this ratio by r, we have, from Eq. (30.34):

where qT is the NTU of the condenser.

$$r = \frac{\left. \frac{dQ}{dA} \right|_{A=0}}{\left. \frac{dQ}{dA} \right|_{A=A_T}} = e^{\frac{UA_T}{W_t C_t}} = e^{\eta_T} \qquad (30.35)$$

where η_T is the NTU of the condenser.

Let us consider a typical condenser with the following data.

$U = 500$ Btu/hr/sq.ft./°F
$A_T = 50,000$ sq. ft.
$W_T = 10 \times 10^6$ lb/hr
$C_T = 1.0$ Btu/lb-°F
We have, for this condenser,

$$\eta_T \frac{(500)(50,000)}{(10 \times 10^6)(1.0)} = 2.5$$

Hence, $r = e^{2.5} = 12.18$

In other words, the rate of steam consumption near the inlet tubesheet (left end) is more than 12 times the rate at the outlet temperature! It is also noted that the skewedness of steam consumption *does not* depend on the steam or cooling water temperatures.

Let us now examine the condenser of Fig. 30.9 equipped with full tube supports which essentially compartmentalizes the shell restricting longitudinal flow of steam between adjacent support plate spaces. The air removal equipment exerts a uniform suction pressure at the bottom which will tend to draw an equal amount of steam across each baffle space. However, the steam requirement rate is quite unequal. As a result, the tube surface located in the left end will be starved of steam, leading to high air concentration in the tube bundle. The air concentration will have the distribution in the manner of Fig. 30.9. The amount of heat transfer in left end

PHOTOGRAPH C: RECTANGULAR CONDENSER WITH NON-SEGMENTAL BAFFLES DURING FABRICATION
(Courtesy Holtec International)

compartments will be reduced due to the pressure of air, resulting in loss of thermal performance.

In addition to the reduction in the thermal performance, the high concentration of oxygen can cause corrosion problems in the condenser, and lead to untimely tube failures. To avoid the undesirable concentration of non-condensibles, designers slot out holes in the support plates to make the compartments as inter-communicative as possible. Another solution is to employ one air removal appara-

tus dedicated to each compartment. This, however, is an expensive option, seldom utilized in the industry. More promising solutions lie in the use of special tube supports which provide for ample longitudinal communication. Known as "non-segmental" tube supports, such a support system has been extensively used in surface condensers provided by Holtec International (Photograph C). A support system of this type permits replenishment of steam in the "cold compartments" by axial migration of steam. The non-segmental support system helps develop a steam flow profile which has the pictorial appearance of Fig. 30.10.

A method to compute the improvement in a condenser's performance by the use of a non-segmental support system is provided in the next section.

30.9 HEAT TRANSFER RELATIONS FOR A CONDENSER WITH NON-SEGMENTAL TUBE SUPPORTS

Assumptions:

(i) We assume that the non-segmental tube support system permits axial migration of non-condensibles and therefore, leads to a uniform overall heat transfer coefficient throughout the condenser. The uniform coefficient U is equal to the coefficient that would obtain in the "drilled tube support unit" at the mid-section $x = \ell/2$ in Fig. 30.10.

(ii) In the drilled tube support unit, the heat transfer coefficient varies in the axial direction, its variation solely caused by the uneven amount of air concentration in the bundle.

(iii) The air concentration at any section in the drilled unit is proportional to the coincident rate of steam condensation.

The expression for U given in the HEI Surface Condenser Standards [2] does not have any dependence on the steam side coefficient.

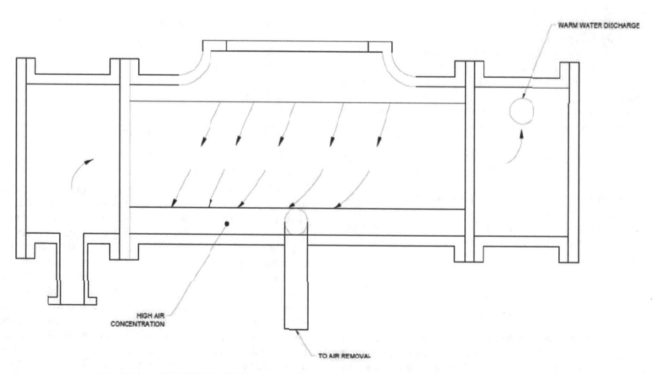

FIG. 30.10 STEAM FLOW IN A NON-SEGMENTAL BAFFLE-EQUIPPED CONDENSER

In a well-deaerated condensing process under vacuum conditions, the condensing coefficient is much larger than waterside coefficient. For this reason, the HEI procedure excludes the shellside film coefficient evaluation from the overall heat transfer computation. We will follow a similar track, and assume that the shellside resistance to heat transfer is solely due to the presence of noncondensibles. Furthermore, following Standiford [17], we will assume that the shellside resistance to heat transfer is proportional to the coincident air concentration. In conjunction with assumption (iii) above, this implies that the shellside resistance is proportional to the derivative of shellside condensation with the tube surface area (dW_s/dA).

The overall heat transfer coefficient depends on the tubeside flow velocity, tube material, I.D. gage, etc. Keeping all these variables constant, but allowing the air concentration to vary leads to the following functional form for U.

$$U = \frac{1}{\alpha \beta \dfrac{dW_s}{dA}} \qquad (30.36)$$

where α lumps the effects of film resistance due to tubeside flow, tubeside metal resistance and tubeside fouling. β contains the proportionality constant which relates the shellside coefficient to the local condensing rate.

We can now construct the necessary relationships to quantify the heat duty of a "drilled unit."

The heat balance between the shellside and tubeside fluids yields

$$W_t C_t \, dt = \lambda \, dW_s \qquad (30.37)$$

Equating the heat transfer through the tube wall to the shellside heat duty gives

$$U \, dA \, (T_c - t) = \lambda \, dW_s \qquad (30.38)$$

Substituting for U from Eq. (30.36) we have

$$\frac{(T_c - t)}{\alpha + \beta \dfrac{dW_s}{dA}} = \lambda \frac{dW_s}{dA} \qquad (30.39)$$

or

$$(T_c - t) = \lambda \frac{dW_s}{dA}\left(\alpha + \beta \frac{dW_s}{dA}\right) = +\frac{dW_s}{dA}\frac{d^2W_s}{dA^2}$$

Noting from Eq. (30.37)

$$\frac{dt}{dA} = \frac{\lambda}{W_t C_t}\frac{dW_s}{dA}$$

or

$$\frac{\lambda}{W_t C_t}\frac{dW_s}{dA} = \lambda\left[\alpha \frac{d^2W_s}{dA^2} + 2\beta \frac{dW_s}{dA}\frac{d^2W_s}{dA^2}\right]$$

or

$$\alpha \frac{d^2W_s}{dA^2} + 2\beta \frac{dW_s}{dA}\frac{d^2W_s}{dA^2} - \frac{1}{W_t C_t}\frac{dW_s}{dA} = 0 \qquad (30.40)$$

The initial conditions are

$$W_s = 0 \, @ \, A = 0 \qquad (30.41)$$

The other condition is obtained from Eq. (39), as follows. Let

$$\frac{dW_s}{dA} = \chi \text{ at } A = 0 \qquad (30.42)$$

Then

$$\frac{T_C - t_i}{\alpha + \beta\chi} = \lambda\chi 1$$

or

$$\alpha\lambda X + \beta\lambda x^2 = (T_c t_i) = 0$$

or

$$(\beta\lambda)\chi^2 + (a\lambda)\chi + (t_i T_c) = 0$$

or

$$\chi = \frac{\alpha\lambda \ + [\alpha\lambda^2 + 4(T_c - t_i)\beta\lambda]^{1/2}}{\alpha\beta\lambda} \qquad (30.43)$$

Equations (30.41) and (30.42) furnish the necessary conditions to solve Eq. (30.40).

30.10 SOLUTION PROCEDURE

Equation (40) gives the governing differential equation which must be solved to the initial conditions (Eq. 30.41 and Eq. 30.42). This equation is readily solved using a numerical method.

Let us subdivide the condensing surface into N small surfaces, each containing a small surface area ΔA, i.e.

$$N \Delta A = A_T$$

or

$$\Delta A = A_T/N$$

The value of the condensed steam to the left of node I ($I = 1$, $2....N + 1$) is denoted by w_i. From Eq. (30.41) we have

$$W_i = 0 \qquad (30.44)$$

Equation (30.42), in finite difference form, can be written as

$$\frac{w_2 w_1}{\Delta A} = \chi$$

or

$$w_2 = \chi \, \Delta A$$

We use the following discretizations for the derivation of w_s

$$\frac{dW_s}{dA}\frac{w_{i+1} + w_i}{\Delta A} \qquad (30.45)$$

$$\frac{d^2W_s}{dA^2} = \frac{w_{i+1} + w_{i-1}}{\Delta A^2}^{-2w_i} \qquad (30.46)$$

Substituting Eqs. (45) and (46) into Eq. (40), we have

$$aw^2_{i+1} + bw_{i+1} + c = 0; i = 2,3....(N-1) \qquad (30.47)$$

where

$$a = \frac{2\beta}{\Delta A^2} \qquad (30.48)$$

$$b = \frac{\alpha}{\Delta A} + \frac{2\beta}{\Delta A^2}(w_{i+1} - 3w_i)\frac{1}{W_t C_t} \qquad (30.49)$$

$$c = \frac{\alpha}{\Delta A}(-2w_i + w_{i-1}) + \frac{2\beta}{\Delta A^2}[2w_i^2 - w_i w_{i-1}]$$
$$+ \frac{w_i}{W t_t C_t} \qquad (30.50)$$

Setting I = 2, the values of a, b and c follow directly from Eqs. (30.48) to (30.50), and Eqs. (30.43) and (30.44). The quadric equation in w_3 (Eq. 47) next gives the value of w_3.

$$w_3 = \frac{\pm b + (b^2 - 4ac)^{1/2}}{2a} \qquad (30.51)$$

The procedure is continued for $I = 3,4...(N-1)$. Thus, the total amount of steam condensed as a function of the surface area coordinate becomes known.

The non-segmental tube support systems, however, lack the in-plane load carrying capability of the classical drilled plate tube support, and as such do not provide the lateral support to the shell plates (in rectangular shell units) to the extent provided by "plate type" support plates.

30.11 OFF-DESIGN CONDITION OPERATION AND NON-CONDENSIBLES REMOVAL

The operating condition in the surface condenser is seldom at the "design point." Changes in the plant load, cooling medium temperature, among other variables, alter operating pressure on the condenser. In the absence of other complicating factors (which we will take up shortly), the determination of the off-design condition condenser pressure is a straightforward process. We recall Section 4 which gives the heat duty for a condenser with an arbitrary number of tube passes;

$$Q_T = W_t C_t (T_c - t_i)(1 - e^{-\eta_T})$$

where

$$\eta_T = \frac{UA_T}{W_t C_t}$$

Furthermore,

$$Q_T = w_S \lambda$$

Therefore,

$$W_S \lambda = w_T c_T (T_c - t_i)(1 - e^{-\eta_T}) \qquad (30.52)$$

The quantities in the above expression refer to any operating condition. In order to perform the calculations without a manual lookup of steam properties in the steam tables, the latent heat λ can be replaced by the following algebraic relationship

$$q = a_1 (705.47 - T_c)^{b_1} \qquad (30.53)$$

where

$$a_1 = 94.26$$

$$b_1 = 0.375$$

In the above, T_c is the condensing temperature in °F, and λ is the latent heat in Btu/lb.

Substituting for λ in Eq. (30.52), we have

$$W_s a_1 (705.47 - T_c)^{b_1} = W_t C_t (T_c = t_i)(1 - e^{-\eta_T}) \qquad (30.54)$$

This equation can be solved for the unknown quantity, say T_c, if any or all of the input quantities are varied.

Example:

The following data defines the design point of a surface condenser.

Surface area, A_T = 50,000 sq. ft.
Overall coefficient, U = 500 Btu/sq.ft./hr°/°F
Cooling water flowrate, W_t, C_t = 1.0 Btu/lb°/°F
Condensing temperature, T_c = 110°F
Cooling water inlet temperature, t_i −80°F
Therefore,

$$\eta_T = \frac{UA_T}{W_t C_t} = \frac{(500)(50,000)}{10 \times 10^6} = 2.5$$

The condensing rate W_s for this condition given by Eq. (30.54):

$$W_s = \frac{(10 \times 10^6)(110 - 80)(1 - e^{2.5})}{(94.26)(705.47 - 110)^{.375}} = 266,000 \text{lb/hr}$$

It is noted that the saturation pressure to correspond to the saturation temperature T_c −110°F is 1.275 psi (2.6 in. of Hg). Let us now determine the condensing temperature and the associated condenser pressure if the cooling water temperature is changed to 50°F. Let T_c' denote the corresponding condensing temperature.

Once again, Eq. (54) yields
$(266,000)(94.26)(705.47 - T_c')^{.375}$
$= (10 \times 10^6)(T_c' - 50)(1 - e^{-2.5})$

or

$$2.371(705.47 - T_c')^{.375} = T_c' - 50$$

Trial and error solution of the above equation yields T_c' = 80.5°F.

The associated saturation pressure can be looked up in the steam table, or evaluated from the curve fit formula

$$p = e^{\left[a_2 + \frac{b_2}{T_c + C_2}\right]} \qquad (30.55)$$

where, $a_2 = 14.47$; $b_2 = -6.99 \times 10^3$, $C_2 = 381.73$

For $T_c = 80.5°F$, Eq. (30.55) yields p = 0.521 psi (1.06 in. of Hg).

The drop in the turbo-generator backpressure to 0.521 psi would result in a welcome improvement in the efficiency of the power cycle. Alas, such improvements do not always occur. The reason lies in the devaporizing zone in the condenser, which is unable to remove the growing quantity of air from the equipment to support efficient heat transfer. This is a core problem in surface condenser design, which warrants a comprehensive exposition.

Condensers are almost invariably equipped with a devaporizer section, also known as air cooling section, wherein the air/vapor mixture gets enriched with air as the water vapor condenses out. The enriched mixture is expelled from the condenser near the cold tubesheet using an evacuation device. Typically, a steam jet ejector or a mechanical vacuum pump (liquid ring pump) is used for this purpose. A look at the performance characteristics of a typical steam jet ejector (Fig. 30.11) reveals the root of the problem. Figure 30.11 shows the evacuation capacity expressed as a percentage of the capacity at 1 in. Hg (abs) pressure.

It is observed that the suction rate drops rapidly with the drop in the condenser backpressure in a steam jet ejector. The drop is similar for a liquid ring pump system. We computed, in the foregoing example problem, that as the cooling water temperature changes from 80°F to 50°F, the condenser operating pressure would drop from 1.275 psi to 0.521 psi. Such a drop in backpressure would cause approximately a 40% reduction in the steam jet ejector system of Fig. 30.11. While the evacuation rate drops, however, the rate of non-condensible ingression into the turbine increases due to the fact that a lower condenser pressure implies that a larger portion of the turbine is under subatmospheric conditions.

The result is a backup of non-condensibles in the condenser resulting in air blanketing of the tubes which in turn reduces the condenser heat duty and raises the backpressure.

To summarize, as the cooling water temperature is reduced, the air removal system sized for the design condition fails to remove all non-condensibles. This results in air blanketing of the condenser tube, causing a less than expected reduction in the condenser pressure.

The deleterious effect of oxygen concentration on the condenser life has been stated before and therefore, does not require further elaboration here.

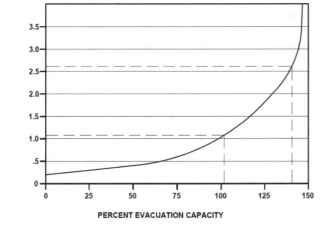

FIG. 30.11   AIR REMOVAL CHARACTERISTICS OF A TYPICAL STEAM JET INJECTOR

TABLE 30.5   AMOUNT OF AIR PER POUND OF MIXTURE EXITING THE CONDENSER

Item	Design condition	Off-Design condition
Backpressure, psi	1.275	0.521
Cooling water inlet temperature, °F	80°	50°
Mixture temperature	85°	55°
Vapor pressure, p_v, psi	0.596	0.214
Air weight fraction from Eq. (30.56)	0.329	0.364

Similar conclusions can be drawn if the steam through-put in the plant is reduced (low load condition).

Finally, we note that the composition of the air-vapor mixture leaving the condenser can be computed in an approximate manner using Dalton's law, which states that the mole count of a gas in a mixture is in direct proportion to its partial pressure. The sum of the partial pressure of the water vapor and air is equal to the backpressure. The partial pressure of the water vapor is the saturation pressure corresponding to the temperature of the mixture leaving the condenser which depends on the air cooler section design. If P_v and P_b, respectively, denote the vapor pressure and backpressure, then Dalton's law gives the weight fraction of air in the mixture, f_a, as

$$f_a = \frac{29}{(29+18)} \frac{p_o - p_v}{p_b} = \frac{29}{47} \frac{p_o - p_v}{p_b} \qquad (30.56)$$

where the molecular weight of the non-condensibles is assumed to be 29.

Let us compute the amount of air per pound of mixture leaving the condenser in the above example problem for the design condition $t_i = 80°F$, and for the off-design condition, $t_i = 50°F$. We will assume that the air-vapor mixture is cooled to within 5°F of the cooling water inlet temperature. The data is presented in Table 30.5.

Clearly, the more the mixture is subcooled, the less is the load on the evacuation apparatus. However, there is an obverse consideration. Subcooled condensate dripping from the air cooler section will carry dissolved oxygen into the hotwell, preventing a more complete purge of this corrosive substance. The situation calls for a design trade-off.

30.12   CATHODIC PROTECTION

In contrast to most power plant heat exchangers, surface condensers use a natural coolant, such as water from a lake, a river, or

TABLE 30.6   PROPERTIES OF COMMON ANODE MATERIALS

Anode material	Drive voltage with respect to steel in the water	Electrochemical efficiency
Magnesium	0.7	40–50%
Zinc	0.25	90%
Aluminum	0.25	80%

TABLE 30.7 IMPRESSED CURRENT ANODIC MATERIALS DATA

Anode name	Metallurgical data	Current (A/m^2)	Max. voltage	Limitations
Platinized titanium	A thin discontinuous layer (2.5 μm) of platinum over titanium substrate	10,000	8	May deplatinize at higher voltages, or if AC ripples in the substrate
Platinized niobium	Thin coat of platinum over niobium substrate	10,000	40–50	Limited due to high cost
Lead silver	98% pb, 2% Ag	100–200 (max.)		Water must have high chlorine content to provide the necessary conductivity.
Lead-silver antimony	94% Pb, 1% Ag, 6% Sb	>30 amp/m^2		Lead peroxide film on the anode, voltage must be >30 V
High silicon cast iron alloys	Duriron 14.5% Si 0.95% C, remaining Fe or modified duriron containing trace amounts of chlorine and molybdenum	>30 A/m^2		Short life due to high rate of consumption

a sea. Exposure of waterbox to the cooling water causes corrosion. The severity of corrosion depends on the materials of construction, dissolved materials, particularly salt in the cooling water, degree of turbulence, etc. Higher salinity in the water tends to promote galvanic corrosion between dissimilar materials present in the waterbox. If not controlled, rapid pitting of tube ends, tubesheet ligaments and waterbox walls may force power plant outage.

An increasingly popular method to arrest corrosion is to apply polymeric coatings on the waterbox walls, pass partition plates, and other exposed ferritic internals. Coatings, however, may have minuscule cracks which provide nucleation sites for corrosion to take hold and propagate. Another method of corrosion control commonly referred to as cathodic protection is sometimes used alone or in conjunction with coatings. In principle, cathodic protection sets up an electrolyte current between the metal surface to be protected and a foreign material submerged in the body of water in the waterbox. The flow of electrons associated with the electrolytic current interferes with the corrosion process. The current can be generated by employing an electrochemically active material, such as zinc, aluminum, or magnesium. The galvanic potential difference between the introduced metal (anode) and the waterbox structure (cathode) causes a gradual loss of the anode. The choice and quantity of the anode depends on the current density requirements, desired galvanic voltage differential and length of service required between anode replacements. Table 30.6 gives drive voltage between common anode materials with respect to steel in sea water, and their electrochemical efficiencies (defined as the percentage of the material that is used up in providing useful current).

The advantage of higher driving voltage of magnesium is largely negated by its low efficiency. Moreover, some protective coatings tend to disbond at high cathodic potentials induced by magnesium anodes. Aluminum's performance as an anode has been quite erratic, presumably caused by small variations in grain size, heat treatment and metallurgical impurities. For these reasons, zinc anodes have received widest acceptance in surface condensers where the water temperatures are invariably below the 140°F threshold, above which zinc begins to become ineffective due to passivation.

Low current densities obtainable with the passive cathodic protection system may not suffice in many situations. In such cases, an "impressed current system" is utilized. A direct current (dc) power source circulates electrolytic current which, depending on the anode utilized, can be as high as 1000 Amperes/m^2 of active anode surface. Table 30.7 lists some commonly used anode materials in impressed current systems and their attributes.

Impressed current systems are coming into increasing use because passive sacrificial anode systems are often inadequate to provide the high current requirements, or are impractical due to the quantity of anode surface required. Designers of impressed current systems must guard against other side effects of excessively high currents. Some tubing materials, notably titanium and certain high chromium molybdenum containing ferritic stainless steels, are susceptible to embrittlement at high cathodic potentials. While an appropriate current density is a prerequisite to a successful cathodic protection system, the distribution of the current "throw" in the waterbox is equally important. Prior experience and good engineering judgment have been sole guides in this matter. However, use of numerical solutions of the current distribution field equations to help place and orient anodes appears to be a feasible technology in the near future.

30.13 NOMENCLATURE

A_T:	Total effective heat transfer area
C_t:	Specific heat of cooling water (assumed constant)
C^*:	HEI coefficient factor
c:	Location of thermal centerline
D:	Diffusion coefficient
d_o:	Tube O.D.
d_i:	Tube I.D.
dQ:	Differential quality of heat transferred on surface area dA
dA:	Differential heat transfer surface area
f_a:	Air weight fraction in the mixture
F_1, F_2, F_3:	HEI Correction factors
f:	Molecular exchange condensation fraction
G_n:	Condensate formation in the n-th tube row
g:	Acceleration due to gravity
h_t:	Tubeside film coefficient
h_o:	Shellside film coefficient

h_f:	Heat transfer coefficient across the liquid condensate boundary layer
h_i:	Vapor–liquid interface heat transfer coefficient
h_L:	Liquid phase forced convection coefficient
h_{sh}:	Shear-controlled coefficient
h_{gr}:	Gravity-controlled coefficient
k:	Cooling water conductivity
k_t:	Tube metal conductivity
l:	Tube length
m:	Mass condensing rate per unit area
N_u:	Nusselt number
Pr:	Prandtl number
p_s:	Steam condensing pressure
p_m:	Pressure of mixture
p_υ:	Partial pressure of steam in the mixture
Q_T:	Total heat duty
R:	Universal gas constant per unit weight of water vapor
r_{fi}:	In-tube fouling resistance
r_{fo}:	Shellside fouling resistance
r_m:	Tube metal resistance
Re:	Reynolds number
t:	Tube wall thickness
T_c:	Condensing temperature
t_i:	Cooling water inlet temperature
t_o:	Cooling water outlet temperature
t:	Cooling water temperature (generic term)
U:	Overall heat transfer coefficient
V:	Specific volume of vapor
W_t:	Cooling water flow rate
W_s:	Steam condensing rate
x:	Area fraction
X_{tt}:	Lockart–Martinelli parameter
x_L:	Liquid weight fraction
ΔT_i:	Temperature drop across vapor/liquid
ΔT_a:	Interface due to kinetic effects
Δt:	Temperature rise of cooling water
θ_o:	Terminal temperature difference
θ_i:	Initial temperature difference
η:	Dimensionless quality
η_T:	NTU of condenser
λ:	Latent heat of condensing at T_c
$(\Sigma G)_n$:	Total flow rate of condensate per unit area running off n-th tube
θ_m:	LMTD
ρ:	Density
μ:	Dynamic viscosity

30.14 REFERENCES

1. Singh, K.P. and Soler, A.I. (1984). Mechanical Design of Heat Exchangers and Pressure Vessels, Arcturus Publishers, Cherry Hill, NJ (also available from holtecinternational.com).

2. Standards for Steam Surface Condensers, 10th Edition (2006), HEI, Inc., Cleveland, Ohio.

3. Singh, K.P. Theory and Practice of Thermal-Hydraulic Design of Heat Exchangers, by K.P. Singh (forthcoming).

4. Silver, R.S., "An Approach to General Theory of Surface Condensers," Proc. Inst. Mech. Eng., vol. 178, part 1, no. 14, pp. 339–357, 1963-1964.

5. Silver, R.S., "Heat Transfer Coefficients in Surface Condensers," Engineering, vol. 161, p. 505, London, 1946.

6. Silver, R.S. and Simpson, H.C., (1961) "Condensation of Superheated Steam," Proc. Conf. NEL, East Kilbride, vol. 39, March.

7. Nusselt, W. Die Oberflachen-Kondensation des Wasserdampfes, VDI Z., vol. 60, pp. 541-546 and 569-575, 1916.

8. Kern, D.Q. (1950). Process Heat Transfer, McGraw-Hill, New York, p. 302.

9. Bow, W.J. (1983) "Space Allotment for Surface Condensers", Symp. On State-of-the-Art Condenser Technology, I.A. Diaz-Tous and R.J. Bell (ed), EPRI, Palo Alto, CA.

10. British Electrical and Allied Manufacturers' Association (1967). Recommended Practice for the Design of Surface Type Steam Condensing Plant, BEAME, London.

11. Marto, P.J. and Nunn, R.H. (1981), Power Condenser Heat Transfer Technology, Hemisphere, New York.

12. Orrok, G.A., (1910). The Transmission of Heat in Surface Condensation, Trans. ASME, vol. 32, pp. 1139–1214.

13. Macnair, E., (1981). Introduction, in Power Condenser Heat Transfer Technology, eds. P.J. Marto and R.H. Nunn, p 5, Hemisphere, New York.

14. Wenzel, L.A., (1981) in Power Condenser Heat Transfer Technology, eds. P.J. Marto and R.H. Nunn, pp. 182–183, Hemisphere, New York.

15. McNaught, J.M. (1982). Two-Phase Forced Convection Heat Transfer during Condensation on Horizontal Heat Tube Bundles, Heat Transfer 1982, vol. 5, pp. 125–131, Munich.

16. Colburn, A.P. and Hougen, O.A. (1934). Design of Cooler Condensers for Mixtures of Vapors with Noncondensing Gases, Ind. Eng. Chem., vol. 26, pp. 1178–1182.

17. Standiford, F.C. (1979). Effect of Non-Condensibles on Condenser Design and Heat Transfer, Chem. Eng. Prog., vol. 75, pt. 2, pp. 59–62.

18. Chisholm, D. (1981). Modern Developments in Marine Condensers: Noncondensable Gases: An Overview, in Power Condenser Heat Transfer Technology, eds. P.J. Marto and R.H. Nunn, pp. 95–142, Hemisphere, New York.

TOWARD ENERGY EFFICIENT MANUFACTURING ENTERPRISES

Kevin W. Lyons, Ram D. Sriram, Lalit Chordia and Alexander Weissman

ABSTRACT

Industrial enterprises have significant negative impacts on the global environment. Collectively, from energy consumption to greenhouse gases to solid waste, they are the single largest contributor to a growing number of planet-threatening environmental problems. According to the Department of Energy's Energy Information Administration, the industrial sector consumes 30% of the total energy and the transportation sector consumes 29% of the energy. Considering that a large portion of the transportation energy costs is involved in moving manufactured goods, the energy consumption of the industrial sector could reach nearly 45% of the total energy costs. Hence, it is very important to improve the energy efficiency of our manufacturing enterprises. In this chapter, we outline several different strategies for improving the energy efficiency in manufacturing enterprises. Energy efficiency can be accomplished through energy savings, improved productivity, new energy generation, and the use of enabling technologies. These include reducing energy consumption at the process level, reducing energy consumption at the facilities level, and improving the efficiency of the energy generation and conversion process. The primary focus of this chapter is on process level energy efficiency. We will provide case studies to illustrate process level energy efficiency and the other two strategies.

31.1 INTRODUCTION

Companies often find it difficult to obtain needed traction on addressing energy efficiency efforts. Company managers view energy use as a necessary cost for conducting business and they have difficulty in competing energy with other core operational needs [1]. Recent public perception and marketplace pressures have companies taking a second look at energy efficiency within the enterprise and with life cycle considerations. The life cycle of a product can have several beginnings. One such cycle that starts with raw material extraction and processing is shown in Fig. 31.1. This cycle continues with the pre-design and fabrication of the relevant semi-finished products, includes manufacturing and assembly of the final product as well as its transportation, use and maintenance, and concludes with the end-of-life operations. This last stage includes recycling of materials and, after adequate treatment, final disposal of waste.

A generalized version of this cycle is shown in Fig. 31.2. The figure shows two cycles. The first cycle depicts the extraction of material from the Earth and putting waste back into the Earth. We would like to minimize this flow and, in particular, achieve zero landfill. The second cycle includes pre-design, production, use, and post-use stages of the product life cycle. The thick arrows represent material and information flow between these stages. The reverse arrow from use stage to production stage denotes the field data from product use into design and manufacturing to improve the design.

The various stages in the above product life cycle have a significant impact on the environment. According to a recent report from the University of Cambridge, "the industrial system can account for 30% or more of greenhouse gas generation in industrialized countries" [2]. This statement reflects only a part of the total impact on the environment. Energy consumption and waste production are other major factors that affect the environment. For example, according to the Department of Energy's Energy Information Administration, the industrial sector consumes 31% of the total energy and the transportation sector consumes 28% of the energy [3]. Considering that a large portion of the transportation energy costs is involved in moving manufactured goods, the energy consumption in the industrial sector could reach nearly 45% of the total energy costs. The U.S. Energy Information Administration's Annual Energy Review 2009 provides an excellent source of information for energy consumption in various sectors [4]. It also shows energy flows from source (e.g., coal, hydroelectric power, renewable energies) to a particular sector (e.g., transportation, industry, etc.).

A product's energy life cycle includes all aspects of energy production [5]. Depending on the type of material and the product, energy consumption in certain stages may have a significant impact on the product energy costs. For example, 1 kg of aluminum requires about 12 kilograms of raw materials and consumes 290 MJ of energy [6]. Several different strategies can be used to improve the energy efficiency of manufacturing enterprises, including reducing energy consumption at the process level, reducing energy consumption at the facilities level, and improving the efficiency of the energy generation and conversion process. While the primary focus of this chapter is on process level energy efficiency, we will briefly discuss energy reduction methods and efficient energy generation processes through case studies.

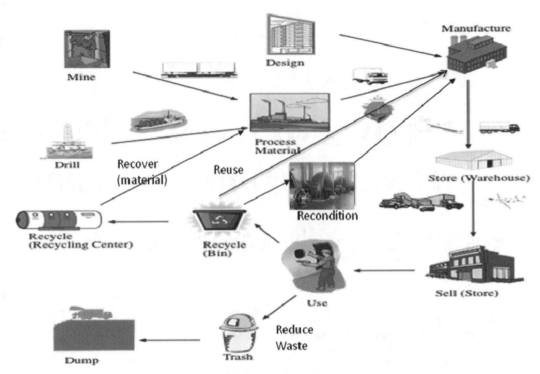

FIG. 31.1   PRODUCT LIFE CYCLE: FROM MINING TO REUSE

31.1.1   Outline of the Chapter

A rationale for energy-efficient manufacturing is provided in the next section (Section 31.2). The concept of unit manufacturing processes is introduced in Section 31.3, followed by a classification of these processes in Section 31.4. Section 31.5 describes the mechanisms used to determine energy consumption. Improving the efficiency of this energy consumption is the realm of Section 31.6. Several case studies are provided in Sections 31.7 to 31.9. Section 31.7 presents a case study on improving energy efficiency in injection molding. A case study of various innovations used by a small manufacturer for energy efficient manufacturing is provided in Section 31.8. In Section 31.9 we discuss a specific technique — using supercritical fluids — that can be effectively used to improve energy efficiency and to improve processes that generate energy

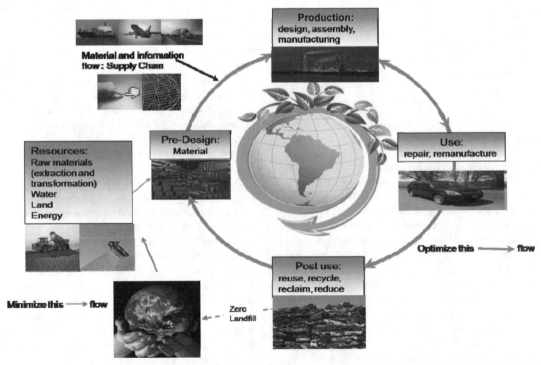

FIG. 31.2   A CLOSED LOOP VIEW OF PRODUCT LIFE CYCLE AND SUSTAINABLE MANUFACTURING

from non-traditional sources. Finally, in Section 31.10, we point out how best practices, regulations, and standards can play an important role in increasing energy efficiency.

31.2 ENERGY EFFICIENT MANUFACTURING

Energy efficiency efforts and the use of renewable energy sources is an essential component that manufacturing enterprises must address to cope with the current global environmental crises [7]. Manufacturer's energy inputs can be described by the following five progressive stages [5]:

1. *primary energy input*, which is the total volume of energy assembled to serve industrial needs.
2. *central generation*, which mainly occurs in powerhouses where fuel is converted to heat and power by a steam plant, power generator, or co-generator.
3. *distribution*, which pipes heat and sends power from central generation to process units.
4. *energy conversion*, which transforms heat and power to usable work, and involves motors, fans, pumps, and heat exchangers.
5. *processes*, where converted energy transforms raw materials and intermediates into final products.

The primary focus of this chapter is on energy conversion and manufacturing processes although aspects of primary energy input, central generation, and distribution are covered when considering life cycle analysis (LCA).

Advanced manufacturing sciences and technologies are necessary to support, promote, and implement energy reduction and renewable efforts. Yet, progress of these efforts is hindered as industry lacks the science-based approaches enabling quantifiable measurement techniques, tools, and data that support objective evaluation of progress against specific aspects of product life cycle. Beyond energy reduction and renewal technologies, manufacturers are starting to look at the entire life cycle of products and services to conserve energy and natural resources, minimize negative environmental impacts, ensure safety for employees, communities, and consumers, and improve economic viability [8].

The U.S. Department of Energy (DoE) and the U.S. Council for Automotive Research (USCAR, www.uscar.org) created a technology roadmap for energy reduction in automotive manufacturing. The goal of the roadmap is to guide decision-making for future research, development, and demonstration projects through identification of potential ways to reduce energy intensity in automotive manufacturing and the associated supply chain. This report is organized around five major operations within automotive manufacturing [9]: 1) Body in White and Components, 2) Paint, 3) Powertrain and Chassis Components, 4) Final Assembly, and 5) Plant Infrastructure. Through the combined data collection, analysis and reporting of energy consumption occurring within these major operations, USCAR members were able to recognize and take appropriate actions for energy efficiency at the major operation level. Figure 31.3 illustrates the major operations, a description of each operation, the flow of materials between processes, and approximate percent of energy that each process requires of the total enterprise energy. Such flows can be developed for other industries.

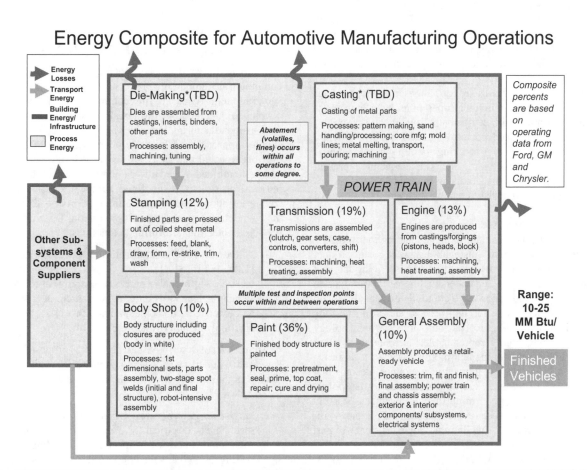

FIG. 31.3 DISTRIBUTION OF ENERGY USE IN MAJOR OPERATIONS IN A TYPICAL AUTOMOBILE MANUFACTURING PLANT [9]

Companies find it difficult to accurately determine energy utilization within a plant or enterprise beyond major manufacturing operations; down to the plant floor equipment level. Companies recognize the need to address energy efficiency practices yet they lack formal descriptions of resources that promote IT automation and the development/ improvement of tools. As a result, current process improvement efforts are likely to address localized problems, potentially failing to address the bigger picture that systems and life cycle approaches would have identified. If companies are able to take into account system and life cycle considerations in regards to addressing energy efficiency issues they must identify and evaluate competing performance attributes or tradeoffs. A process improvement may decrease energy consumption at one stage of manufacturing, but increase consumption at another, or increase waste. Similarly, an overall reduction in manufacturing energy consumption may produce a product that requires more energy to operate, or may result in a less durable product that must be replaced more often. For example, a product that takes 5 MJ to manufacture but lasts for 10 years is more energy-friendly than a product that takes 4 MJ to manufacture but only lasts for 5 years (and therefore must be replaced). Analyzing these tradeoffs and using them to make decisions requires detailed, comprehensive models.

31.3 UNIT MANUFACTURING PROCESSES

As stated in [10], manufacturing, reduced to its simplest form, "involves the controlled application of energy to convert raw materials (typically supplied in simple or shapeless forms) into finished products with defined shape, structure, and properties." Joining and assembly operations can also be viewed similarly. The energy applied during the unit manufacturing operations may be mechanical, thermal, electrical, or chemical in nature and sufficient detail is needed to understand the allocation of energy consumption within each unit process.

Unit manufacturing processes are formal descriptions of manufacturing resources at the individual operations level (e.g., casting, machining, forming, surface treatment, joining, and assembly) required to produce finished goods. Engineers have historically evaluated complex systems by breaking them down into smaller and more manageable parts that together still adequately represent the complete system, and are computational tractable. This approach avoids the difficult task of creating accurate abstractions of the production system.

Unit manufacturing processes provide a science-based methodology for companies. This can aid in understanding their production processes and equipment thus enabling process and product performance improvements. By identifying and defining the unit process, engineers gain a better understanding of the production process performance, thus allowing them to identify, analyze, and improve energy efficiency of the unit process and ultimately the enterprise. Initial applications of the unit process methodology include: 1) providing highly reproducible, accurate positioning of production equipment component motions needed to improve precision levels, 2) developing innovative equipment designs that dampen vibrations so that they are not transmitted to the tooling and workpiece, 3) reducing warm-up time from process start-up to operational steady state to attain minimal energy use and, 4) increasing speed of operation — while achieving consistency of part characteristics, such as dimensional control [10].

The intent of unit processes, and their ultimate utility, is to enable collections of unit processes to define an entire process flow for a component or product. This linking of unit processes, with the output of one process serving as the input for the next process, clearly illustrates the dependency of one process on another for achieving process and product performance. This collection of unit processes also supports calculation of the energy use required to complete specific operations or to produce a product. Unit process models support decisions regarding continuous improvement of the individual unit process as well as the system of unit processes. In addition, social factors, such as rapid response to customer needs or having safe working conditions, can also be addressed. Over the years, the use of unit process methodologies has achieved most success in the chemical sector.

Enhancements to the unit manufacturing process concept have continued to evolve. To increase their utility there was a need for an innovative methodology to incorporate life cycle considerations into the model. A new methodology called Unit Process Life Cycle Inventory (UPLCI) was developed to use the manufacturing unit process as the basis for life cycle inventory [11]. UPLCI involves 1) preparing a process description that includes appropriate supporting information for describing value-add steps and 2) developing process mass loss equations and applicable examples that assist users in applying methods to their work, and references to supporting equations and data. The UPLCI model is further refined through the study of four types of data: 1) time, 2) power, 3) consumables, and 4) emissions [11]. Whereas the initial unit process model applications could be quite complex, UPLCI model development looks for simplicity and tries to minimize information or excessive rigor in deriving estimated energy consumption to maximize productivity and promote wider acceptance of the methodology by industry.

31.4 CATEGORIZATION OF UNIT MANUFACTURING PROCESSES

There are hundreds of unit manufacturing processes. However, common traits among the processes can be used as a basis for organizing them. By identifying common traits, a taxonomy can be developed. One approach to taxonomy construction is to arrange various manufacturing processes according to function [12] while another approach can be descriptive headings that alert companies of expected energy types and use. In all cases, taxonomies add structure and systematic categorization facilitates search and retrieval operations. This allows companies to locate a generalized description of a manufacturing unit process that will serve as the foundation for its unique instantiation. Listed below are two widely adopted manufacturing taxonomies:

1. Allen and Todd's manufacturing processes reference guide [13] (see Fig. 31.4).
2. CO2PE! — "initiative process taxonomy based on the German standard DIN 8580 (Fertigungsverfahren, Begriffe, Einteilung) and extended with some auxiliary processes like compressed air supply, cooling systems, etc." [14].

Another set of descriptive categories was proposed by a National Research Council (NRC) study group [10]. This top-level taxonomy organizes all processes into the five descriptive categories listed below.

1. **mass-change processes**, which remove or add material by mechanical, electrical, or chemical means (included are the traditional processes of machining, grinding, shearing, and plating, as well as such non-traditional processes as water jet, electro-discharge, and electrochemical machining).

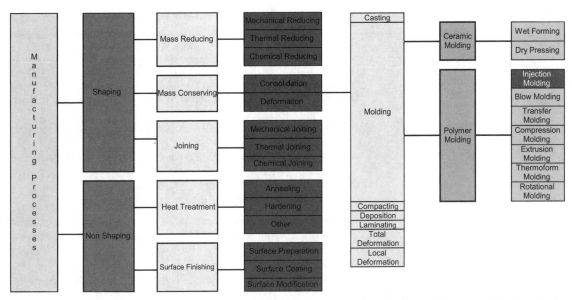

FIG. 31.4 PARTIAL TAXONOMY TO HIGHLIGHT INJECTION MOLDING PROCESS. DATA FROM REF. [13]

2. *phase-change processes*, which produce a solid part from material originally in the liquid or vapor phase (typical examples are the casting of metals, the manufacture of composites by infiltration, and injection molding of polymers).

3. *structure-change processes*, which alter the microstructure of a workpiece, either throughout its bulk or in a localized area such as its surface (shot-peen stress relief, heat treatment and surface hardening are typical processes within this family; the family also encompasses phase changes in the solid state, such as precipitation hardening).

4. *deformation processes*, which alter the shape of a solid work-piece without changing its mass or composition (classical bulk-forming metalworking processes of rolling and forging are in this category, shot-peen forming, as are sheet-forming processes, such as deep drawing and ironing).

5. *consolidation processes*, which combine materials such as particles, filaments, or solid sections to form a solid part or component (powder metallurgy, ceramic molding, and polymer-matrix composite pressing are examples, as are *permanent* joining processes, such as welding and brazing).

The NRC report also suggested a sixth category that recognized the likelihood of innovative configurations of unit processes.

6. *integrated processes*, which combine more than one specific unit process into a single piece of equipment or into a group of work stations that are operated under unified control.

31.5   DETERMINING ENERGY CONSUMPTION

A product's energy life cycle describes its total energy impact, including all stages of its manufacture through the end of its operating life and includes its eventual disposal [5]. This is referred to as cradle-to-grave analysis that captures relevant sustainability data starting with the extraction of raw materials and accounting for all operations until the final disposal of these materials. The total energy should reflect the collective contributions of life cycle factors (such as embodied energy in raw materials, scrap, and dis-

posal) and in-direct manufacturing activities (such as transportation of materials, product packaging, and HVAC).

To determine energy consumption rates at the unit process level requires specific knowledge of the production process (resource inputs, outputs) and the measurement methodology necessary to support reporting and decision support requirements. Typically, there are a number of possible methods that can be used to measure energy rates. Each of the methods is based on various assumptions and has unique precision and accuracy. They also have different costs, depending on the physical and software resources required to take the measurements, the time to take the measurements, the maintenance for continuous operation, and the time for reducing and formatting the data for reporting and decision support. There are new advances in acquiring data from unit processes that make it easier for companies to collect critical measurement data for assessing energy consumption and other sustainability data [15].

A growing number of companies have installed and begun to use Energy Management Systems (EMS) that provide monitoring and reporting capabilities at a sub-station or possibly at the major operation level. Facets of energy management have a notable common requirement: the need for advanced data collection and analytical tools that facilitate energy-efficient practices [16]. Table 31.1 describes some data sources that companies should explore for determining their energy management requirements.

When direct energy consumption data is unavailable, an estimate may be made instead. This may be done when it is desirable to make a prediction of energy consumption before manufacturing processes have been designed and implemented. Oftentimes, this is done using an allocation scheme based on specific energy consumption (SEC) [17]. SEC is defined as the amount of energy used by a specific process for a unit quantity of material. In injection molding, the mass of the part, which can be obtained from a Computer-aided Design (CAD) model, is multiplied by the SEC for the given material, which can be found in an LCA database. From this calculation, an estimate of energy consumption is obtained.

We will illustrate the above concept through an example of an estimation for the energy cost of an injection-molded part, as shown in Figure 31.5.

TABLE 31.1 DATA SOURCES FOR ENERGY MANAGEMENT

Data Source	Data Description
Energy Management Control System Data (EMCS)	Building data from automation systems, HVAC, lighting systems, boiler, chiller, turbine, and other equipment
Energy Meter Data	Data from energy meters and submeters, which can include electricity, chilled water, steam, gas, fuel, water, and other metered resources
Enterprise Resource Planning (ERP)	Enterprise-level business data, such as supply-chain, asset, financial, project, and others
Data Historian	A historical data repository that efficiently stores data from manufacturing process, facility metered, or other types of historical data
Weather Data	Temperature, pressure, humidity, and other weather data
Building Data	Building type, design specifications, and square-footage data
Web Data	Various internet data sources such as Department of Energy (DOE) or U.S. Energy Information Administration (EIA) energy databases or forecasted weather data
Billing Data	Incoming and outgoing utility billing data, which may also come from the ERP system
Scheduling Data	Line schedules, room schedules, facilities schedules, personnel schedules, and others
Excel Data	Historical spreadsheets and reports of Excel data

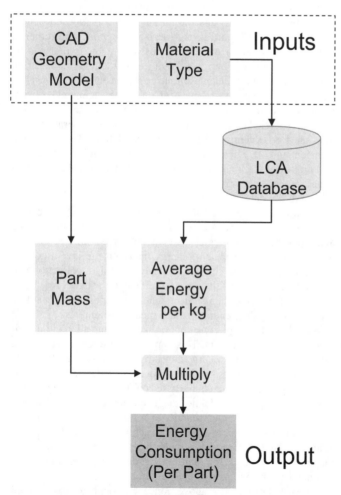

FIG. 31.5 CURRENT METHOD USED TO INVENTORY ENERGY CONSUMPTION FOR INJECTION MOLDING

Unfortunately, the available LCA databases only provide an average over the range of machines used in injection molding. This is inadequate because properties of the specific machine used dramatically influence energy consumption. Larger machines require more thermal energy to maintain the polymer temperature, and more power to move the heavier injection and clamping mechanisms. These generalizations lead to wildly inaccurate energy estimates.

In addition, the allocation scheme based on SEC and part mass do not account for the influence of part geometry and cycle time. Parts having the same volume and therefore the same mass, but different geometry can have significantly different cycle times and therefore require different amounts of energy to manufacture. For example, let us consider the two injection-molded parts shown in Fig. 31.6. Both parts are made using the same material and have the same volume and mass. However, the maximum wall thickness of the smaller, more compact part (a) is twice that of the larger, thinner part (b). The cooling time for an injection molded part is proportional to the square of the maximum wall thickness [18]. Therefore the cooling time for the cup in Fig. 31.6 (a) will be approximately 4 times that of the cup in Fig. 31.6 (b). During the cooling time, the machine is idle and continues to consume energy. Therefore increased cooling time, along with increased cycle time of the operation, also results in increased energy consumption. Studies by Gutowski [19] and Krishnan [20, 21] show that the energy consumed by overhead operations such as maintaining the polymer melt and the mold temperature along with pumping fluids and coolants, can be more than the energy used during each production run. Thick parts may especially require active cooling, which requires use of even more energy to supply coolants.

Gutowski and Krishnan [19–21] have shown that machines with a typically higher throughput tend to consume less energy per part. This can be explained by the influence that cycle time has on energy consumption as described above. Since the baseline idling energy is relatively constant, a machine having lower typical cycle times allocates less idling energy per part.

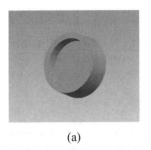 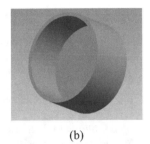

| (a) | (b) |

FIG. 31.6 TWO DIFFERENT PARTS WITH EQUAL VOLUME BUT DIFFERENT WALL THICKNESSES AND COOLING TIMES. PART (A) HAS A WALL THICKNESS OF 1.27 MM, WHILE PART (B) HAS A WALL THICKNESS OF 0.635 MM. BOTH PARTS HAVE A VOLUME OF 54.7327 CM^3

To account for the effects of baseline idling energy, Gutowski divides the specific energy consumption into two components: one component represents the energy used while the machine is idle, and the second component represents the additional energy used to process each unit of material. However, this method still does not account for the variations in power consumption at different stages of the molding cycle. A 2007 study [22] investigating the effects of conformal cooling channels on energy consumption showed that a 40% reduction in cycle time for the same part on the same machine results in only a 20% reduction in energy consumption. This suggests that the portion of the cycle that was shortened consumed power at a rate lower than the average for the entire molding cycle. Therefore, an approach that accounts for a specific part geometry and machine at each stage of the molding cycle could help to achieve a more accurate estimate of energy consumption.

The arrangement of immediate production processes (molding, assembly, finishing) to meet a production schedule can have a dramatic influence on energy consumption in manufacturing processes. This influence is commonly known as energy efficiency [23–25], which is summarized by Kumara as: "Energy efficiency mainly relates to optimizing the ratio of production output to the energy input for the technical building services (heating and cooling) and production machines" [26]. A good model of energy consumption should therefore account for production volume, scheduled and unscheduled maintenance of machines, and the design of the factory, in addition to the energy used during the actual molding process. To determine how this ratio of production to energy input can be optimized, discrete event simulation (DES) can be applied. DES, in combination with LCA data, is one possible approach for analyzing the cause and effect of various scenarios where time, resources, place, and randomness determine the outcome and being sustainable is considered crucial. For such analysis, only a few research publications exist: Solding and Petku [27] and Solding and Thollander [28] both describe how DES can be utilized to lessen the electricity consumption for foundries. Ostergren et al. [29] and Johansson et al. [30] describe how DES can be utilized in combination with LCA for quantifying environmental impacts during food production. DES has also been explored for an automobile paint shop [31] but has not yet been explored for injection molding processes.

Several authors have investigated the interconnected roles in various manufacturing processes of geometry, material, equipment, and production policy in energy consumption and waste production over the product's entire life cycle. Oftentimes, this is rolled into the single metric of "embodied energy," which is the total yielded energy used to create and destroy a product throughout its entire lifecycle [32]. This has been used to compare energy costs of milling marble slabs versus marble tiles [33], for example. Embodied energy has also been considered in the context of re-manufacturing multi-process products, such as engines [34], which are made through die casting followed by machining, and the production of double-glazed windows [35], which are made through the float-glass process and milling. Life cycle analyses of die casting [36, 37] and sand casting [38] have also been performed using embodied energy.

Other authors have examined the relative energy costs of different processes which can be used to make the same product. For example, Cho et al. [39] compare the reduced yielded energy cost of continuous casting with hot extrusion and heat treatment for producing copper wirerods. Other works investigate the relative energy costs of semi-solid forging, traditional forging, and die casting for metal alloys [40, 41].

31.6 IMPROVING ENERGY EFFICIENCY

Energy efficiency refers to technologies and standard operating procedures (SOPs) that reduce the volume of energy per unit of industrial production or energy intensity which is defined as the amount of energy it takes to produce a dollar of goods [6]. The Department of Energy, Office of Energy Efficiency, and Renewable Energy's Industrial Technologies Program (DoE) works to improve the energy intensity of U.S. industry through coordinated research and development, validation, and dissemination of innovative energy efficiency technologies and practices [42]. DoE figures show that industry can achieve energy reductions of nearly 20% to 30% through procedural, behavior, and cultural changes without capital expenditures. The best metrics for energy efficiency have to be clearly measureable, have goals objectively expressed quantitatively and on a time scale, and have status clearly communicated. These goals may be set at the product design stage, at the manufacturing process level, or an approach that spans both design and manufacturing.

Many companies have already begun to improve energy efficiency on the product level by integrating environmental considerations into their product development processes. This effort has come to be part of a paradigm known as Design for Environment (DfE). DfE is defined as "the systematic consideration of design performance with respect to environmental, health, safety, and sustainability objectives over the full product and process life cycle" [43]. Design for Environment goes beyond mere compliance with environmental regulations, in which pollutants are simply cleaned up after manufacturing to the minimum extent required by law. Instead, the potential environmental impact of a product or process throughout its life cycle is considered while it is still being designed. This generates value for companies in several ways such as improved public image, safeguarding of resources vital to the company's continuing productivity, and attenuation of clean-up costs after manufacturing. Oftentimes, a direct savings in energy and resource consumption can be realized as well. Finally, DfE has a trickle-down effect. Companies that demand sustainably sourced materials and components can foster competition between suppliers, no longer just on a basis of cost, but on lower environmental impact as well. Several sets of guidelines for integrating Design for Environment into the design process have been proposed [44–48]. Some of these guidelines have come under criticism [49] because they do not offer a means for quantitatively validating a design decision and ensuring that it does, indeed, result in a net reduction in environmental impact. Therefore, it is clear that quantitative methods for assessing environmental impact are vital if DfE is to be an effective tool.

To improve unit manufacturing processes and energy efficiencies, companies have at their disposal a number of techniques and tools. Companies that have a strong lean and green culture typically have positioned their site positively with regards to energy efficiency. Also, companies that have implemented Lean principles use proven methods such as Value Stream Mapping (VSM) to ensure minimum waste and improved efficiencies [1, 50]. A Lean and Energy Toolkit [1] gives guidance on developing an energy planning and management roadmap. This roadmap starts at an initial assessment followed by design process, opportunity evaluation, and implementation phases. This toolkit goes on to state three techniques for measuring or estimating the energy used by production processes: 1) metering, 2) estimating, and 3) energy studies. These techniques have been discussed in Section 31.5.

31.7 IMPROVING ENERGY EFFICIENCY THROUGH IMPROVED PRODUCT DESIGN: A CASE STUDY IN INJECTION MOLDING

To illustrate energy efficiency methods, a plastic injection molding example will be used. Plastics are used in a vast majority of products produced in the U.S. and globally. One of the major manufacturing processes used to process plastics is injection molding. This is illustrated in the taxonomy of Fig. 31.4. To improve the energy efficiency of manufacturing resources one must understand the process in sufficient detail to identify all aspects on how energy is consumed and how use varies over time during setup and steady state operations. Weissman et al. present a methodology for estimating the energy consumed to injection-mold a part that would enable environmentally conscious decision making during the product design [51]. Table 31.2 conceptually shows that engineering changes can be made to the part design to improve the process energy efficiency without impacting the product functionality.

During injection molding, energy is consumed during the cycles to melt, inject, and pressurize the resin, open and close the mold, and pump water for cooling. The energy requirements during filling, cooling, and resetting can be determined from the cycle times and the power profile of the machine. Weissman et al. note that the main environmental concerns associated with injection molding are energy consumption and waste generation (mostly waste resin).

To develop an accurate method for estimating energy consumption for injection molded parts, an algorithm consisting of five steps was formulated [51]. These steps are:

(1) Determine a surrogate runner arrangement, and its volume, for the mold.
(2) Approximate the parameters of the machine that will be used based on the production requirements.
(3) Estimate various components of the cycle time for molding a part.
(4) Estimate the number of setup operations based on the delivery schedule.
(5) Multiply these times by the appropriate average power used in each stage by the selected machine, and sum to get the total energy consumption.

31.7.1 Surrogate Runner Arrangement

First, the CAD model of the part is analyzed to determine the mold cavity volume. In addition to the volume of the part, the volume of the runner system and sprue must also be considered. The runners and sprue are a system of channels which carry the molten polymer from the injection nozzle to various cavities in the mold. In some parts, especially parts at the small scale, the runner system can be much larger than the part. Hence it is important to carefully select the runner layout for estimating the projected volume of the mold cavity.

Figure 31.7 illustrates eight different sprue/runner layouts for four-cavity molds. These layouts are commonly used layouts which use fishbone and ladder layouts. The most appropriate runner layout is selected based on the critical quality metrics such as shrinkage, shear level, part density, mold machining constraints, etc., while optimizing for the cycle time and the overall runner volume. The geometry of the mold and the sprue location also plays a significant role in the selection of the most appropriate runner layout. Considering the complex nature of this problem, manufacturers currently select the most appropriate runner/sprue layout based on their prior experience.

Based on the selected runner layout, the projected area of the runner system and the runner volume can be computed. This information is then used for selecting the machine for completing the injection molding operation.

31.7.2 Selection of Machine

The next step is to estimate the size of the injection molding machine required to mold the part. Machine size is primarily driven by the clamping force required to hold the mold closed during the injection cycle, the shot size required by the volume of the part and runners, and the stroke length required to clear the maximum depth of the part during part ejection. The part volume and maximum depth of the part can be determined from the geometry model. The required clamping force can then be determined from the relationship between the maximum cavity pressure and the projected area of the cavity.

TABLE 31.2   INJECTION MOLDING PARAMETERS THAT IMPACT ENERGY CONSUMPTION [51]

Design feature	Operational impact	Energy impact
Thicker minimum part thickness	Increased cooling time	More energy used while machine idles longer
Larger projected part area	Greater clamping force needed	More energy used by more powerful hydraulic/servo mechanisms
Greater part depth	Longer stroke needed to eject part	More energy needed to operate clamp mechanism for longer
Higher specific heat of material	Higher temperatures needed to melt	More energy needed by heating unit
Better part finish	Additional post-processing steps	Energy consumed by those steps

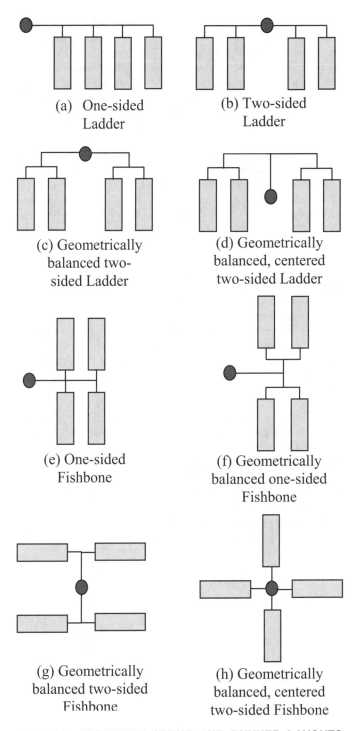

(a) One-sided Ladder

(b) Two-sided Ladder

(c) Geometrically balanced two-sided Ladder

(d) Geometrically balanced, centered two-sided Ladder

(e) One-sided Fishbone

(f) Geometrically balanced one-sided Fishbone

(g) Geometrically balanced two-sided Fishbone

(h) Geometrically balanced, centered two-sided Fishbone

FIG. 31.7 DIFFERENT SPRUE AND RUNNER LAYOUTS FOR FOUR-CAVITY MOLDS [51]

The maximum pressure in the mold can be determined using Moldflow®, given the predicted mold design from the first step and the recommended injection pressure. We then assume that the manufacturer will use the cheapest machine which can provide the necessary clamping force, shot size, and stroke length. The required shot size is equal to the volume of the part, plus the volume of the runners and sprue. The stroke length is typically estimated by a linear relationship with the maximum depth of the part. A machine which meets these criteria can be looked up in a machine database.

31.7.3 Estimation of Cycle Times

Once the machine has been selected, the cycle time for the part is estimated. The molding cycle can be broken down into three stages: injection, packing and cooling, and reset. These stages, as well as their sub-stages and other auxiliary stages in a typical injection molding operation, are shown in the state transition diagram in Fig. 31.8.

During the injection stage, the pressure at the injection nozzle is gradually increased. This is done to maintain a constant volumetric flow rate, as the melt cools and solidifies. The estimated fill time for the mold cavity can be derived based on the maximum flow rate. Fill time is approximated as twice the cavity volume, divided by the maximum flow rate of polymer from the nozzle.

Next, the pressure is held and then gradually dropped as the part cools and contracts in the mold. We assumed that active cooling is not used. Using the first term of the Carslaw and Jaeger solution [18], the cooling time in seconds can be estimated from the maximum wall thickness of the part, the thermal diffusivity of the material, the polymer injection temperature, the recommended mold

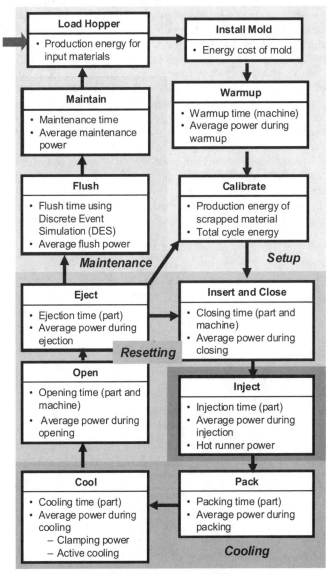

FIG. 31.8 INJECTION MOLDING PROCESS — DECOMPOSITION OF ENERGY CONSUMING STATES [51]

temperature, and the recommended part ejection temperature. The maximum wall thickness can be determined from the part model. The remaining parameters are properties of the material which can be found in the material datasheet provided by the supplier, or derived from those properties. For example, the thermal diffusivity can be computed from the specific heat, thermal conductivity, and density of the material.

Finally, after ejection of the part, the mold is prepared for the next cycle. This time is estimated by applying an overhead to the dry cycle time for the machine. The dry cycle time is typically a measure of the injection molding machine performance that indicates the time for the machine to perform the actions necessary to manufacture a part, without the part actually being produced. The overhead is proportional to the square root of the ratio of the stroke length for our part and the maximum stroke length from the machine.

31.7.4 Estimation of Setup Operations

Before the start of the production, the machine must be set up, which also consumes energy. Setup processes include steps such as warming up the machine, installing the mold, and calibrating the machine. The injection molding machine consumes significant amount of energy during warm up, and then continues to consume energy as it idles during mold installation. Before start of production, the injection molding process needs to be stabilized. This is done to establish process equilibrium to ensure complete filling of the part, avoid jetting, or other undesirable conditions that can compromise product quality. Manufacturers typically reject the first few tens of parts before beginning the production. Therefore the energy consumed during this step is included as part of the machine calibration.

To determine the total energy used during setup processes, it must first be determined how often the machine must be set up during the production schedule of the entire production volume. Typically, the entire production volume will not be completed in a single production run. In general, injection molded parts are produced

based on the production requirement and the delivery schedule. The customer specified delivery schedule involves a request for a certain number of parts at regular time intervals. Thus, to save on the inventory cost before delivery to the customer, the manufacturer makes parts in batches. The batch size should be larger than the number of parts delivery requirement at each time interval. Any remaining parts are stored at the expense of the manufacturer until the next delivery. Since larger batch sizes require fewer setups there is a tradeoff between the setup cost and the inventory cost.

Figure 31.9 shows the relationship between the delivery schedule and the production schedule over the entire production volume. The manufacturer produces a certain number of parts, and delivers to the customer at regular intervals. During this time, undelivered parts remain in storage. When the parts in storage have been depleted, the manufacturer makes a new batch of parts, and continues to ship them out according to the customer's delivery schedule, typically a regular interval of time.

The tradeoff can be formulated as a single variable optimization problem. The solution to this problem is the optimal number of setup operations which minimize the cost to the manufacturer over the entire production volume. For this problem, it is assumed that the batch production period is much larger than the delivery period, and so lead time can be ignored. Furthermore, it is assumed that the manufacturer must pay for a constant amount of storage; even as the manufacturer's inventory is depleted, they must continue to pay for the entire space needed to accommodate a batch of parts.

The solution to the optimization problem shows that the batch period which minimizes cost is a function of the delivery period k, the delivery volume n, and the unit costs associated with machine setup and part storage. Together with the total production volume, the optimal number of batches can then be found.

31.7.5 Computing Total Energy Consumption

The energy used during filling, cooling, and resetting can be determined from the cycle times and the power profile of the machine. Assuming that energy consumption per unit of time on a

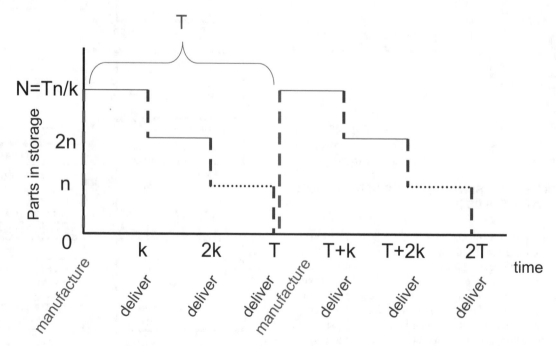

FIG. 31.9 GRAPH SHOWING DELIVERY SCHEDULE AND PRODUCTION SCHEDULE IN TERMS OF PARTS IN STORAGE VERSUS TIME [51]

given machine is constant for a given stage of the cycle, the power required, in watts, during each stage can be looked up. Multiplying the power consumed at each stage by that stage's estimated time, and then dividing by the number of mold cavities, the total energy at each stage is computed. Energy used in these stages can be added up to arrive at a total manufacturing energy cost.

This specific example of energy consumption during injection molding that is shown in Fig. 31.8 does not reflect a complete life cycle analysis (LCA), as the embodied energy of the raw material, transportation, product use, and recycling/disposal energy contribution was not included. Such challenges should for various unit manufacturing processes should be undertaken in the future.

31.8 REDUCING ENERGY CONSUMPTION AT MANUFACTURING FACILITY: A CASE STUDY AT HARBEC PLASTICS, INC

Bob Bechtold is the CEO of HARBEC Plastics, Inc. and is an innovator in implementing sustainable manufacturing practices [52]. His company, which makes high-quality injection-molded parts, has made a considerable commitment to being green. HARBEC Plastics Inc. disproved a common misconception that "Being green is nice but we can't afford it" through eco-economic factors implemented at HARBEC. For example, the CHP (combined heat and power) micro-turbines, which are capable of generating 100% of HARBEC power requirements, provide air conditioning and heat for an injection molding facility, while grid connection provides appropriate back up. HARBEC's air-conditioning system, which uses an absorption chiller, turns exhaust gas waste heat into free air conditioning.

In the area of renewable energy, HARBEC installed a 250-kW wind generator to accomplish wind/microturbine hybrid electricity generation. The projected energy production is 300,000 to 350,000 kWH per year, or about 20% of the total HARBEC annual energy requirements.

HARBEC, over a 7-year time span, replaced all standard hydraulic type equipment with all-electric injection molding machines. The advantage of electric machines is that these machines do not use power when they are in a static state, which is a significant portion of the time; the machines are capable of doing the same or a better job than the hydraulic machines, using as much as 50% less energy.

In the area of lighting, HARBEC replaced every fixture, ballast, and high bay sodium lamps with new T-8 type fluorescent bulbs and reflectors. These sustainable manufacturing practices allowed HARBEC to ensure that the lighting energy consumed was reduced by 48% on average company-wide. HARBEC is a big proponent of Leadership in Energy and Environmental Design (LEED), although not LEED-certified, HARBEC implemented LEED principles wherever it could.

HARBEC has significantly improved their water treatment system by installing a bi-metallic water treatment plant, which does not require any chemicals for water treatment. This enabled them to save thousands of dollars per year on chemicals, and eliminated the need for people to handle them. This new water treatment plant provided 850,000 gallons (3,864,176.5 liters) of fresh water input to their pond, which in turn provides water capacity sufficient for their sprinkler system, and also provides cooling. In a talk at NIST, Bob Bechtold emphasized the overall lesson learned: "If you want to make an environmental impact, and save money, use energy efficiently."

31.9 IMPROVING EFFICIENCY OF ENERGY GENERATION AND CONVERSION WITH EMERGING TECHNOLOGIES: A CASE STUDY IN USE OF SUPERCRITICAL FLUIDS AT THAR TECHNOLOGIES

Emerging technologies provide one way to apply new technology to reduce or remove energy inefficiencies associated with older technologies. New technologies may improve the efficiency of energy generation, or decrease the energy loss associated with converting matter from one phase to another. One example of an emerging technology that can be used to improve efficiency on many different facets is supercritical fluids.

Supercritical fluids (SCFs) exist at temperatures and pressures above the critical point, where the differences between gas and liquid states lose their significance. Various physical properties such as density and dielectric constant change rapidly with pressure and enthalpy and are to some extent "tunable," i.e., they can be manipulated to achieve a processing goal. Such fluids have a variety of energy efficient applications in solvent extraction, laboratory analysis and heat transfer.

Supercritical fluid extraction (SFE) offers multiple advantages compared to traditional extraction technologies such as organic-solvent extraction or distillation. With SFE there is no residual solvent in either the extract or in the raffinate. This translates into lower operating costs because of the reduction in post-processing steps, clean-up, and safety measures. Products extracted with SFE deliver the most natural aromas and flavors because the volatile compounds are not removed as they are in a post-processing step to remove residual solvent. SFE works at low temperatures, resulting in less deterioration of thermally-labile components in the extracts. Since there is no oxygen in the process, the potential for oxidation of the extract is significantly reduced. Supercritical fluids — particularly carbon dioxide ($scCO_2$) — have application in the production or recovery of alternative energy. Several examples of $scCO_2$'s use are provided below.

31.9.1 Food and Fuel: Biodiesel with a Food-Production Bonus

Even though the ability of SFE with carbon dioxide to separate triglycerides from oil seeds is well understood and practiced, several technological problems inhibit the development of a continuous, "green" version of this biofuel process. To date, all known extraction of oils or herbal essences using SFE has been performed with batch processing. Fuel production, expected to occur at feed volumes 10 to 100 times greater than those used in typical herbal extract facilities, demands continuous processing. Various groups have worked on continuous or semi-continuous systems but none have succeeded for a variety of reasons, such as leakages due to carbon dioxide's low viscosity and the sealing of rotating or linear devices under high pressure. Also, the conventional method of cycling CO_2 through a large pressure range of approximately 70 000 000 Pa (700 bar) — for the purpose of changing the dissolving characteristics of the solvent — imposes an energy burden on the process.

Depending on the feedstock, valuable solid co-products, from high-volume animal feed to low-volume, premium-priced de-fatted soy flour, could also be made using SFE. Byproducts include glycerol and soybean hulls (used in animal feed). This process also offers significant environmental advantages. Current biodiesel processing involves the extraction of oil from soybeans by hexane, a hazardous air pollutant (HAP) according to the U.S. Environmental Protection Agency (EPA). SF process eliminates two major sources of pollution from conventional hexane-extraction plants:

1. *Fugitive hexane solvent* itself, which shows up mainly in air emissions, as well as in the meal byproduct that is sold as animal feed.
2. *Solid and aqueous waste products* resulting from de-gumming, de-colorization and de-odorization, which are downstream processes of hexane extraction. Some of the byproducts, such as lecithin, are valuable and can be recovered and re-sold.

31.9.2 Ethanol Extraction

Supercritical fluid extraction of ethanol from aqueous fermentation broth holds promise as a means of cutting the energy consumption that would otherwise go to distillation and subsequent molecular-sieve dehydration. For ethanol derived from corn, distillation and dehydration account for 40% of the energy input, counting cultivation, harvesting, other processing and distribution. Supercritical fluid extraction could possibly cut this figure to 20%. The process is, in its simplest rendition, a drop-in replacement of the distillation and de-hydration units, as shown in Fig. 31.10.

This process first separates water from ethanol in a liquid-phase extractor. Then, in a lower-pressure gas-phase stripper, it condenses the ethanol from the SCF solvent before re-compressing the solvent and returning it to the water separator for a new round of liquid-phase extraction. Sufficient pressure must be maintained in the water separator to guarantee liquid conditions. A temperature gradient is needed across the vertical length of the separator to ensure that water drops to the bottom, while ethanol-rich SCF solvent migrates to the top. The next step of the process, stripping out the ethanol from the solvent, takes place in the gas phase. To do this, latent heat is added in the stripper — mainly by heat exchange with compressed, lean SCF solvent (i.e., depleted of ethanol).

31.9.3 Heating and Cooling Using a Natural Refrigerant

Besides its myriad process applications, scCO$_2$, as well as sub-critical CO$_2$, also act as excellent heat transfer fluids; any fluid below the critical pressure will have a region of vapor-liquid equilibrium that can be put to good use in heat transfer. Carbon dioxide was, in fact, the original refrigerant, dating back more than a century ago. In the first half of the twentieth century, it lost out, first to inorganic refrigerants, sulfur dioxide and ammonia, then to organic chlorofluorcarbons in the years just before World War II. But by the end of the century, concern over ozone depletion brought a renewed focus on CO$_2$, which is environmentally benign and non-toxic, as the carbon dioxide is always contained and recycled. Fugitive emissions, which are possible, would contribute in miniscule amounts to global warming. Fugitive emissions of fluorcarbons, however, pose a significant hazard to the ozone layer. By supplanting fluorocarbons with carbon dioxide, ozone-depletion is minimized with no significant increase in global warming potential.

Supercritical CO$_2$ has more going for it than environmental advantages. It exhibits unique thermodynamic properties that have only recently gained appreciation. For example, the density of scCO$_2$ is greater than that of gaseous freon by a factor of about seven. This translates to smaller equipment, despite the higher pressure required to contain scCO$_2$. Furthermore, the temperatures of evaporation and condensation for carbon dioxide at sub-critical pressures relate to typical air and ground temperatures in such a way as to open applications in geothermal heat rejection and absorption. In other words, sub-critical carbon dioxide is an attractive geothermal heat-transfer fluid for both heating and cooling.

Geothermal heat-transfer technology involves circulating CO$_2$ through the ground, to depths typically less than 150 m. Depending on the mode of operation — cooling or heating — the ground acts as a heat sink to absorb energy (as in cooling) or a heat source to produce energy (as in heating applications). In both cases, the ground temperature is typically 12 centigrade, plus or minus a few degrees, throughout the year. A schematic of this is shown in Fig. 31.11. This technology differs substantially from deep underground recovery of heat from superheated sources, such as geysers.

There are actually two modes of cooling, one employing a compressor to move CO$_2$ in gaseous form and the other relying on a pump to move it in liquid form. A cooling cycle based on liquid pumping satisfies most applications when humidity is already at comfortable levels. For times in a day when humidity rises uncomfortably high, the compressed cycle is started in order to condense water from room air. This combination of pumped/compressed

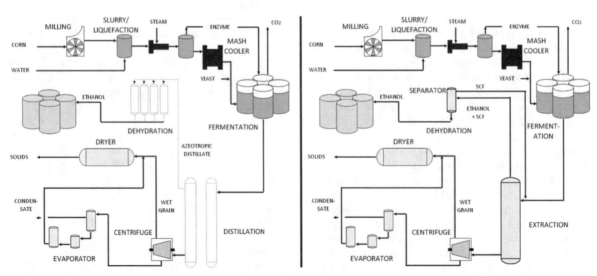

FIG. 31.10 SUPERCRITICAL FLUID REPLACEMENT FOR DISTILLATION AND DEHYDRATION IN CORN ETHANOL PRODUCTION

Thar Geothermal System

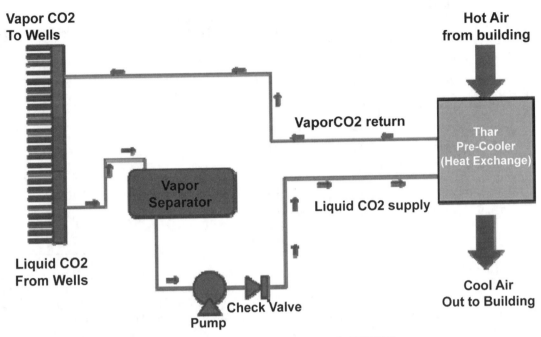

FIG. 31.11 PUMPED CO₂ COOLING SYSTEM

cooling is much more efficient than compressed cooling alone. The heating mode is always a compression cycle. In this case, liquid CO_2 is depressurized to 3,500 kPa to 4,000 kPa (35 bar to 40 bar) so that it evaporates — rather than condensing — underground. Subsequent compression of the evaporated fluid produces temperatures high enough to heat a room. Similar heat-pump systems in use today employ freon-type fluids, but, as noted earlier, gas densities are only one-seventh that of CO_2, and so piping and heat exchange equipment are much larger.

Numerous variations of the scheme are possible. For example, an existing building with a conventional air-conditioning system that employs freon fluid with air-blown heat rejection can be supplemented with a pumped-cooling geothermal system, so as to reduce the energy load overall.

31.9.4   Miniature Refrigeration Systems

Carbon dioxide has even been tried as the heat transfer fluid for a miniature refrigeration system, small enough to fit inside a desktop computer. One of the co-authors (Lalit Chordia, Thar Technologies) designed and built a small compressor capable of moving enough fluid, at sufficient pressure, to dissipate approximately 100 watts of heat from a computer chip. As part of this effort, Chordia's company also developed a microchannel heat exchanger that came in direct contact with a chip-sized heat source.

Heat exchangers were built with channel widths of less than 100 microns. The channels were open at the edges of a metal foil that was only about 0.3-mm thick. Headers attached to the channeled edges ensured even distribution of fluid either into or out of the foil. Supercritical carbon dioxide is a good choice of a heat-transfer fluid with this type of construction because of its low viscosity. Compared to liquid water, and for the same amount of heat capacity, pressure drop through the channels is one-fifth. Furthermore, carbon dioxide poses less of a hazard in the event of mechanical failure, such as a leak. Thermal resistance, on the other hand, ben-

efits from higher flow rates of supercritical CO_2. Tests conducted by Thar showed that pressure drop would likely exceed 140 kPa in order to achieve resistances of less than the 0.5 C/W. Thus, the use of copper rather than stainless-steel is especially important.

The compressor for this application was a reciprocating type, totally enclosed in a casing that included the motor. This way, there was no possibility of leakage of CO_2 from the crankcase. The entire assembly was small enough to fit in the palm of a hand.

31.9.5   Power Generation Using the CO₂ Brayton Cycle

Not all applications for CO_2 as a heat-transfer fluid involve underground operations. One concept captures solar heat in tubes, allowing pressurized fluid to rise to a high enough temperature that heat recovery becomes much more efficient than in conventional water-based systems — and more efficient than photovoltaic systems, for that matter. The same concept can be applied to the recovery of waste heat if such sources exist at temperatures of at least 150 C, which might occur in underground sources starting several hundred meters below the surface.

The cycles shown in Fig. 31.12 operate clockwise, starting with staged liquid pumping, with intercooling, to get to a supercritical pressure for heat absorption. In the case of concentrated

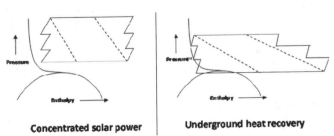

FIG. 31.12   SOLAR HEAT COLLECTION WITH CO₂

TABLE 31.3 BRAYTON CYCLE EFFICIENCIES

Stages	Max fluid T (°C):	Units	149 Underground heat recovery	427 Concentrated solar thermal
Single	Lowest pressure	kPa		20,700
	Highest pressure	kPa		34,500
	Efficiency	%		0.532
Double	Lowest pressure	kPa	90	20,700
	Highest pressure	kPa	275	69,000
	Efficiency	%	19.1%	55.7%
Triple	Lowest pressure	kPa	70	10,300
	Highest pressure	kPa	275	34,500
	Efficiency	%	22.5%	72.6%

solar power, energy is absorbed by fluid running through flow channels onto which light is focused. In waste heat recovery, some type of tubular heat exchanger might be employed. Once the fluid is hot enough it is depressurized through a series of turbines to recover energy as electrical power. The fluid coming out of the last turbine is still hot enough to provide substantial pre-heating to the pressurized fluid by means of a recuperator, the thermodynamic boundaries of which are represented in Fig. 31.12 by the dotted lines. Just how efficient these cycles can be is shown in Table 31.3.

Another application of such water-based systems, that can have potentially huge commercial consequence, is the replacement of steam-generation electrical utilities with ones employing $scCO_2$ instead. Such plants would be smaller and cheaper. Equipment sizes — particularly turbines — would be much smaller, and operating temperatures would be low enough to permit construction with cheaper carbon steel. Carbon dioxide is non-polar and non-corrosive, so high alloy steels are not required.

31.10 BEST PRACTICES, ENVIRONMENTAL REGULATORY POLICIES, AND STANDARDS

Best practices, regulations and standards play an important role in promoting and supporting energy management practices [1]. To be globally competitive companies must be able to confidently manufacture products that conform to regulatory requirements and standards, and most importantly, demonstrate this conformance when required [53]. Companies typically develop specific sets of best practices to achieve competitive advantage. They may also adopt best practices associated with a particular industry sector (aerospace, automotive) or standard (ISO 9000 Quality Management, ISO 14000 Environmental Management, ISO 50001 Energy Management). Representative examples of such practices are discussed below.

- Toyota developed superior value by partnering with suppliers to reduce its cumulative process waste. In 2000, Toyota issued its Green Supplier Guidelines where it asked the suppliers to go beyond legal and social requirements and to undertake activities that support Toyota's environmental goals. This resulted in ninety eight percent of Toyota's North American suppliers becoming ISO 14001 certified/registered (NAM). Toyota also shared its best practices and ideas with its suppliers and years later rolled out Eco-VAS, a comprehensive system used to measure and reduce the environmental impact of a vehicle across its entire life cycle. Toyota promoted energy conservation awareness throughout its supply chain, holding energy reduction events comparable to events conducted within its manufacturing plants for years [6, 54].
- For a number of years, several metal casting companies, backed by industry consortiums and funded through Department of Defense and Department of Energy [55], have been developing new and innovative energy reduction best practices.
- Buildings present one of the best opportunities to economically reduce energy consumption and limit greenhouse gases [56, 57]. A number of companies are establishing procedures to measure the combined impact of manufacturing operations (e.g., plant floor resources and processes) and building services and controls (e.g., HVAC-lighting and delivery) — an important capability for accurately evaluating a company's true performance in meeting sustainability objectives.

Sustainable manufacturing is motivating companies to implement new design and analysis procedures, energy reduction methods, material reduction efforts, and improved materials handling practices. An NIST workshop on sustainable manufacturing identified that current evaluation methods for energy consumption were not sufficient to measure environmental impacts [58]. Evaluation methods and decision-support tools are critical for companies to consider potential investments in energy efficiency. The evaluation methods should be able to calculate the environmental impacts, and these evaluation methods should consider the energy life cycle and source types. Additionally, there is a need for developing energy simulation models and analysis tools for a trade-off analysis between investment and environmental impacts.

31.11 SUMMARY

In order to identify, analyze, and improve energy efficiencies, an enterprise must have a clear understanding of the performance of its production processes and the correlation of its process controlling parameters to the performance of its products. Many processes, methods, and tools currently in use for measuring energy utilization were not built upon or developed using scientific methods. Science-based approaches add credibility by establishing a foundation for the development of energy utilization processes, methods, and tools which will support measurements enabling quantifiable progress toward meeting energy use objectives. Through disciplined research and the pursuit of new resources and technologies founded on well-structured science-based methods, researchers can expect to rapidly discover and increase human knowledge and understanding of energy management and control processes. This will allow for transition and implementation of the technologies into manufacturing areas that can benefit from the structured and formalized approach that science-based approaches provide. Ultimately, engineering methodologies and tools that achieve process energy efficiency without impacting product functionality will be commonplace.

31.12 DISCLAIMER

No approval or endorsement by NIST of any commercial product or service is intended or implied. Certain commercial companies, software tools, and services are identified in this report to facilitate better understanding. Such identification does not imply recommendation or endorsement by NIST nor does it imply those identified are necessarily the best available for the purpose.

31.13 ACKNOWLEDGMENTS

We would like to thank Satyandra Gupta, Mahesh Mani, Rachuri Sudarsan, Bob Bechtold, and John Davis for providing material used in this chapter, and for the National Institute of Standards and Technology for providing financial support.

31.14 REFERENCES

1. U.S. Environmental Protection Agency, *The Lean and Environment Toolkit*, 2007. Available at: http://www.epa.gov/lean/toolkit/Lean EnviroToolkit.pdf.

2. Evans, S., Bergendahl, B., Gregory, M., and Ryan, C., *Towards a Sustainable Industrial System: With Recommendations for Education, Research, Industry, and Policy*, University of Cambridge Institute for Manufacturing, Department of Engineering: Cambridge, 2009. Available at: http://www.ifm.eng.cam.ac.uk/sis/industrial_sustainability_report.pdf.

3. U.S. Energy Information Administration, *Independent Statistics and Analysis*, 2009. Available at: http://www.eia.doe.gov/emeu/aer/consump.html.

4. U.S. Energy Information Administration's Annual Energy Review, pages 408, URL: http://www.eia.doe.gov/aer/, August 2010.

5. National Association of Manufacturers (NAM), *Efficiency and Innovation in U.S. Manufacturing Energy Use*. Available at: http://www.ext.vt.edu/resources/energyguide/energy-management/index.html.

6. Gutowski, T., *Design and Manufacturing for the Environment*, in The Handbook of Mechanical Engineering, K.H. Grote and E.K. Antonson (editors), Springer Verlag, 2009.

7. U.S. Department of Energy. *Energy Efficiency and Renewable Energy*. Available from: http://www.eere.energy.gov/.

8. U.S. Department of Commerce, *How the DOC Defines Sustainable Manufacturing*, 2010; Available from: http://trade.gov/competitiveness/sustainablemanufacturing/how_doc_defines_SM.asp.

9. U.S. Department of Energy, *Technology Roadmap for Energy Reduction in Automotive Manufacturing*, U.S. Department of Energy, Office of Energy Efficiency and Renewable Energy, Industrial Technologies Program and the U.S. Council for Automotive Research (USCAR), September 2008. Available at: http://www1.eere.energy.gov/industry/intensiveprocesses/pdfs/auto_industry_roadmap.pdf.

10. Unit Manufacturing Process Research Committee — Commission on Engineering and Technical Systems — National Research Council, *Unit Manufacturing Processes: Issues and Opportunities in Research*, The National Academy Press, 1995.

11. Kalla, D., Twomey, J., and Overcash, M., *Unit Process Life Cycle Inventory for Product Manufacturing Operations*, in ASME International Manufacturing Science and Engineering Conference (MSEC), 2009.

12. *List of Manufacturing Processes*, in *Wikipedia*, 2010. Available at: http://en.wikipedia.org/wiki/List_of_manufacturing_processes.

13. Todd, H.R. and K.D. Allen, *Manufacturing Processes Reference Guide*, 1994, Industrial Press Inc.

14. Cooperative Effort on Process Emissions in Manufacturing (CO2PE!). *CO2PE! — Taxonomy*, 2010. Available from: http://www.mech.kuleuven.be/co2pe!/taxonomy.php.

15. *MTConnect*. 2010. Available from: http://mtconnect.org.

16. Soplop, J., *An Algorithmic Approach to Enterprise Energy Management: Developing an Integrated Energy Solution Utilizing Real-time Data Collection and Predictive Modeling Capabilities*, In International Conference on Applied Energy, Singapore, 2010. Available at: http://www.rockwellautomation.com/solutions/sustainability.

17. Thiriez, A. and T. Gutowski, *An Environmental Analysis of Injection Molding*, in IEEE International Symposium on Electronics and the Environment, San Francisco, California, USA, May 8-11, 2006.

18. Boothroyd, G., P. Dewhurst, and W. Knight, *Product Design for Manufacture and Assembly*. Second Edition, Marcel Dekker, New York, 2002.

19. Gutowski, T., J. Dahmus, and A. Thiriez, *Electrical Energy Requirements for Manufacturing Processes*, in 13th CIRP International Conference on Life Cycle Engineering, Leuven, Belgium, 2006.

20. Krishnan, S.S., et al., *Machine Level Energy Efficiency Analysis in Discrete Manufacturing for a Sustainable Energy Infrastructure*, in International Conference on Infrastructure Systems, Indian Institute of Technology, Chennai, India, 2009.

21. Krishnan, S.S., Balasubramanian, N., Subramanian, E., Arun Kumar, V., Ramakrishna, G., and Murali Ramakrishnan, A., *Sustainability Analysis and Energy Footprint based Design in the Product Lifecycle*, Designing Sustainable Products, Services and Manufacturing Systems, Chakrabarti, A., Rachuir, S., Sarkar, P., and Kota, S. (editors), Research Publishing Services, 2009.

22. Morrow, W.R., Qi, H., Kim I., Mazumder, J., and Skerlos, S.J., *Environmental aspects of laser-based and conventional tool and die manufacturing*, Journal of Cleaner Production 15(10), pages 932-943, 2007.

23. Lafreniere, E.L., *Energy Conservation in Injection Molding Machines*, in Society of Plastics Engineering (http://www.4spe.org/), ANTEC, 1981.

24. Ames, K., *Saving Energy in Injection Molding Machines*, in Society of Plastics Engineering (http://www.4spe.org/), ANTEC, 1982.

25. Nunn, R.E. and Ackerman, K., *Energy Efficiency in Injection Molding*, in Society of Plastics Engineering (http://www.4spe.org/), ANTEC, 1981.

26. Kumara, S., *Conceptual Foundations of Energy Aware Manufacturing*, in Performance Metrics for Intelligent Systems Workshop, National Institute of Standards and Technology, Gaithersburg, Maryland, 2009.

27. Solding, P. and D. Petku, *Applying Energy Aspects on Simulation of Energy-Intensive Production Systems*, In *Proceedings of the 2005 Winter Simulation Conference*. 2005. Orlando, FL, USA.

28. Solding, P. and P. Thollander. *Increased Energy Efficiency in a Swedish Iron Foundry Through Use of Discrete Event Simulation*, In Proceedings of the 2006 Winter Simulation Conference, Monterey, CA, 2006.

29. Östergren, K. Berlin, J., Johansson, B, Sundström, B., Stahre, J., Tillman, A-M., A Tool for Productive and Environmentally Efficient Food Production Management, In Book of Abstracts, European Conference of Chemical Engineering (ECCE-6), Copenhagen 16-20 September 2007.

30. Johansson, B., Stahre, J., Berlin, J., Östergren, K., Sundtröm, B, and Tillman, A-M., *Discrete Event Simulation with Lifecycle Assessment Data at a Juice Manufacturing System*, In Proceedings of the 5th FOODSIM Conference, University College, Dublin, Ireland, 2008.

31. Johansson, B., Skoogh, A., Mani, M., and Leong, S., *Discrete Event Simulation as Requirements Specification for Sustainable Design of Manufacturing Systems*, In Proceedings of the Performance Metrics for Intelligent Systems (PERMIS) Workshop, National Institute of Standards and Technology, Gaithersburg, MD, 2009.

32. Odum, H.T., *Environmental Accounting: Emergy and Environmental Decision Making*, Wiley, 1995.

33. Traverso, M., Rizzo, G., and Finkbeiner, M., *Environmental Performance of Building Materials: Life Cycle Assessment of a Typical Sicilian Marble*, International Journal of Life Cycle Assessment, 15(1): p. 104-114, 2010.

34. Sutherland, J.W., Adler, D., Haapala, K. R., and Kumar, V., *Comparison of Manufacturing and Remanufacturing Energy Intensities with Application to Diesel Engine Production*, CIRP Annals — Manufacturing Technology, 57(1): p. 5-8, 2008.

35. Weir, G. and Muneer, T., *Energy and Environmental Impact Analysis of Double-glazed Windows*, Energy Conversion and Management, 39(3-4), 1998.

36. Dalquist, S. and Gutowski, T., *Life Cycle Analysis of Conventional Manufacturing Techniques: Die Casting*, Working Paper, Massachusetts Institute of Technology, Cambridge, USA, 2004.

37. Tan, R. and Khoo, H., *Zinc Casting and Recycling*, International Journal of Life Cycle Assessment, 10(3): p. 211-218, 2005.

38. Dalquist, S. and Gutowski, T., *Life Cycle Analysis of Conventional Manufacturing Techniques: Sand Casting*, in ASME International Mechanical Engineering Congress and RD&D Expo, California, USA, 2004.

39. Cho, H., et al., *Estimation of Energy Requirements and Atmospheric Emission in Continuous Casting Process for Copper Wire*, In THERMEC 2003, Madrid, Spain, Trans Tech Publications, 2003.

40. Choe, K.H., Cho, G.,S., Lee, K. W., and Kim, M. H., *Hot Forging of Semi Solidified High Strength Brass*, In *Semi Solid Process of Alloys and Composites*, Busan, Korea, Trans Tech Publications, 2006.

41. Kiuchi, M. and Kopp, R., *Mushy/semi-solid Metal Forming Technology — Present and Future*, CIRP Annals — Manufacturing Technology, 51(2): p. 653-670, 2002.

42. U.S. Department of Energy, *EERE: Industrial Technologies Program Home Page*, 2010. Available from: http://www1.eere.energy.gov/industry/.

43. Fiksel, J., *Design for Environment: A Guide to Sustainable Product Development*, New York, McGraw-Hill, 2009.

44. Telenko, C., Seepersad, C, C., and Webber, M. E., *A Compilation of Design for Environment Principles and Guidelines*, In ASME IDETC Design for Manufacturing and the Lifecycle Conference, Brooklyn, New York, 2008.

45. van Nes, N. and J. Cramer, *Design Strategies for the Lifetime Optimisation of Products*. The Journal of Sustainable Product Design, 3(3-4), pages 101-107, 2003.

46. Anastas, P.T. and Zimmerman, J. B., *Design through the 12 Principles of Green Engineering*, Environmental Science Technology, 37(5): p. 94A-101A, 2003.

47. McDonough, W., Braungart, M., Anastas, P. T., and Zimmerman, J. B., *Applying the Principles of Green Engineering to Cradle-to-Cradle Design*, Environmental Science Technology, 37(23): p. 434A-441A, 2003.

48. Abele, E., Anderl,R., and Birkhofer, H., *Environmentally Friendly Product Development: Methods and Tools*. Springer Verlag, London, 2005.

49. Telenko, C. and Seepersad, C. C., *A Methodology for Identifying Environmentally Conscious Guidelines for Product Design*, Journal of Mechanical Design, 132(September), 2010.

50. Paju M., Leong, S., Heilala, J., Johansson, B., Lyons, K., Heikkila, A., and Hentula, M., *Framework and Indicators for a Sustainable Manufacturing Mapping Methodology*, in *Winter Simulation Conference*, 2010. See http://www.wintersim.org/.

51. Weissman, A., Gupta, S. K., and Sriram, R. D., *A Methodology for Accurate Design-Stage Estimation of Energy Consumption for Injection Molded Parts*, In Proceedings of the ASME 2010 International Design Engineering Technical Conferences & Computers and Information in Engineering Conference (IDETC/CIE), Montreal, Quebec, Canada, 2010.

52. *HARBEC Plastics, Inc.*, 2010; Available from: http://www.harbec.com/.

53. Nidumolu, R., Prahalad, C. K., and Rangaswami, M. R., *Why Sustainability is now the Key Driver of Innovation*, in *Harvard Business Review*, September 2009.

54. Toyota. *North America Environmental Report*, 2008. Available from: http://www.toyota.com/about/enviroreport2008/06_enviro_manage.html.

55. Eppich, R. and Naranjo, R. D., *Implementation of Metal Casting Best Practices*, 2007; Available from: http://www1.eere.energy.gov/industry/metalcasting/pdfs/implementation_final.pdf.

56. National Science and Technology Council Subcommittee for Buildings Technology R&D, *Net Zero Energy High Performance Green Buildings*, 2008. Available from: http://www.bfrl.nist.gov/buildingtechnology/meetings/documents/May2008_Workshop_Material/Hurst-Keynote_1.pdf.

57. NIST Research Program, *Measurement Science for Net-Zero Energy, High-Performance Green Buildings*, 2010. Available from: http://www.nist.gov/public_affairs/factsheet/green_buildings_2010.cfm.

58. Sudarsan, R., Sriram, R.D., Narayanan, A., Sarkar, P., Lee, J., Lyons, K. W., and Kemmerer, S. J., "Sustainable Manufacturing: Metrics, Standards, and Infrastructure: NIST Workshop Report," National Institute of Standards and Technology, Gaithersburg, MD, NISTIR 7683, April 2010.

THE ROLE OF NANO-TECHNOLOGY FOR ENERGY AND POWER GENERATION: NANO-COATINGS AND MATERIALS

Douglas E. Wolfe and Timothy J. Eden

32.1 INTRODUCTION

Chapter 32 discusses a variety of nano-coatings and materials used in the energy and power generation fields. Nano-coatings, nano-composite coatings, nano-layered coatings, functional graded coatings, and multifunctional coatings will be presented. These coatings can be deposited by a wide range of methods and techniques including physical vapor deposition processes (PVD) such as cathodic arc, sputtering, and electron beam evaporation as well as chemical vapor deposition (CVD) and thermal spray. The various types of nano-coatings and their roles in assisting to generate energy and power for the fuel cell, solar cell, wind turbine, coal, and nuclear industries will also be discussed. This chapter provides a brief description of how the past and present state-of-the-art nano-technology within the different industrial areas such as the turbine, nuclear, fuel cell, solar cell, and coal industries is used to improve efficiency and performance. Challenges facing these industries as pertaining to nano-technology and how nano-technology will aid in the improved performance within these industries will also be discussed. The role of coating constitution and microstructure including grain size, morphology, density, and design architecture will also be presented with regards to the science and relationship with processing-structure-performance relationships. This chapter will conclude with a summary of the future role of nano-technology and nano-coatings and materials in the fields of power generation and energy.

32.2 CURRENT STATE-OF-THE-ART

We are surrounded by nano-technology and materials. The number of materials, applications and design architectures on the nano-scale is staggering and continues to grow at a high rate. Nano-technology is defined as the study of matter on the molecular and atomic scale [1]. In general, nano-technology describes components, devices, materials, coatings, etc that are 1 to 100 nm in at least one dimension. For example, 1 nm is to a meter as the size of a marble is to the size of the earth. This section discusses the current state-of-the-art in nano-technology in a few select areas within the solar cell, fuel cell, wind, and turbine industries. The role of nano-technology, nano-coatings and materials for power and

energy varies depending on the working environment, but in general can be classified into the following categories: improved wear resistance, corrosion resistance, erosion resistance, thermal protection, and increased surface area for energy storage. Since not all environments are the same, slight modifications may be required to optimize nano-materials and nano-coatings for a particular application. For example, Holleck [2] summarized a variety of design architectures for improving the performance of hard, wear-resistant coatings as shown in Fig. 32.1. These include composite coatings, functional gradient coatings, superhard coatings, superlattice coatings, metastable multifunctional, solid solution, nano-crystalline, multilayer coatings, and mixed combinations. With all the material choices that we have ranging from binary, ternary, and quaternary nitride, boride, carbide, oxide, and mixed combinations, choosing the optimum coating material and design architecture can be challenging. The approach to materials solutions starts with understanding the system performance, operating environment, maintenance issues, material compatibility, cost, and life cycle. Based on these system requirements, the optimum materials are selected based on intrinsic properties using a weighted design selection methodology, followed by selecting the optimal application technology from one of the over 100 coating and hybrid coating technologies available. After the appropriate coating technology is selected, optimization of the deposition process should result in optimal performance of the nano-coating system. However, this does not always result in a nano-structured coating with improved performance. A brief review of the current state-of-the-art in solar cell, fuel cell, wind energy, and turbine systems will be discussed.

32.3 SOLAR CELL

Typical silicon-based solar cells have efficiencies (η) on the order of 15% to 20% with the m-Si-based cells producing slightly higher efficiencies than p-Si cells. With these efficiencies, the silicon wafer-based photovoltaic (PV) cells are dominant in the market, and this current level of dominance is expected to continue for the next decade until second-generation thin film solar cells such as a-Si, CdTe, CuInSe, CuInS, CIGS, third-generation quantum dots, quantum robes, nano-tubes, and fourth-generation composite solar

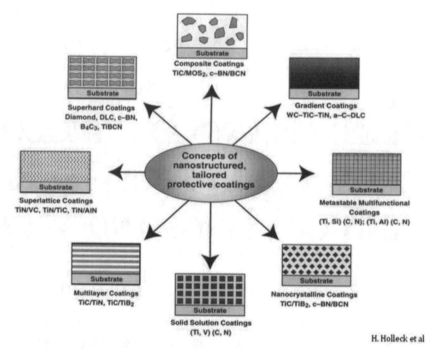

FIG. 32.1 VARIOUS CONCEPTS FOR NANOSTRUCTURED, NANOCOMPOSITE, NANOCRYSTALLINE, SOLID SOLUTION, SUPERLATTICE, AND MULTILAYER COATINGS [ADAPTED FROM HOLLECK [2]]

cells comprised of nano-particles in transparent medium become cheaper and more efficient [3–10]. With maximum efficiencies of only 20%, there is a great deal of room for improvement in the operation of these Si-wafer-based PV cells. A reason for these low efficiencies lies in the reflectivity of bulk silicon, which can reflect 35% of incoming solar radiation [11], radiative losses, sub-bandgap losses, and thermalization losses. To increase efficiency of the cell, reflectivity of the front surface must be as low as possible. Anti-reflective (AR) coatings used on the front surface are tuned to a specific wavelength, based on coating thickness, making them inefficient for collecting broadband solar radiation. Surface texturing is another way to increase absorptivity as shown in Fig. 32.2 [12,13]. Etching can be used to produce random pyramid texture on the surface of m-Si cells, but is not effective for p-Si cells, since etch rate is affected strongly by grain orientation. Other attempts at decreasing front surface reflectivity involve advanced design features such as (1) effective front and rear passivation, (2) effective light trapping via front surface texturing and back surface reflector, (3) reduced shading and contact recombination, (4) selective emitter formation, and (5) higher diffusion length to cell ratio.

On the back side, high reflectivity is desired in order to increase bulk absorption. Techniques have been explored to develop very high efficiency multi-junction cells which preferentially absorb photons at different wavelengths to increase overall absorption, though manufacturing costs are currently prohibitively high [14,15]. On the back side, aluminum back surface field (Al BSF) is typical, but it is not an ideal reflector and causes warping with thin wafers (<150 μm) because of screen printing and 1000°C heat treatment which are part of the processing procedure. New technology is needed to improve the light trapping of the cells, both front- and backside, while limiting warping, and reducing process steps.

Future efforts involve embedding rare earth nano-particles tailored to result in "up-conversion," "down-conversion," and mixed combinations to increase solar cell bandwidth absorption and in-

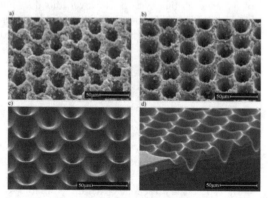

FIG. 32.2 EXAMPLES OF SURFACE TEXTURING VIA STANDARD PULSED LASER (LEFT) AND THE SURFACE OF "BLACK SILICON" PRODUCED WITH A FEMTOSECOND LASER (RIGHT) [12,13]

crease cell efficiencies. This technique offers promise in boosting cell efficiencies to levels above 60% by shifting the wavelengths of unabsorbed photons into high absorption regions of the solar cell to utilize a significantly larger portion of the incident solar radiation.

To increase the efficiency of solar cells, it is desirable to minimize the amount of light reflected off the front surface of the cell. Though etchants are commonly used to texture mono-Si cells, this is only possible because of the regular grain orientations. To address light capturing in polycrystalline-Si cells (random grain orientation), researchers in Poland, Australia, and Germany have investigated laser texturing via scribed lines and individual pock marks. Surface texturing works by varying the angle of incoming light to (a) increase path length within the cell through refraction, and (b) increase the opportunities for multiple reflections [12,16]. As crystalline Si cells become thinner (from 250 microns to <50 microns) due to high material costs, it will become more important to optimize light capture, since conventional designs allow a significant portion of incoming light to pass through, or reflect out of the wafer without freeing up electrons to perform useful work.

Historically, texturing and conventional anti-reflective coatings on polycrystalline-Si wafers are homogenous and optimized to a specific wavelength, which is partially responsible for the noticeable increase in reflectivity at near-UV and near-IR wavelengths reported in the literature [17,18]. However, much higher efficiencies could be realized using conversion processes.

32.4 UP-CONVERSION TECHNOLOGY AND NANO-STRUCTURED MATERIALS

Conventional photovoltaic solar cell materials are typically designed to utilize the standard polychromatic solar spectrum. In contrast, advanced up- and down-photon conversion materials can be tailored to alter the standard solar spectrum in order to better match the absorption characteristics of the solar cell device. By tailoring the absorption characteristics of the solar cell, significant improvements of up to 66% efficiencies can be obtained by reducing sub-bandgap losses, thermalized losses, and radiative losses as shown in Fig. 32.3a [9]. In addition, these photon conversion material systems can be added to existing solar cells or designed into thin film solar cell fabrication for high rate production.

Inherent in the technique is the ability to minimize thermalization and bandgap losses by converting wavelengths that are normally lost into usable energy via luminescent up- and down-conversion processes, resulting in increased solar cell efficiencies.

The primary material candidates are rare earth doped phosphors due to the increased number of potential vibrational and rotational energy bands. These material systems show significant promise in enhancing energy efficiencies, but further understanding of the detailed mechanisms must be understood [3–5]. Additional research in the area of material selection for the up- and down-conversion processes, transparent medium, particle size, morphology, dispersants, etc., is needed to utilize the full potential of these advanced solar cell coatings and possibly lower cost manufacturing methods.

Shalav, et al. have explored the potential of using luminescent layers for enhanced silicon solar cell performance via the up-conversion processes [9]. In theory, up-conversion processes can have efficiencies three times greater than the best current solar cells. Fig. 32.3b shows a schematic of a typical cross section of a three layered up-conversion photovoltaic cell. By using Er^{+3} in $NaYF_4$ host material, Shalav, et al showed an increase in the absorption band width, but it was still too narrow to produce significant increases in cell efficiency.

Proper selection of the host material and the appropriate rare earth dopants (activators and sensitizers) for the up-conversion processes can result in significant enhancements in absorption and energy conversion. Up-conversion phosphorescent materials (materials that emit high energy photons from stimulation due to multiple low energy photons) are unique in the fact that stimulation occurs in a temporary or metastable state in the sensitizer located in the forbidden zone of the activator as illustrated in Fig. 32.3c. When choosing the activator, it is important to choose phosphors doped with materials consisting of many energy bands such as the rare earth ions of Er^{+3} or Tm^{+3}. The sensitizer is typically an ion that is easily excited to a higher state and has enough decay time to stimulate the activator's emission band, thus resulting in increased efficiencies. Since the phosphor materials are contained within the host material of either an oxide or fluoride, no degradation of the conversion system occurs since no material is consumed. The life of such solar cells is expected to be more than 20 years. However, for future solar cells consisting of conducting polymer films, photobleaching could degrade the solar cell performance. However, low cost tape casting methods combining luminescent up-conversion materials with conductive silicon coatings result in lower costs for high efficiency solar cells. In addition, depending on the materials chosen, further enhancements could eventually lead to even thinner solar cells with silicon being deposited onto the surface of the nano-composite up-conversion system with aluminum reflector plates.

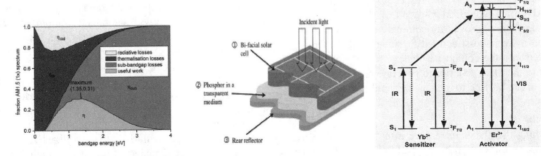

FIG. 32.3   (a) THEORETICAL FRACTIONAL SPECTRAL INTENSITY LOSSES AT ROOM TEMPERATURE, (b) CROSS SECTION OF A THREE LAYERED UP-CONVERSION PHOTOVOLTAIC SOLAR CELL, AND (c) Yb-Er UP-CONVERSION PROCESS [9]

32.5   FUEL CELL TECHNOLOGIES

A fuel cell is defined as an electrochemical cell that converts chemical energy into electricity. The electricity is generated from chemical reactions between the fuel (i.e., hydrogen) and the oxidant (i.e., oxygen, from air). The reactants flow into the fuel cell where they are consumed and release a by-product (water). However, depending on the type of reactants and by-products, degradation (i.e., corrosion, increase in contact resistance, poisoning, etc.) of fuel cells usually occur. There are several different types of fuel cells described elsewhere [19–24]. The role of nano-coatings and nano-particles in fuel cells are generally associated with thin coatings for the bi-polar plates or as gold, platinum, and/or palladium nano-particle catalysts for proton exchange membrane (PEM) fuel cells. Recent efforts have shown promise with Pt and Pd nano-particles supported on graphite nano-tubes or nano-fibers as catalysts. However, efforts are still needed to minimize catalyst poisoning and improve the catalyst efficiency [25–27]. In addition, higher efficiencies and lower cost catalysts are needed to reduce the overall cost of fuel cells. Pt, Pd, Pt-Ru, and their alloys play important roles in the electroreduction of oxygen and electrooxidation of fuel.

For high temperature solid oxide fuel cells (SOFC), research continues in the areas of fabricating cathodes comprised of nano-composite materials of NiO/YSZ, MnYCoO, and $(La,Sr)MnO_3$ (LSM) that exhibit large interconnected porosity and surface area for increased oxygen ion and electron transport [28–31]. In addition, mixed ionic and electronic cathode materials based on the perovskite structure such has $(Sm,Sr)CoO_3$ (SSC), $(La,Sr)CoO_3$ (LSC), $(La,Sr)(Co,Fe)O_3$ (LSCF), and $(Ba,Sr)(Co,Fe)O_3$ (BSCF) have been investigated. However, as the operating temperatures of the fuel cell are increased to increase efficiency, increased grain growth and reaction with the yttria stabilized zirconia (YSZ) electrolyte occurs resulting in greater parasitic losses and poor performances.

32.6   WIND ENERGY

Wind energy is rapidly gaining interest as a renewable energy source. Large scale wind farms are being built on and off the coast lines. As a result, increased challenges occur based on the working environment ranging from icing in cold regions to insects in humid environments.

32.6.1   Icing

Heaters have been built in the blades to assist with de-icing. These are usually carbon or metal fiber elements which generate heat when power is applied. However, if the heater elements fail or get struck by lightening, potential damage to the turbine can result. Datili, et al. provide a good review with respect to the various state of the art methods for de-icing [32]. Only a few "built in" heater elements and designs are approved for commercial use. In the future, we may see more nano-structured thermoelectric materials being incorporated into the composite blades or nano-composite coatings applied to the blade exterior to generate heat during turbine operation that can be used to melt ice on wind turbines [32].

32.6.2   Insects

Although washing the blades to remove insects improves efficiency, it requires the turbines to be stopped, which reduces turbine efficiency. Others wait for the rain to remove insects [32]. The state-of-the-art lies in the application of nano-composite coatings that provide a smooth surface, with low surface energy and low friction coefficient that prevents insects from depositing. These low surface energy coatings result in weak adhesion of the insects on the blades and in insect removal due to shear forces during blade rotation [33].

32.6.3   Super-Hydrophobic Nano-coatings

Superhydrophobic coatings containing nano-particles of silicon dioxide (SiO_2) embedded in epoxies and commercial paints have been shown to repel water which serves two primary benefits in improving the efficiency of wind energy by reducing drag and by reducing damage associated with corrosion. Super-hydrophobic coatings were inspired by the "lotus-leaf effect" in an attempt to replicate the lotus-leaf morphology to repel water. Numerous nano-particles mixed with epoxies and polymers, electrodeposition of copper, gold, and silver followed by wet etching have all been investigated with mixed results [34].

32.6.4   Erosion

Erosion damage does not only affect wind turbine blades, but also solar cells and compressor turbine components reducing efficiency and performance. Unfortunately, due to the complexity of brittle and ductile erosion mechanisms, no simple coating system provides adequate erosion under all conditions. In general, for wind turbine blades, polymeric materials and coatings such as tape are often used to minimize damage associated with erosion. However, these need to be replaced often as they do not provide adequate resistance to hard particle impacts. As a result, the wind turbines must be shut down for repairs affecting the overall efficiency of the turbine. Advanced coatings based on ternary nitrides and nano-composite structures are being investigated for hard particle erosion for turbine compressor components. However, erosion resistance is heavily dependent on the erosive particle type, particle velocity, angle of impingement, particle morphology, working environment, and the material being impacted. In addition, in desert environments, sand and particulate carried by the wind cause erosion or scratch damage of highly reflective solar cells. This damage results in reduced energy efficiency.

32.7   TURBINE

Industrial and government goals for turbine applications are to increase engine efficiency, increase turbine inlet temperature (TIT), reduce NOx and COx emissions, reduce fuel consumption, increase component life and increase durability and reliability. Engines are not 100 percent efficient in converting fuel to energy. Improvements are needed to increase engine efficiency. For the hot section of turbine components, approaches to improve engine efficiency and increase turbine inlet temperature include better cooling path designs in the base alloys, materials development including directionally solidified, single crystal alloys, and ceramic matrix composites (CMC), and by applying thermal barrier coatings (TBCs). Thermal barrier coatings are used to protect metallic components of turbine engines from high gas temperature as illustrated in Fig. 32.4. The TBC forms an insulative layer that reduces the combustion heat flow and decreases the operating surface temperature of the underlying components resulting in increased service life. Ceramic materials, particularly yttria-stabilized zirconia (YSZ), are widely used as TBC materials because of their high temperature capability and low thermal conductivity. Advanced coating formulations based on rare earth oxide dopant cluster formations results

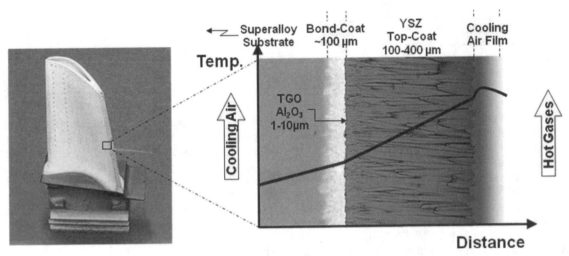

FIG. 32.4 THERMAL BARRIER COATING PROTECTING THE BOND-COATED SUPERALLOY FOR HOT SECTION TURBINE COMPONENTS

in even lower thermal conductivity and higher temperature stability [35]. Future TBCs will include gadolinium zirconate ($Gd_2Zr_2O_7$). To be effective, The TBC must adhere strongly to the component and remain adherent throughout many heating and cooling cycles. The latter requirement is particularly demanding due to the different coefficients of thermal expansion (CTE) between ceramic layer and the metallic substrates (generally super alloys) they protect. The coating system is generally composed of the TBC, a bond coating, a thermally grown oxide layer (TGO), and the base alloy. The oxidation-resistant bond coat is often employed to promote adhesion and extend the service life of the TBC, as well as protect the underlying substrate from damage by oxidation and hot corrosion attack. Bond coats used on superalloy substrates are typically in the form of overlay coatings such as MCrAlX(M= Fe, Co and/or Ni; X= Y, Hf or other rare earth), diffusion aluminide or platinum modified aluminide coatings [36] with advanced systems based on the γ–γ' compositions.

The service life of TBCs is typically limited by a spallation event brought on by thermal fatigue. Spallation occurs more often when the critical thickness (5 to 7 μm) of the TGO at the TBC/bond coat interface is exceeded. In addition to the CTE mismatch between a ceramic TBC and a metallic substrate, spallation can also occur as a result of the TBC structure becoming densified with deposits that form on the TBC during integrated gasification combined cycle (IGCC) turbine operation.

State-of-the-art TBC coatings are usually deposited by electron beam (EB) evaporation physical vapor deposited (PVD) technique. The EB-PVD YSZ is characterized by a columnar grain structure with inter-columnar gaps/porosity. The individual columns also contain microscopic porosity (intracolumnar) that helps in reducing thermal conductivity combined with intracolumnar YSZ nano-grained morphology as shown in Fig. 32.5. The lateral strain compliance with the EB-PVD system results from the columnar structures and intercolumnar gaps. The relatively higher spallation resistance of the EB-PVD coating compared to thermal spray is related to the better strain compliance.

The traditional thermal barrier coating (TBC) system is composed of a nickel-based super-alloy (often with internal cooling passages) that is coated with a metallic bond coating of either nickel-aluminide, Pt-modified aluminide or MCrAlX (where M

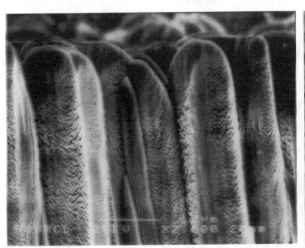

FIG. 32.5 TYPICAL COLUMNAR MICROSTRUCTURE OF YTTRIA STABILIZED ZIRCONIA (YSZ) SHOWING (a) THE COLUMNAR MICROSTRUCTURE AND (b) HIGH MAGNIFICATION SHOWING THE NANO MORPHOLOGY OF THE INTRACOLUMNAR YSZ GRAINS

is either Ni, Co, Fe, or mixed combination, and X is either Y, Hf, or Si), followed by a TBC of which YSZ is the most common [36]. Between the metallic bond coating and ceramic TBC, a thermally grown oxide (TGO), primarily the α-Al$_2$O$_3$ phase, is produced during coating deposition, which leads to better coating adhesion. However, this TGO grows during engine operation and when the thickness exceeds 5 to 7 μm, it eventually leads to spallation, the primary failure mechanism of the TBC. The main deposition methods for applying the nickel-aluminides or Pt-modified aluminide bond coatings are plating and chemical vapor deposition (CVD) [37], whereas MCrAlX coatings are generally applied by low-pressure plasma spray (LPPS) and electron beam-physical vapor deposition (EB-PVD) [38,39]. TBCs are generally applied by either thermal spray or EB-PVD depending on the final service application, with the growing trend towards the latter for higher performance applications [40]. YSZ (ZrO$_2$–8wt.%Y$_2$O$_3$) is an ideal TBC candidate as it has good thermal shock resistance, high thermal stability, low density, and low thermal conductivity [41].

Increases in operating temperatures needed for IGCC turbine operation have required improved high temperature corrosion/oxidation-resistant coatings. Efforts also include bond coat (corrosion or oxidation-resistant coatings) composition and microstructural modifications that are expected to further improve adhesion between the base alloy and ceramic topcoat, and to minimize growth of the thermally grown oxide layer that is often reported as the source of component failure. In addition, some researchers are exploring nano-grained and nano-composite coatings for improved high temperature performance. However, most nano-structured materials will suffer from creep at elevated temperatures. This is discussed in more detail later.

Studies regarding the high temperature corrosion of thermal barrier coatings (TBC) for industrial gas turbine applications in which "dirty" fuels are commonly used have identified several common impurities found in coal and fossil fuels. These impurities include sodium, sulfur, phosphorus, lead, mercury, and especially vanadium. These impurities react with conventional YSZ turbine blade coatings, severely limiting the coating lifetime. Therefore, it is of great interest to better understand the failure mechanisms and develop alternative materials that react less readily with fuel contaminants to increase gas turbine (GT) performance and reliability and differentiate the chemical mechanism from the thermomechanical mechanisms.

Standard YSZ EB-PVD coatings contain 6 to 8 wt% yttria and crystallize in the metastable t′ phase that is derived from a martensitic distortion of the "stabilized" cubic fluorite structure of zirconia. This rapidly cooled t′ structure is the most desirable of all of the possible polymorphs in the yttria-zirconia system for TBC applications. Jones [42] described several mechanisms of chemical attack on 8YSZ coatings. These include chemical reaction, mineralization, bond coat corrosion and physical damage due to molten salt penetration.

32.8 CHEMICAL REACTION AND MINERALIZATION

Acidic species such as SO$_3$ and V$_2$O$_5$ have been shown to react with the yttria stabilizing the t′ phase, destabilizing the Y$_2$O$_3$-ZrO$_2$ by extraction of the Y$_2$O$_3$. Of these, V$_2$O$_5$ has been determined to be the worst offender. Susnitsky and Hertl [43] have studied the reaction mechanism in detail. The reaction:

Equation 32.1:

$$Zr_{1-x}Y_xO_{2-0.5x} (t') + yV_2O_5 \rightarrow 2(1-y) ZrO_2 \text{ (monoclinic)} + 2y\ YVO_4$$

is especially deleterious to the TBC integrity. Vanadium has been shown to leach yttria out of the stabilized zirconia leaving an yttria deficient destabilized monoclinic zirconia phase. The large volume expansion (8% to 9%) caused by this transformation leads to the TBC spalling and exposing the bond coat to further chemical attack.

In contrast, mineralization describes a catalytic process by which a metastable phase (in this case, the t′ phase) is broken into its stable phase assemblages by a catalyst or mineralizer [44]. For example, ceria stabilized zirconia was investigated as a corrosion-resistant coating due to the fact that ceria does not react with vanadium pentoxide:

Equation 32.2:

$$Zr_{1-x}Ce_xO_{2-0.5x} (t') + yV_2O_5 \rightarrow (1-x)ZrO_2 \text{ (monoclinic)} + xCeO_2 + yV_2O_5$$

However, vanadium acts as a mineralizer, destabilizing the t′ phase without reacting to form the vanadate. If no vanadium is present in the coal, but high sulfur concentrations exist, then YSZ is de-stabilized by the following reaction:

Equation 32.3:

$$ZrO_2(Y_2O_3) + 3SO_3(+Na_2SO_4) \rightarrow Y_2(SO_4)_3 \text{ (in a Na}_2SO_4 \text{ solution)}$$

In addition, other impurities such as sodium and phosphorous (derived from coal liquefaction) have been shown to destabilize the zirconia phase as described below:

Equation 32.4:

Sodium destabilization:

$$ZrO_2(Y_2O_3) + P_2O_5 \rightarrow ZrO_2 \text{ (monoclinic)} + 2YPO_4$$

Equation 32.5:

Phosphorous destabilization:

$$8ZrO_2 + 4Na + O_2 + 6P_2O_5(g) \rightarrow 4NaZr_2(PO_4)_3$$

Therefore, the corrosion mechanisms can be very complex and highly dependent on the environment. In addition, due to thermal cyclic oxidation, the thermally grown oxide thickness increases which results in coating spallation. Figure 32.6 shows an uncoated nickel-based alloy after exposure to Type I hot corrosion after 100 hours and 1000 hours showing significant degradation. Figure 32.6c shows the same alloy after 1000 hours exposure with a protective coating applied showing no alloy degradation.

During cyclic oxidation at elevated temperatures, the bond coat or base alloy forms a protective oxide layer. With increasing time, oxygen diffusion inward toward the base metal and outward diffusion of iron, aluminum, chromium, nickel or other transition metal results in a thicker oxide scale. It has been shown in advanced thermal barrier coating systems that primary failure of the thermal protection is due to delamination of the protective thermal insulating ceramic (YSZ) due to the growth of the thermally grown oxide layer between the bond coating and the thermal barrier topcoat.

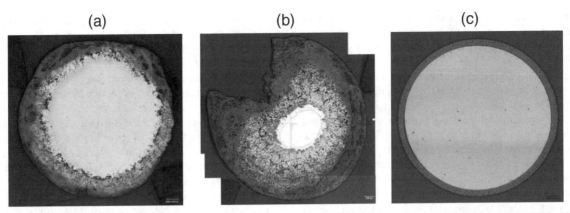

FIG. 32.6 NICKEL BASED ALLOY SHOWING TYPE I HOT CORROSION AFTER (a) 100 HRS, (b) 1000 HRS AND AFTER (c) 1000 HRS WITH A PROTECTIVE COATING SHOWING THE BENEFITS OF COATINGS UNDER HIGH-TEMPERATURE EXTREME ENVIRONMENTS

Delamination of the oxide layer typically occurs due to crack propagation at or near the metallic/oxide interface generally caused by surface defects, bond coat pore relaxation, or grain boundaries acting as sites for future failure. High temperature oxidation is generally associated with oxide scale growth due to the rapid diffusion of oxygen through the oxide layer and outward diffusion of aluminum, chromium or iron from the bond coat or uncoated base alloy. The growth of the oxide scale results in excess stress and modulus mismatch with the underlying metal. At high enough stress, the system fails at the metallic/oxide interface. The other primary failure mechanism results from the non-uniform growth of the oxide layer near grain boundaries where the oxide scale is drawn into the bond coat region at imperfection sites; this is often termed rumpling. However, there is some debate as to whether rumpling is actually observed in fielded components. Lastly, microstructural changes within the coating or base alloy cause tensile stresses, crack initiation, crack propagation due to cycling (known as ratcheting), crack convergence, and finally delamination of the oxide scale exposing a fresh metallic surface. The process repeats until little to no protective layer or base alloy is left.

32.9 IMPACT OF NANO-STRUCTURED MATERIALS AND COATINGS AND THEIR FUTURE IN ENERGY

Isolating the individual contributions to power and energy with regards to nano-coatings and materials is very challenging; as with most applications, several overlapping requirements must be addressed. Therefore, the next sections discuss how nano-structure and nano-composite coatings will play important roles in improving wear resistance using nano-composite and self lubricating smart coatings to reduce energy consumption for machining applications, protective coatings and lubrication for gear assemblies in both gas turbine and wind turbine applications, improved erosion and corrosion resistance, and the concerns due to creep when using nano-structured materials at elevated temperatures.

32.10 HARD COATINGS FOR IMPROVED WEAR RESISTANCE

Hard coatings will continue to play an important role in energy conversion as substantial energy is often lost due to wear and friction of rotating components. Reducing the friction and wear of materials by applying nano-composite coatings will result in improved energy conversion efficiency. Future novel concepts include the development of nano-layered self-lubricated "smart" coatings with embedded sensors for high temperature wear/erosion resistance metallic alloys such as titanium and steel alloys. It has been theorized that nano-layered coatings with multiple coherent or/and semi-coherent interfaces with tailored microchemistry will influence the internal residual stresses and will contribute towards better performance in toughness and wear resistance for coated titanium alloys. It is anticipated that the microchemistry tailored nano-layered coatings will offer superior high temperature wear resistance associated with fine grained microstructure and strong dislocation movement confinement along with low friction coefficient. Embedding sensors into the coating design will assist in better prediction and monitoring of the life of the component. "Chameleon" surface adaptation of materials systems, in general, falls within a category of wear-resistant coatings in which the surface chemistry, structure, and mechanical properties reversibly react with the environmental conditions and applied loads in order to optimize wear characteristics under these conditions.

It has been found that tool materials (wear resistant) suitable for high temperatures (650°C) and high cutting speeds permit machining with lower cutting forces. The application of coating technologies in which a hard layer is deposited on a relatively soft substrate material (titanium) is advantageous in terms of reduced energy in machining. Although there have been great advances in the development of hard coating materials, and wear and abrasion-resistant coatings, not much research has been conducted on nano-layered self-lubricated (NLSL) "smart" coating technology with embedded sensors. It is believed that the strong dislocation confinement in nano-layers will yield much higher high-temperature hardness than in monolithic PVD/CVD coatings, providing significantly improved wear and thermal resistance.

Smart nano-layered self-lubricating coatings will have increased hardness, better toughness, and lower friction coefficients that will allow more heat dissipation and energy absorption for improved wear resistance. Energy enhanced coating processes will provide the ability to tailor the microstructure and control density, hardness, structure (i.e., amorphous, polycrystalline), intrinsic stress, and crystallographic orientation, all of which affect the properties of the coating system.

There have been constant challenges in enhancing the life of components in the tooling, aerospace, and energy industries due to the complexity of the working environment and aggressive conditions. It has been well documented that hard, wear-resistant coatings can play an important role in enhancing component life [45]. The major controlling factors are attributed to: (i) the choice of coating materials, (ii) choice of coating process and thus microstructure, (iii) structure and design of multilayer, nano-layer coating system, (iv) substrate material, geometry, and surface condition, and (v) working environement, etc. First generation coatings primarily consisted of monolayered micrometer thick films of transition metal carbides and nitrides which aided in improving tool life by several hundred percent. Second generation coatings are presently being explored in which materials with different chemical, physical, tribological, and mechanical functions are being combined in multilayer, nano-composite, nano-structure coating systems. An intermediate layer such as titanium (Ti) or titanium nitride (TiN) is generally applied to the substrate to facilitate adhesion of the subsequent coating. The intermediate layers are responsible for increased hardness and toughness, and an outer layer is needed to reduce friction, galling/tool edge build up, and chemical reactivity. This complex requirement of high hardness, high toughness, and good adhesion between the different materials (i.e., coating/substrate and layer/layer interfaces) combined with the need for self-lubricating, wear-resistant properties makes "smart" nano-layered self-lubricating coatings an ideal choice.

Generally, most coatings fail at the coating substrate interface during machining operations as the shear stresses build up at the interface resulting in delamination of the coatings. Stresses at the interface can be altered by changing the thickness and degree of coherency (i.e., coherent, semi-coherent, incoherent interface) of each individual layer within the NLSL coating design. It has been established (Fig. 32.7) that coatings with increased hardness (Fig. 32.7a) and decreased residual stresses (Fig. 32.7b) can be obtained by decreasing the individual layer thickness of multilayer coatings consisting of TiC/Cr$_{23}$C$_6$ and TiC/TiB$_2$. Vicker's hardness measurements of the TiB$_2$/TiC multilayer coatings on WC-6wt.%Co–0.3wt.%TaC were found to be 3294 VHN$_{0.050}$ to 3991 VHN$_{0.050}$ for total layer thickness of 2 to 10 layers as shown in Fig. 32.7a. The hardness increased with increasing number of layers, while keeping approximately the same volume fraction of each phase for each of the coatings [46,47].

The increase in Vicker's hardness number is attributed mainly to the reduction in grain size (which inhibits dislocation mobility) and the microstructure [48]. Dislocations in one layer cannot penetrate the interface unless there is enough energy to force the

dislocation to move across it. As a result, dislocations tend to pile up near interfaces. As dislocations pile up, the trailing dislocations force the leading dislocation across the interface. In addition, when the thickness of the alternating layers is large, dislocation mobility can occur within each individual layer. By reducing the thickness of the individual layers (i.e., increasing the total number of layers) to a nano-scale size, fewer dislocations are formed, resulting in fewer dislocations piling up near the interface. As a result, the amount of applied stress needed to migrate the leading dislocation across the interface increases, since less force generated by dislocation piling up is produced. Similarly, when the individual layer thickness of the multilayer coating is small enough such that only one dislocation exists throughout the thickness of the layer, the dislocation must cross the interface unaided, as there are no other dislocations available to force the leading dislocation across the interface. The only force providing dislocation mobility across the interface is the applied stress (indent). As a result, higher loads are needed to deform the material and are observed by the higher Vicker's hardness numbers as a function of decreasing individual layer thickness.

Differences in the lattice parameter also contribute to the increase in hardness, as well as interfacial stress, as these differences will affect the total number of interfacial dislocations. When the lattice parameter difference between the two materials is large, an increased number of dislocations exist in the incoherent interface region to reduce strain associated with the mismatch [49]. Dislocation movement within and across the interface is again controlled by dislocation pile up. As the individual layer thickness decreases (i.e., increase the total number of layers), the interfacial boundary layer thickness decreases which reduces the total number of available dislocations, and thus increases the required force to cause deformation.

Several researchers have shown that the hardness of a material increases with decreasing grain size according to the following relationship:

Equation 32.6:
$$H = H_o + kd^n$$

Where H = hardness, d = average grain size, n = grain size exponent (typically –1/2), H$_o$ = constant, and k = constant. Figure 32.7c shows the average Vicker's hardness number plotted as a function of the reciprocal square root of the average grain diameter for the TiB$_2$/TiC multilayer coatings and appears to follow the modified Hall-Petch relationship.

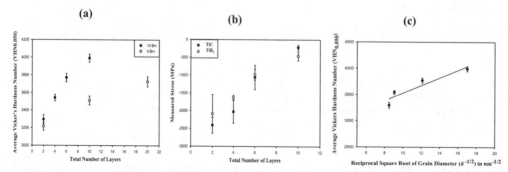

FIG. 32.7 AVERAGE VICKER'S HARDNESS NUMBER (VHN$_{0.050}$) AS (a) A FUNCTION OF TOTAL NUMBER OF INDIVIDUAL LAYERS, (b) MEASURED STRESS AS A FUNCTION OF TOTAL NUMBER OF LAYERS AND (c) RECIPROCAL SQUARE ROOT OF GRAIN SIZE (D$^{-1/2}$) FOR TiB$_2$/TiC MULTILAYER COATINGS DEPOSITED ON WC-6wt.%Co-0.3wt.%TaC BY ARGON ION BEAM ASSISTED, EB-PVD

32.11 NANO-STRUCTURED CONFIGURATION, STRATEGY AND CONCEPTUAL DESIGN

New technology thrusts in tailored coating design include functionally graded coatings, composite coatings, super hard, metastable multifunctional coatings, nano-crystalline coatings, solid solution coatings, multilayer, and superlattice coatings [2]. The coating can be tailored to meet specific material properties. Further improvements may even be obtained by combining one or more of the new technology thrusts, i.e., multilayer design with self-lubricating materials. Of the various new technology concepts, the most common is the multilayer design (Fig. 32.8). There are three major classifications of multilayer coatings, the first being coatings with a limited number of single layers (typically 3 to 13 total layers). The fundamental concept is the combination of different materials with similar crystal structures (TiC and $Cr_{23}C_6$, i.e., cubic) and lattice spacings. The creation of multilayers can interrupt the columnar growth of the grains resulting in a smaller grain size and higher hardness according to the modified Hall-Petch relationship ($H \sim 1/(d)^{1/2}$) [46,47]. The second classification consists of multilayer coatings with a high number of multilayers composed of different materials from different crystal systems as illustrated in Fig. 32.8b with Cr_3C_2 and graphite. The large number of layers corresponds to a high interface volume in addition to interrupting the columnar structure. The third classification is superlattice coatings. Superlattice coatings can only occur when constituents are isostructural with similar lattice spacings, chemical bonding, and atomic radii (NbN/TiN and TiC/TiN) as illustrated in Fig. 32.8c [50].

Properties of multilayer coatings are believed to be a function of the composition, structure and volume of the interface region, but have never been fully investigated. The amount of energy dissipated at the interface heavily depends on whether the interfaces are coherent, semi-coherent, or incoherent. As a crack approaches an interface, it is deflected, which reduces crack propagation. Interfaces also absorb energy from a propagating crack which results in stress relaxation and prevents further crack propagation, thus improving abrasion resistance. Lastly, interfacial defects such as porosity and microcracks serve to relax generated stress. The interface's ability to absorb energy results in improved toughness. Therefore, increasing the total number of layers increases the total number of interfaces (and thus, the total interfacial volume), which should result in increased toughness.

In designing the proper multilayer system, the task can be divided into two areas: structural design and functional design. The structural design can be tailored by changing the geometry and morphology of the grains, grain size, orientation, thickness of the individual layers and the constitution and size of the interfacial boundary layer. Changing these structural components and tailoring them to enhance toughness and hardness should produce more wear-resistant coatings. The functional design aspect includes the selection of the material which is dependent on the type of bonding (i.e., metallic, ionic, and covalent), and the thermal, physical, and mechanical properties of the material. The hardness of a material is often dictated by its intrinsic properties. The intrinsic hardness of a material is heavily dependent on the crystal structure as well as the type of bonding. Materials with a high degree of covalent bonding are generally harder than those with metallic or ionic bonding because the bond lengths are generally shorter for covalent bonding. Chemical stability and inertness occur as a result of ionic bonding, whereas metallic bonds yield better adhesion and toughness. However, few materials consist of entirely one type of bonding. Mixed bonding can be tailored to create a film that exhibits properties indicative of each type of bonding [19]. With increasing demands on material systems, often high hardness and good toughness are required. Materials with high hardness, good chemical inertness and low friction coefficients are good candidate materials for increasing the life of coated tools using the multilayer concept.

It is well documented that the coating microstructure is heavily dependent on the deposition parameters and process [46,47]. As a result, the interfacial constitution, volume, and structure should also be affected by these same factors. For example, deposition at high temperatures should lead to a larger interfacial size as a result of thermal diffusion, especially for systems with high amounts of solubility. In addition, high energy coating processes, such as cathodic arc and ion beam assisted deposition (IBAD), should also yield greater interfacial layers resulting from the intermixing of the underlying material and depositing coating caused by high-energy bombardment. Therefore, deposited multi-nano-layered (MNL) coatings will exhibit smaller grain size and increased toughness with multifunctional capabilities tailored for specific applications.

These multilayer concepts are designed to: (i) increase the hardness and strength of the deposited films, (ii) to obtain wear protective films with low chemical reactivity and low friction, and (iii) to facilitate a strong adhesion between the film and the substrate. High adhesion strength occurs when the energy of the interfaces is small. The formation of an intermetallic phase near the coating/substrate interface (resulting from inter diffusion) generally yields high adhesion (inter diffusion), but often acts detrimentally during machining conditions due to large differences in physical properties near the interface. For example, during CVD deposition of TiC on WC-Co substrates, brittle η phases (Co_6W_6C or Co_3W_3C) are often observed at the TiC/WC-Co interface which are undesirable (decarburization). Formation of the η phase can

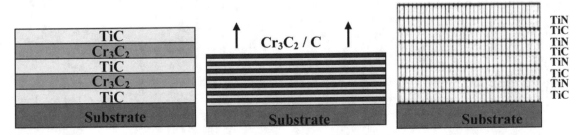

FIG. 32.8 ILLUSTRATION OF MULTILAYER CONCEPTS WITH (a) SIMILAR CRYSTAL STRUCTURES AND LATTICE SPACINGS, (b) DIFFERENT MATERIALS FROM DIFFERENT CRYSTAL SYSTEMS, AND (c) ISOSTRUCTURAL MATERIALS WITH SIMILAR ATOMIC RADII, CHEMICAL BONDING, AND LATTICE SPACINGS

be avoided by applying a thin layer of titanium and reducing the substrate temperature.

32.12 COHERENT, SEMI-COHERENT, INCOHERENT INTERFACES

Depending on the material systems, the degree of lattice mismatch can be quite significant. If the lattice mismatch is too high, excessive stress will result, which can lead to delamination or non-adherence within the NLSL coating. Interfaces are often characterized as being either coherent, semi-coherent, or incoherent and are illustrated in Figs. 32.9a–32.9d. A coherent interface is a boundary between two or more materials with a complete, perfect matching of the lattice spacings. Figure 32.9a illustrates a strain-free coherent interface between two materials with the same crystal structure but different chemical compositions. Similarly, Fig. 32.9b shows a strain-free coherent interface where the lattice is different for each material. It is very rare to observe a perfect coherent interface. Generally, coherent interfaces have a slight mismatch at the interface which leads to coherency strains throughout the adjoining lattices. Typical energies of coherent interfaces are $20-100$ mJ/m^2.

A semi-coherent interface contains an array of interfacial dislocations which accommodate the mismatch in the lattice between the different materials to minimize the interfacial energy (Fig. 32.9c). Typical values of semi-coherent interfaces range from 100 to 500 mJ/m^2 but are highly dependent on the composition and the crystal structure. Lastly, Fig 32.9d illustrates an incoherent interface where the lattices of each material do not match resulting in much higher energies on the order of 500 to 1000 mJ/m^2.

32.13 LATTICE MISMATCH

Sizable differences between the lattice parameters of nano-layered materials can significantly affect the stress state of the deposited coatings. In addition, the lattice mismatch between the various layers will differ depending on the orientation (crystallographic texture) of the coating as a result of different lattice spacings for the various lattice planes for the various structures of the deposited materials. Therefore, controlling the coatings growth orientation, by changing the energy of the vapor cloud through ion bombardment, will help in controlling the stress state of the NLSL coating system.

The orientation of the films with respect to each other, as well as to the substrate, is very important for minimizing residual stresses.

If the mismatch is too great, dislocations will occur which leads to decreased strength. Preventing dislocation motion is one of the strengthening mechanisms in materials. Therefore, changes in the lattice parameters will affect the amount of lattice mismatch between the various layers in a NLSL coating system. This becomes increasingly important for a thin film with a high number of layers. Ion beam assisted deposition can be used to tailor the growth of the films (by controlling the flux to ion ratio), in order to minimize the degree of lattice mismatch. Too high a lattice mismatch could result in a very weak interfacial layer, which in turn will result in lower hardness. IBAD can be used to tailor the orientation of the various layers to increase the lubricious nature of the coating system. For example orienting the (0001) planes of the graphite parallel to the coating surface should increase the lubricity of the multilayer coating design.

32.14 CHALLENGES

Future challenges in nano-technology will include a better understanding of the fundamental mechanisms associated with the interfacial volume, interfacial constitution and structure, and their affect on hardness, crystallographic texture, residual stress, and nano-layered self-lubricating multilayer coatings for high temperature applications.

32.15 COATING MATERIALS SELECTION

There are several different types of coatings available today for providing increased wear resistance, which are briefly discussed, below. Each one has its advantages and disadvantages and is heavily dependent on the environmental conditions. For example, TiN is a good general purpose coating which has increased the life of cutting tool inserts by several hundred percent. Other coatings and their applications are listed below in Table 32.1 with selected properties found in Table 32.2 [51–56].

As with any design problem, the choice of material and process plays an instrumental role in coating performance. The hardness of a material is often dictated by both its intrinsic properties as well as its microstructural features. Wear resistance is often associated with hardness. Therefore, by increasing the hardness through materials selection and multilayer design, improved wear resistance is expected. By combining the above hard materials in a multi-nano-layered coating system, with a composite (CrC/C) or multilayer (Al$_2$O$_3$/TiC, Al$_2$O$_3$/TiAlN) design concept, a coating system with good wear re-

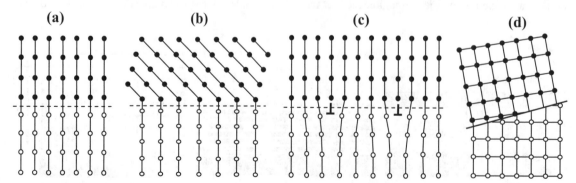

FIG. 32.9 ILLUSTRATION OF A STRAIN-FREE, COHERENT INTERFACE FOR TWO MATERIALS WITH (a) SAME CRYSTAL STRUCTURE BUT DIFFERENT CHEMICAL COMPOSITIONS, (b) SAME CRYSTAL STRUCTURE BUT DIFFERENT LATTICE SPACINGS, (c) OF A SEMI-COHERENT INTERFACE WITH EDGE DISLOCATIONS, AND (d) AN INCOHERENT INTERFACE

TABLE 32.1   COATING MATERIALS AND APPLICATIONS

Coating material	Application	Characteristics
TiN	Iron-based material	General purpose; abrasive and adhesive resistance
TiC	Hard, abrasive materials	High hardness; excellent abrasion and wear resistance
TiCN	Cast iron, brass, aluminum alloys	High hardness, toughness; high feed and speed rates
TiAlN	Irons, nickel- and titanium- alloys	Excellent oxidation; semi-dry and dry machining
CrN	Copper, titanium	Resistance to oxidation, corrosion, and adhesive wear
CrC	Titanium, exotic metals	High hardness and enhanced lubricity
WC/C	Steel	High lubricity and low adhesive wear, dry machining
Al$_2$O$_3$	Interrupted cutting abrasive materials	Resistance to oxidation, high lubricity, chemical and thermal barrier

sistance, improved toughness, increased tool life and chemical resistance with self-lubricating properties can be obtained.

32.15.1   Aluminum Oxide (Al$_2$O$_3$)

Aluminum oxide is one of the more recent materials being investigated for wear applications. Its high diffusion and oxidation-resistant properties make it a material of choice for certain applications. α-Al$_2$O$_3$ is usually deposited at elevated temperatures (~1000 °C) by chemical vapor deposition, but more recent advancements in coating deposition techniques allow for a low temperature PVD α−Al$_2$O$_3$ deposition. The main disadvantages of the CVD process are first, the large grain size of the coating, resulting in lower fracture toughness as compared to PVD coatings deposited at lower temperatures, and second, the process produces hazardous by-products/wastes. The high temperatures are needed to obtain the α-phase which is the hardest (higher wear resistance) and most stable of the various aluminum oxide phases. It is theorized that

α-Al$_2$O$_3$ coatings can be produced under non-equilibrium conditions with relatively low temperatures (<700 °C) and high energies. This can be achieved by changing the low-energy (0.1eV) flux (evaporated material) to high-energy (50 to 1000 eV) by simultaneous ionized gas bombardment and condensing the deposited atoms. The increased energy added to the system allows lower deposition temperatures while still giving the adatoms enough surface mobility to re-arrange themselves in stable lattice positions. As a result of the many different phases of aluminum oxide, if the deposited phase is not stabilized, a phase change can occur which undergoes a volume change resulting in excessive stresses, spallation, and thus coating failure. Al$_2$O$_3$ can also be stabilized in the amorphous structure up to 1100°C by incorporating nitrogen into the coating, and has been reported to have excellent wear-resistant properties. Few materials will adhere to Al$_2$O$_3$ due to its high enthalpy of formation which results in a low friction coefficient. Al$_2$O$_3$ also has excellent abrasion resistance as a result of its high temperature hardness.

32.15.2   Titanium Carbide (TiC)

As with most carbides, titanium carbide (TiC) has a wide range of properties similar to both metallic and ceramic materials [46,47]. For example, TiC has very high hardness (~2800 to 3200 kg/mm^2) and strength, similar to ceramics, but still maintains very good electrical and thermal conductivity associated with the parent metal. The unique properties of TiC are derived from its complex bonding nature. The degree of covalent bonding within TiC is mainly a function of the Ti-C bonding. TiC was selected as a candidate material because of its chemical compatibility with Al$_2$O$_3$ and titanium-based alloys, high hardness and abrasion resistance. In addition, TiC can be deposited at low temperatures resulting in a small grain size which yields better high-temperature fracture resistance.

32.15.3   Chromium Carbide (Cr$_3$C$_2$)

Three main phases for chromium carbide are Cr$_3$C$_2$, Cr$_7$C$_3$, and Cr$_{23}$C$_6$, with orthorhombic Cr$_3$C$_2$ being the most common having a hardness of 1300 to 1800 kg/mm^2. Unlike most carbides, all of the chromium carbides have lower melting temperatures than pure chromium. Within the chromium carbide structure, both Cr-Cr and Cr-C bonds give rise to the unique properties of chromium carbide with the Cr-C bonding being weak and Cr-Cr very strong. As a result, chromium carbide has good chemical resistance and high lubricity during the machining of titanium and does not weld with titanium, resulting in good dimensional tolerances. The reasons for the improved machining characteristics are not fully understood, but are most likely associated with chromium forming a chromium oxide layer at the surface which contributes to increased oxidation protection and diffusion resistance to the underlying

TABLE 32.2   GENERAL PROPERTIES OF VARIOUS CERAMIC MATERIALS USED IN COATING INDUSTRY [51–56]

Property	TiN	TiC	TiAlN	Al$_2$O$_3$	Cr$_3$C$_2$	CrN	WC/C
Hardness (kg/mm^2)	2000–2200	2800–3200	2600–3000	1500–2000	1300–1800	1700–2500	900–1100
Friction coefficient	0.40–0.65	0.40–0.65	0.40–1.7	<0.30	<0.30	<0.30	<0.10
Melting point (°C)	2930	3160	–	2015	1895	1500	2630
Crystal structure	cubic	cubic	cubic	hex	ortho a=5.5	cubic	Hex/amor
Lattice parameter (Å)	4.238	4.327	Composition dependent	4.7592	c=2.83 b=11.49	4.140	a=2.906, c=2.839
Density	5.43	4.938	–	3.965	6.66	6.14	15.67/2.25
Oxidation resistance	low	low	good	excellent	good	good	low

coating material. Similar work has been done by sputtering WC/C multilayers for the machining of steels [57]. A similar concept is being investigated which combines the self-lubricating properties of graphite with the wear-resistant properties and chemical compatibility of CrC in machining titanium-based alloys, making CrC/C an ideal NLSL coating system candidate.

In choosing the multilayer design, close attention to the crystal structures is important to minimize interfacial stresses associated with lattice mismatch between the two coating layer materials. It is believed that the individual layer thickness will greatly affect the crystallographic orientation of the deposited coatings in an effort to minimize the lattice mismatch, thus resulting in better coating performance.

32.15.4   Nitrides

Titanium nitride, chromium nitride, aluminum nitride, and zirconium nitride are all hard, wear-resistant materials that can be deposited by a variety of physical vapor deposition techniques. Titanium nitride has been the most studied, with chrome nitride and aluminum nitride coatings more commonly used when corrosion protection is also required.

32.15.5   Titanium Nitride (TiN)

The use of titanium nitride as a coating material for wear applications has been around for several decades. The hard TiN coating increases tool life by as much as tenfold, metal removal rates for coated drill bits are more than doubled, and the number of regrinds before the tool is consumed is increased substantially over uncoated tools. The properties that make TiN an attractive coating for the tool industry are: high hardness, low coefficient of friction, good chemical/thermal stability, good adhesion, and good corrosion resistance. The main property is its hardness, which results from the rocksalt structure with its high degree of metallic and covalent bonding. TiN gets its strength from the small separation of atoms, large surface energy, and high Young's modulus. The other physical properties that result in TiN being one of the most widely used commercial wear-resistant coatings are: high yield stress (2.0×10^6 psi), high Young's modulus (50×10^6 psi), high specific gravity (5.4 g/cc), high hardness (1200 to 3000 VHN) and high melting point (2950°C.)

32.15.6   Titanium Aluminum Nitride (TiAlN)

The most studied ternary wear-resistant compound is titanium aluminum nitride. As with most nitride structures, TiAlN is a defect structure having a wide range of compositions, and thus properties, including hardness. A peak in TiAlN hardness occurs where the lattice parameter is a minimum and the material is close to transitioning from a NaCl crystal structure to a ZnS crystal structure. The hardness increase of TiAlN (3500 VHN as compared to TiN) is most likely due to complexities of the crystal structure. TiAlN also provides added oxidation and corrosion protection as aluminum can migrate to the surface forming a protective Al_2O_3 layer. TiAlN coatings have exhibited both higher hardness as well as higher erosion resistance. Physical properties of the TiAlN system vary as a function of composition. Typical properties are: E = 434.7 GPa, G = 178.4 GPa, melting temperature of 2930°C, thermal expansion coefficient = 7.5 μm/μm°C, and specific gravity = 4.6 g/cc.

Interestingly, composite materials or composite features already exist in the ternary systems. For example, TiAlN, based on the composition of the bulk coating, can be considered a nanocomposite material of TiN and AlN comprised of the Rocksalt and Wurtzite structures, respectively, as shown in Fig. 32.10.

It is believed that the proper ratio of the Rocksalt/Wurtzite phases is what gives TiAlN its unique properties of increased hardness and increased toughness under certain deposition parameters. Typically, increased hardness results in lower toughness. The nano-composite behavior of TiAlN is also seen with $Ti_{(1-x)}Si_xN$.

32.15.7   Titanium Chromium Nitride (TiCrN)

TiCrN coatings are a class of coatings which typically incorporate Cr into a TiN-based coating system in order to enhance specific characteristics of TiN including high temperature hardness and corrosion resistance. The increased resistance to corrosion and degradation of mechanical properties at higher temperatures are typically attributed to formation of protective chromium oxide compounds. The coating composition and microstructure is very dependent on the deposition technique. A wide variety of coating compositions can be produced including single phase TiCrN, or a mixture of phases such as TiN, CrN, Cr_2N, and the microstructure of these coatings can range from nano-scale columnar grains to

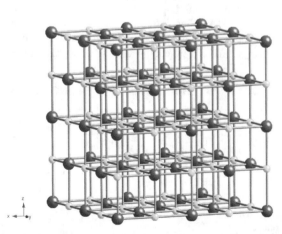
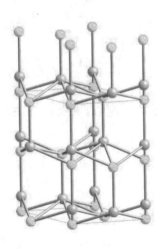

FIG. 32.10   CRYSTAL STRUCTURES OF (LEFT) TITANIUM NITRIDE (ROCKSALT STRUCTURE - TITANIUM ATOMS (BLUE) AND NITROGEN ATOMS (GREEN), AND (RIGHT) ALUMINUM NITRIDE (WURTZITE) — ALUMINUM ATOMS (ORANGE) AND NITROGEN ATOMS (GREEN)

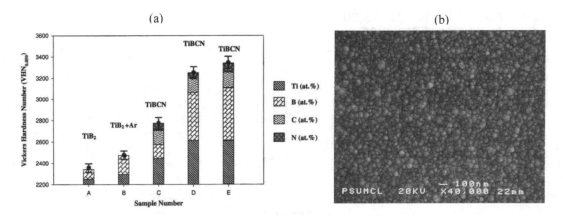

FIG. 32.11 (a) PLOT SHOWS VARIATION IN HARDNESS OF NON-EQUILIBRIUM TiBCN PHASES AS A FUNCTION OF COMPOSITION AND (b) CORRESPONDING SEM MICROGRAPH SHOWING NANO-GRAINED MICROSTRUCTURE [58]

an amorphous-like microstructure. For many PVD techniques, a single phase of TiCrN with the NaCl structure is typically produced, or a mixture of NaCl structured stoichiometric TiN and CrN compounds. The hardness values of the coatings can be tailored from similar to CrN (1200 to 2000 Hv) to greater than that of TiN (>3000 Hv). Other mechanical properties such as wear resistance can also be tailored with deposition parameters, so TiCrN is readily adaptable for a wide range of applications.

32.15.8 Titanium Boron Carbon Nitride (TiBCN)

Super hard TiBCN coatings were successfully deposited by ion beam assisted, electron beam-physical vapor deposition and are shown in Fig. 32.11a [58]. Titanium, titanium diboride, and carbon (through tungsten) were co-evaporated by energetic electron beams while simultaneously bombarding the substrates with varying ionized gas ratios of nitrogen and argon to obtain super hard TiBCN coatings. The hardness of the TiBCN coating was reported to be equivalent to a soft diamond-like carbon film and is attributed in part to the nano-grained microstructure observed in Fig. 32.11b.

32.16 APPROACH IN APPLYING COATINGS

Performance of coatings depends upon the coating properties which are a function of the coating microstructure which is dictated by the coating deposition process as illustrated in Fig. 32.12. There are over 100 coating processes commercially available, each having their own advantages and disadvantages. A brief list of a few of the coating deposition processes is listed in Table 32.3 with most coating processes being related to either physical vapor deposition (PVD) or chemical vapor deposition (CVD) processes.

The term CVD is defined as a process whereby the constituents of a gas or vapor react chemically in a vacuum chamber and deposit on the substrate surface in the form of a thin film, nano-tube, or nano-wire. The CVD process usually takes place between temperatures of 500 to 1100°C, but depends on the material. Various metallic and ceramic coatings can be deposited at a rate of 2 to 10 microns/hour. The residual stresses in a CVD coating are generally tensile. However, in the last 20 years, several variants and modifications to the conventional CVD process have been made which include atmospheric pressure (APCVD), low pressure (LPCVD), metal-organic (MOCVD), Photo-enhanced (PHCVD), Laser-induced (LICVD), and plasma enhanced (PECVD).

Physical vapor deposited coatings are generally applied at much lower temperatures and include evaporation, sputtering, and cathodic arc. In general, most vapor deposited coatings are nano-grained. Figure 32.13 shows a transmission electron micrograph of a nano-composite physical vapor deposited coating with a grain size on the order of 5 nm.

In the EB-PVD process, focused high-energy electron beams generated from electron guns are directed to melt and evaporate ingots as well as to preheat the substrate inside the vacuum chamber as shown in Fig. 32.14. Due to the change in pressure, the vapor rises and traverses the vacuum chamber where it condenses on the substrate forming the coating. To obtain more uniform coatings, like in all PVD processes, the sample is often rotated during the coating process. The depth of the melt and degree of vaporization is controlled by restricting the kinetic energy of the electron beam. This restriction allows the deposition rates to range from a few

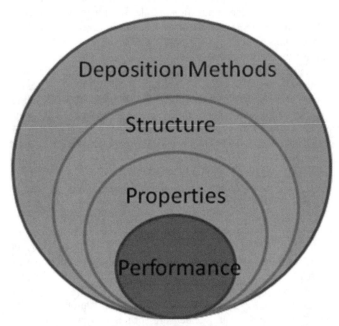

FIG. 32.12 PERFORMANCE–PROPERTY–STRUCTURE–PROCESSING RELATIONSHIP ILLUSTRATION FOR COATINGS

TABLE 32.3   SELECT COATING DEPOSITION PROCESSES FOR NANOCOATINGS

Physical Vapor Deposition (PVD)
- Thermal Evaporation
- Electron Beam Physical Vapor Deposition (EB-PVD)
  o Reactive evaporation
  o Ion assisted evaporation
- Sputtering
  o Diode
  o Reactive
  o Magnetron
  o Unbalanced magnetron
  o Ion Beam
  o High Power Impulse Magnetron Sputtering (HIPIMS)
- Pulsed Laser Deposition (PLD)
- Molecular Beam Epitaxy (MBE)
- Cathodic Arc (CA) Deposition

Chemical Vapor Deposition
- Atmospheric pressure (APCVD)
- Low pressure (LPCVD)
- Metal-organic (MOCVD)
- Photo-enhanced (PHCVD)
- Laser-induced (LICVD
- Plasma enhanced (PECVD).

Electrodeposition Processes
- Electroplating
- Electroless Plating
- Electrophoretic Deposition

Spin Coating

Thermal Spray Deposition

---

angstroms per second to as high as 100 µm/minute, depending on the material [59]. EB-PVD allows coatings to be applied over a wide temperature range from liquid nitrogen to several thousand °C depending on the application and desired microstructure. This is becoming increasingly important as depositing coatings at lower temperatures is desired to maintain the mechanical properties of temperature sensitive substrates.

The EB-PVD process offers many desirable characteristics such as flexible deposition rates (1 nm/min up to 100 µm/minute, depending on material), dense coatings, strong metallurgical bonding, precise composition control, columnar, amorphous, and poly-crystalline microstructure, low contamination, and high thermal efficiency. It should be noted that the microstructure is heavily dependent on the processing parameters. For example, good adhesion is obtained at higher substrate temperatures due to diffusion bonding, but can be facilitated by the use of ion beam assisted deposition (IBAD). IBAD serves two purposes. First, it changes the kinetic energy of the vapor cloud. The average energy of the condensing species for materials directly evaporated is approximately 0.1 eV with a narrow energy distribution. In contrast, the average energy of the condensing species from sputtering processes is 2–10 eV, with typical values being 5 eV. In addition, the energy distribution of the condensing species from sputtering techniques is broader (2 to 5 eV) than that of evaporation (0.1 to 0.2 eV). The energy of the condensing species is very important in film growth in order to produce atom mobility. At lower energies (<10 eV), very little modification to the microstructure occurs as it takes approximately four times the bonding energy (~3 to 5 eV) to

break atomic bonds and result in surface mobility of the atoms. The energy is only high enough to cause local (surface) bond breaking and molecular dissociation. Between energies of 10 eV – 40 eV, not only is there enough energy present to break surface bonds, but also enough to result in an appreciable amount of surface mobility, which affects film growth and density. However, ion beam assisted deposition can be tailored to incorporate ions with much greater energy (over 1000 eV) if desired.

Ion bombardment is also used to preclean the substrates prior to applying the coatings in the EB-PVD chamber (i.e., sputter cleaning) to promote better adhesion. The two major effects occurring during this pre-cleaning step are: (1) removal of adsorbed hydrocarbons and water molecules and (2) increasing the density of nucleation sites for condensation. Energetic ions bombarding the surface remove hydrocarbons and water molecules which are weakly adhered to the surface. Not removing these materials/molecules prior to deposition results in poor adhesion as they serve as weak links for bonding. In addition, energy and current density of the ion beam during sputter cleaning can have a significant impact on the number of nucleation sites. For example, high-energy atoms can cause localized defects which serve as nucleation sites. The increased number of nucleation sites is believed to result in a higher bond density between the substrate and deposited layer. In addition, when bombardment is introduced during deposition, it enhances the surface mobility of the atoms which increases interactions with the incoming molecules causing intermixing of materials at the interfaces which results in better bonding, and thus increased adhesion.

Similarly, cathodic arc falls into the classification of Physical Vapor Deposition (PVD) coating techniques. The term PVD denotes those vacuum deposition processes where the coating material is evaporated or removed by various mechanisms (resistance heating, ablation, high-energy ionized gas bombardment, or electron gun), and the vapor phase is transported to the substrate forming a coating. PVD is often classified as a "line-of-sight" process in

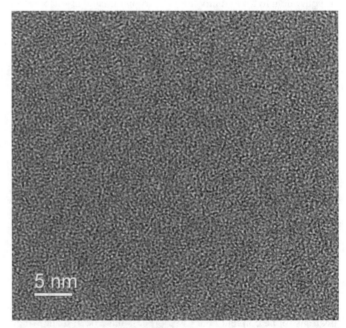

FIG. 32.13   TRANSMISSION ELECTRON MICROGRAPH SHOWING THE NANOGRAINED STRUCTURE OF A PHYSICAL VAPOR DEPOSITED COATING

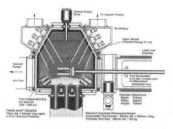

FIG. 32.14   SCHEMATIC DIAGRAM AND PHOTOGRAPHS SHOWING THE EB-PVD PROCESS

which evaporated atoms travel from the source material to the substrate in a straight path. The residual stresses in the PVD coating are generally compressive due to coating bombardment, but can be controlled depending on the deposition parameters. These compressive stresses are often beneficial as they retard the formation and propagation of cracks in the coating. PVD coating processes generally take place between temperatures of 200°C to 500°C to minimize stresses associated with thermal expansion mismatch as compared to the high temperatures (1000°C) of traditional CVD.

In the cathodic arc (CA) deposition process a pulsed or continuous high current-density, low voltage electric current is passed between two separate electrodes (cathode and anode) under vacuum, vaporizing the cathode material while simultaneously ionizing the vapor, forming a plasma. The high current density (usually 10^4 to 10^6 A/cm^2) causes arc erosion by vaporization and melting while ejecting molten solid particles from the cathode surface, with a high percentage of the vaporized species being ionized with elevated energy (50 to 150 eV), and some multiply charged.

Nano-layered coatings can easily be deposited using PVD techniques as shown in Fig. 32.15 in which nano-layers of TiN and CrN were deposited by cathodic arc for improved corrosion and erosion resistance. Figure 32.15 shows a Scanning Transmission Electron Micrograph (STEM) of TiN (dark region) and CrN (light region) of a nano-layered coating with corresponding x-ray energy dispersive spectroscopy (EDS) analysis confirming the elemental composition. The individual layer thickness (i.e., interfacial size, volume, and structure) as well as the residual stress state can be controlled by altering the deposition parameters such as deposi-

tion rate, temperature, pressure, rotation speed, etc. The kinetic energy of the depositing species in cathodic arc are much greater than those of other PVD processes with energies between 50 and 150 eV. Therefore, the plasma becomes highly reactive as a greater percentage of the vapor is ionized. In addition, the cathodic arc process allows tailoring of the interfacial products, especially in multilayer coatings, and does not produce a distinct coating/substrate interface, which may be undesirable. As a result of the high kinetic energy, an intermixed layer of the substrate and coating or between layers of a multilayer coating (10 to 300 Å thick) can be formed which increases the degree of coating adhesion while minimizing residual stresses.

32.17   MICROSTRUCTURE AND PROPERTY ENHANCEMENT

In the last several years, ion beams have gained increased importance during the deposition process to enhance the properties of the depositing film. Ion bombardment of the substrate occurs while the source material is evaporated by either resistance or EB. High-energy bombardment processes have a tendency to create a larger number of coating defects or greater damage. Therefore, low energy bombardment is the preferred method. However, higher energies are needed for ceramic coatings in order to obtain proper surface mobility, and thus the desired microstructure.

The state of the internal stresses developed in the coating can be changed from tensile to compressive by the forcible injection

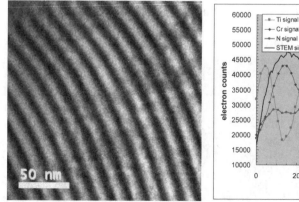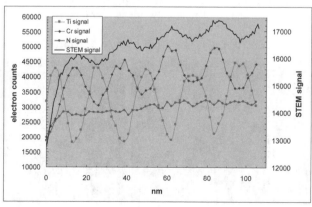

FIG. 32.15   (LEFT) SCANNING TRANSMISSION ELECTRON MICROGRAPH (STEM) OF TiN (DARK REGION) AND CrN (LIGHT REGION) OF A NANOLAYERED COATING WITH CORRESPONDING (RIGHT) X-RAY ENERGY DISPERSIVE SPECTROSCOPY (EDS) ANALYSIS CONFIRMING THE ELEMENTAL COMPOSITION

of high-energy atoms (i.e., ion implantation). Thus the ability to control the stress level is an additional feature of the IBAD process. Chemical vapor deposited coatings generally form with tensile stresses due to the thermal expansion mismatch with the substrate, which often limits the coating thickness before spallation occurs. Ion bombardment during deposition has a tendency to reduce the tensile stress and often changes the intrinsic stress from tensile to compressive. Depending on the energy of the ion beam, texturing or preferred crystal growth orientation can be controlled. Changes in the crystal structure of the film have also been reported with IBAD.

In addition, numerous authors have reported increases in the average hardness of coatings deposited with IBAD. The increase in hardness is obtained by increasing the density, decreasing grain size, changing the stress state, and controlling the crystallographic texture of the coating. Ion implantation (higher degree of energy than IBAD) of Ni and Ti into the surface of high-speed tool steels increases the surface hardness (improved wear resistance), but also introduces large amounts of compressive stress which often leads to premature failure. Improved step coverage (i.e., high surface roughness or complex geometries) has also been reported when using IBAD. This is most likely the result of increased atom mobility under bombardment.

Besides temperature and bombardment as methods of introducing energy into the system, energy can also be added through chemistry (i.e., chemical reactions). During the reactive ion beam assisted (RIBA) deposition process, a beam of ionized species (nitrogen, oxygen, acetylene, methane, argon or mixed combination), is directed towards the substrate where they collide, transfer their kinetic energy to the growing film, and undergo chemical reactions to form the coating. For example, when titanium atoms condense on the substrate's surface, they undergo a chemical reaction with methane or acetylene molecules to form TiC under suitable thermodynamic conditions. The acetylene (or methane) decomposes into carbon and hydrogen; of which the C reacts with titanium forming TiC. The stoichiometry depends on the arrival rate of both constituent species (carbon and titanium). The additional energy resulting from the ions bombarding the surface provides the required energy to initiate the chemical reactions, as well as densification of the film, texturing and stress. These benefits are desirable in many applications including optics and microelectronics, as well as high-wear applications such as cutting tools.

32.17.1 Crystallographic Texturing

Increased awareness has been given to the degree of crystallographic texturing during coating deposition. Crystallographic texture occurs when the crystal lattice of randomly oriented grains within a polycrystalline material preferentially align themselves in a particular direction. This is commonly observed in vapor deposition techniques (growth orientation) and during mechanical rolling. The degree of crystallographic texture is associated with how strongly these grains are oriented in particular directions, and has been correlated with increases in hardness and wear resistance properties, larger piezoelectric effects along the (0001) axes in ZnO, and is being explored in ferromagnetic polycrystalline films with large anisotropic properties used as memory discs [57]. Texturing is very important especially when considering multi-nano-layer coatings in which there is a large difference in the lattice parameters. By tailoring the growth orientation of selective lattice planes, stresses can be minimized especially for NLSL coatings.

Figure 32.16 shows (a–b) TiC/CrC and (c) Al$_2$O$_3$/8YSZ multilayers deposited by co-evaporation of multiple ingots by EB-PVD. By changing the individual layer thickness, the volume fraction of the interface changes. The effects of changing the individual layer thickness on the interfacial composition, structure, and stress state results in unique properties.

32.17.2 Erosion

Depending on the design of the erosion-resistant multilayer coating, the performance can vary depending on the type of environment, i.e., abrasive media (i.e., morphology, type, size). Erosion of materials has long been a concern for structural and design engineers, but it has become increasingly important in the last few years for military and commercial aircraft, especially compressor blisks and helicopter rotors as well as power plants. For example, when aircraft take-off and land, vortexes are formed which often result in the incorporation of hard solid particles of sand, dust and ice with the air flow. In addition, volcanic ash and sandstorms produce hard particles which get ingested into engines. These hard particles impact compressor blades at various angles resulting in severe erosion leading to decreased efficiency and performance. Recent studies have shown that increasing size, temperature, and velocity of the impacting particle generally result in an increase in the erosion rate of the material [60,61].

The mechanism of erosion due to solid particle impingement is much more complex and poorly understood. No single coating material exists which offers erosion resistance properties at both high angle, as well as low angle sand impingement. In general, high hardness (brittle) materials (such as TiN or ternary nitrides) provide excellent erosion resistance against particles with an impingement angle of less than 45° as shown in Fig. 32.17a. However, erosion of the hard brittle materials increases at the higher angles of sand impingement, i.e., 90°. In comparison, metallic (ductile)

(a) TiC/CrC · · · · · · · · · (b) TiC/CrC · · · · · · · · · (c) Al$_2$O$_3$/8YSZ

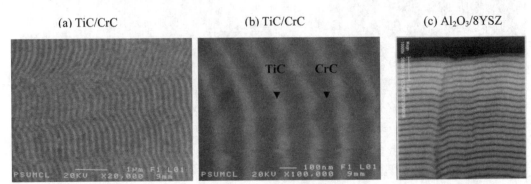

FIG. 32.16   SEM MICROGRAPHS SHOWING (a-b) TiC/CrC AND (c) Al$_2$O$_3$/ZrO$_2$ MULTILAYERED COATINGS PRODUCED BY THE CO-EVAPORATION OF THREE (Ti, Cr, AND GRAPHITE) AND TWO (Al$_2$O$_3$ AND ZrO$_2$) INGOTS, RESPECTIVELY, BY EB-PVD

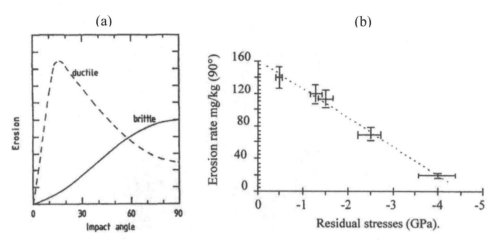

FIG. 32.17 (a) GENERAL EROSION RATES AS A FUNCTION OF IMPACT ANGLE (DEG.) FOR DUCTILE AND BRITTLE MATERIALS, AND (b) EROSION RATE VERSUS RESIDUAL STRESS FOR W/W-N MULTILAYER COATING [60,61]

materials such as nickel-based alloys including Stellite6 and Nu-calloy45 (Fig. 32.17a), show increased erosion resistance at the higher angles (90 deg), but poor resistance at the oblique (20 to 35 deg.) angles of hard particle (i.e., sand) impingement.

The erosion rate from the impact of solid particles is defined as [60]:

Equation 32.7:
$$W = M_p K f(\alpha) V_p^n$$

Where:
W = Erosion rate
M_p = Mass of sand impacting the surface
V_p = The particle velocity on impact
α = The particle impact angle
K, n = Constants determined by the target material and erodent materials, respectively.

According to the above formula, the erosion rate increases with increasing mass and velocity of the abrasive particles. In order to compare erosion rates from different coating materials (i.e. density/porosity), the following formula is often used as it considers the individual volume impacts which normalize for changes in particle size based on fixed volume fractions.

Equation 32.8:
$$W = V_l \pi d^3 / (6 Q_v t C_v)$$

Where:
W = Erosion rate
V_l = Volume loss of the target (mass loss/density)
d = Particle mean diameter
t = Time
C_v = Volume fraction of particles
Q_v = Volumetric flow rate (kg/sec)

There are several issues which arise in developing erosion-resistant coatings. For example, as previously discussed, hard monolithic coatings such as TiN only protect the compressor blisks at low particle impingement angles (Fig. 32.17a) [61,62]. Depending on the coating process and deposition conditions, thin hard TiN

coatings can be limited to 5 to 8 μm in thickness due to increased residual stresses. Often the internal residual stresses of the coating are greater than the adhesive strength of the coating/substrate interface which results in delamination or buckling, and thus coating failure. In contrast, ductile metallic coatings offer excellent erosion resistance against high angle bombardment, but are poor against low angle impingement. Thus the combination of hard coatings (TiN) and ductile metallic coatings in the form of multilayers will offer the best coating properties resistant to erosion at both high and low angle impingement. In addition, multilayers composed of hard and ductile materials will allow application of thicker coatings (20 to 50 μm) which will further increase erosion resistance. Coating technology offers a unique advantage in tailoring the coating with respect to composition, microstructure, and properties depending on the desired application.

Hard particle erosion (i.e., sand) of metallic materials occurs by two main mechanisms, cutting wear and repeated plastic deformation. As sand or other hard particles impact a metallic surface at low impingement angles, the particle plastically deforms the metal surface forming an indent. After the initial indent, the particle will skid across the surface if its velocity is high enough, resulting in gouging or cutting wear/micromachining. The critical velocity is defined as the particle velocity at which the particle will slide or skid across the surface after forming an indent. If the critical velocity of hard particles can be controlled, cutting wear can be significantly reduced. However, controlling the type, size and velocity of hard particles in atmospheric conditions is near impossible. In contrast, hard materials (i.e., ceramics) erode much quicker at higher angles due primarily to Hertzian fracture, or sub-critical crack growth typical of brittle ceramics as schematically shown in Fig. 32.18. For brittle materials, as the particle impacts the surface, Herztian stress fields develop. These stress fields result in sub-surface cracks and crack propagation near localized defects within the coating. With continued particle impaction and stress generation, these cracks grow in length. Eventually, perpendicular and parallel cracks within the coating intersect resulting in coating erosion through spallation from subsequent impaction (coating is eroded), which is often the dominant erosion mechanism for brittle materials. The hard brittle layers absorb the majority of the impact energy from hard particle impingement, minimizing the stress seen by the ductile substrate which aids in preventing debonding. The soft compliant layer minimizes crack propagation within the hard

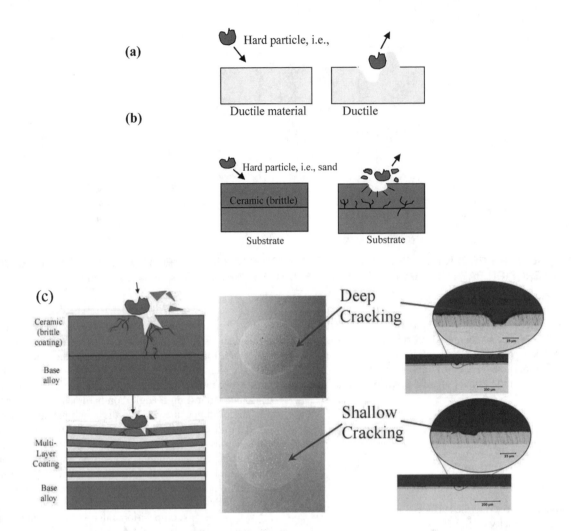

FIG. 32.18   PRINCIPLE EROSION MECHANISM OF (a) DUCTILE METALLIC, (b) BRITTLE CERAMICS AND (c) THE BENEFITS OF CERAMIC-METALLIC MULTILAYER COATINGS

phase which normally leads to coating failure. The most damaging cracks are the radial cracks which initiate beneath the coating surface from coating flexure on the compliant substrate.

Therefore, if the erosion resistance of both metallic and ceramic materials can be incorporated into a multilayer coating of properly selected individual layer thickness, the erosion resistance against hard particle impact can be significantly increased over a wide range of impact angles. Early efforts at examining the erosion resistance of cermet materials (WC-Co and CrC-NiCr) resulted in improved erosion resistance. However, these coatings applied to compressor blisks did not have the erosion resistance that the bulk materials of the same composition had. This was explained due to the microstructure of the thermal sprayed coatings having interlamellar splat boundaries consisting of melted, unmelted, and partially melted particles, and a large volume of porosity. The hard particles would often erode the soft metallic binder (i.e., Co, Ni or Cr) exposing the hard ceramic particles which would eventually be removed through subsequent impacts on the surface.

As a result of these early investigations, multilayer coatings consisting of Cr/Cr-C, W/W-N, W/W-C, Ti/TiB$_2$, Ti/TiN attempted to combine the mechanical and erosion properties of a soft compliant

metallic layer with a hard ceramic layer to increase the erosion resistance against solid particle impact over the entire range of possible impingement angles. In addition, previous studies evaluating the erosion resistance of C/CrC multilayer coatings showed that the erosion resistance increased with increasing compressive stress. These results are quite similar to the increase in hardness (and thus wear resistance) as a function of increasing compressive stress. As the individual layer thickness changes, depending on the material system, the stress state of the coating can be tailored based on lattice mismatch and degree of bombardment. Most PVD coating processes produce high values of compressive stress which have long been known to improve the wear resistance of materials. However, when hard particles impact the surface, they induce increased amounts of compressive stress and deformation. If the induced stresses are too high for the coating to absorb, crack initiation will occur to relieve the stress. Thus, tailoring the multilayer structure and design will improve the erosion resistance.

The incorporation of the soft metallic interlayers results in the required compliance to allow thicker coatings to be deposited without increased residual stresses leading to delamination from the substrate. Multilayer coatings can be deposited much thicker than the standard 5- to 8-μm-thick films by tailoring the stress state to prevent de-bonding. This allows thicker coatings to be deposited, thereby providing further protection against erosion. The performance against erosion is heavily dependent on the choice of material, method of deposition, multilayer configuration and design, i.e., interlayer thickness, number of layers, interfacial boundary, etc.

Erosion resistance can be significantly increased through multilayer design. However, there is no clear understanding as to the optimum individual layer thickness, interfacial boundary effects (i.e., size, structure, composition), or amount of residual stress. Therefore, present efforts are underway to evaluate a variety of multilayer coatings consisting of one of the following hard phase materials: (Ti,Si)N, (Ti,Zr)N, (Ti,Al)N, (Ti,Cr)N, or (Ti,Nb)N in combination with a soft, compliant matrix (Zr, Ti, Cr, Hf, Nb). This type of multilayer configuration is designed for increased toughness and results in a hardness value between the metallic compliant layer and the hard brittle ceramic layer, but primarily depends on the volume fraction of each material and individual layer thickness (lattice mismatch and degree of coherency).

32.18 TIME-DEPENDENT DEFORMATION (CREEP)

The most significant hurdle that nano-structured materials and coatings must overcome at elevated temperatures is creep. Creep is the time-dependent permanent deformation of material at elevated temperature due to long term exposure to high stress levels below a materials' yield point. Creep increases with increasing temperature. In general, ceramic and metallic materials undergo creep at a temperature of 40% to 50% and 30% of the melting temperature of the material, respectively.

Creep deformation is a concern for material applications where materials are subjected to stress at elevated temperature such as in heat exchangers, jet engine components, nuclear power plants, and steam turbine power plants. The mechanisms of creep usually involve grain boundary diffusion, bulk diffusion, climb assisted glide, and thermally activated glide [63,64]. There are three stages of creep for constant stress at elevated temperature over an extended period of time as shown in Fig. 32.19. These stages are transient or primary creep, secondary or steady-state creep, and tertiary

or accelerated creep. In the primary, or initial stage of creep, the strain rate starts out high, but then decreases with increased strain as a result of work hardening. The strain rate eventually reaches a constant, steady-state (secondary) creep in which the stress dependence is strongly dependent on the creep mechanism. Lastly, in the tertiary or accelerated creep stage, the strain rate is increasing and usually results in the formation of cracks, voids, and necking phenomenon which results in failure.

Equation 32.9:
The creep rate is determined by the following equation:
$$d\varepsilon/dt = (C\sigma^m/d^b)\exp[-Q/kT]$$

where:
ε = creep strain
t = time
C = constant which is a function of creep mechanism and material
σ = applied stress
d = grain size
Q = creep mechanism activation energy
T = temperature
k = Boltzmann's constant
m = strain exponential constant
b = inverse grain size exponential constant

However, at elevated stresses, dislocation movement controls the creep where the activation energy (Q) equals the activation energy of self diffusion and the inverse grain size exponent equals zero ($b=0$). As a result, dislocation creep has no grain size dependence, but rather a strong dependence on the applied stress. However, when atoms diffuse through the lattice (bulk diffusion) resulting in grain growth in the direction of the applied stress, we call this diffusional controlled creep or Nabarro-Herring creep. Q again equals $Q_{self\ diffusion}$, but $m = 1$ and $b = 2$. As a result, Nabarro-Herring (N-H) creep is usually categorized as having weak dependence on stress and a moderate grain size dependence. The creep rate increases as the grain size is decreased (i.e., nano-grained materials). N-H creep has a strong dependence on temperature as an atom must overcome the energy barrier to diffuse along the lattice sites. N-H creep is the dominate creep mechanism at higher temperatures (i.e., close to the material's melting temperature).

The second form of diffusional creep is Coble creep in which the atoms diffusion along the grain boundaries in the direction of the applied stress resulting in elongation of the grains. As a result, Coble creep has a much higher dependence on grain size than N-H creep with m and b typically being 1 and 3, respectively. Coble creep causes much greater deformation for nano-structured materials at lower temperatures. Since the activation barrier for Coble creep equals the activation barrier for grain boundary diffusion (which is less than the activation barrier of self diffusion), Coble creep occurs at lower temperatures than N-H. Although Coble creep has the same linear dependence on stress as N-H creep, it does not have as high a temperature dependence as temperature-derived vacancies along grain boundaries are less prevalent (Coble) as compared to the bulk.

Select creep diffusion paths and exponents are listed in Table 32.4 [62] for various creep mechanisms. As observed in Table 32.4, no grain size dependence exists for dislocation creep mechanisms, but there is a strong dependence on grain size for diffusional creep mechanisms such as vacancy flow through grains and along grain boundaries, as well as for interfacial creep mechanism and grain boundary sliding mechanism.

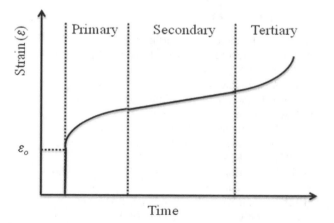

FIG. 32.19 SCHEMATIC ILLUSTRATION OF THE VARIOUS STAGES OF CREEP

38.19 THE FUTURE OF NANO-MATERIALS AND COATINGS

The future of nano-materials and coatings in the energy and power industries is exciting. As we look back over the last 20 years and examine all of the technological advances based on nano-materials, nano-composites, and nano-structures, it is difficult to imagine where we will be in the next 20 years. Nano-tubes like the ones in Fig. 32.20 will play a major role in electronic devices, solar cells, rechargeable batteries and ultracapacitors. Titania nano-tubes have been extensively investigated for use in solar cells as well as hydrogen splitting to produce energy. Future improvements in the efficiency of die-sensitized solar cells (DSSC) will continue to grow the photovoltaic industry [65]. Future research in doping and surface modifications should result in improved photoconversion efficiencies for solar cells.

Nano-structured materials will play important roles in energy storage [66–70], the hydrogen economy, and high efficiency and fast-charging nano-batteries. Companies like Sony and Toshiba have already commercialized nano-batteries that recharge 50 to 60 times faster than the conventional lithium ion batteries [71]. If these could be scaled for the automotive industry, these could revolutionize the energy sources for electric cars. There are several review articles that discuss the use of nano-technology including nano-particles for improving battery efficiency that is attributed to increased electrolyte conductivity by a factor of 6, increased life cycle of nano-composites by reducing volume expansion, and improved nano-structuralization. The future for lithium ion re-

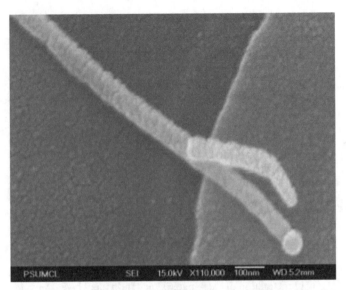

FIG. 32.20 SCANNING ELECTRON MICROGRAPH OF A TYPICAL NANOTUBE. BASED ON PROCESSING, THE DIAMETER AND LENGTH–WIDTH RATIO CAN BE TAILORED TO FORM SINGLE-WALL NANOTUBES (SWNT) OR MULTI-WALLED NANOTUBES (MWNT)

TABLE 32.4 DIFFUSION PATHS AND CREEP EQUATION EXPONENTS FOR VARIOUS CREEP MECHANISMS [62]

Creep mechanism	b	m	Diffusion Path
Dislocation creep mechanisms			
Dislocation glide climb, climb controlled	0	4–6	Lattice
Dislocation glide climb, glide controlled	0	3	Lattice
Dissolution of dislocation loops	0	4	Lattice
Dislocation climb by pipe diffusion	0	5	Dislocation core
Dislocation climb without glide	0	3	Lattice
Diffusional creep mechanisms			
Vacancy flow through grains (Nabarro-Herring Creep)	2	1	Lattice (bulk)
Vacancy flow along grain boundaries (Coble Creep)	3	1	Grain boundary
Interface reaction control	1	2	Lattice/grain boundary
Grain boundary sliding mechanism			
Sliding with liquid	2	1	Lattice (bulk)
Sliding without liquid (diffusion controlled)	3	1	Grain boundary

chargeable batteries will be the nano-ion battery due to higher energy density capacity and faster re-charging. Unlike batteries that store energy chemically, supercapacitors or ultracapacitors physically store energy by separating the negative and positive charges. With future advancements in nano-technology, supercapacitors could overcome the current barriers of low efficiency, long cycle life, and high cost of electrode miniaturization.

Nano-materials and coatings will continue to play a role in fuel cell technology as less expensive and more efficient catalysts are developed [72,73]. In addition, nano-particle additive research to increase the hydrogen release rate will continue to have a positive impact in hydrogen fuel energy storage.

Until the issues of creep are addressed, nano-materials and coatings will continue to be challenged for high temperature energy applications. Nano-materials have been around for several centuries, and it is expected that they will be around for many more.

32.20 ACRONYMS

Al BSF: Aluminum Back Surface Field
APCVD: Atmospheric Pressure Chemical Vapor Deposition
BSCF: $(Ba,Sr)(Co,Fe)O_3$
CA: Cathodic Arc
CIGS: Copper Indium Gallium Selenide
CMC: Ceramic Matrix Composites
CTE: Coefficient of Thermal Expansion
CVD: Chemical Vapor Deposition
DSSC: Die-Sensitized Solar Cells
EB-PVD: Electron Beam – Physical Vapor Deposition
GT: Gas Turbine
HIPIMS: High Power Impulse Magnetron Sputtering
IBAD: Ion Beam Assisted Deposition
IGCC: Integrated Gasification Combined Cycle
LICVD: Laser-Induced Chemical Vapor Deposition

LPPS: Low Pressure Plasma Spray
LPCVD: Low Pressure Chemical Vapor Deposition
LSC: $(La,Sr)CoO_3$
LSCF: $(La,Sr)(Co,Fe)O_3$
LSM: $(La,Sr)MnO_3$
MBE: Molecular Beam Epitaxy
MOCVD: Metal-Organic Chemical Vapor Deposition
near-IR: Near Infrared
near-UV: Near Ultraviolet
NLSL: Nano-layered Self-Lubricating
PECVD: Plasma Enhanced Chemical Vapor Deposition
PEM: Proton Exchange Membrane
PHCVD: Photo-Enhanced Chemical Vapor Deposition
PLD: Pulsed Laser Deposition
PV: Photovoltaic
PVD: Physical Vapor Deposition
RIBA: Reactive Ion Beam Assisted
SOFC: Solid Oxide Fuel Cell
SSC: $(Sm,Sr)CoO3$
STEM: Scanning Transmission Electron Microscopy
TBC: Thermal Barrier Coating
TGO: Thermally Grown Oxide
TIT: Turbine Inlet Temperature
VHN: Vicker's Hardness Number
YSZ: Yttria Stabilized Zirconia

32.21   REFERENCES

1. Mansoori G.A., *Principles of Nanotechnology: Molecular-based study of condensed matter in small systems*, World Scientific, Hackensack, NJ, (2005) 1-30.

2. Holleck H., "Design of Nanostructured Thin Films for Tribological Applications," *Surface Engineering: Science and Technology I, Ed.* by Ashok Kumar, Yip-Wah Chung, John J. Moore, and John E. Smugeresky, The Minerals, Metals & Materials Society, Warrendale, PA (1999) 207-218.

3. Richards B.S., "Enhancing the performance of silicon solar cells via the application of passive luminescence conversion layers," *Solar Energy Materials & Solar Cells,* 90 (2006) 2329-2337.

4. Trupke T, Green M.A., and Wurfel P., "Improving solar cell efficiencies by up-conversion of sub-band-gap light," *Journal of Applied Physics*, 92 [7] (2002) 4117-4122.

5. Trupke T, Shalav A, Richards B.S., Wurfel P., and Green M.A., "Efficiency Enhancement of Solar Cells by Luminescent Up-conversion of Sunlight," *Solar Energy Materials & Solar Cells,* 90 (2006) 3327-3338.

6. Fan Q.H., Chen C., Liao X., Xiang X., Zhang S., Ingler W., Adiga N., Hu Z., Cao X, Du W., Deng X, "High efficiency silicon-germanium thin film solar cells using graded absorber layer," *Solar Energy Materials and Solar Cells*, 94 (2010) 1300-1302.

7. Mishima T., Taguchi M., Sakata H., Maruyama E., "Development status of high-efficiency HIT solar cells," *Solar Energy Materials and Solar Cells*, (2011) in press.

8. Chou C.S., Yang R.Y., Yeh C.K., and Lin Y.J. "Preparation of TiO_2/Nano-metal composite particles and their applications in dye-sensitized solar cells," *Powder Technology*, 194 (2009) 95-105.

9. Shalav A., Richards B.S., Green M.A., "Luminescent layers for enhanced silicon solar cell performance: up conversion," *Solar Energy Materials and Solar Cells*, 91 (2007) 829-842.

10. Wang D.H., Choi D.G., Lee K.J., Jeong J.H., Jeon S.H., Park O.O., and Park J.H., "Effect of the ordered 2D-dot nano-patterned anode for polymer solar cells," *Organic Electronics*, 11 (2010) 285-290.

11. Shockley W. and Queisser H., "Detailed Balance Limit of Efficiency of p-n Junction Solar Cells," *Journal of Applied Physics*, 32 [3] (1961) 510-519.

12. Abbott M. and Cotter J., "Optical and Electrical Properties of Laser Texturing for High-efficiency Solar Cells," *Prog. Photovolt. Res. Appl.*, 14 (2006) 225-235.

13. Mount D., "The Interesting Case of Black Silicon," *Journal of Vacuum Technology and Coating,* (September 2007) 43-47.

14. Cai W., Gong X., and Cao Y., "Polymer solar cells: Recent development and possible routes for improvement in the performance," *Solar Energy Materials and Solar Cells*, 94 (2010) 114-127.

15. Li H.B.T., Franken R.H., Rath J.K., and Schropp R.E.I., "Structural defects caused by a rough substrate and their influence on the performance of hydrogenated nano-crystalline silicon n-i-p solar cells," *Solar Energy Materials and Solar Cells*, 93 (2009) 338-349.

16. Dobrzanski L.A. and Drygala A., "Laser texturization of crystalline silicon for solar cells," *Journal of Achievements in materials and Manufacturing Engineering*, 29[1] (2008) 7-14.

17. Han K.S., Shin J.H., and Lee H., "Enhanced transmittance of glass plates for solar cells using nano-imprint lithography." *Solar Energy Materials and Solar Cells*, 94 (2010) 583-587.

18. Han K.S., Shin J.H., Yoon W.Y., and Lee H., "Enhanced performance of solar cells with anti-reflection layer fabricated by nano-imprint lithography." *Solar Energy Materials and Solar Cells*, 94 (2011) in press.

19. Zhao F, Wang Z, Liu M, Zhang L, Xia C, and Chen F., "Novel nano-network cathodes for solid oxide fuel cells," *Journal of Power Sources*, 185 (2008) 13-18.

20. Kim S.D., Moon H., Hyun S.H., Moon J., Kim J., and Lee H.W., "Nano-composite materials for high-performance and durability of solid oxide fuel cells," *Journal of Power Sources*, 163 (2006) 392-397.

21. Liu Z., Guo B., Huang J., Hong L., Han M., Gan L.M., "Nano-TiO_2-coated polymer electrolyte membranes for direct methanol fuel cells," *Journal of Power Sources*, 157 (2006) 207-211.

22. Jiang S.P., "A review of wet impregnation-An alternative method for the fabrication of high performance and nano-structured electrodes of solid oxide fuel cells," *Materials Science and Engineering A*, 418 (2006) 199-210.

23. Xin X., Wang S., Zhu Q., Xu Y., and Wen T., "A high performance nano-structure conductive coating on a crofer22APU alloy fabricated by a novel spinel powder reduction coating technique," *Electrochemistry Communications*, 12 (2010) 40-43.

24. Kannan A.M., Kanagala P., and Veedu V., "Development of carbon nanotubes based gas diffusion layers by in situ chemical vapor deposition process for proton exchange membrane fuel cells," *Journal of Power Sources*, 192 (2009) 297-303.

25. Kannan A.M. and L. Munukutla L., "Carbon nano-chain and carbon nano-fibers based gas diffusion layers for proton exchange membrane fuel cells," *Journal of Power Sources*, 167 (2007) 330-335.

26. Jian Z., Xia C., and Chen F., "Nano-structured composite cathodes for intermediate-temperature solid oxide fuel cells via an infiltration/impregnation technique," *Electrochemica Acta*, 55 (2010) 3595-3605.

27. Chien C.C. and Jeng K.T., "Noble metal fuel cell catalysts with nano-network structures," *Materials Chemistry and Physics*, 103 (2007) 400-406.

28. Gestel T.V., Sebold D., Meulenberg W.A., and Buchkremer H.P., "Development of thin-film nano-structured electrolyte layers for application in anode-supported solid oxide fuel cells," *Solid State Ionics*, 179 (2008) 428-437.

29. Johnson C., Gemmen R., and Orlovskaya N., "Nano-structured self-assembled LaCrO3 thin film deposited by RF-magnetron sputtering on a stainless steel interconnect material," *Composites: Part B*, 35 (2004) 167-172.

30. Lee J.J., Moon H., Park H.G., Yoon D.I., and Hyun S.G., "Applications of nano-composite materials for improving the performance of anode-supported electrolytes of SOFCs," *International Journal of Hydrogen Energy*, 35 (2010) 738-744.

31. Chockalingam R., Amarakoon V.R.W., and Giesche H., "Alumina/cerium oxide nano-composite electrolyte for solid oxide fuel cell applications," *Journal of the European Ceramic Society*, 28 (2008) 959-963.

32. Dalili N., Edrisy A., and Carriveau R., "A review of surface engineering issues critical to wind turbine performance," *Renewable and Sustainable Energy Reviews*, 13 (2009) 428-438.

33. Makkonen L., Laakso T., Marjaniemi M., Finstad K.J., "Modeling and prevention of ice accertion on wind turbines," Wind Energy, 25 [1] (2001) 3-21.

34. Karmouch R. and Ross G.G., "Superhydrophobic wind turbine blade surfaces obtained by a simple deposition of silica nanoparticles embedded in epoxy," *Applied Surface Science*, 257 (2010) 665-669.

35. Miller R.A., "Thermal Barrier Coatings for Aircraft Engines – History and Directions," *Thermal Barrier Coating Workshop*, NASA CP vol 3312 (1995) 79.

36. Wolfe D.E., Singh J., Miller R.A., Eldridge J.I., Zhu D.M., "Tailored microstructure of EB-PVD 8YSZ thermal barrier coatings with low thermal conductivity and high thermal reflectivity for turbine applications," *Surface and Coatings Technology*, 190 (2005) 132-149.

37. Goward G., Boone D., Pettit F., United States Patent #3,754,903 (1973).

38. Kaden U., Leyens C., Peters M., and Kaysser W.A., *Elevated Temperature Coatings: Science and Technology III*, ed. by J.M. Hampikian and N.B. Dahotre, The Minerals, Metals and Materials Society, Warrendale, PA (1999) 27-38.

39. Cosack T., Pawlowski L., Schneiderbanger S., and Sturlese S., *Trans. ASME*, (1992) 2-GT-319.

40. Gell M., Vaidyanathan K., Barber B., Cheng J., Jordan E., *Metall Trans*, 30A (1999) 427.

41. Ravichandran K.S., An K., Dutton R.E., Semiatin S.L., *Journal of American Ceramics Society*, 82[3] (1999) 673.

42. Jones R.L. and Mess D., "Improved tetragonal phase stability at 1400C with scandia, yttria-stabilized zirconia," *Surface and Coatings Technology*, 86-87 (1996) 94-101.

43. Susnitzky D.W., Hertl W., and Carter C.B., "Vanadia-induced transformations in yttria-stabilized zirconia," *Ultramicroscopy*, 30[1-2] (1989) 233-241.

44. Hill M.D., Phelps D.P., and Wolfe D.E., "Corrosion Resistant Thermal Barrier Coating Materials for Industrial Gas Turbine Applications," *CESP*, 29[4] (2008) 123-132.

45. Wolfe D.E. and Singh J., "Microstructural evolution of titanium nitride (TiN) coatings produced by reactive ion beam-assisted, electron beam physical vapor deposition (RIBA, EB-PVD)", *Journal of materials Science*, 34 (1999) 2997-3006.

46. Wolfe D.E. and Singh J., "Synthesis of titanium carbide/chromium carbide multilayers by the co-evaporation of multiple ingots by electron beam physical vapor deposition," *Surface and Coatings Technology*, 160 (2002) 206-218.

47. Wolfe D.E. and Singh J., "Synthesis and characterization of multi-layered TiC/TiB2 coatings deposited by ion beam assisted, electron beam-physical vapor deposition (EB-PVD)," *Surface and Coatings Technology*, 165 (2003) 8-25.

48. Rao S.I., Hazzledine P.M., and Dimikuk D.M., Materials Research Society Symposium Proceedings, *Materials Research Society*, 362 (1995) 67-77.

49. Koch C.C., Shen T.D., Malow T., and Spaldon O., Mat. *Res. Soc. Symp. Proc.*, 362 (1995) 253-263.

50. Holleck H. and Schier V., *Surface and Coatings Technology*, 76-77 (1995) 328-336.

51. Upadhyaya G.S., *Nature and Properties of Refractory Carbides*, Nova Science Publishers, Inc., 1996.

52. Pierson O.H., *Handbook of Refractory Carbides and Nitrides: Properties, Characteristics, Processing, and Applications*, Noyes Publications, Westwood, New Jersey, 1996.

53. Dubach R., Curtins H., and Rechberger H., *Surface and Coatings Technology*, 94-95 (1997) 622-626.

54. Schulz H., Dorr J., Rass I.J., Schulze M., Leyendecker T., and Erkens G., *Surface and Coatings Technology*, 146-147 (2001) 480-485.

55. Derflinger V., Brandle H., and Zimmermann H., *Surface and Coatings Technology*, 113 (1999) 286-292.

56. Storms E.K., *The Refractory Carbides*, Academic Press, Inc., New York, 1967.

57. Carvalho N.J.M. and DeHosson J.T.M., *Thin Solid Films*, 388 [1-2] (2001) 150-159.

58. Wolfe D.E. and Singh J., "Synthesis and characterization of TiBCN coatings deposited by ion beam assisted, co-evaporation electron beam-physical vapor deposition (EB-PVD)," *Journal of Materials Science*, 37 (2002) 3777-3787.

59. Singh J. and Wolfe, D.E., "Review: Nano and Macro-Structured Component Fabrication by Electron Beam-Physical Vapor Deposition (EB-PVD)," *Journal of Material Science*, 40 (2005) 1-26.

60. Tilly G.P., "Erosion by impact of solid particles," *Treatise on material Science and Technology*, Academic Press, New York, 1979.

61. Haugen K., Kvernvold O., Ronold A., Sandber R., "Sand erosion of wear-resistant materials: Erosion in choke valves," *Wear*,186-187 (1995) 178-188.

62. Gachon Y., Ienny P., Forner A., Farges G., Sainte Catherine M.C., Vannes A.B., "Erosion by solid particles of W/W-N multilayer coatings obtained by PVD process," *Surface and Coatings Technology*, 113 (1999) 140-148.

63. Chokshi A.H., "Unusual stress and grain size dependence for creep in nanocrystalline materials," *Scripta Materialia*, 61(2009) 96-99.

64. Ovidko I.A. and Sheinerman A.G., "Enhanced ductility of nano-materials through optimization of grain boundary sliding and diffusion processes," *Acta Materialia*, 57 (2009) 2217-2228.

65. Lee J.H., Park N.G., and Shin Y.J., "Nano-grain SnO_2 electrodes for high conversion efficiency SnO_2-DSSC," *Solar Energy Materials and Solar Cells*, 2010.

66. Driver D., "Making a material difference in energy," *Energy Policy*, 36 (2008) 4302-4309.

67. Jeon K.J., Theodore A., and Wu C.Y., "Enhanced hydrogen absorption kinetics for hydrogen storage using Mg flakes as compared to conventional spherical powders," *Journal of Power Sources*, 183 (2008) 693-700.

68. Choudhury P., Srinivasan S. S., Bhethanabotla V. R., Goswami Y., McGrath K., and Stefanakos E. K., "Nano-Ni doped Li–Mn–B–H system as a new hydrogen storage candidate," *International Journal of Hydrogen Energy*, 34 (2009) 6325-6334.

69. Chang J.K., Tsai H.Y., and Tsai W.T., "A metal dusting process for preparing nano-sized carbon materials and the effects of acid post-treatment on their hydrogen storage performance," *International Journal of Hydrogen Energy*, 33 (2008) 6734-6742.

70. Srinivasan S.S., "Effects of nano additives on hydrogen storage behavior of the multinary complex hydride LiBH$_4$/LiNH$_2$/MgH," *International Journal of Hydrogen Energy*, (2010) 9646-9652.

71. Peng B., and Chen J., "Functional materials with high-efficiency energy storage and conversion for batteries and fuel cells," *Coordination Chemistry Review*, (2009) 2805-2813.

72. Chen Y., Zhang G., Ma J., Zhou Y., Tang Y., and Lu T., "Electro-oxidation of methanol at the different carbon materials supported Pt nano-particles," *International Journal of Hydrogen Energy*, (2010) 10109-10117.

73. Choi H., Kim H., Hwang S., Choi W., and Jeon M., "Dye-sensitized solar cells using graphene-based carbon nano composite as counter electrode, " *Solar Energy Materials and Solar Cells*, 2010.

INDEX

Page numbers followed by f and t indicate figures and tables, respectively.